FOUNDATIONS OF ELECTRONICS

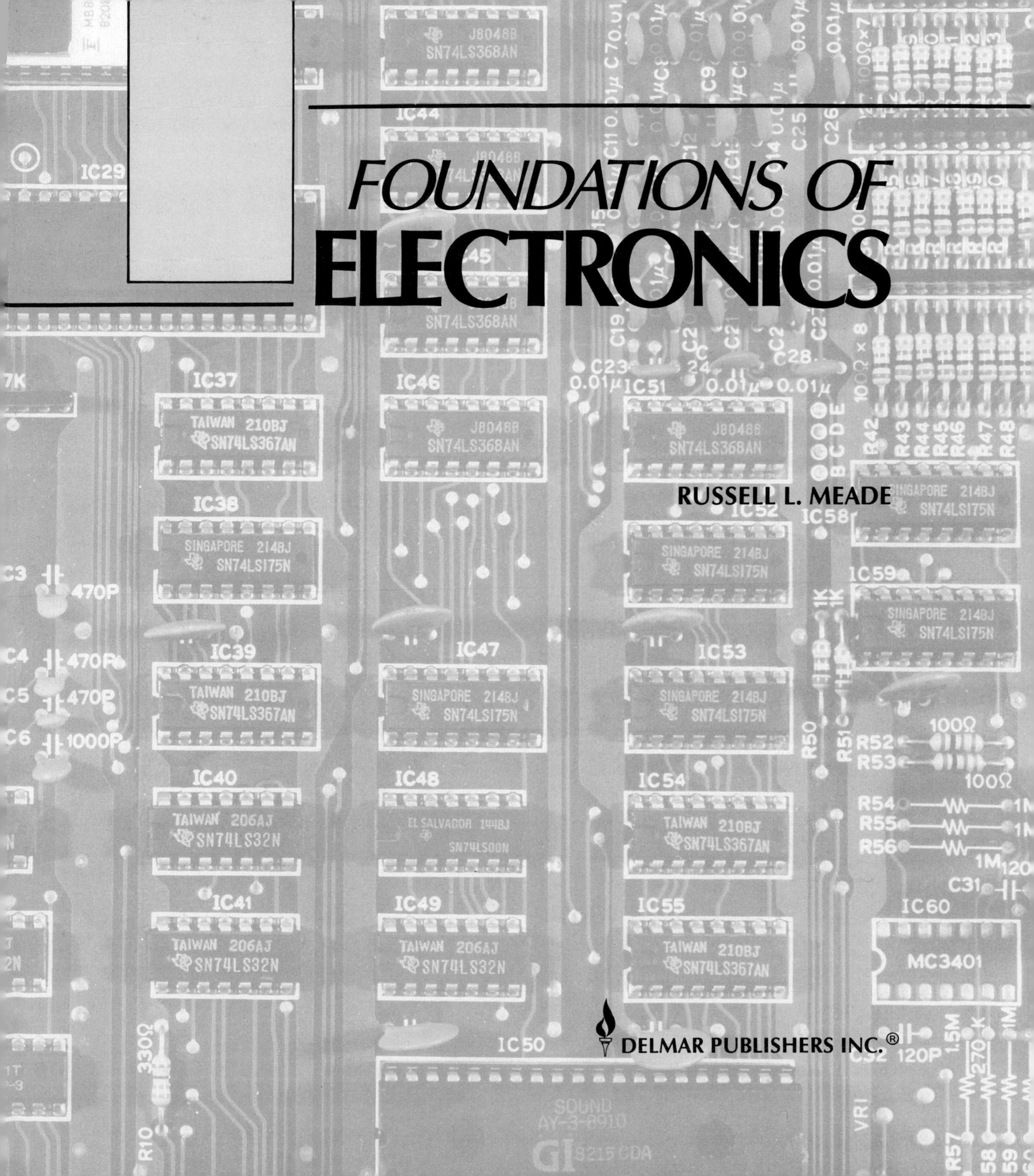
FOUNDATIONS OF
ELECTRONICS
RUSSELL L. MEADE
DELMAR PUBLISHERS INC.®

NOTICE TO THE READER

Publisher does not warrant or guarantee any of the products described herein or perform any independent analysis in connection with any of the product information contained herein. Publisher does not assume, and expressly disclaims, any obligation to obtain and include information other than that provided to it by the manufacturer.

The reader is expressly warned to consider and adopt all safety precautions that might be indicated by the activities described herein and to avoid all potential hazards. By following the instructions contained herein, the reader willingly assumes all risks in connection with such instructions.

The publisher makes no representations or warranties of any kind, including but not limited to, the warranties of fitness for particular purpose or merchantability, nor are any such representations implied with respect to the material set forth herein, and the publisher takes no responsibility with respect to such material. The publisher shall not be liable for any special, consequential or exemplary damages resulting, in whole or in part, from the readers' use of, or reliance upon, this material.

Dedication

To my beloved wife Betty, for her patience, love, support and encouragement throughout this challenging project.

Cover photo depicting VACREL® photopolymer film solder mask, provided courtesy of Du Pont Company, Wilmington, DE

Delmar staff:

Administrative Editor: Wendy Jones
Developmental Editor: Mary E. Ormsbee
Managing Editor: Susan L. Simpfenderfer
Production Supervisor: Larry Main
Art Coordinator: Michael Nelson
Design Supervisor: Susan C. Mathews

For information, address Delmar Publishers Inc.
2 Computer Drive, West, Box 15-015

Printed in the United States of America
Published simultaneously in Canada
by Nelson Canada,
a division of the Thomson Corporation

Library of Congress Cataloging-in-Publication Data

Meade, Russell L.
Foundations of electronics / Russell L. Meade.
p. cm.
ISBN 0-8273-2993-8 (textbook)
1. Electronics. I. Title.
TK7816.M43 1991
621.381—dc20 90-46972
CIP

10 9 8 7 6 5 4 3

CONTENTS

PART I FOUNDATIONAL CONCEPTS

CHAPTER 1 Basic Concepts of Electricity 2

Definition, Physical, and Chemical States of Matter 5 Composition of Matter 7 Structure of the Atom 8 Concept of Electron Theory 12 Ions 13 Energies That Change Electrical Balance 14 Conductors, Semiconductors, and Insulators 14 Sample of an Electrical System 15 Basic Principles of Static Electricity 17 Electrical Potential 21 Charges in Motion 24 Three Important Electrical Quantities 26 Basic Electrical Circuit 28

CHAPTER 2 Ohm's Law 36

Basic Electrical Units and Abbreviations 39 Using the Metric System to Help 40 Diagrams Used for Electronic Shorthand 43 Ohm's Law and Relationships between Electrical Quantities 49 Rearranging to Find the Unknown Quantity 52 Sample Application of Metric Prefixes and Powers of Ten 54 Some Basic Rules Relating to Powers of Ten 55 Direction of Current Flow 58 Polarity and Voltage 58 Work, Energy, and Power 59 Commonly Used Versions of Ohm's Law and Power Formulas 62 Resistors 65 Resistor Color Code 67

PART II BASIC CIRCUIT ANALYSIS

CHAPTER 3 Series Circuits 78

Definition and Characteristics of a Series Circuit 81 Resistance in Series Circuits 81 Voltage in Series Circuits 84 Kirchhoff's Voltage Law 86 Power in Series Circuits 89 Effects of Opens and Troubleshooting Hints 91 Effects of Shorts and Troubleshooting Hints 93 Designing a Series Circuit to Specifications 98 Special Applications 100 Critical Thinking and the SIMPLER Troubleshooting Sequence 107 Introduction to Troubleshooting Skills 107 The SIMPLER Sequence for Troubleshooting 107 Troubleshooting Levels 113 Chapter Challenge 114 A Block-level Troubleshooting Example 115 A Component-level Troubleshooting Example 116

CHAPTER 4 Parallel Circuits 124

Definition and Characteristics of a Parallel Circuit 127 Voltage in Parallel Circuits 128 Current in Parallel Circuits 128

Kirchhoff's Current Law 129 Resistance in Parallel Circuits 132 Methods to Calculate Total Resistance (R_T) 132 Power in Parallel Circuits 140 Effects of Opens in Parallel Circuits and Troubleshooting Hints 142 Effects of Shorts in Parallel Circuits and Troubleshooting Hints 144 Similarities and Differences between Series and Parallel Circuits 147 Designing a Parallel Circuit to Specifications 148 Sources in Parallel 149 Current Dividers 149 Chapter Challenge 154

CHAPTER 5 **Series-Parallel Circuits** **162**

What Is a Series-Parallel Circuit? 165 Approaches to Recognize and Analyze Series and Parallel Circuit Portions 166 Total Resistance in Series-Parallel Circuits 169 Current in Series-Parallel Circuits 173 Voltage in Series-Parallel Circuits 176 Power in Series-Parallel Circuits 178 Effects of Opens in Series-Parallel Circuits and Troubleshooting Hints 180 Effects of Shorts in Series-Parallel Circuits and Troubleshooting Hints 183 Designing a Series-Parallel Circuit to Specifications 185 Loaded Voltage Dividers 187 The Wheatstone Bridge Circuit 195 Chapter Challenge 200

CHAPTER 6 **Basic Network Theorems** **208**

Maximum Power Transfer Theorem 211 Superposition Theorem 215 Thevenin's Theorem 219 Norton's Theorem 222 Converting Norton and Thevenin Equivalent Parameters 227

PART III **ELECTRICAL QUANTITIES**

CHAPTER 7 **Cells and Batteries** **232**

Chemical Action in a Cell 235 The Common Dry Cell 237 Operating Characteristics of Dry Cells 238 The Common Wet Cell 239 Construction and Operational Characteristics of the Lead-Acid Battery 240 Charging the Lead-Acid Battery 244 Methods of Testing Lead-Acid Batteries 245 Care of Lead-Acid Batteries 247 Examples of Other Types of Primary and Secondary Cells 248 Chapter Challenge 252

CHAPTER 8 **Magnetism and Electromagnetism** **256**

Background Information 259 Fundamental Laws, Rules, and Terms to Describe Magnetism 260 Elemental Electromagnetism 264 Important Magnetic Units, Terms, Symbols, and Formulas 270 Practical Considerations about Core Materials 275 The B-H Curve 276 The Hysteresis

Loop 277 Induction and Related Effects 280 Faraday's Law 284 Lenz's Law and Reciprocal Effects of Motors and Generators 284 Summary of Magnetism 288

CHAPTER 9 **DC Measuring Instruments** 296
General Requirements of Basic Meter Movements 299 The d'Arsonval (Moving Coil) Movement 299 Current Meters 302 Current Meter Shunts 306 Voltmeters 315 Calculating the Multiplier Values 316 Ohmmeters 324 VOMs, DMMs, and Other Related Devices 327 Troubleshooting Hints 332 Chapter Challenge 339

CHAPTER 10 **Basic AC Quantities** 346
Background Information 349 Generating an AC Voltage 352 Some Basic Sine-wave Waveform Descriptors 357 Period and Frequency 358 Phase Relationships 360 Important AC Sine-wave Current and Voltage Values 363 The Purely Resistive AC Circuit 367 Other Waveforms 369

CHAPTER 11 **The Oscilloscope** 376
Background Information 379 Key Sections of the Scope 380 Combining Horizontal and Vertical Signals to View a Waveform 386 Measuring Voltage and Current with the Scope 389 Using the Scope for Phase Comparisons 393 Measuring Frequency Ratios with the Scope 395 Chapter Challenge 401

PART IV **REACTIVE COMPONENTS**

CHAPTER 12 **Inductance** 406
Background Information 409 Review of Faraday's and Lenz's Laws 410 Self-inductance 411 Factors That Determine Inductance 413 Inductors in Series and Parallel 414 Energy Stored in the Inductor's Magnetic Field 417 The L/R Time Constant 419 Summary Comments About Inductors 424 Troubleshooting Hints 426

CHAPTER 13 **Inductive Reactance in AC** 432
V and I Relationships in a Purely Resistive AC Circuit 435 V and I Relationships in a Purely Inductive AC Circuit 435 Concept of Inductive Reactance 437 Relationship of X_L to Inductance Value 438 Relationship of X_L to Frequency of AC 439 Methods to Calculate X_L 440 Inductive Reactances in Series and Parallel 443 Final Comments about Inductors

and Inductive Reactance 444 Quality Factor 446 Chapter Challenge 448

CHAPTER 14 RL Circuits in AC 454
Review of Simple R and L Circuits 457 Using Vectors to Describe and Determine Magnitude and Direction 459 Introduction to Common AC Circuit Analyses 461 Fundamental Analysis of Series RL Circuits 466 Fundamental Analysis of Parallel RL Circuits 477 Examples of Practical Applications for Inductors and RL Circuits 482 Chapter Challenge 485

CHAPTER 15 Basic Transformer Characteristics 492
Background Information 495 Coefficient of Coupling 495 Mutual Inductance and Transformer Action 496 Mutual Inductance Between Coils Other Than Transformers 498 Important Transformer Ratios 500 Transformer Losses 508 Characteristics of Selected Transformers 509 Troubleshooting Hints 513

CHAPTER 16 Capacitance 522
Definition and Description of a Capacitor 525 The Electrostatic Field 525 Charging and Discharging Action 526 The Unit of Capacitance 529 Energy Stored in Capacitor's Electrostatic Field 530 Factors Affecting Capacitance Value 531 Capacitance Formulas 533 Total Capacitance in Series and Parallel 535 Finding Voltage When Three or More Capacitors Are in Series 538 The RC Time Constant 540 Types of Capacitors 544 Typical Color Codes 550 Typical Problems and Troubleshooting Techniques 551 Chapter Challenge 555

CHAPTER 17 Capacitive Reactance in AC 560
V and I Relationships in a Purely Resistive AC Circuit 563 V and I Relationships in a Purely Capacitive AC Circuit 563 Concept of Capacitive Reactance 564 Relationship of X_C to Capacitance Value 564 Relationship of X_C to Frequency of AC 565 Methods to Calculate X_C 566 Capacitive Reactances in Series and Parallel 569 Voltages, Currents, and Capacitive Reactances 572 Final Comments about Capacitors and Capacitive Reactance 572 Chapter Challenge 576

CHAPTER 18 RC Circuits in AC 582
Review of Simple R and C Circuits 585 Series RC Circuit Analysis 585 Parallel RC Circuit Analysis 591 Similarities and Differences between RC and RL Circuits 596 Waveshaping

and Non-sinusoidal Waveforms 597 Other Applications of RC Circuits 598 Troubleshooting Hints and Considerations for RC Circuits 601

CHAPTER 19 **RLC Circuit Analysis** **608**
Basic RLC Circuit Analysis and Power in AC Circuits 611 The Series RLC Circuit 613 The Parallel RLC Circuit 620 Power in AC Circuits 623 Rectangular and Polar Vector Analysis 630 Algebraic Operations 637 Application of Rectangular and Polar Analysis 639 Chapter Challenge 644

CHAPTER 20 **Series and Parallel Resonance** **650**
X_L, X_C, and Frequency 653 Series Resonance Characteristics 653 The Resonant Frequency Formula 655 Some Resonance Curves 659 Q and Resonant Rise of Voltage 660 Parallel Resonance Characteristics 663 Parallel Resonance Formulas 664 Effect of a Coupled Load on the Tuned Circuit 666 Q and the Resonant Rise of Impedance 666 Selectivity, Bandwidth, and Bandpass 667 Measurements Related to Resonant Circuits 670 Filter Applications of Nonresonant and Resonant RLC Circuits 674 Chapter Challenge 683 Chapter Challenge 684

PART V **INTRODUCTORY DEVICES AND CIRCUITS**

CHAPTER 21 **Diodes and Power Supply Circuits** **690**
Semiconductor Materials 693 The P-N Junction 698 The Semiconductor Diode 699 Diode Clippers 702 The Power Supply System 703 The Half-wave Rectifier Circuit 705 The Full-wave Rectifier Circuit 708 The Bridge Rectifier Circuit 710 Basic Power Supply Filters 713 Basic Voltage Multiplier Circuits 717 Chapter Challenge 720 Chapter Challenge 721

CHAPTER 22 **Overview of Transistors** **726**
Background Information 729 Transistor Alpha and Beta Parameters 731 Common Amplifier Configurations and Characteristics 733 Overview of Field Effect Transistors 736 Other Semiconductor Devices and Applications 737 Information for the Technician 741

CHAPTER 23 **Overview of Integrated Circuits and Digital Electronics** **748**
Brief History of ICs 751 Some Classifications of ICs 751 Advantages and Disadvantages of ICs 755 Examples of Analog

and Digital Modes of Operation 755 IC Packaging and Basing 755 Common Number Systems Used With Digital ICs 756 Basic Logic Gates 757 Other Common Digital Circuits 763 Memory 763 The Microprocessor 764 The Basic Digital Computer 764

CHAPTER 24 **Introduction to Transistor Amplifiers and Oscillators** **770**

How a Transistor Amplifies 773 Methods of Classifying Amplifiers 774 Notations Regarding Common-emitter (CE) Amplifier 780 Notations Regarding Common-base (CB) Amplifier 784 Notations on the Common-collector (CC) Amplifier 785 Notations on FET Amplifier Configurations 786 Introduction to Oscillator Circuits 787

CHAPTER 25 **Introduction to Operational Amplifiers** **796**

Background Information 799 Linear and Nonlinear Op-amp Applications 799 Elemental Op-amp Information 800 Characteristics of the Ideal Op-amp 803 An Inverting Amplifier 803 A Noninverting Amplifier 804 Basic Op-amp Parameters 805 Sample Op-amp Applications 807

Appendix A **813**

Appendix B **816**

Appendix C **852**

Appendix D **853**

Appendix E **854**

Appendix F **855**

Glossary **856**

Index **881**

INTRODUCTION

The Wonders of Electronics

Today, because of electronics, we have capabilities that could not even be dreamed about just a few years ago!

In the field of *communications,* Samuel F.B. Morse invented the telegraph in the 1830s, Alexander Graham Bell invented the telephone in the 1870s, Heinrich Rudolf Hertz, succeeded in transmitting a radio signal a short distance in the late 1880s, and Guglielmo Marconi invented the first practical "wireless" telegraph in the 1890s. As remarkable as their achievements were, none of these inventors could have imagined today's satellite communications systems, the modern color television, computers "talking" to each other via data transmission systems, or even the facsimile systems (FAX) of today.

In *manufacturing,* who could have foreseen that engineering drawings would be drawn by punching keys on a keyboard of a computer-aided drafting and design (CADD) system. This same CADD system may then be interfaced with a computer-aided manufacturing (CAM) system that automatically takes the computer-developed drawings and controls the systems that create the actual parts drawn. Automation in manufacturing involves everything from robots that make welds in automobiles to machines that automatically drill, punch, mill, shape, bend, and form metals, plastics and other materials into useful products. Computer-controlled "processing" plants develop and produce integrated circuits and semiconductor devices, chemicals, paints, and a myriad of other products.

Today's *service businesses,* such as banks, real estate, insurance, fast food, department stores, and hundreds of others could not work without electronic transfer of funds, computerized property listings, computerized billing of premiums and point of sale (POS). Computerized cash registers can keep a "running inventory," keep track of pricing of every item, print receipts, and even compute the change. Governmental agencies (such as the IRS) could not perform their daily duties without help from electronic systems.

Virtually all that we do in our everyday living is affected by electronics. Home appliances used to cook, wash, clean and provide entertainment are electronic marvels.

Even our automobiles utilize great amounts of electronic wizardry. Imagine getting in your car, punching in your "user" number on a digital panel, and having the automobile's electronic systems then automatically set all mirrors, seat positions and the car's suspension and steering systems to suit the user. Unrealistic? No, not really! Many of these types of amenities are found already in our newer automobiles.

What tomorrow may bring can hardly be imagined. At any rate we know electronics is here to stay. You have chosen a good profession.

The Importance of Learning Fundamentals

Becoming a competent technician in the field of electronics demands that you get a thorough grasp of fundamentals.

In any field of endeavor, whether it be nursing, law, or a craft, there are fundamental terms, symbols, concepts and principles that must be learned.

In electricity and electronics, it is vital that these basic principles be learned for they are the foundation upon which you build an understanding of all that follows in this profession.

Set your mind to work diligently at learning new terms, concepts and the important foundational truths! You will reap large benefits later as you get further into this study.

Mastering the terms, symbols, formulas and basic principles taught in this book will launch you into this stimulating field and will allow you to pursue it as far as your desire will carry you.

We wish you the best of success as you engage in the learning process . . . and beyond!

PREFACE

The Book's Aim

The purpose of this book is to present electrical and electronics fundamentals in an interesting and easy-to-understand style.

Knowledge of fundamentals is *critical* to success in any facet of the electrical or electronics professions. *Foundations of Electronics* is designed to communicate basic facts, concepts and principles in a "reader-friendly" manner. Furthermore, the text fosters your ability to apply this knowledge to "real-life" situations.

This text stresses the development of logical thinking patterns and practical applications of knowledge, not just the presentation of cold theory.

The Book's Approach

Several important strategies are used in the book's presentation. Some of these include:

- Helping you understand rather than just memorize important formulas.
- Giving meaningful overviews to provide a context for content details.
- Moving from the known to the unknown.
- Providing immediate application of knowledge gained.
- Offering clear and frequent examples.
- Making liberal use of pictorial and diagrammatic information.
- Formatting that allows either group or individualized instruction approaches.
- Using frequent practice problems to reinforce concepts.
- Offering strategically located checkpoints.
- Focusing on troubleshooting, safety and providing practical hints.
- Including useful summaries that distill the essence of what should be learned.
- Providing a unique troubleshooting technique, called the SIMPLER Method that sharpens critical thinking skills as it teaches troubleshooting.
- Supplying Chapter Challenge circuit problems to allow practice of troubleshooting with the SIMPLER technique.

The Book's Arrangement

The text begins with a brief overview of the wonders of today's electronic technology and the great opportunities afforded in this field. Next, crucial terms and elemental concepts are presented as foundations for the topics addressed throughout the remainder of the book. Concepts such as the electron theory, basic electrical units and symbols and how electronic circuitry can be illustrated with diagrams provide the stepping stones to what follows.

Essential circuit fundamentals are then presented so that students can develop the ability to analyze various circuit configurations. In the process of learning these fundamentals, several passive devices are studied; including resistors, inductors, and capacitors. Also, important test instruments and their uses are discussed. The book then applies the fundamental concepts, principles and theories, first to DC circuits, and then to AC circuits.

Once these foundations have been laid, you will see several practical applications of these theories in circuits. For example, you will see how rectifiers and filters work to provide a universally significant system called the electronic power supply. Power supplies are found in virtually all electronic equipment.

Finally, you will get a glimpse of some of today's technological marvels. These include semiconductor transistors, integrated circuits and the microprocessor chip.

It is our hope that as you move from the introductory concepts of electronics through the principles of the circuit analysis to practical circuit applications and on to the introduction of modern electronic devices, you will find the book interesting, informative and very useful in your training.

Using the Book

The format of *Foundations of Electronics* is "user-friendly." Each chapter features key terms, an introductory outline, a preview of the chapter's contents, and learning objectives. The body of the chapter includes frequent step-by-step examples and practice problems, and In-Process Learning Checks to reinforce concepts. Each chapter closes with a detailed summary of important points as well as review and "Test Yourself" questions. (Answers to the practice problems, In-Process Learning Checks, and odd-numbered review and test yourself questions may be found in Appendix B.)

A system of icons has been incorporated through the text to quickly identify the features described below:

✔ **IN-PROCESS LEARNING CHECKS** provide an opportunity for you to check your understanding of the subject each step of the way. Answers are provided in Appendix B.

PRACTICE PROBLEMS reinforce concepts learned just a few paragraphs earlier. Answers are provided in Appendix B.

SAFETY HINTS highlight requirements for safety, both while learning, and when applying electronic principles.

TROUBLESHOOTING HINTS offer experienced-based information about circuit problems—what to look for and how to fix them.

CHAPTER CHALLENGE problems appear throughout the text. Using the SIMPLER troubleshooting method, you will be able to guide yourself through each Chapter Challenge circuit and think critically about the problem you need to solve. The SIMPLER sequence and accompanying challenges are explained fully in the next section. Answers are in Appendix C.

Color Insert

One possible approach to solving each Chapter Challenge is provided in step-by-step illustrations in the book's color insert.

The illustrations on the following pages show a sample Chapter Challenge and give instructions on how to complete the challenge. We hope that you and your instructor will have great success with the SIMPLER sequence technique and that these Chapter Challenges will make learning and teaching troubleshooting skills easier and more enjoyable.

Accuracy and Development

Great care has been taken to ensure that this text meets the needs of the instructors and students who have chosen the ever-changing field of electronics. Through an exhaustive review process, we have tailored the book's organization and teaching tools to our demanding audience. At the onset of the process, a detailed national survey of electronics instructors was conducted to determine the best textbook approach to introductory electronics. Our sales representatives conducted in-depth one-on-one interviews to confirm the methods and teaching tools used in this text. More than 65 electronics instructors and professionals have helped shape this book throughout the development process, providing their ideas, performing accuracy checks, and sharing their experiences. Due to the extensive reviewing and accuracy testing used during development, this book is both an efficient and effective teaching and learning tool.

Troubleshooting With the SIMPLER Sequence

To become logical troubleshooters, technicians must develop good critical thinking skills. This book introduces a useful troubleshooting approach called the SIMPLER sequence. This successful troubleshooting method integrates critical thinking all along the way and encourages you to solve problems logically and efficiently. Instructors may use and teach other sequences, and the concepts outlined in the SIMPLER sequence are easily used with a wide variety of approaches.

② Following the instructions, work through the SIMPLER sequence to find the problem with the circuit.

① Study the challenge circuit diagram and the starting point information.

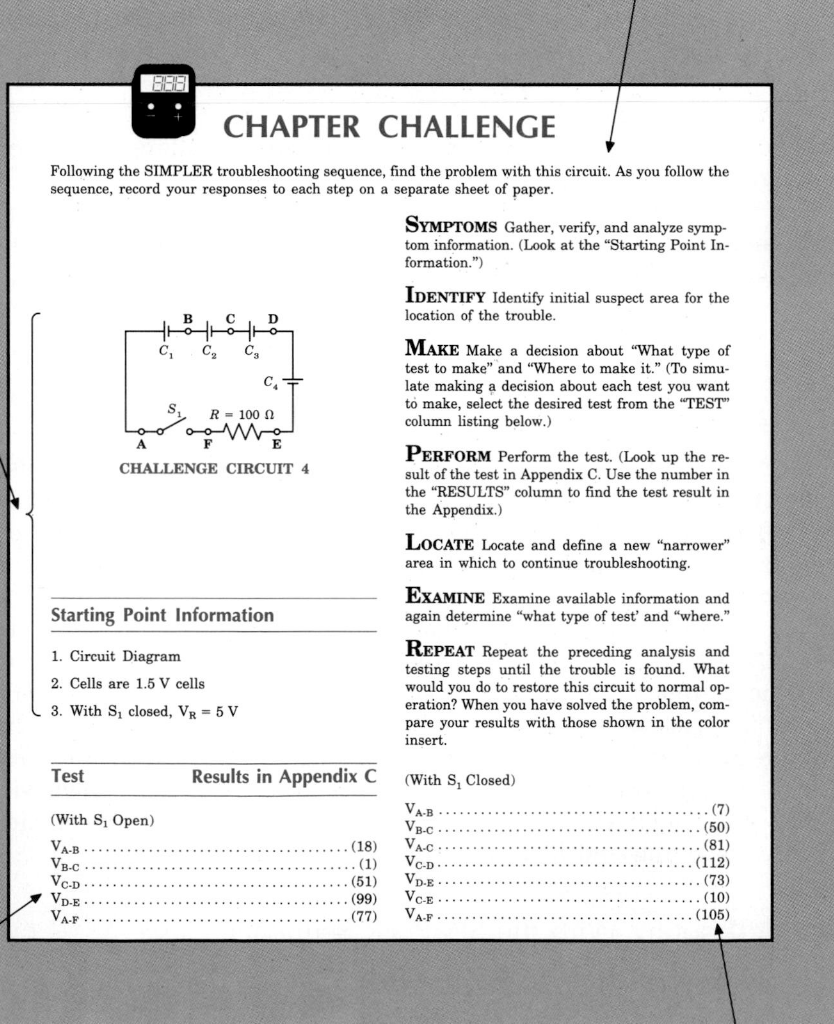

CHAPTER CHALLENGE

Following the SIMPLER troubleshooting sequence, find the problem with this circuit. As you follow the sequence, record your responses to each step on a separate sheet of paper.

CHALLENGE CIRCUIT 4

Starting Point Information

1. Circuit Diagram
2. Cells are 1.5 V cells
3. With S_1 closed, V_R = 5 V

SYMPTOMS Gather, verify, and analyze symptom information. (Look at the "Starting Point Information.")

IDENTIFY Identify initial suspect area for the location of the trouble.

MAKE Make a decision about "What type of test to make" and "Where to make it." (To simulate making a decision about each test you want to make, select the desired test from the "TEST" column listing below.)

PERFORM Perform the test. (Look up the result of the test in Appendix C. Use the number in the "RESULTS" column to find the test result in the Appendix.)

LOCATE Locate and define a new "narrower" area in which to continue troubleshooting.

EXAMINE Examine available information and again determine "what type of test' and "where."

REPEAT Repeat the preceding analysis and testing steps until the trouble is found. What would you do to restore this circuit to normal operation? When you have solved the problem, compare your results with those shown in the color insert.

Test	Results in Appendix C
(With S_1 Open)	
$V_{A\text{-}B}$	(18)
$V_{B\text{-}C}$	(1)
$V_{C\text{-}D}$	(51)
$V_{D\text{-}E}$	(99)
$V_{A\text{-}F}$	(77)
(With S_1 Closed)	
$V_{A\text{-}B}$	(7)
$V_{B\text{-}C}$	(50)
$V_{A\text{-}C}$	(81)
$V_{C\text{-}D}$	(112)
$V_{D\text{-}E}$	(73)
$V_{C\text{-}E}$	(10)
$V_{A\text{-}F}$	(105)

③ As you work through the SIMPLER sequence, choose tests from the list that you think might lead you to the problem.

④ Look at the number across from the test you chose in the Results column.

APPENDIX C

Chapter Challenge Test Results

Find the number listed next to the test you chose and record the result.

1. 1.4 V
2. signal normal
3. 2 kΩ
4. 2 V
5. 1.5 V
6. slightly high
7. 1.5 V
8. 0 V
9. 0 V
10. 3 V
11. 0 V
12. 7 V
13. ∞ Ω
14. 0 Ω
15. 0 V
16. 1 kΩ
17. 400 Ω
18. 1.5 V
19. ?
20. 1.5 V
21. ∼
22. 10 V
23. no signal
24. circuit operates normally
25. 10 kΩ
26. slightly high
27. 0 V
28. 0 V
29. 0 Ω
30. greatly below max value
31. 5 V
32. 400 Ω
33. ∞ Ω
34. 6.4 V
35. noticeably high
36. 4 kΩ
37. 10 kΩ
38. 3.3 V
39. 14 V
40. 0 V
41. ≈1.5 mH
42. 3.0 V
43. 7 V
44. 1.5 V
45. high
46. 0 Ω
47. 0 V
48. 2.3 V
49. 14 V
50. 0.5 V
51. 1.5 V
52. 10 V
53. ∿ (14 VAC)
54. 10 kΩ
55. max value
56. 10 V
57. signal normal
58. ∞ Ω
59. 1.5 kΩ
60. 275 pF
61. 14 V
62. 1 kΩ
63. ∞ Ω
64. 1.5 V
65. ≈6 VDC
66. 0 V
67. 7.5 V
68. 10 kΩ
69. slightly low
70. signal normal
71. no change in operation
72. 0 Ω
73. 1.5 V
74. ∿ (120 VAC)
75. 10 V
76. slightly low
77. 5.9 V
78. 3.1 V
79. 7.5 V
80. 7 V
81. 2 V
82. 0 V
83. 10 V
84. 0 Ω
85. slightly below max value
86. 1.5 V
87. 5 Ω
88. 14 V
89. 2.3 V
90. ∞ Ω
91. 0 V
92. normal
93. 10 kΩ
94. 5 V
95. 10 V
96. ≈11 V
97. 0 V
98. slightly low
99. 1.5 V
100. ∞ Ω
101. 0 V
102. 0 V
103. 0 V
104. ≈5 kΩ
105. 0 V
106. no signal
107. 50 pF
108. ∞ Ω
109. 12 kΩ
110. 0 V
111. no change in operation
112. 1.5 V
113. signal normal
114. 297 Ω
115. 0 V
116. ≈5 V
117. within normal range

⑤ Turn to Appendix C and find that number from the results column. Next to the number in the Appendix you will find the result this test would yield. As you locate the number of the test you chose, record the result listed next to it on a sheet of paper.

⑥ When you have completed the challenge and pinpointed the circuits problem, check your work against the data in the color insert. The insert offers one possible sequence of steps that might have been used to get to the solution, along with step-by-step color photographs.

CIRCUIT 5

SYMPTOMS: Information. When the ohmmeter test probes are touched to zero the ohmmeter, there is not meter response. This indicates a bad meter, a lack of circuit continuity somewhere, or a dead battery.

IDENTIFY initial suspect area: The total circuit, since a discontinuity can be caused anywhere in the circuit.

MAKE test decision: Since the battery is the only item in the circuit having a definite shelf life, let's first check the battery. Measure V between points A and B with the probes open.

PERFORM 1st Test: The result of measuring $V_{A\text{-}B}$ is 1.5 V, which is normal. NOTE: With the test probes open, there is no load current drawn from the battery, thus it might measure normal without load and not completely good when under load. However, this tells us the battery is probably not totally dead, which it would have to be to match our symptom information.

LOCATE new suspect area: The remainder of the circuit, since discontinuity can be anywhere along the current path.

EXAMINE available data.

REPEAT analysis and testing: By checking continuity between points C and E (with test leads not touching), we can check the continuity of the two circuit elements, R_1 and the test-lead/test-probe.

CHAPTER 9 CHALLENGE

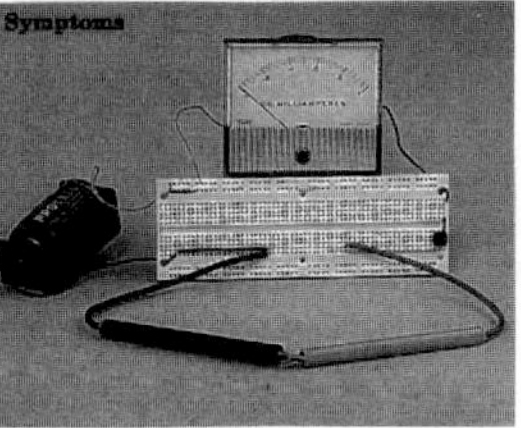

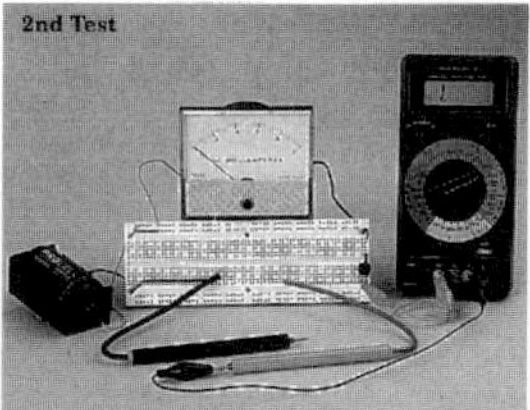

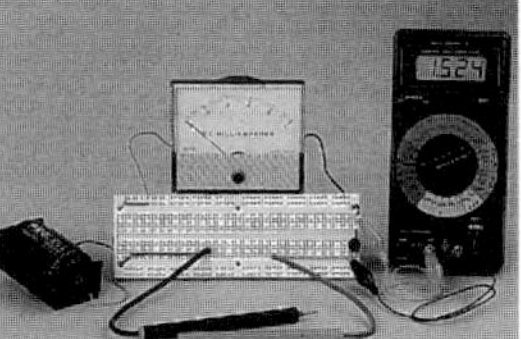

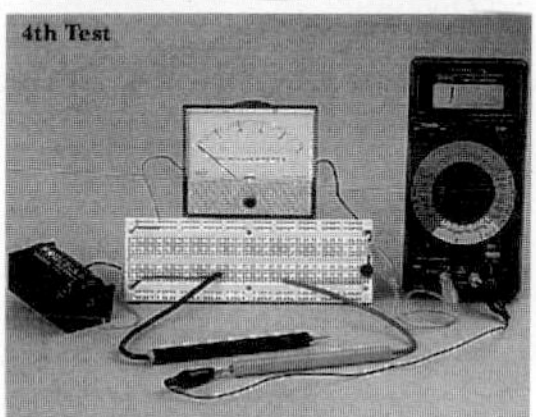

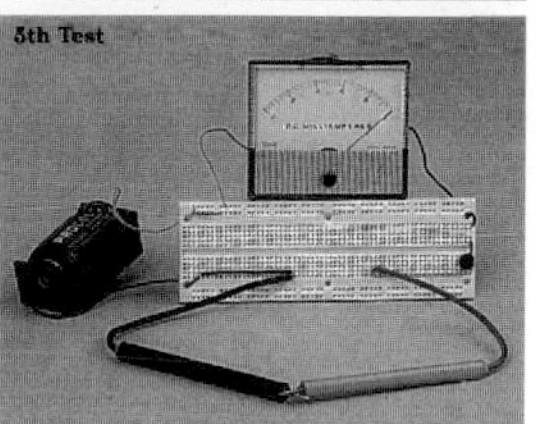

2nd Test: Check $R_{C\text{-}E}$. R is infinite ohms, indicating a problem. (NOTE: This meter shows "1" when measuring infinite ohms.)

3rd Test: Check $R_{C\text{-}D}$ to verify that R_1's resistive element or wiper arm is not causing the discontinuity. R is about 1500 ohms. However, moving the wiper arm causes R to change, which implies everything is normal.

4th Test: Check $R_{D\text{-}E}$. R is infinite ohms, indicating the discontinuity is between these points. A further check shows the test-lead wire is broken inside the test-probe plastic handle.

5th Test: Repair the wire connection to the test probe. Then, try to zero the ohmmeter. Now it works.

Chapter Challenge. As you move through this book, you will encounter a series of troubleshooting problems called Chapter Challenge. Each Chapter Challenge includes a challenge circuit schematic and starting point symptoms information. Using the SIMPLER sequence, you will step through and solve the simulated troubleshooting problem.

Selecting Tests and Finding the Results. To simulate making a parameter test on a particular portion of the challenge circuit, you may select a test from the test listing provided with the Chapter Challenge. You may then look up the results of the test you selected by looking in Appendix C. Look in the appendix for the identifier number assigned to the test on the Chapter Challenge page. Next to that number, you will see the parameter value or condition that would be present if you make that same test on the actual circuit simulated by the Chapter challenge.

The Learning Package

Foundations of Electronics has a well-correlated lab manual that can multiply the effectiveness of the learning process. The manual has 71 easy-to-use projects that provide hands-on experience with the concepts and principles you study in the theory text.

The format of the manual ensures that you will not miss the forest through the trees. You will not perform a whole series of cook-book steps, or the whole project, only to find out at the end you did not learn what you should have in performing the project. Instead, you can check your results each step of the way.

The projects are arranged as stand-alone, single-concept projects; however, they are grouped logically to allow maximum instruction flexibility. Each project can be performed individually or several related projects can be performed during one session.

Again, we wish you success as you have opportunity to use the two books as partners in your training. A Student Study Guide is available to help reinforce the concepts in the text along with a student computer software work disk.

For the instructor, transparency masters, an instructor's guide for the text and lab manual, a testbank, and a computerized testbank are available from Delmar Publishers Inc.

Scheduling Instructions

Ideally, your program can use all the material provided in the text and the correlated laboratory manual to maximize learning and practice opportunities.

The following suggestions are provided as possible scenarios for those whose time is more limited:

- For a one-year (three-quarter or two-semester) program; Chapters 1–21 may generally be covered in their entirety.
- Time may be saved by deleting or minimizing reading assignments in the section of Chapter 2 covering powers of ten if you already have covered this topic elsewhere.
- Certain portions of the Network Theorems in Chapter 6 may also receive light treatment, as appropriate to your course.
- Also, some programs prefer to cover cells and batteries in briefer fashion as simply a lecture, which does not include all the data in Chapter 7.
- Additionally, not all programs require the exhaustive treatment on analog meters provided in Chapter 9. For example, the section on the Ayrton (Universal) shunt might be deleted or minimized in assignments to save time.
- If vector algebra and manipulation of complex numbers are not required in your program, the last half of Chapter 19 may be omitted.
- Further customizing of the program to your program's needs may be done by appropriate selection of the specific laboratory experiments provided in the *Foundations of Electronics* laboratory manual.
- Chapters 22 through 25 provide introductions to transistor and integrated circuit (IC) devices and select circuit applications. As time permits, your program may use these chapters to introduce these topics.
- For programs that must cover the dc and ac fundamentals in a two-quarter timespan, use the "trimming" ideas listed.
- Additionally, limit assignments and discussions in Magnetism and Electromagnetism (Chapter 8) to the "fundamental principles" portions, deleting the discussion of the many SI magnetic units and presentation of "motor" and "generator" affects.

ACKNOWLEDGEMENTS

Many individuals have helped in shaping this textbook package. We wish to thank the reviewers for their time and valuable insights.

Phillip D. Anderson
Muskegon Community College

Robert L. Arndt
Belleville Area College

Tom Birum
Lima Senior High School

William Blanton
Delta Quachita Vocational
Technical Institute

Stanley N. Brown
Vincennes University

Phillip J. Chiarelli
Electronics Institute

John Ephraim
East Tennessee State University

Louis Feldman
ENY High School

Bill Frazier
Gwinnett Technical School

Arnie Gugarty
Wentworth Technical Institute

A.E. Hall
Central Texas College

Thomas C. Harrison
Pickens Technical Institute

John J. Hatch
ITT Technical Institute

Fred Kiehle
Georgia Department of Technical and
Adult Education

William Earl Armistead, Jr.
ITT Technical Institute

Issa Batarseh
Olive-Harvey College

Leelan C. Blackmon
Wallace State Community College

Teresa G. Bowen
Waycross-Ware Technical Center

William A. Campbell
Bay Area Vocational Technical
Center

Derek Elliott
ITT Technical Institute

Robert Evans
Anoka Technical College

Elwood Fennimore
Spokane Community College

Victor Gibson
Interactive Learning Systems

Walter Ray Gunter
Shreveport—Rossier Regional
Technical Institute

Jill A. Harlamert
DeVry Institute of Technology

Floyd Hastings
Francis Tuttle Area Vocational-
Technical Center

Charles Hollins
Cerritos College

Stanley W. Lawrence
Salt Lake Community College

C.J. Lemmon
Renton Vocational Technical Institute

Al Mikula
Johnson and Wales University

John L. Morgan
DeVry Institute of Technology

Dan Nemanich
ITT Technical Institute

Dennis M. Patton
Washington—Holmes Area Vocational Technical Center

Cheryl M. Reed
ECPI Computer Institute

Bill Robertson
ITT Technical Institute

Frank Shannon
Coleman College South Bay

Charles F. Solomon
Texas State Technical Institute

Bill White
Georgia Department of Technical and Adult Education

George S. McInturff
L.H. Bates Vocational Technical Institute

Donald J. Montgomery
ITT Technical Institute

Rodney F. Oakley
Hennepin Technical College

Roy L. Ray
Isothermal Community College

Alan Reinmiller
Southeast Community College

Bruce Sargent
Middlesex Community College

Arlyn L. Smith
Alfred State College

David E. Tester
ITT Technical Institute

Richard Thomson
Northwestern Electronic's Institute

Glenn B. Wilkins
Hillwood Comprehensive High School

We are particularly grateful to Fred Kerr and the students at DeVry-Atlanta who offered tremendous help in producing the color insert:

Paul Zecchino, James Grimmett, Mike LeCoq, Dale Zenga, Rob Crane, Anthony Rivers, and Rick Sherman.

Ronald J. Coler and Clifford Scott Miller of LMC-Cadex, Inc. proved to be very helpful in generating ideas and serving as a professional resource. Thank you.

We also wish to thank ACK Radio supply of Atlanta, Georgia, for its generous loan of equipment.

Finally, we want to recognize all who responded to our survey and the one-on-one interviews. You no doubt will see your influence as you read through the pages of *Foundations of Electronics.*

To the staff at Delmar who put in long hours and lived out their commitment to excellence—thank you! This book would not exist as it is today were it not for the professional expertise of Wendy Jones, Mary Ormsbee, Susan Simpfenderfer, Susan Mathews, Larry Main, and Mike Nelson.

PART I

FOUNDATIONAL CONCEPTS

Basic Concepts of Electricity

Key Terms

Ampere
Atom
Circuit
Compound
Conductor
Coulomb
Current
Electrical charge
Electron
Element
Energy
Force
Free electrons
Insulator
Ion
Load
Matter
Molecule
Neutron
Ohm
Polarity
Potential
Proton
Resistance
Semiconductor
Source
Valence electrons
Volt

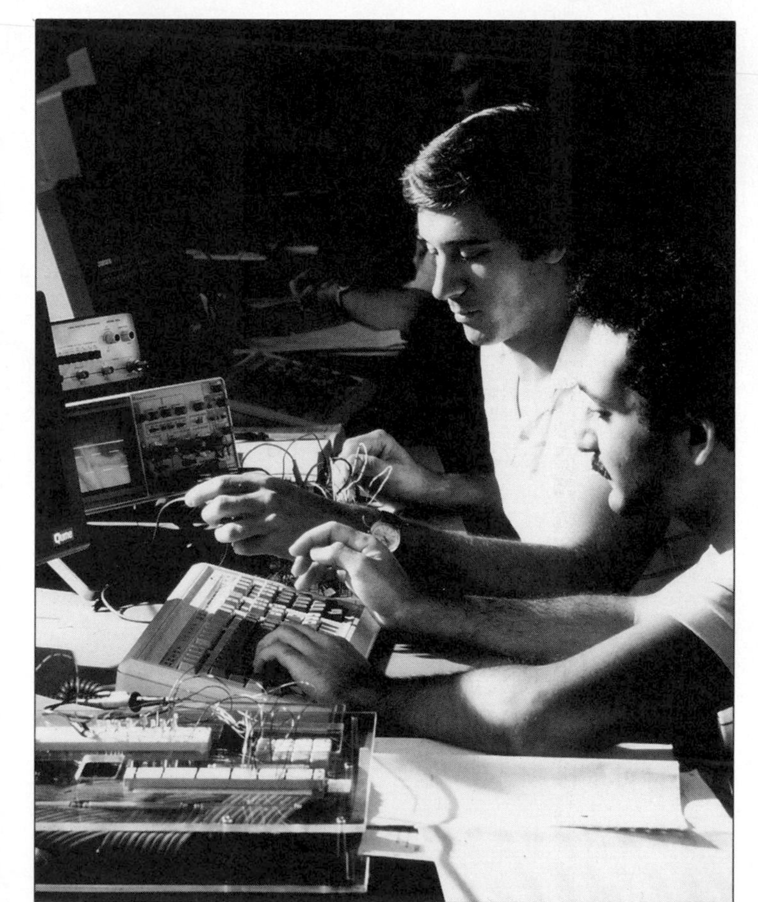

Courtesy of GE Fanuc Automation

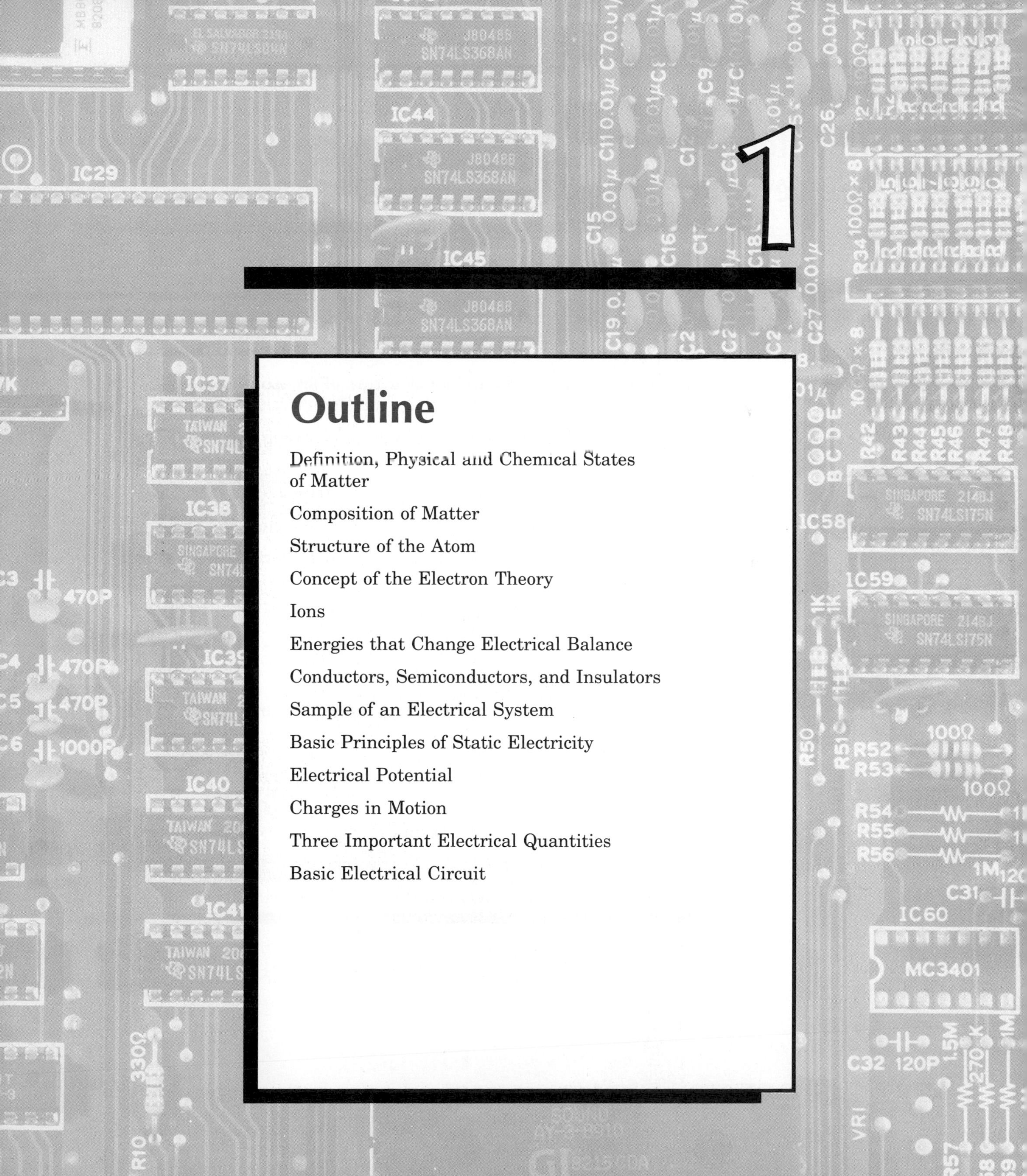

1

Outline

Definition, Physical and Chemical States of Matter

Composition of Matter

Structure of the Atom

Concept of the Electron Theory

Ions

Energies that Change Electrical Balance

Conductors, Semiconductors, and Insulators

Sample of an Electrical System

Basic Principles of Static Electricity

Electrical Potential

Charges in Motion

Three Important Electrical Quantities

Basic Electrical Circuit

Chapter Preview

This chapter will introduce you to concepts and theories proven useful as "thinking tools" when studying electricity and electronics. It also provides the understanding you will need as a technician when analyzing, installing, designing, and troubleshooting electronic circuitry.

Objectives

After studying this chapter, you will be able to:

- Define the term **matter** and list its physical and chemical states
- Describe the difference between **elements** and **compounds**
- Discuss the characteristics and structure of an **atom, molecule,** and **ion**
- Define the electrical characteristics of an **electron, proton,** and **neutron**
- Explain **valence electrons** and **free electrons**
- List the methods used to create electrical imbalances
- Describe the characteristics of **conductors, semiconductors,** and **insulators**
- State the law of electrical charges
- Discuss the terms **polarity** and **reference points**
- Define charge and its unit of measure **coulomb**
- Define **potential** (emf) and give its unit of measure
- Define **current** and explain its unit of measure
- Calculate current when magnitude and rate of charge motion is known
- Define **resistance** and give its unit of measure
- List the typical elements of an electrical circuit
- Describe the difference between closed and open circuits

DEFINITION, PHYSICAL AND CHEMICAL STATES OF MATTER

Definition of Matter

Everything that we see, touch, or smell represents some form of matter. In Figure 1–1 the hills, lake, trees, and air represent **matter** in various forms. Although many definitions have been used, **matter** may be defined as anything that has weight and occupies space. We can also say that matter is what all things are made of and what our senses can perceive.

The basic building block of all matter is the atom. Later in this chapter, atoms will be thoroughly discussed because of their importance to the understanding of electronics.

Physical States of Matter

Matter exists in three *physical* states. Matter is either a **solid** (for example, the chair you are sitting on or the bridge and trees in Figure 1–1), a **liquid** (water), or a **gas,** such as oxygen or hydrogen.

FIGURE 1–1 A scene full of matter (Photo by Ruby Gold)

When matter exists in the liquid or gaseous state, its dimensions are determined by the container.

Chemical States of Matter

The chemical states of matter are **elements, compounds,** and **mixtures.** An element, Figure 1–2, is a substance that cannot be chemically broken into simpler substances. In fact, an element has only one kind of atom. Examples of chemical elements are gold, iron, copper, silicon, oxygen, and hydrogen.

FIGURE 1–2 Some familiar metallic elements

A compound is formed by a chemical combination of two or more elements. In other words, compounds are built from two or more atoms that combine into molecules. Also, a compound has a definite structure and characteristic (i.e., same weight and atomic structure). Figure 1–3 shows two examples of compounds. Examples of compounds are water, which is the chemical union of hydrogen and oxygen, and sugar, which contains carbon (a black, tasteless solid) plus hydrogen and oxygen (two gases).

Mixtures are a combination of substances where the individual elements possess the same properties as when they are alone. There is no chemical change resulting from the combination as there is with compounds. In Fig-

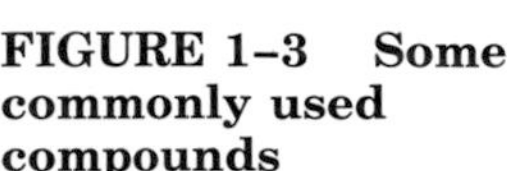

FIGURE 1–3 Some commonly used compounds

ure 1–4, mixing gold dust and sand would not yield a new, chemically different entity or compound. The substances would simply be mixed or combined.

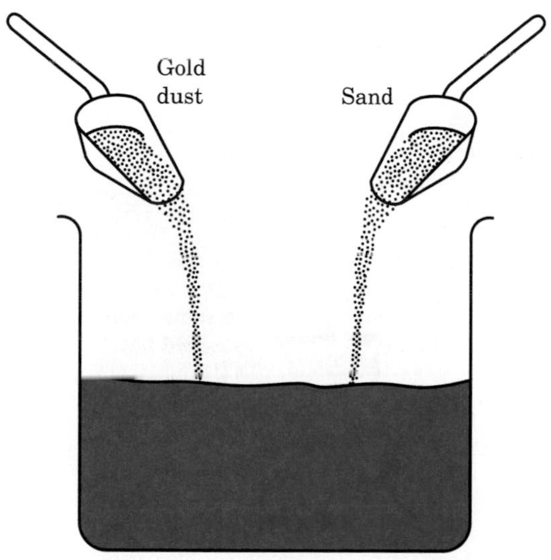

FIGURE 1–4 Mixtures are a combination of substances where the individual elements possess the same properties as when they are alone. A mixture is different than a compound.

COMPOSITION OF MATTER

The building blocks of all matter are atoms. Atoms combine chemically to form molecules, and the new compound (matter) created is different than each element that went into the compound's makeup, Figure 1–5. Also, a specific substance's molecules are different than the molecules of other types of matter.

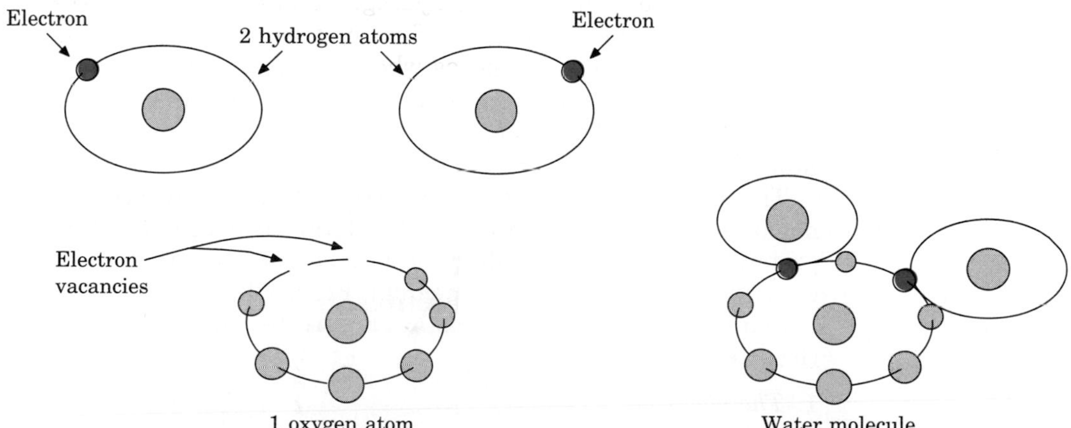

FIGURE 1–5 A molecule of water is formed from oxygen and hydrogen atoms.

In summary, *all matter is composed of atoms and molecules.* The smallest particle into which a compound can be divided but retain its physical properties is the molecule. The smallest particle into which an element can be divided but retain its physical properties is the atom, Figure 1–6.

FIGURE 1–6 The smallest particle of a compound that has the same characteristics as the compound is a molecule. The smallest particle of an element that will show the same characteristics as the element is an atom.

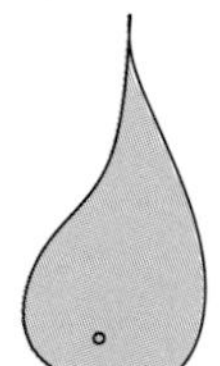

A *molecule* of water will exhibit the same physical characteristics as a drop of water.

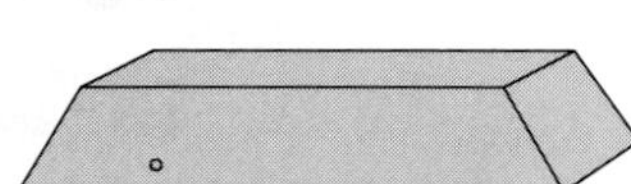

An *atom* of gold will show the same chemical properties as a gold bar.

STRUCTURE OF THE ATOM

The Particles

The particles of the atom that interest electronics students are the **electron, proton,** and **neutron.** Although science has identified other particles (mesons, positrons, neutrinos), studying these particles is not necessary to understand the electron theory.

The Model

A Danish scientist, Niels Bohr, developed a model of atomic structure that explains the electron theory. In his model the atom consists of a nucleus in the center of the atom with electrons orbiting the nucleus. The nucleus has two particles called protons that are electrically positive in charge and neutrons that are neutral in electrical charge. The orbiting electrons are negatively charged particles, Figure 1–7. The common analogy for this model is our solar system: Planets orbit the sun like electrons orbit the nucleus of the atom.

The net charge of the atom is neutral because the orbiting electrons' total negative charge strength equals the total positive charge strength of the protons in the nucleus. Also, the number of protons and electrons in an electrically balanced atom are equal, Figure 1–8.

Some interesting characteristics, Figure 1–9, of these three atomic particles are:

1. *The electron:*
   - has a negative electrical charge
   - has small mass or weight (9×10^{-28} grams)

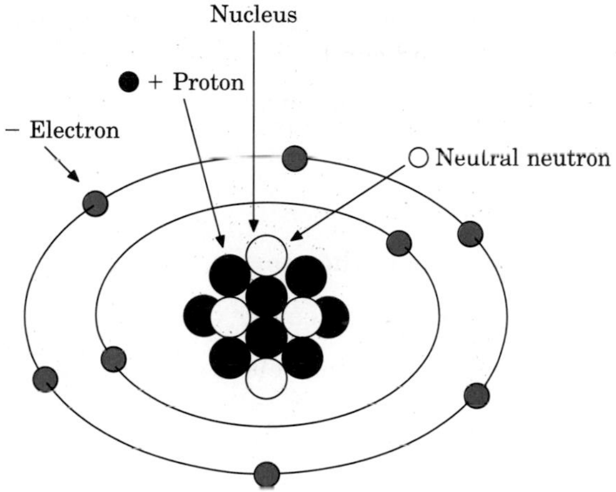

FIGURE 1–7 Bohr's model of the atom. The nucleus (center) contains protons and neutrons. Electrons orbit the nucleus.

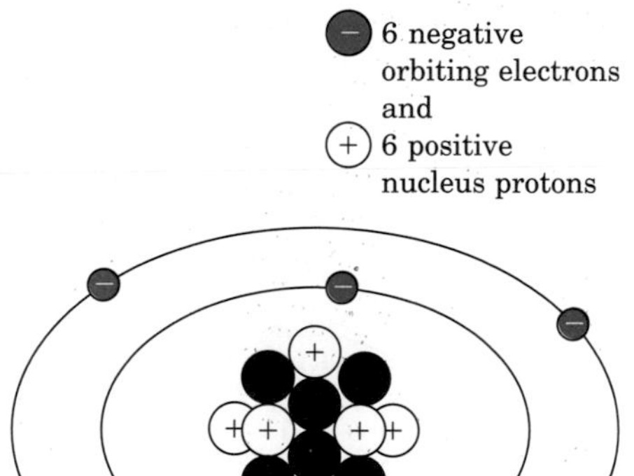

FIGURE 1–8 Typically atoms are electrically balanced.

ELECTRONS	PROTONS	NEUTRONS
Negative charge	Positive charge	No charge
Small mass / weight (about 9×10^{-28} grams)	1,836 times heavier than electron	Similar in mass / weight to proton
Travel in orbits outside of nucleus	Located in the nucleus	Located in the nucleus
Rates of speed (trillions of orbits per sec.)	Equal in number to atom's electrons	

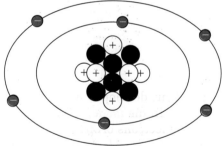

FIGURE 1–9 Some facts about atomic particles

- travels around the nucleus at unbelievable speed (trillions of times a second)
- helps determine the atom's chemical characteristics

2. *The proton:*
   - has a positive electrical charge
   - is located in the nucleus of the atom
   - is approximately 1,800 times heavier than an electron
   - is equal in number to the atom's electrons
3. *The neutron:*
   - has no electrical charge
   - is located in the nucleus of the atom
   - is about the same mass or weight as a proton
   - may vary in number for a given element to form different "isotopes" of the same element; for example, hydrogen has three isotopes: protium, deuterium, and tritium, Figure 1–10

FIGURE 1–10
What atomic difference is there between these isotopes?

Deuterium
(heavy hydrogen)
nucleus =
1 proton
1 neutron

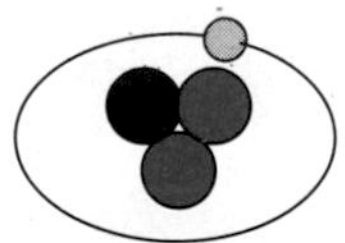

Tritium
(heavy, heavy hydrogen)
nucleus =
1 proton
2 neutrons

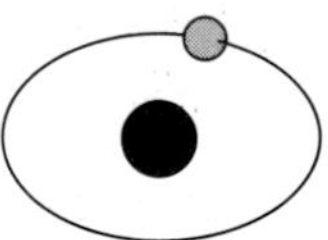

Protium
(light hydrogen)
nucleus =
1 proton

Atomic Number and Weight

Although it is not necessary to thoroughly understand atomic numbers and atomic weights, it is appropriate to mention them.

The atomic number of an element is determined by the number of protons in each of its atoms. For example, the atomic number for copper is 29 and the atomic number for carbon is 6, Figure 1–11.

The atomic weight of an element is determined by comparing the weight of its atoms to an atom of carbon-12. For example, hydrogen, the simplest atom, has an atomic weight of about 1.007 "atomic mass units," and copper's atomic weight is 63.54.

IN-PROCESS LEARNING CHECK I

It will be a good idea at this point to initiate you into a technique that will be used from time to time throughout the book: the "In-Process Learning Check." Rather than wait until the end of the chapter to find out whether you're missing any of the key points, these special "checkups" will provide you with a chance to assure that you're learning the key points. Fill in the blanks for the following statements. If you have trouble with any of them, simply move back to that topic and refresh your memory. Answers are in Appendix B.

1. Matter is anything that has ______ and occupies ______.
2. Three physical states of matter are ______, ______, and ______.
3. Three chemical states of matter are ______, ______, and ______.
4. The smallest particle that a compound can be divided but retain its physical properties is the ______.
5. The smallest particle that an element can be divided into but retain its physical properties is the ______.
6. The three parts of an atom, which interest electronics students, are the ______, ______, and ______.
7. The atomic particle having a negative charge is the ______.
8. The atomic particle having a positive charge is the ______.
9. The atomic particle having a neutral charge is the ______.
10. The ______ and ______ are found in the atom's nucleus.
11. The particle that orbits the nucleus of the atom is the ______.

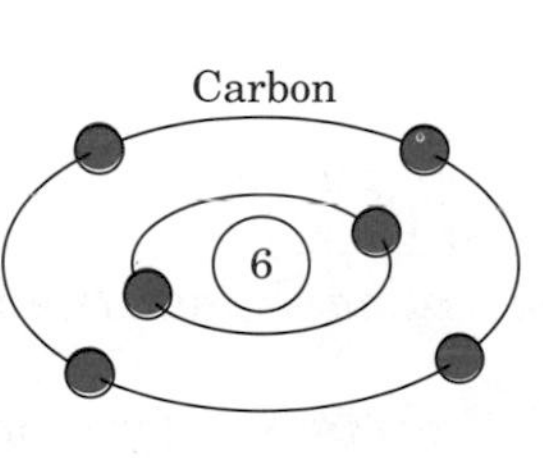

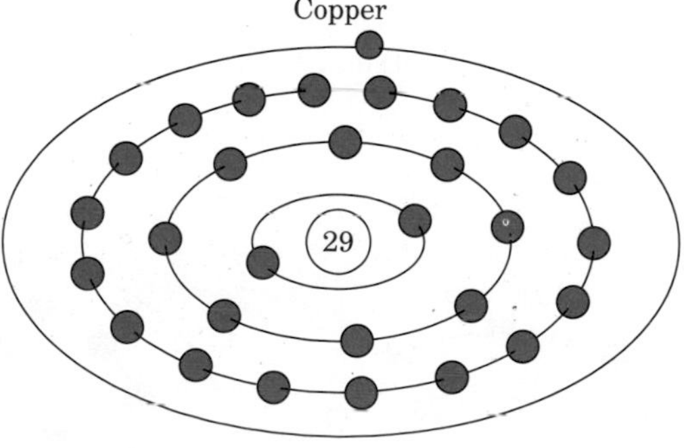

FIGURE 1–11 Carbon and copper atoms

Concept of Atomic Shells

As you have already seen, all electrons traveling around the nucleus of the atom do not travel in the same path or at the same distance from the nucleus. Electrons align themselves in a structured manner, Figure 1–12.

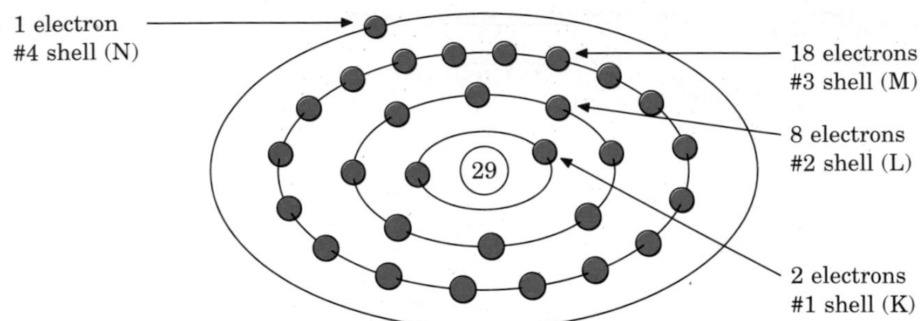

FIGURE 1–12 Electrons align at different distances from the nucleus.

Each ring, or shell, of orbiting electrons has a maximum number of electrons that can locate themselves within that shell if the atom is stable. The formula to use is $2n^2$, where n equals the number of the shell. For example, the innermost shell (closest to the nucleus) may have a maximum of 2 electrons; the second shell, 8 electrons; the third shell, 18 electrons; and the fourth shell, a maximum of 32 electrons. The outermost shell, whichever number shell that is for the given atom, can never contain more than 8.

CONCEPT OF ELECTRON THEORY

Purpose

The electron theory helps visualize atoms and electrons as they relate to electrical and electronic phenomena, and the previous descriptions of atomic structure are part of this electron theory. The next discussion of valence and free electrons illustrates some practical aspects of the electron theory.

Valence Electrons

Valence electrons are those electrons in the outermost shell of the atom. The number of valence electrons in the atom determines its stability or instability, both electrically and chemically, Figure 1–13.

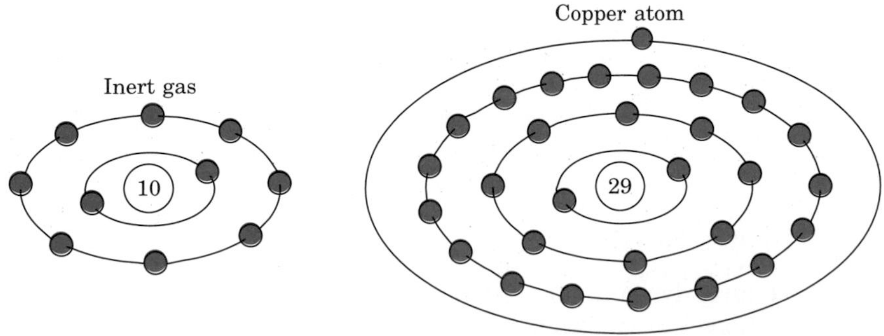

FIGURE 1–13 The number of valence electrons affects the chemical and/or electrical stability of an atom.

For all atoms, this outermost shell is *full* when it has *8 electrons.* If there are 8 outermost ring electrons, then the material is stable and does not easily combine chemically with other atoms to form molecules. Also, these electrons are not easy to move from the atom. Examples of stable atoms are inert gases, such as neon and argon.

If the material's atoms have fewer than 8 outermost ring electrons, the material is chemically and electrically active. Often the material will chemically combine with other atoms to gain stability and to form molecules and/or atomic bonds. Electrically, these valence electrons can be easily moved from their original home atom and are sometimes referred to as free electrons. Examples of materials are copper, gold, and silver.

Also, there are materials that have 4 outermost ring electrons. They are halfway between being stable and very unstable. Germanium and silicon are two examples. These materials are used in most of today's solid-state, "semiconductor" devices, such as transistors and integrated circuits.

In summary, the concepts of the electron theory are based on atomic structure and help explain the various electrical phenomena discussed in this book.

IONS

When an electron leaves its original home atom because of chemical, light, heat, or other types of energy, it leaves behind an atom that is no longer electrically neutral. An **ion** is any atom that is not electrically balanced (or neutral) and that has gained or lost electrons. A positive ion is an atom having fewer electrons than protons (a deficiency of electrons). A negative ion is an atom having more electrons than protons (an excess of electrons), Figure 1–14.

When an electron is torn from a neutral atom, leaving a positive ion, or when an electron is added to a neutral atom, leaving a negative ion, the process is called ionization. Later in your study of electronics, you will see how this process can be useful in various electronic devices.

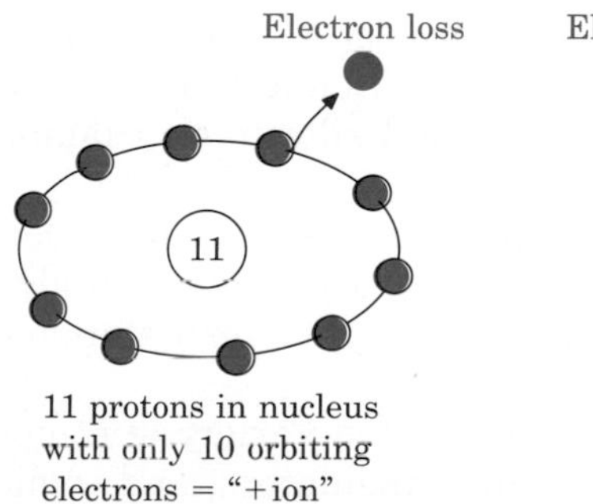

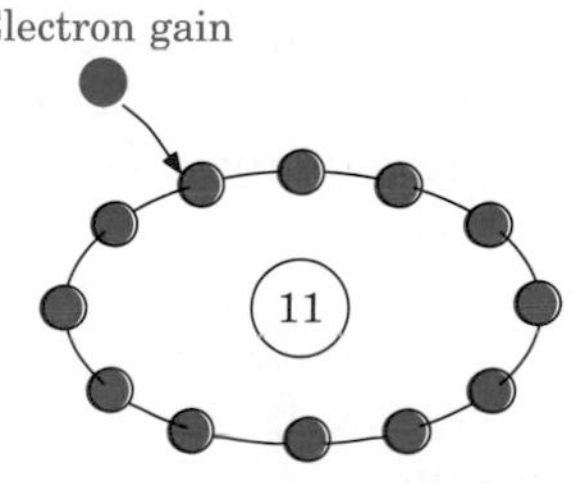

FIGURE 1–14 Ions are atoms that have lost or gained electrons.

ENERGIES THAT CHANGE ELECTRICAL BALANCE

Why should anyone want to change the electrical balance of atoms, or control electron movement? As your study of electronics continues, you will see that the ability to control the movement of electrons, or electron flow, is the basis of electronics.

Figure 1–15 displays some common sources of energy causing electron movement and/or separation of charges. These sources are:

- Friction (static electricity)
- Chemical energy (batteries)
- Mechanical energy (a generator or alternator)
- Light energy
- Heat energy

By appropriately using one or more of these energies, sources of electrical energy may be developed. These sources develop electrical potential, or the potential to move electrons through whatever circuits are electrically connected to these sources. Electrical **circuits** are closed paths designed to carry, manipulate, or control electron flow for some purpose. Later in this chapter, electrical potential and movement of electrons are discussed to show how the concepts presented thus far are foundations for the study of electronics.

CONDUCTORS, SEMICONDUCTORS, AND INSULATORS

Conductors have many free electrons, such as copper, gold, and silver. These materials conduct electron movement easily because their outermost ring electrons are loosely bound to the nucleus. In other words, their outermost ring contains 1, 2, or 3 electrons rather than 8 electrons needed for atomic stability, Figure 1–16a.

Semiconductors, Figure 1–16b, are materials that are halfway between the conductors' characteristic of few outermost shell electrons and the stable, inert materials with 8 valence electrons. Semiconductor materials have 4 outermost ring electrons. Germanium and silicon are examples of semiconductor materials.

Insulator materials, Figure 1–16c, do not easily allow electron movement because their outermost shell electrons are tightly bound to the atom. Insulators have few free electrons. Examples of insulator materials are glass and ceramic.

Electrical and electronic circuits comprise a variety of components and interconnections consisting of conductors, insulators, and semiconductors.

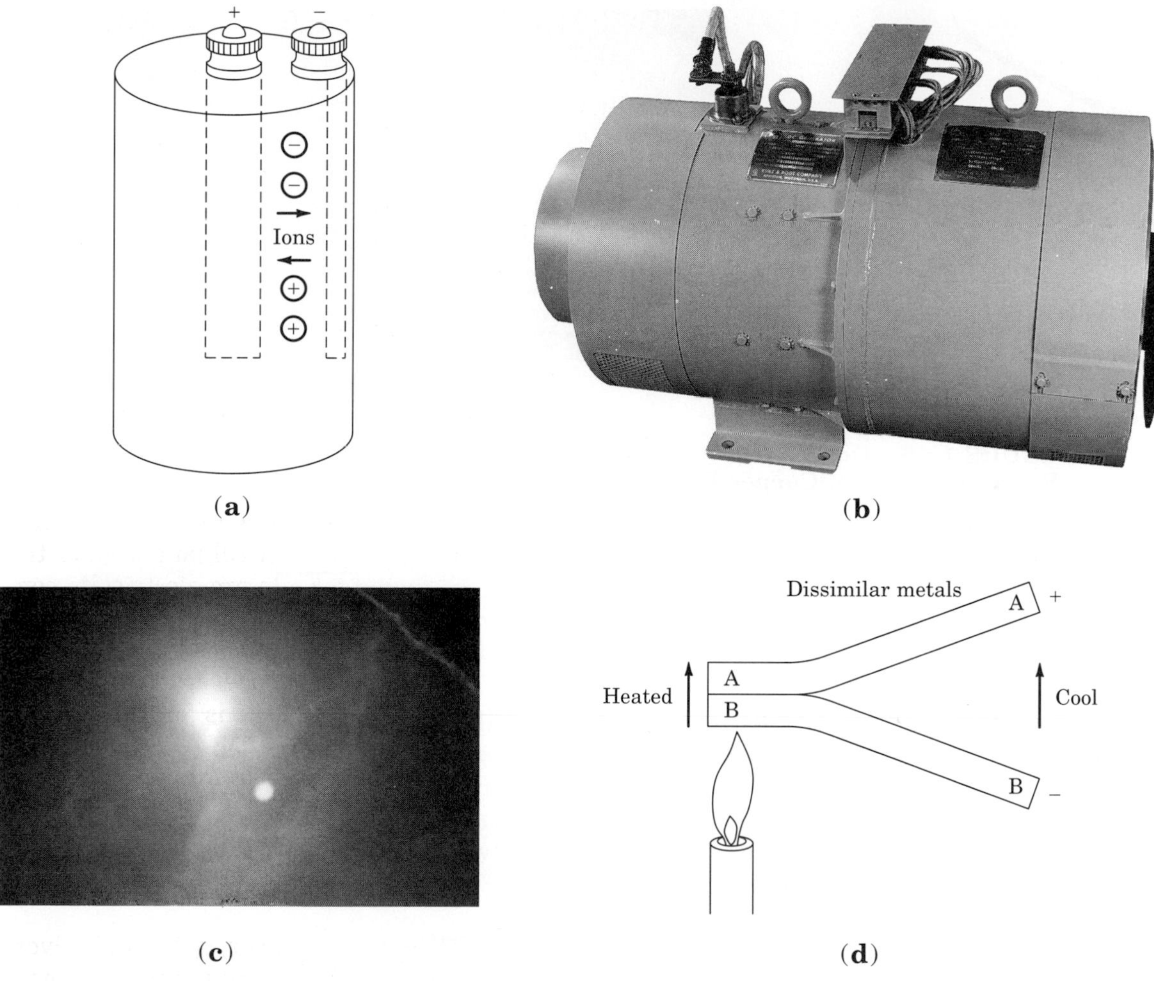

FIGURE 1–15 Some energy sources that create useful electrical energy
a. A battery uses the chemical energy in its cell to create voltage.
b. A generator converts mechanical energy into electricity. (Courtesy of Kurz & Root Company)
c. The sun is a source of light energy.
d. A thermocouple uses heat energy to produce an electric current.

SAMPLE OF AN ELECTRICAL SYSTEM

An electrical system typically has a source of electrical energy, a way to transport that electrical energy from one point to another, and an electrical load, Figure 1–17.

The source supplies energy through the transporting means to the load.

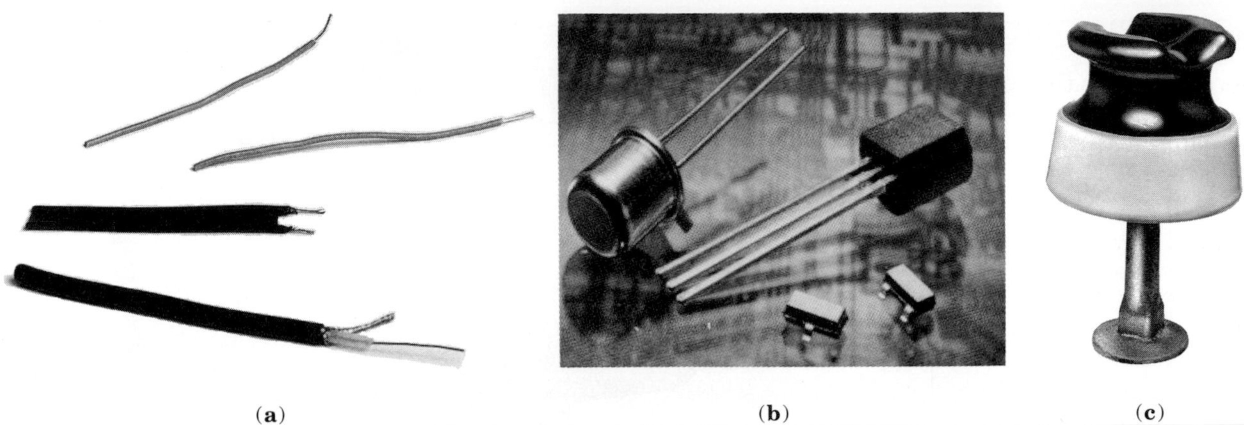

FIGURE 1–16 Conductors (a), semiconductors (b), and insulator (c) (Photos courtesy of Cooper Power Systems)

FIGURE 1–17 Parts of a basic electrical system

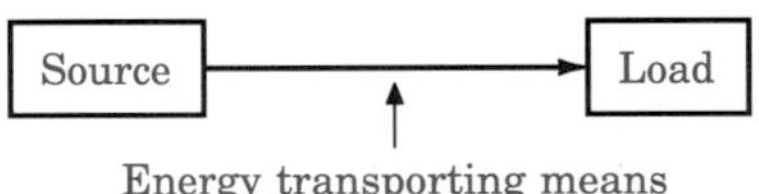

The load may convert that electrical energy into another form of electrical energy or into another type of energy, such as heat, light, or motion.

Figure 1–18 illustrates an electrical lighting circuit. Observe that the method to transport the electrical energy to the light bulb (the load) is the conductor wires. The load then converts the electrical energy into another useful form, light.

FIGURE 1–18 Components of a commonly used electrical system

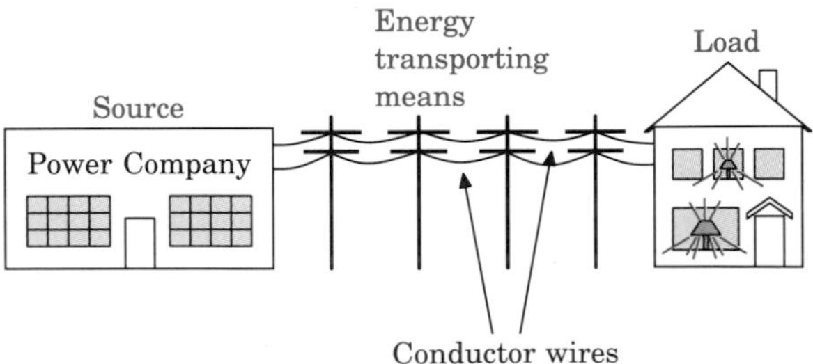

The rest of this chapter will discuss basic electrical laws and principles to help you in understanding what is happening in this basic light circuit. You should seek answers to the following questions: How does the source cause electrical energy or electricity to move through the conductors to the load? In what manner and/or form does the electricity move from one point to another? What determines the quantity of electricity moved from the source to the load?

BASIC PRINCIPLES OF STATIC ELECTRICITY

Before studying the movement of electrons in electrical circuits, it will be helpful to learn some basic principles about charges and static electricity as a frame of reference.

What Is Static Electricity?

Everyone has been electrically shocked after walking on a thick pile rug and touching a door knob or after sliding across an automobile seat and touching the door handle. Where did that electricity come from? Sometimes this electricity is called static electricity. This term is somewhat misleading because static implies stationary and electrons are in constant motion around the nucleus of the atoms. Think of this type of electricity as being associated with nonconductors or insulator materials.

The Basic Law of Electrical Charges

In many science courses, the experiment of rubbing a rubber rod with fur or a glass rod with silk to develop charges is frequently performed, Figure 1–19. After doing this experiment, paper or other light materials are sometimes attracted to the charged body or object. What is meant by a charged body? This means that the object has more or less than its normal number of electrons. In the case of the rubber rod, it gained some electrons from the fur, and the fur lost those electrons to the rod. Thus, the net charge of the rubber rod is now negative, since it has an excess of electrons. Conversely, the fur is positively charged since it lost some of its electrons but retains the same number of protons in its nuclei.

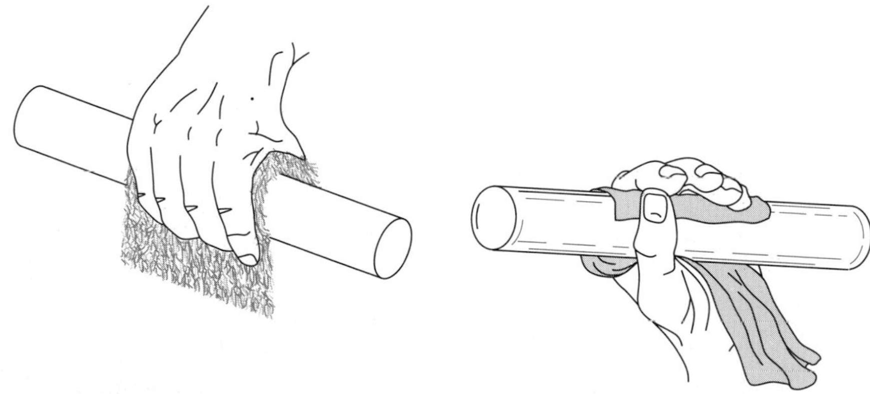

FIGURE 1–19 Static electricity is usually associated with nonconductor materials.

With the silk and glass rod experiment, the glass rod loses electrons and becomes positively charged. Furthermore, another experiment would show there is some attraction between the charged rubber rod (negative) and the charged glass rod (positive). This leads to a conclusion that has been accepted as the basic electrical law: Unlike charges attract each other, and like charges repel each other. Remember this important electrical law, Figure 1–20.

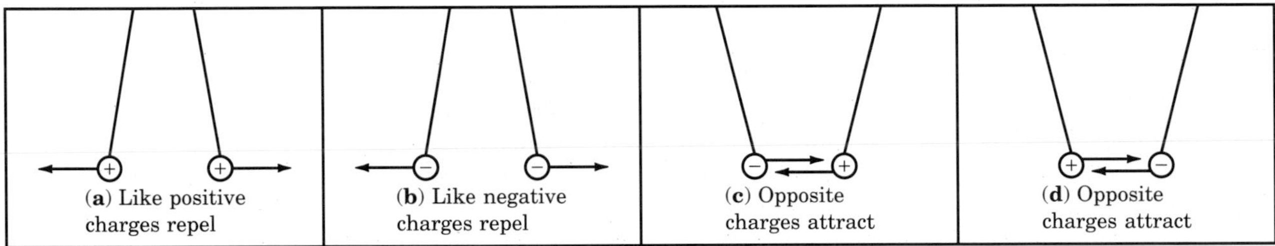
(a) Like positive charges repel
(b) Like negative charges repel
(c) Opposite charges attract
(d) Opposite charges attract

FIGURE 1–20 Basic law of charges is that like charges repel and unlike charges attract.

Polarity and Reference Points

You might have noticed that a common way of showing the difference between charges is by identifying them with a minus sign for negative or a plus sign for positive. Using positive (+) or negative (−) signs frequently indicates electrical **polarity.** Polarity denotes the relative electrical charge of one point in an electrical circuit "with reference to" another point, Figure 1–21. The notion of something "with reference to" something else is not really mysterious. Everyone has heard statements such as "John is taller

FIGURE 1–21 In a flashlight cell, point A is positive with respect to point B, and point B is negative with respect to point A.

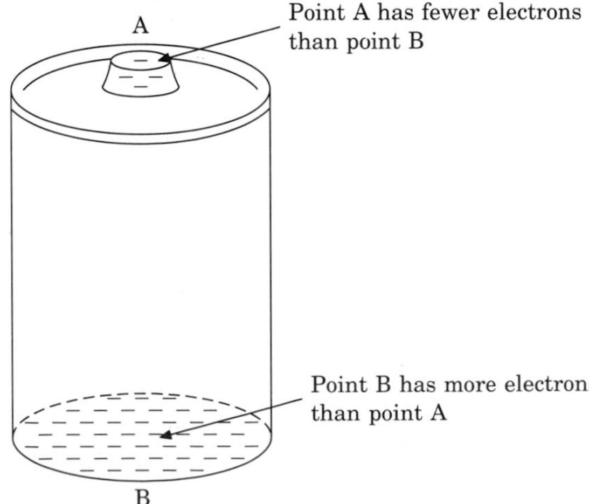
A
Point A has fewer electrons than point B
Point B has more electrons than point A
B

than Bill," or "Bill is shorter than John." In the first statement, Bill is the reference point. In the second statement, the reference point is John.

Another familiar example of polarity and reference points is that the upper tip of the earth's axis is called the North Pole and the lower tip, the South Pole. These terms refer to their geographic locations.

Review some key facts:

1. Electrons are negatively charged particles and protons are positively charged.
2. Any substance or body that has excess electrons is negatively charged.
3. Any substance or body that has a deficiency of electrons is positively charged.
4. Unlike charges attract and like charges repel.

Refer to Figures 1–22a, b, and c and answer the following questions:

1. In Figure 1–22a, which ball has a deficiency of electrons?
2. In Figure 1–22b, will the balls attract or repel each other?
3. In Figure 1–22c, if ball B has an equal number of electrons and protons, what is the polarity of ball A with respect to B?

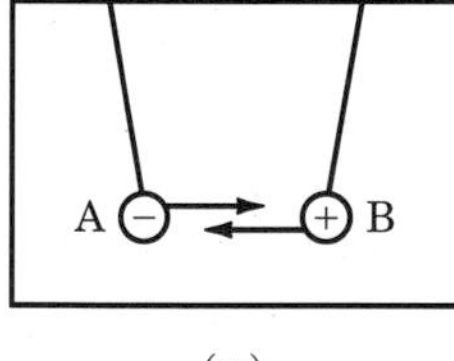

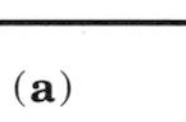

(a)

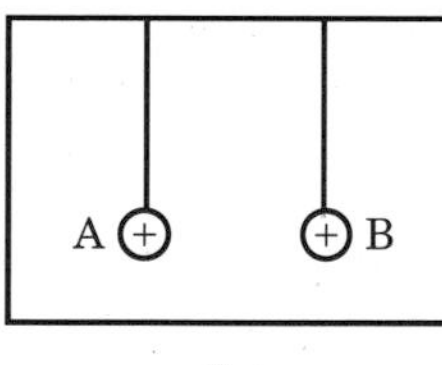

(b)

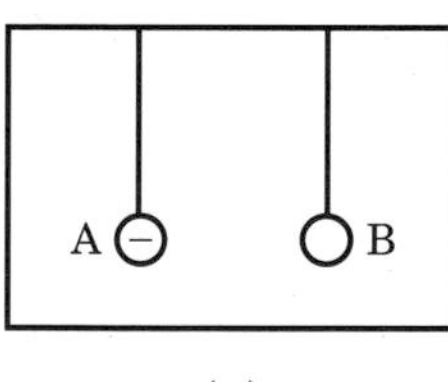

(c)

FIGURE 1–22 Some arrangements of electrical charges

The answers are 1) **B;** 2) **repel;** and 3) **negative.** In number one, the positive polarity indication on ball B indicates a deficiency of electrons. In number two, both balls have the same polarity charge, so they repel each other. In number three, if ball B has an equal number of electrons and protons, it is electrically neutral. However, ball A remains negative with respect to ball B, since ball A is not electrically neutral. In this case, ball A is negatively charged with respect to anything that does not have an equal or greater negative charge.

Coulomb's Formula Relating to Electrical Charges

Examining the forces of attraction or repulsion (repelling force) between charged bodies, Charles Coulomb, a French physicist, determined that the amount of attraction or repulsion between two charged bodies depended on the amount of charge of each body and the distance between the charged bodies. He expressed this relationship with the formula:

$$\text{Force} = \frac{\text{Charge on body \#1} \times \text{Charge on body \#2}}{\text{Distance between them squared}}$$

Restated:

Formula 1–1	$F = \frac{q_1 \times q_2}{d^2}$

where F = force (newtons),
q = charge in units of charge (coulombs),
and d = distance between charged bodies (meters)

Coulomb's law of charges states that the force of attraction or repulsion between two charged bodies is directly related to the product of their charges and inversely related to the square of the distance between them. In other words, the force equals the product of the charges divided by the square of the distance between them.

Unit of Charge

In honor of Charles Coulomb, the unit of charge is designated the "**coulomb**." Many electrical units that define quantity or magnitude of electrical parameters are named after famous scientists who perform related experiments.

Electrical units of measure are powerful reference values used every day in electronics. Even as someone had to determine liquid quantities as a gallon, quart, pint, or cup, scientists define units of measure for electrical parameters so that each has its own "reference point" with respect to quantity or value.

In the case of electrical charge, the unit is the coulomb. This is the amount of electrical charge represented by 6.25×10^{18} electrons, or 6,250,000,000,000,000,000 electrons. Now you know the unit for charge used for the "qs" in the previous formula:

$$F = \frac{q_1 \times q_2}{d^2}$$

Fields of Force

You are familiar with several types of "fields," for example, gravitational fields and magnetic fields. Fields of force are represented by imaginary lines, which represent the field of influence for the force involved. Figure 1–23 shows electrostatic fields between unlike charges, Figure 1–23a, and between like charges, Figure 1–23b. This only shows a visualization of the nature of these fields, as represented by the lines. It does not attempt to

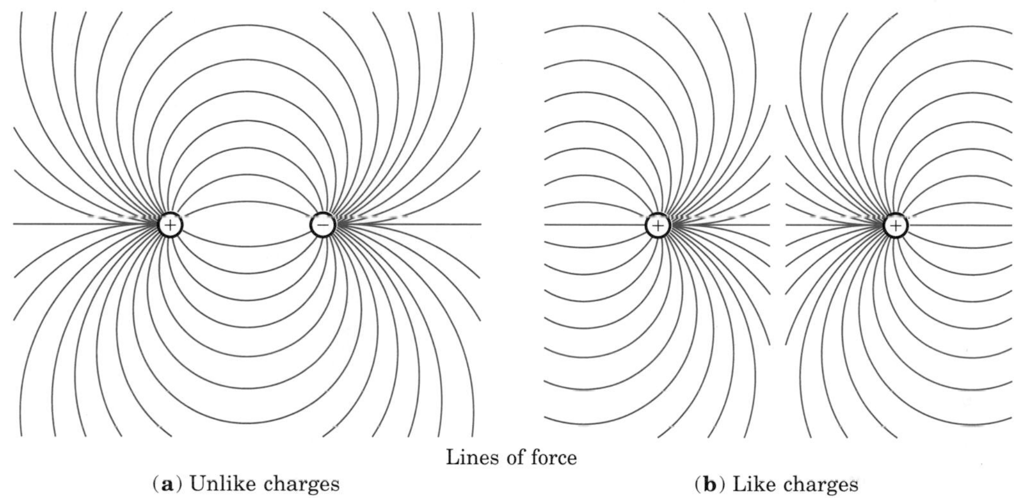

FIGURE 1–23 Electrostatic fields represented by lines of force

indicate the strength of the charges, the strength of the field of force, or the distance between the charges. This representation helps us visualize the tangible forces between electrical charges.

ELECTRICAL POTENTIAL

What Is Electrical Potential?

The implication drawn from these tangible forces is that they can be harnessed to do work. In this case, the difference of charge levels at two points has the potential to move electrons from a point of excess electrons to a point of electron deficiency, if a suitable path is provided. (Remember, the protons are in the nucleus and are not free to move). This difference between two points having different charge levels is often termed a **potential** *difference.* This force, which moves electrons from one place to another, is called an **electromotive force** (emf). Remember these important terms because you will use them often. See Figure 1–24 for another example of potential energy and the potential to do work.

The Unit of Electrical Potential or Electromotive Force

Scientists have established a measure for the unit of charge (the coulomb), and the unit of measure for this difference of potential. They have named this unit of potential difference the **volt** in honor of Alessandro Volta who invented the electric battery and the electric capacitor. Often, this potential difference between two points, which is measured in volts, is called voltage.

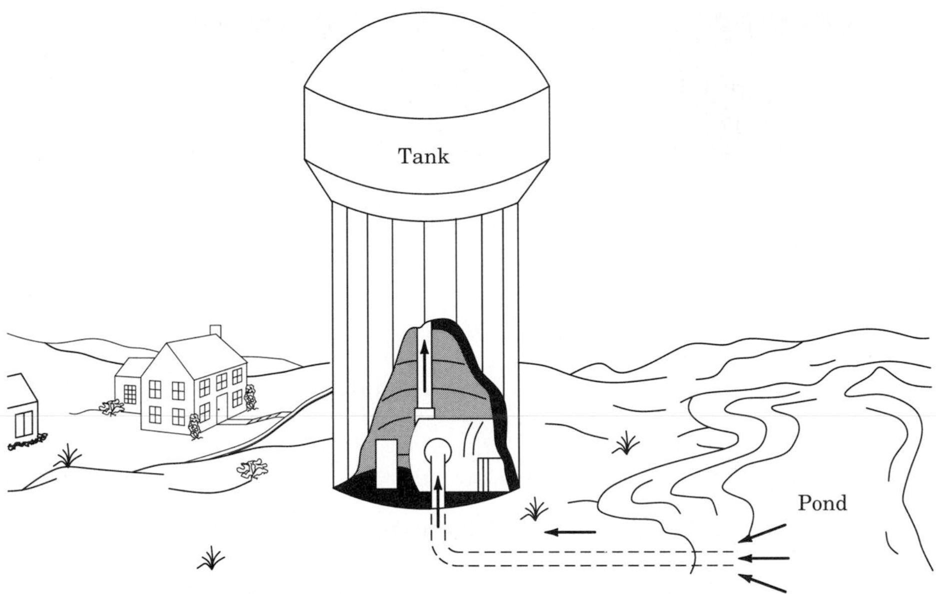

FIGURE 1–24 The water in this tank has potential energy due to its height above the ground; it has energy to move through pipes to homes in a nearby town. Likewise, electrical potential can move electrons, provided a path for such movement exists.

An in-depth discussion will come later. For now mentally picture that charges and/or different points are positive or negative relative to each other (that is, polarity); that there are different quantities of charge depending on the number of excess electrons or electron deficiency; and that there is a potential difference between such points, which provides an electromotive force capable of causing electron movement from one point to another point.

What Means Can Produce Electrical Potential?

Since electrical potential difference (electromotive force), or voltage, performs the electrical work of moving electrons from place to place, what establishes and maintains this voltage?

Earlier in this chapter, we discussed various types of energy that can cause "electrical imbalance," which you now know as electromotive force, or voltage. Static electricity, chemical energy, mechanical energy, light energy, and heat energy were all mentioned. Look at Figure 1–25 to see some methods of using these various energies to establish and maintain differences of potential (voltage) between two points.

(a)

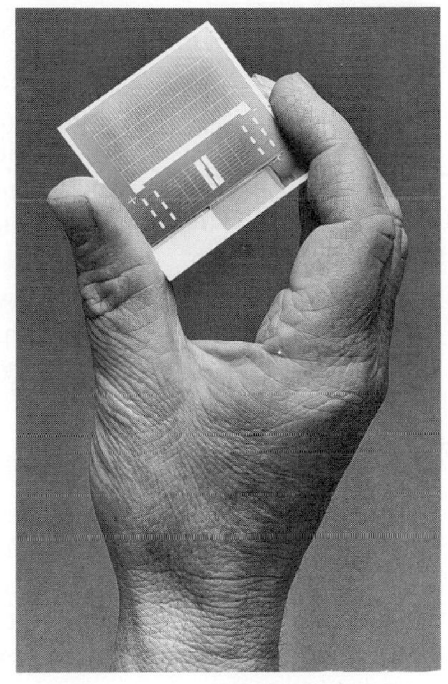

(b)

(c)

(d)

FIGURE 1–25 Several methods of producing voltage or emf
a. A battery converts chemical energy into voltage. (Courtesy of Eveready Battery Company)
b. A solar cell harnesses light energy. (Courtesy of The Energy Technology Visuals Collection)
c. Heat energy allows the thermocouple to produce useful electric current. (Courtesy of CGS Thermodynamics)
d. Generators turn mechanical energy into electricity. (Courtesy of Kurz & Root Company)

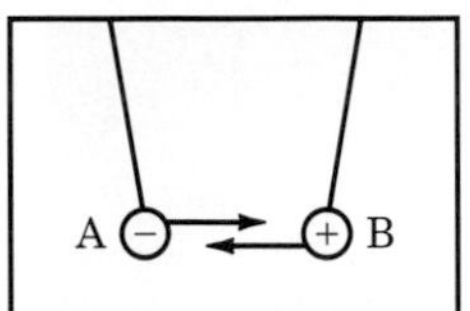

FIGURE 1–26 Two different charges

CHARGES IN MOTION

What happens if the two balls in Figure 1–26 touch each other? If you said that the excess electrons on ball A move to ball B and attempt to overcome the deficiency of electrons on ball B, you are right. In fact, electrons would continue to move until ball A and ball B had equal charges, or were neutral with respect to each other.

If the balls do not touch each other but are connected with a copper conductor wire, Figure 1–27, what happens? If you surmise that some electrons move through the conductor wire until the charges on ball A and ball B were equal, you are right.

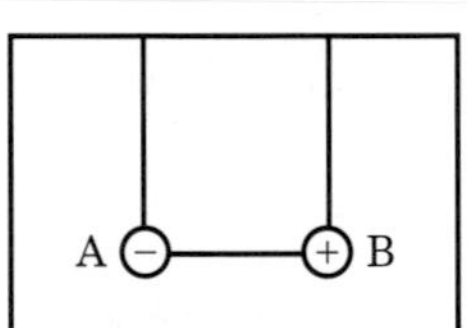

FIGURE 1–27 If a conductor wire is connected to ball A and ball B, what happens?

Current

Are the electrons that move to ball B the same electrons that left ball A? Probably not. As you have already studied, the conductor wire has many free electrons moving within it. Having a positive charge at one end of the conductor and a negative charge at the other end (a potential difference between its ends) causes the electrons to move from the negative end to the positive end. This movement of electrons is known as electrical **current flow,** or **current.**

An Analogy of Current Flow

Figure 1–28 illustrates the movement of electrons (current flow) from atom to atom within the conductor (from negative to positive ends). Bin A has a large quantity of rubber balls (representing electrons) and bin B has a few. The row of people between the two bins represent the conductor. They pass the balls from person to person, similar to the conductor material passing electrons from atom-to-atom when a current flows through the conductor. The progressive movement of balls from bin A to bin B simulates the movement of electrons from ball A through the conductor to ball B. In the conductor, as a free electron leaves its original atom and moves to an adjacent atom, it is replaced by another electron from another atom. This process is multiplied millions of times when there is current through a conductor. As

FIGURE 1–28 An analogy relating to electron movement

an electron leaves one end of the conductor (attracted by ball B's positive charge), it is replaced by an electron entering the conductor from ball A. Figure 1–29 also illustrates current flow.

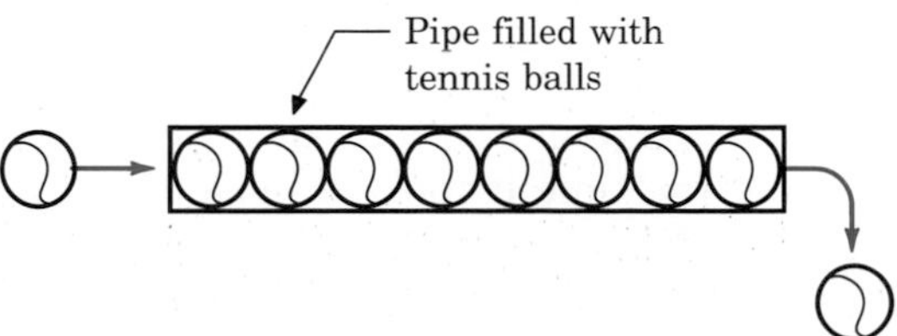

FIGURE 1–29 Another analogy relating to electron movement

Obviously, these analogies are oversimplifications of the physics involved in current flow, but they will aid your understanding as we discuss current flow and the specific electrical phenomena involved.

The Unit of Current

Recall that the unit of charge, the coulomb, was established in honor of Charles Coulomb, and the unit of potential difference, the volt, was named after Alessandro Volta. In like manner, the unit of measure for current was named in honor of the French mathematician and physicist, Andre Ampere.

An **ampere** of current is the quantity of electron movement represented by a flow rate of one coulomb of charge per second. Restated, *a flow of one coulomb per second = one ampere.*

A Formula That Relates Charge Movement, Time, and Current

A common formula that relates coulombs of charge, time (in seconds), and current (in amperes) is:

Formula 1–2	$Q = I \times T$

where Q = charge in coulombs,
I = current in amperes,
and T = time in seconds.

This formula allows you to calculate the amount of charge moved, if you know the current in amperes and the time in seconds.

Also, this formula can be transposed to calculate current, if the charge and time are known. This transposition yields $I = \frac{Q}{T}$. For example, if 10 coulombs of charge move from one point to another point in an electrical circuit over 2 seconds, current equals 5 amperes $\left(I = \frac{10}{2}\right)$. If at this point you do not understand how the formula was transposed, be encouraged by knowing that some techniques of transposing will be discussed again later in the book. Look at this formula and the examples, and use your knowledge about direct and inverse relationships to understand the formula. Is current *directly* or *inversely* related to the quantity of charge moved? Is current *directly* or *inversely* related to the duration of time taken to move the given charge? Your answers should have indicated a *direct* relationship between current and quantity of charge moved and an *inverse* relationship between current in amperes and the time it took to move the given charge. Recall that the direct relationship between quantities means that as one quantity increases, the other quantity increases. The inverse relationship indicates that as one quantity increases, the other quantity decreases, or vice versa.

IN-PROCESS LEARNING CHECK II

Fill in the blanks as appropriate.

1. An electrical system (circuit) consists of a source, a way to transport the electrical energy, and a ________.
2. Static electricity is usually associated with ________ type materials.
3. The basic electrical law is that ________ charges attract each other and ________ charges repel each other.
4. A positive sign or a negative sign often shows electrical ________.
5. If two quantities are *directly* related, as one increases the other will ________.
6. If two quantities are *inversely* related, as one increases the other will ________.
7. The unit of charge is the ________.
8. The unit of current is the ________.
9. An ampere is an electron flow of one ________ per second.
10. If two points have different electrical charge levels, there is a difference of ________ between them.
11. The volt is the unit of ________ force, or ________ difference.

THREE IMPORTANT ELECTRICAL QUANTITIES

This chapter has discussed the magnitude of electrical charges (amount or value of charge); polarity (either negative or positive); and the difference of charge between two points that creates a potential difference (electromotive force or voltage) which can move electrons from one point to another (a current flow).

Also, you have learned that scientists have defined units of measure for the quantities of charge, potential difference, and current. Recall that the unit of charge is the coulomb (6.25×10^{18} electrons); the unit of electromotive force, or potential difference, is the volt; and the unit of current flow, or current, is the ampere. Of the electrical parameters described, you will frequently use the units for current and electromotive force (i.e., the ampere and the volt).

A third electrical quantity you will often use is **resistance.** If you have tried to sandpaper a piece of wood, you have experienced *physical resistance.*

One analogy to electrical resistance is the friction to fluid, like water passing through a pipe, Figure 1–30. The resistance encountered by the water flow is related to the smoothness or roughness of the pipe's surface and the diameter of the pipe. It takes a source of pressure, such as a pump or gravity, to cause water to flow through the pipe against the pipe's frictional resistance.

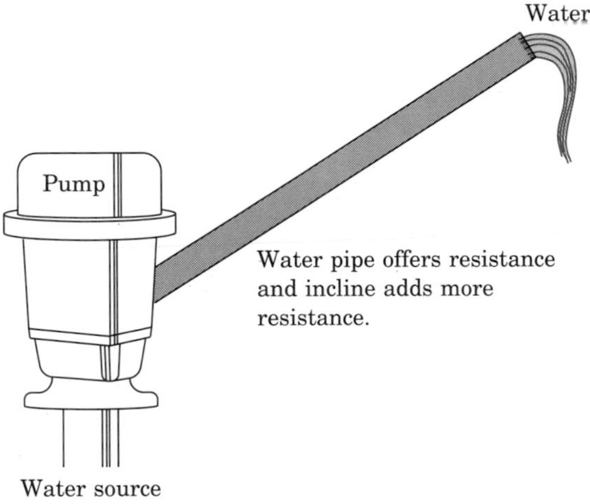
Water
Pump
Water pipe offers resistance and incline adds more resistance.
Water source

FIGURE 1–30 Water pipes and/or gravity show resistance to water flow.

Current flow through a conductor encounters molecular resistance to its flow. The amount of resistance to current flow depends on a number of factors. One factor is the type of material through which the current is flowing. Other factors are the dimension of the conductor wire, including its cross-sectional area and its length, and temperature. An electrical pump (emf or voltage) causes current to flow through the resistance in its path.

As you have probably reasoned, electrical resistance is the opposition shown to current flow. Again, an electrical unit has been named after a scientist, Georg Simon Ohm; thus, the unit of resistance is the **ohm.**

The ohm of resistance has been defined in several ways. One definition states an ohm is the resistance of a column of pure mercury with a cross-sectional dimension of 1 millimeter squared and a length of 106.3 centimeters at a temperature of zero degrees centigrade. Another definition is an ohm is the amount of resistance that develops 0.24 calories of heat when one ampere of current flows through it. The definition that is most useful is an ohm of resistance is the amount of electrical resistance that limits the current to one ampere when one volt of electromotive force is applied.

BASIC ELECTRICAL CIRCUIT

Generalized Description

Earlier in the chapter, a simple description was given of an electrical circuit. Recall that an electrical circuit has three basic ingredients: the source, the means to carry electricity from one point to another point, and a load. Before thoroughly examining this basic circuit, we will discuss the difference between a closed and an open circuit.

The Closed Circuit

You have learned that if a path is provided and when a potential difference exists between two points, electrons move from the point of excess electrons (negative polarity point) to the point of electron deficiency (positive polarity point). One method of providing this path is to connect a conductor wire between the points, thus providing a route through which electrons move and establish current flow. A closed circuit is a complete, unbroken path through which electrical current flows whenever voltage is applied to that circuit, Figure 1–31.

FIGURE 1–31 A closed circuit provides an unbroken path for current flow.

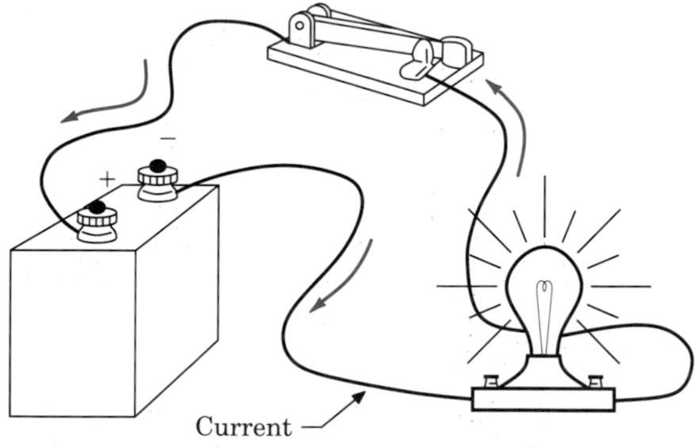

Closed circuit

The Open Circuit

You probably already understand that if a closed circuit is an unbroken path for electron flow, then an open circuit has a break in the path for current flow, Figure 1–32. This break may be either a desired (designed) break or an

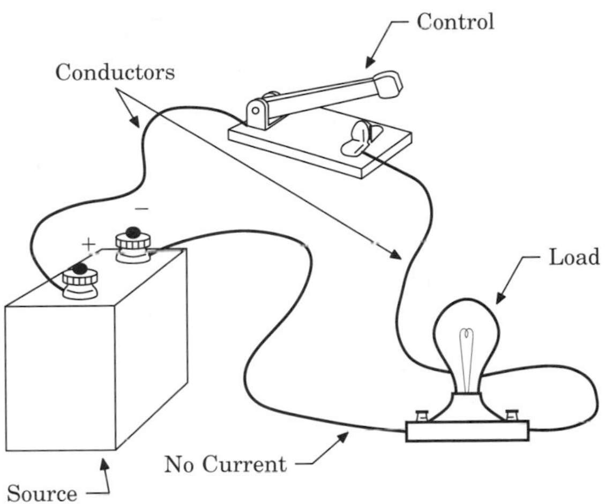

FIGURE 1–32 An open circuit has a break somewhere in the current path.

undesired (unplanned) break. The most common method for purposely opening a circuit is an electrical switch, such as the one you use to turn the lights on or off. The switch is a fourth element of the basic electrical circuit, and many circuits contain such a control component, or circuit. In Figure 1–32 the switch is the control component.

The Basic Electrical Circuit Summarized

The basic electrical circuit has a source, a means of conducting electron movement (current), a load, and may have a control element, such as a switch.

In order for current to flow (electron movement) through the circuit, there must be a voltage source and a closed circuit or a complete current-conducting path. If the conductor wire(s) were accidentally connected across the source with no other load component, the wire (with its small resistance) becomes the load. Depending on the resistance of the wire and the amount of source voltage, this undesired low resistance condition (sometimes called a short circuit) would probably result in the wire overheating and melting, thus breaking the circuit.

Applying a voltage (electromotive force) to a closed circuit causes current to flow through the circuit conductors and load from the source's negative side through the circuit and back to the source's positive side, Figure 1–33. The amount of current (amperes) that flows depends on the value of voltage (volts) applied to the circuit and how much resistance (ohms) the circuit offers to current flow, Figure 1–34. Remember these points because they will become more important as your studies proceed.

FIGURE 1–33
Direction of current flow through a circuit

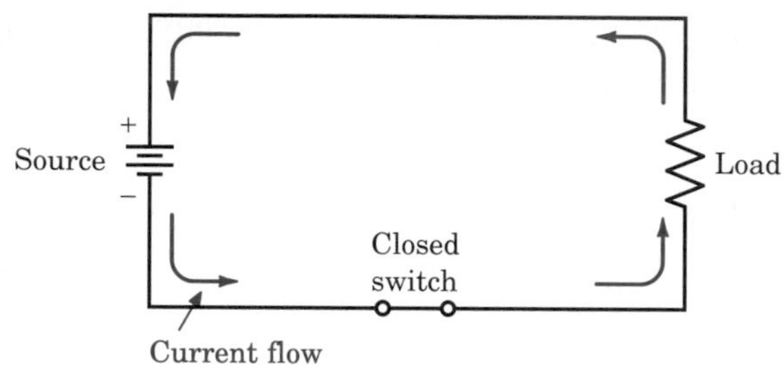

PARTS OF A CIRCUIT	EXAMPLE OF EACH
SOURCE	POWER COMPANY GENERATORS
TRANSPORTING MEANS	CONDUCTOR WIRES
LOAD	LIGHT BULB/S
(CONTROL)	(SWITCH)

IMPORTANT QUANTITIES	
EMF	(VOLTS)
RESISTANCE	(OHMS)
CURRENT	(AMPERES)

FIGURE 1–34 Important facts about a basic electrical circuit. As emf increases, current increases. As resistance increases, current decreases.

SUMMARY

- Matter occupies space, has weight, and sometimes can be tangible to one or more of our five senses.
- Matter's physical states include solids, liquids, and gases.
- Matter's chemical states include elements, compounds, and mixtures.
- Elements are comprised of only one type of atom. But compounds are unique combinations of different elements' atoms that result in a new matter or substance.
- A mixture does not cause chemical or physical changes in the elements being mixed.
- The smallest particle into which a compound can be divided and retain its basic characteristics is a molecule. Elements can be divided to the atom level and retain their inherent characteristics.
- The atom is the fundamental building block of matter. Special combinations of atoms, called molecules, are the building blocks of compounds.
- Bohr's model of an atom pictures the central portion of the atom, the nucleus, being comprised of protons and neutrons. Around this nucleus is a particle called the electron, which orbits the nucleus like planets orbit the sun.
- The electron theory helps illustrate electricity and behavior of electrons in matter and relates to the atom model with its protons, neutrons, and electrons.
- The atom's proton has a positive electrical charge, while the orbiting electrons have a negative electrical charge. The neutrons in the center of the atom add mass to the atom but are electrically neutral.

- Protons and neutrons are approximately 1800 times heavier than electrons.
- The net charge of an atom is typically neutral, since there are equal numbers of electrons and protons and their electrical charges offset each other.
- The atomic number of an element is the number of protons in the nucleus of each of its atoms.
- The atomic weight of an element is a comparison of its weight to that of carbon-12.
- Electrons travel around the nucleus of an atom in shells that are at different distances from the nucleus. The first ring (shell) carries a maximum of 2 electrons; the second shell outward a maximum of 8; the next shell, 18, and the fourth shell, 32. The outermost electron ring, valence electrons, determines if the atom is electrically and chemically active or stable and is full when it has 8 electrons.
- If the outer shell of an atom contains 8 electrons, it is stable. If the outer ring has less than 4 electrons, it is usually a conductor. If the valence electrons number 4, it is a semiconductor material.
- Materials, such as copper, that have only 1 outermost ring electron per atom are good conductors. These electrons, sometimes called free electrons, are easily moved within the material from atom to atom.
- An atom that has lost or gained electrons is called an ion. If it has lost electron(s), it is a positive ion. If it has gained electron(s), it is a negative ion.
- Ions and/or movement of electrons within a material are caused by various external and internal energies or forces. Several examples of energy that can electrically unbalance atoms or move electrons are friction, chemical, heat, and light.
- The law of electrical charges is that unlike charges attract and like charges repel each other.
- Static electricity is usually associated with nonconductor materials and is often caused by friction between them.
- Polarity designates differences between two points or objects relative to each other, or opposites. For example, electrical polarities of "−" or "+" show which point has excess electrons and electron deficiency relative to each other. Magnetic polarity indicates differences between magnetic poles (North and South). Also, electrical polarity shows direction of current flow through an electrical component or circuit. For example, electron movement (current) moves from the source's negative side to the source's positive side through the components and circuitry connected to that source.
- Coulomb's law of charges states, the force of attraction or repulsion between two charges is directly related to the product of the two charges and is inversely related to the square of the distance between them.
- The unit of charge is the coulomb, which represents 6.25×10^{18} electrons of charge.
- Electrical potential is the potential to perform the electrical work of moving electrons. It is sometimes termed electromotive force.
- The unit of measure for potential difference (electromotive force) is the volt.
- Several types of energy that establish and maintain potential differences between two points (i.e., create a voltage source) are chemical, heat, light, magnetic, and mechanical energies.
- Organized electron movement, or electron flow, is known as current flow or current.
- Current flow caused by electron movement is from the negative side of the voltage source to

the positive side of the voltage source through the circuit connected to the source, if the circuit is a closed circuit.

- ▶ The unit of current is the ampere, which represents a rate of electron flow of one coulomb per second.
- ▶ A formula that relates amount of charge (coulombs), current (amperes), and time (seconds) is $Q = I \times T$, where Q = coulombs, I = amperes, and T = seconds. Therefore, the formula may be transposed to read $I = \frac{Q}{T}$.
- ▶ Three important electrical quantities of measure are:
  1. volt, the unit of potential difference, or amount of potential difference causing one ampere of current to flow through a resistance of one ohm.
  2. ampere, the unit of current flow, or amount of current flow represented by a rate of charge flow of one coulomb per second.
  3. ohm, the unit of resistance, or amount of electrical resistance limiting the current to one ampere with one volt applied.
- ▶ A basic electrical circuit has an electrical source (emf), a way to transport electrons from one point to another point, and a load, which uses the electrical energy. Another part of many basic circuits is a control device, such as a switch.
- ▶ A closed circuit has an unbroken path for current flow: from one side of the source, through the circuitry and back to the other side of the source.
- ▶ An open circuit does not provide a continuous path for current flow. The path is broken, either by design (using a switch) or by accident. The open circuit presents an infinite resistance path for current.
- ▶ A short circuit is an undesired, low resistance path for current flow. If the short is across both terminals of a voltage source, it frequently causes undesired effects, such as melted wires, fire, and other damage to the voltage source.
- ▶ In the basic electrical circuit, the voltage source provides the electromotive force that moves electrons and causes current flow. The circuit conductors and components provide resistance to current, thereby limiting the current. The various techniques of controlling electrons and electron flow to perform desired effects form the basis of electronic science.

REVIEW QUESTIONS

1. Define the term matter.
2. List the three physical states of matter and give examples of each.
3. List the three chemical states of matter and give examples of each.
4. Define the term element and give two examples of elements.

5. Define compound and give two examples of compounds.

6. Sketch a hydrogen atom. Identify each particle.

7. On your hydrogen atom sketch, designate the electrical charge of each particle.

8. a. Which has more weight or mass, the electron or the proton?
   b. Approximately how many times heavier is the one compared to the other?

9. Define the term free electron.

10. a. Define the term ion.
    b. What causes an ion to be classified as a positive ion?

11. Name four energy types that can be used to purposely create electrical imbalances in atoms of materials.

12. How many valence electrons do each of the following materials have?
    a. A conductor.
    b. A semiconductor.
    c. An insulator.

13. Describe the basic law of electrical charges.

14. Describe Coulomb's law of electrical charges.

15. If the distance between two charges is tripled and the charge strength of each charge remains the same, what happens to the force of attraction or repulsion between them?

16. In an electrical circuit, what is potential difference?

17. What unit of measure expresses the amount of electromotive force present between two identified points in a circuit?

18. In an electrical circuit, what is current flow?

19. a. What basic unit of measure expresses the value of current in a circuit?
    b. What is the value of current flow if 20 coulombs of charge move past a given point in 5 seconds?

20. Describe the term resistance in terms of voltage and current, and give the unit of measure.

21. Describe the difference between a closed circuit and an open circuit.

22. List three or more elements of a basic electrical circuit.

TEST YOURSELF

1. Draw a diagram illustrating an electrically balanced atom that has 30 protons in its nucleus and whose first three shells are full, with the fourth (last) shell having two electrons.
2. Label all electrical charges in your diagram.
3. What kind of material is pictured in your diagram?
4. Is this material a conductor, semiconductor, or insulator?
5. What is its atomic number?
6. Define the term polarity as used in electrical circuits.
7. Draw an electrical circuit showing voltage source, conductor wires, and load. Show the polarity of that source, and, by using an arrow, show the direction of current flow through the circuit. (Assume a closed circuit.)

Ohm's Law

Key Terms

Ampere (A; mA; μA)

Battery (B)

Block diagrams

Conductance (G)

Current (I)

Ohm (Ω; kΩ; MΩ)

Ohm's law

Polarity

Resistance (R)

Resistor (R)

Resistor color code

Schematic diagrams

Siemens (S; mS)

Switch (S)

Volt (V; mV; μV)

Voltage (V)

Watt (W; mW)

Flexible circuits courtesy of NASA

2

Outline

Basic Electrical Units and Abbreviations
Using the Metric System to Help
Diagrams Used for Electronic Shorthand
Ohm's Law and Relationships of Electrical Quantities
Rearranging to Find the Unknown Quantity
Sample Application of Metric Prefixes and Powers of Ten
Some Basic Rules Relating to Powers of Ten
Direction of Current Flow
Polarity and Voltage
Work, Energy, and Power
Commonly Used Versions of Ohm's Law and Power Formulas
Resistors
Resistor Color Code
Safety Hints

Chapter Preview

This important chapter will provide a thorough knowledge of Ohm's law. Ohm's law provides the fundamental background you need to understand and apply in your career as a technician. It is the basis for calculations of electrical quantities in all types of electrical and electronic circuits.

You will have opportunity to combine the knowledge you have already learned with some new and practical information. For example, units and symbols for electrical quantities and components, metric prefixes, the powers of ten, schematic diagrams, and other useful information are in this chapter. Although proficiency in these areas comes through practice, you can immediately obtain useful experience as you integrate this knowledge with the study of Ohm's law. This chapter equips you with the remaining "fundamental tools of thinking" needed in your studies to become an electronics technician.

Objectives

After studying this chapter, you will be able to:

- List the units of measure for charge, potential (emf), **current, resistance,** and **conductance** and give the appropriate abbreviations and symbols for each
- Use metric system terms and abbreviations to express subunits or multiple units of the primary electrical units
- Recognize and/or draw the diagrammatic representations for conductors that cross and electrically connect, and that cross and do not connect
- Recognize and/or draw the diagrammatic symbols for seven elemental electronic components or devices
- Use **block** and **schematic diagrams**
- Explain the relationships among current, **voltage,** and resistance
- Use **Ohm's law** to solve for unknown circuit values
- Illustrate the direction of current flow and **polarity** of voltage drops on a schematic diagram
- Use metric prefixes and powers of ten to solve Ohm's law problems
- Explain power dissipation
- Use appropriate formulas to calculate values of power
- Give the characteristics of several types of **resistors**
- Use the **resistor color code**
- List key safety habits to be used in laboratory work

BASIC ELECTRICAL UNITS AND ABBREVIATIONS

Charge

Recall that the basic unit of charge is the coulomb and that the letter "Q" (or q) represents charge. However, you need to know that the abbreviation for the coulomb is **C.** (Remember that one C equals 6.25×10^{18} electrons.) Notice that this is a capital "C" and not a lower case "c." If the unit abbreviation or symbol for electronic measurement is named after an individual, such as coulomb, volt, ampere, or ohm, the abbreviation is a capital letter and not a lower case letter. Keep this in mind and get in the habit of correctly writing unit abbreviations or symbols.

Potential

The unit for potential difference, or electromotive force, is the **volt,** named after Alessandro Volta. The abbreviation, or symbol, for this unit is **V. Voltage** is expressed in volts. Recall that one volt equals the amount of electromotive force (emf) that moves a current of one ampere through a resistance of one ohm.

Current

The unit of measure for current flow is the ampere. The abbreviation, or symbol, for this basic unit of measure is **A.** Don't confuse the letter "I" (thought of as intensity of electron movement) that represents the general term current with the "A" used for the unit of current. Remember that one ampere equals an electron flow of one coulomb per second past a given point.

Resistance

Resistance is another electrical parameter that uses two letters: "R" represents the general term resistance and the Greek letter omega (**Ω**) represents the unit of resistance, the **ohm**. Remember that one Ω equals the resistance that limits the current to one ampere with one volt applied. [NOTE: Figure 2–1 shows some "multimeters" used to measure V, A and Ω, the units of measure we have just discussed.]

Conductance

Another electrical parameter is **conductance.** It sometimes is defined as the comparative ease with which current flows through a component or circuit. Conductance is the opposite of resistance. The unit of conductance is the **siemens** (**S**) named after Ernst von Siemens. The abbreviation for the general term conductance is **G,** and the unit of measure is the siemens (S).

(a)

(b)

FIGURE 2–1 Multimeters used to measure V, A, and Ω
a. An analog volt-ohm-milliammeter (VOM) (Courtesy of Simpson Electric Company)
b. A digital multimeter (DMM) (Courtesy of Simpson Electric Company)

Conductance, in siemens, is defined as the reciprocal of resistance in ohms, that is; $G(S) = \frac{1}{R}$ or $G = \frac{1}{R}(S)$.

An alternate unit for conductance is the "mho." This is ohm backward and is used due to the inverse relationship of resistance and conductance. The formula $G(\mho) = \frac{1}{R}$ or $G = \frac{1}{R}(\mho)$ and symbol ($\mho$) are the same.

These units of measure and abbreviations, or symbols, are important, so a chart has been provided, Figure 2–2. Study this chart until you have memorized all the units and their abbreviations.

USING THE METRIC SYSTEM TO HELP

Some Familiar Metrics

Everyone has been exposed to the metric system. The meter (approximately 39 inches), centimeter (one one-hundredth of a meter), millimeter (one one-thousandth of a meter), and kilometer (1,000 meters) are not uncommon terms. As you can see, the prefix to meter identifies a multiple or subdivision of the basic unit. In other words, the meter is the basic unit and the prefix centi equals $\frac{1}{100}$, milli equals $\frac{1}{1{,}000}$, and kilo equals 1,000 times the basic unit.

This technique of using metric prefixes is also useful in identifying multiples or subdivisions of basic electrical units. For example, in electronic circuits it is common to use thousandths or millionths of an ampere, or thou-

Brief Definition of Quantity	Electrical Quantity or Parameter	Basic Unit of Measure	Abbreviation or Symbol for Unit
Excess or deficiency of electrons	Charge (Q)	Coulomb	C (6.25×10^{18} electrons)
Force able to move electrons	Potential Difference (EMF)	Volt (Force that moves one coulomb of charge per second through one ohm of resistance)	V
Progressive flow of electrons	Current (I)	Ampere (An electron flow rate of one coulomb per second)	A
Opposition to current flow	Resistance (R)	Ohm (A resistance that limits current to a value of one ampere with one volt applied)	Ω
Ease with which current can flow through a component or circuit	Conductance (G)	Siemens (The reciprocal of resistance, or, $\frac{1}{R}$)	S

FIGURE 2–2 Important electrical units' abbreviations and symbols

sands of ohms of resistance. Although these subunits and multiple-units do not represent all prefixes encountered in electronics, they do represent those you will be immediately applying. Study Figure 2–3 until you learn each of the metric terms and their abbreviations or symbols.

There is one other bit of related knowledge that can be of great help to you in practical work. That is mathematical ways to represent each of these metric prefixes. These mathematical representations are helpful in studying electronic circuits, because simple calculations can be used to verify, or predict, circuit behavior.

FIGURE 2–3 Some commonly used metric units

METRIC TERM	SYMBOL	MEANING	TYPICAL USAGE WITH ELECTRONIC UNITS
pico	p	one millionth of one millionth of the unit	picoampere (pA)
nano	n	one thousandth of one millionth of the unit	nanoampere (nA) nanosecond (ns)
micro	μ	one millionth of the unit	microampere (μA) microvolt (μV)
milli	m	one thousandth of the unit	milliampere (mA) millivolt (mV)
kilo	k	one thousand times the unit	kilohms (kΩ) kilovolts (kV)
mega	M	one million times the unit	megohm(s) (MΩ)

Since the subunit and multiple-unit prefixes, Figure 2–3, are based on a decimal system (multiples or submultiples of 10), it is convenient to express these prefixes in powers of ten. Recall that:

$$10^0 = \text{(ten to the zero power)} = 1$$
$$10^1 = \text{(ten to the 1st power)} = 10$$
$$10^2 = \text{(ten to the 2nd power)} = 10 \times 10 = 100$$
$$10^3 = \text{(ten to the 3rd power)} = 10 \times 10 \times 10 = 1{,}000$$
$$10^{-2} = \text{(ten to the negative 2nd power)} = \frac{1}{100} = 0.01$$
$$10^{-3} = \text{(ten to the negative 3rd power)} = \frac{1}{1{,}000} = 0.001$$

Figure 2–4 shows how powers of ten relate to the electrical units and prefixes discussed earlier. **Learn these because you will use them frequently.**

Study Figure 2–4. You can surmise that powers of ten can simplify writing very large or very small numbers. For example, instead of writing 3.5 microamperes (A) as 0.0000035 amperes, you can state it as 3.5×10^{-6}

FIGURE 2–4 Powers of ten related to metric and electronic terms

NUMBER	PWR OF 10	TERM	SAMPLE ELECTRONIC TERM
0.000000000001	10^{-12}	pico	pA (1×10^{-12} ampere)
0.000000001	10^{-9}	nano	nA (1×10^{-9} ampere)
0.000001	10^{-6}	micro	μA (1×10^{-6} ampere)
0.001	10^{-3}	milli	mA (1×10^{-3} ampere)
1,000	10^{3}	kilo	kΩ (1×10^{3} ohms)
1,000,000	10^{6}	mega	MΩ (1×10^{6} ohms)

amperes. The power of ten tells us how many decimal places and which direction the decimal moves to state the number as a unit. Thus, 3.5×10^{-6} indicates the decimal moves six places to the left when stated as the basic unit (i.e., 0.0000035 amperes). This is three and one-half millionths of an ampere.

Another example is 5.5 megohms (MΩ) of resistance simplified as 5.5×10^{6} ohms. How would this be stated as a number representing the basic unit? If you said the power of ten indicates moving the decimal six places to the right, you are correct. This yields 5,500,000 ohms.

Obviously, it is easier to manipulate numbers using powers of ten rather than using the bulky multiple decimal places involved in the small and large numbers typically associated with electrical/electronic units. You will get some practice using powers of 10 later in the chapter. (See In-Process Learning Check I.)

DIAGRAMS USED FOR ELECTRONIC SHORTHAND

Some Basic Symbols

Electronic circuits are frequently represented by either block diagrams, schematic diagrams, or both. In order to interpret and use schematic diagrams, it is necessary to learn some basic electronic symbols used in schematics. Because diagrams can be made to represent an infinite variety of circuits and configurations in a concise manner, schematic diagrams are the electronic technician's or engineer's shorthand.

Although you may not yet know the characteristics and applications of all the components introduced at this point, these components and symbols are frequently used in electronic circuits, Figure 2–5. Become familiar with the symbols; the details of their operation and applications will unfold as your learning continues.

Examine Figure 2–5 and observe a few of the important electronic symbols. Don't be worried about committing these to rote memory at this time. You will automatically learn these by repeated usage as you continue in your studies.

During your experiences in electronics you will encounter various types of diagrams which can be helpful. Several of these types include **block diagrams, schematic diagrams,** chassis layout diagrams, and others. For now, we'll focus on the block diagram and the schematic diagram since these are the kind you will be using most.

Block Diagrams

Block diagrams are useful because they are simple and conveniently portray the "functions" relating to each block. Recall that we used a simple block diagram in an earlier chapter when a simple electrical system was discussed.

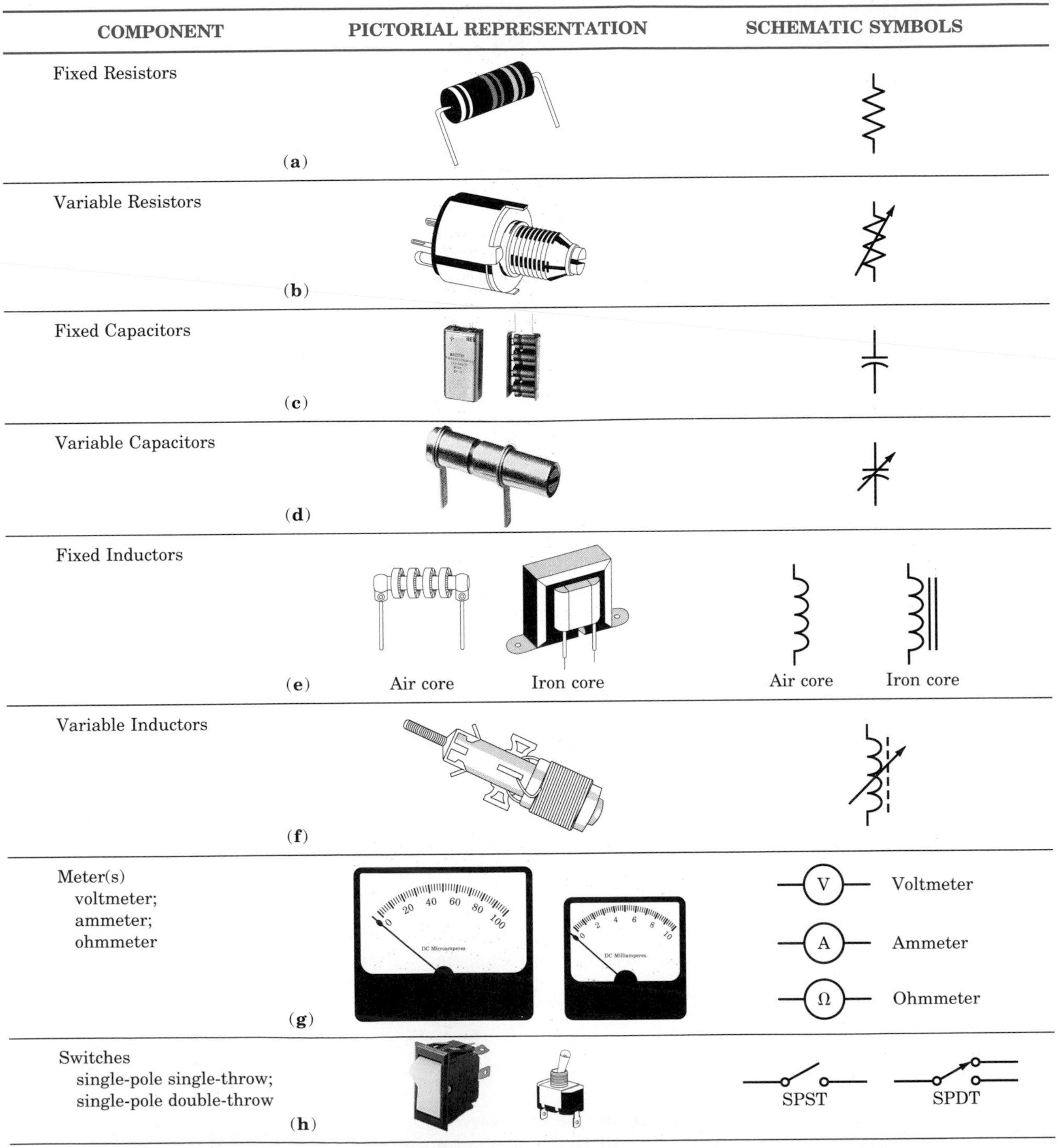

COMPONENT		PICTORIAL REPRESENTATION	SCHEMATIC SYMBOLS
Fixed Resistors	(a)		
Variable Resistors	(b)		
Fixed Capacitors	(c)		
Variable Capacitors	(d)		
Fixed Inductors	(e)	Air core; Iron core	Air core; Iron core
Variable Inductors	(f)		
Meter(s) voltmeter; ammeter; ohmmeter	(g)		V Voltmeter; A Ammeter; Ω Ohmmeter
Switches single-pole single-throw; single-pole double-throw	(h)		SPST; SPDT

FIGURE 2–5 Various components and their schematic symbols (Photo c courtesy of Mallory Capacitor Co.; photo d courtesy of Johanson General Electric; photos h, i, j; and k courtesy of Grayhill, Inc.)

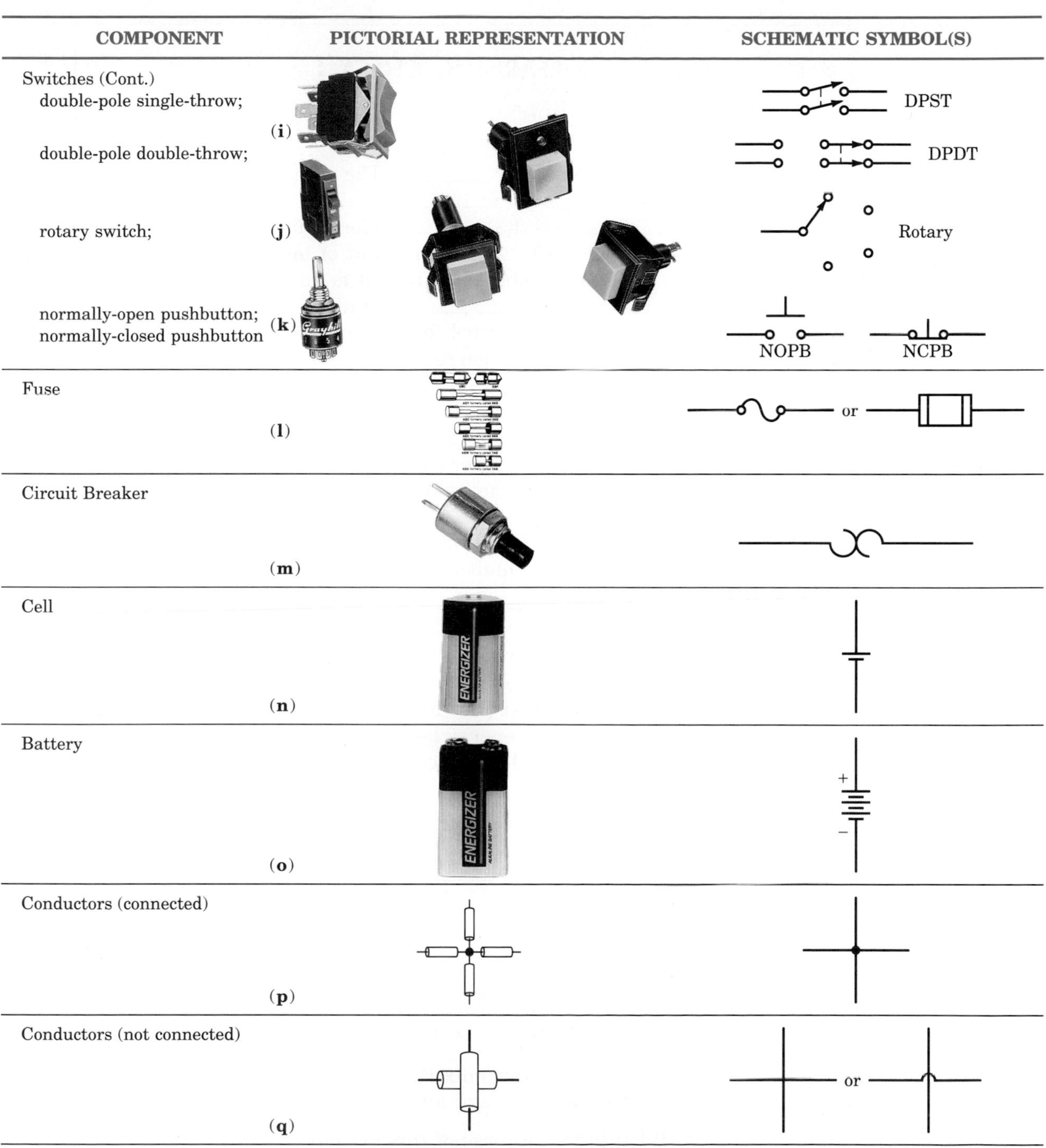

COMPONENT
PICTORIAL REPRESENTATION
SCHEMATIC SYMBOL(S)
Switches (Cont.)
double-pole single-throw;
(i)
DPST
double-pole double-throw;
DPDT
rotary switch;
(j)
Rotary
normally-open pushbutton;
normally-closed pushbutton
(k)
Grayhill
NOPB
NCPB
Fuse
(l)
or
Circuit Breaker
(m)
Cell
ENERGIZER
(n)
Battery
ENERGIZER
(o)
+
−
Conductors (connected)
(p)
Conductors (not connected)
(q)
or

FIGURE 2–5 cont. Various components and their schematic symbols (Photo l courtesy of Bussman Division, Cooper Industries; photo m courtesy of Square D Co.; photos n and o courtesy of Eveready Battery Co.)

IN-PROCESS LEARNING CHECK I

Fill in the blanks as appropriate.

1. Charge is represented by the letter ________. The unit of measure is the ________, and the abbreviation is ________.
2. The unit of potential difference is the ________. The symbol is ________.
3. The abbreviation for current is ________. The unit of measure for current is the ________, and the symbol for this unit is ________.
4. The abbreviation for resistance is ________. The unit of measure for resistance is the ________, and the symbol for this unit is ________.
5. Conductance is the ________ with which current can flow through a component or circuit. The abbreviation is ________. The unit of measure for conductance is the ________, and the symbol for this unit is ________.
6. How many microamperes does 0.0000022 amperes represent? ________ How is this expressed as a whole number times a power of 10? ________
7. What metric prefix represents one-thousandth of a unit? ________ How is this expressed as a power of 10? ________
8. Using a metric prefix how would you express 10,000 ohms? ________ How is this expressed as a whole number times a power of 10? ________

Refer to Figure 2–6 and see if you can understand the concept of a block diagram. Note that the block diagram quickly conveys what the system does (amplifies audio signals). Also it graphically illustrates the direction of signal flow and electrical energy between blocks. Most important, it conveys the primary function of the circuitry within each block.

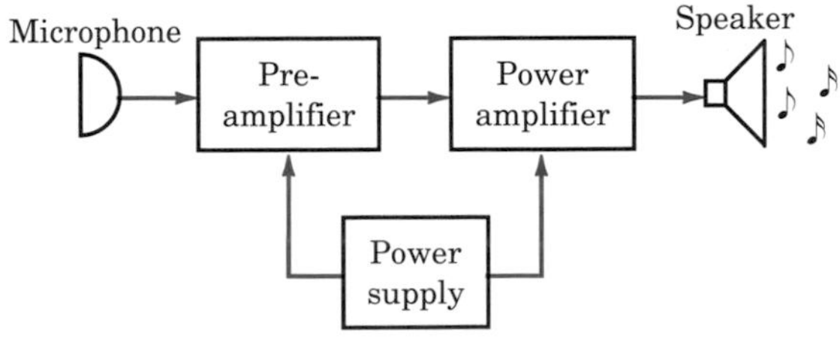

FIGURE 2–6 Sample block diagram (audio amplifier)

PRACTICE PROBLEM I

Answers are in Appendix B.
Look at Figure 2–7 and answer the following questions:

1. What is the purpose of this system?
2. Does signal flow from oscillator to RF amplifier, or vice versa?

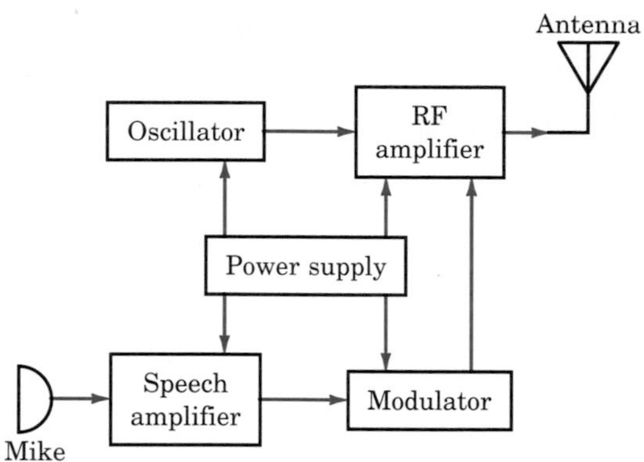

FIGURE 2–7 Sample block diagram (simple AM transmitter)

3. Does signal flow from RF amplifier to modulator, or vice versa?
4. Does electrical energy flow from the speech amplifier to the power supply, or vice versa?

Schematic Diagrams

Now that you have seen the usefulness of block diagrams, let's move on to the schematic diagrams.

In schematic diagrams you apply the knowledge of electrical symbols. Because the symbols are standardized, it is easy for others working in electronics to "decode" the information that the original designer of the schematic is conveying. The information is concise and useful in communicating the detailed electrical connections in the circuitry, the types and values of components, and other important information.

PRACTICE PROBLEM II

Look at Figure 2–8 and answer the following questions:

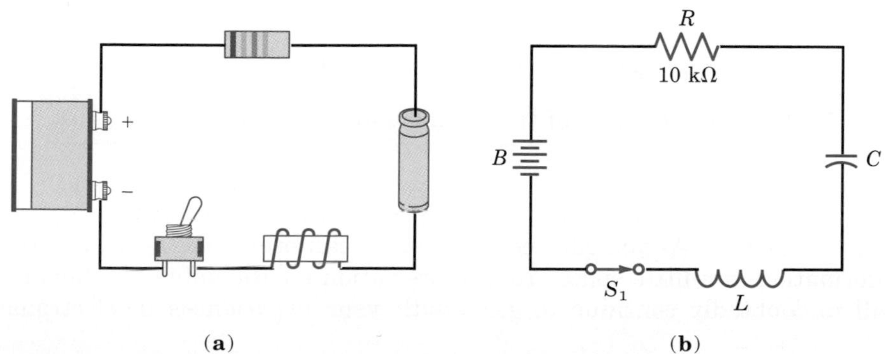

FIGURE 2–8a Sample pictorial diagram

FIGURE 2–8b Sample schematic diagram

1. How many different kinds of components are there in the circuit? Name them.
2. What is the value of the resistor?
3. What kind of switch is used, a SPST or a SPDT?
4. Is the voltage source a cell or a battery?

PRACTICE PROBLEM III

Look at Figure 2–9 and answer the following questions:

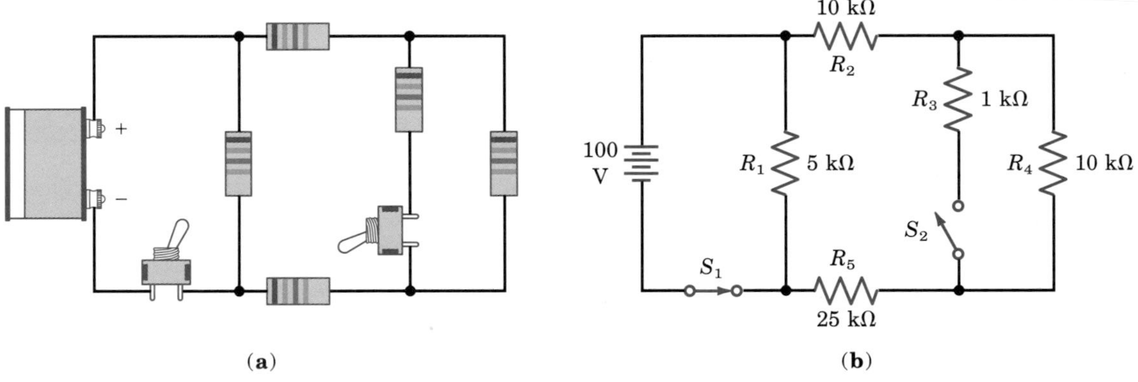

FIGURE 2–9a Sample pictorial diagram

FIGURE 2–9b Sample schematic diagram

1. Without counting the conductor wires, how many different kinds of components are in the circuit?
2. How many resistors are in the circuit?
3. Which resistor has the highest resistor value? What is its value?
4. How many switches are in the circuit?
5. What is the value of the circuit applied voltage?

Did you do well? As you can see, schematics can provide a large amount of information in a little space. Your appreciation for the value of schematics will undoubtedly continue to grow with your experiences in electronics.

OHM'S LAW AND RELATIONSHIPS BETWEEN ELECTRICAL QUANTITIES

Georg Simon Ohm, after whom the unit of resistance is named, observed and documented important relationships between three fundamental electrical quantities. He verified a relationship that indicated that electrical current in a circuit was *directly related* to the voltage applied and *inversely related* to the circuit's resistance. Recall that directly related means as one item increases or decreases, the other item also increases or decreases in step with the first item. Inversely related means as one item increases, the other item decreases, and vice versa. Also recall that in defining some of the electrical quantities, the relationships in **Ohm's law** were implied in the definitions. For example, an ohm is that amount of resistance that limits current to one ampere when one volt is applied.

The Relationship of Current to Voltage with Resistance Constant

These relationships are stated in a formula known as Ohm's law, where:
Current (amperes) = Potential difference (volts) / Resistance (ohms)

OR

Formula 2–1	$I = \frac{V}{R}; \left(\text{Amperes} = \frac{\text{Volts}}{\text{Ohms}}\right)$

EXAMPLE
If a circuit has 10 volts applied voltage and a resistance of 5 ohms, what is the value of current in amperes?

Answer: $I = \frac{V}{R}$, where $I = \frac{10\text{ V}}{5\ \Omega} = 2$ amperes. (See Figure 2–10)

As you can see from the formula, if V increases and R remains unchanged, then V divided by R is a larger number and the value of I increases, Figure 2–11.

For another example of this concept, assume V is 20 volts instead of 10V (in Figure 2–10) and R remains at 5 ohms. Current becomes 4 amperes because 20 divided by 5 equals 4. As V increases, I increases, showing a direct relationship of I and V.

The Relationship of Current to Resistance with Voltage Unchanged

In Figure 2–10, if voltage remains at 10 volts but the resistance increases to 10 ohms, what is the new current value? If you said one ampere, you are right because $I = \frac{V}{R}$, where $I = \frac{10\text{ V}}{10\ \Omega} = 1$ ampere. As R increases, I decreases, showing an inverse relationship of I and R, Figure 2–11.

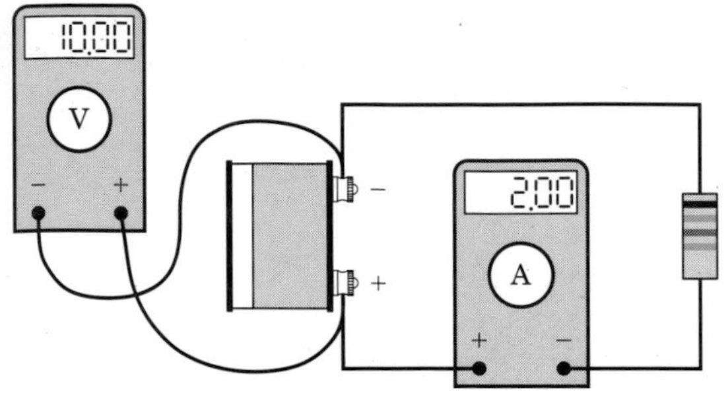

$$I = \frac{V}{R}$$

$$I = \frac{10\ V}{5\ \Omega} = 2\ A$$

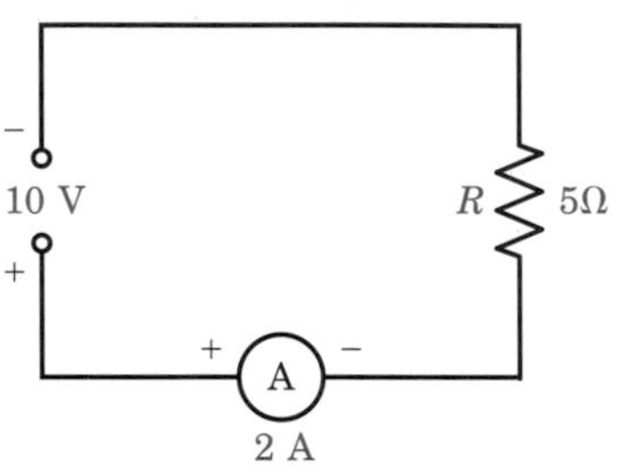

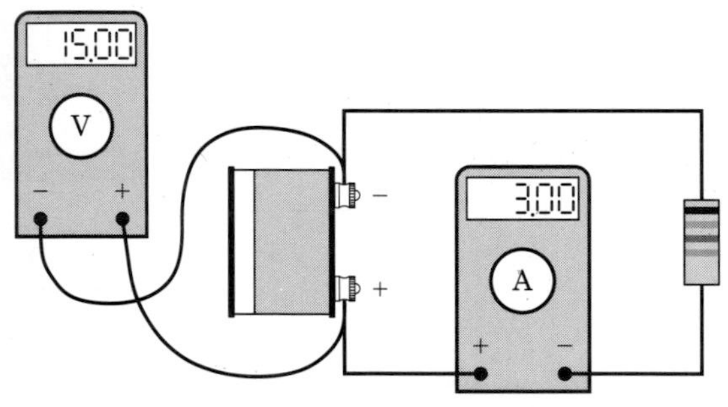

$$I = \frac{V}{R}$$

$$I = \frac{15\ V}{5\ \Omega} = 3\ A$$

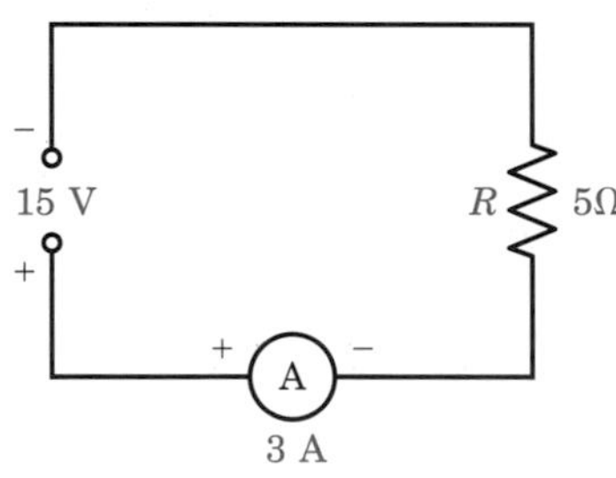

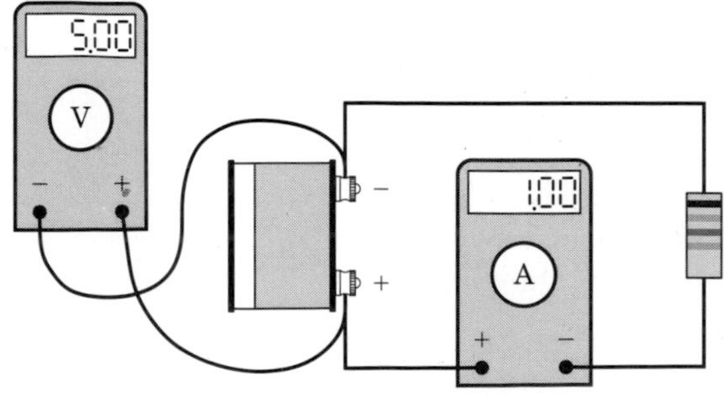

$$I = \frac{V}{R}$$

$$I = \frac{5\ V}{5\ \Omega} = 1\ A$$

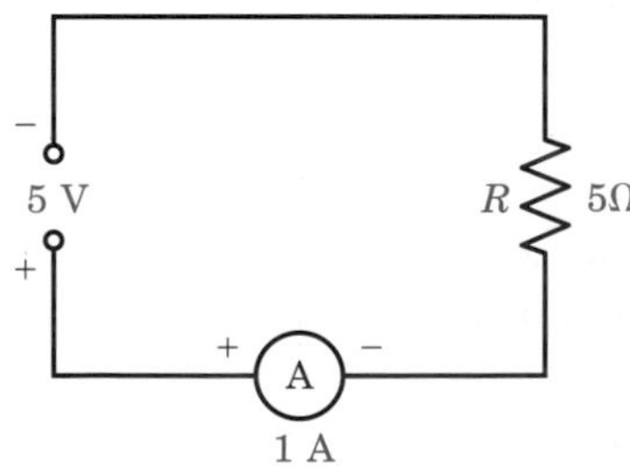

FIGURE 2–10 Examples of how to use $I = \frac{V}{R}$ Ohm's law formula

If	$\frac{V\uparrow}{R\rightarrow}$	then I↑	(I directly related to V. If V doubles, I doubles, if R is not changed)
If	$\frac{V\downarrow}{R\rightarrow}$	then I↓	(I directly related to V. If V halves, I halves, if R is not changed)
If	$\frac{V\rightarrow}{R\uparrow}$	then I↓	(I inversely related to R. If R doubles, I halves, if V is not changed)
If	$\frac{V\rightarrow}{R\downarrow}$	then I↑	(I inversely related to R. If R halves, I doubles, if V is not changed)

FIGURE 2–11 Direct and inverse relationships

PRACTICE PROBLEM IV

Look at Figure 2–12 and determine the current for the circuit shown.

FIGURE 2–12 Practice problem

PRACTICAL NOTES

To solve problems, get in the habit of writing down the appropriate formula, substituting the knowns into the formula, then solving for the unknown. Also, if the problem is stated in words (without a diagram), draw a simple diagram illustrating the circuit in question. It will help you think clearly!

So far, you have looked at only one arrangement of Ohm's law: $I = \frac{V}{R}$. Now, a quick review of transposing an equation will be given because this process isolates the unknown quantity of interest on one side of the equation in order to simplify finding its value.

Recall that an equation is a statement of equality between two quantities, or sets of quantities. The equality is expressed by the "=" sign. For example, the Ohm's law equation of $I = \frac{V}{R}$ says that the current is equal to the voltage divided by the resistance. Notice that the I is shown on one side of the equation, isolated from the other factors. To find its value, it is only necessary to carry out the operations on the other side of the equation. If both V and R are known, then their values are substituted into the equation to solve for I.

REARRANGING TO FIND THE UNKNOWN QUANTITY

How can the $I = \frac{V}{R}$ equation be transposed to solve for V or R? Before doing that, let's look at two basic premises relating to equations.

1. *Multiplying* or *dividing* both sides of an equation by the *same* factor does not change the equality.
2. By using the correct value factor, cancellation on one side of the equation can be made to provide the desired change in the equation by isolating the desired unknown.

EXAMPLE

Dividing and cancelling technique

$A = B \times C$ Solve for C

To isolate "C" (that is, to get it by itself on one side of the equation), divide *both* sides of the equation by B and cancel factors, as appropriate.

$$\frac{A}{B} = \frac{\cancel{B} \times C}{\cancel{B}} \quad \text{or} \quad \frac{A}{B} = C \quad \text{or} \quad C = \frac{A}{B}$$

EXAMPLE

Multiplying and cancelling technique

$I = \frac{V}{R}$ Solve for V

To isolate "V," multiply both sides of the equation by R, and cancel factors, as appropriate.

Therefore, $I \times R = \frac{V}{R} \times R$ or $I \times R = \frac{V\cancel{R}}{\cancel{R}}$

Then, $IR = V$ or

Formula 2–2	$V = I \times R$

How can we transpose our original $I = \frac{V}{R}$ formula to solve for R? First, multiply both sides of the equation by R so that the R can cancel out on one side.

Therefore, $I = \frac{V}{R}$ Then, $I \times R = \frac{V \times \cancel{R}}{\cancel{R}}$, and $I \times R = V$

Now, I and R are on one side of the formula and R is no longer a denominator in a fraction.

How can we isolate R?

If both sides are divided by I, then the I on the left side of the equation will cancel because I divided by I = 1.

Therefore, IR = V. Then, $\frac{\cancel{I} \times R}{\cancel{I}} = \frac{V}{I}$

Formula 2–3	$R = \frac{V}{I}$

As you can see, either multiply or divide *by the factor you are trying to cancel,* depending on whether that factor is in the numerator or the denominator of the equation. Multiply both sides of the equation by that factor, if you are trying to cancel the factor in a denominator. Divide both sides of the equation by the factor you are trying to cancel, if the factor is in the numerator.

You have now been exposed to all three variations of Ohm's law formula: $I = \frac{V}{R}$, $V = I \times R$, and $R = \frac{V}{I}$. This review of transposing formulas to isolate unknowns will help as other similar occasions arise. A useful memory aid often helps students remember these three variations.

The simple visual aid to memory that has been devised which has proven very useful for many years in helping beginning students to remember the three variations of the Ohm's law formula is shown in Figure 2–13.

FIGURE 2–13 Ohm's law memory aid

Simply look at (or cover up) the desired quantity (I, V, or R) and observe the relationships of the other two quantities to remember the correct version of the formula:

NOTE: If looking at I, then the aid shows $\frac{V}{R}$; thus $I = \frac{V}{R}$

If looking at V, then the aid shows $I \times R$; thus $V = I \times R$

If looking at R, then the aid shows $\frac{V}{I}$; thus $R = \frac{V}{I}$

Restating these three Ohm's law variations:

$$I = \frac{V}{R} \text{ or Amperes} = \frac{\text{Volts}}{\text{Ohms}}$$

$$V = I \times R \text{ or Volts} = \text{Amperes} \times \text{Ohms}$$

$$R = \frac{V}{I} \text{ or Ohms} = \frac{\text{Volts}}{\text{Amperes}}$$

Knowing any two of the three electrical quantities, you can solve for the unknown value by using the appropriate version of the Ohm's law formula. This capability is important to know.

In the practical world of electronics, you usually will not be dealing with just the basic units of voltage, amperes, and ohms. You will frequently work with millivolts, milliamperes, microamperes, kilohms, and megohms.

SAMPLE APPLICATION OF METRIC PREFIXES AND POWERS OF TEN

Suppose that a circuit has 10,000 ohms of resistance with 150 volts of applied voltage, what is the current?

Solution

$$I = \frac{V}{R}$$

$$I = \frac{150 \text{ V}}{10{,}000\ \Omega} \quad \left(\text{or } \frac{150 \text{ V}}{10 \text{ k}\Omega}\right)$$

$$I = 0.015 \text{ amperes} \quad (\text{or } 15 \text{ mA})$$

It is cumbersome to divide units by thousands of units (kilounits), isn't it? It turns out that when you divide units (volts) by kilounits (kΩ), the answer is in milliunits (mA). That is a convenient thing to remember, and you will use it often.

However, how would this work if we used powers of ten?

Solution

$$I = \frac{V}{R}$$

$$I = \frac{150 \text{ V}}{10 \times 10^3\ \Omega}$$

$$I = 15 \times 10^{-3} \text{ amperes (or 15 milliamperes, i.e., 15 mA)}$$

Did you notice that when 10^3 is brought from the denominator to the numerator, it becomes 10^{-3}? Suppose we had divided one by one thousand. How many times would one thousand go into one? The answer is one one-thousandth time. Recall that one thousand equals 10^3 and one one-thousandth equals 10^{-3}. Therefore, when 150 volts are divided by 10,000 ohms, the answer is in thousandths of an ampere.

Stated mathematically:

$$\frac{1}{1{,}000} = 0.001 \text{ or one thousandth (or } 1 \times 10^{-3}\text{)}$$

So:

$$\frac{150}{10{,}000} = 0.015 \text{ or 15 thousandths (or } 15 \times 10^{-3}\text{)}$$

SOME BASIC RULES RELATING TO POWERS OF TEN

Since you will be using powers of ten frequently, we will summarize some rules.

Rules for Moving Decimal Places Using Powers of Ten

1. To express a decimal as a whole number times a power of ten, move the decimal point to the right and count the number of places moved from the original point. The number of places counted (or moved) is the appropriate negative power of ten.

EXAMPLE

The number 0.00253 is expressed using powers of ten as follows:

(1) 253×10^{-5} (2) 2530×10^{-6} (3) 25.3×10^{-4} and so on.

---

2. To express a large number as a smaller one times a power of ten, move the decimal point to the left and count the number of places moved from the original point. The number of places counted is the appropriate positive power of ten.

EXAMPLE

The number 42,853 is expressed using powers of ten as follows:

(1) 4.2853×10^4 (2) 42.853×10^3 (3) 4285.3×10^1 and so on.

Rule for Multiplication Using Powers of Ten

When multiplying powers of ten, add the exponents (algebraically). (NOTE: Laws of exponents are applicable when using powers of ten.)

EXAMPLE

$10^2 \times 10^2 = 10^4$

$2 \times 10^4 \times 2 \times 10^{-3} = 4 \times 10^1$

Rule for Division Using Powers of Ten

When dividing powers of ten, subtract the exponents (algebraically).

EXAMPLE

$$\frac{4 \times 10^{-2} \text{ (numerator)}}{2 \times 10^{-4} \text{ (denominator)}} = 2 \times 10^{-2} \times 10^4 = 2 \times 10^2$$

$$\frac{5 \times 10^3}{2.5 \times 10^6} = 2 \times 10^3 \times 10^{-6} = 2 \times 10^{-3}$$

NOTE: Change the sign of the denominator power of ten and add algebraically to the numerator power of ten exponent. If the power of ten exponent in the denominator is the same sign and same value as the numerator, they cancel.

Rules for Multiplication and Division Combined

The rules are: a. Multiply all factors in the numerator.
b. Multiply all factors in the denominator.
c. Divide.

EXAMPLE

$$\frac{15 \times 10^2 \times 10 \times 10^4}{2 \times 10^{-2} \times 12.5 \times 10^{-2}} = \frac{150 \times 10^6}{25 \times 10^{-4}} = 6 \times 10^{10}$$

Rule for Addition of Powers of Ten

Change all factors to the same power of ten and add the coefficients. The answer retains the same power of ten as the added factors.

EXAMPLE

$$(2 \times 10^2) + (5 \times 10^3) = (0.2 \times 10^3) + (5 \times 10^3) = 5.2 \times 10^3$$

Examples of coefficients

Rule for Subtraction of Powers of Ten

Change all powers of ten to the same power of ten and subtract the coefficients.

EXAMPLE

$(5 \times 10^3) - (2 \times 10^2) = (5 \times 10^3) - (0.2 \times 10^3) = 4.8 \times 10^3$

See if you can apply Ohm's law, metric prefixes, and powers of ten.

IN-PROCESS LEARNING CHECK II

1. Refer to the diagram below and determine the circuit current. I = ________ amperes, or ________ mA.

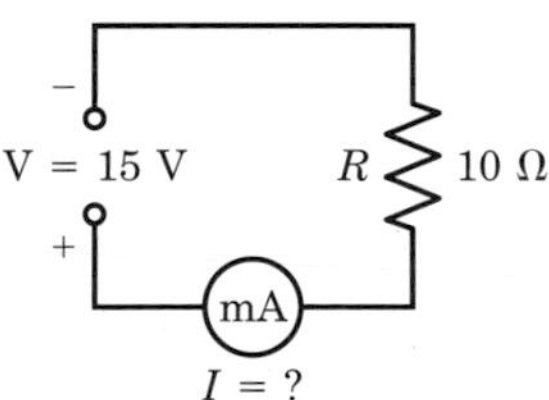

2. Refer to the circuit below and determine the circuit applied voltage. V = ________ volts.

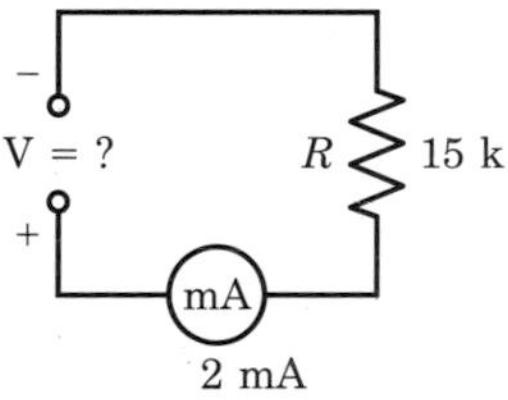

3. Refer to the circuit below and determine the value of circuit resistance. R = ________ ohms, or ________ k ohms.

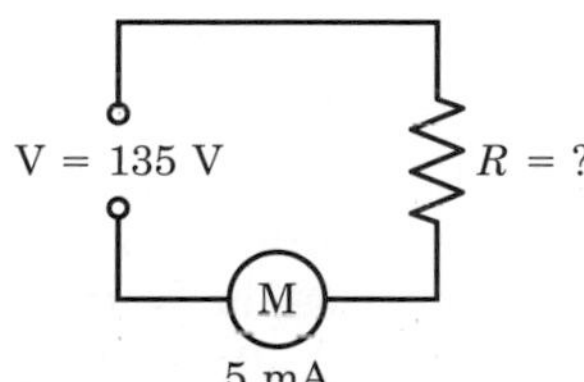

4. a. If R increases and V remains the same in Problem 1, does current increase, decrease, or remain the same?
   b. If I increases and R remains the same in Problem 2, would the voltage applied be higher, lower, or the same as stated in the problem?
   c. In Problem 3 if current doubles but the voltage remains the same, what must be true regarding the circuit resistance?

DIRECTION OF CURRENT FLOW

In Chapter 1, there was a brief discussion about electrons (negative in charge) moving to a positively charged point in a circuit. When circuit diagrams are shown, it is quite common to show the direction of current flow through the circuit by means of an arrow. In this case, we are referring to electron current flow. Figure 2–14 shows how arrows can be used to show direction of electron current flow.

FIGURE 2–14 Method of showing direction of current flow

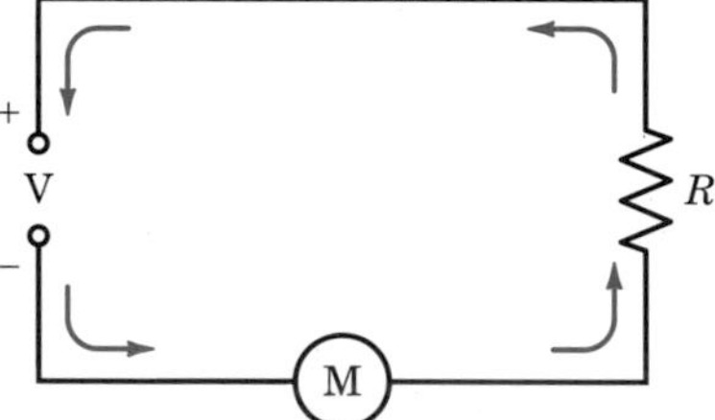

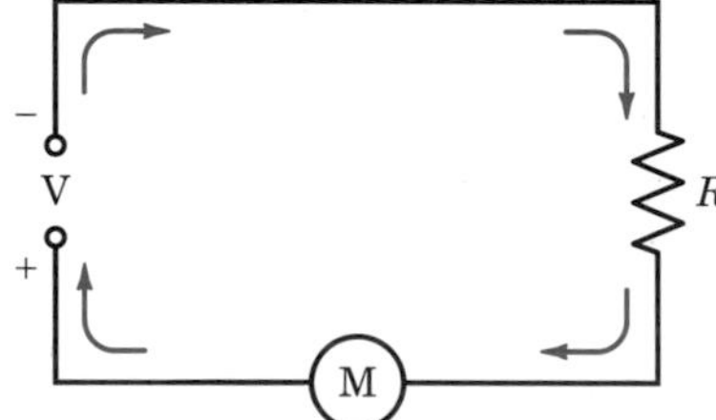

POLARITY AND VOLTAGE

The concept of polarity, one point in a component or circuit being considered more negative or positive than another point, was discussed earlier. Recall that the more negative point was labeled with a negative sign, and the less negative (or positive point) was labeled with a positive sign. This technique indicates there is a potential difference, or voltage, between the two points.

PRACTICAL NOTES

You should have observed throughout our discussions that current flows through a component and/or circuit. Voltage does **not** "go" or "flow." It is a difference of potential between points, so think of voltage as being **across,** or **between** two points.

To label the polarity of voltages across components in a circuit diagram, note the direction of current flow, then label the point where the current enters the component as the negative end. The other end of the component (where the current exits) is labeled as the positive point. This process is true for all components in the circuit, except the voltage source. In that case, the negative side of the source is the source of the electron flow, and the positive end of the source is the end where the electrons return after traveling through the external circuit, Figure 2–15.

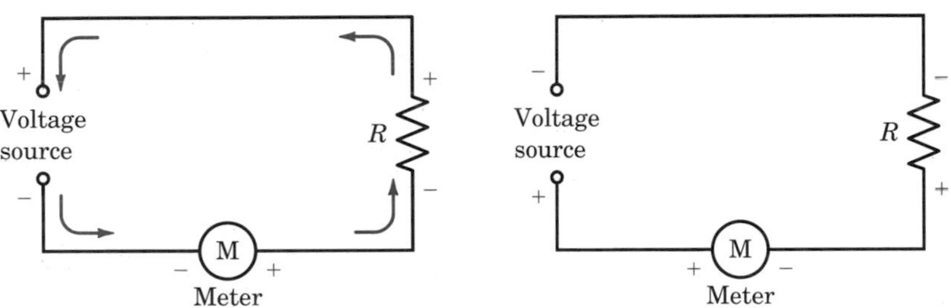

FIGURE 2–15 Illustrating polarity of voltages

WORK, ENERGY, AND POWER

Many students may remember facts regarding work, energy, and power from high school science courses; however, a brief review will be given for those who are unfamiliar with these terms, or those who may need a refresher.

Work is the expenditure of energy.

Formula 2–4	work = force × distance

The basic unit of work is the foot poundal. At sea level, 32.16 foot poundals equals 1 foot pound.

Energy is the ability to do work, or that which is expended in doing work. The basic unit of energy is the "erg," and 980 ergs equals one gram centimeter. This is a small unit so the unit joule (pronounced jewel) is often used, and one joule equals 10^7 ergs.

Mechanical power is the rate of doing work. For example, the "horsepower" equals a rate of 550 foot pounds per second.

Therefore, electrical energy is the ability to do electrical work. The unit of energy used for electrical energy is, again, the joule. For electrical applications, the joule is defined as the energy required to move one coulomb of electrical charge between two points having a difference of potential of one volt between them.

The unit of electrical power is the **watt,** named after the scientist, James Watt. The watt is electrical work at the rate of one joule per second. Since joules can be related to the basic unit of energy, the erg, and the erg is related to force multiplied by distance, it can be determined that 746 watts is equivalent to one horsepower of mechanical power, or 746 watts equals one horsepower, Figure 2–16.

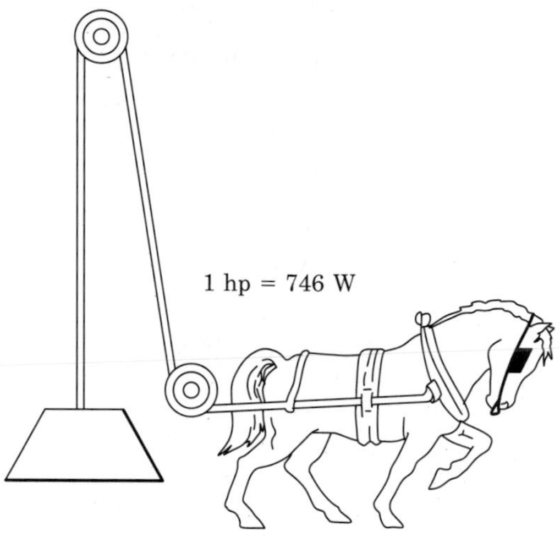

FIGURE 2–16 Concept of the unit of "horsepower"

The Basic Power Formula

Relating electrical power to practical circuits:

- A joule is the energy used to move one coulomb of charge between two points with a one volt potential difference.
- An ampere is the rate of charge movement of one coulomb per second.
- A watt is the rate of doing work at one joule per second.

Thus, one watt of power equals one ampere times one volt. The formula is:

Power (in watts) = Voltage (in volts) × Current (in amperes)

or

Formula 2–5	$P = V \times I$

EXAMPLE

If a circuit has 50 mA of current flow and an applied voltage of 10 volts, how much power does the circuit dissipate? Using the basic formula for power:

$P = V \times I$
$P = 10\ V \times 50\ mA$
$P = 10 \times 50 \times 10^{-3} = 500 \times 10^{-3}$ watts, or 500 milliwatts (mW)

Variations of the Basic Power Formula

Just as the Ohm's law formula can be rearranged, so can the power formula. The formula we started with was $P = V \times I$. Two other arrangements are $V = \frac{P}{I}$ and $I = \frac{P}{V}$. Incidentally, the same style of visual aid used for the Ohm's law formula will work here, Figure 2–17.

FIGURE 2–17 Power formula memory aid

Again, by looking at the desired parameter (or covering it up) and seeing the arrangement of the other two parameters; the correct formula is evident. For example, if you know V and I, to find power take $V \times I$. If you know the power and the current, simply divide P by I to get V, and so forth.

Other practical variations of the power formula are developed by combining knowledge of Ohm's law and the power formulas. For example, since $V = I \times R$ (Ohm's law) and $P = V \times I$, the formula can be combined to show that $P = I^2 \times R$ or I^2R. You have substituted for V in the Ohm's law formula $P = V \times I$; therefore $P = (I \times R) \times I$.

Formula 2–6	$P = I^2R$

Using this same idea, you can derive the formulas that state:

$P = \frac{V^2}{R}$, $P = VI$, $P = V \times \left(\frac{V}{R}\right)$, or

Formula 2–7	$P = \frac{V^2}{R}$

You have simply substituted the Ohm's law formula for I into the "I" part of the $P = V \times I$ power formula.

In summary, three ways to solve for power include:

1. $P = V \times I$
2. $P = I^2 \times R$
3. $P = \frac{V^2}{R}$

Each of these formulas can be rearranged using the transposition and cancelling techniques learned earlier in this chapter. For example, you can solve for R in the second formula ($P = I^2R$) with $R = \frac{P}{I^2}$, and you can solve for $I^2 = \frac{P}{R}$ or $I = \sqrt{\frac{P}{R}}$. Also, you can use the third formula $\left(P = \frac{V^2}{R}\right)$ and get the expressions $R = \frac{V^2}{P}$ and $V = \sqrt{PR}$.

COMMONLY USED VERSIONS OF OHM'S LAW AND POWER FORMULAS

The most important equations to remember are $I = \frac{V}{R}$, $V = IR$, and $R = \frac{V}{I}$ for the Ohm's law formulas and $P = VI$, $P = I^2R$, and $P = \frac{V^2}{R}$ for the power formulas. By understanding and knowing these Ohm's law and power formulas and by becoming adept at applying them to practical electronic problems, you will be on the way to becoming a good technician.

For a change of pace from simply reading about these formulas, we now want you to see how these formulas can be applied to practical problems. Follow the steps carefully as several sample problems and solutions are shown. You will be doing this sort of thing on your own throughout the remainder of your training.

EXAMPLE

Refer to Figure 2–18 for the following.

1. In Diagram 1, find the voltage (V). Ohm's law formula for voltage is $V = I \times R$.

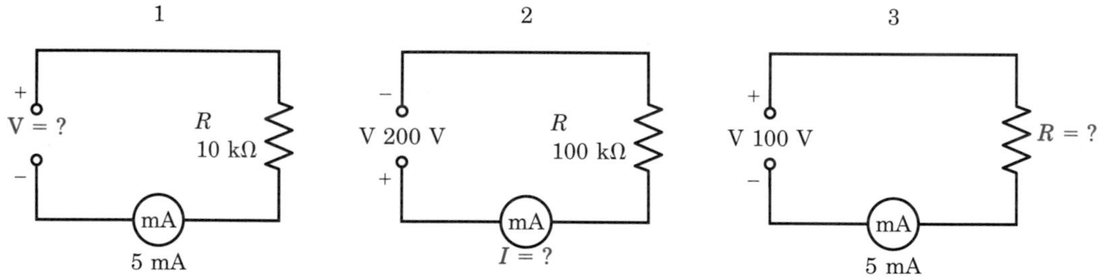

FIGURE 2–18 Ohm's law problems

Substitute the known values for I and R.
Therefore, $V = 5 \times 10^{-3}$ amperes $\times 10 \times 10^3$ ohms $V = 50$ volts

Remember: When multiplying powers of ten, add their exponents algebraically:
10^{-3} and 10^3 add to 10^0, which equals 1. Also, milliunits multiplied by kilounits result in units, and, in this case, mA × k results in volts.

2. In Diagram 2, find the circuit current (I). Ohm's law formula for current is $I = \frac{V}{R}$.

Substitute the known values for V and R.
Therefore, $I = \frac{200 \text{ V}}{100 \text{ k}}$ or $I = \frac{200 \times 10^0}{100 \times 10^3} = 2 \times 10^{-3} = 2\text{mA}$

Remember: When dividing powers of ten, subtract the exponents algebraically. Since the voltage did not show a power of ten, show it as 200×10^0 [$10^0 = 1$]. To subtract the denominator 10^3 from the numerator 10^0, change the sign of the denominator power of ten, then add 10^{-3} to 10^0, which equals 10^{-3}. Also, units divided by kilounits result in milliunits. In this case, $\frac{\text{V}}{\text{k}\Omega}$ results in milliamps (mA).

3. In Diagram 3, find the resistance (R). Ohm's law formula for resistance is $R = \frac{V}{I}$.

   Substitute the known values for V and I.

   Therefore, $R = \frac{100\text{ V}}{5\text{ mA}}$ or $R = \frac{100}{5 \times 10^{-3}}$; $R = 20\ \text{k}\Omega$

   Remember: When moving a power of ten in the denominator by dividing it into the numerator, change its sign and add the exponents as appropriate. The 10^{-3} for the milliamps becomes the 10^3 for kilohms in the answer. Also, in this problem units divided by milliunits result in kilounits; thus, volts divided by milliamps result in the answer as kilohms.

In summary, when applying Ohm's law and power formulas:

- look at the circuit to determine the knowns and unknown(s)
- select and write the appropriate formula to use
- substitute the knowns into the formula
- use metrics and powers of ten to manage large and small numbers
- remember that units divided by milliunits equals kilounits and that units divided by kilounits equals milliunits
- remember that milliunits multiplied by kilounits equals units
- recall various expressions for Ohm's law

EXAMPLE

Use Figure 2–19 for the following.

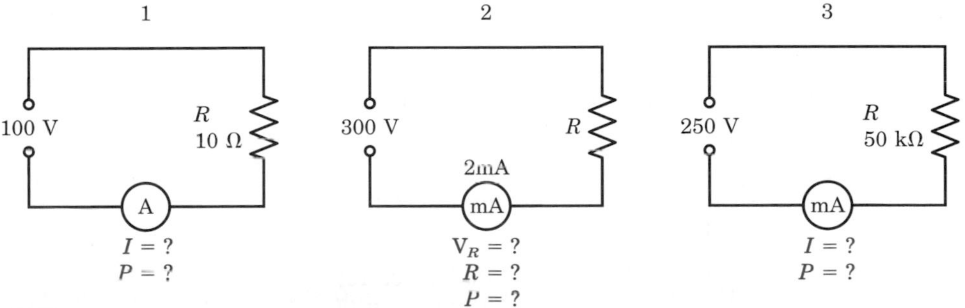

FIGURE 2–19 Combination Ohm's law and power formulae problems

1. In Diagram 1, to solve for current, which formula should we use first? Observe that values for V and R are known.

   Therefore, $I = \frac{V}{R}$

   $$I = \frac{100\ V}{10\ \Omega} = 10\ A$$

   Knowing the current, you can solve the remaining unknowns. Therefore, $P = V \times I = 100\ V \times 10\ A = 1{,}000$ watts (or 1 kilowatt)
   Also, since V, I and R are known, you can use the other power formulas, as well: $P = I^2R$ or $P = \frac{V^2}{R}$

2. In Diagram 2, which formula might be used first? In this case, voltage and current are known values, but resistance is unknown. If the current meter does not drop any appreciable voltage, you could assume that V_R equalled the applied voltage and first solve for power. In many problems, you can start at more than one place, depending on the known value(s).
   Let's start with the power formula

   $$P = V \times I$$

   $$P = 300\ V \times 2 \times 10^{-3}\ A = 600 \times 10^{-3} \text{ watts (or 600 mW)}$$

   Now solve for the other unknowns:

   Assume 300 V across R, where $V_R = 300$ V

   Therefore, $R = \frac{V}{I}$

$$R = \frac{300\ V}{2 \times 10^{-3} A}$$

$R = 150 \times 10^3$ ohms (or 150 kΩ)

3. In Diagram 3, a good starting point is to first solve for I since you know the values of V and R.

   Therefore, $I = \frac{V}{R}$

   $$I = \frac{250\ V}{50\ k\Omega} = \frac{250}{50 \times 10^3} = 5 \times 10^{-3} \text{ amperes (or 5 mA)}$$

   Now that voltage and current are known, it is easy to solve for the other unknown.

   Therefore, $P = V \times I$

   $P = 250\ V \times 5 \times 10^{-3}\ A = 1{,}250 \times 10^{-3}$ watts or 1.25 watts

In summary, now you should know:

- how to find the known value(s) from the given diagram or facts
- where there are several starting places it is best to look for the formula that might be the easiest to use and start there
- how to combine Ohm's law knowledge with power formulas

PRACTICE PROBLEM V

Try the problems in Figure 2–20 to see if you can apply logic and appropriate formulas. Don't cheat yourself by looking ahead. You can find the answers in Appendix B.

Now check yourself to see how well you did! You will be getting more practice, both in lab and within this text, as you proceed in your training.

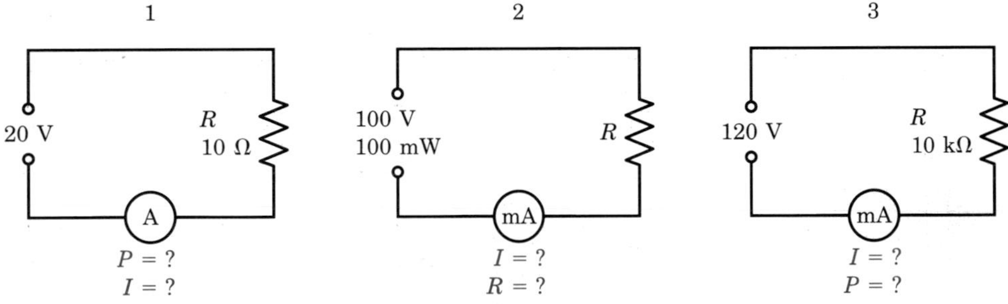

FIGURE 2–20

If you have difficulty in solving these, take a few minutes to review appropriate parts of this chapter to find the formulas and the needed guidance or hints.

RESISTORS

Uses

Resistors limit and/or help determine current values for a given circuit. This current controlling function is one of the important roles that resistors perform in electronics. Another function, which you'll study later, is dividing voltage. Both of these purposes are important.

Types

Resistors are designed to perform these functions under a variety of operating conditions, including various voltage, current, and power levels. Each situation demands certain characteristics of this component. Due to these demands, several basic types of resistors have been designed, Figure 2–21, and include:

1. Carbon/composition
2. Wire wound
3. Special purpose
4. Fixed, semi-fixed, and variable

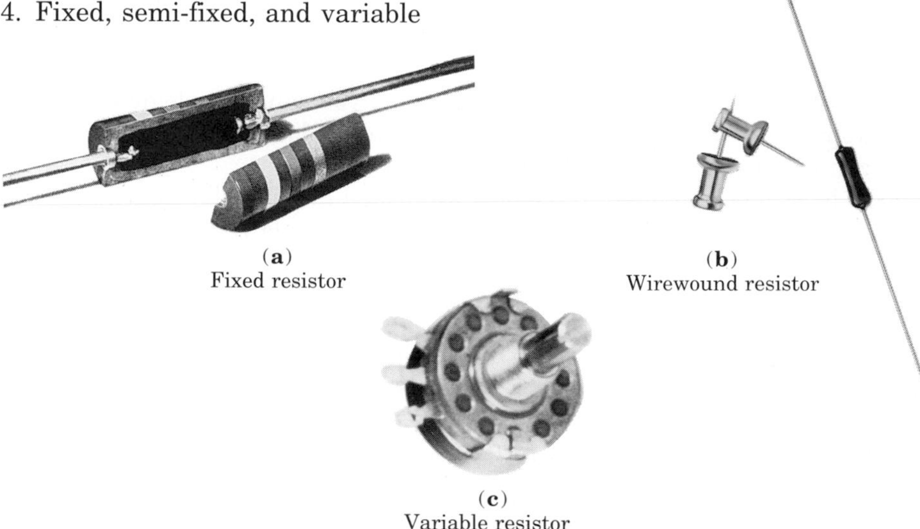
(a) Fixed resistor

(b) Wirewound resistor

(c) Variable resistor

FIGURE 2–21 Various types of resistors (Photos a and c courtesy of Allen Bradley Co., Inc.; photo b courtesy of IRC, Inc.)

Basic Construction

Carbon, or composition resistors, are composed of carbon or graphite mixed with powdered insulating material. The mixture controls the amount of resistance exhibited for a given dimension of the material. These carbon resistors appear either in a tubular shape or in a circular, flat disc form for special resistive control elements, as shown in Figure 2–21.

Wire wound resistors generally comprise special resistance wire wrapped around a ceramic, or insulator core. The amount of resistance is controlled by the size of the wire, the type of material in the wire, and the length of the wire wrapped around the form, Figure 2–22.

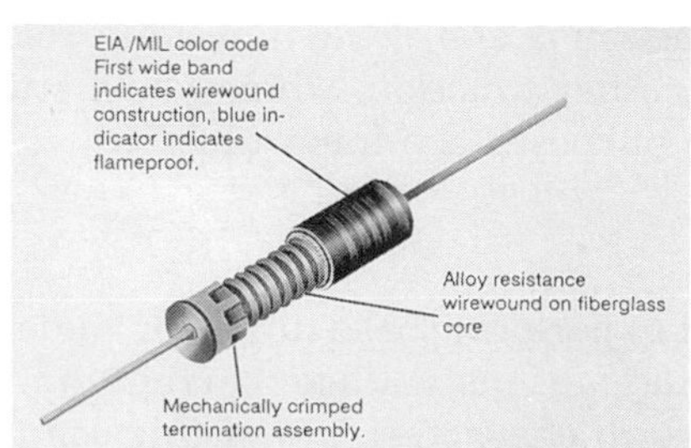

FIGURE 2–22 Wire wound resistors (Courtesy of IRC, Inc.)

The power-handling capability of any resistor is a function of its physical size and ability to dissipate heat. Typically, power ratings for carbon resistors range from $\frac{1}{4}$ watt to 2 watts, and wire wound resistors range from about 5 watts to over 100 watts or more.

Familiarity with these resistors will be gained in laboratory applications. Important characteristics to consider when selecting or replacing resistors include value in ohms, wattage rating, voltage rating, accuracy of value or tolerance rating, physical dimensions, and other similar factors.

RESISTOR COLOR CODE

A skill used continuously by electronic technicians is being able to decode the **resistor color code.** Because many resistors are so physically small, it is difficult to print their value in a readable size. A system of color coding has been adopted and standardized by the Electronics Industries Association (EIA); thus, technicians everywhere understand the code.

The band system of color coding consists of three or four bands, or stripes, of color around the tubular-type carbon resistors. Special film-type resistors, which have a thin layer of resistive material coating the outside of a ceramic form, may have five stripes. The bands important to you now are the first four bands.

Look at Figure 2–23 and observe the placement of these color bands. The first color band is the one closest to the end of the resistor. The other color bands progress toward the center of the resistor. The key is to learn the significance of *each color* and of *each band position.*

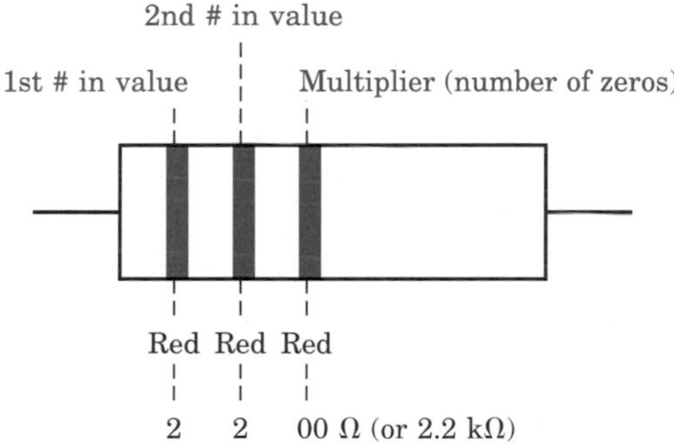

FIGURE 2–23
Color band positions

Significance of Colors

A good starting point is to learn the meaning of each color. Note that the color code starts with dark colors and progresses toward lighter colors. One method used to memorize a sequence of items is a memory aid sentence. Figure 2–24 shows how the sentence *"**B**ig **B**oys **R**ace **O**ur **Y**oung **G**irls, **B**ut **V**iolet **G**enerally **W**ins"* can be used to remember the color code.

0	1	2	3	4	5	6	7	8	9
BLACK	BROWN	RED	ORANGE	YELLOW	GREEN	BLUE	VIOLET	GRAY	WHITE
BIG	BOYS	RACE	OUR	YOUNG	GIRLS	BUT	VIOLET	GENERALLY	WINS

FIGURE 2–24 Color values chart

The colors represent the sequential numbers zero through nine. The first letter of each word in the sentence is the same as the first letter of each color. Learn these!

Significance of Band Positions

Refer to Figures 2–23 and 2–25 to see the significance of the color band positions and how the first three bands provide the resistance value. The first band represents the first number of the resistor's value in ohms. The second band provides the second number of the value. The third band (the *multiplier* band) indicates how many zeros are added to the first two numbers to decode the resistor's value.

The fourth band is the *tolerance* band. It indicates that the actual value of the resistor might fall in a range of values of plus or minus some percentage of the color-coded value. If there is no fourth band color, the tolerance is 20%. A silver fourth band color band signifies a tolerance of plus or minus 10%. A gold fourth band indicates a tolerance of 5%, Figure 2–26. Notice that a red fourth band represents a tolerance of 2%.

FIGURE 2–25 The fourth band shows tolerance rating.

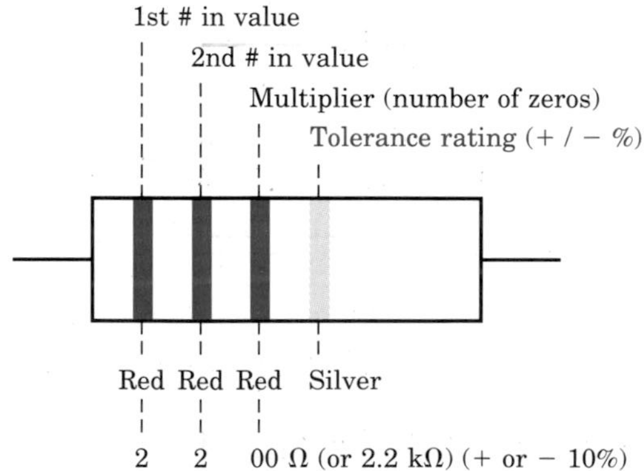

Special Use of Gold and Silver Colors

Very low value resistors sometimes require a *multiplier* value for the third band that indicates the numbers decoded from the first two color bands are multiplied by either 0.1 or 0.01 to find the color-coded value for the resistor.

A *gold third band* indicates the multiplier to be one-tenth (0.1). For example, if the first two color bands are red and violet (27), then a gold third band would indicate 0.1 × 27 = 2.7 ohm resistor value.

A *silver third band* indicates a multiplier of one-hundredth (0.01). This means if the resistor is color-coded with brown-black-silver bands (in that order), its value is 0.1 ohm.

You will get practice problems from this book and in laboratory work; however, several examples, Figure 2–26, are provided to help you understand the system.

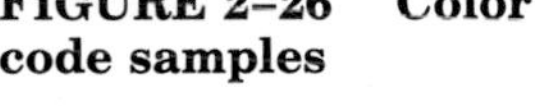

FIGURE 2–26 Color code samples

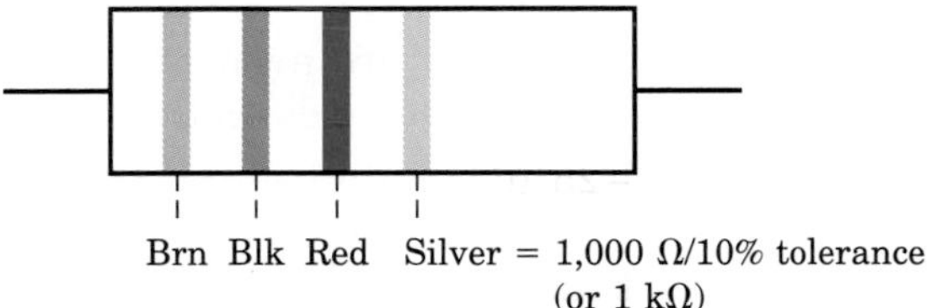

(NOTE: at 10% tolerance the actual resistance could be between 900 Ω and 1100 Ω and still be within its rating; i.e. 1,000 Ω + or − 100 Ω.)

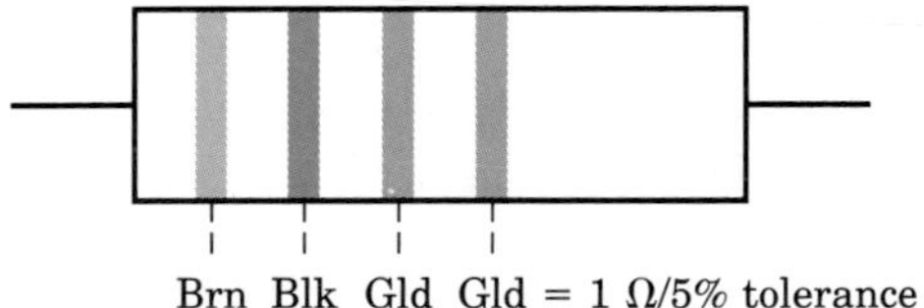

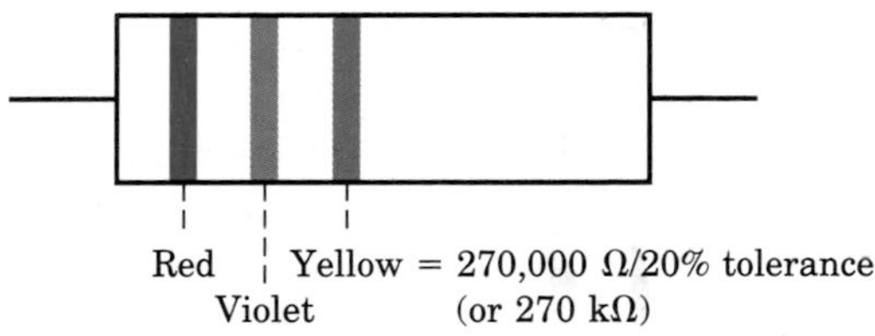

SAFETY HINTS

You may have already been exposed to some safety hints about electricity and electronics. Nevertheless, here are a few facts and tips that can serve either as information or as reinforcement of earlier learning.

Background Facts

Electrical shock occurs when your body provides a path for current between two points having a difference of potential. Generally, the amount of injury a person receives from a shock depends on the amount of current forced through the path of current. A range of effects from minor discomfort, severe muscular contractions, severe burns, ventricular fibrillation (erratic heart action), and death can occur. Many times the injury caused by involuntary jerking or muscular contractions is worse than the shock. It has been stated that currents as low as 150 milliamperes can be fatal under the right conditions.

Having studied Ohm's law, you are aware that the amount of current through any current path depends on the voltage value and the resistance of the path. A higher voltage causes more current to flow through the circuit. The higher the resistance path, the lower the current. Therefore, anything that lowers the voltage on those circuits or any means of increasing resistance of possible current paths through the body is helpful.

Some Safety Tips

1. *Remove power* from circuits before working on them, as appropriate.

2. If you must work on a circuit with power applied, *work with only one hand at a time.* This will prevent touching two points in the circuit simultaneously and providing a current path from one hand, through your body, to the other hand. Also, avoid touching two points with any part of your body. For example, your hand touching a point in the circuit, and your arm leaning against the circuit chassis, ground, or another point in the circuit can provide "shocking results."

3. *Insulate your body* from ground so a path is not provided from one point (through your body) to ground.

4. Use only *tools with insulated handles.*

5. *Know where the closest switch or main breaker switch is located* that will turn off power to the system. Make sure everyone working near you also knows this.

6. *Remove metal jewelry.* Dangling jewelry can not only make electrical contact, but, if you are working on "rotating machinery," can entangle you with the machinery.

7. When working on circuits with capacitors that store charge even after the power has been removed, *discharge the capacitors* prior to working on the circuit.
8. *Ground all equipment that should be grounded,* including equipment chassis, cabinets, and so forth.

Safety When Working With Hot Soldering Irons

1. *Position hot soldering irons properly* to prevent accidentally laying your arm, hands, or other objects on them.
2. *Wear safety glasses* when soldering.
3. *Be careful when shaking or wiping hot solder off a soldering iron.* Hot or molten solder splatters easily and causes serious burns.

NOTE: Other safety hints will be provided at strategic times during your training. Don't just read these safety hints; *follow them!*

SUMMARY

- Charge (Q or q) is measured in coulombs. The abbreviation for coulomb is C. A typical subunit of the coulomb is the μC, or microcoulomb, meaning one-millionth of a coulomb.
- Potential difference, or voltage, is represented by the letter "V." The unit of measure is the volt, also abbreviated "V." The submultiples and multiple of voltage commonly found include mV (millivolt), μV (microvolt), and kV (kilovolt).
- Current (I) is measured in amperes. The abbreviation for the unit of measure is A. Typical submultiples of this unit are mA (milliampere), μA (microampere), and occasionally pA (picoampere).
- Conductance uses the letter G and the unit of measure is the siemens (S). The commonly found submultiples are mS (millisiemens) and μS (microsiemens).
- Resistance uses the letter R and the unit of measure is the ohm (Ω). Typical multiples of this unit are the kΩ (kilohm) and the MΩ (megohm).
- Several components found in many circuits are resistors, capacitors, inductors, switches, batteries (or voltage sources), and the conductor wires. Know the standard symbols used for each.
- Commonly used metric subunit prefixes are pico or micro micro meaning one millionth of one millionth (one trillionth) of a unit; nano meaning one thousandth of a millionth (a billionth) of a unit; micro meaning one millionth of a unit; and milli meaning one thousandth of a unit.
- Powers of ten for the above subunit prefixes are 10^{-12} for pico (or micro micro), 10^{-9} for nano, 10^{-6} for micro, and 10^{-3} for milli units.

- Common metric multiple-unit prefixes are kilo (1,000 times the unit) and mega (1,000,000 times the unit).
- Symbols for all of the above metric prefixes are p (pico), n (nano), μ (micro), m (milli), k (kilo), and M (mega).
- Block diagrams illustrate the major functions within a circuit or system and convey the flow of signals, electrical power, and so forth. General flow is typically shown from left to right.
- Schematic diagrams provide more information than block diagrams about types of components in the circuit/s, their values, and specific connections and related electrical parameter values.
- Block diagrams provide an overview of the system, whereas schematic diagrams provide the information needed to enable circuit analysis, circuit repairs or maintenance, and/or desired circuit modifications.
- Current is directly related to voltage and inversely related to the resistance. Ohm's law states this as $I = \frac{V}{R}$.
- If the other factors in the equation are known, the techniques of transposition allow rearrangement of the equation to solve for an unknown factor. For example, by rearranging Ohm's law formula where $I = \frac{V}{R}$, two useful equations are derived where $V = I \times R$ and $R = \frac{V}{I}$.
- Powers of ten are useful when working with large or small numbers in formulas. When using powers of ten, the rules relating to exponents are used.
- The direction of current flow through a circuit, external to the voltage source, is from the source's negative side, through the circuit, and back to the source's positive side. Arrows are frequently drawn on schematic diagrams to illustrate this.
- When labeling the polarity of voltage across a given component in a circuit, the component end where the current enters is labeled the negative end and the end where the current exits is labeled the positive end.
- Electrical power is the rate of performing electrical work and is generally dissipated in a circuit or in a component in the form of heat. A watt of power is the performance of electrical work at a rate of 1 joule per second.
- A joule is the energy moving one coulomb of charge between two points that have a difference of potential of one volt between them.
- Several common formulas to calculate electrical power, or power dissipation, are $P = V \times I$, $P = I^2R$, and $P = \frac{V^2}{R}$.
- Resistors perform several useful functions in electronic circuits: They limit current and divide voltage.
- Several types of resistors are carbon or composition resistors, film resistors, and wire wound resistors. Another way to categorize resistors is fixed, semi-fixed, and variable resistors.
- Carbon or composition resistors normally are manufactured with power dissipation ratings from $\frac{1}{4}$ to 2 watts. Wire wound resistors generally are rated to handle power levels from 5 watts to over 100 watts.
- Because carbon resistors are typically very small, a color-coding system has been devised to indicate resistor values.
- The sentence *"Big Boys Race Our Young Girls, But Violet Generally Wins"* can help you remember the ten colors in the code: Black, Brown, Red, Orange, Yellow, Green, Blue, Violet, Gray, and White.
- Safety rules relating to electrical circuits, hot soldering irons, and rotating machinery should always be followed.

REVIEW QUESTIONS

1. Convert 3 mV to an equivalent value in V.

2. If a value of current is 12 μA, this can be stated as a value of 12×10 to what power amperes? What is this same value of current when stated as a decimal value of amperes?

3. What submultiple of a milli unit is a micro unit?

4. How many kΩ are there in a MΩ?

5. Draw a schematic diagram illustrating a voltage source (battery), two resistors, and a SPST switch all connected end-to-end. Illustrate the first resistor connected to the source's negative side, the other resistor connected to the source's positive side, and the switch between the resistors.

6. Draw a block diagram illustrating a power supply, a phonograph pickup, an audio amplifier, and a speaker. Show the direction of flow of signal(s) and/or electrical power.

7.  is the symbol for a fixed ______________.

8.  is the symbol for a variable ______________.

9. 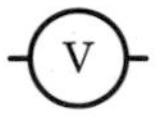 (V) is the symbol for a ______________.

10.  + − is the symbol for a ______________.

11. State the three forms of:
    a. the Ohm's law formula.
    b. the formula for electrical power.

12. Convert the following units:
    a. 15 milliamperes to amperes.
    b. 15 milliamperes to 15 times a power of ten.
    c. 5,000 volts to kilovolts.
    d. 5,000 volts to 5.0 times a power of ten.
    e. 0.5 megohms to ohms.
    f. 100,000 ohms to megohms.
    g. 0.5 ampere to mA.

13. If V is doubled and R halved for a circuit, what is the relationship of the new current to the original current?

14. If current triples in a circuit but R remains unchanged, what happens to the power dissipated by the circuit?

15. a. Draw a schematic diagram showing a voltage source, a current meter, and a resistor.
    b. Assume the voltage applied is 30 volts and the current meter reads 2 mA. Calculate the circuit resistance and power dissipation.
    c. Label all electrical parameters, the polarity of voltage across the resistor and across the voltage source, and show the direction of current flow.

16. Define work, energy, and power and give the units for electrical energy and power.

17. Name two or more types of resistors and discuss their physical construction.

18. Give the resistance value and tolerance rating for the following color-coded resistors:
    (NOTE: The first color shown is the first band; the second color represents the second band, and so forth.
    a. Yellow, violet, yellow, and silver.
    b. Red, red, green, and silver.
    c. Orange, orange, black, and gold.
    d. White, brown, brown, and gold.
    e. Brown, red, gold, and gold.

19. Name at least three important parameters, or factors, to consider when selecting a resistor for a given circuit.

20. List five or more safety rules to follow when working on electrical circuits and/or around hot soldering irons or rotating machinery.

TEST YOURSELF

1. Draw a block diagram illustrating a simple intercom system. Use at least three blocks, and label all blocks and diagram features.
2. Draw a schematic diagram showing two resistors, one capacitor, one inductor, two switches, and a battery.
3. a. Draw a schematic diagram showing a source, a resistor, and a current meter.
   b. Assume the power dissipation is 1,000 watts and resistance is 10 ohms. Transpose the $P = I^2R$ formula and solve for I (I = ________ amperes).
   c. Calculate the circuit applied voltage (V_A = ________volts).
   d. Label all parameters on the circuit diagram, including the direction of current flow and the polarity of voltages on each component.
4. List the color coding for the following resistors:
   a. 1,000 ohms, 10% tolerance.
   b. 27k ohms, 20% tolerance.
   c. 1 megohm, 10% tolerance.
   d. 10 ohms, 5% tolerance.
5. Using powers of ten, change the following as directed:
   a. $\dfrac{10^7 \times 10^{-4}}{10^5 \times 10^2 \times 10^{-2}}$ to a whole number.
   b. $10^3 \times 10^5 \times 10^6 \times 10^{-2}$ equals ten to what power?
   c. Add: $(7.53 \times 10^4) + (8.15 \times 10^3) + (225 \times 10^1)$ equals?
   d. Subtract: $(6.25 \times 10^3) - (0.836 \times 10^2)$ equals?

PERFORMANCE PROJECTS CORRELATION LISTING

Suggested performance projects that correlate with topics in this chapter are:

CHAPTER TOPIC	PERFORMANCE PROJECT	PROJECT NUMBER
Potential	Voltmeters (Use & Care of Meters series)	1
Current	Current Meters " "	2
Resistance	Ohmmeters " "	3
Relationship of Current to Voltage with Resistance Constant	Relationship of I & V with R Constant (Ohm's law series)	5
Relationship of Current to Resistance with Voltage Unchanged	Relationship of I & R with V Constant (Ohm's law series)	6
The Basic Power Formula	Relationship of Power to V with R Constant (Ohm's law series)	7
Variations of the Basic Power Formula	Relationship of Power to I with R Constant (Ohm's law series)	8
Resistor Color Code	Resistor Color Code (Ohm's law series)	4

PART II

BASIC CIRCUIT ANALYSIS

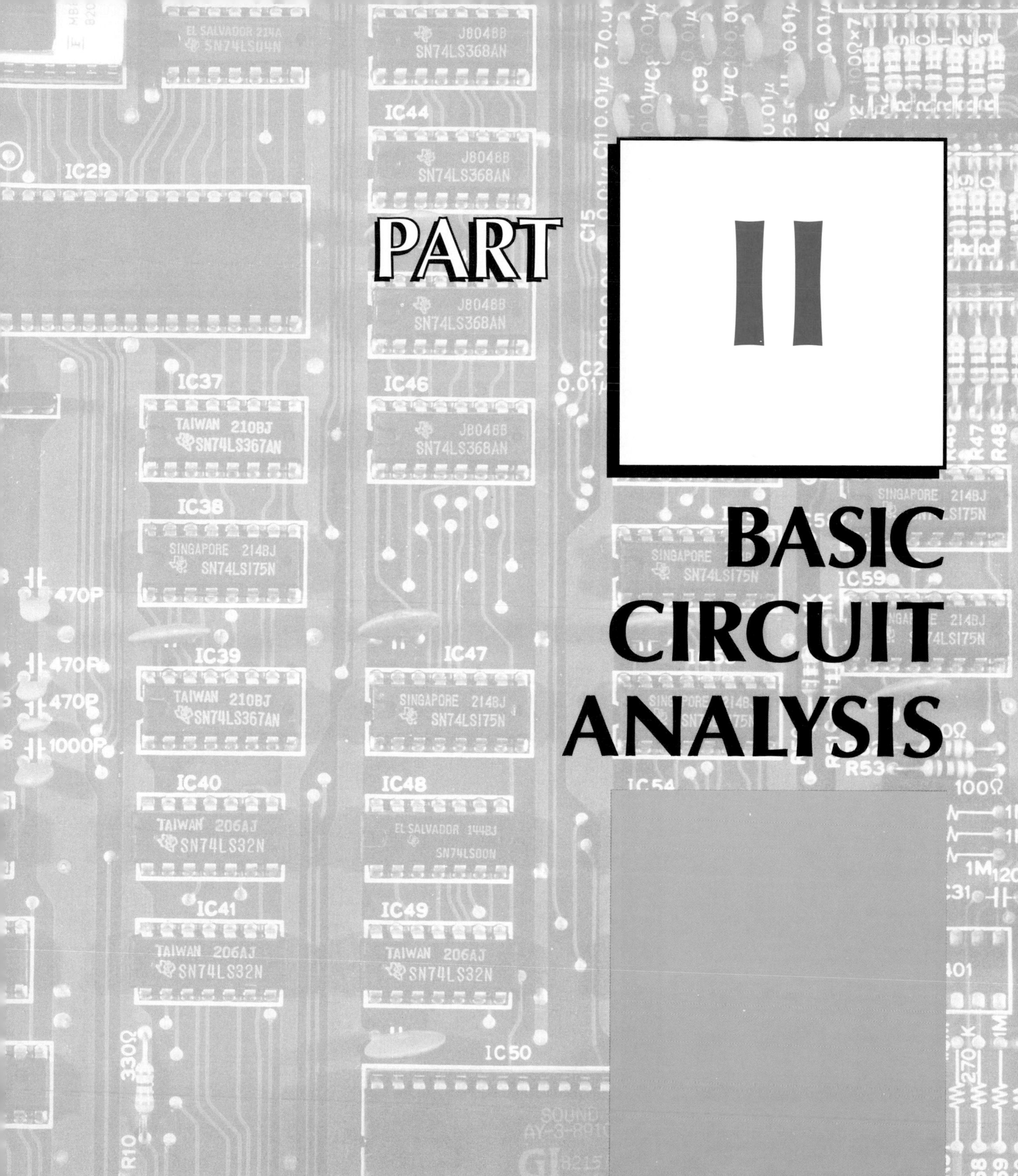

Series Circuits

Key Terms

Dropping resistor

Kirchhoff's voltage law

Open circuit

Series circuit

Short circuit

Voltage-divider action

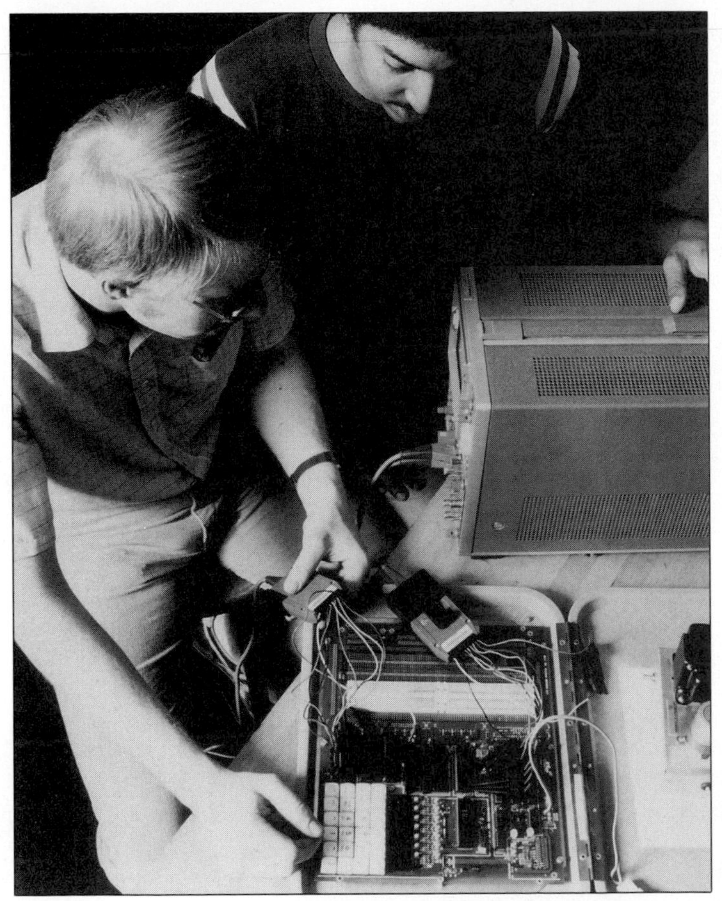

Courtesy of DeVry Institutes

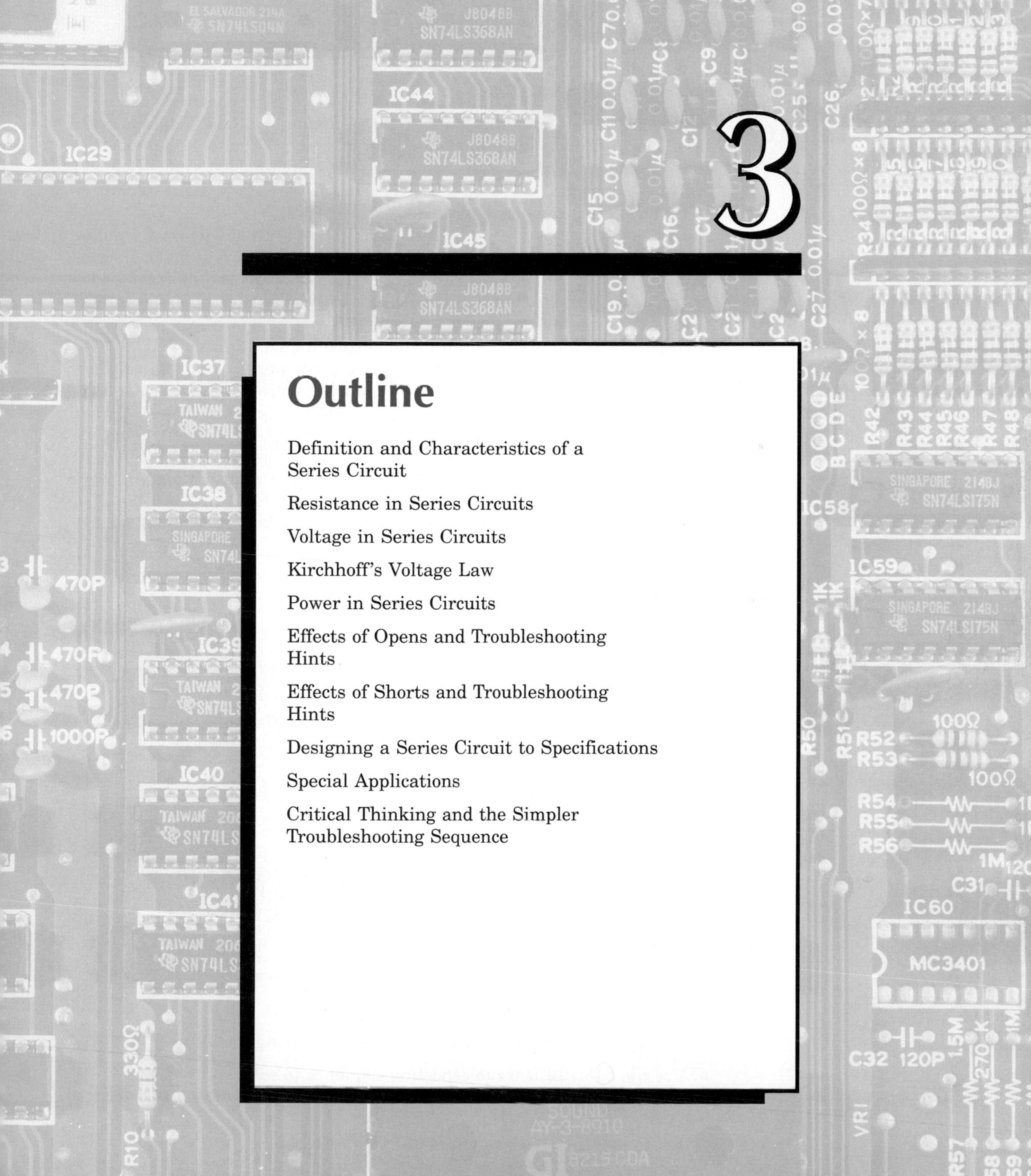

3

Outline

Definition and Characteristics of a Series Circuit

Resistance in Series Circuits

Voltage in Series Circuits

Kirchhoff's Voltage Law

Power in Series Circuits

Effects of Opens and Troubleshooting Hints

Effects of Shorts and Troubleshooting Hints

Designing a Series Circuit to Specifications

Special Applications

Critical Thinking and the Simpler Troubleshooting Sequence

Chapter Preview

Series circuits are found in many systems and subsystems. Examples are light switch circuits and automobile ignition switch circuits. It is important that you learn the characteristics of the series circuit. That knowledge will aid you in analyzing series circuits whether they are stand-alone circuits or portions of more complex circuits.

In studying series circuits, you will apply Ohm's law and power formulas. Also, you will learn how Kirchhoff's voltage law helps you analyze series circuits. Insight will be gained about the effects of opens and shorts in series circuits. You will see how voltage sources are used in series circuits. Also, voltage dividers and the concept of reference points will be examined. The final portion will present a useful troubleshooting method that will be valuable to you throughout your training, and more importantly, in your career.

You will be introduced to a special troubleshooting sequence called the SIMPLER sequence. By solving a troubleshooting problem called Chapter Challenge, you will gain experience using the SIMPLER sequence. The system for performing the procedure and solving the Chapter Challenge will be discussed at the point where the Chapter Challenge is introduced. Answers to the Chapter Challenges are located in Appendix C.

Objectives

After studying this chapter, you will be able to:

- Define the term **series circuit**
- List the primary characteristics of a series circuit
- Calculate the total resistance of series circuits using two different methods
- Calculate and explain the voltage distribution characteristics of series circuits
- State and use **Kirchhoff's voltage law**
- Calculate power values in series circuits
- Explain the effects of **opens** in series circuits
- Explain the effects of **shorts** in series circuits
- List troubleshooting techniques for series circuits
- Design series circuits to specifications
- Series-connect voltage sources for desired voltages
- Analyze a **voltage divider** with reference points
- Calculate the required value of a series-**dropping resistor**
- Use the SIMPLER troubleshooting sequence to solve a problem

DEFINITION AND CHARACTERISTICS OF A SERIES CIRCUIT

A **series circuit** is any circuit having only one path for current flow. In other words, two or more electrical components or elements are connected so the same current passes through all the connected components.

This situation exists when components are connected end-to-end in the circuit external to the source, Figure 3–1. This is termed a two-component series circuit. Also, notice there is only *one possible path* for current flow from the source's negative side, through the external circuit and back to the source's positive side.

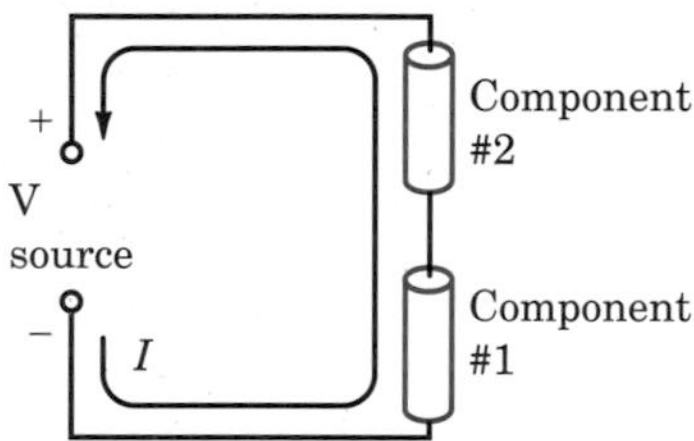

FIGURE 3–1 Components connected end-to-end in a two-component series circuit

Key Characteristics

Excluding the source, how many components are in series in Figure 3–2? If you said four, you are correct! The circuit in Figure 3–3 is a six-component series circuit. In all these cases, the important characteristics to remember are there is only one path for current flow and the current is the same through all parts of a series circuit. As a technician, you will have many opportunities to apply this knowledge.

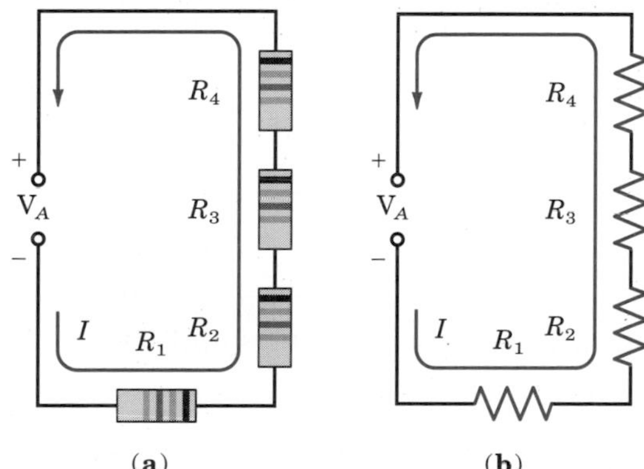

FIGURE 3–2 Four-resistor series circuit

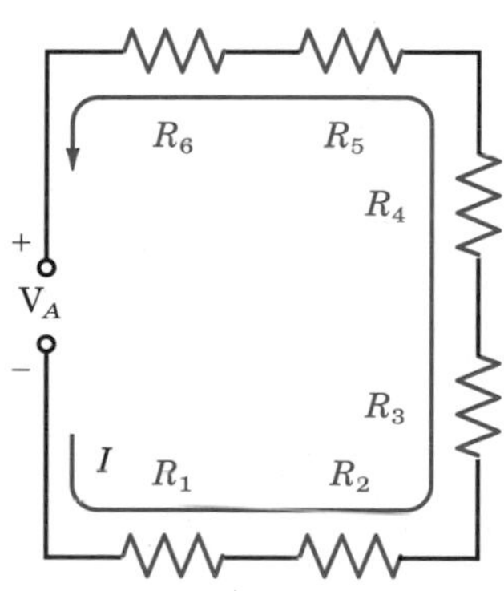

FIGURE 3–3 Six-resistor series circuit

RESISTANCE IN SERIES CIRCUITS

Knowing the characteristics that identify a series circuit helps you analyze the important parameters of these circuits.

First, let's analyze the series circuit in terms of its resistance to current flow. In previous chapters we have limited circuits to a single resistor, and the circuit's resistance to current flow was obvious. Now, we are connecting two or more resistors in tandem or in series. How does this affect the circuit resistance?

Series Circuit Total Resistance Formula

If the circuit current must sequentially flow through all the resistors since there is only one path for current, then the total resistance to current flow will equal the *sum* of all the resistances in series. Once again, in series circuits, the total resistance (R_T or R total) equals the sum of all resistances in series.

Formula 3–1	$R_T = R_1 + R_2 \ldots + R_n$

R_n represents each of the remaining resistor values. Refer to Figure 3–4 for an application of this formula. This illustrates that to find a series circuit's total resistance, simply add all the individual resistance values.

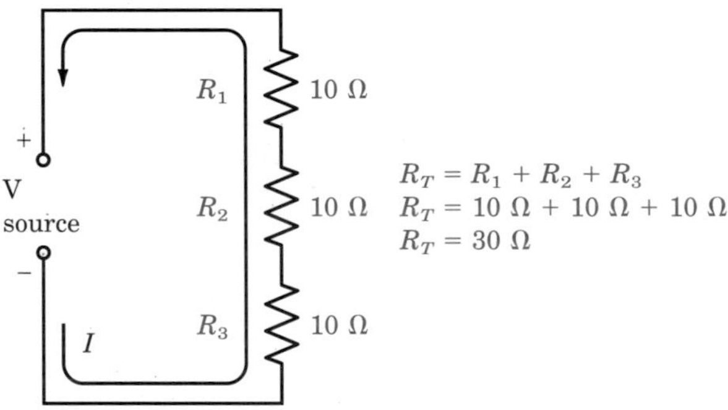

FIGURE 3–4 Application of series-circuit total resistance formula

PRACTICE PROBLEM I

Answers are in Appendix B.

Referring back to Figure 3–2, what is the circuit's total resistance if R_1 and R_2 are 10 kΩ resistors and R_3 and R_4 are 27 kΩ resistors?

The Ohm's Law Approach

Another important method to determine total resistance *in any circuit,* including series circuits, is Ohm's law. Recall that the Ohm's law formula to find resistance is $R = \frac{V}{I}$. To find total resistance, this formula becomes:

Formula 3–2	$R_T = \frac{V_T}{I_T}$

If there is any means of determining the value of total voltage applied to the series circuit *and* the total current (which is the same as the current through any part of the series circuit), Ohm's law can be used to find total resistance by using $R_T = \frac{V_T}{I_T}$.

EXAMPLE
Look at Figure 3–5 and use Ohm's law to find total resistance. Your answer should be 30 ohms because 60 volts divided by 2 amperes equals 30 ohms. Does this show the value of each of the resistors in the circuit? No, it only reveals the circuit's total resistance.

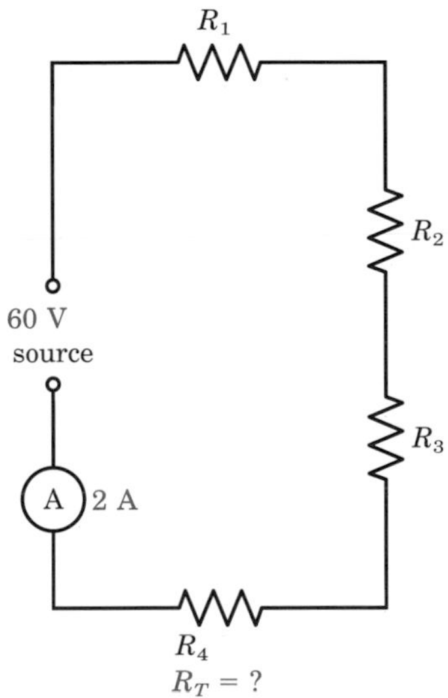

FIGURE 3–5 Finding R_T by Ohm's law

EXAMPLE
Now look at Figure 3–6. Again, it is possible to use Ohm's law to determine the circuit's total resistance. Because we know the voltage drop across R_1 and know that the current through any part of a series circuit is the same, we can determine the value of R_1.

Therefore, R total $= \frac{70 \text{ V}}{2 \text{ A}} = 35$ ohms. And R_1 is found by dividing the voltage across it by the current through it. Thus, R_1 equals 30 V divided by 2 amperes equals 15 ohms.

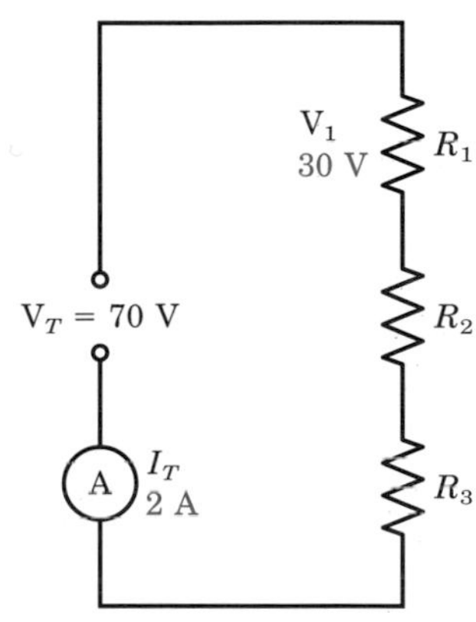

FIGURE 3–6 Finding one R value by Ohm's law

PRACTICE PROBLEM II

Again refer to Figure 3–6, if V_T = 100 V, I_T = 1 A, and V_I = 47 V, what is the value of R_1? What is the value of R_T?

We have discussed and illustrated two important facts about series circuits. First, *current* is the same throughout all parts of a series circuit. Second, the total circuit *resistance* equals the sum of all the resistances in series, which indicates total R must be greater than any one of the resistances. We will now study voltage, a third important electrical parameter in series circuits.

VOLTAGE IN SERIES CIRCUITS

To help you understand how voltage is distributed throughout a series circuit, refer to Figure 3–7.

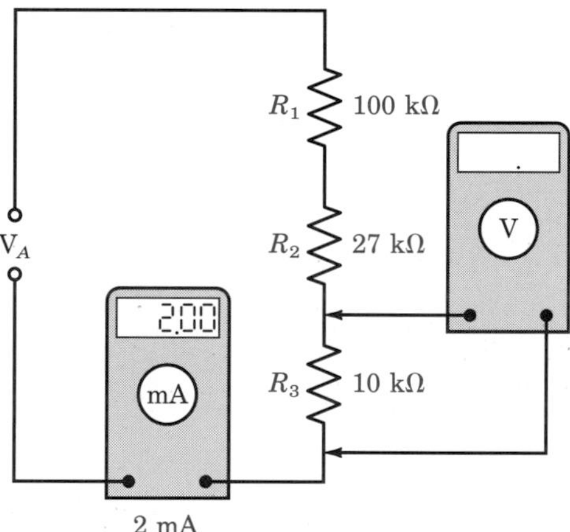

FIGURE 3–7 Voltage distribution in series circuits

Individual Component Voltages

Since there is only one path for current, the current (I) through each resistor (R_1, R_2, and R_3) must have the same value since it is the same current, Figure 3–7. From the study of Ohm's law, you know the value of voltage dropped by R_1 must equal its I times its R, or

Formula 3–3	$V_1 = I_{R_1} \times R_1$

EXAMPLE
In this case, the current is 2 mA and the resistance of R_1 is 100 kΩ; therefore, V_1 = 200 volts. (Remember that milliunits times kilounits equals

units). Voltage dropped by R_2 must equal 2 mA times 27 kΩ, or 54 volts. The voltmeter, which is measuring the voltage across R_3, would indicate 20 volts because 2 mA × 10 kΩ = 20 volts.

PRACTICE PROBLEM III

In the circuit, Figure 3–7, if the voltmeter reads 30 V, what are the values of V_A, I_T, V_{R_2}, and V_{R_1}?

PRACTICAL NOTES

In review, several observations are made.

1. The largest value resistor in a series circuit drops the largest voltage and the smallest value resistor drops the least voltage. These drops occur because the current is the same through each of the resistors and each resistor's voltage drop equals I times its R.
2. Since the same current passes through all components in series, a given component's voltage drop equals *the same percentage or fraction* of the total circuit applied voltage as its resistance value is of the total circuit resistance.
3. Relating the above concepts, when comparing two specific components' voltage drops, the ratio of voltage drops is the same as the ratio of their individual resistances.

Refer back to the voltages calculated for the circuit in Figure 3–7. Did the largest resistor value drop the highest voltage? Yes, R_1 dropped 200 volts compared to 54 volts for R_2 and 20 volts for R_3. Is the comparative ratio of voltage drops between resistors the same as the ratio of their resistances? Yes, R_1 resistance value is 10 times that of R_3 and V_1 is 10 times V_3. Also note that R_2 voltage drop is 2.7 times that of R_3, which is the same as their resistance ratios.

The concept of ratios of voltage drops equaling the ratios of resistances can be used to compare the voltage drops of any two components in the series circuit. For example, if the R values and applied voltage are known, it is possible to find all of the voltages for each component throughout the circuit using the proportionality technique. Naturally, you can also determine individual voltages around the circuit by solving for total resistance, then total current $\left(\frac{V_T}{R_T}\right)$, and then determining each I × R drop with $I \times R_1$ for V_1, $I \times R_2$ for V_2, and so forth.

Finding the Value of Applied Voltage

Again referring to Figure 3–7, let's examine some ways to find the circuit applied voltage. One obvious way is the Ohm's law expression where $V_T = I_T \times R_T$. In this circuit, V total (or V applied) = 2 mA times R_T. As you have previously learned, R_T in this circuit equals 100 kΩ + 27 kΩ + 10 kΩ = 137 kΩ total resistance. Therefore, circuit applied voltage equals 2 mA × 137 kΩ = 274 volts.

Another way to find applied voltage is to add all the individual voltage drops to find total voltage, just as we add all the individual resistances to find total resistance. This yields the same answer where 200 volts + 54 volts + 20 volts = 274 volts. This method suggests an important concept called Kirchhoff's voltage law.

KIRCHHOFF'S VOLTAGE LAW

Kirchhoff's voltage law states the *arithmetic sum* of the voltages around a single circuit loop (any complete closed path from one side of the source to the other) *equals V applied (V_A).* It also says the *algebraic sum* of all the loop voltages, including the source or applied voltage, must *equal zero.* That is, if you observe the polarity and value of voltage drops by the circuit elements and the polarity and value of the voltage source, and add the complete loop's values algebraically, the result is zero. For our purposes, the arithmetic sum approach will be most frequently used. It illustrates how Kirchhoff's voltage law helps determine unknown circuit parameters. Refer to Figure 3–8 as you study the following example.

EXAMPLE

If the V_A is 50 volts and V_2 is 20 volts, how can Kirchhoff's voltage law help determine V_1? If the sum of voltages (*not* counting the source) must equal V applied, then:

$$V_A = V_1 + V_2$$

Since we know V_A and V_2, transpose to solve for V_1.

$$\text{Therefore, } V_1 = V_A - V_2 = 50\text{V} - 20\text{V} = 30\text{V}$$

If individual voltage drops were known but not the applied voltage, find V applied by adding the individual voltage drops. In this case 30 V + 20 V = 50 V. This agrees with Kirchhoff's voltage law that the arithmetic sum equals the applied voltage. Kirchhoff's voltage law can help you find un-

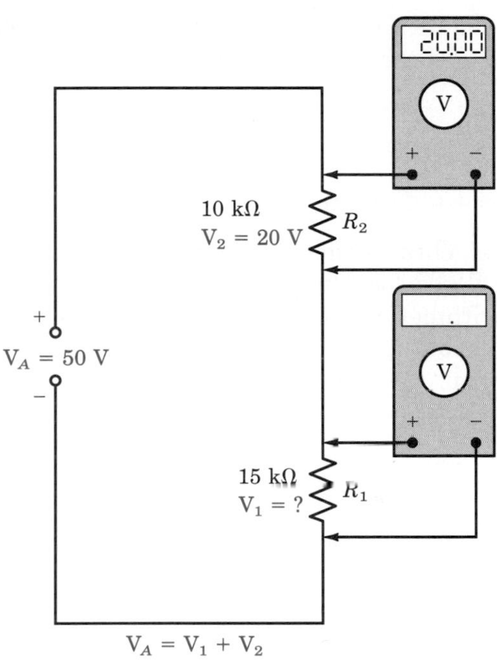

FIGURE 3–8 Using Kirchhoff's voltage law where $V_A = V_1 + V_2$

known voltages in series circuits through either addition or subtraction, as appropriate. And if there are more than two series components:

Formula 3–4	$V_T = V_1 + V_2 \ldots + V_n$

where V_n represents each of the remaining voltage values.

In Figure 3–9, observe the polarity of voltages around the closed loop. To indicate polarities used in Kirchhoff's voltage law, trace the loop from the source's positive side (Point A), through the resistors and back to the source's negative side (Point B). Consider any voltage a positive voltage whose + point is reached first, and vice versa. In this case, the first voltage

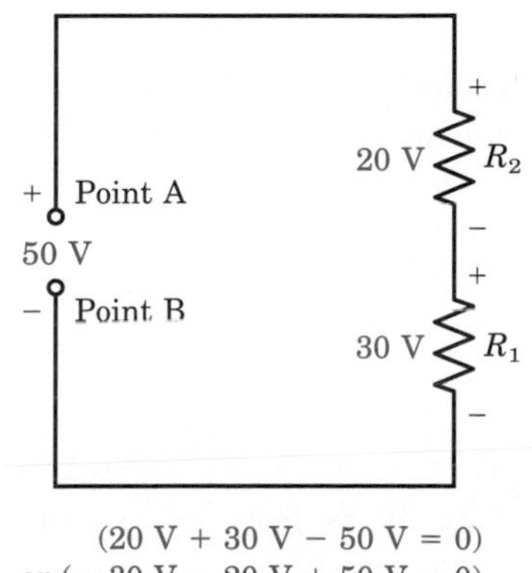

FIGURE 3–9 Kirchhoff's algebraic sum example

reached is +20 V; the next voltage is +30 V, and the source negative terminal is −50 V. Adding these voltages yields:

$$(+20) + (+30) + (-50) = 0; \quad \text{or } 20 + 30 - 50 = 0$$

Tracing the other direction through the circuit gives:

$$(-30) + (-20) + (+50) = 0; \quad \text{or } -30 - 20 + 50 = 0$$

In either case, the algebraic sum of the voltage drops *and* the voltage source around the entire closed loop equals zero.

Likewise, the arithmetic sum of the *voltage drops* around a given loop must equal the value of the *applied voltage,* regardless of which direction is used to trace the loop.

PRACTICE PROBLEM IV

Refer again to Figure 3–8. Using the proportionality of voltage drops concept and Kirchhoff's voltage law, if V_1 is 45 V, what are the values of V_2 and V_A?

IN-PROCESS LEARNING CHECK I

Fill in the blanks as appropriate.

1. The primary identifying characteristic of a series circuit is that the ________ is the same throughout the circuit.
2. The total resistance in a series circuit must be greater than any one ________ in the circuit.
3. In a series circuit, the highest value voltage is dropped by the ________ value resistance, and the lowest value voltage is dropped by the ________ value resistance.
4. In a two-resistor series circuit, if V applied is 210 volts and one resistor's V drop is 110 volts, what must the voltage drop be across the other resistor? ________ V.
5. Answer each part of this question with increase, decrease, or remain the same. In a three-resistor series circuit, if the resistance value of one of the resistors increases, what happens to the circuit total resistance? ________ To the total current? ________ To the adjacent resistor's voltage drop? ________
6. What is the applied voltage in a four-resistor series circuit where the resistor voltage drops are 40 V, 60 V, 20 V, and 10 V, respectively? ________

A Quick Review

Current is the same throughout a series circuit; total resistance equals the sum of all the individual resistances in series, and voltage distribution around a series circuit is directly related to the resistance distribution, since I is the same through all components and each component's V drop = I × its R. Kirchhoff's voltage law states that the arithmetic sum of voltage drops equals the voltage applied, or the algebraic sum of all the voltage drops and the voltage source equal zero.

CALCULATOR CORNER

Try solving this problem with a calculator, following these instructions:

Problem: Find R_T, I_T and V_T for a series circuit containing three 20 kΩ resistors where $V_1 = 10$ V.

Step 1: Turn calculator on and clear.

Step 2: Find R_T by inputting 20,000 + 20,000 + 20,000, then push the (=) button. The answer should be 60,000. (Of course, this is 60,000 Ω, or 60 kΩ.)

Step 3: Find I by finding the I through R_1, where $I_1 = \frac{V_1}{R_1}$. Clear calculator by pushing the (C) button. Input as follows: Punch in 10 (for V_1), then the (÷) symbol button, then 20,000 (for R_1), then the (=) button. The answer should be 0.0005. (Of course, this is 0.0005 amperes or 0.5 mA.)

Step 4: Find V_T by using the formula $V_T = I_T \times R_T$. You may clear the calculator (or leave the 0.0005 you had just calculated, since you need it in this calculation). At any rate, be sure that 0.0005 is showing, then push the (×) symbol button, then input 60,000 (for the R_T value). Push the (=) button. The answer should be 30, representing 30 volts.

NOTE: This approach is appropriate for calculators that do not use reverse Polish notation. Some brands, depending on their notation approach, will not use the same sequence of key strokes.

Try this one on your own, using a calculator. Answers are in Appendix B. For the circuit shown to the right, find the parameters called for:

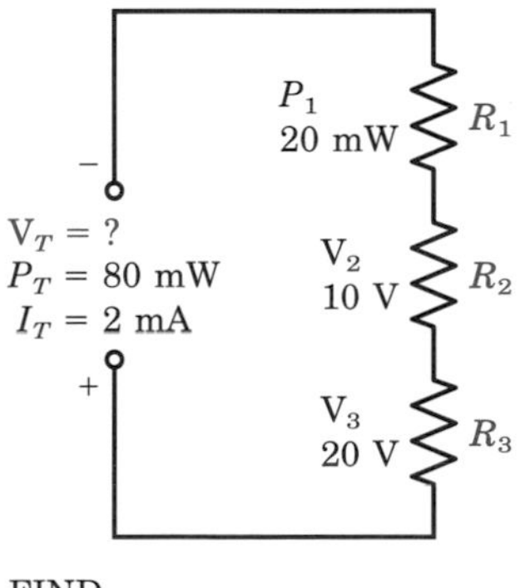

FIND
V_T = ______
P_3 = ______
R_1 = ______
R_2 = ______
R_3 = ______

POWER IN SERIES CIRCUITS

The last major electrical quantity to be discussed in relation to series circuits is power. Recall that power dissipated by a component or circuit (or supplied by a source) is calculated with the formulas P = V × I, $P = I^2R$, or $P = \frac{V^2}{R}$. You have learned that in series circuits the current (I) is the same through all components. Therefore, the I^2 factor in the $P = I^2R$ formula is the same for each component in series and the largest R value dissipates the most power. Conversely, the component with the least R value dissipates the least power (I^2R), Figure 3–10.

Individual Component Power Calculations

Individual component power dissipations can be found if any two of the three electrical parameters for the given component are known. For example, if the component's resistance and the current through it are known, use the $P = I^2R$ formula. If the component's voltage drop and the current are known, use the P = V × I formula. Knowing the voltage and resistance allows you to use the $P = \frac{V^2}{R}$ version of the power formula.

As implied above, the individual power dissipations around a series circuit are *directly related* to the resistance of each element, just as the voltage distribution is directly related to each element's resistance. A specific resistor's power dissipation is the same percentage of the circuit's total power dissipation as its resistance value is of the total circuit resistance (R_T). For example, if its resistance is one-tenth $\left(\frac{1}{10}\right)$ the total circuit resistance, it dissipates 10% of the total power. If it is half the R_T, it dissipates 50% of the total power, and so forth, Figure 3–10.

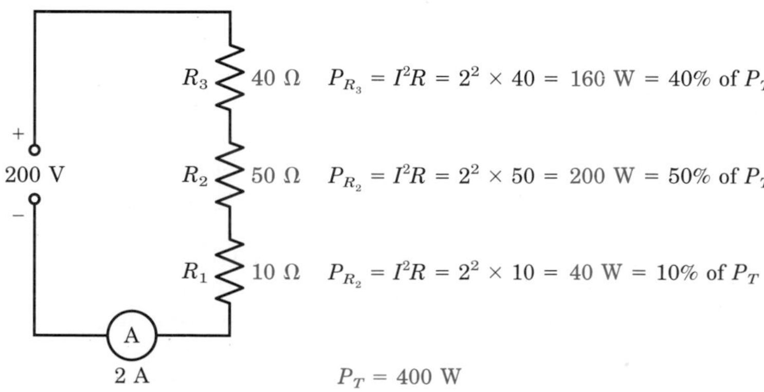

FIGURE 3–10 Largest *R* dissipates most power, and smallest *R* dissipates least power.

Total Circuit Power Calculations

Total power dissipation (or power supplied to the circuit by the power source) is determined by adding all the individual power dissipations, if they are known. See Formula 3–5.

Formula 3–5	$P_T = P_1 + P_2 \ldots + P_n$

PRACTICE PROBLEM V

Refer to Figure 3–11. Assume the source voltage is 90 volts.

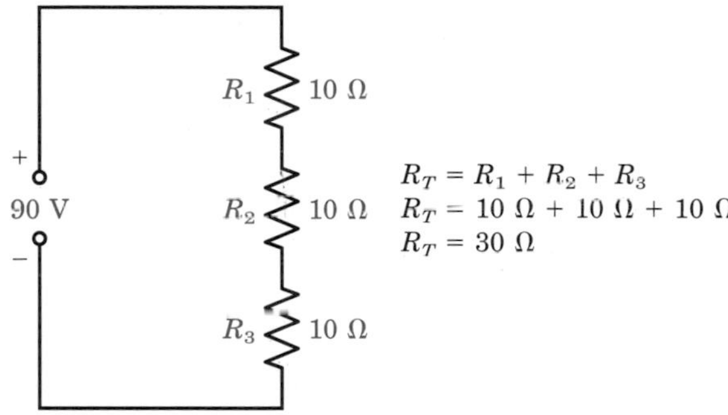

FIGURE 3–11

Find the values for:

P_1 = ________ watts V_1 = ________ volts

P_2 = ________ watts V_2 = ________ volts

P_3 = ________ watts V_3 = ________ volts

P_T = ________ watts R_3 = ________ % of R_T

I_T = ________ amperes P_3 = ________ % of P_T

EFFECTS OF OPENS AND TROUBLESHOOTING HINTS

Opens

Thus far, you have studied how to analyze current, resistance, voltage, and power in series circuits. At this point we will introduce some practical situations that might occur in *abnormal* series circuits. The two extreme conditions are opens and shorts.

First, what are the effects of an open in a series circuit? An **open circuit** occurs anytime a break in the current path occurs somewhere within the series circuit. Since there is only one path for current flow, if a break (or open) occurs *anywhere* throughout the current path, there is no current flow. In effect, the circuit resistance becomes "infinite" and I_T (circuit current) must decrease to zero.

If an open causes circuit resistance to increase to infinity and circuit current to decrease to zero, how does this affect the circuit voltages and power dissipations?

EXAMPLE

In Figure 3–12, assume that R_3 is physically broken in half, causing an open or break in the current path. Since there is no complete path for current flow, what would the current meter read? The answer is zero mA.

FIGURE 3–12 An open in a series circuit

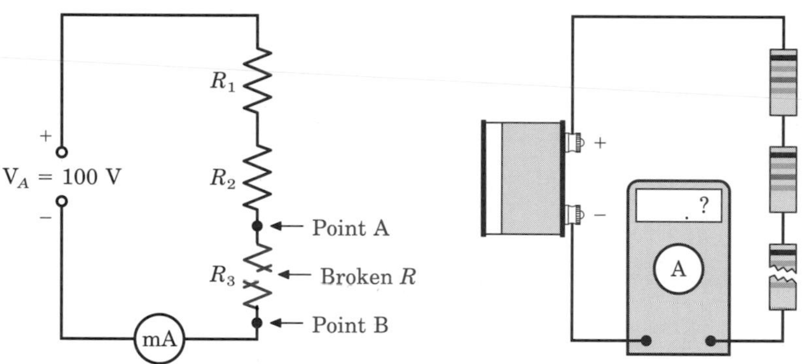

If there is no current, what is the I × R drop of R_1? The answer is zero volts since zero times any value of R equals zero. And the answer for V_2 is also zero for the same reason. This means that the difference of potential across each of the good resistors (R_1 and R_2) is zero. That is, R_1 has zero voltage drop and R_2 has zero voltage drop. In effect, the positive side of the source potential is present anywhere from the source itself around the circuit to point A. The negative side of the source potential is present at every point along the circuit on the opposite side of the break all the way to point B.

What is the potential difference across the break, or open? If your answer is 100 volts (or V_A), you are correct! This indicates there will be no I × R drops across the *good* components in the circuit, and the circuit applied voltage appears across the open portion of the circuit, regardless of where the open appears in the circuit.

An example of a purposely opened circuit is the light switch circuit in your home. When you turn the light on, you *close* an open set of contacts on the switch, which is in series with the light. When you turn the light off, you *open* those switch contacts in series with the light; therefore, you open the path for current flow, causing the applied voltage to appear across the opened switch contacts and zero volts to appear across the light element(s). If you measure the voltage across the switch with the lights *off,* you will measure the source voltage, or V applied.

PRACTICE PROBLEM VI

1. In a series circuit having a 100-volt source and comprised of four 27-kΩ resistors, indicate whether the following parameters would increase, decrease, or remain the same if one of the resistors opened.
   a. Total resistance would ______________.
   b. Total current would ______________.
   c. Voltage across the unopened resistors would ____________.
   d. Voltage across the opened resistor would ______________.
   e. Total circuit power dissipation would ______________.
2. Would the conditions you have described for question 1, above, be true, no matter which resistor opened? ________________

TROUBLESHOOTING HINTS

When you suspect an open in a series circuit, you can measure the voltage across each component. The component, or portion of the circuit across which you measure V applied, is the opened component, or circuit portion. *Caution:* To measure these voltages, power must be applied to the circuit. Use all safety precautions possible.

An alternative to this "power-on" approach is to use *"power-off"* resistance measurements. The two points in the circuit at which you measure "infinite ohms" (with the circuit completely disconnected from the voltage/power source) is the opened portion of the circuit. Resistance of the good components should measure their rated values.

EFFECTS OF SHORTS AND TROUBLESHOOTING HINTS

Shorts

What are the effects of a short in a series circuit? First, what is a short? A **short circuit** may be defined as an undesired very low resistance path in or around a given circuit. An example is when someone drops a metal object across two wires connected to the output terminals of a power source. In this case you short all the other circuits connected to the power source terminals and provide a *very low* resistance path for current flow through the new metal current path.

A short may occur across one component, several components, or the complete circuit (as in our dropped metal object example). Regardless, consider a short to be a virtually zero resistance path that disrupts the normal operation of the circuit.

EXAMPLE

Observe Figure 3–13. Notice that the undesired low resistance path is a piece of bare wire with a virtually zero resistance that has fallen across leads of R_1. *Without* the short, the circuit total resistance would equal 20 kΩ ($R_1 + R_2$). Therefore, with 20 volts applied voltage, the current is 1 mA $\left(\frac{20\text{ V}}{20\text{ k}\Omega}\right)$. *With* the short, the resistance from point A to point B is close to zero ohms since R_1 is shorted). As far as the voltage source is concerned, "it sees" only 10 kΩ resistance of R_2 as total circuit resistance. Thus, current is $\frac{20\text{ V}}{10\text{ k}\Omega}$, or 2 mA (double what it would be without the short). The short causes R_T to decrease to half its normal value and current to double since the V applied is unchanged.

This circuit analysis points to some important generalizations. 1) If any portion of a series (or any other type circuit) is shorted (or simply decreases in resistance), the total circuit resistance decreases, and 2) The circuit's total current increases, assuming the applied voltage remains unchanged.

What happens to the other circuit parameters if a short (or decreased resistance) occurs in a series circuit? Using our sample circuit in Figure 3–13: Let's think through it.

1. R_1 becomes zero ohms.

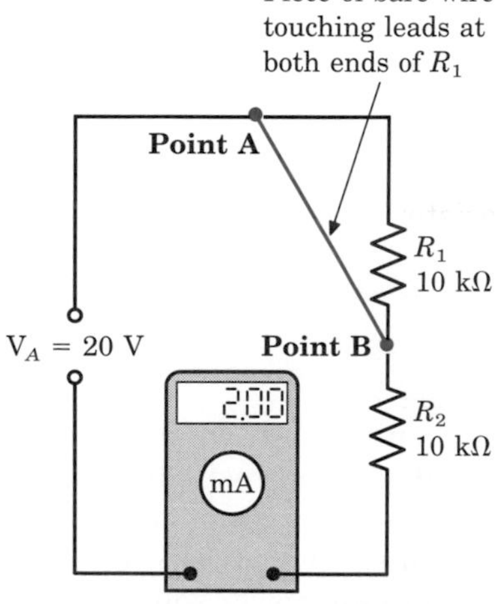

Circuit current with R_1 shorted

2 mA

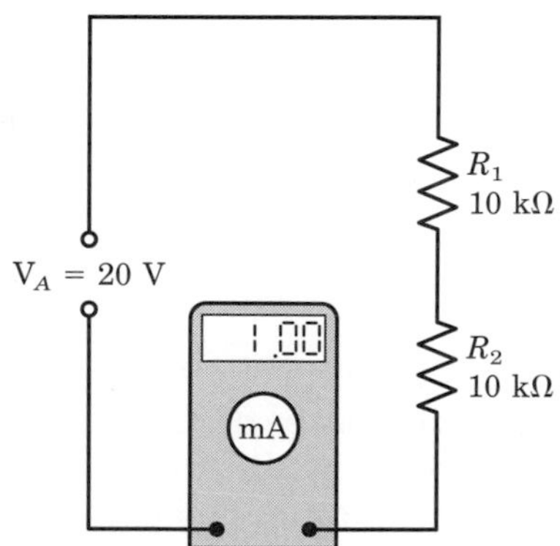

Circuit current with R_1 NOT shorted

1 mA

FIGURE 3–13 Effects of a short in a series circuit

2. R_T decreases, causing circuit current (I_T) to increase.
3. Since I has increased and R_2 is still the same value, the voltage across R_2 increases. In this case, it will increase to V applied (20 V) since there is no other resistor in the circuit to drop voltage.
4. V_1 (the voltage across R_1) decreases to zero volts since any value of I times zero ohms equals zero volts.
5. The total circuit power dissipation increases, since I has doubled and V applied has remained unchanged ($P = V \times I$).
6. Power dissipated by the good resistor (R_2) increases, since its R is the same, but the current through it has doubled ($P = I^2R$).
7. Of course, the power dissipated by R_1 decreases to zero.

What happens if the *total* circuit is shorted instead of part of it? This means both R_1 and R_2 are shorted and the short is across the power supply terminals. The voltage source is looking at zero ohms, and it would try to supply "infinite" current. In reality this cannot happen. Probably the power supply fuse would blow, and/or the power supply and circuit wires/conductors might be damaged. However, the generalizations for the partially shorted circuit would also apply. That is, circuit resistance goes down, total current increases (until a fuse blows or the power supply "dies"), and the voltage across the shorted portion of the circuit decreases to zero.

PRACTICE PROBLEM VII

1. If a series circuit having a 200-volt source is composed of a 10-kΩ, a 27-kΩ, a 47-kΩ and a 100-kΩ resistor, and the 47-kΩ resistor shorts, what will the voltages be across each of the resistors listed? (Don't forget to use the "draw-the-circuit" hint to help you in your analysis.)
2. With the short present across the 47-kΩ resistor, do the voltage drops across the other resistors increase, decrease, or remain the same?

TROUBLESHOOTING HINTS

If circuit fuses blow, components are too hot, or smoke appears, there is a good chance that a component, circuit, or portion of a circuit may have acquired a short.

Again, measuring resistances helps locate the problem that requires removing power from the circuit to be tested. It is best to remove power from the circuit until the fault can be cleared; therefore, voltage measurement techniques must be used judiciously, if at all. Many times, applying power to circuits of this type causes other components or circuits to be damaged because excessive current may pass through the non-shorted components or circuits.

With power removed from the circuit, it is possible to use an ohmmeter to measure the resistance values of individual components, or of selected circuit portions. If the normal values are known, it is obvious which component(s), or portion(s) of the circuit have the very low resistance value. Often, there will be visual signs of components or wires that have overheated, leaving a trail that indicates where the excessive current flowed as a result of the short.

In special cases, if voltage measurements are used, voltage drops across the good components are higher than normal (due to the higher current), and voltage drops across the shorted component(s) or circuit portion(s) are close to zero.

If there are current meters or light bulbs in the circuit, the meter readings are higher than normal and/or the unshorted light bulbs glow brighter than normal.

A Special Troubleshooting Hint

A simple but useful technique used in troubleshooting circuits, which have sequential "in-line" components or subcircuits, is called the divide-and-conquer approach, or split technique. This technique can save many steps in series situations where one of many components or subcircuits in the sequential line is the malfunctioning element.

The technique is to make the first test *in the middle* of the circuit. The results tell the technician which half of the circuit has the problem. For example, Figure 3–14 represents a twelve-light series Christmas tree circuit with an open bulb. Recall, if one light becomes open, all the lights go out since the only path for current has been interrupted. Note in Figure 3–14, the first test is made from one end of the "string" to the "mid-point", in this case between bulbs 6 and 7. If making a voltage test with power on and the meter indicates V applied, then the open bulb must be between bulb 1 through bulb 6. If V measures 0 volts, then the problem is in the other half of the circuit between bulb 7 through bulb 12. With *one* check, half of the circuit has been eliminated as the possible area of trouble.

Incidentally, this technique can be used with the power-off resistance measurement approach. With the circuit *disconnected* from the power source, if the first R measurement shows infinite resistance between bulbs 1 and 6, the problem is in that half of the circuit. If the reading is a low resistance, then the problem is in the other half of the circuit between bulb 7 through bulb 12.

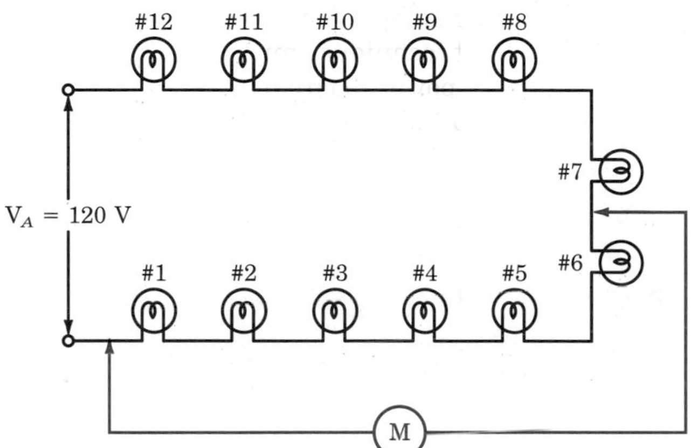

FIGURE 3–14 Divide-and-conquer troubleshooting technique—1st step NOTE: Meter "M" may be either a voltmeter if voltage is applied to the circuit under test or, an ohmmeter if the circuit is disconnected from the power source.

After dividing the abnormally operating circuit's suspect area in half by making the first measurement made at the circuit mid-point, the next step is to divide the remaining suspect area of the circuit in half again by making the second check from one end of the suspect section to its mid-point; thus, narrowing down to a quarter of the circuit the portion that remains in question, Figure 3–15. This splitting technique can continue to be used until only two components or subcircuits remain. Then each is checked, as appropriate, to find *the* malfunctioning component.

The divide-and-conquer approach can be used in any in-line (linear) system of components or circuits where current flow, power, fluids, or signals must flow sequentially from one component or subcircuit to the next and is useful in troubleshooting electrical, electronic, hydraulic and many other systems having in-line condition.

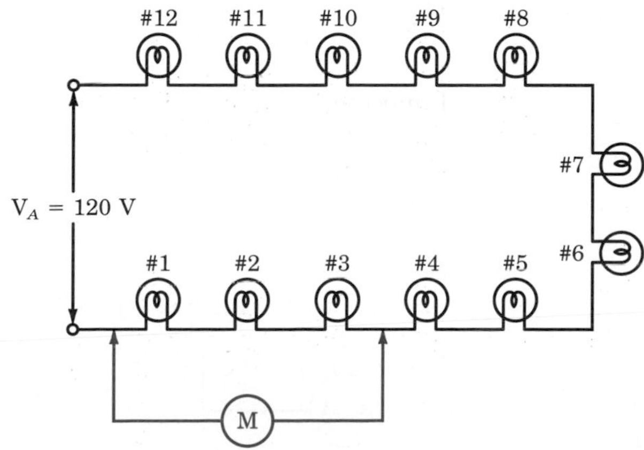

FIGURE 3–15 Divide-and-conquer troubleshooting technique—2nd step NOTE: Meter "M" may be either a voltmeter if voltage is applied to the circuit under test or, an ohmmeter if the circuit is disconnected from the power source.

PRACTICAL NOTES

In review, solve problems with the following steps:

1. Collect all known values of electrical parameters.
2. If no circuit diagram is given, draw a diagram and write known quantities on the diagram.
3. Solve the first part of the problem at a point where you have, or can easily find, sufficient knowns to solve for an unknown. Then proceed by solving the remaining portions of the problem, as appropriate.

DESIGNING A SERIES CIRCUIT TO SPECIFICATIONS

Let's apply this in a sample design problem, then you can try one to see whether you have learned the process.

EXAMPLE

Design a three-resistor series circuit where:

- Two of the resistors are 10 kΩ
- Total circuit current equals 2 mA
- V applied equals 94 volts

Answer: The first step is to collect the knowns, which were given to you. The next step is to draw the circuit and label the knowns.

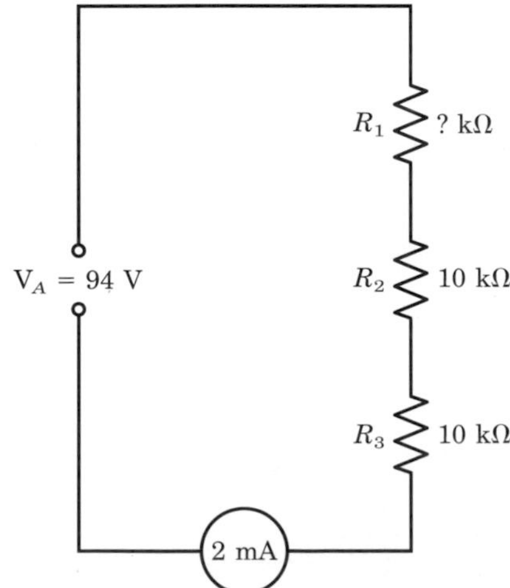

The third step is to start at a point where sufficient knowns are available to solve for a desired unknown.

To apply the third step, observe that in the circuit we drew, we know both the I and the R values for both R_2 and R_3. This makes it easy to find their voltage drops.

Thus, using the $V = I \times R$ formula, $V_2 = I \times R_2 = 2\text{ mA} \times 10\text{ k}\Omega = 20$ volts. And, $V_3 = I \times R_3 = 2\text{ mA} \times 10\text{ k}\Omega = 20$ volts.

The unknown is the value of R_1, which completes our design problem.

If the voltage drop of R_1, can be found, it is easy to find its resistance value. This is because we know the current must be 2 mA and its resistance must equal its voltage drop divided by 2 mA. It is convenient to use Kirchhoff's voltage law to find V_1. According to Kirchhoff, V_A must equal $V_1 + V_2 + V_3$. Thus, transposing to solve for the unknown V_1:

$$V_1 = V_A - (V_2 + V_3)$$
$$V_1 = 94\text{ V} - (20\text{ V} + 20\text{ V})$$
$$V_1 = 54\text{ volts}$$

Therefore, $R_1 = \dfrac{54\text{ V}}{2\text{ mA}} = 27\text{ k}\Omega$.

Another way to solve R_1's value is to find R_T, where $R_T = \dfrac{V_T}{I_T}$. Thus, $R_T = \dfrac{94\text{ V}}{2\text{ mA}} = 47\text{ k}\Omega$. Knowing total resistance equals the sum of all individual resistances in a series circuit and R_2 plus R_3 equals 20 kΩ, R_1 must make up the difference of 27 kΩ to provide a total resistance of 47 kΩ. The more formal way to illustrate this is to write the R_T formula, then transpose it to solve for R_1. Thus,

$$R_T = R_1 + R_2 + R_3$$

Therefore,

$$R_1 = R_T - (R_2 + R_3) = 47\text{ k}\Omega - (10\text{ k}\Omega + 10\text{ k}\Omega) = 27\text{ k}\Omega$$

Does our design meet the specifications? Yes, it is a three-resistor series circuit where two of the resistors are 10 kΩ, the total circuit current is 2 mA, and the applied voltage equals 94 volts.

PRACTICE PROBLEM VIII

Try this next design problem yourself. Complete it on a separate sheet of paper, then check your results with the answers in Appendix B.

Problem: Design a three-resistor series circuit where R_1 is 20 kΩ and drops two-fifths of V applied, R_2 drops 1.5 fifths of V applied, and R_3 has a voltage drop equal to V_2. Assume V applied is 50 V. Draw the circuit and label all V, R, P, and I parameter values.

SPECIAL APPLICATIONS

Voltage Sources in Series

Voltage sources can be connected in series to provide a higher or lower total (resultant) voltage than one of the sources provides alone. The *resultant* voltage of more than one voltage in series depends on the values of each voltage *and* whether they "series-aid" or "series-oppose" each other. Figure 3–16a is a series-aiding type connection. Figure 3–16b is an example of a series-opposing arrangement.

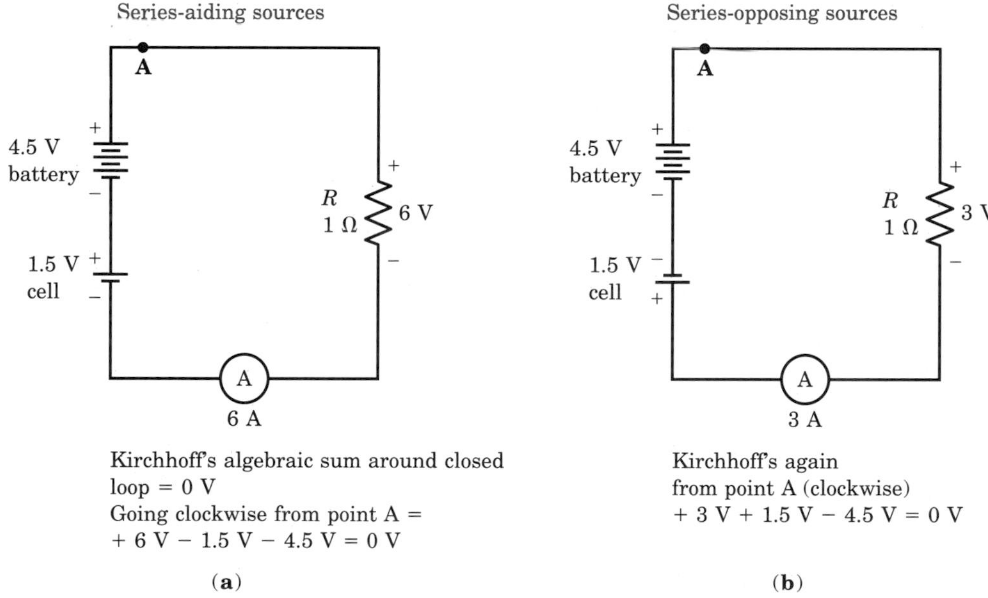

FIGURE 3–16 Series-connected voltage sources

Note, in Figure 3–16a, that the "resultant" voltage applied to the circuit from the series-aiding sources equals the sum of the two sources. In this case 4.5 V + 1.5 V = 6 volts. The resulting current through the resistor is 6 amperes, since $I = \frac{V}{R} = \frac{6\text{ V}}{1\ \Omega} = 6\text{ A}$. Ways to know these sources are series-aiding are 1) the negative terminal of one source is connected to the positive terminal of the next, and 2) both sources try to produce current in the same direction through the circuit.

In Figure 3–17, what is the total voltage applied to the circuit? What is the current value? If you answered 27 volts for the resultant voltage and 0.5 mA for the current, you are correct.

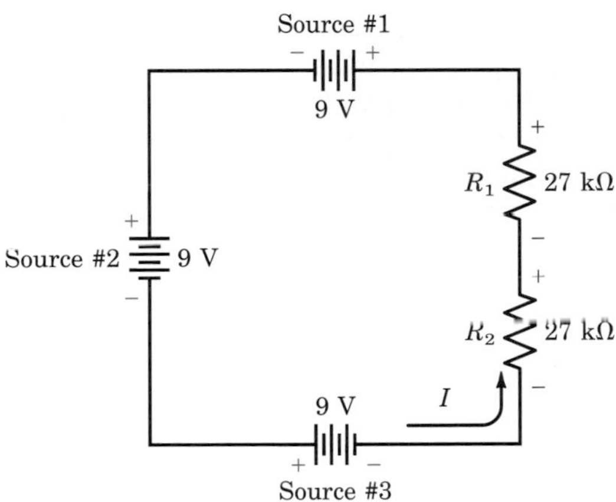

FIGURE 3–17 Series-aiding sources have the following features:
a. Sources are in series.
b. Negative terminal of one source connects to positive terminal of next, and so on.
c. Both sources try to produce current in the same direction through the circuit.
d. Verifiable using Kirchhoff's law.

As you can see in Figure 3–18, series-opposing sources are connected so that: 1) the negative terminal of one source is connected to the negative terminal of the next, or positive-to-positive; and 2) the sources try to produce current through the circuit in opposite directions.

To determine the resultant or equivalent voltage of series-opposing sources, *subtract* the smaller voltage from the larger voltage.

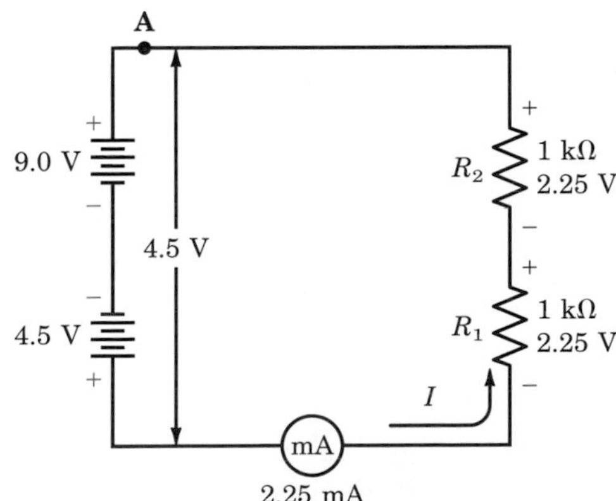

FIGURE 3–18 Series-opposing sources have the following features:
a. Sources are in series.
b. Negative terminal of one source connects to negative terminal of next (or positive connects to positive), and so on.
c. All three sources try to produce current in opposite directions through the circuit.
d. Verifiable using Kirchhoff's law.

EXAMPLE

In Figure 3–18, 9.0 V minus 4.5 V equals a resultant of 4.5 V applied to the circuit connected to the series-opposing sources.

A final point about the concept of series-aiding and series-opposing voltages is that voltage drops can be series-aiding or series-opposing just like voltage sources are series-aiding or series-opposing. If the voltage drops are such that series-connected polarities are − to +, or + to −, *and* the current through the series components is in the same direction, they are series-aiding voltage drops. For example, V_{R1} and V_{R2} in Figure 3–18 are series-aiding. Conversely, if the polarities of series component voltage drops are − to −, or + to +, they are series-opposing voltage drops.

Simple Voltage Dividing Action and Reference Points

Voltage Dividing. As you have learned, voltage drops around a series circuit are proportional to the resistance distribution since the current is the same through all components. You can *select* the values of resistors in series to "distribute" or "divide" the applied voltage in any desired fashion. Therefore, combining this concept with the series-aiding voltage drop idea, we can create a simple **voltage-divider action** using a series circuit, Figure 3–19. Notice the polarity of voltage drops, and the difference of potential between the various points identified in the circuit.

Equal resistance values are used for R_1, R_2, and R_3, and the voltage source is equally divided by the three resistors. Also, since the voltage drops are series-aiding, the voltages at the various points are cumulative.

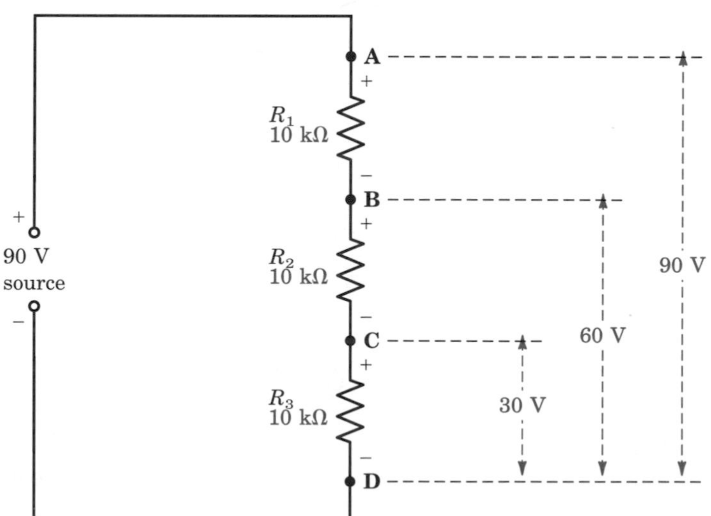

FIGURE 3–19 Example of voltage-divider action

PRACTICE PROBLEM IX

Referring again to Figure 3–19, assume the parameters are changed so source voltage = 188 V, R_1 = 20 kΩ, R_2 = 27 kΩ, and R_3 = 47 kΩ. Indicate the voltages between the following points: D to C; C to B; B to A; and D to B.

Reference Point(s). Recall in an earlier chapter, we discussed the idea of something "with respect to" or "in reference to" something else. The example used was "John is taller than Bill", or "Bill is shorter than John."

In electronic circuits it is common to relate electrical parameters, like voltage, to some common reference point in the circuit. The point frequently used as an electrical reference point in circuits is called "chassis ground," and the electrical schematic symbol is ⏚, Figure 3–20.

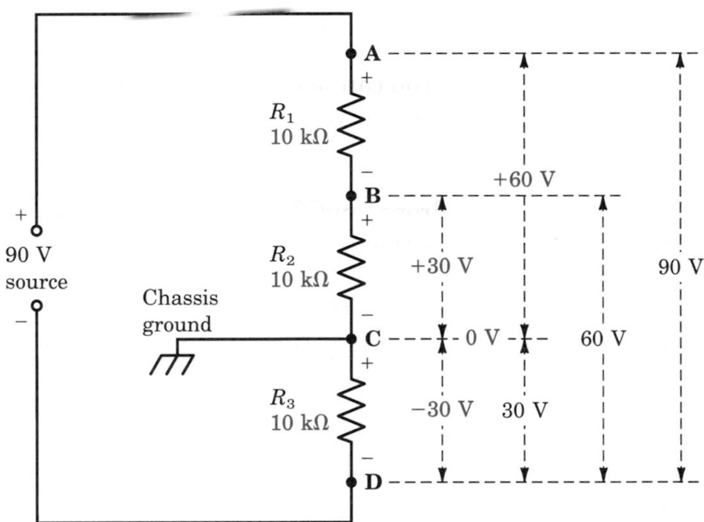

FIGURE 3–20
Chassis ground as a reference point at point C. The 30 V, 60 V, and 90 V indicators show the cumulative V drops. The −30 V, +30 V and +60 V indicators show voltages with respect to point C, where ground reference is connected.

In Figure 3–20, point C is where the chassis ground is connected. How is this used as a reference point to describe the voltages along the "divider"? The voltages with respect to this ground reference point can be described as follows:

Point D with respect to chassis ground = −30 V
Point C with respect to chassis ground = 0 V
Point B with respect to chassis ground = +30 V
Point A with respect to chassis ground = +60 V

Suppose that the chassis ground reference point is moved from point C to point B, Figure 3–21. How are the voltages described with respect to the reference point?

Point D with respect to chassis ground = −60 V
Point C with respect to chassis ground = −30 V
Point B with respect to chassis ground = 0 V
Point A with respect to chassis ground = +30 V

Polarity

Again, the polarity and value of voltage are described in terms of its polarity and value with respect to the designated reference point. For example, in

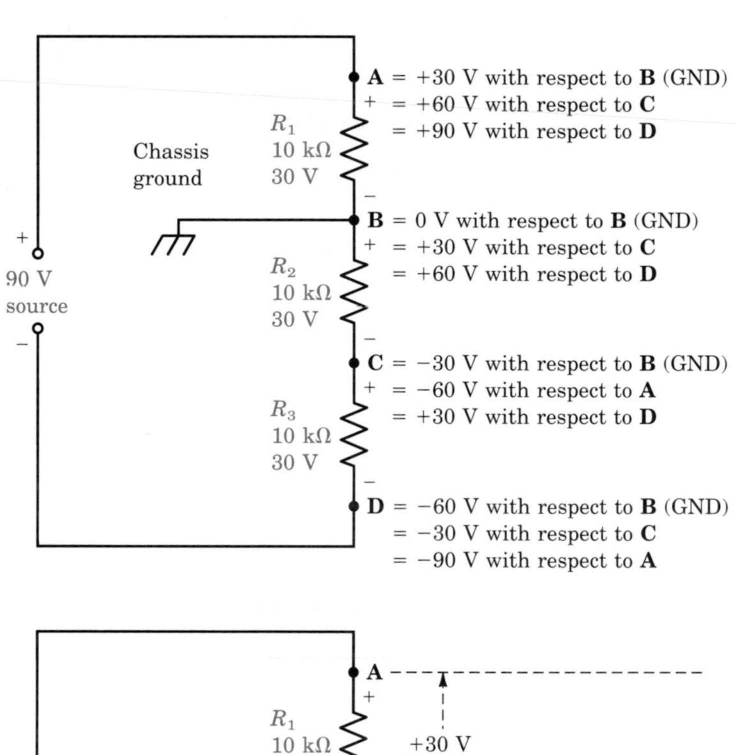

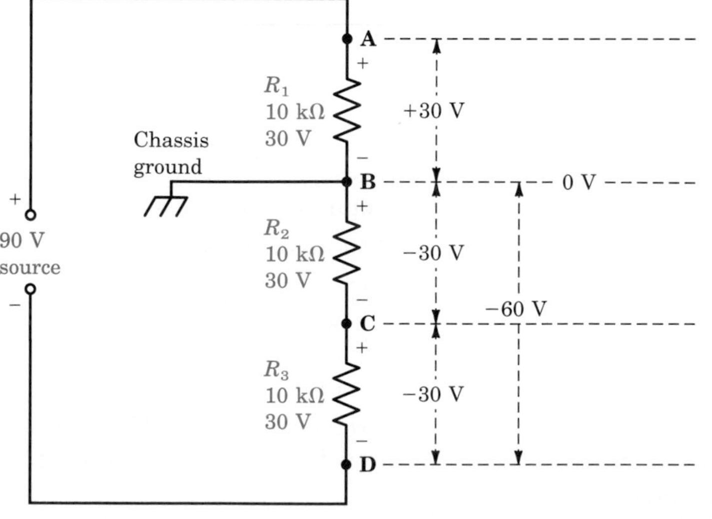

FIGURE 3–21 Chassis ground as a reference point at point B

Figure 3–21, the voltage at point C with respect to the voltage at point D is +30 V. Conversely, the voltage at point D with respect to the voltage at point C is −30 V. (Note the direction and amount of current is the same in both of the examples.)

An in-depth study of various types of voltage dividers will be in the chapter on series-parallel circuits. However, the idea of reference points will be used throughout your studies and your career as a technician.

The Voltage-Dropping Resistor

One final topic relating to some special applications of series circuits is the series-**dropping resistor.**

Often, there is a need to supply a specific voltage at a specific current to an electrical load. If the "fixed" power supply voltage available is higher than the desired voltage value, a series-dropping resistor is used that drops the voltage to the desired level, Figure 3–22.

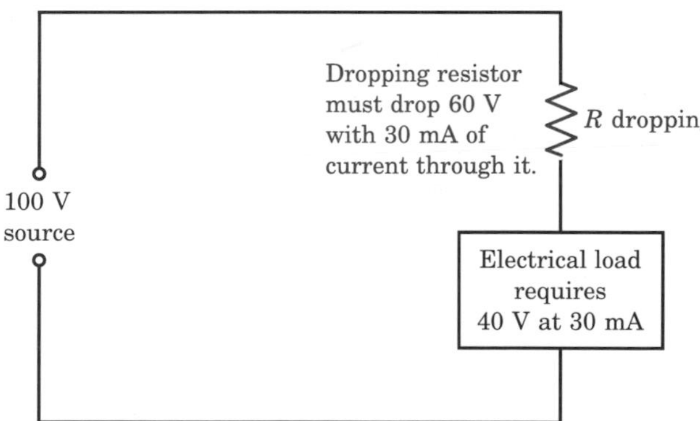

FIGURE 3–22 Example of a series-dropping resistor

EXAMPLE

The current through the electrical load and the series-dropping resistor must be the same since they are in series. To calculate the value of the dropping resistor, the value of voltage it must drop, and the value of current the load requires, provide the knowns to solve the problem. In this case, the dropping resistor must drop 60 V, and the load requires 30 mA of current. Therefore, the dropping resistor must have 30 mA passing through it *when* it is dropping 60 V. Ohm's law indicates the resistor value must be $R = \frac{V}{I}$, or the dropping resistor value $= \frac{60\text{ V}}{30\text{ mA}} = 2\text{ k}\Omega$. Naturally, this design calculation technique can be used to select resistor values for an infinite variety of dropping-resistor applications and is another example of the many special applications for series circuits.

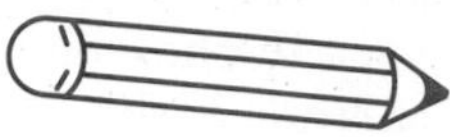

PRACTICAL NOTES

The power rating of the dropping resistor must be greater than it is required to dissipate. Usually, a rating *of at least two times* the actual dissipation is chosen.

It should be pointed out that a voltage drop is the difference in voltage between two points caused by a loss of pressure (or emf) as current flows through a component offering opposition to current flow. An example is the potential difference caused by a voltage drop across a resistor.

PRACTICE PROBLEM X

1. What is the minimum power rating for the dropping resistor, Figure 3–22?
2. What should the value of the dropping resistor be if the applied voltage were 150 volts and the load requirements remained 40 V at 30 mA? What is the minimum power rating for the dropping resistor?

CRITICAL THINKING AND THE SIMPLER TROUBLESHOOTING SEQUENCE

Your most valuable asset as a technician is the ability to think critically and logically. The ability to mentally move from a general principle or general case to a specific situation (deductive reasoning) is valuable. Likewise, moving from a specific case to a probable general case is worthwhile (inductive reasoning).

Throughout the remainder of this text, we will help you develop and enhance your logical reasoning skills. To get started, study the following section called "Introduction to Troubleshooting Skills."

INTRODUCTION TO TROUBLESHOOTING SKILLS

All electronic technicians and engineers perform troubleshooting to some degree throughout their careers! Whether designing, installing, testing, or repairing electronic circuits and systems, the skill of locating problem areas in the circuit or system by a *logical* narrowing down process becomes a valuable asset.

To enhance your skills of logical troubleshooting, we will introduce a simple sequence of troubleshooting steps you can easily learn. We'll call it the "SIMPLER" sequence (or method) for troubleshooting.

In many of the remaining chapters, you will have an opportunity to practice this sequence and solve "Chapter Challenge" applications problems. By the time you finish the book, this SIMPLER sequence will be second nature; thus, enhancing your skills and making you an even more valuable technician.

THE SIMPLER SEQUENCE FOR TROUBLESHOOTING

S—YMPTOMS .... Gather, verify, and analyze symptom information

I—DENTIFY .... Identify initial suspect area for location of trouble

M—AKE .... Make decisions about: "What type of test to make" and "Where to make it"

P—ERFORM .... Perform the test

L—OCATE .... Locate and define new "narrower" area in which to continue troubleshooting

E—XAMINE .... Examine available information and again determine "what type test" and "where"

R—EPEAT .... Repeat the preceding analysis and testing steps until the trouble is found

Further Information About Each Step in the Sequence

1. **S**YMPTOMS may be provided or collected from a number of sources, such as:
   a. What *the user* of the circuit or system tells you.
   b. *Your five senses* frequently can give you strong clues to the trouble. Sight and hearing allow evaluation of the circuit or system output indicators, e.g., normal or abnormal picture, normal or abnormal sound, system meter readings. Sight also allows you to see obviously overheated/burned components. Smell helps find overheated or burned components. (Once you have smelled certain types of overheated or burned components, you will never forget the smell!) Touch can also be used but should be used *very cautiously*! Bad burns result from touching a component that is running exceedingly hot! (Some components run that way normally, others do not.)
   c. By using your evaluation of the *easy-to-check indicators.* For example, if a TV has sound but no picture, or a radio receiver has hum, but you can't tune in radio stations. Most systems have easy-to-get-at switches and controls that aid in narrowing down possible trouble areas before you begin in-depth troubleshooting. For example, TVs have channel selectors, radios have tuning and volume controls, and electronic circuits frequently have switches and variable resistances.
   d. By *comparing actual operation to normal operation and "norms."* "Norms" are normal characteristics of signals, voltages, currents, etc., that are known by experience or determined by referring to diagrams and documentation about the circuit or system being checked. This means you, as a technician, should know how to read and thoroughly understand block and schematic diagrams. (In specific cases, you may have to trace the actual circuit and draw the diagrams yourself.)

2. **I**DENTIFY the initial suspect area by:
   a. Analyzing all the symptom data.
   b. From the analysis of the symptom data, determining all the possible sections of a system, or components within a circuit, that might cause or contribute to the problem.
   c. Circle, or in some way mark, the initial suspect area.

3. **M**AKE a decision about "what kind of test" and "where" by:
   a. Looking at the initial suspect area, and determining where a test should be made to narrow down the suspect area most efficiently. And, determining what type of test would be most appropriate to make. (NOTE: Many times the "where" will dictate what type of test is to be made. At other times, the easiest type of test to make,

which yields useful information, is the determinant of both the "where" and "what kind" of tests.)

b. Typically, you begin with general tests, such as looking at obvious indicators, manipulating switches and controls, and similar measures. As you narrow down the suspect area, your tests become more precise, such as voltage, current, or resistance measurements. Finally, component substitution with a known good component for the suspected bad one confirms your analysis. (In some cases, eradicating the problem is done by soldering, or removing a short, or moving a wire rather than changing a component.)

4. **P**ERFORM what your "where-and-what-kind-of-test" analysis has indicated (i.e., what kind of test and where), and you are provided with new information and insight to use in your quest to narrow down and find the problem.

5. **L**OCATE the questionable area of the circuit or system using information you have. Again, you will identify this new smaller area by circling or bracketing it so you won't make needless checks outside the logical area to be analyzed.

6. **E**XAMINE the collected symptom information, test results, and other data you have. Now make a decision about the next test that will provide further meaningful information.

7. **R**EPEAT the analysis, testing, and narrowing down process as many times as required, and you will eventually find the fault in the circuit or system. At this point, you have successfully used the SIMPLER sequence!

Using the SIMPLER Sequence: An Example

The information given to you is:

A basic circuit is comprised of a voltage source, conducting wires, a switch, and a light bulb (with its socket). It is known that the voltage source and conductor wires have been previously tested and they are good.

The complaint (which reveals symptom information) is the light bulb does not light, even when the switch is flipped to the "on" position.

1. The **symptom** information gathered is that the light bulb doesn't light! You verify this by connecting the circuit. Analysis of the symptom information can be done by mentally visualizing the circuit, or drawing the circuit diagram, Figure 3–23. It is then apparent that if the source and the conductor wires are good, the initial suspect areas are the switch, light bulb, and socket.

2. Now you mentally (or on paper) **identify** the portion of the circuit containing the switch, the light bulb, and the socket (by circling or

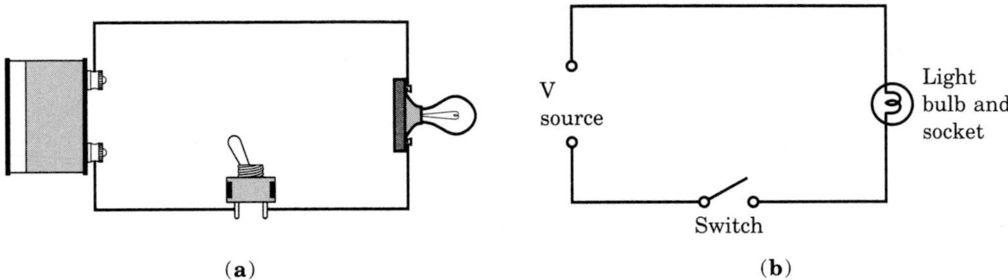

FIGURE 3–23 Basic circuit diagram

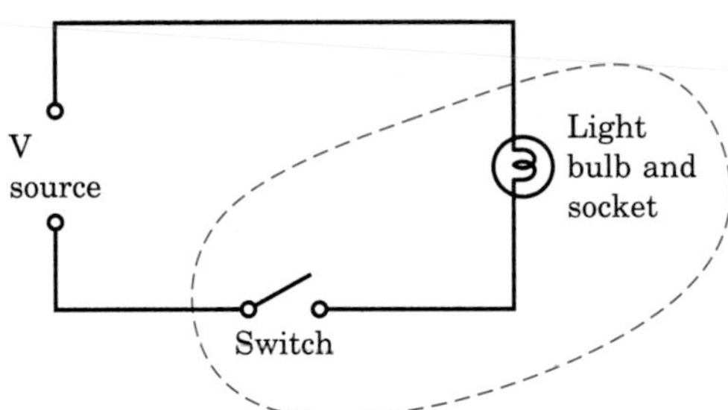

FIGURE 3–24 Initial suspect area

bracketing), Figure 3–24. This becomes the "initial suspect area" for locating the trouble.

3. You now **make** a decision about "what type of test to make" and "where to make it." Some of the possibilities are to: a) make a voltage test with a voltmeter across the light bulb socket; b) turn the switch to the "off" position and replace the bulb with a known good bulb, then turn switch to the "on" position to see if the replacement bulb lights; c) disconnect the power source, unscrew the bulb, and make an ohmmeter test of the suspect bulb to see if it has continuity; d) measure the voltage across the switch terminals with the switch open and with it closed to see if the voltage changes; and e) disconnect the power source and use an ohmmeter to check the switch in both the "off" and "on" positions to see if it is operating properly.
   Since changing the bulb would be reasonably easy to do and bulbs do fail frequently, you decide to make the light bulb test first, Figure 3–25.

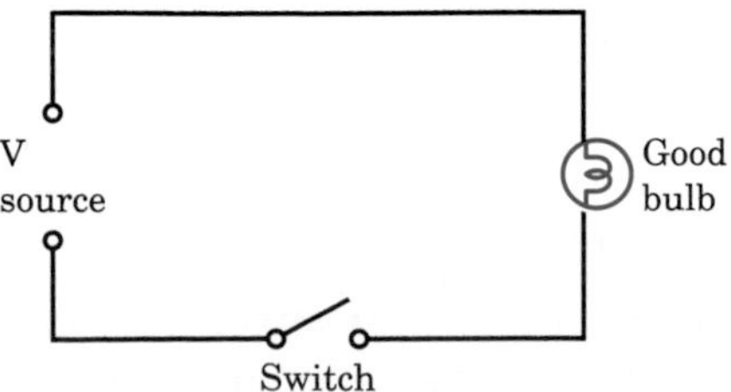

FIGURE 3–25 Checking the light bulb first

4. When you **perform** this test, you find that the "good" replacement bulb does not light, even with the power source connected to the circuit and the switch in the "on" position. Or if you tested the "suspect" bulb using the ohmmeter check approach, you would find the bulb checks good, Figure 3–26.

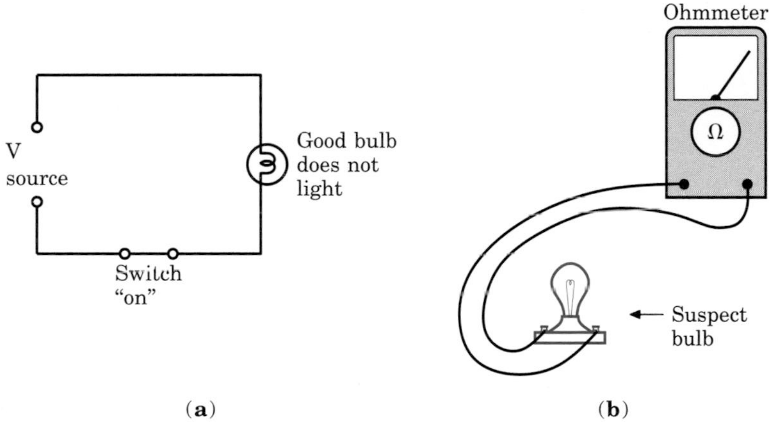

FIGURE 3–26 Results of test a. Good bulb does not light. b. Suspect bulb has continuity.

5. You can now **locate** the new smaller suspect area that includes only the socket and switch, Figure 3–27.

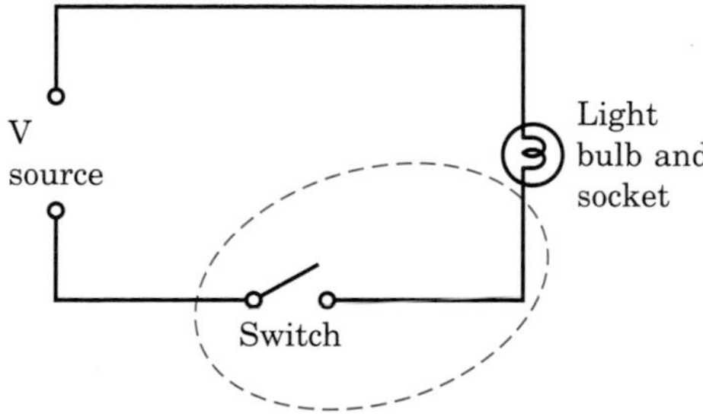

FIGURE 3–27 Narrowed suspect area

6. When you **examine** all the accumulated information, you decide the most likely suspect between the socket and switch might be the switch since it is somewhat of a mechanical device which fails more frequently. Three ways you can check into this possibility are to: a) disconnect the power source from the circuit and use an ohmmeter to check the switch continuity in both the "off" and "on" positions; b) measure the voltage at the light socket to see if voltage is present with the switch in the "on" position; and c) measure voltage across the switch in both the "off" and "on" positions to see if there is volt-

age across the switch when in the "off" position and zero volts present when the switch is in the "on" position. After examining the information your decision about "what type of test" and "where" is to check voltage at the light socket since this is an easy check to make, Figure 3–28.

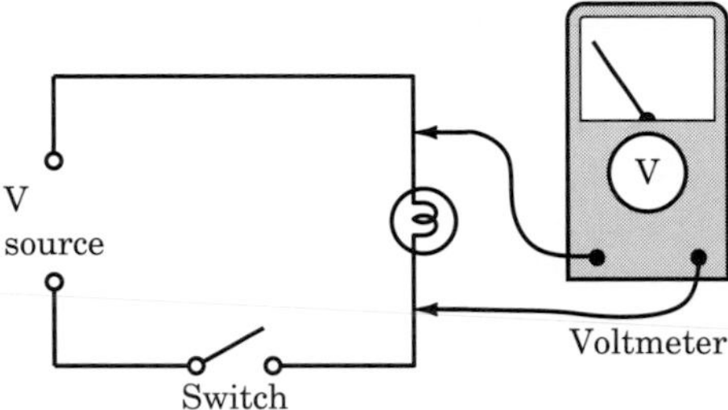

FIGURE 3–28 Voltage check at light socket

7. **Repeating** the analysis *and* testing procedures, you make the test at the bulb socket with the switch "on." Voltage is present, Figure 3–29. This indicates the switch is good. With this analysis, your remaining suspect is the light socket. In testing the socket (either by ohmmeter continuity testing with the power off or by replacing with a known good socket), you find the socket is indeed bad. You have found the problem! The solution is obvious: Replace the socket. After replacing the socket, you verify the accuracy of your troubleshooting by testing the circuit. Your reward is that the circuit works properly, Figure 3–30.

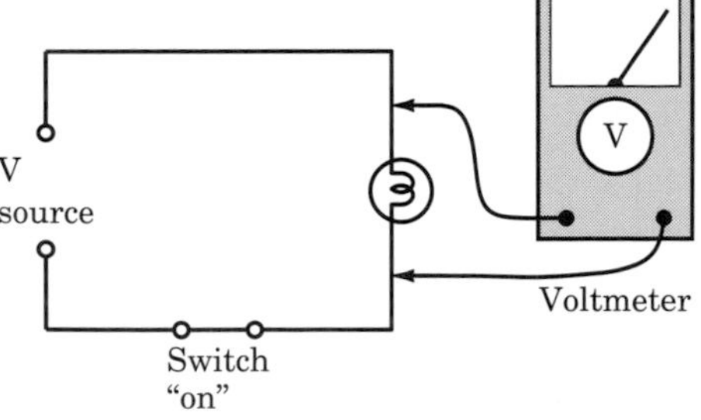

FIGURE 3–29 Voltage present at socket

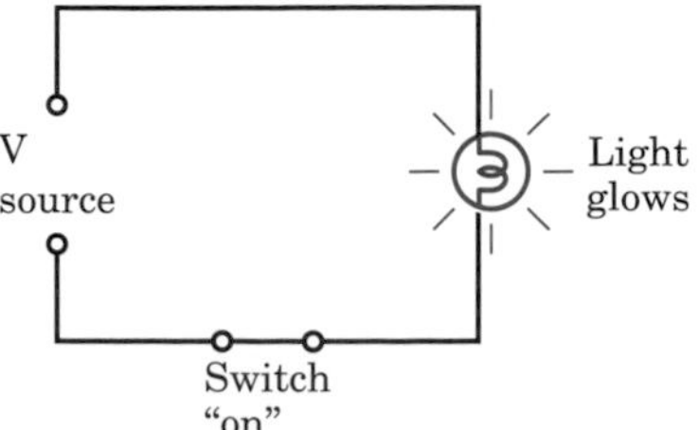

FIGURE 3–30 The circuit works

Although this hypothetical case is simple, it illustrates the concepts of the SIMPLER troubleshooting sequence. Practice using the sequence because it will help you become a logical troubleshooter!

TROUBLESHOOTING LEVELS

There are two basic levels of troubleshooting technicians may have to perform. One is troubleshooting to the "block" or "module" level. The other is troubleshooting to the single component level.

The block level requires knowledge of the normal "inputs" and "outputs" used for each block or module in the total system. These inputs and/or outputs may be audio or video signals, certain voltage levels, certain current values, and so forth.

To troubleshoot to the block level, we'll refer to the "goz-inta/goz-outa" theory, where the "goz-inta" refers to block inputs and the "goz-outa" refers to the block output(s). If the input quantities (signals, voltages, etc.) check normal for a given block but the output quantities are abnormal, the problem is probably in that block.

On the other hand, if the input quantities are not normal, you trace backward to where the input is supposed to come from until the place of abnormality is found. In some cases, the abnormality can be caused by the input circuitry of the block being tested, rather than from blocks feeding this block. Your job is to isolate the block causing the problem and to replace it. See a block-level troubleshooting example on page 115.

Single-component level troubleshooting requires knowledge of normal parameters throughout the circuitry within each block of the system. By isolating and narrowing down (using the SIMPLER sequence approach), you will eventually narrow the problem to the bad component(s) within a module or block. Replacing the bad component(s) solves the problem. See an example of component-level troubleshooting on page 116.

The SIMPLER Sequence and Troubleshooting Levels

The SIMPLER sequence can be applied to both the block and component levels of troubleshooting. You will have a chance to try both troubleshooting levels in the Chapter Challenge problems you will find throughout this book.

Now, try the Chapter Challenge here on series circuits to begin practicing these concepts.

CHAPTER CHALLENGE

Following the SIMPLER troubleshooting sequence, find the problem with this circuit. As you follow the sequence, record your responses to each step on a separate sheet of paper.

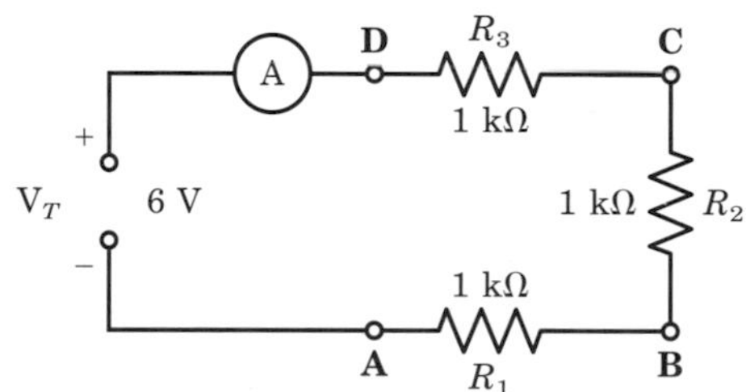

CHALLENGE CIRCUIT 1

SYMPTOMS Gather, verify, and analyze symptom information. (Look at the "Starting Point Information.")

IDENTIFY Identify initial suspect area for the location of the trouble.

MAKE Make a decision about "What type of test to make" and "Where to make it." (To simulate making a decision about each test you want to make, select the desired test from the "TEST" column listing below.)

PERFORM Perform the test. (Look up the result of the test in Appendix C. Use the number in the "RESULTS" column to find the test result in the Appendix.)

LOCATE Locate and define a new "narrower" area in which to continue troubleshooting.

EXAMINE Examine available information and again determine "what type of test' and "where."

REPEAT Repeat the preceding analysis and testing steps until the trouble is found. What would you do to restore this circuit to normal operation? When you have solved the problem, compare your results with those shown in the color insert.

Starting Point Information

1. Circuit diagram
2. V_T = 6 V
3. Measured I_T = 1.5 mA

Test	Results in Appendix C
$V_{A\text{-}B}$	(5)
$V_{B\text{-}C}$	(86)
$V_{C\text{-}D}$	(42)
Res. of R_1	(16)
Res. of R_2	(62)
Res. of R_3	(3)
R_T	(36)

A BLOCK-LEVEL TROUBLESHOOTING EXAMPLE

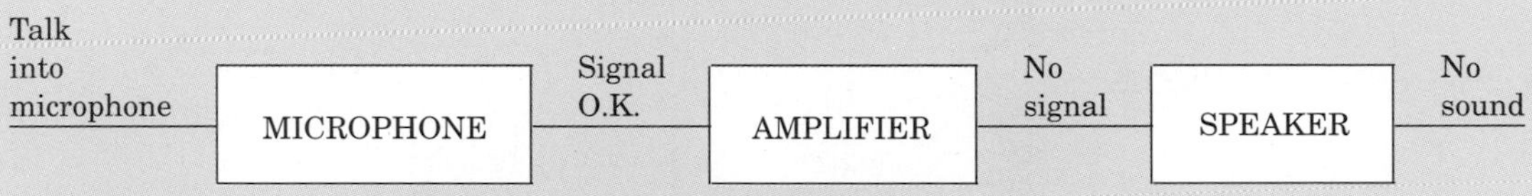

FIGURE 3–31

If you were told there was no sound from the speaker, Figure 3–31, even though someone were talking into the microphone, block-level troubleshooting scenario might be as follows:

Symptoms No sound from speaker

Identify initial suspect area anywhere in the system for this situation. The trouble could be in the speaker block, amplifier block, or microphone block.

Make decisions about test (what kind of test and where). Probably make test about in the middle of the system (at the amplifier input). That way you have cut the area of possible trouble in half with your first test. (Remember the "divide-and-conquer" technique explained with Figure 3–15.)

Perform the test: As you can see from the diagram, the signal into the amplifier is OK (the "goz-inta" is all right for this block).

Locate the new smaller suspect area. Your first test has eliminated the microphone as a possible trouble spot. The remaining possible trouble area is the amplifier block and the speaker block.

Examine available information and make new test decisions. Now see if the amplifier output is normal. The "type" of check is a signal check; the "where" is at the amplifier output.

Repeat testing and analysis procedure. A check at the amplifier output shows no signal. (The "goz-outa" signal is bad.) If the input to the amplifier is good and the output of the amplifier is bad, the chances are the trouble is in the amplifier block. This is troubleshooting to the block level. If we were to go inside the block to find the component in the block causing the trouble, that would be troubleshooting to the component level.

A COMPONENT-LEVEL TROUBLESHOOTING EXAMPLE

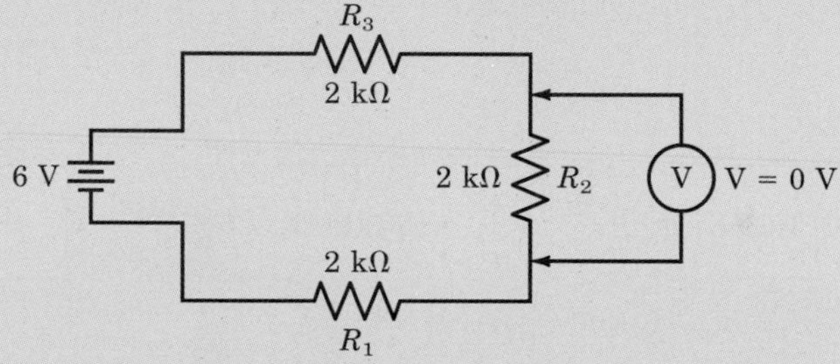

FIGURE 3–32

If you were given the facts in Figure 3–32, and told the voltage source had been checked as being O.K., component-level troubleshooting might be done as follows:

Symptoms   No voltage across R_2, even though circuit input voltage is normal.

Identify   initial suspect area. The trouble could be with any one of the resistors or with the interconnecting wires in the circuit, since the symptom could be caused by R_2 being shorted or by an open elsewhere in the circuit.

Make   decisions about test (what type of test and where). Since voltage tests are easy and yield much information for small effort, the "what kind of test" is answered with a voltage test. Since you had been told that the voltage source is O.K., you decide to check the voltage across either one of the two remaining resistors to determine if current is flowing anywhere in the circuit.

Perform   the test. You decide to make the voltage test across R_1. The result of the test is that V_{R_1} is found to be 6 volts. This is abnormal for this circuit!

Locate   the new smaller suspect area. The voltage test across R_1 leads you to believe it may be the suspect! (Recall in a series circuit V applied appears across the open portion of the circuit, and zero volts appear across the good components).

Examine   available information and make new test decisions. The logical new test might be to disconnect the power source and use an ohmmeter to check the value of resistance exhibited by R_1. The "what kind of test" is a resistance check. The "where to make the test" decision is across R_1.

Repeat   testing and analysis procedure. When R_1 is tested with the ohmmeter, it measures infinite resistance, or "open." R_1 is indeed the culprit. This is verified by replacing R_1 and energizing the circuit. When this is done, the voltage drops measured across each of the resistors is 2 volts. This is troubleshooting down to the component level.

SUMMARY

- The definition and characteristics of a series circuit state that the circuit components are connected so that there is only one current path; therefore, the current is the same through all components in series.
- Current in a series circuit is calculated by Ohm's law, using either total voltage divided by total resistance $\left(I = \frac{V_T}{R_T}\right)$ or the voltage drop across one of the components divided by its resistance $\left(\text{for example, } I = \frac{V_1}{R_1}\right)$.
- Total resistance in a series circuit equals the sum of the individual resistances ($R_T = R_1 + R_2 \ldots + R_n$). R_T can also be calculated if V_T and I are known using $R_T = \frac{V_T}{I_T}$.
- Because I is the same throughout all components in series, each component's voltage drop ($I \times R$) is proportional to its resistance compared to the total resistance. The largest value resistance drops the highest percentage or fraction of V applied; the smallest value resistance drops the smallest proportion of voltage.
- V applied must equal the sum of the voltage drops around the closed loop. That is, if there are three components around the series string, V applied must equal the sum of the three components' voltage drops. If there are four components, V applied equals the sum of the four individual voltage drops.
- Kirchhoff's voltage law states the arithmetic sum of the voltages around a single circuit loop equals V applied. Also, Kirchhoff's voltage law can be stated as the algebraic sum of all the loop voltages, including the source, must equal zero. It is important when utilizing Kirchhoff's law(s) to notice the voltage polarities.
- The power distribution throughout a series circuit is directly related to the resistance distribution. For example, if a component's R is 25% of the total R, it dissipates 25% of the total power supplied by the source.
- Total power dissipated by a series circuit (or any other type circuit) equals the sum of all the individual power dissipations. Total power can be calculated by $P_T = V_T \times I_T$, or $P_T = P_1 + P_2 \ldots + P_n$.
- A break or open anywhere along the path of current in a series circuit causes R_T to increase to infinity; I_T to decrease to zero; voltage drops across unopened components or circuit portions to decrease to zero; and circuit applied voltage to be felt across the open portion of the circuit. P_T decreases to zero since I equals zero.
- A short (very low resistance path) across any component, or portion, of a series circuit causes R_T to decrease; I_T to increase; voltage drops across the normal portions of the circuit to increase; and voltage dropped across the shorted portion of the circuit to decrease to virtually zero. A short also causes P_T to increase and the power dissipated by the remaining unshorted components to increase.
- When designing a circuit or solving for unknown quantities in an electrical circuit, it is good to: 1) collect all known values; 2) draw a diagram and label known value(s); 3) solve at any point where enough knowns are available to solve for an unknown. (NOTE: Usually, Ohm's law and/or Kirchhoff's law(s) will be used in the solution).
- When troubleshooting a series circuit with an open condition, voltage measurements will reveal zero voltage drop across good components (or circuit portions) and V applied across the open points. Take resistance measurements only with the circuit disconnected from the source. Then normal components will reveal normal readings, and the opened component/portion will reveal an infinite ohms reading.

- When troubleshooting a series circuit with a short condition, use *resistance* measurements with the power source disconnected from the circuit. Normal components (or circuit portions) will measure normal values of R, while the shorted component (or circuit portion) will measure very low, or zero resistance.
- Series circuits are useful for several special applications, such as series-aiding and series-opposing voltage sources, simple voltage-dividing action, and series-dropping an available voltage down to a desired level.

REVIEW QUESTIONS

NOTE: While working on these questions, don't forget to draw and label diagrams, as appropriate. It is a good habit to form.

1. If a series string of two equal resistors draws 5 amperes from a 10-volt source, what is the value of each resistor?

2. What is the current through a series circuit comprised of one 10-kΩ, one 20-kΩ, and one 30-kΩ resistor? Assume the circuit applied voltage is 40 volts.

3. Draw a diagram of a series circuit containing resistors having values of 50 Ω, 40 Ω, 30 Ω, and 20 Ω, respectively. Label the 50 ohm resistor as R_1, the 40 ohm resistor as R_2 and so forth. Calculate the following and appropriately label your diagram:
   a. What is the value of R_T?
   b. What is V applied if I equals 2 amperes?
   c. What is the voltage drop across each resistor?
   d. What is the value of P_T?
   e. What value of power is dissipated by R_2? By R_4?
   f. What fractional portion of V_T is dropped by R_4?
   g. If R_3 increases in value while the other resistors remain the same, would the following parameters increase, decrease, or remain the same?
      (1) Total resistance

(2) Total current
(3) V_1, V_2, and V_4
(4) P_T

h. If R_2 shorted, would the following parameters increase, decrease, or remain the same?
(1) Total resistance
(2) Total current
(3) V_1, V_3, and V_4

4. According to Kirchhoff's voltage law, if a series circuit contains components dropping 10 V, 20 V, 30 V, and 50 V respectively, what is V applied? What is the algebraic sum of voltages around the complete closed loop, including the source?

5. Draw a diagram showing how you would connect three voltage sources to acquire a circuit applied voltage of 60 volts, if three sources equaled 100 V, 40 V, and 120 V, respectively.

6. a. Calculate the P_T and P_1 values for the circuit shown in Figure 3-33.
   b. What is the V_T?
   c. What is the R_T?
   d. How many times greater is P_4 than P_1?
   e. If R_4 shorted what would the value of P_T become? Of P_1?

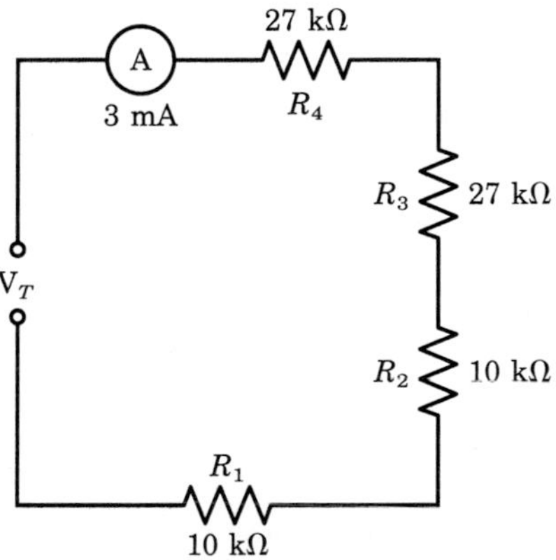

FIGURE 3–33

7. What precaution should be taken when troubleshooting a series circuit suspected of having a short? Briefly explain why this precaution should be taken.

8. If a series circuit's current suddenly drops to zero, what has happened in the circuit?

9. If certain components are overheating, circuit current has increased, and the voltage drops across R_2 and R_3 have drastically decreased, which of the following has occurred?

a. The total circuit has been shorted.
b. The total circuit has been opened.
c. R_1 has shorted.
d. R_2 and R_3 have been shorted out.
e. None of the above.

10. Find the value of V_A for the series circuit shown in Figure 3–34.
    a. Assume that R_3 and R_4 are equal. What is the value of R_3?
    b. What is the value of R_2?

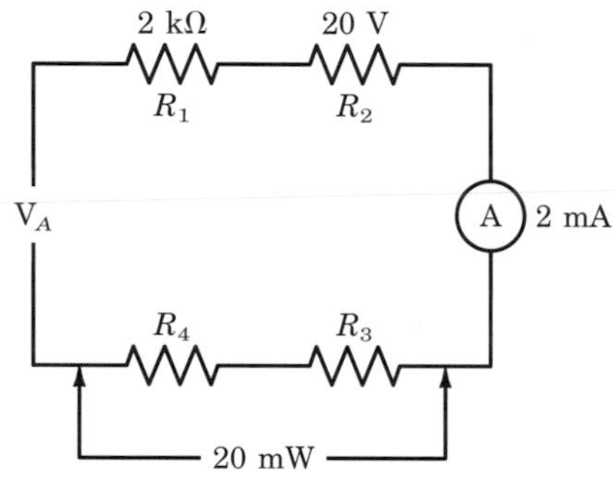

FIGURE 3–34

11. Draw a simple three-resistor series voltage divider that provides equal voltage division of a 180-volt source and draws 2 mA of current.

12. In a three-resistor series circuit, if V applied is 200 volts, V_1 is 40 V, and V_2 is 90 V, what is the value of V_3?

13. In a three-resistor series circuit, if V_1 is 25 V, V_2 is 50 V, and V_3 is twice V_1, what is the value of V applied?

14. What basic concept is often used in troubleshooting a defective series circuit that has a number of components in series?

15. For the conditions shown in the circuit in Figure 3–35 to what value is R_1 adjusted?
    a. What is the value of I_T?
    b. What percentage of total power is dissipated by R_1?
    c. If R_1 were set to the middle of its R range, what would be the value of I_T?
    d. If R_1 were set at the least value possible, what value of voltage would be indicated by the voltmeter in the circuit?

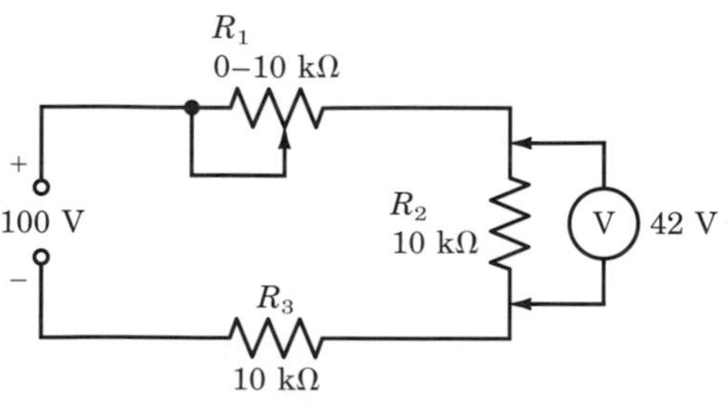

FIGURE 3–35

16. a. Find R_T and V_T for the circuit shown in Figure 3–36.
    b. What value will meter M_1 indicate?
    c. What value will meter M_3 indicate?
    d. What is the value of I_T for this circuit?
    e. If all R values remain the same, but P_T doubles, what is the new value of V_T? What is the new I_T?

17. Refer to Figure 3–36. If all R values are the same, but P_{R_5} equals 45 mW:
    a. What is the value of P_{R_2}?
    b. What is the value of P_T?
    c. What is the value of I_T?

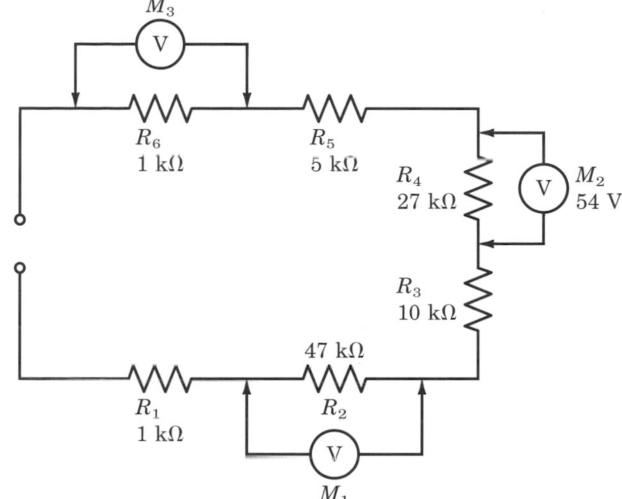

FIGURE 3–36

TEST YOURSELF

1. Draw a circuit diagram of a series circuit where V_3 is three times the value of V_1 and V_2 is twice V_1. Assume V applied is 60 V, R_T is 120 kΩ and calculate:
   a. Value of R_1.
   b. Values of V_1, V_2, and V_3.
   c. Values of I and P_T.

2. Determine the value of series-dropping resistor needed to drop a source voltage of 100 volts to the appropriate value for a load rated as 50V at 5mA. Draw the circuit and label all components and electrical parameters.

PERFORMANCE PROJECTS CORRELATION CHART

Suggested performance projects in the Laboratory Manual that correlate with topics in this chapter are:

CHAPTER TOPIC	PERFORMANCE PROJECT	PROJECT NUMBER
Resistance in Series Circuits	Total Resistance in Series Circuits	9
Definition and Characteristics of a Series Circuit	Current in Series Circuits	10
Voltage in Series Circuits	Voltage Distribution in Series Circuits	11
Power in Series Circuits	Power Distribution in Series Circuits	12
Effects of Opens in Series Circuits and Troubleshooting Hints	Effects of an Open in Series Circuits	13
Effects of Shorts in Series Circuits and Troubleshooting Hints	Effects of a Short in Series Circuits	14

NOTE: It is suggested that after completing the above projects, the student should be required to answer the questions in the "Summary" at the end of this section of projects in the Laboratory Manual.

Parallel Circuits

Key Terms

Assumed voltage method

Current divider circuit

Equivalent circuit resistance

Kirchhoff's current law

Parallel branch

Parallel circuit

Product-over-the-sum method

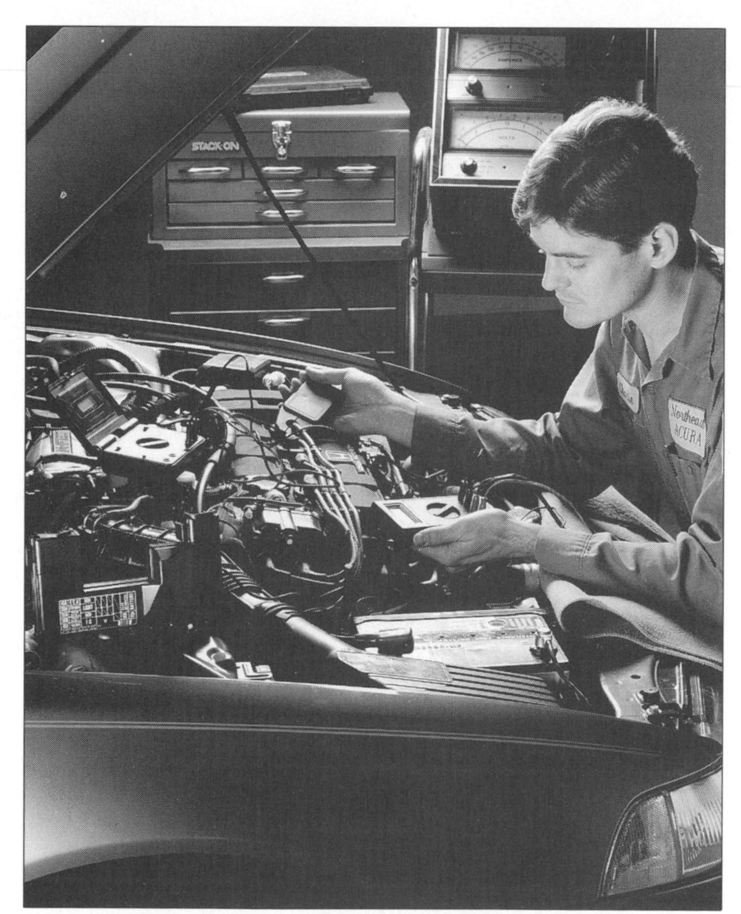

Courtesy of Cleveland Institute of Electronics, Inc.

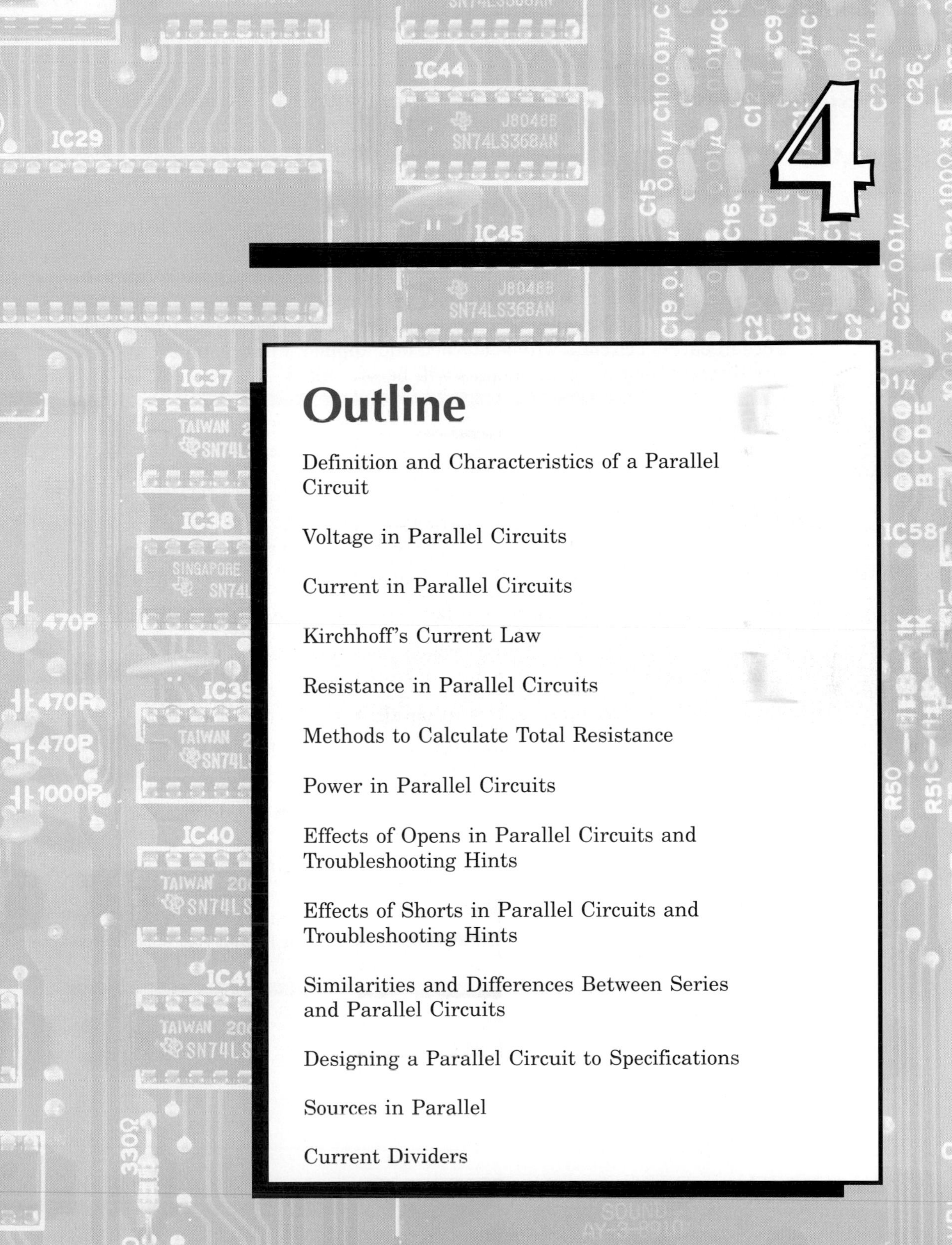

4

Outline

Definition and Characteristics of a Parallel Circuit

Voltage in Parallel Circuits

Current in Parallel Circuits

Kirchhoff's Current Law

Resistance in Parallel Circuits

Methods to Calculate Total Resistance

Power in Parallel Circuits

Effects of Opens in Parallel Circuits and Troubleshooting Hints

Effects of Shorts in Parallel Circuits and Troubleshooting Hints

Similarities and Differences Between Series and Parallel Circuits

Designing a Parallel Circuit to Specifications

Sources in Parallel

Current Dividers

Chapter Preview

Most electrical and electronic circuits, no matter how simple or complex, contain portions that can be examined using parallel circuit analysis. Of course, some circuits contain only parallel circuit arrangements. For example, the lights and wall outlets in your home generally use parallel circuitry to connect several lights or outlets on one circuit, Figure 4–1. Many automobile accessories (heaters and radios) are connected to the battery using parallel circuit methods, Figure 4–2.

In this chapter, you will again apply Ohm's law and the power formulas. Also, Kirchhoff's current law and some *vital contrasts* between parallel and series circuits will be examined. Several approaches to solve for total resistance in parallel circuits will be learned and applied. Finally, information on circuit troubleshooting techniques will be presented.

Objectives

After studying this chapter, you will be able to:

- Define the term **parallel circuit**
- List the characteristics of a parallel circuit
- Determine voltage in parallel circuits
- Calculate the total current and branch currents in parallel circuits
- Compute total resistance and branch resistance values in parallel circuits using at least three different methods
- Determine conductance values in parallel circuits
- Calculate power values in parallel circuits
- List the effects of opens in parallel circuits
- List the effects of shorts in parallel circuits
- Describe troubleshooting techniques for parallel circuits
- Use current divider formulas
- Apply the SIMPLER troubleshooting sequence to a Chapter Challenge problem

Lights in a home are connected in parallel.

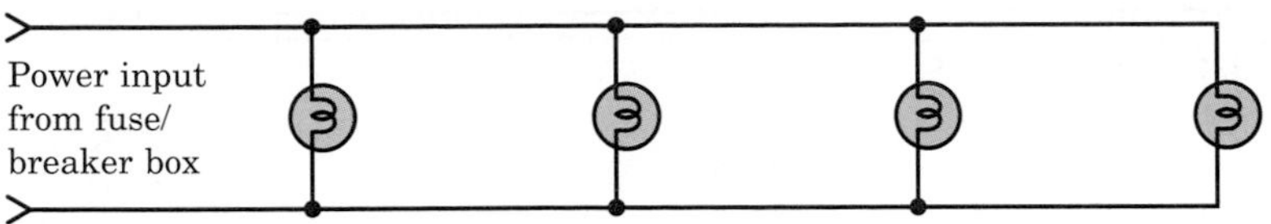

FIGURE 4–1 Parallel circuits are common in a home.

Wall outlets in a home are connected in parallel.

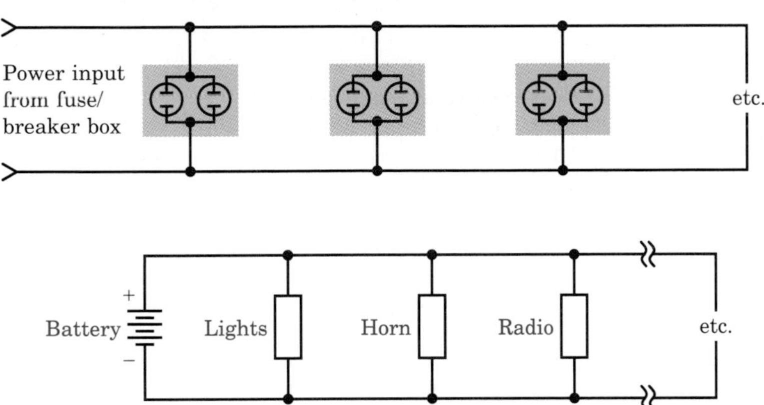

FIGURE 4–2 Automotive accessories are often connected in parallel across the battery. NOTE: Series switches in circuits are not shown.

DEFINITION AND CHARACTERISTICS OF A PARALLEL CIRCUIT

Recall from the last chapter the important characteristic defining a series circuit is only one path for current exists. Therefore, the current through all components is the same. Because of this fact, the voltage distribution throughout the circuit (voltage across each of the components) depends on each component's resistance value (i.e., the highest R value dropping the highest voltage or the lowest R value dropping the lowest voltage).

The significant features of **parallel circuits** are: 1) *the voltage across all parallel components must be the same,* Figure 4–3; and 2) *there are two or more paths (branches) for current flow,* Figure 4–4. Because the voltage is

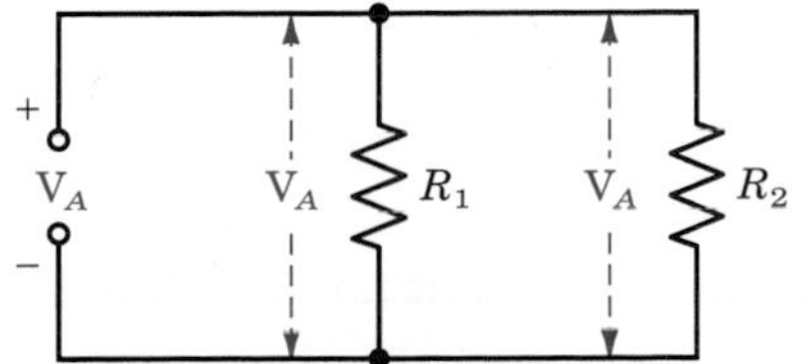

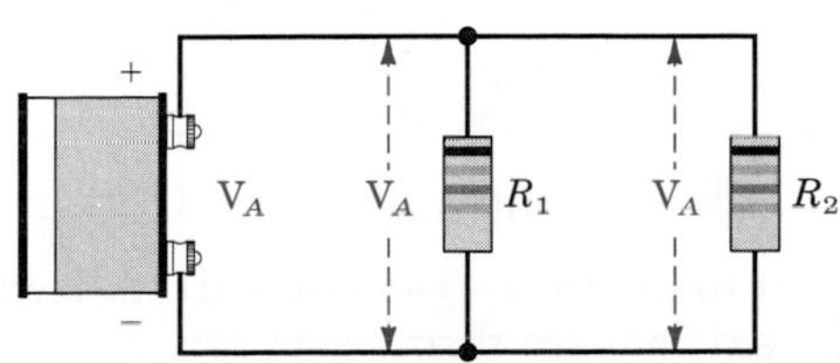

FIGURE 4–3 Voltage in parallel circuits is the same across all components, where $V_{R_1} = V_A$ and $V_{R_2} = V_{R_1} = V_A$.

FIGURE 4–4 A basic difference between series and parallel circuits

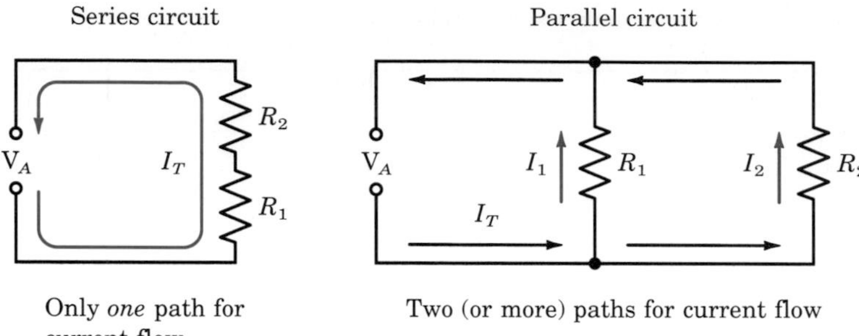

the same across each **parallel branch,** the current distribution throughout the circuit depends on each branch's resistance value. Each branch current is *inverse* to its resistance value, which means the higher the resistance value, the lower the branch current, Figure 4–5.

FIGURE 4–5 In parallel circuits, current through each branch is *inversely* proportional to each branch's resistance.

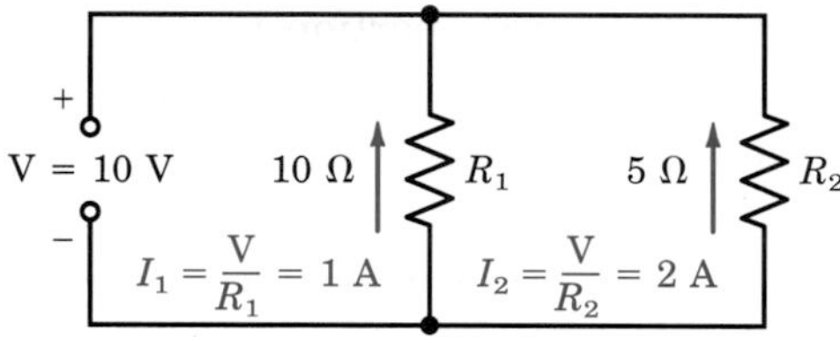

Thus, a parallel circuit is defined as one where the voltage across all (parallel branch) components is the same and where there are two or more branches for current flow.

VOLTAGE IN PARALLEL CIRCUITS

If you know the voltage applied to a parallel circuit or the voltage across any one of the branches, you automatically know the voltage across any one or all of the parallel branches. This voltage is the same across all branches of a parallel circuit, Figure 4–3.

PRACTICE PROBLEM I

Answers are in Appendix B.

Again refer to Figure 4–3. If R_2 is 27 kΩ and current through R_2 is 2 mA, what is the value of V_{R1}? Of V_A?

CURRENT IN PARALLEL CIRCUITS

Branch Currents and Total Current

Refer to Figure 4–6 and note the following points as current and Kirchhoff's current law are discussed:

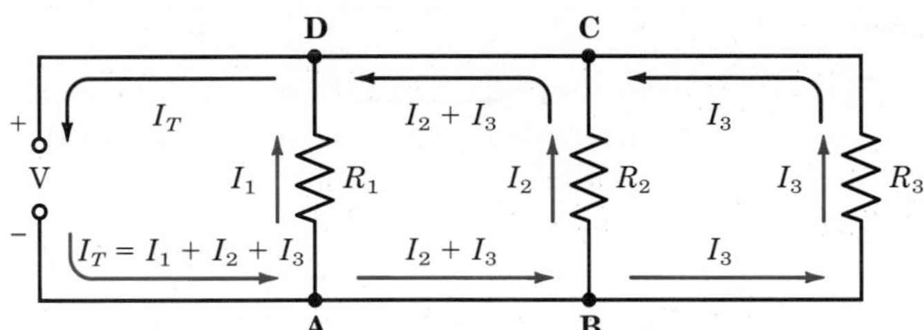

FIGURE 4–6 Kirchhoff's current law states current coming to a point must equal the current leaving that same point.

- Total circuit current leaves from the source's negative side and travels along the conductor until it reaches the circuit "junction" at point A.
- At point A, a portion of the total current travels through branch resistor R_1 (current path I_1) on its way back to the source's positive side. The remaining portion travels on to the circuit junction at point B.
- At point B, current "splits" and a portion travels through R_2 (current path I_2) and the remainder through R_3 (current path I_3) on its way back to the source's positive side.
- At point C, currents I_2 and I_3 rejoin and travel to junction D. Thus, current value from point C to D must equal $I_2 + I_3$.
- At point D, current $I_2 + I_3$ join current I_1 and this total current travels back to the source's positive side. NOTE: *The total current equals the sum of the branch currents.* This is an important fact; remember it! Mathematically:

Formula 4–1	$I_T = I_1 + I_2 + I_3 \cdots + I_n$

PRACTICE PROBLEM II

Assume a five-branch parallel circuit. What is the formula for I_T? Which resistor passes the most current? Which resistor passes the least current?

KIRCHHOFF'S CURRENT LAW

The previous detailed statements express **Kirchhoff's current law.** Simply stated, this law is *the value of current entering a point must equal the value of current leaving that same point.*

In parallel circuit branch currents, since V across each branch is the same, the current divides through the branches in an inverse relationship to the individual branch resistances. That is, the highest R value branch allows the least current and the lowest R value branch allows the highest branch current.

EXAMPLE

In using Ohm's law to solve for branch currents, this fact becomes apparent, Figure 4–7. It suggests an interesting proportionality approach in thinking about branch currents in parallel circuits, Figure 4–8. You can see that if branch 1 has twice the resistance of branch 2, it will have half the current value through it. Conversely, branch 2 will have twice the current value of branch 1. If branch 1 were to have one-fourth the R value of branch 2, then it would have four times the current of branch 2, and so forth.

FIGURE 4–7 Current is inverse to branch *R* value.

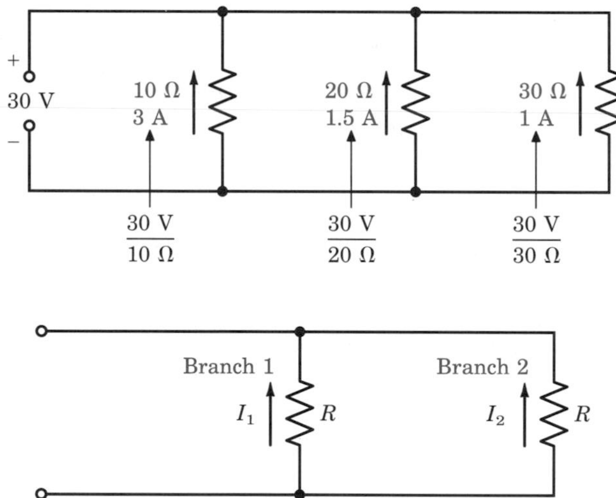

FIGURE 4–8 Looking at branch 1 via the proportionality approach: If branch 1 has twice the *R* of branch 2, then I_1 = half of I_2 (or $I_2 = 2 \times I_1$). If branch 1 has one-fourth the *R* of branch 2, then $I_1 = 4 \times I_2$ (or I_2 is one-fourth of I_1).

Now you know voltage is the same across each parallel branch, current to and from any given circuit point must be the same, branch currents are inverse to their branch resistances, and total current equals the sum of the branch currents. It's time to see if you can apply this knowledge and Ohm's law and Kirchhoff's law to solve some problems.

PRACTICE PROBLEM III

Look at Figure 4–9 and, without looking ahead, determine the values for V_2, V_1, V_A, I_1, and I_T. Read the following explanation.

To find V_2 use Ohm's law, where $V_2 = I_2 \times R_2 = 1\text{ A} \times 50\ \Omega = 50\text{ V}$. Since the voltage must be the same across all parallel branches, then V_1 and V_A must also equal 50 V in this circuit. Since you know V_1 and R_1, then Ohm's

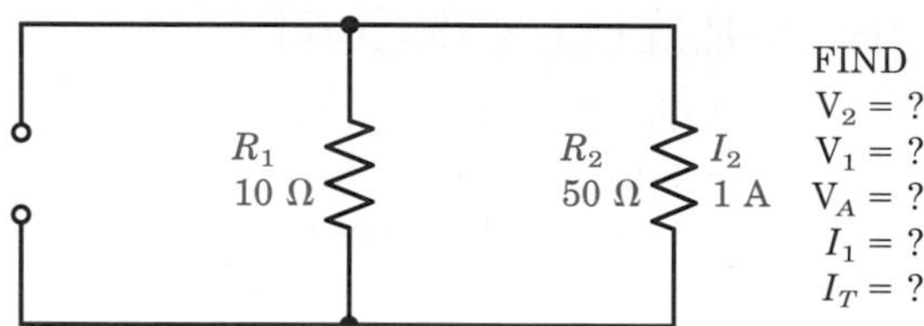

FIGURE 4–9 Parallel circuit problem

law will again enable you to find the answer for I_1. That is, I_1 must equal $\frac{V_1}{R_1}$, or 5 amperes. And I_T equals the sum of the branch currents; thus, 1 A + 5 A = 6 A.

Using Ohm's law, what must the value of R_T be? The answer is $R_T = \frac{V_T}{I_T} = \frac{50 \text{ V}}{6 \text{ A}} = 8.33\ \Omega$. It is interesting to note the total circuit resistance is *less* than either branch's resistance. With respect to the power source, the two branch resistances of 10 Ω and 50 Ω could be replaced by one resistor with a value of 8.33 Ω and it would still supply the same total current and circuit power. Later in the chapter you will study several methods to determine total circuit resistance for parallel circuits. However, at this point, remember that total resistance of parallel circuits is always less than the least branch resistance.

PRACTICE PROBLEM IV

Refer to Figure 4–10 and with your knowledge of parallel circuits, solve for the unknowns. Don't cheat yourself by looking up the answers prior to working the problems.

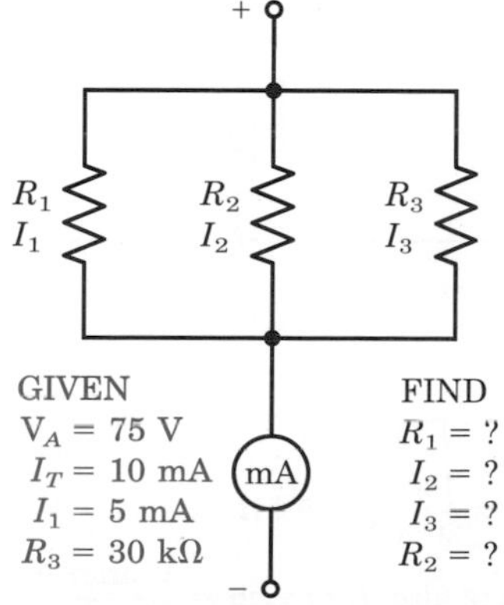

FIGURE 4–10 Parallel circuit problem

RESISTANCE IN PARALLEL CIRCUITS

Again, total resistance of parallel circuits is always *less than the least branch resistance.* Let's examine the reason for this and learn several useful ways to solve for total resistance in parallel circuits.

Since each parallel branch provides another current path where the voltage/power source must supply current, then the source must provide more current for each additional branch in a parallel circuit. Since the V is the same from the source, but the I supplied by the source is higher for every branch added in parallel, then Ohm's law $\left(R = \frac{V}{I}\right)$ tells us that the total or **equivalent circuit resistance** must decrease every time another branch (or current path) is added to the parallel circuitry. For example, if a circuit comprises only the lowest R value branch, the total resistance equals that value. If *any* value R branch is added in parallel with that branch, another current path has been provided. Therefore, total current supplied by the source increases, source voltage remains constant, and total or equivalent circuit resistance decreases. Moving forward from that thought, let's look at several methods available to determine parallel circuit total resistance.

METHODS TO CALCULATE TOTAL RESISTANCE (R_T)

The Ohm's Law Method

Recall the formula $R_T = \frac{V_T}{I_T}$ to find total resistance.

EXAMPLE

Look at Figure 4–11. The total resistance according to Ohm's law is $R_T = \frac{100\ \text{V}}{20\ \text{mA}} = 5\ \text{k}\Omega$. A logical extension of this law for parallel circuits is that

FIGURE 4–11 Ohm's law method to find R_T

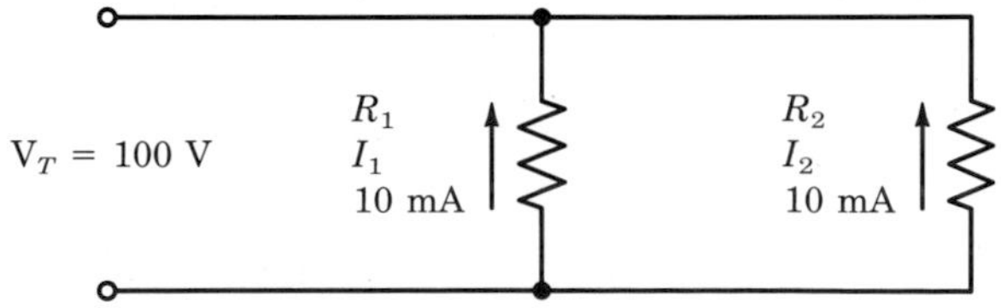

$V_T = 100$ V; R_1, I_1 10 mA; R_2, I_2 10 mA

$$R_T = \frac{V_T}{I_T}$$

$$R_T = \frac{V_T}{(I_1 + I_2)}$$

$$R_T = \frac{100\ \text{V}}{20\ \text{mA}} = 5\ \text{k}\Omega$$

$$\left(\text{Note also: } R \text{ branch} = \frac{\text{V branch}}{I \text{ branch}}\right)$$

each branch resistance can be solved by knowing branch voltage and branch current. This is stated as $R \text{ branch} = \frac{V \text{ branch}}{I \text{ branch}}$.

Again, refer to Figure 4–11. Since we know the voltage is the same across all branches and must equal V applied, then $R_1 = \frac{100 \text{ V}}{10 \text{ mA}} = 10 \text{ k}\Omega$. $R_2 = \frac{100 \text{ V}}{10 \text{ mA}} = 10 \text{ k}\Omega$. And, $R_T = \frac{100 \text{ V}}{20 \text{ mA}} = 5 \text{ k}\Omega$.

Note this result of $R_T = 5 \text{ k}\Omega$ agrees with our earlier statement that total circuit R must be less than any one of the branch resistances in parallel.

PRACTICE PROBLEM V

Again, refer to Figure 4–11 and apply Ohm's law. If $I_1 = 15$ mA and $R_1 = 2 \text{ k}\Omega$, what are the values of V_{R1} and V_T? If $R_2 = 10 \text{ k}\Omega$, what is the value of R_T?

The Reciprocal Method

Since total current in a parallel circuit equals the sum of the branch currents, we can derive a reciprocal formula for resistance using this fact plus Ohm's law. Since $I_T = I_1 + I_2 \ldots + I_n$, and Ohm's law states $I_T = \frac{V}{R_T}$ and each branch current equals its V over its R, then:

$$\frac{V}{R_T} = \frac{V}{R_1} + \frac{V}{R_2} + \frac{V}{R_3} \ldots + \frac{V}{R_n}$$

Recall R_n represents each of the remaining resistor values. Since V is common in parallel circuits and is the same across each of the resistances, we divide by V in each case and derive:

Formula 4–2	$\frac{1}{R_T} = \frac{1}{R_1} + \frac{1}{R_2} + \frac{1}{R_3} \cdots + \frac{1}{R_n}$

This formula is used for any number of branch resistances to find total or equivalent resistance of the circuit (R_T or R_e).

EXAMPLE

If a three-resistor circuit has a 10-Ω branch, a 15-Ω branch, and a 20-Ω branch, the total, or equivalent resistance, is found as follows:

$$\frac{1}{R_T} = \frac{1}{10} + \frac{1}{15} + \frac{1}{20} = \frac{6}{60} + \frac{4}{60} + \frac{3}{60} = \frac{13}{60}$$

Therefore, $R_T = \dfrac{60}{13} = 4.62\ \Omega$

Again, R equivalent is less than the least branch. Another variation of the reciprocal method is:

Formula 4–3	$R_T = \dfrac{1}{\dfrac{1}{R_1} + \dfrac{1}{R_2} + \dfrac{1}{R_3} \cdots + \dfrac{1}{R_n}}$

PRACTICAL NOTES

The reciprocal method is easy with calculators. If you do not have a calculator, methods other than the reciprocal method may be easier to use.

PRACTICE PROBLEM VI

Assume a parallel circuit has branch resistances of 10 kΩ, 27 kΩ, and 20 kΩ, respectively. Use the reciprocal method and find R_T.

The Conductance Method

Recall resistance is the opposition shown to current flow, and conductance, conversely, is the ease with which current passes through a component or circuit. To calculate conductance use the reciprocal of resistance, $G = \frac{1}{R}$, where G is conductance in siemens (S) and R is resistance in ohms (Ω).

EXAMPLE

If a circuit has a resistance of 10 Ω, it exhibits one-tenth S of conductance. Conversely, if a circuit has one-tenth S of conductance, it must have 10 ohms of resistance. That is, $R = \frac{1}{G}$, where $R = \frac{1}{0.1} = 10\ \Omega$. Total conductance of parallel branches is found by adding all conductances of parallel branches, Figure 4–12.

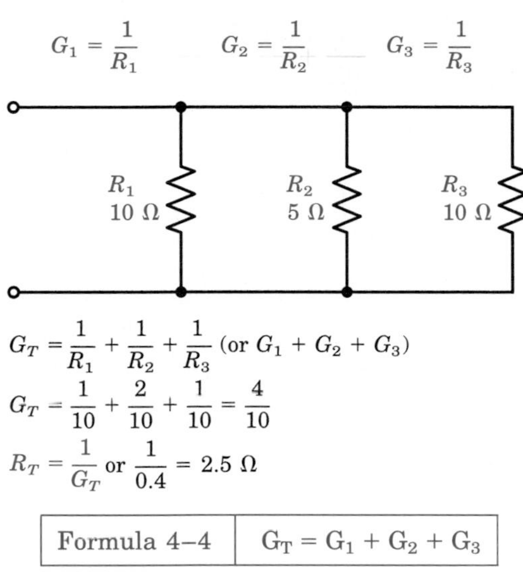

$G_1 = \frac{1}{R_1}$
$G_2 = \frac{1}{R_2}$
$G_3 = \frac{1}{R_3}$
R_1 10 Ω
R_2 5 Ω
R_3 10 Ω
$G_T = \frac{1}{R_1} + \frac{1}{R_2} + \frac{1}{R_3}$ (or $G_1 + G_2 + G_3$)
$G_T = \frac{1}{10} + \frac{2}{10} + \frac{1}{10} = \frac{4}{10}$
$R_T = \frac{1}{G_T}$ or $\frac{1}{0.4} = 2.5\ \Omega$

FIGURE 4–12 The conductance method to find R_T

Formula 4–4	$G_T = G_1 + G_2 + G_3$

In other words, $\frac{1}{R_T} = \frac{1}{R_1} + \frac{1}{R_2} + \frac{1}{R_3}$.

PRACTICE PROBLEM VII

Refer to Figure 4–13, and use the conductance method to find the circuit total resistance.

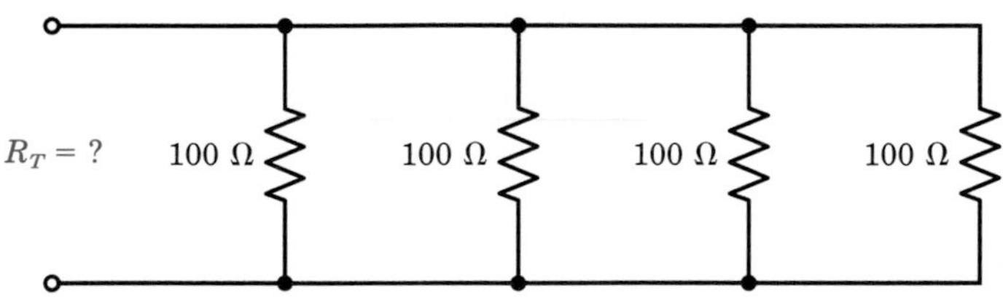

FIGURE 4–13 Conductance method problem

The Product-Over-The-Sum Method

One popular method to solve for total resistance in parallel circuits is the **product-over-the-sum method.** The mathematical formula is derived from the basic reciprocal resistance formulas. That is $\frac{1}{R_T} = \frac{1}{R_1} + \frac{1}{R_2}$. Using algebra this becomes:

$$R_T = \frac{1}{\frac{1}{R_1} + \frac{1}{R_2}}$$

. . . and:

Formula 4–5	$R_T = \frac{R_1 \times R_2}{R_1 + R_2}$

EXAMPLE

An illustration of this formula is in the two-resistor parallel circuit of Figure 4–14.

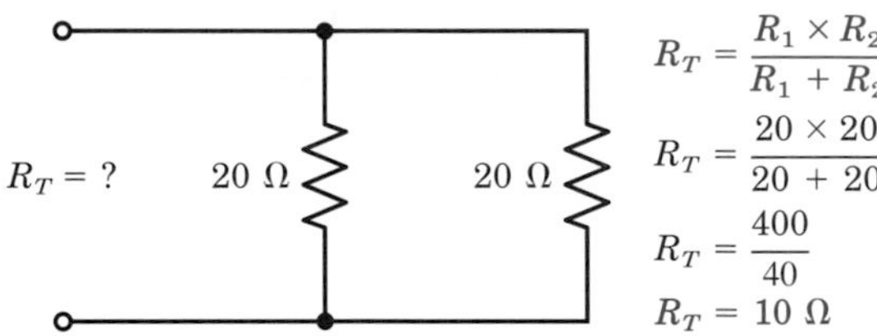

FIGURE 4–14 The product-over-the-sum method to find R_T

What would be the total circuit resistance if $R_1 = 100$ ohms and $R_2 = 50$ ohms? In this case, the product is $100 \times 50 = 5000$, and the sum is $100 + 50 = 150$. Therefore, $R_T = \frac{5000}{150}$, and total resistance = 33.33 ohms. This product-over-the-sum method works for any two branch values.

PRACTICE PROBLEM VIII

If the circuit in Figure 4–14 has resistances of 30 Ω and 20 Ω, respectively, solve for R_T using the product-over-the-sum method.

This method works well for two-branch circuits. How might a similar technique be used for a three-branch circuit? The answer is simple: Use the formula on two branches, *then use these results* from that calculation as being one equivalent resistance in parallel with the third resistor, and repeat the process. Of course, you can also use the reciprocal or conductance approach where $\frac{1}{R_T} = \frac{1}{R_1} + \frac{1}{R_2} + \frac{1}{R_3}$.

EXAMPLE

In Figure 4–15 three branch resistances are in parallel. Since each branch has a resistance of 3 Ω, what is the circuit total resistance?

Using the product-over-the-sum formula for the first two branches yields: $\frac{R_1 \times R_2}{R_1 + R_2} = \frac{3 \times 3}{3 + 3} = \frac{9}{6} = 1.5\ \Omega$, indicating that the equivalent resistance of branches 1 and 2 = 1.5 ohms. However, there is another branch involved! The equivalent resistance of 1.5 Ω is in parallel with branch R_3 (3 Ω). Repeating the product-over-the-sum formula for this situation gives:

$$\frac{R_e \times R_3}{R_e + R_3} = \frac{1.5 \times 3}{1.5 + 3} = \frac{4.5}{4.5} = 1\ \Omega.$$

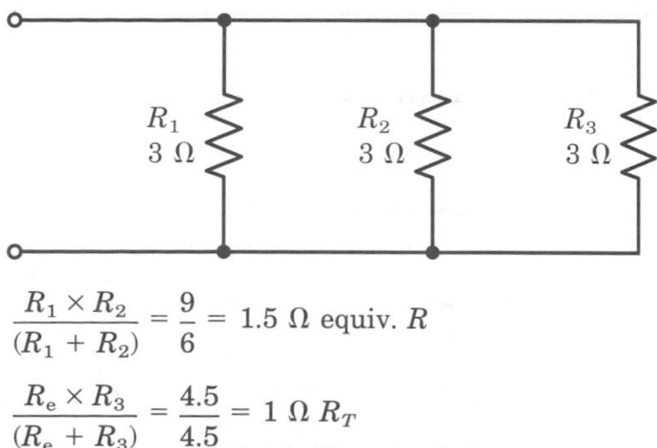

FIGURE 4–15 Using product-over-the-sum method with more than two branches

$$\frac{R_1 \times R_2}{(R_1 + R_2)} = \frac{9}{6} = 1.5\ \Omega \text{ equiv. } R$$

$$\frac{R_e \times R_3}{(R_e + R_3)} = \frac{4.5}{4.5} = 1\ \Omega\ R_T$$

Note that using the reciprocal approach would also yield:

$$\frac{1}{R_T} = \frac{1}{3} + \frac{1}{3} + \frac{1}{3} = \frac{1}{1} = 1\ \Omega.$$

PRACTICE PROBLEM IX

If the resistances in Figure 4–15 are 50 Ω, 100 Ω, and 200 Ω, respectively, what is the value of R_T? Use the product-over-the-sum approach to calculate.

What would happen if there were four branches? To find the equivalent resistance of any two branches, use the product-over-the-sum formula. Then find the equivalent resistance of the other two branches in the same way, and use the product-over-the-sum formula for the two equivalent resistances to find total circuit resistance.

A Useful Simplifying Technique

You have undoubtedly observed in Figure 4–14 described earlier, which had two 20-Ω resistors in parallel, that the total R = 10 Ω. This number is half the value of one of the equal branch resistances. What do you suppose the total resistance of a three-branch parallel circuit is if the circuit has three equal branches of 30 Ω each? If you said 10 Ω again, you are right. That is, two equal parallel branches are equivalent to half the R value of one of the equal branches, three equal branches are equivalent to one-third the value of one of the equal branches, and so on. The idea is to mentally *combine* equal resistance branches to find their equivalent resistance. Doing this mental process reduces the complexity of the problem.

Look at Figure 4–16. By realizing that R_2 and R_4 can be combined as an equivalent resistance of 10 Ω, then combined with the two remaining 10-Ω branches, you can see the total circuit resistance will be one-third of 10 Ω, or 3.33 Ω.

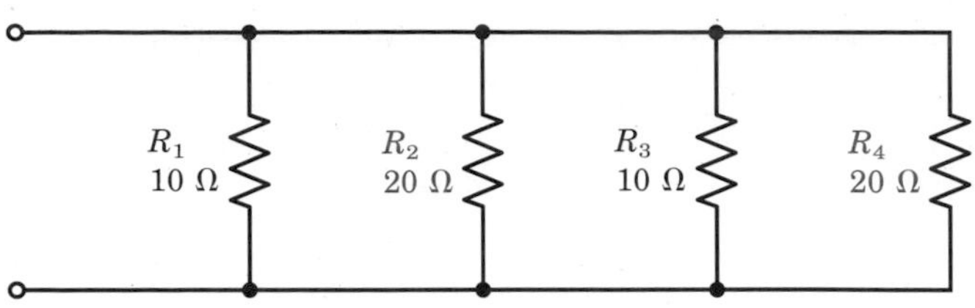

FIGURE 4–16 A useful simplifying technique

Step 1 = Combine R_2 and R_4 to yield 10 Ω.
Step 2 = Knowing that three 10 Ω resistors in parallel yields $\frac{1}{3}$ of 10 Ω as their R_T. Then R_T = 3.33 Ω.

This same concept can be used whenever branches of a multi-branched parallel circuit have equal values. Combine equal branches in a way to simplify to the fewest remaining branches. Then use the most appropriate method to solve for total circuit resistance, such as product-over-the-sum method, reciprocal method, Ohm's law, etc.

The Assumed Voltage Method

One technique to solve for total resistance of parallel circuits containing miscellaneous "mismatched" resistance values is the **assumed voltage method.** You'll frequently find problems of this sort on FCC radio licensing tests.

The technique uses a two-step approach:

1. Assume an uncomplicated or easy circuit applied voltage value to determine each branch current, should the "assumed voltage" actually be applied to the circuit. For example, choose a value that is an even multiple of all the branch resistor values.
2. Determine the branch currents by dividing the assumed voltage by each branch R value. Then add branch currents to find total current. Now *divide "assumed voltage" by calculated total current to find total circuit resistance.*

EXAMPLE

Refer to Figure 4–17. Following the above steps, the solution for this problem is as follows:

1. Assume 300 volts applied voltage since it is easily divided by each of the branch R values.
2. Solve each branch current using Ohm's law. Branch currents are 12 A, 6 A, 3 A, and 2 A, respectively.
   Total current equals sum of branch currents = 23 A

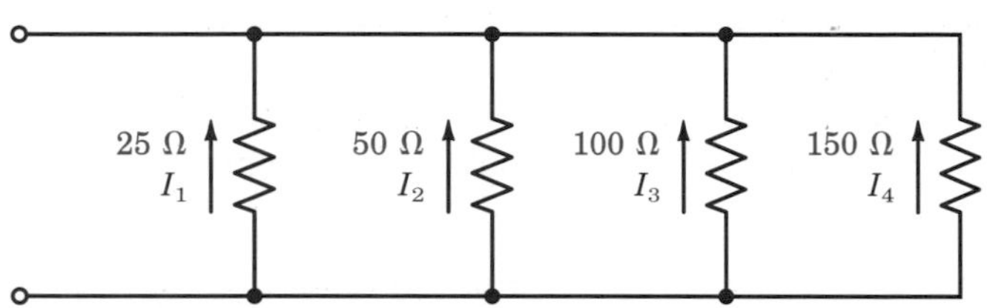

A good assumed voltage is 300 V due to R values.

$I_1 = 12$ A $I_2 = 6$ A $I_3 = 3$ A $I_4 = 2$ A

I_T = sum of branch currents = 23 A

$$R_T = \frac{300\text{ V}}{23\text{ A}} = 13.04\ \Omega$$

FIGURE 4–17 The assumed voltage method to find R_T

$$\text{Thus, } R_T = \frac{300\text{ V}}{23\text{ A}} = 13.04\ \Omega$$

In essence, this technique often allows you to mentally perform the first step without paper and pencil to find the total current. All that remains is one division problem, $\frac{V_T}{I_T}$, to find total circuit resistance.

In Figure 4–17, if the R values were changed to 10, 13, 20, and 26 ohms respectively, what would be the total circuit resistance? Try using the assumed voltage method to find the answer.

Your answer should be 3.77 Ω. One value of voltage that could have been assumed is 260 V, since it is evenly divisible by all the R values in the circuit. NOTE: It does not matter if you assumed another value of voltage applied, your answer will be the same!

A Useful Parallel Circuit Resistance Design Formula

As a technician, you will probably want to purposely decrease the resistance of an existing circuit by placing a resistor in parallel with it. Or you may need a resistor of some unavailable specific value. In either case, you have a number of resistors, but none are the right value. This happens frequently! How can we determine the unknown resistor value to be paralleled with a given (known) value to achieve the desired resultant resistance?

The answer is a formula derived from the product-over-the-sum method:

Formula 4–6	$R_u = \frac{R_k \times R_e}{R_k - R_e}$

Where R_u = R unknown

R_e = equivalent R of paralleled resistors

R_k = known R value that will be paralleled by R unknown achieving the desired R equivalent.

EXAMPLE

What value of R is needed in parallel with a 10-ohm resistor to achieve a total resistance of 6 ohms? See the solution in Figure 4–18.

FIGURE 4–18 A practical circuit design formula

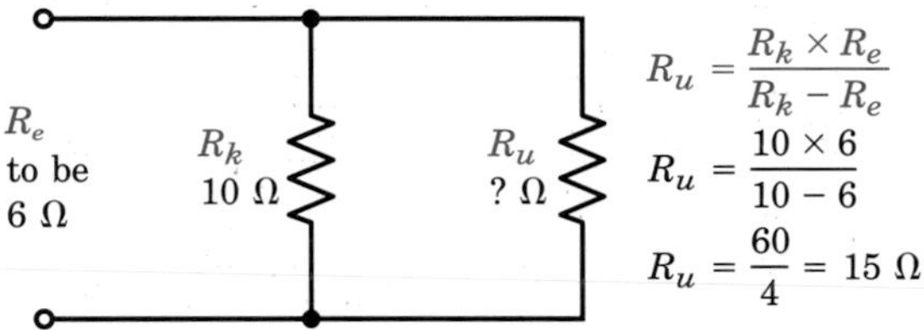

PRACTICE PROBLEM X

Use this technique and solve the problem in Figure 4–19.

FIGURE 4–19 Problem for applying design formula

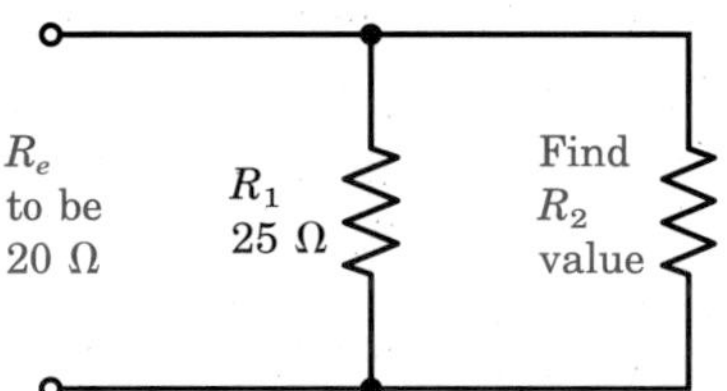

POWER IN PARALLEL CIRCUITS

Total Power

Total power for *any* resistive circuit (series, parallel, or combination) is computed by adding all individual power dissipations in the circuit, or by using one of the power formulas, $P_T = V_T \times I_T$; $P_T = I_T^2 \times R_T$; or $P_T = \frac{V_T^2}{R_T}$. Likewise, power dissipated by any one resistive component is calculated using that particular component's parameters. That is, its voltage drop, current through it, and its resistance value.

Look at the circuit in Figure 4–20 as you read the next section about power in parallel circuits.

Power Dissipated by Each Branch

To find a branch power dissipation, use the parameters applicable to that branch. In the case of branch 1 (R_1), $P = 100\ V \times I_1$. To find I_1, divide branch

$P_1 = V_1 \times I_1$ $P_2 = V_2 \times I_2$ $P_3 = V_3 \times I_3$
$P_1 = 100\text{ V} \times 1\text{ mA}$ $P_2 = 100\text{ V} \times 4\text{ mA}$ $P_3 = 100\text{ V} \times 5\text{ mA}$
$P_1 = 100\text{ mW}$ $P_2 = 400\text{ mW}$ $P_3 = 500\text{ mW}$

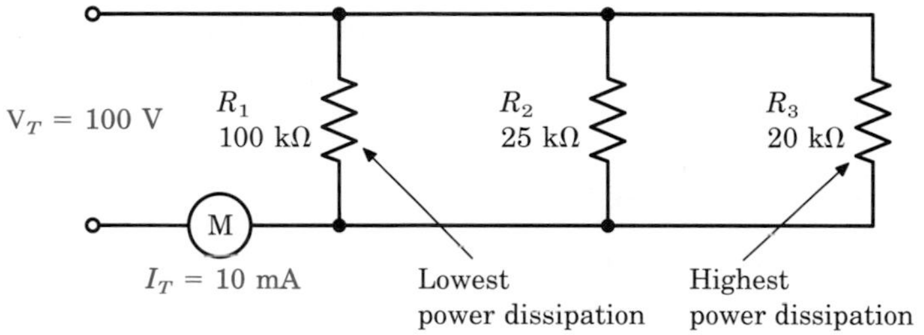

$P_T = 100\text{ mW} + 400\text{ mW} + 500\text{ mW} = 1000\text{ mW (or 1 W)}$
or
$P_T = 100\text{ V} \times 10\text{ mA} = 1000\text{ mW (or 1 W)}$

FIGURE 4–20 Power in parallel circuits

V (100 V) by branch R (100 kΩ). The answer is 1 mA. Thus, $P_1 = 100\text{ V} \times 1\text{ mA} = 100\text{ mW}$.

$$\text{Branch 2: Branch I} = \frac{100\text{ V}}{25\text{ k}\Omega} = 4\text{ mA}$$
$$\text{Branch P} = 100\text{ V} \times 4\text{ mA} = 400\text{ mW}$$
$$\text{Branch 3: Branch I} = \frac{100\text{ V}}{20\text{ k}\Omega} = 5\text{ mA}$$
$$\text{Branch P} = 100\text{ V} \times 5\text{ mA} = 500\text{ mW}$$

EXAMPLE

If we added all the individual power dissipations in the circuit in Figure 4–20, the total power = 100 mW + 400 mW + 500 mW = 1 W. Using the power formula, since V_T equals 100 volts and I_T equals 10 mA, the total power is 100 V × 10 mA = 1000 mW (or 1 W). In either case, the total power supplied to and dissipated by the circuit is calculated as 1 watt.

PRACTICE PROBLEM XI

What is the total power in the circuit of Figure 4–20 if V_T = 150 V?

Relationship of Power Dissipation to Branch Resistance Value

Refer again to Figure 4–20 and note that the *lowest* branch resistance dissipates the *most* power; the *highest* value R branch dissipates the *least* power. This is because V is the same for all branches, and current through each branch is *inverse* to its R value. **This is the opposite of what you learned about series circuits.** In *series circuits,* the largest value R dissipates the most power because its I is the same as all others in the circuit; however its V is directly proportional to its R value.

Soon we will look at other interesting similarities and differences between series and parallel circuits that you should learn. However, before we do that, let's look at some variations from the norm when parallel circuits develop opens or shorts.

EFFECTS OF OPENS IN PARALLEL CIRCUITS AND TROUBLESHOOTING HINTS

Opens

Analysis of one or more opened branches in a parallel circuit is simple. Refer to Figure 4–21.

FIGURE 4–21 Effects of an open in a parallel circuit

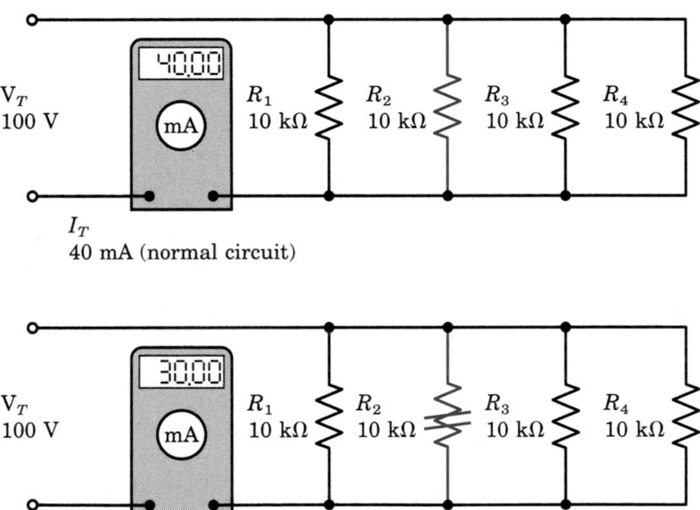

1. Normal operation for this circuit is: V_T = 100 V; I_T = 40 mA; P_T = 4000 mW (4 W), each branch I = 10 mA, and each branch P = 1000 mW (1 W).
2. If branch 2 opened (R_2), then total current decreases to 30 mA, since there is no current through branch 2. Under these conditions, the following is true:
   - Total current decreases (30 mA rather than 40 mA).
   - Total voltage and branch voltages remain unchanged.
   - Total power decreases (100 V × 30 mA = 3000 mW, or 3 W).

- Current through the good (unopened) branches remains unchanged since both their V and R values stay the same.
- Also, since voltage and current through unopened branches is unaffected, individual branch power dissipations stay the same, *except* for the opened branch or branches.
- Since current decreases to zero through the opened branch, its power dissipation also drops to zero. The voltage across the opened branch, however, stays the same.

PRACTICAL NOTES

Some generalizations worth noting about an open in *any* circuit (series, parallel, or combination) are:

1. An open causes total circuit resistance to increase.
2. An open causes total circuit current to decrease.
3. An open causes total circuit power to decrease.

Incidentally, opens can be caused either by *components* opening, or by *conductor paths* opening. On printed circuit boards, a cracked conductor path or poor solder joint can cause an open. On "hard-wired" circuitry, a broken wire (or poor solder joint) can cause an open conductor path, Figure 4–22.

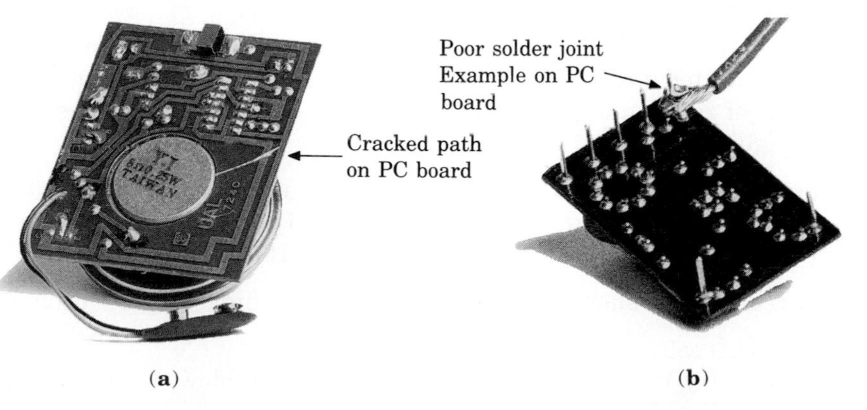

FIGURE 4–22 Open circuits can be caused by broken conductor paths or poor solder joints.

TROUBLESHOOTING HINTS

Since an open in a parallel branch causes total resistance to increase, total current to decrease, and current through the opened branch to decrease to zero, troubleshooting hints help isolate and identify these facts.

With power removed from the circuit, you can measure its total resistance with an ohmmeter. If the measured value indicates a higher than normal value for the circuit, then there is a good chance an open has occurred or at least some part of the circuit has opened or increased in R value. If it is convenient to electrically "lift" (electrically isolate) one end of each branch one at a time from the remainder of the circuit as you watch the meter, you can spot the troubled branch. For example, if lifting one end of a branch from the circuit causes a change in R (increase), the branch just lifted is probably good. If lifting of the branch causes no change in the R reading, you have probably found the branch with the problem. Since it was already opened, opening it again didn't change anything!

If you prefer using "power-on" current readings, the first indication of trouble you will see is a lower-than-normal total circuit current. If you can conveniently connect the meter to measure each branch's current individually, the branch with zero current is the open branch. Other branches should measure normal values.

EFFECTS OF SHORTS IN PARALLEL CIRCUITS AND TROUBLESHOOTING HINTS

Shorts

A shorted branch in a parallel circuit is a serious condition. Since each branch is connected directly across the voltage source, a shorted branch puts a resistance of zero ohms, or close to zero ohms, directly across the source. Refer to Figure 4–23.

1. Normal operation for this circuit is the same as described earlier for Figure 4–21, where $V_T = 100$ V; $I_T = 40$ mA; $P_T = 4000$ mW (4 W); each branch I = 10 mA; and each branch P = 1000 mW (1 W).
2. If branch 2 (R_2) shorted, then the following things would probably happen:
   - I_T would try to drastically increase until a fuse blew, or the power supply "died"!
   - V_T would drastically decrease! Due to greatly increased $I \times R$ drop across the source's internal resistance, the output terminal voltage

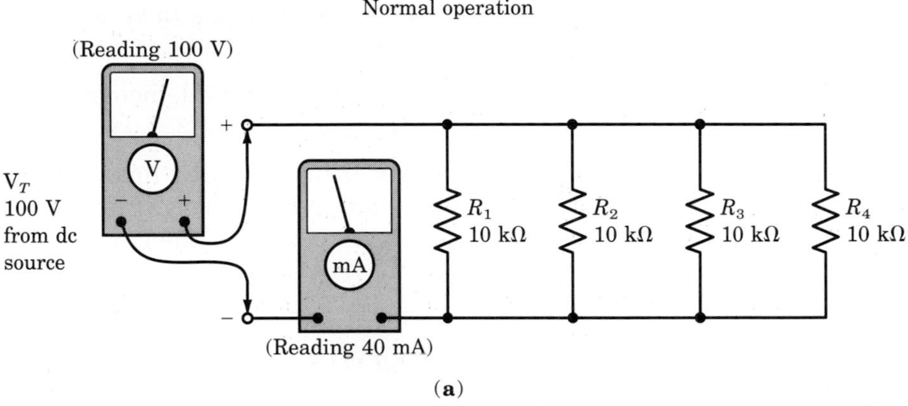

(a)

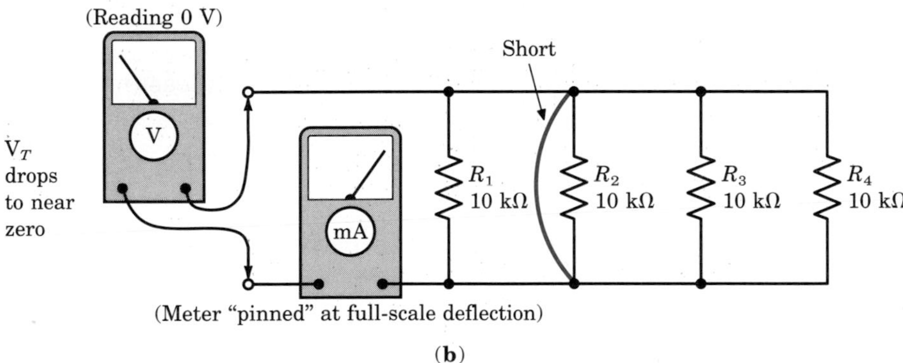

(b)

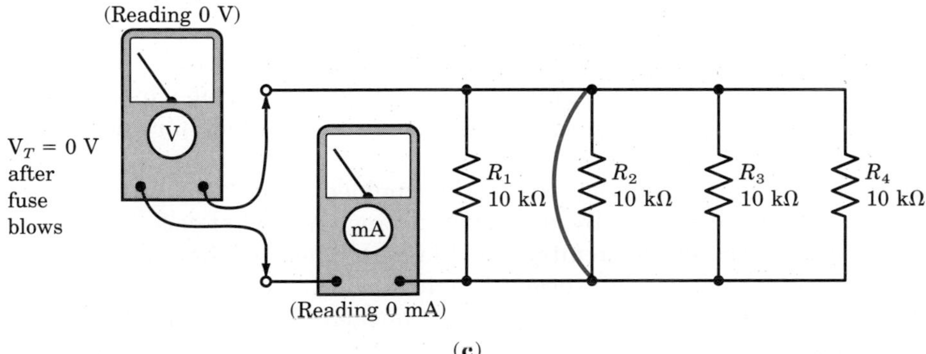

(c)

FIGURE 4–23 Effects of a short in a parallel circuit. R_T decreases and I_T increases until the fuse blows or the power source quits.

would decrease. It would probably drop to nearly zero volts with most power sources.

- P_T would drastically increase until the power source quit, or a fuse blew!

- Current through the other branches would decrease to zero, or close to it, because the V has dropped to almost zero, or zero.
- Total circuit resistance would drastically decrease! Remember the total resistance will be less than the least branch R.
- If there were any power supply fuses or house circuit fuses associated with the faulty circuit, they would probably blow. Also, there might be some smoke rising from some sections of the circuit or power supply.

PRACTICAL NOTES

Some generalizations about a short in *any* type of circuit (series, parallel, or combination) are:

1. A short causes total circuit resistance to decrease.
2. A short causes total circuit current to increase. (At least momentarily!)
3. A short causes total circuit power to increase. (At least momentarily!)
4. A short likely causes circuit voltage to decrease. (Probably to zero volts!)

TROUBLESHOOTING HINTS

First *disconnect* the circuit in question from the power source! Resistance checks can then be made on the circuitry without further damage either to the circuitry, power source, connecting wires, and so forth. Power-on troubleshooting should *not* be done for shorted conditions.

Again, the best method to find the troubled area is to isolate each branch one at a time. By electrically lifting one end of each branch from the remaining circuitry, an ohmmeter check of the branch resistance can be made. Normal R values will be found for the good branches. Close to zero ohms will be measured across the shorted branch or component.

Once the branch is located, a visual inspection of components, wires, printed circuit paths, and so on generally reveals the problem. If there is an internal short within the component, an R measurement of the component reveals this problem.

SIMILARITIES AND DIFFERENCES BETWEEN SERIES AND PARALLEL CIRCUITS

Some interesting and important similarities and differences between series and parallel circuits that you should know and remember are in the chart below. Use the following chart as an In-Process Learning Check for this chapter.

COMPARING SERIES AND PARALLEL CIRCUITS

SERIES CIRCUIT CHARACTERISTICS	PARALLEL CIRCUIT CHARACTERISTICS
R_T = Sum of Rs	R_T = Less than lowest R in parallel
$I_T = \frac{V_T}{R_T}$	$I_T = \frac{V_T}{R_T}$
I_T = Same current throughout the circuit	I_T = Sum of branch currents, where branch current $(I_b) = \frac{V_b}{R_b}$; branch I is inverse to branch R
$V_T = I_T \times R_T$	$V_T = I_T \times R_T$
V_T = Sum of all IR drops	V_T = Same as across each branch
$V_{R_x} = I_{R_x} \times R_X$; Any given R's voltage drop is related to the ratio of its R to R total: $V_{R_x} = \left(\frac{R_X}{R_T}\right) \times V_T$	Voltage across all resistors is same value and equals the circuit applied voltage
P dissipated by a given R is *directly related* to its R value. Largest R in circuit dissipates most power; smallest R dissipates least power.	P dissipated by a given R is *inverse* to its R value. Largest R dissipates least power, smallest R value dissipates most power.
$P_T = V_T \times I_T$; or sum of all Ps	$P_T = V_T \times I_T$; or sum of all Ps
Opens cause 0 current; 0V across good components; and V_T (applied circuit voltage) to appear across open	*Opens* cause no change in good branches; opened branch has zero current, but V across all branches stays the same
Shorts cause I to increase; V to decrease across short; V to increase across remaining components; and total circuit power to increase.	*Shorts* cause I_T and P_T to increase (at least momentarily); V to decrease across all branches; and current through good branches to decrease.

DESIGNING A PARALLEL CIRCUIT TO SPECIFICATIONS

EXAMPLE

Assume you want to create a four-branch parallel circuit so that branches 1 and 3 have equal current values, and branches 2 and 4 also have equal current values but their current values are twice that of branches 1 and 3. Further assume that branch 1 has an R value of 10 kΩ and its current value is 10 mA.

The design problem involves determining all the circuit parameters and drawing the schematic for this circuit. Label all component values and all V, R, and I electrical parameters for each branch and for the total circuit.

Recall the way to begin is to draw a diagram and label all the knowns. Then proceed in the most appropriate manner to solve the unknowns based on the knowns.

See Figure 4–24a for this first step, which is drawing diagram and labeling all knowns.

Step two uses the knowns to solve for the unknowns, then labeling diagram with appropriate parameters.

- Since branch 1's R and I values are known, it's easy to find its voltage value. Thus, V = I × R = 10 mA × 10 kΩ = 100 V.
- Since this is a parallel circuit, 100 V is the voltage across all the branches as well as the V applied value.
- Because branches 3 and 1 have the same current value, and since we know the V applied to both branches is the same, it is obvious branch 3 resistance value must be the same as branch 1. So we now know branch 3 R_3 = 10 kΩ, I_3 = 10 mA, and V_3 = 100 V.

Step 1: Draw diagram. Label and/or list "knowns."

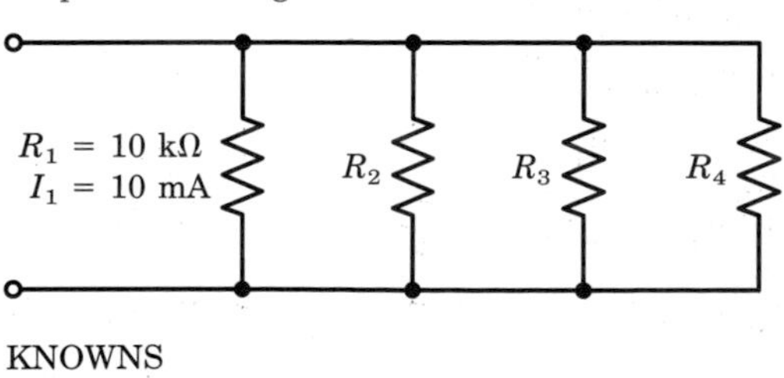

KNOWNS
$I_1 = I_3$
$I_2 = 2 \times I_1$
$I_4 = 2 \times I_1$
I_1 = 10 mA

(a)

Step 2: Use knowns to solve unknowns. Label diagram.

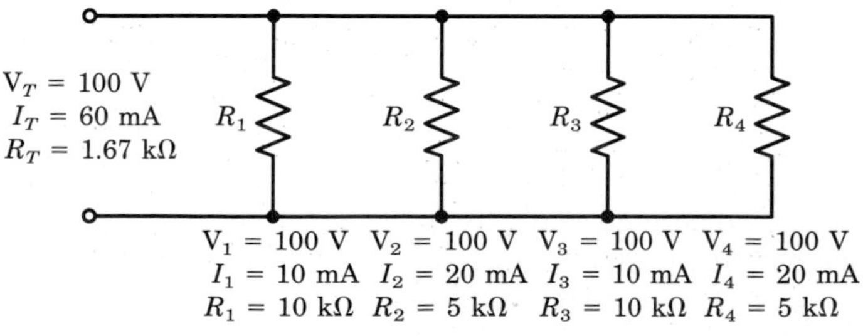

(b)

FIGURE 4–24 Circuit design problem

- Because branches 2 and 4 have twice the current value as branches 1 and 3, it is obvious their current values must be 20 mA each. Since the V across each branch is 100 V, each branch's R value is $\frac{100\text{ V}}{20\text{ mA}} =$ 5 kΩ.
- Solving the total circuit current and resistance are simple, and you already know the V applied value. Total current (I_T) equals the sum of the branch currents = 60 mA. Total resistance (R_T) = $\frac{V_T}{I_T} = \frac{100\text{ V}}{60\text{ mA}} = 1.67\text{ k}\Omega$.

  Incidentally, you can use the R values and solve for the total resistance. The two 10 kΩ branches are the equivalent of 5 kΩ. This 5 kΩ in parallel with two other 5 kΩ branches is equivalent to 5 kΩ divided by 3, or 1.67 kΩ. Remember the trick with equal value branches?
- The last task is to label all appropriate parameters on the diagram, Figure 4–24b.

PRACTICE PROBLEM XII

Try this problem yourself. Design and draw a three-resistor parallel circuit so that:

1. V applied to the circuit is 120 volts.
2. The total circuit power dissipation is 1.44 W.
3. Branch 1 has half the current value of branch 2 and one-third the current value of branch 3.

Label your diagram with all V, I, R, and P values for the circuit and for each branch.

SOURCES IN PARALLEL

Voltage and power sources may be connected in parallel to provide higher current delivering capacity (and power) to circuits or loads (at a given voltage) than can be delivered by only one source. The sources should be equal in output terminal voltage value when using this technique. A common application of this method is the "paralleling" of batteries or cells, Figure 4–25.

CURRENT DIVIDERS

You are aware that parallel circuits cause the total current to divide through that circuit's branches, called **current divider circuits.** You also know the

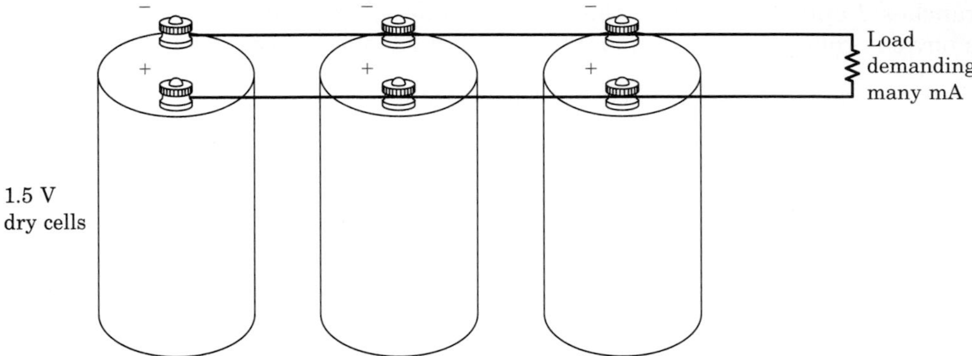

FIGURE 4–25 Equal voltage sources in parallel can deliver more current and power than one source alone.

current divides in inverse proportion to each branch resistance value. These characteristics of parallel circuits offer the possibility of current dividing "by design," where one purposefully controls the ratios of branch currents by choosing desired values of resistances. Also, this current-dividing feature gives another method to find current through branches if only the total resistance, branch resistances, and total current are known.

General Formula for Parallel Circuit with Any Number of Branches

A general formula to find any given branch's current is:

Formula 4–7	$I_X = \left(\frac{R_T}{R_X}\right) \times I_T$

where I_X is a specified branch's current value; R_T is the parallel circuit total resistance; and R_X is the value of resistance of the specified branch. Refer to Figure 4–26 for example applications of this formula.

PRACTICE PROBLEM XIII

Refer again to Figure 4–26. Assume that $R_1 = 27\ k\Omega$, $R_2 = 47\ k\Omega$, $R_3 = 100\ k\Omega$, $R_T = 14.64\ k\Omega$, and $I_T = 3.42$ mA. Find each of the branch currents using the divider formula, as appropriate.

Simple Two-Branch Parallel Circuit Current Divider Formula

If there are only two branches, Figure 4–27, a formula is available to solve for each branch current, knowing I_T and branch R values:

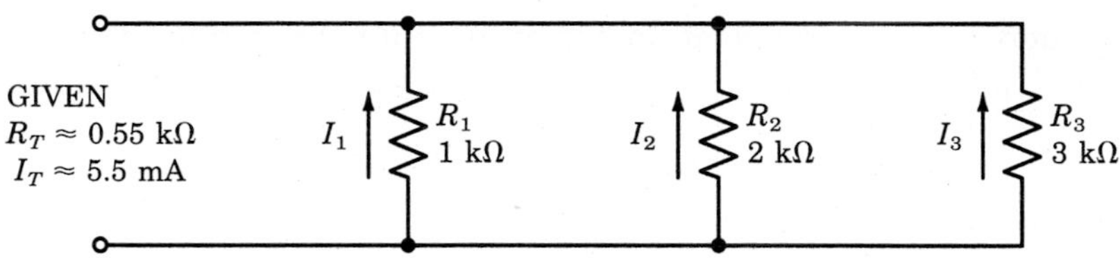

FIGURE 4–26 A general current-divider example

Find approximate values for I_1; I_2 and I_3

$$I_1 \approx \left(\frac{R_T}{R_1}\right) \times I_T = \left(\frac{0.55\ \text{k}\Omega}{1\ \text{k}\Omega}\right) \times 5.5\ \text{mA} \approx 3.0\ \text{mA}$$

$$I_2 \approx \left(\frac{R_T}{R_2}\right) \times I_T = \left(\frac{0.55\ \text{k}\Omega}{2\ \text{k}\Omega}\right) \times 5.5\ \text{mA} \approx 1.5\ \text{mA}$$

$$I_3 \approx \left(\frac{R_T}{R_3}\right) \times I_T = \left(\frac{0.55\ \text{k}\Omega}{3\ \text{k}\Omega}\right) \times 5.5\ \text{mA} \approx 1.0\ \text{mA}$$

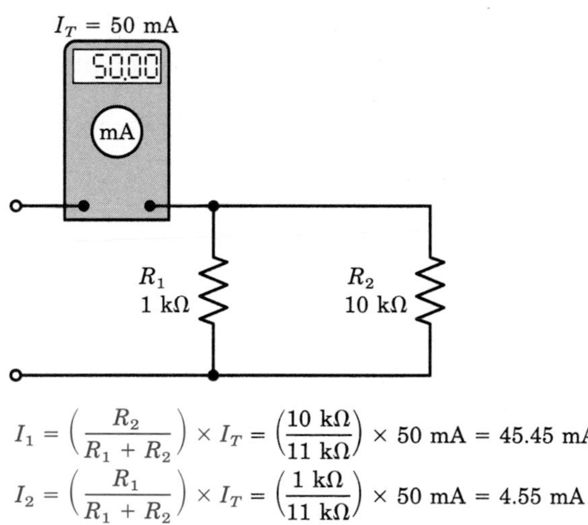

$$I_1 = \left(\frac{R_2}{R_1 + R_2}\right) \times I_T = \left(\frac{10\ \text{k}\Omega}{11\ \text{k}\Omega}\right) \times 50\ \text{mA} = 45.45\ \text{mA}$$

$$I_2 = \left(\frac{R_1}{R_1 + R_2}\right) \times I_T = \left(\frac{1\ \text{k}\Omega}{11\ \text{k}\Omega}\right) \times 50\ \text{mA} = 4.55\ \text{mA}$$

FIGURE 4–27 Two-resistor current divider

Formula 4–8	$I_1 = \left(\frac{R_2}{R_1 + R_2}\right) \times I_T$

$$I_1 = \frac{10\ \text{k}\Omega}{11\ \text{k}\Omega} \times 50\ \text{mA} = 0.909 \times 50\ \text{mA} = 45.45\ \text{mA}$$

AND

Formula 4–9	$I_2 = \left(\frac{R_1}{R_1 + R_2}\right) \times I_T$

$$I_2 = \frac{1\ \text{k}\Omega}{11\ \text{k}\Omega} \times 50\ \text{mA} = 0.0909 \times 50\ \text{mA} = 4.55\ \text{mA}$$

NOTE: There is an inverse relationship of current to branch Rs. Thus, when solving for I_1, R_2 is in the numerator; when solving for I_2, R_1 is in the numerator.

PRACTICE PROBLEM XIV

Refer again to Figure 4–27 and assume the values of R_1 and R_2 change to 100 kΩ and 470 kΩ, respectively. Also, assume the total current is 2 mA. Find the values of current through each branch using the appropriate two-branch divider formula. Check your answers against those at the end of the chapter.

PRACTICAL NOTES

You have learned that whenever *any* resistance value is placed in parallel with an existing circuit, it alters the circuit resistance and current because another branch has been added; thus providing another current path.

In a later chapter, we will discuss how this feature impacts the types and uses of measurement devices, such as voltmeters. For now, we only want to alert you to an important observation about parallel circuits.

As already stated, no matter what value of R is paralleled with (or "bridged across") an existing circuit, the circuit R will decrease. But the question is *how much* change will there be?

Look at Figure 4–28. Note the impact (change) that adding a 1 megohm resistor branch has on the circuit resistance. As you can see, adding a *high* R in parallel causes a small change in total circuit R.

FIGURE 4–28 Change in R_T of a given circuit when adding 1 MΩ branch

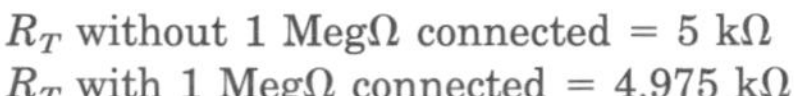

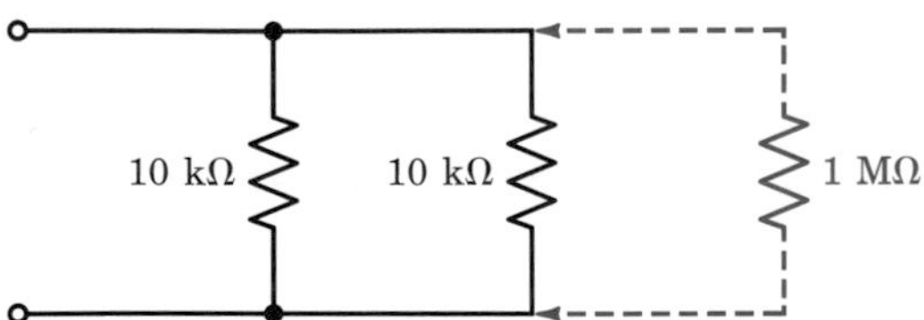

Look at Figure 4–29. Note the change in R_T caused by introducing a 1 kΩ resistor in parallel with the same original circuit. Adding a *low* R in parallel with an existing circuit causes a much more significant change in the circuit total resistance (and total current) than paralleling an existing circuit with a high R value.

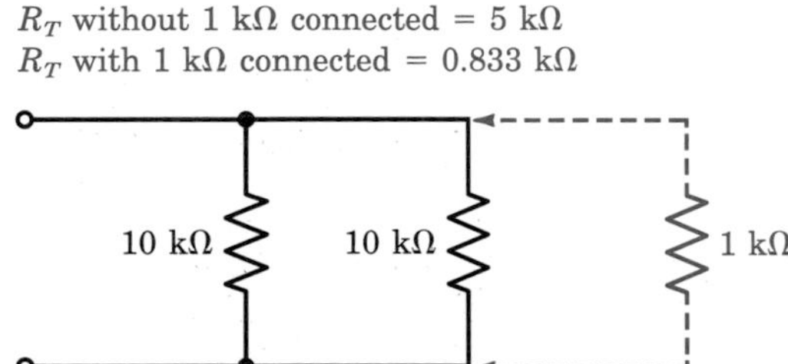

FIGURE 4–29 Change in R_T of a given circuit when adding 1 kΩ branch

The amount of change caused by adding a new branch R is related to *both* the R value of the "new" R branch *and* to the total circuit resistance of the original circuit. If the original circuit's equivalent resistance is low initially, adding a low value R in parallel will not have nearly the effect that it does when the original circuit has a high R_T.

SUMMARY

- A parallel circuit is one in which there are two or more paths (or branches) for current flow, and all the branches have the same voltage applied across them.
- Total circuit current divides through the parallel branches in an inverse ratio to the branch resistances.
- Total circuit current in a parallel circuit equals the sum of the branch currents.
- Kirchhoff's current law states the value of current entering a point must equal the value of current leaving that same point in the circuit.
- The total resistance of a parallel circuit must be less than the least resistance branch R value.
- Several methods to find the total resistance of parallel circuits are:

1. *Ohm's law* $\left(R_T = \frac{V_T}{I_T}\right)$
2. *reciprocal resistance method* $\left(\frac{1}{R_T} = \frac{1}{R_1} + \frac{1}{R_2} \cdots + \frac{1}{R_n}\right)$
3. *conductance method* $\left(R_T = \frac{1}{G_T}\right)$
4. *product-over-the-sum method* $\left(R_T = \frac{R_1 \times R_2}{R_1 + R_2}\right)$
5. *assumed voltage method.*

- One technique to solve for parallel circuit total resistance is to combine equal branch R values into their equivalent resistance value, whenever possible. For example, two 10 kΩ resistors in parallel equal 5 kΩ equivalent resistance (or half of one equal branch R value). Three 10 kΩ resistors in parallel equal 3.33 kΩ equivalent resistance (or one-third of one equal branch R value).
- A useful formula to find the R value needed in parallel with a given R value to obtain a de-

CHAPTER CHALLENGE

Following the SIMPLER troubleshooting sequence, find the problem with this circuit. As you follow the sequence, record your responses to each step on a separate sheet of paper.

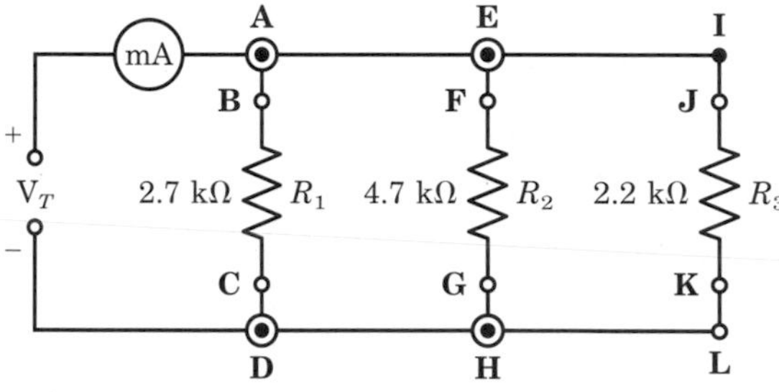

CHALLENGE CIRCUIT 2

Starting Point Information

1. Circuit diagram
2. $V_T = 10$ V
3. Measured $I_2 = 2.1$ mA
4. Measured $I_T = 6.7$ mA
5. Rs measured with remainder of circuit isolated from points being tested

SYMPTOMS Gather, verify, and analyze symptom information. (Look at the "Starting Point Information.")

IDENTIFY Identify initial suspect area for the location of the trouble.

MAKE Make a decision about "What type of test to make" and "Where to make it." (To simulate making a decision about each test you want to make, select the desired test from the "TEST" column listing below.)

PERFORM Perform the test. (Look up the result of the test in Appendix C. Use the number in the "RESULTS" column to find the test result in the Appendix.)

LOCATE Locate and define a new "narrower" area in which to continue troubleshooting.

EXAMINE Examine available information and again determine "what type of test" and "where."

REPEAT Repeat the preceding analysis and testing steps until the trouble is found. What would you do to restore this circuit to normal operation? When you have solved the problem, compare your results with those shown in the color insert.

Test	Results in Appendix C
$V_{A\text{-}D}$	(83)
$V_{B\text{-}D}$	(11)
$V_{E\text{-}H}$	(22)
$V_{F\text{-}G}$	(95)
$V_{I\text{-}L}$	(52)
$V_{J\text{-}K}$	(75)
$V_{A\text{-}B}$	(8)
$V_{C\text{-}D}$	(56)
$V_{I\text{-}J}$	(103)
$R_{A\text{-}B}$	(46)
$R_{C\text{-}D}$	(90)
$R_{I\text{-}J}$	(14)
$R_{K\text{-}L}$	(29)

sired equivalent (total resistance) value is $\frac{R_k \times R_e}{R_k - R_e}$ where R_u is the unknown R value being calculated; R_e is the resultant total resistance desired from the "paralleled" resistors, and R_k is the known value resistor across which the unknown value resistor is paralleled to achieve the desired equivalent, or total parallel circuit resistance. (Also, you can use $\frac{1}{R_e} - \frac{1}{R_k} = \frac{1}{R_u}$).

▶ Total power supplied to, or dissipated by, a parallel circuit is found with the power formulas $P_T = V_T \times I_T$, $I_T^2 \times R_T$; or $\frac{V_T^2}{R_T}$ or by summing the power dissipations of all the branches.

▶ Power dissipated by any branch of a parallel circuit is computed with the power formulas by substituting the specified branch V, I, and/or R parameters into the formulas, as appropriate.

▶ Branch power dissipations are such that the highest R value branch dissipates the least power, and the lowest R value branch dissipates the most power.

▶ An open branch in a parallel circuit causes total circuit current (and power) to decrease; total circuit R to increase; current through the unopened branches to remain the same, and current through the opened branch to decrease to zero. The circuit voltage remains the same.

▶ A shorted branch in a parallel circuit may cause damage to the circuit wiring and/or to the circuit power supply and/or may cause fuse(s) to blow. Until a fuse blows or the circuit power source fails, the short causes total circuit current to drastically increase; circuit R to decrease to zero ohms; and power demanded from the source to greatly increase. When a fuse blows or the source fails, circuit V, I, and P decrease to zero.

▶ When troubleshooting parallel circuits with an open branch problem, isolate the bad branch by power-off resistance checks or by power-on branch-by-branch current measurements. Another technique is to lift each branch electrically from the remainder of the circuit while monitoring circuit resistance or circuit current. The branch that causes no change is the bad branch.

▶ When troubleshooting parallel circuits with a shorted branch problem, *turn the power off and remove the circuit from the power source terminals before performing power-off checks to isolate the problem.* Branch-by-branch resistance measurements usually locate the problem. Also, you can electrically lift and reconnect each branch sequentially while monitoring the total circuit resistance.

▶ There are several significant *differences* between parallel and series circuits. Current is *common* in series circuits; voltage is *common* in parallel circuits. *Voltage* divides around a series circuit in *direct* relationship to the R values. *Current* divides throughout a parallel circuit in *inverse* relationship to the branch R values. In series circuits, the largest R dissipates the most power; in parallel circuits, the smallest R dissipates the most power. In series circuits, the total resistance equals the sum of all resistances. In parallel circuits, the total circuit resistance is *less* than the smallest branch R value.

▶ Voltage/power sources can be connected in parallel to provide higher current/power at a given voltage level.

REVIEW QUESTIONS

1. Refer to Figure 4–30 and find R_T.

   Circle the method you used to find the answer:

   Ohm's law
   Reciprocal method
   Product-over-the-sum method
   Assumed voltage method

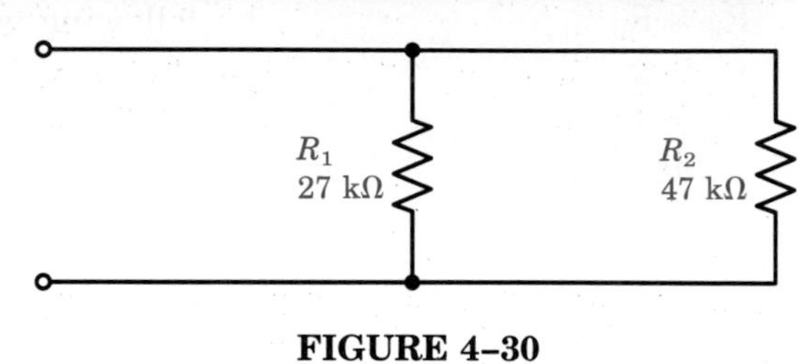

FIGURE 4–30

2. Refer to Figure 4–31 and find R_T.

   Describe the method(s) you used to solve this problem.

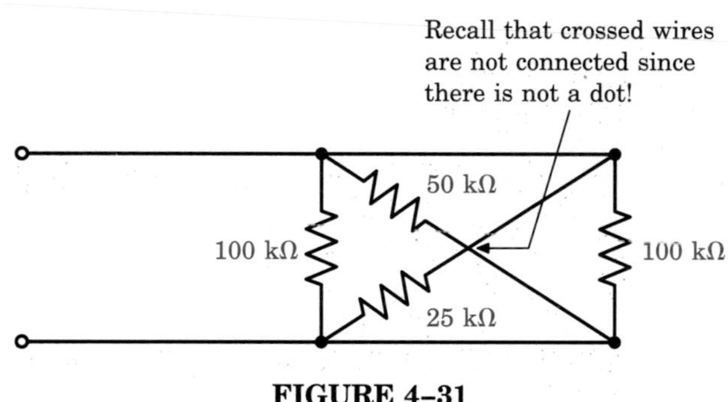

FIGURE 4–31

3. Refer to Figure 4–32 and find I_2.

   Describe the method(s) you used to solve this problem.

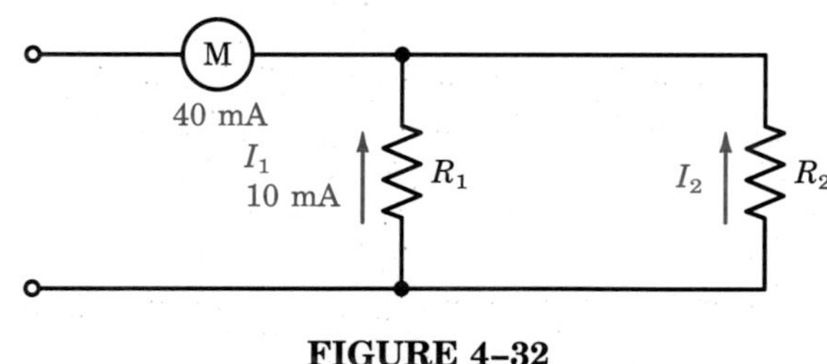

FIGURE 4–32

4. Assume in Figure 4–33 that each of the six resistors is 150 kΩ and determine the following circuit parameters:

   R_T, I_T, and P_T;
   Current reading of meter 1
   Current reading of meter 2

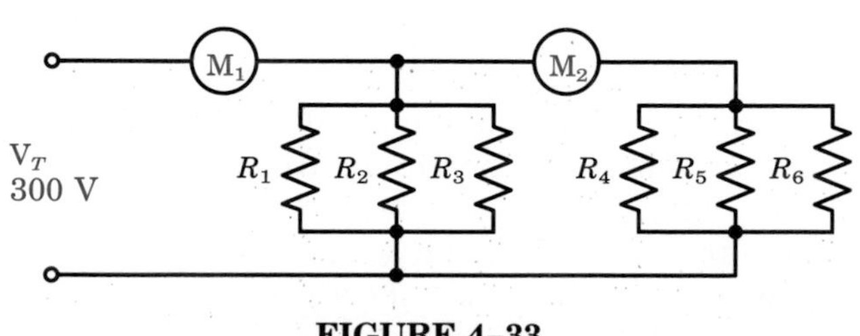

FIGURE 4–33

5. Refer to Figure 4–34 and using the information provided, answer the following with:
   "I" for increase
   "D" for decrease
   "RTS" for remain the same
   a. If R_2 opens,
      I_1 will ________ P_T will ________
      I_2 will ________ V_1 will ________
      I_3 will ________ V_T will ________.
      I_T will ________
   b. Assume that R_3 shorts internally. Prior to fuse or power supply failure,
      I_1 will ________ I_3 will ________
      I_2 will ________ total circuit current will ________.

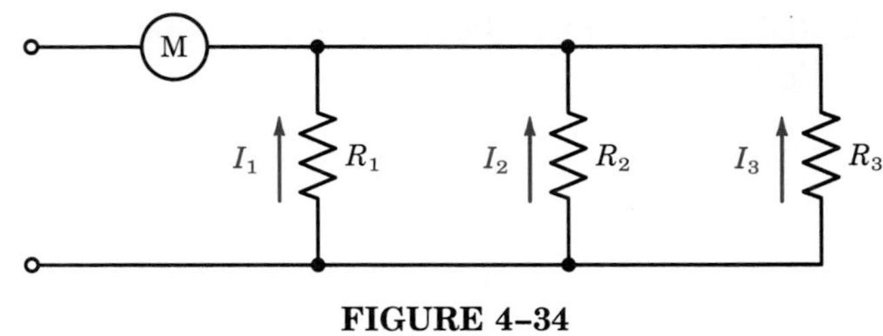

FIGURE 4–34

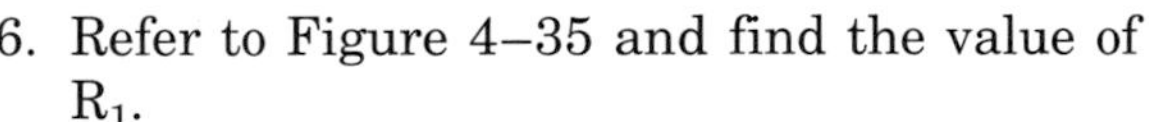

6. Refer to Figure 4–35 and find the value of R_1.

   R_1 = ________ kΩ

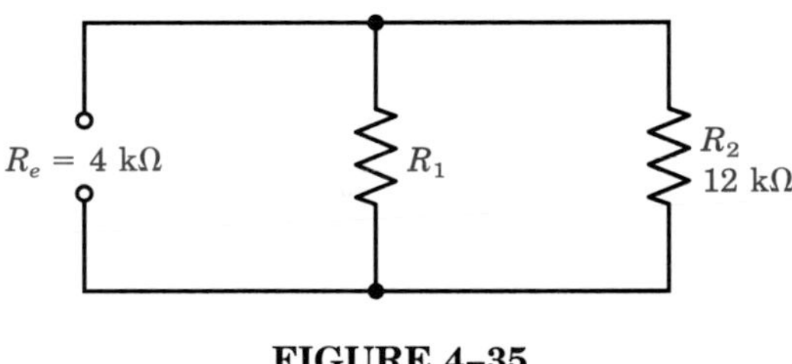

FIGURE 4–35

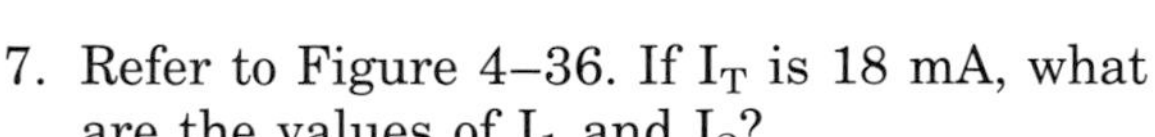

7. Refer to Figure 4–36. If I_T is 18 mA, what are the values of I_1 and I_2?

   I_1 = ________ mA
   I_2 = ________ mA

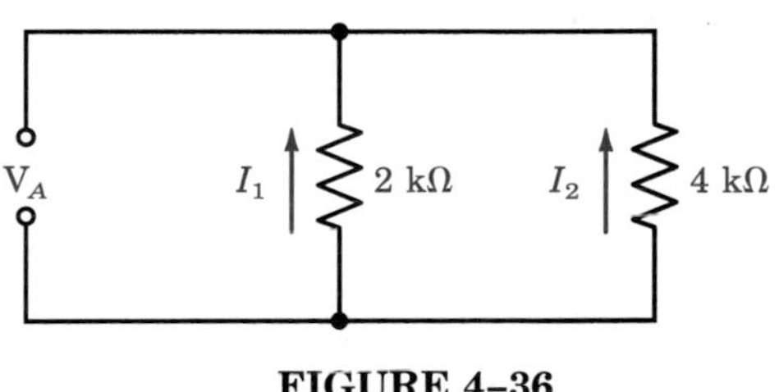

FIGURE 4–36

8. Refer to Figure 4–37. What is the value of R_1?

   R_1 = ________ kΩ

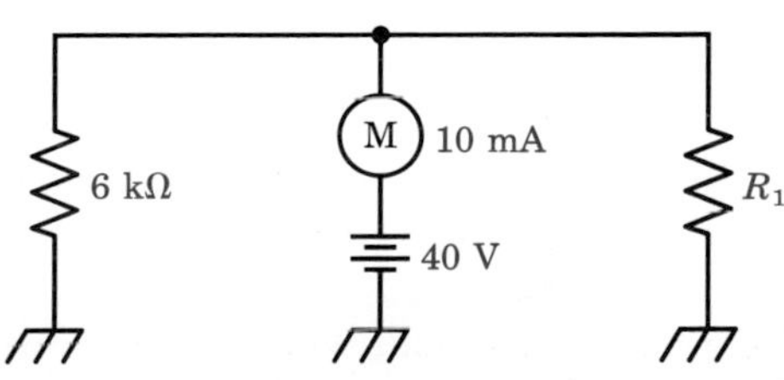

FIGURE 4–37

9. Refer to Figure 4–38. What is the value of P_{R_1}?

   P_{R_1} = ________W.

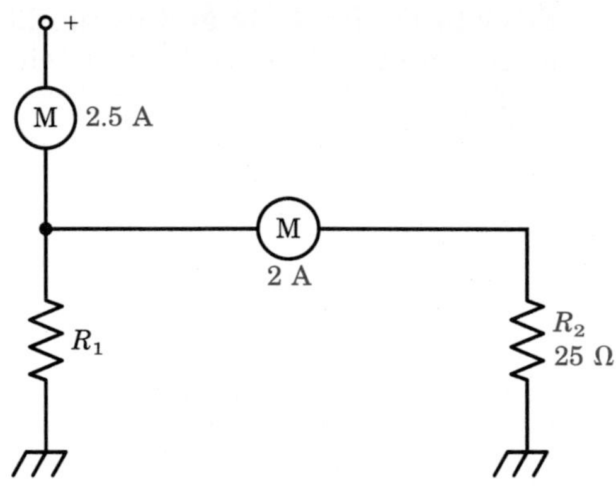

FIGURE 4–38

10. Refer to Figure 4–39 and find the following parameters:

    R_T = ________
    I_T = ________
    P_T = ________
    I_1 = ________
    I_2 = ________
    I_3 = ________
    I_4 = ________
    I_5 = ________
    P_1 = ________
    P_2 = ________
    P_3 = ________
    P_4 = ________
    P_5 = ________

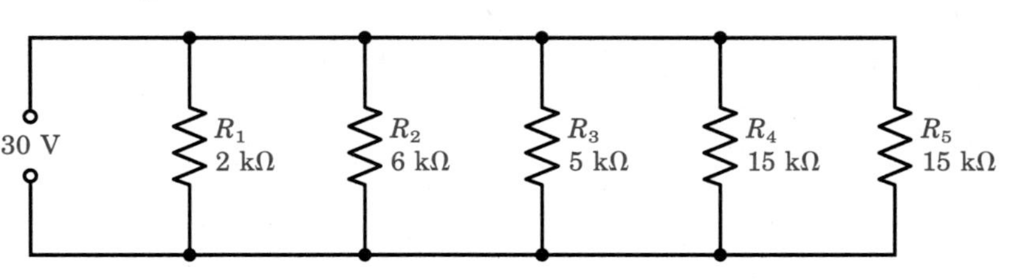

FIGURE 4–39

11. Refer to Figure 4–40 and answer the following:
    a. What is current reading of M_1?
    b. What is current reading of M_2?
    c. What is current reading of M_3?
    d. What is current reading of M_4?
    e. What is current reading of M_5?
    f. What is the power dissipation of R_1?
    g. What is the power dissipation of R_2?
    h. What is the power dissipation of R_3?
    i. What is total power dissipation of the circuit?
    j. If R_2 opened, what is new value of P_T?

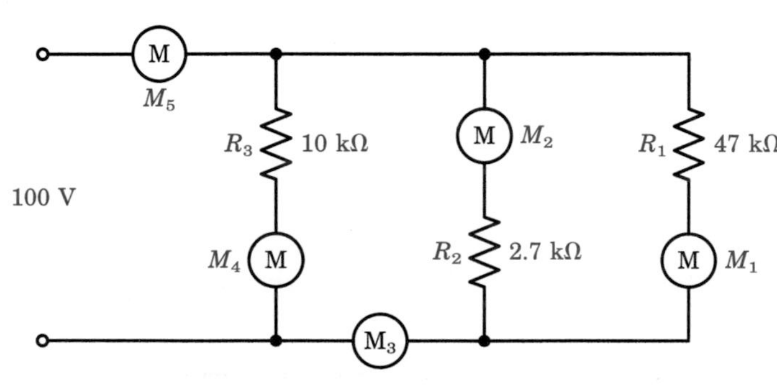

FIGURE 4–40

Refer to this circuit, Figure 4–41, when answering questions 12 through 16.

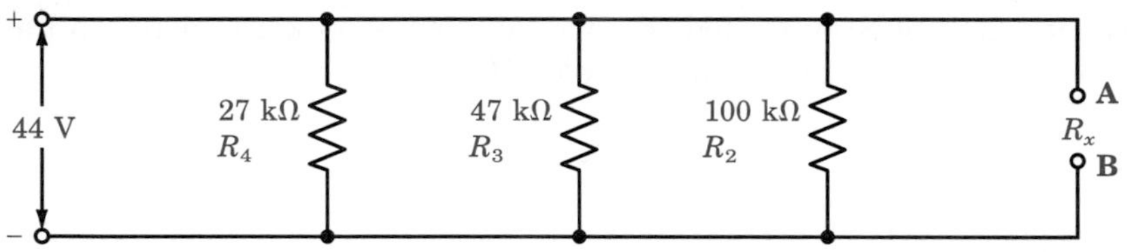

FIGURE 4–41

12. With nothing connected between points A and B, show your "step-by-step" work and solve for R_T using:
    a. The reciprocal method
    b. The conductance method
    c. The assumed voltage method
    d. The Ohm's law method

13. What resistance (R_X) needs to be connected between points A and B for the circuit total resistance to equal 10 kΩ?

14. Assume the R_X of the preceding problem is left connected to the circuit, what happens to the circuit R_T if R_4 is doubled in value and R_2 is halved in value? Would the answer be the same if R_X is removed?

15. With the R_X of problem 13 still connected:
    a. Which resistor in the circuit dissipates the most power?
    b. What is the value of power dissipated by this resistor?

16. With the R_X of problem 13 *disconnected:*
    a. Which resistor in the circuit dissipates the least power?
    b. What is the value of power dissipated by this resistor?

TEST YOURSELF

1. Draw a circuit diagram of a three-branch parallel circuit with the largest R having a value of 100 kΩ. This current divider circuit should have values that cause the total circuit current to divide as follows:
   Branch 1's current is half of branch 2 and branch 2's current is one-fifth of branch 3.
   Label all R values on your diagram.

2. What value of R is needed in parallel with a 50 kΩ resistor to have a resultant equivalent resistance of 10 kΩ?

PERFORMANCE PROJECTS CORRELATION CHART

Suggested performance projects in the Laboratory Manual that correlate with topics in this chapter are:

CHAPTER TOPIC	PERFORMANCE PROJECT	PROJECT NUMBER
Voltage in Parallel Circuits	Voltage in Parallel Circuits	17
Current in Parallel Circuits & Kirchhoff's Current Law	Current in Parallel Circuits	16
Resistance in Parallel Circuits	Equivalent Resistance in Parallel Circuits	15
Power in Parallel Circuits	Power Distribution in Parallel Circuits	18
Effects of Opens in Parallel Circuits & Troubleshooting Hints	Effects of an Open in Parallel Circuits	19
Effects of Shorts in Parallel Circuits & Troubleshooting Hints	Effects of a Short in Parallel Circuits	20

NOTE: It is suggested that after completing the above projects, the student should be required to answer the questions in the "Summary" at the end of this section of projects in the Laboratory Manual.

Series-Parallel Circuits

Key Terms

Bleeder current

Bleeder resistor

Bridge circuit

Bridging or shunting

Ground reference

In-line

Load

Loaded voltage divider

Loading effect

Series-parallel circuit

Subcircuit

Wheatstone bridge circuit

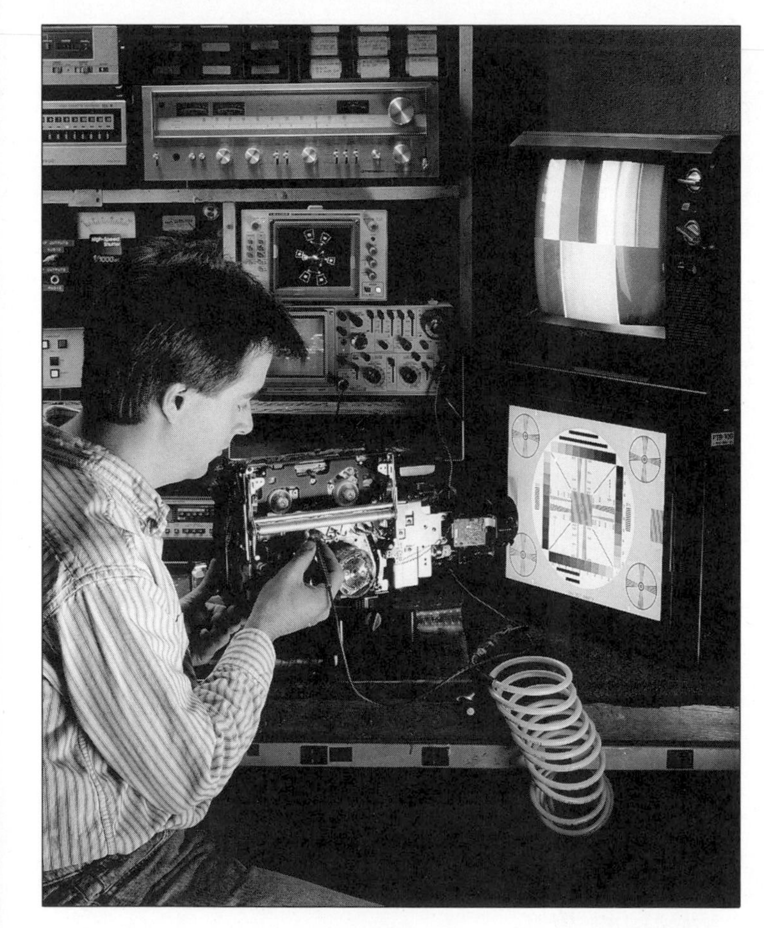

Courtesy of Cleveland Institute of Electronics, Inc.

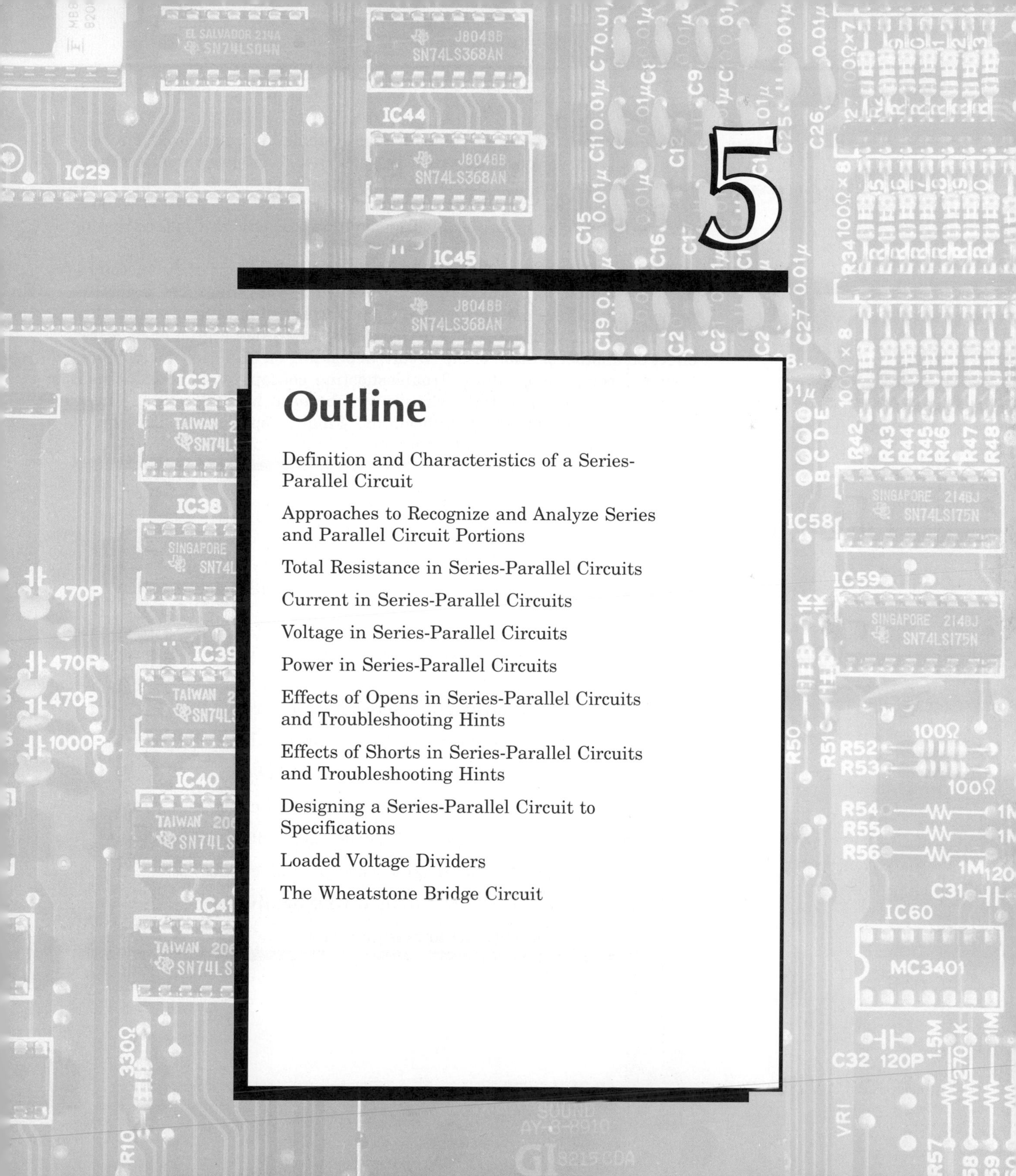

5

Outline

Definition and Characteristics of a Series-Parallel Circuit

Approaches to Recognize and Analyze Series and Parallel Circuit Portions

Total Resistance in Series-Parallel Circuits

Current in Series-Parallel Circuits

Voltage in Series-Parallel Circuits

Power in Series-Parallel Circuits

Effects of Opens in Series-Parallel Circuits and Troubleshooting Hints

Effects of Shorts in Series-Parallel Circuits and Troubleshooting Hints

Designing a Series-Parallel Circuit to Specifications

Loaded Voltage Dividers

The Wheatstone Bridge Circuit

Chapter Preview

Most electronic devices, equipment, and systems are comprised of series-parallel circuitry. Consumer products, such as high fidelity amplifiers, televisions, computers, and virtually all electronic products are filled with series-parallel circuits. A combination of both series and parallel connections form circuits that are coupled, or combined to perform a desired task or function.

In this chapter, you will examine various configurations of these "combinational" circuits and will become familiar with analyzing them, using the principles you have already learned about series and parallel circuits. Also, you will see how a change in one part of a circuit has a greater or lesser effect on another portion of the circuit, depending on total circuit configuration and component values. Troubleshooting concepts will be examined. Your understanding of these circuits will be used by designing a simple series-parallel circuit to specifications. Finally, special applications of series-parallel circuitry will be studied.

Objectives

After studying this chapter, you will be able to:

- Define the term **series-parallel circuit**
- List the primary characteristic(s) of a series-parallel circuit
- Determine the total resistance in a series-parallel circuit
- Compute total circuit current through any given portion of a series-parallel circuit
- Calculate voltages throughout a series-parallel circuit
- Determine power values throughout a series-parallel circuit
- Analyze the effects of an open in a series-parallel circuit
- Analyze the effects of a short in a series-parallel circuit
- Design a simple series-parallel circuit to specifications
- Explain the **loading effects** on a series-parallel circuit
- Calculate values relating to a **loaded voltage divider**
- Make calculations relating to **bridge circuits**

WHAT IS A SERIES-PARALLEL CIRCUIT?

Definition

A **series-parallel circuit** contains a combination of both series-connected and parallel-connected components. There are both **in-line** series current paths and "branching-type" parallel current paths, Figure 5–1.

FIGURE 5–1
A basic series-parallel circuit

(**a**) Pictorial

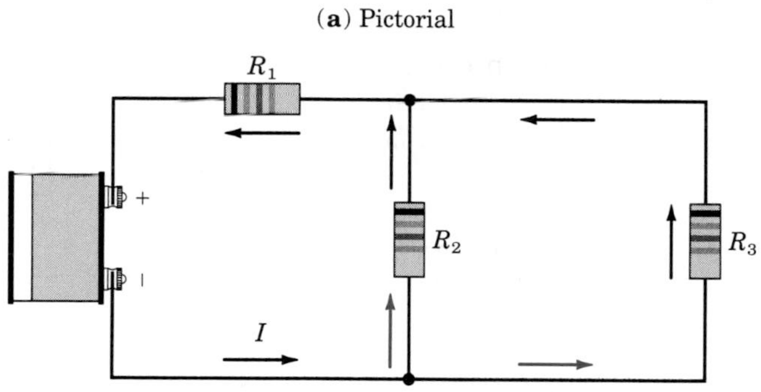

(**b**) Schematic

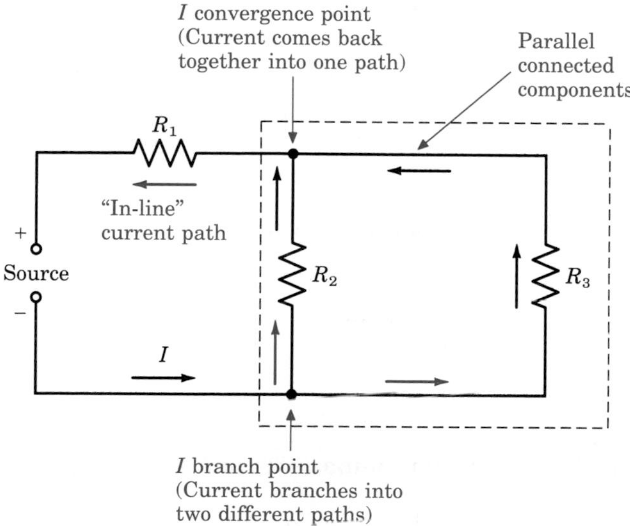

Characteristics

Series components in a series-parallel circuit may be in series with other *individual* components, or with other *combinations* of components, Figure 5–2.

Parallel components in a series-parallel circuit may be in parallel with other *individual* components, or with other *combinations* of components, Figure 5–3.

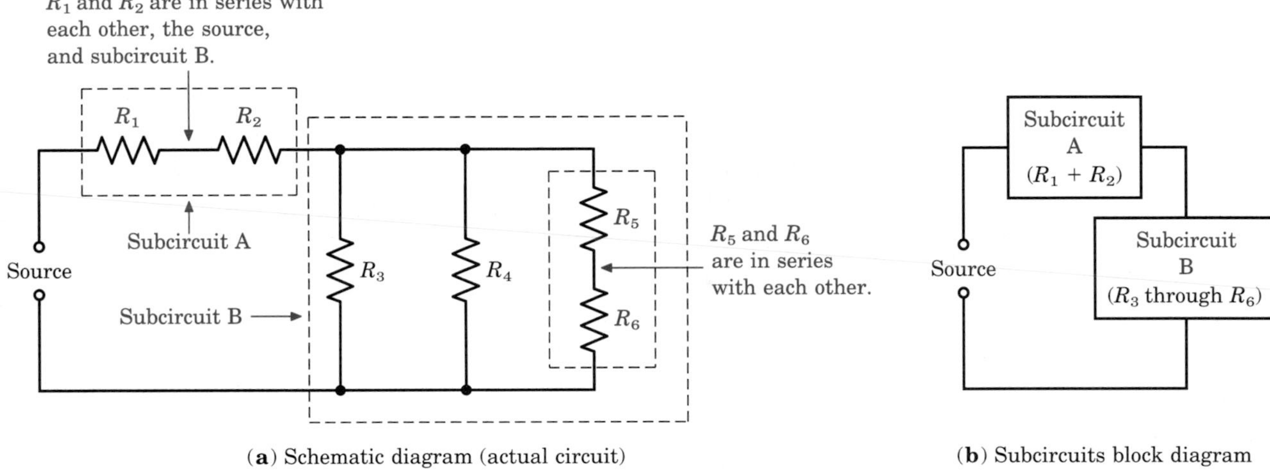

(**a**) Schematic diagram (actual circuit) (**b**) Subcircuits block diagram

FIGURE 5–2 Series components in a series-parallel circuit

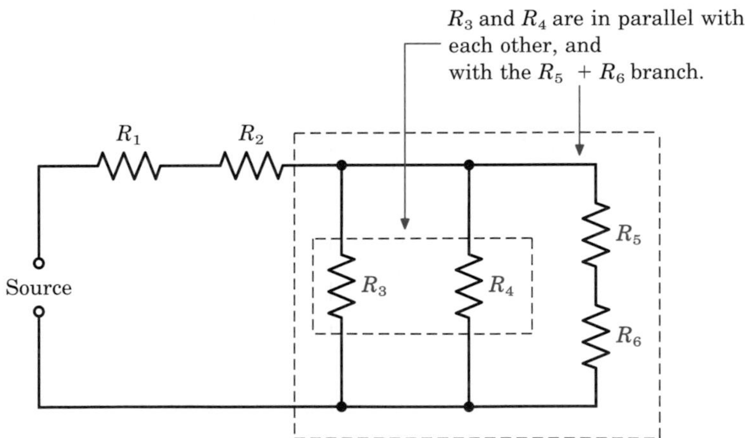

FIGURE 5–3 Parallel components in a series-parallel circuit

APPROACHES TO RECOGNIZE AND ANALYZE SERIES AND PARALLEL CIRCUIT PORTIONS

1. Start analysis at the portion of the circuit farthest from the source and work back toward the source to identify those components or circuit portions that are in series and in parallel.

2. Trace common current paths to identify components in series, Figure 5–4. Components, or combinations of components with common current are in series with each other.

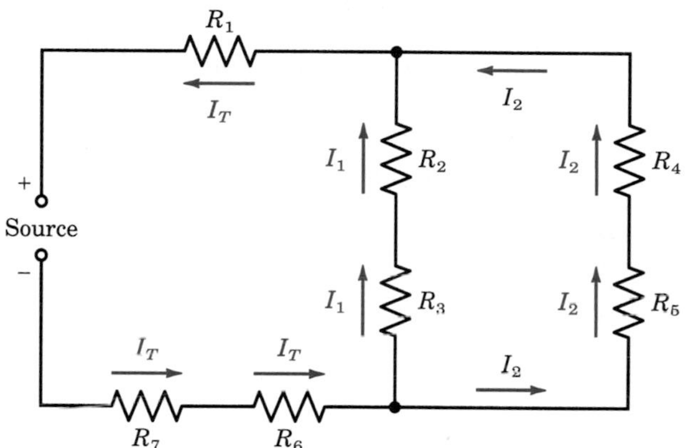

Starting "away from the source" and working back:
1. R_4 and R_5 have common current (I_2) and are in series with each other.
2. R_2 and R_3 have common current (I_1) and are in series with each other.
3. R_1, R_6, and R_7 have common current (I_T) and they are in series with each other, and with the total parallel combination of the $R_2 + R_3$ branch in parallel with the $R_4 + R_5$ branch.

FIGURE 5–4 Tracing common current paths to identify series circuit portions

3. Observe voltages shared (in common) to identify components in parallel, Figure 5–5. Components, or combinations of components with the

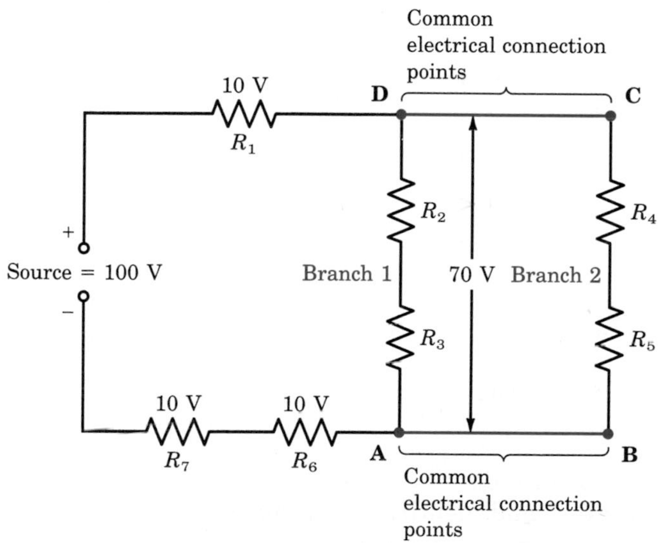

NOTE: Points **A** and **B** are electrically the same point and points **C** and **D** are electrically the same point. Therefore, the voltage from point **B** to point **C** equals the voltage from point **A** to point **D**.

Branch 2 ($R_4 + R_5$) is across the same voltage points as branch 1 ($R_2 + R_3$). Therefore, branch 1 and branch 2 are in parallel with each other.

FIGURE 5–5 Observing common voltage points to identify parallel circuit portions

same voltage connection points (at both ends) are in parallel with each other.

4. Observe current branching and converging points to identify components, or combinations of components that are in parallel with each other, Figure 5–6.
   a. A point where current splits, or branches, is one end of a parallel combination of components, Figure 5–6, points A and B.
   b. The point where those same currents converge (or rejoin) is the other end of that same parallel combination, Figure 5–6, points C and D.

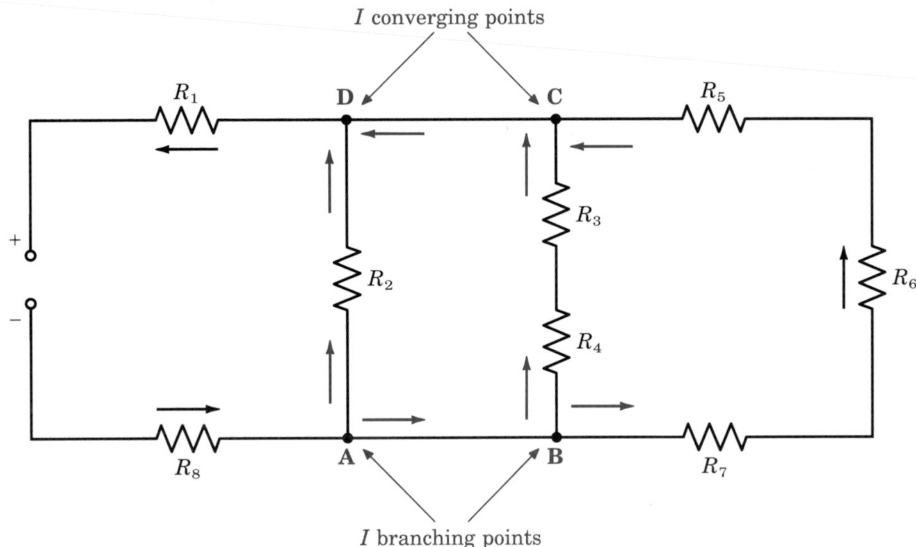

FIGURE 5–6 Observing current branching and converging points to identify parallel circuit portions

5. In summary, in the series portions of the circuit, current is common and the voltages (and resistances) are additive. In the parallel portions of the circuit, voltage is common and the branch currents are additive.

PRACTICE PROBLEM I

Answers are in Appendix B.

Refer to Figure 5–7 and identify *individual* or single components.

1. Carry total current.
2. Are in parallel with each other. That is, have the same voltage across them.

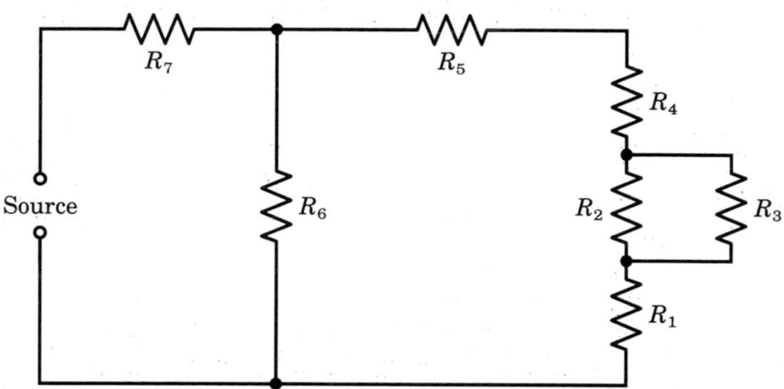

FIGURE 5–7 Practice problems

3. Are in series with each other. That is, have the same current through them.

Let's move to some analysis of series-parallel circuit electrical parameters.

TOTAL RESISTANCE IN SERIES-PARALLEL CIRCUITS

If total circuit current and circuit applied voltage are known, then solving for total circuit resistance is easy.

Using Ohm's Law

Simply use Ohm's law, where $R_T = \dfrac{V_T}{I_T}$.

If source voltage (V_S) and total current are unknown, then apply the concepts discussed in the preceding section.

Using the "Outside Toward the Source" Approach

1. Start at the end of the circuit farthest from the source and work toward the source, identifying and solving the series, the parallel, and the series-parallel combinations, Figure 5–8. You will apply series circuit rules for series portions, and parallel circuit rules for parallel portions in the circuitry.

2. First, identify and solve parallel component segments. The common voltage and the current branching and converging concepts are true for R_2 and R_3; therefore, these two components are in parallel, Figure 5–8a.

   As you recall, the equivalent resistance of two equal value resistors in parallel is equal to half the value of one equal branch R. Thus, R_c

FIGURE 5–8
Analyzing a circuit to find R_T

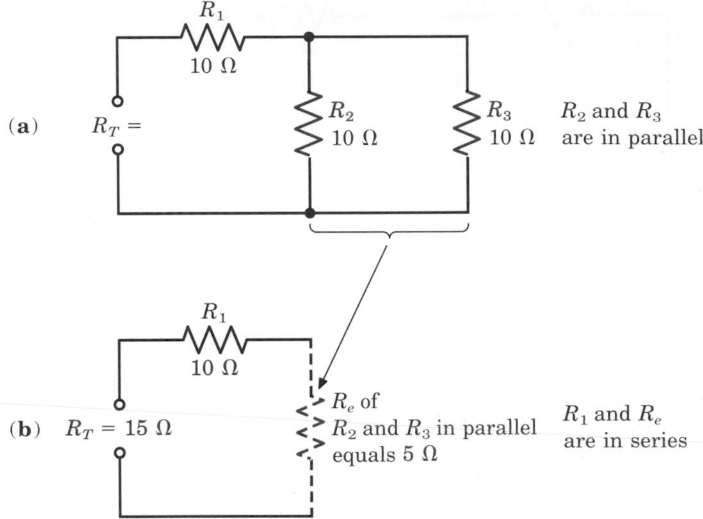

of R_2 and R_3 = 5 Ω. Naturally, you can also solve it with the product-over-the-sum formula. That is:

$$R_e = \frac{R_2 \times R_3}{R_2 + R_3} = \frac{(10\ \Omega) \times (10\ \Omega)}{(10\ \Omega) + (10\ \Omega)} = \frac{100\ \Omega}{20\ \Omega} = 5\ \Omega.$$

3. Next, work toward the source to solve the series and/or series-parallel combinations.

   Total current splits and rejoins as it flows through the $R_2 - R_3$ parallel combination. Total current also flows through R_1. Thus, R_1 and the parallel combination of R_2 and R_3 are in series with each other, Figure 5–8b.

   Therefore, total resistance of R_1 in series with the $R_2 - R_3$ parallel combination equals 10 Ω + 5 Ω = 15 Ω. And the circuit total resistance is 15 Ω.

The Reduce-and-Redraw Approach

A useful method to analyze series-parallel circuits is the reduce-and-redraw approach. This technique simplifies circuit analysis.

EXAMPLE

Look at Figures 5–9a–d as R_T is discussed for this more complex series-parallel combination circuit.

1. R_3 and R_4 are in parallel with each other and have an equivalent resistance that we'll call R_{e_1}. Thus, R_{e_1} = 5 kΩ.

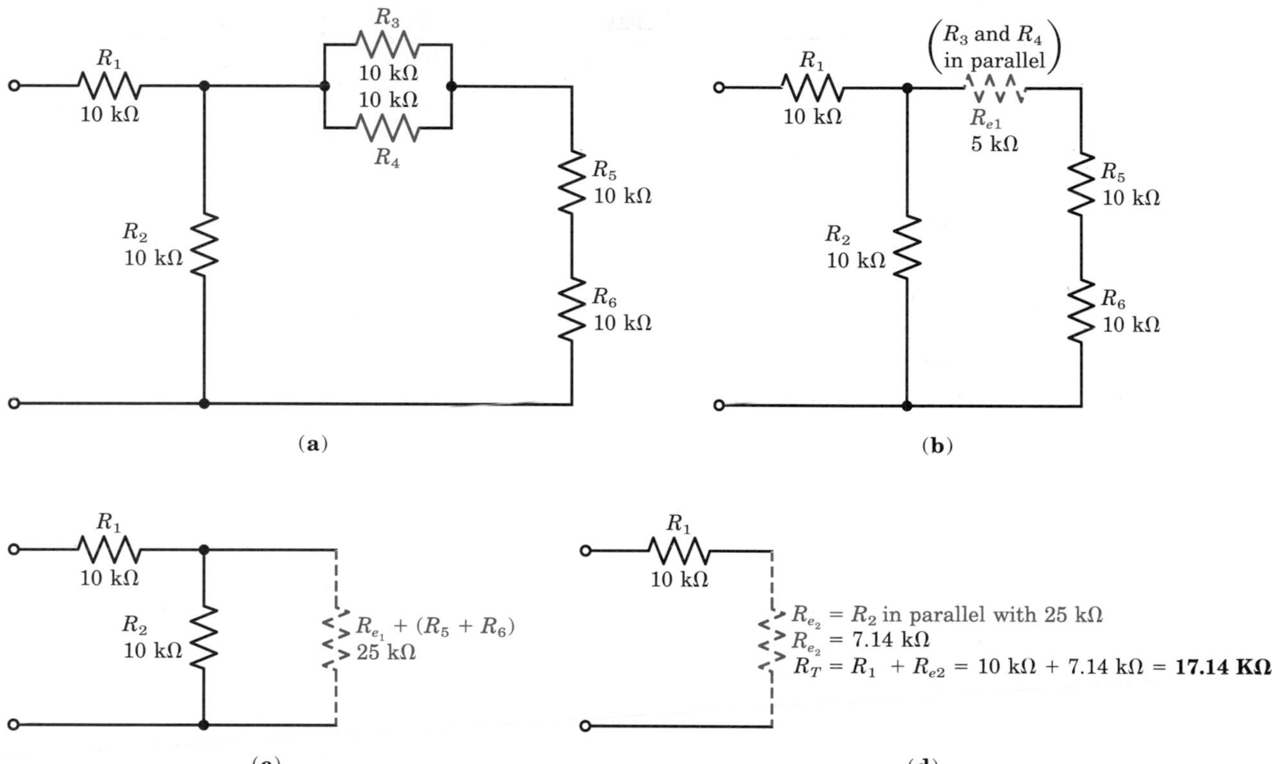

FIGURE 5–9 Simplifying by the "reduce and redraw" technique

$$R_{e_1}\text{, the resistance of } R_3 \text{ and } R_4 \text{ in parallel} = \frac{(10\text{ k}\Omega)(10\text{ k}\Omega)}{(10\text{ k}\Omega) + (10\text{ k}\Omega)} =$$

$$\frac{100 \times 10^6}{20 \times 10^3} = 5 \times 10^3 = 5\text{ k}\Omega. \quad \text{(Figure 5–9b)}$$

Of course, a simpler way is to remember that two equal branch resistances in parallel have an equivalent resistance equal to half of one of the branch resistances.

2. R_5 and R_6 are in series with each other and with R_{e_1} of parallel resistors R_3 and R_4. See Figure 5–9c and the calculations below.

   The resistance of $R_{e_1} = 5\text{ k}\Omega$
   The resistance of $R_5 + R_6$ in series $= 10\text{ k}\Omega + 10\text{ k}\Omega = 20\text{ k}\Omega$
   The resistance of $R_{e_1} + R_5 + R_6 = 25\text{ k}\Omega$

3. R_2 (10 kΩ) is in parallel with the total 25 kΩ combination of components R_3 through R_6 ($R_{e_1} + R_5 + R_6$). See Figures 5–9c and Figure 5–9d and the following calculations.

$$R_{e_2} = \frac{(10\text{ k}\Omega)(25\text{ k}\Omega)}{(10\text{ k}\Omega) + (25\text{ k}\Omega)} = \frac{250 \times 10^6}{35 \times 10^3} = 7.14 \times 10^3 = 7.14\text{ k}\Omega.$$

4. The overall combination of components R_2 through R_6 (R_{e_2} is in series with R_1. See Figure 5–9D and the following calculations.

   Total circuit resistance (R_T) = $R_1 + R_{e_2} = 10\text{ k}\Omega + 7.14\text{ k}\Omega =$ $17.14\text{ k}\Omega$

IN-PROCESS LEARNING CHECK I

Try solving the total resistance problems in Figures 5–10a and 5–10b. Use the outside toward the source approach to solve Figure 5–10a and use the reduce and redraw technique to solve the circuit in Figure 5–10b. You can then compare your answers with those given in Appendix B.

FIGURE 5–10 Solve for R_T values in each circuit.

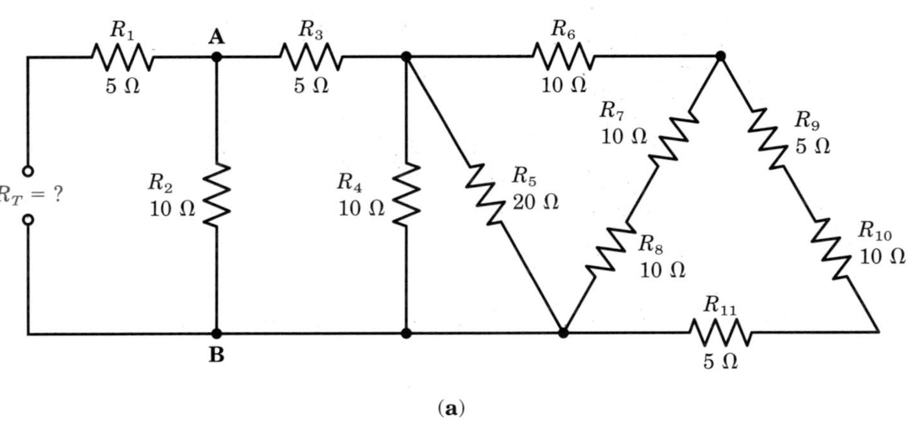

(a)

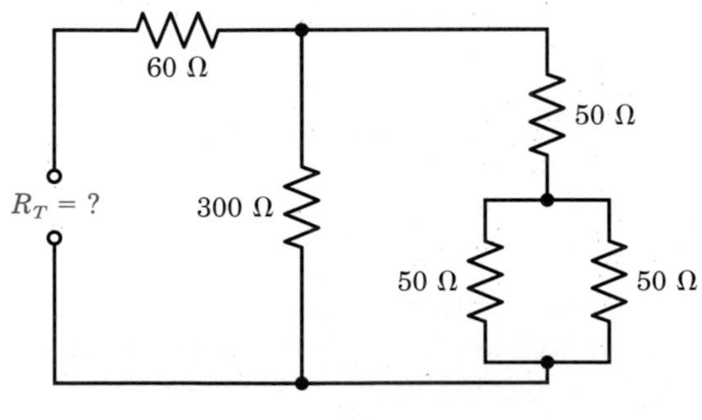

(b)

CURRENT IN SERIES-PARALLEL CIRCUITS

Examine Figure 5–11 as we now discuss current in a series-parallel circuit. One can start by finding the total circuit current, then work from that point to determine current distribution throughout the circuit using Ohm's law, Kirchhoff's laws, current divider formulas and/or other techniques.

It is easier if Ohm's law is used to find total current, where $I_T = \frac{V_S}{R_T}$. In this circuit, the source voltage (V_S) is 175 volts and the total resistance is 175 Ω. For practice, verify the R_T value using the techniques you learned in the preceding section.

Now that R_T and V_S are known, it's easy to solve for I_T.

$$I_T = \frac{V_S}{R_T} = \frac{175\ \text{V}}{175\ \Omega} = 1\ \text{A}$$

What about the distribution of current throughout the circuit? For this circuit, distribution of current is easy to find because of the values used. One example of a thought process that can be used is as follows:

1. R_1 and R_7 must have 1 ampere through them, since they are in series with the source and the remaining circuitry. In other words, I_T passes through both these resistors, Figure 5–11.

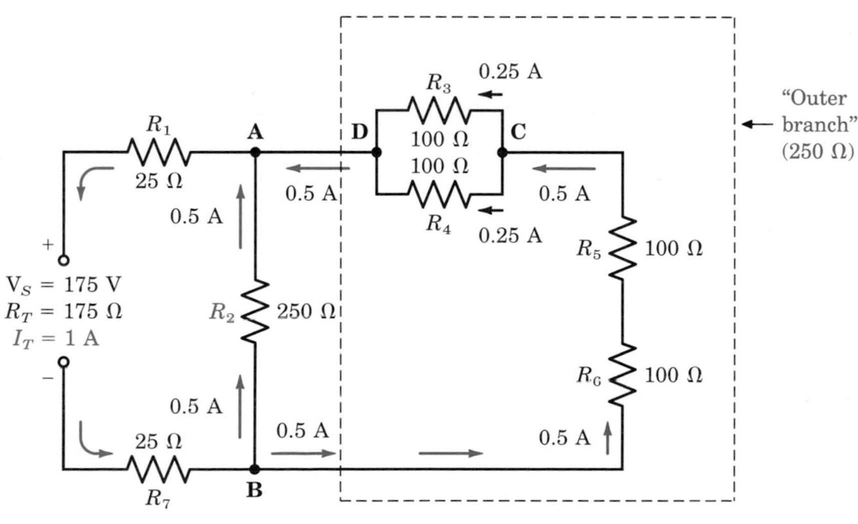

FIGURE 5–11 Solve for current values throughout a series-parallel circuit.

2. Since R_2 (250 Ω) and the outer branch (250 Ω) are equal resistance value branches, the total current equally divides between them. Thus, current through R_2 = 0.5 ampere, and current through the outer branch must equal 0.5 ampere, Figure 5–11.

3. It is apparent in tracing current through the outer branch that all of the 0.5 ampere must pass through both R_5 and R_6. However, this 0.5 ampere evenly splits between R_3 and R_4 because they are equal value parallel resistances. This means current through R_3 must be 0.25 ampere, and current through R_4 is also 0.25 ampere. In other words, the 0.5 ampere evenly splits at point C into two 0.25 ampere branch currents and rejoins at point D, Figure 5–11.

To find circuit resistance, you frequently start at a point farthest from the source and work back toward the source. *But when analyzing circuit current, you often start at the source with total current, then work outward through the circuit, analyzing the current distribution.* In this case, the circuit currents are determined to be:

I_T; I through R_1; and I through R_7 = 1 ampere.
I through R_2; I through R_5; and R_6 = 0.5 ampere.
I through R_3 and I through R_4 each = 0.25 ampere.

Of course, several other methods, or combinations of methods can be used to provide solutions, such as Ohm's law, Kirchhoff's voltage and current laws, and current-divider formulas. In those cases where the values are not as convenient, you will probably have to use other methods.

EXAMPLE

Having solved for total current (1 ampere), you can refer to Figure 5–11 and apply known values to solve for the unknown parameters.

1. I_T passes through both R_1 and R_7. We can use Ohm's law to find the voltage drops. Thus,

$$V_1 = I_T \times R_1 = 1\text{ A} \times 25\ \Omega = 25\text{ V}$$
$$V_7 = I_T \times R_7 = 1\text{ A} \times 25\ \Omega = 25\text{ V}$$

2. According to Kirchhoff's voltage law, the sum of the voltages around any closed loop must equal V applied. This means the sum of V_1 + V_{A-B} + V_7 must equal 175 V.
Thus, 25 V + V_{A-B} + 25 V = 175 V

$$50\text{ V} + V_{A-B} = 175\text{ V}$$
$$V_{A-B} = 175\text{ V} - 50\text{ V} = 125\text{ V}$$

3. Kirchhoff's current law indicates the current leaving point A must equal the current entering point A. Stated another way, the total of the currents through R_2 and the outer branch must equal I_T in this case.

a. Use Ohm's law to find current through R_2.

$$\text{I through } R_2 = \frac{V_2}{R_2} = \frac{125\ \text{V}}{250\ \Omega} = 0.5\ \text{A}$$

b. Using Kirchhoff's current law:

$$\text{I through outer branch} = I_T - I\ (\text{through } R_2);$$
$$= 1\ \text{A} - 0.5\ \text{A} = 0.5\ \text{A}$$

4. Ohm's law and Kirchhoff's voltage law are now used to solve the voltage drops across the resistors in the outer branch and the currents through R_3 and R_4.

Knowing the outer branch current is 0.5 A:

a. $V_6 = 0.5\ \text{A} \times R_6 = 0.5\ \text{A} \times 100\ \Omega = 50\ \text{V}$

b. $V_5 = 0.5\ \text{A} \times R_5 = 0.5\ \text{A} \times 100\ \Omega = 50\ \text{V}$

Kirchhoff's voltage law indicates that:

c. $V_{C-D} = V_{A-B} - (V_5 + V_6) = 125\ \text{V} - (50\ \text{V} + 50\ \text{V}) = 25\ \text{V}$

Ohm's law can now be used to solve for the currents through R_3 and R_4.

d. $\text{I through } R_3 = \dfrac{V_3}{R_3} = \dfrac{25\ \text{V}}{100\ \Omega} = 0.25\ \text{A.}$

e. $\text{I through } R_4 = \dfrac{V_4}{R_4} = \dfrac{25\ \text{V}}{100\ \Omega} = 0.25\ \text{A.}$

PRACTICE PROBLEM II

For practice in solving for current in a series-parallel circuit, refer to Figure 5–12. Solve for I_T, I_1, I_2, and I_3.

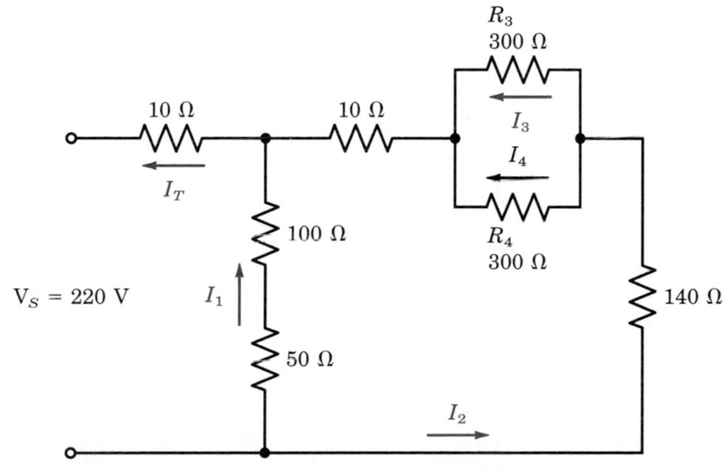

FIGURE 5–12 Practice problems—Finding current in series-parallel circuits

VOLTAGE IN SERIES-PARALLEL CIRCUITS

Voltage distribution throughout a series-parallel circuit is such that: 1) voltage across components, or circuit portions that are in series with each other are additive, 2) voltage across components, or circuit portions that are in parallel must have equal values.

As you would expect, a given component's voltage drop in a series-parallel circuit is contingent on its circuit location and its resistance value. The location dictates what portion of the circuit's total current passes through it. Consequently, location of any given component within the series parallel circuit has direct influence on its I × R drop.

Refer to Figure 5–13 as the voltage distribution characteristics are analyzed with two methods: the Ohm's law method and resistance and voltage divider techniques.

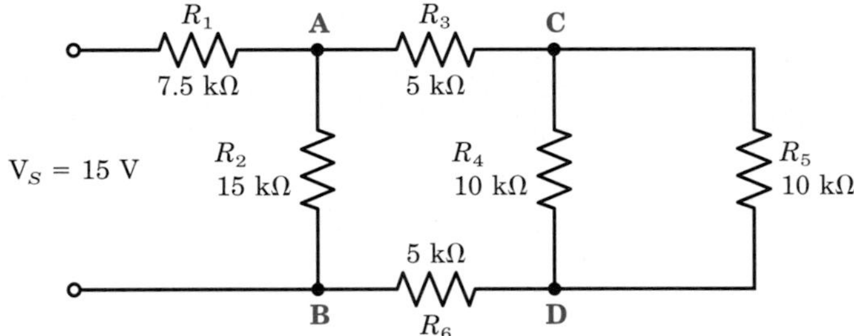

FIGURE 5–13 Analyzing voltages in a series-parallel circuit

Ohm's Law Method

Solve circuit R and I values, then use Ohm's law to compute voltages.

1. $R_{C-D} = \dfrac{R_4 \times R_5}{R_4 + R_5} = \dfrac{(10 \times 10^3) \times (10 \times 10^3)}{(10 \times 10^3) + (10 \times 10^3)}$
$= \dfrac{100 \times 10^6}{20 \times 10^3} = 5 \times 10^3 = 5\ k\Omega$

2. $R_3 + R_{C-D} + R_6 = 5\ k\Omega + 5\ k\Omega + 5\ k\Omega = 15\ k\Omega.$

3. $R_{A-B} = \dfrac{(R_2)(R_3 + R_{C-D} + R_6)}{(R_2) + (R_3 + R_{C-D} + R_6)}$
$= \dfrac{(15 \times 10^3) \times (15 \times 10^3)}{(15 \times 10^3) + (15 \times 10^3)} = \dfrac{225 \times 10^6}{30 \times 10^3} = 7.5 \times 10^3 = 7.5\ k\Omega$

4. $R_T = R_1 + R_{A-B} = 7.5\ k\Omega + 7.5\ k\Omega = 15\ k\Omega$.

5. $I_T = \frac{V_T}{R_T} = \frac{15\ V}{15\ k\Omega} = 1\ mA$.

6. $V_1 = I_T \times R_1 = 1\ mA \times 7.5\ k\Omega = 7.5\ V$

7. Current splits evenly between branches R_2 and the R_3 through R_6 branch since each branch R equals 15 kΩ. This means that:

   $V_2 = 0.5\ I_T \times R_2 = 0.5\ mA \times 15\ k\Omega = 7.5\ V$
   $V_3 + V_{C-D} + V_6$ equals 7.5 V.

8. $V_3 = 0.5\ I_T \times R_3 = 0.5\ mA \times 5\ k\Omega = 2.5\ V$

9. Current splits evenly between R_4 and R_5; thus, each carries 0.25 mA of current.

   $V_4 = 0.25\ I_T \times R_4 = 0.25\ mA \times 10\ k\Omega = 2.5\ V$
   $V_5 = 0.25\ I_T \times R_5 = 0.25\ mA \times 10\ k\Omega = 2.5\ V$

10. $V_6 = 0.5\ I_T \times R_6 = 0.5\ mA \times 5\ k\Omega = 2.5\ V$

Resistance and Voltage Divider Techniques

1. Observing the circuit, you can see that:
   a. R_4 and R_5 are in parallel; thus, $V_4 = V_5$.
   b. R_e of $R_{4-5} = 5\ k\Omega$.
   c. R_3, R_{4-5}, and R_6 are in series and together form a 15-kΩ branch of the circuit. Furthermore, each of these drops one-third the voltage present from point A to point B, since each of the three entities represents 5 kΩ.
   d. This 15-kΩ branch is in parallel with R_2, another 15 kΩ branch. Thus R_e from point A to point B equals 7.5 kΩ.
   e. R_1 is also 7.5 kΩ and is in series with the equivalent 7.5 kΩ found from point A to point B. Thus V_1 and V_{A-B} are equal, and each equals half the voltage applied (Kirchhoff's voltage law).

2. Applying the principles of voltage dividers:

   a. $V_{A-B} = \frac{R_{A-B}}{R_T} \times V_T = \frac{7.5\ k\Omega}{15\ k\Omega} \times 15\ V = 7.5\ V$

   b. $V_{C-D} = \frac{R_{4-5}}{R_3 + R_{4-5} + R_6} \times V_{A-B} = \frac{5\ k\Omega}{15\ k\Omega} \times 7.5\ V = 2.5\ V$

   c. $V_3 = V_{4-5} = V_6$

d. $V_2 = V_{A-B}$

e. $V_1 = \frac{R_1}{R_T} \times V_T = \frac{7.5\ k\Omega}{15\ k\Omega} \times 15\ V = 7.5\ V$

PRACTICE PROBLEM III

Analyze the voltage and current parameters throughout the circuit of Figure 5–14. List the voltage dropped by each resistor and the current through each resistor.

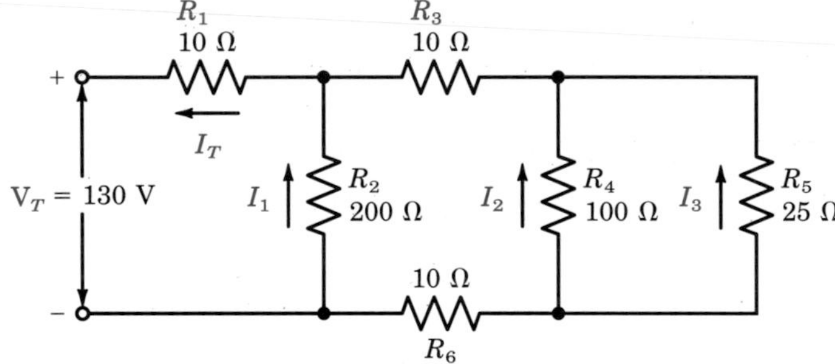

FIGURE 5–14 Voltage and current practice problem for a series-parallel circuit

POWER IN SERIES-PARALLEL CIRCUITS

Analysis of power dissipations throughout a series-parallel circuit employs the same techniques used in series or parallel circuits. Important points to remember are:

1. Total power $= V_T \times I_T \left(\text{or } I_T{}^2 \times R_T\text{; or, } \frac{V_T{}^2}{R_T}\right)$
2. Total power also equals the sum of all individual power dissipations.
3. Individual component power dissipations are calculated using the individual component's parameters. That is, $V \times I$; I^2R; or $\frac{V^2}{R}$ values.

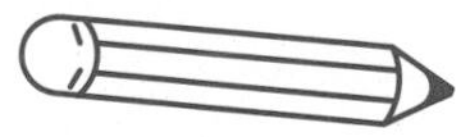

PRACTICAL NOTES

Note that:

1. In series circuits, the largest value resistor dissipates the most power, and the lowest value resistor dissipates the least power.
2. In parallel circuits, the smallest resistance branch dissipates the most power, and the largest resistance branch dissipates the least power.
3. In series-parallel circuits, the resistor dissipating the most power *may or may not* be the highest value resistor in the circuit. The resistor dissipating the lowest power *may or may not* be the lowest value resistor in the circuit.

The power dissipated is determined by the position of the given resistor in the total circuit configuration. An illustration of this is in Figure 5–15. Note that:

- R_7 is twice the value of R_1 but dissipates only half the power.
- R_1 and R_2 are equal value resistors but each dissipates a different value of power.

To make sure you are learning to analyze the key parameters for a series-parallel circuit, perform the following In-Process Learning Check.

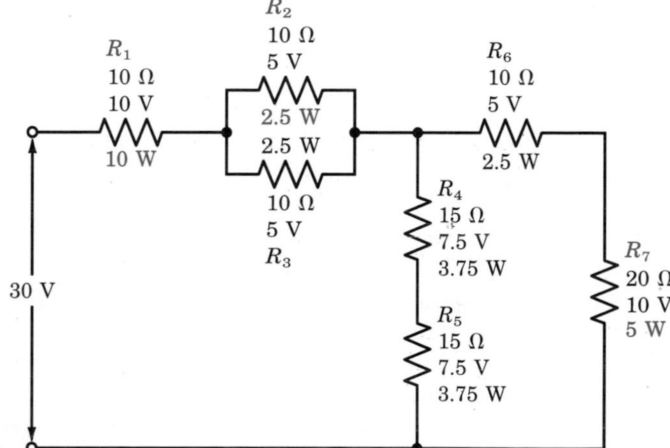

FIGURE 5–15 Effect of location on power, current, and voltage

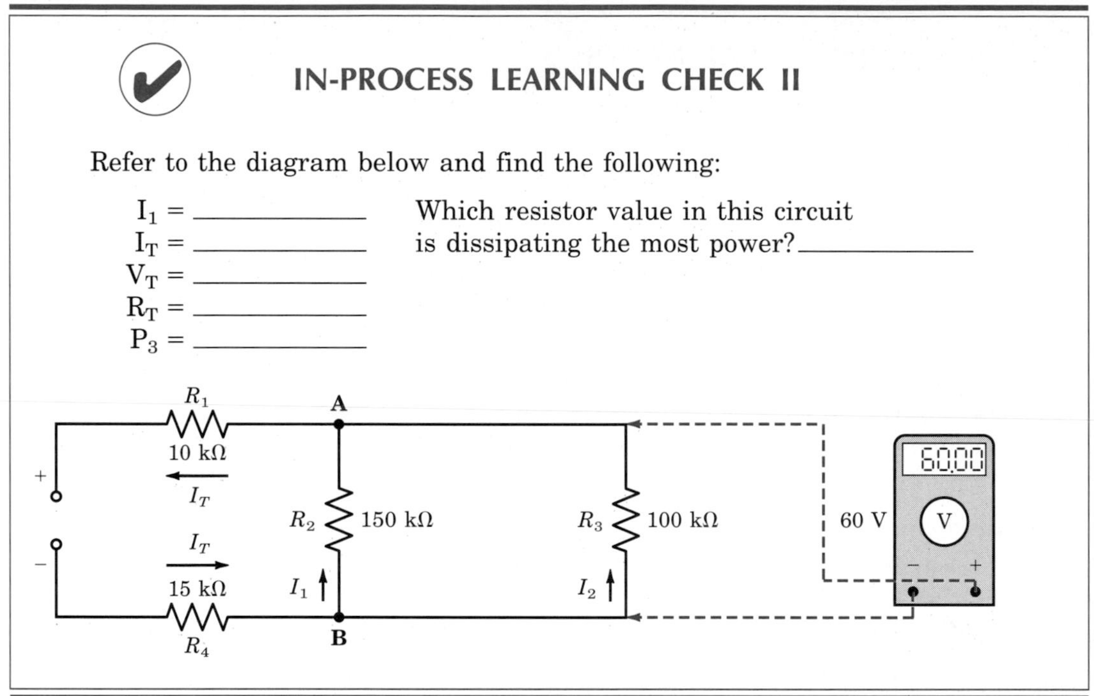

IN-PROCESS LEARNING CHECK II
Refer to the diagram below and find the following:
I1 = ______
IT = ______
VT = ______
RT = ______
P3 = ______
Which resistor value in this circuit is dissipating the most power? ______
R1
10 kΩ
A
IT
R2
150 kΩ
I1
R3
100 kΩ
I2
60 V
60.00
V
15 kΩ
R4
B

EFFECTS OF OPENS IN SERIES-PARALLEL CIRCUITS AND TROUBLESHOOTING HINTS

Effects

As you have seen, the electrical location of a given component in a series-parallel circuit has great influence on all its electrical parameters. The location of an open in the circuit configuration also has a strong influence on all the parameters throughout the circuit.

Looking at the circuit used in In-Process Learning Check II, you can see that if either R_1 or R_4 open, then the path for total current is broken. Thus, all other components' voltage drops, currents, and power dissipations decrease to zero!

On the other hand, if R_3 opened, total current decreases, since the total resistance increases from 85 kΩ to 175 kΩ. But current does not decrease to zero. Thus, individual component voltages, currents, and power dissipations change; however V applied remains the same.

Here is a **Special thinking exercise** for you. Again refer to the In-Process Learning Check circuit and assume R_2 opens. Will the following parameters increase, decrease, or remain the same? See Appendix B for answers.

V_T will ______ I_T will ______

V_1 will ______ I through R_1 will ______

V_2 will ______________ I through R_2 will ______________

V_3 will ______________ I through R_3 will ______________

V_4 will ______________ I through R_4 will ______________

P_T will______________ P_3 will______________ P_4 will______________

Total power and total current should decrease because total resistance increases while V applied remains the same.

Voltage drops across R_1 and R_4 also decrease, since they now have less current through them and their resistances remain unchanged.

Voltage across R_2 and R_3 increases because the series components (R_1 and R_4) are dropping less of the applied voltage, leaving more voltage to be dropped across the parallel combination of R_2 and R_3, Points A–B.

Current decreases through R_1 and R_4 (remember I_T decreased); however, current through the unopened parallel branch, R_3, increases since current is no longer splitting between R_2 and R_3. Obviously, current through R_2 decreases to zero, since it is an open path.

Finally, power dissipations change. P_T, P_1, and P_4 decrease due to the drop in I_T, and P_3 increases because its current increases.

By now you should realize logical thinking is required when analyzing the sequence of events and the effects of an open in a series-parallel circuit. There is not just one simple approach. Rather, think of the interaction between the various portions of the circuit on each other and on the total circuit parameters.

TROUBLESHOOTING HINTS

Because the component's position in the circuit is important due to the interaction it has on the remaining circuitry, the hints given here are somewhat general.

If an open occurs in *any* component in the circuit, total resistance increases; consequently, total current decreases. If you measure (or monitor) total current and it is lower than normal with normal applied voltage, an open, or at least an increased resistance value, has occurred somewhere in the circuit.

NOTE: Two typical ways to determine total current in a circuit are shown in Figure 5–16.

a. Insert current meter in series with source and the circuitry.

b. Measure voltage across a known resistor value that carries total current, and use Ohm's law $\left(\frac{V}{R}\right)$ to determine current.

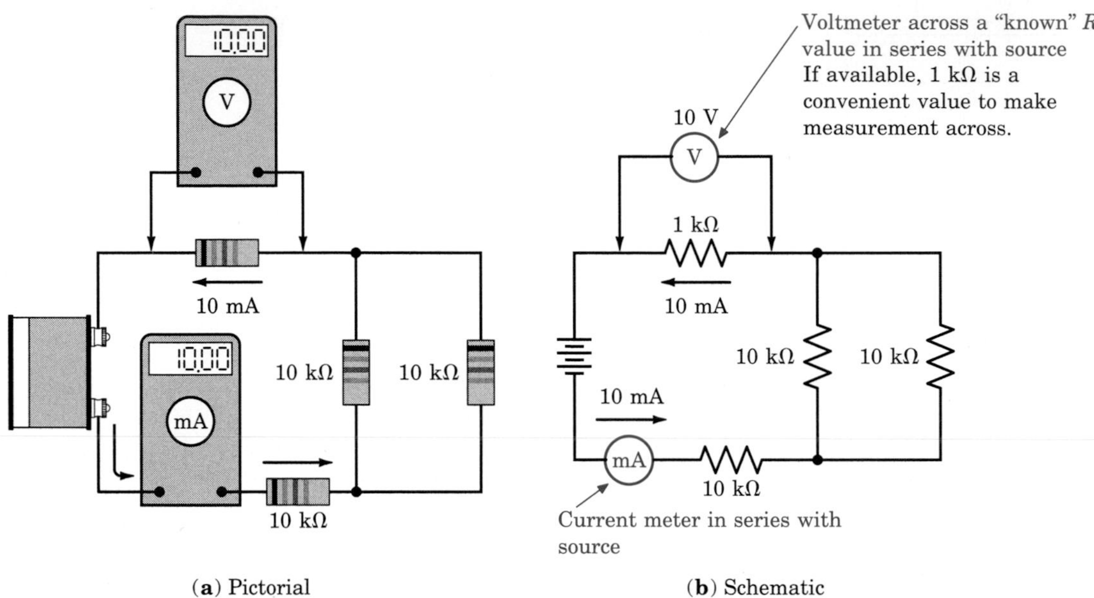

(a) Pictorial (b) Schematic

FIGURE 5–16 Two ways of checking total current

PRACTICAL NOTES

A convenient resistance value for this approach is a 1 kΩ resistor, if available, because its voltage drop equals the number of mA through it. In other words, $\frac{10\text{ V}}{1\text{ k}\Omega} = 10\text{ mA}$, Figure 5–16.

1. If an open occurs in any component in series with the source, total R increases to ∞, and total I decreases to zero. As you learned in series circuits, V applied appears across the opened series component and zero volts are dropped across the good components.
2. If an open occurs in any component other than those in series with the source, current measurements and/or *common sense* will isolate the problem.
3. If current measurements cannot be easily performed, you can make a number of voltage checks throughout the circuit. Observe where voltages are higher and lower than normal, then logically determine where the open is located.

EFFECTS OF SHORTS IN SERIES-PARALLEL CIRCUITS AND TROUBLESHOOTING HINTS

Effects

Again, the location of the faulty component or circuit portion in the total circuit configuration controls what happens to the parameters of the remaining good components.

Refer to Figure 5–17. If the shorted component is in a line that carries total current, that is, in series with the source, then: 1) total current increases, 2) voltage across the shorted component drops to zero, and 3) all other voltages throughout the circuit increase (except V_T).

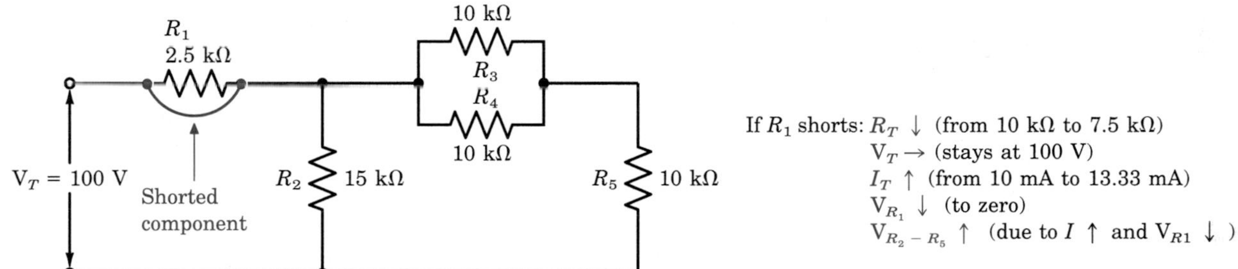

FIGURE 5–17 Effects of a shorted series (in-line) element

Refer to Figure 5–18. If the shorted component is elsewhere in the circuit, then: 1) total current still increases since total R decreases; however, 2) voltage across the shorted component *and* components *directly* in parallel with the shorted component decreases to zero, since the shorted component is shunting the current around components in parallel with it; and 3) other components' voltage drops may increase or decrease, depending on their location.

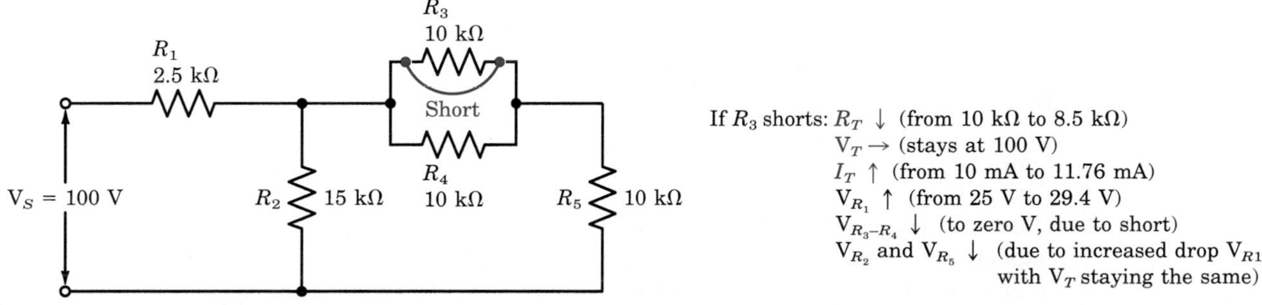

FIGURE 5–18 Effect of shorted element in the combinational portion of a series-parallel circuit

TROUBLESHOOTING HINTS

Some general thoughts about troubleshooting shorts in series-parallel circuits are:

1. A short of *any* component in the circuit causes total resistance to decrease and total current to increase.
2. A higher than normal total current (or lower than normal R_T) implies that either a short or a lowered R has occurred somewhere in the circuit.
3. Voltage measurements across the circuit components reveal close to zero volts across the shorted component and those directly in parallel with it. The good components may have either higher or lower than normal voltage levels, depending on their individual positions within the circuit. Voltage measurement is one simple way to locate the component or combination of components that are shorted.
4. Once the shorted portion of the circuit is located, resistance measurements will isolate the exact component or circuit portion containing the short. Resistance measurement involves turning the power off during measurement and lifting one end of each branch from the circuit.

Let's see if your analysis skills are developing. Try the troubleshooting problems in the following In-Process Learning Check.

IN-PROCESS LEARNING CHECK III

1. Refer to Figure 5–19. Determine the defective component and the nature of its defect.
2. Refer to Figure 5–20. Indicate which component or components might be suspect and what the trouble might be.
3. Refer to Figure 5–21. Indicate which voltages will change and in which direction, if R_2 increases.

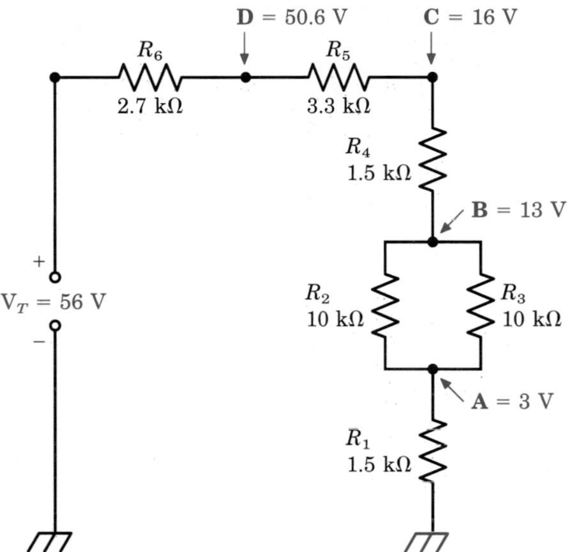

FIGURE 5–19
In-Process Learning Check: Troubleshooting problem #1

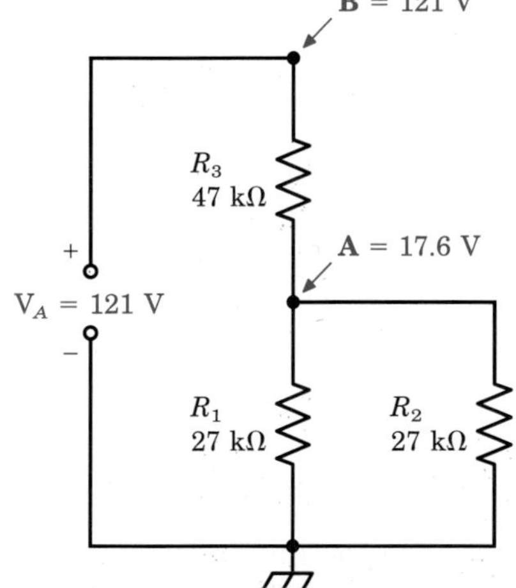

FIGURE 5–20
In-Process Learning Check: Troubleshooting problem #2

DESIGNING A SERIES-PARALLEL CIRCUIT TO SPECIFICATIONS

For a design problem, let's assume we are designing and drawing the circuit diagram for a simple lighting circuit that contains four light bulbs, each nor-

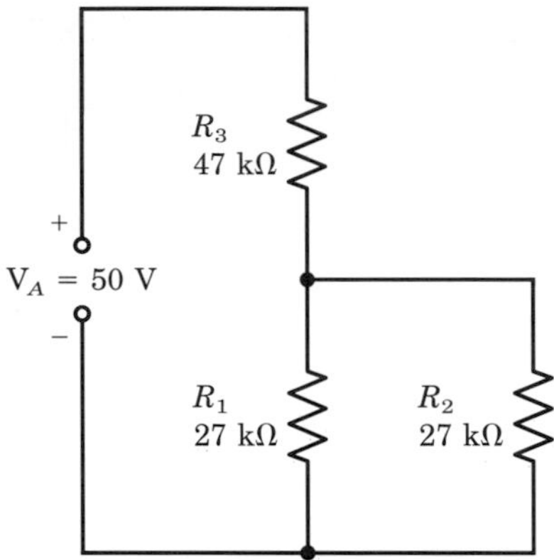

FIGURE 5–21
In-Process Learning Check: Troubleshooting problem #3

mally operates at 10 volts. Using four lights and three SPST switches, configure the circuit so switch 1 acts as an on-off switch for all the lights; switch 2 controls two of the lights' on-off modes (assuming switch 1 is on), and switch 3 controls the remaining lights' on-off conditions (assuming switch 1 is on). The source voltage is 20 volts.

EXAMPLE

The thought processes to solve this problem are as follows:

1. Since the source equals 20 volts and the bulbs are 10-V bulbs, it is logical to assume two bulbs are in series across the 20-volt source.
2. Since there are four lights, it is also logical to assume there are two, two-bulb branches connected to the 20-volt source voltage.
3. Since switch 1 controls the on-off condition of the total circuit connected to the source, it is logical that this switch is in series with the source and connected between the source and the remaining circuitry.
4. Since the circuit specification is to control each of the two bulb branches independently, switches 2 and 3 must be placed in series with the bulbs in each branch, respectively.

Using the above knowns, it is possible to draw the circuit diagram, Figure 5–22.

PRACTICE PROBLEM IV

See if you can design a series-parallel circuit that meets the following conditions:
Given: six 10-kΩ resistors and a 45-volt source, design and draw the circuit diagram where:
V_1 = 20 V; V_2 = 5 V; V_3 = 5 V; V_4 = 5 V; V_5 = 5 V; V_6 = 20 V

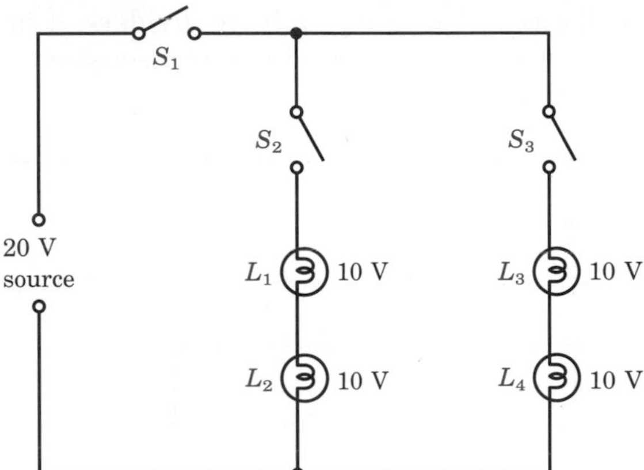

FIGURE 5–22 Sample design problem circuit

LOADED VOLTAGE DIVIDERS

Two applications of series-parallel circuits, commonly encountered by electronic technicians are **loaded voltage dividers** and the **Wheatstone bridge circuit.**

Frequently in electronic circuits, a requirement is to provide different levels of voltage for various portions of the circuit or system. Also, these different portions may demand different load currents. Voltage dividers are often used with power supplies to provide and distribute these different levels of voltages at the different current values demanded.

Refer to Figure 5–23. Recall in the Series Circuits chapter we learned voltage values at different points along the series circuit are controlled by the ratios of the resistances throughout the series string. These voltage dividers are called unloaded voltage dividers because no external circuit loads (current-demanding components or devices) are connected to the voltage divider circuit.

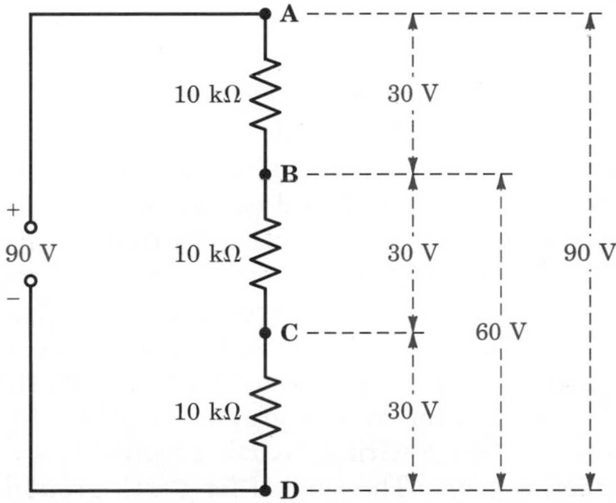

FIGURE 5–23 Unloaded voltage divider

To help in our discussion of loaded voltage dividers, it is appropriate to define some important terminology used in our discussion. Refer to Figure 5–24 as you study these definitions.

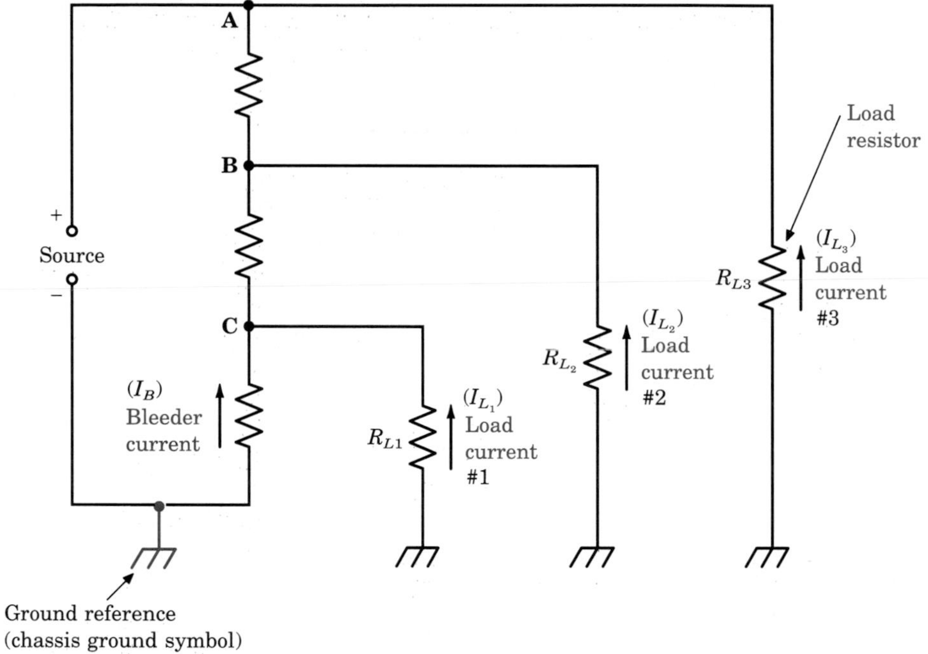

FIGURE 5–24 Terms related to loaded voltage dividers

1. **Load** is a component/device or circuit drawing current from a power source.
2. **Load current** is the current required by the components or circuits that are connected to the power source and/or its output voltage divider.
3. $\mathbf{R_L}$ is the resistance value of the load resistor or component.
4. **Ground reference** is an electrical *reference point* used in defining or measuring voltage values in a circuit.

   Frequently the electrical ground reference point in a circuit is chassis ground that is the common metal conducting path through which many components or circuits are electrically connected to one side of the power supply. In printed circuits, it is a common metal conductor *path* where many components are electrically connected, or "commoned." In circuits built on a metal chassis, the metal chassis is often the "return" path, or common conductor path; hence, the term "chassis ground." The symbol for chassis ground is ⏊. In electrical wiring circuits, such as house wiring, "earth ground" is a common reference and connecting point. The symbol for earth ground is ⏚.

5. **Bleeder current** is considered the fixed current through a resistive network, or a bleeder resistor connected across the output of the power supply. Many power supplies use a bleeder resistor that draws a fixed minimum current from the supply to help regulate the power supply output voltage; that is, keeping it more constant under varying load conditions.

6. **Bleeder resistor** is the resistor, or resistor network, connected in parallel with the power supply circuitry, that draws the bleeder current from the source. Another common function of a bleeder resistor is to discharge power supply capacitors after the circuit is turned off, for safety purposes. You will learn more about this in later chapters that discuss power supply circuits.

7. **Potentiometer** (Figure 5–25) is a three-terminal resistive device used as a voltage-dividing component. One terminal is at one end of the resistive element, a second contact is at the other end of the resistive element, and the third contact is a "wiper arm" that can move to any position along the resistive element. The position of the wiper arm along the resistive element determines voltage from it to either end contact.

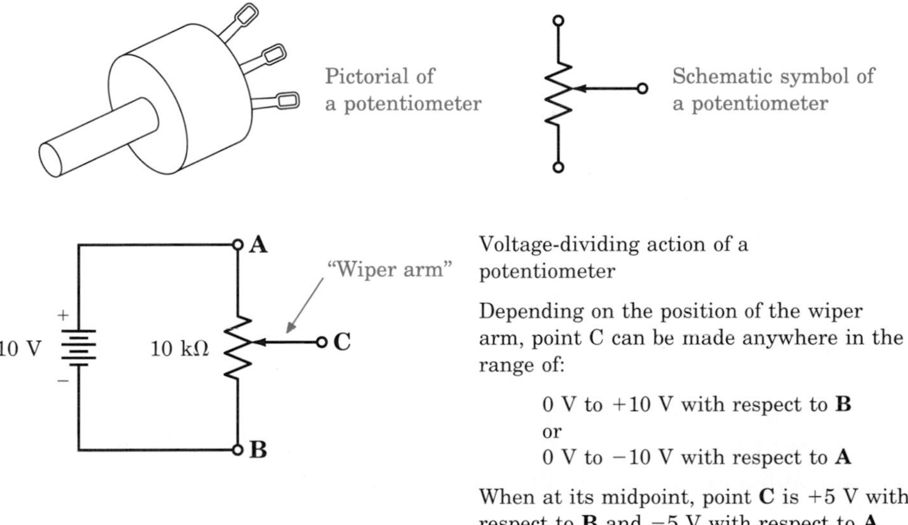
Pictorial of
a potentiometer
Schematic symbol of
a potentiometer
A
"Wiper arm"
+
10 V
−
10 kΩ
C
B
Voltage-dividing action of a
potentiometer
Depending on the position of the wiper
arm, point C can be made anywhere in the
range of:
0 V to +10 V with respect to B
or
0 V to −10 V with respect to A
When at its midpoint, point C is +5 V with
respect to B and −5 V with respect to A.

FIGURE 5–25 The potentiometer

With these terms in mind, let's look at a brief analysis of a typical loaded voltage divider circuit.

Two-Element Divider with Load

Observe Figures 5–26a and 5–26b as you study the following discussion. Let's think through the circuit.

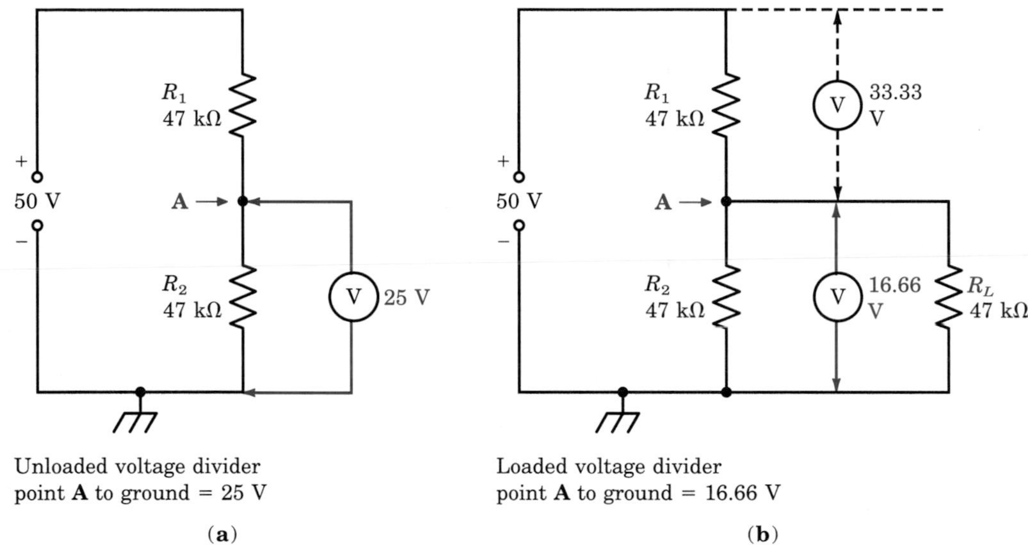

FIGURE 5–26 Two-element divider circuit

What causes voltage between point A and ground reference to decrease from 25 volts to about 16.7 volts when load resistor R_L is connected? Before the load was connected, the two-series-connected equal value resistors divided the input voltage equally (25 V each) since they were passing the same current.

When the load resistor was connected, a new branch path for current was created through R_L. The current through R_L must also pass through R_1 to return to the source's positive side. Now R_1 has the original unloaded voltage divider current through it, *plus* the load current of R_L.

If the current through R_1 increases and its R value is unchanged, then its I × R drop must increase. In this case it increased from 25 volts to about 33.33 volts. Thus, only 16.66 volts remain to be dropped from point A to ground compared to 25 volts previously dropped before the load was connected.

Three-Element Divider with Multiple Loads

Observe in Figure 5–27a and 5–27b the changes caused by connecting a load to a three-element voltage divider are similar to those changes that occur in a two-element divider.

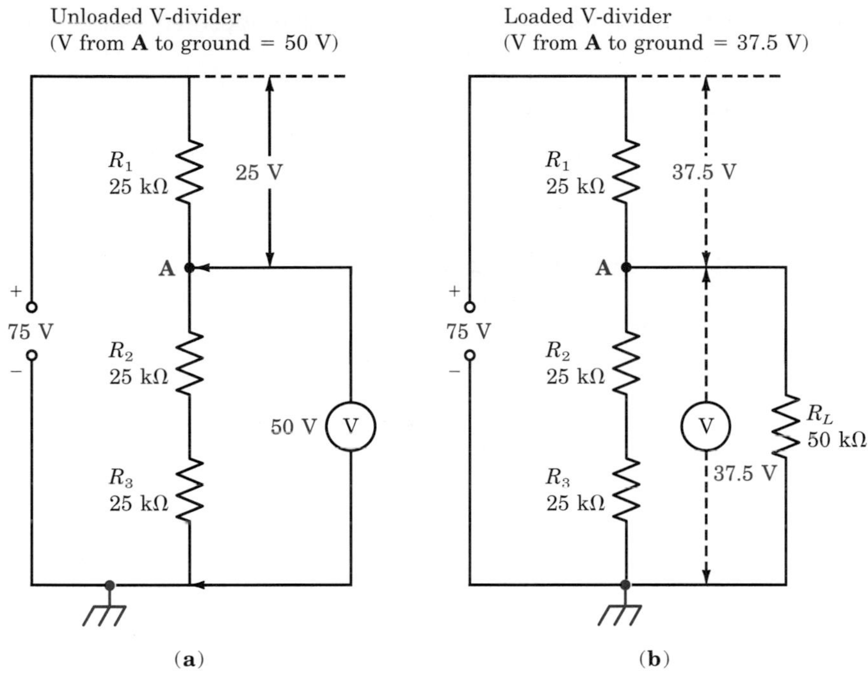

FIGURE 5–27 Comparison of loaded and unloaded voltage dividers (Effect of connecting a load to a voltage divider circuit)

Let's look at the three-element loaded voltage divider in Figure 5–27b from the perspective of Ohm's law and Kirchhoff's voltage law.

R_T is solved by using your knowledge of series and parallel circuits. R_L is in parallel with the series combination R_{2-3}. This indicates the R_L value of 50 kΩ is in parallel with 50 kΩ so the equivalent resistance from point A to ground equals 25 kΩ. This 25 kΩ is in series with the R_1 value of 25 kΩ yielding a total circuit resistance of 50 kΩ.

Solve I_T using Ohm's law, where $I_T = \dfrac{V_T}{R_T} = \dfrac{75\text{ V}}{50\text{ k}\Omega} = 1.5\text{ mA}.$

Since I_T passes through R_1, $V_1 = I_T \times R_1 = 1.5\text{ mA} \times 25\text{ k}\Omega = 37.5$ volts.

Since point A to ground is composed of two equal 50 kΩ branches (R_L and R_{2-3}), total current evenly divides between the branches. This means 0.75 mA passes through R_L and 0.75 mA passes through R_{2-3}. The R_L voltage drop $= 0.75\text{ mA} \times 50\text{ k}\Omega = 37.5$ V; R_2 voltage drop $= 0.75\text{ mA} \times 25\text{ k}\Omega = 18.75$ V; R_3 voltage drop also $= 0.75\text{ mA} \times 25\text{ k}\Omega = 18.75$ V; therefore, $V_{2-3} = 37.5$ V.

Of course, by Kirchhoff's voltage law (the sum of the voltage drops around a closed loop must equal V applied), you know the voltage from point A to ground has to be 37.5 volts. Since the calculated R_1 drop is 37.5 volts, the remainder of V applied has to drop from A to ground reference.

Let's move one step further in studying loaded voltage dividers by performing a design and analysis problem to achieve specified voltage levels and load currents.

EXAMPLE

Refer to Figure 5–28 for the desired parameters of the circuit to be designed. Note the voltage-dividing system provides voltage and current levels as follows:

Load 1 (R_{L_1}) = 25 V at 12.5 mA
Load 2 (R_{L_2}) = 75 V at 25 mA
Load 3 (R_{L_3}) = 225 V at 50 mA

Bleeder current (I_B) = 12.5 mA

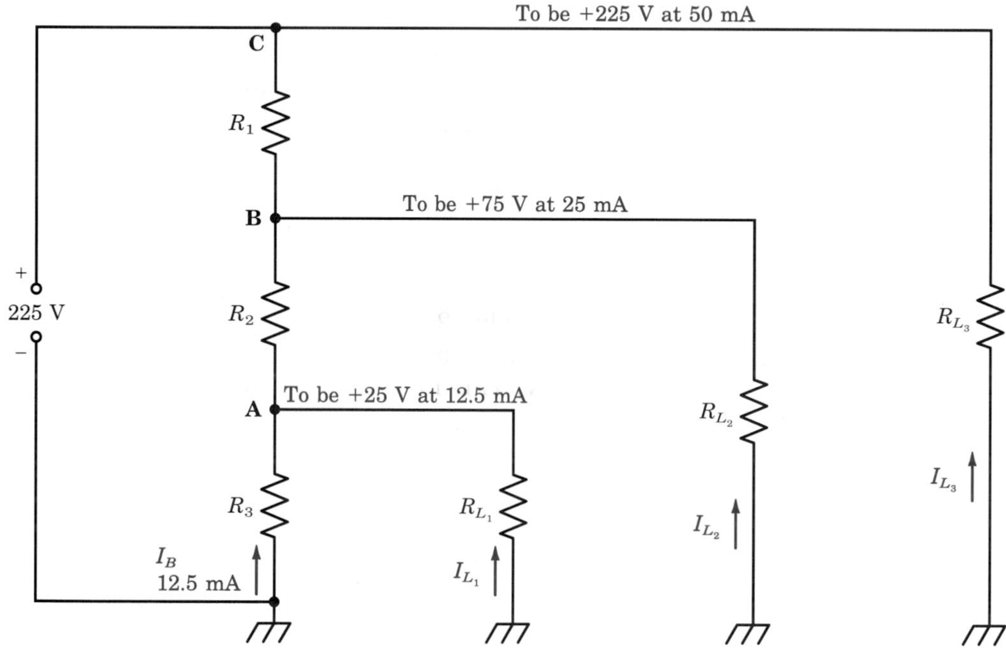

FIGURE 5–28 Analyzing and designing a loaded voltage divider circuit

For additional information we'll make one task to determine the resistance values of loads 1, 2, and 3. Using the parameters provided, this is not critical to the design tasks but will be of interest.

Since you know the desired voltage across each load and the currents passing through them, simply apply Ohm's law to find load resistance values.

$$R_{L_1} = \frac{25\text{ V}}{12.5\text{ mA}} = 2\text{ k}\Omega.$$

$$R_{L_2} = \frac{75\text{ V}}{25\text{ mA}} = 3\text{ k}\Omega.$$

$$R_{L_3} = \frac{225\text{ V}}{50\text{ mA}} = 4.5\text{ k}\Omega.$$

The second task, a design task, is to determine the values of voltage divider resistors R_1, R_2, and R_3 that will provide the appropriate voltages to the loads.

Again refer to Figure 5–28 as we examine the design task as follows:

1. Looking at the circuit, you can see that none of the load currents pass through R_3. This means only the given bleeder current (I_B) of 12.5 mA is passing through R_3. Because its voltage must equal the load 1 voltage of 25 V since they are in parallel, then $R_3 = \frac{25\text{ V}}{12.5\text{ mA}} = 2\text{ k}\Omega$.

2. The R_2 current must equal the 12.5-mA bleeder current, *plus* the 12.5-mA load 1, or 25 mA. The voltage dropped by R_2 must equal 75 V minus the voltage dropped by R_3. Thus, V_{R_2} equals 50 V.

   Therefore, $R_2 = \frac{50\text{ V}}{25\text{ mA}} = 2\text{ k}\Omega$.

3. The R_1 current must equal the bleeder current plus load 1 current plus load 2 current. Therefore, the current through R_1 equals 12.5 mA (bleeder current) plus 12.5 mA (I_{L_1}) plus 25 mA (I_{L_2}) equals 50 mA. Thus, R_1 voltage drop must equal 225 V minus the drop from point B to ground, or 225 V − 75 V = 150 V. This means the value of R_1 must $= \frac{150\text{ V}}{50\text{ mA}} = 3\text{ k}\Omega$. Look at Figure 5–29 for the completed design.

A Practical Voltage Divider Variation

Look at the circuit in Figure 5–30. With the potentiometer's wiper arm at the top (position 1), the 20 kΩ R_L is in parallel with total potentiometer resistance of 30 kΩ, yielding an equivalent resistance of 12 kΩ from point A to ground. The circuit voltage-dividing action is dividing the 200 volts across the 40-kΩ total circuit resistance consisting of the 28-kΩ resistor and the 12-kΩ parallel circuit equivalent resistance in series. Therefore, R_L drops twelve-fortieths of 200 V, or 60 V. The 28-kΩ resistor drops the remaining 140 V.

On the other hand, if the wiper arm is at position 2 (or ground), both ends of R_L will connect to ground. Therefore, the voltage across R_L is zero V. This means that with the potentiometer, we can vary the voltage across R_L from zero to sixty volts. As you will see throughout your electronics experience, this "variable-voltage-divider" capability of the potentiometer is convenient.

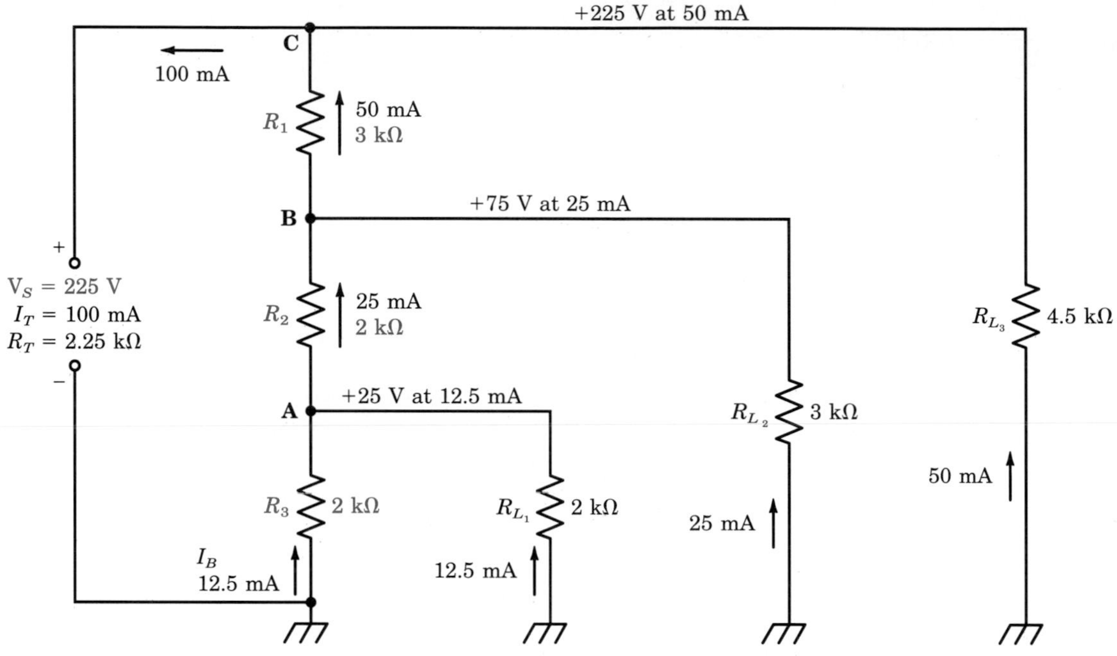

FIGURE 5–29 Voltage divider problem solution

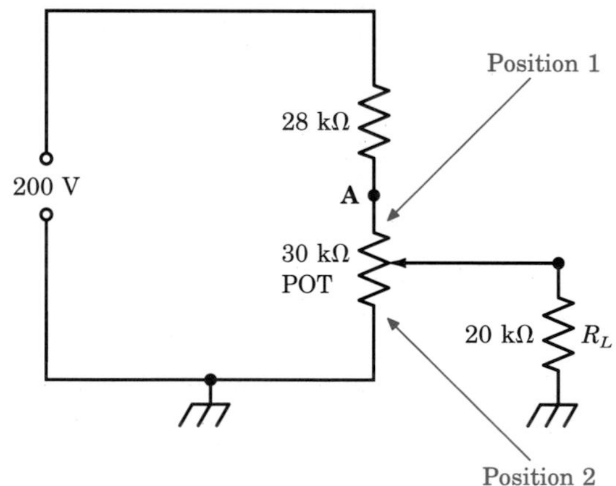

FIGURE 5–30 Example of potentiometer varying voltage output

The potentiometer enables varying V across R_L from 0 to 60 V

PRACTICE PROBLEM V

Refer to Figure 5–31 and answer the following:

1. What is the value of voltage across R_L? How much power is dissipated by R_L?

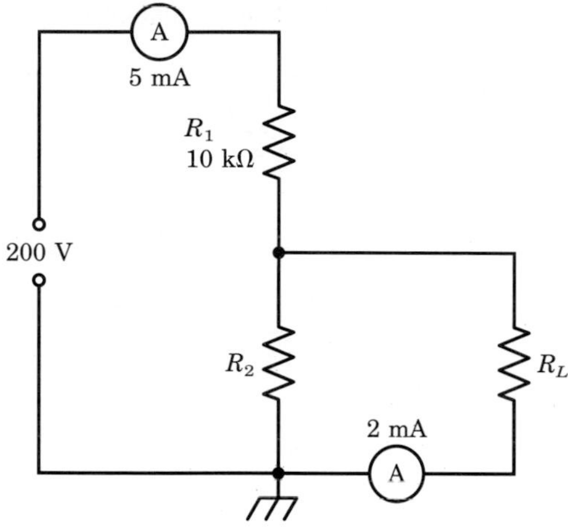

FIGURE 5–31

2. What is the value of R_L?
3. What is the value of R_2?
4. If V_{RL} suddenly decreases due to a change in the R_2 value, does this indicate R_2 has increased or decreased in value?
5. If the decrease in V_{RL} in question 4 was caused by a change in R_1 value, would it indicate R_1 has increased or decreased in value?
6. If R_L is disconnected from the voltage divider in the circuit, what is the new value of V_{R_2}?
7. What happens to V_{RL} if R_1 opens? If R_2 shorts?

THE WHEATSTONE BRIDGE CIRCUIT

Various types of special series-parallel circuits, called **bridge circuits,** are used to make measurements in electronic circuits, Figure 5–32.

Note that zero volts appear between points A and B when the bridge is balanced. You will see this balance *only* when the ratio of the resistances in the left arm and in the right arm of the bridge are equal.

For example, observe Figure 5–32. In the left arm of the bridge, R_1 = 10 kΩ and R_2 = 20 kΩ. In the right arm of the bridge, R_3 = 50 kΩ and R_4 = 100 kΩ. If V applied is 150 volts, then R_1 drops one-third the voltage across the left arm of the bridge, or 50 V (since it is one-third of the left arm's total resistance). R_2 drops the other two-thirds of the 150 volts, or 100 V.

What about the other arm of the bridge? The same 150 volts is applied across that arm's 150 kΩ, with R_3 (50 kΩ) dropping one-third the applied

FIGURE 5–32 A balanced bridge circuit exists when $\frac{R_1}{R_2} = \frac{R_3}{R_4}$.

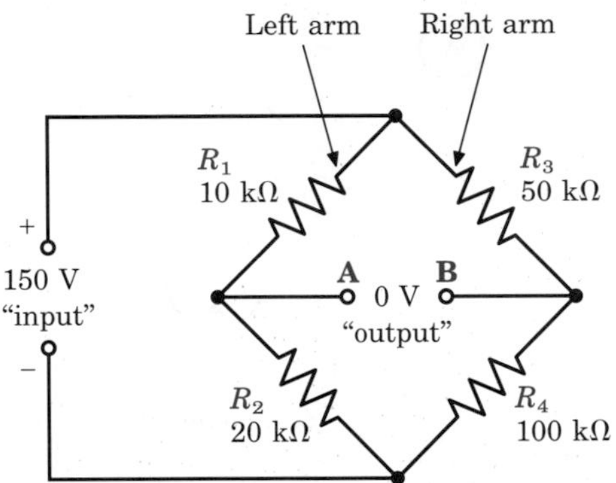

voltage, or 50 V, and R_4 dropping two-thirds the applied voltage (R = two-thirds the right arm's total R), or 100 V.

Note the voltage differential between points A and B is zero. This is because point A is at a potential of +100 volts with respect to the power supply's negative terminal. Point B is also at a +100 volts with respect to the power supply's negative terminal. Thus, the difference of potential between point A and B is zero volts, since they are at the same potential.

This balanced condition (zero volts between points A and B) always exists whenever the resistance ratios of the top resistor to the bottom resistor in the left and right arm of the bridge are the same. That is,

Formula 5–1	Balanced when: $\frac{R_1}{R_2} = \frac{R_3}{R_4}$

NOTE: The values of the resistors do not matter. It's the *ratio* of the resistances in each arm that determines the voltage distribution throughout that arm. Recall the series circuit concepts. If both arms have equal voltage distribution, then the midpoint of each arm is at the same potential and the bridge is balanced.

The **Wheatstone bridge circuit,** a special application of the bridge circuit, measures unknown resistance values. When the bridge circuit is balanced, the output terminals (points A and B) have zero potential difference between them. On the other hand, if the bridge circuit is unbalanced, there is a difference of potential between the output terminals. Refer to Figure 5–33 as you read the following discussion.

One method of using the Wheatstone bridge circuit to determine an unknown resistance is to:

1. Carefully select R_1 and R_2 to be matched (equal) values.
2. Place a sensitive "zero-center-scale" current meter between points A and B, which are the output terminals of the bridge circuit.
3. Use a calibrated variable resistance, such as a "resistance decade

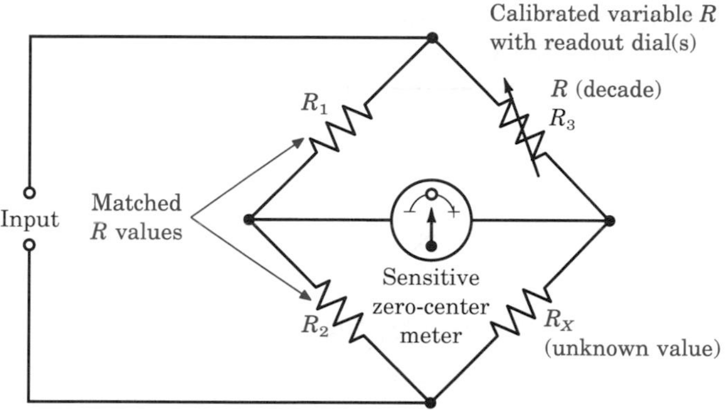

FIGURE 5–33 Concepts of the Wheatstone bridge circuit

box," as R_3. NOTE: This type variable resistance has calibrated dials to help you read its resistance setting, Figure 5–34.

4. Place the unknown resistor (R_X) in the same arm of the circuit as the calibrated variable resistance.
5. Since R_1 and R_2 are equal value in the left arm, then the bridge is balanced when the calibrated variable resistance and the unknown resistance equal each other in the right arm. In this case, the sensitive current meter's pointer is pointing at "0" in the center of the scale. If the bridge is unbalanced, the pointer points to the left of center, indicating current through the meter in one direction, or to the right of center, indicating current in the other direction. The pointer's direction is dependent on the polarity of the voltage difference between points A and B, as determined by the comparative voltages at points A and B in the bridge network.

FIGURE 5–34 Decade resistor box used as a calibrated R value

6. Once the calibrated variable resistance is adjusted for zero current reading (balanced), the value of the unknown resistor is read directly from the calibrated variable resistance's dial(s).

Another method to determine the unknown resistance value is to use the known resistance values and the concepts of the balanced bridge circuit.

1. Recall balance occurs when $\frac{R_1}{R_2} = \frac{R_3}{R_4}$.
2. With these relationships, it is mathematically shown that:

   $R_1 \times R_4 = R_2 \times R_3$ (cross-multiplying factors in formula)
3. If R_4 is the unknown resistance, (R_X), the formula can be rearranged to solve R_X as follows:

Formula 5–2	$R_X = \frac{R_2 \times R_3}{R_1}$

4. Look at Figure 5–35. Let's assume some values to see if the steps described above will work.

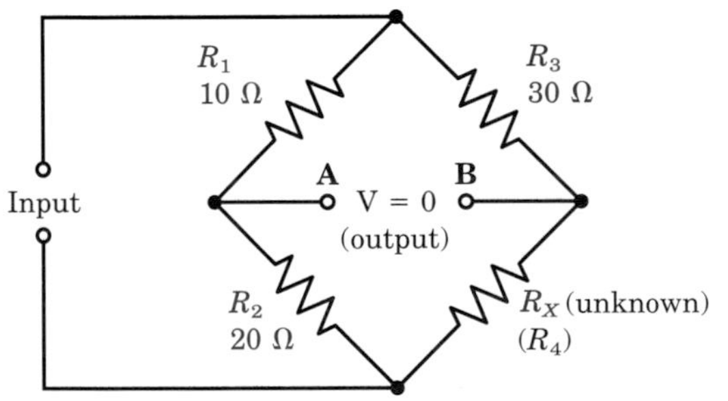

1. Balance: $\frac{R_1}{R_2} = \frac{R_3}{R_4}$; therefore $\frac{10}{20} = \frac{30}{R_X}$
2. Rearranging: $R_1 \times R_4 = R_2 \times R_3 \therefore 10\,R_X = 600\ \Omega$
3. Solving for R_X: $R_X = \frac{R_2 \times R_3}{R_1} = \frac{20 \times 30}{10} = \frac{600}{10} = 60\ \Omega$

FIGURE 5–35 The calculation approach may be used to solve for the R_X value in a bridge circuit.

a. Balance: $\frac{R_1}{R_2} = \frac{R_3}{R_X} = \frac{10}{20} = \frac{30}{R_X}$

b. Rearranging: $R_1 \times R_X = R_2 \times R_3 = 10 \times R_X = 20 \times 30$

c. Solving for R_X: $R_X = \frac{R_2 \times R_3}{R_1} = \frac{600}{10} = 60\ \Omega$

5. Verifying results by substituting our answer into the original balanced bridge formula: $\frac{R_1}{R_2} = \frac{R_3}{R_X} = \frac{10}{20} = \frac{30}{60} = \frac{1}{2}$. (It checks out!)

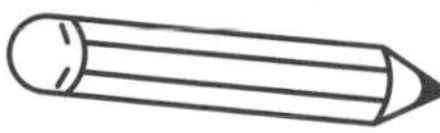

PRACTICAL NOTES

Because series-parallel circuits are combinational circuits that are connected in an infinite variety of configurations, no single analysis technique is applicable. Rather, it is necessary to break the circuit down into its series and parallel portions for analysis.

In studying this chapter, you have learned the series portions of series-parallel circuits are analyzed using series-circuit concepts, where 1) current is the same through series components and/or circuit portions, and 2) voltage drops on series components are proportional to their resistance ratios.

Also you have learned that the parallel portions of the series-parallel circuits abide by the rules of parallel circuits, where 1) voltage is the same across parallel components or combinations, and 2) current divides in inverse proportion to the parallel branch resistances.

Furthermore, you have observed that the position (location) of a given component or circuit portion in the circuit influences all the parameters throughout the circuit. Therefore, to troubleshoot series-parallel circuits, be sure to isolate individual circuit portions using concepts appropriate to that circuit portion. Then move to the next portion, and so forth, until the faulty circuit portion is found. Isolation techniques are then used in that portion to isolate the specific faulty component or device within that portion.

SUMMARY

- A series-parallel circuit is a combination of series and parallel connected components and/or circuit portions.
- Series components, or circuit portions are analyzed using the rules of series circuits.
- Parallel components, or circuit portions are analyzed using the rules of parallel circuits.
- Points throughout the circuit where current divides and rejoins help identify parallel portions.
- Points throughout the circuit that share common voltage also help identify components, or circuit portions that are in parallel.
- Paths in the circuitry that carry the same current help identify the series circuit elements.
- Analysis of total circuit resistance is approached by starting away from the source, then solving and combining results as one works back toward the source. (NOTE: If V_T

CHAPTER CHALLENGE

Following the SIMPLER troubleshooting sequence, find the problem with this circuit. As you follow the sequence, record your responses to each step on a separate sheet of paper.

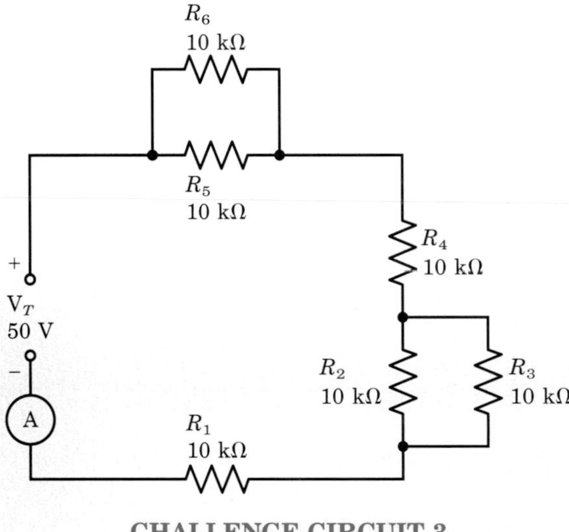

CHALLENGE CIRCUIT 3

Starting Point Information

1. Circuit diagram
2. I_T measures 1.4 mA
3. V_{R_1} measures 14 V
4. Rs measured with remainder of circuit being isolated from points being tested.

Test	Results in Appendix C
V_{R1}	(49)
V_{R2}	(80)
V_{R3}	(12)
V_{R4}	(39)
V_{R5}	(61)
V_{R6}	(88)
Res. of R_1	(25)
Res. of R_2	(54)
Res. of R_3	(93)
Res. of R_4	(37)
Res. of R_5	(68)
Res. of R_6	(33)

SYMPTOMS Gather, verify, and analyze symptom information. (Look at the "Starting Point Information.")

IDENTIFY Identify initial suspect area for the location of the trouble.

MAKE Make a decision about "What type of test to make" and "Where to make it." (To simulate making a decision about each test you want to make, select the desired test from the "TEST" column listing below.)

PERFORM Perform the test. (Look up the result of the test in Appendix C. Use the number in the "RESULTS" column to find the test result in the Appendix.)

LOCATE Locate and define a new "narrower" area in which to continue troubleshooting.

EXAMINE Examine available information and again determine "what type of test" and "where."

REPEAT Repeat the preceding analysis and testing steps until the trouble is found. What would you do to restore this circuit to normal operation? When you have solved the problem, compare your results with those shown in the color insert.

and I_T are known, this is unnecessary. Use $R_T = \frac{V_T}{I_T}$ formula).

- ▶ Analysis of current throughout the circuit is approached by finding the total current value. Then working outward from the source, analyze currents through various circuit portions.
- ▶ Power dissipations throughout the circuit are analyzed by finding the appropriate I, V, and R parameters. Then use the appropriate power formulas, such as $V \times I$, I^2R, or $\frac{V^2}{R}$.
- ▶ A short, or decreased value of resistance, *anywhere* in the circuit causes R_T to decrease and I_T to increase.
- ▶ An open, or increased value of resistance *anywhere* in the circuit causes R_T to increase and I_T to decrease.
- ▶ Two special applications of series-parallel circuits are voltage dividers and bridge circuits used for various purposes.

REVIEW QUESTIONS

1. Refer to Figure 5–36 and determine R_T.

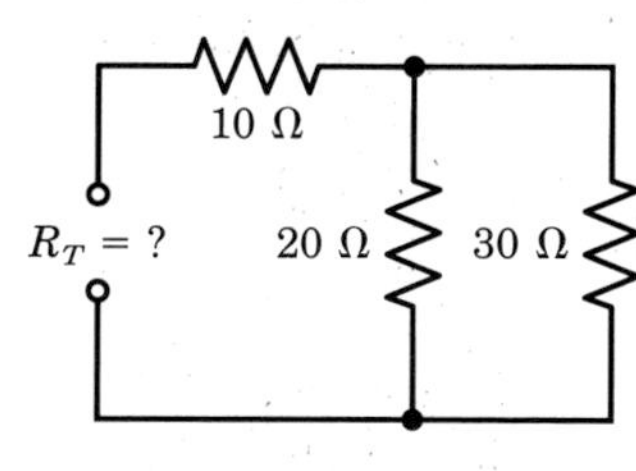

FIGURE 5–36

2. Refer to Figure 5–37 and find:

I_T ________ V_3 ________ I_2 ________

V_2 ________ V_5 ________ I_3 ________

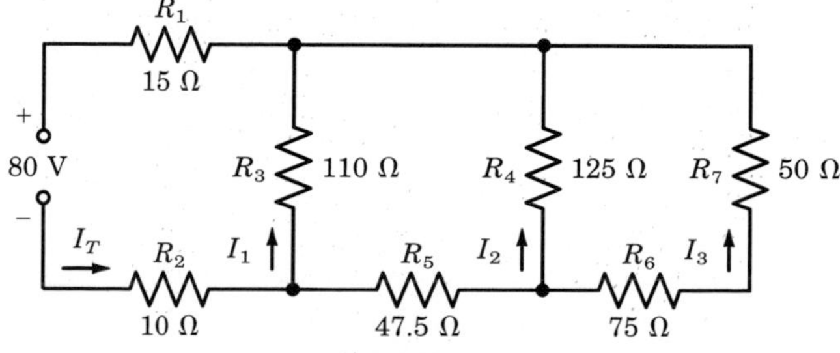

FIGURE 5–37

3. Refer to Figure 5–38 and answer the following with I, for increase; D, for decrease; or RTS, for remains the same. Assume that R_5 shorted:

   R_T will ________

   R_4 will ________

   V_1 will ________

   V_3 will ________

   P_T will ________

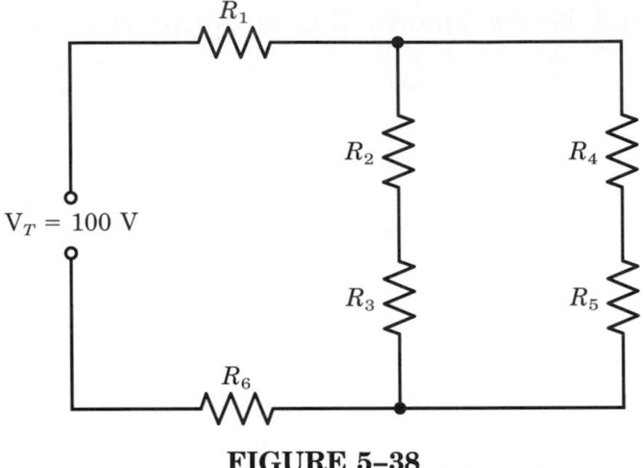

FIGURE 5–38

4. In Figure 5–39, what is the maximum total current possible?

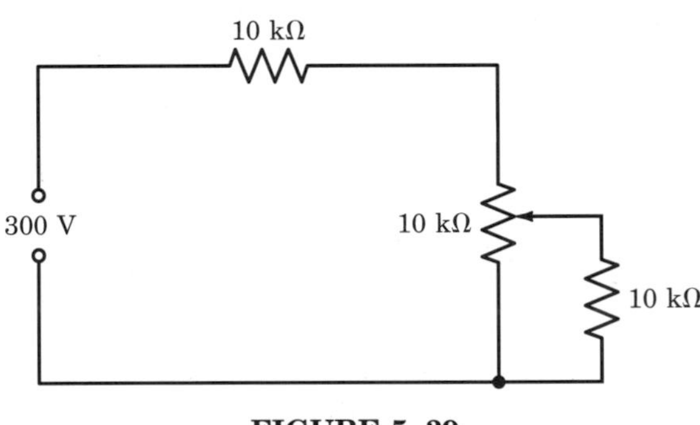

FIGURE 5–39

5. In Figure 5–40, what are the minimum and maximum voltages possible from point A to ground, if the setting of the potentiometer is varied? From ________ V to ________ V

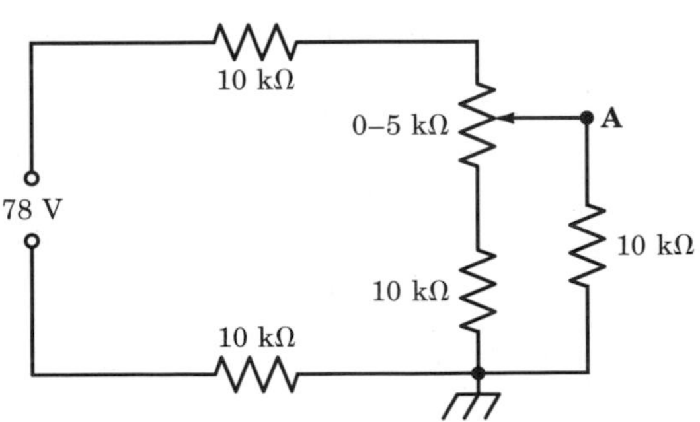

FIGURE 5–40

6. In Figure 5–41, I_T is 10 amperes. Find R_1 and P_{R2}.

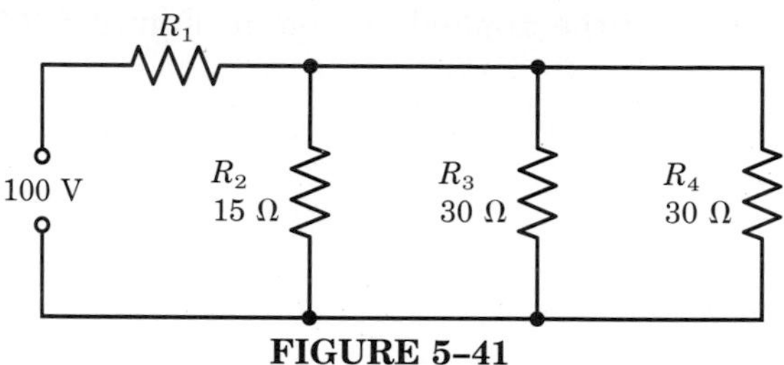

FIGURE 5–41

7. In Figure 5–42, find the value of R_4 and I_2.

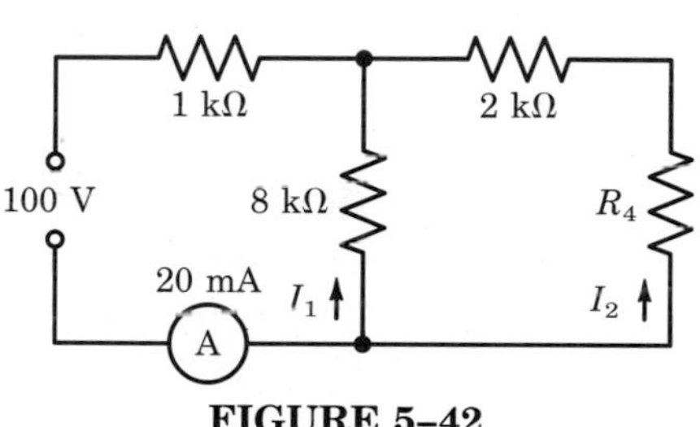

FIGURE 5–42

8. In Figure 5–43, find the resistance between points A and B.

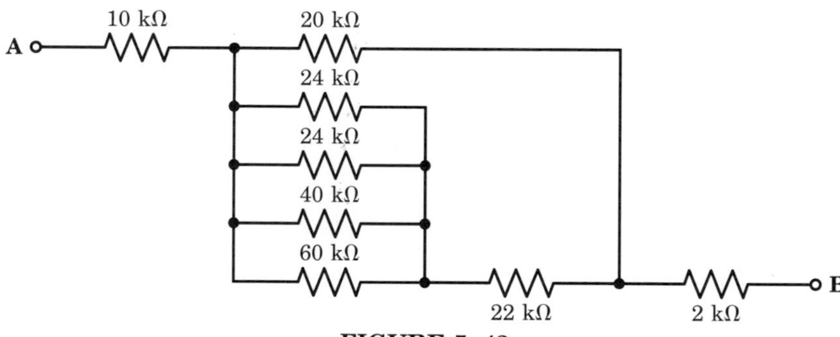

FIGURE 5–43

9. In Figure 5–44, will the light get brighter, dimmer, or remain the same after the switch is closed?

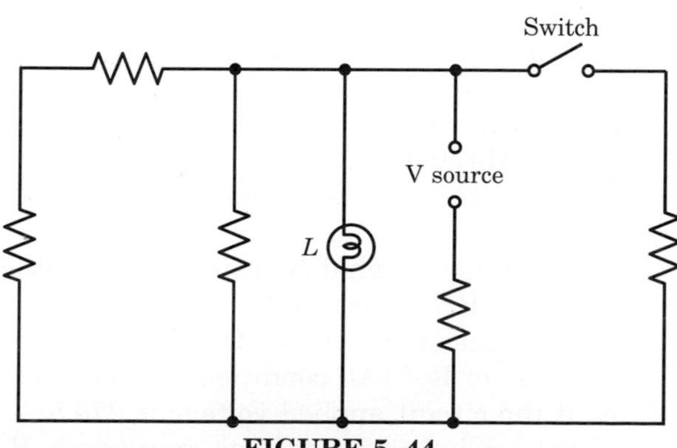

FIGURE 5–44

10. Find the applied voltage in Figure 5–45.

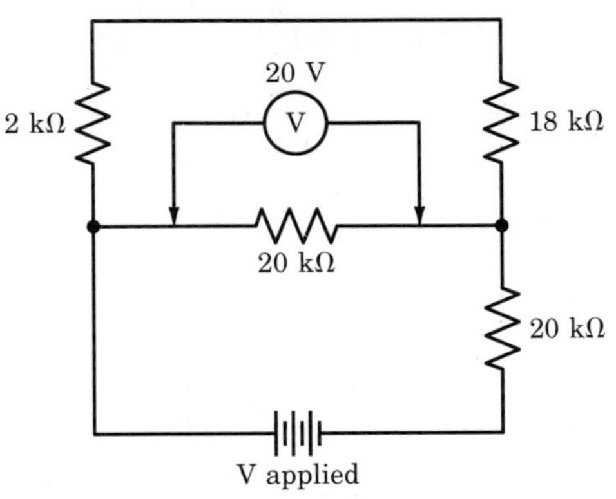

FIGURE 5–45

11. Accurately copy the diagram, Figure 5–46, below on a separate sheet of paper. When you are done, draw arrows on the diagram showing the direction of current through each resistor. Using this circuit, assume all resistors have a value of 100 k ohms. Answer the following questions. (Round answers to whole numbers.)
    a. Find R_T.
    b. Find I_T.
    c. Find V_{R_1}.
    d. Find V_{R_2}.
    e. Find V_{R_3}.
    f. Find V_{R_4}.
    g. Find V_{R_5}.
    h. Find V_{R_6}.
    i. Find V_{R_7}.
    j. Find V_{R_8}.

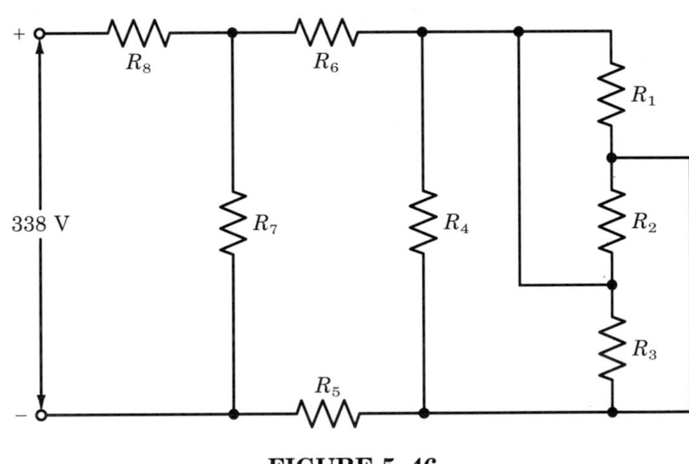

FIGURE 5–46

12. Refer to the circuit in Figure 5–47 on the next page and answer the following questions:
    a. If R_1 increases, what happens to the current through R_5?
    b. If R_7 decreases, what happens to the voltage across R_9?
    c. If R_4 shorts, what happens to the voltage across R_2? Across R_8?
    d. If all resistors are 10 kΩ, what is the value of R_T? (All conditions are normal!)
    e. If the circuit applied voltage is 278.5 volts, what is the voltage drop across R_1?

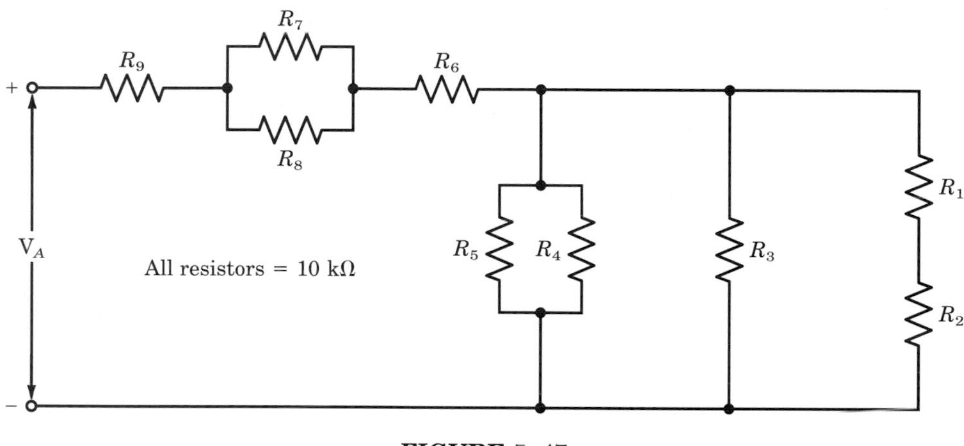

FIGURE 5–47

f. If the circuit applied voltage is reduced to half the value shown in question e, what is the power dissipation of R_9?
g. If the circuit applied voltage is 557 volts, which resistor(s) dissipate the most power?
h. Regardless of the applied voltage, which resistor(s) dissipate the least power?
i. For the conditions defined in question g, what power is dissipated by the resistor dissipating the most power? What power is dissipated by the resistor dissipating the least power?

TEST YOURSELF

1. Determine the values of R_1, R_2, R_3 and R_4 for the circuit, Figure 5–48.

2. Determine the value of R_X in the circuit, Figure 5–49: Assume the bridge is balanced by the variable R set at 100 kΩ.

FIGURE 5–48

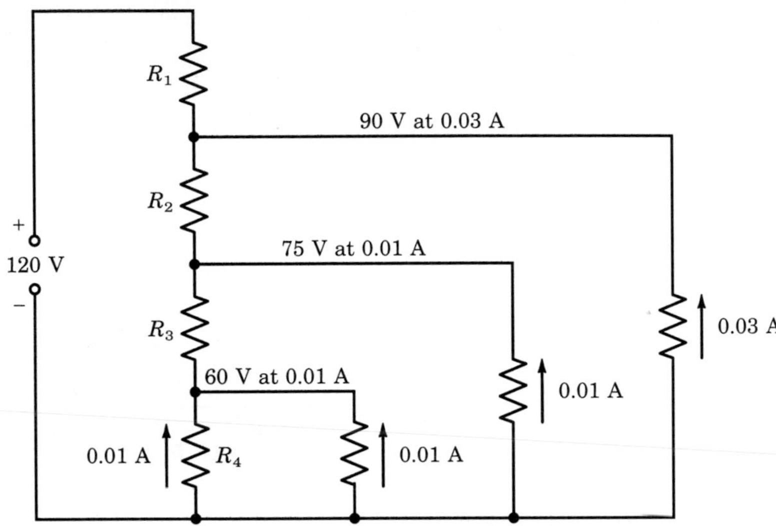

FIGURE 5–49

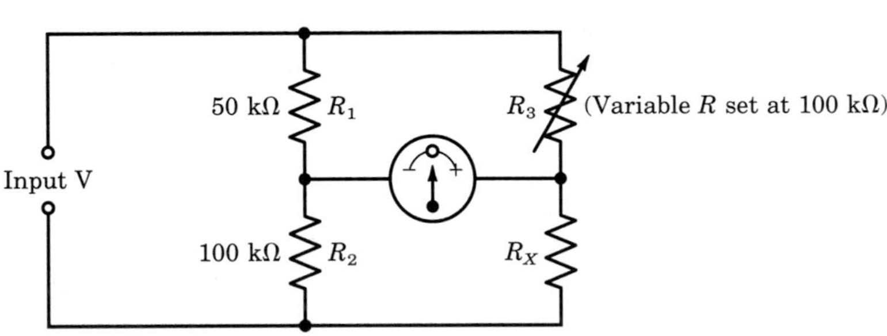

PERFORMANCE PROJECTS CORRELATION CHART

Suggested performance projects in the Laboratory Manual that correlate with topics in this chapter are:

CHAPTER TOPIC	PERFORMANCE PROJECT	PROJECT NUMBER
Resistance in S-P Circuits	Total Resistance in S-P Circuits	21
Current in S-P Circuits	Current in S-P Circuits	22
Voltage in S-P Circuits	Voltage Distribution in S-P Circuits	23
Power in S-P Circuits	Power Distribution in S-P Circuits	24
Effects of Opens in S-P Circuits	Effects of an Open in S-P Circuits	25
Effects of Shorts in S-P Circuits	Effects of a Short in S-P Circuits	26

NOTE: After completing above projects, perform Summary at end of Series-Parallel Circuits performance projects section in the Laboratory Manual.

Basic Network Theorems

Key Terms

Bilateral resistance

Constant current source

Constant voltage source

Efficiency

Impedance

Linear network

Maximum power transfer theorem

Multiple-source circuit

Network

Norton's Theorem

Steady-state condition

Superposition theorem

Thevenin's Theorem

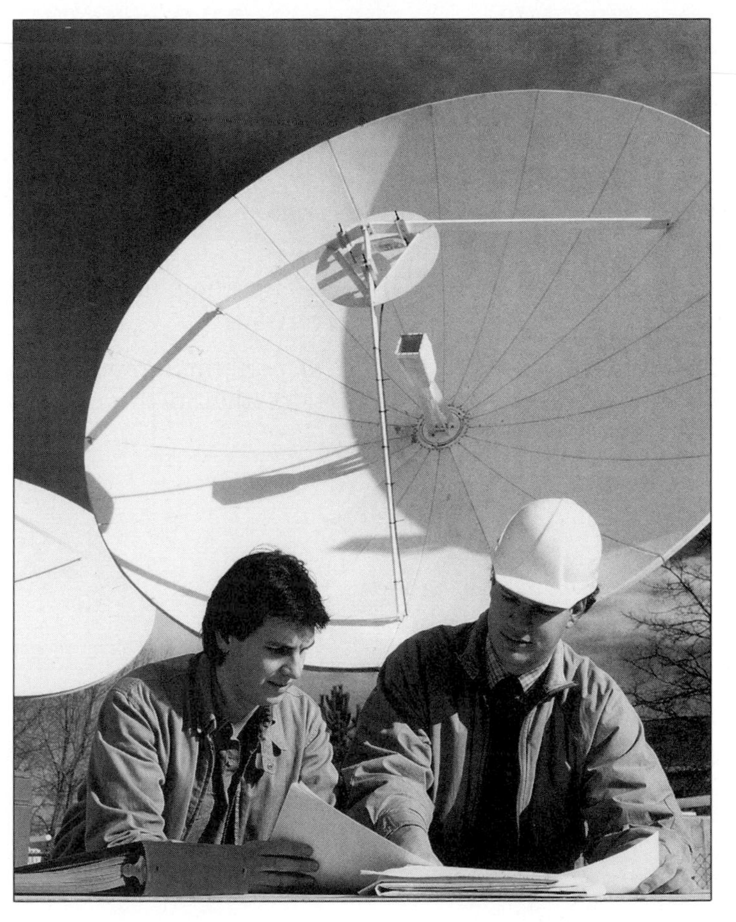

Courtesy of Cleveland Institute of Electronics, Inc.

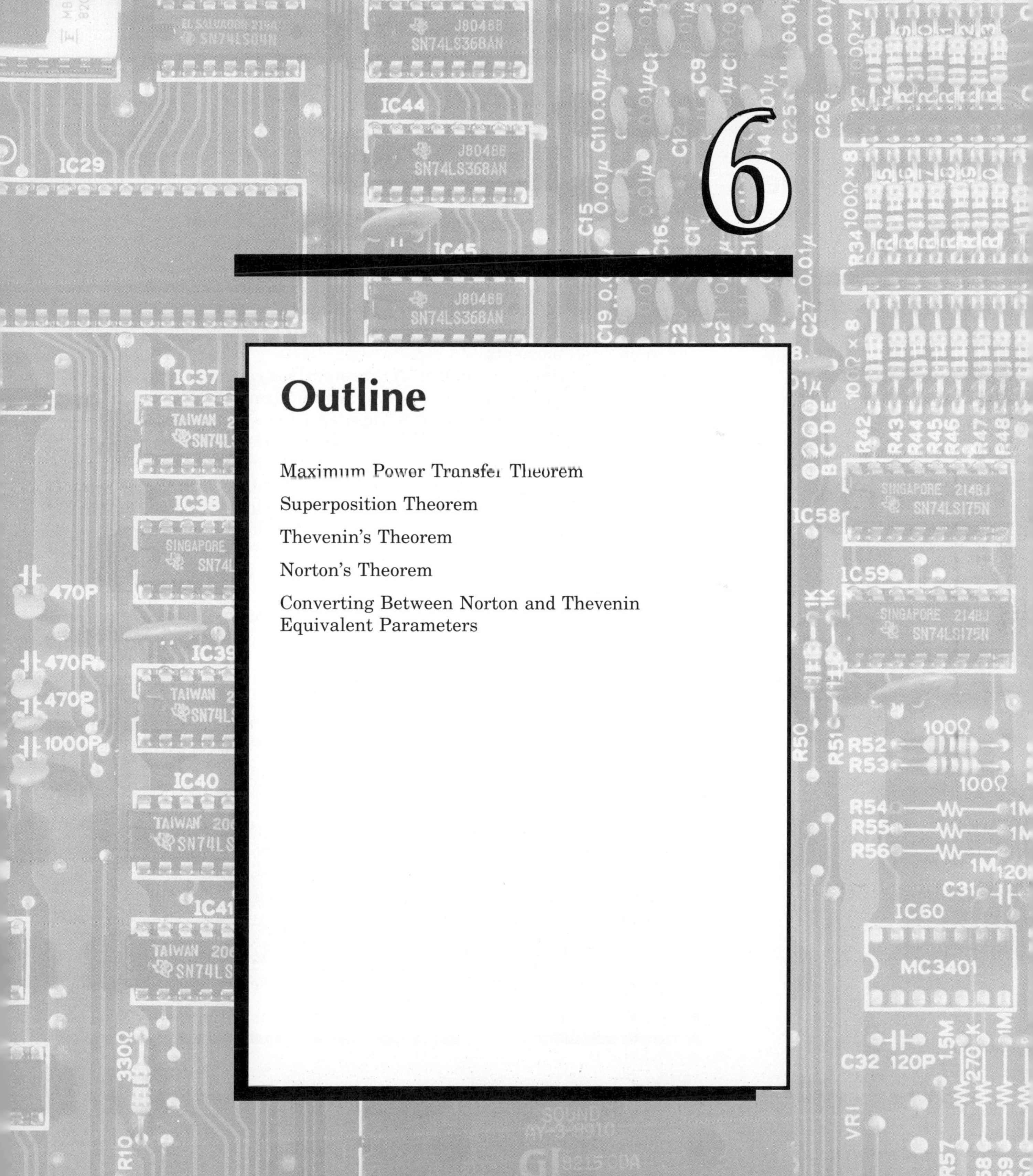

6

Outline

Maximum Power Transfer Theorem

Superposition Theorem

Thevenin's Theorem

Norton's Theorem

Converting Between Norton and Thevenin Equivalent Parameters

Chapter Preview

In this chapter, several commonly used electrical network theorems will be discussed. These include **maximum power transfer, superposition, Thevenin's** and **Norton's** theorems. For our purposes, a network is defined as a combination of components that are electrically connected. Theorems are ideas or statements that are not obvious at first, but that we can prove using some more accepted premises. When the theorems are proved this way, they also become laws. The theorems you will learn about in this chapter can be proven in theory and in practice.

The theorems in this chapter deal with **bilateral resistances** that are in **linear networks** under **steady-state conditions,** not during transient switch on or off moments.

Using the maximum power transfer theorem, you can know what value of load resistance (or impedance) allows maximum transfer of power from a source to its load, (e.g., what speaker impedance in your stereo allows transfer of maximum power from amplifier to speaker). Using the superposition theorem, you can readily analyze circuits having more than one source.

One of the advantages in using Thevenin's theorem is once the Thevenin equivalent circuit has been determined, you can easily predict electrical parameters for a variety of load resistances (R_L) without having to recalculate the whole network each time.

Norton's theorem provides another approach to simplify complex networks and make circuit parameter analysis easier.

Objectives

After studying this chapter, you will be able to:

- State the **maximum power transfer theorem**
- Determine the R_L value needed for maximum power transfer in a given circuit
- State the **superposition theorem**
- Solve circuit parameters for a circuit having more than one source
- State Thevenin's theorem
- Determine V_L and I_L for various values of R_L connected across specified points in a given circuit or network
- State **Norton's theorem**
- Apply Norton's theorem in solving specified problems
- Convert between Norton and Thevenin equivalent parameters

MAXIMUM POWER TRANSFER THEOREM

In many electronic circuits and systems, it is important to have maximum transfer of power from the source to the load. For example, in radio or TV transmitting systems, it is desired to transfer the maximum power possible from the last stage of the transmitter to the antenna system. In your high fidelity amplifier, you want maximum power transferred from the last amplifier stage to the speaker system. This is usually accomplished by proper "**impedance** matching," which means the load resistance or impedance matches the source resistance or impedance. NOTE: In dc circuits, such as you have been studying, the term for opposition to current flow is resistance. In ac circuits, which you will soon study, the term for opposition to current flow is impedance.

If it were possible to have a voltage source with zero internal resistance but able to maintain constant output voltage (even if a load resistance of zero ohms is placed across its output terminals), the source theoretically could supply infinite power to the load, where $P = V \times I$. In this case, V would be some value and I would be infinite. Thus, P = V value × infinite current = infinite power in watts.

Since these conditions are not possible, the practical condition is that there is an optimum value of R_L where maximum power is delivered to the load. Any higher or lower value of R_L causes less power to be delivered to the load.

The Basic Theorem

The **maximum power transfer** theorem states maximum power is transferred from the source to the load (R_L) when the resistance of the load *equals* the resistance of the source. In other words, it is a series circuit. In practical circuits, the internal resistance of the source causes some of the source's power to be dissipated within the source, while the remaining power is delivered to the external circuitry, Figure 6–1.

Refer to the circuits in Figure 6–2a, b, and c. Assuming the same source for all three circuits, notice the power delivered to the load in each case.

Observe in the first circuit, Figure 6–2a, the load resistance (R_L) equals the internal resistance of the source (r_{int}). The power delivered to the load is 10 watts.

In the circuit of Figure 6–2b, the load resistance is less than the source's internal resistance. The power delivered to the load is 8.84 watts.

In the third circuit, Figure 6–2c, R_L is greater than r_{int}, and the power delivered to the load is 7.5 watts. You can make calculations for a large number of R_L values to confirm that maximum power is delivered when $R_L = r_{int}$.

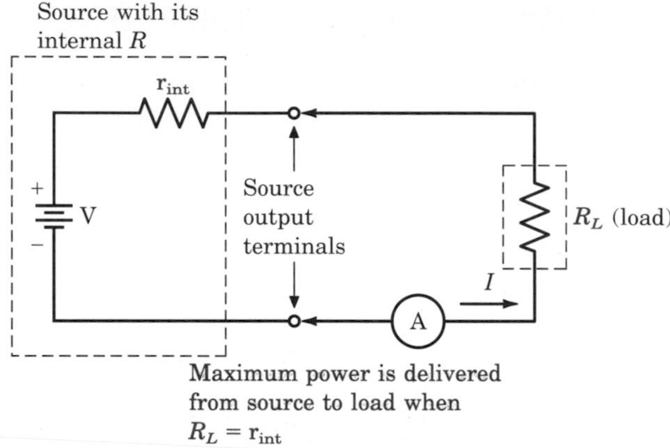

FIGURE 6–1 Maximum power transfer theorem (NOTE: R_T seen by source is the series total of r_{int} + R_L.)

Efficiency Factor

Formula 6–1	$\text{Efficiency (\%)} = \frac{P_{out}}{P_{in}} \times 100$

If you calculate the **efficiency** factor, based on what percentage of total power generated by the source is delivered to the load, you will find efficiency equals 50% when R_L equals r_{int}.

In Figure 6–2a, V_T = 20 V and I_T = 1 A. Thus, P_T (or P_{in}) = 20 V × 1 A = 20 W. The load is dissipating $(1\text{ A})^2(10\ \Omega)$ = 10 W, or half of the 20 watts.

In Figure 6–2b, V_T = 20 V and I_T = 1.33 A. Thus, P_T = 20 V × 1.33 A = 26.6 W but only 8.84 W is delivered to the load. Therefore, efficiency = $\left(\frac{8.84}{26.6}\right) \times 100 = 33.2\%$.

On the other hand, in Figure 6–2c, P_T = 10 W and P_L = 7.5 W. Thus, 75% of the power generated by the source is delivered to the load. However, the smaller value of total power produced by the source causes P_L to be less than when there was 50% efficiency (R_L matched r_{int}).

Where is the remainder of the power consumed in each case? If you said by the internal resistance of the source, you are right! In other words, when there is maximum power transfer, half the total power of the source is dissipated by its own internal resistance; the other half is dissipated by the external load. When the efficiency is about 33%, Figure 6–2b, 67% of the total power is dissipated within the source. When efficiency is 75%, Figure 6–2c, 25% of the total power is dissipated within the source, while 75% is delivered to the load.

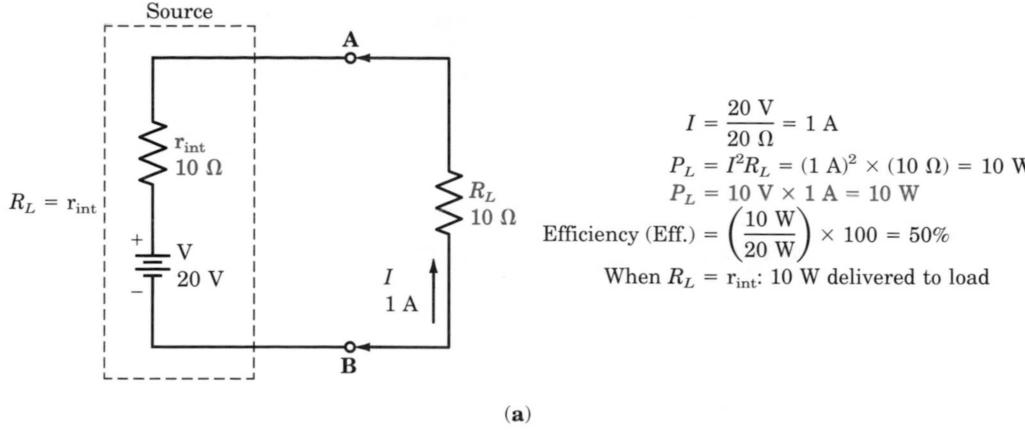

(a)

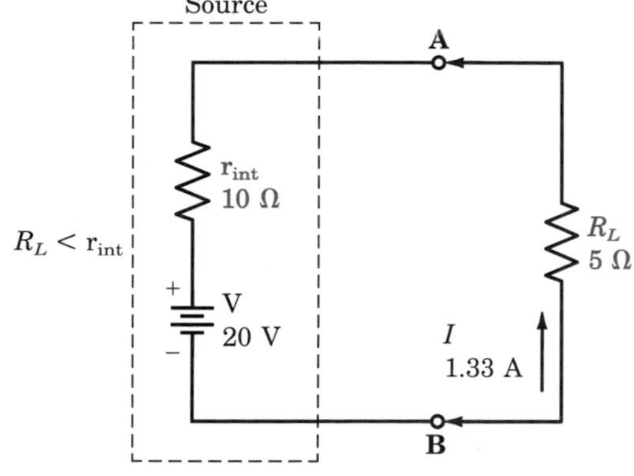

$$I = \frac{20\text{ V}}{15\ \Omega} = 1.33\text{ A}$$
$$P_L = I^2R_L = (1.33)^2 \times (5\ \Omega) = 8.84\text{ W}$$
$$P_L = 6.65\text{ V} \times 1.33\text{ A} = 8.84\text{ W}$$
$$\text{Eff.} = \left(\frac{8.84\text{ W}}{26.6\text{ W}}\right) \times 100 = 33.2\%$$

When $R_L < r_{int}$: less than 10 W delivered to load

(b)

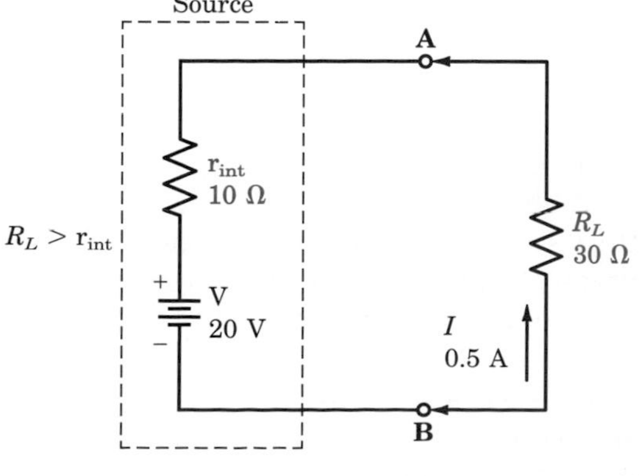

$$I = \frac{20\text{ V}}{40\ \Omega} = 0.5\text{ A}$$
$$P_L = I^2R_L = (0.5)^2 \times (30\ \Omega) = 7.5\text{ W}$$
$$P_L = 15\text{ V} \times 0.5\text{ A} = 7.5\text{ W}$$
$$\text{Eff.} = \left(\frac{7.5\text{ W}}{10\text{ W}}\right) \times 100 = 75\%$$

When $R_L > r_{int}$: less than 10 W delivered to load

(c)

FIGURE 6–2 Verification of maximum power transfer theorem

Summary of the Maximum Power Transfer Theorem

- Maximum power transfer occurs when the load resistance equals the source resistance.
- Efficiency at maximum power transfer is 50%.
- When R_L is greater in value than r_{int}, efficiency is greater than 50%.
- When R_L is less than the source internal resistance, efficiency is less than 50%.

PRACTICAL NOTES

You may or may not have noticed the amount of voltage at the output terminals of the source varies with load current. This is due to the internal IR drop across r_{int}. That is, higher currents result in greater internal IR drop and less voltage available at the output terminals of the source. In fact, one way to obtain an idea of a source's internal resistance is to measure its open circuit output terminal voltage, then put a load that demands current from the source across its terminals. The change in output terminal voltage from no-load (zero current drain) to loaded conditions allows calculation of r_{int}.

Formula 6–2	$r_{int} = \dfrac{V\text{ (no load)} - (V\text{ loaded})}{I(\text{load})}$

If V no-load is 13 V and V loaded is 12 V with 100 A load current, then: $\dfrac{13-12}{100} = 0.01\ \Omega$

A practical example of this approach to determine if a source's internal resistance is higher than it should be is the automotive technician's technique of checking a battery's output voltage at no-load, then under heavy load. If the terminal voltage of the battery drops too much under load, its internal resistance is too high, and the battery is either defective or needs charging.

PRACTICE PROBLEM I

Answers are in Appendix B.
Try solving the maximum power transfer problem in Figure 6–3.

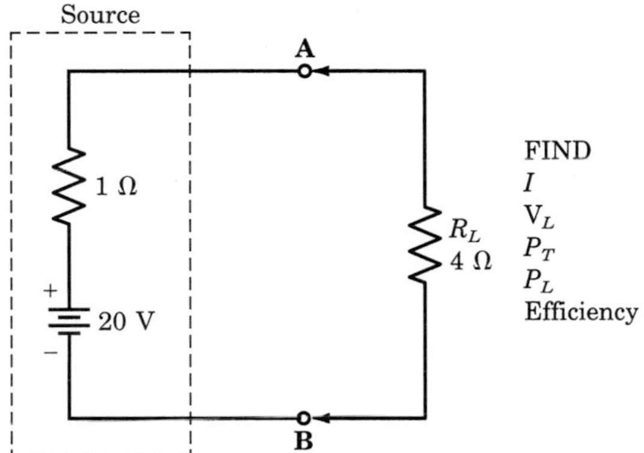

FIGURE 6–3 Maximum power transfer problem

SUPERPOSITION THEOREM

Frequently, you will analyze circuitry having more than one source. The superposition theorem is a useful tool in these cases. The **superposition theorem** states in linear circuits having more than one source, the voltage across or current through any given element equals the algebraic sum of voltages or currents produced by each source acting alone with the other sources disabled, Figure 6–4. NOTE: To disable voltage sources for purposes of calculations, consider them shorted. To disable current sources, consider them opened.

EXAMPLE

A circuit containing two voltage sources—**a multiple-source circuit**—is illustrated in Figure 6–4a. In Figures 6–4b through 6–4d, the step-by-step application of the superposition theorem is shown. ***Important:*** **Noting the direction of current flow and polarity of voltage drops is important for each step.**

In Figure 6–4b, the 60-V source is considered disabled (shorted out). Notice this effectively places R_3 in parallel with R_2. This means the R_1 value of 2 kΩ is in series with the parallel combination of $R_2 - R_3$ across the 30-V source. Thus, total R is 3 kΩ. Total I is 10 mA. R_1 passes the total 10 mA, whereas, the 10 mA is equally divided through R_2 and R_3 with 5 mA through each one. Using Ohm's law, the voltages are computed as: $V_1 = +20$ V with respect to point A; $V_2 = -10$ V with respect to point A; and $V_3 = +10$ V with respect to point B.

In Figure 6–4c, the 30-V source is considered disabled. In this case, R_1 and R_2 are in parallel with each other and in series with R_3 across the 60-V source. Notice the voltage drops are such that $V_1 = +20$ V with respect to point A; $V_2 = +20$ V with respect to point A; and $V_3 = +40$ V with respect to point B.

In Figure 6–4d, you see the algebraically computed results. Recall V_1 is +20 V with respect to A with the 60-V source disabled and +20 V with the

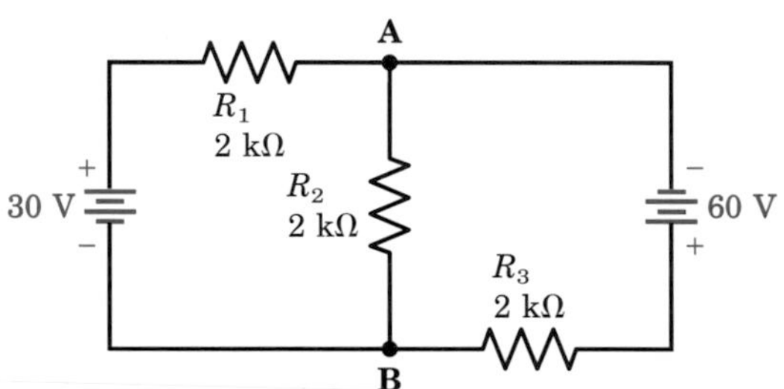

(a)

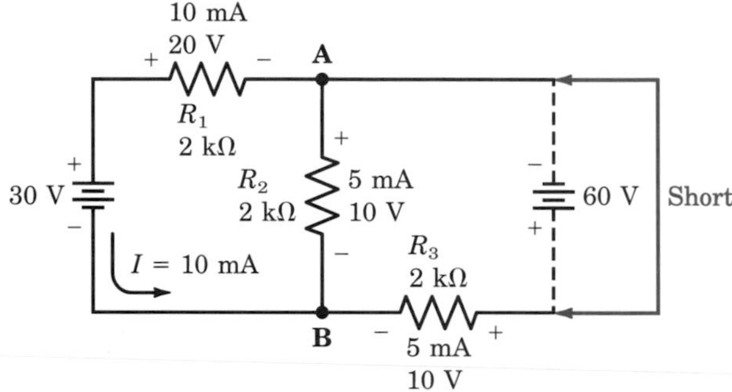

NOTE: If you were going to physically do this, you would *remove* the 60-V source and replace it with a conductor.

(b)

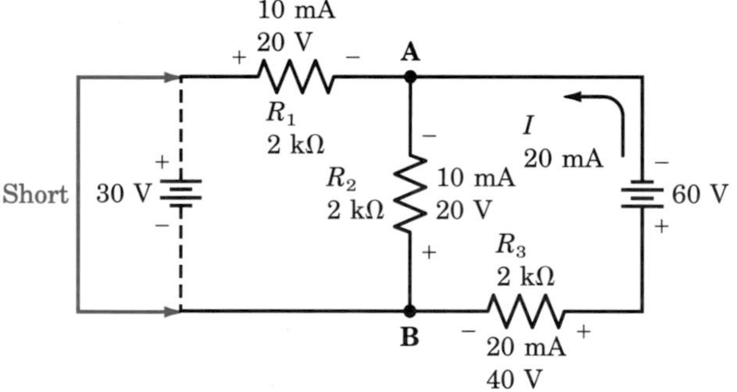

NOTE: To physically perform this step, you would *remove* the 30-V source and replace it with a "short."

(c)

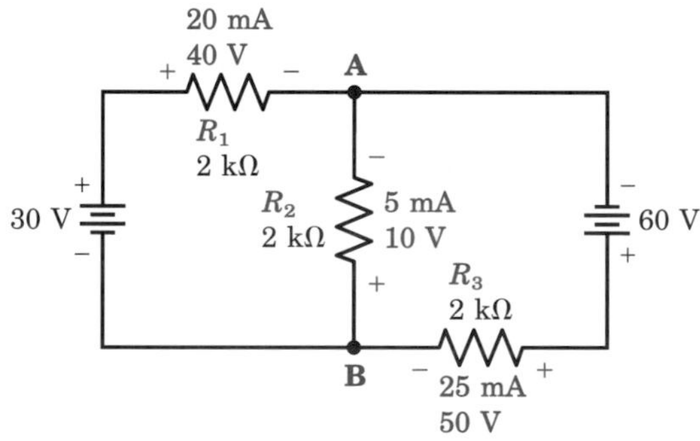

(d)

FIGURE 6–4 Verification of superposition theorem

30-V source disabled. This means the voltages are additive and the resultant voltage across R_1 equals +40 V with respect to point A. This same type addition also determines the resultant value of V_2 equals +10 V with respect to point A, and the resultant value of V_3 equals +50 V with respect to point B.

In a similar manner, directions and values of the currents through each component are determined, using the algebraic summation technique. For example, note in Figure 6–4b, current flows from point B toward point A through resistor R_2 and has a value of 5 mA. In Figure 6–4c, current flows from point A toward point B through R_2 and has a value of 10 mA. This means the resultant current through R_2 with both sources active is 5 mA traveling from point A toward point B. You get the idea! Refer to Figure 6–4d to see the other resultant values, as appropriate.

EXAMPLE

If the polarity of each of the sources in Figure 6–4 is reversed (while the circuit configuration remains the same), the results are like those in Figure 6–5.

The same analysis is used. That is, disable each voltage source, one-at-a-time by viewing them as shorted. Then determine the currents and voltages throughout the circuit, using Ohm's law and so on.

With the 60-V source assumed as shorted, V_1 equals −20 V with respect to point A; V_2 equals +10 V with respect to point A; and V_3 equals −10 V with respect to point B, Figure 6–5b.

With the 30-V source assumed as shorted, V_1 equals −20 V with respect

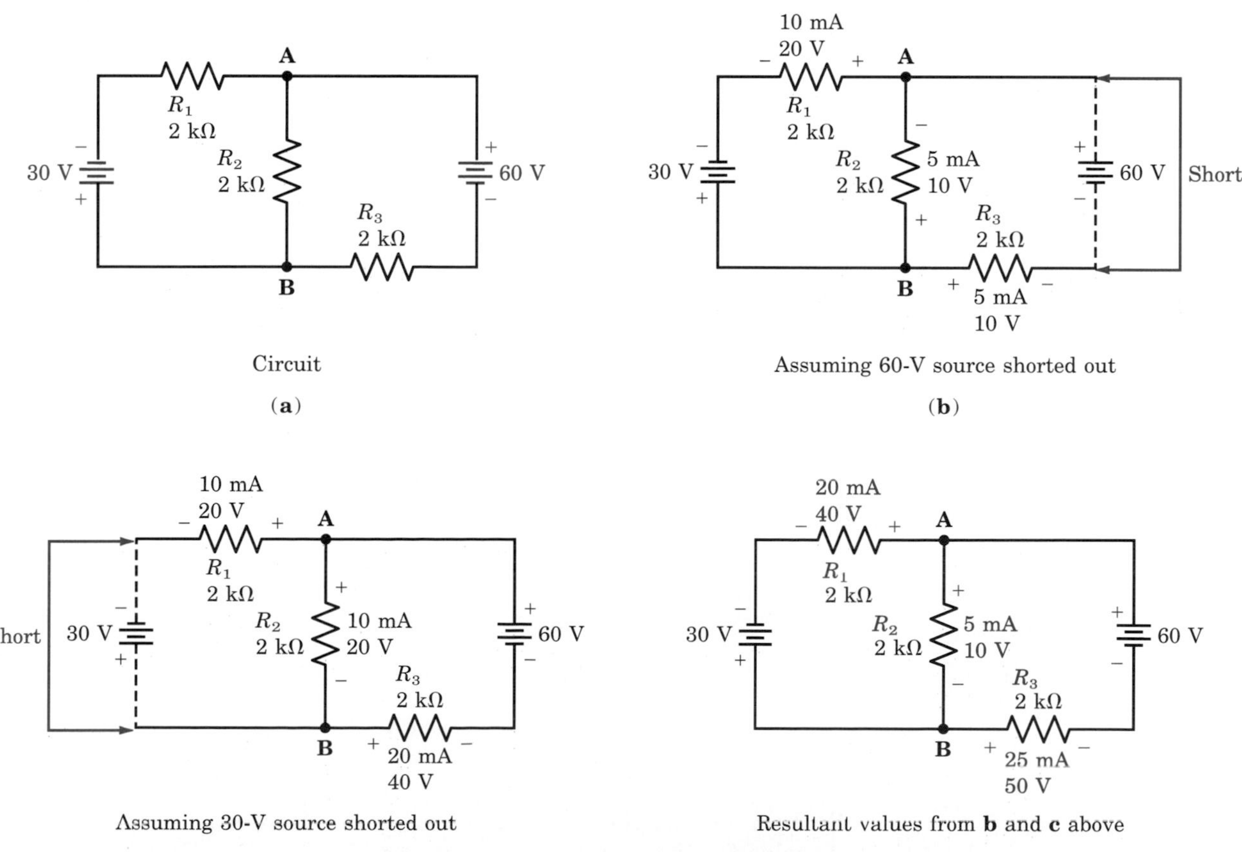

FIGURE 6–5 Superposition theorem example: reversed polarities

to point A; V_2 equals −20 V with respect to point A; and V_3 equals −40 V with respect to point B, Figure 6–5c.

Combining the results of the two assumed conditions, V_1 equals −40 V with respect to point A; V_2 equals −10 V with respect to point A; and V_3 equals −50 V with respect to point B, Figure 6–5d.

As you can see, when only the polarity of the sources changes, the resultant direction of currents and polarities of the voltage drops reverses but their values remain the same. Compare Figure 6–4d and 6–5d.

Summary of the Superposition Theorem

- Ohm's law is used to analyze the circuit, using one source at a time.
- Then the final results are determined by algebraically superimposing the results of all the sources involved.

PRACTICE PROBLEM II

For practice in using the superposition theorem, try the following problems:

1. Solve for the resultant current(s) and voltage(s) for the circuit in Figure 6–6.

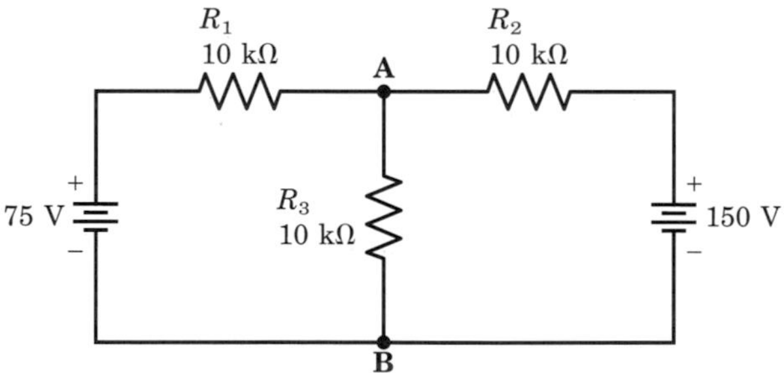

FIGURE 6–6 Superposition theorem problem

FIND
I through R_1
V across R_1, with respect to Point **A**
I through R_2
V across R_2, with respect to Point **A**
I through R_3
V across R_3, with respect to Point **A**

2. Assume only the 150-volt source is reversed in polarity and solve for the resultant current(s) and voltage(s). **Caution: Only one source is being reversed!**

IN-PROCESS LEARNING CHECK I

Fill in the blanks as appropriate.

1. For maximum power transfer to occur between source and load, the load resistance should be ___________ the source resistance.
2. The higher the efficiency of power transfer from source to load, the ___________ the percentage of total power is dissipated by the load.
3. Maximum power transfer occurs at _______% efficiency.
4. If the load resistance is less than the source resistance, efficiency is ___________ the efficiency at maximum power transfer.
5. To analyze a circuit having two sources, the superposition theorem indicates that Ohm's law ___________ be used.
6. What is the key method in using the superposition theorem to analyze a circuit with more than one source?

   ____
7. Using the superposition theorem, if the sources are considered "voltage" sources, are these sources considered shorted or opened during the analysis process? ___________.
8. When using the superposition theorem when determining the final result of your analysis, the calculated parameters are combined, or superimposed, ___________.

THEVENIN'S THEOREM

Thevenin's theorem is used to simplify complex **networks** so the calculation of voltage(s) across any two given points in the circuit (and consequently, current(s) flowing between the two selected points) is easily determined.

Once the Thevenin equivalent circuit parameters are known, solving circuit parameters for *any value* of R_L merely involves solving a two-resistor series circuit problem. A **constant voltage source** is assumed. This is a source whose output voltage remains constant under varying load current (or current demand) conditions.

As you recall, voltage distribution was considered when you studied series circuits. Thevenin's theorem deals with a simple equivalent series circuit to simplify voltage distribution analysis. Let's see how this is applied.

The Theorem Defined

One version of **Thevenin's theorem** is stated as any **linear** two-terminal **network** (of resistances and source[s]) can be replaced by a simplified equivalent circuit consisting of a single voltage source (called V_{TH}) and a single series resistance (called R_{TH}), Figure 6–7.

FIGURE 6–7
Thevenin's theorem equivalent circuit

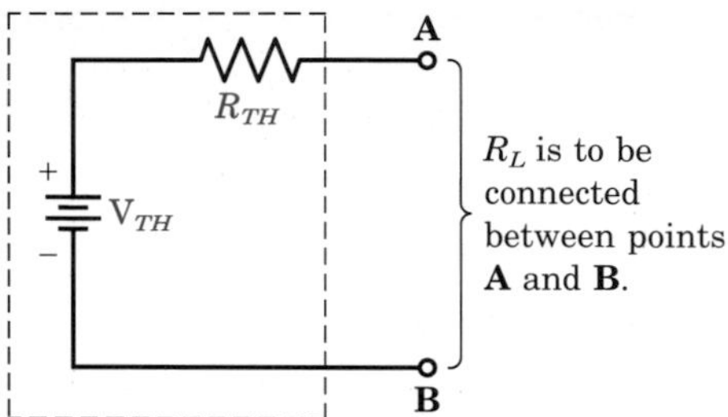

V_{TH} is the open-terminal voltage felt across the two load terminals (A and B) with the load resistance removed. R_{TH} is the resistance seen when looking back into the circuit from the two load terminals (without R_L connected) with voltage sources replaced by their internal resistance. For simplicity, we assume them shorted.

Refer to Figures 6–8a, b, c, and d. Figure 6–8a shows the original circuit

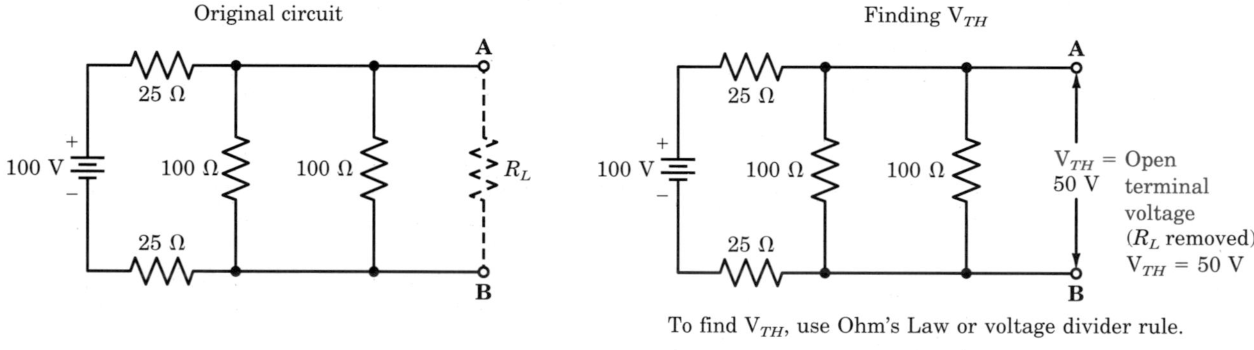

(a) (b)

Finding R_{TH}

A
25 Ω
Short
100 Ω
100 Ω
R_{TH} = R "looking back" into circuit with V source shorted out.
R_{TH} = 25Ω
25 Ω
B

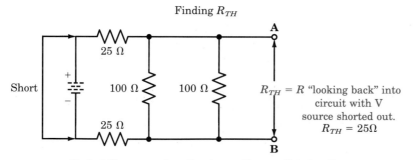

To find R_{TH}, use rules of series and/or parallel circuits.

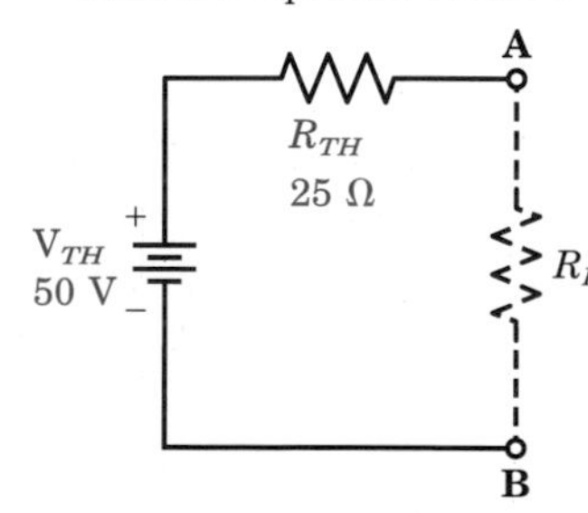

If: R_L = 25 Ω
Then: I_L = 1 A
V_L = 25 V

(c) (d)

FIGURE 6–8 Thevenin's theorem illustration

that we will use to determine the Thevenin equivalent circuit, a process called "Thevenizing" the circuit.

Figure 6–8b illustrates how the value for V_{TH} is determined. Notice R_L is considered opened (between points A and B). V_{TH} is considered the value of voltage appearing at network terminals A and B (with R_L removed). For example, if Ohm's law is used, R_T is 100 Ω; V_T is 100 V; thus $I_T = 1$ A. Each of the 25-Ω series resistors drops 25 V (1 A × 25 Ω). The 1 A evenly splits between the two 100-ohm parallel resistors. Thus, voltage between points A and B = 0.5 A × 100 Ω, or 50 V. Therefore, $V_{TH} = 50$ V.

Figure 6–8c shows the process to find R_{TH}. Mentally short the source and determine the circuit resistance by looking toward the source from the two network points A and B. With the source shorted, the two 100-Ω parallel resistors are effectively in parallel with 50 Ω (25 Ω + 25 Ω). The equivalent resistance of two 100-ohm resistors is 50 Ω, and that 50 Ω in parallel with the remaining 50-Ω branch yields an R_T of 25 Ω. This is the value of R_{TH}.

Figure 6–8d shows the Thevenin equivalent circuit of a 50-volt source (V_{TH}) and the series R_{TH} of 25 Ω. It is thus possible to determine the current through any value of R_L and the voltage across R_L by analyzing a simple two-resistor series circuit.

EXAMPLE

If R_L has a value of 25 Ω, what are the values of I_L and of V_L? If you answered 1 ampere and 25 volts, you are correct!

Using formula 6–3, solve this way:

$$I_L = \frac{50\ \text{V}}{25\ \Omega + 25\ \Omega} = \frac{50\ \text{V}}{50\ \Omega} = 1\text{A}$$

Formula 6–3	$I_L = \dfrac{V_{TH}}{R_{TH} + R_L}$

V_L is calculated by using Ohm's law ($I_L \times R_L$), or the voltage divider rule where:

Formula 6–4	$V_L = \dfrac{R_L}{R_L + R_{TH}} \times V_{TH}$

Using formula 6–4, solve this way:

$$V_L = \frac{25\ \Omega}{25\ \Omega + 25\ \Omega} \times 50\ \text{V} = \frac{25\ \Omega}{50\ \Omega} \times 50\ \text{V} = .5 \times 50\ \text{V} = 25\ \text{V}.$$

Summary of Thevenin's Theorem

The Thevenin equivalent circuit consists of a source, V_{TH}, (often called the Thevenin generator) and a single resistance, R_{TH}, (often called the Thevenin internal resistance) in series with the load (R_L), which is across the specified network terminals. Therefore, this equivalent circuit is useful to simplify calculations of electrical parameters for various values of R_L since it reduces the problem to a simple two-resistor series circuit analysis.

PRACTICAL NOTES

Obviously, in a practical circuit, you can remove the load resistor (R_L) and measure the open-terminal voltage at those two points with a voltmeter to find V_{TH}. Then you can remove the power supply from the circuit and short the points where its output terminals are connected. Next, you can use an ohmmeter to measure the circuit resistance (at the open-terminal points) by looking at where the source points are shorted. Thus, the measured value of R is R_{TH}.

One example of a practical application of Thevenin's theorem is in analyzing an unbalanced-bridge circuit. Try applying your Thevenin's theorem knowledge to the following practice problems.

PRACTICE PROBLEM III

1. Refer again to Figure 6–8. What are the values of I_L and V_L, if R_L is 175 Ω.
2. In Figure 6–8, what value of R_L must be present if I_L is 125 mA?
3. In Figure 6–9, what are the values of V_{TH} and R_{TH}? What are the values of I_L and V_L, if R_L equals 16 kΩ?

NORTON'S THEOREM

Norton's theorem is another method to simplify complex networks into manageable equivalent circuits for analysis. Again, the idea is to devise the equivalent circuit so parameter analysis at two specified network terminals (where R_L is connected) is easy.

Thevenin's theorem uses a voltage-related approach with a voltage source/series resistance equivalent circuit where voltage divider analysis is applied. On the other hand, Norton's theorem uses a current-related approach with a current source/shunt resistance equivalent circuit where a current-divider analysis approach is applied.

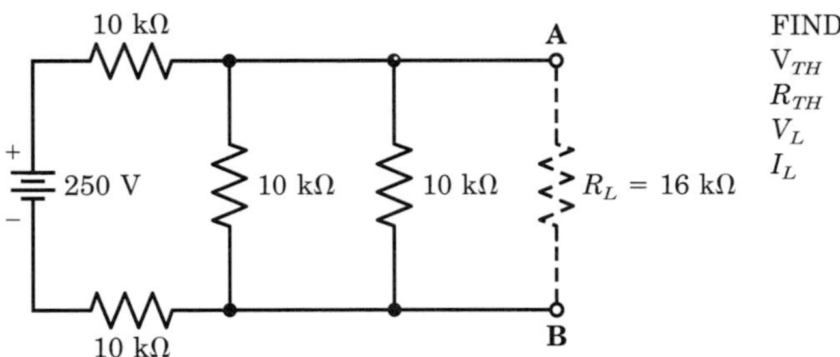

FIGURE 6–9 Thevenin's theorem practice problem

The Theorem Defined

Norton's theorem says any linear two-terminal network can be replaced by an equivalent circuit consisting of a single (**constant**) **current source** (called I_N) and a single shunt (parallel) resistance (called R_N), Figure 6–10.

Refer to Figure 6–11, points A and B. I_N is the current available from the Norton equivalent current source. I_N is determined as the current flowing between the load resistance terminals, if R_L were shorted (i.e., if R from points A to B is zero ohms). R_N is the resistance of the Norton equivalent shunt resistance. This value is computed as the R value seen when looking in the network with the load terminals open (R_L not present) and sources of *voltage* replaced by their R_{int}, Figure 6–12.

This method is the same one used to calculate R_{TH} in Thevenin's theorem.

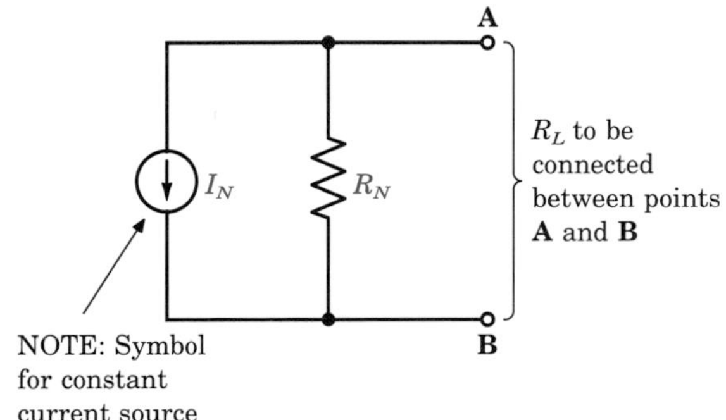

FIGURE 6–10 Norton's theorem equivalent circuit

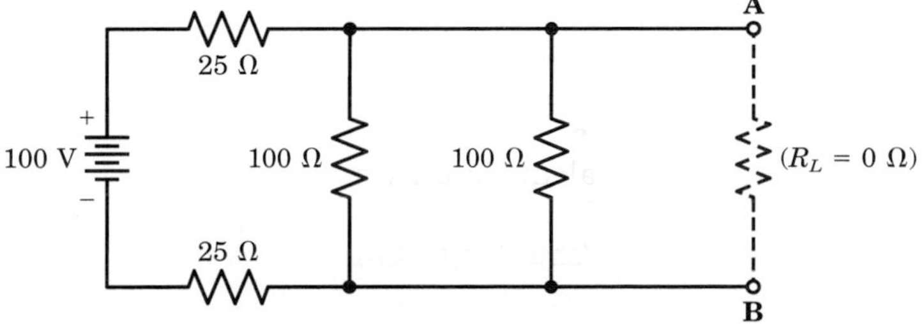

FIGURE 6–11 Norton's I_N = Current from A to B, if R_L shorted (0 Ω). (In this case, I_N would equal $\frac{100\text{ V}}{50\ \Omega}$ = 2 A.)

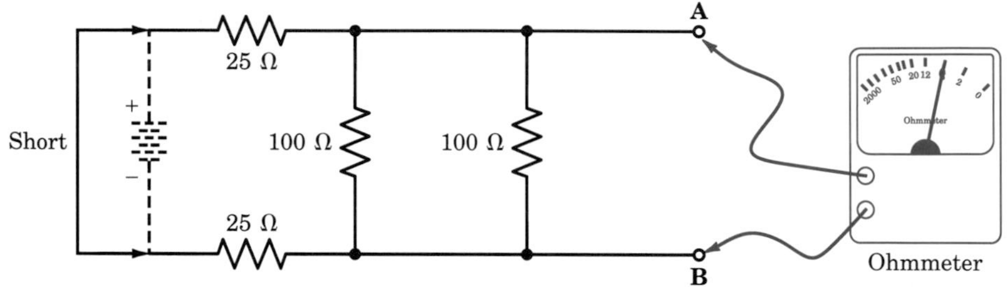

FIGURE 6–12 Norton's $R_N = R_{A\text{-}B}$ with R_L removed and source considered as a short. (In this case, $R_N = 25\ \Omega$.)

In fact, the R_{TH} and R_N values are the same. The only difference is that in Norton's theorem, this resistance value is considered in parallel (or shunt) with the I_N equivalent current source. In Thevenin's theorem, this resistance value is considered in series with the V_{TH} equivalent voltage source.

Observe the similarities and differences between the two types of equivalent circuits in Figure 6–13b and 6–13c. Note these equivalent circuits are derived from the same original circuit used to explain Thevenin's theorem earlier in the chapter.

Correlating the prior thinking of the Thevenin analysis to the Norton

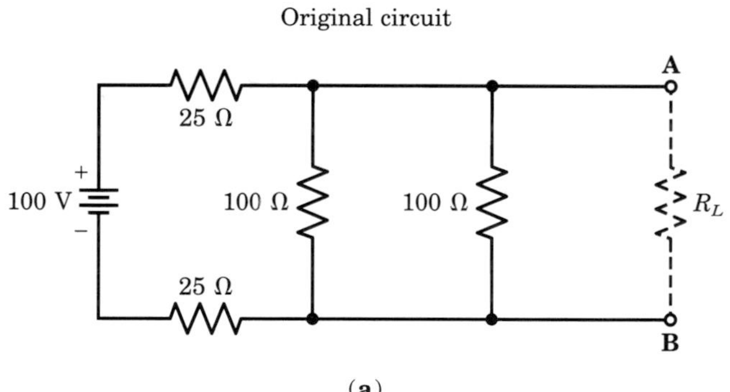

Thevenin equivalent circuit

A, R_{TH} 25 Ω, V_{TH} 50 V, R_L, B

(b)

Norton equivalent circuit

A, I_N 2 A, R_N 25 Ω, R_L, B

(c)

FIGURE 6–13 Comparisons of Thevenin and Norton equivalent circuits

analysis, let's assume the same 25 Ω R_L (load resistance) used in Figure 6–8, and apply the Norton equivalent circuit approach.

First, let's see how the Norton equivalent circuit parameters, Figure 6–13c, are derived. Analysis starts with the original circuit, Figure 6–14a.

I_N equals current between points A and B with R_L shorted. This short also shorts the two 100-Ω parallel resistors, so the source sees a total circuit R of 50 Ω (the two 25-Ω resistors in series, through the short). Thus, current between points A and B is $\frac{100\text{ V}}{50\ \Omega}$, or 2 amperes ($I_N = 2$ A), Figure 6–14b.

R_N equals R looking back into the network with points A and B open and with *voltage* source(s) shorted. R_N thus equals the equivalent resistance of the two 100-ohm parallel resistors (or 50 Ω) in parallel with the two 25-Ω series resistors' branch (or 50 Ω). This yields a resultant circuit equivalent resistance of 25 Ω (R_N = 25 Ω), again see Figure 6–14b.

Now, refer to Figure 6–14c to see how this equivalent circuit is used to determine electrical parameters associated with various values of R_L.

FIGURE 6–14
Application sample of Norton's theorem

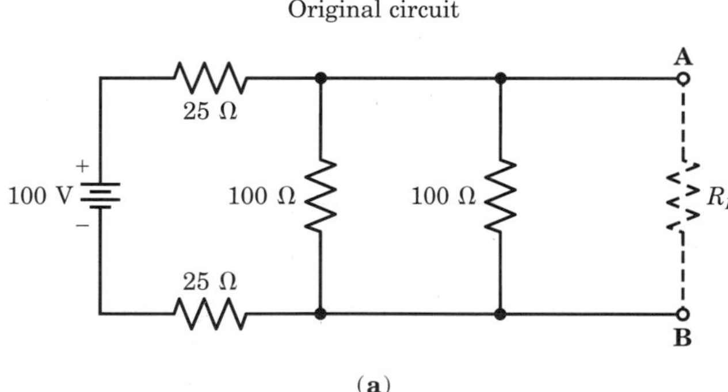

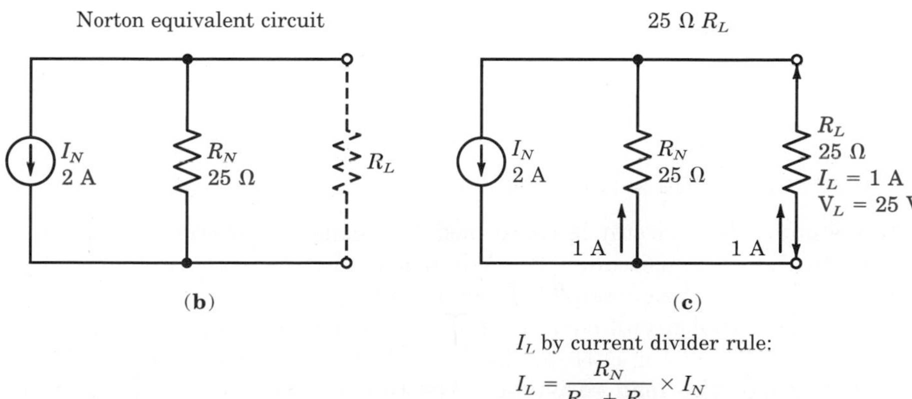

I_L by current divider rule:

$$I_L = \frac{R_N}{R_N + R_L} \times I_N$$

$$I_L = \frac{25}{(25 + 25)} \times 2 = 0.5 \times 2 = 1\text{ A}$$

If an R_L value of 25 Ω is placed from terminal A to B, it is apparent the Norton equivalent circuit current of 2 amperes equally divides between R_N and R_L, since they are equal value parallel branches. This means $I_L = 1$ A. V_L is easily calculated using Ohm's law where, $I_L \times R_L = 1\text{ A} \times 25\ \Omega = 25$ V. It is interesting to note the I_L and V_L computed using Norton's theorem are exactly the same values as those calculated for a load resistance of 25 Ω when using Thevenin's theorem for the same original circuit, Figure 6–8. It makes sense the same R_L connected to the same circuit has the same operating parameters, regardless of the method used to analyze these parameters!

EXAMPLE

Suppose the R_L value changes to 75 Ω. The Norton equivalent circuit is the same. That is, $I_N = 2$ A and $R_N = 25$ Ω. With the higher value of R_L, I_L obviously is a smaller percentage of the I_N value.

Using the current divider rule, which is derived from the fact current divides inverse to the resistance of the branches, we can determine I_L from the following formula:

Formula 6–5	$I_L = \frac{R_N}{R_N + R_L} \times I_N$

With the 75 Ω R_L and the 25 Ω R_N:

$$I_L = \frac{25}{25 + 75} \times 2 = .25 \times 2 = 0.5\text{ A}$$

To find V_L, again use Ohm's law where $V_L = .5\text{ A} \times 75\ \Omega = 37.5$ V

PRACTICE PROBLEM IV

In Figure 6–14a, assume R_L is 60 Ω. What are the values of I_L and V_L?

Summary of Norton's Theorem

Norton's equivalent circuit is comprised of a constant current source (I_N) and a shunt resistance called R_N. When a load (R_L) is connected to this equivalent circuit, the current (I_N) divides between R_N and R_L according to the rules of parallel circuit current division (i.e., inverse to the resistances). The simple two-resistor current divider rule is applied to determine the current through the load resistance. Try the following Norton's theorem problems.

PRACTICE PROBLEM V

1. Use Norton's theorem and determine the values of I_L and V_L for the circuit in Figure 6–14a, if R_L has a value of 100 Ω.
2. Refer again to Figure 6–14a. Will I_L increase or decrease if R_L changes from 25 Ω to 50 Ω? Will V_L increase or decrease?

CONVERTING NORTON AND THEVENIN EQUIVALENT PARAMETERS

Since we have been discussing the similarities and differences between the two theorems, it is appropriate to show you some simple conversions that can be done between the two.

You already know R_{TH} and R_N are the same value for a given network or circuit. You may have noticed if you multiply $I_N \times R_N$, the answer is the same as the value of V_{TH}.

Let's summarize these thoughts to show you the conversions possible.

Finding Thevenin parameters from Norton parameters:

$$R_{TH} = R_N$$
$$V_{TH} = I_N \times R_N$$

Finding Norton parameters from Thevenin parameters:

$$R_N = R_{TH}$$
$$I_N = \frac{V_{TH}}{R_{TH}}$$

SUMMARY

- The theorems studied apply to two-terminal linear networks that have bilateral resistances or components.
- The maximum power transfer theorem indicates maximum power is transferred from source to load when the load resistance equals the source resistance. Under these conditions the efficiency is 50%.
- The superposition theorem says electrical parameters are computed in linear circuits with more than one source by analyzing the parameters produced by each source acting alone, then superimposing each result by algebraic addition to find the circuit parameters with multiple sources.
- Thevenin's theorem simplifies any two-terminal linear network into an equivalent circuit representation of a voltage source (V_{TH}) and series resistance (R_{TH}).
- Steps for Thevenizing a circuit are:
  1. Open (or remove R_L) from the network circuitry.

2. Determine the open-circuit V at points where R_L is to be connected. This is the value of V_{TH}.
3. Mentally short the circuit voltage source and determine resistance by looking into the network (from R_L connection points, with R_L disconnected). This is the value of R_{TH}.
4. Draw the Thevenin equivalent circuit with source = V_{TH} and series resistor of the R_{TH} value.
5. Make calculations of I_L and V_L for the desired values of R_L (where R_L is assumed to be across the equivalent circuit output terminals) by using two-resistor series circuit voltage analysis.

- ▶ Voltage distribution in a Thevenin equivalent circuit is computed using Ohm's law and/or the voltage divider rule.
- ▶ Norton's theorem simplifies any two-terminal linear network into an equivalent circuit representation of a current source (I_N) and a parallel or shunt resistance (R_N).
- ▶ Steps for Nortonizing a circuit are:

1. Determine the value for I_N by finding the value of current flowing through R_L if it were zero ohms.
2. Find the value for R_N by assuming the R_L is removed from the circuit (R_L = infinite ohms) and any voltage sources are zero ohms. Calculate the circuit resistance by looking from the R_L connection points toward the source(s).
3. Draw the Norton equivalent circuit as a current source of a value equal to I_N and a resistance in parallel with the source equal to R_N.
4. Compute R_L parameters, as appropriate, by assuming the load resistance at the output terminals of the equivalent circuit, using concepts of two-resistor parallel circuit current division analysis.

- ▶ Current division in a Norton equivalent circuit is computed using Ohm's law and/or the current divider rule.

REVIEW QUESTIONS

1. Name at least two applications for the maximum power transfer theorem.
2. In Figure 6–15, use the superposition theorem, and determine the voltage across R_1 and R_2.
3. Refer again to Figure 6–15. Assume the polarity of V_{S2} is reversed. What is the value of V_{R1} and V_{R2}?

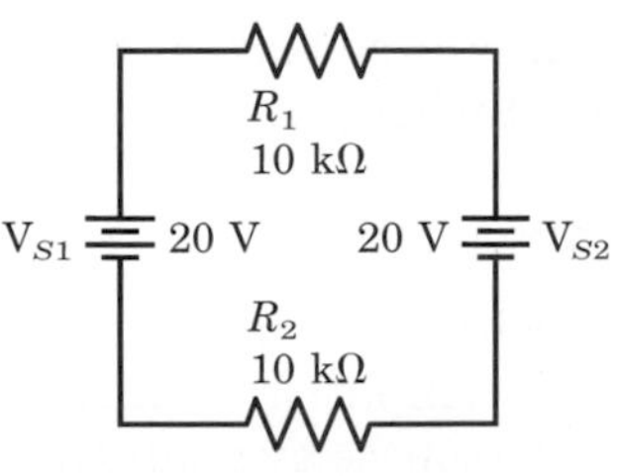

FIGURE 6–15

4. In Figure 6–16, use Thevenin's theorem and determine V_{TH}, R_{TH}, I_L and V_L. Assume R_L is 10 kΩ.

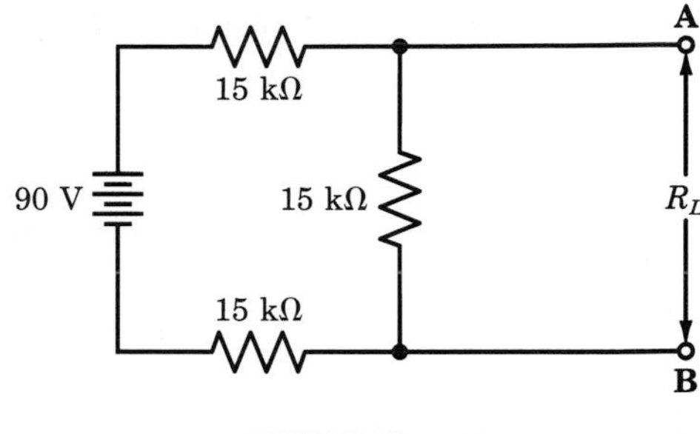

FIGURE 6–16

5. Refer again to Figure 6–16. Assume R_L changes to 25 kΩ. What are the values of V_{TH}, R_{TH}, I_L, and V_L?

6. Draw the Norton equivalent circuit for the circuit in question 4.

7. Draw the Norton equivalent circuit for the circuit in question 5.

8. Define the term "a bilateral linear network."

9. State the sequence of steps used in "Thevenizing" a circuit.

10. State the sequence of steps used in "Nortonizing" a circuit.

11. If the efficiency of power transfer from source to load is 60%, should the load resistance be increased or decreased to enable maximum power transfer?

12. If the voltage at the source terminals of a circuit is 100 volts without load and 90 volts when 10 amperes are drawn from the source, what value is the load resistance to afford maximum power transfer?

13. For the circuit in Figure 6–17, what are the values of R_{TH} and V_{TH}?

14. For the circuit in Figure 6–18, what are the values of R_N and I_N?

15. Convert Norton parameters to Thevenin parameters, given $I_N = 3$ A and $R_N = 10$ Ω.

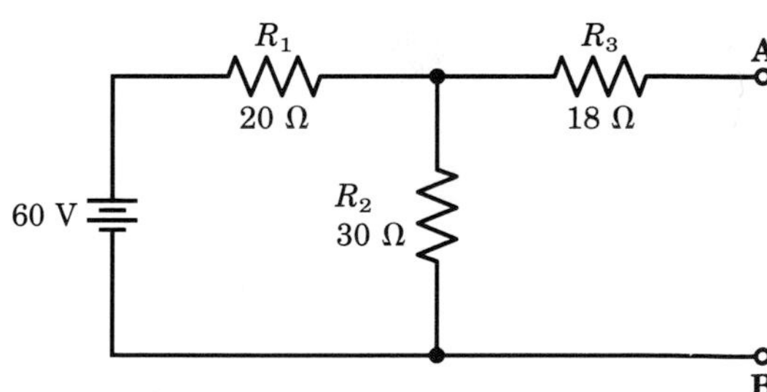

FIGURE 6–17 Find R_{TH} and V_{TH}.

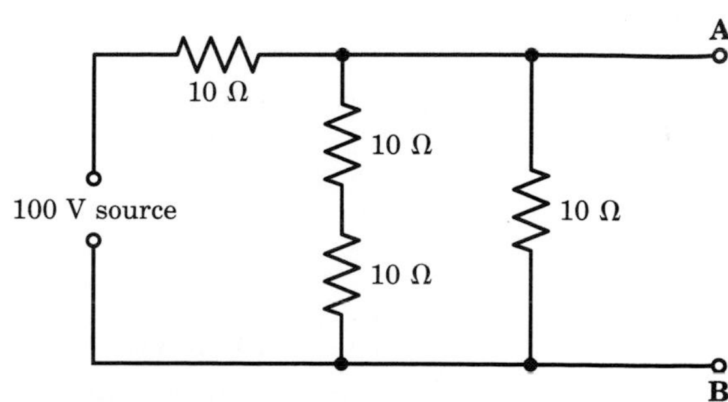

FIGURE 6–18 Find R_N and I_N.

16. Convert Thevenin parameters to Norton parameters, given $R_{TH} = 15\ \Omega$ and $V_{TH} = 25$ V.

17. Describe the most important application of the superposition theorem.

18. Which theorem relates most closely to voltage divider analysis techniques?

19. Which theorem relates most closely to current divider analysis techniques?

TEST YOURSELF

1. Perform all the computational Review Exercise Questions using a computer and the *Basic* (immediate mode) approach. (Ask your instructor for information on how to perform resistive circuit analysis with the computer.)
2. Perform all the computational Review Exercise Questions using a computer and the *Spreadsheet* approach.

PERFORMANCE PROJECTS CORRELATION CHART

Suggested performance projects in the Laboratory Manual that correlate with topics in this chapter are:

CHAPTER TOPIC	PERFORMANCE PROJECT	PROJECT NUMBER
Maximum Power Transfer	Maximum Power Transfer Theorem	29
Thevenin's Theorem	Thevenin's Theorem	27
Norton's Theorem	Norton's Theorem	28

NOTE: After completing above projects, perform the Summary at end of performance projects section in the Laboratory Manual.

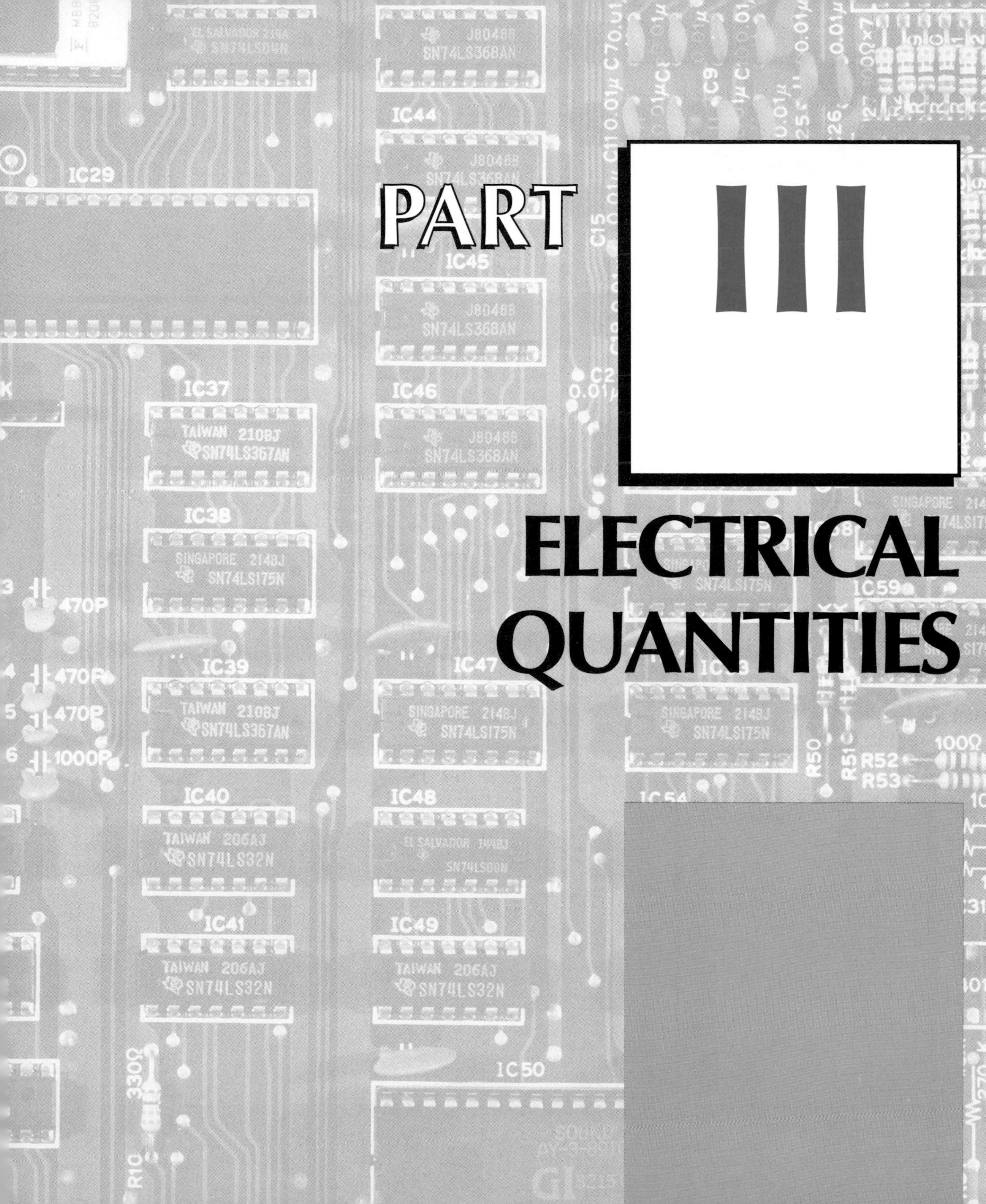

PART III

ELECTRICAL QUANTITIES

Cells and Batteries

Key Terms

Amalgamation

Ampere-hour rating

Battery

Cell

Depolarizer

Electrode

Electrolyte

Local action

Polarization

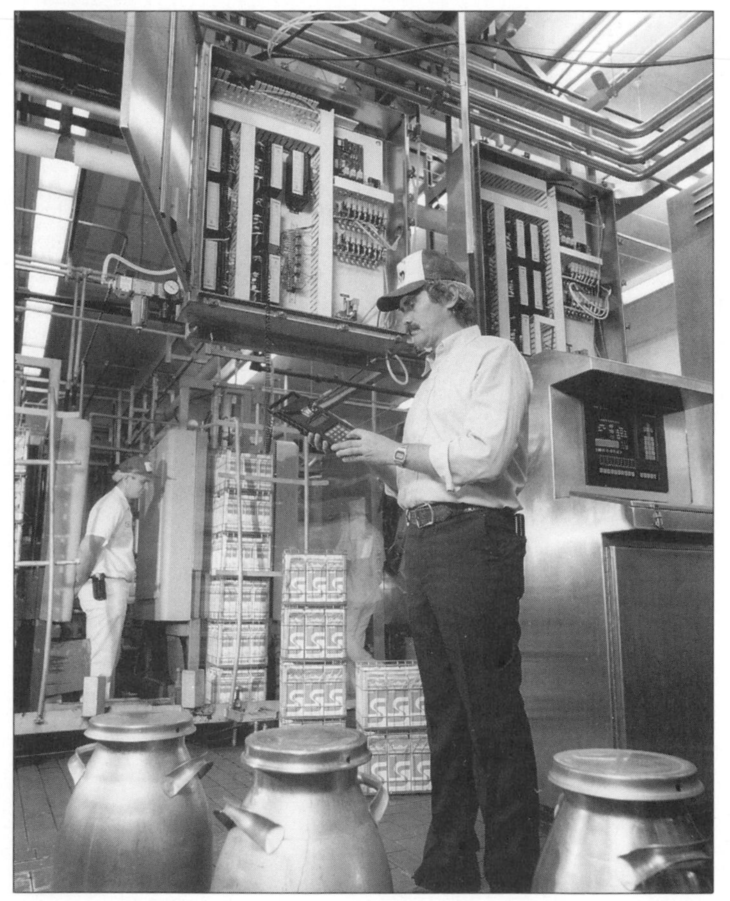

Courtesy of GE Fanuc Automation

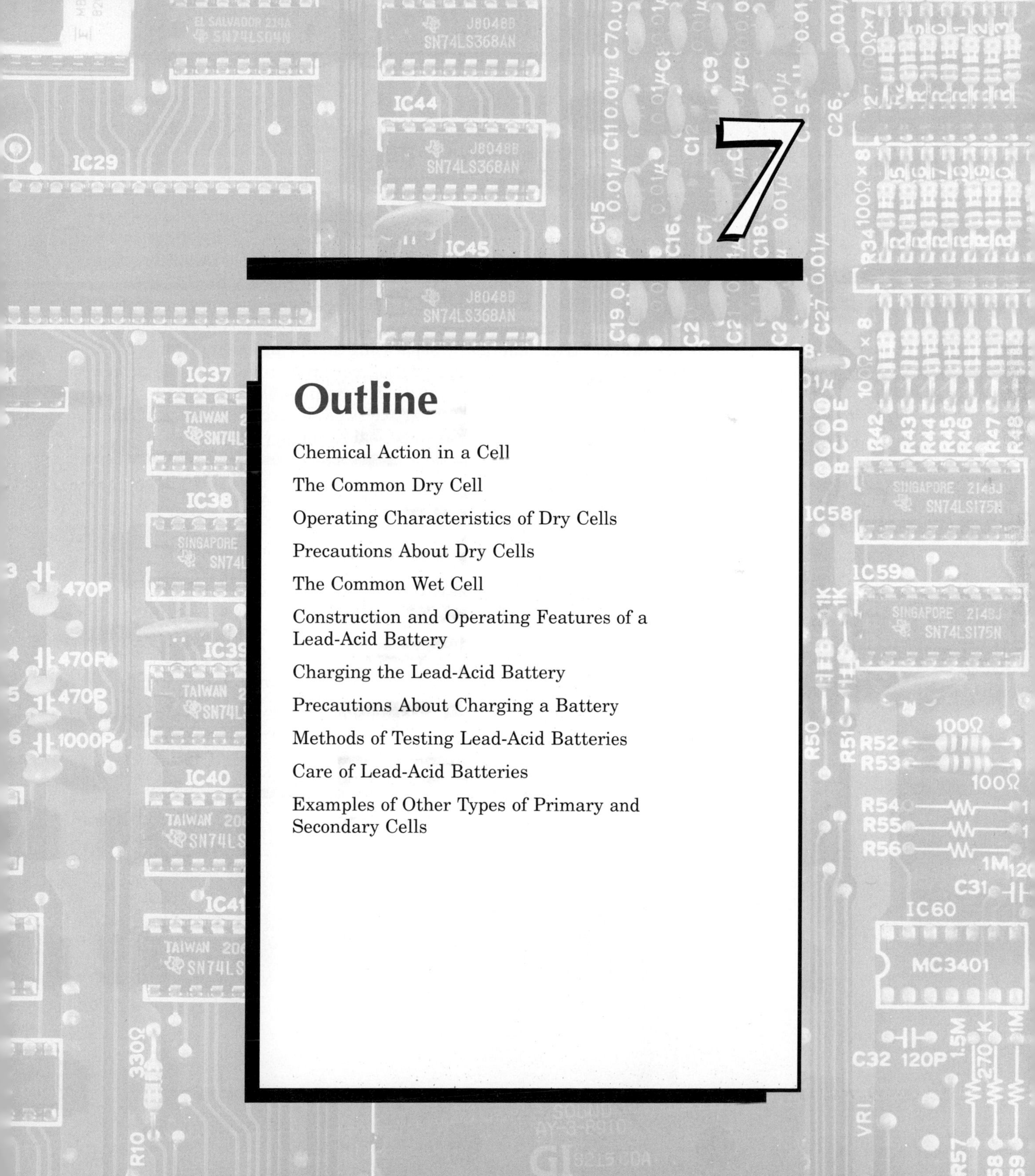

7

Outline

Chemical Action in a Cell

The Common Dry Cell

Operating Characteristics of Dry Cells

Precautions About Dry Cells

The Common Wet Cell

Construction and Operating Features of a Lead-Acid Battery

Charging the Lead-Acid Battery

Precautions About Charging a Battery

Methods of Testing Lead-Acid Batteries

Care of Lead-Acid Batteries

Examples of Other Types of Primary and Secondary Cells

Chapter Preview

In this chapter you will study common types of cells and batteries; learn about their operation; and examine some of their basic features. Also, some important facts about their care and usage are presented.

Cells and batteries are important sources of dc voltage and current that operate a variety of electrical and electronic circuits and devices. Everything, from automated cameras, burglar alarms, emergency lighting systems, to portable radio and TV receivers, is made possible by various types of cells and batteries, Figure 7–1.

A battery is two or more electrically connected cells usually contained in a single unit. Cells and batteries depend on the process of changing chemical energy into electrical energy by separating and/or producing charges.

Two general categories of these dc sources are nonrechargeable primary cells and rechargeable secondary cells. The common flashlight battery is an example of a nonrechargeable primary cell and the rechargeable automobile storage battery (made of several cells) is a secondary cell.

Objectives

After studying this chapter, you will be able to:

- Describe the difference between primary and secondary **cells**
- Briefly explain chemical action in a cell
- List physical and operating features of dry cells
- List physical and operating features of lead-acid cells
- Describe precautions to take with dry cells and wet cells
- List a variety of primary and secondary cells
- List typical electrical parameters of at least three primary and secondary cells

(a)

(b)

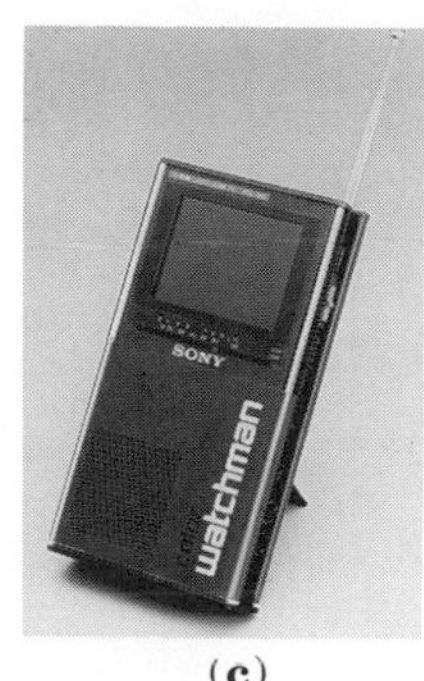

(c)

FIGURE 7–1 Some common items that use batteries as a power source: (a) 35 mm camera; (b) portable radio; (c) portable television (Photo a courtesy of Pentax; photo b courtesy of Philips Cons Elect; photo c courtesy of Sony, Inc.)

CHEMICAL ACTION IN A CELL

An early version of a cell that converts chemical energy to electrical energy is called a voltaic cell, named in honor of Alessandro Volta.

A voltaic cell consists of two different metal plates (copper and zinc) immersed in a diluted acid solution (like a weak sulfuric acid solution), Figure 7–2. These metal plates are the **electrodes** and the acid solution is the **electrolyte.**

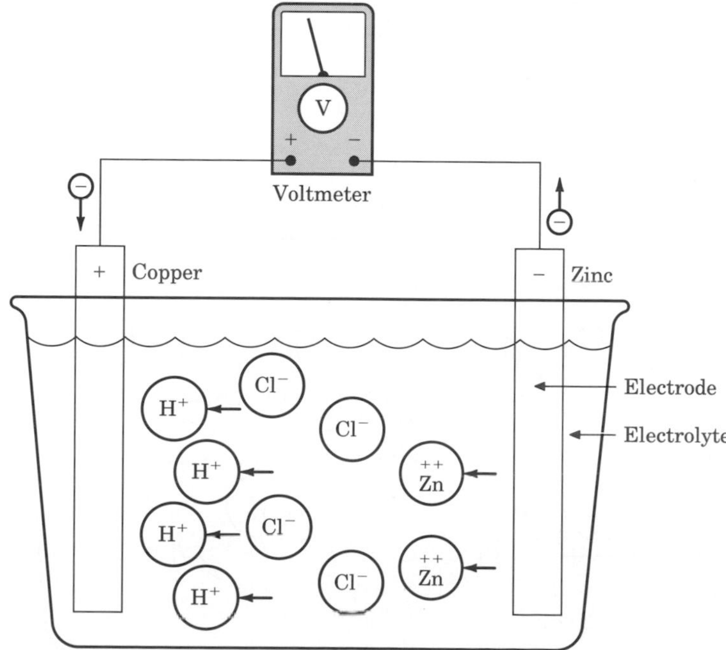

FIGURE 7–2 A simple voltaic cell consists of two different metal plates (electrodes) immersed in a diluted acid solution (electrolyte).

The chemical action occurring under these conditions is briefly described as follows:

1. When sulfuric acid dissolves in water, it splits into positive hydrogen ions and negative sulfate ions. (Recall from Chapter 1 that positive ions are atoms that have lost electrons and are positively charged entities. Negative ions are atoms that have gained electrons and are negatively charged entities).
2. When the zinc plate is immersed in the acid solution, the acid attacks the zinc. This action causes zinc ions to separate from the plate, leaving the zinc plate with an excess of electrons. The zinc plate has a net negative charge. NOTE: When the zinc plate wears away, the cell can no longer operate, and the cell is dead.
3. The positive hydrogen ions draw electrons from the copper plate, leaving the copper plate positively charged.
4. Because of the chemical actions described, the zinc plate is negative with respect to the copper plate (and/or the copper plate is positive with respect to the zinc plate). Therefore, a difference of potential exists between the electrodes. This electromotive force (emf) causes current to flow between the electrode terminals, if a conducting path is connected between them, Figure 7–3.

FIGURE 7–3 A cell produces current through the conducting path between electrodes.

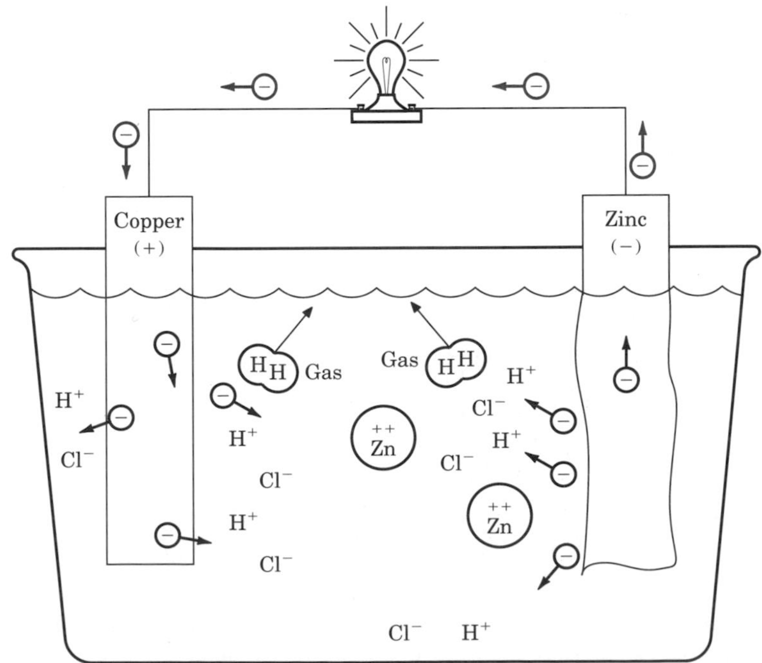

5. The open circuit difference of potential in primary cells depends on the metals used as the electrodes.

In summary, the concepts discussed show that chemical reactions produce electromotive force. This principle of changing chemical energy to electrical energy is fundamental to the operation of all cells and batteries.

THE COMMON DRY CELL

Using the materials previously described, the useful life of the voltaic cell is very short. Therefore, it is not a practical type of cell.

A more useful primary cell is the carbon-zinc cell (also known as the LeClanche cell). This cell is commonly termed a dry cell.

Construction

The basic construction of a dry cell is shown in Figure 7–4. The key components are:

1. the positive electrode, which is carbon;
2. the negative electrode, which is zinc; and
3. the electrolyte, which is a solution of ammonium chloride called sal ammoniac that forms a paste-like substance.

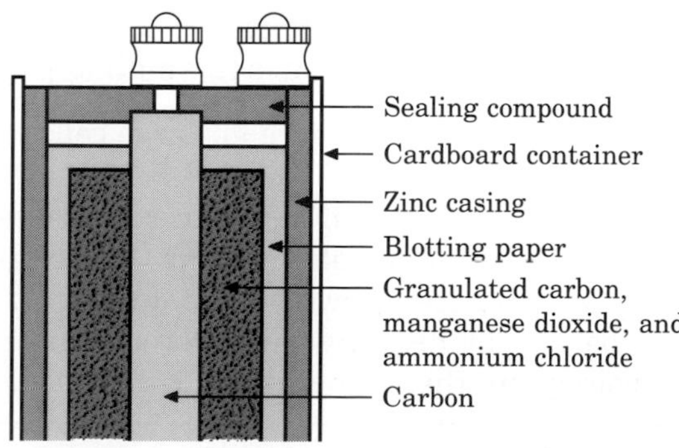

FIGURE 7–4 The construction of a carbon-zinc (LeClanche) cell, commonly termed a dry cell

Polarization

As the zinc electrode dissolves during the chemical process, hydrogen bubbles form on the positive carbon electrode that is detrimental to the cell's operation. This process is called **polarization.**

To overcome polarization, a depolarizer agent is introduced that combines its oxygen with the hydrogen bubbles to form water. This prevents the buildup of the undesired bubbles. A common depolarizing agent is manganese dioxide.

Local Action

Another operational problem confronted in the manufacture of dry cells is **local action.** Due to the manufacturing process, most zinc has impurities (such as carbon). Such impurities are suspended in the electrolyte close to the zinc and create independent cell actions. This undesired "voltaic" action dissolves the zinc.

To reduce local action, a thin coating of mercury is put on the zinc by a process called **amalgamation.** Amalgamation seals the impurities, separating them from the electrolyte to prevent local action.

Shelf Life

An unused dry cell has an indeterminant shelf life, since some local action occurs and the electrolyte paste will eventually dry up, depending on storage conditions.

The cell becomes worn-out when the zinc, electrolyte, or depolarizer is consumed. This is observed when output voltage, under normal load current conditions, is appreciably lower than it should be.

OPERATING CHARACTERISTICS OF DRY CELLS

During a dry cell's useful life, voltage output ranges from about 1.4 volts to 1.6 volts with 1.5 volts the nominal voltage.

Current capability typically ranges from a few milliamperes for small cells with small electrodes to $\frac{1}{8}$ ampere or more for large cells, such as a #6 dry cell.

As a cell or battery ages, its internal resistance increases. Consequently, at a given current level, its terminal voltage decreases accordingly. Recall in the previous chapter we described how you can determine the internal resistance of a cell or battery by dividing the difference between no-load and full-load terminal voltage by the load current at full load.

PRACTICE PROBLEM I

Answers are in Appendix B.

1. What is the internal resistance of a dry cell having a no-load terminal voltage of 1.5 volts and a full-load terminal voltage of 1.25 volts with a load current of 100 mA?
2. According to our nominal normal range values discussed earlier, is this cell considered good, fair, or bad in performance?

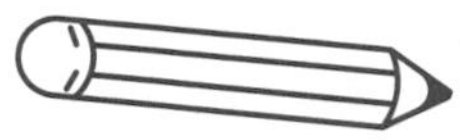

PRACTICAL NOTES

As indicated earlier, drying of the sal ammoniac electrolyte paste and local action shorten the shelf life of dry cells. Also, trying to draw too much current and/or drawing high current for extended periods will shorten operating life.

To lengthen shelf life, dry cells and batteries should not be stored in hot environments and should not be operated at currents that are not within normal specified operating ranges.

If terminal voltage appreciably drops from the nominal rated value, the internal resistance of the cell has increased due to its deteriorated operating condition. To insure proper operation of the circuit(s) using the cell as a source, replace the cell when this condition is apparent.

IN-PROCESS LEARNING CHECK I

Fill in the blanks as appropriate.

1. The carbon-zinc type dry cell is composed of a negative electrode made of ____________ and a positive electrode made of ____________. The electrolyte solution in this type cell is ____________ ____________, called sal ____________.
2. The process which causes hydrogen bubbles to form on the positive electrode, and which is detrimental to the cell's operation is called ____________.
3. Manganese dioxide is often added to the cell to prevent the process mentioned in question 2. This agent is sometimes known as a ____________ agent.
4. Another problem that can occur in dry cells is that of little batteries being formed due to impurities suspended in the cell's electrolyte near the zinc electrode. This type activity is known as ____________ ____________. To reduce this undesired activity, mercury is put on the zinc by a process called ____________.
5. The shelf life of dry cells is (enhanced, hurt) ____________ by storing the cells in hot conditions.
6. Nominal cell terminal voltage of dry cells is about ____________ V.

THE COMMON WET CELL

The lead-acid storage cell is the most common secondary cell. Desirable features are that it is rechargeable and can deliver high currents. For example, automobile batteries deliver several hundred amperes on a short-term basis.

Often lead-acid batteries are used in other applications in addition to being a power source in automobiles. In fact, there are many other practical uses of lead-acid batteries. One use is to provide power for mobile or portable operation of radio transmitters. Actually, anytime dc power is needed in a portable or mobile operation, this battery is often used.

In permanent installations, telephone companies often use glass-encased, stationary lead-acid cells because of their advantages.

CONSTRUCTION AND OPERATIONAL CHARACTERISTICS OF THE LEAD-ACID BATTERY

The key components are illustrated in Figure 7–5 and include:

1. the positive plate(s), which is lead peroxide (PbO_2) (brown);
2. the negative plate(s), which is spongy lead (Pb) (gray);

FIGURE 7–5 The construction of a lead-acid (secondary) battery

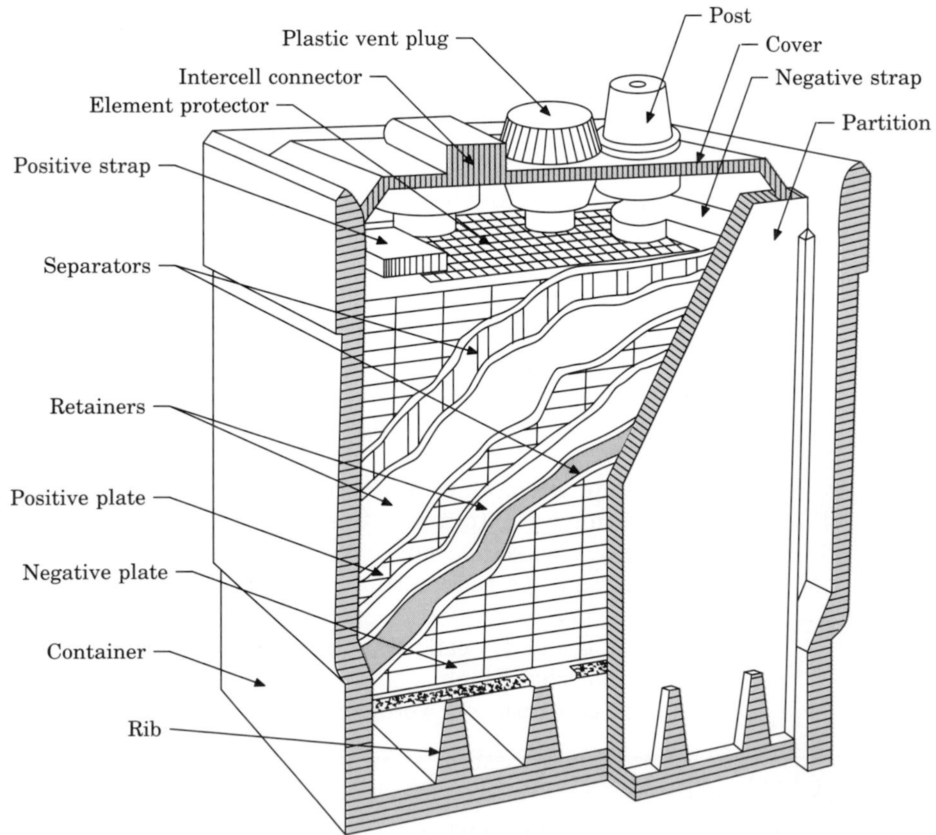

3. the electrolyte, which is dilute sulphuric acid (H_2SO_4);
4. nonconducting separators; and
5. hard rubber or plastic enclosure.

Charged Condition(s)

When the automotive lead-acid battery is fully charged, the conditions include:

1. Positive plate(s) are lead peroxide and dark brown.
2. Negative plate(s) are spongy lead and gray.
3. The electrolyte specific gravity is between 1.275 and 1.30.
4. Closed-circuit (normal load) voltage per cell is about 2 volts. (Open-circuit voltage per cell is about 2.2 volts without load when the battery is fully charged).

Discharging Condition(s)

As the lead-acid battery is discharging, Figure 7–6, the battery's chemical actions cause:

1. The positive plates to deteriorate, resulting in formulation of lead sulfate and water, which further dilutes the electrolyte;
2. The negative plates to deteriorate, resulting in formulation of lead sulfate; and
3. The electrolyte to weaken and decrease in specific gravity.

Discharged Condition(s)

When the lead-acid battery has fully discharged, again see Figure 7–6, typical conditions are:

1. Both positive and negative plates are changed to lead sulfate.
2. The creation of water weakens the sulfuric acid electrolyte so the specific gravity (measured with a hydrometer) drops below 1.150.
3. Closed-circuit voltage per cell drops to about 1.75 volts.
4. If the battery remains in an uncharged condition for too long, the sulfate hardens and the chemical reversal (via charging action) becomes impossible. This is called sulfation.

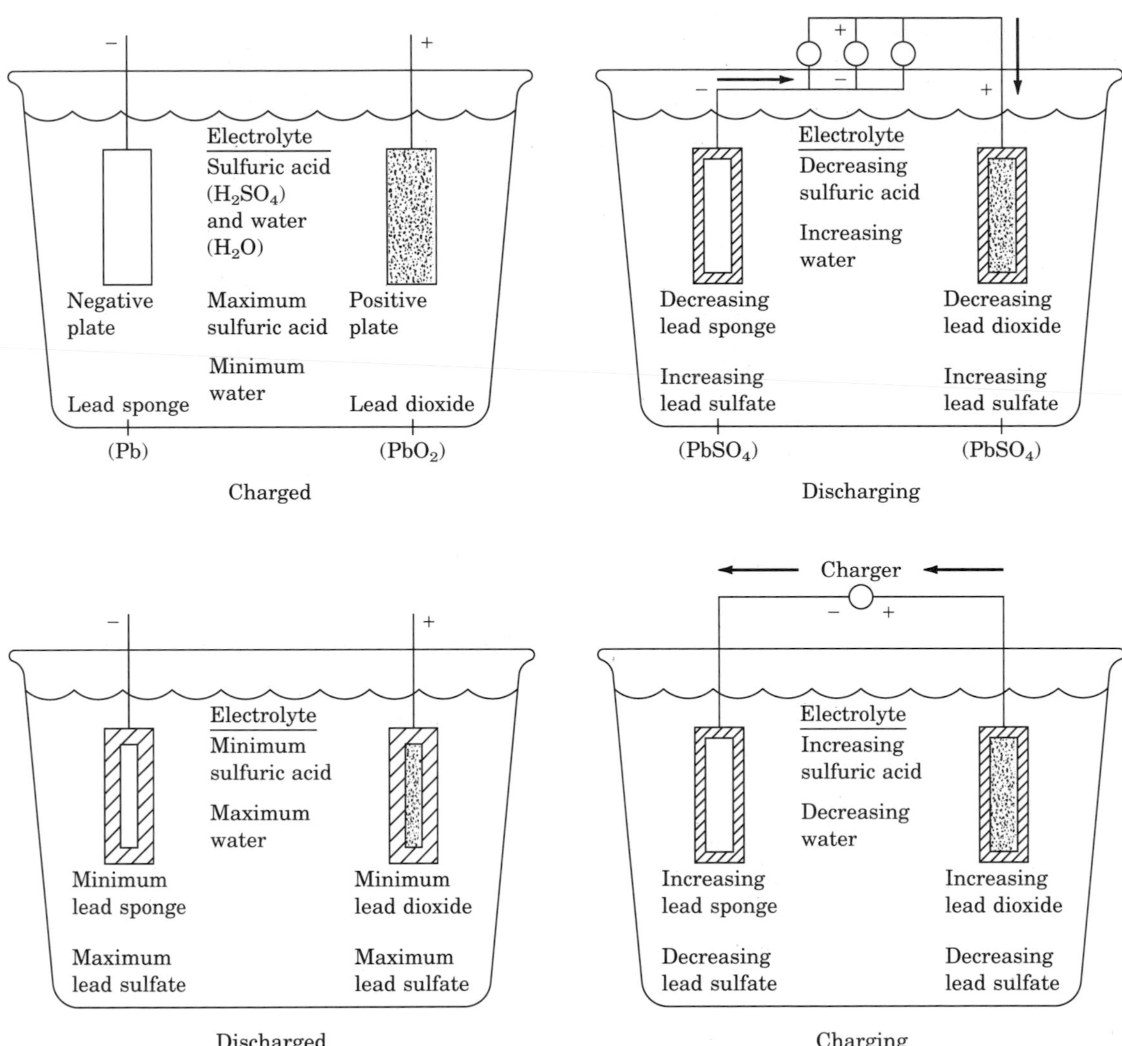

FIGURE 7–6 Conditions of cell from charged to discharged to charging states

Charging Condition(s)

During the charging process, chemical reactions reverse and the plates regain their original chemical properties. For example, the sulfuric acid becomes less dilute.

Examine Figure 7–7 to see the changes in chemical properties during the charging and discharging processes.

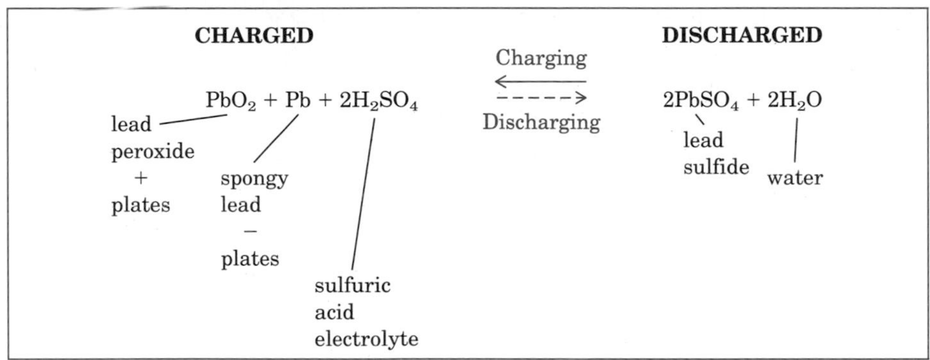

FIGURE 7–7 Changes in chemical properties occur during charging and discharging actions.

Battery Ratings

Typically batteries are rated based on an ampere-hour basis. The batteries we have been discussing are rated based on how many amperes they deliver over a specified time without their output voltage falling below some specified value. A rule of thumb frequently used is the number of amperes continuously delivered for eight hours.

EXAMPLE

A battery rated at 160 ampere-hours can deliver $\frac{160}{8}$, or 20 amperes for eight hours without terminal voltage falling below the specified value. On the other hand, if this same battery is rated on the 20-hour system, it will deliver 8 amperes for 20 hours since $\frac{160}{20} = 8$.

Different rating systems can be used. However, the catalog information or a specification sheet for the battery indicates the hours involved in the ampere-hour or milliampere-hour rating.

PRACTICE PROBLEM II

1. What continuous level of current can a battery with a 150 ampere-hour rating provide? Assume a 20-hour rating system.
2. What continuous level of current can this battery provide over eight hours?

CHARGING THE LEAD-ACID BATTERY

Overview

Charging any partially or fully discharged secondary battery involves forcing current through it in the reverse direction (from the normal current flow) when it is acting as a source.

Look at Figure 7–8. Notice the difference in current flow directions when the battery is a source (providing current to a load) and when the battery is the load for the charging device or circuit. NOTE: To accomplish the charging action, the charger voltage must be greater than the battery voltage; the negative output terminal of the charger is connected to the negative battery terminal; and the positive charger output terminal is connected to the battery positive terminal. That is, the connections are *negative-to-negative* and *positive-to-positive.*

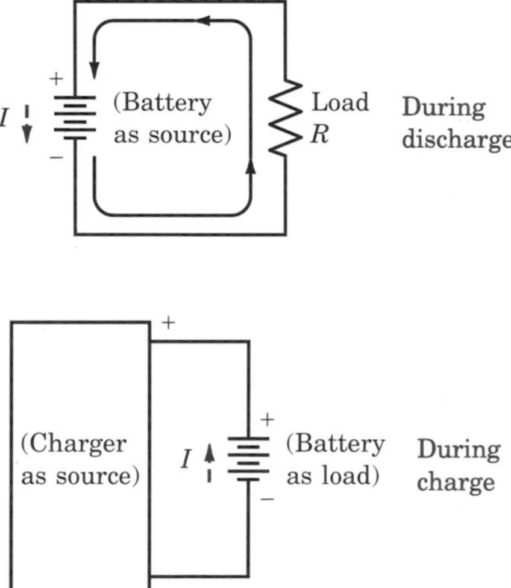

FIGURE 7–8
Directions of current through battery during charge and discharge

Two Charging Methods

The constant voltage method, Figure 7–9, applies a constant charging voltage to the battery. When the battery voltage, which is series-opposing to the charger voltage, is low, the charging current is high. As the battery increases its charge, thus its terminal voltage, the battery voltage almost equals the charging source. Thus, the charging current's value tapers downward as the battery charges and its series-opposing voltage nears that of the charger voltage.

The constant current method, Figure 7–10, essentially charges the battery at a constant current rate (usually the battery's normal discharge rate)

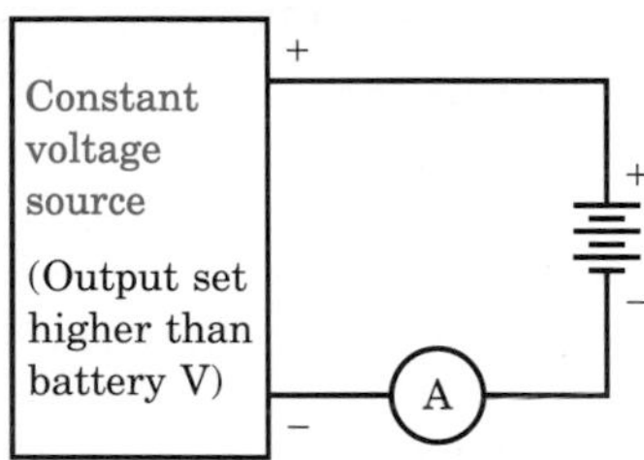

Higher charging current initially when battery's series-opposing voltage is lower.

Charging current tapers downward in value as battery charges, and its series-opposing voltage nears that of the charger voltage.

NOTE: A constant voltage source is one that has very low internal R; thus, varying values of load current won't cause source output voltage to change much.

FIGURE 7–9 Constant voltage method of charging a battery

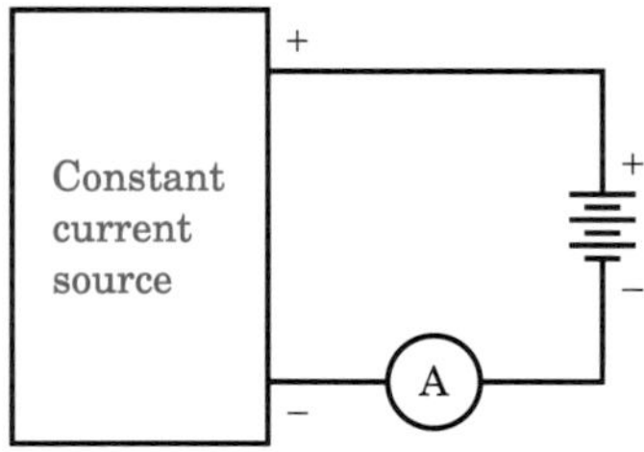

Charged at near normal discharge rate for about 90% of charging cycle, then reduced to about 40% of normal discharge rate for the remainder of the charging time.

NOTE: A constant current source has a high internal *R* compared to *R* of load; thus, load has little effect on circuit current.

FIGURE 7–10 Constant current method of charging a battery

for about 90% of the charging time. During the remaining time, the rate reduces to about 40% of normal discharge rate. For example, a 100-ampere-hour battery initially charges at a rate of about 12 amperes, then reduces to about 5 amperes for the remaining period of charge.

METHODS OF TESTING LEAD-ACID BATTERIES

Two common methods to test the condition of lead-acid batteries are to measure the specific gravity of the electrolyte solution and to make a short-term heavy-load test.

SAFETY HINTS

Don't charge at too fast a rate! This action damages the plates in the battery (buckling) and causes excessive gassing (produces hydrogen gas) and/or overheating. Avoid doing this! The longer the charging time and the lower the charging rate, the more safely the task can be performed. Generally, charging at less than a 10-ampere level for 24 hours is good. A faster rate can be used with the constant voltage method, starting at 20 to 40 amperes which will taper down with charging time. However, ***caution*** should be taken to assure the battery is not overheating during the charging cycle.

As the battery approaches full charge and begins producing hydrogen gas that escapes through open caps or vents, ***it is dangerous to allow any flames or sparks near the charging battery since this gas is volatile!*** Many people have been seriously hurt when exploding batteries throw acid on them! *Observe* and *remember* the cautions in Figure 7–11!

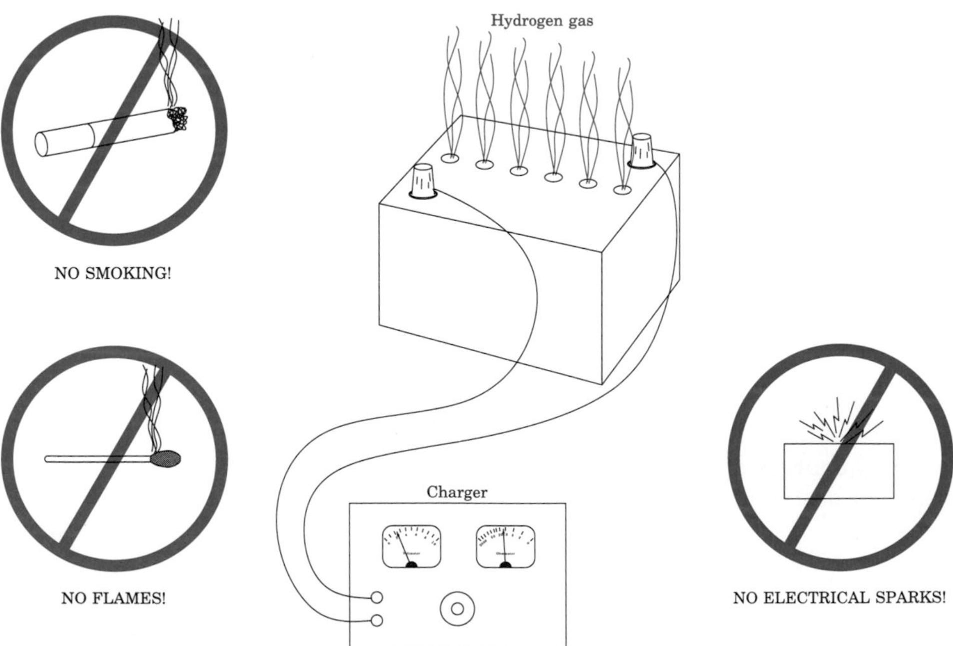

FIGURE 7–11 Charging batteries can explode due to gassing. Avoid flames and sparks at all times!

Hydrometer Test

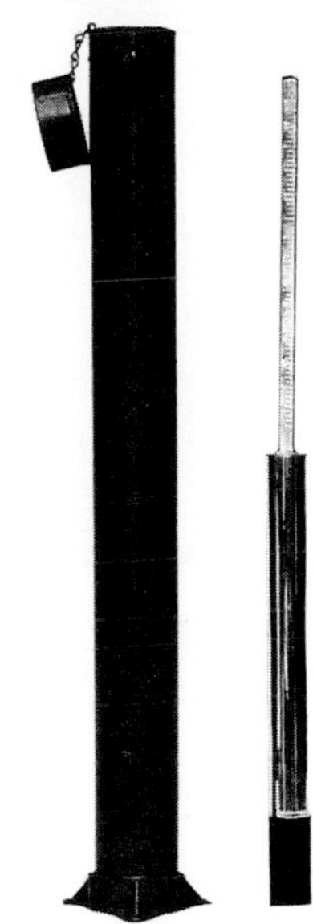

FIGURE 7–12 Hydrometer checks specific gravity of electrolyte. (Courtesy of Bico Braun Int.)

The hydrometer is a device containing a float that hovers evenly with the level of the liquid in the device when the liquid is water. If the liquid suctioned into the hydrometer has a specific gravity (s.g.) that is heavier than water, the float does not displace its volume and floats above the liquid level, Figure 7–12. Incidentally, specific gravity compares the weight of a liquid to an equal volume of water at 39.2 degrees Fahrenheit (F).

Because the float changes position with respect to the top of the liquid with fluids of different specific gravities, the float is calibrated to register specific gravity at different positions on the float.

Since a fully charged cell (electrolyte) has a s.g. of about 1.280 and a discharged s.g. of about 1.150, the hydrometer indicates the cell's state of charge. The s.g. of an electrolyte is inversely proportional to temperature; thus, as temperature increases s.g. decreases. The standard procedure is to correct all readings to 80 degrees F. This procedure requires a correction factor of plus or minus 0.0004 points per degree F. That is, if s.g. reads 1.270 at 110 degrees, then the correction factor yields a reading of $1.2700 + [(110 - 80) \times 0.0004] = 1.282$. A temperature correction scale and thermometer are common and integral parts of the hydrometer.

Short-Term Heavy-Load Test

The tester used for the short-term heavy-load test is a voltmeter with a special scale, indicating the cell's condition, Figure 7–13. When this device is connected across a cell, it provides a load of about 200 amperes on the cell. If the voltage under load falls significantly below the nominal 2 volts per cell value, the pointer on the meter scale indicates a bad condition.

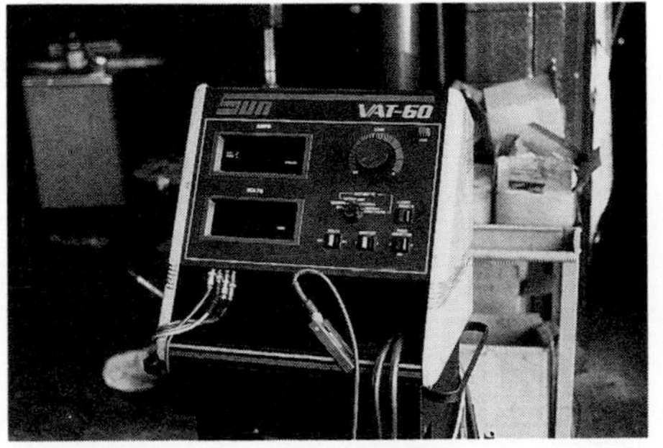

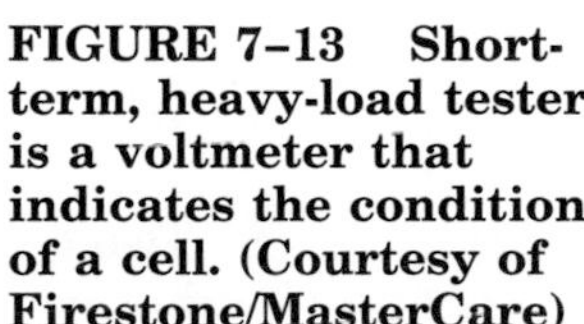

FIGURE 7–13 Short-term, heavy-load tester is a voltmeter that indicates the condition of a cell. (Courtesy of Firestone/MasterCare)

CARE OF LEAD-ACID BATTERIES

Basic rules to care for lead-acid batteries include the following:

1. Don't let the battery remain in an uncharged condition.

2. If the battery is not a completely enclosed type, make sure the level of liquid covers the plates.
3. If the battery is not a completely enclosed type, make sure that no dirt or foreign substances enter the battery when the cap(s) are removed.
4. Keep negative and positive terminals clean.
5. Keep the battery in a charged condition.

IN-PROCESS LEARNING CHECK II

1. The lead-acid battery comprises (primary, secondary) ________ cells which (are, are not) ________ rechargeable.
2. Lead peroxide makes up the (negative, positive) ________ plates of a lead-acid battery. Spongy lead makes up the (negative, positive) ________ plates of a lead-acid battery. The electrolyte of a lead-acid battery is dilute ________________.
3. The normal closed-circuit voltage per cell in a lead acid battery is approximately ________ volts.
4. Does the electrolyte in a lead-acid battery become more or less dilute as the battery discharges? ________.
5. Does the specific gravity of the electrolyte in a lead-acid battery increase or decrease as the battery becomes more discharged? ________. A fully discharged battery has a specific gravity of approximately ________.
6. What method is used to rate the current delivering capabilities of a lead-acid battery? ________.
7. Why are sparks and/or flame dangerous near a charging battery? ________.
8. Two types of tests that can be used to check the condition of a lead-acid battery are: a ________ test and the short-term ________ test.

EXAMPLES OF OTHER TYPES OF PRIMARY AND SECONDARY CELLS

Electronic parts catalogs list a variety of primary and secondary cells and batteries. We will briefly mention several common types whose application depend on inherent features, such as size, weight, current capacity, voltage, longevity, and durability.

Examples of Rechargeable Secondary Cells and Batteries

The Edison Cell. The Edison (nickel-iron) cell is frequently used in heavy-duty industrial applications. This cell is lighter and more durable than the lead-acid cell. Although it endures many rechargings, it is more expensive than the lead-acid cell.

FIGURE 7–14 The nickel-cadmium battery

The Nickel-Cadmium Battery. The nickel-cadmium (ni-cad) battery is the most frequently used, all-purpose rechargeable dry battery available today, Figure 7–14. Catalog listings for this battery indicate capacities in the range of 110 to over 1,000 milliampere-hour ratings. This battery has a long shelf life and is typically used in applications where there is cyclic, rather than continuous use. It is usually recharged using a "trickle" (slow) charge method.

Other Types. Other rechargeable secondary cells and batteries are lead-calcium, silver-zinc, and silver-cadmium.

Examples of Nonrechargeable Primary Cells and Batteries

Primary cells and batteries are used in a variety of applications, such as watches, cameras, computer memory protection circuits, and radios.

Alkaline Cell. The alkaline cell has a longer shelf life than the standard carbon-zinc dry cell, Figure 7–15.

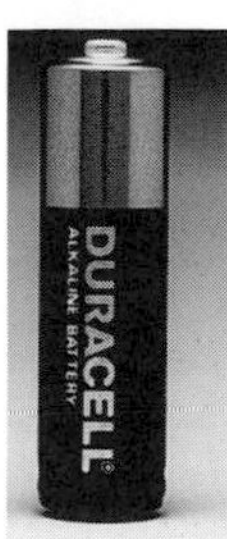

FIGURE 7–15 The alkaline cell (Courtesy of Duracell, Inc.)

Mercury Cell. The mercury is often used in voltage reference applications because of the stable output voltage (typically 1.35 volts per cell) during its lifetime, Figure 7–16.

Zinc-Chloride Cell. The main feature of the zinc-chloride cell is high current capacity.

Lithium Cell. Figure 7–17 shows examples of lithium cells. Because of their long life, lithium cells are commonly used in watches, cameras, calculators, computers, and similar devices. Although they are expensive, their long life and energy density make them practical.

Solar Cell. The solar cell is a special category of primary cells used in space vehicles, calculators, and so forth. It has a voltage of about 0.26 volts per cell and a very low current rating.

Refer to Figure 7–18 for a listing of primary and secondary cells and batteries and their nominal output voltage per cell.

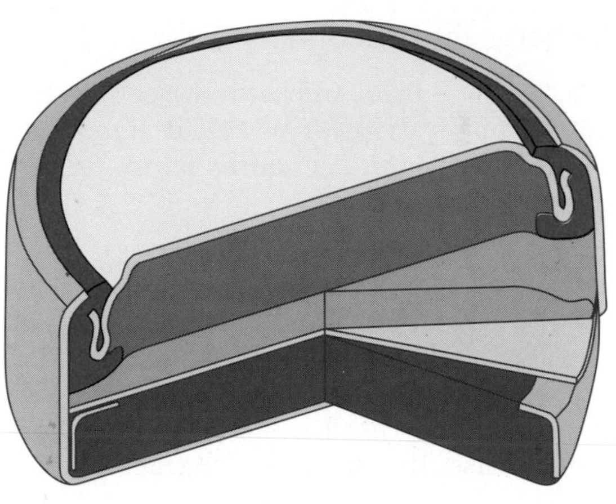

FIGURE 7–16 The mercury cell

Final Comments About Cells and Batteries

Selection of the right cell or battery for the job depends on the requirements of the application relative to *size, current drains, voltage, frequency of replacement,* and *cost.*

Recall from an earlier chapter the possibilities of connecting cells or batteries, Figure 7–19a, in series to increase voltage, Figure 7–19b, and of connecting like-value voltage sources in parallel to increase current delivery, Figure 7–19c. When performing either process, it is recommended that the cells and batteries have the same manufacturer's electrical ratings.

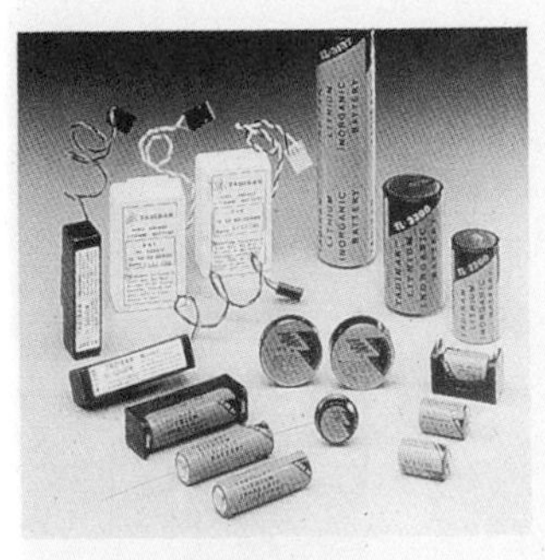

FIGURE 7–17 The lithium cell (Courtesy of Tadiran Electronics)

Also, recall that as a cell or battery deteriorates, its internal resistance increases. This means that for a given current drain, its output terminal voltage decreases due to the increased internal I × R drop at the given current value. NOTE: Some *newer types* of batteries typically *maintain* their *voltage* output then *suddenly die.* Examples are mercury (primary) cells and nickel-cadmium (secondary) cells.

Shelf life of cells is greatly affected by temperature. Most dry cells operate near 70 degrees F, so long exposure to higher temperatures drastically

FIGURE 7–18 Chart of primary and secondary cells/batteries with nominal voltage values

PRIMARY CELLS/BATTERIES (Non-rechargeable)		SECONDARY CELLS/BATTERIES (Rechargeable)	
TYPE	NOMINAL V	TYPE	NOMINAL V
Carbon-Zinc	1.5	Lead-Acid	2.2
Alkaline	1.5	Nickel-Iron	1.4
Mercury	1.35	Nickel-Cadmium	1.2
Silver-Oxide	1.5		
Lithium	3.0		

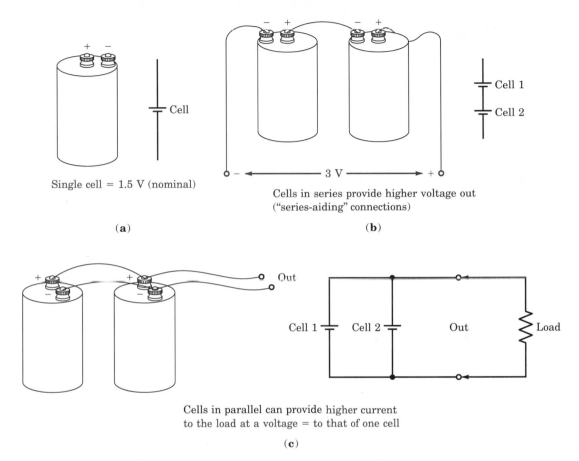

Single cell = 1.5 V (nominal)

(a)

Cells in series provide higher voltage out ("series-aiding" connections)

(b)

Cells in parallel can provide higher current to the load at a voltage = to that of one cell

(c)

FIGURE 7–19 Advantages of multiple-cell connections

shortens their life. Conversely, batteries stored for long periods at low temperatures (e.g., 0 degrees F) have a good shelf life.

PRACTICAL NOTES

Caution: *Do not leave dry cells or batteries in equipment for long periods without checking them!* With age, most cells and batteries leak electrolyte or acid that damages battery holder contacts, circuitry, or equipment. Therefore, remove a dead or degenerated cell or battery, even if a replacement is not available. Do not leave a dead cell or battery in the equipment.

Finally, do not recharge batteries at too high a charging rate; be cautious with sparks or flames near a charging battery; and, of course, be careful to prevent spillage of the electrolyte acid! Review Figure 7–11 one more time!

CHAPTER CHALLENGE

Following the SIMPLER troubleshooting sequence, find the problem with this circuit. As you follow the sequence, record your responses to each step on a separate sheet of paper.

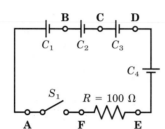

CHALLENGE CIRCUIT 4

Starting Point Information

1. Circuit Diagram
2. Cells are 1.5 V cells
3. With S_1 closed, $V_R = 5$ V

Test	Results in Appendix C
(With S_1 Open)	
$V_{A\text{-}B}$	(18)
$V_{B\text{-}C}$	(1)
$V_{C\text{-}D}$	(51)
$V_{D\text{-}E}$	(99)
$V_{A\text{-}F}$	(77)

SYMPTOMS Gather, verify, and analyze symptom information. (Look at the "Starting Point Information.")

IDENTIFY Identify initial suspect area for the location of the trouble.

MAKE Make a decision about "What type of test to make" and "Where to make it." (To simulate making a decision about each test you want to make, select the desired test from the "TEST" column listing below.)

PERFORM Perform the test. (Look up the result of the test in Appendix C. Use the number in the "RESULTS" column to find the test result in the Appendix.)

LOCATE Locate and define a new "narrower" area in which to continue troubleshooting.

EXAMINE Examine available information and again determine "what type of test" and "where."

REPEAT Repeat the preceding analysis and testing steps until the trouble is found. What would you do to restore this circuit to normal operation? When you have solved the problem, compare your results with those shown in the color insert.

(With S_1 Closed)	
$V_{A\text{-}B}$	(7)
$V_{B\text{-}C}$	(50)
$V_{A\text{-}C}$	(81)
$V_{C\text{-}D}$	(112)
$V_{D\text{-}E}$	(73)
$V_{C\text{-}E}$	(10)
$V_{A\text{-}F}$	(105)

SUMMARY

- Cells and batteries (combinations of cells) are important sources of dc voltage and current, especially for portable equipment (e.g., emergency back-up equipment).
- Primary cells are *not rechargeable,* have a variety of voltage ratings, current capabilities, expected shelf life ratings, and costs, and are produced in many sizes and shapes.
- Secondary cells and batteries *are rechargeable.* Secondary cells or batteries (such as lead-acid batteries and ni-cad batteries) are available in a variety of ratings, sizes, shapes, and costs.
- The common dry cell is composed of a positive carbon electrode, a negative zinc electrode, and an electrolyte of ammonium chloride (and water), called sal ammoniac.
- During a dry cell's normal operating chemical process, hydrogen bubbles form around the positive electrode. This polarization degenerates the battery's output. To reduce polarization, a depolarizing agent is placed in the dry cell during manufacture. A common depolarizing agent is manganese dioxide.
- Local action also causes degeneration in dry cells. Impurities in the negative zinc electrode suspend in the electrolyte near the zinc, creating their own voltaic action and eating away the zinc. To reduce local action, the inside of the zinc container (negative electrode) is covered with a thin coating of mercury; thus preventing most potential for local action.
- Shelf life of cells and batteries depends on storage conditions and chemical properties.
- Voltage ratings of cells and batteries depend on the electrode material. Current ratings largely depend on the size (active area) of the electrodes.
- When using dry cells, care should be taken with their storage. Also, frequently check their condition, and make replacements, as appropriate.
- One common secondary (rechargeable) cell is the lead-acid cell. In this cell, the positive plates are lead peroxide, the negative plate(s) are spongy lead, and the electrolyte is diluted sulfuric acid.
- Typical specific gravity (s.g.) of a fully charged lead-acid cell's electrolyte is about 1.280. As the cell discharges, its specific gravity declines, and when the cell has fully discharged, the s.g. is about 1.150.
- Another test to check the condition of a lead-acid battery (and other types) is a short-term heavy-load test. During this test, if the terminal voltage falls below a specified value, this indicates a weak or bad cell or battery.
- Batteries are rated according to the amount of current they deliver for a specified time. For example, a battery rated at 120 ampere-hours can nominally deliver 15 amperes for 8 hours, or 6 amperes for 20 hours. Small batteries and cells are rated in milliampere-hours.
- Rechargeable batteries are charged at high rates for short periods, or at lower rates for longer periods to restore them to full charge. The preferred method is to charge at lower rates for longer periods, decreasing the risk of cell or battery damage as well as preventing excessive gassing.
- In addition to the common carbon-zinc primary cell, many other types exist, such as alkaline, mercury, zinc-chloride and lithium cells.
- Common types of rechargeable batteries are the lead-acid, nickel-iron (Edison) and nickel-cadmium.

REVIEW QUESTIONS

1. Briefly define positive ion.
2. In a carbon-zinc cell, which element is the positive electrode and which is the negative electrode?
3. Briefly describe the process of amalgamation.
4. What term describes the undesired collection of hydrogen bubbles around a dry cell's positive electrode?
5. Name a common depolarizing agent.
6. Name a common rechargeable cell used for many small portable electronic devices.
7. What type of cell is frequently a reference voltage source due to its stable output voltage?
8. Name two advantages and one disadvantage of the lead-acid cell.
9. What is the ampere-hour rating of a battery that delivers 12.5 amperes for 8 hours without its output voltage excessively dropping?
10. Using the constant voltage method, describe the process and features of charging action when a battery is recharged.

TEST YOURSELF

1. Explain the polarity relationships between charger and battery when recharging a battery.
2. Using the constant current method, explain the process when a battery is recharged.
3. Refer to an electronic parts catalog and find the typical ampere-hour or milliampere-hour ratings for:
   a. Ni-cad cells
   b. Alkaline dry cells
4. What is the internal resistance of a battery whose no-load voltage is 12 volts; however, the terminal voltage drops to 11 volts when a load current of 20 amperes is demanded?

5. Refer to an electronic parts catalog. Find and record the voltage and current for the following cells:
   a. AAA
   b. AA
   c. A
   d. C
   e. D
6. What primarily determines the voltage delivering capability of a single cell?
7. What primarily determines the current delivering capability of a single cell?
8. Why does a cell's output voltage decrease under load, particularly when the cell is going bad?
9. What are some common applications of the modern lithium cell?
10. What makes the lithium cell so useful for the applications you named in question 9?

Magnetism and Electromagnetism

Key Terms

AT
D
Electromagnetism
Faraday's law
Flux
Flux density
H
Induction
Left-hand rules
Lenz's law
Lines of force
Magnetic field
Magnetic field intensity
Magnetic polarity
Magnetism
Magnetomotive force
μ
μ_r
Permeability
Relative permeability
Reluctance
Residual magnetism
Saturation
Tesla
Weber

Courtesy of Hewlett-Packard Company

8

Outline

Background Information

Fundamental Laws, Rules, and Terms To Describe Magnetism

Elemental Electromagnetism

Important Magnetic Units, Terms, Symbols, and Formulas

Practical Considerations About Core Materials

The B-H Curve

The Hysteresis Loop

Induction and Related Effects

Faraday's Law

Lenz's Law and Reciprocal Effects of Motors and Generators

Summary Comments and Observations About Magnetism

Chapter Preview

Of all the phenomena in the universe, one of the most fascinating is magnetism and its unique relationship to electricity.

Knowledge of magnetism is important to the understanding of important devices, such as electrical meters, motors, generators, inductors, and transformers. You will study these devices as you proceed in your goal to become a technician.

In this chapter you will study basic principles about magnetism; some units of measure used with magnetic circuits; and important relationships between magnetic and electrical phenomena.

Objectives

After studying this chapter, you will be able to:

- Define **magnetism, magnetic field, magnetic polarity,** and **flux**
- Draw representations of magnetic fields related to permanent magnets
- State the magnetic attraction and repulsion law
- State at least five generalizations about magnetic lines of force
- Draw representations of fields related to current-carrying conductors
- Determine the polarity of electromagnets using the **left-hand rule**
- List and define at least five magnetic units of measure, terms, and symbols
- Draw and explain a B-H curve and its parameters
- Draw and explain a hysteresis loop and its parameters
- Explain motor action and generator action related to magnetic fields
- List the key factors related to induced emf
- Briefly explain **Lenz's law**

BACKGROUND INFORMATION

Early History

Early discovery of magnetism happened by people noticing the unique characteristics of certain stones called lodestones. These stones (magnetite material), which were found in the district of Magnesia in Asia Minor, attracted small bits of iron. It was also noted that certain materials stroked by lodestones became magnetized.

Materials or substances that have this unique feature are called magnets, and the phenomenon associated with magnets is called **magnetism.** Materials that are attracted by magnets are called magnetic materials.

As early as the eleventh century, the Chinese used magnets as navigational aids, since magnets that are suspended or free to move align approximately in a north-south geographic direction.

Today we have materials, called permanent magnets, that retain their magnetism for long periods. Examples of permanent magnets include hard iron or special iron-nickel alloys.

Magnetic materials that lose their magnetism after the magnetizing force is removed are termed temporary magnets.

Later Discoveries

Centuries later the relationship between electricity and magnetism was discovered.

In the 1800s several scientists made important observations. Hans Oersted noted that a free-swinging magnet would react to a current-carrying wire, if they were in close proximity. As long as current passed through the wire the reaction occurred. But when there was no current, there was no interaction between the magnet and wire. This indicated a steady current caused a steady magnetic effect near the current-carrying conductor. This "field of influence" would eventually be called a **magnetic field.**

Later in the same century, Joseph Henry and Michael Faraday discovered interesting phenomena that occurred when changing current levels or moving magnetic fields were present.

Henry determined when current changed levels in a conductor, current was induced in a nearby conductor (having a closed current path).

Faraday determined a permanent magnet moving in the vicinity of a conductor induced current in the wire (if there existed a complete path for current through the conductor).

All of these discoveries are very important! You will see these scientists' names again since several units of magnetic measure are named in their honor.

FUNDAMENTAL LAWS, RULES, AND TERMS TO DESCRIBE MAGNETISM

When a magnet is created, there is an area of magnetic influence near the magnet called a **magnetic field.** A magnetic field is established either by alignment of internal magnetic forces within a magnetic material, Figure 8–1, or by organized movement of charges (current flow) through conductor materials. This magnetic field is composed of magnetic **lines of force,** sometimes called **flux lines.** Thus, magnetic flux refers to all the magnetic lines of force associated with that magnet. The symbol for flux is the Greek letter Phi ϕ. Each magnetic field line is designated by a unit called the maxwell (Mx). That is, one magnetic field line equals one maxwell, or 50 magnetic field lines equal 50 maxwells.

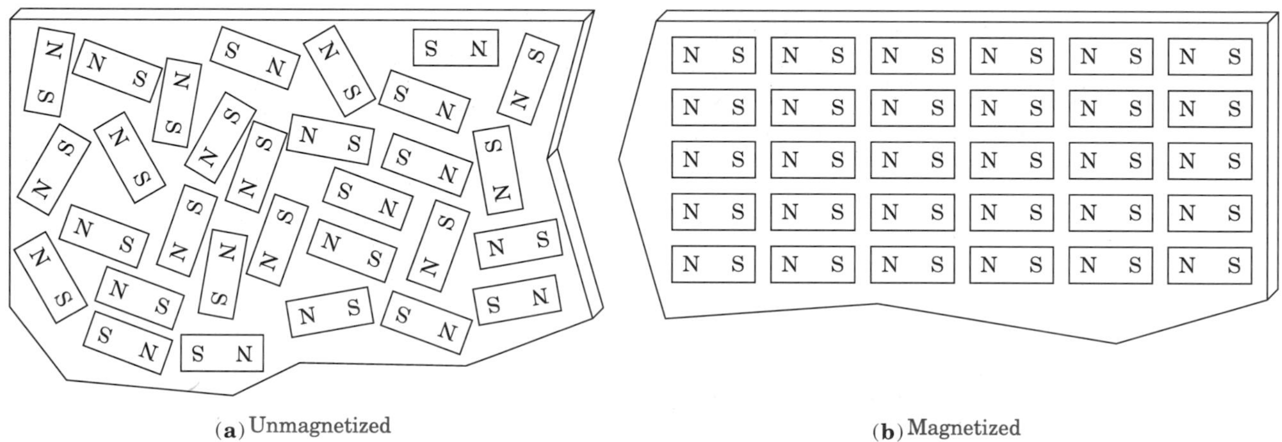

(a) Unmagnetized (b) Magnetized

FIGURE 8–1 Magnetized and unmagnetized conditions

Another basic unit associated with flux is the weber. Whereas the maxwell equals one line of flux, the weber (Wb) equals 10^8 lines of flux, or 100 million flux lines. Thus, a microweber equals 100 lines, or 100 Mx.

Another fundamental term related to flux is flux density that refers to the number of lines per given unit area. We'll discuss flux density later in the chapter.

If a nonmagnetic material, like cardboard, is laid on top of a bar magnet (or magnets), sprinkled with small iron filings then is tapped or vibrated, the iron filings position themselves in a pattern illustrating the effect of the magnetic field around the magnet, Figure 8–2.

A diagram frequently used to show this effect is in Figure 8–3. Notice that at each end of the bar magnet, the lines of force are concentrated, but as they get farther from the ends, they spread. As you can see, the ends of the bar magnets (where the lines of force are concentrated) are labeled "N" and "S" and are the "poles" of the magnet. That is, the N and S indicate North-seeking and South-seeking poles.

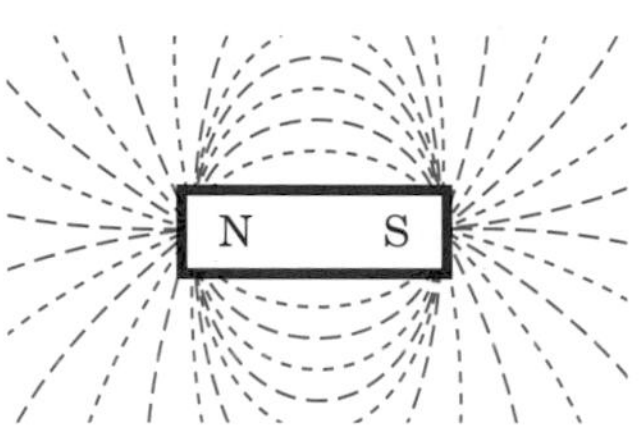

FIGURE 8–2 Iron filings displaying magnetic field pattern of a bar magnet

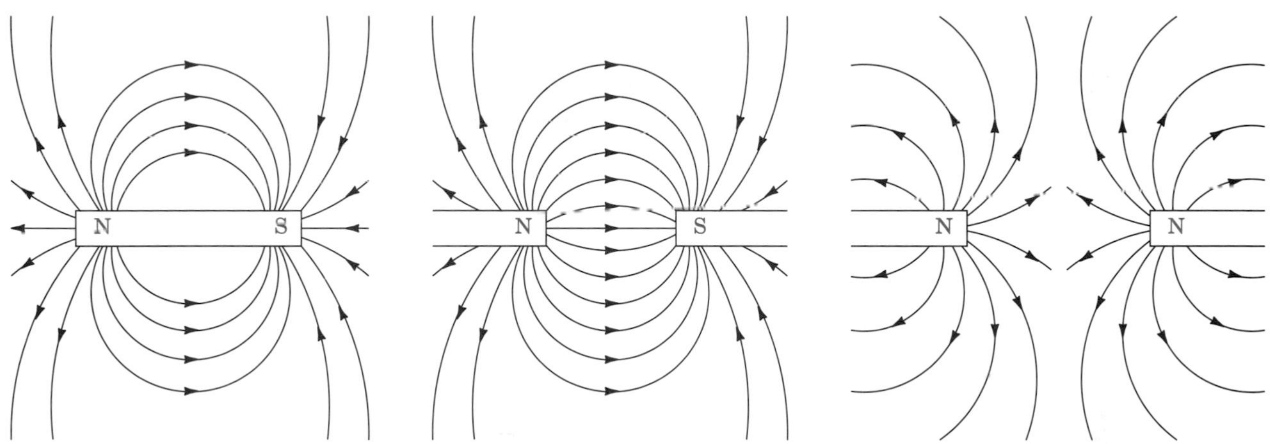

FIGURE 8–3 Lines of force

To represent directions of the lines of force, it is customary to say the flux lines exit the North pole and reenter the South pole of the magnet. Inside the magnet, of course, the flux lines travel from the South pole to the North pole. Each flux line is an unbroken loop, or ring.

In Figure 8–4 you can see two bar magnets used to illustrate that *like*

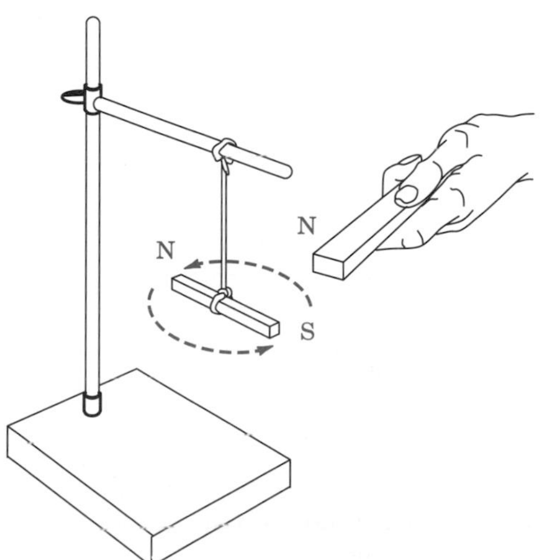

FIGURE 8–4 Like poles repel each other, and unlike poles attract each other.

poles repel each other, and unlike poles attract each other. This is an important law so remember it!

Rules Concerning Lines of Force

Along with the law defined above, you need to know the following important generalizations related to *lines of force* (flux lines), Figure 8–5.

1. Magnetic lines are continuous, Figure 8–5a.
2. Magnetic lines flow from North to South Pole outside the magnet and from South to North inside the magnet, Figure 8–5a.

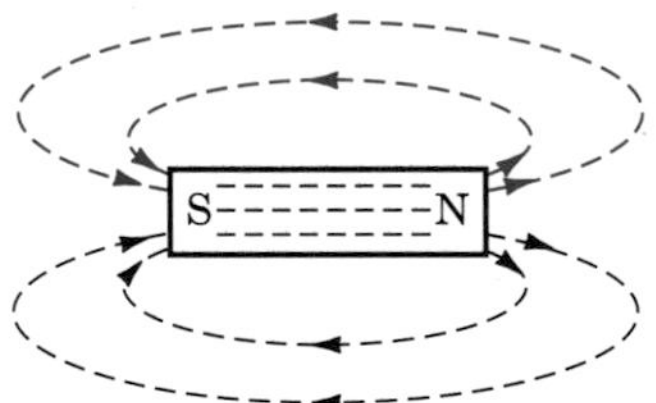

Magnetic lines are continuous and are assumed to flow from North to South poles external to the magnet and from South to North poles inside the magnet.

(a)

S

N

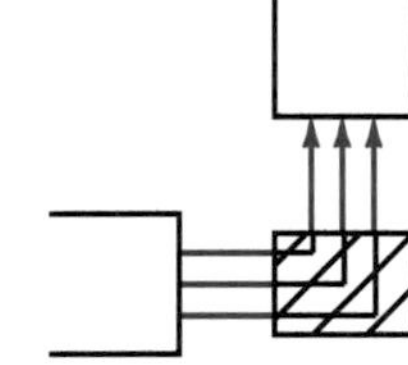

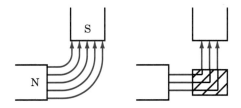

Magnetic lines tend to take the shortest or easiest path.

(b)

Lines in same direction repel each other, strengthening overall field.

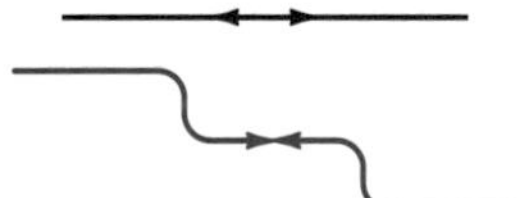

Lines in opposite directions attract and cancel each other, weakening overall field.

(c)

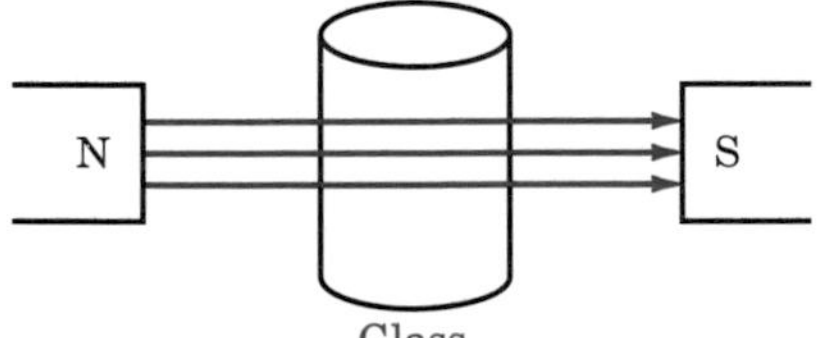

Magnetic lines penetrate nonmagnetic materials.

(d)

Lines of force don't cross each other.

(e)

FIGURE 8–5 Rules about lines of force

3. Magnetic lines take the shortest or easiest path, Figure 8–5b.
4. Lines in the same direction repel each other but are additive, strengthening the overall field. Lines in opposite directions attract and cancel each other, weakening the overall field, Figure 8–5c.
5. Magnetic lines penetrate nonmagnetic materials, Figure 8–5d.
6. Lines of force do not cross each other, Figure 8–5e.

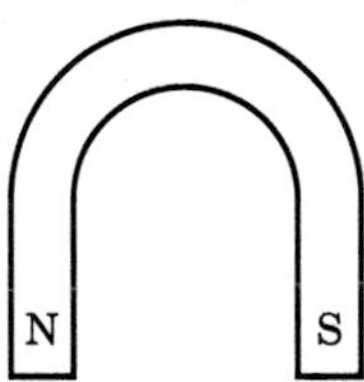

FIGURE 8–6 Horseshoe magnet

Some Practical Points About Permanent Magnets

Before discussing electromagnetism, here are some practical points about permanent magnets.

1. They are not really permanent, because they lose their magnetic strength over time, or if heated to a high temperature, or if physically pounded upon, and so on.
2. They can be configured into different shapes in addition to bar magnets. One common shape is termed a horseshoe magnet due to its appearance, Figure 8–6.
3. The field lines' shape depends on the easiest path for flux, Figure 8–7.

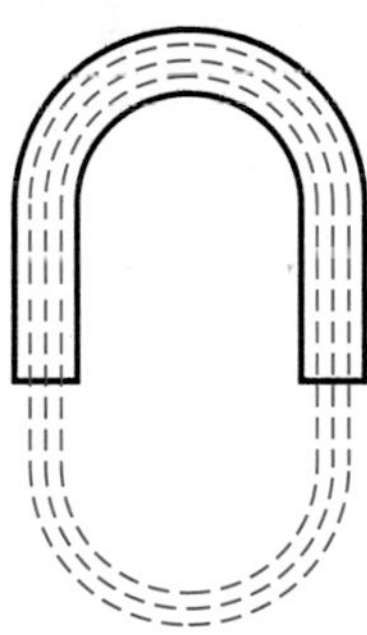

FIGURE 8–7 Shape of field lines depends on easiest path for flux.

4. Appropriate methods to store magnets help preserve their field strengths for longer periods. For example, store bar magnets so opposite poles are together, or use a "keeper bar" of magnetic material across the ends of a horseshoe magnet, Figure 8–8.

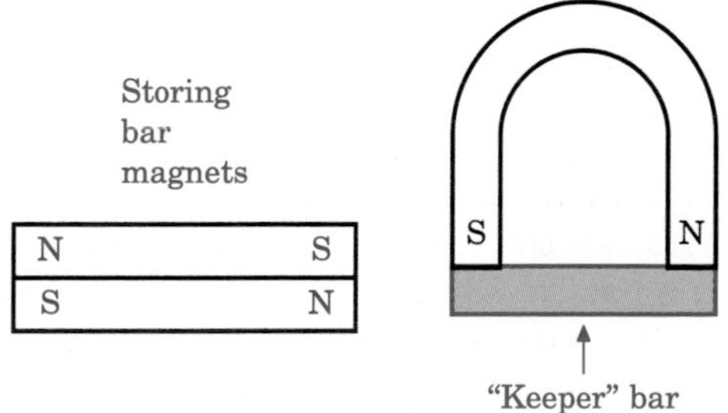

FIGURE 8–8 Preserving the strength of magnets

5. A magnet induces magnetic properties in a nearby object of magnetic material, if the object is in the path of the original magnet's field lines, Figure 8–9.

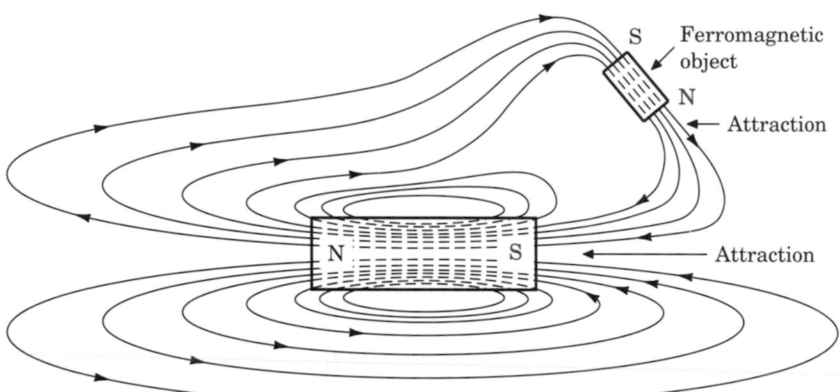

NOTE: Polarity of "induced" magnetism places "unlike" poles next to each other. This explains how magnets attract metal objects.

FIGURE 8–9 Induced magnetism

IN-PROCESS LEARNING CHECK I

Fill in the blanks as appropriate.

1. Small magnets that are suspended and free to move align in a ________ and ________ direction.
2. Materials that lose their magnetism after the magnetizing force is removed are called ________ ________. Materials that retain their magnetism after the magnetizing force is removed are called ________ ________.
3. Can a wire carrying dc current establish a magnetic field? If it does, is this field stationary or moving?
4. A law of magnetism is like poles ________ each other and unlike poles ________ each other.
5. The maxwell is a magnetic unit that represents how many lines of force?
6. The weber is a magnetic unit that represents how many lines of force?
7. Are magnetic lines of force continuous or noncontinuous?
8. Lines of force related to magnets exit the ________ pole of the magnet and enter the ________ pole.
9. Do nonmagnetic materials stop the flow of magnetic flux lines through themselves?
10. Can magnetism be induced from one object to another object?

ELEMENTAL ELECTROMAGNETISM

Direction of Field Around a Current-Carrying Conductor

Recall that Hans Oersted discovered a current-carrying wire has an associated magnetic field. A relationship between the current direction through the wire and the magnetic field direction around the current-carrying conductor is established while moving the compass around the conductor and noting the effect on the compass needle.

Note in Figure 8–10a you are looking at the end of wire (blue dot), and the current is coming toward you. In Figure 8–10b you get an overview of the relationship of current direction with the magnetic field around the current-carrying conductor. Both of these figures illustrate one of the **left-hand rules.**

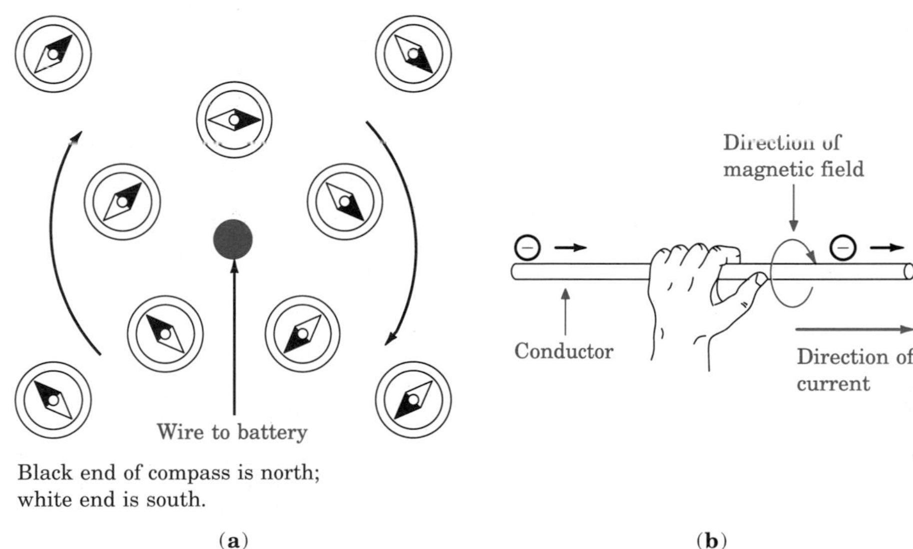

FIGURE 8–10 (a) Field around current-carrying conductor; (b) Left-hand rule for current-carrying conductor

The Left-Hand Rule for Current-Carrying Conductors

Grasping the wire with your left hand so your extended thumb points in the direction of current flow, the magnetic field around the conductor is in the direction of the fingers that are wrapped around the conductor.

EXAMPLE

Another way to illustrate this rule is shown in Figure 8–11. In this illustration the current (and your thumb) is coming toward you from the conductor with the dot, Figure 8–11a. The dot is like looking at the tip of an arrow coming toward you. In this case, the magnetic field is in a clockwise direction.

In the other picture, Figure 8–11b, the conductor has a plus sign, indicating current is going away from you into the page. This is similar to looking at the tail feathers of an arrow. In this case, the direction of magnetic field is counterclockwise.

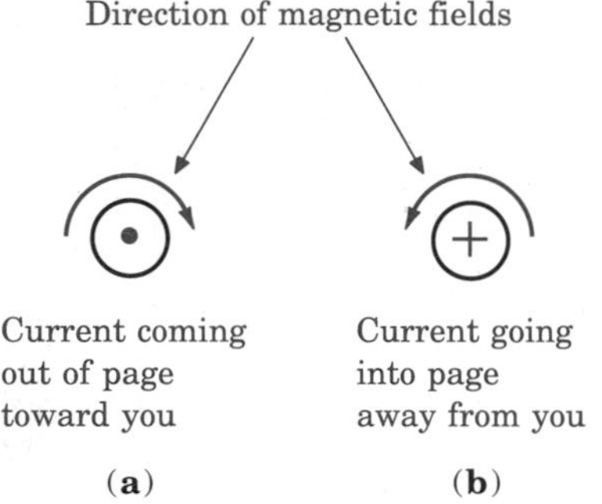

FIGURE 8–11 Magnetic fields around current-carrying conductors

PRACTICE PROBLEM I

Answers are in Appendix B.

1. Draw a picture of two parallel conductors with the direction of current going away from you in both cases.
2. Using arrows, show the directions of the magnetic fields around each conductor.

Force Between Parallel Current-Carrying Conductors

Look at Figure 8–12a. Notice that when two parallel current-carrying wires have current in the same direction, the flux lines between the wires are in opposite directions. This situation causes the flux lines to attract and cancel each other, and the magnetic field pattern is modified since lines of force cannot cross. This indicates the two adjacent wires move toward the weakened portion of the field or toward each other.

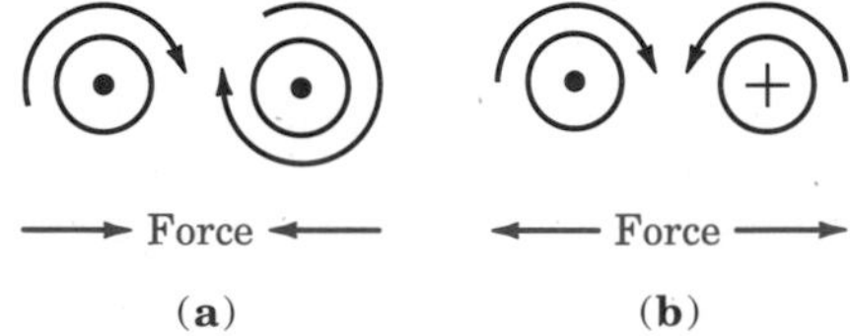

FIGURE 8–12 Force between adjacent current-carrying conductors

On the other hand, if the two wires have current in opposite directions, Figure 8–12b, the lines of force between the two conductors move in the same direction. Hence, the flux lines repel each other and the wires move apart.

PRACTICE PROBLEM II

Indicate whether the conductors move together (toward each other) or apart in Figure 8–13a, Figure 8–13b, and Figure 8–13c.

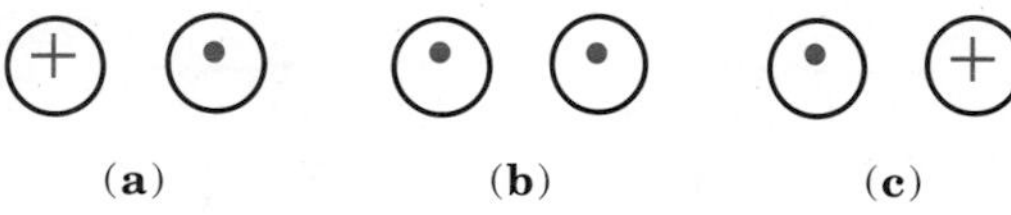

FIGURE 8–13 Current-carrying conductor problems

Field Around a Multiple-Turn Coil

Individual fields of conductors carrying current in the same direction modify to a larger and stronger field, as shown in Figure 8–14.

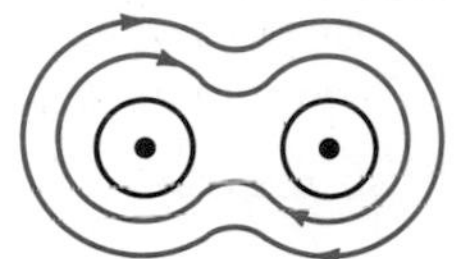

FIGURE 8–14 Stronger "combined" field

When a single current-carrying wire forms multiple loops (such as in a coil), there are a number of fields that combine from adjacent turns of the coil, producing a large field through and around the coil, Figure 8–15.

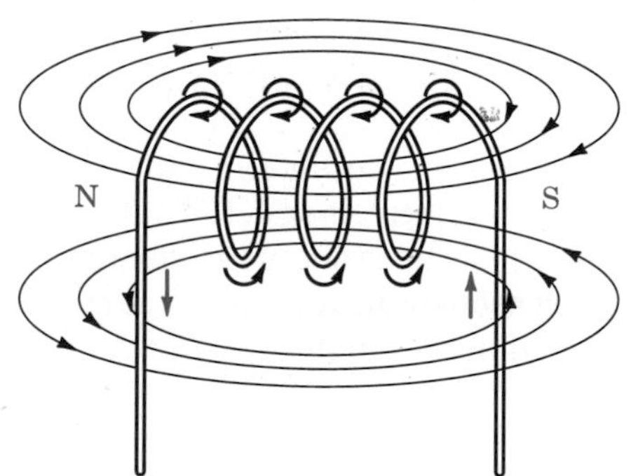

FIGURE 8–15 Strong field of coil and establishment of magnetic polarity

Also notice in Figure 8–15 the magnet established by this coil has North and South magnetic poles, like a permanent bar magnet. The magnetic polarity of this electromagnet is determined by another left-hand rule.

Left-Hand Rule To Determine Polarity of Electromagnets

Grasping the coil so your fingers are pointing in the same direction as the current passing through the coil, the extended thumb points toward the end of the coil that is the North pole, Figure 8–16.

PRACTICE PROBLEM III

See if you can determine the magnetic polarity of the coils in Figure 8–17a, 8–17b, and 8–17c.

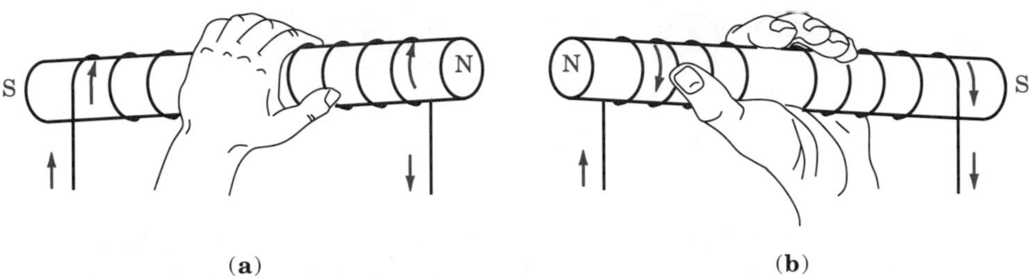

FIGURE 8–16 Left-hand rule to determine magnetic polarity of an electromagnet

FIGURE 8–17 Polarity-determination problems

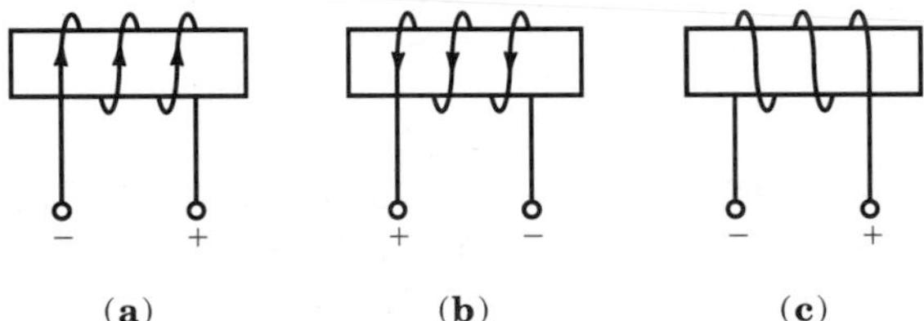

Factors Influencing the Field Strength of an Electromagnet

From the previous discussions, you should surmise the number of turns influences the strength of the electromagnet. That is, the higher the number of turns per unit length of coil, the greater the strength of the magnetic field.

Also, you might have reasoned the amount of current through the conductor directly effects the magnetic strength, since a higher current indicates more moving charge per unit time, Figures 8–18a and 8–18b. These are correct assumptions.

Another factor influencing strength is the characteristics of the material in the flux-line path. If the material in the flux path is magnetic, the strength is greater compared with a path using nonmagnetic material, Figure 8–18c.

The length of the magnetic path also influences strength. For example, suppose we have a 100-turn coil with a given length of "X." Now if we spread that 100 turns over a length twice as great (2X), the strength reduces because the turns are farther apart, thus the effective magnetizing force is less, Figure 8–18d.

To summarize, field strength is influenced by:

1. number of turns on the coil;
2. length of the coil (number of turns per unit length);
3. value of current through coil; and
4. type of core (magnetic characteristics of material in flux path).

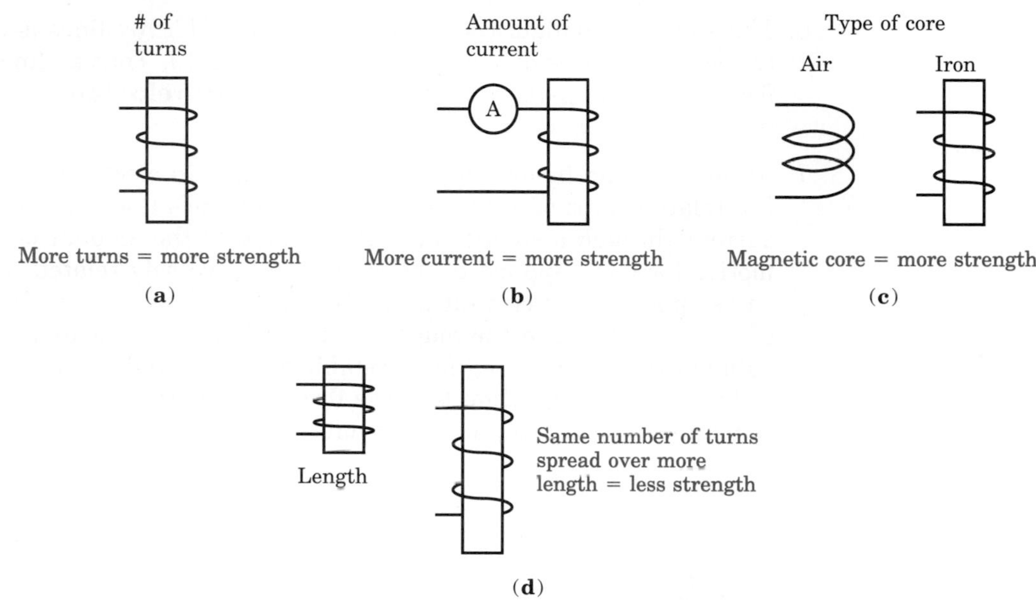

FIGURE 8–18 Some factors that determine strength of an electromagnet

Highlighting Magnetism and Electromagnetism Facts Discussed Thus Far:

1. Like poles repel, and unlike poles attract, again see Figure 8–4.
2. Amount of repulsion or attraction depends on the strength and proximity of the poles involved. Repulsion or attraction is directly related to the poles' strength and inversely related to the *square* of their separation distance.

Formula 8–1	$\text{Force} = \dfrac{\text{Pole 1 Strength} \times \text{Pole 2 Strength}}{d^2}$

3. Strength of magnet is related to number of flux lines in its magnetic field per unit cross-sectional area.
4. Magnetic lines outside of the magnet travel from North pole to South pole, again see Figure 8–5a.
5. Characteristics of the flux path determine how much flux is established by a given magnetizing force. Therefore, if the path is air, vacuum, or any nonmagnetic material, less flux is established. If the path is a magnetic material flux lines are more easily established. These concepts indicate when air gaps are introduced into a magnetic path, the magnetic strength of the field is diminished.

6. The shape or configuration of the magnetic field flux lines is controlled by features of the materials in their path. That is, lines of force follow the path of least opposition, called **reluctance** in magnetic circuits, again see Figure 8–5b.
7. An analogy can be drawn between Ohm's law in electrical circuits and the relationships of a magnetic circuit. In Ohm's law, the amount of current through a circuit is directly related to the amount of electromotive force (V) applied to the circuit and inversely related to the circuit's opposition to current flow (R). In a magnetic circuit, the number of flux lines through the magnetic circuit is directly related to the value of the magnetizing force establishing flux and inversely related to the circuit's opposition to those flux lines. This opposition is the reluctance of the path, and the symbol is R, Figure 8–19b.

FIGURE 8–19
Highlights

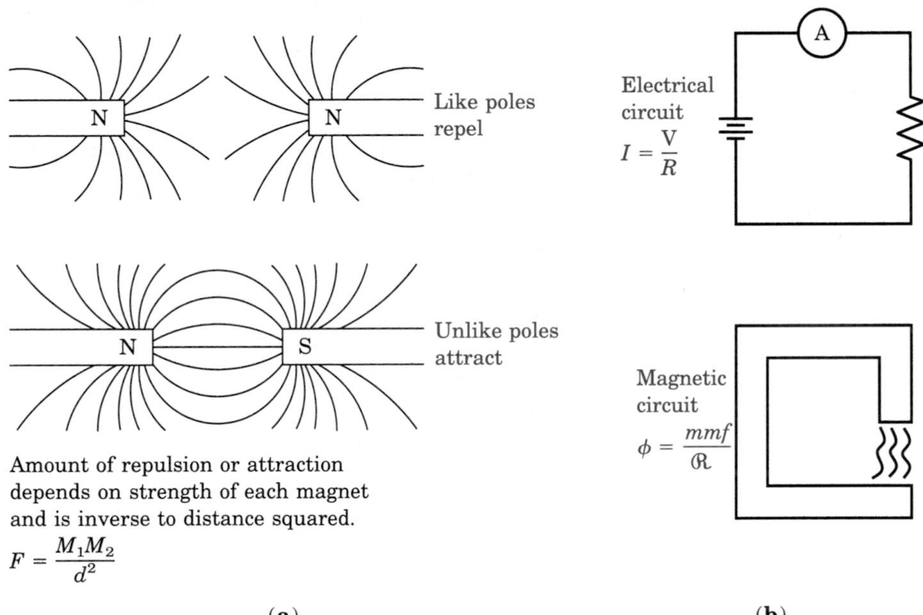

So we can further examine these factors with respect to common magnetic units of measure, it is now appropriate to discuss magnetic units.

IMPORTANT MAGNETIC UNITS, TERMS, SYMBOLS, AND FORMULAS

Recall a single flux line (or magnetic line of force) is a maxwell, and 10^8 flux lines equal a Weber (Wb). Also, although magnetizing force was presented, we did not specifically define it.

Before thoroughly discussing magnetic units, some background about systems of units would be helpful. Most of us daily use units of measure pertaining to dimensions, volume, weight, speed, and so on. In the field of electronics, it is critical that some meaningful system of units exists.

Before 1960 there were several systems of units related to electricity, electronics, and magnetism. One system was the common unit of the volt, ampere, and watt; another system dealt with electrostatics; and a third system explained magnetics. Both electrostatics and magnetics used a system called the "cgs" system, where c was distance in centimeters, g was weight in grams, and s was time in seconds.

IN-PROCESS LEARNING CHECK II

1. For a given coil dimension and core material, what two factors primarily affect the strength of an electromagnet? ________ and ________.
2. The left-hand rule for determining the polarity of electromagnetics states that when the fingers of your left hand point (in the same direction, in the opposite direction) ________ as the current passing through the coil, the thumb points toward the (North, South) ________ pole of the electromagnet.
3. Adjacent current-carrying conductors which are carrying current in the same direction tend to (attract each other, repel each other) ________.
4. If you grasp a current-carrying conductor so that the thumb of your left hand is in the direction of current through the conductor, the fingers "curled around the conductor" (will indicate, will not indicate) ________ the direction of the magnetic field around the conductor.
5. When representing an end view of a current conductor pictorially, it is common to show current coming out of the paper via a (dot, a cross, or plus sign) ________.

However, since the 1960s, a system has been accepted that uses the *Meter* for distance, the *Kilogram* for weight, the *Second* for time, and the *Ampere* (necessary for electrical systems) for current measure. This system is abbreviated as the *MKSA.* Since this system has been adopted internationally, it is called the International System, abbreviated "SI," *Système internationale d'unités*.

Examples of the units, symbols, and formulas are in Figure 8–20. You will need to refer to this chart several times as you study the next section.

TERM OR QUANTITY	SYMBOL OR ABBREVIATION	SI UNIT AND FORMULA	CGS UNIT AND FORMULA
Flux (lines)	ϕ	weber (Wb) $= \frac{\text{number lines}}{10^8}$	Maxwell (Mx) number lines in field
Flux Density (Magnetic flux per unit cross-sectional area at right angles to the flux lines).	B	$\frac{\text{webers}}{\text{sq meter}}$ = tesla (T) $B = \frac{\phi\ (\text{mks})}{A\ (\text{mks})}$ or teslas $= \frac{\text{Wb}}{\text{sq mtr}}$	$\frac{\text{lines}}{\text{sq cm}}$ $B = \frac{\phi\ (\text{cgs})}{A\ (\text{cgs})}$ or gauss $= \frac{\text{Mx}}{\text{sq cm}}$
Magnetomotive Force (That which forces magnetic lines of force through a magnetic circuit).	MMF	Ampere-turns or AT = NI	F (cgs) = 0.4πNI or F (gilberts) = $1.26 \times N \times I$
Magnetic Field Intensity (Magnetomotive force per unit length)	H	Ampere-turns per meter $\frac{\text{NI}}{\text{length}}$ OR $\frac{\text{AT}}{\text{meter}}$	H (oersteds) $= \frac{0.4\pi \text{NI}}{\text{length}}$ or $\frac{1.26\ \text{NI}}{1\ \text{cm}}$
Permeability (Ability of a material to pass, conduct or concentrate magnetic flux; analogous to conductance in electrical circuits).	μ	Vacuum or air: $\mu_o = \frac{B}{H}$ B is teslas; H is $\frac{\text{AT}}{\text{meter}}$ where free space is considered to have an "absolute" permeability of: $4\pi \times 10^{-7}$ or 12.57×10^{-7}	Free space: $\mu_o = \frac{B}{H}$ B is in gauss; H is in oersteds; In cgs system, $\mu_o = 1$.
Relative Permeability (Not constant because it varies with the degree of magnetization).	μ_r	Relative permeability of a material is a ratio. Thus, $\mu_r = \frac{\text{Flux density with core material}}{\text{Flux density with vacuum core}}$ Where, flux density in the core material is: $B = \mu_o\mu_r H$ teslas, and absolute permeability of core materials is: $\mu = \frac{B}{H} = \mu_O\mu_r$ (SI units)	Same concept in both systems

FIGURE 8–20 Magnetic units, symbols, and formulas chart

The Basic Magnetic Circuit

Since technicians do not typically have to deal with all the systems of units presented in Figure 8–20, we will not dwell on this subject. However, you should be aware of these units and have an understanding of the magnetic circuit.

The magnetic equivalent of Ohm's law, sometimes called Rowland's law, is shown in Figure 8–21.

OHM'S LAW	ROWLAND'S LAW	EXPLANATION
$R = \frac{V}{I}$, where: Resistance in ohms equals electromotive force, in volts divided by current in amperes. R Reluctance	$R = \frac{F(mmf)}{\phi}$, where: Reluctance (the magnetic circuit's opposition to the establishment of flux) equals the magnetomotive force (F) divided by the flux (ϕ). Also, it can be shown that: $R = \frac{l}{\mu A}$ where l = length of the path in meters; μ = permeability of the path A = cross-sectional area of the path perpendicular to flux, in meters.	Reluctance equals magnetomotive force divided by flux. Flux equals magnetomotive force divided by reluctance. F (mmf) = $\phi \times$ R or flux times reluctance

FIGURE 8–21 Rowland's Law

Relationships Among B, H, and Permeability Factors

From our previous discussion, it is possible to note H (magnetic field intensity or mmf per unit length) is the factor producing flux per unit area (or flux density, B) within a given magnetic medium.

It is further observed the amount of B a given H produces directly relates to the permeability of the flux path. For a vacuum only it is stated that:

Formula 8–2	$\mu_o = \frac{B}{H}$

Formula 8–3	$B = \mu_o \times H$

Formula 8–4	$H = \frac{B}{\mu_o}$

Referring again to Figure 8–20 and other related discussions, we will now apply this information to solve some problems dealing with magnetic parameters. This will give you a better understanding of the relationships.

EXAMPLES

Problem 1: What is the **magnetomotive force,** if a current of 5 amperes passes through a coil of 100 turns?

Answer:

Formula 8—5	mmf = NI

Thus, mmf = 100 × 5 = 500 ampere-turns

Problem 2: If the length of the coil in problem 1 is 0.1 meters, what is the **magnetic field intensity?**

Answer:

Formula 8–6	$H = \frac{AT}{meter}$

$$\text{Thus, } H = \frac{500}{0.1} = 5000 \text{ ampere-turns per meter}$$

Problem 3: What is the **flux density** (SI units) if there are 10,000 lines of force coming from a North magnetic pole face (surface) having a cross-sectional area of 0.002 square meters?

Answer:

Formula 8–7	$B \text{ (in Tesla)} = \frac{\phi}{A} = \frac{Wb}{m^2}$

$$\text{Thus, } B = \left[\frac{\frac{10^4}{10^8}}{2} \times \left(10^{-3}m^2\right)\right]$$

$$= \frac{10^{-4}}{0.002} = \frac{0.0001}{0.002} = 0.05 \text{ tesla}$$

NOTE: Stated in micro units, $\frac{100\ \mu Wb}{2000\ \mu m} = 0.05$ tesla

Problem 4: What is the **"absolute" permeability** (SI) of a core material if its relative permeability (μ_r) is equal to 2000?

Answer:

Formula 8–8	$\mu = \mu_o \times \mu_r$

Thus, $\mu = (12.57 \times 10^{-7}) \times 2000 = 2514 \times 10^{-6}$

Problem 5: What is the **flux density** (B) in tesla (T) if there are 100 webers per square meter?

Answer:

Formula 8–9	$\text{Tesla} = \dfrac{\text{Wb}}{\text{sq meter}}$

Thus, T = 100

Problem 6: If a magnetic circuit has an air gap with a length of 2 mm and a cross-sectional area of 5 cm^2, what is the gap's **reluctance?**

Answer:

Since $\mu_r = 1$ for air, then:

$$R = \frac{l}{\mu A} = \left[\frac{2 \times 10^{-3}}{12.57 \times 10^{-7} \times 5 \times 10^{-4}}\right]$$

$$= \frac{2 \times 10^{-3}}{62.85 \times 10^{-11}}$$

$$= (2000 \times 10^{-6})\,(62.85 \times 10^{-11})$$

$$= 31.82 \times 10^{5} \text{ SI units}$$

Problem 7: If a given coil has a flux density of 3000 webers per square meter, using unknown core material and it has a flux density of 1000 webers per square meter using air as its core, what is the **relative permeability** of the "brand x" material?

Answer:

$$\mu_r = \frac{\text{flux density with unknown core}}{\text{flux density with air core}}$$

$$= \frac{3000}{1000}$$

$$= 3$$

PRACTICAL CONSIDERATIONS ABOUT CORE MATERIALS

Since many of the components and devices you will be studying use principles related to electromagnetism, we will now consider the practical aspects of the core materials used in these components.

You know that for a given magnetizing force, a coil's magnetic field strength is weaker if its core is air than if its core has a ferromagnetic substance, Figure 8–22.

Also, we have indicated that certain magnetic properties of a material are not constant but vary with degrees of magnetization. Therefore, there is not a constant relationship between the magnetizing force field intensity (H) and the flux density (B) produced.

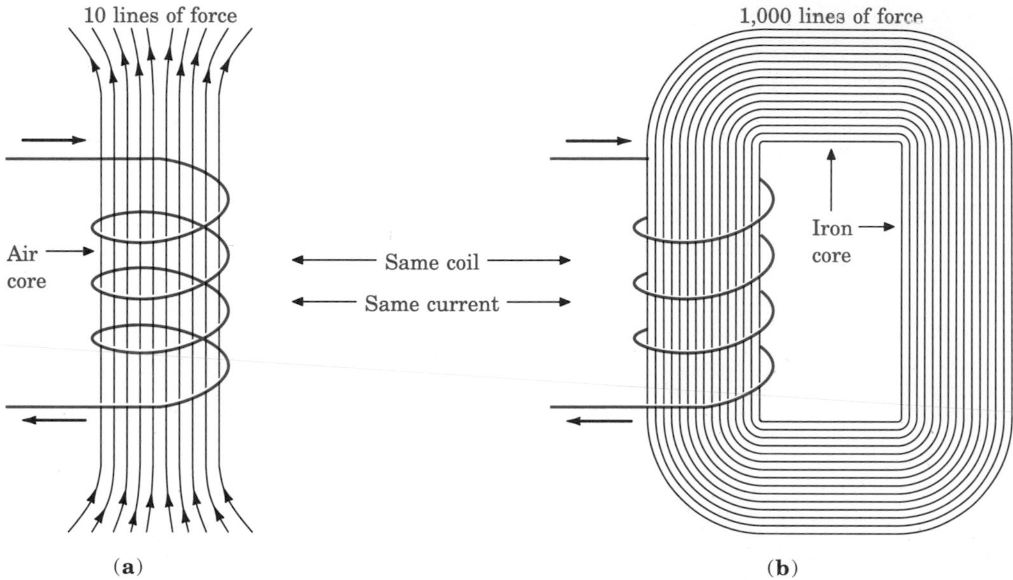

FIGURE 8–22 Effects of core material(s)

THE B-H CURVE

Manufacturers of magnetic materials often provide data about their materials to help select materials for a given purpose. One presentation frequently used is a B-H curve, or a magnetization curve, Figure 8–23. Notice in Figure 8–23 that as the magnetizing intensity (H) increases, the flux density (B) increases but not in a complete linear fashion. This graph provides several important points.

1. During magnetization (the solid line), a point is reached when increasing H (by increasing current through the coil) does not cause a further significant flux density increase. This is due to the core reaching "saturation," represented at point S on the graph. At this point it can't contain more flux lines.

FIGURE 8–23 B-H curve

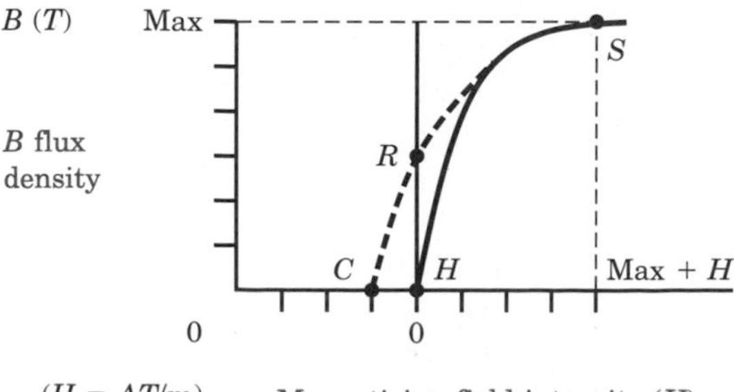

2. During demagnetization (with current decreasing to zero through the coil), the dotted line indicates some other interesting characteristics:
   a. When the H reaches zero (no current through coil), some flux density (B) remains in the core due to "residual" or "remanent" magnetism. The amount is indicated by the vertical distance from the graph baseline to point R.
   b. To remove the residual magnetism and reduce the magnetism in the core to zero, it takes a magnetizing intensity in the opposite direction equal to that represented horizontally from point 0 to point C. This value indicates the amount of "coercive" force required to remove the core's residual magnetism.

An example of how the B-H curve is used to illustrate important magnetic features of different materials is shown in Figure 8–24. For example, notice different materials saturate at different levels of flux density (B). Saturation occurs when a further increase in H does not result in an appreciable increase in B.

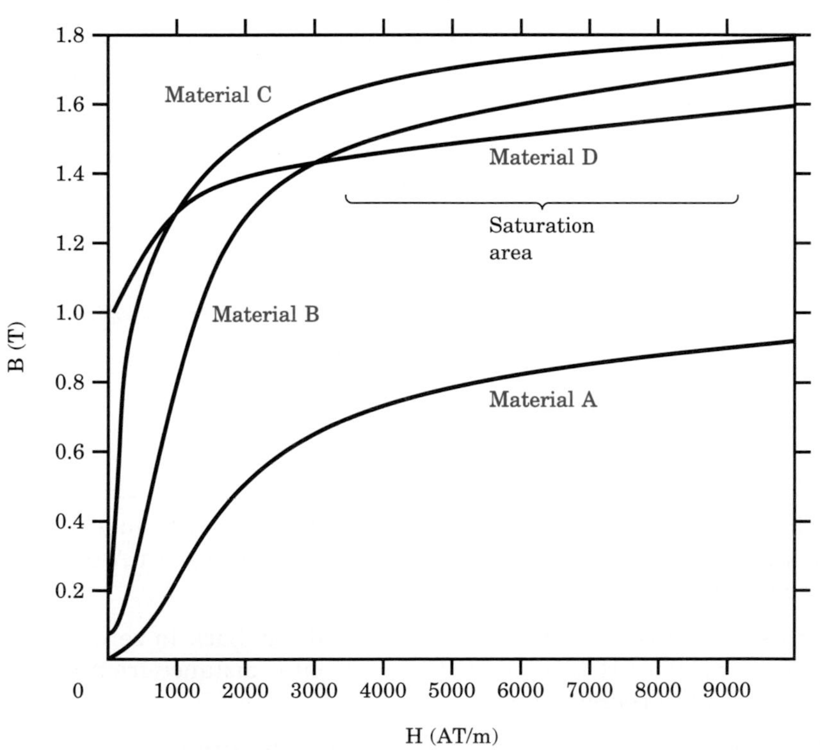

FIGURE 8–24 Magnetization curves for different metals

THE HYSTERESIS LOOP

Another useful graph is the hysteresis loop. It portrays what happens when a core is saturated with current in one direction through its coil, then the current direction is reversed and increased until the core is saturated in the opposite polarity. If this is continually repeated, a hysteresis loop graph, similar to that in Figure 8–25, can be created that enables meaningful analysis of the material in question. This is typically done by application of an alternating current source to the coil. NOTE: "Hysteresis" means "lagging behind." As this graph demonstrates, the flux buildup and decay lags behind the changes in the magnetizing force.

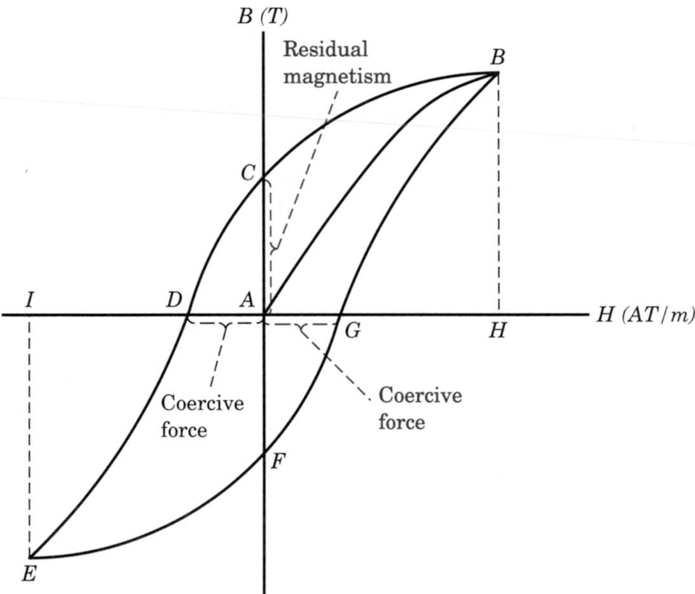

FIGURE 8–25
Hysteresis loop

Important information provided by this graph includes:

1. Points A to B represent the initial magnetization curve from zero H until the core reaches saturation.
2. Points B to C represent the flux density decay as H changes from its maximum level back to zero. The vertical distance from A to C is the amount of residual magnetism in the core.
3. Points C to D represent the flux density decay back to zero as the opposite polarity H is applied. The horizontal distance from A to D represents the coercive force.
4. Points D to E represent the buildup of flux density to core saturation in the opposite polarity from the initial magnetization. Of course, the total horizontal distance from point A to I shows the H required for this saturation. The vertical axis of the graph shows the flux density (B) involved.

5. Points E to F show decay of flux density as H changes from its maximum level of this polarity back to zero. The residual magnetism is represented from points A to F.
6. Points F to G illustrate the decay of flux density back to zero due to the coercive force of the reversed H overcoming the residual magnetism. The coercive force is from points A to G.
7. Points G back to B complete the loop, displaying the increase of B to saturation in the original direction. For this segment of the loop, the vertical distance from point B to H is the flux density value at saturation.
8. The area inside the hysteresis loop represents the core losses from this cycle of magnetizing, demagnetizing, magnetizing in the opposite polarity, and demagnetizing the core. These losses are evidenced by the heating of the core as a result of the "magnetic domains" continuously having to be realigned with alternating current (ac) applied to the coil. We'll discuss core losses in greater detail at a later point in the book.

Incidentally, you may wonder how material that has been magnetized can be demagnetized. One method is to put the material through this hysteresis loop cycle while gradually decreasing the H swing. The result would be shown as a shrinking area under the hysteresis loop. One practical way to accomplish a decrease in H swing is to use a large ac demagnetizing coil so the material is placed within the coil. Gradually pull the material away from the ac energized coil, Figure 8–26. You may have seen a TV technician use a "degaussing" coil on color picture tubes that have become magnetized, which degrades the picture quality.

IN-PROCESS LEARNING CHECK III

1. A "B-H" curve is also known as a __________ curve.
2. B stands for flux __________.
3. H stands for magnetizing __________.
4. The point where increasing current through a coil causes no further significant increase in flux density is called __________.
5. The larger the area inside a "hysteresis loop" the (smaller, larger) __________ the magnetic losses represented.

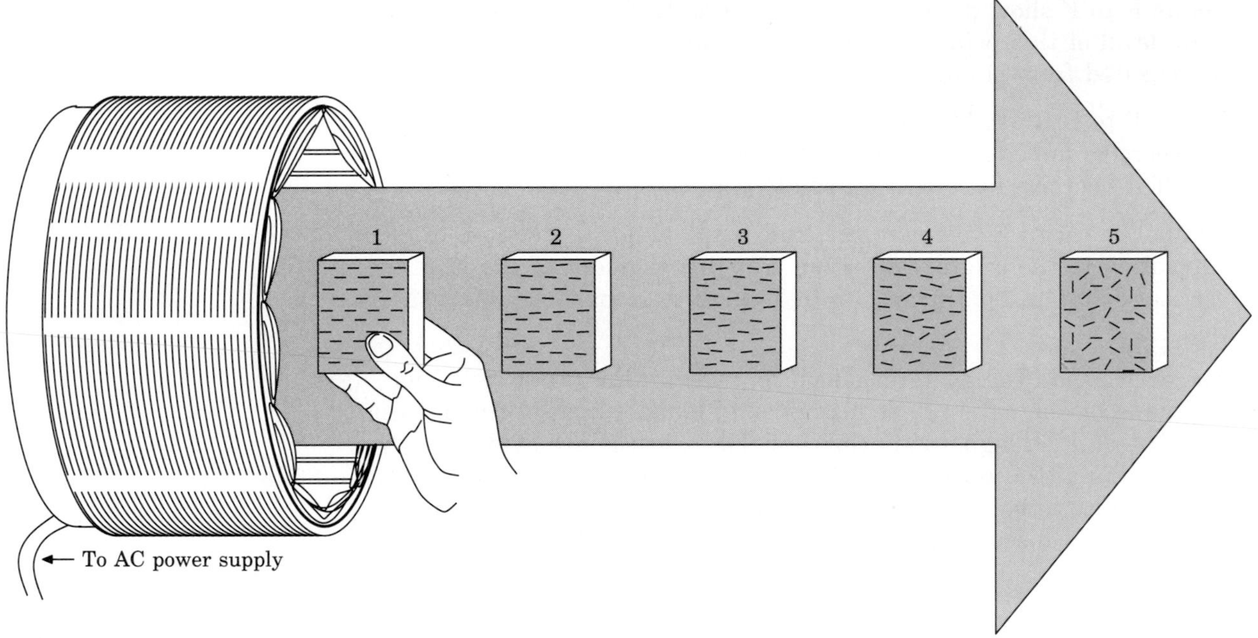

FIGURE 8–26 Example of a demagnetizing coil

INDUCTION AND RELATED EFFECTS

Motor Action

When talking about motor action, we refer here to converting electrical energy to mechanical energy. You have studied the law that states lines of force in the same direction repel each other, and lines of force in opposite directions attract and cancel each other. One practical application of this phenomenon is motor action. The simplest example of this is a current-carrying conductor placed in a magnetic-field environment.

Look at Figure 8–27 and note the following:

1. The direction of the magnet's flux lines—North to South.
2. The direction of the current-carrying conductor's flux lines (Use the left-hand rule for current-carrying conductors.)

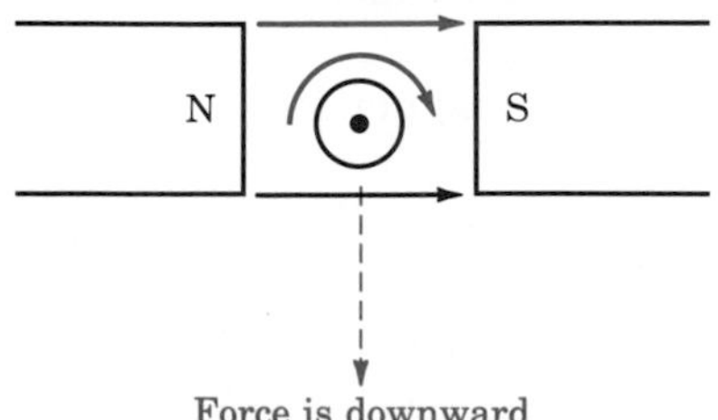

FIGURE 8–27 Motor action on a single conductor

3. The interaction between the conductor's flux lines and the magnet's lines of force is such that the lines at the top of the conductor are repelled (lines in the same direction repel); the lines at the bottom of the conductor are attracted (lines in opposite directions attract); and the field is weakened at the bottom. Therefore, the conductor will be pushed down. In other words, the lines at the top aid each other and the lines at the bottom cancel each other.
4. The amount of force that pushes the conductor down depends on the strength of both fields. The higher the current through the conductor, the stronger its field. And the higher the flux density of the magnet, the stronger its field.
5. Another method to cause stronger motor action is to make the single conductor into a coil (in the form of an armature) with many turns, thus increasing the ampere-turns and the strength of its field, Figure 8–28.

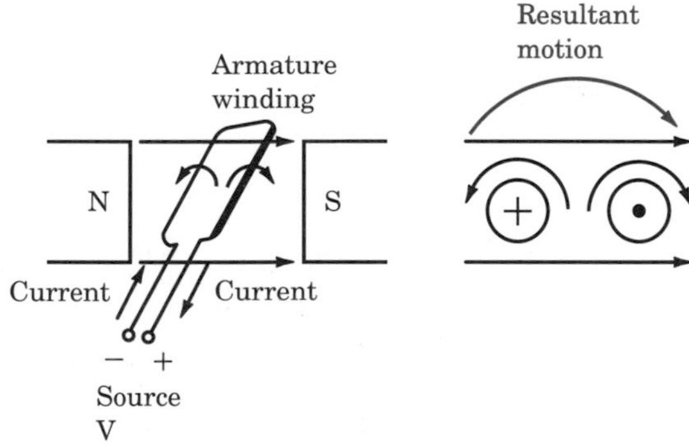

FIGURE 8–28 Concept of stronger motor action with an armature coil

6. Refer to Figure 8–29b. When the current through the conductor is reversed or the magnetic poles of the magnet are switched, the force (motor action) is in the opposite direction of that in Figure 8–29a. When both current and poles are simultaneously reversed, Figure 8-29c, the force remains in the original direction, prior to being reversed, as that shown in Figure 8–29a.

Generator Action

When talking about generator action, we refer to converting mechanical energy to electrical energy. Throughout the chapter, we have alluded to a relationship between electricity and magnetism. Electromagnetic induction clearly illustrates this linkage. As you will recall, Faraday observed there is an induced emf (causing current if there is a path) when a conductor cuts

FIGURE 8–29 Effects of changing direction of current or field

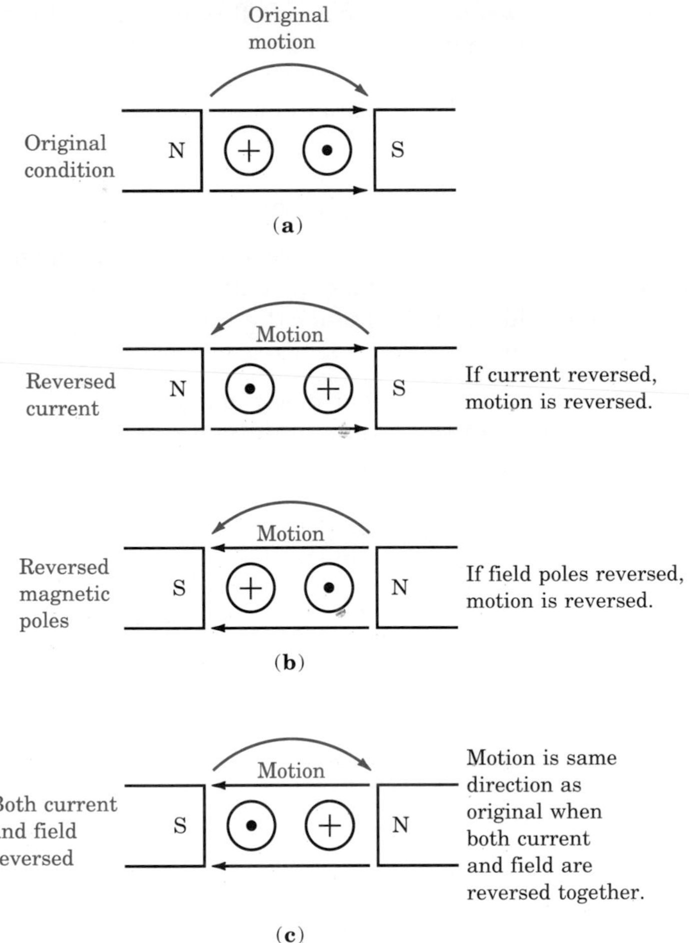

across magnetic flux, or conversely, when the conductor is cut by lines of force. In essence, *when there is relative motion* between the two (i.e., the conductor and the magnetic lines of flux), an emf is induced.

Observe Figure 8–30 and note the following:

1. The direction of the magnet's field: North to South
2. The physical direction of the conductor movement with respect to the magnetic field of the magnet: Upward.
3. The fact that when the conductor cuts this field in the direction as indicated, the direction of induced current is toward you, Figure 8–30a.

   In fact, the left-hand rule for current-carrying conductors (studied earlier) would indicate the direction of current, as shown. For example, if you aim your fingers in the direction of the flux (N to S) and

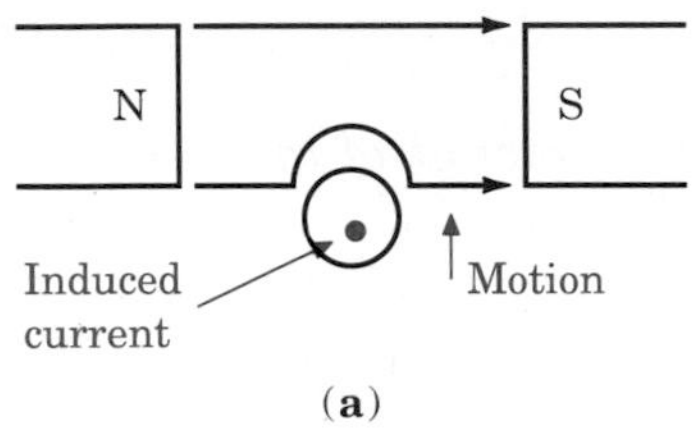

(a)

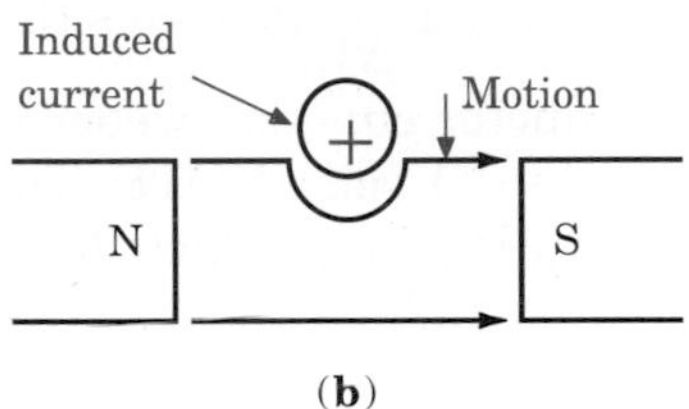

(b)

FIGURE 8–30
Converting mechanical energy into electrical energy

bend your fingers in the same direction the magnet's field is bent, then the induced current is in the same direction as your extended thumb.

4. If the conductor moves down into the field, rather than up, the induced current is in the opposite direction, Figure 8–30b. (Try the left-hand rule again!)
5. The faster we move the conductor relative to the field, the higher the induced current (and/or emf), since more lines are being cut per unit time.

You may be wondering how cutting flux lines with a conductor causes induced emf, thus producing current flow (assuming a closed current path). Conceptually, it is briefly explained by indicating the magnetic fields involved with each of the electrons in the wire are caused to align, causing electron movement, which is current flow. When the conductor is open (with no complete path for current flow), one end of the conductor will be caused to have an excess of electrons (negatively charged); the other end has a deficiency of electrons (positively charged). Thus, we have an induced emf. If a closed path for current is provided, this induced emf causes current to flow.

What factors affect the amount of induced voltage? The key factors are:

1. amount of flux
2. number of turns linked by the flux
3. angle of cutting the flux
4. rate of relative motion

All of these factors relate to the number of flux lines cut per unit time.

FARADAY'S LAW

Faraday's law states *the amount of induced emf depends on the rate of cutting the flux (ϕ).* The formula is:

Formula 8–10	$V_{ind} = \frac{\Delta\phi}{\Delta T}$

where, V_{ind} is the induced emf, and $\frac{\Delta\phi}{\Delta T}$ indicates the *rate* of cutting the flux. Using the SI system, a conductor cuts one weber of flux (10^8 lines) in one second, so the induced voltage is 1 volt. (NOTE: The symbol Δ is the fourth letter in the Greek alphabet called delta, which, for our purposes, represents the amount of change in, or a small change in). Sometimes, the lower case letter "d" is used to indicate delta rather than the triangular symbol. Thus:

Formula 8–11	$V_{ind} = N\ (\#\text{ of turns}) \times \frac{d\phi\ (Wb)}{dt\ (sec.)}$

EXAMPLE

If a 100-turn coil is cutting 3 webers per second ($\frac{d\phi}{dt}$ in formula), then the induced V is $100 \times 3 = 300$ volts. In essence, 3×10^8 lines of flux are cutting the 100-turn coil each second, resulting in 300 volts of induced voltage.

Although we will not thoroughly discuss generators, it is important you know that if a loop of wire is rotated within a magnetic field, Figure 8–31, the maximum voltage is induced when the conductor(s) cut at right angles to the flux lines. Voltage will incrementally decrease as the loop rotates at less than right angles, reaching zero when the loop conductor moves parallel to the lines of flux (in the vertical position), Figure 8–32. The key point is when the conductor moves parallel to the lines of flux (an angle of 0 degrees), no voltage is induced; when the conductor moves at right angles to the flux lines, maximum voltage is induced; and when the conductor moves at angles between zero and 90 degrees, some value between zero and the maximum value is induced.

LENZ'S LAW AND RECIPROCAL EFFECTS OF MOTORS AND GENERATORS

Lenz's Law

As you have already noted, the law of energy conservation is you don't get something for nothing! In the case of motor action in the generator, to generate electricity with the generator we have to supply mechanical energy that

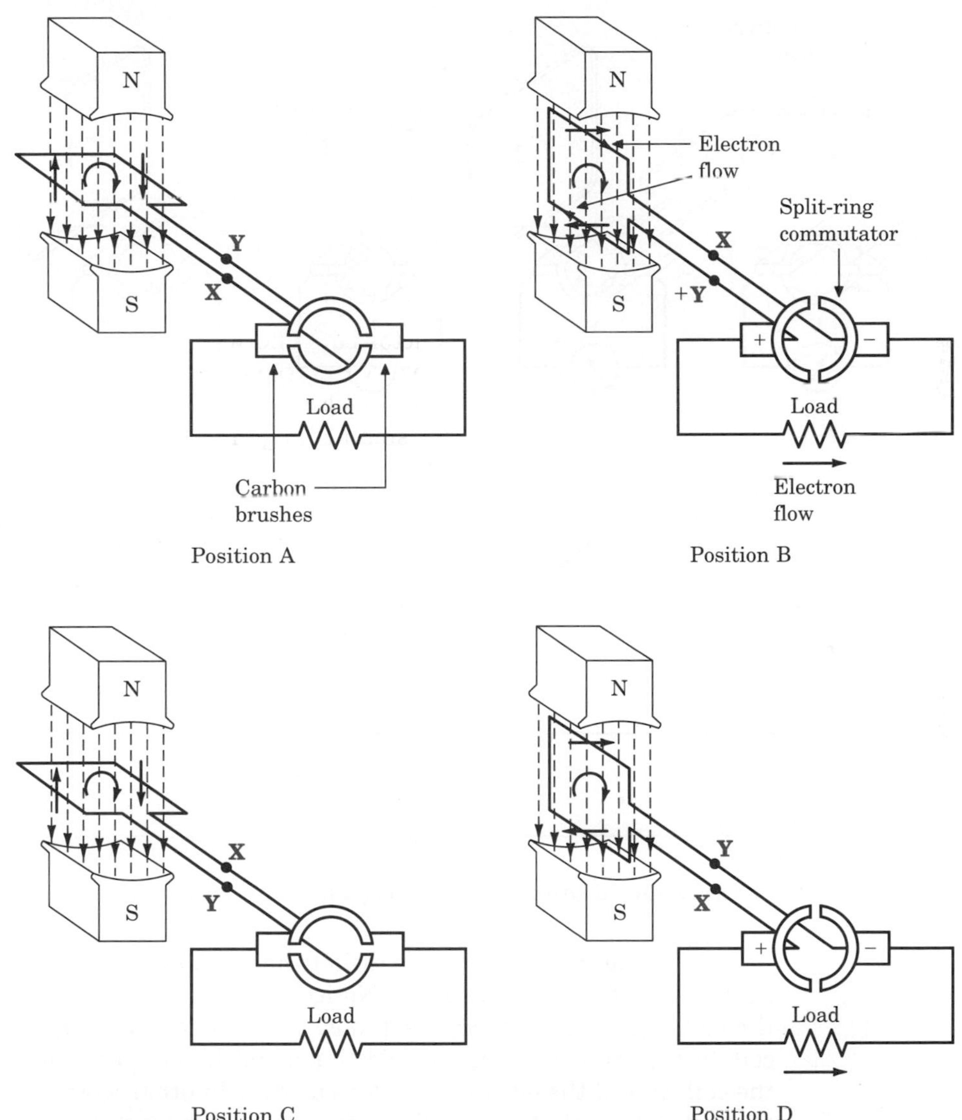

FIGURE 8–31 Regions of maximum and minimum induced voltage

turns the armature shaft and overcomes the force of the opposing motor effect that results from the induced current and its resultant field(s).

Lenz's law states *the direction of an induced voltage, or current, is such that it tends to oppose the change that caused it.*

Another way of stating this law is when there is a change in the flux linking a circuit, an emf/current is induced that establishes a field that opposes the change.

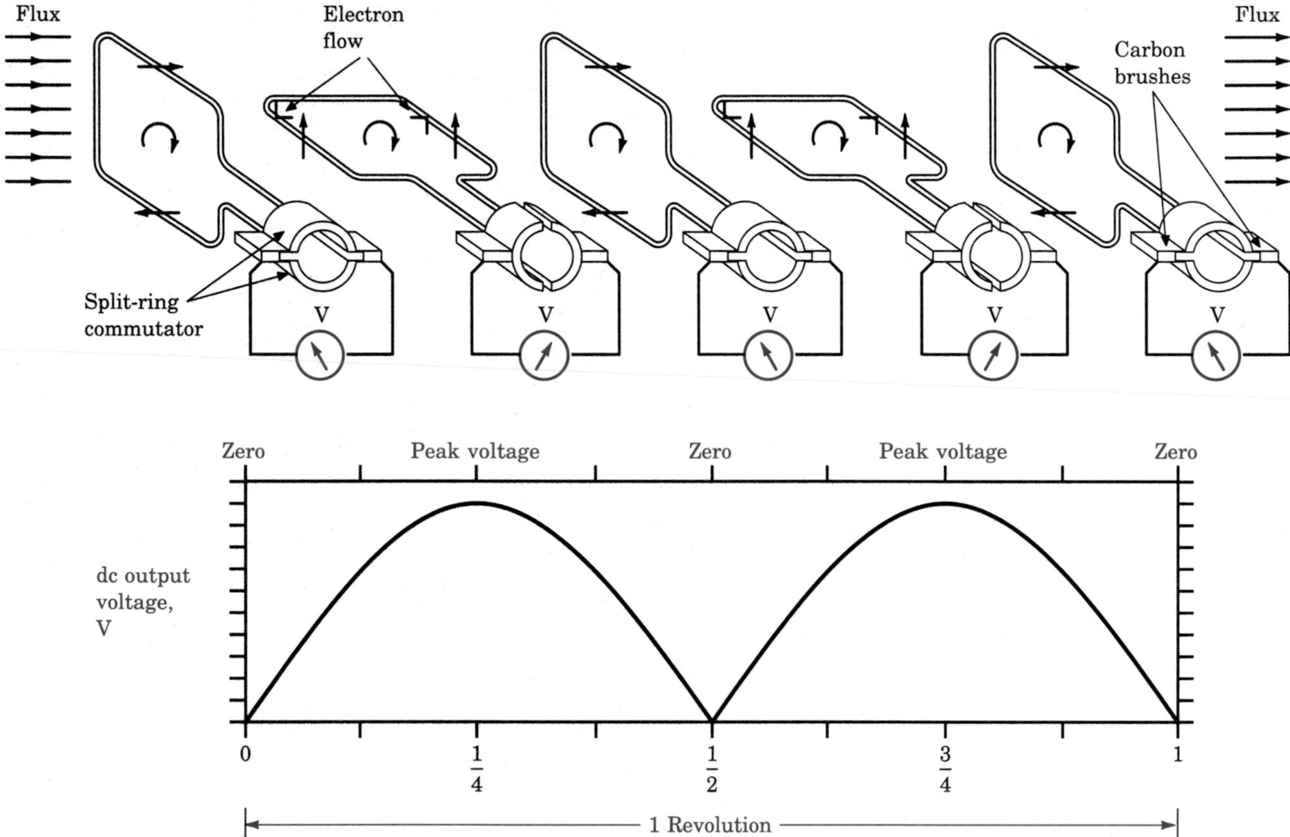

FIGURE 8–32 Variation in output voltage through one rotation of loop

Look at Figure 8–33 and use the left-hand rules for current-carrying conductors *and* for electromagnets. Notice when the magnet is moved *down* in the coil, the induced current establishes a North pole at the top end of the coil, facing (and repelling) the North pole of the magnet, whose flux linking the coil caused the original induced current. In other words, as we force the magnet down into the coil, the induced current produces a field to oppose the motion, like Lenz said. Conversely, when we pull the magnet out of the coil, the induced current is in the reverse direction, establishing a South pole at the top of the coil and tending to oppose retraction of the magnet. Again, Lenz's law is proven correct!

Motor Effect in a Generator

The generator converts mechanical energy (armature shaft rotation) into electrical energy (generator's output voltage). As you know, when the generator's conductors (armature windings) move through the magnetic field of

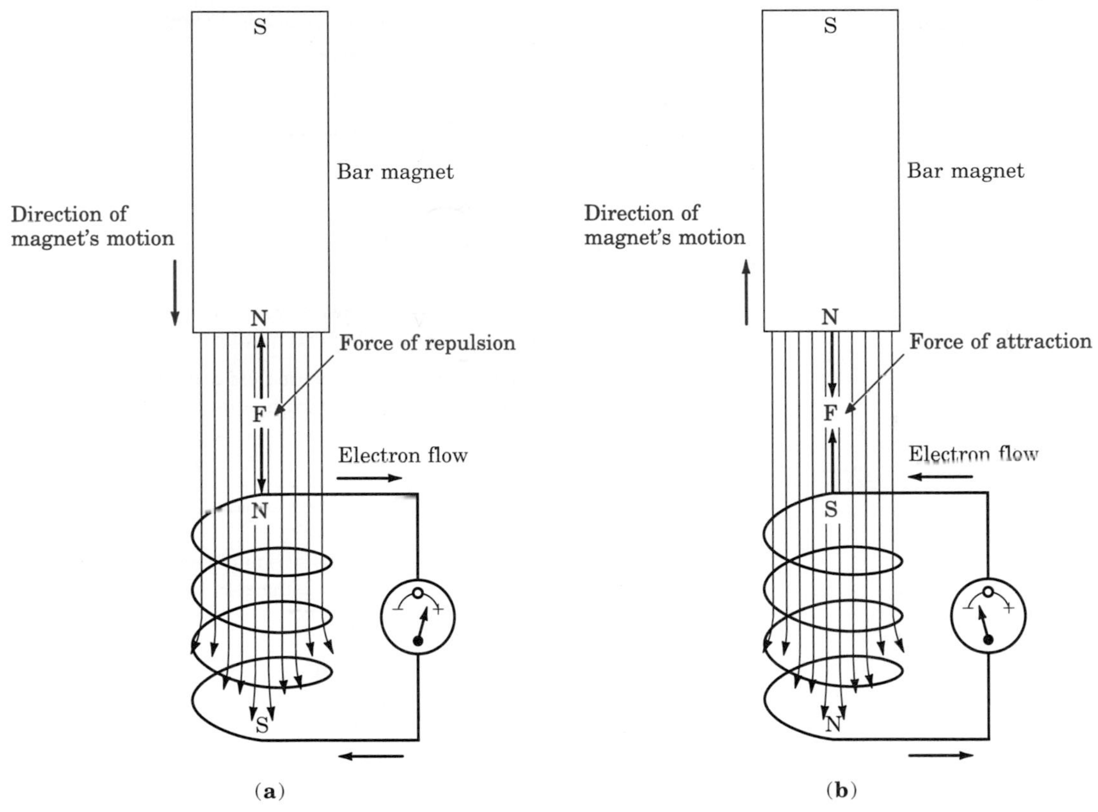

FIGURE 8–33 Lenz's Law

the generator's field magnets, voltage is induced according to Faraday's law. This induced voltage causes current through the conductors, assuming a closed path for current. The *direction* of the resulting induced current is such that the magnetic field established around the conductor(s), resulting from the induced current, *opposes* the motion causing it, Figure 8–34.

Generator Effect in a Motor

In Figure 8–35 you can see the direction of current we are passing through the conductor causes the loop to rotate clockwise due to appropriate motor action.

Also, you can observe when the conductor rotates clockwise, the wire loop's left side moves up into the flux and the wire loop's right side moves down into the flux. Since we have movement between a conductor and flux lines, there is an induced voltage/current.

This induced current is in the opposite direction from the current that is causing the motor's armature to move clockwise. This phenomenon is often termed generator effect in a motor and is related to Lenz's law.

FIGURE 8–34 Motor effect in a generator

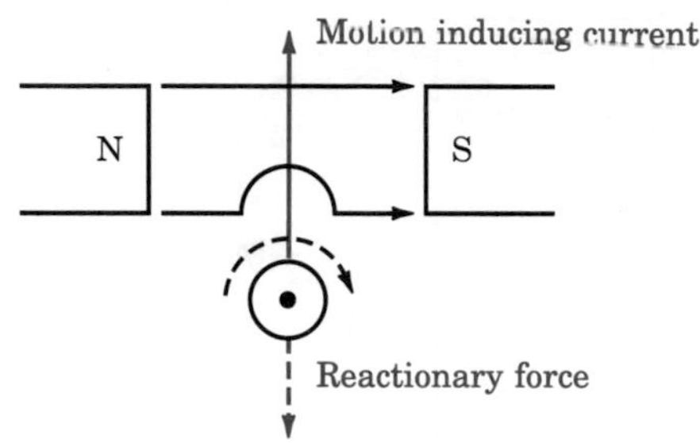

Induced current (coming toward you) results from upward mechanical movement.

Induced current sets up magnetic field around conductor which opposes upward motion that induced it in the first place.

FIGURE 8–35 Generator effect in a motor is such that the induced current due to cutting flux will be in opposite direction of current causing motor to rotate.

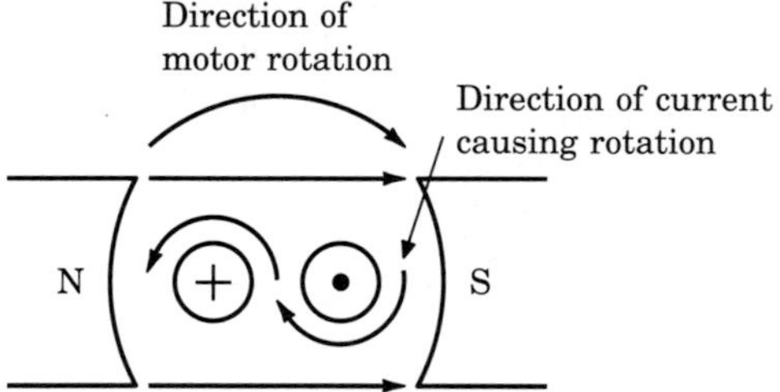

SUMMARY OF MAGNETISM

Several topics not previously discussed are now presented since you now have an understanding of magnetism. These topics are:

1. the toroidal coil form (closed magnetic path);
2. magnetic shields;
3. shaping fields across a gap via pole pieces;
4. the Hall effect; and
5. special classifications of materials.

Toroidal Coil Form

Notice in Figure 8–36 the toroidal coil form resembles a doughnut and is made of soft iron, ferromagnetic, or other low-reluctance, high-permeability material. The key feature is this coil effectively confines the lines of flux within itself (because there are no air gaps); therefore, it is a very efficient carrier of flux lines. It allows little "leakage flux" (flux outside the desired

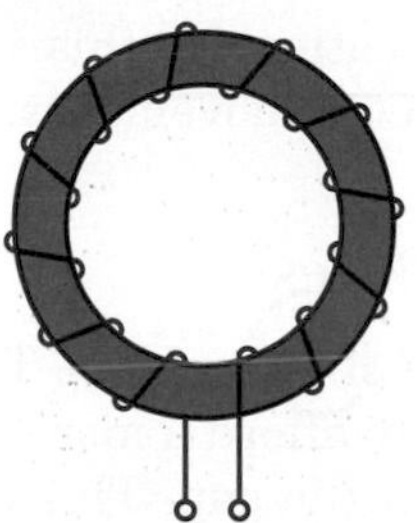

Flux is virtually contained within core
with very little "leakage" flux

FIGURE 8–36
Toroid coil

path, generally in the surrounding air), thus does not create magnetic effects on nearby objects. At the same time, its internal field is affected very little by other magnetic fields near it.

It is interesting to note there are no poles, as such, in the toroid configuration. However, if one takes a slice out of the doughnut, creating an air gap, North and South poles are established at opposite sides of the air gap.

Magnetic Shields

Remember magnetic lines of force penetrate nonmagnetic materials. To protect components, devices, and so forth from stray or nearby magnetic fields, a magnetic shield is used, Figure 8–37. A high-permeability, soft iron enclosure is useful for the shielding function. Actually, it is simply diverting the lines of flux through its low-reluctance path, thus protecting anything con-

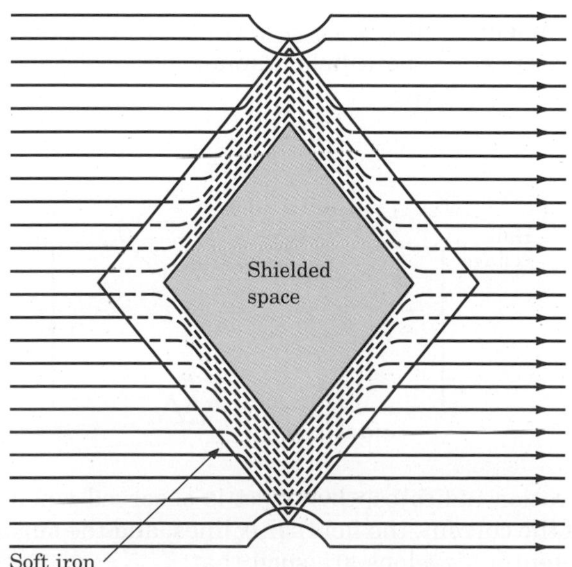

FIGURE 8–37
Concept of the magnetic shield

tained inside of it. One application for this type shielding is in electrical measuring instruments or meter movements.

Shaping Fields Via Pole Pieces

Notice in Figure 8–38 the use of pole pieces. These pieces control the shape of the field across the air gap to afford a more linear field with almost equal flux density across the area of the gap. This control is beneficial for many applications.

FIGURE 8–38
Concept of pole pieces linearizing a field

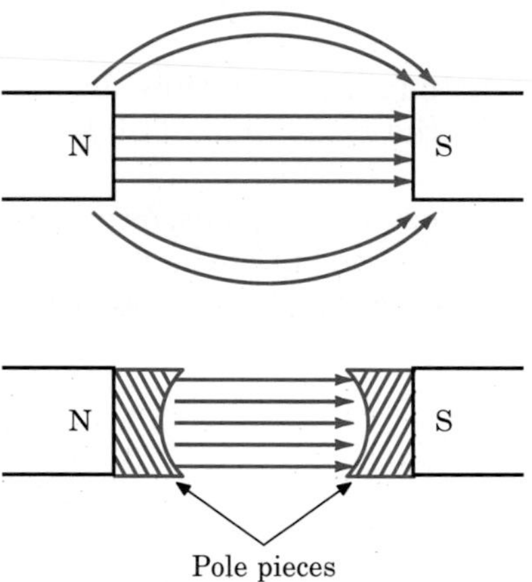

The Hall Effect

Figure 8–39 illustrates the Hall effect. Named after its discoverer, E. H. Hall, the phenomenon states when a current-carrying material is in the presence of an external magnetic field, a small voltage develops on two opposite surfaces of the conductor.

FIGURE 8–39
The Hall effect

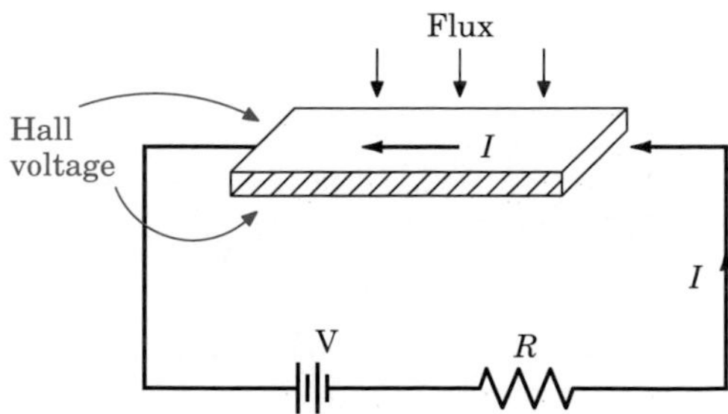

As the illustration indicates, this effect is observed when the positional relationships of the current, the field's flux lines, and the surfaces where the difference of potential develops are such that:

1. Flux is perpendicular to direction of current flow through the conductor. (NOTE: The amount of voltage developed is directly related to the flux density, B).
2. If current is traveling the length of the conductor, the Hall voltage (v_H) develops between the sides or across the width of the conductor.

Indium arsenide is a semiconductor material used in devices, such as gaussmeters, to measure flux density via the Hall effect. This material develops a relatively high value of v_H per given flux density. Since the amount of Hall-effect voltage developed directly relates to the flux density where the current-carrying conductor is immersed, a meter connected to this probe can be calibrated to indicate the flux density of the probed field, Figure 8–40.

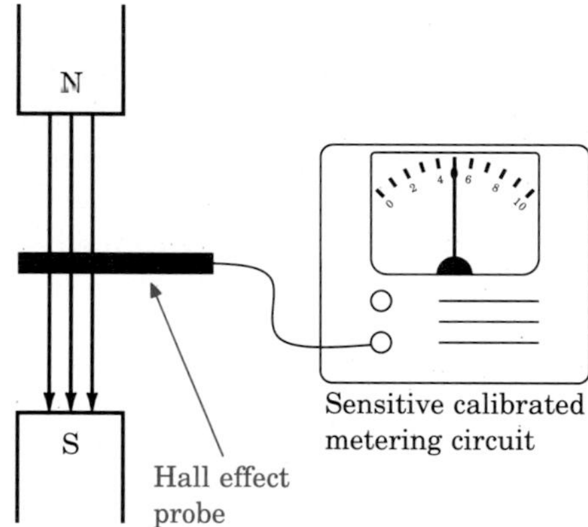

FIGURE 8–40 Using the Hall effect to measure flux density

Special Classifications of Materials

In this chapter, we have primarily been discussing ferromagnetic materials that typically have high permeabilities and can be strongly magnetized. Examples include iron, steel, nickel, and various alloys. Other classifications of materials worth noting are:

1. *Paramagnetic* materials that have weak magnetic properties. Examples are aluminum and chromium.
2. *Diamagnetic* materials that have permeabilities of less than one and align perpendicular to the direction of magnetic fields. Examples are mercury, bismuth, antimony, copper, and zinc.
3. *Ferrites* that are powdered and compressed materials with high permeabilities but high electrical resistance that minimizes I^2R/eddy current losses when used in high frequency applications. Examples are nickel ferrite, nickel-cobalt ferrite, and yttrium-iron garnet. Note that eddy currents are those induced into the core material that cause undesired I^2R losses within the core.

SUMMARY

- A magnet is an object that attracts iron, steel, or certain other materials. It can be natural (magnetite or lodestone) or produced in iron, steel, and other materials in a form called a permanent magnet. Also, a magnet can exist in soft iron and other materials as a temporary magnet that requires a current-carrying coil around it.
- Magnetic materials, such as iron and nickel, are attracted by magnets and can be made to have magnetic properties.
- Magnets that are suspended and free to move align in an approximate North/South direction. The end of the magnet seeking the earth's North pole is the North-seeking pole. The other end of the magnet is the South-seeking pole.
- A magnetic field composed of magnetic lines of force called flux surrounds a magnet. This region is where the magnetic forces act.
- Lines of force are continuous; travel outside the magnet from the magnet's North pole to the magnet's South pole; and flow inside the magnet from the South pole to the North pole. They travel the shortest path or the path with least opposition to the flow of flux, and when traveling in the same direction, repel each other. When traveling in opposite directions, they attract each other. Furthermore, lines of force do not cross each other and will penetrate nonmagnetic materials.
- A current-carrying conductor or wire is surrounded by a magnetic field. The strength of the magnetic field is directly related to the value of current passing through it.
- A coil of wire carrying current develops a magnetic field that is stronger than a straight conductor with the same current. The magnetic field becomes stronger if an iron core is inserted in the coil, making it an electromagnet.
- A basic law of magnetism is like poles repel each other and unlike poles attract each other.
- Permanent magnets come in a variety of shapes, for example, bar-shaped (rectangular), horseshoe-shaped, and disc-shaped. Permanent magnets are not really permanent since they lose their magnetism over time.
- Induction of magnetism occurs when a magnetized material is brought in proximity to or rubbed across some magnetic material.
- The left-hand rule to determine direction of the magnetic field surrounding a current-carrying conductor states when the wire is grasped in the left hand with the extended thumb in the direction of current flow through the wire, then the fingers are wrapped around the conductor in the same direction as the magnetic field.
- The left-hand rule to determine the polarity of electromagnets (i.e., which end is the North pole and which end is the South pole) states when the coil is grasped with the fingers pointing in the same direction as the current, the extended thumb points toward the end of the coil that is the North pole.
- Factors influencing the strength of an electromagnet are: the number of turns on the coil; the coil's length; the coil's value of current; and the type of core.
- Several important terms and units relating to magnetism include: flux; weber; tesla; magnetomotive force; magnetic field intensity; permeability; and relative permeability.
- Comparing magnetic circuit parameters with electrical circuits and Ohm's law can be performed by stating that the amount of flux flowing through a magnetic circuit is directly related to the magnetomotive force and inversely related to the path's reluctance (opposition) to flux.

- Core materials differ in their magnetic and electrical features. For example, some materials saturate at lower levels of magnetization.
- An ideal core material has very high permeability; loses all its magnetism when there is no current flow in the coil; does not easily saturate; and has low I^2R loss due to eddy currents.
- An important relationship exists between flux density produced by a given magnetic field intensity and the permeability of the material involved. This relationship stated for the permeability of a vacuum is:

$$\mu_o = \frac{B}{H}, \text{ and } B = \mu_o \times H \text{ and } H = \frac{B}{\mu_o}$$

where: μ_o is the absolute permeability of a vacuum
B is flux density and
H is the magnetic field intensity

- When an ac passes through an electromagnet's coil, there is a lag between the magnetizing force and the flux density produced. This is illustrated by the B-H curve and the hysteresis loop. These graphs also illustrate residual magnetism and the coercive force needed to overcome it.
- Faraday's law states the amount of voltage induced when there is relative motion between conductor(s) and magnetic flux lines is such that when 10^8 lines of flux are cut by one conductor (or vice versa) in one second, one volt is induced. In other words, induced voltage relates to the number of lines of force cut per unit time.
- Lenz's law states when there is an induced voltage (or current), the direction of the induced voltage or current opposes the change causing it.
- A motor converts electrical energy into mechanical energy.
- A generator converts mechanical energy into electrical energy.

REVIEW QUESTIONS

NOTE: Redraw illustrations on separate paper to answer questions involving diagrams.

1. Briefly define magnetism, flux, field, pole, and magnetic polarity.
2. Draw a typical field pattern for the magnetic situations in Figure 8–41. Indicate direction(s) of flux lines with arrows.

N S S S

FIGURE 8–41

3. State the law regarding magnetic attraction and repulsion.
4. Draw magnetic field lines around the conductors in Figure 8–42 and indicate direction with arrows.

FIGURE 8–42

5. Indicate the North and South poles of the current-carrying coil in Figure 8–43.

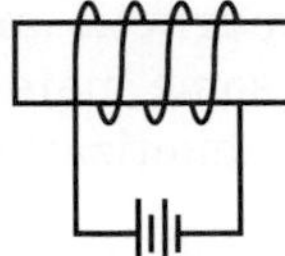

FIGURE 8–43

6. List five statements about the behavior of magnetic lines of force.
7. What term indicates the ease with which a material passes, conducts, or concentrates flux?
8. What term indicates the opposition to magnetic lines of force?
9. Using SI units, define the terms, expressions, or units for flux and flux density. Show symbols, if appropriate.
10. Using SI units, define the terms, expressions, or units for magnetomotive force, magnetic field, and field intensity. Show symbols, if appropriate.
11. Using SI units, define the terms, expressions, or units for permeability and relative permeability. Show symbols, if appropriate.
12. How many lines of force are represented by the weber? (NOTE: Express as a whole number, not a power of ten).
13. What is the difference between magnetomotive force and magnetic field intensity?
14. If the permeability of a given flux path is unchanged and H doubles, what happens to B?

15. What happens to the reluctance of a given flux path if its length triples while the magnetic circuit's μ and A remain unchanged?

16. If current doubles through a given electromagnet's coil, does reluctance increase, decrease, or remain the same? (Assume the magnet is in its linear operating range.)

17. For the situation in question 16, will flux density increase, decrease, or remain the same?

18. What is meant by saturation in a magnetic path?

19. How does a magnetic shield give the effect of shielding something from magnetic lines of force?

20. What does a gaussmeter measure?

TEST YOURSELF

1. Draw a typical hysteresis loop and explain how saturation, residual magnetism, and coercive force are shown on the graph.
2. If an air gap is 2 mm in length and has a cross-sectional area of 10 cm^2, what is the air gap's reluctance?
3. If a pole piece has a cross-sectional area of 0.005 m^2 and is emitting 1 μWb of flux:
   a. How many lines of flux are involved?
   b. What is the flux density, expressed in teslas?
4. How much induced emf results if a coil of 500 turns is cut at a rate of 10 μWb per second?

DC Measuring Instruments

Key Terms

Backoff ohmmeter scale
Core
Damping (electrical)
Frame/movable coil
Full-scale current
$I_m(I_{fs})$
I_s
Loading effect
Meter shunt (current meter)
Moving-coil (d'Arsonval) movement
Multimeter
Multiplier resistor
Ohms-per-volt rating
Pointer
Pole pieces
R_m
R_{mult}
Range selector (switch)
Scale
Sensitivity rating
Series-type ohmmeter circuit
Springs
Stops
Universal (Ayrton or ring) shunt
Zero-adjust control

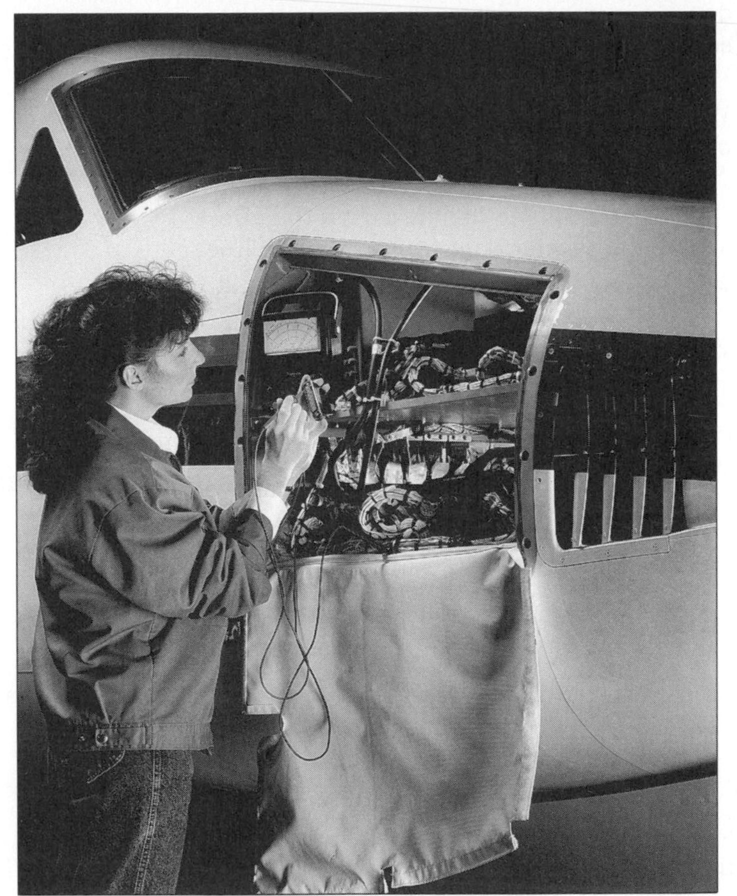

Courtesy of Cleveland Institute of Electronics, Inc.

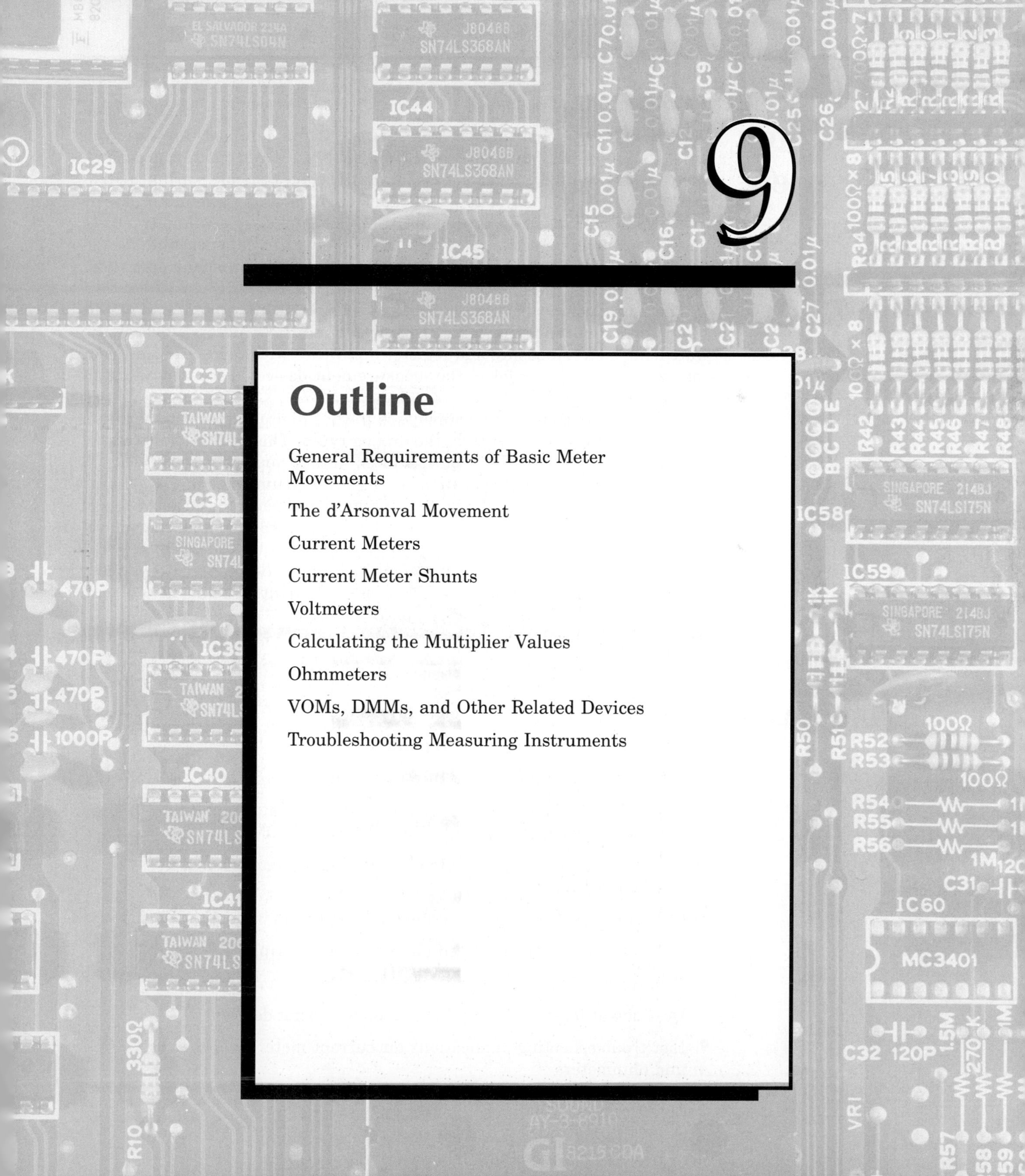

9

Outline

General Requirements of Basic Meter Movements

The d'Arsonval Movement

Current Meters

Current Meter Shunts

Voltmeters

Calculating the Multiplier Values

Ohmmeters

VOMs, DMMs, and Other Related Devices

Troubleshooting Measuring Instruments

Chapter Preview

Lord Kelvin can be paraphrased as saying, "If you can measure what you are speaking about, you know something about it. If you can't, your knowledge is very limited." Someone else has indicated that measuring instruments are like special eyes to technicians, engineers, and scientists. The dc measuring instruments you will study in this chapter are invaluable aids to people working in electronics. These instruments are indispensable for designing, troubleshooting, servicing, and monitoring equipment operation. Some samples of these instruments are in Figure 9–1.

There is a technical distinction between the terms instrument and meter. Instruments measure the value of present quantities. Meters, such as the watt-hour meter, measure and register quantities with respect to time. For convenience, we will refer to the measurement devices discussed using either term.

Although the trend in instruments is toward digital meters, rather than the analog types, we will study the analog types. There are a multitude of analog instruments still in use. At a later time in your education, you will study the principles and operations of digital instruments.

In this chapter, you will examine the most frequently used moving-coil movement, called the d'Arsonval movement. You will see how this movement is incorporated in measuring instrument circuits that measure current, voltage, and resistance. Also, you will study the proper use of these instruments; the instruments' effects on the circuit(s) being tested; and some hints about their practical applications.

Objectives

After studying this chapter, you will be able to:

- Describe the **d'Arsonval movement**
- List key parts and functions of a movement
- Define linear deflection, full-scale sensitivity, shunt, multiplier, **sensitivity rating, loading effect,** VOM, and DMM
- Explain the proper methods to connect and use current meters, voltmeters, and ohmmeters
- Calculate the values required for **current meter shunts**
- Calculate the values required for voltmeter multipliers
- Briefly explain calibration of series-type ohmmeters
- Describe at least two special-purpose measuring devices
- List troubleshooting applications for current meters, voltmeters, and ohmmeters

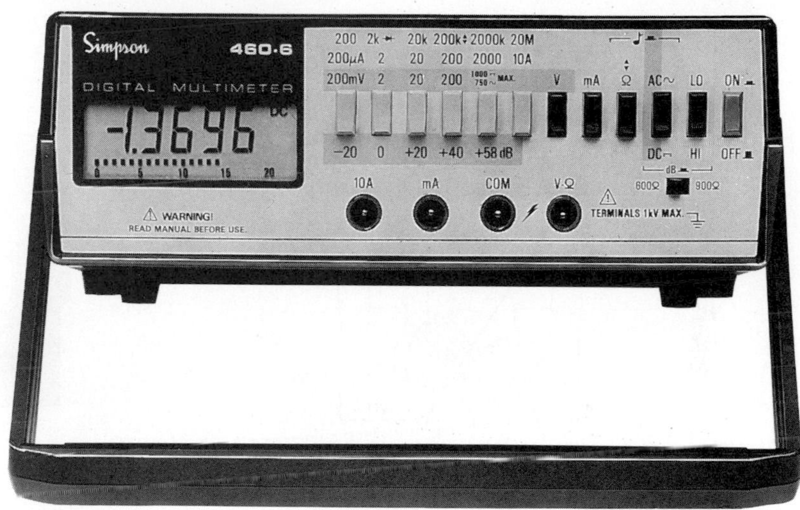

(a) (b)

FIGURE 9–1 Examples of measuring instruments. (Photos courtesy of Simpson Electric Company)

GENERAL REQUIREMENTS OF BASIC METER MOVEMENTS

Three essentials most analog meter movements contain are:

1. A moving element that reacts to the magnitude of the electrical parameter being measured.
2. An indicating function that shows how much reaction has occurred.
3. A damping function that aids the moving element to reach its appropriate resting place without undue overshoot or back and forth mechanical oscillation.

THE D'ARSONVAL (MOVING COIL) MOVEMENT

The **d'Arsonval movement** contains parts meeting the above requirements and is a device that reacts to the value current through it.

Refer to Figure 9–2 and note the following key parts and their functions in this movement.

1. *Permanent magnet*—establishes magnetic field.
2. *Pole pieces*—linearize and strengthen magnet's field
3. *Core*—soft iron core concentrates, strengthens magnetic field and, along with pole pieces, produces a uniform air gap between the permanent magnet's pole pieces and itself.

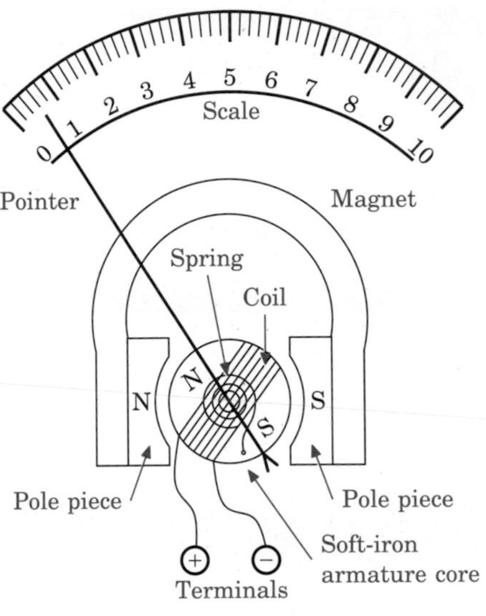

(**a**) Moving-coil movement

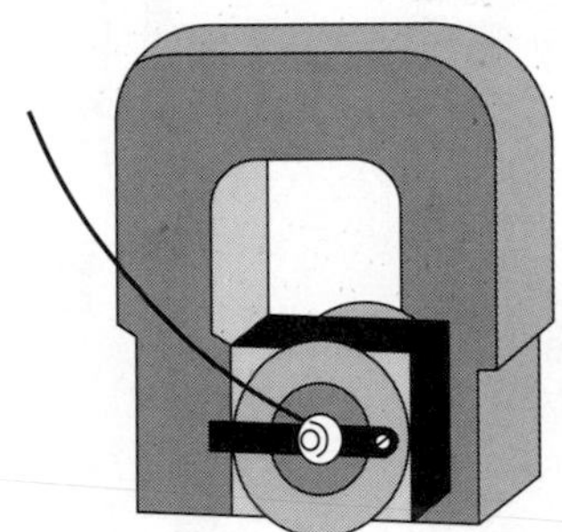

(**b**) "Phantom" view of commercial permanent-magnet moving-coil movement

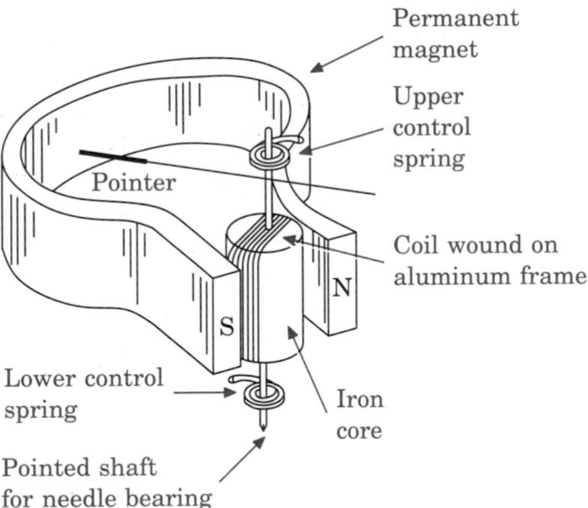

(**c**) Moving element with pointer and springs

FIGURE 9–2 Key parts of a moving-coil movement

4. *Movable coil and support frame*—light aluminum frame provides support and damping action for coil; movable coil creates electromagnetic field when current passes through it.
5. *Spindle and jewel bearings*—spindle supports movable coil and coil frame; jewel bearings provide a low-friction bearing surface to allow turning action of movable coil mechanism. (NOTE: Sapphire bearings are not shown.)

6. *Springs*—provide electrical contact or current path to coil and mechanical torque, restoring coil to starting position when current is not present; also provide a force against which the movable coil acts.
7. *Pointer and counterbalance*—provide means of indicating amount of deflection.
8. *Scale*—calibrated in appropriate units gives meaning to pointer deflection.
9. *Stops*—provide left and right limits to pointer deflection. (NOTE: Stops are not shown.)

Principle of Operation

The operation of the **moving-coil movement** is based on the motor action occurring between the permanent magnet's field and the field established by the movable coil when current passes through it, Figures 9–2 and 9–3, as appropriate.

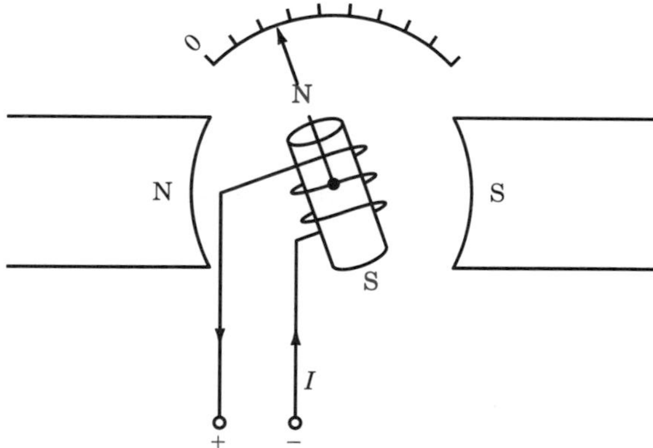

FIGURE 9–3 Polarity required for proper direction of motor action

The moving element function, described earlier, involves many parts—the moving coil, the light aluminum coil frame, the **pole pieces,** the magnetic **core** piece, the spindle, the **springs,** and the sapphire bearings on which the spindle rotates. You will recall the strength of an electromagnet depends on the number of turns involved (per given length), the current through those turns, and the permeability features of the flux path. Since the number of turns and the flux path are known values, the only variable, in this case, is the amount of current through the movable coil. The higher current produces a stronger motor action; thus, the moving coil moves further.

Observe in Figures 9–2a and 9–3 the polarity of the electromagnetic field set up by current passing through the moving coil is such that the electromagnet's North pole is adjacent to the permanent magnet's North pole, and the electromagnet's South pole is adjacent to the permanent magnet's South pole. This means there is a force of repulsion. Thus, the movable coil moves in a clockwise direction when current passes in the correct direction.

It also is evident that current in the opposite direction through the coil reverses the polarity of the coil's field, causing attraction rather than repulsion. Thus, the movable coil moves in a counterclockwise direction, possibly damaging the fragile meter movement. Note that proper operation of this movement depends on dc current (current in only one direction) and proper polarity, or direction of that current. The indicating function is performed by the combination of the **pointer** and calibrated **scale.**

The **damping** function is really an electrical damping provided by the light aluminum coil frame. Recall that when a conductor moves within or links a magnetic field and has relative motion, a current-producing emf is induced. The direction of the induced current establishes a magnetic field that opposes the motion causing the induced current in the first place (Lenz's law). This action causes a mechanical damping effect on the movement of the movable-coil-pointer assembly. The result is the pointer does not significantly overshoot its appropriate resting place, nor does it oscillate back and forth or above and below the appropriate deflection location for long periods.

Advantages and Disadvantages of the Moving-Coil Movement

Primary advantages of the moving-coil movement are:

1. Linear scale, including scale divisions, is easy to read.
2. Construction provides a self-contained magnetic shielding effect.
3. It can be very accurate.
4. It can be quite sensitive since it has a large number of turns on coil and a strong permanent magnet.

Some disadvantages of the moving-coil movement are:

1. It is quite fragile so will not stand rough treatment.
2. It measures only dc parameters.
3. It is very expensive.

CURRENT METERS

How Current Meters are Connected

The moving-coil movement is primarily a current indicator. Its basic action depends on the magnitude of the current through its moving coil. Thus, to

IN-PROCESS LEARNING CHECK I

Fill in the blanks as appropriate.

1. Three essential elements in an analog meter movement include a __________ element, an __________ means, and some form of __________.
2. The purpose of the permanent magnet in a d'Arsonval meter movement is to establish a fixed __________ __________.
3. The movable element in a moving coil meter moves due to the interaction of the fixed __________ __________ and the magnetic field of the movable coil when current passes through the coil.
4. Electrical conduction of current in a movable coil meter movement is accomplished through the __________, which also provides a desired (mechanical magnetic) __________ force against which the movable coil must act.
5. The element in a moving-coil movement producing electrical damping is the __________.

measure current, the circuit to be tested must be broken, and the current meter inserted in series with the circuit path where it is desired to measure current value (being sure to observe proper polarity), Figure 9–4. As you already know, current is the same throughout a series circuit path. Hence, the meter indicates the current through that portion of the circuit where it is inserted.

1. Turn power off.
2. Break part of circuit to be tested.
3. Insert current meter in series.
4. Observe polarity.

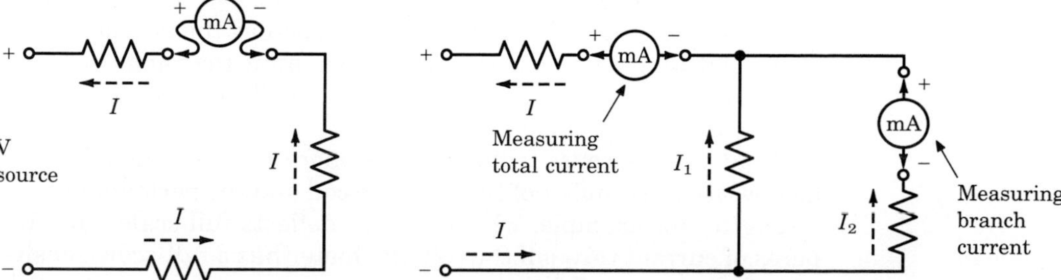

FIGURE 9–4 Proper connection of current meters

Sensitivity

Current meters come in a variety of measurement ranges, Figure 9–5. Two common methods to define the **sensitivity** of current meters are the amount of current required to cause full-scale deflection and the number of millivolts dropped by the coil at a current equal to full-scale deflection. Sensitive basic movements (for example, the laboratory galvanometer) measure current values in the range of microamperes; other movements measure current in milliamperes.

(a)

(b)

FIGURE 9–5 Meters come in different sizes, shapes, and sensitivities. (Photos courtesy of Simpson Electric Co.)

However, there is a limit to the range of measurement that a basic movement can be designed. Physical factors, such as the size of wire, the weight of the coil frame, and the size of springs, limit the range when designing a basic movement. As you will see in a later discussion, using meter shunts helps solve this problem.

What factors cause a meter to be more or less sensitive? The primary factors are the number of turns on the coil and the permanent magnet's field strength. For example, when a meter deflects full-scale with 10 microamperes of current through the coil, the meter has a full-scale sensitivity (fs) of 10 microamperes. When it takes 100 microamperes to cause full-scale deflection, the full-scale (fs) sensitivity rating is 100 microamperes and so on.

Another physical factor influencing meter sensitivity is the force of the hairsprings. In order to have a very sensitive movement (one taking a small current value causing full-scale deflection of the pointer), it takes a strong permanent magnet and a coil with many turns of fine wire, and hairsprings without much force.

On the other hand, to have a less sensitive meter, a movable coil using less turns of larger diameter wire is used. Also, using a weaker permanent magnet and/or hairsprings with more force produces a less sensitive meter.

Typically, basic movements are produced in ranges from a few microamperes to a few milliamperes. When measuring currents higher than this, the use of a parallel (current bypass) path, called a **meter shunt,** is necessary.

The number of millivolts dropped by the meter movement at full-scale current adds yet another useful rating because it involves *both* the **full-scale current (I_m)** and the resistance of the wire in the moveable coil (R_m) of the meter movement. For example, if the meter is a 1 mA meter (fs = 1 mA) and the meter resistance (R_m) is 10 Ω, then when the meter deflects a full-scale value of 1 mA, the meter movement is dropping a voltage equal to $I_m \times R_m = 1 \text{ mA} \times 10\ \Omega = 10$ millivolts. Therefore, this movement is rated as a 10 millivolt, 0–1 mA movement.

Trade-Offs

What are the trade-offs of basic meter movements? Two important trade-offs are the desired sensitivity and the desire not to disturb the tested circuit more than necessary.

A more sensitive meter has a smaller wire and greater number of turns on the movable coil. Small diameter wire has a higher resistance per unit length than larger wire. This means that the more sensitive the meter the higher is its own internal resistance called $\mathbf{R_m}$. Consequently, a meter with higher resistance has more effect on the tested circuit, since its R_m is a larger percentage of the test circuit's total resistance. Ideally, a current meter should have zero internal resistance so as not to disturb the circuit tested. But this zero resistance is impossible.

PRACTICAL NOTES

A common rule of thumb about current meter resistance versus the resistance of the circuit tested is the current meter should not be more than about 1% of the resistance of the tested circuit.

EXAMPLE

For example, if the circuit through which you are trying to measure current has a resistance of 10,000 ohms, the current meter should not have a resistance greater than 100 ohms to ensure that the tested circuit is not altered too much when the current meter is inserted.

CURRENT METER SHUNTS

The Basic Shunt

Current meter shunts extend the range of current that a given meter movement measures. The shunt is simply a low-resistance current path connected in parallel (or shunt) with the meter movement (the moving coil). The shunt effectively causes some percentage of the current to bypass the meter through the parallel path with the remaining percentage passing through the meter to cause deflection, Figure 9–6.

FIGURE 9–6 Concept of a "shunt" resistor path to extend meter range

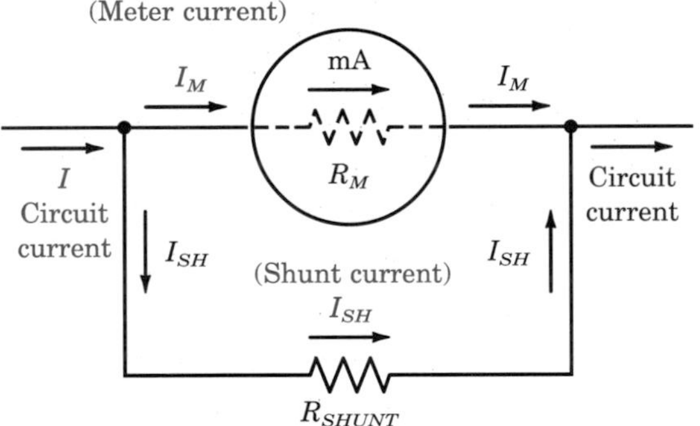

Circuit current "splits." A percentage of current passes through meter while remaining percentage passes through shunt. Thus, $I_M + I_{SH}$ = circuit current.

Important Parameters. To calculate the required shunt resistance value that extends the meter's range, several electrical parameters are of importance. These parameters are:

1. the full-scale current for the meter movement ($\mathbf{I_{fs}}$, or I_m);
2. the resistance of the meter movement (R_m);
3. the desired maximum current, or the extended meter range value to be indicated by full-scale deflection (I);

4. the current that passes through the shunt (I_s or $I - I_m$); and
5. the resistance of the shunt (R_s).

Methods Used to Calculate R_{shunt} Value. More than one approach can be used to calculate the R value needed to shunt a given meter movement to increase its measurement range. However, all the techniques evolve from variations of Ohm's law, Figure 9–7.

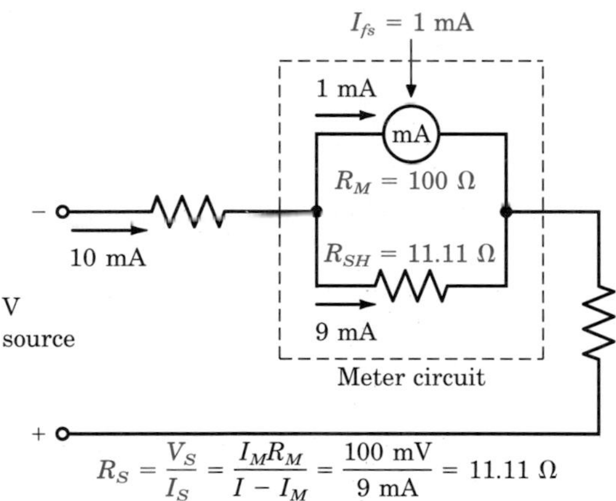

FIGURE 9–7 Basic shunt resistance calculation

One method is to use *Ohm's law* in its basic form, where:

Formula 9–1	$R_s = \frac{V_s}{I_s}$

NOTE: V_s is the same as the parallel voltage V_m (or I_mR_m).

$\mathbf{I_s}$ is the difference between the highest current to be measured (the extended range full-scale value) and the movement's full-scale current value ($I_s = I - I_m$).

Substituting in Formula 9–1:

Formula 9–2	$R_s = \frac{I_mR_m}{(I - I_m)}$	OR	Formula 9–3	$R_s = \frac{I_mR_m}{I_s}$

Incidentally, the I_mR_m product is the same as the meter's millivolt rating spoken of earlier, assuming full-scale current value in milliamperes. The $(I - I_m)$ portion in Formula 9–2 is really the current value that must pass through the shunt (I_s) in Formula 9–3. Therefore, if you know the millivolt rating of the meter and the current value passing through the shunt, the value of $R_s = \frac{\text{millivolt rating}}{I_s}$.

EXAMPLE

Given a 0–1 mA movement with a meter resistance of 100 Ω, what value of shunt R is needed to make the meter act as a 0–10 mA instrument?

Answer: $R_s = \dfrac{I_m R_m}{(I - I_m)}$

$$R_s = \frac{(1 \times 10^{-3}) \times 100}{(10 \times 10^{-3}) - (1 \times 10^{-3})}$$

$$R_s = \frac{(100 \times 10^{-3})}{(9 \times 10^{-3})}$$

$$R_s = 11.11\ \Omega$$

Another approach is to use the *inverse proportionality concept.* That is, branch currents are inverse to branch resistances.

For example, if it is desired to have nine times the current through branch 2 as there is current through branch 1, then make the R of branch 2 equal to one-ninth that of branch 1. The inverse relationship for our case is stated as:

Formula 9–4	$\dfrac{R_s}{R_m} = \dfrac{I_m}{I_s}$

NOTE: The ratio of the *shunt* resistance *to* the *meter* resistance is the same as the ratio of the *meter* current *to* the *shunt* current. Notice the inverse relationship on the two sides of the equation.

Another formula expressing the ratio concept is:

Formula 9–5	$R_s = \dfrac{R_m}{N - 1}$

where R_s = R of shunt, R_m = R of meter, and N = number of times range is being multiplied. In other words, 1 mA meter whose increased range is 10 mA, so N equals 10.

EXAMPLE

Using the same meter and desired current range as in the previous problem, substitute values in the inverse relationship equation as follows:

$$\frac{R_s}{R_m} = \frac{I_m}{I_s}$$

$$\frac{R_s}{100\ \Omega} = \frac{1\text{ mA}}{9\text{ mA}} \quad \text{or} \quad R_s = \text{one-ninth of } 100\ \Omega$$

$$R_s = 11.11\ \Omega$$

Using this same concept, if you want to double the basic range of a given meter, what is the shunt resistance value? If you said the shunt resistance (R_s) equals the meter resistance (R_m), you are correct! In this case, we want the total current to equally split between the shunt path and the meter movement path; thus, the R values must be equal. Since you now understand the inverse proportionality method, try the following problem.

PRACTICE PROBLEM I

Answers are in Appendix B.

What shunt resistance value is needed to extend the range of a 0–1 mA meter to 0–100 mA. Assume the meter movement's R is 100 Ω.

A Basic Multiple-Range Meter Circuit

Notice in Figure 9–8 we have several shunts that can be selectively switched and connected with the meter. Obviously, since each shunt is a different value, the meter circuit can measure different values of current, depending on which shunt is connected to the circuit.

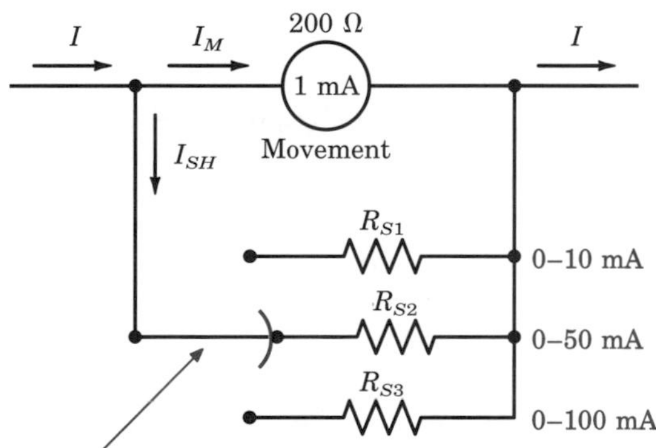

FIGURE 9–8 Switchable shunts provide multiple ranges with one meter.

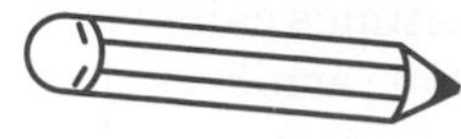

PRACTICAL NOTES

1. *A special shorting-type switch is used to switch ranges.* This type of switch, with its very long wiper contact, assures the meter is never without a shunt! In other words, when changing ranges, this switch makes contact with the next switch contact before it breaks contact with the contact it is leaving. Of course, the reason we don't want to leave the meter without a shunt when switching

between ranges is that when measuring current in a circuit having current greater than the basic movement's full-scale current rating, the meter is damaged or ruined by excessive current through the movement during that moment when it is unshunted.

2. *The shunt value for each range is determined in the same manner as you determined the single shunt,* Figure 9–9 for sample calculations.

FIGURE 9–9 Sample calculation of shunts for three ranges

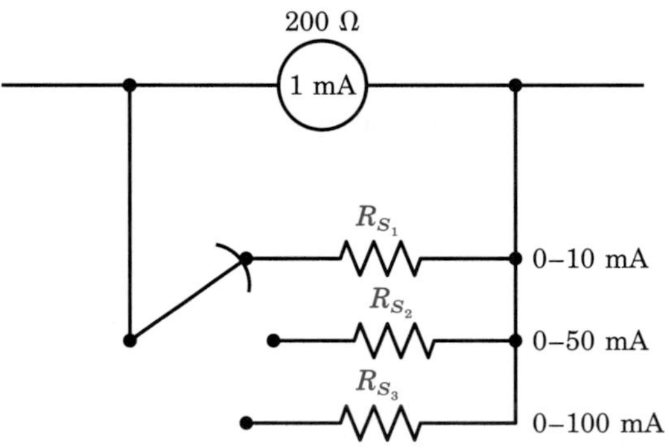

$$R_{S1} = \frac{200\ \text{mV}}{9\ \text{mA}} = 22.22\ \Omega$$

$$R_{S2} = \frac{200\ \text{mV}}{49\ \text{mA}} = 4.08\ \Omega$$

$$R_{S3} = \frac{200\ \text{mV}}{99\ \text{mA}} = 2.02\ \Omega$$

The Universal (Ayrton or Ring) Shunt

General Information. The **universal shunt,** sometimes called the **Ayrton** or ring shunt, is another popular approach that achieves multiple ranges with a single meter movement. It is a variation from the single-shunt concept, using several series resistors with electrical contacts (or "taps") between them, Figures 9–10a and 9–10b.

In essence, we create a series-parallel circuit where the R values for each path are varied by switching between the contacts or taps. By switching the meter test leads to the various taps, different ranges are selected. NOTE: In this case, a shorting-type switch can be used but is not required since the meter is never without a shunt. With any measuring instrument, however, it is always *safest* to leave or start the meter in its highest range setting. This prevents too high a current through the meter movement itself. If the reading provides too little deflection, then you can continue switching down to the next lower range until the meter deflects an appropriate amount for easy reading and best accuracy. It is best to have a reading at mid-scale or higher whenever possible, since the meter is most accurate at its full-scale deflection.

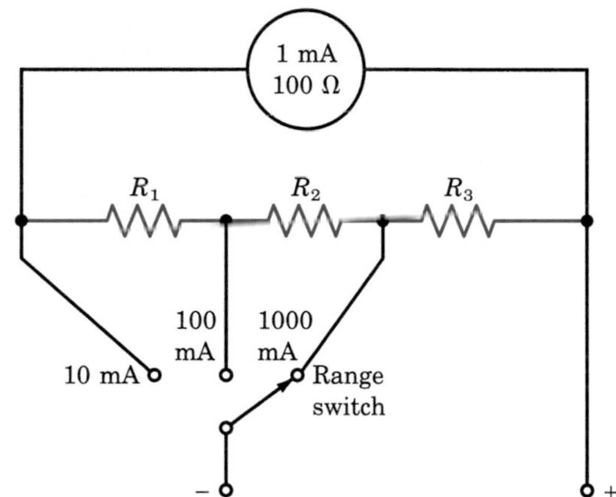

FIGURE 9–10a Example of the universal (Ayrton or ring) shunt

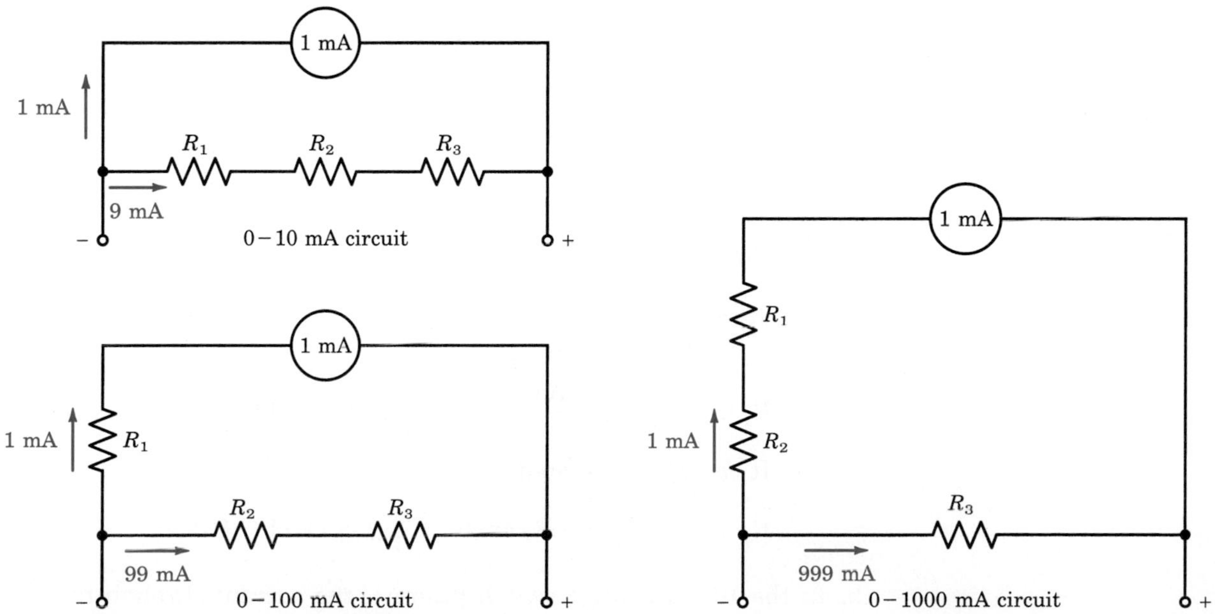

FIGURE 9–10b Equivalent circuits of Ayrton shunt ranges

Ayrton Shunt Calculations. Refer to Figure 9–11 as you study the following discussion.

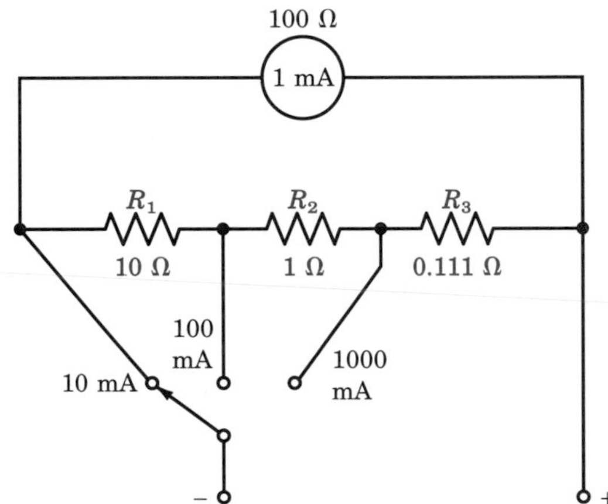

FIGURE 9–11 Sample universal shunt circuit calculation results

1. The resistive network, R_1, R_2, and R_3 form the universal shunt in this circuit.
2. The position of the range selector switch determines which resistances are in the meter circuit and which are in the shunt circuit.
3. The basic meter movement is a 0–1 mA movement with an R_m value of 100 Ω.
4. Using the inverse proportionality concept:
   a. In the 10-mA range switch position, the current through the meter path is 1 mA and the current through the shunt path is 9 mA.

   Therefore, $\dfrac{R_s}{R_m} = \dfrac{I_m}{I_s}$

   $$\frac{R_1 + R_2 + R_3}{100\ \Omega} = \frac{1\ \text{mA}}{9\ \text{mA}}$$

   $$R_1 + R_2 + R_3 = \frac{100}{9} \text{ or } R_s = \frac{1}{9} \times 100 = 11.11\ \Omega$$

   It can also be stated that:

   $$R_2 + R_3 = \frac{100}{9} - R_1 \text{ or } (R_2 + R_3) = 11.11 - R_1$$

   b. In the 100-mA range switch position, the current through the meter path must still be 1 mA. But this time, the current through the shunt path must equal 99 mA. Notice in this position of the switch, R_1 and R_m are in the meter path; however, R_2 and R_3 are in the shunt path. Substituting in our formula:

$$\frac{R_2 + R_3}{R_1 + R_m} = \frac{1 \text{ mA}}{99 \text{ mA}}$$

Therefore, R_s or $(R_2 + R_3) = \dfrac{R_1 + 100}{99}$

Using the expressions in parts a and b above, for $R_2 + R_3$ an equation is written that states:

$$\frac{100}{9} - R_1 = \frac{R_1 + 100}{99}$$

$$1100 - 99R_1 = R_1 + 100$$

$$R_1 = 10\ \Omega$$

It can also be stated that:

$$R_2 + R_3 = \frac{100}{9} - 10 = 1.11\ \Omega$$

Therefore, $R_2 = \dfrac{10}{9} - R_3$

c. In the 1000-mA range switch position, the current through the meter path must still be 1 mA. But this time, the current through the shunt path must equal 999 mA. In this position of the switch, $R_2 + R_1 + R_m$ are in the meter path and R_3 is in the shunt path. Again, substituting in our formula:

$$\frac{R_3}{R_2 + R_1 + R_m} = \frac{1 \text{ mA}}{999 \text{ mA}}$$

$$999R_3 = R_2 + 110$$

$$999R_3 - R_2 = 110$$

Substituting the $\dfrac{10}{9} - R_3$ expression for R_2 then:

$$999R_3 = \frac{10}{9} + 110 = 111.11$$

$$R_3 = 0.111\ \Omega$$

Therefore, $R_2 = 1.11 - 0.11 = 1\ \Omega$

Collecting our information, then:

$R_1 = 10\ \Omega$
$R_2 = 1\ \Omega$
$R_3 = 0.11\ \Omega$

Using the fact that the voltages across the meter circuit and across the shunt circuit must be equal in each case, we can verify our results as follows (NOTE: Minor discrepancies are due to rounding-off factors):

In the 10-mA range

- Current through the meter portion of the circuit is 1 mA and resistance of meter circuit is 100 Ω. Therefore, V_m = 100 mV.
- Current through the shunt portion of the circuit is 9 mA and resistance is 11.111 Ω. Therefore, V_s = 100 mV.

In the 100-mA range

- Current through the meter portion of the circuit is 1 mA and resistance of the meter circuit is 110 Ω. Therefore, V_m = 110 mV.
- Current through the shunt portion of the circuit is 99 mA and resistance is 1.111 Ω. Therefore, V_s = 110 mV.

In the 1000-mA range

- Current through the meter portion of the circuit is 1 mA and resistance of the meter circuit is 111 Ω. Therefore, V_m = 111 mV.
- Current through the shunt portion of the circuit is 999 mA and resistance is 0.111 Ω. Therefore, V_s = 111 mV.

There are other ways of solving for these values, such as simultaneous equations. Most technicians will probably never have to design universal shunts unless they work at a test instrument manufacturing company; therefore, we won't pursue the topic further. Simply be aware that many multi-range commercial instruments use this approach to achieve multiple current ranges with a given basic movement.

Current Meter Key Points Capsule

1. Moving-coil movements are essentially dc current indicators.
2. The minimum current measured is based on the sensitivity of the basic meter movement.
3. The sensitivity of a basic meter movement is determined by the number of turns in the moving coil, the strength of the magnetic field, and other related mechanical factors, such as the springs, the weight of the coil frame, and so forth.

4. Currents higher than the basic movement's full-scale current rating can be measured if appropriate values of resistance are connected in shunt (or parallel) with the basic movement.
5. Current meters are generally linear devices. That is, when 50% of full-scale current value flows through the meter movement, deflection is mid-scale on the meter. When 25% of fs current is measured, the pointer moves one-fourth full-scale distance on the scale.
6. To measure current, meters must be connected in series, observing proper polarity, and have adequate range.
7. Because current meters are connected in series, the lower the meter resistance (R_m) used in making the measurement, the less the actual circuit operating conditions will be changed and the more useful the reading will be.

VOLTMETERS

The Purpose of the Multiplier Resistor

Another extremely useful application for which movements can be adapted is that of measuring voltage, Figure 9–12. Remember that the basic meter movement is a current indicator. By adding select value resistors, called **multiplier resistors,** in series with the meter movement, it is possible to measure voltage. The multiplier resistor limits the current through the movement. The amount of meter deflection is directly related to the voltage applied to the meter circuit. For example, a voltage equaling one-third the voltage range causes one-third scale deflection. As you will soon see, different multiplier values provide different voltage ranges.

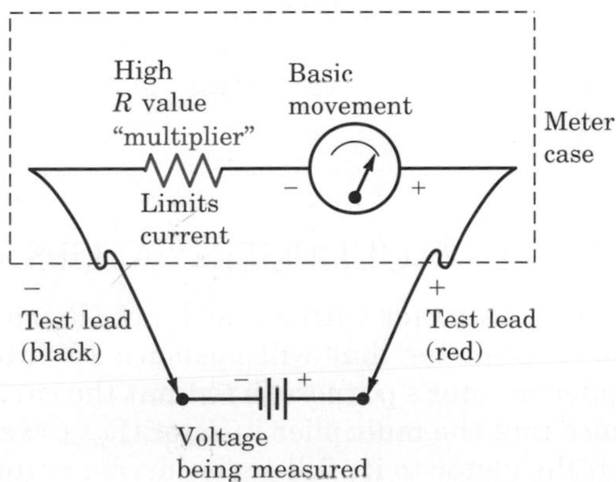

FIGURE 9–12 The multiplier resistor

Connecting the Voltmeter to the Circuit Being Tested

Refer to Figure 9–13. Notice that although the voltmeter circuit is a series circuit ($R_m + R_{mult}$), the voltmeter is always connected *in parallel* with the two points across where it is desired to measure the potential difference, or voltage. Because the voltmeter is connected in parallel with the portion of the circuit being tested, it is ***not*** *necessary to break (open) the circuit,* as is done when inserting a standard current-measuring instrument for measuring current.

FIGURE 9–13 Connecting a voltmeter to make voltage measurements

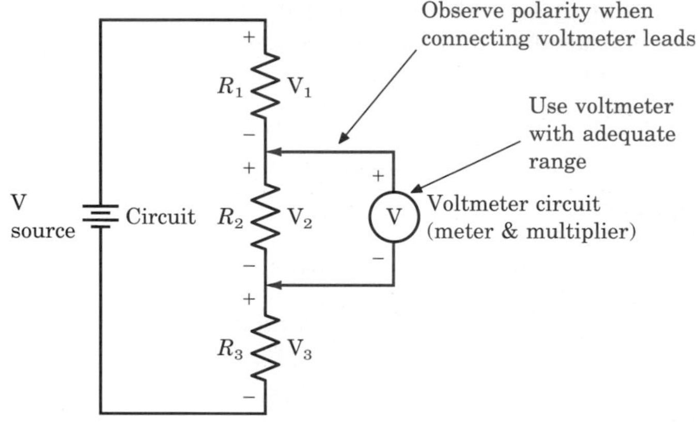

It is necessary, however, to *observe polarity* when measuring dc voltages, or the meter deflection will be in the wrong direction, as it is for the basic current meter.

As indicated earlier in the chapter, *a sufficiently high range* is needed to assure no more than full-scale deflection of the meter; otherwise, damage will result. Put another way, you are asking for bent pointers and maximum smoke if an adequate range is not used! Again, look at Figure 9–13 to see these points illustrated.

CALCULATING THE MULTIPLIER VALUES

Using our familiar 0–1 mA meter with a resistance (R_m) of 100 Ω, let's see how we can design a voltmeter that will measure 10 V, 50 V, and 100 V.

Since the multiplier resistor's purpose is to limit the current through the meter, it makes sense that the multiplier resistor (R_{mult}) value should limit the current through the meter to its full-scale current rating at the voltage range we want to design. That is, to have a 10-volt range, the series value of

$R_m + R_{mult}$ should limit current to 1 mA when it is measuring 10 volts. This is a calculation for a two-resistance series circuit using Ohm's law. As you read the following examples, refer to Figure 9–14.

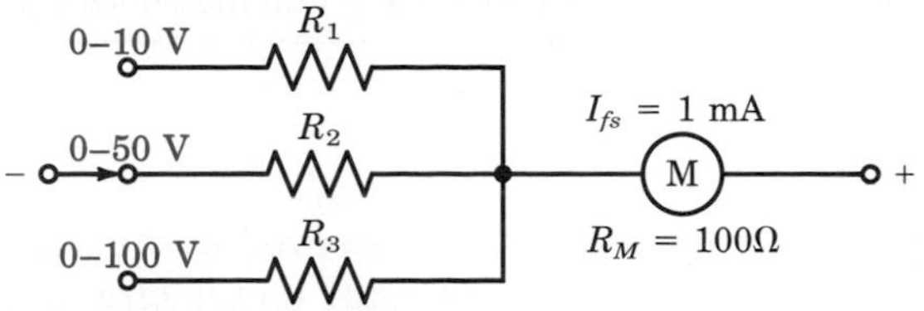

0–10 V R_1
0–50 V R_2
0–100 V R_3
I_{fs} = 1 mA
M
R_M = 100Ω
− +

$$R_1 = \frac{10\text{ V}}{1\text{ mA}} - 100\ \Omega = 10{,}000\ \Omega - 100\ \Omega = 9{,}900\ \Omega$$
$$R_2 = \frac{50\text{ V}}{1\text{ mA}} - 100\ \Omega = 50{,}000\ \Omega - 100\ \Omega = 49{,}900\ \Omega$$
$$R_3 = \frac{100\text{ V}}{1\text{ mA}} - 100\ \Omega = 100{,}000\ \Omega - 100\ \Omega = 99{,}900\ \Omega$$

FIGURE 9–14
Examples of multiplier calculations

EXAMPLES

For the 10-volt range, then:

(Total R of meter circuit)

Formula 9–6	$R_{mult} + R_m = \frac{\text{V (range)}}{\text{I (full-scale)}}$

(Multiplier R)

Formula 9–7	$R_{mult} = \frac{\text{V (range)}}{\text{I (fs)}} - R_m$

$$\text{Therefore, } R_{mult} = \frac{10\text{ V}}{1\text{ mA}} - 100\ \Omega = 10{,}000 - 100 = 9{,}900\ \Omega$$

For the 50-volt range:

(Total R of meter circuit) $R_{mult} + R_m = \frac{V}{I}$

(Multiplier R) $R_{mult} = \frac{V}{I} - R_m$

$$R_{mult} = \frac{50\text{ V}}{1\text{ mA}} - 100\ \Omega = 50{,}000 - 100 = 49{,}900\ \Omega$$

For the 100-volt range:

(Total R of meter circuit) $R_{mult} + R_m = \frac{V}{I}$

(Multiplier R) $R_{mult} = \frac{V}{I} - R_m$

$$R_{mult} = \frac{100\text{ V}}{1\text{ mA}} - 100\ \Omega = 100{,}000 - 100 = 99{,}900\ \Omega$$

PRACTICE PROBLEM II

Given a 50-μA movement with R_m value of 2,000 Ω, what value of multiplier resistor is needed to create a voltmeter range of 250 V?

Approximating Multiplier Values

An interesting point you have probably observed is that the meter's resistance is a very small percentage of the total meter circuit R ($R_m + R_{mult}$). Therefore, not much error is introduced if we indicate that R_{mult} is essentially equal to $\frac{\text{V (desired range)}}{\text{I(fs)}}$.

Range Selector Switch

Observe in Figures 9–15a and 9–15b that a **range selector switch** selects the desired voltage measurement range by switching in the appropriate multiplier resistor(s). No matter what range is selected, the multiplier resistance limits current to the meter's full-scale current value when the voltage applied to the total metering circuit (meter plus multiplier) equals the voltage range value.

Also note that either the individual R multiplier or the cumulative R multipliers approach can be used. Most commercial multi-range meters use the cumulative R type circuitry. Refer again to Figure 9–15b.

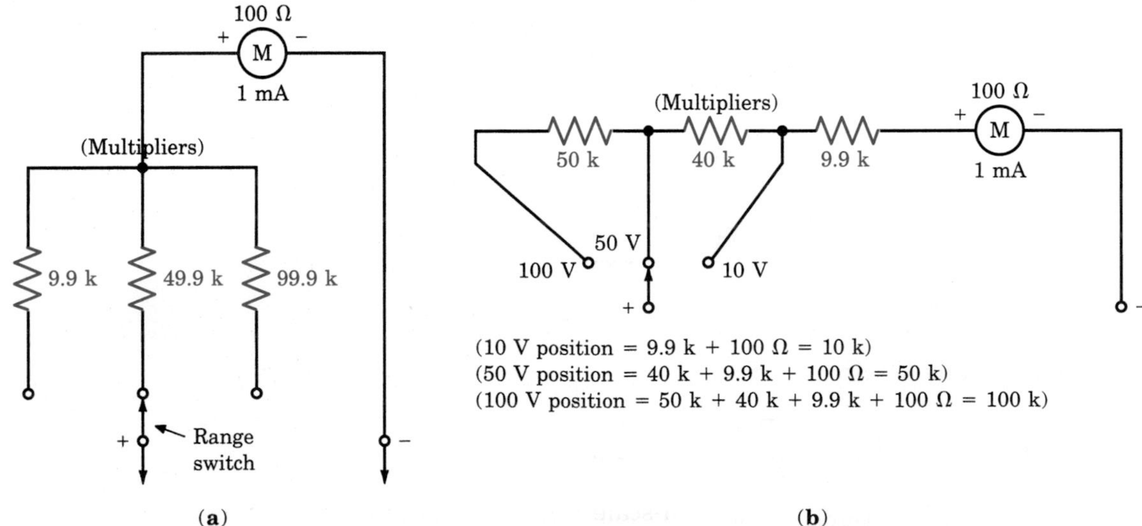

FIGURE 9–15 The range selector switch

SAFETY HINTS

Caution About Accuracy

There are many times when measurement accuracy is *critical!* In certain sensitive circuits, a percentage point of inaccuracy can make a meaningful difference—the difference between operating or not operating, in some cases. For that reason, we want to remind you: ***Always be aware of the test instruments' accuracy as related to the acceptable level of accuracy needed to measure the circuit being tested!***

Meter scales have calibrations that either directly match or are a manageable multiple or submultiple of the various selectable ranges. Refer to Figure 9–16.

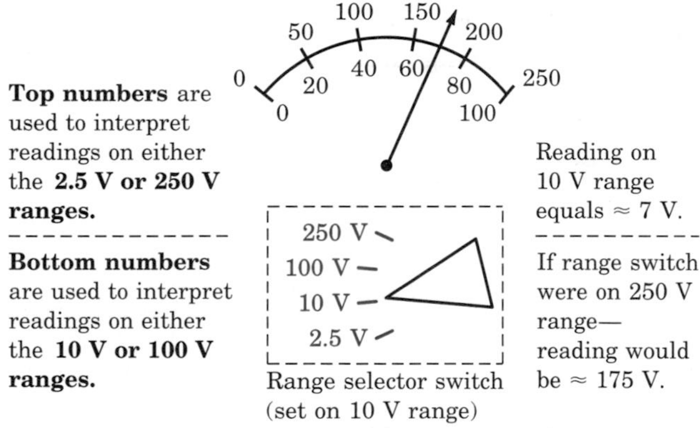
0
50
100
150
200
250
0
20
40
60
80
100
Top numbers are used to interpret readings on either the **2.5 V or 250 V ranges.**
Bottom numbers are used to interpret readings on either the **10 V or 100 V ranges.**
250 V
100 V
10 V
2.5 V
Range selector switch (set on 10 V range)
Reading on 10 V range equals ≈ 7 V.
If range switch were on 250 V range— reading would be ≈ 175 V.

FIGURE 9–16 Meter scales correlate to selectable ranges.

Voltmeter Ohms-Per-Volt $\left(\frac{\Omega}{V}\right)$ Ratings

Previously we discussed the full-scale current- and millivolt-type ratings used with the basic current meter movements.

Another useful rating, related particularly to voltmeters, is the **ohms-per-volt rating** $\left(\frac{\Omega}{V}\right)$, Figure 9–17. This rating indicates how many ohms of meter circuit resistance ($R_m + R_{mult}$) must be present to limit current to the meter's full-scale current rating value *for each volt* applied. In other words, $R = \frac{1\ V}{I(fs)}$, or the reciprocal of the full-scale current rating provides the ohms-per-volt rating.

EXAMPLE

A 1-mA movement has an $\frac{\Omega}{V}$ rating of $\frac{1}{1 \times 10^{-3}}$, or 1,000-ohms per volt.

A sensitive 50-μA movement has an ohms-per-volt rating of $\frac{1}{50 \times 10^{-6}}$, or 20,000 ohms per volt.

A 10-mA movement has a rating equal to the reciprocal of 0.010, or $\frac{1}{10 \times 10^{-3}} = 100$ ohms per volt.

FIGURE 9–17 The ohms-per-volt rating

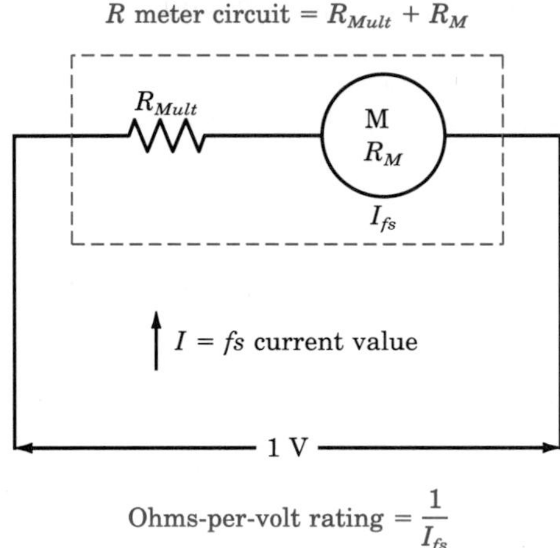

Greater sensitivity of the basic movement means a higher ohms-per-volt rating. Two observations are worth noting related to this rating.

One, if you know the ohms-per-volt rating, it's easy to calculate the multiplier R value for any desired voltmeter range, neglecting the small value of R_m. However, *be careful!* If a 20,000-$\frac{\Omega}{V}$ movement has a 2,000-Ω meter resistance, which is typical, a 10% error is introduced on the 1 V range by this technique. One example of using this ballpark or approximating technique is as follows. If the meter rating is 10,000-$\frac{\Omega}{V}$, and you want to create a 100-V range, multiply the rating by 100. The answer is 100 × 10,000, or 1 megohm. You can see the trend!

Another more important fact is this rating gives you a good idea of how much meter loading effect will occur when you make voltage measurements. The next section discusses this concept in more detail. It is important, so study it!

PRACTICE PROBLEM III

1. What is the ohms-per-volt rating of a meter that uses a 20-microampere movement?
2. What is the full-scale current of a meter rated at 2,000-ohms per volt?

Voltmeter Loading Effect

Since a voltmeter is connected in parallel with the component, or the circuitry where it is measuring voltage, it would be ideal if the voltmeter circuitry had *infinite* resistance. This would mean none of the normal circuit operating conditions is altered with the voltmeter connected. But this is not the case! The voltmeter circuit has some finite value of resistance, and it alters circuit conditions! This situation is called **loading effect,** Figure 9–18.

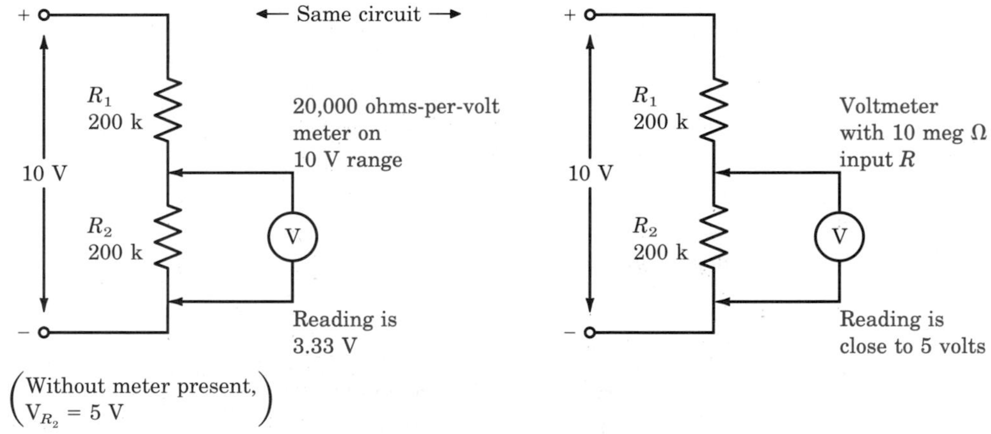

FIGURE 9–18 Voltmeter loading effect

It is apparent the higher the voltmeter circuit's resistance, the less loading effect it has on the tested circuit. That is, the higher the ohms-per-volt rating of the voltmeter circuit *or* the higher the voltage range, the less the loading effect. Again greater sensitivity of the basic movement means a higher ohms-per-volt rating. This situation correlates with the voltmeter causing less loading effect on the tested circuit.

Also, a higher voltage range results in a higher multiplier resistance, which causes less voltmeter loading effect, Figure 9–19.

The greater the voltmeter circuit resistance ratio to the component or circuit resistance being tested, the more accurate the measurement reflects

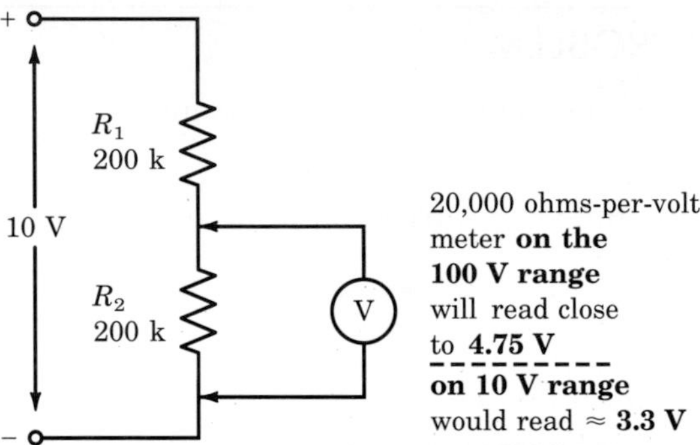

+
R1
200 k
10 V
R2
200 k
V
−
20,000 ohms-per-volt
meter on the
100 V range
will read close
to 4.75 V
on 10 V range
would read ≈ 3.3 V

FIGURE 9–19 Sample conditions of loading effect

circuit conditions. That is, the less the circuit conditions change with the voltmeter attached.

Stated another way, not only is the meter circuit resistance important, *but* the tested circuit's resistance characteristics are equally important! A higher circuit resistance prior to voltmeter connection results in greater disturbance of normal operating parameters when the meter is connected! In other words, low resistance circuits are disturbed less by the meter connection than high resistance circuits, Figure 9–20.

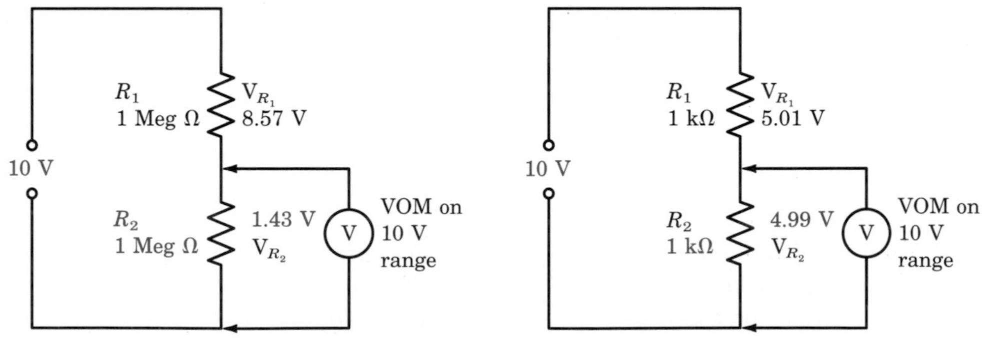

Using a 20,000 ohms-per-volt VOM—loading effect
changes parameters as shown.
R1
1 Meg Ω
VR1
8.57 V
10 V
R2
1 Meg Ω
1.43 V
VR2
V
VOM on
10 V
range
R1
1 kΩ
VR1
5.01 V
10 V
R2
1 kΩ
4.99 V
VR2
V
VOM on
10 V
range
Without the voltmeter, VR1 and VR2 = 5 V each.
Note the loading effect difference between the
two circuits!
Large loading effect is on high R circuit.
Small loading effect is on low R circuit.

FIGURE 9–20 Variation in loading effect due to differences in circuits being tested

PRACTICAL NOTES

One practical rule of thumb is to be sure the resistance of the voltmeter circuit (on the range being used) is at least ten times the resistance of the circuit portion being tested by the voltmeter. For practical purposes, commercial voltmeters range from 20,000-$\frac{\Omega}{V}$ sensitivity ratings for common multimeters, or VOMs (volt ohm milliammeters), to 10 megohms, or have higher input circuit resistances for the voltmeter circuitry common in digital measuring devices, Figure 9–21. Obviously, meters with multi-megohm input resistances cause negligible loading effect on lower-resistance circuits to be tested.

FIGURE 9–21 Higher meter circuitry resistance means less loading effect.

Voltmeter Key Points Capsule

1. A voltmeter is a combination of a current meter and a series current-limiting resistor, called a multiplier. The multiplier resistance limits current through the voltmeter when voltage is applied to the circuit.
2. A common sensitivity rating system for voltmeters is the ohms-per-volt rating that expresses how much resistance is needed to limit current through the meter to full-scale value when one volt is applied to the voltmeter circuitry.
3. The multiplier resistance value needed to achieve a given range voltmeter is determined by using Ohm's law and simple series-circuit techniques. $(R_m + R_{mult}) = \frac{V \text{ desired range}}{I_{fs}}$
4. Linear scales are usually present on dc voltmeters.
5. To measure voltage, the voltmeter is connected in parallel with the component or circuit portion where the difference of potential is to be measured. When measuring dc voltage, it is necessary to observe polarity. Also, care must be taken to assure that the voltage range of the metering circuit is sufficient to handle the voltage to be measured.
6. Because the voltmeter is connected in parallel, the higher the metering circuit resistance, the less it changes the circuit conditions, and the more accurately voltmeter reading shows the actual circuit operating conditions.

OHMMETERS

NOTE: Although **series-type ohmmeter circuits** are becoming less common, due to the availability of digital multimeter devices, we will study the series-type ohmmeter briefly so you can gain a basic understanding of ohmmeter concepts.

The ohmmeter is yet another application for dc meter movements. An ohmmeter circuit has three main elements—a battery (generally housed in the meter case), the meter movement, and some current-limiting resistance, Figure 9–22.

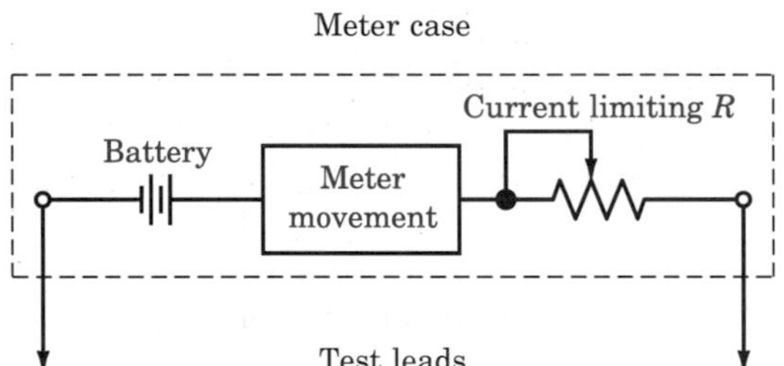

FIGURE 9–22 Typical ohmmeter elements

Because resistance measurements are made with the power off on the tested component or circuit, the internal battery becomes the current source during testing. Obviously, the meter reacts to the amount of current flowing, and the current-limiting resistance controls the amount of current flowing through the meter.

Adjusting, Using, and Reading the Ohmmeter

The most common ohmmeter (used in analog instruments) uses the series-type ohmmeter circuit. Refer to Figure 9–23 and notice that when the test leads are not touching each other or are not connected to an external current path, there is an "open" circuit. Therefore, there is no current through the meter, and the **pointer** is at its left-hand resting place.

On the ohmmeter scale, zero current indicates infinite resistance between the test probes. The scale is marked with the infinity symbol (∞) on the left end of the scale. When the test leads are touched together, there is zero ohms between the test leads, and there is a complete current path. That is, current flows through the meter. The amount of current is controlled by adjusting the variable resistor (called the **zero-adjust control**) so there is exactly full-scale deflection on the current meter. The right end of the ohm-

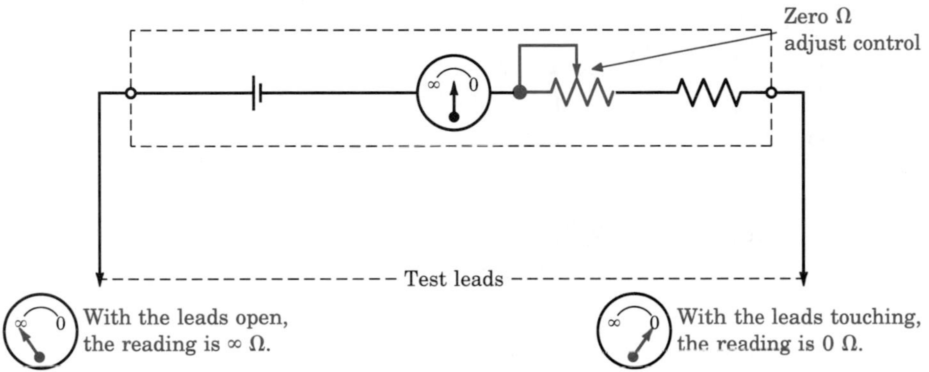

FIGURE 9–23 Series-type ohmmeter circuit characteristics

meter scale is marked with a "0" calibration mark, indicating that when full-scale current is flowing, there is zero ohms between the test probes.

It is apparent when a resistance value between 0 and ∞ ohms is placed between the test leads, some amount of current flows that is less than full-scale current value and more than zero current. Therefore, the scale is calibrated to show this resistance value. Series circuit analysis is used to show what percentage of full-scale deflection occurs for various resistance values, Figure 9–24.

Note the ohmmeter scale is a nonlinear scale! This nonlinear scale contrasts with most ammeter and voltmeter scales that have evenly spaced or linear scales. A nonlinear ohmmeter scale is sometimes termed a **backoff ohmmeter scale,** since it is a "right-to-left" scale. In other words, zero is on the right end of the scale, and higher resistance readings are to the left. Again, see Figure 9–23.

Typically, commercial multimeters (volt ohm milliammeters or VOM) have more than one range for measuring resistance that enables reading both low and high R values. Notice in Figure 9–25 the ohmmeter range

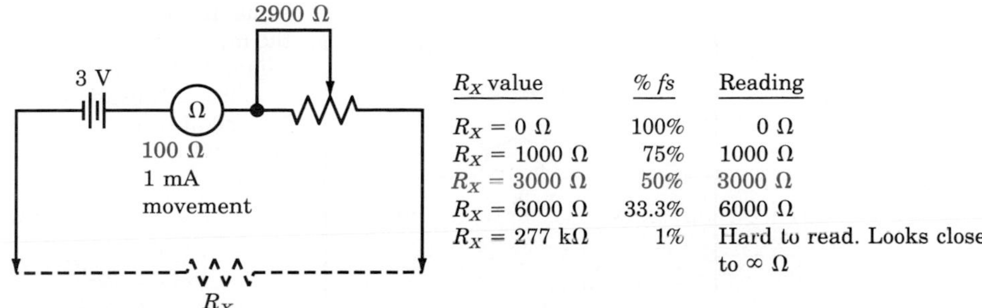

R_X value	% *fs*	Reading
$R_X = 0\ \Omega$	100%	0 Ω
$R_X = 1000\ \Omega$	75%	1000 Ω
$R_X = 3000\ \Omega$	50%	3000 Ω
$R_X = 6000\ \Omega$	33.3%	6000 Ω
$R_X = 277\ \text{k}\Omega$	1%	Hard to read. Looks close to ∞ Ω

FIGURE 9–24 Ohmmeter deflection for various R_X values

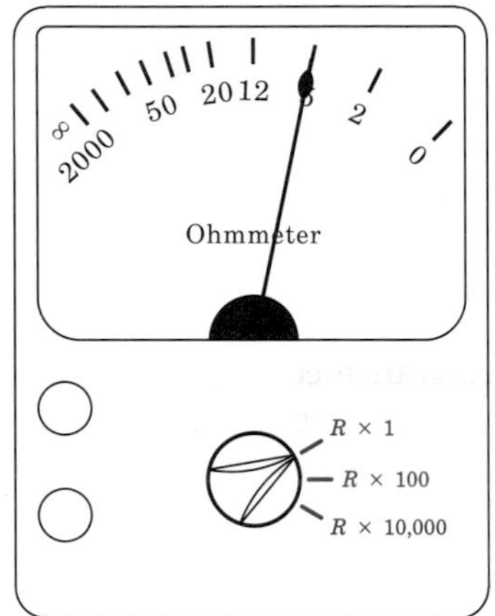

FIGURE 9–25 Switchable ohmmeter ranges

If the pointer is pointing at "5" on the scale, then
$R \times 1 = 5\ \Omega$
$R \times 100 = 500\ \Omega$
$R \times 10{,}000 = 50{,}000\ \Omega$

switch shows ranges expressed in terms of R times some factor. For example, in the R × 1 range, the ohmmeter scale is read directly. In the R × 100 range position, the technician should multiply the scale reading by 100 to interpret the value of resistance being measured. The method for multiple ranges involves switching to different metering circuit shunt resistances, which is necessary because when measuring high resistances, little current flows; thus high meter sensitivity is needed. When measuring low resistances, much current flows; thus, the meter can be shunted, as appropriate. Also, it is common to find a switching of internal voltage source(s), so when measuring high resistances the voltage supplied to the circuit is higher.

We have been discussing the common series-type ohmmeter circuit. There is also a shunt-type circuit where the meter, its internal V source, and the resistance under test are in parallel with each other, Figure 9–26. We won't discuss this in detail because this type of ohmmeter is not frequently

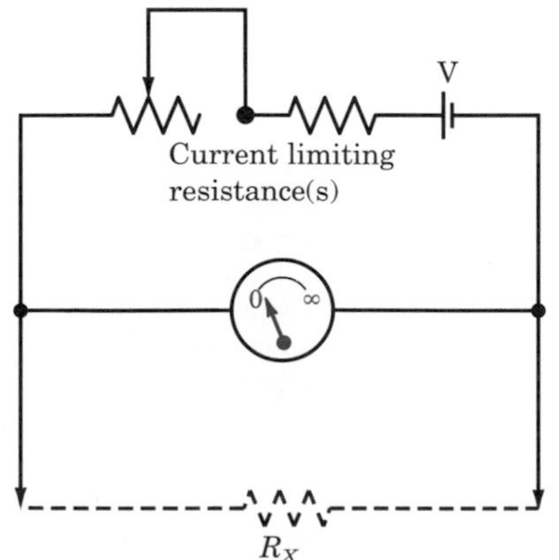

FIGURE 9–26 Shunt-type ohmmeter circuit

R_X shunts the current around the meter. Therefore, the lower the R_X value, the less the current through the meter and the higher the R_X value, the higher the current through the meter. Thus, zero ohms is on the left end of scale and $\infty\ \Omega$ is on the right end.

used. However, it is worth mentioning that this type of circuitry is useful for measuring very low resistance values and uses a normal left-to-right scale, rather than a backoff scale.

Ohmmeter Key Points Capsule

1. The series-type ohmmeter is a dc movement, an internal dc source, and a series current-limiting resistance connected in series. The test probes, when connected to the external component(s) being tested, place the unknown value of R being measured in series with the metering circuit.
2. The shunt-type ohmmeter places the movement, voltage source circuitry, and resistance under test in parallel with each other.
3. The commonly used series-type ohmmeter uses a backoff (right-to-left) scale that is nonlinear. The lower value resistance scale calibrations (on the right end of the scale) are spread apart more than the higher resistance value calibrations, which appear on the left end of the scale.
4. Ohmmeters should never be used to measure components or circuit portions that have power applied. (Most of the time, if that is done, the meter will be damaged). The ohmmeter's internal battery supplies the necessary voltage and current to make the measurement(s).
5. When using the series-type ohmmeter, the test probes are touched together and the zero-adjust control is adjusted so there is full-scale current through the meter and the ohmmeter scale reads "zero" resistance. After "zeroing" the meter, the test probes are then connected across the component or components being tested to read their resistance value(s).
6. Most ohmmeters have more than one resistance range to measure a wider range of resistance values. The range selector switch indicates R times some factor at each switch position. For example, if the meter indicates 100 on the scale and the range selector switch is in the R × 10 position, the actual resistance being measured is 10 × 100, or 1,000 Ω.

VOMs, DMMs AND OTHER RELATED DEVICES

Background Information

As indicated throughout the chapter, most test instruments are designed so they have multiple ranges for the electrical quantity they are measuring. Figure 9–27 shows an example of a panel meter designed to measure one

FIGURE 9–27 Single-range panel meters (Courtesy of Simpson Electric Company)

type of electrical quantity. This type of meter is found in electronic equipment, such as radio transmitters and power supplies, and frequently has only one range.

As an electronics technician, you will probably use devices called **multimeters** in your troubleshooting and designing tasks. These meters not only have multiple ranges for each type of electrical quantity to be measured, but can measure more than one of the basic electrical quantities (I, V, or R). Let's take a brief look at several types of popular multimeters.

VOMs

The VOM (volt ohm milliammeter) is a popular multimeter used for many years by technicians, Figure 9–28. The advantages and disadvantages of the VOM are:

ADVANTAGES	DISADVANTAGES
• Compact and portable • Does not require an external power source • Measures dc current, dc voltage resistance, and ac voltage • Typically has multiple ranges for each electrical quantity it measures • Is analog in nature, so is *good* for adjusting, tuning, or peaking tuned circuits	• Can cause significant meter loading effects on circuit being tested • Less accurate than a DMM • Analog readout scale can introduce "parallax" reading errors if pointer is viewed from an angle

FIGURE 9–28 The popular VOM (volt ohm milliammeter) (Courtesy of Simpson Electric Company)

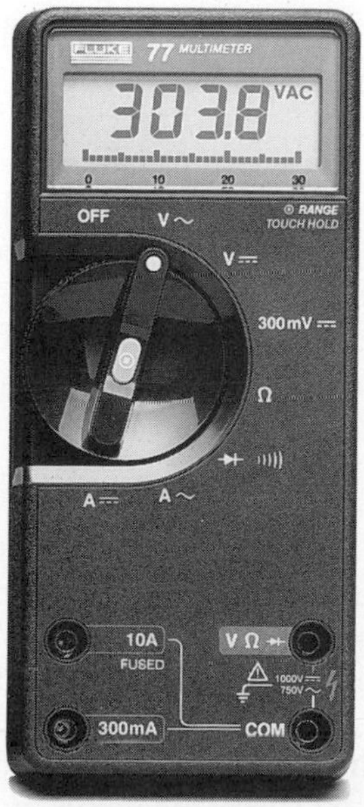

FIGURE 9–29 The DMM (digital multimeter): a portable DMM (Courtesy of the John Fluke Mfg. Co., Inc.)

DMM

The DMM (digital multimeter), Figure 9–29, has become very popular because it offers the capabilities of a VOM in addition to other distinct advantages. The advantages and disadvantages of the DMM are:

ADVANTAGES	DISADVANTAGES
• Digital readout usually prevents a reading error • Can have "autoranging" that prevents destroying a meter by having it on the wrong range • Can have "autopolarity" that prevents problems from connecting meter to test circuit with wrong polarity • Can have excellent accuracy (plus or minus 1%) • Portable and uses internal batteries	• Can be physically larger than equivalent VOM (not in all cases) • Can be more costly to buy and operate • Due to "step-type" digital feature rather than "smooth" analog action, are *not as good* for adjusting "tuned" circuits or "peaking" tunable responses

Related Devices

There are special probes used with multimeters and special self-contained metering devices that you should know about.

High Voltage Probe. Basic multimeters cannot generally measure very high voltages in the multikilovolt range; however, a high-voltage probe will enable this to be done, Figure 9–30. In essence, the probe consists of a high-value resistance that acts as a multiplier resistance, external to the meter. The probe's resistor(s) are designed to withstand high voltages. One typical application of this probe is to measure high voltages on TV picture tubes.

Clamp-On Current Probe. Another accessory found on some multimeters is the clamp-on current probe that measures ac current, Figure 9–31. When ac current levels are in the ampere ranges, this device reacts to the magnetic fields surrounding the current-carrying conductors and translates this activity into current readings. The key advantage to this device is that it measures current *without having to break the circuit* when inserting the metering circuit. The clamp-on probe simply surrounds the current-carrying conductor without breaking the circuit. NOTE: This device *does not measure dc* current!

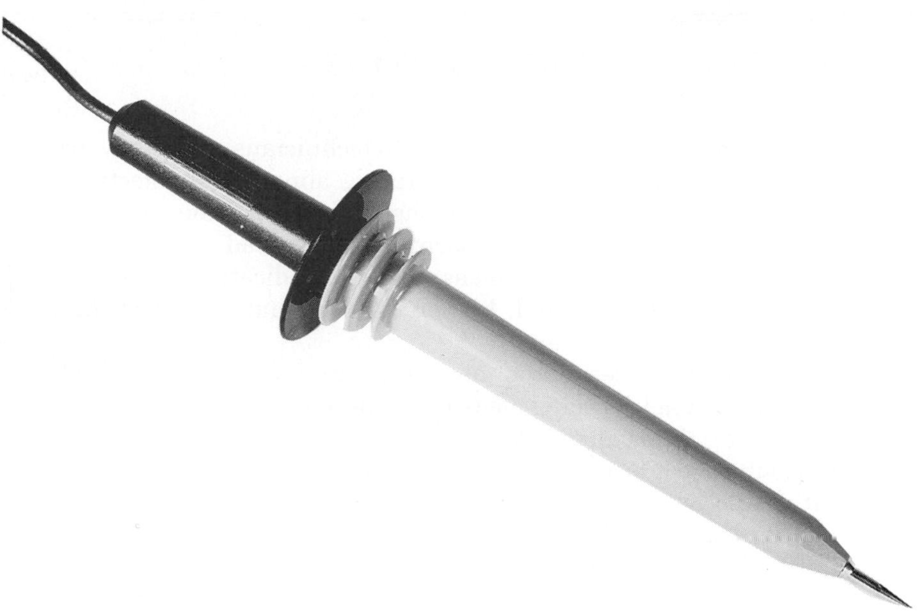

FIGURE 9–30 The high-voltage probe (Photo reprinted with permission from Perozzo, *Practical Electronics Troubleshooting*, ©1985 by Delmar Publishers Inc., Albany, NY)

Other Devices. There are a variety of special clamp-on measurement devices described in many electronic distributor catalogs. You will find the clamp-on meters we've been describing, as well as clamp-on volt-ohm-milliam-

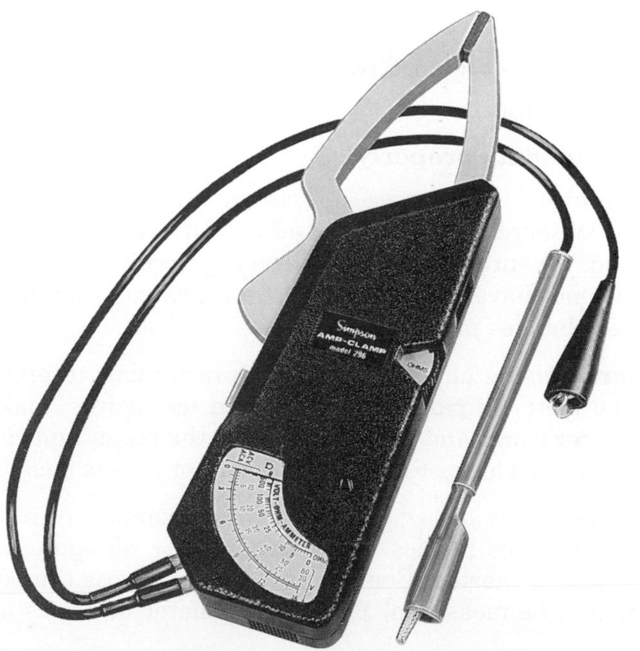

FIGURE 9–31 The clamp-on current probe (Courtesy of Simpson Electric Company)

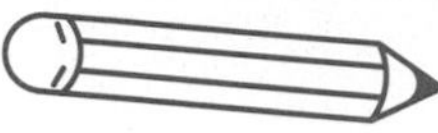

PRACTICAL NOTES

Measuring instruments are a vital tool to technicians when they are troubleshooting circuits that are operating abnormally. Electrical parameters indicate whether a component, circuit portion, or a total circuit is operating normally, abnormally, or is dead.

In a *qualitative* sense, the measurements indicate whether the values measured are normal, higher than normal, or lower than normal.

In a *quantitative* sense, the measurements indicate specific values for the parameters being measured. These values are then interpreted as normal or abnormal.

To troubleshoot any circuit or system, the technician must know "what should be." That is, he or she *must know the norms.* This is gained through experience, knowledge, or reference to appropriate technical documentation, such as schematics.

Although the following discussion is not exhaustive or comprehensive, the hints and techniques can be useful to you during your training and throughout your career as a technician. Study and apply these items, as appropriate.

meters, clamp-on digital wattmeters, and "power factor" meters. (You will learn about "power factor" meters in later chapters).

TROUBLESHOOTING HINTS

General Information

Use test instruments properly:

1. **Current meters**—*Turn power off circuit.* Insert meter in series, making sure to observe proper range and polarity. Turn power on circuit to be tested. Be sure the current metering circuit (meter or meter and shunt) does not have too high a resistance, causing undue or unaccounted for changes in circuit conditions.

2. **Voltmeters**—Using appropriate safety precautions to prevent shock, connect the meter in parallel with desired test points, making sure to observe proper range and polarity. Be sure the meter and range do not cause unaccounted for meter loading because of too low a meter resistance.

3. **Ohmmeters**—*Turn power off circuit* where component or circuit portion to be measured resides. Make appropriate zero-adjust calibration with test probes shorted. Connect test probes across component or circuit portion to be measured. Be sure ohmmeter batteries are good.

Techniques for Voltmeter Measurements

1. **Connecting one lead to common or ground**—Many electronic circuits have a common or ground reference point, such as the chassis or a common conductor "bus" where all voltages are referenced. By connecting one lead of the voltmeter (frequently the black or negative lead) to this ground reference, measurements are made throughout the circuit by moving the other lead to the various points where voltage measurements are desired, Figure 9–32. Frequently, the black lead on a meter has an alligator clip, making it easy to attach the lead to an electrically common point in the circuit.

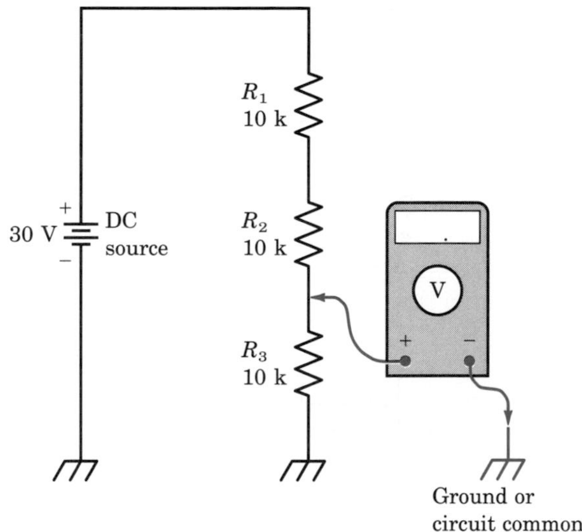

FIGURE 9–32 Making voltage measurements with respect to ground reference

2. **Using voltage checks to determine current**—Because it is often awkward and time consuming to break a circuit when installing a current meter, technicians frequently use voltage measurements to determine current.

   For example, if the measured voltage drop across a 1 kΩ resistor is 15 volts, it is simple (Ohm's law) to deduce the current through that resistor must be 15 mA. That is, $I = \frac{V}{R} = \frac{15\text{ V}}{1\text{ k}\Omega} = 15\text{ mA}$. Obviously, Ohm's law enables you to determine current regardless of the R value where you are measuring the voltage. This technique prevents you from having to a) turn off the circuit; b) break the circuit; c) insert a current meter; d) make the measurement, then turn circuit off again; and e) reconnect the circuit when the current meter is removed.

3. **Using voltage measurements to find opens**—Recall from a previous chapter that when an open is in a series circuit, the applied voltage appears across the open. This means the voltmeter is a very effective means to find the circuit's open portion. The series circuit applied voltage appears across the open component(s), and zero volts are measured across the remaining component(s), Figure 9–33.

FIGURE 9–33 Voltage measurements can find "opens."

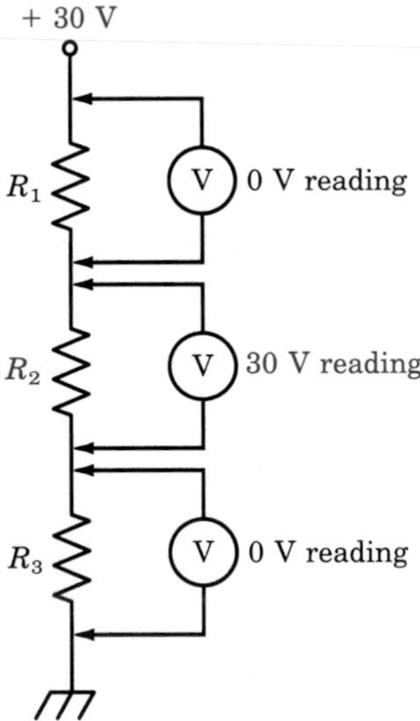

4. **Using voltage measurements to find shorts**—Again recall that zero volts IR drop appear across a shorted component or circuit portion. Voltage measurements quickly show this.

Techniques for Current Meter Measurements

1. **Total current measurement as a circuit condition indicator**—Total current value gives a general indication of circuit condition. If the total current is higher than normal . . . a lower-than-normal total circuit resistance is indicated. If the total current is lower than normal, a higher-than-normal total circuit resistance is indicated. Additional isolation techniques are then used to find the problem.

2. **Special jacks for current measurement**—Some circuits or systems have built-in closed-circuit jacks where the current meters are connected to measure current in the circuit, Figure 9–34.

When the current meter is not present, the jack is shorted, providing a closed-current path. But when the current meter is present, the current must pass through the meter. (NOTE: This current-jack situation is uncommon.) More typically, voltage measurements determine current(s). In transmitters and other similar circuits where monitoring current is important, permanently mounted and installed current meters are used.

3. **Clamp-on metering for ease of measurement**—Previously we discussed the advantage of not breaking the circuit when using this type of current measurement. The limitations are available ranges and physical situations where it is impossible to connect such a meter.

(a)

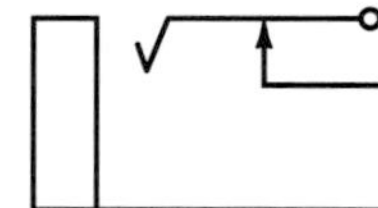

Schematic symbol for this type phone jack

(b)

FIGURE 9–34 Shorting-type jack (Photo reprinted with permission from Perozzo, *Practical Electronics Troubleshooting*, ©1985 by Delmar Publishers Inc., Albany, NY)

Techniques for Ohmmeter Measurements

1. **Continuity checks to find opens**—One frequently used check is the continuity check. The ohmmeter indicates ∞ ohms resistance if there is no "continuity," or an open between the points where the probes are connected. Of course, an open is one extreme condition that can be measured.

   A continuity check is very useful for checking fuses, light bulbs, switches, circuit wires, and circuit board paths. For example, all of these items will show very low or almost zero ohms resistance across their terminals, if they are normal. If the measurement shows infinite ohms, the item is obviously open or blown.

   In fact, sometimes visual checks are used to confirm where the open is occurring. You can see the broken filament in an automobile light bulb and a broken copper path on the circuit board!

2. **Ohmmeter checks to find shorts**—Ohmmeter checks are valuable for finding shorted components or circuit portions. It is apparent when measuring a component or circuit resistance that when the resistance is zero ohms, there is a short. In the case of a closed switch, this is the normal situation. However, in cases where there should be a measurable resistance value, you have located a problem. By electrically lifting one end of each component or circuit portion from the remaining circuitry and making appropriate measurements, you can determine which component is shorted.

3. **Checking continuity of long cables or power cords**—Frequently, it is handy to verify whether a cable has a break in one of its conductors, or a short between conductors. Examples are TV twin leads, coaxial cables for antenna installations, and cables in computer network systems. Note Figure 9–35 and see one popular technique used to verify continuity in such cables. Rather than having an ohmmeter with 200-foot-long test leads, the technician creates temporary shorts on

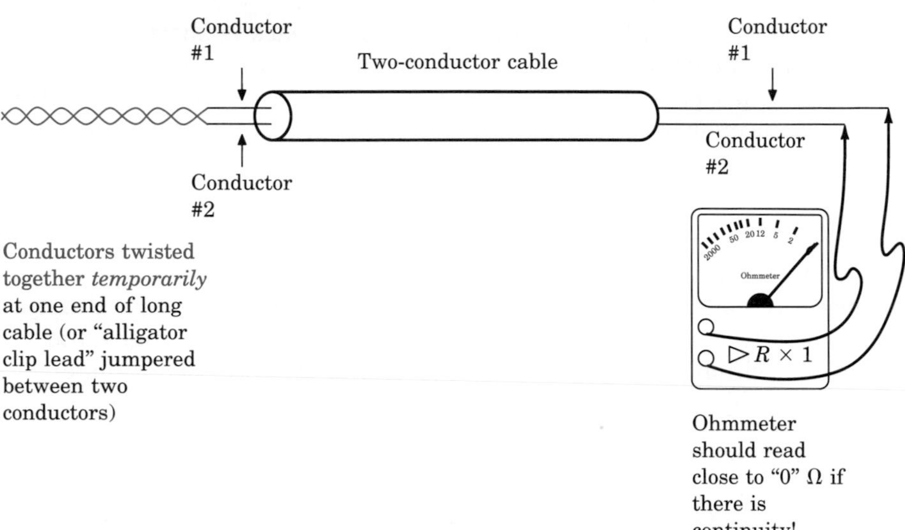

FIGURE 9–35 Checking continuity of long cables

the conductors at one end of the cable. Then, using the ohmmeter at the other end, determines if there is continuity throughout the cable length for both conductors. After making the measurement, the technician removes the temporary short to establish normal operation. Incidentally, when the temporary short is removed, there should *not* be continuity. If there is, this indicates an undesired low resistance, or shorted path between conductors.

4. **Measuring resistances with the ohmmeter**—Reiterating, *never* connect an ohmmeter to a circuit with power on it! Remove the power plug!

Watch out for "sneak paths"! One frequent error made by technicians is not sufficiently isolating the component or circuit portion to be tested from the remaining circuitry. Any unaccounted current path that remains electrically connected to the specific path you want to measure can change the measured R values from the "expected" to the "unexpected." Where possible, it is best to completely remove, or at least lift one end of the component being tested from the remaining circuit.

Another common "sneak path" occurs when the technician inadvertently allows his or her fingers to touch the test probes. This puts their body resistance in parallel with whatever is being measured. Where it is impossible to isolate the component or circuit portion to be tested, but "in-circuit" tests are required, have a schematic diagram, or good knowledge about the circuitry so you can accurately interpret the measurements.

Naturally, when measuring resistances, use all of the normal operating procedures. That is, perform the zero-adjust calibration for the range to be used, remove power from the circuit or component being tested, and use the proper range.

If resistance measurements indicate improper values, proceed with additional isolating procedures, as appropriate.

Closing Comments About Measuring Instruments

In this chapter, we have focused on dc measurement instruments. We did mention, however, that most multimeters have additional built-in circuitry enabling the measurement of ac voltages.

Also you should know there are several other types of meter movement designs that react to ac quantities, which were not discussed in this chapter. These movements include the attraction-type moving-iron meter movement, the repulsion-type moving-iron movement, and the electrodynamometer movement (frequently used in wattmeters). It is beyond the scope of this chapter to discuss how each of these types operate, but we did want you to be aware of them.

As indicated throughout the chapter, meter movements are very fragile instruments. It is important that you give them the proper care to prevent physical abuse, and that you use proper operating procedures to prevent electrical damage.

Many instruments are manufactured with electrical protection circuitry built in. A variety of semiconductor circuits, from simple diode protection to more elaborate circuits, are examples of these circuits.

We have mentioned that technicians should be aware of factors affecting accuracy when making measurements. Meter loading effects were mentioned as one of these factors. Use simple series-parallel circuit analyses to determine the voltmeter's effect on the tested circuit. Use series circuit analyses to assess the current meter's effect on a given circuit.

Another error is that of parallax when reading scales on analog meters, Figure 9–36. Some instruments have a mirror on the scale so the technician will not read the pointer from an angle that might cause a misinterpreted reading, Figure 9–37. By making sure the real pointer is superimposed over the reflection, the technician will be reading the scale at the angle that yields the best accuracy.

Another point is the basic accuracy rating of the instrument itself. Typically, meters are rated from 1–3% accuracy at full-scale deflection for analog instruments and from 0.1–1% (or plus/minus one digit) on digital meter readouts. It is essential technicians understand the limitations of the instruments used and have knowledge of the circuits or systems they are testing to properly interpret what is being measured. As a technician, you will find that *properly used* and *properly interpreted* circuit parameter mea-

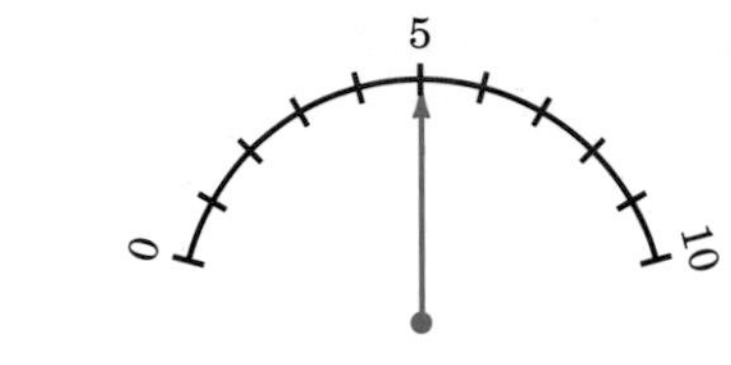

Viewed straight on

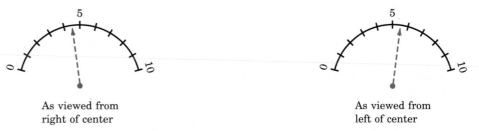

As viewed from right of center

As viewed from left of center

FIGURE 9–36 "Parallax" error

surements (enabled by the test instruments we have been discussing) are essential to the performance of your job. This is true whether the task be in designing, troubleshooting, servicing, or monitoring the operation of circuits and systems.

FIGURE 9–37 Mirrored scale for minimizing parallax error

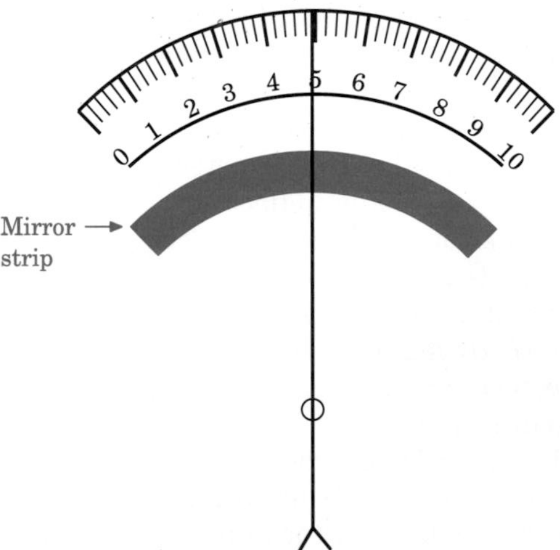

The technician makes sure the "real pointer" is superimposed over the reflection of the pointer when reading the meter to minimize the parallax reading error.

CHAPTER CHALLENGE

Following the SIMPLER troubleshooting sequence, find the problem with this circuit. As you follow the sequence, record your responses to each step on a separate sheet of paper.

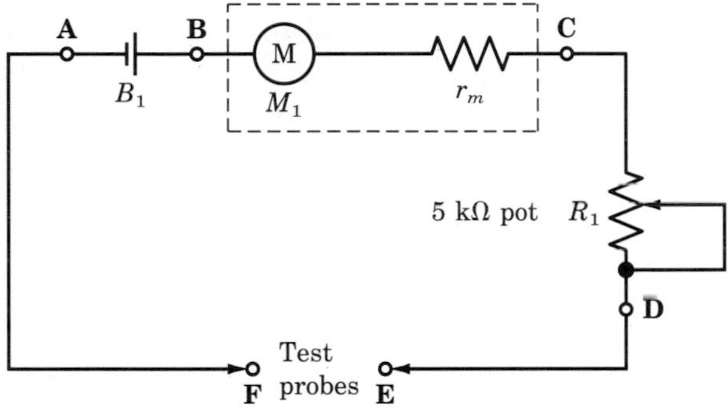

CHALLENGE CIRCUIT 5

Starting Point Information

1. Circuit diagram
2. B_1 is a 1.5 V cell
3. Meter is a 1 mA movement
4. There is no meter response when the test probes are touched together to "zero" on the ohmmeter

Test	Results in Appendix C
(With test probes not touching)	
V_{A-B}	(44)
V_{B-C}	(115)
V_{C-D}	(9)
V_{D-E}	(91)
V_{E-F}	(40)
V_{F-A}	(101)
R_{C-E}	(13)
R_{C-D}	(59)
R_{D-E}	(108)
R_{F-A}	(84)
(With test probes touching)	
V_{A-B}	(64)
V_{B-C}	(27)
V_{C-D}	(97)
V_{D-E}	(20)
V_{E-F}	(110)
V_{F-A}	(47)

SYMPTOMS Gather, verify, and analyze symptom information. (Look at the "Starting Point Information.")

IDENTIFY Identify initial suspect area for the location of the trouble.

MAKE Make a decision about "What type of test to make" and "Where to make it." (To simulate making a decision about each test you want to make, select the desired test from the "TEST" column listing below.)

PERFORM Perform the test. (Look up the result of the test in Appendix C. Use the number in the "RESULTS" column to find the test result in the Appendix.)

LOCATE Locate and define a new "narrower" area in which to continue troubleshooting.

EXAMINE Examine available information and again determine "what type of test' and "where."

REPEAT Repeat the preceding analysis and testing steps until the trouble is found. What would you do to restore this circuit to normal operation? When you have solved the problem, compare your results with those shown in the color insert.

SUMMARY

- A moving-coil meter movement (often called the d'Arsonval movement) contains several key elements—permanent magnet, pole pieces, soft iron core, movable coil and support frame, spindle and bearings, springs, pointer and counterbalance, calibrated scale, and stops.
- The moving-coil movement depends on a force of repulsion between the permanent magnet's fixed magnetic field and the movable coil's magnetic field when current passes through the movable coil.
- Electrical damping prevents the instrument's pointer from oscillating above and below the actual reading; develops from the effects of induced current in the movable coil frame; and is analyzed by Lenz's law.
- Moving-coil meters are current indicators and can be manufactured with sensitivities ranging from microamperes to milliamperes.
- Movement sensitivity relates to the strength of the permanent magnet, the number of turns on the moving coil, and other factors.
- Current meter sensitivities are usually expressed in full-scale current, but are also expressed in the amount of voltage dropped by the coil with full-scale current value passing through it, (e.g., 1 mA, 50 mV movement).
- When using the moving-coil movement as a current meter, the instrument is connected in series with the circuit portion where it is desired to measure current flow. Thus, breaking or opening the circuit, inserting the meter, and observing proper polarity and range are required. (*Caution:* Turn power off circuit before breaking circuit).
- The minimum current value an instrument measures is determined by its basic movement full-scale current sensitivity. An instrument can measure higher currents by using current meter shunts. A shunt is a parallel-resistance path that divides the circuit current. Thus, a percentage of the current passes through the meter and the remainder passes through the shunt.
- Parallel circuit analysis (current-divider) techniques determine the shunt resistance value for a desired current range. Inverse proportionality of current dividing in parallel circuits is a useful technique for this analysis, where $\frac{R_s}{R_m} = \frac{I_m}{I_s}$.
- When multiple ranges are available on the instrument for current measurements (selectable shunts), the switch is normally a shorting-type switch to assure the basic meter movement is never without a shunt between switch positions.
- The universal, or Ayrton shunt is a frequently used series-parallel circuit and also provides multiple current ranges. This shunt does not require a shorting-type switch to protect the movement, as there is always a shunt across the meter.
- A moving-coil meter movement can measure voltage by adding a series multiplier resistance. The multiplier R value and the meter's resistance must limit the current to the movement's full-scale current value for the voltage range desired. Ohm's law and series-circuit techniques determine the multiplier value. (For all practical purposes, the value of $R_{mult} = \frac{\text{V (desired range)}}{\text{I (full-scale)}}$. To be accurate, subtract the meter's resistance value from this answer.)
- Multiple ranges for voltage measurement are obtained by using a switch and contacting different multiplier resistance values at each switch position.
- Voltmeters are often rated by ohms-per-volt sensitivity. This rating indicates how many

ohms must be in the metering circuit to limit current to the meter's full-scale current value when one volt (1 V) is applied. The formula for this is $\frac{\Omega}{V}$ (sensitivity) $= \frac{1\text{ V}}{I_{fs}}$.

- ▶ Because voltmeters are connected in *parallel* with the portion of circuitry under test, the meter circuit's resistance alters the test circuit's operating conditions. This is called meter loading. The higher the meter circuit's input resistance, or $\frac{\Omega}{V}$ rating, the less loading effect it will cause.
- ▶ Most analog ohmmeters use series-type circuitry, whereby the meter's R, the internal voltage source, and the resistance under test are all in series. Therefore, the ohmmeter scales read from right to left, with low R values appearing on the right side and higher R values appearing on the left side. This scale is sometimes called a backoff scale. Also, the scale calibrations for ohms are nonlinear, but the scale calibrations for current and voltage are linear.
- ▶ Factors to be observed when using dc instruments include: use the proper range; connect the meter to the circuit in a proper fashion; observe proper polarity; and, with the ohmmeter, make sure power is removed from the component or circuit being tested.
- ▶ Multimeters measure more than one type of electrical parameter, such as voltage, current, and resistance.
- ▶ A VOM typically has lower R input than a DMM. Therefore, a VOM causes more meter loading effects when measuring voltage.
- ▶ A number of special probes and devices are available for electrical measurements. Some of these include clamp-on current probes, thermocouples (measure high frequency currents), clamp-on multimeters, and high-voltage probes.
- ▶ Using dc measuring instruments to troubleshoot actually means using common sense, properly applying Ohm's law, and having some knowledge of how the tested circuit or system operates.

REVIEW QUESTIONS

1. A current meter is rated as a 2-mA, 200-mV meter movement. What is its full-scale current rating? What is the resistance of the movement?
2. It is desired to increase the range of a 5-milliampere movement to 500 mA. What value of shunt resistance should be used? (Assume $R_m = 20\ \Omega$)
3. Briefly explain the action of a shorting-type switch. Under what circumstances should it be used in dc measuring instruments?

4. Should a current meter have a high or low resistance? Why?

5. Draw a three-resistor parallel circuit with a dc source. Show where and how current meters are connected to measure I_T and the current through the middle branch. Indicate polarities, as appropriate.

6. A current meter is rated at 3 mA, 150 mV. What multiplier resistance value is needed to use this meter as a 0–30-V meter? Draw and label the circuit.

7. If the full-scale current rating of a meter is 40 μA, what is its ohms-per-volt sensitivity rating as a voltmeter? What is the value of $R_m + R_{mult}$ if the voltage range selected is the 500-volt range?

8. Refer to Figure 9–38 and indicate what the voltmeter reading will be if the meter has $\frac{\Omega}{V}$ sensitivity rating of 20,000 $\frac{\Omega}{V}$. (Assume that you are using the 0–10-V range). What would the R_2 voltage drop be without the voltmeter present?

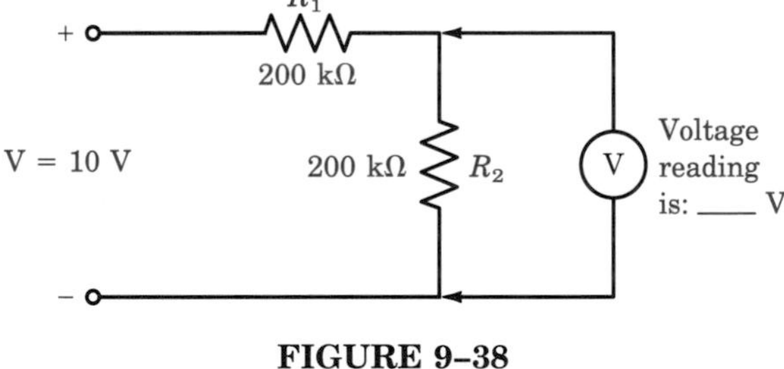

FIGURE 9–38

9. What is the value of the metering circuit resistance for a current meter having a 1 mA, 100 mV rating that is shunted to measure 0–2 mA?

10. If a voltmeter is reading 22 volts on the 0–50-volt range, what percentage of full-scale current is flowing through the meter movement? What would the reading be if 75% of full-scale current were flowing?

11. Should a voltmeter circuit have a high or low resistance? Why?

12. Draw the basic circuit of a series-type ohmmeter. Show the test probes and an unknown value resistor to be measured. (R_x)

13. For the circuit in question 12, if the meter is a 1 mA, 100-Ω movement and the ohmmeter voltage source is a 1.5-V cell, what would be the calibration value at mid-scale on the ohmmeter scale? At quarter scale?

14. Current meters are connected in __________ with the component or circuit to be tested. Voltmeters are connected in __________ with the component or circuit to be tested. For ohmmeters using the backoff scale, the resistance under test is in __________ with the meter, its current limiting resistor, and the internal voltage source.

15. Why is it easier to use voltmeter checks, rather than current meter checks when troubleshooting?

16. Why are pole pieces used in dc moving-coil movements?

17. What is meant by damping relative to dc moving-coil meter movements?

18. How is damping achieved in a moving-coil instrument?

19. What are two distinct advantages of a DMM compared with an analog VOM?

20. For what situation is an analog-type instrument easier to use than a digital-type instrument?

TEST YOURSELF

1. Draw a diagram illustrating a basic movement and a two-range Ayrton shunt. Assume the meter to be a 50-μA, 2,000-Ω metering circuit. The desired current ranges are: 0–1 mA and 0–10 mA. Label all component values and show polarity of all terminals.

2. If a voltmeter with a sensitivity rating of 1,000 $\frac{\Omega}{V}$ that is set on the 0–50-volt range is connected across one resistor of a series circuit comprised of two 100-kΩ resistors, what will be the voltmeter reading? (Assume a 100-volt dc source.)

3. In question 2, what would be the reading if the meter had a sensitivity of 100,000 $\frac{\Omega}{V}$ and the 0–100-volt range were used?

4. In either question 2 or 3, if two identical meters, each set on the same voltage range were simultaneously connected across R_1 and R_2 respectively (i.e., one meter measuring V_1, the other measuring V_2), what voltage reading would each show? (Still assume 100 volts applied voltage).

PERFORMANCE PROJECTS CORRELATION CHART

Suggested performance projects that correlate with topics in this chapter are:

CHAPTER TOPIC	PERFORMANCE PROJECT	PROJECT NUMBER
Current Meter Shunts	The Basic Current Meter Shunt	30
Voltmeters	The Basic Voltmeter Circuit	31
Ohmmeters	The Series Ohmmeter Circuit	32

After completing the above projects, perform the "Summary" checkout at the end of the performance projects section.

Basic AC Quantities

Key Terms

ac
alternation
average value
coordinate system
coordinates
cosine
effective value (rms)
frequency (f)
Hertz (Hz)
instantaneous value
magnitude
peak value (V_{pk} or maximum value, V_{max})
peak-to-peak value (pk-pk)
period (T)
phase angle (θ)
phasor
quadrants
sine
sine wave
sinusoidal quantity
vector

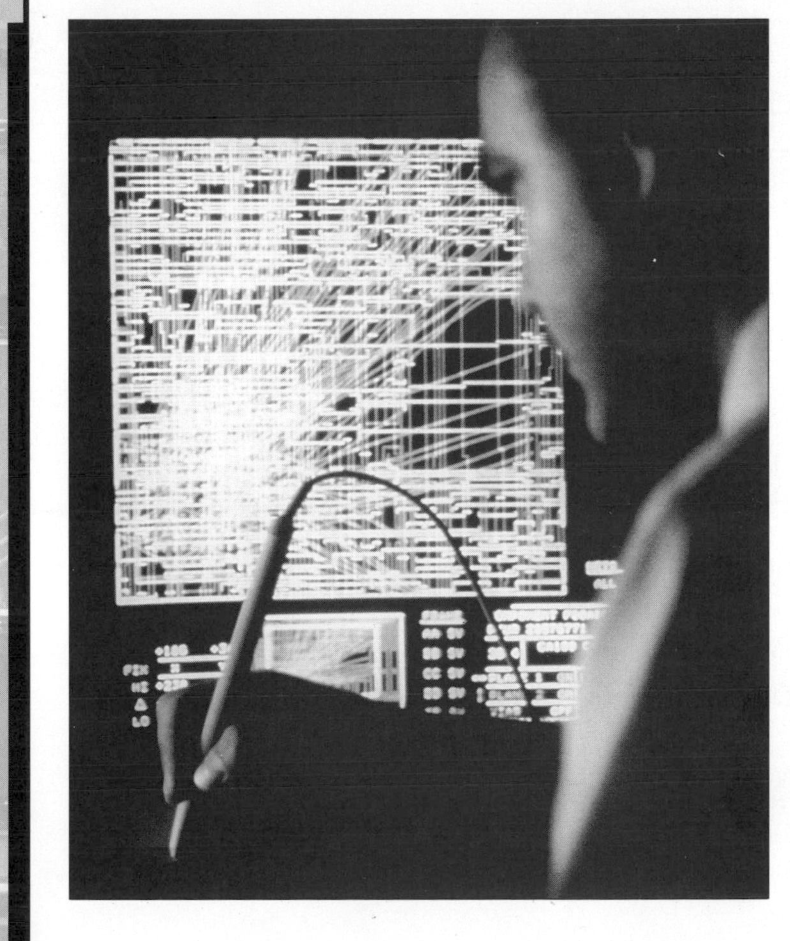

Courtesy of International Business Machines Corporation.

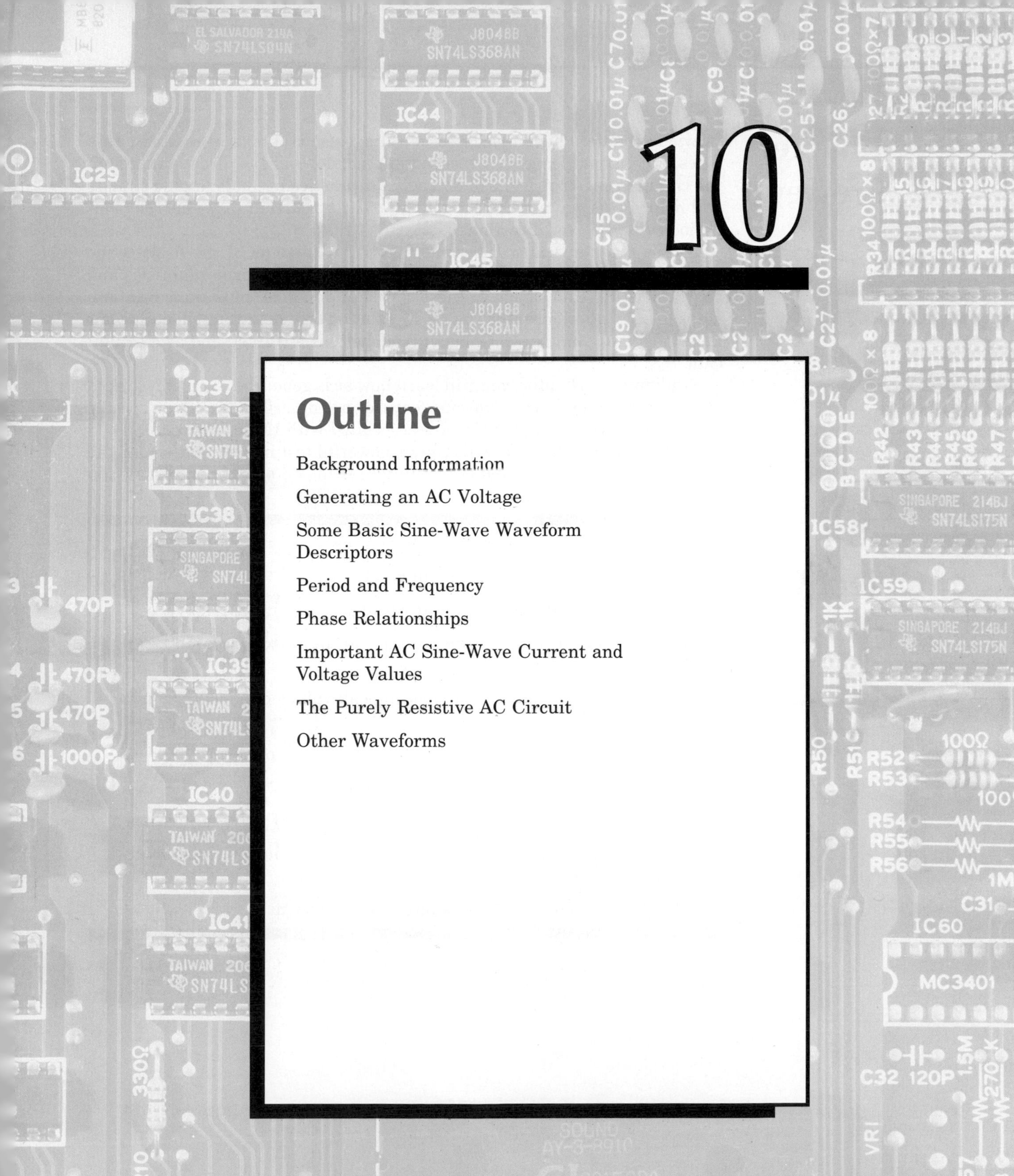

10

Outline

Background Information

Generating an AC Voltage

Some Basic Sine-Wave Waveform Descriptors

Period and Frequency

Phase Relationships

Important AC Sine-Wave Current and Voltage Values

The Purely Resistive AC Circuit

Other Waveforms

Chapter Preview

Previous chapters have discussed dc voltages, currents, and related quantities. As you have learned, direct current (dc) flows in one direction through the circuit. Your knowledge of dc and your practice in analyzing dc circuits will help you as you now investigate alternating current (ac)—current that periodically reverses direction.

In this chapter, you will become familiar with some of the basic terms used to define important ac voltage, current, and waveform quantities and characteristics. Knowledge of these terms and values, plus their relationships to each other, form the critical foundation on which you will build your knowledge and skills in working with ac, both as a trainee and as a technician.

Studying this chapter, you will learn how ac is generated; how it is graphically represented; how its various features are described; and how specific important values, or quantities are associated with these features.

In the next chapter, you will study a powerful test instrument, called the oscilloscope, which will enable you to visualize the ac characteristics you will study in this chapter.

Objectives

After studying this chapter, you will be able to:

- Draw a graphic illustrating an **ac** waveform
- Define **cycle, alternation, period, peak, peak-to-peak,** and **effective value (rms)**
- Compute effective, peak, and peak-to-peak values of ac voltage and current
- Explain **average** with reference to one-half cycle of ac
- Define and calculate **frequency** and period
- Draw a graphic illustrating phase relationships of two **sine waves**
- Describe the phase relationships of V and I in a purely resistive ac circuit
- Label key parameters of non-sinusoidal waveforms

BACKGROUND INFORMATION

The Difference Between DC and AC

Whereas direct current is unidirectional, alternating current, as the name suggests, alternates in direction, and thus is bidirectional, Figure 10–1. The sources of dc you are familiar with include batteries and dc power supplies. Batteries cannot produce alternating current.

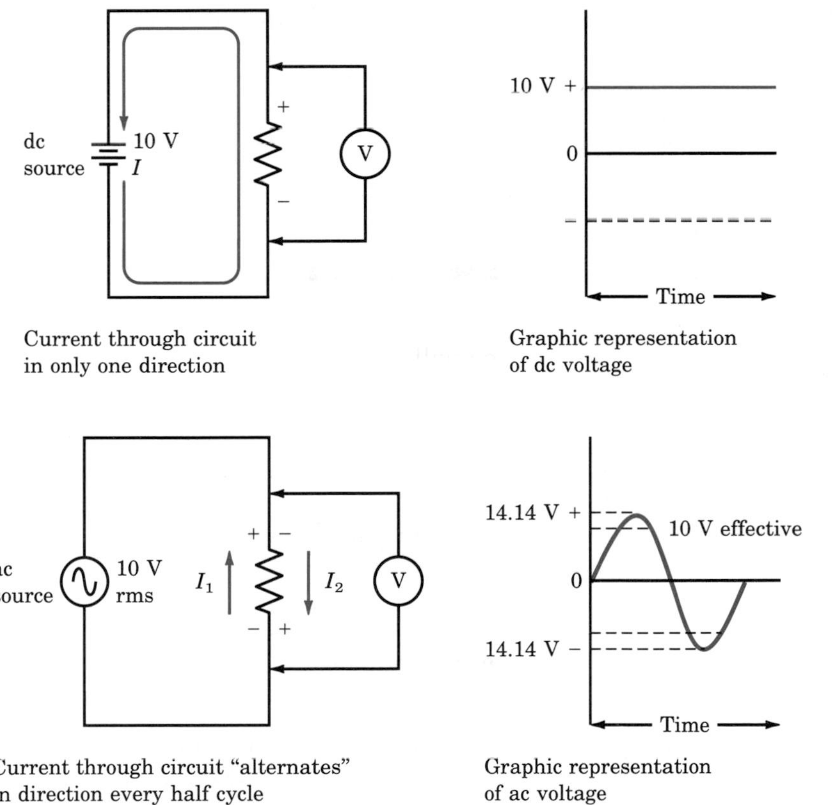

FIGURE 10–1 Graphic comparing dc and ac waveforms

Alternating Current Used for Power Transmission

Alternating current **(ac)** has certain advantages over dc when transmitting electrical power from one point to another point. These advantages are why power is brought to your home in the form of ac rather than dc.

Some interesting contrasts between dc and ac are shown in Figure 10–2. This ac electrical power is produced by alternating-current generators. A simple discussion about generating ac voltage will be covered shortly. However, some groundwork will now be presented.

FIGURE 10–2 Similarities and differences between dc and ac power

CHARACTERISTIC	IS CHARACTERISTIC OF: AC	DC
Current in one direction only	No	Yes
Current periodically alternates in direction one-half the time in each direction	Yes	No
V can be "stepped up" or "stepped down" by transformer action	Yes	No
Is most efficient power to transport over long distances	Yes	No
Useful for all purposes	Yes	No

Defining Angular Motion and the Coordinate System

As we deal with ac quantities, it is helpful to understand how angular relationships are commonly depicted. Refer to Figures 10–3a (angular motion) and 10–3b (the coordinate system) as you study the following discussions.

Mathematical and Graphic Concepts About AC

1. Angular motion is defined in terms of a 360° circle or a four-quadrant coordinate system.
2. The zero degree (starting or reference plane) is a horizontal line or arrow extending to the right, Figure 10–3a. NOTE: Other terms you might see for this line or arrow are a rotating radius and a vector called a **phasor.** A **vector** identifies a quantity that has (or illustrates) both magnitude *and* direction.
3. Angles increase in a positive, counterclockwise (CCW) direction from the zero reference point. In other words, the horizontal phasor to the right is rotated in a CCW direction until it is pointing straight upward, indicating an angle of 90° (with respect to the 0° reference point).
   - When the phasor continues in a CCW rotation until it is pointing straight to the left, that point represents 180° of rotation.

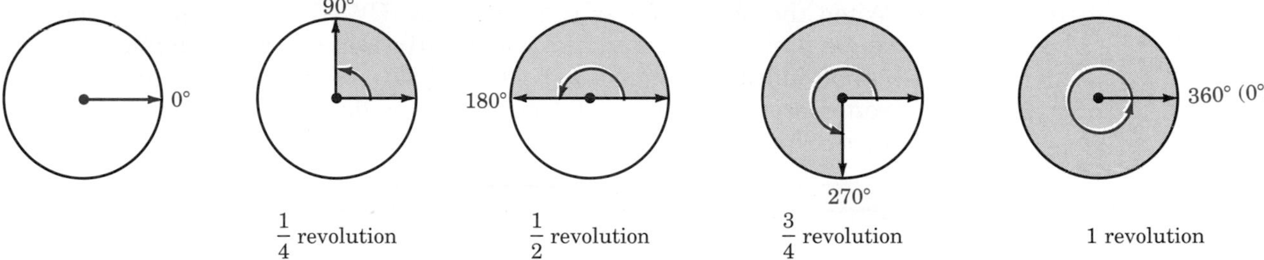

FIGURE 10–3a Describing angular motion

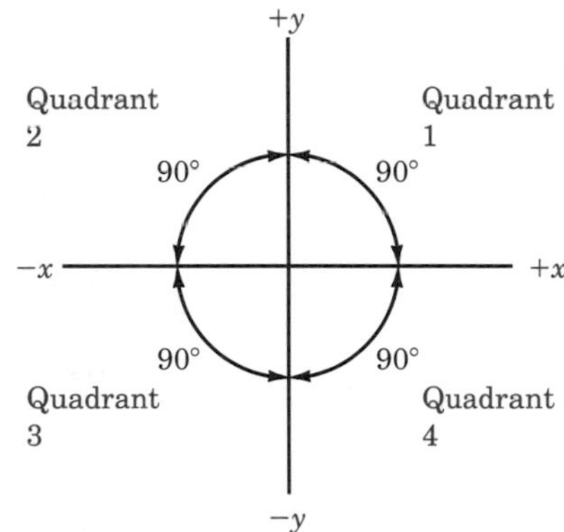

FIGURE 10–3b The coordinate system

- When rotation continues until the phasor is pointing straight downward, the angle at that point is 270°.
- When the phasor completes a CCW rotation, returning to its starting point of zero degrees, it has rotated through 360°.
- When the phasor is halfway between 0° and 90°, the angle is 45°. When the phasor is halfway between 90° and 180°, the angle is 135°.

4. The **coordinate system,** Figure 10–3b, consists of two perpendicular lines, representing one of the two axes. The horizontal line is the x-axis. The vertical line is the y-axis. It is worth your effort to learn the names and directions of these two axes—they will be referenced often during your technical training. Remember the x-axis is horizontal and the y-axis is vertical.

5. There are four **quadrants** associated with this coordinate system. Again, look at Figure 10–3b and note that the angles between 0° and 90° are in quadrant 1. Angles between 90° and 180° are in quadrant 2. Angles between 180° and 270° are in quadrant 3; and angles between 270° and 360° (or 0°) are in quadrant 4.

6. When the phasor moves in a clockwise (CW) direction, the various angle positions are described as negative values. For example, the location halfway between 360° (0°) and 270° (−90°) in quadrant 4 (the same location as the +315° point) can be described as either −45° or +315°.

IN-PROCESS LEARNING CHECK I

1. Define the difference between dc and ac.
2. Which direction is the reference, or zero degree position when describing angular motion?
3. Is the y-axis the horizontal or vertical axis?
4. Is the second quadrant between 0 and 90 degrees, 90 and 180 degrees, 180 and 270 degrees, or 270 and 360 degrees?
5. Define the term vector.
6. Define the term phasor.

Keeping in mind these thoughts about defining angles in degrees and a reference point on the coordinate system, let's move to a brief explanation of how an ac sine-wave voltage is generated.

GENERATING AN AC VOLTAGE

Refer to Figure 10–4a and note the elementary concepts of an ac generator. The simplified generator consists of a magnetic field through which a conductor is rotated. As you would expect when the maximum number of magnetic field flux lines are being cut (or linked) by the rotating conductor, maximum voltage is induced. Recall that if 10^8 flux lines are cut by a single conductor in one second, the induced voltage is one volt. Conversely, when minimum (or zero) flux lines are cut, minimum (or zero) voltage is induced.

Also observe the slip rings and brushes that connect the external circuit load to the rotating loop so each end of the load is always connected to the same end of the rotating conductor. If we monitor or measure the voltage across the load 360-degrees (one complete rotation of the conductor loop), the graphic display of the varying magnitude and polarity of the generator's output voltage would look similar to that shown in Figure 10–4b. This graphic representation of the changing amplitude and direction of the output voltage over time is called a waveform.

The Sine Wave

This unique, single-frequency waveform reflects a quantity that constantly changes in amplitude and periodically reverses in direction. This quantity is

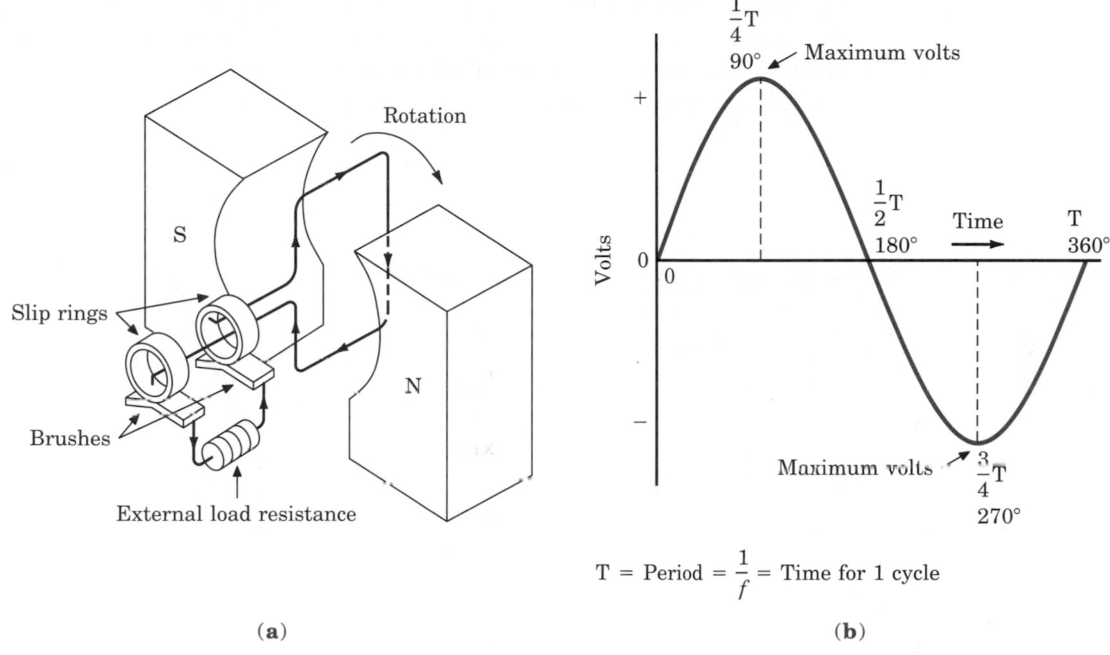

FIGURE 10–4 Basic concepts of an ac generator

termed **sine wave.** Refer to Figure 10–5 to see how the sine-wave's amplitude relates to the various points as the radius line rotates through 360°. This waveform is called a sine wave because the amplitude (amount) of voltage at any given moment is directly related to the sine function, used in trigonometry. That

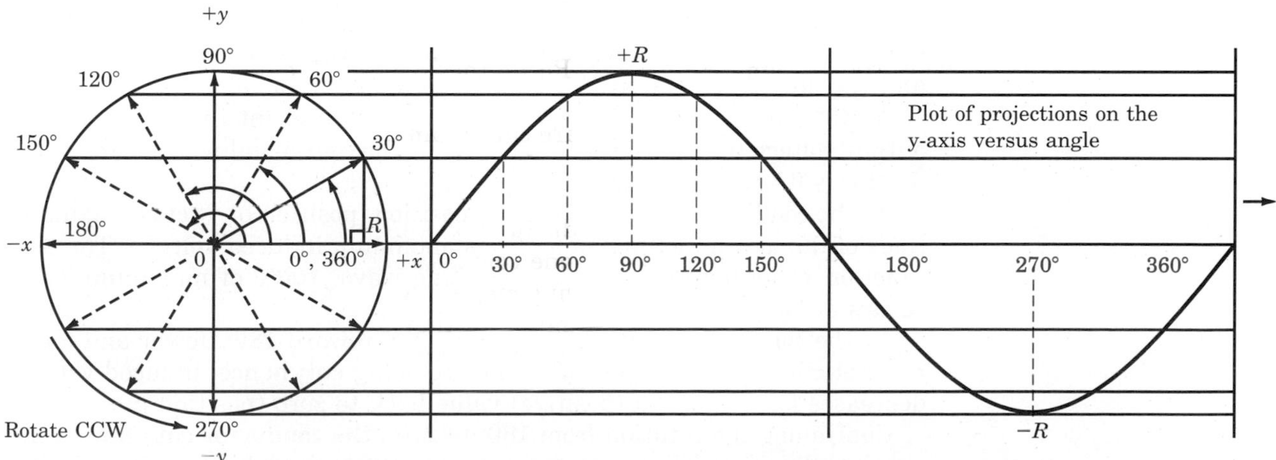

FIGURE 10–5 Projections of sine-wave amplitudes in relation to various angles

is, the trigonometric sine (sin) value for any given angle, represented along the sine wave, relates directly to the voltage amplitude at that same angle. NOTE: The sine of a given angle, using a right triangle as the basis of analysis, is equal to the ratio of the opposite side to the hypotenuse, Figure 10–6.

For angle A:

1. The side opposite the right angle is called the *hypotenuse* (c).
2. The side opposite angle A is called the *opposite side* (a).
3. The side of angle A which is not the hypotenuse is called the *adjacent side* (b).

Sine (sin) of angle $= \dfrac{\text{opposite side}}{\text{hypotenuse}}$; $\sin A = \dfrac{a}{c}$

Cosine (cos) of angle $= \dfrac{\text{adjacent side}}{\text{hypotenuse}}$; $\cos A = \dfrac{b}{c}$

Tangent (tan) of angle $= \dfrac{\text{opposite side}}{\text{adjacent side}}$; $\tan A = \dfrac{a}{b} = \dfrac{\sin A}{\cos A}$

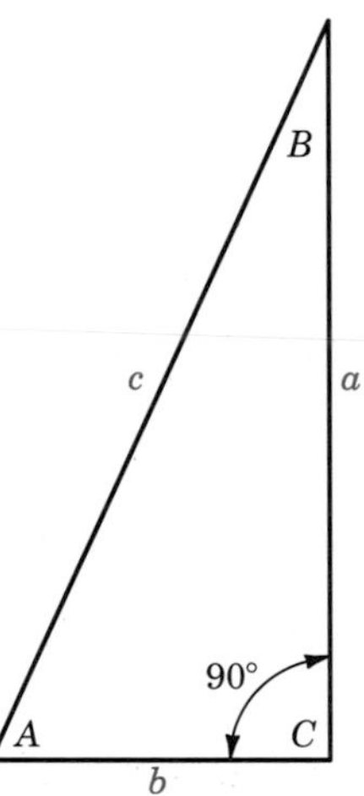

FIGURE 10–6 The right triangle

Refer to Figure 10–7 as you read the following discussion. For simplicity, we will equate the angle of rotation from the starting point of the ac generator's mechanical rotation (where no flux lines are being cut by the rotating conductor) to the angle of the sine wave output voltage waveform. For example, at the 0 degree point of rotation, no flux lines are being cut and the output voltage is zero (position a). If you find the sine value of 0° on a "trig" table, or use the "sin" function on a calculator, you would find it to be zero, and the output voltage at this time is zero volts.

At 45° into its rotation, the conductor cuts sufficient flux lines causing output voltage to rise from 0 to 70.7% of its maximum value. The sine value for 45° is 0.707.

As the conductor rotates to the 90° position (position b), it cuts maximum lines of flux per unit time; hence, has maximum induced voltage. The sine function value for 90° is 1.00. For a sine wave, 100% of maximum value occurs at 90°.

As the conductor rotation moves from 90° toward 180°, fewer and fewer magnetic lines of force per unit time are being cut; hence, induced voltage decreases for maximum (positive) value back to zero (position c).

Continuing the rotation from 180° to 270°, the conductor cuts flux in the opposite direction. Therefore, the induced voltage is of the opposite polarity. Once again, voltage starts from zero value and increases in its opposite

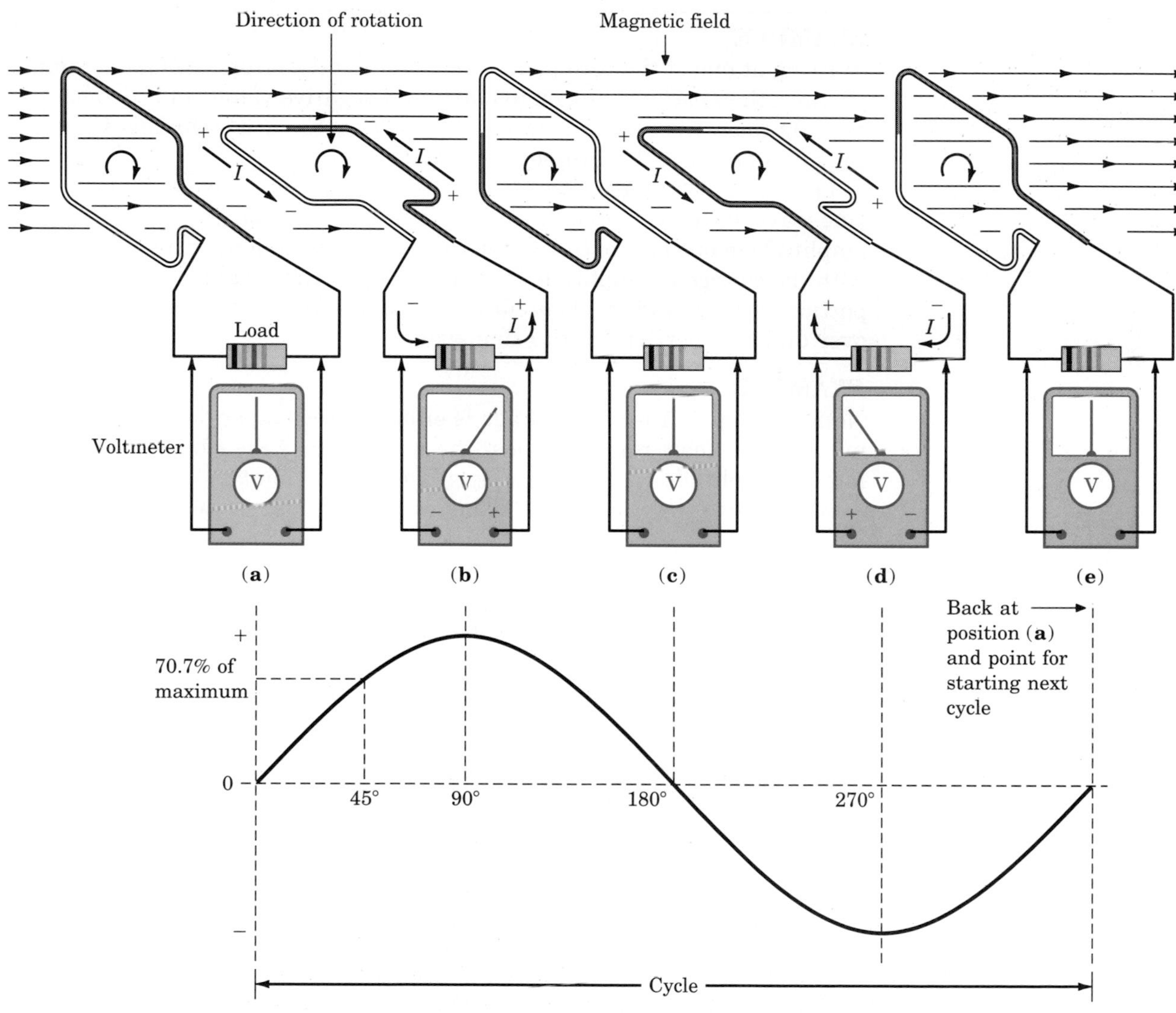

FIGURE 10–7 Relationship of flux-lines cut in relation to output voltage

polarity magnitude until it reaches maximum negative value at 270° of rotation (position d).

From 270° degrees to 360°, the output voltage decreases from its maximum negative value back to zero (position e). This complete chain of events of the sine wave is a **cycle.** If rotation is continued, the cycle is repeated.

(NOTE: A cycle can start from *any* point on the waveform. That particular cycle continues until that same polarity and magnitude point is reached again. At that point, repetition of the recurring events begins again, and a new cycle is started.)

EXAMPLE

Starting at one of the zero points, voltage increases from zero to maximum positive, then back to zero, on to maximum negative value, and then back to zero. As you can see, the sine wave of voltage continuously changes in amplitude and periodically reverses in polarity.

Let's look at another interesting aspect of the sine wave of voltage. The sine-wave waveform is nonlinear. That is, there is not an equal change in amplitude in one part of the waveform for a given amount of time, compared with the change in amplitude for another equal time segment at a different portion of the waveform. Therefore, the *rate of change* of voltage is different over different portions of the sine wave.

EXAMPLE

Notice in Figure 10–8 the voltage is either increasing or decreasing in amplitude at its most rapid rate near the zero points of the sine wave, (i.e., the largest amplitude *change* for a given amount of time).

Also, notice that near the maximum positive and negative areas of the

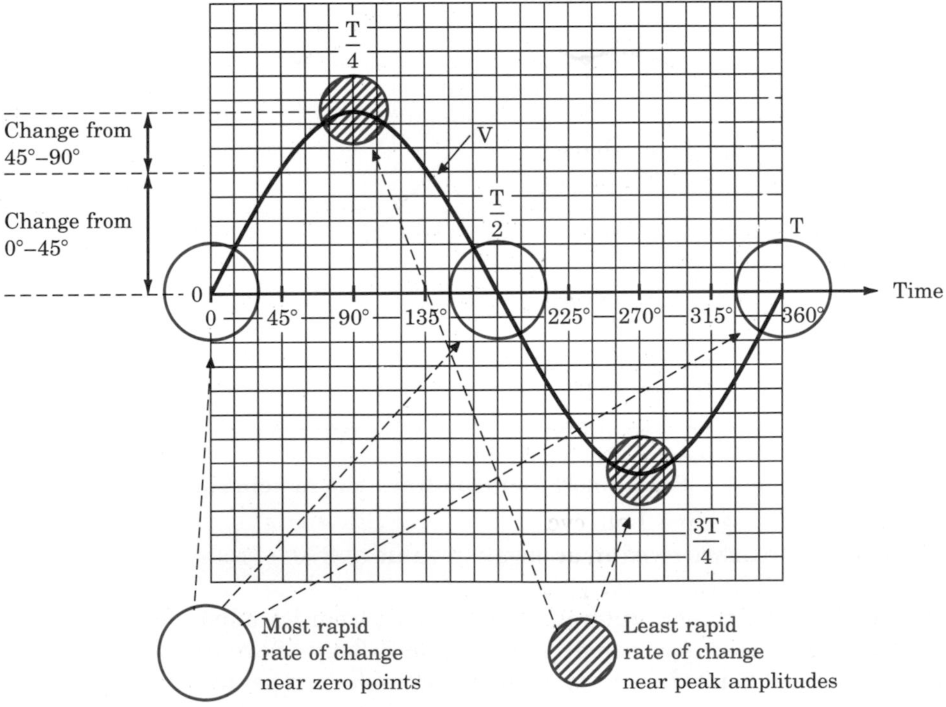

T = time for one cycle. This is known as the "period" of the waveform.

FIGURE 10–8 Different rates of change of voltage at various parts of sine wave

sine wave, specifically, near the 90° and 270° points, the rate-of-change of voltage is minimum. This rate of change has much less *change* in amplitude per unit time. Knowledge of this rate-of-change (with respect to time) will be helpful to you in some later discussions, so try to retain this knowledge.

PRACTICE PROBLEM I

Answers are in Appendix B.

1. Use a trig table or a calculator and determine the sine of 35 degrees.
2. Indicate on the following drawing at what point cycle 1 ends and cycle 2 begins, if the starting point for the first cycle is point A.

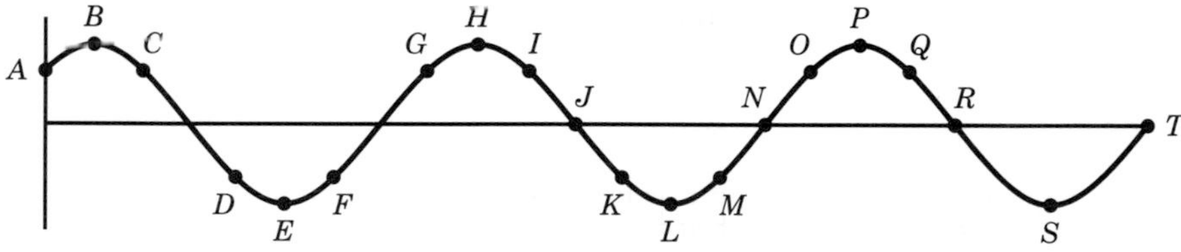

3. Which points on the drawing above represent areas of maximum rate of change of the sine wave(s)?

SOME BASIC SINE-WAVE WAVEFORM DESCRIPTORS

Refer to Figure 10–9. The horizontal line (x-axis) divides the sine wave into two parts—one above the zero reference line, representing positive values, and one below the zero reference line, representing negative values. Each half of the sine wave is an **alternation.** Naturally, the upper half is the "positive alternation," or positive half cycle. The lower half of the wave is the "negative alternation," or negative half cycle. With zero as the starting point, an alternation is the variation of an ac waveform from zero to a maximum value, then back to zero (in either polarity).

Examine Figure 10–9 again, and learn what is meant by positive peak, negative peak, and peak-to-peak values of the sine wave. Also, note in this illustration how a cycle is depicted.

Generalizing, a cycle, whether in electronics or any other physical phenomena, is one complete set of recurring events. Examples of this are one complete revolution of a wheel, the 24-hour day (representing one complete revolution of the earth), and the 11-year sunspot cycle.

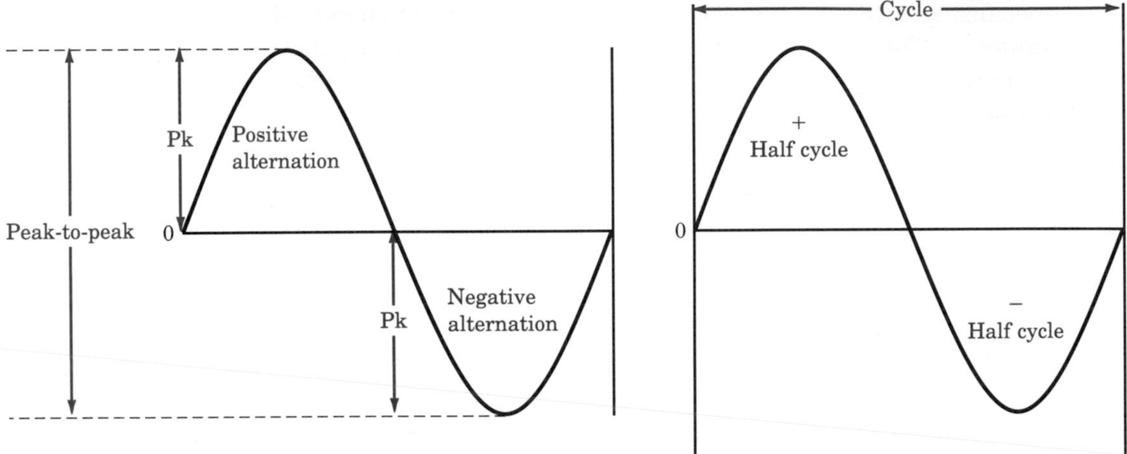

FIGURE 10–9 Several important descriptors of sine wave

PERIOD AND FREQUENCY

Period

The ac sine-wave waveform has been described in terms of amplitude and polarity and has been referenced to various angles. It should be emphasized that the sine-wave occurs over a period of time—it does not instantaneously vary from zero to maximum to zero to opposite-polarity maximum back to zero. The household ac, with which you are familiar, takes $\frac{1}{60}$ of a second to complete a cycle. Look at Figure 10–10 and observe how the sine-wave illustrations have been labeled. Notice that time is along the horizontal axis. In Figure 10–10a, a one-cycle-per-second signal is represented. If this waveform represented one cycle of the 60-cycle ac house current, the time, or **period (T)** of a cycle would be $\frac{1}{60}$ of a second, rather than the one second shown for the one-cycle signal. Learn this term period! Again, period is the time required for one cycle.

Frequency

The number of cycles occurring in one second is the **frequency (f)** of the ac, refer again to Figure 10–10. The period of one cycle is inversely related to frequency. That is, the period equals the reciprocal of frequency, or $\frac{1}{f}$. The letter "T" represents period; that is, "T" is "time" of one cycle. Therefore:

Formula 10–1	$T\,(\text{sec}) = \frac{1}{f\,(\text{Hz})}$

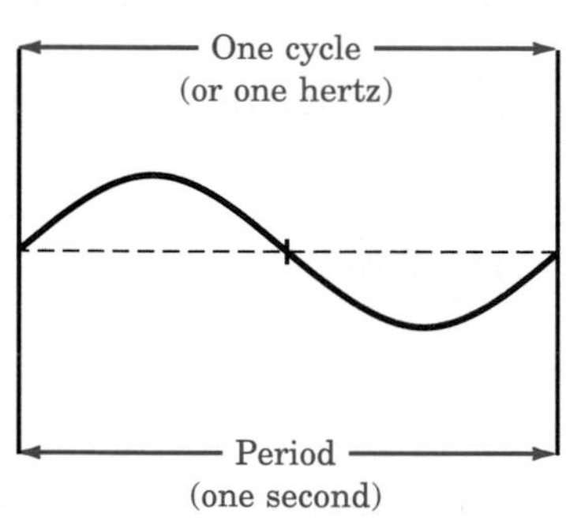

One cycle in one second equals a "frequency" of one hertz

(a)

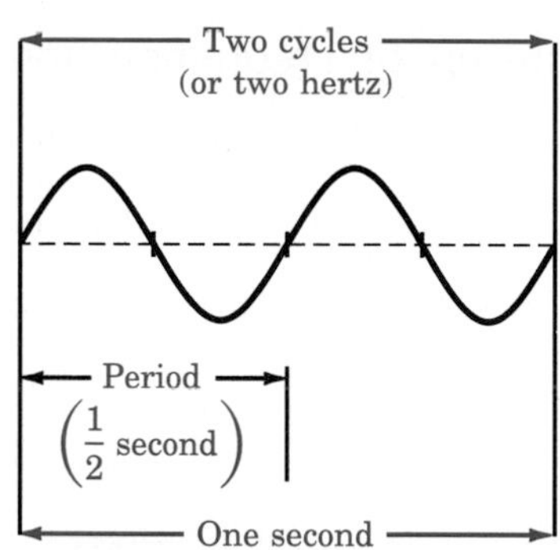

Two cycles in one second equals a "frequency" of two hertz

(b)

FIGURE 10–10 Comparing two frequencies

Restated, the period of one cycle (in seconds) equals 1 divided by the frequency in cycles per second. Rather than saying "cycles per second" when talking about frequency, we use a unit called the **Hertz (Hz),** named for the famous scientist Heinrich Hertz. For example, one Hz equals one-cycle-per-second; 10-Hz equal 10 cycles-per-second; 1 kHz equals 1,000 cycles-per-second; and 1 MHz equals 1,000,000 cycles-per-second.

EXAMPLE

1. The period of a 100-Hz sine wave is $\frac{1}{100}$ of a second, $\left(T = \frac{1}{f}\right)$.

2. The period of a 1000-Hz signal is $\frac{1}{1000}$ of a second, $\left(T = \frac{1}{f}\right)$.

Observe the frequency spectrum chart in Figure 10–11. Notice the abbreviation for Hertz (Hz) *starts* with a capital H and is abbreviated Hz to prevent confusion with another electrical quantity you will study later—the Henry, which is abbreviated with a capital H.

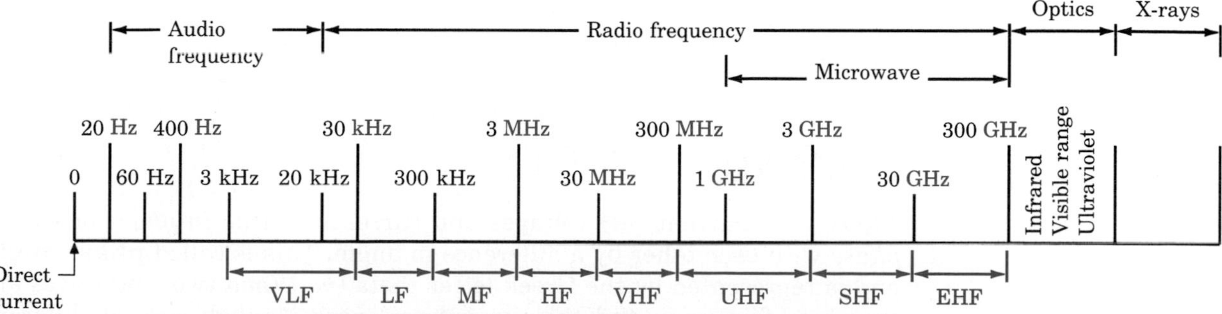

FIGURE 10–11 Frequency spectrum chart

Relationship of Frequency and Period

Converse to calculating the period of one cycle by taking the reciprocal of the frequency, $\left(\text{i.e., } T = \frac{1}{f}\right)$, the formula is transposed to find the frequency, if you know the time it takes for one cycle. The formula is:

Formula 10–2	$f\,(\text{Hz}) = \frac{1}{T\,(\text{sec})}$

This indicates the period is inverse to the frequency, and vice versa. That is, the higher (or greater) the frequency, the less time required for one cycle (the shorter the period), or the longer (or greater) the period, the lower the frequency must be. In other words, if $T\uparrow$, then $f\downarrow$; and if $f\uparrow$, then $T\downarrow$.

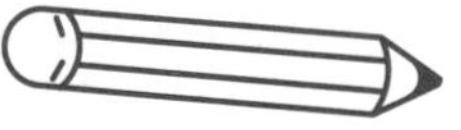

PRACTICAL NOTES

You may be interested in knowing that radio waves travel at 300 million meters per second in "free space." The "wavelength" of a radio wave is defined by how far it travels during one period. (The time for one cycle of the given frequency). This means the higher the frequency, the shorter its wavelength. One formula used for this is $\text{wavelength (in meters)} = \frac{300}{f\,(\text{in MHz})}$.

EXAMPLE

1. A signal with a period of 0.0001 seconds indicates a signal with a frequency of 10,000 Hz, $\left(f = \frac{1}{T}\right)$.
2. A signal with a period of 0.005 seconds indicates a signal with a frequency of 200 Hz, $\left(f = \frac{1}{T}\right)$.

PHASE RELATIONSHIPS

Alternating current (ac) voltages and currents can be *in-phase* or *out-of-phase* with each other by a difference in angle. This is called **phase angle** and is represented by the Greek letter theta (θ). When two sine waves are the same frequency and their waveforms pass through zero at different

IN-PROCESS LEARNING CHECK II

1. Determine the frequency of an ac signal whose period is 0.0001 second.
2. What is the period for a frequency of 400-Hertz?
3. What time does one alternation for a frequency of 10-kHz take?
4. Does the y-axis represent time or amplitude in waveforms?
5. For waveforms, at what *angular points* does the amplitude equal 70.7% of the positive peak value? At what angular points is the sine wave at a 70.7% level during the negative alternation?
6. In the coordinate system, in which quadrants are all the angles represented by the positive alternation of a sine wave?
7. What is the period of a frequency of 15-MHz?
8. What frequency has a period of 25-msec?
9. As frequency increases, does T increase, decrease, or remain the same?
10. The longer a given signal's period, the ____________ the time for each alternation.

times, and when they do not reach maximum positive amplitude at the same time, they are out-of-phase with each other.

On the other hand, when two sine waves are the same frequency, and when their waveforms pass through zero and reach maximum positive amplitude at the same time, they are in-phase with each other.

Look at Figure 10–12 and note the following:

1. Sine waves a and b are in-phase with each other because they cross the zero points and reach their maximum positive levels at the same time (shown on the x-axis). Zero points at times T_0, T_2, and T_4, and the positive peak at time T_1.

2. Sine waves b and c are out-of-phase with each other because they do not pass through the zero points and reach peak values at the same time. In fact, sine wave b reaches the peak positive point (at T_1) of its waveform 90 degrees before sine wave c does (at T_2). It can be said that sine wave c "lags" sine wave b (and also sine wave a) by 90°. Conversely, it can be said that sine wave b "leads" sine wave c by 90°. In other words, when viewing waveforms of two *equal-frequency* ac signals, the signal reaching maximum positive first is leading the other signal, in phase. "First" refers to the peak that is closest to the y-axis; thus, showing an occurrence at an earlier time.

Obviously, ac signals can be out-of-phase by different amounts. This is illustrated by sine-wave waveforms and phasor diagrams, Figure 10–13. Figure 10–13a shows two signals that are 45° out-of-phase. Which signal is leading

FIGURE 10–12 Phase relationships

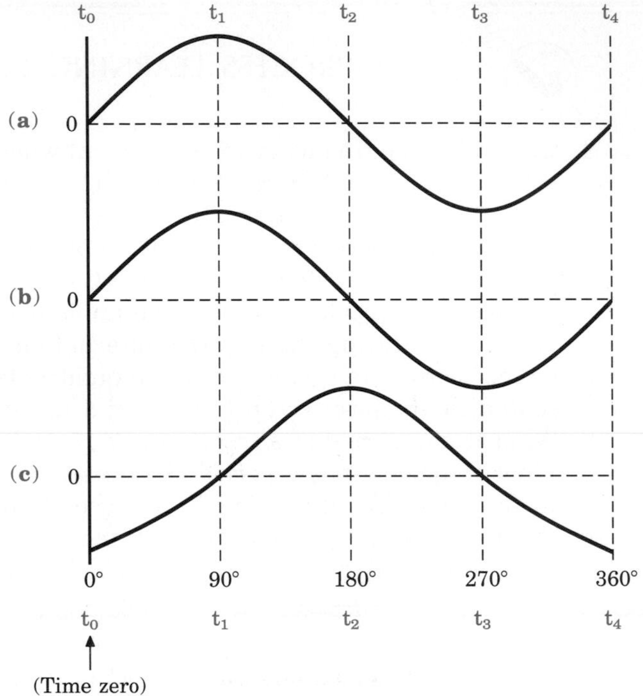

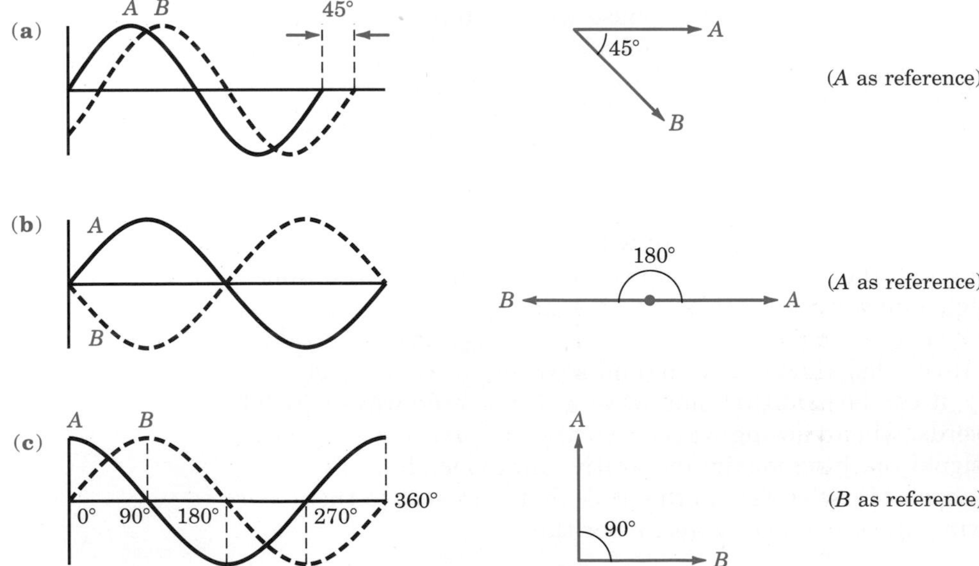

FIGURE 10–13 Diagrams depicting out-of-phase signals

the other one by 45°? The answer, of course, is a is leading b by 45°.

In Figure 10–13b, waveforms a and b are 180° out-of-phase. If these signals were of equal amplitude, they would cancel each other out.

In Figure 10–13c, you see two signals that are 90° out-of-phase.

Notice how the phasors in Figure 10–13 illustrate both magnitude and relative phase for several conditions. Observe that the *length* of the phasors can represent *amplitude* values of the two waveforms. Both phasors must represent the same relative point or quantity for each of the two waveforms being compared.

PRACTICE PROBLEM II

1. Illustrate two equal-frequency sine waves that are 60° out-of-phase with each other. Label the leading wave as A, and the lagging wave as B. Also, show the A waveform starting at the zero degree point!
2. Assume the signals in problem 1 are equal in amplitude. Draw and label a diagram using phasors to show the conditions described in problem 1. Also, show the B phasor as the horizontal vector!

IMPORTANT AC SINE-WAVE CURRENT AND VOLTAGE VALUES

The importance of learning the relationships between the various values used to measure and analyze ac currents and voltages cannot be overemphasized! These values are defined in this section. Learn these relationships well, since you will use them extensively in your training and work from this point on.

Brief Review of Terms

Look at Figure 10–14 and recall the maximum positive or negative amplitude (height) of one alternation of a sine wave is called **peak value.** For voltage this is called V_{pk}, or V_M; for current this is called I_{pk}, or I_M.

Also remember the total amplitude from the peak positive point to the peak negative point is called **peak-to-peak** value, sometimes abbreviated as $V_{pk\text{-}pk}$ when speaking of voltage values.

In other words, for a symmetrical waveform:

$$V_{pk} = \frac{1}{2} \times V_{pk\text{-}pk}, \text{ and}$$

$$V_{pk\text{-}pk} = 2 \times V_{pk}$$

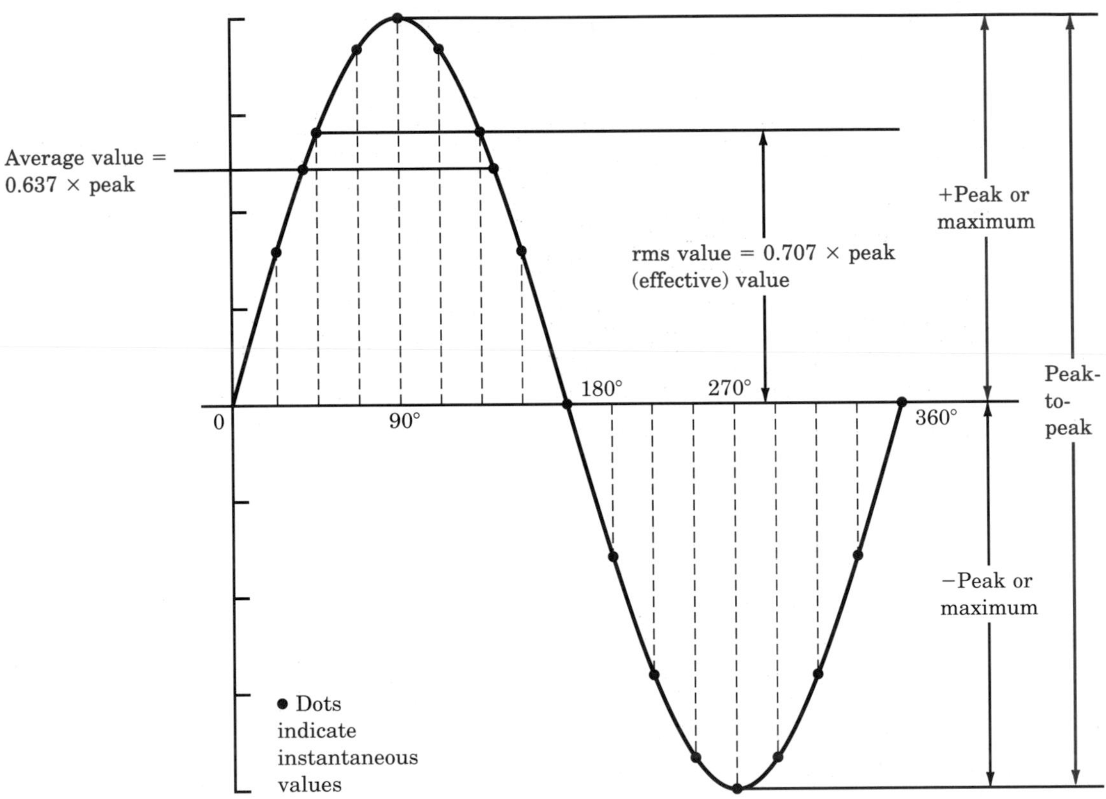

FIGURE 10–14 Important values related to sine-wave waveforms

Some New Terms and Values

Refer again to Figure 10–14 as you study the following discussions:

1. Since a given amount of current through a component in either direction causes the same amount of voltage drop, power dissipation, heating effect, and so on, the effects of an ac voltage or current are analyzed on a single alternation. For example, the positive alternation causes the same amount of current through a component (like a resistor) as the negative alternation, and vice versa.

2. If you compute the average amplitude or level under the waveform outline of *one-half cycle* or one alternation of an ac waveform, the result is the **average value** equals .637 (63.7%) of the maximum (or peak) value. The reason the average value is more than 50% of the peak value is because the rate of change is slower near peak value than it is near the zero level. That means the voltage level is above 50% of maximum level for a longer time than it is below 50% of maximum level.

In other words, the sine wave is a nonlinear waveform. Therefore, the average is not half of the peak value. *Incidentally, when you "average" over the entire cycle, rather than for one alternation, the result is zero,* since the positive and negative alternations are symmetrical, equal in amplitude and time, and opposite in polarity. However, computing only one alternation or half-cycle, the formula is:

Formula 10–3	Average value = 0.637 × Peak value

EXAMPLE

If the peak value in Figure 10–14 were 1 volt, V_{ave} (one-half cycle) = $0.637 \times 1.0 = 0.637$ V

---

3. The most frequently used value related to ac is called the **effective value (rms).** This is *that value of ac voltage or current producing the same heating effect as an equal value of dc voltage or current.* That is, it produces the same power $\left(I^2R \text{ or } \frac{V^2}{R}\right)$. For example, an effective value of ac current of two amperes through a given resistor produces the same heating effect as two amperes of dc current through that same resistor.

   Effective value is also commonly termed the root-mean-square value. The root-mean-square term comes from the mathematical method used to find its value. It is derived from taking the square root of the mean (average) of all the squares of the sine values. Thus, effective value (rms) is 0.707 times maximum or peak value. Again, the reason effective value is greater than half of the maximum value relates to the nonlinearity of the rate of change of voltage or current throughout the alternation.

Formula 10–4	Effective value (rms) = 0.707 × Peak value

EXAMPLE

Again, assuming 1-volt peak value for the waveform in Figure 10–14, $V_{rms} = 0.707 \times 1 = 0.707$ V

Summary of Relationships for Common AC Values:

1. Average value = 0.637 times peak value. (NOTE: it also = 0.9 times effective value, since 0.637 is approximately $\frac{9}{10}$ of 0.707.)
2. Effective value (rms) = 0.707 times peak value, or 0.3535 times peak-to-peak value. (NOTE: it also equals 1.11 times average value, since 0.707 is 1.11 times 0.637.)

PRACTICAL NOTES

Most voltmeters and current meters used by technicians are calibrated to measure effective, or rms, ac values.

In ac, average power (heating effect) is computed in purely resistive components or circuits by using the effective voltage or current values. That is, $P = V_{rms} \times I_{rms}$, or $P = I^2R$, or $P = \frac{V^2}{R}$.

3. Peak value = maximum amplitude within one alternation. Since effective value (rms) is 0.707 of peak value, it follows that peak value equals $\frac{1}{0.707}$, or is 1.414 times effective value. That is, $V_{pk} = 1.414 \times$ rms value. (NOTE: it also equals 0.5 times peak-to-peak value.)
4. Peak-to-peak value = two times peak value and is the total sine wave amplitude from positive peak to negative peak. It follows then, that effective value (rms) is 0.3535 of the peak-to-peak value (or half of 0.707). That is, $V_{pk\text{-}pk}$ equals $\frac{1}{0.3535}$, or peak-to-peak value equals 2.828 times effective value.

With the knowledge of these various relationships, try the practice problems below.

PRACTICE PROBLEM III

1. The peak value of a sine wave voltage is 14.14 volts. What is the rms voltage value?
2. For the voltage described in question 1, what is the V_{ave} value?
3. Approximately what percentage of effective value computed in question 1 is the average value computed in question 2 above?
4. Does frequency affect the rms value of a sine-wave voltage?
5. Are the results from calculating ac values from the negative alternation the same as ac values calculated from the positive alternation?
6. The peak-to-peak voltage of a sine-wave signal is 342 volts. Calculate the following:

a. ac rms (effective) voltage value ___________
b. peak voltage value ___________
c. instantaneous voltage value at 45° ___________
d. instantaneous voltage value at 90° ___________
e. ac average voltage value (over one-half cycle) ___________

THE PURELY RESISTIVE AC CIRCUIT

Earlier, we discussed in-phase and out-of-phase ac quantities. In purely resistive ac circuits, the ac voltage across a resistance is in-phase with the current through the resistance, Figure 10–15. It is logical that when maximum voltage is applied to the resistor, maximum current flows through the resistor. Also, when the instantaneous ac voltage value is at the zero point, zero current flows and so forth. Ohm's law is just as useful in ac circuits as it has been in dc circuits. (NOTE: **instantaneous values** are expressed as "e" or "v" for instantaneous voltage and "i" for instantaneous current.)

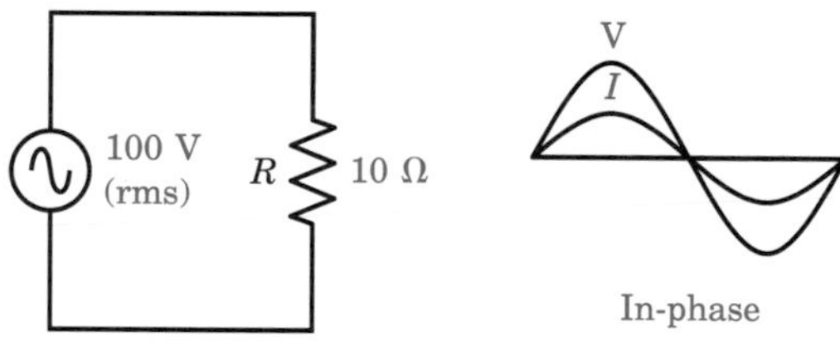

FIGURE 10–15 Basic ac resistive circuit

EXAMPLE

Applying Ohm's law to the circuit in Figure 10–15, if the rms ac voltage is 100 V, what is the effective value of ac current through the 10-Ω resistor? If you answered 10 amperes, you are correct! That is, $I = \frac{V}{R} = \frac{100\text{ V}}{10\ \Omega} = 10\text{ A}$.

In ac circuits where voltage and current are in-phase, the calculations to find V, I, and P are the same as they are in a dc circuit. Generally, effective values are used in these computations, unless otherwise indicated.

If you want to determine the instantaneous voltage at the 30° point of the sine wave, the value would equal the sine of 30° times the maximum value. In our sample case, the sine of 30° = 0.5. Therefore, if the maximum voltage (V_{max} or V_{pk}) is 141.4 volts, the instantaneous value at 30° = 0.5 × 141.4 = 70.7 V.

The analyses you have been using for dc circuits also applies to purely resistive circuits with more than one resistor. The only difference is that you typically will deal with ac effective values (rms). When you know only peak or peak-to-peak values, you have to use the factors you have been studying to find the rms value(s). In later chapters, you will study what happens when out-of-phase quantities are introduced into circuits via "reactive" components.

Try to apply your new knowledge of ac quantities and relationships to the following practice problems.

R_1 10 kΩ
100 V (rms)
R_2 10 kΩ

FIGURE 10–16 Purely resistive series ac circuit

PRACTICE PROBLEM IV

1. Refer to the series circuit in Figure 10–16 and determine the following parameters:
   a. Effective value of V_1
   b. Effective value of V_2
   c. I_{pk}
   d. Instantaneous peak power
   e. P_1

2. Refer to the parallel circuit in Figure 10–17 and determine the following parameters:
   a. Effective value of I_1
   b. Effective value of I_2
   c. I_{pk}
   d. $V_{pk\text{-}pk}$

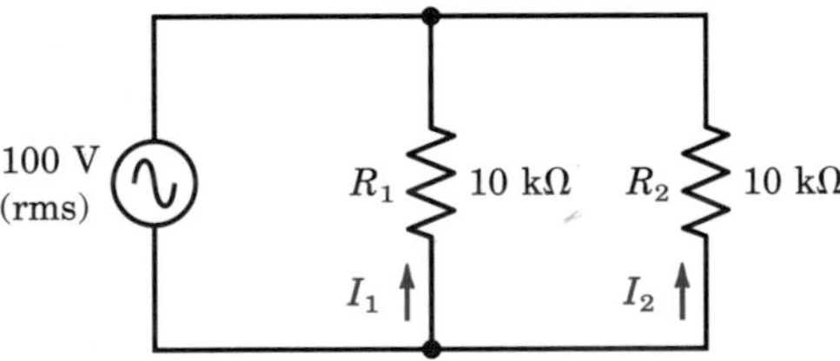

FIGURE 10–17 Purely resistive parallel ac circuit

3. Refer to the series-parallel circuit in Figure 10–18 and determine the following parameters:
   a. Effective value of V_1
   b. Effective value of V_2
   c. Effective value of V_3
   d. P_1 (NOTE: Use appropriate rms values.)

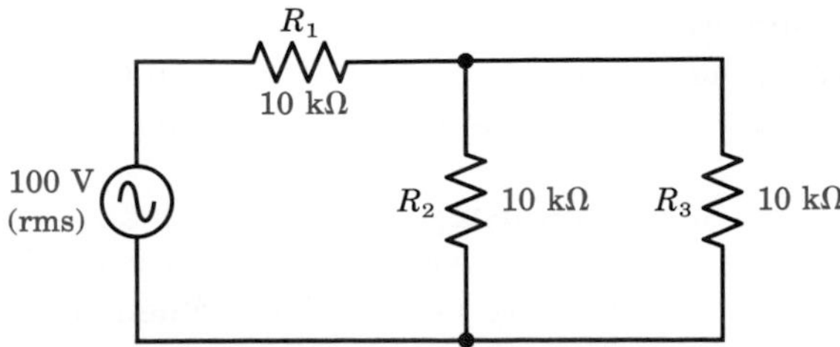

FIGURE 10–18 Purely resistive series-parallel ac circuit

OTHER WAVEFORMS

Although the sine wave is a fundamental ac waveform, there are other non-sinusoidal waveforms that are important for the technician to know and understand. Modern digital electronics systems, such as computers, data communications, radar, and pulse systems, and circuits requiring ramp waveforms, require that the technician become familiar with their features.

It is not within the scope of this chapter to deal in-depth with these various waveforms, but it is important to at least introduce you to them.

The most important waveforms include the square wave, the rectangular wave, and the sawtooth (ramp-shaped) wave. Refer to Figure 10–19 and note the comparative characteristic of each waveform with the sine wave.

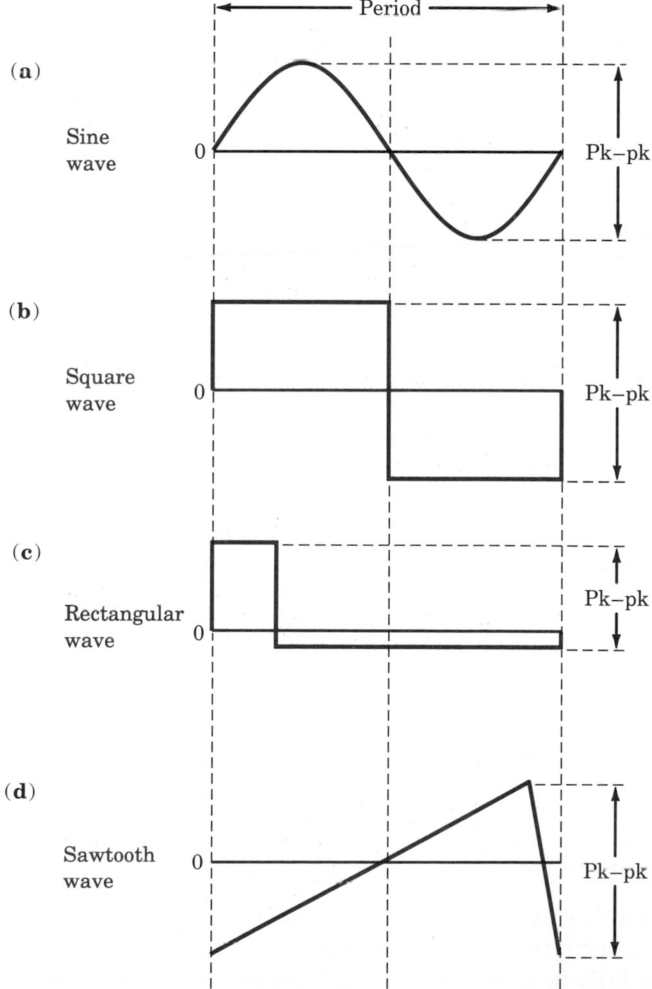

FIGURE 10–19 Comparison of sinusoidal and non-sinusoidal waveforms

The Square Wave

1. The period and frequency of the square wave, Figure 10–19b, is the same as the period and frequency of the sine wave, Figure 10–19a.
2. A square wave cycle is the same as the cycle of the sine wave. That is, when the wave is at the same point in amplitude and polarity and is varying in the same direction as the starting point, the waveform is beginning the next cycle of its recurring characteristic. (This is true for any periodic waveform.)
3. The peak value of the square wave is from zero to the maximum point. However, the average and effective values are the same as the peak value for one alternation.
4. Peak-to-peak value for the square wave is from the maximum positive point of the waveform to the maximum negative point of the waveform. (The same is true for any waveform.)
5. The square wave is symmetrical in positive and negative alternations in both amplitude and time required for the alternation.

The Rectangular Wave

1. The period and frequency of the rectangular wave, Figure 10–19c, is the same as the period and frequency of the sine wave, Figure 10–19a.
2. A rectangular wave cycle is the same as the cycle of the sine wave.
3. Peak-to-peak value for the rectangular wave is from the maximum positive point of the waveform to the maximum negative point of the waveform. (The same is true for any waveform.)
4. The rectangular wave is nonsymmetrical in positive and negative alternations in both amplitude and time required for the alternation.

The Sawtooth Wave

1. The period and frequency of the sawtooth wave, Figure 10–19d, is the same as the period and frequency of the sine wave, Figure 10–19a.
2. A sawtooth wave cycle is the same as the cycle of the sine wave.
3. Peak-to-peak value for the sawtooth wave is from the maximum positive point of the waveform to the maximum negative point of the waveform. (The same is true for any waveform.)
4. The sawtooth wave may or may not be symmetrical in positive and negative alternations in time required for the alternation. It generally rises and falls in a linear fashion from one extreme to the other.

SUMMARY

- Alternating current (ac) varies continually in magnitude and periodically reverses in direction (polarity).
- Whereas ac is bidirectional, dc is unidirectional. Also, ac alternates in polarity, but dc has only one polarity. Furthermore, ac can be transformed or stepped up or down via transformer action, whereas dc cannot.
- The sine wave is related to angular motion in degrees, and any given point's value throughout the sine wave relates to the trigonometric sine function. Maximum values occur at 90° and 270° (−90°) points of the sine wave. Zero values occur at 0°, 180°, and 360° points of each cycle. (Assuming the wave is started at the zero level.)
- A sine-wave voltage is generated by relative motion between conductor(s) and a magnetic field, assuming the conductor(s) links the flux lines via a rotating motion. The ac generator output is typically delivered to the load via slip rings and brushes.
- Quantities having magnitude and direction are frequently represented by lines or arrows called vectors. A vectors' length and relative direction illustrate the quantity's value and angle.
- When representing angles relative to ac quantities, the zero reference point is horizontal and to the right. Angles increase positively from that reference plane in a counterclockwise (CCW) direction .
- When representing angles relative to the coordinate system, angles between 0° and 90° are in quadrant 1; angles between 90° and 180° are in quadrant 2; angles between 180° and 270° are in quadrant 3; and angles between 270° and 360° are in quadrant 4. These positive angles apply with CCW rotation of the phasor.
- A cycle represents one complete set of recurring events or values. In ac, the completion of a positive alternation (half-cycle) and a negative alternation (or vice-versa) completes a cycle.
- A period is the time it takes for one cycle. It is calculated by the formula $T = \frac{1}{f}$.
- Frequency is the number of cycles per second, and Hertz (Hz) indicates cycles per second. Frequency is calculated using the formula $f = \frac{1}{T}$, if the time for one cycle (period) is known. Frequency and period are inversely related.
- Referring to a sine wave, rate-of-change indicates how much amplitude changes for a given amount of time. This rate of change is maximum near the zero level points and minimum near the maximum level points.
- In purely resistive ac circuits, voltage and current are in-phase with each other.
- When ac quantities are out-of-phase, they do not reach the same relative levels at the same time. A generalization is the waveform that reaches maximum positive level first is leading the other quantity in-phase.
- Phase angle is the difference in angle (time difference) between *two equal-frequency* sine-waves. If sine wave A and B are in-phase, there is zero phase angle between them. If sine wave A reaches maximum positive level $\frac{1}{4}$ cycle before sine wave B, sine wave A is leading sine wave B by 90°. Thus, the two signals are out-of-phase by that amount. If sine wave A reaches maximum positive level $\frac{1}{8}$ cycle prior to sine wave B, sine wave A is leading sine wave B in-phase by 45°.

- Average value of ac voltage or current is the average level under one alternation's waveform. Average value is equal to 0.637 times peak value. (NOTE: Average value of a symmetrical waveform over a complete cycle equals zero.)
- Effective value (rms) of ac voltage or current is the value that causes the same heating effect as an equal value of dc voltage or current. Effective value is computed as 0.707 times peak value, or 0.3535 times peak-to-peak value, or 1.11 times average value.
- Peak value (maximum positive or negative value) of ac voltage or current is 1.414 times the effective value. It can also be calculated as being half the peak-to-peak value.
- Peak-to-peak value is the total difference in voltage or current between the positive and negative maximum values. Peak-to-peak value is calculated as two times peak value, or 2.828 times effective value.
- In ac resistive circuits, average power is calculated with the effective values (rms) of I and V.
- Several non-sinusoidal waveforms are the square wave, rectangular wave, and sawtooth wave.

REVIEW QUESTIONS

1. What are two differences between ac and dc?
2. What is the period of a 2-kHz voltage?
3. What is the frequency of an ac having a period of 0.2 μsec?
4. How much time does one alternation of 60-Hertz ac require?
5. Draw an illustration showing one cycle of a 10-volt (rms), 1 MHz ac signal. Identify and appropriately label the positive alternation; the period; the peak value; the peak-to-peak value; the effective value; and points on the sine wave representing angles of 30°, 90°, and 215°.
6. The peak value of a sine wave is 169.73 volts. What is the rms value?
7. What is the frequency of a signal having one-fourth the period of a signal whose T equals 0.01 seconds?
8. Draw a phasor diagram showing voltage A being approximately twice as great as voltage B and leading B by about 45°. (Use B as reference.)
9. What is the peak-to-peak voltage across a 10-kΩ resistor that dissipates 50 mW?

10. Draw an illustration showing one cycle of a 30V pk-pk square wave. Assume the signal has a frequency of 100-Hz. Label the illustration indicating the period and the number of seconds in the positive alternation.

11. If the rms value of an ac voltage doubles, the peak-to-peak value must ____________.

12. The ac effective voltage across a given R doubles. What happens to the power dissipated by the resistor?

13. The period of an ac signal triples. What must have happened to the frequency of the signal?

14. The rms value of a given ac voltage is 75 volts. What is the peak value? What is the peak-to-peak value?

15. What time is needed for one alternation of a 500-Hz ac signal?

16. Refer to Figure 10–20 and answer the following:
    a. Which waveform illustrates the signal with the longest period?
    b. What is the peak-to-peak voltage value of waveform B?

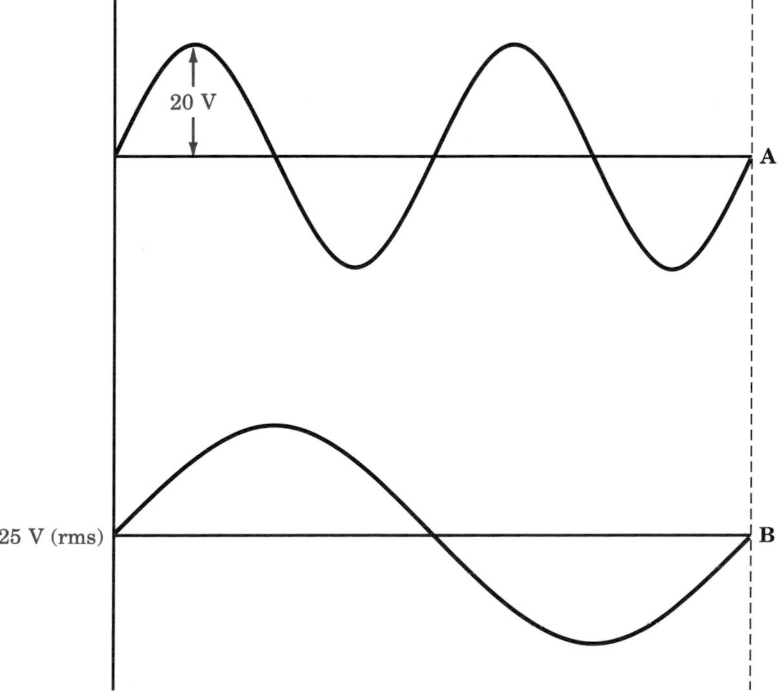

FIGURE 10–20
Review question 16

c. What is the effective value of voltage waveform A?
d. If the period of waveform A equals 0.002 milliseconds, what is the frequency represented by waveform B?
e. If the peak-to-peak value of waveform A changes to 60 volts, what is the new effective voltage value?

TEST YOURSELF

1. Refer to a trig table or calculator, and draw one cycle of a sine wave and a cosine wave on the same "baseline," assuming each maximum value is 10 units. Appropriately label the diagram to show the following information:
   a. Angles of 0, 45, 90, 135, 180, 225, 270, 315, and 360 degrees, marked on the x-axis.
   b. Amplitude values of the sine wave at 45 and 90 degrees, respectively.
   c. Amplitude values of the cosine wave at 0, 45, and 90 degrees, respectively.
   d. Indicate which wave is leading in phase and by how much.
2. One alternation of a signal takes 500 μs. What is its frequency? What is the time of two periods?
3. Draw the diagram of a three-resistor series-parallel circuit having a 10-kΩ resistor (labeled R_1) in series with the source, and two 100-kΩ resistors (labeled R_2 and R_3) in parallel with each other and in series with R_1. Assume a 120-V ac source. Calculate the following parameters; and appropriately label them on the diagram.
   a. V_{R_1}
   b. I_{R_1}
   c. P_{R_1}
   d. I_{R_2}
   e. V_{R_3}
   f. θ between V applied and I total

The Oscilloscope

Key Terms

Attenuator

CRT

Focus control

Gain control(s)

Horizontal (X) Amplifier

Intensity control

Lissajous pattern(s)

Oscilloscope

Position control(s)

Sweep circuit(s)

Sweep frequency control(s)

Synchronization

Synchronization control(s)

Vertical volts/cm control(s)

Vertical (Y) Amplifier

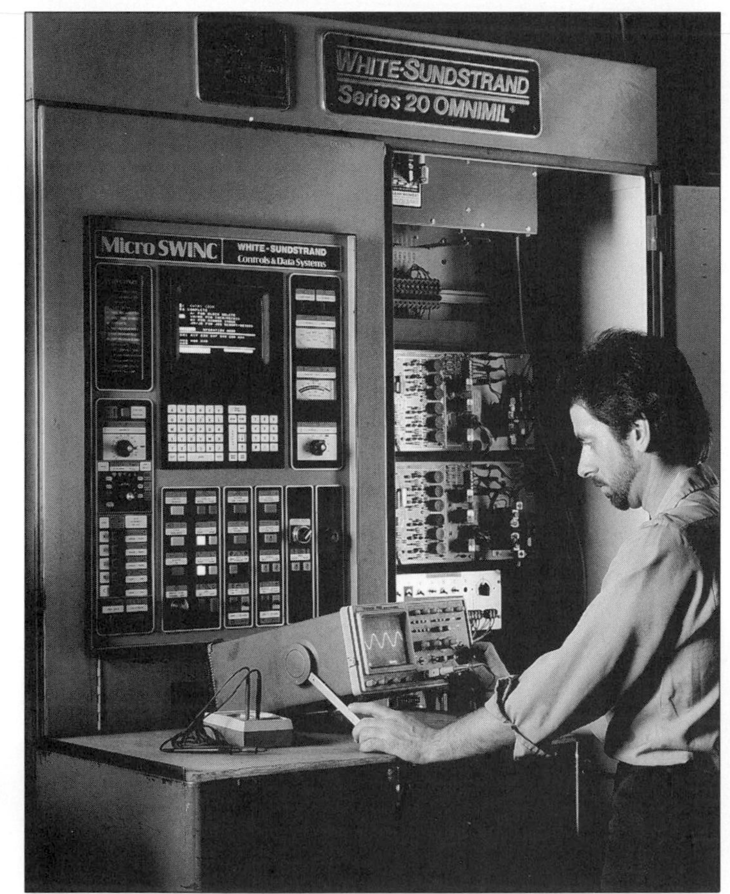

Courtesy of Cleveland Institute of Electronics, Inc.

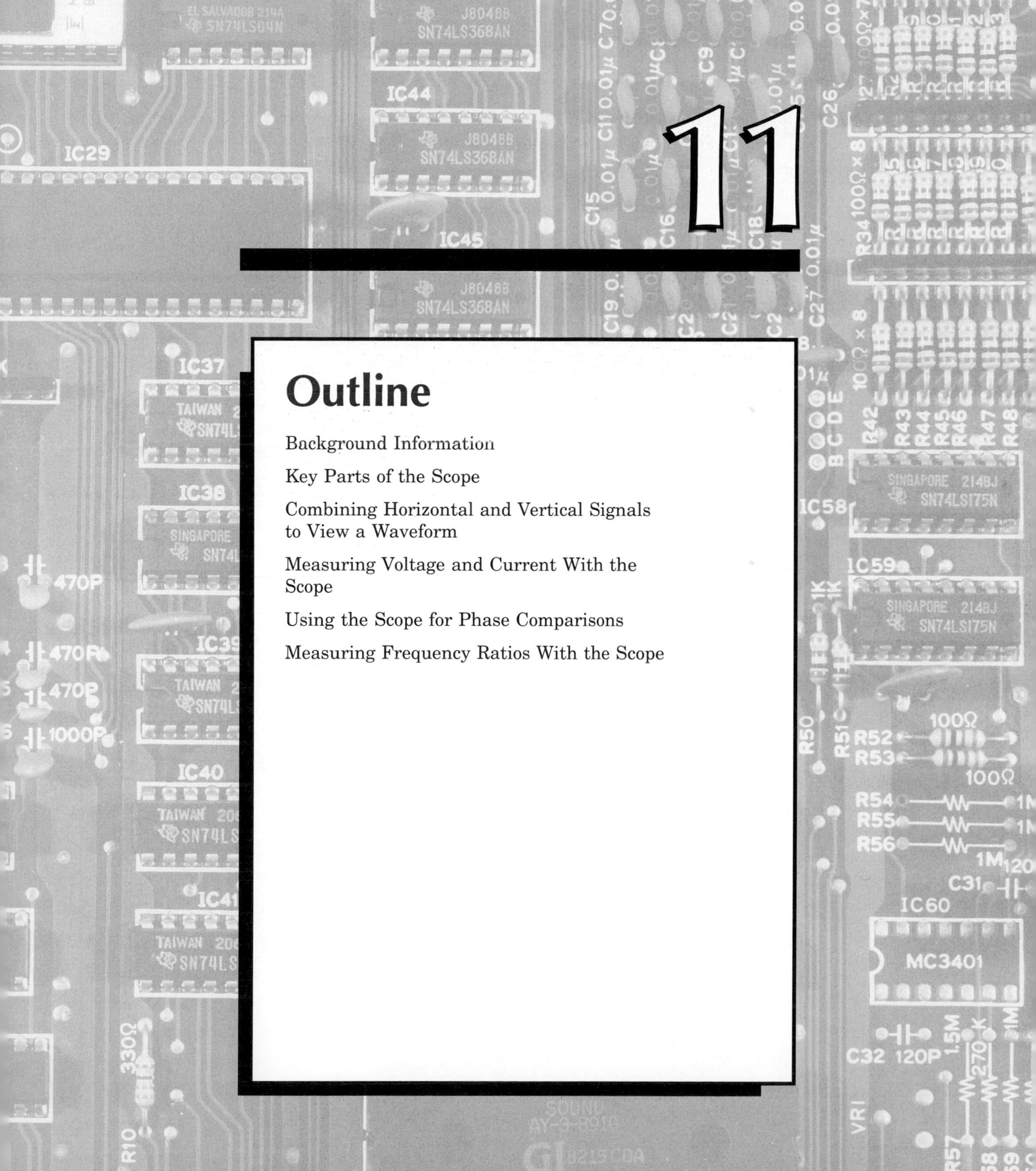

11

Outline

Background Information

Key Parts of the Scope

Combining Horizontal and Vertical Signals to View a Waveform

Measuring Voltage and Current With the Scope

Using the Scope for Phase Comparisons

Measuring Frequency Ratios With the Scope

Chapter Preview

The oscilloscope, frequently called a scope, is without doubt one of the most versatile tools available to technicians and engineers. It can display waveforms, measure voltages and currents, and measure and compare phase relationships and measure and compare frequencies.

In your studies, you have been introduced to other measuring instruments and to some important ac quantities. This chapter will increase your knowledge about measuring techniques. This chapter will also reinforce your knowledge about the important ac quantities you will measure and analyze throughout your remaining studies and in your career as a technician.

The purpose of this chapter is not to discuss the details of circuitry involved in making a scope function, nor to instruct you on subjects that are best learned with actual experience. However, this chapter will, perhaps, inspire you to become familiar with this valuable tool and will be a good background for you to become proficient in using this device.

Objectives

After studying this chapter, you will be able to:

- List the key parts of the **oscilloscope**
- List precautions when using scopes
- List procedures when measuring voltage with a scope
- List procedures to display and interpret waveforms
- Define **Lissajous pattern**
- List procedures relating to phase measurement
- List procedures when measuring frequency with a scope

BACKGROUND INFORMATION

Scopes come in a variety of brands and complexity. Two common categories of scopes include those used for general purpose applications, and the more complex (and expensive) laboratory quality scopes, Figure 11–1. Both categories have certain common features and applications. Knowing these commonalities will enable you to quickly use almost any kind of scope.

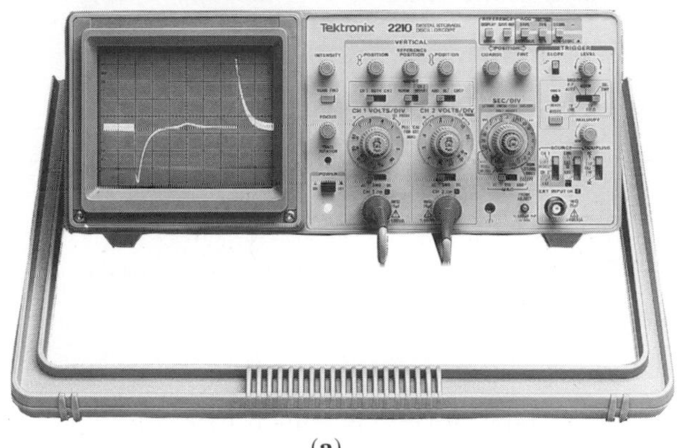

(a)

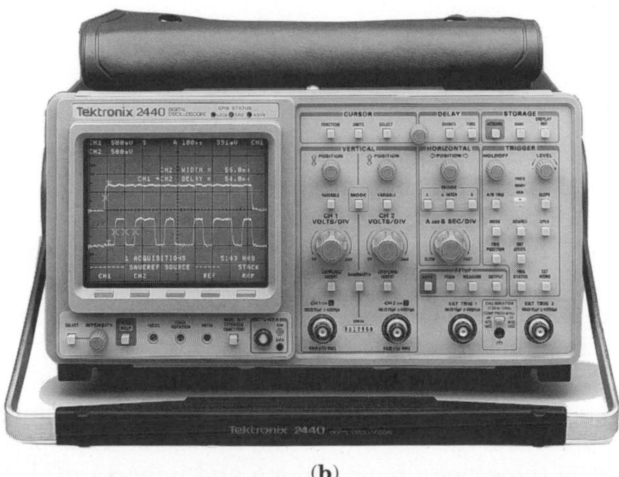

(b)

FIGURE 11–1 General purpose (a) and lab quality (b) oscilloscopes (Photos courtesy of Techtronix, Inc.)

Two general areas of commonality between scopes include:

1. All scopes have a cathode-ray tube (CRT) where the displays are viewed.

2. All scopes have controls and circuits that adjust the display to help you analyze voltage, time, and frequency parameters of the signal(s) under test.

Let's take a brief look at these two critical areas related to scopes.

KEY SECTIONS OF THE SCOPE

The key sections of a scope that will be discussed are:

1. Cathode-ray tube
2. Intensity and focus controls
3. Position controls
4. Sweep frequency controls
5. Vertical section
6. Horizontal section
7. Synchronization controls

A typical scope front panel layout is shown in Figure 11–2. Use this figure while you read the following descriptions of the various controls and circuits.

The CRT

All oscilloscopes have a **cathode-ray tube** (**CRT**), which has a "screen." The CRT is a special tube where one end has some elements that cause an electron stream toward the screen of the tube. Voltages applied to various elements in the CRT control the direction, intensity, and shape of this electron beam, which strikes the screen material on the inside face of the CRT. (For informational purposes, Figure 11–3 shows some of these electrode control elements within the CRT. NOTE: It is not necessary for you to learn the names and functions of these elements at this point.)

The screen material is a luminescent material. That is, the material emits light when struck by electrons. By virtue of this light-emitting feature and the ability to control the electron beam, the oscilloscope produces visual waveforms that is the scope's most unique and important ability.

Intensity and Focus Controls

As implied by their names, the **intensity** and **focus controls** adjust the brightness and focus the sharpness of the spot (or trace) on the CRT screen. The sharper the spot, the clearer and more distinct the traces on the CRT screen. In Figure 11–2, you can see how these controls are labeled on the typical scope's front panel.

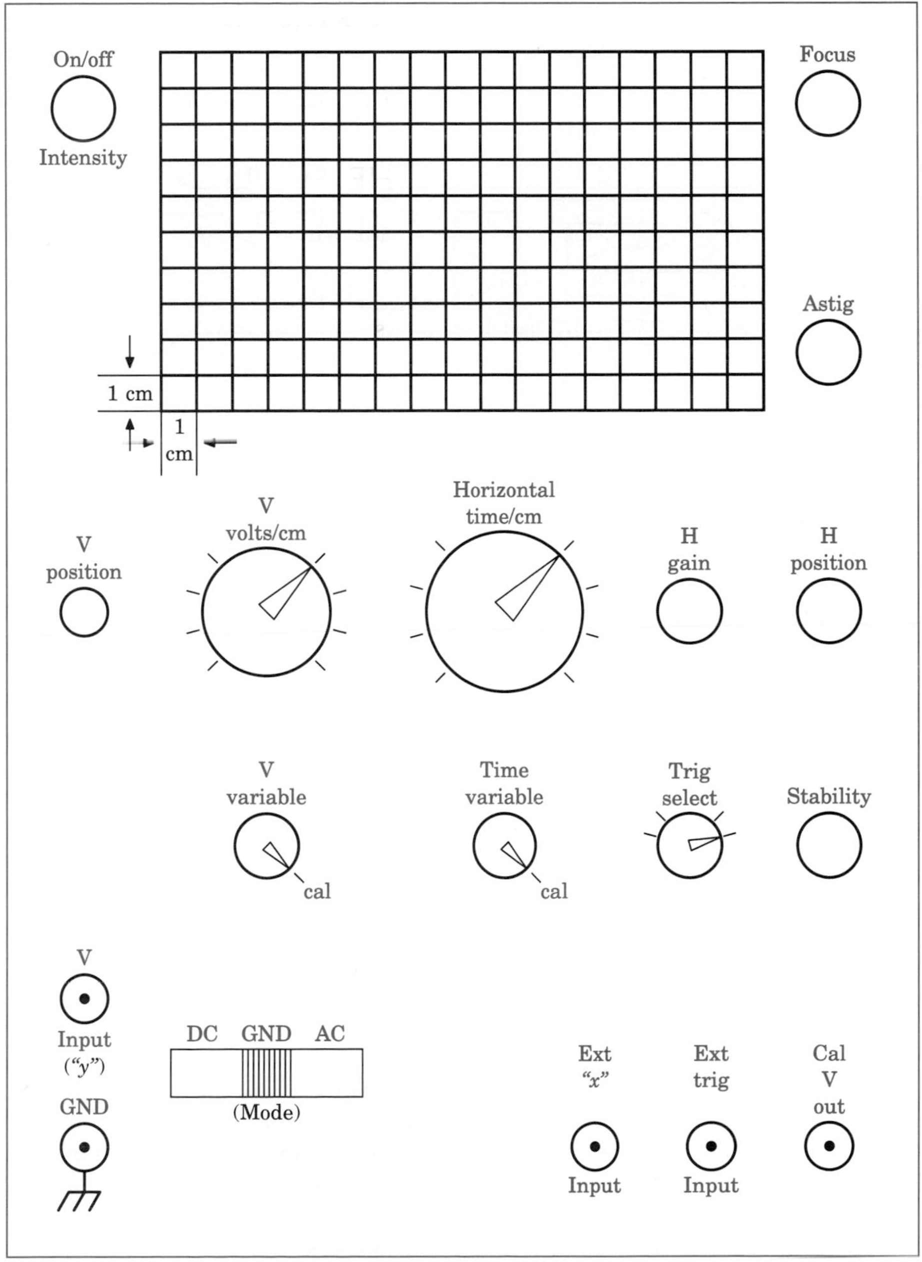

FIGURE 11–2 Typical controls on a single-trace general-purpose scope

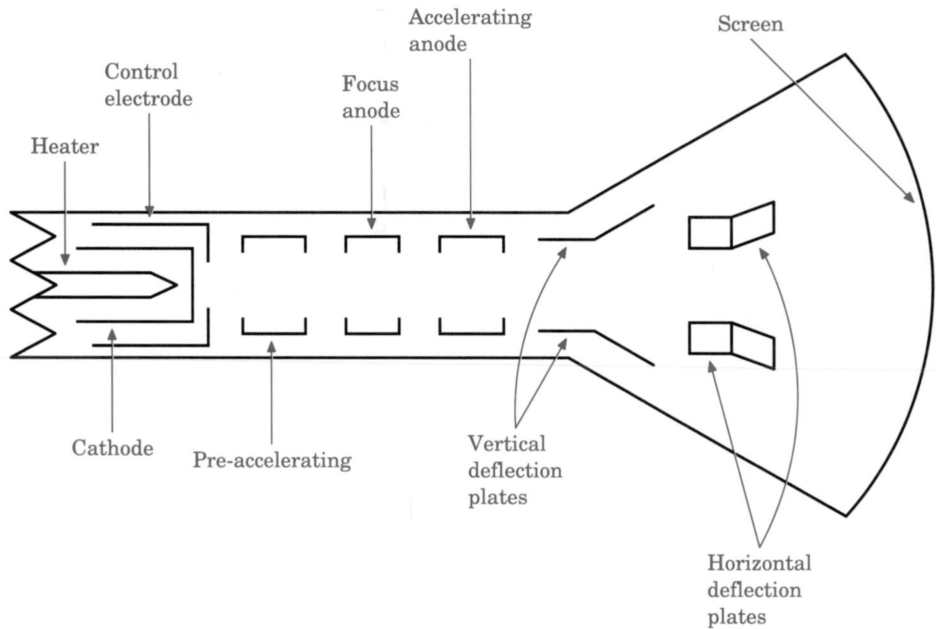

FIGURE 11–3
Elements in a typical cathode-ray tube (CRT)

PRACTICAL NOTES

Caution! One thing you should learn when operating a scope is that it is not good to leave a bright spot at one position on the CRT. The CRT screen material can be burned or damaged if this occurs for any great length of time. Never have the trace bright enough to cause a "halo" effect on the screen.

Position Controls

In Figure 11–3 you can see this CRT has "deflection plates" that control the position and movement of the electron beam coming from the opposite end of the tube and striking the back of the screen. The position where this electron beam strikes the back of the screen (CRT) is controlled by the electrostatic field set up between the plates when there is a difference of potential. (It is worth mentioning that positional control of the electron beam [or stream] in televisions is normally accomplished through *electromagnetic* fields, rather than *electrostatic* fields used by most oscilloscopes.)

The reactions of the electron beam to these fields are illustrated in Figure 11–4. Notice the electron beam is attracted to the plate(s) having a positive charge and repelled from any deflection plates having a negative charge, or

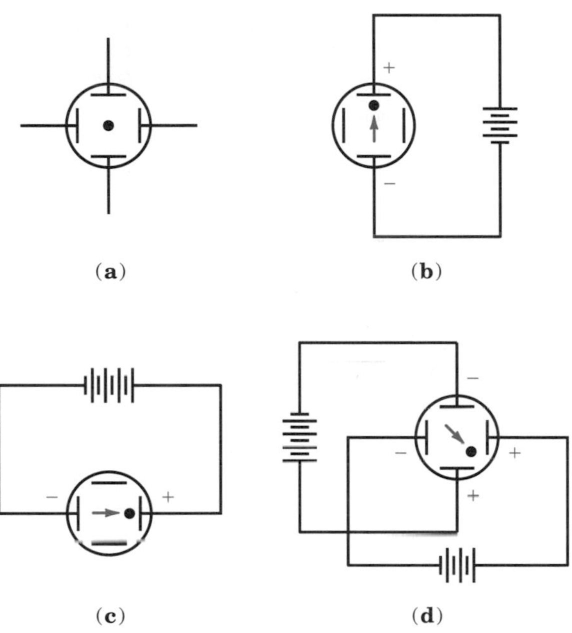

FIGURE 11–4 Electron-beam movement caused by dc voltages on deflection plates

potential. Therefore, the position where the electron beam strikes the screen can be controlled by the voltages present at the deflection plates. Knowing that deflection of the electron beam is toward positive deflection plates and away from negative deflection plates, you can understand the **position controls** simply make the appropriate deflection plates either more positive or more negative, depending on which way we want the electron beam to move on the CRT face.

Look again at Figure 11–2. You can see these front-panel controls that influence both the vertical and horizontal position of the trace. Notice what happens when we apply ac to the deflection plates, Figures 11–5a, b, and c.

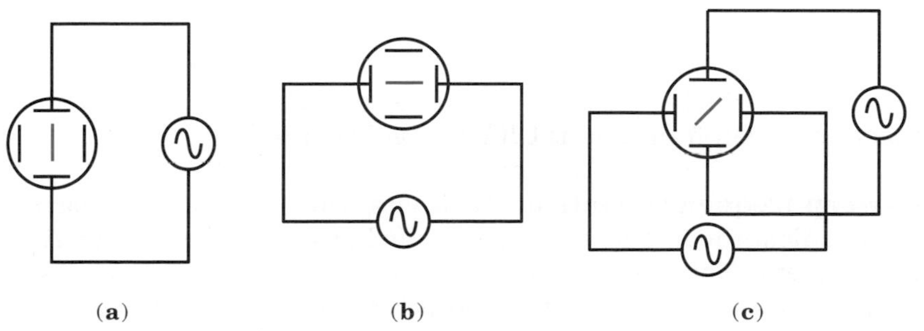

FIGURE 11–5 Electron-beam movement caused by ac voltages on deflection plates

Horizontal Sweep Frequency Control(s)

A sawtooth (or ramp-type) waveform applied to the horizontal plates, creates a horizontal trace on the screen, Figure 11–6. The voltage applied to

FIGURE 11–6 Linear trace caused by sawtooth voltage on horizontal (H) plates

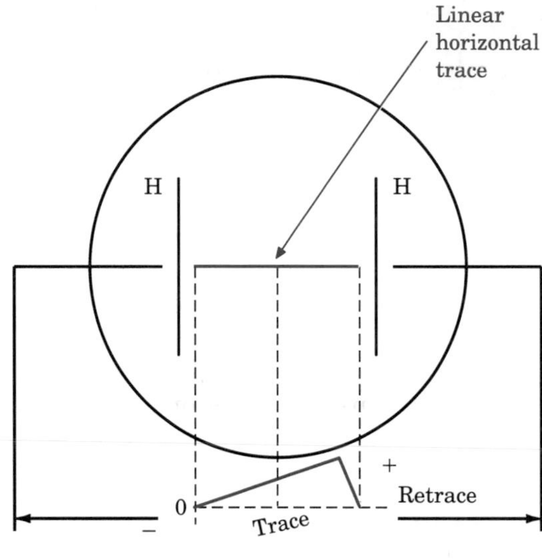

the horizontal plates is sometimes called the horizontal "sweep voltage," because the voltage causes the electron beam to horizontally sweep across the screen, creating a horizontal trace, or line.

The electron beam moves at a constant rate (linear speed), across the screen from left-to-right, then "flies back" to its starting point on the left. This rapid process is continually repeated. The number of times it traces across the screen in a second is determined by the frequency of the sweep voltage. You see a horizontal line on the screen when this happens at a rapid rate. Two reasons for this are:

1. The CRT screen material has a persistence that causes continual emission of light from the screen for a short time after the electron beam has stopped striking that area.
2. The retinas in our eyes also have a persistence feature. This is why you don't see flicker between frames as a movie film is projected, or flicker between frames in the TV "scanning" process.

The **sweep frequency controls** (horizontal time variable and horizontal time/cm) adjust the frequency output of the horizontal **sweep circuitry** in the oscilloscope. Naturally, this determines the number of times per second the electron beam is traced horizontally across the CRT screen. The horizontal frequency control allows us to view signals of different frequencies.

Having accumulated this information, let's now see how the electron beam is further controlled, amplified, or attenuated to provide meaningful displays on the CRT screen.

Vertical Section

Key elements in the vertical section are the vertical input jack(s) (sometimes called Y input), and the **vertical attenuator** and **amplifier,** with related controls. When the scope is a single-trace scope, there is only one vertical input jack. When the scope is a dual-trace scope, there are two vertical input jacks.

The vertical attenuator and amplifier circuitry, and associated controls, decrease or increase the amplitude of the signals to be viewed. That is, the signals which are applied to the vertical section via the vertical input jack(s). View Figure 11–2 and note the calibrated **vertical volts/cm** and the **variable control(s),** which are used for signal level control. Using these controls allows you to adjust for larger or smaller vertical deflection of the CRT trace with a given vertical input signal amplitude.

Horizontal Section

We have briefly discussed some of the important horizontal deflection elements and controls. One of these controls adjusts the frequency of the internally generated **horizontal** sweep signal.

Other important elements related to the horizontal trace include the horizontal **gain control** that changes the length of the horizontal trace line; and a jack (often marked "Ext X" input) that inputs a signal from an external source to the horizontal deflection system in lieu of using the internally generated sweep signal.

Synchronization Controls

The **synchronization controls** synchronize the observed signal with the horizontal trace; thus, the waveform appears stationary. The effect is much like a strobe light that provides "stop-action" when timing an automobile, or the effect of making spinning fan blades appear to be standing still, using a light which is blinking at the appropriate rate. Without going into the technical details, suffice it to say that by starting the trace (left-to-right horizontal sweep) of the electron beam in proper time-relationship with the signal to be observed on the scope (the signal fed to the vertical deflection plates), a stable waveform is displayed. If the horizontal trace and the vertical system signals are not synchronized (i.e., do not start from a given waveform point at a specific time), the waveform is a "moving target" and hard to view.

Some of the jacks and controls typically associated with this synchronizing function are also shown in Figure 11–2, and include:

1. the "external trigger" jack, (for inputting external synchronizing signals),

2. "trig" selector switches or buttons to select whether the source of the synchronizing "trigger" signal is from the scope's internal circuitry or from some external signal source, and
3. "trigger stability" control, which helps set the appropriate level for the triggering signal to work best.

IN-PROCESS LEARNING CHECK I

1. The part of the oscilloscope producing the visual display is the ________ ________ ________.
2. The scope control that influences the brightness of the display is the ________ control.
3. A waveform is moved up or down on the screen by using the ________ ________ control.
4. A waveform is moved left or right on the screen by using the ________ ________ control.
5. For a signal fed to the scope's vertical input, the controls that adjust the number of cycles seen on the screen are the horizontal ________ and the horizontal ________ control.
6. The control(s) that help keep the waveform from moving or jiggling on the display are associated with the ________ circuitry.

COMBINING HORIZONTAL AND VERTICAL SIGNALS TO VIEW A WAVEFORM

Having an electron beam cause a horizontal line across the CRT screen by applying voltage(s) to the horizontal deflection plates alone is not useful for displaying and analyzing waveforms. Causing an electron beam to vertically move by applying voltage(s) to the vertical deflection plates alone also is not useful for displaying and analyzing waveforms. However, "magic" occurs when appropriate voltages are simultaneously applied to both the horizontal and vertical deflection plates.

Look at Figure 11–7. Notice how a sinusoidal waveform is displayed when a sine-wave voltage is applied to the vertical deflection plates, and when a sawtooth voltage of the same frequency (from internal sweep circuitry) is simultaneously applied to the horizontal deflection plates of the CRT. In this case, the frequency is 60 Hertz.

Again refer to Figure 11–7 and study the following discussion. Look at the conditions at times t_0, t_1, t_2, t_3, and t_4.

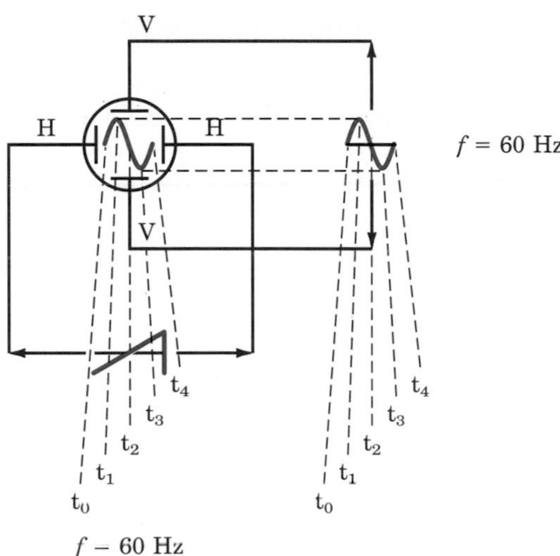

FIGURE 11–7 Waveform with sawtooth on horizontal (H) plates and sine wave on vertical (V) plates

1. The sawtooth voltage is applied to the horizontal deflection plates so that at the beginning of the sawtooth, the electron beam is at the left side of the screen (viewed by the person looking at the front of the scope). Time is t_0.
2. As the sawtooth voltage increases in a positive direction, the right-hand deflection plate becomes increasingly positive. Thus, the beam moves from the screen's left side to its right side. Time is t_1.
3. The 60-Hz sine-wave voltage is applied to the vertical deflection plates. At time zero (t_0) the beam is horizontally on the screen's left side. Since no vertical voltage is at t_0 of the sine wave, a spot appears on the screen's left side.
4. As the horizontal sawtooth signal causes the beam to begin its horizontal trace to the right, the voltage on the vertical deflection plates causes the beam to simultaneously move to its maximum upward position. Time is t_1.
5. By the time the vertical signal has decreased from maximum positive value back to zero (180° into the sine wave), the horizontal signal has caused the beam to arrive at the center of the screen. In other words, the positive alternation of the sine wave has been traced and displayed on the screen at t_2.
6. The same rationale is used when tracing the negative alternation's waveform. That is, as the beam continues its horizontal trek across the screen, the signal applied to the vertical deflection plates is causing the beam to be sequentially moved to the maximum downward position, then back to the zero voltage position, vertically (center of

the screen vertically). During this sequence, the negative alternation of the sine wave has been displayed.

7. Throughout the cycle, one cycle of the sawtooth waveform applied to the horizontal plates has caused one horizontal trace. At the same time, the electron beam has been moved up and down, corresponding to the voltage throughout one sine-wave signal cycle applied to the vertical plates. The result is a sine-wave waveform viewed on the CRT.

8. Because these events have occurred 60 times in one second, our eyes perceive a single sine-wave waveform that traces over itself 60 times in one second.

EXAMPLE

A horizontal sweep frequency of 60 Hz and a vertical signal of 120 Hz result in two cycles of vertical deflections occurring at the same period that one sweep occurred. The display shows two full cycles of a sine wave, Figure 11–8.

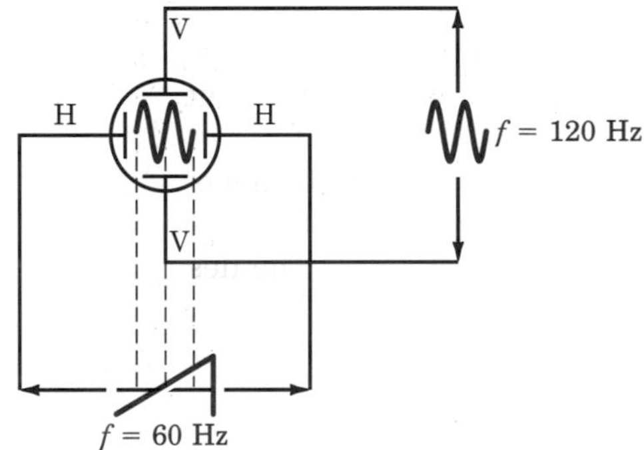

FIGURE 11–8 Result of two cycles of sine-wave signal per one horizontal sweep

PRACTICE PROBLEM I

Answers are in Appendix B.

Assume the vertical frequency is four times the horizontal frequency. What does the waveform look like?

By looking at the electron beam position at a number of instantaneous values of the applied horizontal and vertical signals, you can determine what type of waveform will appear on the scope screen, Figure 11–9. These concepts help interpret what we see on the CRT screen. Later you will see how these concepts help us measure frequency and phase of ac signals.

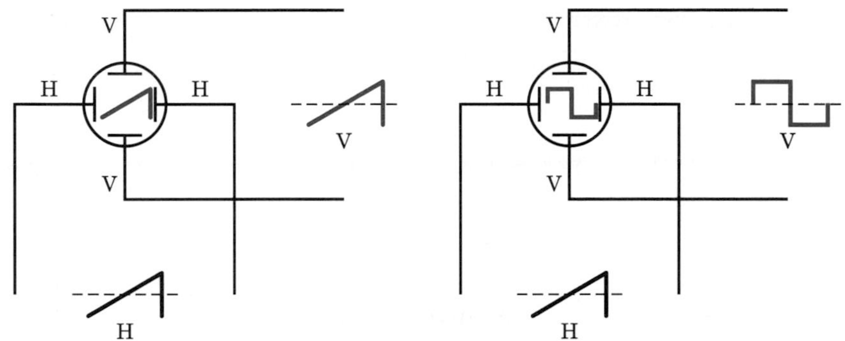

FIGURE 11–9 Illustration of how different waveforms are traced

MEASURING VOLTAGE AND CURRENT WITH THE SCOPE

General Information

The dc and ac voltage measurements made with oscilloscopes are normally not as accurate as those measurements made with good quality multimeters. However, it is often convenient to know how to make these measurements and to interpret waveforms in terms of voltage amplitudes.

As indicated earlier, learning to use the scope in a practical way *only comes through "hands-on" experience.* The descriptions in this chapter provide only conceptual knowledge. Therefore, you will have to expand on these concepts by actual laboratory experiences—be sure you do!

Calibration Voltage

Many scopes have built-in calibrating voltage sources used to set up the scope for interpretation of voltage levels. Typically, the calibrating voltage output is 0.5 V or 1 V peak-to-peak value. Also, this calibrating voltage is often a square wave, Figure 11–10.

Let's assume a 1-volt peak-to-peak calibrating voltage and see how calibration enables us to measure voltages. Refer to Figure 11–11, as appropriate, when you read the following discussion.

1. With *no* signals connected to the vertical input jack, the horizontal sweep and gain controls, vertical and horizontal positioning controls and "sync" controls are adjusted to produce a horizontal line trace across the screen, which is centered both vertically and horizontally on the screen, Figure 11–11a.

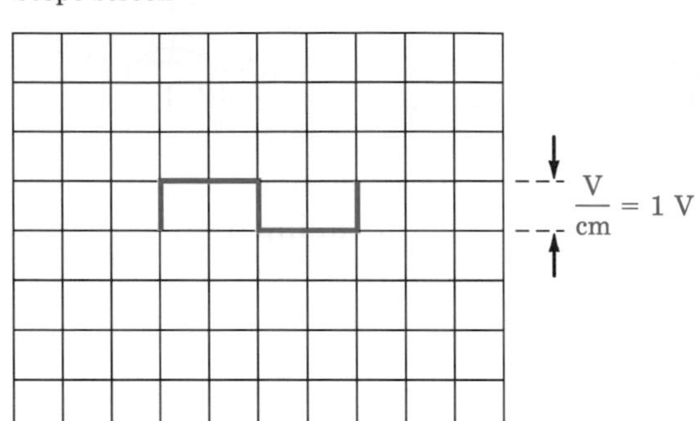

FIGURE 11–10
Typical square-wave calibration voltage

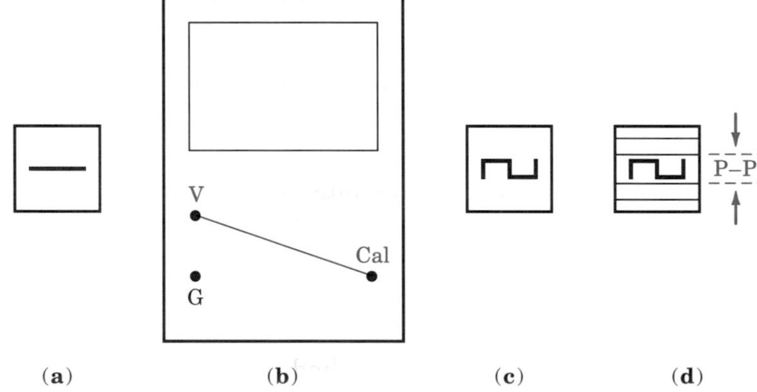

FIGURE 11–11
Calibrating for voltage measurement with calibration signal

2. A "jumper" wire connects the calibrating voltage output terminal on the scope to the vertical input jack, Figure 11–11b.
3. Horizontal frequency controls are adjusted so one cycle of the calibration square wave is displayed on the screen, Figure 11–11c.
4. Vertical "variable" level control is set to the "Cal" position. The volts/cm selector switch is then adjusted so the waveform is displayed with an "easy-to-read" amount of deflection on the screen calibrations (graticule). For example, one major division (1 cm) of vertical deflection appears, Figure 11–11d.
5. When this is done, each vertical square, or division of deflection caused by a voltage applied to the vertical input jack represents one volt (1 V) of input voltage.
6. The scope settings are now adjusted for one division to equal 1 V (peak-to-peak). Let's see how measurements are made.

AC Voltage Measurement

After removing the jumper wire from the vertical input jack (i.e., removing the calibrating voltage signal from the vertical input), there is a vertically and horizontally centered trace across the scope.

Let's now assume we want to measure a small ac (sine-wave) voltage, which is connected between the vertical input jack and ground of the scope. Let's further assume a sine wave appears on the face of the scope, which is four divisions high, from top to bottom (peak-to-peak). What is the peak-to-peak value of the voltage being measured?

If you said four volts peak-to-peak, you are correct. Since each division in height represents one volt (peak-to-peak), then four-divisions-high voltage must be four volts peak-to-peak.

EXAMPLE

What is the rms value of the voltage just measured? (Here is where you apply your knowledge gained from the last chapter!)

1. If the voltage is 4 V peak-to-peak, then it must be 2 V-peak.
2. If it is 2 V peak, then its rms value must be 0.707×2 V, or 1.414 volts.

PRACTICE PROBLEM II

Suppose the voltage in our example had measured 10 volts peak-to-peak on the scope. What would the rms value be? What would be the peak value of this voltage?

When measuring ac voltage with a scope, what you see is the peak-to-peak value. You must then use the appropriate factors to determine the value being viewed. Refer to Figure 11–12 as you read the following discussion.

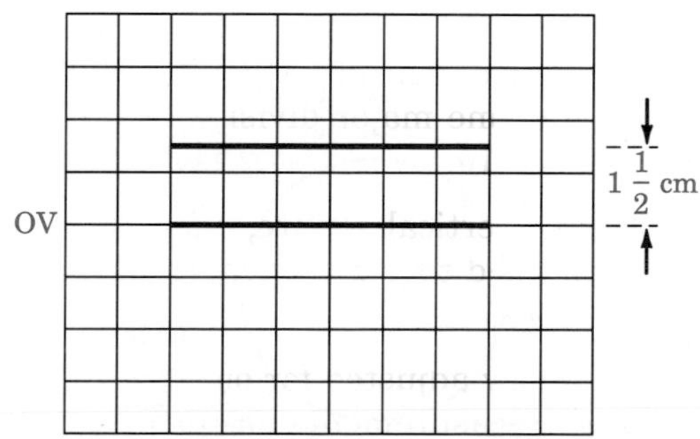

FIGURE 11–12 Direct current (dc) voltage measurement

DC Voltage Measurement

First, the "DC-GND-AC" input mode switch is set to dc. For simplicity, assume the scope is still calibrated so the one V peak-to-peak calibrating square wave causes a one-division-high waveform on the screen. (In other words, the scope is calibrated for one volt causing one division of deflection.)

With no voltage applied between the vertical input jack and ground, the scope shows a centered horizontal trace like before.

Assume when you connect the dc voltage to be measured to the vertical input, the trace jumps up one and one-half divisions from its starting point. What is the dc voltage being measured? If you said 1.5 volts dc, you are right!

Furthermore, you might have indicated the end of the source connected to the vertical input jack was positive at the jack side and negative at the ground connection side, because the trace moved up. Obviously, if you reversed the connections, the trace would move down from center by one and one-half divisions.

Measuring Voltages Greater or Less than the Calibrating Voltage

Obviously, the measured voltages will not always be within a reasonable range of the calibrating voltage value that we have used for explanation purposes. When the voltage to be measured is much greater or less than the calibrating voltage, how is this handled?

One method is to observe the vertical switch setting (calibrated attenuator or deflection factor). When the switch is in the 500-mV position (assuming the variable level control is set at "Cal" position), then for each centimeter (or square) of deflection, 500-mV of voltage is indicated. When the switch is set at the 1 V/cm position, then each division of deflection equals 1 volt and so forth, Figure 11–13.

Another technique often used with many scopes is a test probe that attenuates the signal by a given factor, for example ten times, Figure 11–14.

FIGURE 11–13 Using the calibrated volts/cm switch setting to determine voltage

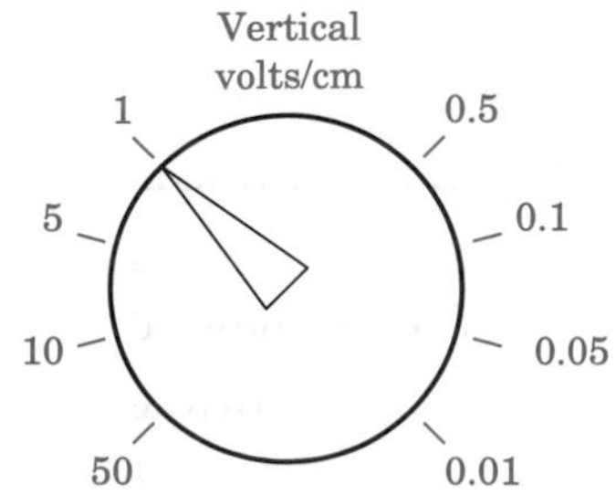

FIGURE 11–14 An example of a "times ten" test probe (Courtesy of SPC Technology)

Thus, you multiply by ten times what you are reading on the scope to find the actual value of voltage applied to the probe. This probe often has another purpose than simply attenuating the signal—it prevents the circuit being tested from being disturbed by the scope's input circuitry, or test cable.

Summary of Voltage Measurement

To measure voltage, a calibration is performed, using the scope's built-in calibrating voltage, or by properly adjusting the vertical input level controls. The voltage can then determined by using the volts-per-division switch setting of the vertical volts/cm control.

Measuring Current

Current is measured by the known-resistor/Ohm's law technique you learned in previous chapters. If you use the scope to measure the voltage across a 1-kΩ resistor, for example, you automatically know the number of volts measured equals the number of milliamperes through the 1-kΩ resistor. That is, $I = \frac{V}{R}$, and volts divided by kΩ equals mA. Obviously, Ohm's law applies to any value resistor. The 1-kΩ value is merely convenient to use, where possible.

USING THE SCOPE FOR PHASE COMPARISONS

In addition to exhibiting waveforms and measuring voltage (or current), the scope can compare the phase of two signals. Two methods for doing this are:

1. An overlaying technique where the two waveforms are visually superimposed on each other. This is accomplished by a dual-trace scope, or

by an electronic switching device. Therefore, the displacement of the two signals positive peaks, or the points where they cross the zero reference axis determines the phase difference.

EXAMPLE

Adjust the horizontal sweep frequency so exactly one cycle of the signal applied to the A (or #1) input covers the screen. (NOTE: ten divisions are typical).

The second signal is applied to the scope B (or #2) input. Since the screen is divided into ten divisions, and ten divisions represent 360 degrees, then each division represents 36 degrees. By noting the amount of offset between the two waveform traces and knowing that each division difference equals 36 degrees, you can easily interpret the phase difference between the two signals.

PRACTICE PROBLEM III

Suppose the horizontal frequency in our example were adjusted so there were exactly 2 cycles across the 10 divisions of the screen. How many degrees would each major division on the screen represent?

---

2. Another method used to compare the phase of two signals is the **Lissajous** (pronounced liz-a-ju) pattern(s) technique. With this method, one signal is applied to the horizontal plates and the other signal is applied to the vertical plates. The resulting waveform (known as a Lissajous pattern) enables the technician to determine the phase differential between the two signals. (NOTE: This technique is not a frequently used method for phase measurement.)

The Overlaying Technique

If a dual-trace scope is used:

1. Two horizontal traces are obtained on the scope. They are then superimposed at the center of the screen by using the positioning controls, as appropriate.
2. Signal #1 is applied to one of the vertical input jacks. Scope sweep frequency is adjusted for one sine wave. Vertical level controls are adjusted to obtain a convenient size waveform for viewing purposes.
3. Signal #2 is connected to the other vertical input jack. Again, appropriate adjustments are made to appropriately display signal #2. The signals should be adjusted to the same vertical size and should be the same frequency.

ANSWERS TO CHAPTER CHALLENGES

The troubleshooting steps illustrated on these pages are examples of just one way the **SIMPLER** sequence might be used to solve each Chapter Challenge.

NOTE: The tolerances of components and of test instruments will create differences between the theoretical values and those actually shown. The theoretical values would be achieved only if all components were precisely rated and the test instruments had zero percent error.

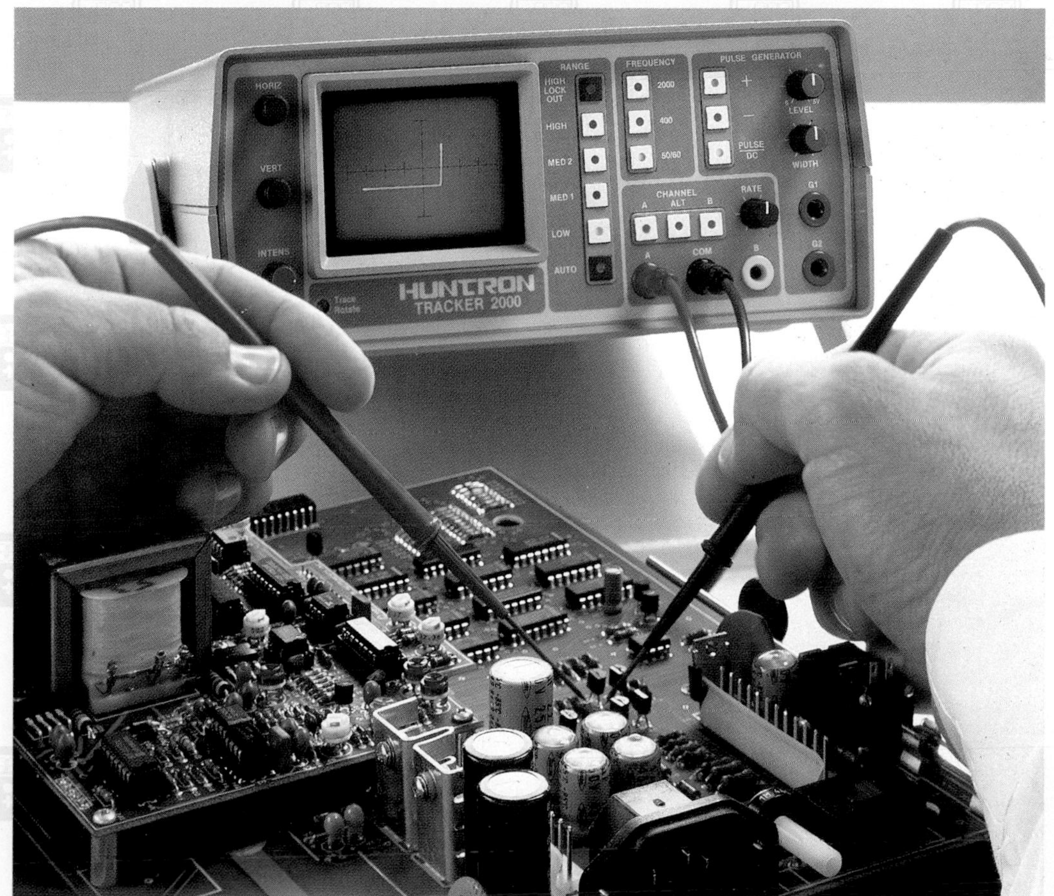

(Courtesy of Huntron Electronics)

CIRCUIT 1

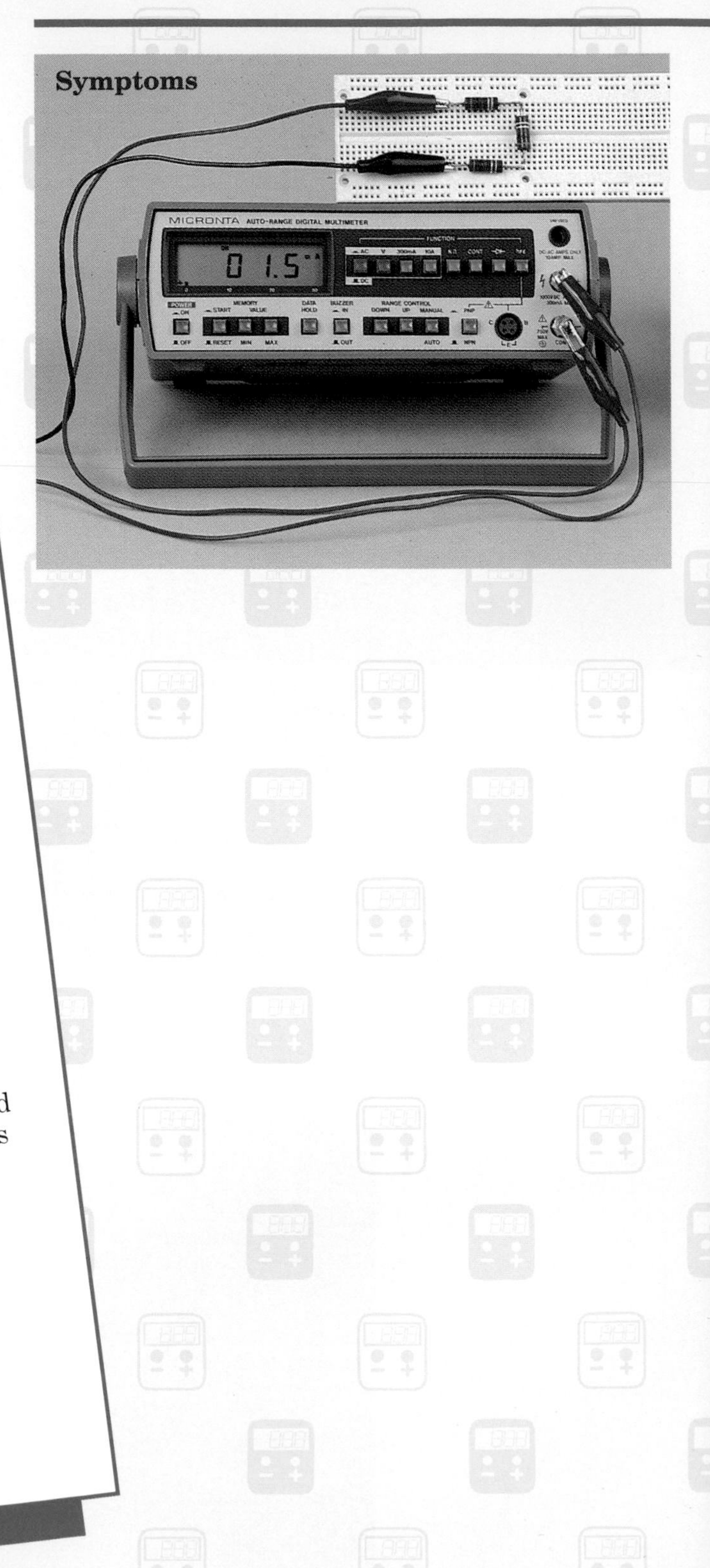

SYMPTOMS: The total current is too low for voltage applied. This implies that the total resistance has increased, meaning that one or more resistor(s) has changed value.

IDENTIFY initial suspect area: R_1, R_2, and R_3 (i.e., total circuit).

MAKE test decision: Check voltage across R_2 (middle of circuit).

PERFORM 1st Test: Look up the test result. $V_{B\text{-}C}$ is 1.5 V.

LOCATE new suspect area: R_1 and R_3. NOTE: $V_{B\text{-}C}$ would be greater than 2 V if R_2 had increased and the other resistors had not.

EXAMINE available data.

REPEAT analysis and testing:

2nd Test: Check voltage across R_1. $V_{A\text{-}B}$ is 1.5 V.
3rd Test: Check voltage across R_3. $V_{C\text{-}D}$ is 3 V.
4th Test: Disconnect the V source and check resistance of R_3. The result is that R_3 is 2 kΩ, which is abnormal.
5th Test: Replace R_3 with a good 1-kΩ resistor and note the current. When this is done, the circuit checks out normal. Each resistor drops two volts, and the circuit current measures 2 mA.

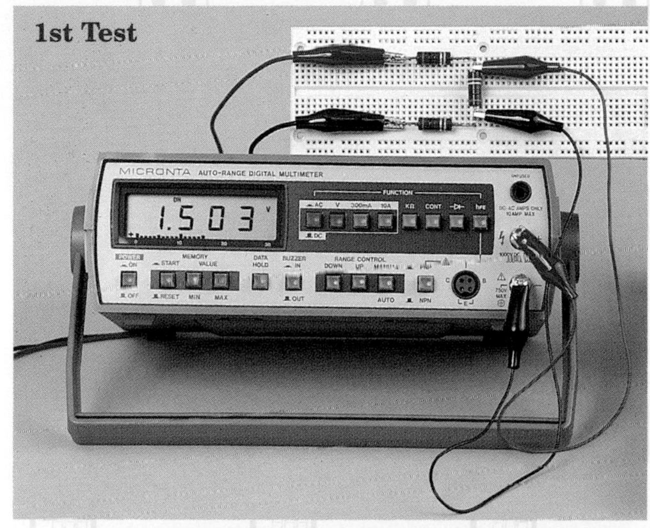
1st Test
MICRONTA AUTO-RANGE DIGITAL MULTIMETER
1.503 V

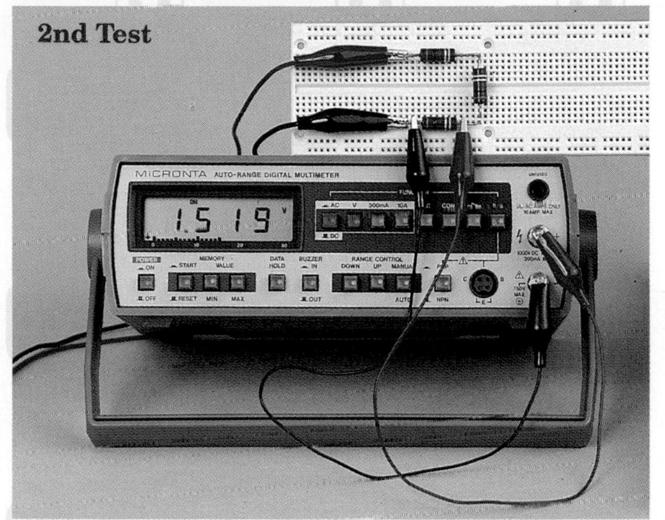
2nd Test
MICRONTA AUTO-RANGE DIGITAL MULTIMETER
1.519 V

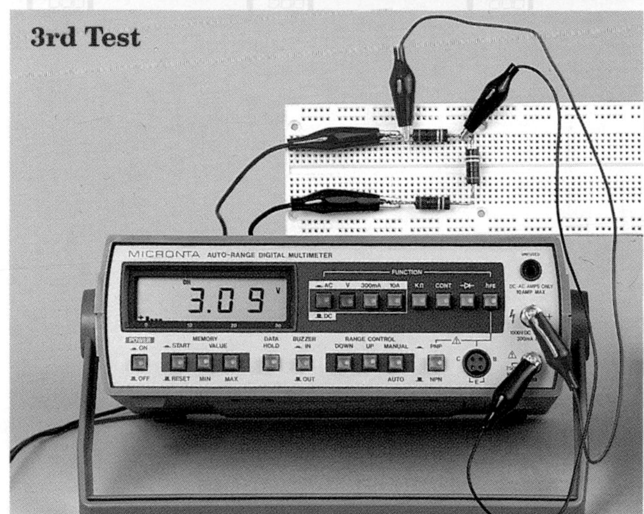
3rd Test
MICRONTA AUTO-RANGE DIGITAL MULTIMETER
3.09 V

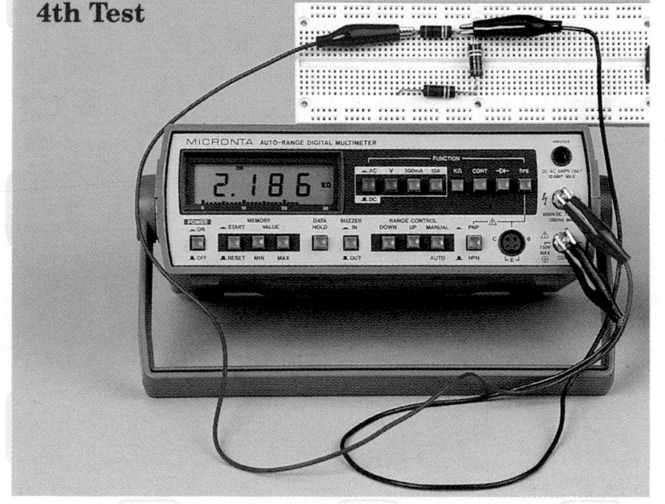
4th Test
MICRONTA AUTO-RANGE DIGITAL MULTIMETER
2.186 KΩ

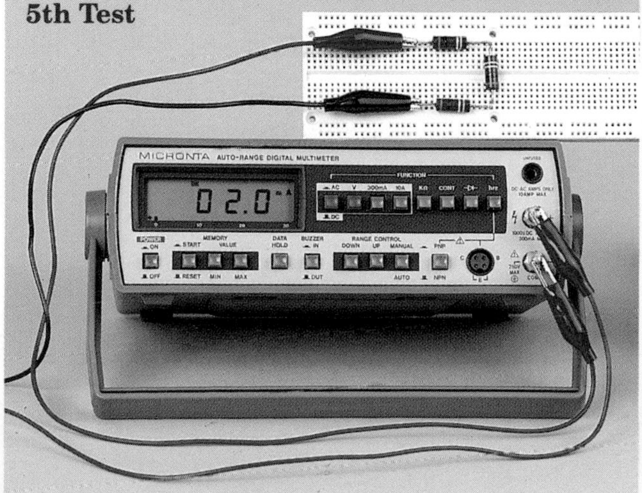
5th Test
MICRONTA AUTO-RANGE DIGITAL MULTIMETER
02.0

CIRCUIT 2

SYMPTOMS: The current through R_2 is normal, but the total current is low for this circuit. This suggests that the total R is higher than normal.

IDENTIFY initial suspect area: Branch 1 (points A to D) and branch 3 (points I to L) are suspect areas since branch 2 seems to be operating normally.

MAKE test decision: Check the voltage between points A and D (branch 1).

PERFORM 1st Test: Look up the test result. $V_{A\text{-}D}$ is 10 V, which is normal.

LOCATE new suspect area: Points A to B and C to D in branch 1 and all of branch 3 are the new suspect areas.

EXAMINE available data.

REPEAT analysis and testing:

2nd Test: The $V_{B\text{-}D}$. $V_{B\text{-}D}$ is 0 V, which is abnormal. This could result from a bad connection between points A and B or between points C and D.

3rd Test: Check $V_{A\text{-}B}$. $V_{A\text{-}B}$. It is 0 V, which is normal.

4th Test: Check $V_{C\text{-}D}$. $V_{C\text{-}D}$. It is 10 V, which is abnormal.

5th Test: Isolate branch 1 and check the resistance between points C and D. $R_{C\text{-}D}$ is infinite ohms, indicating an open. (NOTE: This meter shows a blinking 30.00 to indicate infinite ohms. Other digital multimeters may use different readouts to indicate infinite ohms.)

6th Test: Make a solid connection between points C and D and note the total current reading. It should be 10.4 mA if all the branches are operating properly. That is, the result of a good connection is that total current is approximately 10.4 mA. When this is done, the circuit operates normally.

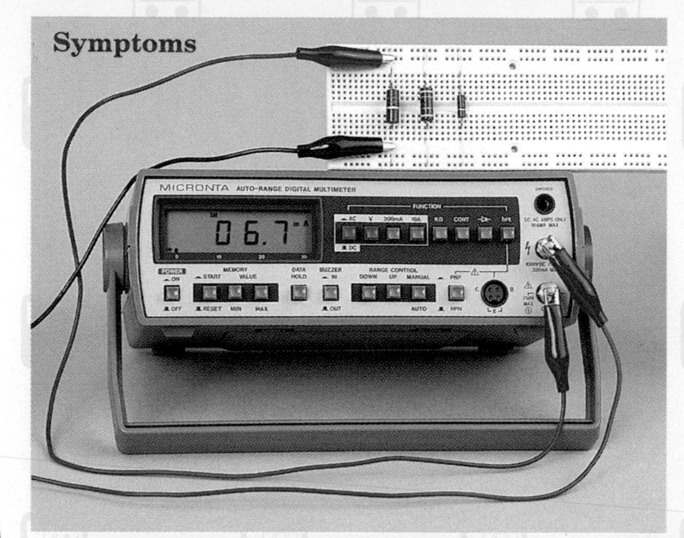

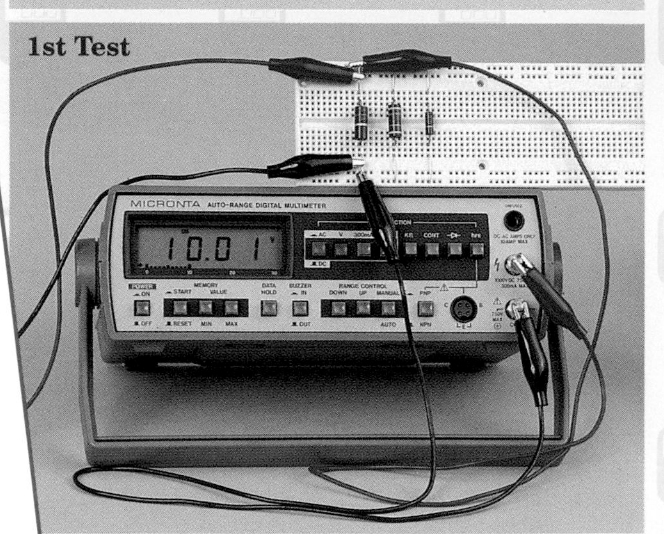

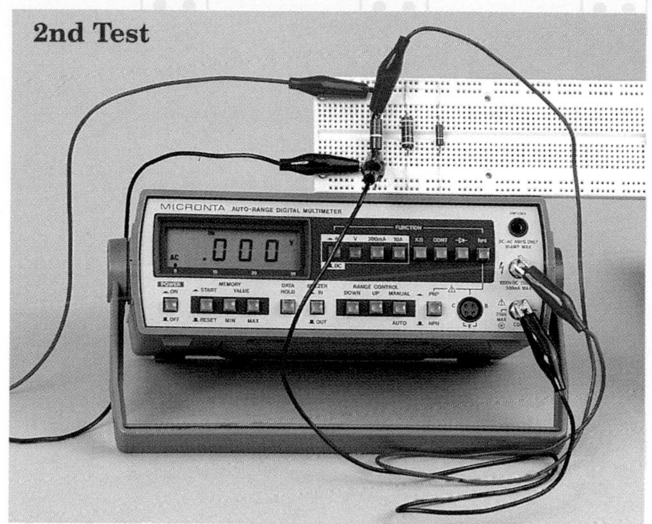

SPECIAL NOTE: You could have found the problem more quickly by calculating what the total current should be, then determining the amount of current missing was equal to the value branch 1 should have. Then you would have immediately known the problem was with branch 1. You still would have to perform isolation tests within that branch to locate the open.

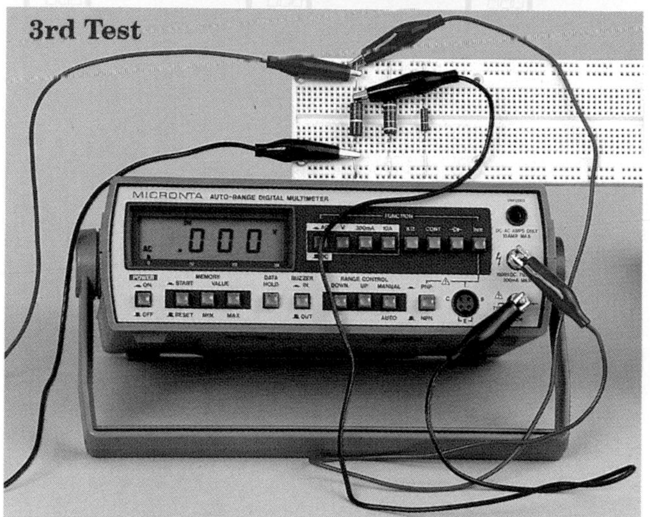

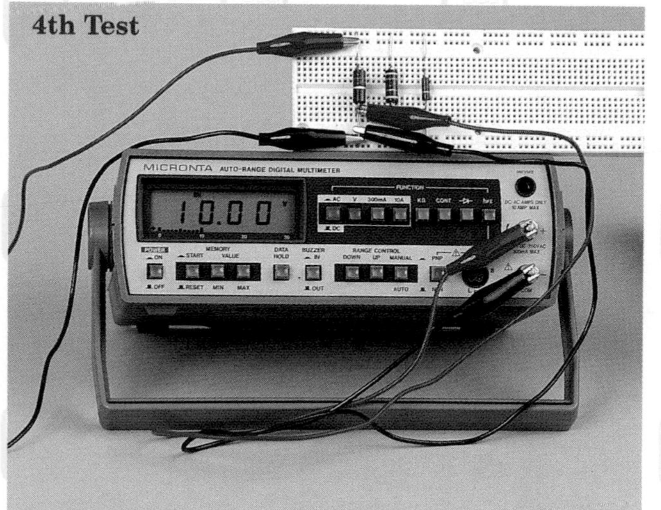

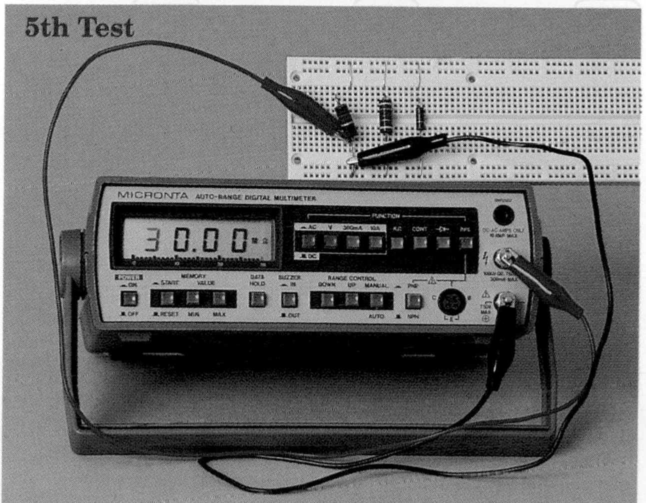

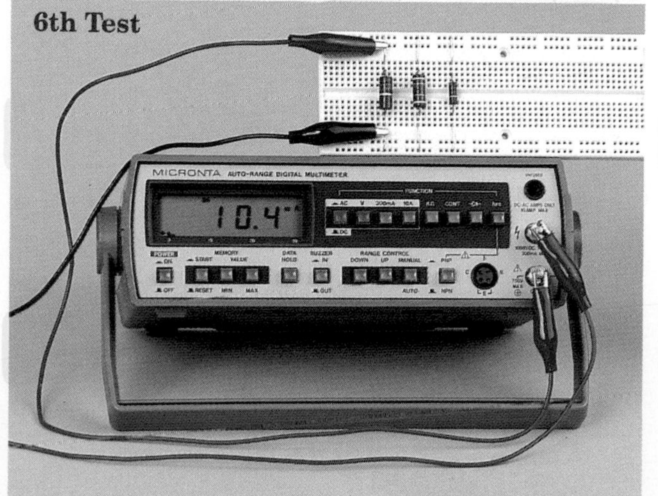

CIRCUIT 3

SYMPTOMS: The total current is a little low. Since current should be close to 1.7 mA, R_T is higher than it should be.

IDENTIFY initial suspect area: All resistors are initial suspect areas since if one or more resistor increased in value, the symptom would be present. However, the most likely suspect resistors and associated circuitry are related to R_2 and R_3, plus R_5 and R_6. This is because an open condition for any resistor branch would change the circuit resistance just enough to cause the symptom.

MAKE test decision: Check the resistance of R_2 (with it isolated).

PERFORM 1st Test: Look up the test result. R_2 is 10 kΩ, which is normal.

LOCATE new suspect area: R_3, R_5, and R_6.

EXAMINE available data.

REPEAT analysis and testing:

2nd Test: Measure the resistance of R_3. R_3 is 10 kΩ, which is normal.
3rd Test: Measure the resistance of R_5. R_5 is 10 kΩ, which is normal.
4th Test: Measure the resistance of R_6. R_6 is infinite ohms.
5th Test: Replace R_6 and note the circuit parameters. When this is done, the circuit operates properly.

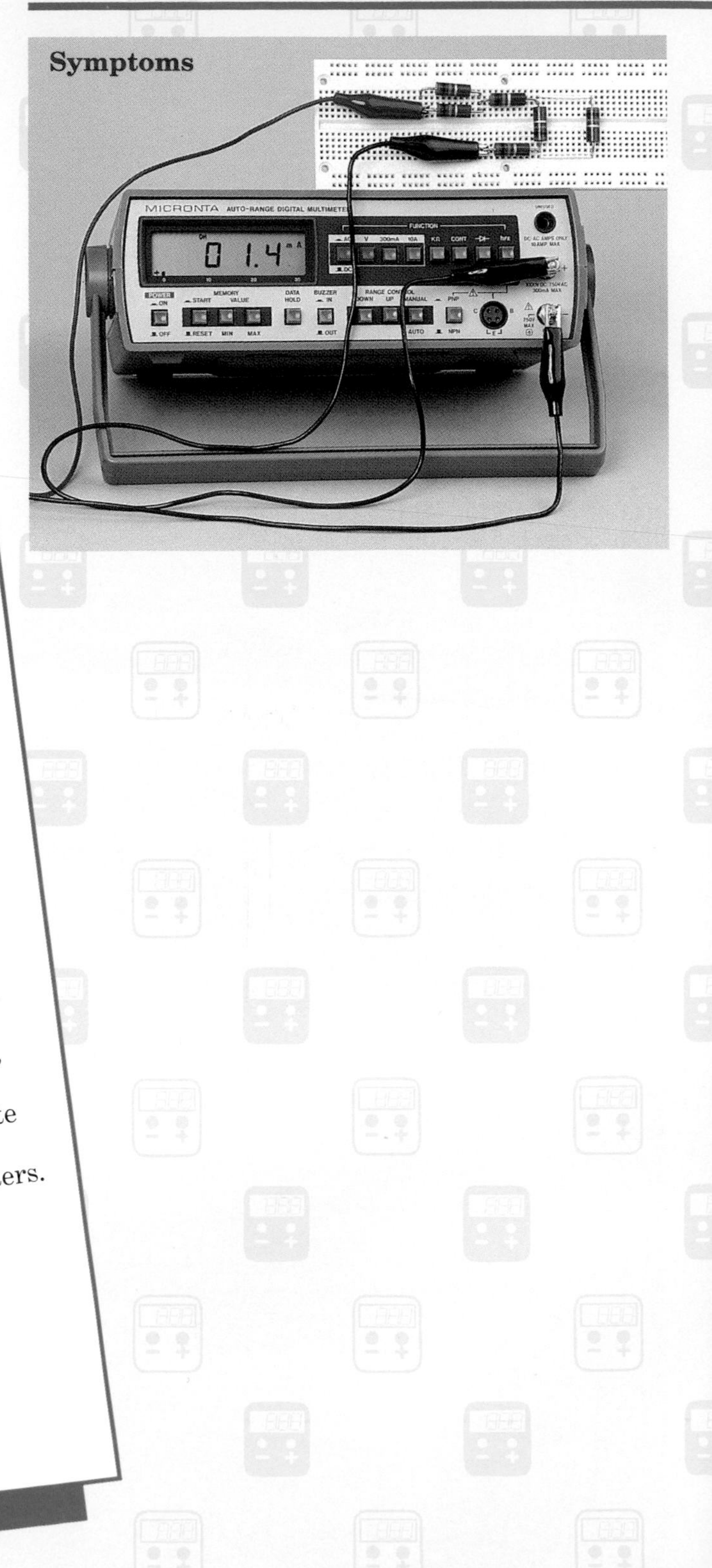

1st Test

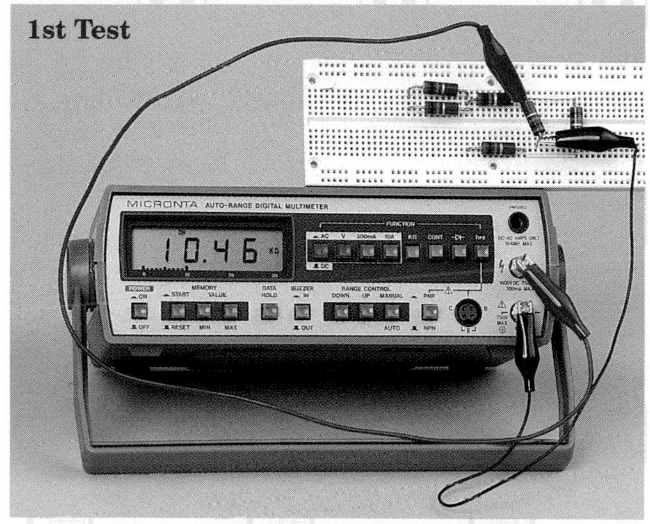

2nd Test

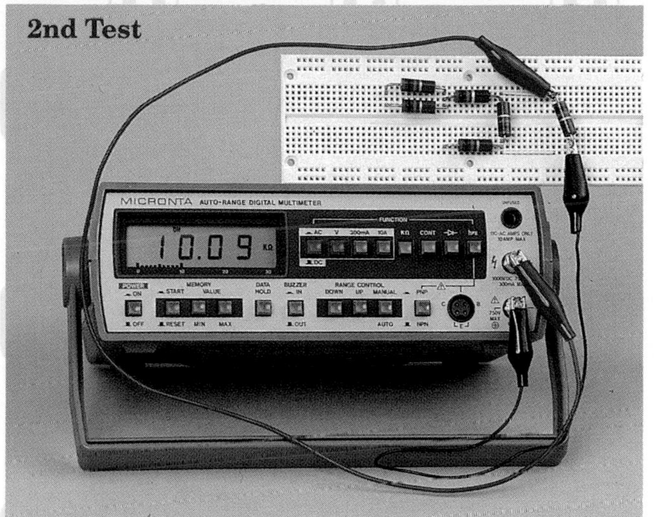

3rd Test

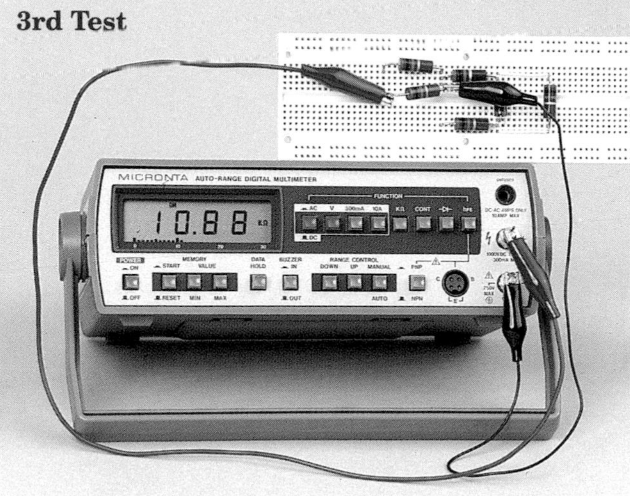

4th Test

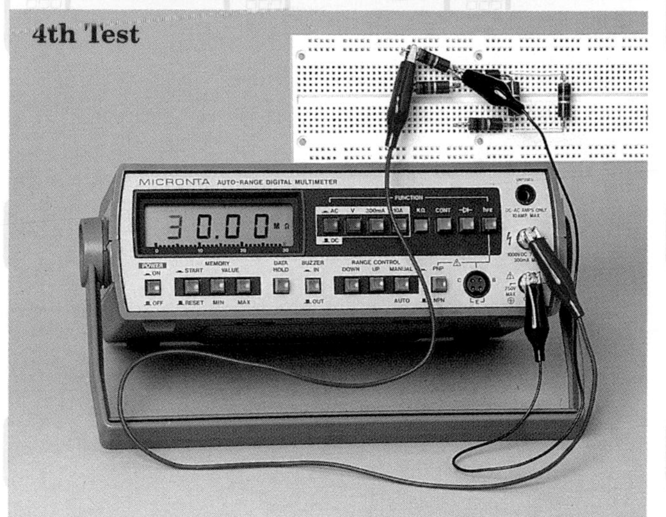

5th Test

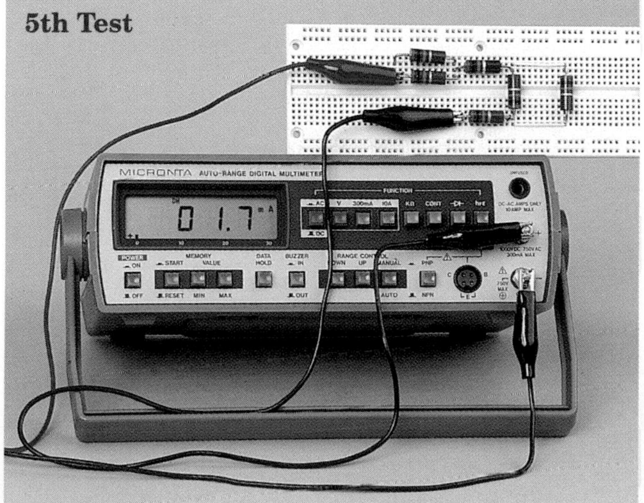

SPECIAL NOTE: Another way to check is to lift one end of each resistor and observe whether total current changed. The resistor you lift without having the total current change is the open one, since opening an open does not change the circuit conditions.

CIRCUIT 4

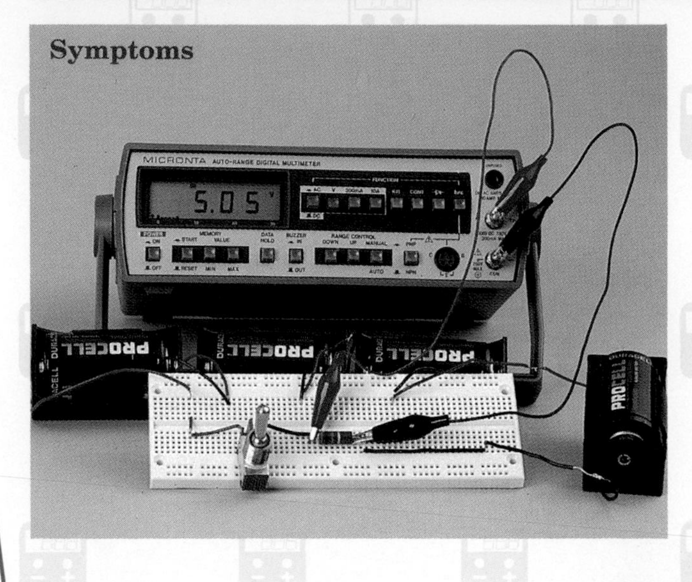

SYMPTOMS: With the switch closed, V_R is 5 volts. However, if all the 1.5 V cells were operating properly, it should be 6 volts. Since the resistor drops 5 volts with the switch closed there is circuit current and circuit continuity. This implies a worn cell in the group.

IDENTIFY initial suspect area: All the cells.

MAKE test decision: If we use the divide-and-conquer (splitting) technique, we can check the voltage from A to C or from C to E. This narrows the suspect area to two cells, either C_1 and C_2, or C_3 and C_4. Let's start with the first pair by checking voltage from point A to C with S_1 closed to cause circuit current. That is, the batteries are being tested under load.

PERFORM 1st Test: The result of measuring $V_{A\text{-}C}$ is 2 V, which is an abnormal reading for two 1.5 V cells in series.

LOCATE new suspect area: Cells C_1 and C_2.

EXAMINE available data.

REPEAT analysis and testing: Check the voltage across either C_1 or C_2 with S_1 closed.

2nd Test: Check voltage across C_1. V is 1.5 V, which is normal.

3rd Test: Check voltage across C_2. V is 0.5 V, which is abnormal, suggesting that cell C_2 is probably the bad cell.

4th Test: Replace cell C_2 with a good cell, and observe voltage across R with S_1 closed. V_R is 6 V, which is normal.

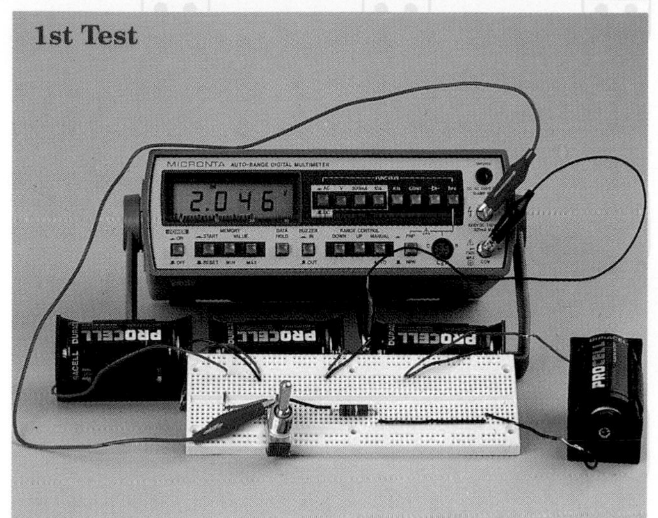
1st Test
MICRONTA
2.046
PROCELL

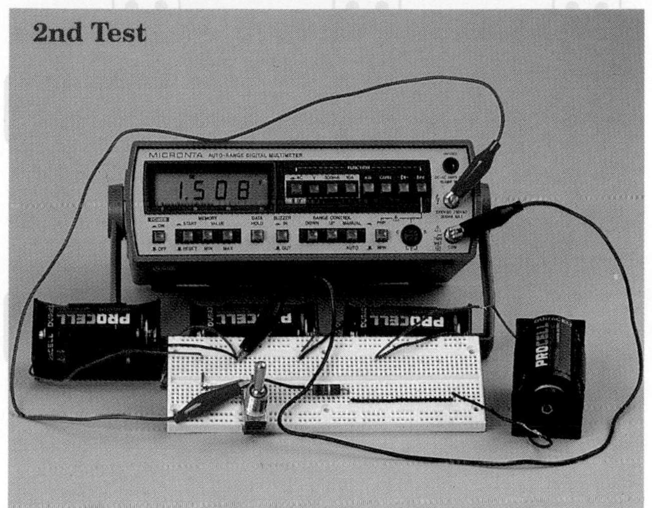
2nd Test
MICRONTA
1.508
PROCELL

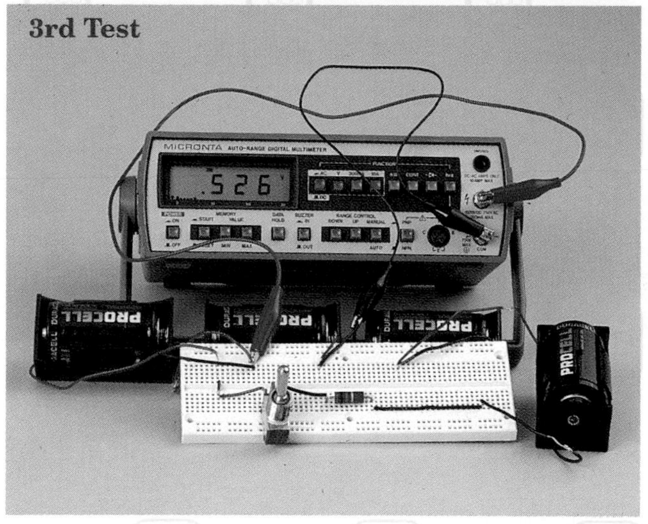
3rd Test
MICRONTA
.526
PROCELL

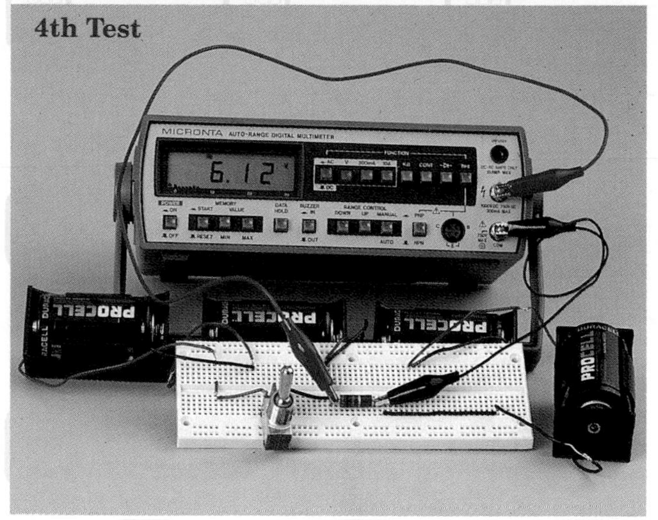
4th Test
MICRONTA
6.12
PROCELL

CIRCUIT 5

SYMPTOMS: Information. When the ohmmeter test probes are touched to zero the ohmmeter, there is not meter response. This indicates a bad meter, a lack of circuit continuity somewhere, or a dead battery.

IDENTIFY initial suspect area: The total circuit, since a discontinuity can be caused anywhere in the circuit.

MAKE test decision: Since the battery is the only item in the circuit having a definite shelf life, let's first check the battery. Measure V between points A and B with the probes open.

PERFORM 1st Test: The result of measuring $V_{A\text{-}B}$ is 1.5 V, which is normal. NOTE: With the test probes open, there is no load current drawn from the battery, thus it might measure normal without load and not completely good when under load. However, this tells us the battery is probably not totally dead, which it would have to be to match our symptom information.

LOCATE new suspect area: The remainder of the circuit, since discontinuity can be anywhere along the current path.

EXAMINE available data.

REPEAT analysis and testing: By checking continuity between points C and E (with test leads not touching), we can check the continuity of the two circuit elements, R_1 and the test-lead/test-probe.

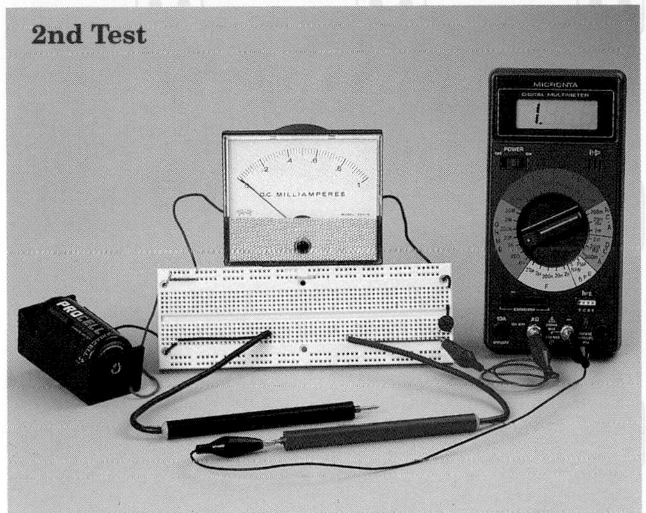

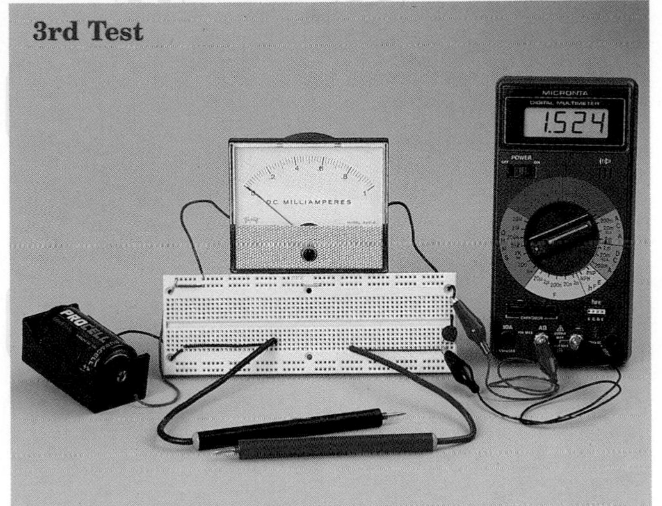

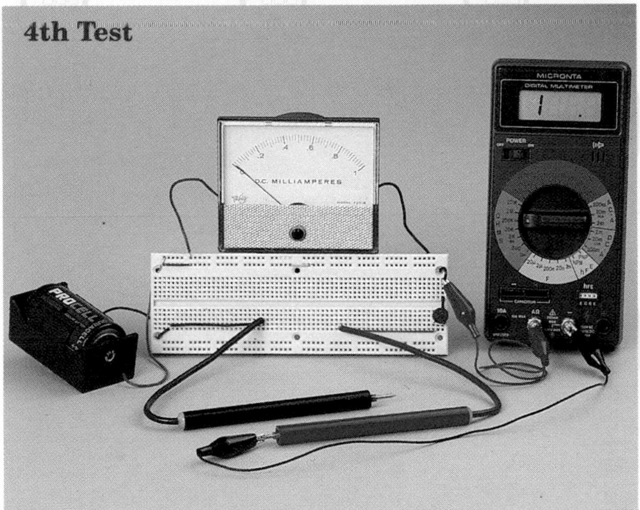

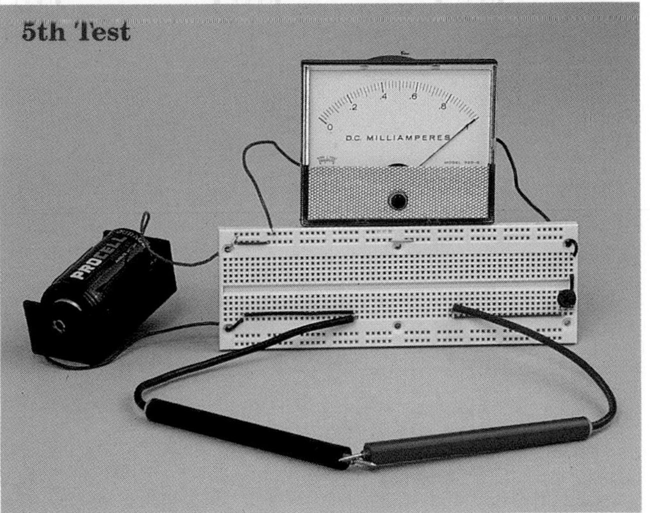

2nd Test: Check $R_{C\text{-}E}$. R is infinite ohms, indicating a problem. (NOTE: This meter shows "1" when measuring infinite ohms.)

3rd Test: Check $R_{C\text{-}D}$ to verify that R_1's resistive element or wiper arm is not causing the discontinuity. R is about 1500 ohms. However, moving the wiper arm causes R to change, which implies everything is normal.

4th Test: Check $R_{D\text{-}E}$. R is infinite ohms, indicating the discontinuity is between these points. A further check shows the test-lead wire is broken inside the test-probe plastic handle.

5th Test: Repair the wire connection to the test probe. Then, try to zero the ohmmeter. Now it works.

CIRCUIT 6

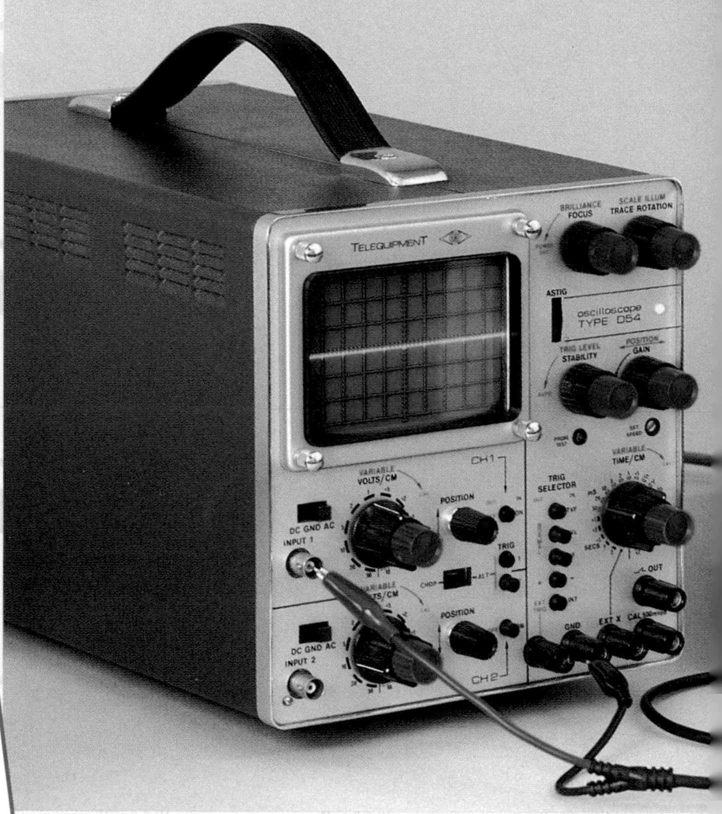

SYMPTOMS: Although there should be a waveform on the CRT screen, there is only a straight horizontal line. This implies there is a horizontal signal being fed to the CRT horizontal plates, but no vertical signal is reaching the CRT vertical plates.

IDENTIFY initial suspect area: The initial suspect area includes all elements involved in the vertical signal path to the CRT vertical plates. That is, the signal input to the vertical amplifier, the vertical amplifier, and the path from the vertical amplifier output to the CRT vertical plates.

MAKE test decision: We'll use the "gozinta-gozouta" technique and first check the input of the vertical amplifier to see if an input signal is present.

PERFORM 1st Test: Check the presence of an input signal for the vertical amplifier at point D. The input signal is present.

LOCATE new suspect area: The new suspect area is the remainder of the vertical signal path to the CRT plates.

EXAMINE available data.

REPEAT analysis and testing:

2nd Test: Check the signal "out of" the vertical amplifier at point E to ground. No signal is present, indicating a problem in the vertical amplifier circuitry. We have performed block-level troubleshooting. At this point, the vertical amplifier would require component-level troubleshooting.

3rd Test: After component-level troubleshooting has been completed and the needed repairs performed, test the system with the same signals. The result is that two sine-wave cycles are seen on the CRT screen.

Symptoms

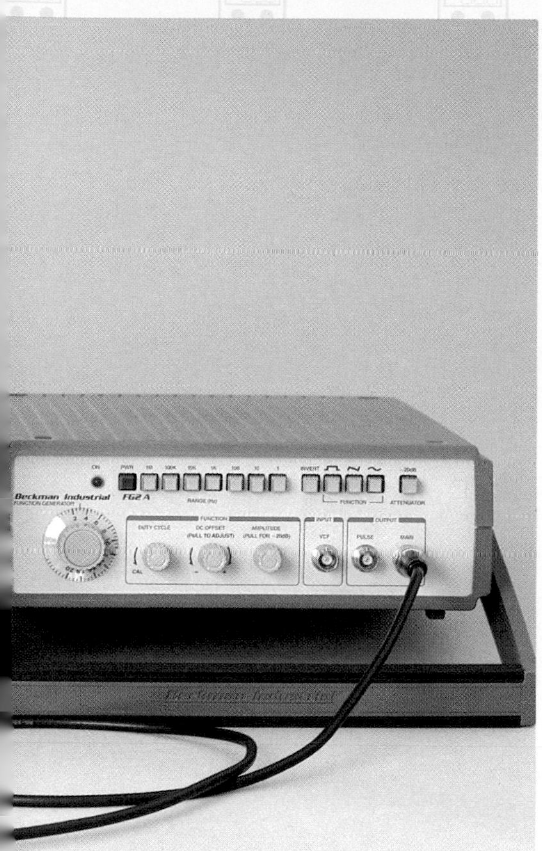

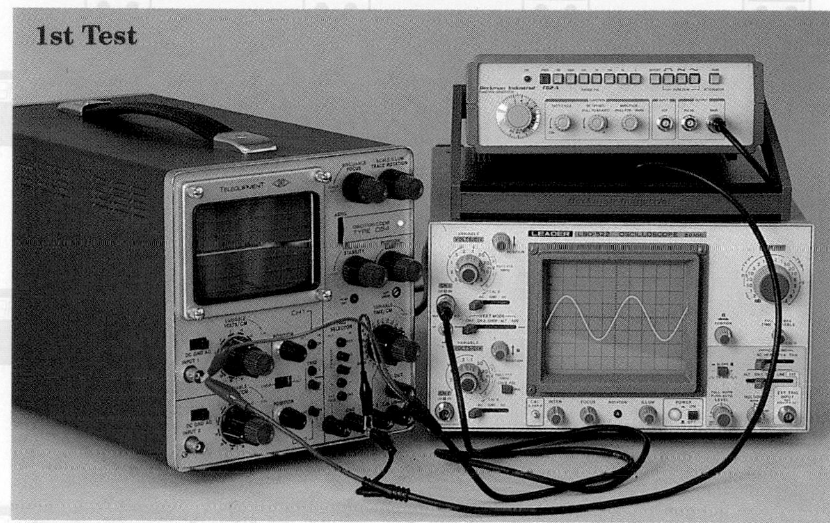

1st Test

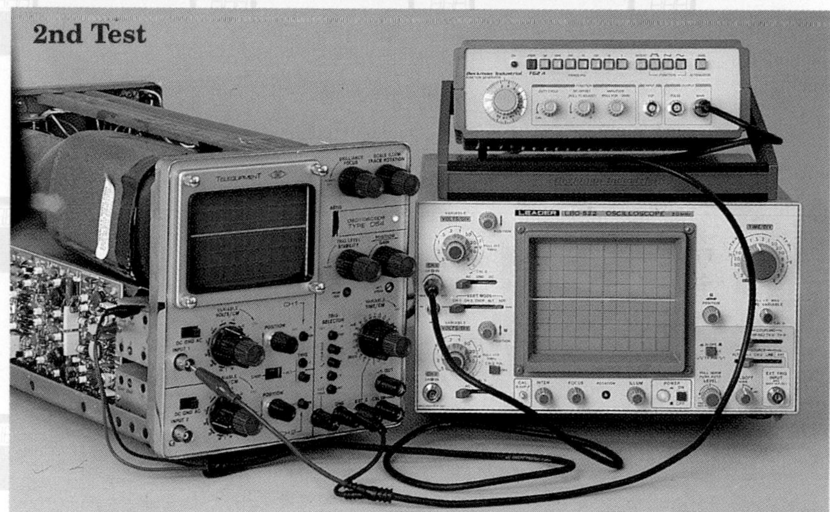

2nd Test

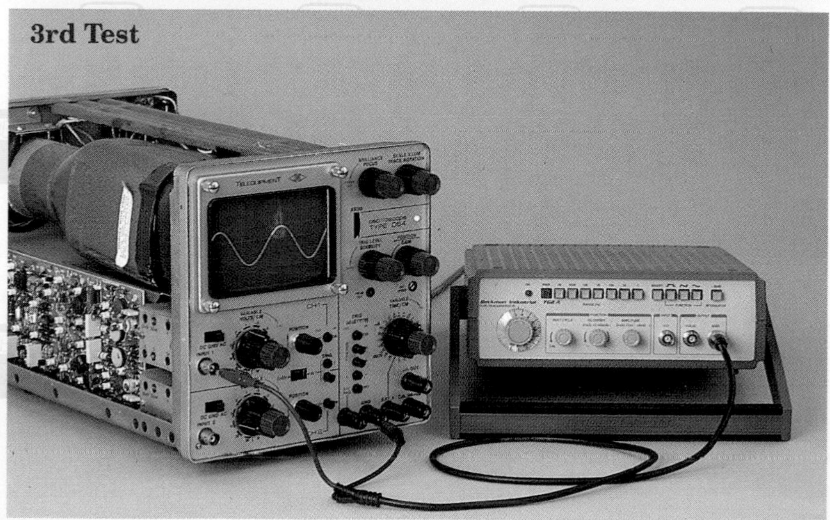

3rd Test

CIRCUIT 7

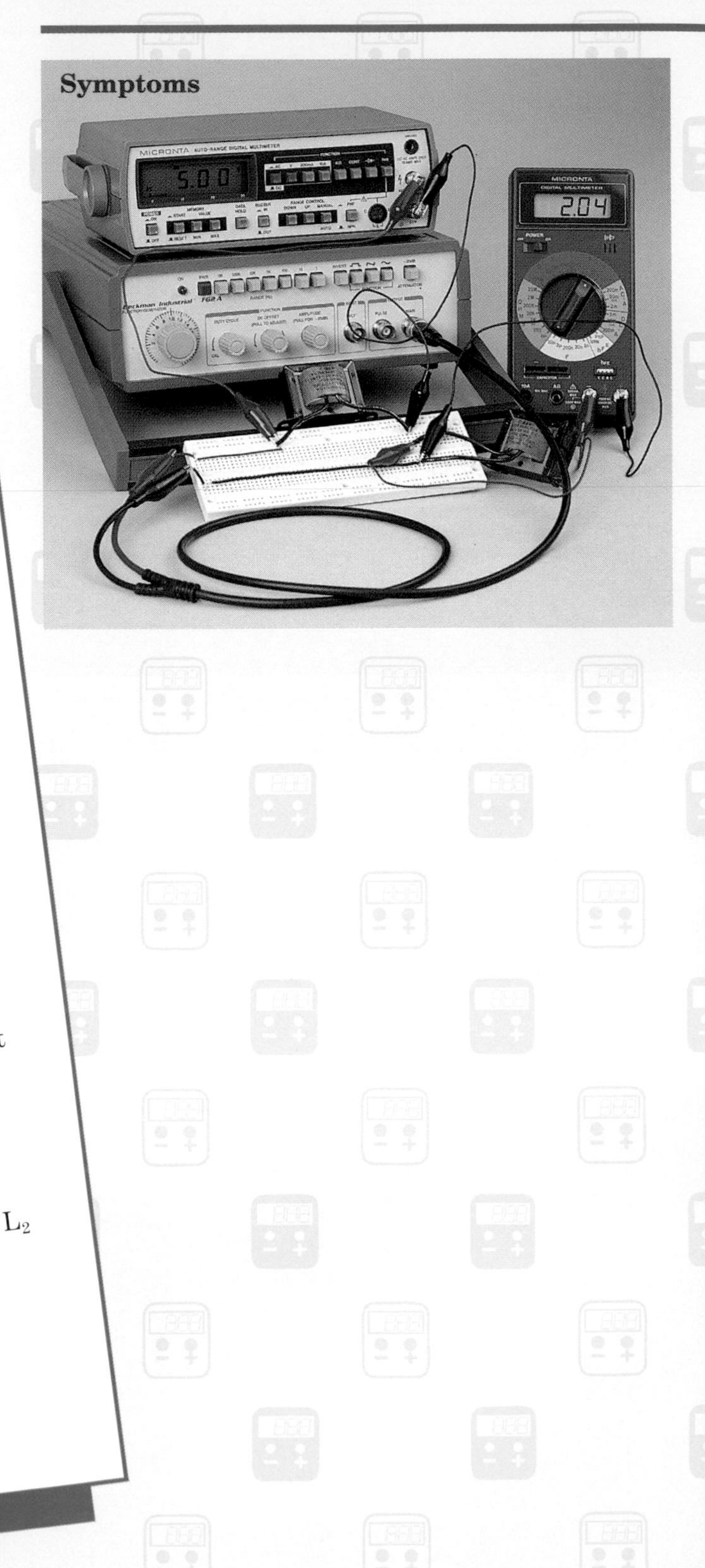

SYMPTOMS: The series circuit inductors are rated at the same value, and in a series circuit the current through both inductors should be the same. However, there is a significant difference in the I x X_L drops across the two inductors. This implies one inductor has a problem.

IDENTIFY initial suspect area: Both inductors until further checks are made.

MAKE test decision: It is less likely that an inductor would increase rather than decrease in inductance (due to shorted turns or change in magnetic path). So let's start by examining the inductor with the lowest voltage drop (L_1). Our assumption is L_2 has remained at the rated value of L, while L_1 may have decreased. This would fit the symptom information regarding L_1's voltage drop being low.

PERFORM 1st Test: Check the dc resistance of L_1 versus its rated value to see if there might be a hint of shorted turns. (NOTE: Source is removed from the circuit.) R is 297 Ω and the rated value is 400 Ω, indicating the possibility of some shorted turns.

LOCATE new suspect area: L_1 is now a strong suspect area. But we still need to check L_2's parameters to compare and verify our suspicions about L_1.

EXAMINE available data.

REPEAT analysis and testing: A resistance check of L_2 might be necessary.

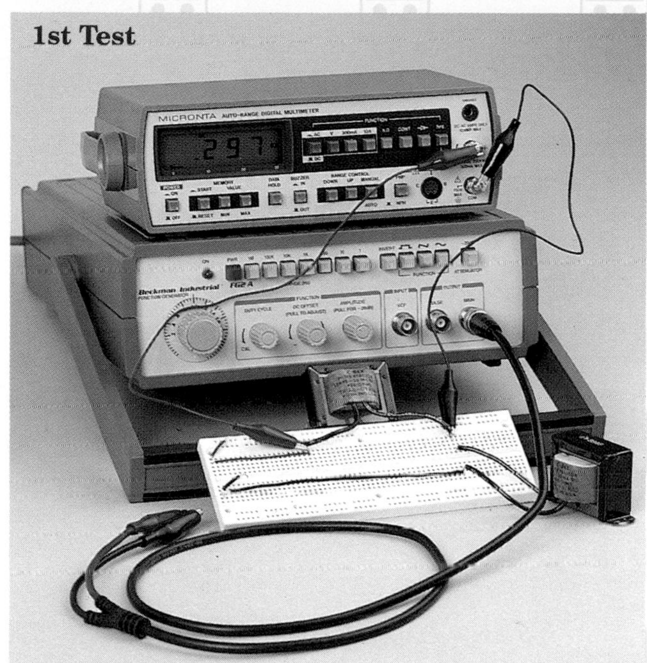

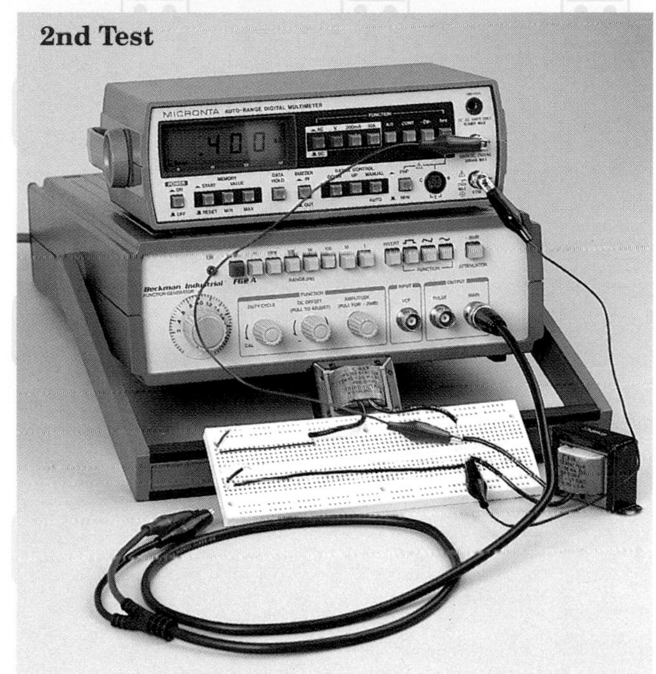

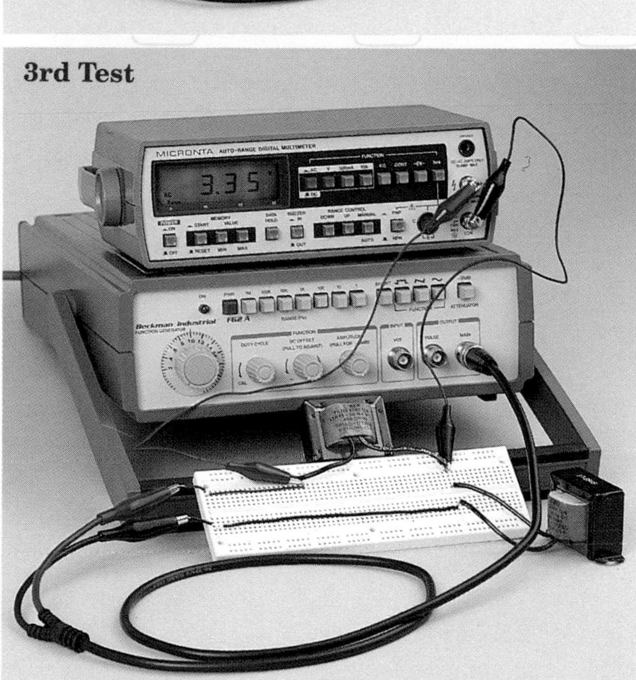

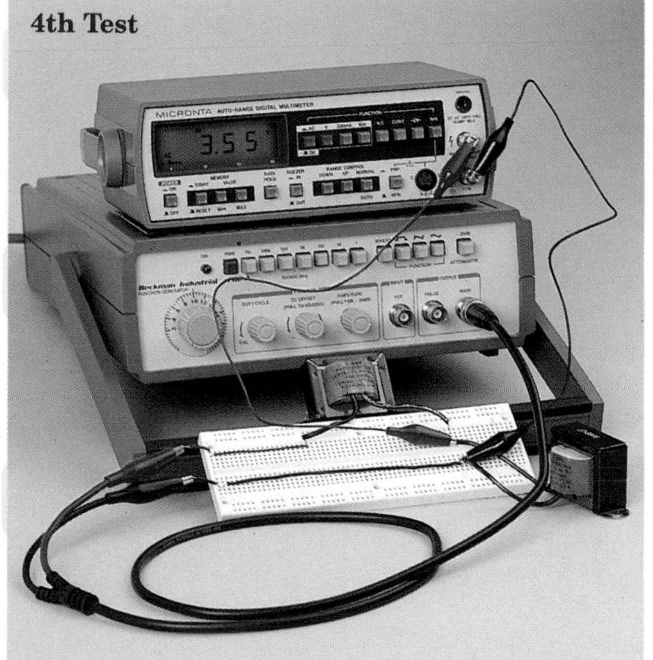

2nd Test: Check $R_{B\text{-}C}$. $R_{B\text{-}C}$ is 401 Ω, which agrees with the rating.

3rd Test: Replace L_1 with the appropriate rated inductor. Then check the voltage across L_1. $V_{C\text{-}D}$ is now close to 3.5 V, which is half of the source voltage. This is normal for this circuit with equal inductor values.

4th Test: Check the voltage across L_2 to verify that it is approximately equal to the voltage across L_1. When checked, the result is also about 3.5 V. The trouble was L_1 has shorted turns. Replacing L_1 restored the circuit to normal.

CIRCUIT 8

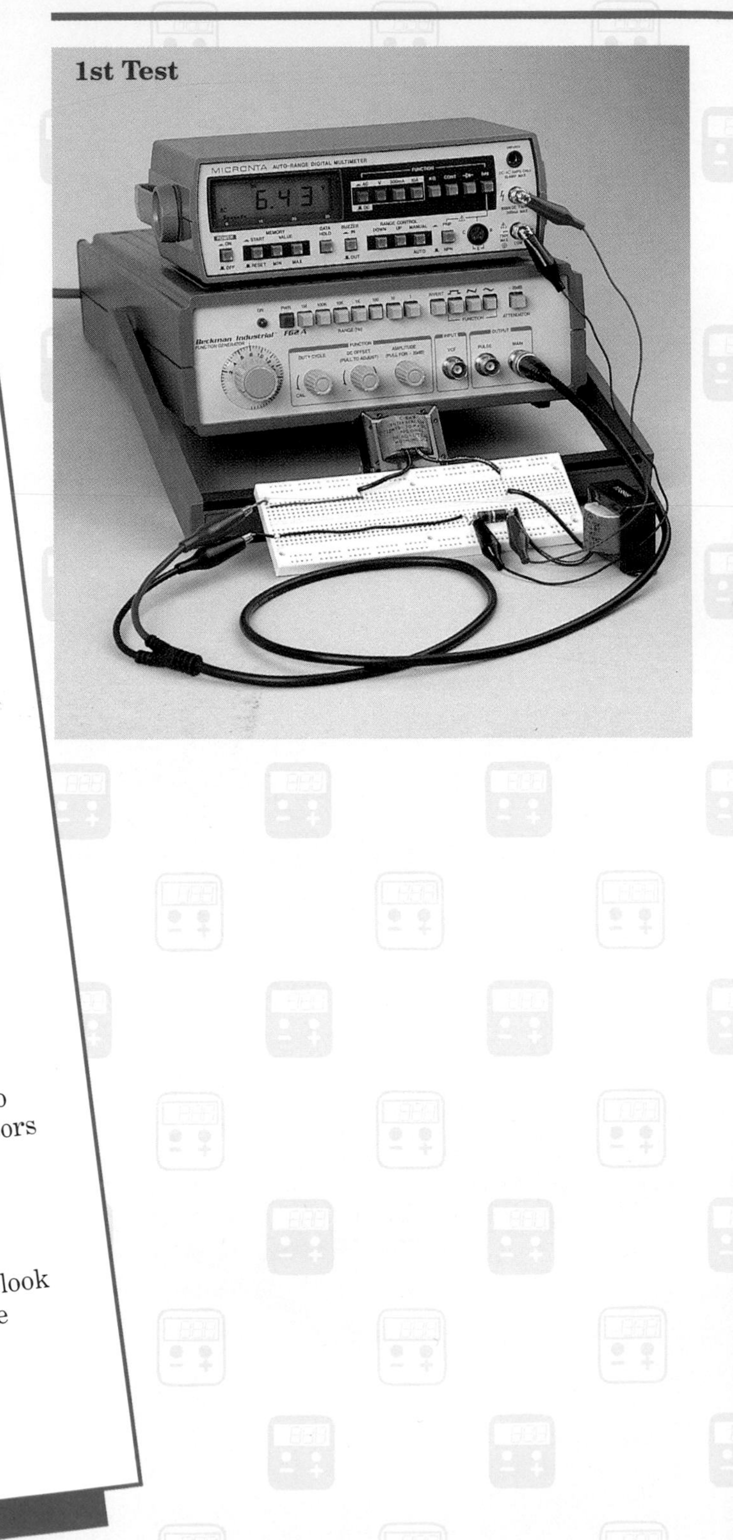

SYMPTOMS: The voltage drop across the resistor is slightly higher than it should be. This could be due to an inductor that has decreased in value due to shorted turns or an R that has increased or opened since, at this point, we don't know how high the voltage is across the resistor.

IDENTIFY initial suspect area: All components (resistor and two inductors) are in the initial suspect area.

MAKE test decision: Since inductors typically open more frequently than they short and since we don't know how high the voltage is across the resistor, it makes sense to check the voltage across R_1.

PERFORM 1st Test: V_{R_1} is 6.4 V. If everything were normal, the resistor would drop 5.9 V, and each inductor would drop about 2.6 V, since their inductive reactances at this frequency would be close to 4.35 kΩ each. The total X_L is about 8.7 kΩ and the rated R is 10 kΩ. Z would then equal about 13.3 kΩ, and circuit current would be approximately 0.59 mA. Thus, the vector sums of the resistive and inductive voltage drops would equal approximately 7.8 V.

LOCATE new suspect area: The new suspect area stays the same as the initial suspect area since we do not have enough information to eliminate the inductors or the resistor.

EXAMINE available data.

REPEAT analysis and testing: It would be good to look at the inductor voltage drops so we can draw some conclusions.

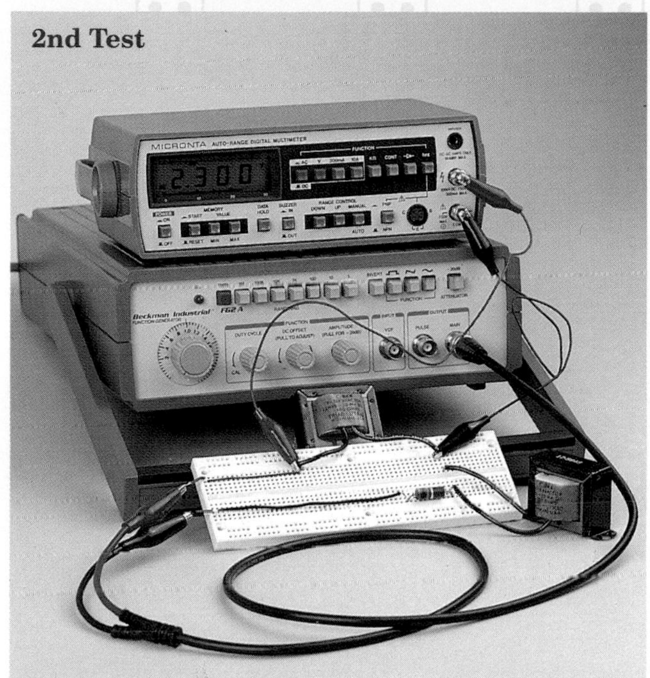

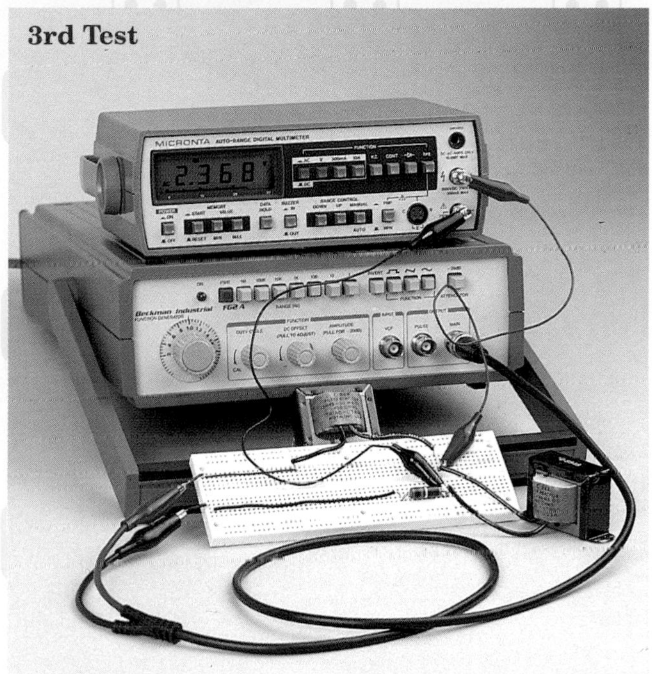

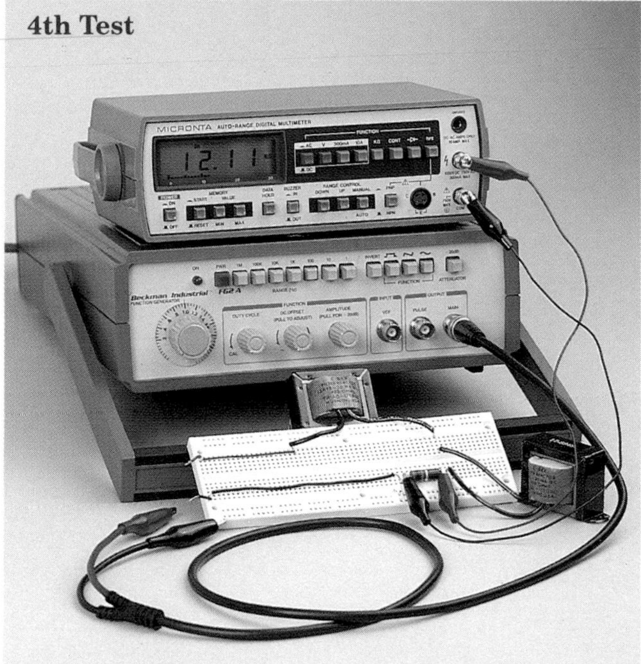

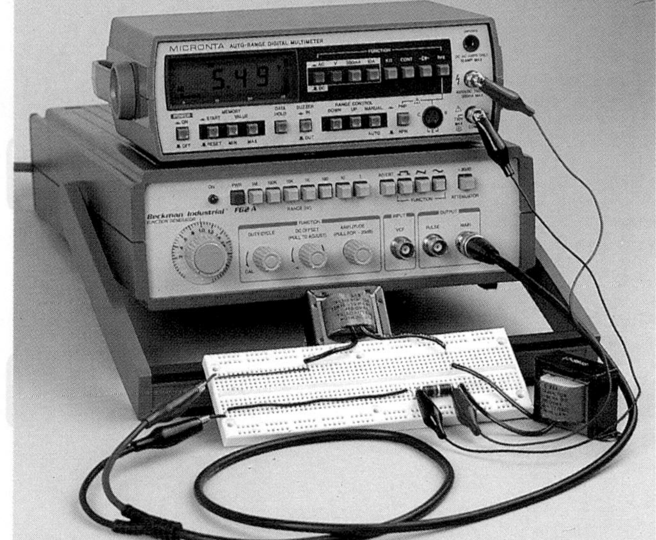

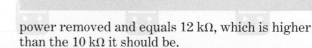

2nd Test: Measure V_{L_1}. V_{L_1} is close to 2.3 volts, which is slightly lower than normal.

3rd Test: Measure V_{L_2}. V_{L_2} is close to 2.36 volts, which is slightly lower than normal.

4th Test: Measure R_1. R_1 is measured with power removed and equals 12 kΩ, which is higher than the 10 kΩ it should be.

5th Test: Since R_1 resistance is high replace R_1 with a new 10-kΩ resistor. After replacing the resistor, all the circuit parameters become normal.

CIRCUIT 9

SYMPTOMS: All we know is that the voltage across C_2 is lower than it should be. If things were normal, voltage across C_1 would be 3.3 volts; the voltage across C_2 would be 3.3 volts; and the voltage across C_3 would also be 3.3 volts. Since Q is the same for all capacitors, the voltage distributions should be as described above. Also, we could predict that if C_2 were opened, its V would be 10 V (which does not agree with our symptom information); if C_2 were shorted or one of the other capacitors were opened, V_{C_2} would be 0 V; and if V_{C_2} were leaky, V would also be close to 0 volts.

IDENTIFY initial suspect area: C_2 is the initial suspect area. However, we can't disregard the other capacitors since we have limited information.

MAKE test decision: Let's first check C_2's voltage to see where we are.

PERFORM 1st Test: Check V_2. V_{C_2} is close to 0 volts, which is lower than normal.

LOCATE new suspect area: If either one of the other capacitors were open, the voltage for C_2 would still be 0 V. If either one of the other two capacitors were leaky or shorted, C_2's voltage would be higher than normal.

EXAMINE available data.

REPEAT analysis and testing: A check of V_{C_1} or V_{C_3} might be necessary.

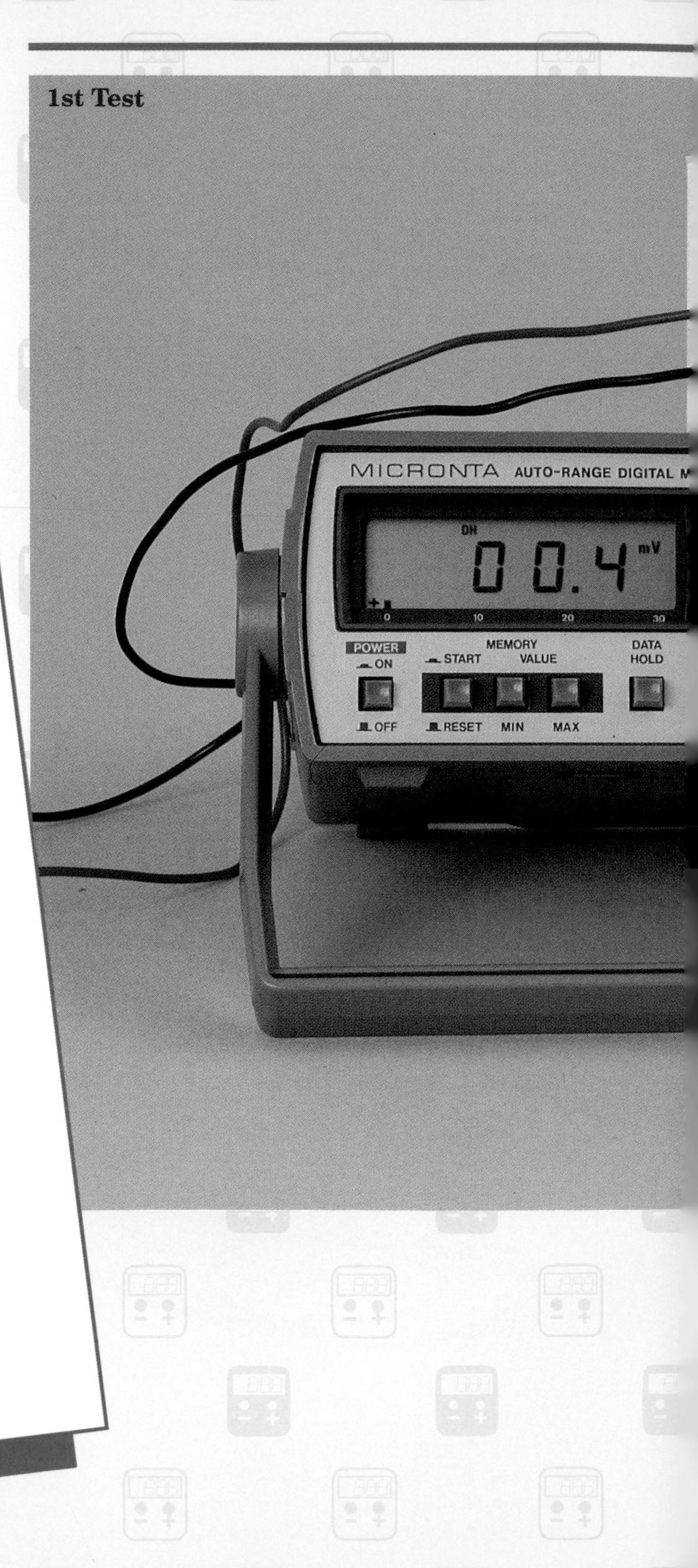

1st Test

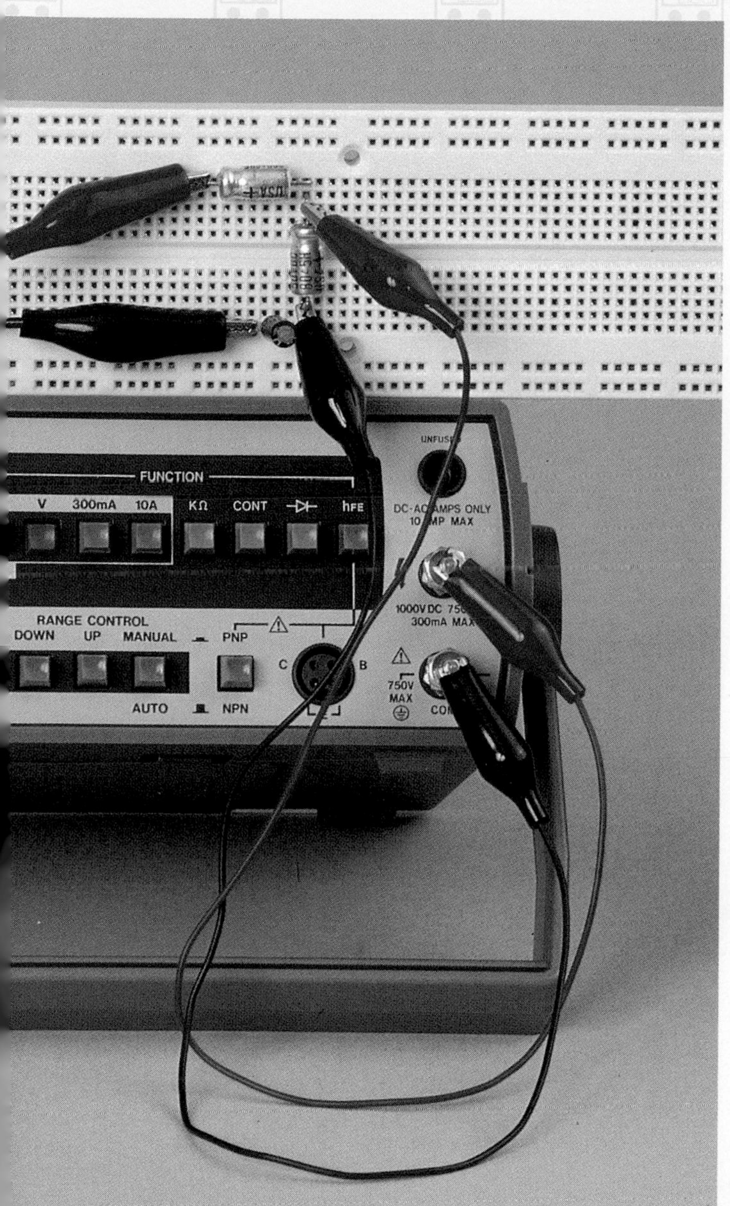

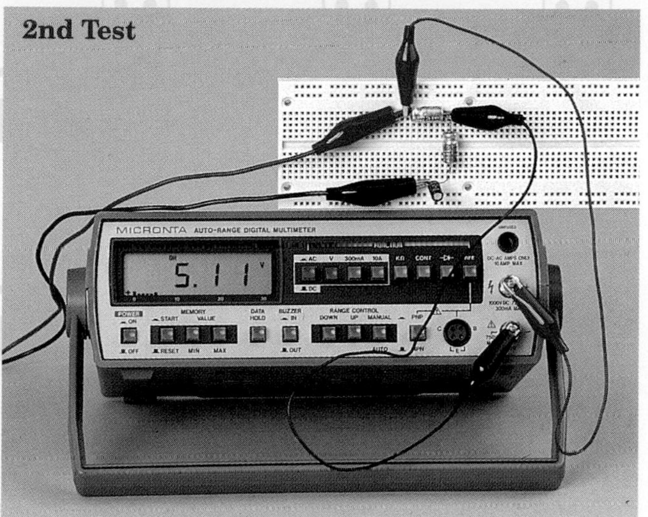

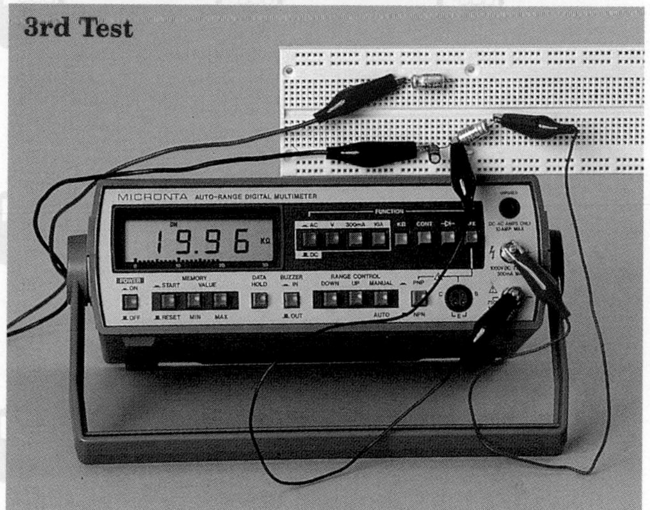

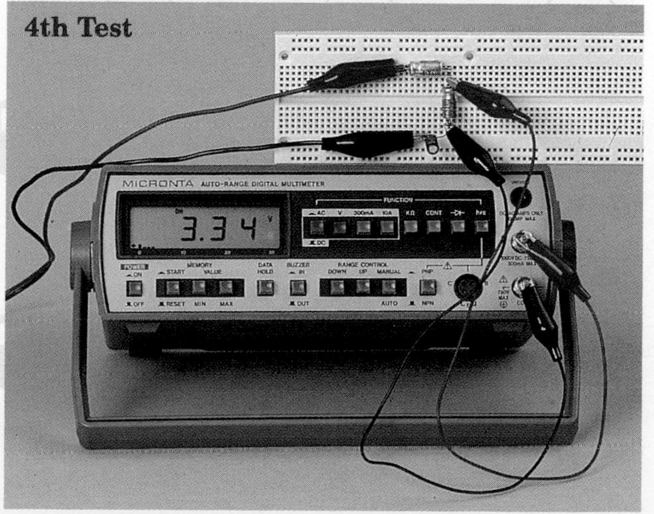

2nd Test: Check V_{C_3}. V_{C_3} is about 5 V, which is higher than normal, but C_3 is certainly not open or shorted. It looks like C_2 might be leaky. Let's take a look at C_2.

3rd Test: Lift one of C_2's leads and check the capacitor's dc resistance. The test reveals a lower than normal R.

4th Test: Replace C_2 and check the circuit operation. The circuit checks out normal. C_2 was leaky.

CIRCUIT 10

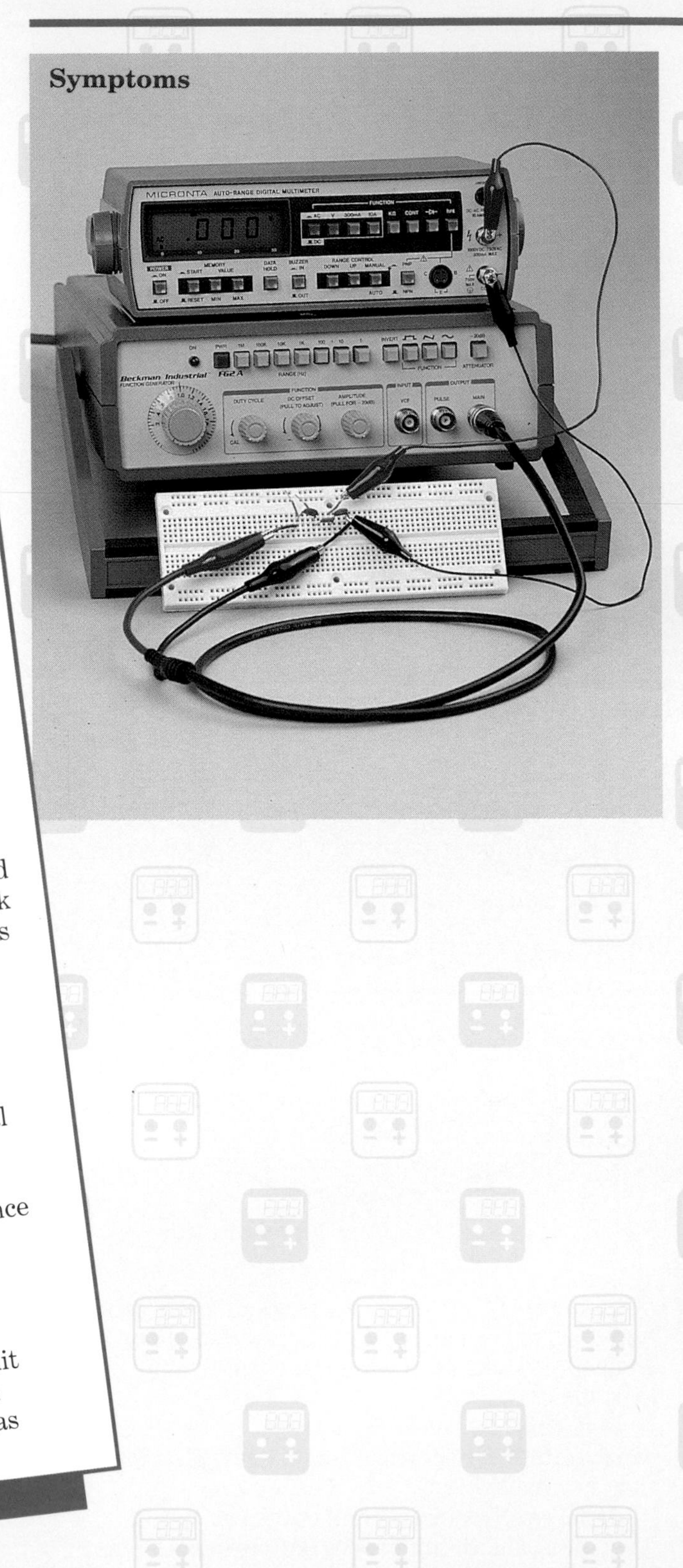

SYMPTOMS: The capacitive voltage divider is not working correctly. The X_C's of the equal value capacitors should be equal. The I x X_C drops should all be approximately equal. At test point 1, the voltmeter should read about 2.5 V; at test point 2 about 5.0 V; and at test point 3 about 7.5 V.

IDENTIFY initial suspect area: If C_1 were shorted, the V at test point 1 would be 0 V. The V at test point 2 would be 3.75 V rather than 2.5 V. If C_3 were shorted, the V at switch position 1 would be 3.75 V not 0 V. If C_1 were open, the V at switch position 1 would be 7.5 V. If C_3 were open, the V at switch position 2 would be 0 V. The normal voltage at switch position 3 is 7.5 V (the source voltage), which is what is measured. Without making our first test, we have narrowed the initial suspect area to C_2. If it were open the circuit would operate similar to the symptom descriptions.

MAKE test decision: Bridge the suspected open capacitor with a known good one of the same ratings and note any changes in circuit operation. We'll try the quick bridging technique. Caution: Make bridging connections with the power off. Then turn power on and make the necessary observations.

PERFORM 1st Test: Bridge C_2 while monitoring voltage at switch position 2. When this is done, V measures about 4.8 V, close to the nominal 5 V normal operation.

LOCATE new suspect area: This is not necessary since we've found our problem.

EXAMINE available data.

REPEAT analysis and testing: Verify that the circuit operates normally by checking the voltage at each switch position **(2nd, 3rd, and 4th Tests)**. C_2 was open.

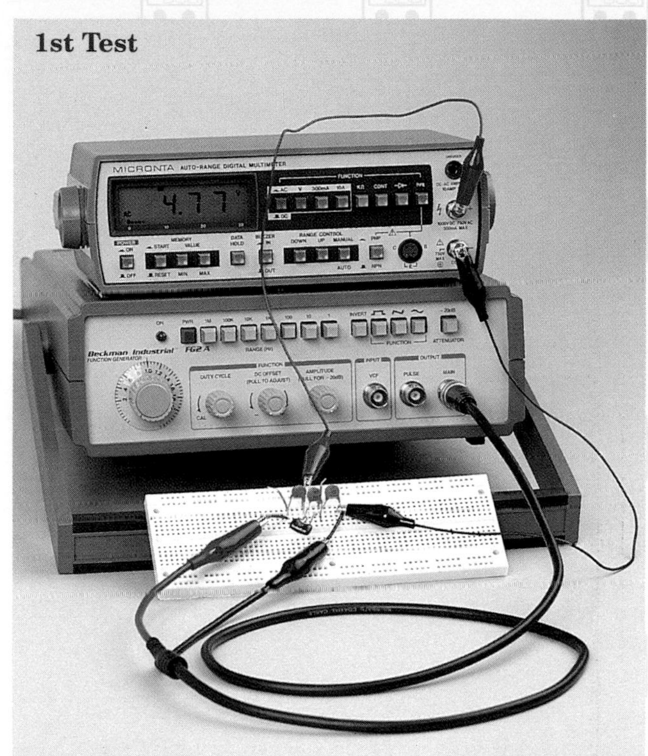

1st Test

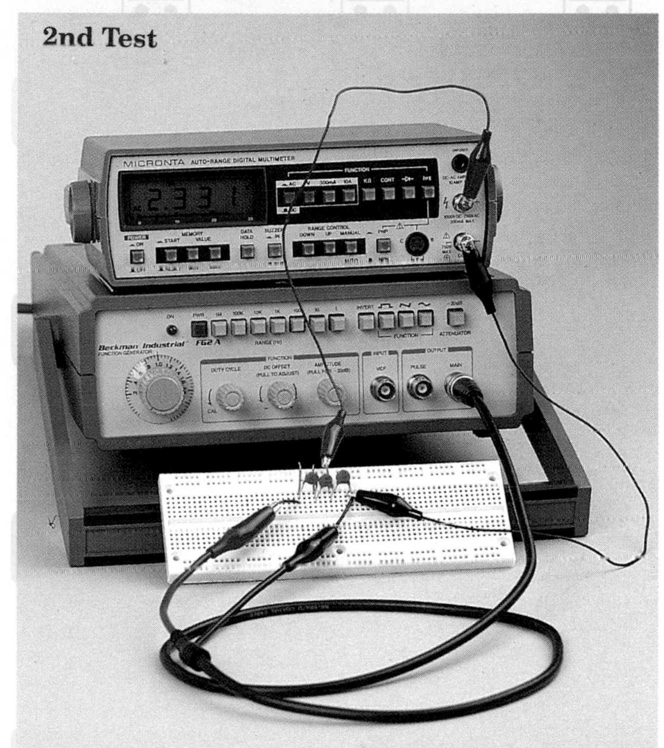

2nd Test

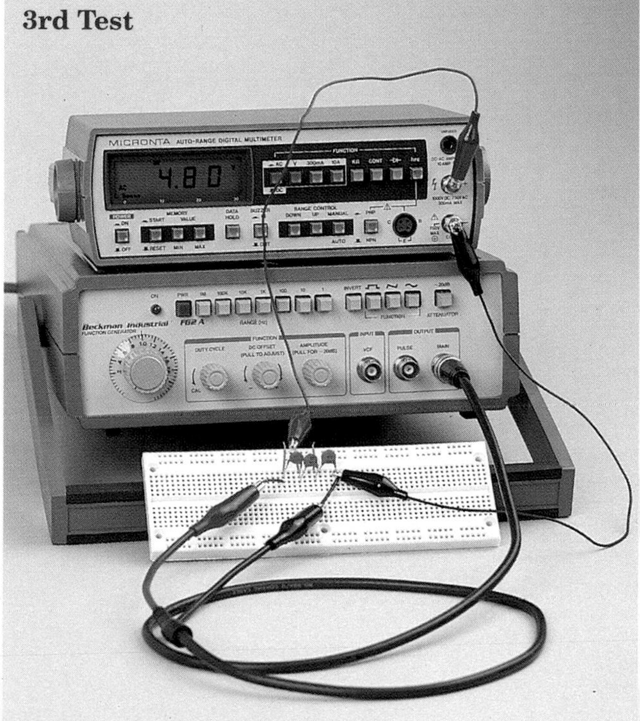

3rd Test

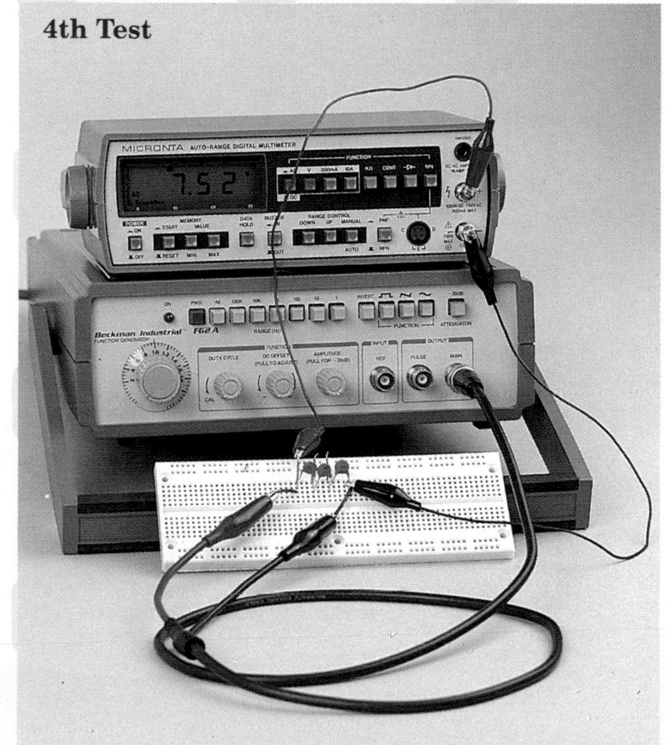

4th Test

CIRCUIT 11

SYMPTOMS: The voltmeter measures approximately 0.057 volts. If the circuit were operating as prescribed by the diagram, the voltmeter should measure about 0.112 V. This implies circuit current is lower than it normally would be. Thus, circuit impedance is higher than normal.

IDENTIFY initial suspect area: Between points C and D, the total inductive reactance of the parallel inductors should be about 9.5 kΩ. Between points B and C, the net capacitive reactance of the parallel capacitors should be about 5.8 kΩ. The net resistance between points A and B should be about 5 kΩ. If the circuit were operating correctly, the total impedance would be about 6.2 kΩ. However, it's acting like the circuit has about 14 kΩ total impedance.

MAKE test decision: The upward shift in Z could be caused by any R, C, or L opening or any C or L shorting. We'll first look at the resistors. Let's look at V_{A-B}.

PERFORM 1st Test: Test V_{A-B}. V_{A-B} is about 3.11 volts, again indicating lower than normal circuit current. We know the Rs are not shorted or the voltage would read zero volts. Also, since we know the circuit current is about 0.57 mA, then 0.57 mA times the normal equivalent R of approximately 5 kΩ comes close to that value. The resistors appear to be acting normally.

LOCATE new suspect area: The Rs have been eliminated. Thus, the reactive components are the new suspect area.

EXAMINE available data.

REPEAT analysis and testing: Let's look at the inductive branches isolating, as appropriate.

Symptoms

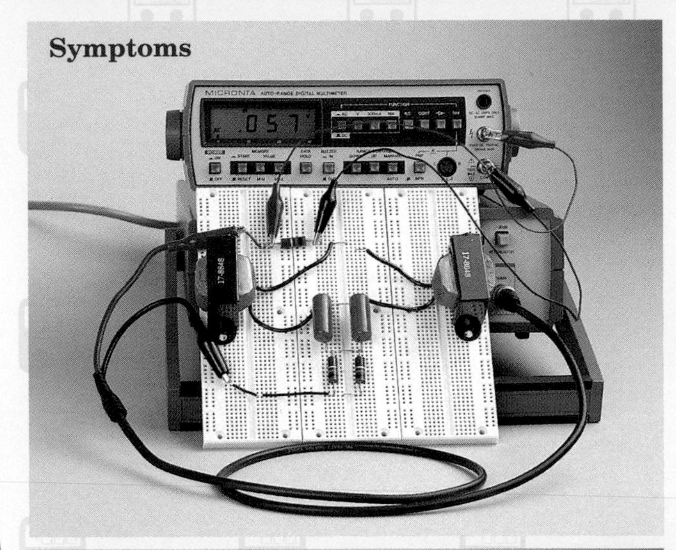

1st Test

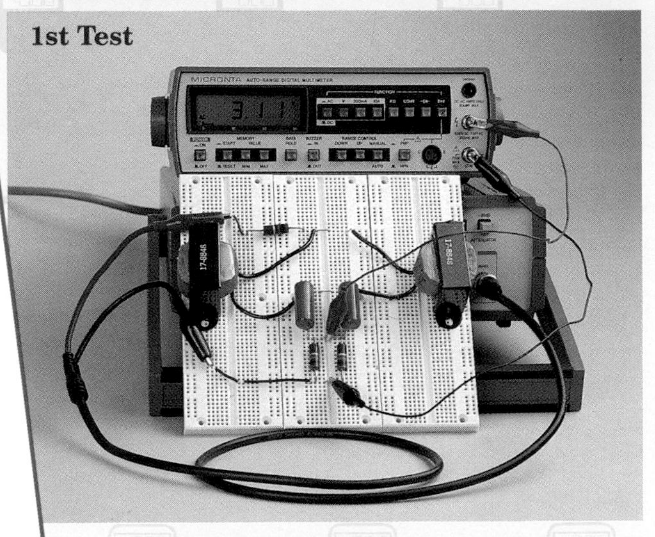

2nd Test: Check $V_{C\text{-}D}$. $V_{C\text{-}D}$ is about 11 volts. An open inductor could cause this situation so let's check that out.

3rd Test: Check $R_{C\text{-}D}$. $R_{C\text{-}D}$ is about 400 Ω. This should be about 200 Ω with the parallel circuit situation.

4th Test: Check R_{L_1}. R_{L_1} is about 400 Ω, which is normal.

5th Test: Check R_{L_2}. R_{L_2} is infinite ohms, which is an open coil.

6th Test: Change L_2 and check the circuit operation. It's normal.

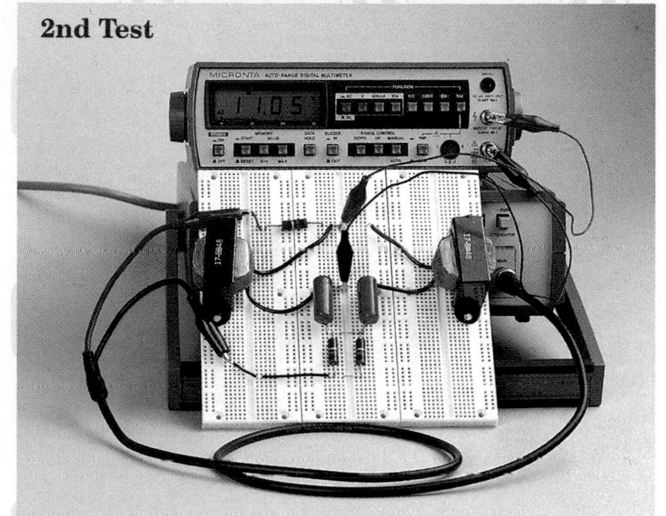
2nd Test

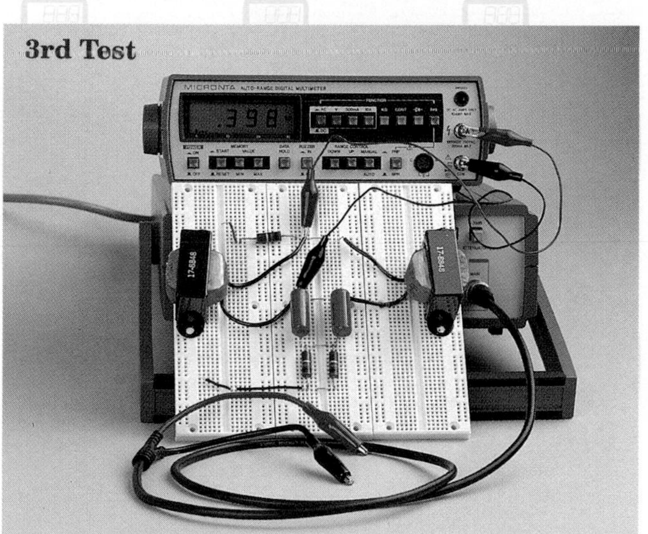
3rd Test

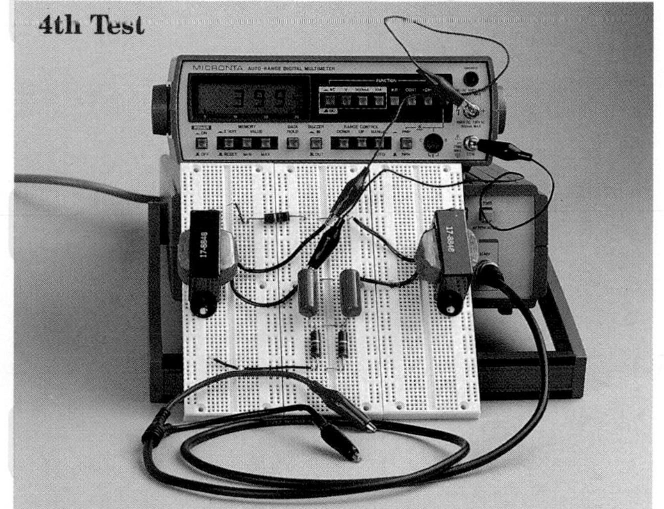
4th Test

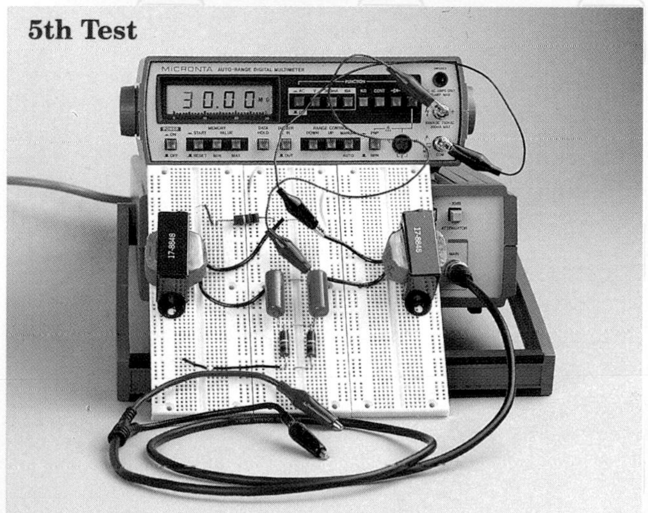
5th Test

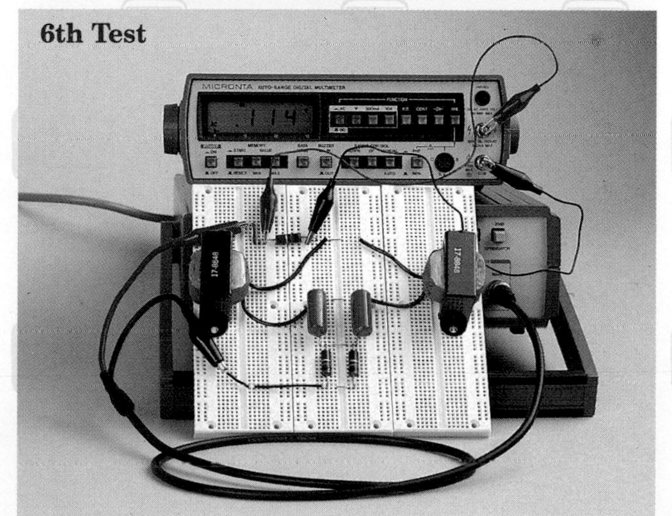
6th Test

CIRCUIT 12

SYMPTOMS: When the variable C is adjusted so the LC circuit is resonant at about 318 kHz source signal (i.e., C is about 100 pF), the scope shows maximum deflection due to maximum I_R drop at series resonance (Photo 1). When the variable C is adjusted so the LC circuit resonance is slightly higher in frequency (plates more unmeshed), the V_R (and scope deflection) slightly decrease as would be expected (Photo 2). When the capacitor is adjusted so the LC resonance is slightly lower (plates more meshed), the V_R and scope deflection drop drastically (Photo 3). This indicates some drastic change in the circuit. The off-resonance impedance must have increased at an abnormal rate for this circuit.

IDENTIFY initial suspect area: Since the only item being varied is the capacitance (because of our tuning efforts), the variable tuning capacitor is the initial suspect area.

MAKE test decision: We can check the tuning capacitor's operation with an ohmmeter to make sure that it isn't shorting at some point in its tuning range. The ohmmeter check is an easy check to find any possible shorts.

PERFORM 1st Test: Perform a resistance check throughout the tuning range. As the plates approach being more fully meshed a short between the stator and rotor plates becomes evident. This appears to be our problem.

LOCATE new suspect area: There is no new area because the variable capacitor has proven to be the likely culprit.

EXAMINE available data.

REPEAT analysis and testing: Let's visually inspect the capacitor to see if we can see some shorted plates.

Symptoms (1)

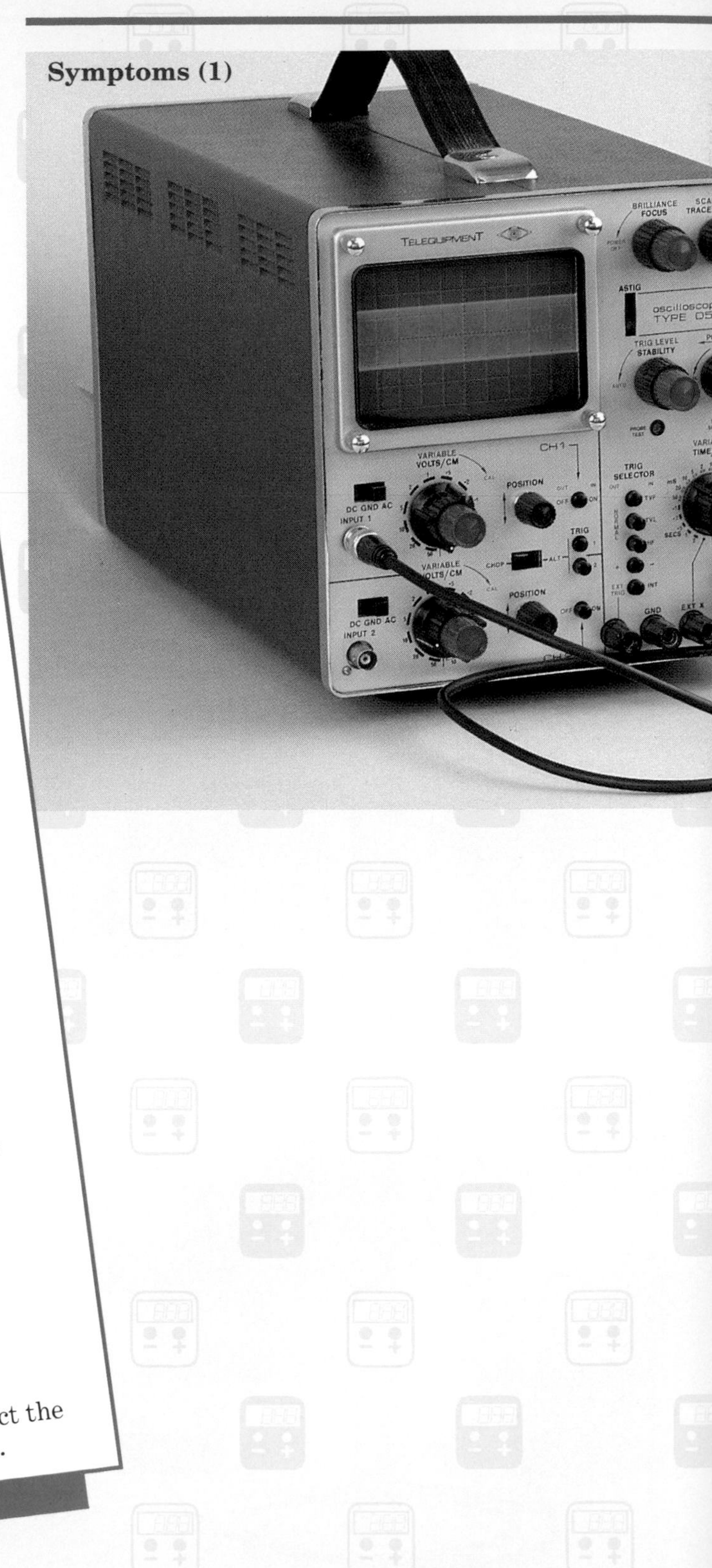

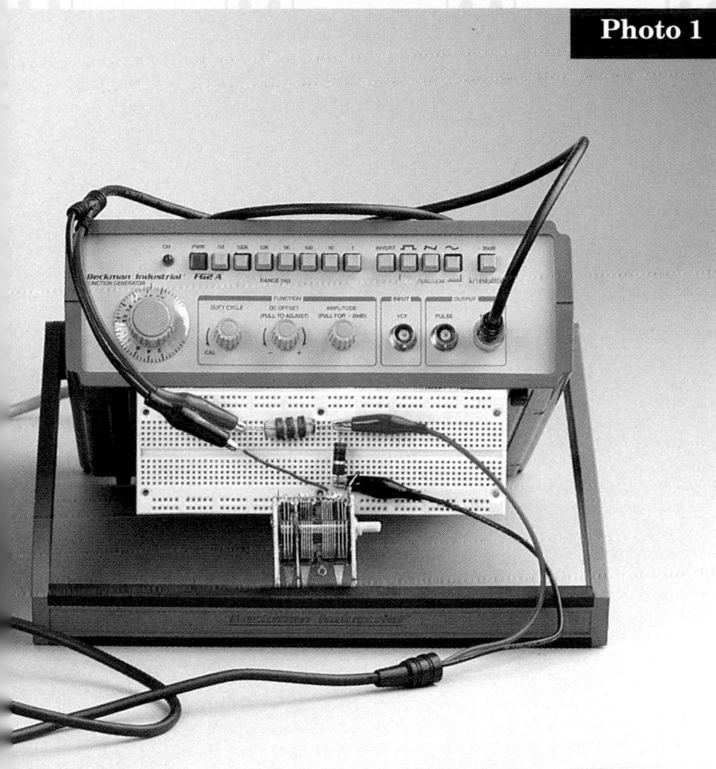

Photo 1

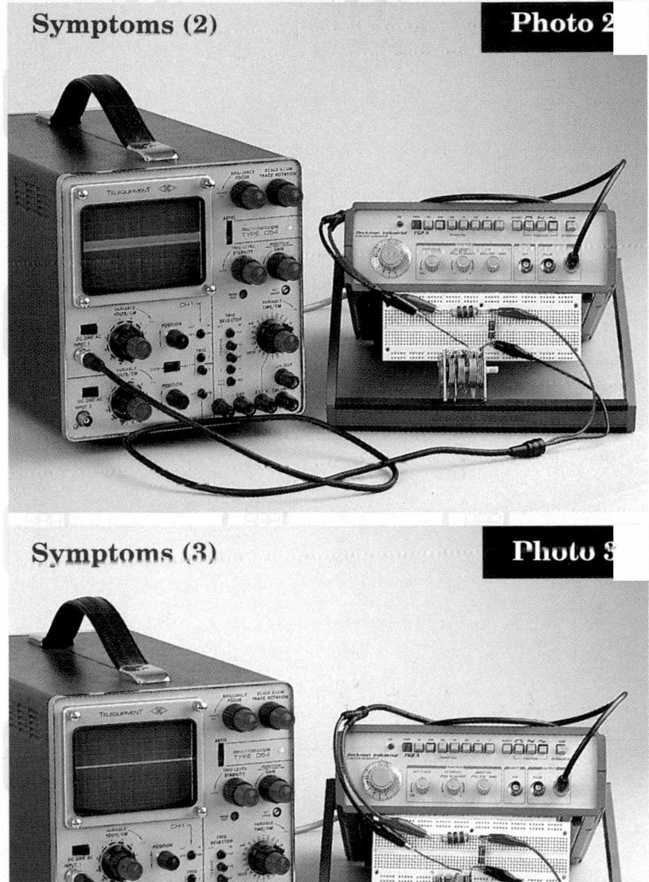

Symptoms (2) — Photo 2

Symptoms (3) — Photo 3

2nd Test: Inspect the capacitor with strong light. The result is you see two plates that are rubbing as the plates begin to more fully mesh. The problem is shorting plates on the tuning capacitor.

3rd Test: Bend the plates to eliminate the rubbing. Then, check with the ohmmeter again, while turning the rotor plates through their range. The result is everything is normal (Photos 4 and 5).

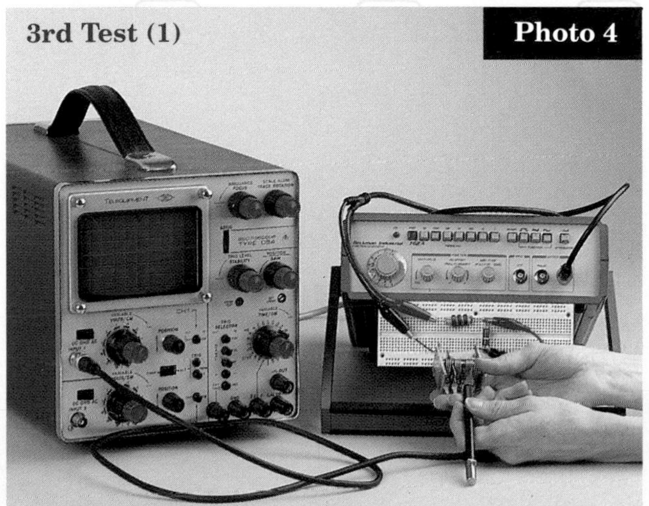

3rd Test (1) — Photo 4

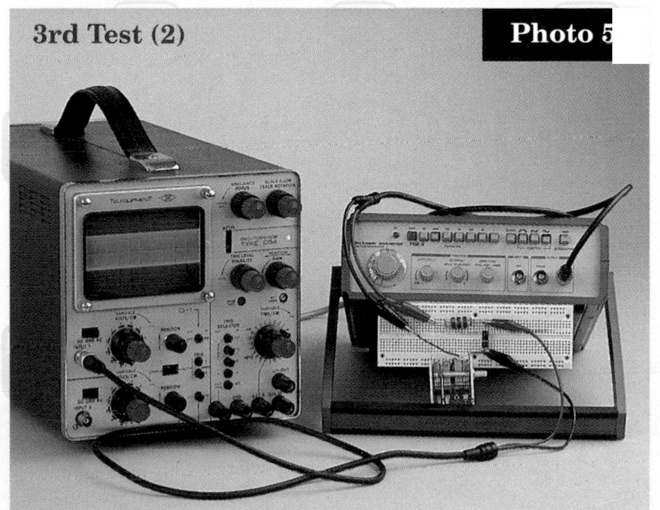

3rd Test (2) — Photo 5

CIRCUIT 13

Symptoms: One would normally expect the tuning range of the LC circuit to be from about 190 kHz to 450 kHz as the C is varied between 50 and 275 μμF. It is actually tuning from about 220 kHz (Photo 1) to 580 kHz (Photo 2).

Identify initial suspect area: Since the frequencies are higher than expected, it looks like the LC product must be less than the rated values. Either the capacitor has decreased by losing plates (which is highly unlikely) or the inductor has decreased for some reason. We can't be sure at this point. The inductor is the initial suspect. However, we'll keep both the C and L in the initial suspect area.

Make test decision: What could make the inductor have less inductance than anticipated? Possibly, shorted turns which would make L decrease. Since the rated dc resistance of this coil is about 7 Ω, an ohmmeter check (with the inductor isolated from the rest of the circuit) might reveal if enough turns are shorted to make a difference in coil resistance.

Perform 1st Test: With L isolated, measure its resistance (Photo 3). R measures about 5 Ω. Even this little resistance change could be significant, since we're dealing with such a small rated resistance in the first place. A number of turns could be shorted.

Locate new suspect area: The hot zone is still the inductor. However, we need to make another inductor check so we can exclude the capacitor.

Examine available data.

Repeat analysis and testing: Since the coil looks suspicious, let's substitute a known good coil of like ratings and see how the circuit operates.

Symptoms (1) Tuned at resonance

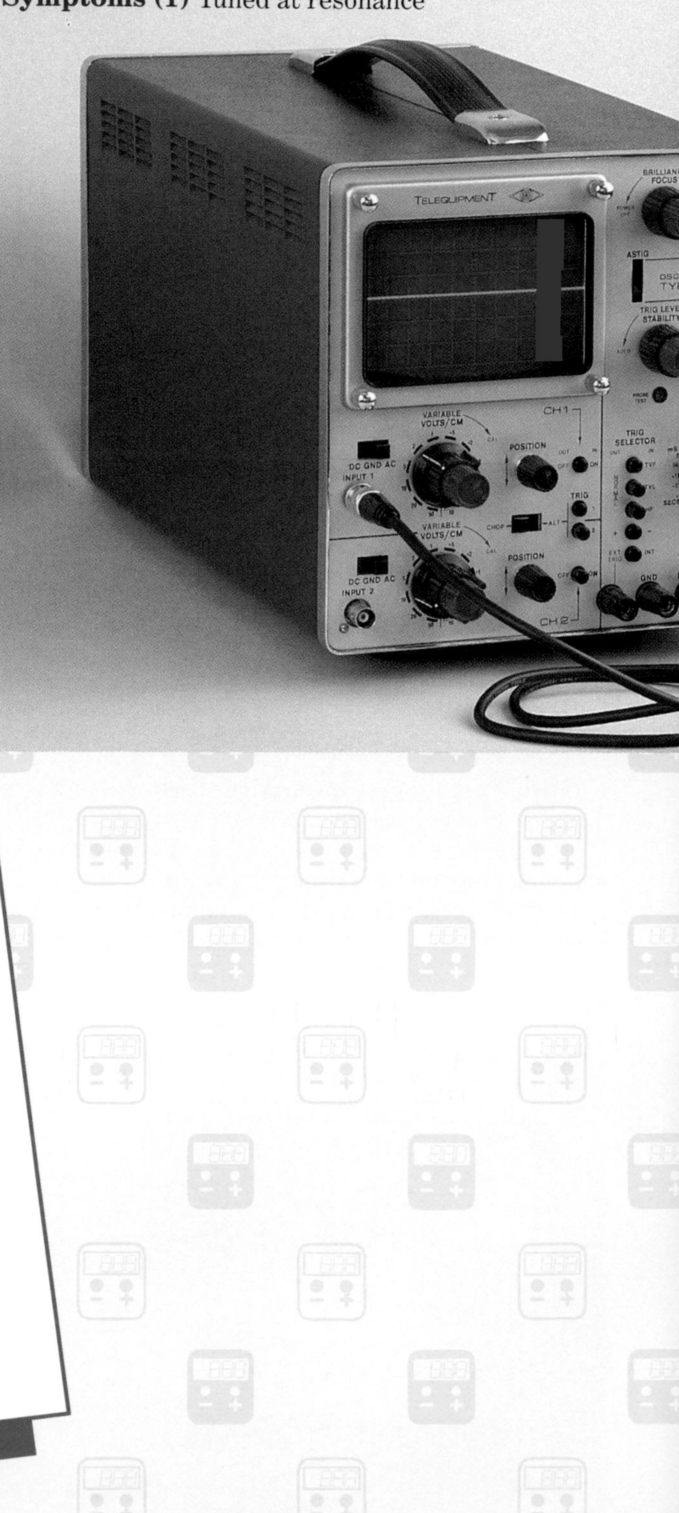

Photo 1

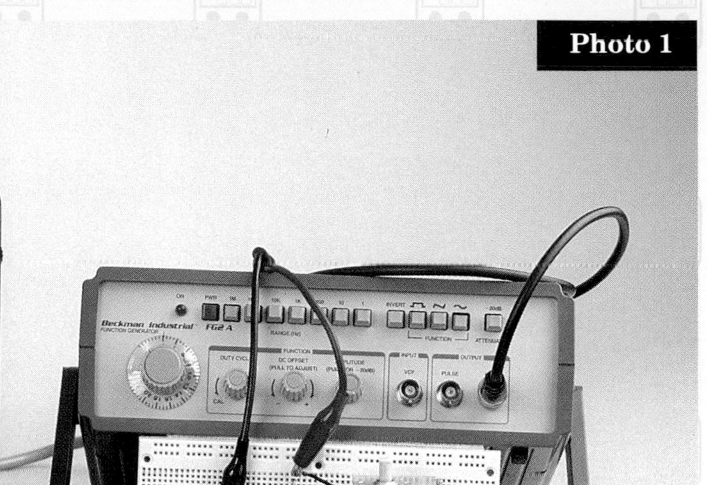

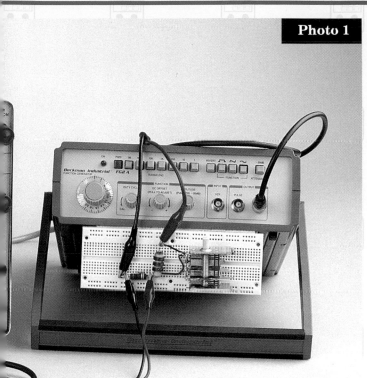

2nd Test: Substitute an inductor of the same ratings. Then check the circuit tuning characteristics The result is the component substitution worked. The circuit tunes with the expected range of frequencies, 190 kHz (Photo 4) to 450 kHz (Photo 5). The trouble was shorted turns on the inductor.

2nd Test (1)
Bending plates

Photo 4

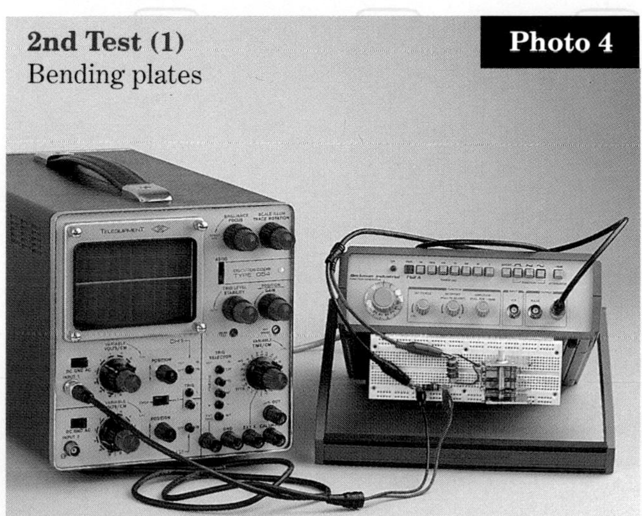

Symptoms (2)
Tuned above resonance

Photo 2

1st Test
Tuned below resonance

Photo 3

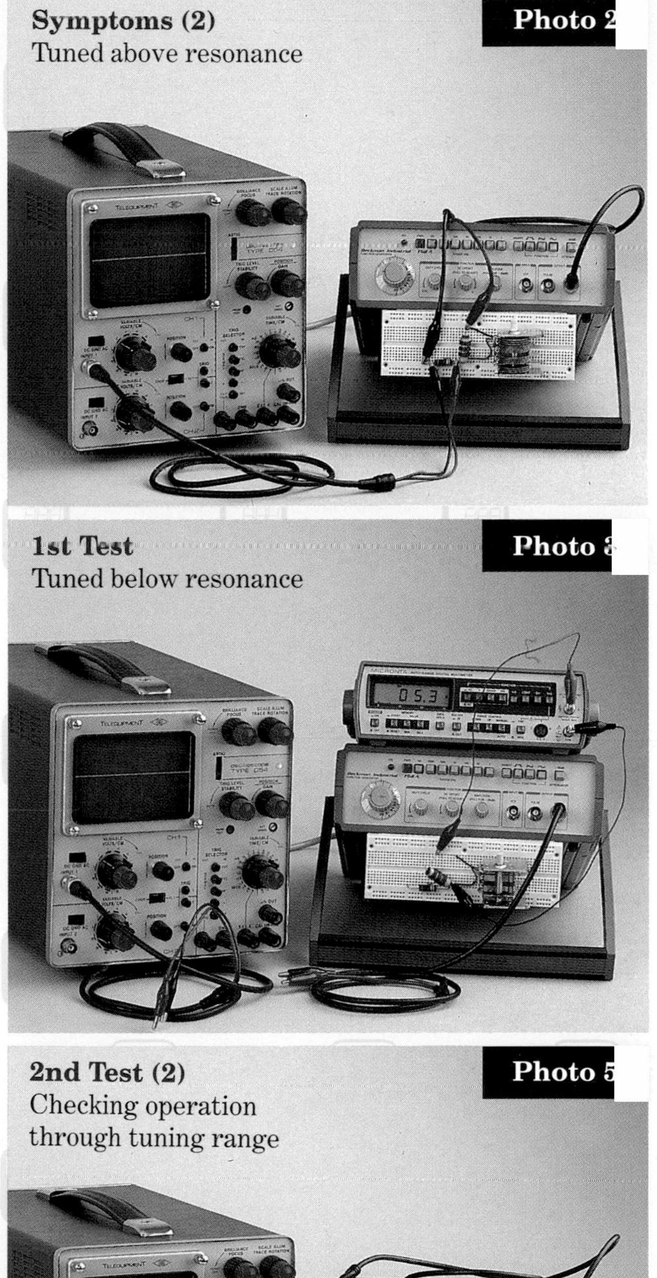

2nd Test (2)
Checking operation through tuning range

Photo 5

CIRCUIT 14

SYMPTOMS: The dc output voltage is considerably lower than it should be.

IDENTIFY initial suspect area: The low voltage can be caused by a transformer problem, a rectifier problem, a filter problem, or an excessively high current drain by the load. For now, let's keep all of these possibilities in the initial suspect area.

MAKE test decision: The magnitude of change in the output voltage is large, and we would probably see smoke if the transformer were that faulty or if the load were that heavy. Therefore we'll make our first check at the midpoint between the rectifier circuit and the filter circuit. That is, at point C or the rectifier output.

PERFORM 1st Test: Using a scope, check the waveform at point C. The result is only one pulse per ac input cycle. Half of the full-wave rectifier output is missing.

LOCATE new suspect area: It looks like we may have a rectifier problem. But, using our "gozinta gozouta" technique, we'll have to verify that the ac input into the rectifier is normal before we exclude the transformer. Thus, the new suspect area includes the transformer and rectifier circuits.

EXAMINE available data.

REPEAT analysis and testing: The next test is to look at the ac input to the rectifier circuit at point B.

2nd Test: Check the waveform at point B. The scope check indicates a sine-wave voltage with peak voltage of about 39 volts. This is what it should be. So we have isolated the trouble to the rectifier block. When we restore the circuit by replacing the faulty component(s), the circuit works fine.

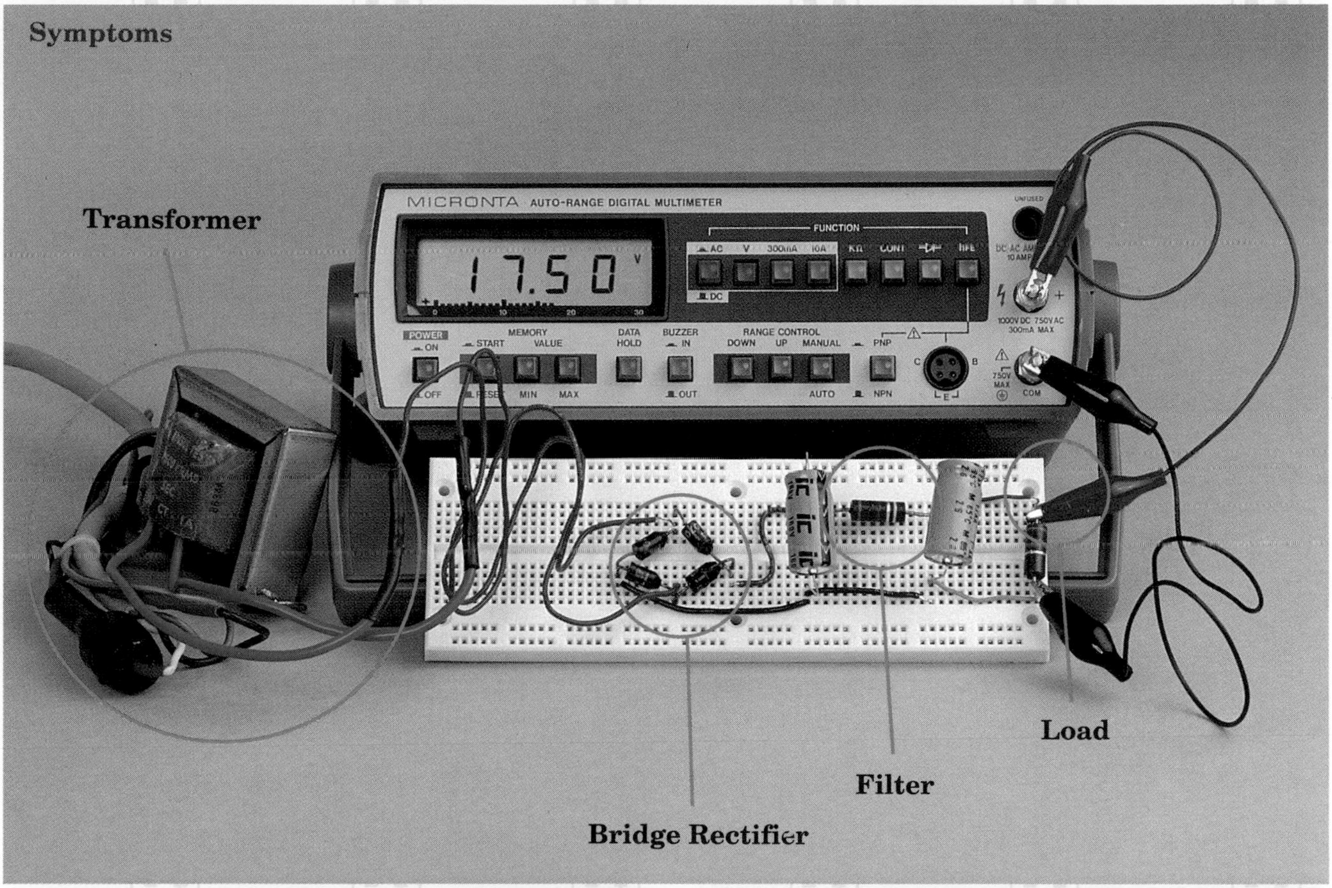

Symptoms
Transformer
MICRONTA AUTO-RANGE DIGITAL MULTIMETER
17.50 V
Bridge Rectifier
Filter
Load

1st Test
17.59

2nd Test
17.56

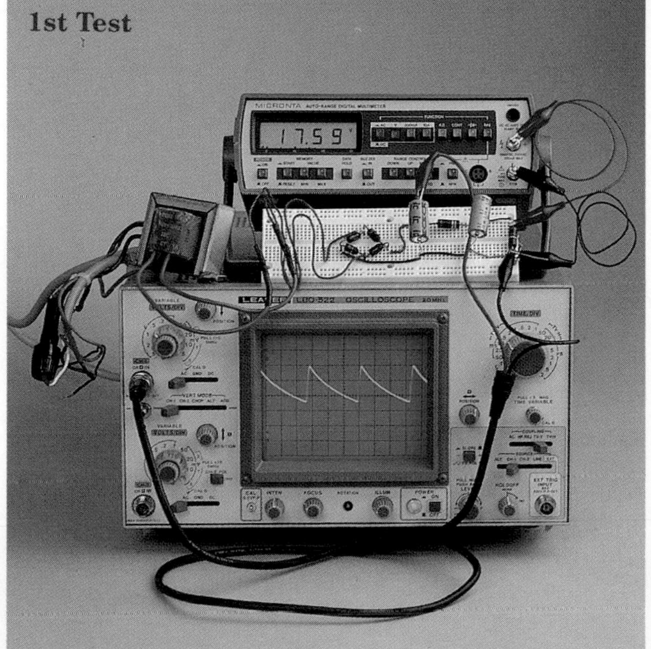

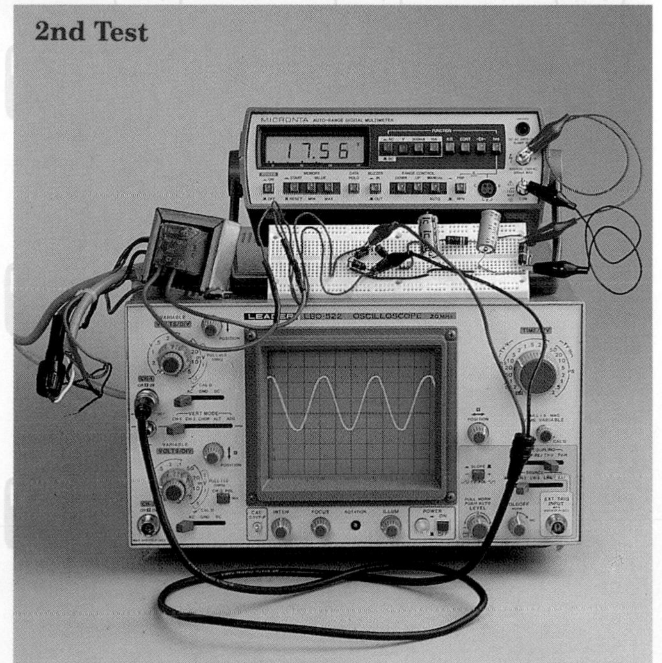

CIRCUIT 15

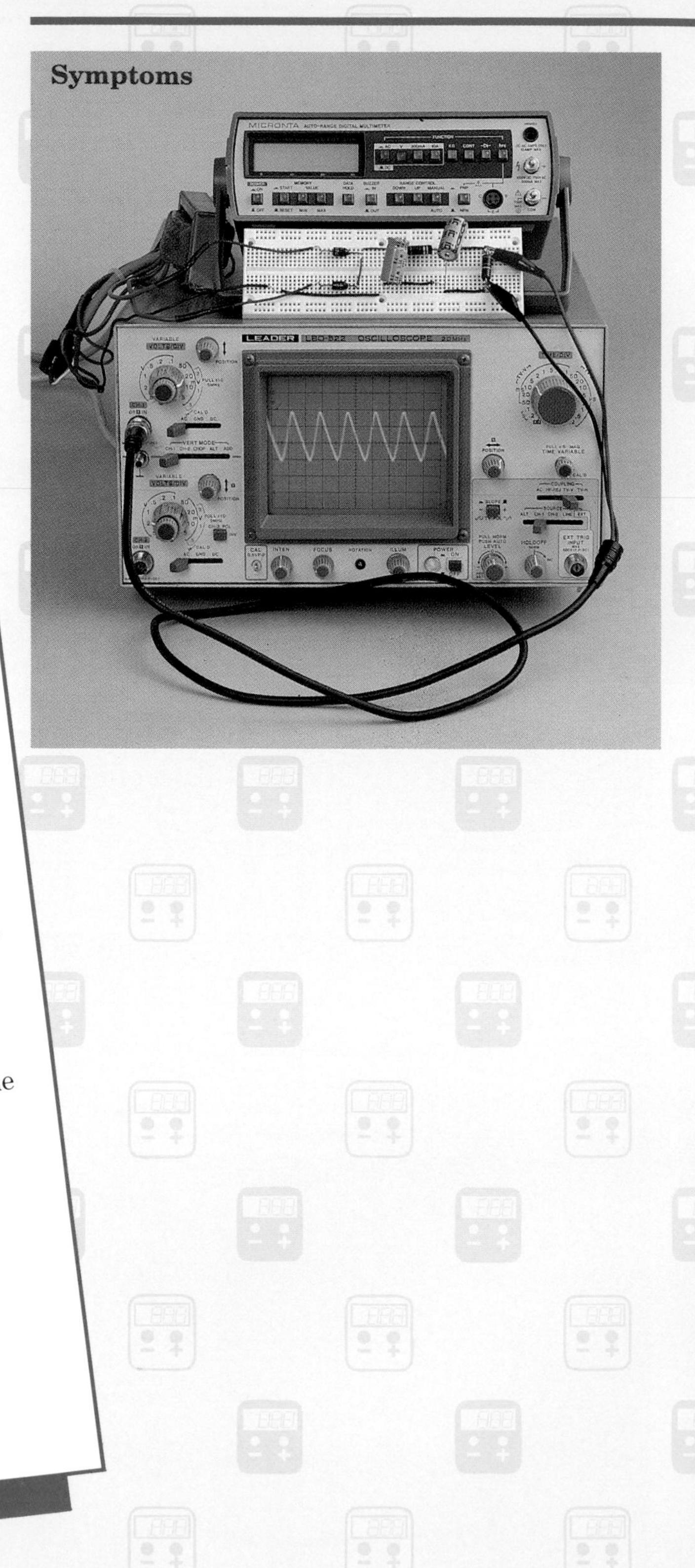

SYMPTOMS: The dc voltage out seems low, and excessive ac ripple is present.

IDENTIFY initial suspect area: The ripple symptom indicates the filter components are involved. If the inductor were shorted, the dc output might be slightly higher than normal. If the inductor were open, there would be no output. So this narrows our initial suspect area to the two filter capacitors.

MAKE test decision: If either capacitor were shorted, the output voltage would be drastically lower than normal (virtually zero). So we need to see if a capacitor might be open, which would agree with our symptom information. Typically, an aging electrolytic capacitor that has dried out and become virtually open will show much more ac than it would if it were normal. First, let's check the ac voltage across the output filter capacitor (C_2).

PERFORM 1st Test: Check the ac voltage across C_2. It shows higher than the normal ac component. However, this could be caused by the input filter capacitor (C_1) not feeding a normal signal to the inductor and this capacitor.

LOCATE new suspect area: The first test is not conclusive, therefore the suspect area remains the same at this point.

EXAMINE available data.

REPEAT analysis and testing.

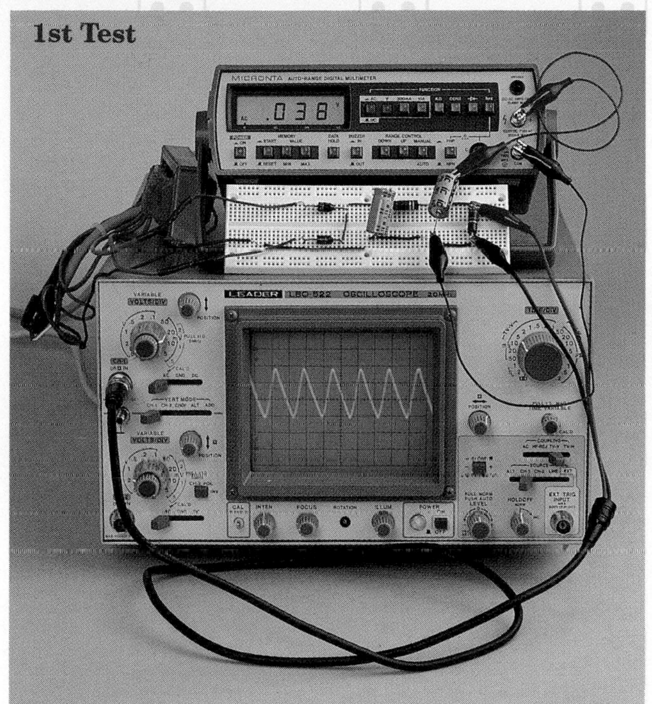

1st Test

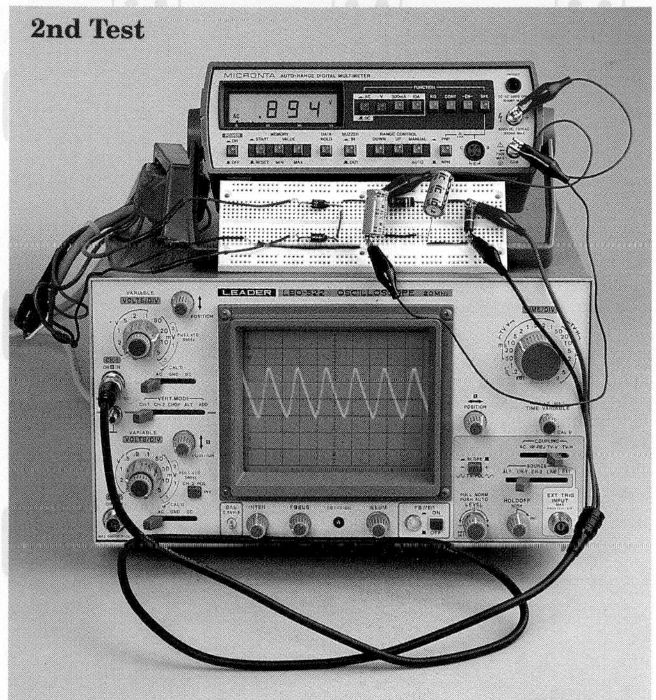

2nd Test

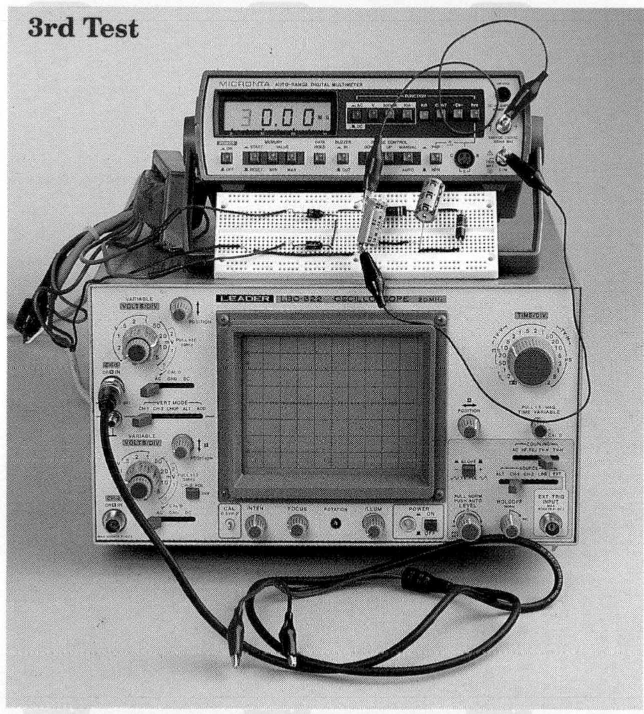

3rd Test

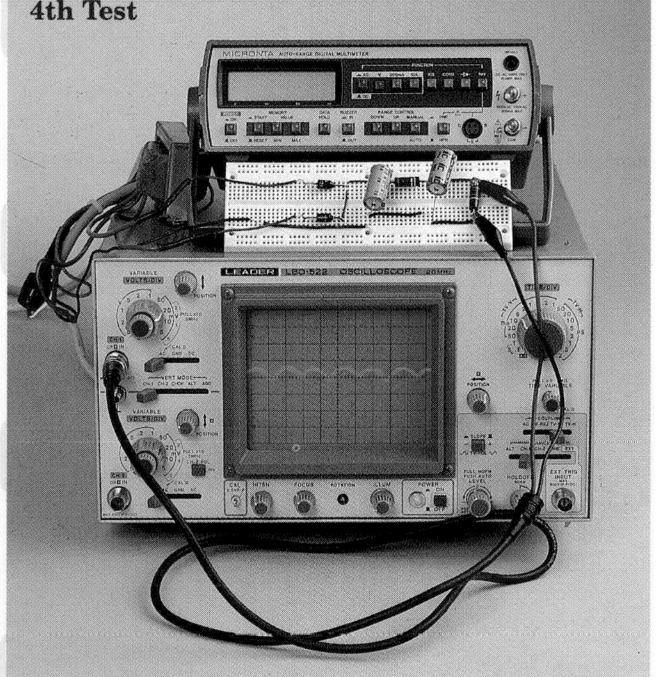

4th Test

2nd Test: Measure ac voltage across C_1. The result shows a much higher than normal ac component.

3rd Test: Make a front-and-back resistance check of capacitor C_1, with power removed. Since the capacitor checks open, we have probably found our problem - an open filter capacitor (C_1).

4th Test: Replace C_1 with a capacitor of appropriate ratings and restore normal operation.

THE SIMPLER SEQUENCE FOR TROUBLESHOOTING

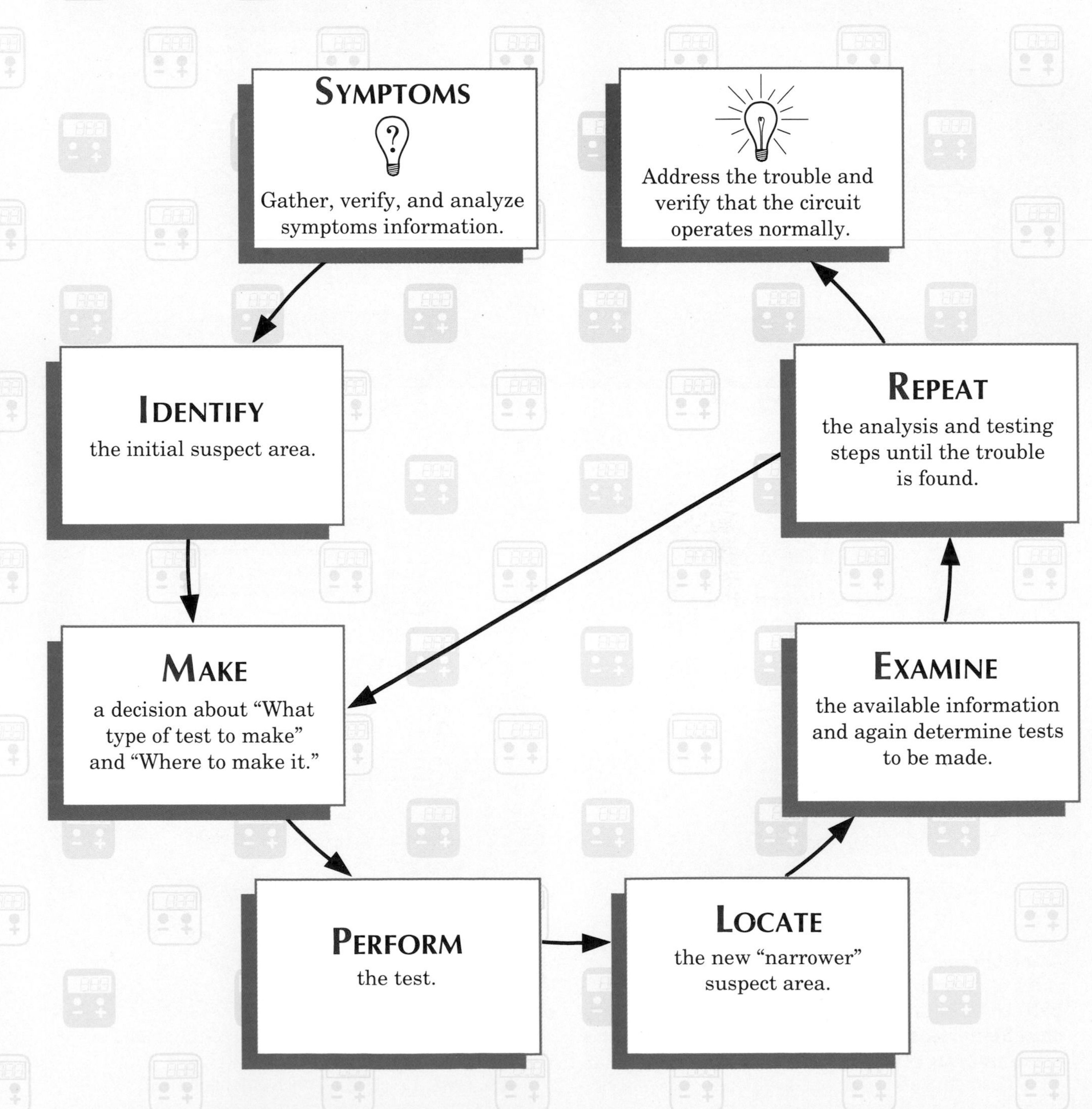

SYMPTOMS
?
Gather, verify, and analyze symptoms information.
IDENTIFY
the initial suspect area.
MAKE
a decision about "What type of test to make" and "Where to make it."
PERFORM
the test.
LOCATE
the new "narrower" suspect area.
EXAMINE
the available information and again determine tests to be made.
REPEAT
the analysis and testing steps until the trouble is found.
Address the trouble and verify that the circuit operates normally.

4. Look at Figure 11–15 to see one method to interpret the display in terms of phase difference between the signals.

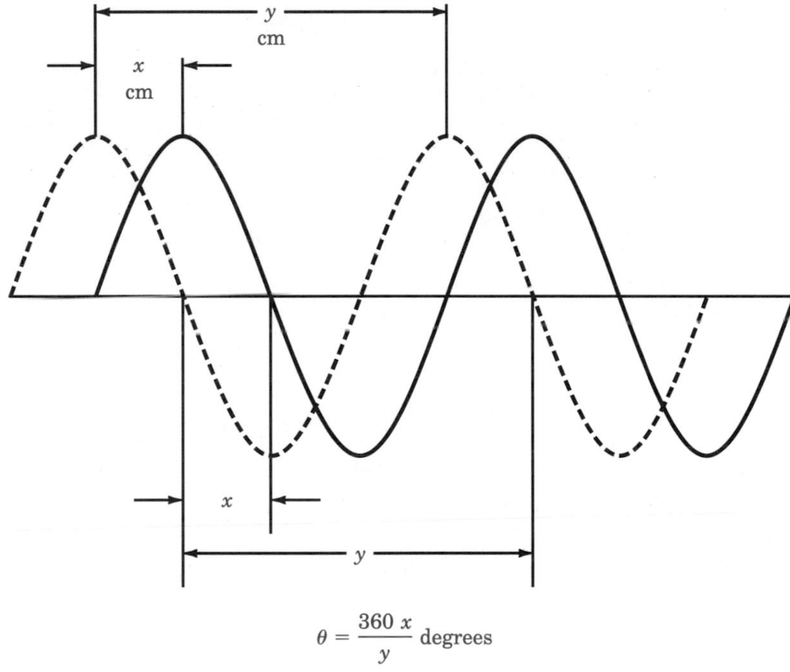

$$\theta = \frac{360\,x}{y} \text{ degrees}$$

x and y are measured in centimeters from the scope graticule.

FIGURE 11–15
Overlay technique for phase measurement

The Lissajous Pattern Technique

This technique is used only with sine-wave signals. Although **Lissajous patterns** are more convenient for determining frequency ratios between two signals than for measuring their phase differences, Lissajous patterns can interpret phase difference between two equal-frequency signals. (Assume the gain controls have been previously adjusted so each signal causes an equal amount of deflection). It is not the intention of this chapter to provide an explanation and formula for this interpretation. You can, however, get a good idea of how this works by looking at Figure 11–16.

MEASURING FREQUENCY RATIOS WITH THE SCOPE

Although there are several methods of using the scope to look at two signals ratio of frequencies, we'll briefly discuss only one of them—the Lissajous pattern technique. This technique is used for frequencies that are multiples, or submultiples of each other.

A Lissajous pattern forms on a scope screen when sine-wave voltages are

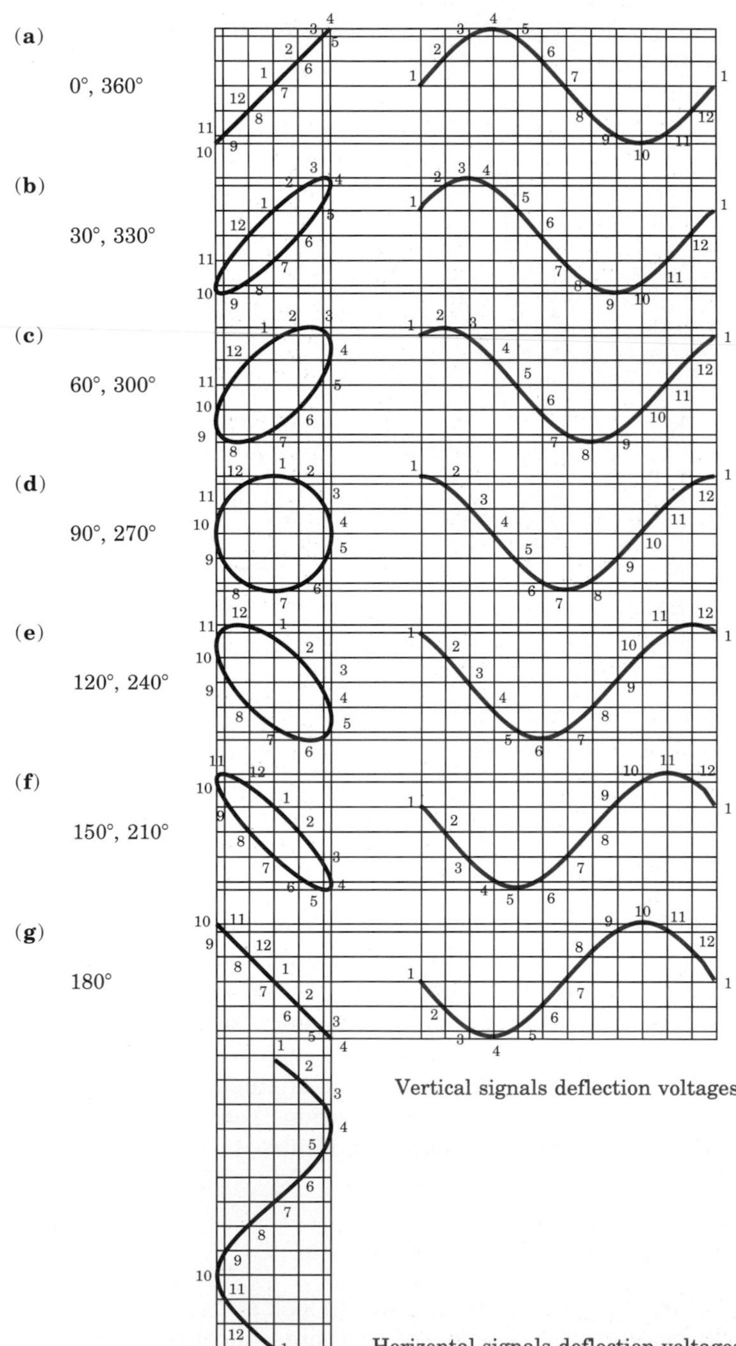

FIGURE 11–16 Lissajous pattern technique for determining phase with H and V signals at same frequency but at different phase relationships

simultaneously applied to the horizontal and vertical deflection plates. The resulting pattern indicates the ratios of the two signals. When the frequency of one signal (the reference signal) is known, it is easy to determine the unknown frequency by interpreting the ratio of frequencies. Without belaboring the point, recall we said that by looking at resultant deflection for a number of instantaneous points, the technician can perform some meaningful interpretations about the signals causing the deflection(s). Figure 11–17 illustrates a point-by-point interpretation of the waveform.

In Figure 11–17, the pattern illustrates the resulting waveform for two sine waves where the vertical signal's frequency is twice that of the signal applied to the horizontal plates. All but pattern C are closed-loop patterns. Pattern C is obviously an open-loop pattern. To interpret the closed-loop patterns, the number of loops tangent to one of the pattern sides (e.g., the horizontal deflection peak(s) on one side) is compared to the number of loops tangent to the top or bottom of the pattern (the vertical deflection peak[s]).

In Figure 11–18 you see the ratio of horizontal to vertical frequency is 1:2. Knowing the horizontal reference frequency is 60 Hz, you automatically also know the vertical frequency must be two times that, or 120 Hz.

EXAMPLE

In Figure 11–19, you see a 2:3 frequency ratio of horizontal-to-vertical signals. What is the vertical signal's frequency when the horizontal signal has a frequency of 100 Hz? The answer is 150 Hz: $\frac{3}{2} = 1.5$; therefore, vertical signal = 1.5 × horizontal signal.

Note the reference signal (or known frequency signal) is usually applied to the horizontal deflection system; whereas, the unknown frequency signal is applied to the vertical deflection system. Although you will only become proficient at interpreting Lissajous patterns by actual practice, we wanted to at least discuss the idea of Lissajous for interpreting frequency comparisons.

PRACTICAL NOTES

You probably will not use the scope for frequency ratio comparisons as much as you will for direct frequency measurement of a single signal. As you continue in your studies, you will learn to make direct frequency measurement by calibrating time on the horizontal axis of the scope display, and then determining frequency by the $f = \frac{1}{T}$ formula.

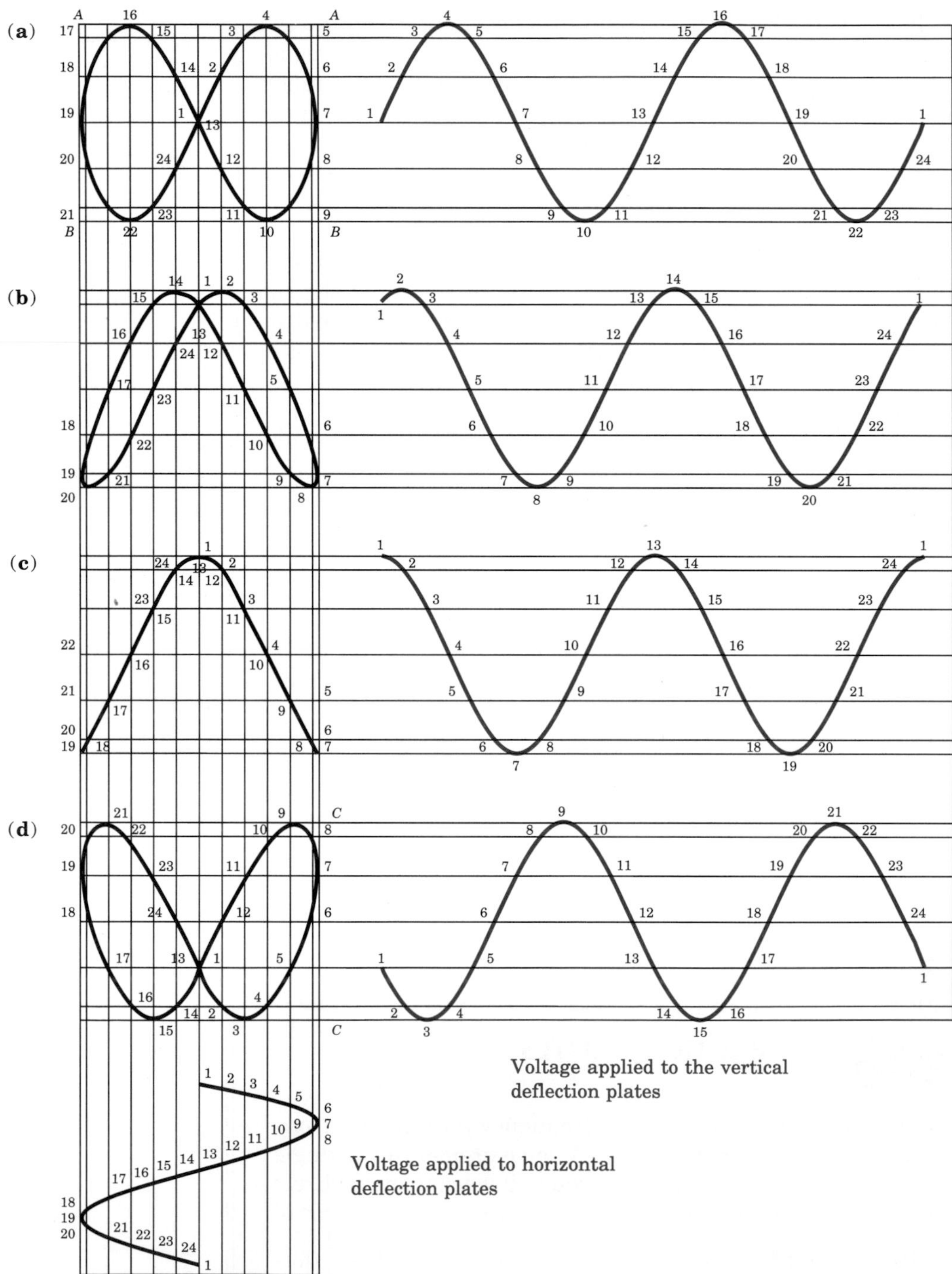

FIGURE 11–17 Determining frequency ratio with Lissajous pattern(s), where V signal is two times H signal frequency

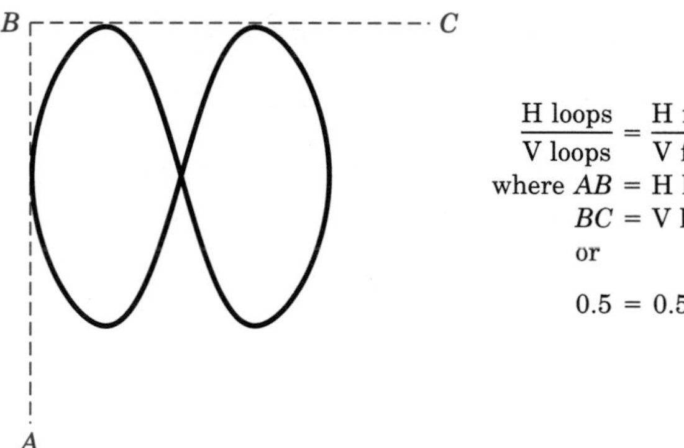

$$\frac{\text{H loops}}{\text{V loops}} = \frac{\text{H freq}}{\text{V freq}}$$

where AB = H loops

BC = V loops

or

$$0.5 = 0.5\,\frac{\text{H}}{\text{V}}$$

FIGURE 11–18 Interpreting a closed-loop Lissajous pattern

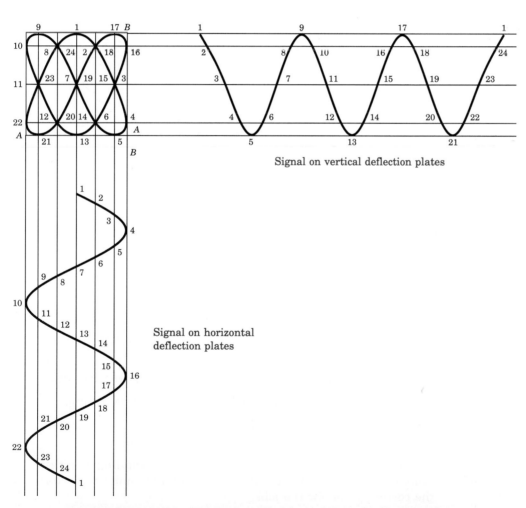

Signal on vertical deflection plates

Signal on horizontal deflection plates

FIGURE 11–19 A 2:3 horizontal-to-vertical frequency ratio

Final Comments About the Oscilloscope

Certainly, reading this chapter has not made you a professional oscilloscope user! We only hope that by briefly discussing the remarkable versatility of the oscilloscope, you will desire to become knowledgeable and practiced in applying this instrument throughout your studies and career. The ability of the scope to display waveforms so that increments of time are measured and interpreted; and its ability to measure relative amplitudes makes this instrument one of the technician's most valuable assets, Figure 11–20. Learn to use it!

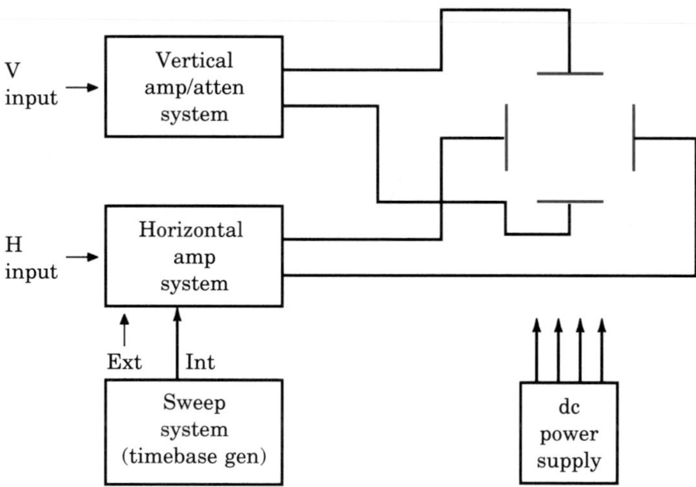

FIGURE 11–20 Simple oscilloscope functional block diagram

PRACTICAL NOTES

Some precautions: When using a scope, the technician should use common sense!

1. Do not leave a "bright spot" on the screen for any length of time.
2. Do not apply signals that exceed the scope's voltage rating.
3. Do not try to make accurate measurements on signals whose frequency is outside the scope's frequency specifications.
4. Be aware that the scope's input circuitry (including the test probe) can cause loading effects on the circuitry under test—use the correct probe for the job!

CHAPTER CHALLENGE

Following the SIMPLER troubleshooting sequence, find the problem with this circuit. As you follow the sequence, record your responses to each step on a separate sheet of paper.

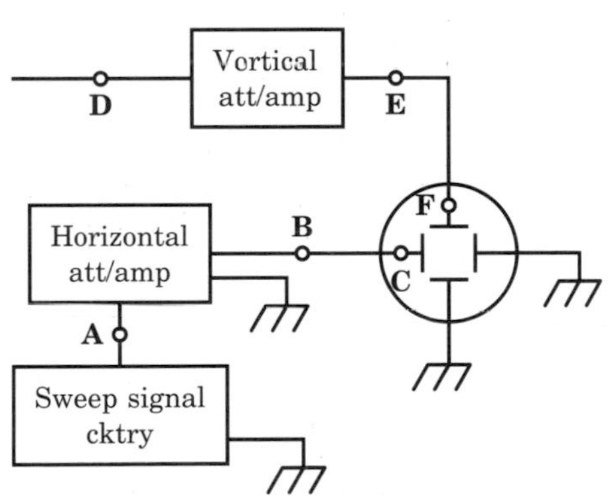

CHALLENGE CIRCUIT 6

SYMPTOMS Gather, verify, and analyze symptom information. (Look at the "Starting Point Information.")

IDENTIFY Identify initial suspect area for the location of the trouble.

MAKE Make a decision about "What type of test to make" and "Where to make it." (To simulate making a decision about each test you want to make, select the desired test from the "TEST" column listing below.)

PERFORM Perform the test. (Look up the result of the test in Appendix C. Use the number in the "RESULTS" column to find the test result in the Appendix.)

LOCATE Locate and define a new "narrower" area in which to continue troubleshooting.

EXAMINE Examine available information and again determine "what type of test" and "where."

REPEAT Repeat the preceding analysis and testing steps until the trouble is found. What would you do to restore this circuit to normal operation? When you have solved the problem, compare your results with those shown in the color insert.

Starting Point Information

1. Simplified block diagram
2. Vertical input signal is 300-Hz ac sine-wave
3. Output of sweep signal circuitry is a 150-Hz sawtooth waveform
4. Horizontal sweep = 150 Hz
5. Trace seen on CRT is a horizontal line from left to right.

Test	Results in Appendix C
Signal/Voltage Checks	
Point A to Ground	(70)
Point B to Ground	(2)
Point C to Ground	(113)
Point D to Ground	(57)
Point E to Ground	(23)
Point F to Ground	(106)

SUMMARY

- The oscilloscope (scope) is an instrument providing a visual presentation of electrical parameters with respect to time and amplitude.
- Intensity and focus controls adjust brightness and clarity of the visual presentation.
- Horizontal controls adjust the frequency of the internally generated horizontal sweep signal; select outside or inside sources for this sweep signal; horizontally position the visual presentation; and adjust the horizontal trace length.
- Vertical controls adjust the vertical amplitude and the vertical position of signals displayed on the screen.
- Synchronization control(s) adjust the start time of the horizontal sweep signal so the visual presentation of a signal is stable, or standing still.
- Scope presentations display the peak-to-peak values of the waveform. Other significant values are interpreted from the wave form by appropriate interpretation factors. For example, rms value = $0.707 \times$ peak, or $0.3535 \times$ peak-to-peak value.
- To measure voltage with an oscilloscope, the appropriate vertical controls are adjusted so the user knows the vertical sensitivity of the scope. That is, how many volts are represented by each division of vertical deflection on the calibrated screen.
- To measure the phase difference between two signals with an oscilloscope, a dual-trace scope is used so each signal is applied to a separate vertical input. Then, by appropriate adjustment, the two signals are superimposed on the visual presentation. Thus, the technician can interpret the phase differential in terms of the two signals.
- Lissajous patterns can also be used to measure phase difference. One signal is applied to the horizontal deflection system and the other to the vertical deflection system. The technician then interprets the resulting screen pattern in forms of phase differential of the two signals.
- To measure frequency (ratio) with an oscilloscope, Lissajous patterns are created by applying a known frequency signal to the horizontal deflection system, while the unknown frequency signal is applied to the vertical deflection system. Then the technician interprets the visual presentation by using the frequency ratio of the two signals.

REVIEW QUESTIONS

1. What is the name and abbreviation of the oscilloscope component that displays signals on its screen?
2. What are four practical uses of the oscilloscope?
3. Why should the intensity control be set to provide the minimum brightness, allowing good readability of the presentation?

4. What control(s) set the sweep frequency, therefore determining the number of cycles of a given signal viewed?

5. To what input terminal(s) is the signal to be viewed normally applied?

6. To what terminal(s) is an external sweep voltage applied?

7. To what terminal(s) is an external synchronizing signal applied?

8. The time/cm control and horizontal gain control settings are set so the horizontal sweep line represents 0.1 ms per centimeter. What is the frequency of a signal if one displayed cycle uses four horizontal centimeters on the horizontal sweep line?

9. Along what axis is deflection caused by a signal applied to the vertical deflection system?

10. Why should the setting of the scope vertical gain control and vertical input switch remain unchanged after initial calibration, if using the scope for voltage measurement?

11. Refer to Figure 11–21 and answer the following questions:
    a. What is the peak-to-peak value of voltage indicated by the waveform display?
    b. What is the value of T for the signal displayed?
    c. If you want the display to show only one cycle of the signal, which control(s) do you adjust on the scope?
    d. What is the time-per-centimeter setting to accomplish a one-cycle display?
    e. The vertical sensitivity is set for 5-volts per centimeter (calibrated). What are the rms, peak, and peak-to-peak values of voltage of the signal?

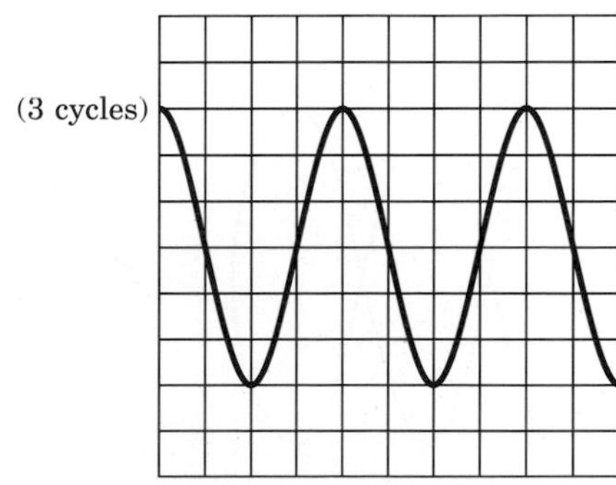

Vertical sensitivity is set at 5 V/cm
Horizontal time/cm is set at 0.001 sec/cm

FIGURE 11–21

TEST YOURSELF

1. Refer to Figure 11–22 and for each pattern shown (except for the model pattern) list the oscilloscope controls used to cause the waveform shown to match the model waveform. (NOTE: A = model waveform.)
2. Refer to Figure 11–23 and for each waveform display, list the unknown vertical frequency based on the known horizontal frequencies and the pattern shown. (NOTE: Use the closed-loop tangent technique.)

FIGURE 11–22

Model waveform

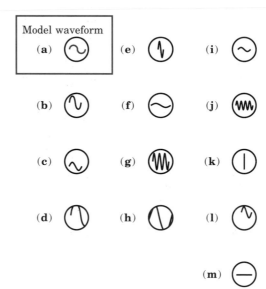

(a) (b) (c) (d) (e) (f) (g) (h) (i) (j) (k) (l) (m)

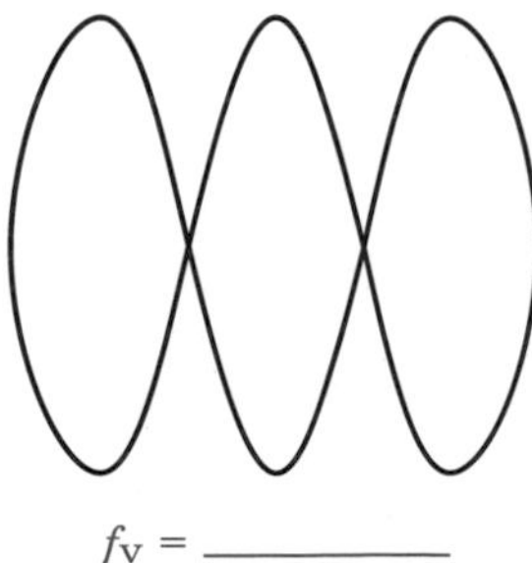

f_V = ________

(a)

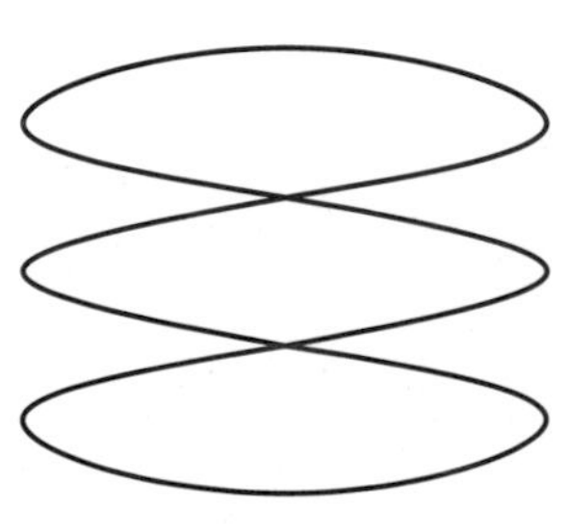

f_V = ________

(b)

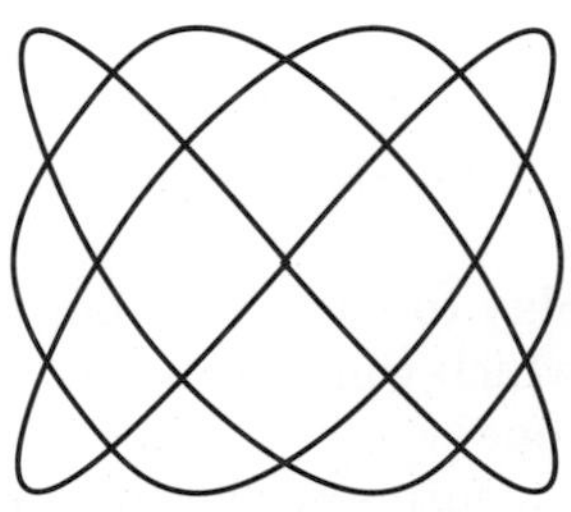

f_V = ________

(c)

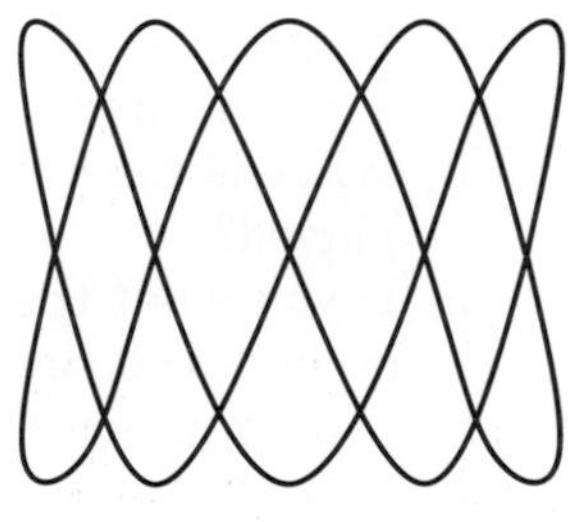

f_V = ________

(d)

Pattern recognition—horizontal frequency: 60 Hz

FIGURE 11–23

PERFORMANCE PROJECTS CORRELATION CHART

Suggested performance projects that correlate with topics in this chapter are:

CHAPTER TOPIC	PERFORMANCE PROJECT	PROJECT NUMBER
Key Sections of the Scope	Basic Operation	33
Measuring Voltage and Current with the Scope	Voltage Measurements	34
Using the Scope for Phase Comparisons	Phase Comparisons	35
Measuring Frequency with the Scope	Frequency Measurement	36

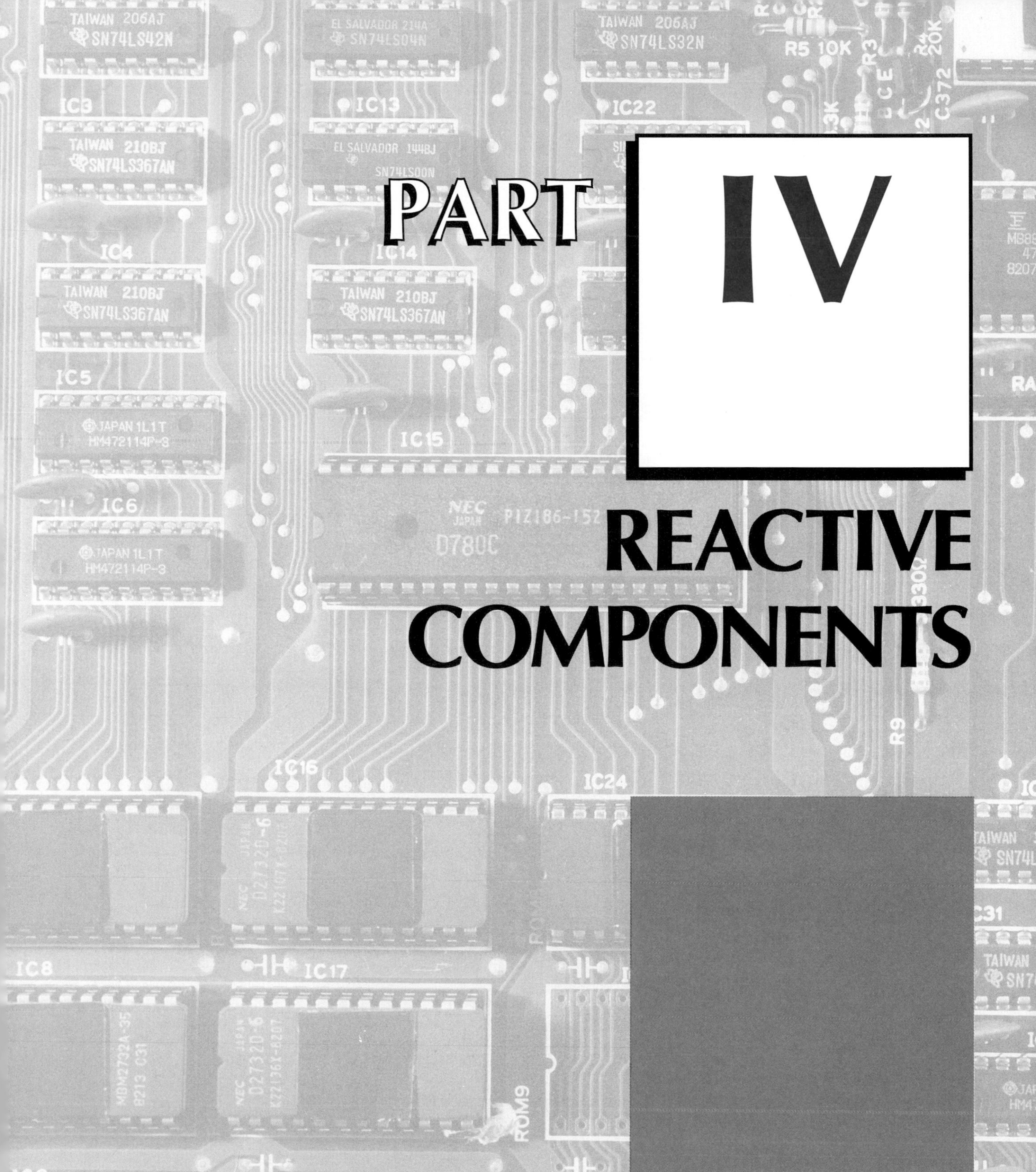

PART IV

REACTIVE COMPONENTS

Inductance

Key Terms

cemf

ϵ

Faraday's Law

H

Henry

Inductance (self-inductance)

Joule

Lenz's Law

L, L_T

L/R time constant

TC

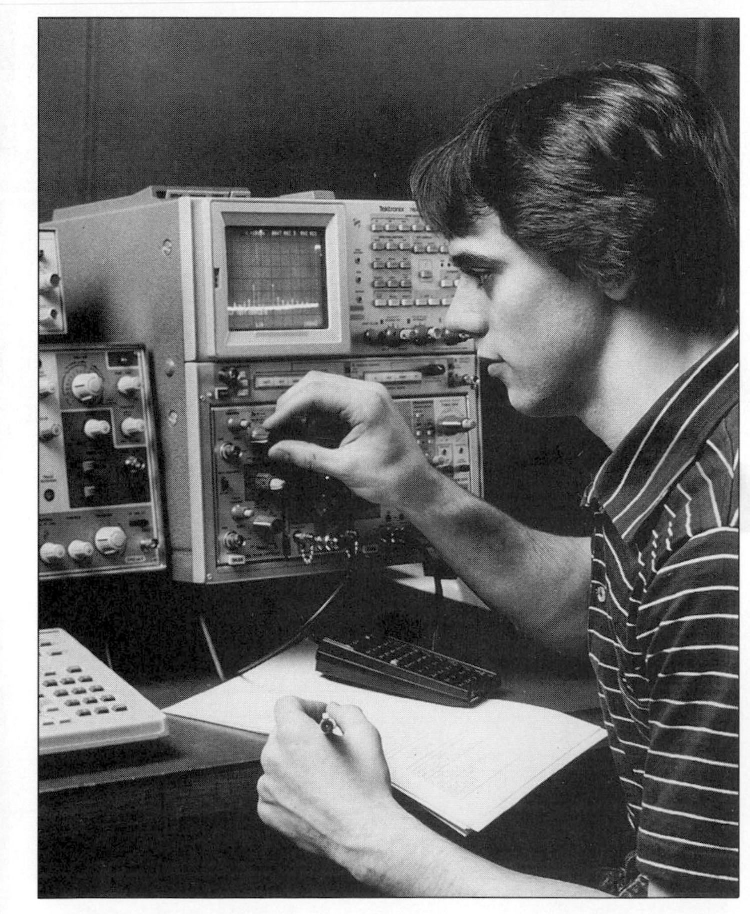

Courtesy of DeVry Institutes

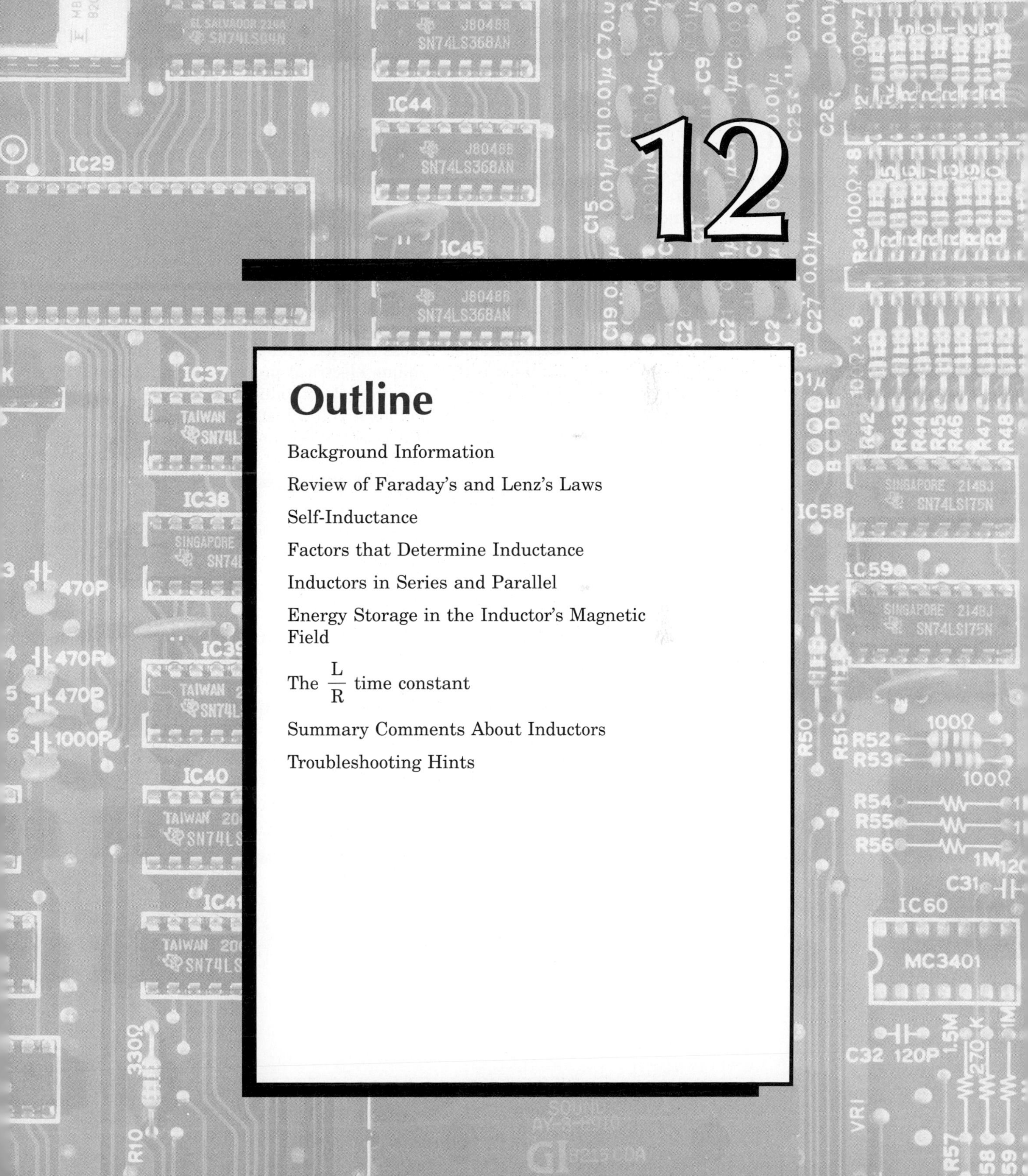

12

Outline

Background Information

Review of Faraday's and Lenz's Laws

Self-Inductance

Factors that Determine Inductance

Inductors in Series and Parallel

Energy Storage in the Inductor's Magnetic Field

The $\frac{L}{R}$ time constant

Summary Comments About Inductors

Troubleshooting Hints

Chapter Preview

The electrical property of resistance and its component, a resistor, are important elements in your understanding of electronics. Likewise, the property of inductance and its related component, an inductor, are equally important.

Inductors have many uses in electronics. For example, they are used in "tuned" circuits, which enable selection of a desired radio or television station. They are used as filter elements to smooth waveforms or to suppress unwanted frequencies. Inductance is also used in transformers, which are devices that increase or decrease ac voltages, currents, and impedances. The inductor (or coil) in automobiles has been used for years to help develop high voltages, which create the spark across the spark plug gaps. These applications and many others use this unique electrical property of inductance.

In this chapter, we will discuss the property of inductance as it relates to dc circuit action. In the following chapters, you will study the action of this electrical property under ac conditions. An understanding of both will be very useful to you!

Objectives

After studying this chapter, you will be able to:

- Define **inductance** and **self-inductance**
- Explain **Faraday's** and **Lenz's laws**
- Calculate induced **cemf** values for specified circuit conditions
- Calculate inductance values from specified parameters
- Calculate inductance in series and parallel
- Determine energy stored in a magnetic field
- Draw and explain time-constant graphs
- Calculate time constants for specified circuit conditions
- List common problems of inductors

BACKGROUND INFORMATION

From previous chapters, you know that electrical resistance (R) is that property in a circuit that opposes current flow. *Inductance* (L) is *the property in an electrical circuit that opposes a **change** in current.* Typical schematic symbols to represent air-core inductors and iron-core inductors are shown in Figure 12–1. Again, remember that inductance does not oppose current flow, but opposes a *change* in current flow.

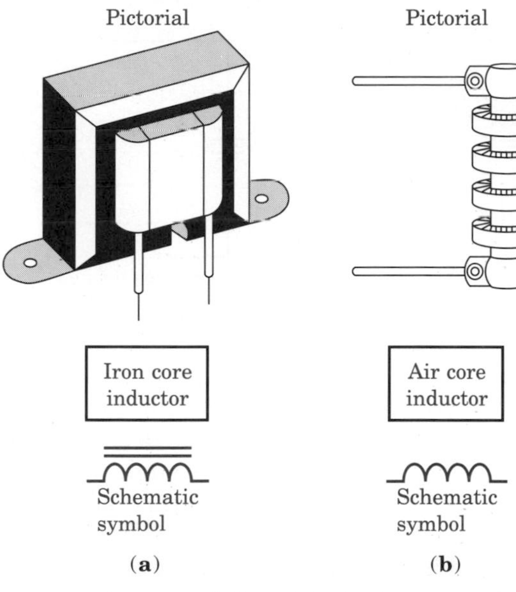

FIGURE 12–1 Inductors and their schematic symbols (a) Iron core (b) Air core

Another important characteristic of inductance and inductors is they store electrical energy in the form of the magnetic field surrounding them. As current increase through the inductor, a resulting magnetic field expands, surrounding the coil, 12–2a. When current tries to decrease, the expanded field collapses and tries to prevent the current from decreasing, Figure 12–2b. (Recall Lenz's law!) As the field expands, it absorbs and

Expanding field

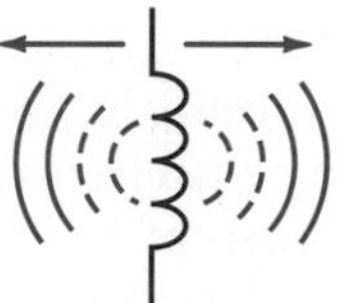

As current increases through the coil, the expanding magnetic field induces cemf, which prevents an instantaneous increase in current.

(a)

Collapsing field

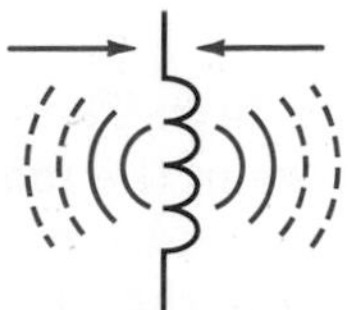

As current decreases through the coil, the collapsing field induces cemf, which prevents an instantaneous decrease in current.

(b)

FIGURE 12–2 Expanding and collapsing fields

stores energy. When the field collapses, it effectively returns that stored energy to the circuit.

Another feature of inductance and inductors is that an emf (electromotive force or voltage) is induced when a current *change* occurs through them. This induced emf is due to the expanding or collapsing magnetic field as it cuts the conductors of the coil. This self-induced emf is sometimes called **counter-emf,** or **cemf,** because its polarity opposes the change that induced it, Figure 12–3. The term **self-inductance** is frequently used to refer to inductors that create this situation.

FIGURE 12–3
The concept of cemf

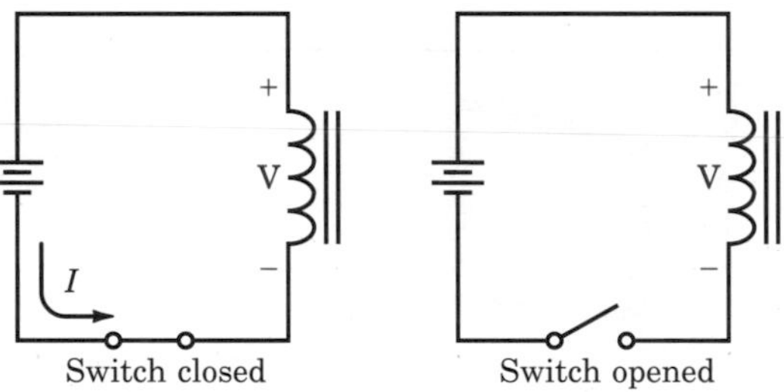

The polarity of the induced cemf is opposite to that of the source.

In a previous chapter, we discussed magnetism and electromagnetism. Additionally, you learned the factors determining the strength of this electromagnetic field around the coil—number of turns; coil length (number of turns per unit length); amount of current through the coil; and the type of core material (characteristics involved in the flux path). Keep these factors in mind, as there is a relationship between these items and the property of inductance, which is addressed in this chapter.

In the earlier discussion of electromagnetism, we examined a constant level dc current through the coil. In this chapter, we will begin to examine what happens during the switch-on/switch-off "transient" (temporary condition) periods when dc is applied or removed from a circuit containing inductance.

REVIEW OF FARADAY'S AND LENZ'S LAWS

Faraday's Law

Recall **Faraday's law:** The emf induced in a circuit is proportional to the time-rate of change in the magnetic flux linking the circuit. In other words, the amount of induced emf depends on the rate of cutting the flux (ϕ).

Also, recall that in the SI system of units, a weber is 10^8 lines of flux. The formula to determine the induced voltage due to cutting flux lines (or flux linkages) is defined as:

$$V_{ind} = \frac{d\phi}{dt}$$

where $\frac{d\phi}{dt}$ indicates rate of cutting or linking flux lines.

For a coil then:

Formula 12–1	$V_{ind} = N \text{ (number of turns)} \times \frac{d\phi \text{ (Wb)}}{dt \text{ (seconds)}}$

EXAMPLE
When a coil has 500 turns and the rate of flux change is 2 webers per second, $V_{ind} = 500 \times \frac{2}{1} = 1000$ volts.

PRACTICE PROBLEM I

Answers are in Appendix B.
What is the induced voltage (per turn) when 2×10^8 lines of flux are cut in 0.5 second?

Lenz's Law

Recall **Lenz's law:** The direction of an induced voltage (or current) opposes the change that caused it. That is, when the amount of magnetic flux linking an electric circuit changes, an emf is induced that produces a current opposing this flux change.

To summarize, Faraday's law describes how much voltage is induced by the parameters affecting it, and Lenz's law describes the polarity and nature of this induced voltage and the current produced. Let's now begin to correlate these laws with some new information as we examine inductance and its dc characteristics.

SELF-INDUCTANCE

In addition to inductance being defined as the property of a circuit that opposes a change in current, another definition for self-inductance (or inductance) is *the property of an electric circuit where a voltage is induced when the current flowing in the circuit changes.* You have learned the symbol for inductance, **L.** The **unit of inductance** is the **henry.** Named in honor of Joseph Henry, this unit of inductance is abbreviated **H.**

As you might suspect, the amount of induced emf produced is related to the amount of inductance (L) in henrys (H) and the rate of current change. A formula expressing this is:

Formula 12–2	$V_L = -L \frac{di \text{ (change in current, in amperes)}}{dt \text{ (change in time, in seconds)}}$

where L = inductance in henrys

$\frac{di}{dt}$ = rate of change of current $\left(\frac{A}{s}\right)$

The minus sign in front of the L indicates the polarity of the induced voltage is opposite to the source voltage that caused the current through the coil.

From the above, a definition for a henry of inductance states that *a circuit has an inductance of one henry when a rate of change of one ampere per second causes an induced voltage of one volt.* The induced voltage is in opposite direction to the source voltage; and therefore, is called **counter-emf** or **back-emf,** again see Figure 12–3.

EXAMPLE

When the rate of current change is five amperes per second through an inductance of ten henrys, what is the induced emf?

$$V_L = -L\frac{di\ \text{(change in current, in amperes)}}{dt\ \text{(change in time, in seconds)}}$$

$$V_L = -10 \times 5 = -\ 50\ \text{volts (Or, induced cemf} = 50\ \text{volts.)}$$

PRACTICE PROBLEM II

What is the counter-emf produced by an inductance of five henrys if the rate of current change is three amperes per second?

From these definitions and formulas, a useful formula to calculate inductance value is:

Formula 12–3	$L = \dfrac{V_L}{\frac{di}{dt}}$ henrys

where L = inductance in henrys

$\frac{di}{dt}$ = rate of current change $\left(\frac{A}{s}\right)$

EXAMPLE

What is the circuit inductance if a current change of 12 amperes in a period of 6 seconds causes an induced voltage of 5 volts?

$$L = \frac{V_L}{\frac{di}{dt}}\ \text{henrys}$$

$$L = \frac{5}{2} = 2.5\ \text{henrys}$$

FACTORS THAT DETERMINE INDUCTANCE

As implied earlier, some of the same factors that affect the strength of a coil's electromagnetic field are also related to the amount of coil inductance.

Look at Figure 12–4. The physical properties of an inductor that affect its inductance include:

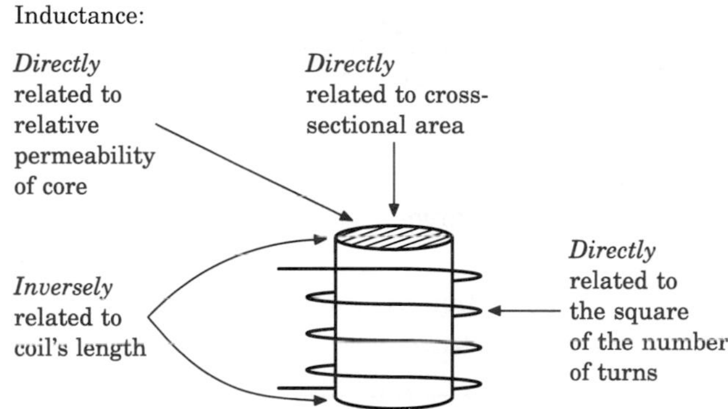
Inductance:
Directly related to relative permeability of core
Directly related to cross-sectional area
Inversely related to coil's length
Directly related to the square of the number of turns

FIGURE 12–4 Factors that affect the amount of inductance

1. *Number of turns.* The greater the number of turns, the greater the inductance value. In fact, inductance is proportional to the square of the turns. That is, when you double the number of turns in a coil of a given length and diameter, the inductance quadruples.
2. *Cross-sectional area (A) of the coil.* This feature relates to the square of the coil diameter. The greater the cross-sectional area, the greater the inductance value. When the cross-sectional area doubles, the inductance (L) doubles.
3. *Length of the coil.* The longer the coil for a given diameter and number of turns, the *less* the inductance value. This is due to a less concentrated flux, thus, less flux cut per unit time and the less the induced voltage produced.
4. *Relative permeability of the core.* For air-core inductors, the relative permeability (μ_r) is virtually one. The higher the permeability of the inductor's core material, the greater the inductance value. Recall that permeability relates to the magnetic path's ability to concentrate magnetic flux lines.

As a technician you probably will not design inductors. However, the generalized formula below illustrates the relationship of inductance to the various factors we have been discussing. In SI units, the formula is:

Formula 12–4	$L = 12.57 \times 10^{-7} \times \frac{\mu_r N^2 A}{l}$

where L = inductance in henrys
μ_r = relative permeability
N = number of turns
A = cross-sectional area in square meters
l = length of coil in meters
12.57×10^{-7} = the absolute permeability of air

IN-PROCESS LEARNING CHECK I

Fill in the blanks as appropriate.

1. All other factors remaining the same, when an inductor's number of turns increases four times, the inductance value ___________ by a factor of ___________.
2. All other factors remaining the same, when an inductor's diameter triples, the inductance value ___________ by a factor of ___________.
3. All other factors remaining the same, when an inductor's length increases, the inductance value ___________.
4. When the core of an air-core inductor is replaced by material having a permeability of ten, the inductance value ___________ by a factor of ___________.

EXAMPLE

An air-core coil that is 0.01 meters in length with a cross-sectional area of 0.001 meters has 2000 turns. What is its inductance in henrys?

$$L = 12.57 \times 10^{-7} \times \left(\frac{1 \times (2000)^2 \times 0.001}{0.01}\right)$$

$$L = 12.57 \times 10^{-7} \times \left(\frac{(40 \times 10^5) \times 1 \times 10^{-3}}{10 \times 10^{-3}}\right)$$

$$L = 12.57 \times 10^{-7} \times 4 \times 10^5 = \text{about } 0.5\text{ H, or } 500\text{ mH}$$

Since the μ_r of air is one, the formula is modified for *air-core* inductors and stated as:

Formula 12–5	$L = 12.57 \times 10^{-7} \times \frac{N^2A}{l}$ Henrys

INDUCTORS IN SERIES AND PARALLEL

For our purposes, we will assume the inductors in this section do not have any *coupling* between them. That is, there are no flux lines from one inductor linking the turns of another inductor. In technical terms, no *mutual inductance* exists between coils.

Inductance in Series

Finding the total inductance of inductors in *series,* Figure 12–5a, is quite simple. Since they are in series, the current change through each inductor is the same. Thus, when they are equal value inductors, the induced voltage in each is equal, and the total inductance is two times the value of either inductor. This implies that when the inductors are not equal in value, the total inductance (**L_T**) is the sum of the inductance values in series. That is, $L_T = L_1 + L_2$.

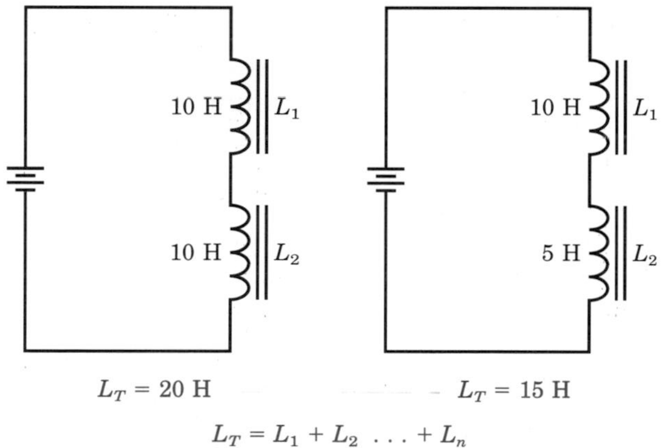

FIGURE 12–5a Total inductance of series inductors

Stated another way: *Inductors in series add inductances in the same way that resistors in series add resistances!*

Formula 12–6	$L_T = L_1 + L_2 \ldots + L_n$

EXAMPLE

When the inductors in Figure 12–5a are tripled in value, the total inductance of the circuit on the left is 60 H, rather than 20 H, and the total inductance of the circuit on the right is 45 H, rather than 15 H.

PRACTICE PROBLEM III

Determine the total inductance of the circuit, Figure 12–5b.

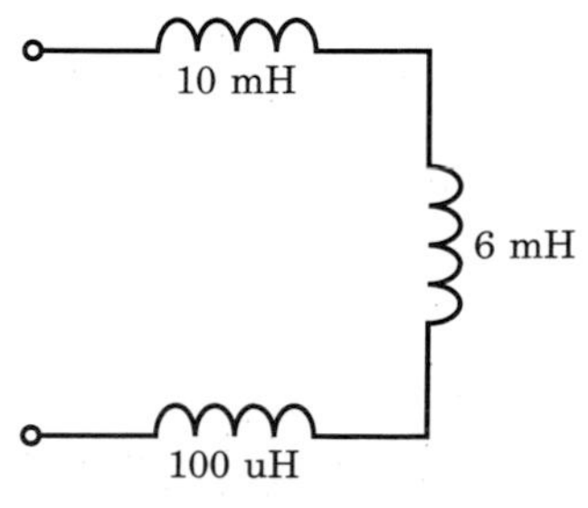

FIGURE 12–5b

Inductance in Parallel

As you might expect, the total inductance of inductors in parallel is less than the least value inductor in parallel. This is analogous to resistors in parallel.

FIGURE 12–6 Finding L_T using the reciprocal formula

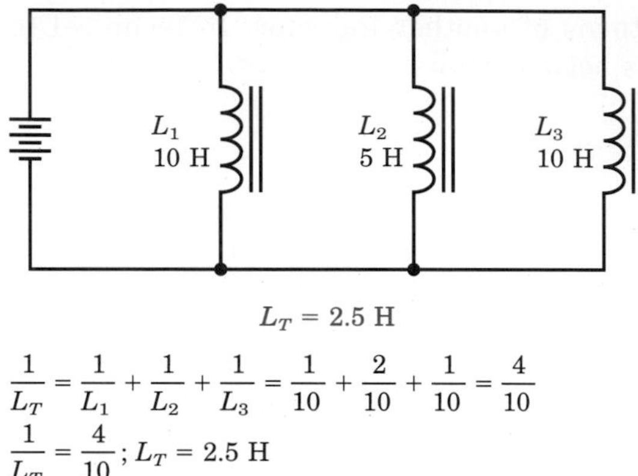

$L_T = 2.5$ H

$$\frac{1}{L_T} = \frac{1}{L_1} + \frac{1}{L_2} + \frac{1}{L_3} = \frac{1}{10} + \frac{2}{10} + \frac{1}{10} = \frac{4}{10}$$

$$\frac{1}{L_T} = \frac{4}{10};\ L_T = 2.5 \text{ H}$$

The general formulas, used in Figure 12–6, used to compute total inductance for inductors in parallel are the reciprocal formulas:

Formula 12–7	$\frac{1}{L_T} = \frac{1}{L_1} + \frac{1}{L_2} + \frac{1}{L_3} \ldots + \frac{1}{L_n}$	OR	Formula 12–8	$L_T = \cfrac{1}{\frac{1}{L_1} + \frac{1}{L_2} + \frac{1}{L_3} \ldots + \frac{1}{L_n}}$

FIGURE 12–7 Finding L_T using the product-over-the-sum formula

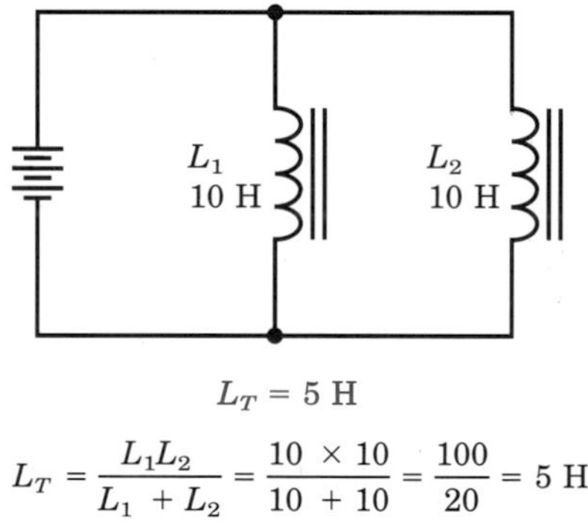

$L_T = 5$ H

$$L_T = \frac{L_1 L_2}{L_1 + L_2} = \frac{10 \times 10}{10 + 10} = \frac{100}{20} = 5 \text{ H}$$

(NOTE: If all units are in henrys, the answer is in henrys.) Also, the product-over-the-sum approach, used in Figure 12–7, works for two inductors in parallel:

Formula 12–9	$L_T = \frac{L_1 \times L_2}{L_1 + L_2}$ Henrys

EXAMPLE

You have probably already surmised that the same time-saving techniques can be used for inductance calculations as you used in resistance calculations. For example, if the two inductors in parallel are of equal value, the total inductance equals half of one inductor's value. This is similar to two 10-Ω resistors in parallel having an equivalent total resistance of 5 Ω.

PRACTICE PROBLEM IV

What is the total inductance of three 15-H inductors that are connected in parallel and have no mutual inductance?

ENERGY STORED IN THE INDUCTOR'S MAGNETIC FIELD

The magnetic field surrounding an inductor when current passes through the inductor is a form of stored energy, Figure 12–8. The current producing the electromagnetic field is provided by the circuit's source. Hence, this stored energy is supplied by the source to the magnetic field.

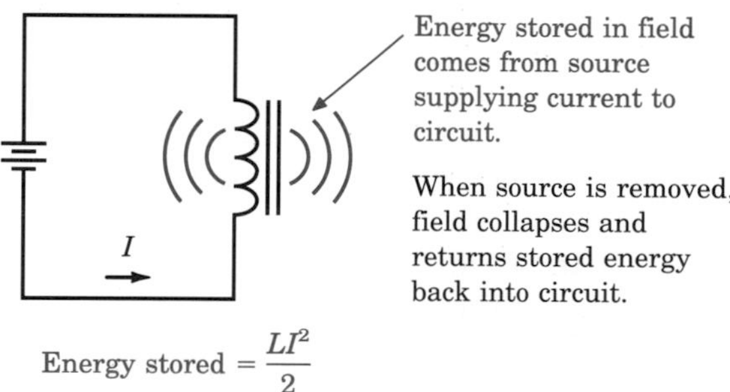

Energy stored in field comes from source supplying current to circuit.
When source is removed, field collapses and returns stored energy back into circuit.
I
Energy stored = $\frac{LI^2}{2}$

FIGURE 12–8 Concept of energy being stored in inductor's magnetic field

When the source is removed from the circuit, the field collapses and returns the stored energy to the circuit in the form of induced emf. This induced emf tries to keep the current through the coil from decaying.

If the inductor were a perfect inductor with no resistance, no I^2R loss would occur; therefore, no power would be dissipated in this energy storing-and-returning process. NOTE: In reality, inductors have some resistance (the resistance of the wire), and they dissipate power; however, for the present discussion, we won't analyze the details of this fact. Note also that in dc circuits, the inductor's resistance (plus any other resistance in series with the inductor) and the amount of voltage applied determine the amount of current through the coil (Ohm's law).

Notice in Figure 12–9, if a coil has 10 ohms of resistance, and a dc voltage of 20 volts is applied, the dc current through the coil will be $\frac{20\text{ V}}{10\ \Omega}$, or 2 amperes.

Computing the energy stored in an inductor's field is not a daily calculation for most technicians. Many technicians, however, take FCC examina-

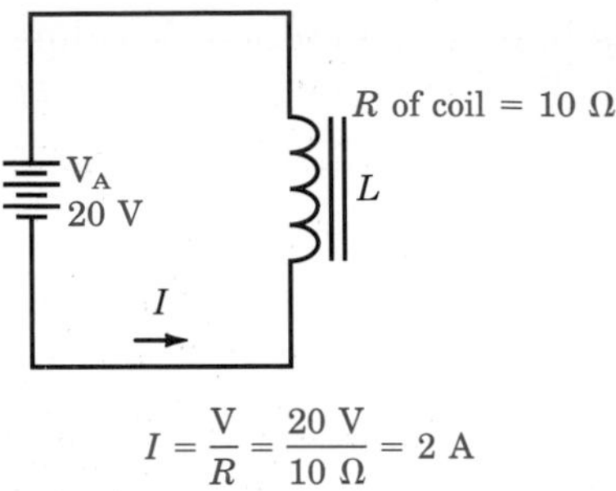

$$I = \frac{V}{R} = \frac{20\ V}{10\ \Omega} = 2\ A$$

FIGURE 12–9 Direct current (dc) through inductor found by using Ohm's law

tions to obtain commercial radio licenses. Since this computation is required for these tests, we will show you how this calculation is performed.

In an earlier chapter on Ohm's law, we discussed the unit of electrical energy, the **joule.** A joule is sometimes defined as the amount of electrical energy required to move one coulomb of electrical charge between two points having a potential difference of one volt. Also remember the watt of power relates to the joule since the performance of electrical work at a rate of one joule per second equals one watt.

Relating all of these factors, it can be shown that the stored energy correlates to the average power supplied to the inductor circuit and the time involved. The amount of energy stored, in turn, relates to the amount of inductance and current involved.

Accumulating these facts, then:

Formula 12–10	Energy (stored in the magnetic field) = 0.5 LI^2 joules

where L = inductance in henrys
I = current in amperes
Energy = joules (or watt-seconds)

EXAMPLE

What is the energy stored by an inductor of 20 henrys that has a current of 5 amperes passing through it?

Substituting into the formula:

$$\text{Energy} = 0.5\ LI^2 \text{ joules}$$

$$= \frac{20 \times (5)^2}{2}$$

$$= \frac{20 \times 25}{2} = \frac{500}{2}$$

$$= 250\ J$$

PRACTICE PROBLEM V

A dc voltage of 100 volts is applied to a 1-H inductor that has 20 Ω of dc resistance. What is the dc current through the coil? How many joules of energy are stored in the inductor's magnetic field?

THE $\frac{L}{R}$ TIME CONSTANT

We earlier established that an inductor opposes a change in current flow due to the expanding or contracting magnetic field as it cuts the turns of the coil. The induced voltage opposes the change that originally induced it (Lenz's law). This implies that even when dc voltage is first applied to a coil, the current cannot instantly rise to its maximum value (the $\frac{V}{R}$ value), which is true!

Let's examine the transient conditions, or the temporary state caused by a sudden change in circuit conditions in a simple inductive circuit when dc is first applied. Refer to Figures 12–10 and 12–11 as you study the following discussion.

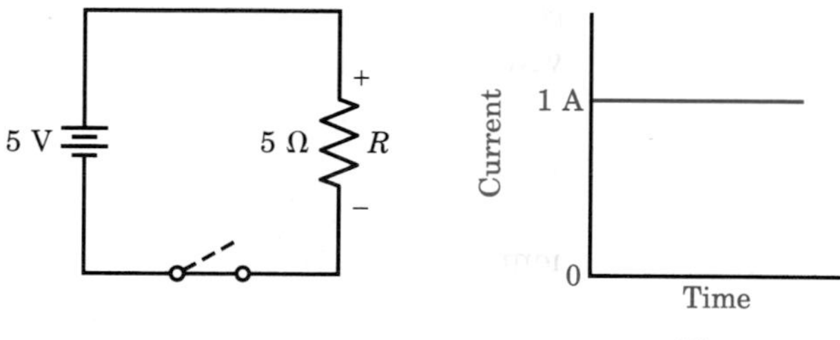

When switch is closed, I immediately rises to $\frac{V}{R}$ value.

FIGURE 12–10 Direct current (dc) change in purely resistive circuit

1. In Figure 12–10, you see the conditions that exist in a purely resistive circuit. When the switch is closed, the current immediately reaches its dc (Ohm's law) value determined by the voltage applied and the resistance involved.

2. In Figure 12–11a, the inductor in the resistor-inductor (RL) circuit prevents current from rising to its maximum value during the first moments after the switch is closed. This is because the current increases from zero to some value, causing an expansion of the magnetic field in the inductor. The expanding field cuts the turns of the coil, inducing a back-emf (cemf) that opposes the current change. Therefore, the current does not abruptly rise from zero to the final (stable) dc value.

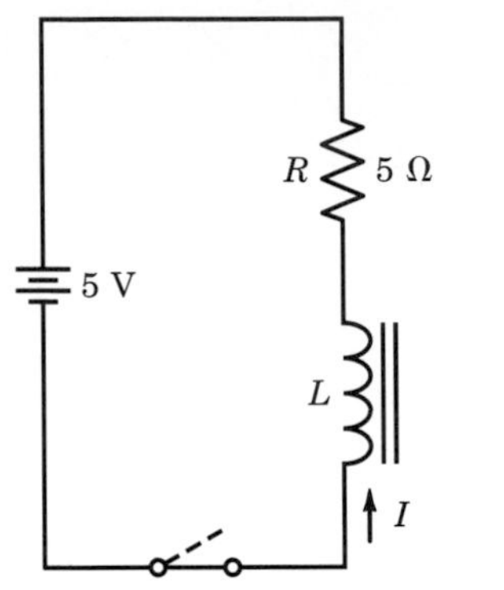

When switch is closed, current rises "exponentially" rather than instantly to a value $= \frac{V}{R}$.

(a)

100%
86.5%
63.2%

t_0 t_1 t_2 t_3 t_4 t_5

$\frac{L}{R}$ $2\frac{L}{R}$ $3\frac{L}{R}$ $4\frac{L}{R}$ $5\frac{L}{R}$

It takes $5\frac{L}{R}$ time constants for current to complete the change.

(b)

FIGURE 12–11 Current change in circuit containing inductance

3. In Figure 12–11b, at time t_1, the current has arrived at only 63.2% of its final value. The time from t_0 to t_1 is equal to the value of $\frac{L}{R}$, where L is in henrys and R is the circuit resistance value in ohms. This value $\left(\frac{\mathbf{L}}{\mathbf{R}}\right)$ is known as one **time constant,** abbreviated as 1 **TC.** NOTE: The symbol for the time constant is the Greek letter tau (τ). In a circuit containing L and R, the formula for time constant is $\tau = \frac{L}{R}$.
4. In Figure 12–11b, at time t_2 (where the time from t_1 to t_2 represents a second time constant), the current has risen 63.2% of the remaining distance to the final dc current level. That is, current rises 63.2% of the remaining 36.8% to be achieved, or 63.2% of 36.8% equal an additional 23.3%. Thus, in two time constants $\left(2 \times \frac{L}{R}\right)$, the current achieves 86.5% of its final maximum dc value, or 63.2% + 23.3% = 86.5%.
5. As time continues, the current rises another 63.2% of the remaining value for each time constant's worth of time. After *five time constants,* the current has virtually arrived at its final dc value. In this case, $\frac{5\text{ V}}{5\ \Omega} = 1$ ampere. Therefore, the amount of time required for current

to rise to maximum $\frac{V}{R}$ value is directly related to the value of inductance (L) and inversely related to the circuit R, as indicated by the $\frac{L}{R}$ relationship.

The curve represented on the time-constant graph is sometimes called an exponential curve. Many things in nature follow this exponential curve.

PRACTICAL NOTES

The exponential curve we are discussing is a plot that illustrates the raising of a given number to successive negative powers, or exponents. The number that relates to this natural growth (or decay) rate of many things in nature (including the plot of time constants being discussed) is the number 2.71828. This "magic" number is often referred to by the Greek letter epsilon (ϵ). That is, ϵ, for many mathematical purposes is considered to be the number 2.71828. For example, $\epsilon^0 = 1.0000$, $\epsilon^{-1} = 0.3679$, $\epsilon^{-2} = 0.1353$, and so on.

The inductor's expanding magnetic field and the resulting induced voltage, cause current to exponentially increase (in a nonlinear fashion), rather than instantly as it would in a purely resistive circuit.

What happens when the inductor's field collapses, that is, when the source is removed after the field has fully expanded? In Figure 12–12, when the switch is in position A, the source is connected to the RL circuit.

When the switch is in position B, the source is disconnected from the RL circuit. However, a current path is provided for the RL circuit, even when the source is disconnected.

Assume the switch was in position A for at least five time constants and the current has reached its maximum level. Therefore, the magnetic field is fully expanded. When the switch is moved to position B, the expanded field collapses, inducing an emf into the coil, which attempts to keep the current from changing. The inductor's collapsing field and induced emf have now, in effect, become the source. Obviously, as the field fully collapses, current decays and eventually becomes zero. The following comments describe what happens during this declining current time. Figure 12–13 shows a graphic display of both the current build-up and decay. Refer to it, as appropriate, while reading the following discussion.

1. The current decays 63.2% from maximum current value toward zero in the first time constant. In other words, it falls to 36.8% of maximum value.

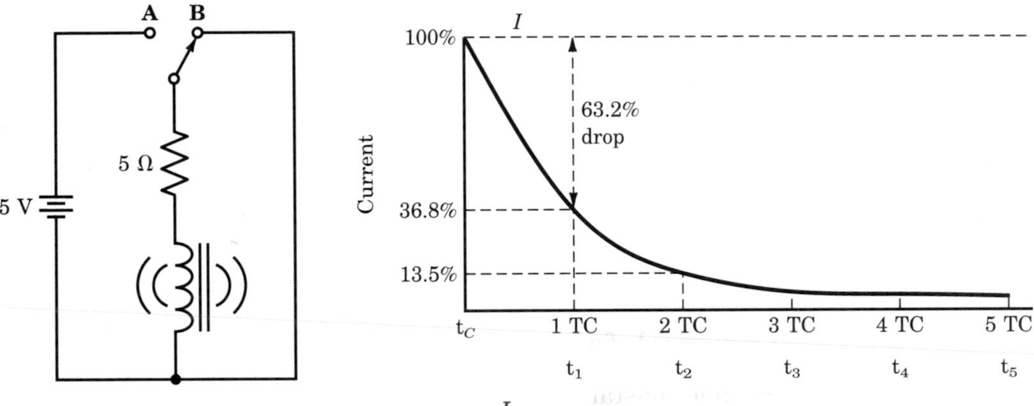

It takes $5\frac{L}{R}$ time constants for current to complete change.

FIGURE 12–12 Circuit with same $\frac{L}{R}$ time constant for current decay as for current buildup

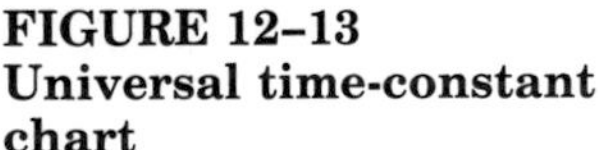

FIGURE 12–13 Universal time-constant chart

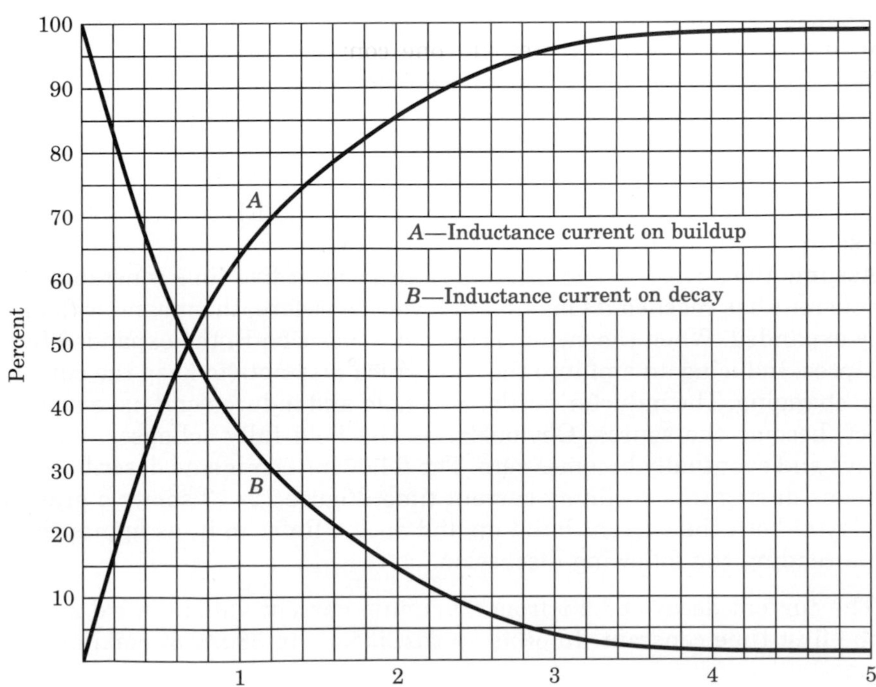

2. During the second TC, it falls 63.2% of the remaining distance, or another 23.3% for a total of 86.5% decay. In other words, the current is now at 13.5% of its maximum value.

3. This continues, until at the end of the five $\frac{L}{R}$ time constants, the current is virtually back at the zero level.

4. Given the same $\frac{L}{R}$ conditions, the current decay is the mirror image (or the inverse) of the current-rise condition, described earlier.

Let's summarize. In a dc circuit comprised of an inductor and a series resistance (RL circuit), it takes five $\frac{L}{R}$ time constants for current to change *from one static level to a "changed" new level,* whether it be an increase or a decrease in current levels. (NOTE: This is true whether the change is from 0 to 1 ampere, from 1 to 5 amperes, or from 4 amperes to 2 amperes, and so forth.) This is an important concept, so remember it!

Observe Figure 12–14 and notice that in a *dc* RL circuit, the sum of V_L and V_R must equal V applied. This means as current increases through the RL circuit, V_R is increasing and V_L is decreasing proportionately until V_R equals approximately V applied and V_L equals approximately 0 V when the current has finally reached a steady-state value. (Assuming the R of the inductor is close to zero ohms). Conversely, when current decays, V_R also decays at the same rate to the steady-state minimum value the circuit is moving toward. The TC curves are one convenient method to determine what values V_R and V_L have after specified times.

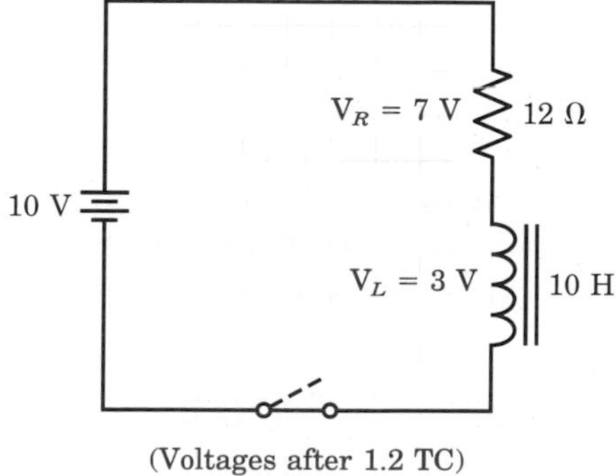

FIGURE 12–14 Simple dc, *RL* circuit

EXAMPLE

Refer to Figure 12–14. What is the value of V_R one second after the switch closed?

Answer: First, compute how many time constants are represented by one second of time: $\tau = \frac{L}{R} = \frac{10\ H}{12\ \Omega} = 0.833$ seconds. Therefore, 1 second = $\frac{1}{0.833}$ time constants, or 1 second = 1.2 TC.

Next, refer to the TC curves and note that at 1.2 TC, the current rises to 70% of maximum value. Therefore, $V_R = 0.7 \times V$ applied.

Thus, V_R at that instant equals $0.7 \times 10\ V = 7$ volts and V_L is 3 volts.

PRACTICE PROBLEM VI

Using the same circuit as in Figure 12–14, determine the V_R and V_L values two seconds after the switch is closed.

A Rapidly Collapsing Field Induces Very High Voltage(s)

In a circuit, such as in Figure 12–14, when the source voltage is disconnected, there is no low resistance current path, as was shown earlier when explaining the current decay events.

For purposes of explanation, let's assume the resistance across the open-switch contacts is extremely high. What happens to the total R in the current path when the switch opens? The R drastically increases (approaching infinity). What impact does this have on the $\frac{L}{R}$ time constant? Since R drastically increases and L remains the same, the time constant must greatly decrease!

The result of all this activity is the expanded field collapses almost instantly. This means many flux lines cut the turns of the inductor in a very short period of time; thus a very high voltage is induced in the coil. (NOTE: This phenomenon produces high voltage and high voltage "arcs" jumping the gaps of spark plugs in your automobile.) In Figure 12–14, it would be possible that the induced voltage is high enough to cause an arc across the open-switch contacts. The stored magnetic energy represented by the collapsing field is thus dissipated by the energy used in producing the arc.

SUMMARY COMMENTS OF INDUCTORS

You have studied that inductors have properties of inductance (that oppose a *change* in current) and resistance in the wire from which the inductor is wound (that opposes current flow). The amount of inductance, L (measured in henrys), relates to the physical properties of the coil, including the length, cross-sectional area, number of turns, and type of core. Also, there is a relationship between the amounts of inductance and induced cemf that develop for any given rate of current change through the coil.

Also, you know when a dc voltage is first applied to a circuit with an inductor present, current does not instantly rise to the value dictated by the dc resistance of the circuit divided by the applied voltage. This resistance is due to the inductor's opposing any current change. It takes five $\frac{L}{R}$ time constants for the complete change from one current level (even if it is zero amperes) to a new current level. Under opposite conditions (i.e., if the dc source is removed from the inductor circuit) current will not abruptly decrease to zero, since the inductor opposes the current change and the collapsing field induces an emf in the coil to attempt to prevent the current from changing. When the time constant is very small during the decay sequence, the field quickly collapses, inducing a very high voltage in the coil. Wise circuit designers, therefore, make sure that current is not abruptly interrupted through the inductors in general dc circuits. When it must be abruptly interrupted, use preventive measures to avoid unwanted voltage spikes. (NOTE: There are some special applications where high voltage is desired, and the component ratings and circuit design are appropriately chosen for this purpose.)

Inductors come in a variety of physical and electrical sizes, Figure 12–15. Laminated iron-core inductors, such as those used in power supplies and audio circuitry, are often in henrys of inductance, Figure 12–15a. Powdered iron-core and variable ferrite-core inductors can have inductance ranges in millihenrys (mH), Figure 12–15b. Air-core inductors are typically in the microhenrys (μH) range, Figure 12–15c.

(a)

(b)

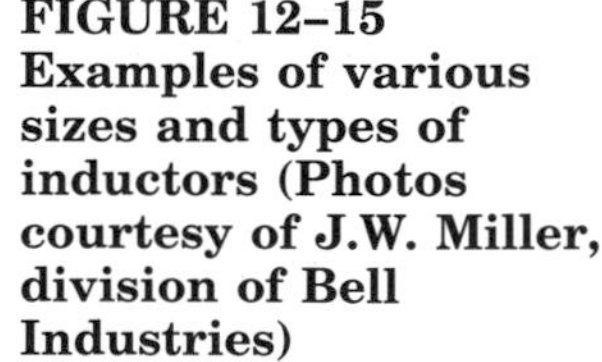

FIGURE 12–15 Examples of various sizes and types of inductors (Photos courtesy of J.W. Miller, division of Bell Industries)

(c)

Also, inductors are wound with various sizes of wire and in varying numbers of turns that affect the dc resistance of the coil. Later, you will study how the resistance to inductance ratios affect inductors' operation in certain ac circuit applications.

Excellent information is available about the details of winding coils to desired specifications in *The Radio Amateur's Handbook,* published by the American Radio Relay League (ARRL). Also, there are numerous inexpensive, special slide rules that allow you to establish required parameters and to read the number of turns, coil length, coil diameter, and so on, needed for the desired results.

TROUBLESHOOTING HINTS

Typical Troubles

Troubles in inductors are usually associated with either opens or shorts. For this reason, the ohmmeter is a useful instrument to check the condition of an inductor—*provided* you know the normal resistance for the coil being checked.

Usual values range from 1 Ω to several hundred ohms for iron-core "chokes" and inductors with many turns and inductance values in the henrys range. Inductors in the millihenrys and microhenrys ranges typically have dc resistances from less than 1 Ω to something less than 100 Ω.

Open Coil. If an inductor is open, it is obvious with an ohmmeter check that it measures infinite ohms. *Careful!* Be sure there are no sneak paths when you measure. If it is easy to do, it is good to isolate the component being measured from the rest of the circuitry when checking it. Probably the most common trouble is this open condition. Either there is a broken turn in the coil, or a bad connection to the coil, and so forth.

Shorted Turns. Because each coil turn must be electrically insulated from all other coil turns for the inductor to operate properly, when the varnish or insulation on adjacent turns breaks down, these turns can short together. This is more difficult to detect with an ohmmeter test. The key is to know what R value the given coil should read. If there are shorted turns, the R value will read lower than normal.

(NOTE: If only one or two turns are shorted, it may be impossible to detect by simple ohmmeter measurements!)

Obviously, the more turns that are shorted, the greater is the difference between the normal value and that which is measured. It is also possible in iron-core inductors to have turns shorting to the core, or being shorted by the core.

Another possibility is the whole inductor is being shorted from one end to the other. The reading will then be zero Ω. This condition, however, is less apt to happen than simply having shorted turns. Many times, an inductor with shorted turns will give the technician a clue by the burned insulation smell or visual effects shown by an extremely overheated component.

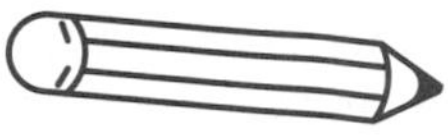

PRACTICAL NOTES

When selecting inductors for replacement or new designs, be sure to notice the important parameter ratings. Four important ratings you should note are:

1. the inductance value
2. the current rating
3. the dc resistance of the inductor
4. any voltage and insulation ratings or limitations related to the inductor's usage.

SUMMARY

- Inductance or self-inductance is the property of a circuit that opposes any *change* in current flow. Any change in current creates moving magnetic flux lines that, when linked (or cutting conductor[s]), induces an emf opposing the current change that initially produced it (Lenz's law).
- An inductor is a component that has the property of inductance. Any current-carrying conductor possesses some amount of inductance. However, most inductors are created by forming a conductor into loops of wire wound around some sort of structure. The amount of inductance created depends on the number of turns, the cross-sectional area, the length of the coil, and the permeability (magnetic characteristics) of the core material.
- Inductors come in a variety of physical and electrical sizes. Iron-core inductors typically have inductances in the henrys ranges. Powdered iron-core and variable ferrite-core inductors have values in the millihenrys range. Air-core inductors are typically in the microhenrys range.
- The amount of back-emf induced in an inductor relates to the amount of inductance involved and the rate of current change. That is, $V_L = -L\frac{di}{dt}$, where L is inductance in henrys and $\frac{di}{dt}$ equals the rate of current change in amperes per second.
- Inductance value is calculated when the inductor physical characteristics and the core magnetic properties (permeability) are known. That is, $L = (12.57 \times 10^{-7}) \times \frac{(\mu_r N^2 A)}{\text{length}}$, where 12.57×10^{-7} is the absolute permeability of air, μ_r is the relative permeability of the core, N is the number of turns, and A is the core's cross-sectional area in square meters.
- Total inductance of inductors in series (with no mutual coupling) is calculated by adding the individual inductance values. That is, $L_T = L_1 + L_2 \ldots + L_n$ This method is similar to finding total resistance of series resistors.
- Total inductance of inductors in parallel (with no mutual coupling) is calculated in a similar

fashion to that used when calculating equivalent resistance of parallel resistors. That is, either the reciprocal formula or the product-over-the-sum formula may be used. These methods are $\frac{1}{L_T} = \frac{1}{L_1} + \frac{1}{L_2} \quad \ldots \quad + \frac{1}{L_n}$, and $L_T = \frac{L_1 \times L_2}{L_1 + L_2}$.

- Energy is stored in the inductor magnetic field when there is current flowing through the inductor. The amount of energy stored is calculated in joules. That is, $E = \frac{LI^2}{2}$ joules, where E is energy in joules, L is inductance in henrys, and I is current in amperes.

- Current through an inductor does not instantly rise to its maximum value. Neither does it instantly fall from the value it is at to the minimum value where it decays. Current takes five time constants to change from the level it is at to a new stable level. A time constant (τ) is equal to the inductance in henrys, divided by the R in ohms, or $\tau = \frac{L}{R}$. (NOTE: The $\frac{L}{R}$ time for the increasing current condition may be different than the $\frac{L}{R}$ time for the decaying current condition depending on the circuitry involved).

- The universal time-constant chart is a graph of the exponential growth and decay of many things in nature. Looking at the curve of increasing values, after one time constant (TC), the quantity reaches 63.2% of maximum value. After two time constants, the quantity has reached 86.5% of maximum value and so on. After five time constants, the quantity has virtually reached 100% of the maximum value to be achieved. Looking at the decay curve, after one time constant, the value has fallen from 100% of maximum to 36.8% of maximum (or a decay of 63.2%). After two time constants, the quantity has fallen to 13.5% of the original maximum value, and so forth. After five time constants, the quantity has fallen to the minimum value involved.

- When an inductor's magnetic field rapidly collapses, a high voltage is induced. Therefore, in dc circuits, precautions should be taken to prevent abrupt interruptions of current through large inductances. Also, circuit design should minimize the effects of such changes.

- Typical problems with inductors are opens or shorts. The ohmmeter is useful to check for these problems, provided the technician has knowledge of normal resistance values for the component being tested. Shorted turns are more difficult to find with ohmmeter checks.

REVIEW QUESTIONS

1. Define inductance.

2. In what form is electrical energy stored by an inductor?

3. Define the four important physical elements that affect the amount of inductance.

4. Explain Lenz's law.

5. Explain Faraday's law.

6. State the formula for one time constant. Define terms.

7. If the iron core is removed from an inductor so that it becomes an air-core inductor, will the coil's inductance increase, decrease, or remain the same?

8. When the current through an inductor has reached its steadystate condition and the current through the coil remains the same, is there an induced back-emf?

9. When additional inductors are connected in series with a circuit's existing inductors, will the total inductance of the circuit increase, decrease, or remain the same? (Assume no mutual coupling exists between inductors.)

10. When additional inductors are connected in parallel with a circuit's existing inductors, will the total inductance of the circuit increase, decrease, or remain the same? (Assume no mutual coupling exists between inductors.)

11. When the number of turns on an inductor, which has an inductance of 100 mH, doubles, while keeping its length, cross-sectional area, and core material constant, what is its new inductance value? (Express your answer in three ways: μH, mH, and H.)

12. Current through an inductor is changing at a rate of 100 mA/s and the induced voltage is 30 mV. What is the value of inductance?

13. What is the new inductance for a 100-μH air-core coil whose core is replaced with one having a relative permeability of 500?

14. How much energy is stored in the magnetic field of a 15-H inductor that is carrying 2 amperes of current?

15. A 100 mH, a 0.2 H, and a 1000 μH inductor are connected in series. Assuming no mutual coupling exists between them, what is the circuit's total inductance? (Express your answer in millihenrys.)

16. A 10-H and a 15-H inductor are connected in parallel. Assuming no mutual coupling exists between them, what is the resultant total inductance?

17. A series RL circuit consists of a 250-mH inductor and a resistor having a resistance of 100 Ω. What is the value of one time constant? How long will it take for current to change completely from one level to another?

18. Refer to the Universal Time Constant (UTC) chart, Figure 12–13, page 422, and determine what percentage of applied voltage appears across the resistor in a series RL circuit after four time constants. What percentage of the applied voltage is across the inductor at this same time?

19. A 500-turn inductor has a total inductance of 0.4 H. What is the inductance when connections are made at one end of the coil and at a tap point connection at the 250th turn?

20. How long will it take the voltage across the resistance in a series RL circuit consisting of 5 H of inductance and 10 Ω of resistance to achieve 86.5% of V applied? (Assume the inductor has negligible resistance.)

TEST YOURSELF

1. If a 1000-turn inductor has an inductance of 4 H, how many additional turns does it take to increase the inductance to 6 H?
2. Draw a graph of the V_R and V_L voltages with respect to V applied for an inductor circuit consisting of a 10-H inductor, a series 10 Ω resistance, and a source voltage of 20 volts. Identify the specific voltages across L and R at the 1TC, 2TC, 3TC, and 4TC points on your graph.
3. Explain why a 1000-turn coil, consisting of wire that has a varnish insulation with a voltage breakdown rating of only 100 volts, can safely handle an applied voltage across the coil at 2000 volts.

PERFORMANCE PROJECT CORRELATION CHART

Suggested performance projects in the Laboratory Manual that correlate with topics in this chapter are:

CHAPTER TOPIC	PERFORMANCE PROJECT	PROJECT NUMBER
Self-Inductance $\frac{L}{R}$ Time Constant	L Opposing Change of Current	37
Inductors in Series and Parallel	Total Inductance in Series and Parallel	38

After completing the above projects, perform "Summary" checkout at the end of the performance projects section in the Laboratory Manual.

Inductive Reactance in AC

Key Terms

Inductive reactance

Leading in phase

Phase angle

Q (figure of merit)

Courtesy of DeVry Institutes

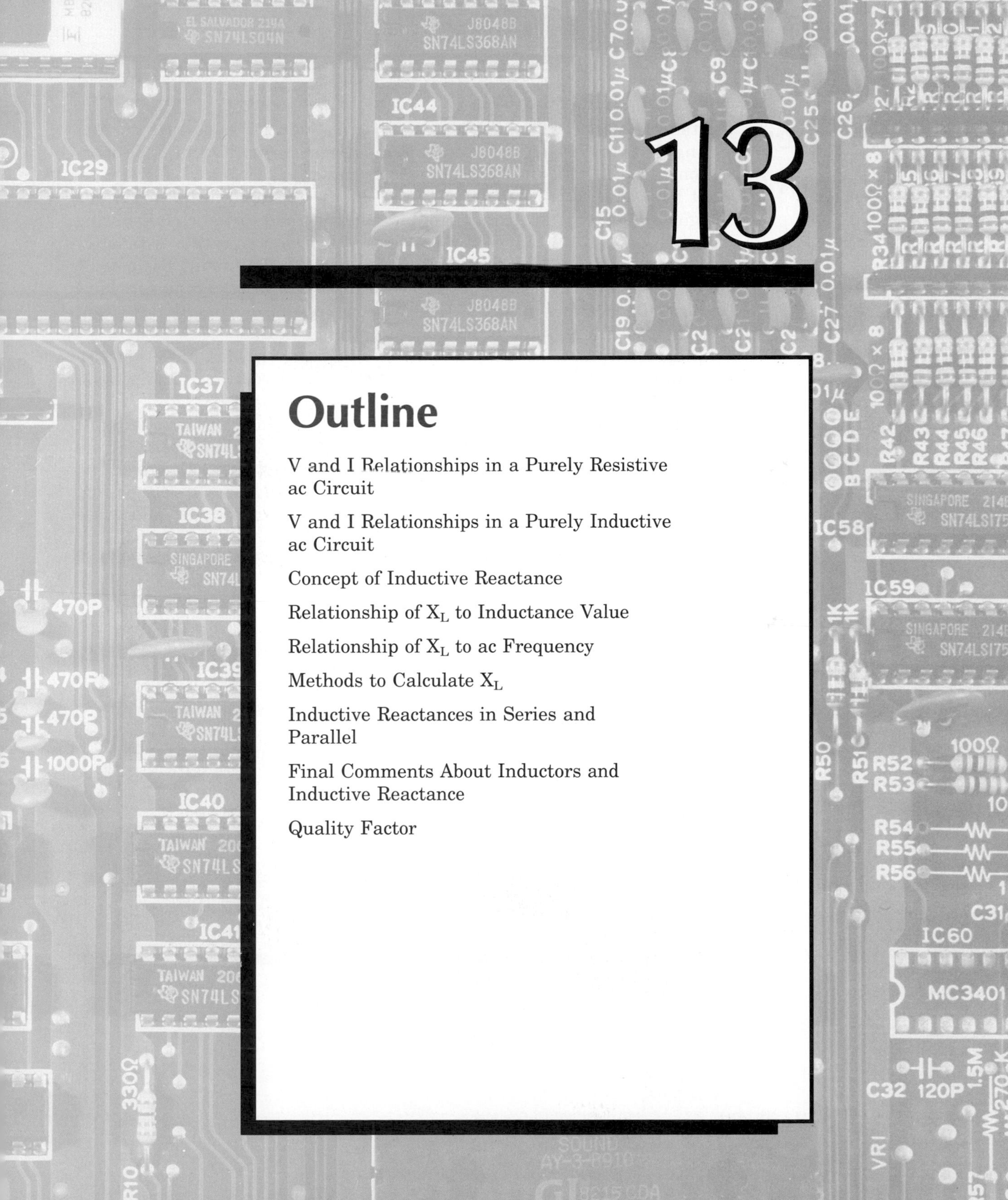

13

Outline

V and I Relationships in a Purely Resistive ac Circuit

V and I Relationships in a Purely Inductive ac Circuit

Concept of Inductive Reactance

Relationship of X_L to Inductance Value

Relationship of X_L to ac Frequency

Methods to Calculate X_L

Inductive Reactances in Series and Parallel

Final Comments About Inductors and Inductive Reactance

Quality Factor

Chapter Preview

In the previous chapter, we learned that an inductor opposes a current change but provides very little opposition to a steady (dc) current. This opposition to the current change is because an inductor reacts to changing current by producing a back-emf. You saw energy is stored in the inductor's magnetic field during current buildup and returned to the circuit during current decay. If the inductor were "perfect" (without dc wire resistance or other losses), it would return all the stored energy to the circuit, dissipate no power, and have no effect on dc circuit parameters. Practical inductors have some resistance and dissipate some power, even in dc circuits. However, these losses are not due to the property of inductance.

Inductors react to ac current, which is constantly varying. Also, they produce an opposition to ac current known as inductive reactance or reactance that will be studied in this chapter.

Important concepts discussed earlier were the inductance value and the rate of current change relationships to the induced voltage appearing across the coil. In this chapter, we will relate these concepts to the ac circuit and explore what happens when inductance is present in a sine-wave ac environment. We will show how these characteristics translate into how inductance influences parameters in ac circuits.

Objectives

After studying this chapter, you will be able to:

- Illustrate V-I relationships for a purely resistive ac circuit
- Illustrate V-I relationships for a purely inductive ac circuit
- Explain the concept of **inductive reactance**
- Write and explain the formula for inductive reactance
- Use Ohm's law to solve for X_L
- Use the X_L formula to solve for inductive reactance at different frequencies and with various inductance values
- Use the X_L formula to solve for unknown L or f values
- Determine X_L, I_L, and V_L values for series and parallel-connected inductances

V AND I RELATIONSHIPS IN A PURELY RESISTIVE AC CIRCUIT

For a quick review, observe Figure 13–1. Note in a purely resistive circuit, the voltage across a resistor is in-phase with the current through it. Both the voltage and current pass through the zero points and reach the maximum points of the same polarity at the same time.

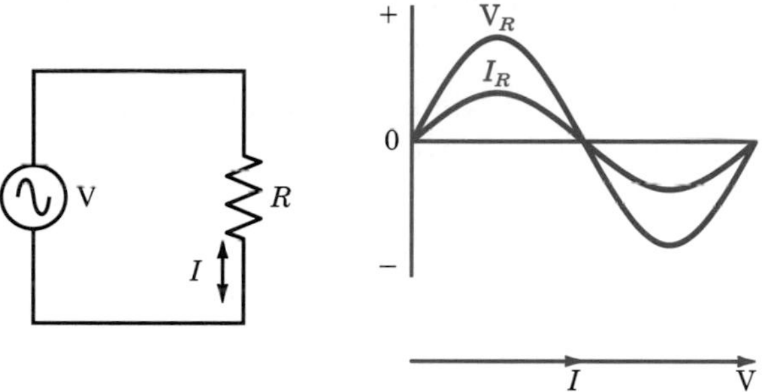

FIGURE 13–1 V and *I* relationships in a purely resistive ac circuit

V AND I RELATIONSHIPS IN A PURELY INDUCTIVE AC CIRCUIT

For purposes of explanation, let's assume the inductor is an ideal inductor, having only inductance and no resistance. What phase relationship will the current through the coil and the voltage across the coil have with respect to each other? Examine Figure 13–2 as you study the following explanations and comments.

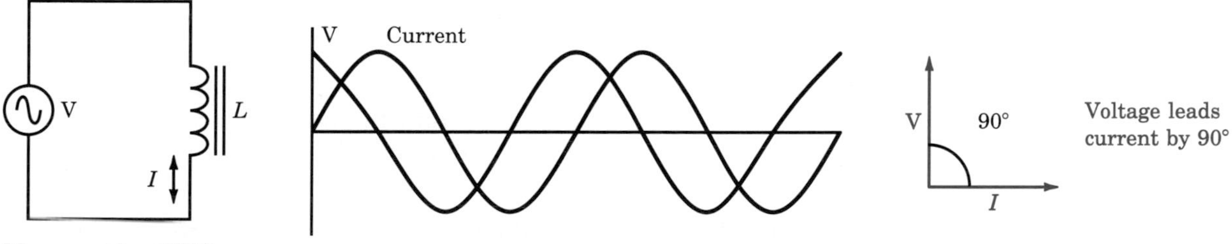

FIGURE 13–2 V and *I* relationships in a purely inductive ac circuit

1. In the circuit shown, the inductor is in parallel with the source output terminals. Therefore, recalling the rule about voltage in parallel circuits, you know the impressed voltage across the coil and the source applied voltage are the same.
2. In the circuit shown, there is only one current path. Therefore, recalling the rule about current in series circuits, you know the cur-

rent through all circuit parts is the same. That is, the circuit current (I) and the current through the inductor (I_L) are the same.

3. Recall the induced voltage (cemf) across the inductor is maximum when the rate of current change is maximum and minimum when the rate of current change is minimum.
4. Also, remember for a sine wave the maximum rate of change occurs as the waveform crosses the zero points. The minimum rate of change occurs at the maximum positive and negative peaks of the sine wave.
5. The implication is that maximum induced V occurs across the inductor when the current sine wave is either increasing from (or through) zero in the positive direction or decreasing from (or through) zero in the negative direction.
6. The *amplitude* of the induced voltage is determined by the rate of current change and the value of L.
7. Two factors determine the *polarity* of the induced voltage:
   a. the direction of the current flow through the coil, and
   b. whether the current is trying to increase or decrease.
8. Since the induced voltage opposes any current change, when the current tries to increase, the cemf will oppose it. When the current tries to decrease, the cemf will again oppose it (Lenz's law).
9. Figure 13–2 illustrates the results of these actions and reactions. In essence, because the induced back-emf is maximum when the rate of current change is maximum and minimum when the rate of current change is minimum, the current lags behind the impressed voltage. The time-lag, or delay of current behind voltage has obviously been caused by the inductance opposition to changes in current flow. The current and voltage are out of phase. The 90° **phase angle** is because the moment when the current increases or decreases, the inductor instantly develops the cemf, which opposes the change in current.
10. Since the V and I are one-fourth cycle apart, they are 90° out of phase with each other. A complete cycle represents 360° or 2π radians, where a radian equals 57.3°, or the angle described by an arc on the circle's circumference that is equal in length to the radius.
11. Using our rule of thumb from a previous chapter, you can see the voltage is at maximum positive value prior to the current reaching maximum positive value. Hence, the voltage is **leading** the current by 90°. Incidentally, it is also accurate to say current is lagging voltage by 90°. Recall the reference point concept—John is taller than Bill, or Bill is shorter than John.

At this point we will introduce a helpful phrase that will serve as a memory aid and that will help throughout your studies in ac and in your career. The understanding of the complete expression will come only after you have studied capacitors and inductors. However, the phrase can also be helpful to you now.

Here's the phrase: **Eli the ice man.** It's a short saying, so you should not have trouble memorizing it, particularly after you understand the implications!

You have just learned in a sine-wave ac circuit environment voltage across an inductor leads the current through the inductor by 90° (in a pure inductance). The word "Eli" helps you remember that voltage leads current since the E (emf or voltage) is before the I (current). Obviously the "L" indicates we are talking about an inductor. (You have probably already surmised that "ice" in this special phrase refers to current (I) leading voltage (E) in a capacitor (C), but we'll discuss this in a later chapter.)

CONCEPT OF INDUCTIVE REACTANCE

Since an inductor reacts to the constantly changing current in an ac circuit by producing an emf that opposes current changes, inductance displays opposition to ac current, Figure 13–3. Recall the inductor did not show opposition to dc current, except for the small resistance of the wire. This lack of opposition is because current must be changing to cause a changing flux that induces the back-emf we've been discussing. In dc the current is not changing, except for the temporary switch-on and switch-off conditions discussed earlier.

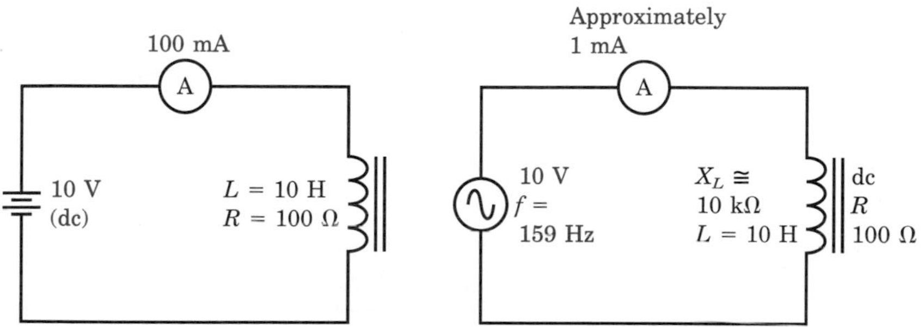

With ac applied, current is much less than with dc applied to same circuit. The additional opposition to current in the ac circuit is due to X_L of the inductor.

FIGURE 13–3 Concept of inductive reactance to ac

This special opposition an inductor displays to ac current is **inductive reactance.** The symbol "X" denotes reactance. Since it is inductive reactance, the correct abbreviation or symbol is $\mathbf{X_L}$. Because X_L is an opposition to current, the *unit of measure for inductive reactance (X_L) is the ohm!*

RELATIONSHIP OF X_L TO INDUCTANCE VALUE

We have already stated that the amount of opposition inductance shows to ac current is directly related to the amount of back-emf induced. For a given rate of current change, the amount of back-emf induced is directly related to the value of coil inductance (i.e., the greater the L, the greater the back-emf produced). Therefore, the higher the value of L, the higher will be the inductive reactance (X_L) value. In other words, if the L value doubles, the number of ohms of X_L doubles; if the L in the circuit decreases to half its original value, X_L decreases to half, and so on, Figure 13–4.

FIGURE 13–4
Relationship of X_L to L

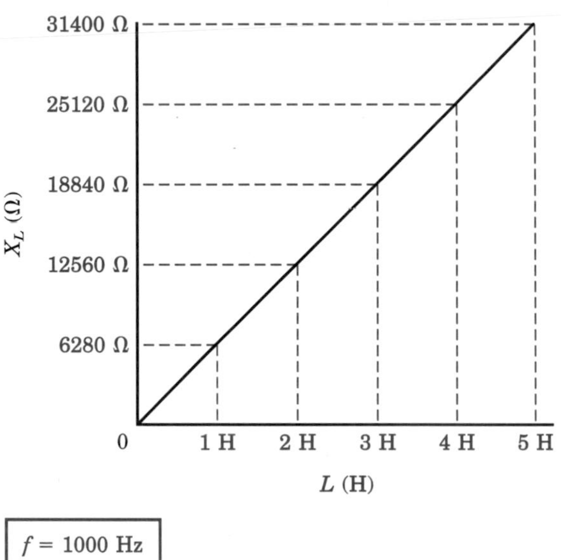

EXAMPLE

If a purely inductive circuit has a 5-H inductor with a second 5-H inductor connected in series, what happens to the circuit current? (Assume all other circuit parameters remain the same.)

Answer: Current decreases to half the original value, since the total opposition to current (X_{LT}) has doubled when inductance was doubled.

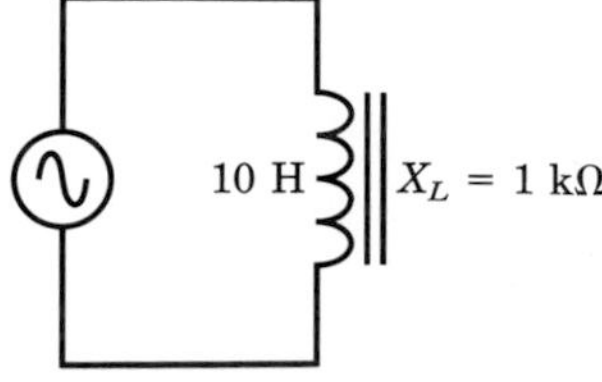

FIGURE 13–5

PRACTICE PROBLEM I

Answers are in Appendix B.

1. For the circuit in Figure 13–5, what is the value of total X_L when two more equal value inductors are added in series with the existing inductor thus making L_T 30 H? (Assume all other circuit conditions remain the same.)

2. If the three 10-H inductors in question 1 are parallel-connected rather than series-connected, what is the total X_L?

RELATIONSHIP OF X_L TO FREQUENCY OF AC

Rate of Change Related to Frequency

It has been previously established that the amount of back-emf produced in an inductor relates to both the value of inductance and to the rate of current change. As you know, the faster the rate of change, the greater the induced emf produced. And the greater the back-emf, the higher the value of opposition or X_L.

Let's take a look at how changing the frequency of an ac signal affects this rate of change. Refer to Figure 13–6 as you study the following discussion.

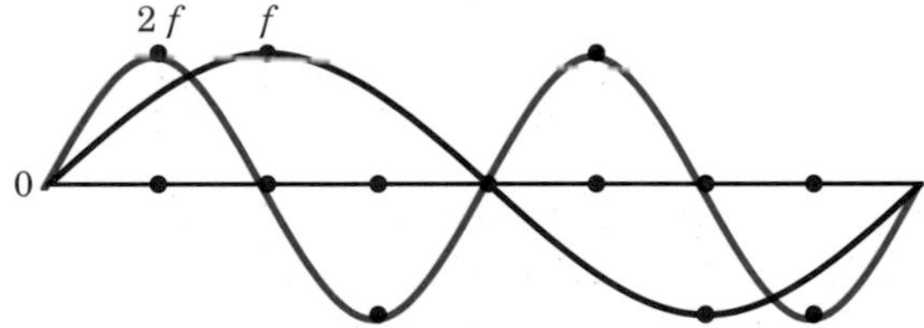

Rate of change = higher for higher frequency

FIGURE 13–6 Comparison of rate of change for two frequencies

1. Only one cycle is shown for the lower of the two frequency signals displayed. Two cycles of the higher frequency signal are shown.
2. The two signals have the same maximum (peak) amplitude. You can see the higher frequency signal must have a higher rate of change of current to reach its maximum amplitude level in half the time. Carrying this thought throughout the whole cycle, it is apparent that the higher frequency signal has twice the rate of change compared to the lower frequency signal.
3. Since the back-emf is directly related to the rate of change, when frequency doubles (which doubles the rate of change), the inductive reactance also doubles. When frequency decreases, rate of current change decreases and, consequently, the inductive reactance decreases proportionately, Figure 13–7.

Rate of Change Related to Angular Velocity

In an earlier chapter, a rotating vector to describe a sine wave was introduced. At this point, we want to indicate a relationship exists between frequency and the angular velocity of the quantity represented by a rotating vector.

FIGURE 13–7
Relationship of X_L to f

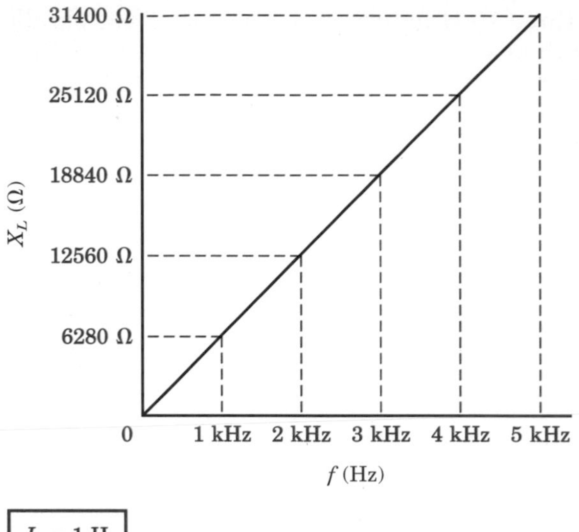

Angular velocity is the speed that a vector rotates about its axis. The symbol representing angular velocity is the Greek letter omega, ω. Thus, the relationship of angular velocity to frequency is ω equals $2\pi f$ (radians per second).

(Recall that 2π radians equal 360 degrees, or one complete cycle). We introduce angular velocity and its mathematical representation at this point to acquaint you with the expression "$2\pi f$" that is used in formulas dealing with ac quantities. You will see this expression as you continue in your studies and in your work as a technician.

METHODS TO CALCULATE X_L

The X_L Formula

Two common methods are used to calculate the value of X_L. One method is to use Ohm's law. The other method is the inductive reactance formula,

Formula 13–1	$X_L = 2\pi f L$

where X_L = inductive reactance in ohms
2π = 6.28 (approximately)
f = frequency in hertz
L = inductance in henrys

You can see from the formula and previous discussions that inductive reactance is directly related to both frequency and inductance. That is, when *either* frequency *or* inductance increases, inductive reactance increases. When *either* frequency *or* inductance decreases, X_L decreases. That is, when

IN-PROCESS LEARNING CHECK I

Fill in the blanks as appropriate.

1. When current increases through an inductor, the cemf (helps, hinders) ____________ the current increase.
2. When current decreases through an inductor, the cemf (helps, hinders) ____________ the current decrease.
3. In a pure inductor, the ____________ leads the ____________ by 90 degrees.
4. What memory aid helps to remember the relationship described in Question 3? ____________
5. The opposition that an inductor shows to ac is termed ____________.
6. The opposition that an inductor shows to ac (increases, decreases, remains the same) ____________ as inductance increases.
7. The opposition that an inductor shows to ac (increases, decreases, remains the same) ____________ as frequency decreases.
8. X_L is (directly, inversely) ____________ related to inductance value.
9. X_L is (directly, inversely) ____________ related to frequency.

one factor doubles while the other one halves, the reactance stays the same, and so on. You get the idea!

The X_L formula can also be transposed to solve for L or f, if the other two factors are known.

Formula 13–2	$L = \frac{X_L}{2\pi f}$

That is, $L = \frac{X_L}{6.28\,f}$ henrys

and

Formula 13–3	$f = \frac{X_L}{2\pi L}$

That is, $f = \frac{X_L}{6.28\,L}$

EXAMPLE

What is the inductive reactance of an iron-core choke having 5 henrys of inductance that is in a 60-Hz ac circuit?

Answer: $X_L = 2\pi f L = 6.28 \times 60 \times 5 = 1{,}884\ \Omega$.

PRACTICE PROBLEM II

1. What is the X_L of a 10-mH inductor that is operating at a frequency of 100 kHz?
2. What is the inductance in a purely inductive circuit, if the frequency of the applied voltage is 400 Hz and the inductive reactance is 200 ohms?
3. What frequency causes a 100-mH inductor to exhibit 150Ω of inductive reactance?

Using the Ohm's Law Approach

The ac current through an inductor is also calculated by using the Ohm's law expression: $I_L = \frac{V_L}{X_L}$. We can transpose this formula to find any one of the three factors, provided we know the other two factors. (We have previously done this using Ohm's law expressions.) This means that:

$$X_L = \frac{V_L}{I_L}$$

$$V_L = I_L X_L$$

$$I_L = \frac{V_L}{X_L}$$

PRACTICAL NOTES

When using Ohm's law, as indicated, it is common to use the (rms) or effective values of voltage and current in your calculations, unless otherwise indicated!

PRACTICE PROBLEM III

Use Ohm's law and answer the following.

1. What is the inductive reactance of an inductor whose voltage measures 25 volts and through which an ac current of 5 mA is passing?
2. What is the voltage across a 2-henry inductor passing 3 mA of current? Assume the frequency of operation is 400 Hz. (Hint: Calculate X_L using the reactance formula first.)

3. What is the current through an inductor having 250 Ω of inductive reactance if the voltage across the inductor measures 10 volts?

INDUCTIVE REACTANCES IN SERIES AND PARALLEL

You are aware that inductances in series are additive just like series resistances are additive. Likewise, you have already learned that inductive reactance, or X_L is directly related to the amount of inductance. Therefore, the ohmic values of inductive reactances of series inductors add like the ohmic values of series resistances add. In other words, $X_{LT} = X_{L_1} + X_{L_2} \ldots + X_{L_n}$, Figure 13–8.

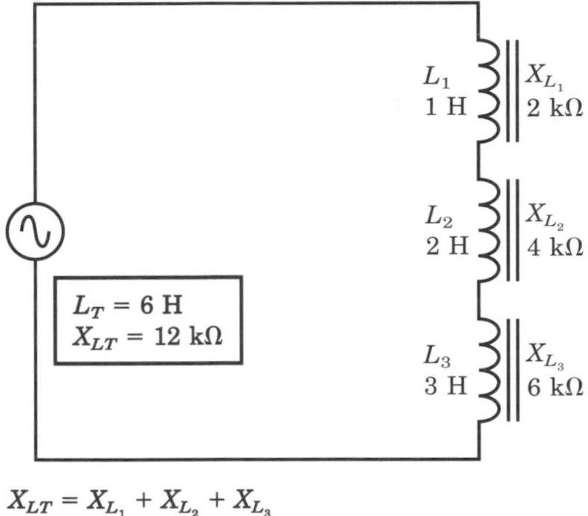

FIGURE 13–8 Inductive reactances in series

Inductance values in parallel are also treated in the same manner that parallel resistance values are treated. That is, the reciprocal or product-over-the-sum approaches are used to find the total or equivalent value(s). Since inductive reactance is directly related to inductance value, total X_L is again correlated to the total inductance of parallel-connected inductors like it was to the total inductance of series-connected inductors, Figure 13–9.

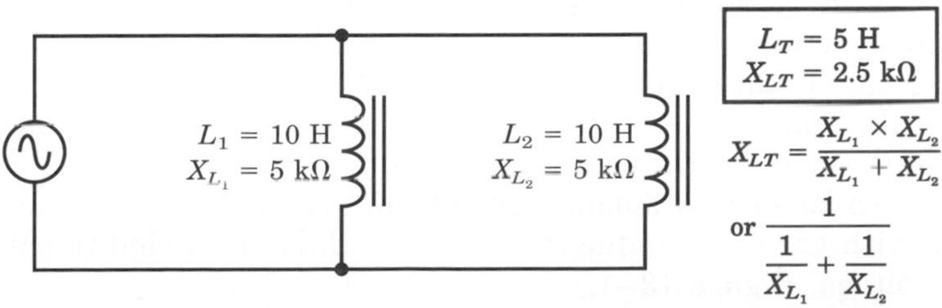

FIGURE 13–9 Inductive reactances in parallel

Useful formulas to find the total inductive reactance of parallel-connected inductors are:

Formula 13–4	$\frac{1}{X_{LT}} = \frac{1}{X_{L_1}} + \frac{1}{X_{L_2}} \ldots + \frac{1}{X_{L_n}}$

or

Formula 13–5	$X_{LT} = \frac{1}{\frac{1}{X_{L_1}} + \frac{1}{X_{L_2}}}$, etc.

For two inductors in parallel, you can use:

Formula 13–6	$X_{LT} = \frac{X_{L_1} \times X_{L_2}}{X_{L_1} + X_{L_2}}$

You can use all the "shortcuts" you have previously learned about parallel circuits—two 25-ohm reactances in parallel yield a total equivalent reactance of 12.5 ohms, or half of either reactor's value and so forth.

FINAL COMMENTS ABOUT INDUCTORS AND INDUCTIVE REACTANCE

Collecting what you now know about inductance and reactance, the following comments can be made:

1. The inductive reactance of an inductor limits or opposes the ac current flow through the inductor. That is, given an inductor in a circuit with a specified value of dc applied voltage, if that same inductor is put into a circuit with the same ac voltage (rms value), the current would be less in the ac circuit than in the dc circuit. This situation results from the opposition of the inductive reactance to ac current, refer to Figure 13–3.

2. Connecting inductances in series causes total inductance to increase, total inductive reactance to increase, and current to decrease (assuming a constant V applied), Figure 13–10.

3. Series-connected inductors may act as a series, inductive ac voltage divider with the greatest voltage across the largest inductance and the smallest voltage across the smallest inductance, and so on. NOTE: In a series circuit containing *nothing but inductors,* the total voltage equals the sum of the individual voltages across the inductors. Even though the current through the inductors is 90° out-of-phase with the voltage across each inductor, each of the inductor's voltages is in-phase with the other inductors' voltages, thus are added to obtain the total voltage, Figure 13–11.

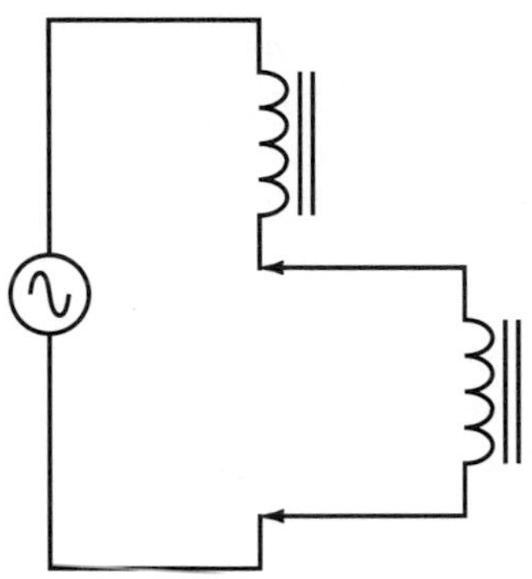

FIGURE 13–10 Effects of connecting inductances in series

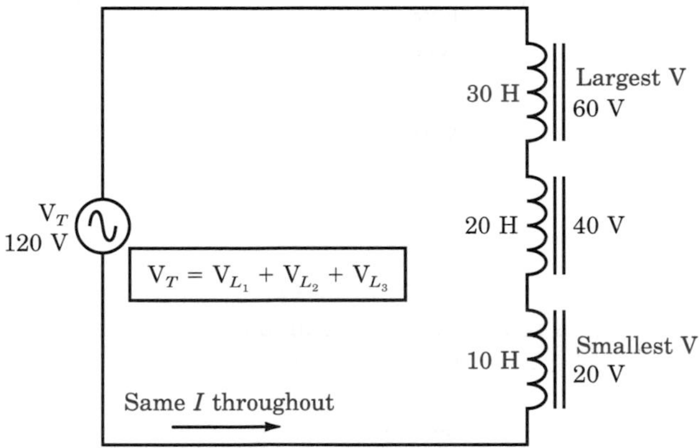

FIGURE 13–11 Voltage division across series-connected inductors

4. Connecting inductances in parallel causes total inductance to decrease, total inductive reactance to decrease, and current to increase (assuming a constant V applied), Figure 13–12.

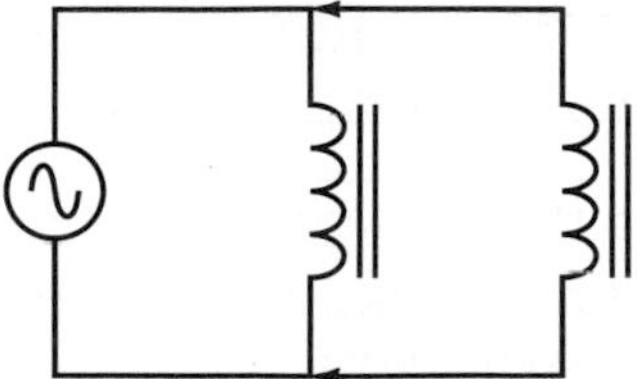

FIGURE 13–12 Effects of connecting inductors in parallel

5. Parallel-connected inductors can act as a parallel inductive ac current divider with the highest current through the lowest inductance and the lowest current through the highest inductance and so on. NOTE: In a parallel circuit containing *nothing but inductors,* the total current equals the summation of the individual inductor branch currents. Again, there is a 90° phase difference between the current through each inductor and the voltage across that inductor. But since all branch currents are lagging the circuit voltage by 90°, the branch currents are in-phase with each other, thus are added to find total current, Figure 13–13.

FIGURE 13–13 Current division through parallel-connected inductors

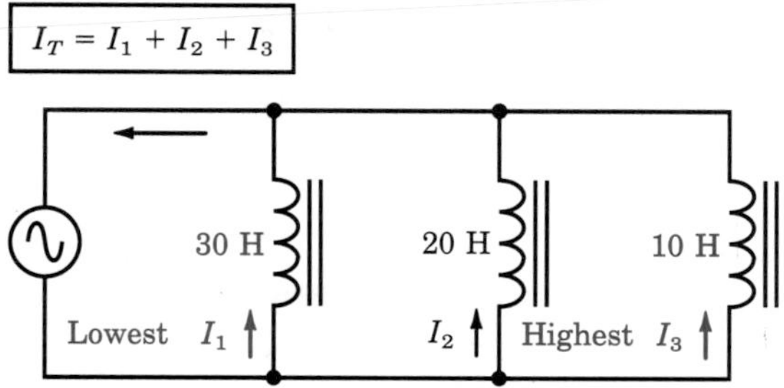

QUALITY FACTOR

We have already implied that a perfect inductor would have zero resistance, and that the only opposition this inductor would show to ac current flow would be the inductive reactance. It has also been stated that every coil has some resistance in the wire making up the coil. In many cases, this resistance is so small that it has negligible effect on the circuit operation. In other cases, this is not true. In fact, under certain conditions it may have considerable effect. For this reason, inductors sometimes have a "quality" rating that is a ratio describing the energy it stores in its field compared to the energy it dissipates due to I^2R losses.

In essence, the ratio of the X_L the inductor shows (at a given frequency) to the R it has, depicts this ratio. That is:

Formula 13–7	$Q = \frac{X_L}{R}$

EXAMPLE

The Q of a coil having 100-mH inductance and 50 Ω resistance that is operating at a frequency of 5 kHz would be:

$$Q = \frac{X_L}{R}$$

$$X_L = 6.28 \times 5 \times 10^3 \times 100 \times 10^{-3} = 3140\ \Omega$$

$R = 50\ \Omega$

$$Q = \frac{X_L}{R} = \frac{3140}{50} = 62.8$$

$Q = 62.8$ (Q has no units since it is a ratio.)

The above discussion indicates that the lower the R that a given inductor (L) has, the higher is its **figure of merit** or **Q.** (Also, the higher the frequency of operation, the greater the X_L and the higher the Q for a given inductor.)

It is generally desirable to use high Q coils, particularly at radio frequencies. A coil is considered to be high Q if its Q value is greater than approximately 20–25. Of course, this is an arbitrary value. It does, however, give some indication of the inductance to resistance ratio of the inductor. Therefore, the value is the ratio of energy stored to energy dissipated or the $\frac{X_L}{R}$ ratio we have been examining.

SUMMARY

- Inductive reactance is the opposition that inductance displays to ac (or pulsating) current.
- Inductive reactance is associated with the amount of back-emf generated by the inductor, which is associated with the value of inductance (L) and the rate of change of current.
- Inductive reactance is measured in ohms. The symbol for inductive reactance is X_L. X_L is calculated by the formula $X_L = 2\pi f L$, where X_L is in ohms, f is in hertz, and L is in henrys. The 2π is the constant equal to two times pi or approximately 6.28.
- Voltage leads current by 90° in a purely inductive circuit. (Voltage and current are in-phase for a purely resistive circuit).
- The first part of the memory-aid phrase helps you remember the phase relationship of voltage and current relative to inductors. The phrase is *Eli the ice man.* The word "*Eli*" indicates that E comes before I in an inductor (L).
- Inductive reactance is *directly proportional* to both inductance (L) *and* to frequency (f).
- Quality (or the figure of merit) of an inductor is sometimes described as its Q. This factor equates the ratio of the inductor's X_L to its R. Thus, $Q = \frac{X_L}{R}$.
- Inductive reactances in series add just as resistances in series add. That is, $X_{LT} = X_{L_1} + X_{L_2} \ldots + X_{Ln}$
- Total (equivalent) inductive reactance of inductances in parallel is computed using similar methods to those used to find total or equivalent resistance of resistors in parallel. That is, the reciprocal formula or the product-over-the-sum formula are used.

  The reciprocal formula(s) include:

  $$\frac{1}{X_{LT}} = \frac{1}{X_{L_1}} + \frac{1}{X_{L_2}} + \ldots \frac{1}{X_{L_n}} \text{ or}$$

  $$X_{LT} = \frac{1}{\frac{1}{X_{L_1}} + \frac{1}{X_{L_2}}}$$

  The product-over-the-sum formula is:

  $$X_{LT} = \frac{X_{L_1} \times X_{L_2}}{X_{L_1} + X_{L_2}}$$

CHAPTER CHALLENGE

Following the SIMPLER troubleshooting sequence, find the problem with this circuit. As you follow the sequence, record your responses to each step on a separate sheet of paper.

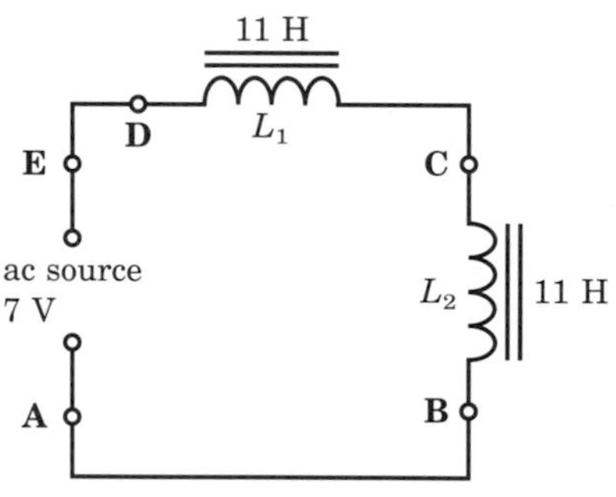

CHALLENGE CIRCUIT 7

SYMPTOMS Gather, verify, and analyze symptom information. (Look at the "Starting Point Information.")

IDENTIFY Identify initial suspect area for the location of the trouble.

MAKE Make a decision about "What type of test to make" and "Where to make it." (To simulate making a decision about each test you want to make, select the desired test from the "TEST" column listing below.)

PERFORM Perform the test. (Look up the result of the test in Appendix C. Use the number in the "RESULTS" column to find the test result in the Appendix.)

LOCATE Locate and define a new "narrower" area in which to continue troubleshooting.

EXAMINE Examine available information and again determine "what type of test" and "where."

REPEAT Repeat the preceding analysis and testing steps until the trouble is found. What would you do to restore this circuit to normal operation? When you have solved the problem, compare your results with those shown in the color insert.

Starting Point Information

1. Circuit diagram
2. Each inductor is rated as 11 H, 400 Ω dc resistance
3. The ac voltage measured across L_1 is significantly lower than the voltage measured across L_2

Test	Results in Appendix C
$V_{A\text{-}B}$	(66)
$V_{B\text{-}C}$	(31)
$V_{C\text{-}D}$	(4)
$V_{D\text{-}E}$	(82)
$V_{E\text{-}A}$	(43)
$R_{B\text{-}C}$	(17)
$R_{C\text{-}D}$	(114)

REVIEW QUESTIONS

1. What is the inductance of an inductor that exhibits 376.8 Ω reactance at a frequency of 100 Hz? What is inductive reactance at a frequency of 450 Hz?

2. What is the X_L of a 5-henry inductor at 120 Hz? What is the value of X_L at a frequency of 600 Hz?

3. If circuit conditions are changed so that frequency triples and the amount of inductance doubles, how many times greater will be the new X_L compared to the original value?

4. Assume circuit conditions change for an inductive circuit so that frequency is halved and inductance is tripled. What is the relationship of the new X_L to the original value?

5. In a circuit containing inductance, when frequency decreases, will circuit current increase, decrease, or remain the same?

6. When the frequency of applied voltage for a given inductive circuit doubles and the circuit voltage increases by three times, will the circuit current increase, decrease, or remain the same?

7. When the frequency applied to a given inductor decreases, will the Q of the coil change? If so, in what way?

8. Does the R of a given inductor change with frequency?

9. What is the total inductance of the circuit in Figure 13–14? When the voltage applied is 100 volts and the current through L_1 is 100 mA, what is the frequency of the applied voltage? (Use one decimal place in your calculations.)

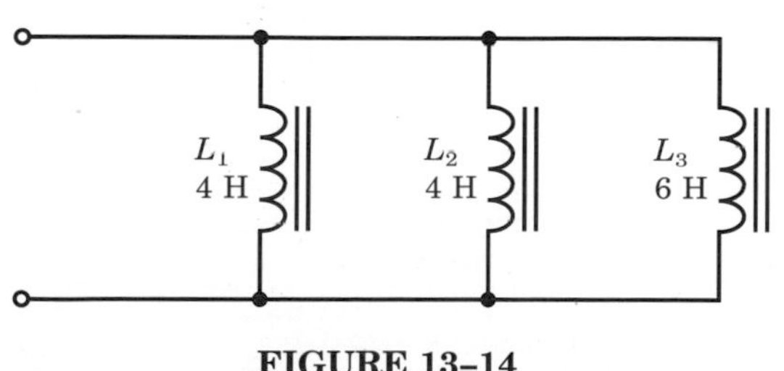

FIGURE 13–14

10. Find L_T for the circuit in Figure 13–15. At approximately what frequency does X_{LT} equal 5,000 ohms?

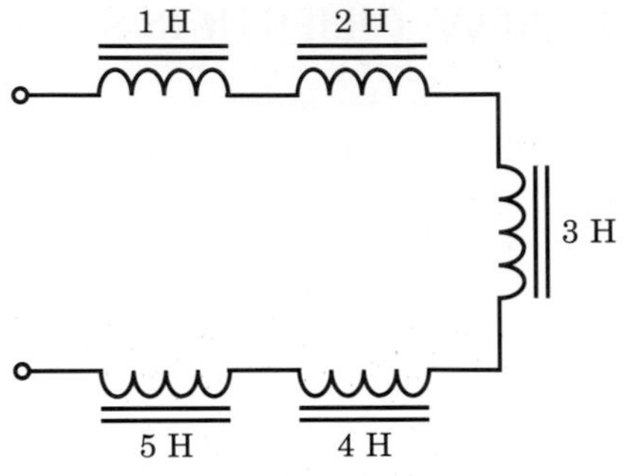

FIGURE 13–15

11. Find the Q of the coil in the circuit shown in Figure 13–16. When the frequency triples and the R of the inductor doubles, does Q increase, decrease, or remain the same? If the applied voltage were dc instead of ac, would Q change? Explain.

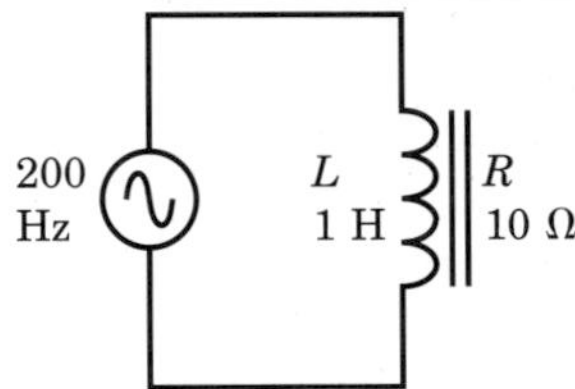

FIGURE 13–16

12. Draw a waveform showing two cycles that illustrates the circuit V and I for the circuit in Figure 13–17. Appropriately label the peak values of V and I, assuming the source voltage is shown in rms value.

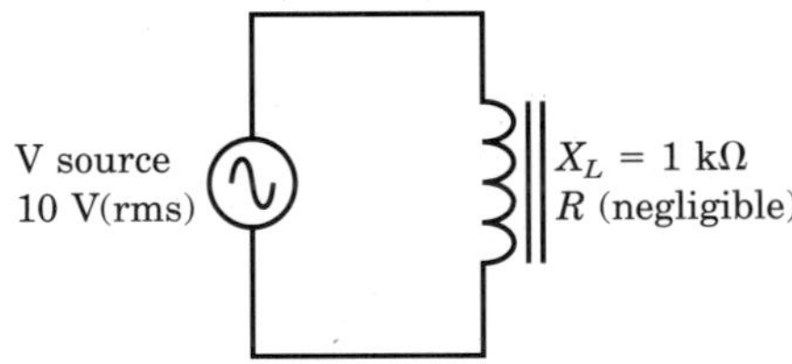

FIGURE 13–17

13. Find V_T, V_{L_1}, and I_T for the circuit in Figure 13–18.

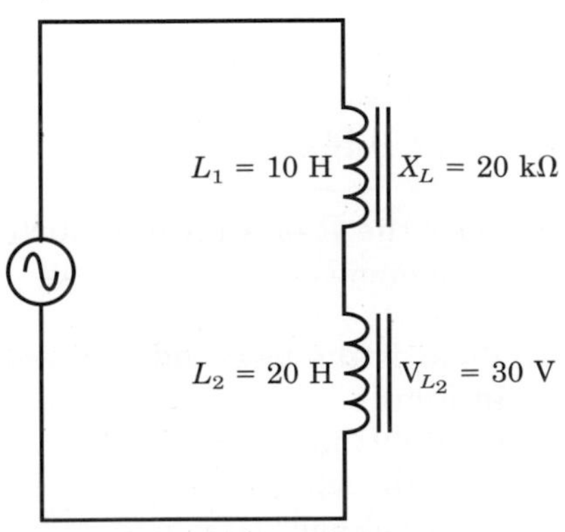

FIGURE 13–18

14. What is the branch current through L_2 for the circuit in Figure 13–19?

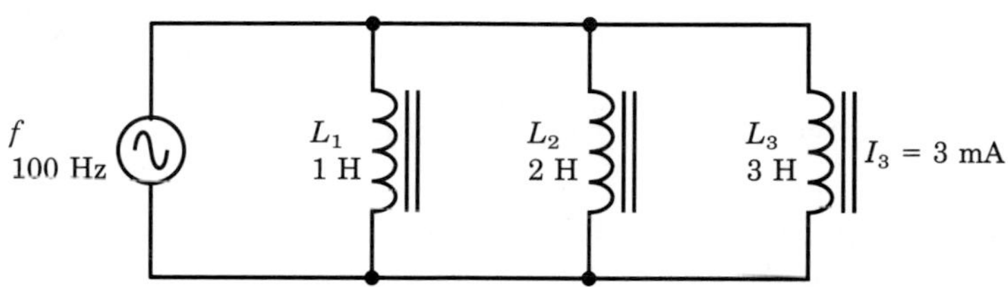

FIGURE 13–19

TEST YOURSELF

1. Determine the total inductance of the circuit in Figure 13–20.

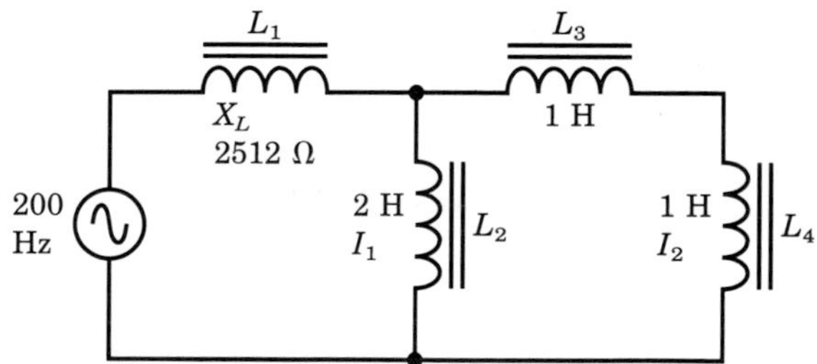

FIGURE 13–20

2. Refer again to Figure 13–20 and answer the following with "I" for increase, "D" for decrease, and "RTS" for remain the same.
   a. If frequency doubles and V applied stays the same,
      L_T will ______;
      X_{LT} will ______;
      I_T will ______;
      V_{L_4} will ______; and
      I_1 will ______.
   b. If L_2 changes to a 1-H inductor,
      L_T will ______;
      X_{L_1} will ______;
      I_T will ______;
      V_{L_1} will ______; and
      f will ______.

PERFORMANCE PROJECT CORRELATION CHART

Suggested performance projects in the Laboratory Manual that correlate with topics in this chapter are:

CHAPTER TOPIC	PERFORMANCE PROJECT	PROJECT NUMBER
Concept of Inductive Reactance	Induced Voltage	39
Relationship of X_L to L Relationship of X_L to f	Relationship of X_L to L and frequency	40
Methods to Calculate X_L	The X_L Formula	41

After completing the above projects, perform "Summary" checkout at the end of the performance projects section in the Laboratory Manual.

RL Circuits in AC

Key Terms

Cosine function

Impedance

Phase angle (θ)

Pythagorean theorem

Scalar

Sine function

Tangent function

Vector

Courtesy of International Business Machines Corporation

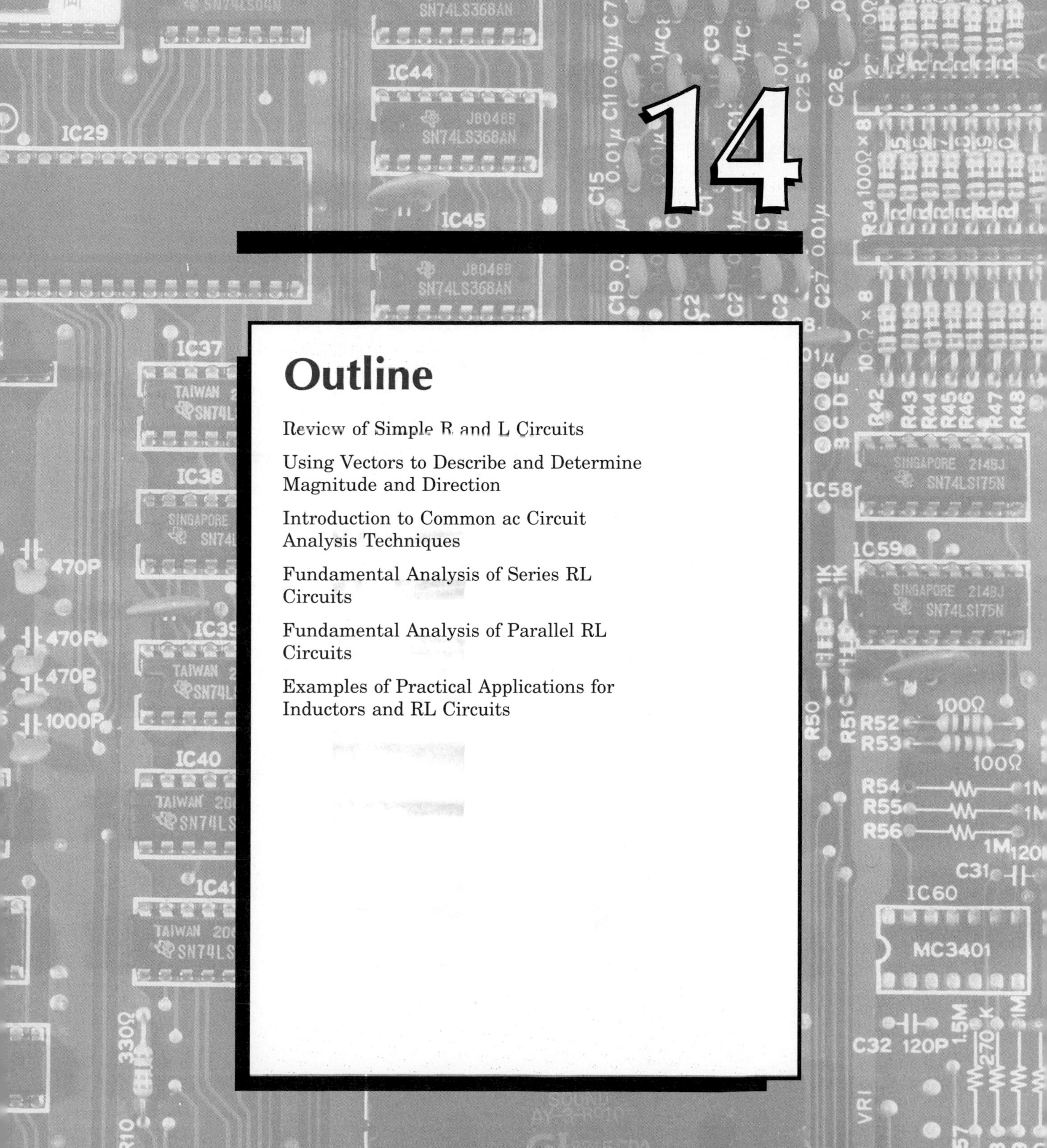

14

Outline

Review of Simple R and L Circuits

Using Vectors to Describe and Determine Magnitude and Direction

Introduction to Common ac Circuit Analysis Techniques

Fundamental Analysis of Series RL Circuits

Fundamental Analysis of Parallel RL Circuits

Examples of Practical Applications for Inductors and RL Circuits

Chapter Preview

In the preceding chapters, you have studied the circuit characteristics of both dc and ac circuits containing only inductance. You have learned basic concepts related to inductors, such as inherent opposition to current changes, the self-induced back-emf, phase relationships of V and I, and other characteristics of inductance.

The purpose of this chapter is to introduce you to basic facts and circuit analysis techniques used to examine ac circuits containing both resistance and inductance (RL circuits). Impedance, a combination of resistance and reactance, will be discussed and analyzed. Further considerations about **phase angle** will be given. Primary methods of using **vectors** and other simple math techniques to analyze RL circuit parameters will also be introduced. Later in the text you will have the opportunity to examine these ac circuit analysis methods in more depth.

The concepts discussed in this chapter are important for you to learn, because they provide valuable knowledge for all your future ac circuit analyses endeavors.

Objectives

After studying this chapter, you will be able to:

- Use **vectors** to determine magnitude and direction
- Determine circuit impedance using the **Pythagorean theorem**
- Determine V_T and I_T using the Pythagorean theorem
- Determine ac circuit parameters using trigonometry
- Calculate ac electrical parameters for series RL circuits
- Calculate ac electrical parameters for parallel RL circuits
- List at least three practical applications of inductive circuits

REVIEW OF SIMPLE R AND L CIRCUITS

Simple R Circuit

The ac current through a resistor and the ac voltage across a resistor are in-phase, as you have already learned. You know this is shown by a waveform display or phasor diagram, Figure 14–1.

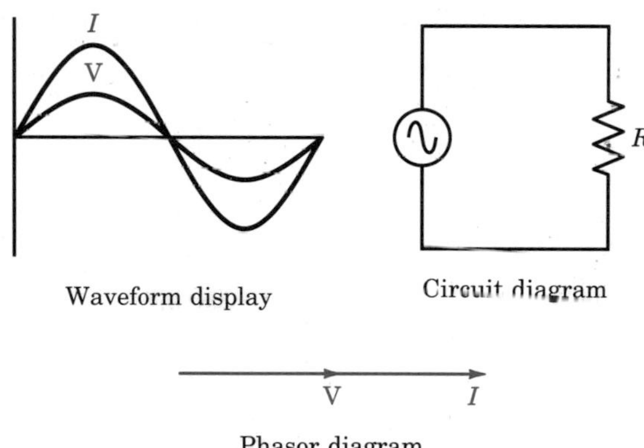

FIGURE 14–1 V and *I* in a simple *R* circuit

In the purely resistive *series* ac circuit, Figure 14–2a, you simply add the individual series resistances to find total circuit opposition to current. Or you add voltage drops across resistors to find total circuit voltage, since all voltage drops are in-phase with each other.

In purely resistive *parallel* circuits, Figure 14–2b, total resistance is

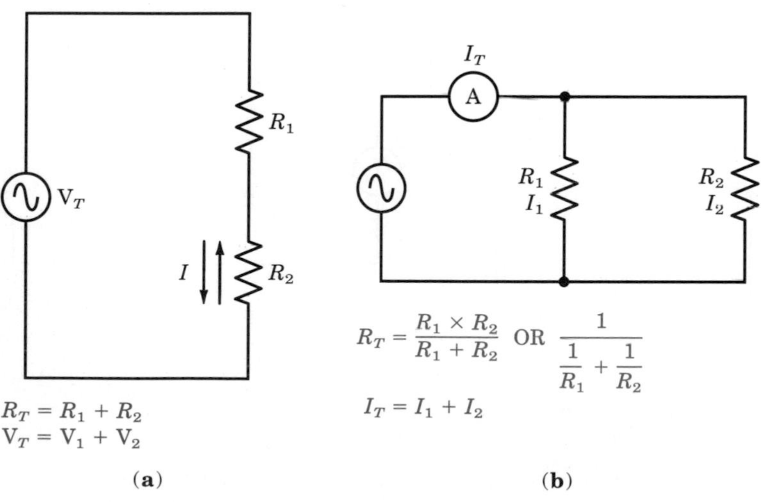

FIGURE 14–2 Purely resistive circuits

found by using the product-over-the-sum or reciprocal methods. The total circuit current is found by adding the branch currents, since all branch currents are in-phase with each other.

Simple L Circuit

In a purely inductive ac circuit, you know current through the inductor and voltage across the inductor are 90° out-of-phase with each other. Again, this is illustrated by a waveform display, Figure 14–3a or a phasor diagram, Figure 14–3b.

FIGURE 14–3 V and *I* in a simple *L* circuit

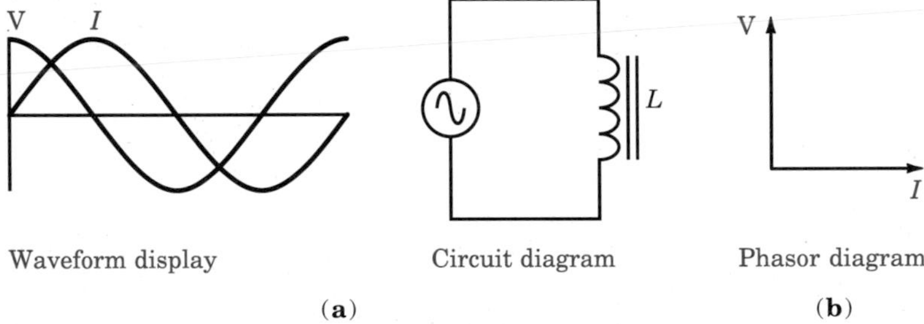

You know that in purely inductive series circuits, total inductance is the sum of the individual inductances; total inductive reactance (X_{LT}) is the sum of all X_Ls; and total voltage equals the sum of all V_Ls, since each V_L is in-phase with the other V_Ls although they are not in-phase with the current, Figure 14–4a. In purely inductive parallel circuits, total circuit opposition to current (X_{LT}) is found by using reciprocal or product-over-the-sum approaches. Also, total circuit current equals the sum of the branch currents,

FIGURE 14–4 Purely inductive circuits

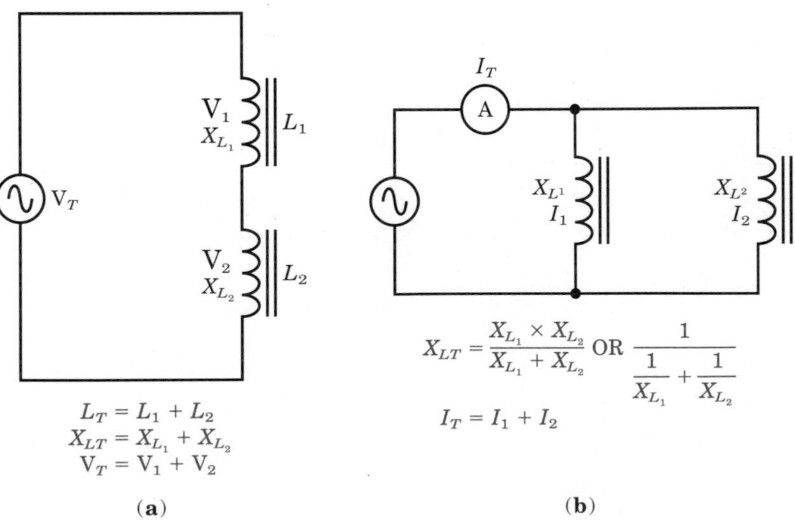

since branch currents are in-phase with each other although not in-phase with the circuit voltage, Figure 14–4b.

USING VECTORS TO DESCRIBE AND DETERMINE MAGNITUDE AND DIRECTION

Background

When ac circuits contain reactances and resistances, their voltage(s) and current(s) are not in-phase with each other. In this case we cannot add these quantities together to find resultant totals as we did with dc circuits or ac circuits that are purely resistive or reactive. Thus, other methods are used to analyze ac circuit parameters.

One of these methods is to represent circuit parameters using **vectors.** Recall *a vector quantity expresses both magnitude and direction.* Examples of quantities having both magnitude and direction that are illustrated using vectors include mechanical forces, the force of gravity, wind velocity and direction, magnitudes and relative angles of electrical voltages and currents in ac.

Scalars, on the other hand, define quantities that exhibit only magnitude. Units of measure, such as the henry or the ohm, are scalar quantities. Even dc circuit parameters, or instantaneous values of ac electrical quantities are treated as scalar quantities, since there are no time differences or angles involved. Simple addition is used to add scalar quantities, but this is not true for vector quantities.

Plotting and Measuring Vectors

Vectors can be plotted to scale to determine *both* magnitude and direction of given quantities, Figure 14–5. If a force of 10 pounds is exerted on Rope 1 and a force of 10 pounds is exerted on Rope 2, what is the total force exerted

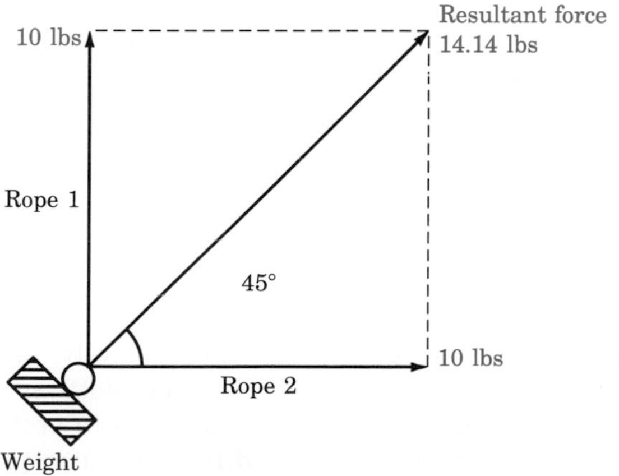

FIGURE 14–5 Vectors show both magnitude and direction.

on the weight, and in which direction is this total force? The resultant force is 14.14 pounds in a direction that is 45° removed from either rope's direction.

Look at Figure 14–6. Notice these vectors are plotted to scale so you can then *measure* both the magnitude and direction of the resultant force. *Measure the length* of the resultant vector *to find magnitude,* and *measure the angle* between the original forces and the resultant force by a protractor *to find direction.*

FIGURE 14–6 Plotting vectors to scale

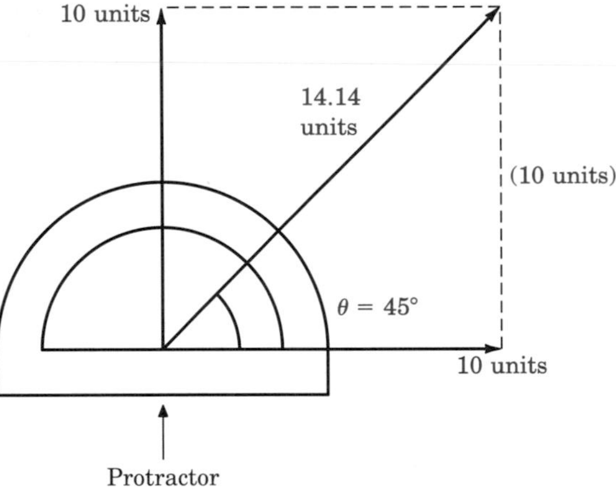

Observe Figure 14–7a that illustrates how vectors are laid end-to-end or head-to-tail to determine resultant values. Also, notice in Figure 14–7b that a parallelogram is drawn that projects the same magnitudes as the end-to-end approach. Although either method can be used to find resultant values, we will use the parallelogram approach most frequently.

FIGURE 14–7 Methods of formatting vectors

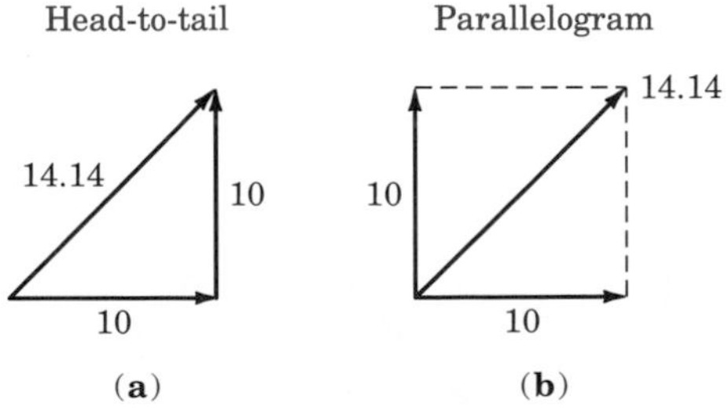

Later in the chapter you will see how this plotting and measuring technique is used to analyze RL circuit parameters. Although it is not always convenient to have graph paper and protractor handy, or to draw circuit parameters to scale when analyzing a circuit, the plotting approach can be used. Also, it is a useful tool to help introduce you to vectors.

Even when not plotted to scale for the direct measurement method, vectors are useful to illustrate ac circuit parameters and show relative magnitudes and angles. Sketching this diagram helps you understand the situation and verify your computations.

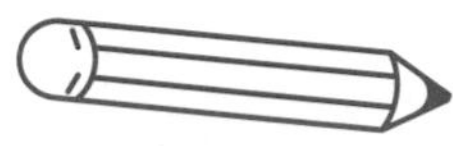

PRACTICAL NOTES

Even when not precisely measuring vectors to determine resultant values, get in the habit of approximating the *relative* lengths and directions of vectors when you draw them to solve ac problems. This allows you to visually determine if your computations are making sense! A good practice is to draw simple vector diagrams whenever you are analyzing ac circuits—they help visualize the situation.

PRACTICE PROBLEM I

Answers are in Appendix B.

Assume the forces on the ropes in Figure 14–5 are 3 pounds for Rope 1 and 4 pounds for Rope 2. What is the magnitude and direction of the resultant force on the weight? (NOTE: Use the plotting-to-scale approach and remember to keep things relative to the scale you use.)

IN-PROCESS LEARNING CHECK I

Fill in the blanks as appropriate.

1. In a purely resistive ac circuit, the circuit voltage and current are ____________-phase.
2. In a purely inductive ac circuit, the current through the inductance and the voltage across the inductance are ____________ degrees out-of-phase. In this case, the ____________ leads the ____________.
3. A quantity expressing both magnitude and direction is a ____________ quantity.
4. The length of the vector expresses the ____________.

INTRODUCTION TO COMMON AC CIRCUIT ANALYSES

Most ac circuit analyses solve for quantities that are illustrated with vectors in right-triangle form. Recall from our previous discussions that a right triangle has one angle forming 90° and two angles that total another 90°, Figure 14–8. Many electrical parameters have this 90° phase difference between quantities. For example, in perfect inductors and capacitors, the voltages and currents are 90° out-of-phase. Also, ac circuit resistive and reactive voltage drops or currents are often represented as 90° out-of-phase with each other.

FIGURE 14–8 Examples of right triangles

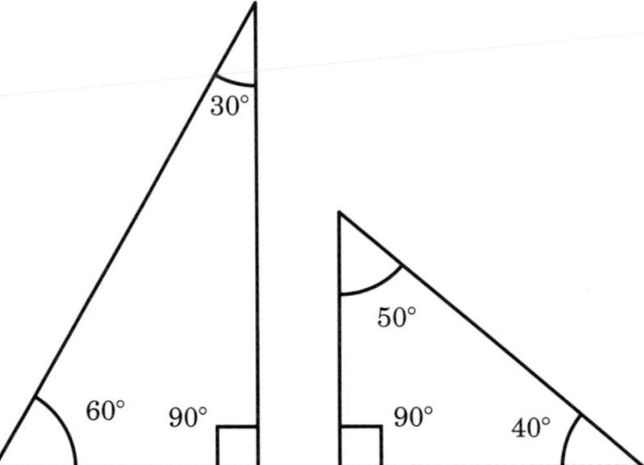

We will discuss two basic techniques used to analyze these right-triangle problems. The first method is the **Pythagorean theorem** that is useful to determine the *magnitude* of quantities. The other method briefly addressed in this chapter is trigonometric functions that determine the circuit phase angles. Also, trigonometry can be used to find the magnitude of various electrical quantities, if the phase angle is known.

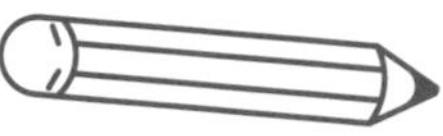

PRACTICAL NOTES

NOTE: The calculations are easy if you have a calculator with square, square root, and "trig" function keys. You probably own or have access to such a calculator. If not, you might want to learn to use these functions as you proceed through the remainder of the chapter. Also, these computations can be performed on a computer if one is available. If these devices are not available, paper and pencil will work just fine! NOTE: *Be aware* that some calculators and most computers only accept radian measure! Therefore, to convert degrees to number of radians, divide the number of degrees by 57.3; and to convert radians to number of degrees, multiply the number of radians by 57.3. Recall that 2π radians equal 360°; that is, 6.28 times 57.3° equals 360°.

Pythagorean Theorem

The Pythagorean theorem states the square of the hypotenuse of a right triangle equals the sum of the squares of the other two sides.

Formula 14–1	$c^2 = a^2 + b^2$

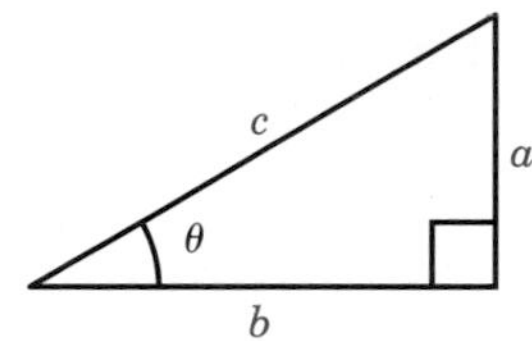

$$c^2 = a^2 + b^2$$
$$\sqrt{c^2} = \sqrt{a^2 + b^2}$$
$$c = \sqrt{a^2 + b^2}$$

FIGURE 14–9 Pythagorean theorem

Refer to Figure 14–9 and review the equations related to the right triangle and Pythagorean theorem.

PRACTICE PROBLEM II

Refer to Figure 14–10. Use the Pythagorean theorem and solve for the magnitude of side V_T.

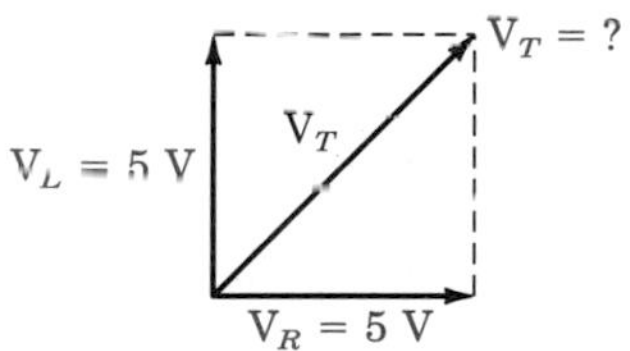

FIGURE 14–10 Problem using Pythagorean theorem

Trigonometric Functions

Trigonometric functions result from special relationships existing between the sides and angles of right triangles. "Trig" is a common shortened term for trigonometry or trigonometric. We will only discuss three basic trig functions used for your study of ac circuits—sine, cosine, and tangent functions.

Notice the side opposite (opp), the side adjacent (adj), and the hypotenuse (hyp) in Figure 14–11. As you can see, the relationship of each side to the "angle of interest" θ determines whether it is called the opposite or the adjacent side. The hypotenuse is *always* the longest side of the triangle.

Refer to Figures 14–11 and 14–12, as appropriate, while you study the following.

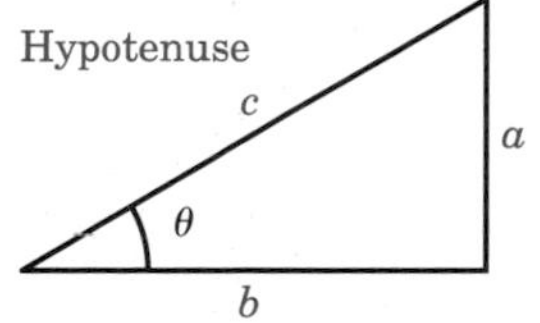

Side opposite θ

Side adjacent to θ

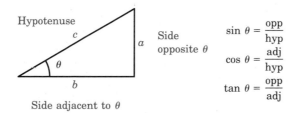

$$\sin\theta = \frac{\text{opp}}{\text{hyp}}$$
$$\cos\theta = \frac{\text{adj}}{\text{hyp}}$$
$$\tan\theta = \frac{\text{opp}}{\text{adj}}$$

$$\sin\theta = \frac{a}{c};\ a = c\sin\theta;\ c = \frac{a}{\sin\theta}$$
$$\cos\theta = \frac{b}{c};\ b = c\cos\theta;\ c = \frac{b}{\cos\theta}$$
$$\tan\theta = \frac{a}{b};\ a = b\tan\theta;\ b = \frac{a}{\tan\theta}$$

FIGURE 14–11 Labeling sides of a right triangle

1. **Sine** of the angle ($\sin\theta$) is the ratio of the side opposite to the hypotenuse.

Formula 14–2	$\sin\theta = \dfrac{\text{opposite}}{\text{hypotenuse}}$

FIGURE 14–12 Partial trigonometry table

ANGLE	SINE	COSINE	TANGENT
0 °	0.0000	1.0000	0.0000
15 °	0.2588	0.9659	0.2680
30 °	0.5000	0.8660	0.5773
45 °	0.7071	0.7071	1.0000
60 °	0.8660	0.5000	1.7321
75 °	0.9659	0.2588	3.7321
90 °	1.0000	0.0000	infinity

NOTE: In Figure 14–11, we have labeled the opposite side to the angle of interest as "a" and the hypotenuse as "c." Thus, $\sin \theta = \frac{a}{c}$ in this case.

2. **Cosine** of the angle ($\cos \theta$) is the ratio of the side adjacent to the hypotenuse.

Formula 14–3	$\cos \theta = \frac{\text{adjacent}}{\text{hypotenuse}}$

NOTE: In Figure 14–11, $\cos \theta = \frac{b}{c}$.

3. **Tangent** of the angle ($\tan \theta$) is the ratio of the side opposite to the side adjacent.

Formula 14–4	$\tan \theta = \frac{\text{opposite}}{\text{adjacent}}$

NOTE: In Figure 14–11, $\tan \theta = \frac{a}{b}$.

Take time to learn and understand the formula for each of these three functions now! It will be worth your time! To help remember these functions, you can use the sentence, *"Oscar had a heap of apples, some came today"!* Arrange the words as follows:

This translates to:

$$\frac{\text{Oscar}}{\text{Had}} \quad \text{Some} \quad \frac{O}{H} = S \quad \text{OR} \quad \frac{\text{Opp}}{\text{Hyp}} = \text{Sine}$$

$$\frac{\text{A}}{\text{Heap}} \quad \text{Came} \quad \frac{A}{H} = C \quad \text{OR} \quad \frac{\text{Adj}}{\text{Hyp}} = \text{Cosine}$$

$$\frac{\text{Of}}{\text{Apples}} \quad \text{Today} \quad \frac{O}{A} = T \quad \text{OR} \quad \frac{\text{Opp}}{\text{Adj}} = \text{Tangent}$$

If we assume a rotating vector with a unit length of one starts at 0° and rotates counterclockwise to the 90° position, the values of sine, cosine, and tangent vary within limits. That is:

$\sin\theta$ increases from 0 to 1.0 (with angles from 0° to 90°)
$\cos\theta$ decreases from 1.0 to 0 (with angles from 0° to 90°)
$\tan\theta$ increases from 0 to ∞ (with angles from 0° to 90°)

See if you can relate this to the sine and cosine waves in Figure 14–13. Assume the maximum (peak) value is one volt for each waveform. Relate the zero points and the maximum points to the angles. Do they compare?

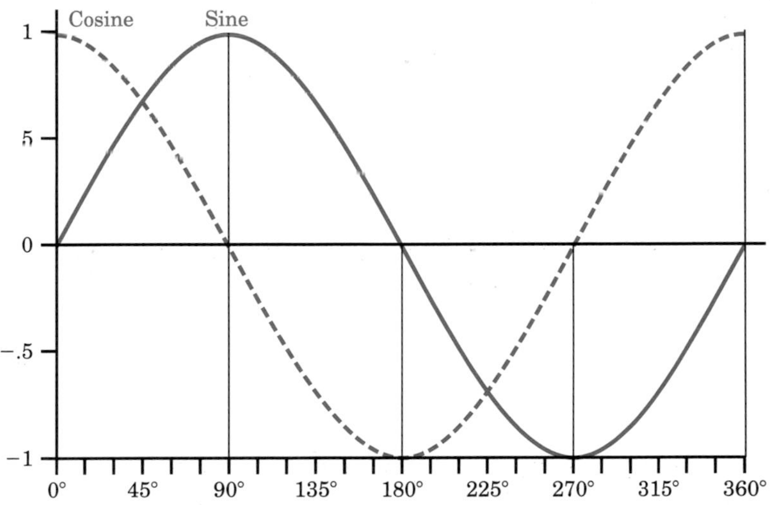

FIGURE 14–13 Sine and cosine waves

Figure 14–14 carries this idea one step further. Notice how these functions vary in the four quadrants of the coordinate system we discussed in an earlier chapter. Again, try relating the sine and cosine waves in Figure 14–13 to the statements in Figure 14–14.

QUADRANT	Sin θ from	Sin θ to	Cos θ from	Cos θ to	Tan θ from	Tan θ to
I (0° to 90°)	0	1.0	1.0	0	0	∞
II (90° to 180°)	1.0	0	0	−1.0	−∞	0
III (180° to 270°)	0	−1.0	−1.0	0	0	∞
IV (270° to 360°)	−1.0	0	0	1.0	−∞	0

FIGURE 14–14 Variations in values of trig functions

EXAMPLE

What is the phase angle if the side opposite is 2 units and the hypotenuse is 2.828 units?

Answer: Referring to the partial trig table in Figure 14–12, the trig formula containing the side opposite and the hypotenuse is the formula for the sine of the angle. Therefore, $\sin \theta = \frac{2}{2.828} = 0.707$. From the table in Figure 14–12, the angle whose sine equals 0.707 is 45°.

PRACTICE PROBLEM III

1. What is the phase angle if the side adjacent has a magnitude of 5 units and the hypotenuse has a magnitude of 10 units?
2. What is the tangent of 45°?
3. What is the sine of 30°?

FUNDAMENTAL ANALYSIS OF SERIES RL CIRCUITS

Now we will take the knowledge you have acquired and apply it to some practical situations and problems, so you can see the benefit of this knowledge!

Refer to Figure 14–15. Notice if you added the voltages across the series resistor and the inductor to find total applied voltage, you would get 20 volts. This answer is wrong! The actual applied voltage is 14.14 volts.

FIGURE 14–15 Series *RL* circuit voltages

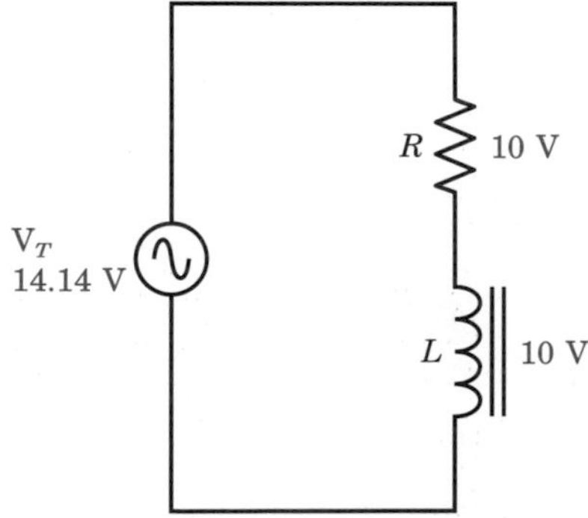

Let's see why 20 volts is the wrong value! Recall in series circuits, the current is the same through all components. Again referring to Figure 14–15 and from our previous discussions, we know the voltage across the resistor is in-phase with the current. Also, we know the voltage across the inductor leads the current by about 90°, which means the voltages across the resistor and the inductor are *not* in-phase with each other. Since the current through both is the same current, one voltage is in-phase with the current (V_R), and one voltage is not in-phase with the current. Therefore, we cannot add the voltages across R and L to find total voltage, as we would do in a simple dc circuit. Rather, we must use a special method of addition called vector addition to find the correct answer. The voltages must be added "vectorially" not "algebraically."

Let's try the two methods previously discussed to find total voltage in series RL circuits.

Using the Pythagorean Theorem to Analyze Voltage in Series RL Circuits

Figure 14–16 illustrates how V_R, V_L, and V_T are typically represented in a right triangle by phasors or vectors.

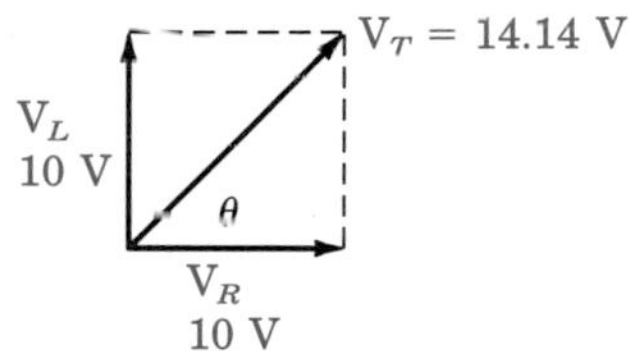

FIGURE 14–16
Voltage vector diagram

EXAMPLE

Let's transfer one Pythagorean theorem formula in Figure 14–9, $c = \sqrt{a^2 + b^2}$, to this situation. As you can see, the longest side is V_T. This is equivalent to the c in the original Pythagorean formula. The other two quantities (V_R and V_L) are the other quantities in the formula. Thus, the formula to solve for V_T becomes:

Formula 14–5	$V_T = \sqrt{V_R^2 + V_L^2}$

Therefore, $V_T = \sqrt{10^2 + 10^2} = \sqrt{200} = 14.14$ volts.

EXAMPLE

Assume the voltage values in Figure 14–15 change so that V_R is 30 volts and V_L is 40 volts. What is the new V applied value?

Answer: The V_T is 50 volts, Figure 14–17.

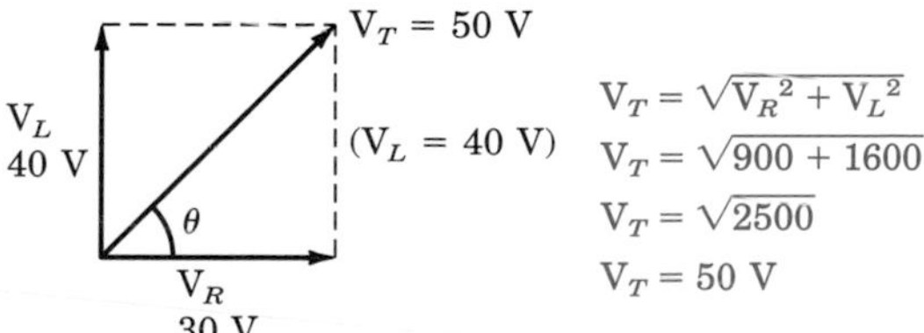

(NOTE: V_T is *not* equal to the sum of V_R and V_L. It is greater than either one alone, but not equal to their sum.)

FIGURE 14–17 Series *RL* circuit voltage analysis

PRACTICE PROBLEM IV

In a simple series RL circuit, V_R is 120 V and V_L is 90 V. What is the V applied value?

You can see that using the Pythagorean theorem, it is easy to find the magnitude of total voltage in a series RL circuit—simply substitute the appropriate values in the formula. We are not able to find the resultant angular direction by this method. However, the trig functions help us.

Using Trig to Solve for Angle Information in Series RL Circuits

Look at Figure 14–18 as you study the following statements.

FIGURE 14–18 Using trig to find angle information

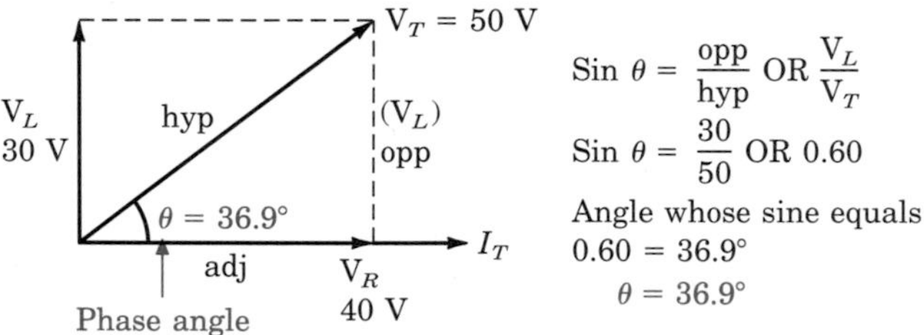

NOTE: Phase angle is considered the angular difference in phase between circuit total voltage and circuit total current. Get in the habit of looking at the sketch to see if the answer is logical. It is obvious that the angle in this case must be less than 45° because the vertical leg is shorter than the horizontal leg. The answer looks and seems logical!

1. To find the desired angle information using trig functions, identify the sides of the right triangle in terms of the circuit parameters of interest. Note in Figure 14–18 that V_L is in the upward (+90°) direction, and V_R is at the 0° reference direction. In *series* circuits, the current is the same throughout the circuit and is in-phase with the resistive voltage. Therefore, I is also on the 0° axis.

PRACTICAL NOTES

For series circuits since current is common to all parts, *current is commonly the reference phasor* and is plotted on the reference phasor 0 degree axis.

Since these directions are important, let's restate them! They are general rules of thumb that you will use from now on! V_L is vertical (up), V_R is horizontal to the right, and I is in-phase with V_R.

2. Referring to the values in Figure 14–18, let's find the *circuit phase angle* that is defined as *the phase angle between circuit applied voltage and circuit total current.*
   a. Consider the angle between V_T and I_T the angle of interest. Thus, the side opposite to this angle is the V_L value and the hypotenuse is V_T.
   b. Use the information relating these two sides in the sine function formula to find the angle. $\left(\text{Recall that } \sin\theta = \frac{\text{opp}}{\text{hyp}}\right)$. Therefore, in our example, $\sin\theta = \frac{30}{50} = 0.60$.
   c. Using a trig table, calculator, or computer to find the angle having a sine function value of 0.60, we find that it is 36.9°.

PRACTICAL NOTES

In ac circuits containing both resistance and reactance, the phase difference in degrees between applied voltage and total current must fall between 0° (indicating a purely resistive circuit) and 90° (indicating a purely reactive circuit). Since both resistance and reactance are present, the circuit cannot act purely resistive or purely reactive.

   d. In this example, since we know all sides, we can also use the other two trig functions. That is, $\cos\theta = \frac{\text{adj}}{\text{hyp}} = \frac{V_R}{V_T} = \frac{40}{50} = 0.8$. The angle with a cosine value of 0.8 is 36.9°. And $\tan\theta = \frac{\text{opp}}{\text{adj}} = \frac{V_L}{V_R} = \frac{30}{40} = 0.75$. The angle with a tangent value of 0.75 is 36.9°.
   e. Use the sine, cosine, or tangent functions to solve for the unknown angle depending on the information available. Obviously, when you know the values for the opposite side and the hypotenuse, use the sine function. When you know the values for side adjacent and the hypotenuse, use the cosine function. When you know only the values for side opposite the angle of interest and the side adjacent to the angle of interest, use the tangent function.

Stated another way, select and use the formula containing the two known values and the value you are solving for.

PRACTICE PROBLEM V

1. Use trig and solve for the phase angle of the circuit in Figure 14–19.

FIGURE 14–19

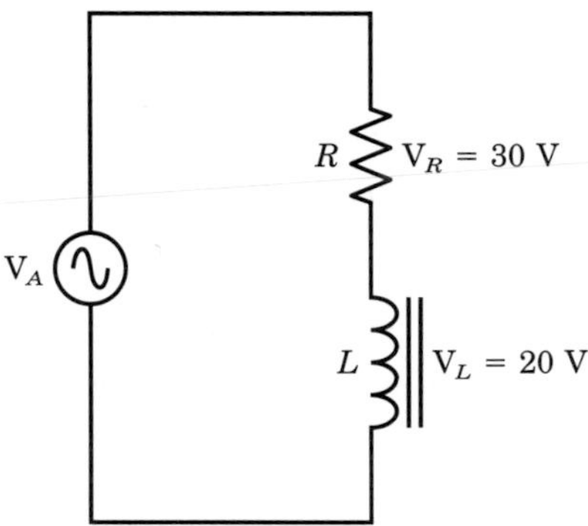

2. Which trig function did you use in this case?
3. If the voltages had been reversed, i.e., V_R = 20 volts and V_L = 30 volts, would the phase angle have been greater, smaller, or the same?

Using Trig to Solve for Voltages in Series RL Circuits

You have seen how the trig helps find angle information. You also learned the Pythagorean theorem is useful to find magnitude or voltage value in series RL circuits. Can trig also be used to find magnitude? Yes, it can! Let's see how!

EXAMPLE

If the angle information is known, determine the sine, cosine, or tangent values for that angle. Next, use the appropriate, available sides information, and find the value of the unknown side. Look at Figure 14–20 and let's see how to do this.

1. Start by making sure the formula for the trig function used includes the unknown side of interest, plus a known side. This allows us to solve for the unknown by using equation techniques.
2. To solve for V_L, use the sine function since the sin θ uses $\frac{\text{opposite}}{\text{hypotenuse}}$. V_L is the unknown side of interest and is the side oppo-

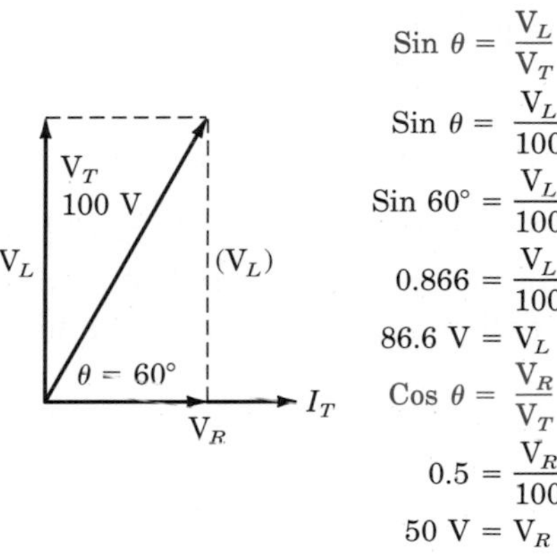

$$\text{Sin } \theta = \frac{V_L}{V_T}$$
$$\text{Sin } \theta = \frac{V_L}{100}$$
$$\text{Sin } 60° = \frac{V_L}{100}$$
$$0.866 = \frac{V_L}{100}$$
$$86.6\text{ V} = V_L$$
$$\text{Cos } \theta = \frac{V_R}{V_T}$$
$$0.5 = \frac{V_R}{100}$$
$$50\text{ V} = V_R$$

FIGURE 14–20 Using trig to solve for voltages in a series *RL* circuit (knowing *θ* and one side)

site the known angle. Since we know the hypotenuse value, we have our starting point.

3. Sin $\theta = \frac{\text{opp}}{\text{hyp}}$, thus sin $60° = \frac{V_L}{100}$. Now we determine the sine of 60° and substitute in the equation. Use the trig table, calculator, or computer and find the sine of 60° is 0.866. (NOTE: round-off to 0.87.) Put this in the formula, thus $0.87 = \frac{V_L}{100}$. Transpose and solve: $V_L = 100 \times 0.87 = 87$ volts.
4. To solve for V_R, use either the cosine or tangent functions with the information now known. Cosine uses the $\frac{\text{adjacent}}{\text{hypotenuse}}$ and tangent uses the $\frac{\text{opposite}}{\text{adjacent}}$. In either case, a known and the desired unknown (the adjacent) values appear in the formulas.
5. Assuming we did not know the V_L value, we use the cosine function. Cos $\theta = \frac{\text{adj}}{\text{hyp}}$, thus cos $60° = \frac{V_R}{100}$. Next, determine the cosine of 60°, which equals 0.5. Thus, substitute in the formula, $0.5 = \frac{V_R}{100}$. Transpose and solve: $V_R = 100 \times 0.5 = 50$ volts.

Now, let's see if these results make sense with respect to our vector diagram in Figure 14–20. Since the angle is greater than 45°, we know V_L must be greater than V_R and it is!

Now we can use the tangent function to check our results, since it uses the $\frac{\text{opposite}}{\text{adjacent}}$ sides. Insert the values found for V_L and V_R into the

$\tan\theta = \frac{\text{opp}}{\text{adj}}$ formula to find that $\frac{86.6}{50} = 1.732$. The angle whose tangent is 1.732 is 60°. It checks out! Now, you try one.

PRACTICE PROBLEM VI

A series RL circuit has one resistor and one inductor. What is the V_R value if the applied voltage is 50 volts and the circuit phase angle is 45°?

Analyzing Impedance in Series RL Circuits

You have learned how to analyze various voltages in series RL circuits by measuring vectors, using the Pythagorean theorem, and applying basic trig functions. Now, let's look at another important ac circuit consideration, impedance.

Impedance is the total opposition an ac circuit offers to current flow at a given frequency. This means impedance is the combined opposition of all **resistances and reactances** in the circuit. Since impedance is sensitive to frequency, there must be some reactance—*Pure* resistance is not sensitive to frequency. The symbol for impedance is "Z", and the unit of impedance is the ohm.

Before discussing impedance, let's review some important facts you already know, Figure 14–21.

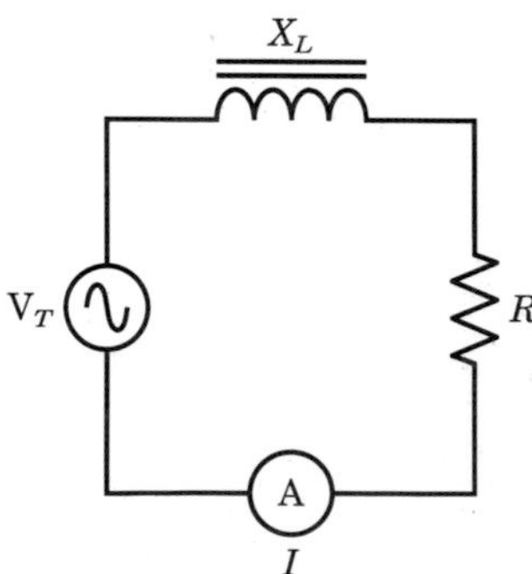

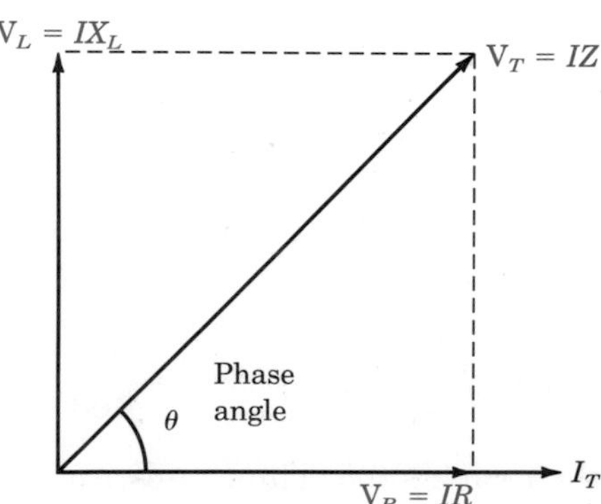

FIGURE 14–21 Series *RL* circuit quantities

1. The voltage drop across the resistor is in-phase with the circuit current through the resistor, and V_R equals IR. Note that this is a series circuit. Since current is the common factor in series circuits, the current is the reference vector and is at the 0° position of the coordinate system. Also V_R is at 0° since it is in-phase with I.

2. The voltage drop across the inductor leads the circuit current by 90° (assuming a perfect inductor with no resistance), and V_L equals IX_L. Since V_L leads I by 90°, the IX_L vector is leading the I vector by 90°.

3. As you know, total circuit voltage is the *vector resultant* of the individual series voltages. Since the circuit is not purely resistive, V_T cannot be at 0°. Since the circuit is not purely inductive, V_T cannot be at 90°. Remember in circuits containing both resistance and reactance, the resultant circuit phase angle between circuit V and circuit I is between 0° and 90°. As you can see from the diagram, total circuit voltage equals IZ, where Z is the total opposition to ac current from the combination of resistance and reactance. Since I is at 0° and IZ is at an angle between 0° and 90°, when we plot Z, it also will be between 0° and 90°.

Let's pursue these thoughts. Refer to Figure 14–22 as you study the following discussion about how to plot a diagram of R, X_L, and Z (impedance diagram) for series RL circuits.

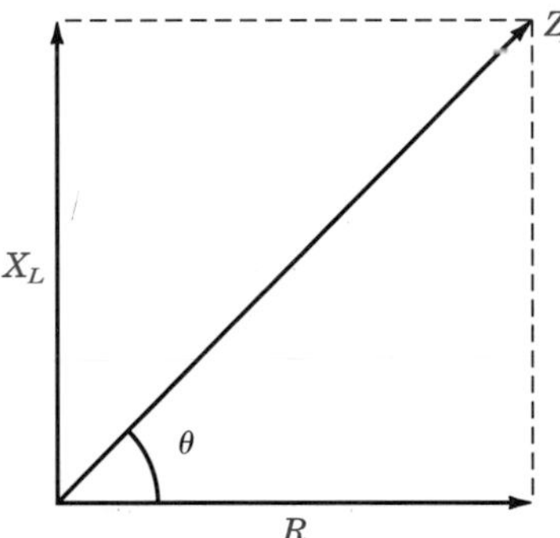

FIGURE 14–22 An impedance diagram

1. The V_R equals IR, and I is plotted at the 0° reference vector position. Also, since the resistor voltage is in-phase with the current through the resistor, R is plotted at the 0° reference position, when plotting an impedance diagram of R, X_L, and Z.
2. Since V_L equals IX_L and V_L is plotted at 90° from the reference vector position, it is logical that X_L is plotted at the 90° position when plotting R, X_L, and Z on an impedance diagram. Thus, the result of I times X_L is plotted at 90°.
3. The V_T equals IZ, and V_T is plotted at an angle (depending on the vector resultant of V_R and V_L) between 0° and 90°. Then it is logical to plot Z at the same angle in an impedance diagram where V_T is plotted in the voltage diagram of the same series RL circuit. In fact, the angle between Z and R is the same as the circuit phase angle between V_T and I_T.

Knowing all this, how can the techniques to analyze voltages in series RL circuits be applied to impedance? As you can see by comparing the diagrams of 14–21 and 14–22, the layout of the impedance diagram looks just like the diagram used in the voltage vector diagram analysis. This implies the techniques used to find impedance (Z) are those techniques used to find V_T. Recall these methods include:

1. Draw to scale and measure to find magnitude and direction.
2. Apply the Pythagorean theorem to find magnitudes.
3. Apply trig functions to find angles or magnitudes.

EXAMPLE

Figure 14–23 illustrates the drawing-to-scale approach. Figure 14–24 shows the Pythagorean theorem version used for impedance solutions. That is:

Formula 14–6	$Z = \sqrt{R^2 + X_L^2}$

FIGURE 14–23 The drawing-to-scale approach

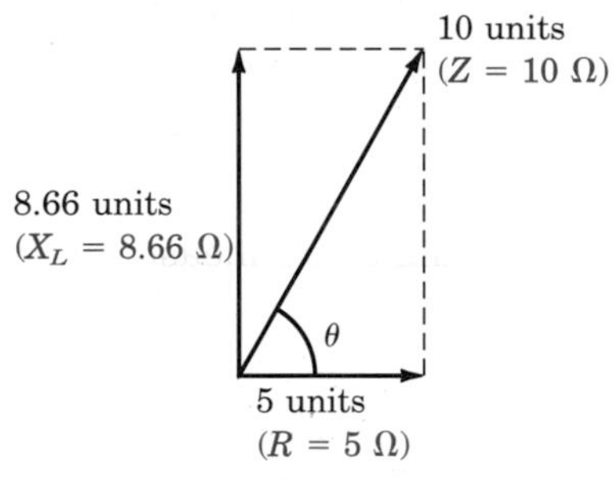

FIGURE 14–24 The Pythagorean theorem approach

$$Z = \sqrt{R^2 + X_L^2}$$
$$Z = \sqrt{5^2 + 8.66^2}$$
$$Z = \sqrt{25 + 75}$$
$$Z = 10\ \Omega$$

(NOTE: Z is *not* equal to the sum of R and X_L. It is greater than either one alone but not equal to their sum.)

Finally, Figure 14–25 solves for the angle using the simple trig technique. Now you try applying these techniques in the following situation.

PRACTICE PROBLEM VII

1. Draw the schematic diagram of a series RL circuit containing 15 Ω of resistance and 20 Ω of inductive reactance.

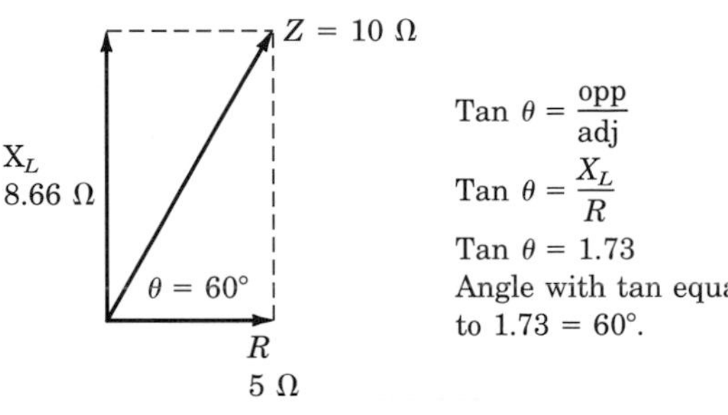

FIGURE 14–25 Using trig to solve for angle information

2. Draw an impedance diagram for the circuit described in question 1 (to scale). Measure and record the impedance value. Use a protractor and measure the angle between R and Z. Label this angle on the diagram.
3. Use the Pythagorean theorem formula where $Z = \sqrt{R^2 + X_L^2}$ and calculate and record the Z value. Does the answer agree with your measurement results?
4. Use appropriate trig formula(s) and solve for the angle where Z should be plotted. Show your work. Does this answer agree with the angle you measured using the protractor in Question 2?

Summary of Analyses of Series RL Circuits

1. The larger the R value compared to the X_L value, the more resistive the circuit will act. Conversely, the larger the X_L compared to the R, the more inductive the circuit will act.
2. The more resistive the circuit, the closer the circuit phase angle is to 0°. The more inductive the circuit, the closer the circuit phase angle is to 90° because circuit voltage is leading circuit current, which is the 0° reference vector in series circuits. Recall in a purely resistive series circuit, phase angle equals 0°. In a purely inductive series circuit, phase angle equals 90°. Refer to Figure 14–26 for some examples of R, X_L, Z, V_R, V_L, V_T, and phase angle (θ) relationships.
3. Since current (I) is the same through all parts of a series circuit, the ratio of R and X_L is the same as the relationship (or ratio) of V_R and V_L. That is, $V_R = IR$ and $V_L = IX_L$. For example, when the R value is two times the X_L value, the V_R value is twice the V_L value.
4. Total circuit impedance is the result of both factors, resistance and reactance. Impedance is greater than each factor but *not* equal to their sum. Right-triangle methods are used to find its value. Refer again to Figure 14–24.

FIGURE 14–26 Examples of parameter relationships in series RL circuits

Assume an I_T value of one ampere in each case. (This means V applied is changing.)

R Ω	V_R (IR)	X_L Ω	V_L (IX_L)	Z Ω ($\sqrt{R^2 + X_L^2}$)	V_T (IZ) volts	θ (Phase Angle) $\left(\text{Tan} = \frac{X_L}{R}\right)$ $\left(\text{or } \frac{V_L}{V_R}\right)$
10	10 V	5	5 V	11.18	11.18	26.56°
10	10 V	10	10 V	14.14	14.14	45°
5	5 V	10	10 V	11.18	11.18	63.43°
100	100 V	10	10 V	100.5	100.5	5.71°
10	10 V	100	100 V	100.5	100.5	84.29°

5. Total circuit voltage is the vector resultant of the resistive and reactive voltages in series. This voltage is greater than each factor but *not* equal to their sum. Right-triangle methods are used. Refer again to Figure 14–17.
6. Phase angle for series RL circuits is found using sine, cosine, or tangent functions. For example:

$$\sin\theta = \frac{V_L}{V_T}$$

$$\cos\theta = \frac{V_R}{V_T}$$

$$\tan\theta = \frac{V_L}{V_R}$$

NOTE: The tangent function is useful for many ac circuit problems because it does not require knowing the hypotenuse value.

For the impedance diagram:

$$\sin\theta = \frac{X_L}{Z}$$

$$\cos\theta = \frac{R}{Z}$$

$$\tan\theta = \frac{X_L}{R}$$

7. A greater R or X_L value results in greater circuit impedance and lower current value for a given applied voltage.
8. The circuit current value is found by using Ohm's law, where $I_T = \frac{V_T}{Z_T}$.

9. *Impedance diagrams can only be drawn and used to analyze series ac circuits!* Also, they can be used to analyze series portions of more complex circuits but *should not* be drawn to analyze parallel circuits.
10. An impedance diagram is not a phasor diagram, since the quantities shown are not sinusoidally varying quantities. However, because Z is the vector resulting from R and X, it can be graphed.

FUNDAMENTAL ANALYSIS OF PARALLEL RL CIRCUITS

Basically, the same strategies we have been using in the series circuit solutions can be applied to parallel circuits. That is, we can use vectors, the Pythagorean theorem, and trig functions. The differences are in *which* parameters are used on the diagrams and in the formulas.

Because *voltage* is the common factor in parallel circuits, it becomes the reference vector at the 0° position on our diagrams. Since I_T equals the vector sum of the branch currents in parallel circuits, our diagrams illustrate the branch currents in appropriate right-triangle form. (Recall that in series circuits, current was the reference and voltages are plotted in right-triangle format.)

Another contrast between parallel RL circuit analysis and series RL circuit analysis is that we *do not* use impedance diagrams to analyze parallel RL circuits.

With these facts in mind, let's apply the principles you have learned to some parallel RL circuit analysis. Refer to Figure 14–27 as you study the following comments.

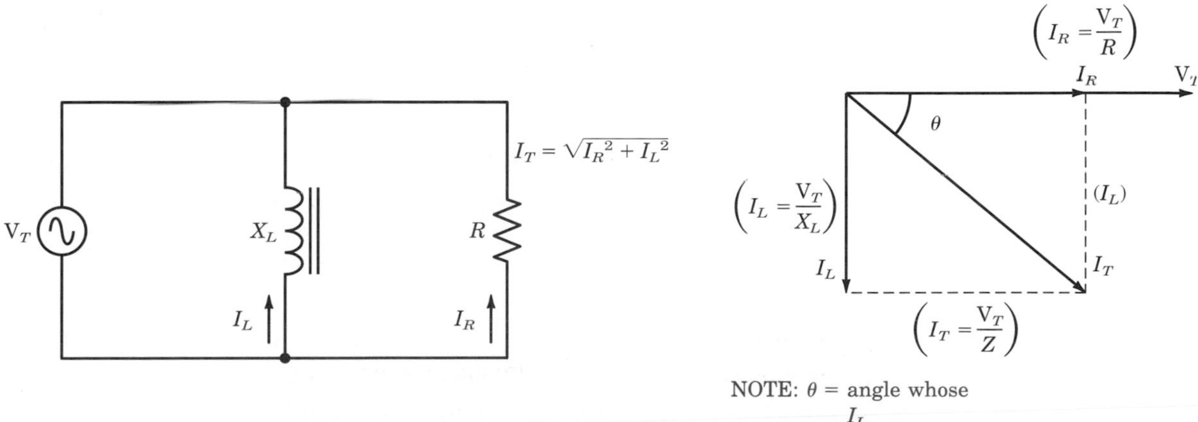

FIGURE 14–27 Parallel *RL* circuit analysis

1. V_T is at the reference vector position, because it is common to both the resistive and the reactive branches.
2. Since I_R (resistive branch current) is in-phase with V_R (or V_T), it is also at the 0° position.
3. We know the current through a coil lags the voltage across the coil by 90°. Thus, I_L is lagging V_T (or V_L) by 90°. (Remember in the four-quadrant system that −90° is shown as straight down.)
4. The total current is the vector resultant of the resistive and the reactive branch currents. Thus, it is between 0° and −90°.
5. The higher the resistive branch current relative to the reactive branch current, the more resistive the circuit will act. This means the smaller phase angle is between V_T and I_T. Also, the greater the inductive branch current relative to the resistive branch current, the more inductive the circuit will act. This means the phase angle is greater between V_T and I_T.
6. In a parallel RL circuit total impedance is the resultant opposition offered to current by the combined circuit resistance and reactance. That is, $Z = \frac{V_T}{I_T}$. Because of the phase relationships in a circuit with both resistance and reactance, the impedance is less than the impedance of any one branch. However the shortcut methods used in purely resistive or reactive circuits will not work. For example, if there are 10 Ω of resistance in parallel with 10 Ω of reactance, the impedance will *not* be 5 Ω but 7.07 Ω. This is because the total current is not the simple arithmetic sum of the branch currents, as it is in purely resistive or reactive circuits. Rather, total current is the vector sum.
7. Branch currents are found using Ohm's law, where $I_R = \frac{V_R}{R}$ and $I_L = \frac{V_L}{X_L}$.
8. Total current is found using the Pythagorean formula, where:

Formula 14–7	$I_T = \sqrt{I_R^2 + I_L^2}$

EXAMPLE

Parameters for the circuit in Figure 14–28 are $V_T = 150$ V, $R = 50$ Ω, and $X_L = 30$ Ω. Find I_T, Z, and the circuit phase angle (θ).

1. Label the known values on the diagram, Figure 14–29.

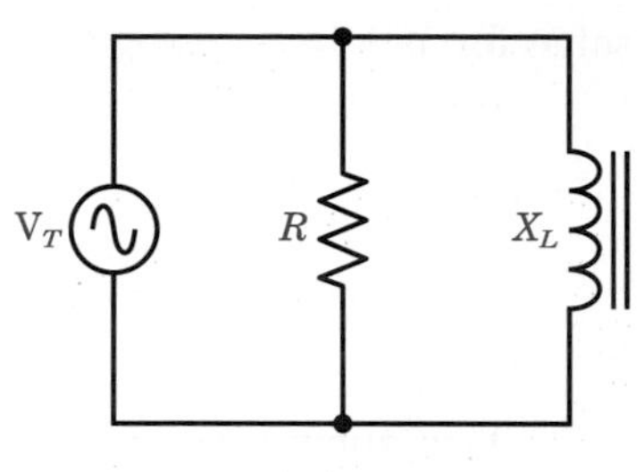

FIGURE 14–28

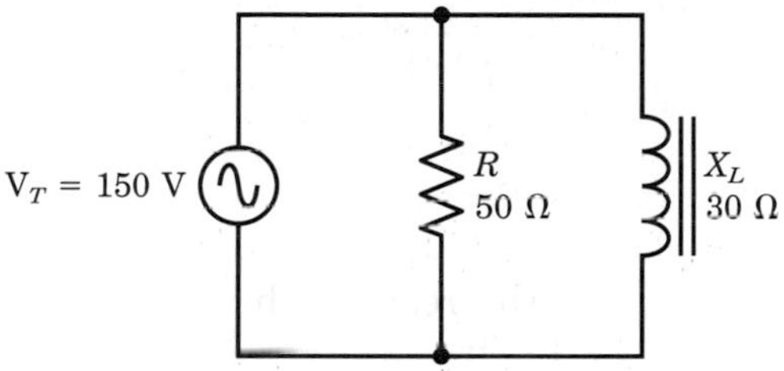

FIGURE 14–29

2. Determine the branch currents:

$$I_R = \frac{V_T}{R} = \frac{150\text{ V}}{50\ \Omega} = 3\text{ A}$$

$$I_L = \frac{V_T}{X_L} = \frac{150\text{ V}}{30\ \Omega} = 5\text{ A}$$

3. Draw a current vector diagram and label knowns, Figure 14–30.

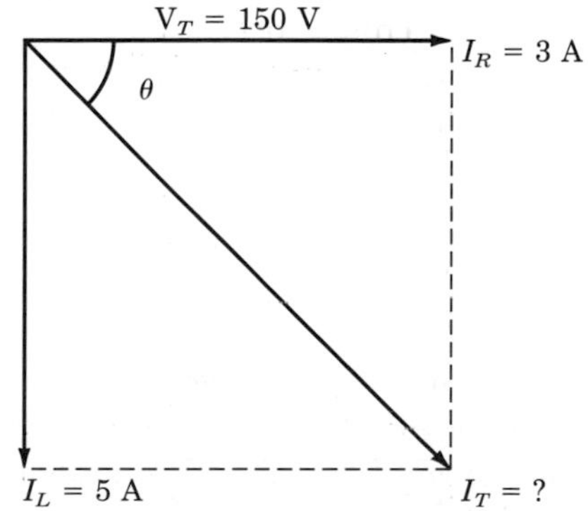

FIGURE 14–30

4. Use Pythagorean theorem and solve for I_T:

$$I_T = \sqrt{I_R^2 + I_L^2}$$
$$I_T = \sqrt{9 + 25}$$
$$I_T = \sqrt{34}$$
$$I_T = 5.83\text{ A}$$

5. Use Ohm's law and solve for Z:

$$Z = \frac{V_T}{I_T}$$

$$Z = \frac{150\ V}{5.83\ A}$$

$$Z = 25.73\ \Omega$$

6. Use trig formula and solve for phase angle:

$$\text{Tan}\ \theta = \frac{I_L}{I_R}$$

$$\text{Tan}\ \theta = 1.666$$

Angle whose tangent = 1.666 is 59°

Answers: $I_T = 5.83$ A, $Z = 25.73\ \Omega$, and $\theta = -59°$

(NOTE: The angle is considered a negative angle because the circuit current is lagging the reference circuit voltage by 59°.)

Now you try the following practice problem.

PRACTICE PROBLEM VIII

A circuit similar to the one in Figure 14–28 has the following parameters: $V_T = 300$ V, $R = 60\ \Omega$, and $X_L = 100\ \Omega$. Perform the same steps previously shown and find I_T, Z, and the phase angle (θ).

Summary of Analysis of Parallel RL Circuits

1. The greater the resistive branch *current* compared to the inductive branch *current,* the more resistive the circuit acts. Conversely, the greater the inductive branch *current* compared to the resistive branch current, the more inductive the circuit acts.

2. The more resistive the circuit is, the closer the circuit phase angle is to 0°. The more inductive the circuit is, the closer the circuit phase angle is to −90°. These situations result from circuit current lagging behind the circuit voltage, which is the 0° reference vector in parallel circuits. Recall in a purely *resistive* parallel circuit $\theta = 0°$, but in a purely inductive *parallel* circuit $\theta = -90°$. Refer to Figure 14–31 to see some examples of the relationships of R, X_L, Z, I_R, I_L, I_T, and phase angle (θ) with selected circuit parameters purposely varied.

3. Branch current is *inverse* to the resistance or reactance of a given branch. For example, when R is two times X_L, current through the *resistive* branch is half the current through the *reactive* branch.

FIGURE 14–31 Examples of parameter relationships in parallel RL circuits

Assume a voltage applied (V_T) of 50 V for each case. (This means I_T will be changing.)						
R Ω	I_R $\left(\frac{V_T}{R}\right)$	X_L Ω	I_L $\left(\frac{V_T}{X_L}\right)$	I_T $(\sqrt{I_R^2 + I_L^2})$	Z $\left(\frac{V_T}{I_T}\right)$	θ (Phase angle) $\left(\frac{\text{Tan } I_L}{I_R}\right)$
10	5 A	5	10 A	11.18 A	4.47 Ω	−63.43°
10	5 A	10	5 A	7.07 A	7.07 Ω	−45°
5	10 A	10	5 A	11.18 A	4.47 Ω	−26.56°
100	0.5 A	10	5 A	5.02 A	9.96 Ω	−84.29°
10	5 A	100	0.5 A	5.02 A	9.96 Ω	−5.7°
						NOTE: The larger the (−) phase angle, the more inductive the circuit is acting.

4. Total circuit impedance of a given parallel RL circuit is less than the opposition of any one branch. However, impedance cannot be found with product-over-the-sum or reciprocal formulas used in dc circuits. Neither can impedance be found using an impedance diagram approach. Ohm's law or other methods are used.

5. Total circuit current is the vector resultant of the resistive and reactive branch currents. Current is greater than any one branch current but *not* equal to their sum. Right-triangle methods are used. Refer again to Figure 14–27.

6. Phase angle for parallel RL circuits is found using sine, cosine, or tangent trig functions. For example:

$$\sin \theta = \frac{I_L}{I_T}$$

$$\cos \theta = \frac{I_R}{I_T}$$

$$\tan \theta = \frac{I_L}{I_R}$$

   NOTE: The tangent function is useful for many ac circuit problems because it does not require knowing the hypotenuse value.

7. A greater R or X_L value results in greater circuit impedance and lower circuit current value for a given applied voltage.

8. The circuit impedance value is found by using Ohm's law, where $Z_T = \frac{V_T}{I_T}$.
9. *Impedance diagrams can only be drawn and used to analyze series ac circuits!* Also, they can be used to analyze series portions of more complex circuits but *should not* be drawn to analyze parallel circuits. Obviously, since the impedance of a parallel RL circuit is smaller than any one branch, the hypotenuse cannot be shorter than one of the sides when using the right-triangle techniques. Therefore, an impedance diagram cannot be drawn for parallel RL circuits.

EXAMPLES OF PRACTICAL APPLICATIONS FOR INDUCTORS AND RL CIRCUITS

Inductors and inductances have a variety of applications. Some of these applications use the property of inductance or mutual inductance. Other applications involve using inductance with resistance or capacitance, or both. Inductance can be used in circuits operating at power-line (low) frequencies; circuits operating in the audio-frequency range; and circuits operating at higher frequencies, known as radio frequencies.

The following is a brief list of applications.

1. Power-frequency applications, Figure 14–32.
   a. Power transformers
   b. Filter circuits (both power line and power supply filters)
   c. Solenoids, electromagnets, relays
2. Audio-frequency applications, Figure 14–33.
   a. Audio transformers (used for coupling and Z-matching)
   b. Audio chokes (as loads for amplifiers)
   c. Filter circuits
3. Radio-frequency applications, Figure 14–34.
   a. Tuning circuit applications
   b. Radio chokes/filter
   c. Waveshaping applications (using L/R time constant characteristics)

Typically, inductance values used at power-line frequencies are in the henrys range. For audio frequencies, values are often in the tenths of a henry, or millihenry ranges. For radio frequency (rf) applications, values are typically in the microhenrys range.

One key characteristic of inductors allowing them to be used in these applications is their sensitivity to frequency. (Recall, X_L varies with frequency.) Also, their relationship to creating and reacting to magnetic fields is another very useful characteristic.

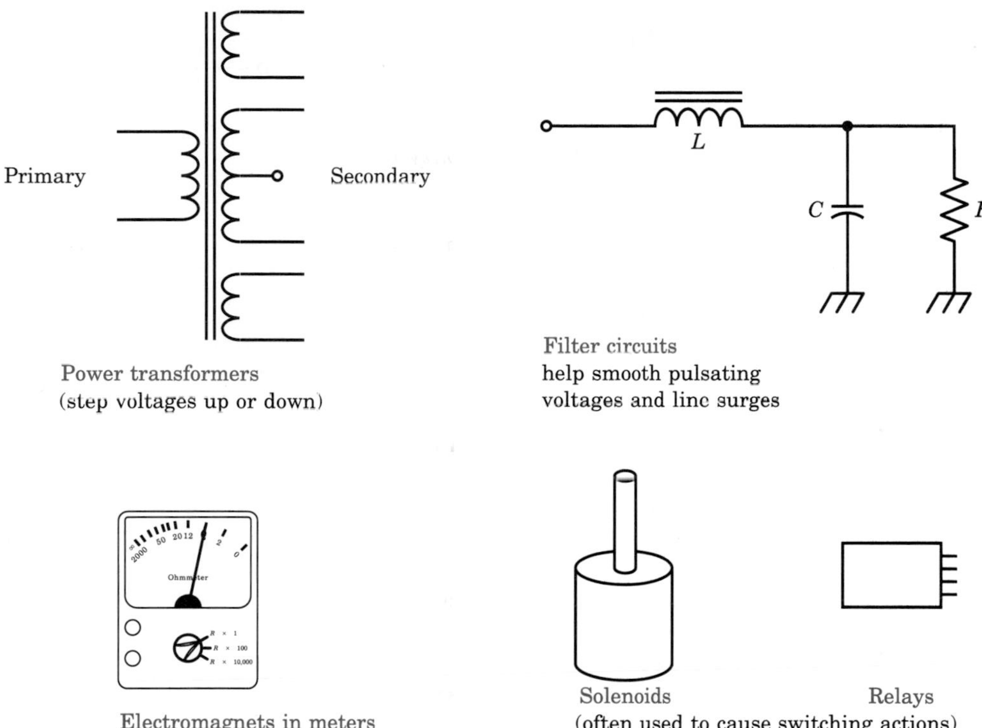

FIGURE 14–32 Examples of applications at low ac (or pulsating dc) frequencies

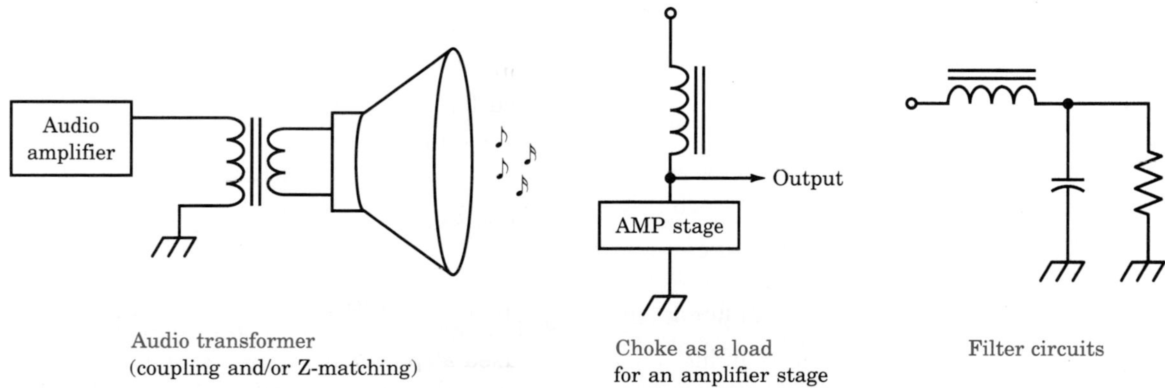

FIGURE 14–33 Examples of applications at audio frequencies

It is beyond the scope of this book to explain all of these applications. However, in this text you already know or will eventually learn all the basic principles involved with these applications.

For example, you know inductive reactance varies with frequency. The

FIGURE 14–34 Examples of radio frequency *(RF)* and other applications

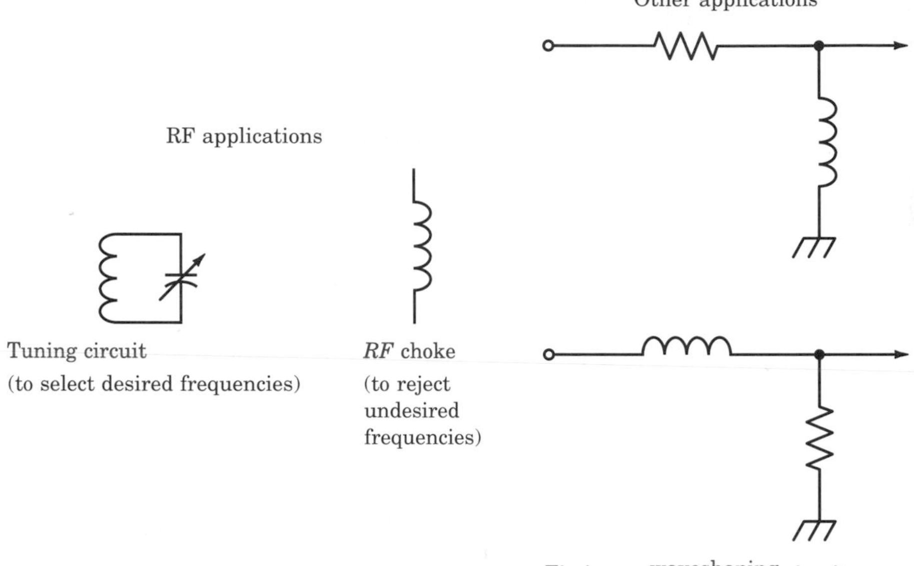

application of inductors in filter and tuning circuits depends on this principle. Also, you studied the principle for waveshaping when you studied $\frac{L}{R}$ time constants.

As you can see, inductors, the property of inductance, and inductive reactance are valuable tools in the field of electronics. Your understanding of inductors will be of great help in your career.

SUMMARY

- Phase angle is the difference in phase between the circuit applied voltage (V_T) and the circuit current (I_T).
- Circuit current and voltage are in-phase in purely resistive ac circuits; are 90° out-of-phase in purely inductive circuits; and are between 0° and 90° out-of-phase in circuits containing both resistance and inductance.
- For *series* ac circuits containing both resistance and inductive reactance, total circuit voltage and impedance cannot be found by adding voltage values or oppositions to current, as is done in dc circuits. Rather, values must be determined by vectors or right-triangle techniques.
- For voltage-current vector diagrams relating to *series* RL circuits, *current* is the *reference vector* positioned at the 0° location since it is common throughout the circuit. Resistive voltage drop(s) are also at the 0° position, since they are in-phase with current. Inductive voltages are shown at 90° (since V_L leads I by 90°). Total circuit voltage is plotted as the vector resultant of the resistive and inductive voltages.

CHAPTER CHALLENGE

Following the SIMPLER troubleshooting sequence, find the problem with this circuit. As you follow the sequence, record your responses to each step on a separate sheet of paper.

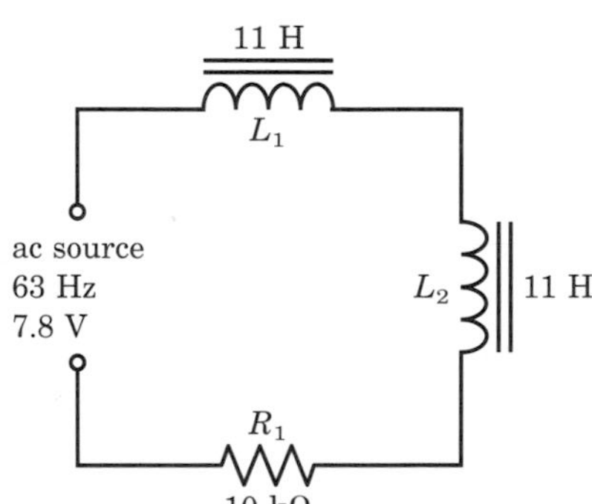

CHALLENGE CIRCUIT 8

SYMPTOMS Gather, verify, and analyze symptom information. (Look at the "Starting Point Information.")

IDENTIFY Identify initial suspect area for the location of the trouble.

MAKE Make a decision about "What type of test to make" and "Where to make it." (To simulate making a decision about each test you want to make, select the desired test from the "TEST" column listing below.)

PERFORM Perform the test. (Look up the result of the test in Appendix C. Use the number in the "RESULTS" column to find the test result in the Appendix.)

LOCATE Locate and define a new "narrower" area in which to continue troubleshooting.

EXAMINE Examine available information and again determine "what type of test" and "where."

REPEAT Repeat the preceding analysis and testing steps until the trouble is found. What would you do to restore this circuit to normal operation? When you have solved the problem, compare your results with those shown in the color insert.

Starting Point Information

1. Circuit diagram
2. V_{R_1} is higher than it should be
3. V Source checked as being 7.8 V

Test	Results in Appendix C
V_{R_1}	(34)
V_{L_1}	(89)
V_{L_2}	(48)
R_1	(109)

- For impedance diagrams (used *only* for *series* circuits), the R values are plotted at the 0° reference position. The X_L value is plotted vertical (up) at 90° (assuming a perfect inductor). Impedance is plotted at a position that is the vector resultant of the R and X_L vectors.

- The Pythagorean theorem applied to *series* RL circuits is used to calculate the total voltage value where $V_T = \sqrt{V_R^2 + V_L^2}$. It is also useful to determine the impedance (Z) value where $Z = \sqrt{R^2 + X_L^2}$.

- The trig functions of sine, cosine, and tangent are used to determine phase angle, if two sides are known. Also, these functions can be used to determine voltage and impedance values of *series* RL circuits, if the phase angle and one side are known.

- For *parallel* ac circuits containing both resistance and inductive reactance, total circuit current cannot be found by adding branch currents, as is done in dc circuits. Right-triangle methods are again used.

- For voltage-current vector diagrams relating to *parallel* RL circuits, *voltage* is the *reference vector* at the 0° location since it is common throughout the circuit. Resistive branch current(s) are also at the 0° position, since they are in-phase with the voltage. Inductive branch currents are at −90° since I_L lags V by 90°. Total circuit current is plotted as the vector resultant of the resistive and inductive branch currents.

- The Pythagorean theorem applied to *parallel* RL circuits is used to calculate the total current value where $I_T = \sqrt{I_R^2 + I_L^2}$.

- The trig functions (sine, cosine, and tangent) are also used to determine phase angle in parallel circuits, as well as series circuits, if two sides are known. They can be used to determine current values in *parallel* RL circuits, if the phase angle and one side are known.

- Plotting vectors to scale and then measuring lengths and angles is one method to determine parameters of interest in RL circuits. For series RL circuits, the resistive and inductive voltage values are plotted to find V_T and θ. For parallel RL circuits, the resistive and inductive branch current values are plotted to find I_T and θ.

- Impedance is the total opposition offered to ac current by a combination of resistance and reactance. The symbol for impedance is "Z", and the unit of impedance is the ohm.

- Inductors, either alone or in conjunction with other components (R or C), are used in many applications. General uses include transformers, tuning circuits, filter circuits and waveshaping circuits.

REVIEW QUESTIONS

1. A series RL circuit contains a resistance of 20 Ω and an inductive reactance of 30 Ω. Find the following parameters using the specified techniques.
   a. Use the plot and measure technique to find Z. (Round-off to the nearest whole

number.) Measure the θ with a protractor. (Round-off to the nearest degree.)
b. Use Pythagorean theorem to find Z.
c. Use trig to find the angle between R and Z.

2. A parallel RL circuit has one branch consisting of a 200-Ω resistance, and another branch consisting of an inductor with X_L equal to 150 Ω. The applied voltage is 300 volts. Find the values for the following parameters. Draw and label the schematic and vector diagrams.
   a. I_R
   b. I_L
   c. I_T
   d. θ
   e. Z

3. Draw the circuit diagram, voltage-current vector diagram and impedance diagram for a series RL circuit having one R and one L. Voltage drop across the 10 kΩ resistor is 5 volts, and circuit phase angle is 45°. Label all components and vectors, and show all calculations.

4. Draw the circuit diagram and voltage-current vector diagram of a parallel RL circuit having one R and one L. Assume a total current of 12 amperes and a phase angle of −75°. Label all components and vectors in the diagrams, and show all calculations. (Round answers to the nearest whole number and the nearest degree values.)

Answer the following with "I" for increase, "D" for decrease, and "RTS" for remain the same.

5. If frequency of the applied voltage in Question 2 doubles,
   a. R will ___________
   b. L will ___________
   c. X_L will ___________

d. Z will ____________
e. θ will ____________
f. I_T will ____________
g. V_T will ____________

6. If the inductance in Question 3 doubles while R remains the same,
a. R will ____________
b. L will ____________
c. X_L will ____________
d. Z will ____________
e. θ will ____________
f. I_T will ____________
g. V_T will ____________

7. If the R value in Question 3 doubles while X_L remains the same,
a. R will ____________
b. L will ____________
c. X_L will ____________
d. Z will ____________
e. θ will ____________
f. I_T will ____________
g. V_T will ____________

8. If the inductance in Question 4 doubles while R remains the same,
a. R will ____________
b. L will ____________
c. X_L will ____________
d. Z will ____________
e. θ will ____________
f. I_T will ____________
g. V_T will ____________

9. If the R value in Question 4 doubles while X_L remains the same,
a. R will ____________
b. L will ____________
c. X_L will ____________
d. Z will ____________
e. θ will ____________
f. I_T will ____________
g. V_T will ____________

10. If both R and X_L values in Question 4 are halved:
    a. R will ___________
    b. L will ___________
    c. X_L will ___________
    d. Z will ___________
    e. θ will ___________
    f. I_T will ___________
    g. V_T will ___________

11. Refer to Figure 14–35 and solve for the following:
    a. Z
    b. V_R
    c. θ

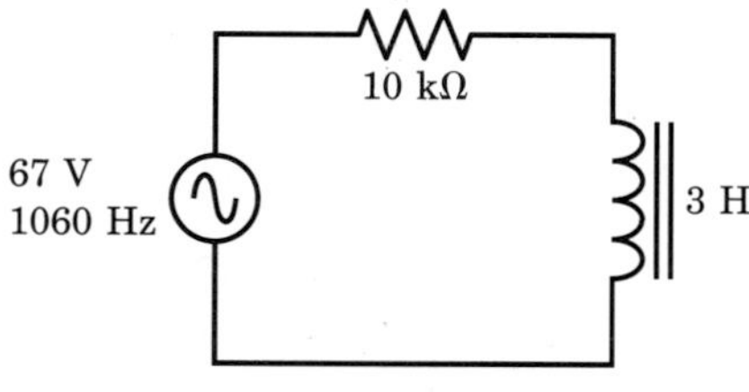

FIGURE 14–35

12. Refer to Figure 14–36 and solve for the following:
    a. X_L
    b. V_L
    c. V_R
    d. θ

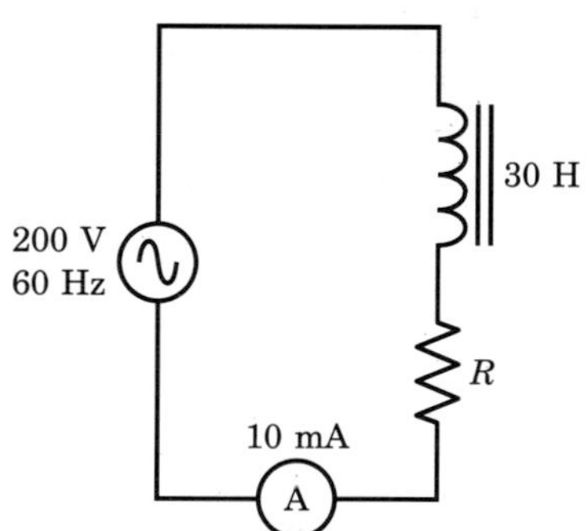

FIGURE 14–36

13. Refer to Figure 14–37 and solve for the following:
    a. I
    b. Z
    c. V_L
    d. X_L
    e. L

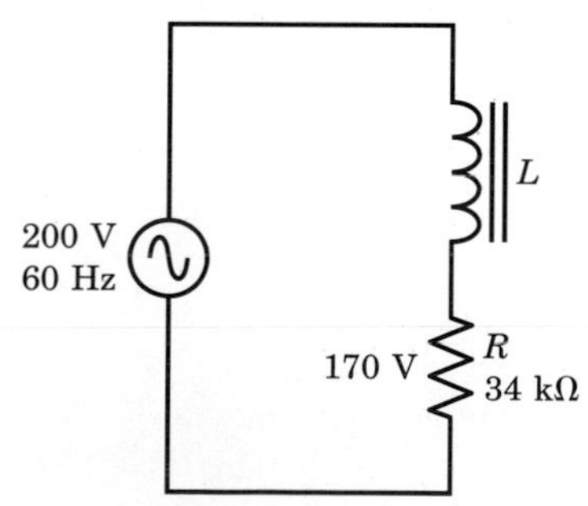

FIGURE 14–37

TEST YOURSELF

1. Draw the circuit diagram and all vector diagrams of the series RL circuit described below to find and illustrate all voltage, current, impedance, and phase angle parameters. Show all calculations.
   Given: Inductance = 10 H
   Voltage leads current by 60°
   Frequency = 1 kHz
   Total voltage (V_T) = 72.5 V

2. Draw the circuit diagram and all vector diagrams for the parallel RL circuit described below to find and illustrate all voltage, current, and phase angle parameters. Show all calculations.
   (NOTE: Rounding-off is acceptable for this problem.)
   Given: Inductance = 1.99 H
   Frequency = 400 Hz
   Resistance = 7 kΩ
   V_T = 21 V

3. Draw the circuit diagram for a series RL circuit containing two 4 henry inductances and two 5-kΩ resistors. The circuit has a frequency of 200 Hz. Find I_T if the applied voltage is 140 volts. (Round-off to the nearest kΩ, volt, and milliampere, as appropriate.)

PERFORMANCE PROJECT CORRELATION CHART

Suggested performance projects in the Laboratory Manual that correlate with topics in this chapter are:

CHAPTER TOPIC	PERFORMANCE PROJECT	PROJECT NUMBER
Fundamental Analysis of Series RL Circuits	Review of V, I, R, Z, and θ Relationships in a Series RL Circuit	42
	V, I, R, Z, and θ Relation-ships in a Series RL Circuit	43
Fundamental Analysis of Parallel RL Circuits	Review of V, I, R, Z, and θ Relationships in a Parallel RL Circuit	44
	V, I, R, Z, and θ Relation-ships in a Parallel RL Circuit	45

After completing the above projects, perform "Summary" checkout at the end of the performance projects section in the Laboratory Manual.

Basic Transformer Characteristics

Key Terms

Autotransformer

Coefficient of coupling (k)

Copper loss

Core loss

Current ratio

Impedance ratio

Isolation transformer

Mutual inductance (M or L_M)

Transformers

Turns ratio

Voltage ratio

Courtesy of Heath Products

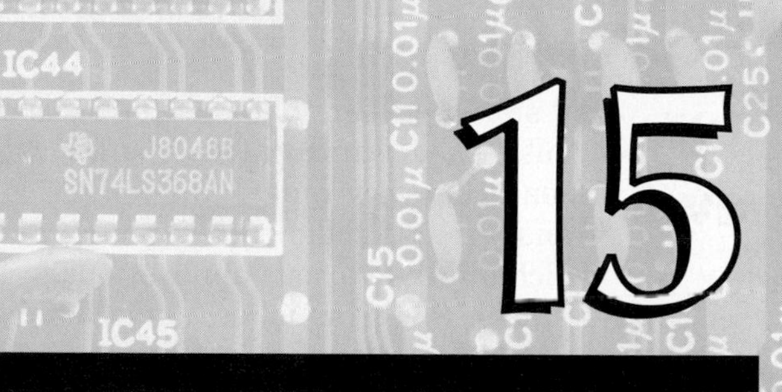

Outline

Background Information

Coefficient of Coupling

Mutual Inductance and Transformer Action

Mutual Inductance Between Coils Other Than Transformers

Important Transformer Ratios

Transformer Losses

Characteristics of Various Transformers

Troubleshooting Hints

Chapter Preview

We stated earlier that one application of electromagnetic induction is the transformer. Furthermore, you know these devices have application in different frequency ranges and for various purposes. That is, there are power transformers, audio transformers, and radio-frequency (rf) transformers, which all find various applications in many types of electronic circuits and systems. Power supplies, audio amplifiers, radio and TV receivers, transmitters, and other systems and subsystems use the unique characteristics of these devices.

In this chapter, you will study the basic features and types of transformers that a technician needs to understand. Practical information, such as schematic symbols and color codes, will be provided. Turns ratios, voltage ratios, current ratios, impedance ratios, and other major aspects about transformers will be studied. Finally, typical malfunctions and troubleshooting techniques will be examined.

Objectives

After studying this chapter, you will be able to:

- Define **mutual inductance**
- Calculate mutual inductance values
- Calculate **coefficient of coupling** values
- Calculate **turns, voltage, current,** and **impedance ratios**
- List, draw, or explain physical, magnetic, electrical, and schematic characteristics of various **transformers**
- List common transformer color codes
- Define at least two types of **core losses**
- List common problems found in transformers
- List troubleshooting procedures

BACKGROUND INFORMATION

You know when a changing current passes through a conductor wire or a coil, an expanding or collapsing magnetic field is produced. This expanding or collapsing field, in turn, induces a counter-emf (cemf). This process of producing voltage through a changing magnetic field is known as electromagnetic induction. When the induced voltage is created across the current-carrying conductor or coil, it is called self-induction. As you know, the symbol for this self-inductance is L.

Another important kind of induction is **mutual induction.** The unit of mutual inductance is the henry and the symbol is M. (NOTE: Sometimes you will also see the abbreviation L_M.)

One definition for mutual inductance is the property of inducing voltage in one circuit by varying the current in another circuit. Refer to Figure 15–1 and note that Coil A is connected to an ac source. Coil B is not connected to any source. However, it is located close to coil A so that most of the flux produced in Coil A by current from the source cuts or "links" Coil B. Since a *conductor, magnetic field,* and *relative motion* will induce voltage between the conductor and field, a voltage is induced in Coil B because of the changing current in Coil A. By definition, there must be mutual inductance. These two circuits (Coil A and Coil B) are positioned so that energy is transferred by *magnetic linkage.* This coupling between circuits is also sometimes called *inductive coupling.* Inductive coupling is due to flux linkages.

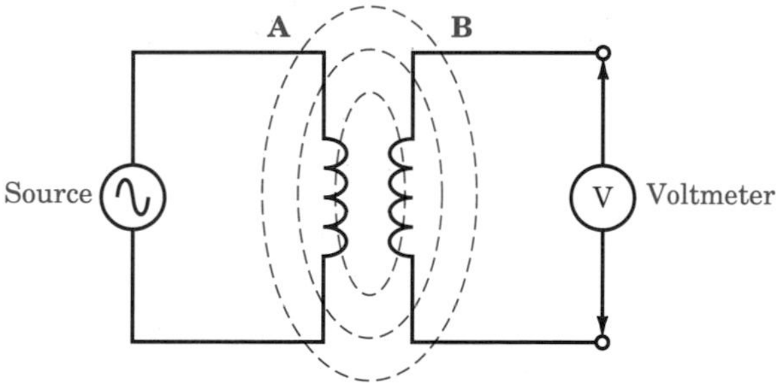

FIGURE 15–1 Mutual induction by flux linkage

COEFFICIENT OF COUPLING

In Figure 15–2 you can see smaller and greater amounts of flux linkage between coils. A term expressing this relationship by the fractional amount of total flux linking the two coils is **coefficient of coupling,** represented by the letter **k.** Coefficient of coupling and mutual inductance are *not* the same. However, they are related. Figure 15–2b shows the coils with the higher coefficient of coupling. The coefficient of coupling is therefore computed as the fractional amount of the total flux linking the two circuits. When *all* the flux from one circuit links the other, coefficient of coupling (k) equals one. When two-thirds the flux links both circuits, k equals 0.66, and so forth.

FIGURE 15–2
Coefficient of coupling related to degree of flux linkage

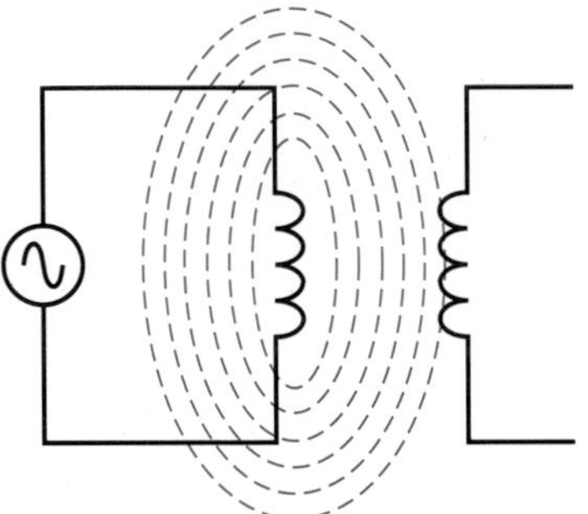

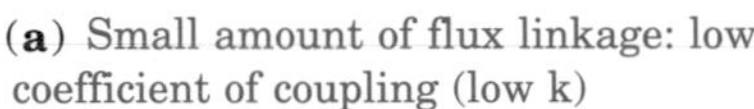

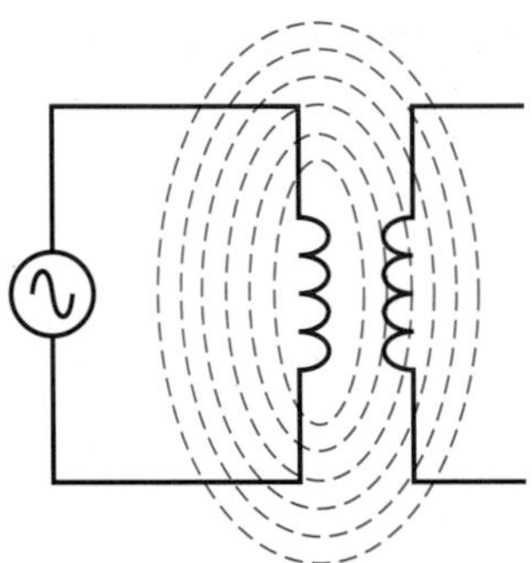

(**a**) Small amount of flux linkage: low coefficient of coupling (low k)

(**b**) Increased amount: higher coefficient of coupling (high k)

See Figure 15–3 for examples of factors that influence the coefficient of coupling. These factors include the proximity of coils (the closer they are the higher the k); the relative positions of coils with respect to each other (parallel coils at a given distance have greater k than perpendicular coils); and other factors, such as when the coils are wound on the same core. Coils wound on the same iron core have a coefficient of coupling almost equal to unity (1). Virtually all the flux produced by one coil links the other coil, with little leakage flux occurring. Air-core coils (often used in rf circuits) have a k that indicates a percentage of coupling from less than 5% (k = 0.05) to about 35% (k = 0.35). Obviously, air-core coils have more leakage flux.

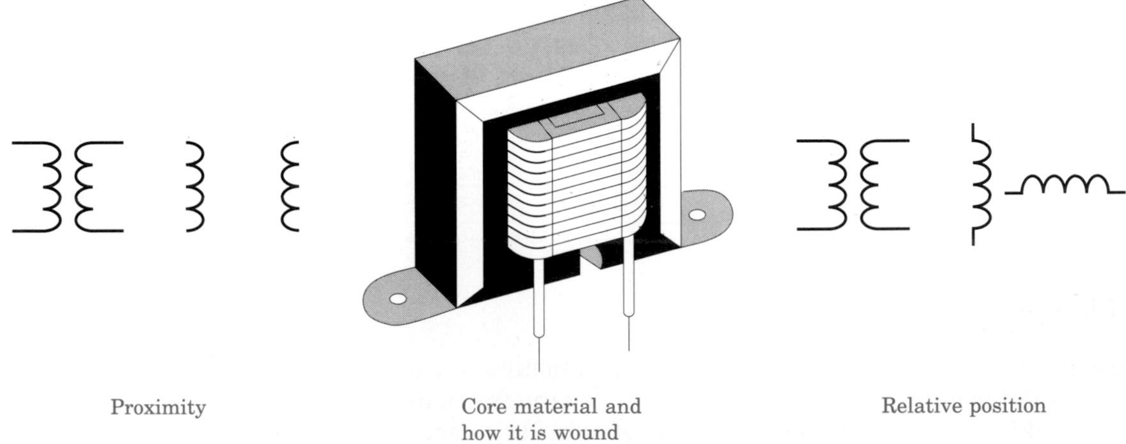

FIGURE 15–3 Factors influencing coefficient of coupling (k)

MUTUAL INDUCTANCE AND TRANSFORMER ACTION

As stated, when two circuits are coupled magnetically, they have mutual inductance. Such is the case for transformers. **Transformers** are devices

IN-PROCESS LEARNING CHECK I

Fill in the blanks as appropriate.

1. Producing voltage via a changing magnetic field is called electromagnetic ___________.
2. Inducing voltage in one circuit by varying current in another circuit is called ___________ ___________.
3. The fractional amount of the total flux that links two circuits is called the ___________ of coupling, which is represented by the letter ___________. When 100% of the flux links the two circuits, the ___________ of coupling has a value of ___________.
4. The closer coils are, the ___________ the coupling factor produced. Compared to parallel coils, perpendicular coils have a ___________ degree of coupling.

that transfer energy from one circuit to another by electromagnetic induction (mutual induction). Typically, transformers have their *primary* and *secondary* windings wound on a single core.

Refer to Figure 15–4. Notice the winding connected to the source is the primary winding. The winding connected to the load is the secondary winding. Some transformers can have more than one secondary winding. Also, observe the schematic symbol(s) used for various transformers.

Another feature shown on some diagrams, Figure 15–4, are sense dots. Sense dots indicate the ends of the windings that have the same polarity at any given moment. When sense dots are *not* shown, there is 180-degree difference between the primary and secondary voltage polarity. That is, if the top of the primary is positive with respect to a given reference point, the top of the secondary is negative, at that same moment.

Two circuits have a mutual inductance of 1 henry when a rate of current change of one ampere per second in the first circuit induces a voltage of one volt in the second circuit.

Mutual inductance between windings is directly related to the inductance of each winding and the coefficient of coupling. The formula is:

Formula 15–1	$M = k\sqrt{L_1 \times L_2}$

where M = mutual inductance in henrys
L_1 = self-inductance of the primary and L_2 = self-inductance of the secondary (for transformers) where Ls are in Henrys

EXAMPLE

A transformer has a k of 1 (100% coupling of flux lines), and the primary and secondary each have inductances of 3 H. The mutual inductance (M or L_M)

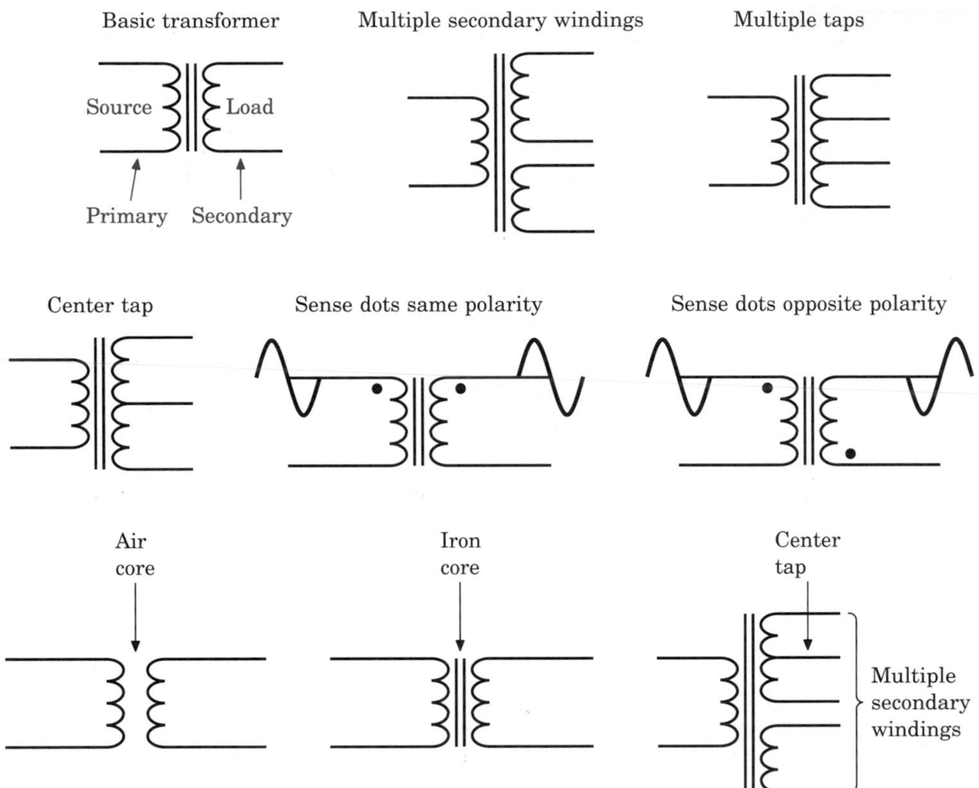

FIGURE 15–4 Primary and secondary transformer windings

is computed as $M = 1\sqrt{3 \times 3} = 1\sqrt{9} = 3$ H. From our definition, this indicates a current change of one ampere per second in the primary induces 3 volts in the secondary.

PRACTICE PROBLEM I

Answers are in Appendix B.
What is the mutual inductance of a transformer where the coefficient of coupling is 0.95, the primary inductance is 10 H, and the secondary inductance is 15 H?

MUTUAL INDUCTANCE BETWEEN COILS OTHER THAN TRANSFORMERS

We have been discussing mutual inductance in transformers. How does mutual inductance (coupled inductance) affect the total inductance of coils other than transformers that are located where their fields can interact?

Series Coils

When two coils are connected in *series* so their fields *aid* each other, total inductance is calculated from the formula:

Formula 15–2	$L_T = L_1 + L_2 + 2M$

where L_T = total inductance (H)
L_1 = inductance of one coil (H)
L_2 = inductance of other coil (H)
M = mutual inductance (H)

When two coils are connected in *series* so their fields oppose each other, the formula becomes:

Formula 15–3	$L_T = L_1 + L_2 - 2M$

Therefore, the general formula for series inductors is:

Formula 15–4	$L_T = L_1 + L_2 \pm 2M$

EXAMPLE

Two inductors are series connected so their fields aid one another. Inductor 1 has an inductance of 8 H. Inductor 2 also has an inductance of 8 H. Their mutual inductance is 4 H. What is the total inductance of the circuit?

Answer:

$$L_T = L_1 + L_2 + 2M = 8 + 8 + 8 = 24 \text{ H}$$

PRACTICE PROBLEM II

What is the total inductance of two series-connected 10-henry inductors whose mutual inductance is 5 H and whose magnetic fields are opposing?

Parallel Coils

When two inductors are in *parallel* with their fields *aiding* each other, the formula is:

Formula 15–5	$L_T = \dfrac{1}{\dfrac{1}{L_1 + M} + \dfrac{1}{L_2 + M}}$

When two inductors are in *parallel* with their fields *opposing* each other, the formula becomes:

Formula 15–6	$L_T = \dfrac{1}{\dfrac{1}{L_1 - M} + \dfrac{1}{L_2 - M}}$

PRACTICE PROBLEM III

1. What is the total inductance of two parallel coils whose fields are aiding, if $L_1 = 8$ H, $L_2 = 8$ H, and $M = 4$ H?
2. What is the total inductance of two parallel coils whose fields are opposing, if $L_1 = 8$ H, $L_2 = 8$ H, and $M = 4$ H?

Relationship of k and M

The coefficient of coupling and mutual inductance are related by the following formulas for k and M.

Formula 15–7	$k = \dfrac{M}{\sqrt{L_1 \times L_2}}$

Recall Formula 15–1 is $M = k\sqrt{L_1 \times L_2}$. As you can see from these two formulas, mutual inductance and coefficient of coupling are directly related!

IMPORTANT TRANSFORMER RATIOS

We will begin this section by making the following assumptions.

1. With *no load* connected to the secondary winding, the primary winding acts like a simple inductor.
2. Current flowing in the primary depends on applied voltage and inductive reactance of the primary. That is, $I_P = \dfrac{V}{X_P}$.
3. The primary current produces a magnetic field that cuts both the primary and secondary turns.
4. A cemf is produced in the primary. This cemf almost equals the applied voltage due to self-inductance.
5. A cemf is produced in *each turn* of the secondary, which equals the voltage induced in *each turn* of the primary due to mutual inductance. This means when there are equal turns on the secondary and the primary, their voltages are equal.

Using these assumptions, let's look at the transformation ratio. That is, the ratio between electrical output and input parameters under specified conditions. When 100% of the primary flux links the secondary winding, the ratio of the induced voltages (i.e., the ratio of secondary voltage to the primary voltage) is the same as the ratio of turns on the secondary compared to the turns on the primary. The symbol for transformation ratio is the Greek letter alpha (α). Thus, $\alpha = \frac{V_S}{V_P} = \frac{N_S}{N_P}$, where V_S is secondary induced voltage; V_P is primary induced voltage; N_S is the number of turns on the secondary winding; and N_P is the number of turns on the primary winding.

Observe Figure 15–5 and notice how the terms step-up and step-down relate to transformers. When the secondary (output) voltage is higher than the primary (input) voltage, the transformer is a step-up transformer. Conversely, when the secondary voltage is lower than the primary voltage, the transformer is a step-down transformer.

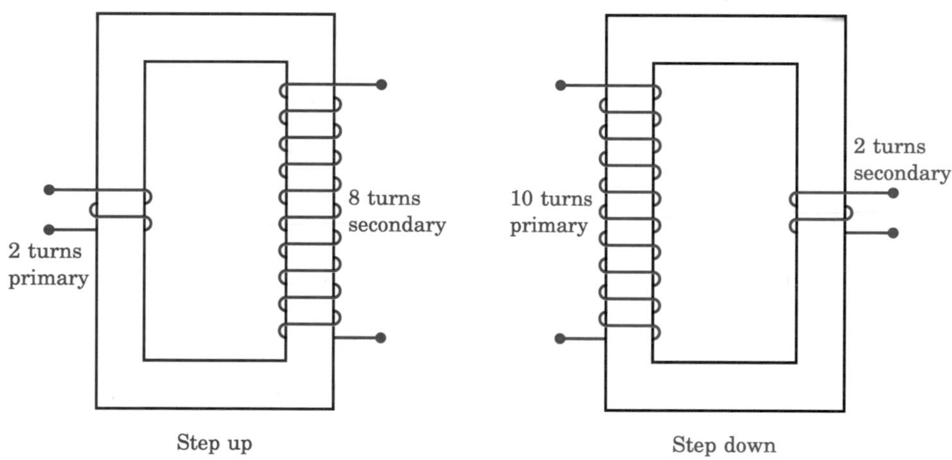

FIGURE 15–5 Step-up and step-down transformers

Turns Ratio

It is often stated the ratio of secondary turns to primary turns $\left(\frac{N_S}{N_P}\right)$ is the **turns ratio** (TR). For example, when the secondary has 1000 turns and the primary has 250 turns, the turns ratio is $\frac{1000}{250} = 4{:}1$. This is called the secondary-to-primary (s-p) turns ratio, Figure 15–6a.

Voltage Ratio

As you learned with the ideal transformer, the **voltage ratio** has the same ratio as the turns ratio, Figure 15–6b. That is:

Formula 15–8	$\frac{V_S}{V_P} = \frac{N_S}{N_P}$ or $\frac{N_S}{N_P} = \frac{V_S}{V_P}$

Secondary-to-primary turns ratio = 4:1
Primary-to-secondary turns ratio = 1:4

250 t 1000 t

Primary Secondary

(a)

Turns ratio

Laminated core

A

V

Input 240 V

Primary winding 300 turns

Stepdown ratio 2:1 (P-S)

Flux

120-V output

V

Secondary winding 150 turns

(b)

Voltage ratio

$$\frac{N_S}{N_P} = \frac{V_S}{V_P} \qquad \frac{N_P}{N_S} = \frac{V_P}{V_S}$$

$$\frac{150}{300} = \frac{120}{240} \text{ OR } \frac{300}{150} = \frac{240}{120}$$

$$\frac{1}{2} = \frac{1}{2} \qquad \frac{2}{1} = \frac{2}{1}$$

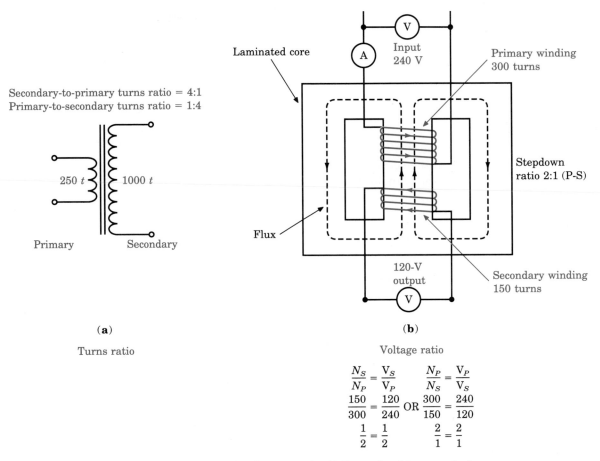

FIGURE 15–6 Turns and voltage ratio (Adapted with permission from Duff and Herman, *Alternating Current Fundamentals*, 3rd edition, © 1986 by Delmar Publishers Inc., Albany, New York)

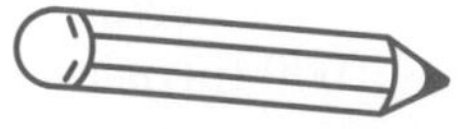

PRACTICAL NOTES

Some people use the opposite approach, i.e., primary-to-secondary ratio. In any case, the technician should be careful in making transformer calculations. Be *consistent* in the way you set up the various ratios. We will use both ratios, as appropriate.

EXAMPLE

1. What is the secondary-to-primary turns ratio of a transformer having eight times as many turns on the primary as on the secondary?

Answer:

$$TR = \frac{N_S}{N_P} = \frac{1}{8} \text{ or } 1{:}8$$

2. What is the secondary voltage on a transformer having a turns ratio of 14:1 (s-p), if the applied voltage is 20 volts?
Answer:

$$\frac{V_S}{V_P} = \frac{N_S}{N_P}; \frac{V_S}{20} = \frac{14}{1} = 20 \times 14 = 280 \text{ volts}$$

PRACTICE PROBLEM IV

1. What is the s-p turns ratio of an ideal transformer whose secondary voltage is 300 volts when 100 volts is connected to the primary?
2. What is the s-p turns ratio of an ideal transformer that steps-up the input voltage six times?
3. What is the secondary voltage of an ideal transformer whose s-p turns ratio is 5:1 and whose primary voltage is 50 volts?
4. What is the voltage ratio of an ideal transformer whose s-p turns ratio is 1:4? Is this a step-up or a step-down transformer?

Current Ratio

Up to now, our discussions have been about transformers with no load connected to the secondary(ies). Notice in Figure 15–7 that when a load is connected to the secondary, current flows through the load and the second-

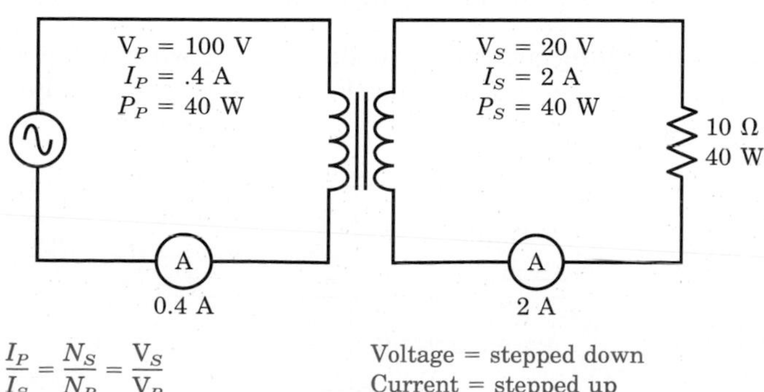

FIGURE 15–7
Current ratio

ary. This current produces a magnetic field related to the secondary current. The polarity produced by the secondary current cancels some of the primary field. Hence, the cemf is reduced in the primary, which reduces the impedance of the primary to current flow. This reduced impedance causes primary current to increase. The interaction between the secondary and the primary is termed reflected impedance. In other words, the impedance reflected across the input or primary results from current in the output or secondary.

In Figure 15–7, the power available to the load from the transformer secondary must come from the source that supplies the primary, since a transformer does not *create* power, but *transfers* power from the primary circuit to the secondary circuit. Transformers are not 100% efficient. Therefore, the power delivered to the primary by its source is slightly higher than the power delivered by the secondary to its load. Since transformers are highly efficient (generally ranging from 90–98% efficient), Figure 15–7 has assumed 100% efficiency to illustrate this principle. Later in the chapter, we will look at some losses that prevent actual 100% efficiency.

Also notice in Figure 15–7 the power delivered to the load is 40 W, where $V_S = 20$ V, $R_S = 10\ \Omega$, and $I_S = 2$ A. The power delivered to the primary is 40 W, where $V_P = 100$ V and $I_P = 0.4$ A. This is a step-down transformer where the voltage has been stepped down five times. That is, $\frac{N_S}{N_P} = 1{:}5$ and $\frac{V_S}{V_P} = 1{:}5$.

What about the **current ratio?** Observe the current in the secondary is five times higher than the current in the primary. Using our secondary-to-primary principle, the current ratio is 5:1, or the inverse of the voltage ratio.

This makes sense, if we assume the power in the secondary is the same as the power in the primary. In other words, $V_SI_S = V_PI_P$. If voltage was decreased by five times, the current must be increased by five times so the V times I (or P) will be the same in both primary and secondary.

(NOTE: Power supplied to the primary must equal power delivered to the secondary plus power lost.)

Now let's relate current ratio to the turns and voltage ratios. Thus, $\frac{I_P}{I_S} = \frac{N_S}{N_P} = \frac{V_S}{V_P}$. Note the inversion in the current ratio. Since quantities equaling the same thing are equal to each other, we can list three useful equations from the above information. First remember that Formula 15–8 says:

Formula 15–8	$\frac{N_S}{N_P} = \frac{V_S}{V_P}$

Formula 15–9	$\frac{N_S}{N_P} = \frac{I_P}{I_S}$

Formula 15–10	$\frac{V_S}{V_P} = \frac{I_P}{I_S}$

If you know any three of the quantities in any of these equations you can solve for the unknown quantity, assuming 100% efficiency.

EXAMPLE

1. A transformer has 1000 turns on the secondary, 250 turns on the primary, and the primary voltage is 120 volts. What is the secondary voltage value?
   Answer:

$$\frac{N_S}{N_P} = \frac{V_S}{V_P}$$

$$\frac{1000}{250} = \frac{V_S}{120}$$

$$250\ V_S = 120{,}000$$

$$V_S = \frac{120{,}000}{250} = 480 \text{ volts}$$

2. A transformer has a secondary-to-primary turns ratio of 1:5. What is the primary current when the secondary load current is 0.5 ampere?
   Answer:

$$\frac{N_S}{N_P} = \frac{I_P}{I_S}$$

$$\frac{1}{5} = \frac{I_P}{0.5}$$

$$5\ I_P = 1 \times 0.5$$

$$I_P = \frac{0.5}{5} = 0.1 \text{ ampere}$$

3. The primary-to-secondary current ratio of a transformer is 4:1, and the secondary voltage is 60 volts. What is the voltage on the primary?
   Answer:

$$\frac{V_S}{V_P} = \frac{I_P}{I_S}$$

$$\frac{60}{V_P} = \frac{4}{1}$$

$$4\ V_P = 60 \times 1$$

$$V_P = \frac{60}{4} = 15 \text{ volts}$$

Impedance Ratio

The **impedance ratio** is related to the turns ratio, Figure 15–8. Also, since inductance and thus the inductive reactance of a coil, relates to the square of the turns, the impedance also has a similar relationship. A load of a given impedance connected to the transformer secondary is transformed to a different value by what the source sees "looking into" the primary (unless the turns ratio is 1:1). In fact, for an ideal (zero loss) transformer, the impedance transformation (secondary-to-primary) is proportional to the square of the secondary-to-primary turns ratio. That is,

Formula 15–11	$\left(\frac{N_S}{N_P}\right)^2 = \frac{Z_S}{Z_P}$

where Z_S = the impedance of the load connected to the secondary
Z_P = the impedance looking in the primary from source
$\frac{N_S}{N_P}$ = the secondary-to-primary turns ratio

FIGURE 15–8
Impedance ratio

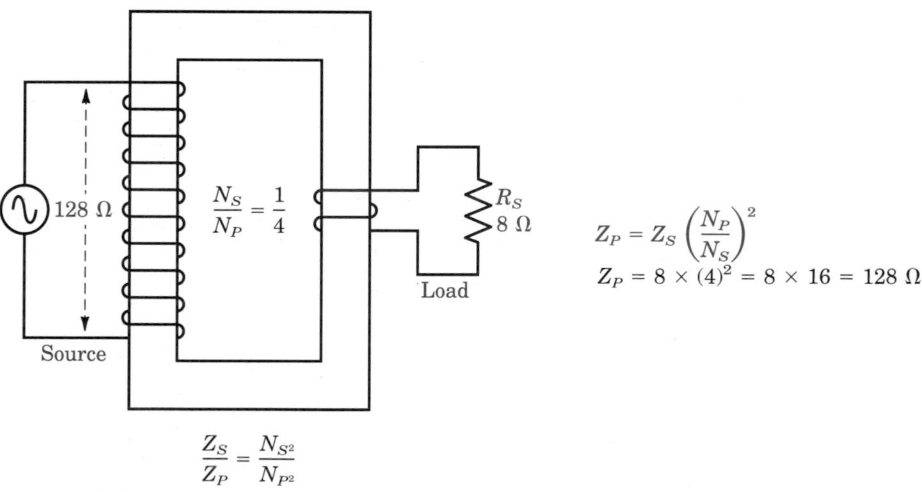

EXAMPLE

When the $\frac{N_S}{N_P}$ turns ratio of a transformer is 1:4 and the impedance connected to the secondary is 8 Ω, the primary impedance is related to the square of the turns ratio. Hence, Z_P is 16 times Z_S. That is $Z_P = 8 \times 16 = 128\ \Omega$, Figure 15–8. A practical formula expressing this in terms of *primary-to-secondary* turns ratio is:

Formula 15–12	$Z_P = Z_S \left(\frac{N_P}{N_S}\right)^2$

EXAMPLE

A transformer has a secondary-to-primary turns ratio of 0.5, and a load resistance of 2000 ohms is connected to the secondary. What is the apparent primary impedance?

Answer:

Since the secondary-to-primary turns ratio is 0.5, the secondary has half as many turns as the primary, or the primary has twice as many turns as the secondary. Appropriately substitute in the formula:

$$Z_P = Z_S\left(\frac{N_P}{N_S}\right)^2$$

$$Z_P = 2000\ \Omega \times (2)^2$$

$$Z_P = 2000\ \Omega \times 4 = 8{,}000\ \Omega$$

PRACTICE PROBLEM V

In the prior example, suppose the turns ratio reverses. In other words, the secondary has twice as many turns as the primary. What is the apparent primary impedance?

These discussions indicate the primary impedance of a very efficient transformer (as it looks to the power source feeding the primary) is determined by the load connected to the secondary and the turns ratio. Because this is true, transformers are often used to match impedances. That is, a transformer is used so that a secondary load of given value is transformed to the optimum impedance value into which the primary source should be looking. (The source may be an amplifier stage, another ac signal, or power source.) Therefore, a useful formula is:

Formula 15–13	$\frac{N_P}{N_S} = \sqrt{\frac{Z_P}{Z_S}}$

where $\frac{N_P}{N_S}$ = the required *primary-to-secondary* turns ratio

Z_P = the primary impedance desired (required)

Z_S = the impedance of the load connected to the secondary

EXAMPLE

An amplifier requires a load of 1000 ohms for best performance. The amplifier output is to be connected to a loudspeaker having an impedance of 10 ohms. What must the turns ratio *(primary-to-secondary)* be for a transformer used for impedance matching?

$$\frac{N_P}{N_S} = \sqrt{\frac{Z_P}{Z_S}}$$

$$\frac{N_P}{N_S} = \sqrt{\frac{1000}{10}} = \sqrt{100} = 10$$

This indicates the primary must have 10 times as many turns as the secondary.

PRACTICE PROBLEM VI

1. What is the primary-to-secondary impedance ratio of a transformer that has 1000 turns on its primary and 200 turns on its secondary?
2. What impedance would the primary source see looking in a transformer where the primary-to-secondary turns ratio is 16 and the load resistance connected to the secondary is 4 Ω?

TRANSFORMER LOSSES

Because of energy losses, practical transformers do not meet all the criteria for the ideal, 100% efficient devices we have been discussing. These transformer losses consist primarily of two categories. One category is **copper loss,** due to the I^2R copper wire losses that make up the primary and secondary windings. Current passing through the resistance of the wire produces energy loss in the form of heat.

The other category of transformer losses is termed **core loss.** Core losses result from two factors. One factor is called *eddy current* losses. Due to the changing magnetic fields, and since the iron core conducts electricity, current(s) produced in the iron core does not aid transformer output. These eddy currents cause an I^2R core loss related to the core material resistance and the current amount. To reduce the current involved, the iron core is usually constructed of thin sheets of metal called *laminations.* These laminations are electrically insulated from each other by varnish or shellac. Reducing the cross-sectional area of the conductor by laminating significantly increases the core resistance, which decreases the current, and greatly reduces the I^2R loss.

The other core loss is called *hysteresis loss.* Recall magnetic domains in the magnetic core material require energy to rearrange. (Recall in an earlier chapter, the lagging effect of the core magnetization behind the magnetizing force was illustrated by the hysteresis loop). Since ac current constantly changes in magnitude and direction, the tiny molecular magnets within core magnets are constantly being rearranged. This process requires energy that causes some loss.

Copper losses, core losses, leakage flux, and other factors prevent a transformer from being 100% efficient. Efficiency percentage is calculated as output power divided by input power times 100.

Formula 15–14	Efficiency % $= \frac{P_{out}}{P_{in}} \times 100$

To minimize these various losses, the physical features of coils and transformers used for applications in power, audio, and rf frequency ranges slightly differ. We will investigate some of these characteristics in the next section.

CHARACTERISTICS OF SELECTED TRANSFORMERS

Power Transformers

Power transformers, Figure 15–9, typically have the following features:

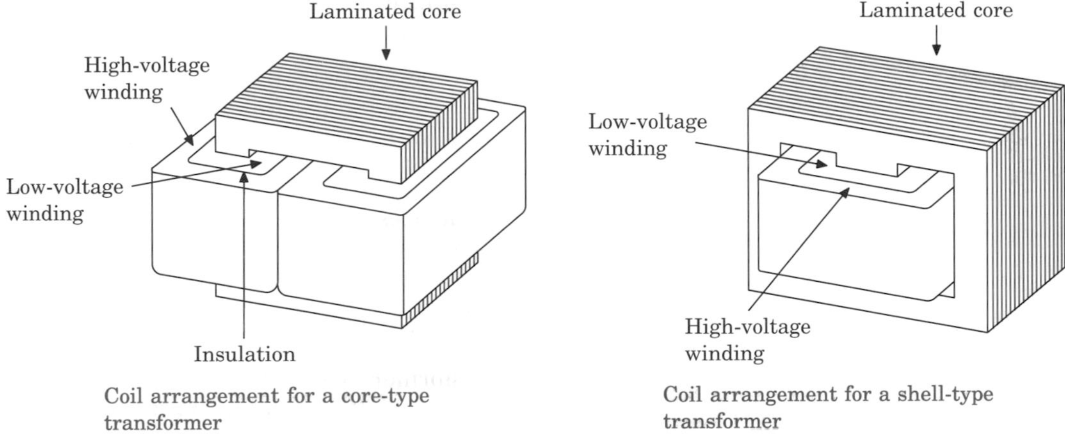

FIGURE 15–9 Power transformers (Adapted with permission from Perozzo, *Practical Electronics Troubleshooting*, © 1985 by Delmar Publishers Inc., Albany, New York)

1. Heavy (due to iron core) and can be very large, like the transformers on telephone poles. However, power transformers in electronic devices are much smaller and vary considerably in size.
2. Laminated iron core to reduce eddy-current losses.
3. Use as short a flux path as possible to reduce leakage flux and minimize magnetizing energy needed.
4. Typically use one of two core shapes—core type (primary and secondary windings on separate legs of the core and shell type (both primary and secondary windings on the center leg), again see Figure 15–9.
5. Can have one or more primary and secondary windings.

6. Can be *one* tapped winding, such as the **autotransformer,** Figure 15–10. The autotransformer is a special single-winding transformer.

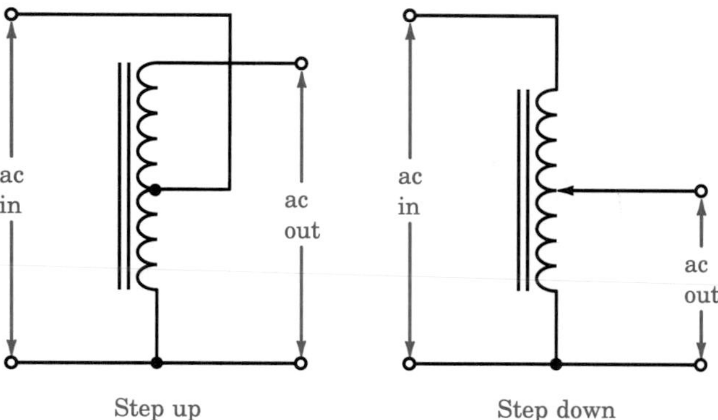

FIGURE 15–10
Autotransformer

When the input source is connected from one end to the tap while the output is across the whole coil, the voltage is stepped up. If the input voltage is applied across the whole coil and the output taken from the tap to one end, the voltage is stepped down.

Variacs or *Powerstats* are trade names for autotransformers having a movable wiper-arm tap that allows varying voltage from zero volts to the maximum level. They are often used as a variable ac source in experimental setups.

SAFETY HINTS

Caution: Bad shock can occur if no **isolation transformer** is used on the autotransformer input side, because there is no isolation from the raw line voltage and ground connections in the autotransformer circuit. For this reason, do *not* use autotransformers with a movable wiper-arm tap unless you plug the autotransformer into an isolation transformer output, which isolates it from the raw ac power source.

7. Can be used to *isolate* the circuitry connected to its secondary from the primary ac source (e.g., the ac powerline). This is accomplished by using a 1:1 turns ratio transformer.

When using this special 1:1 turns ratio transformer, isolation transformer, the load on the secondary has no *direct* connection to the primary source. This provides a safety measure for people working with the circuitry connected to the isolation transformer secondary.

8. Cannot operate at high frequency due to excessive losses that would be created.
9. An example of how power transformer leads are typically color-coded is shown in Figure 15–11a.

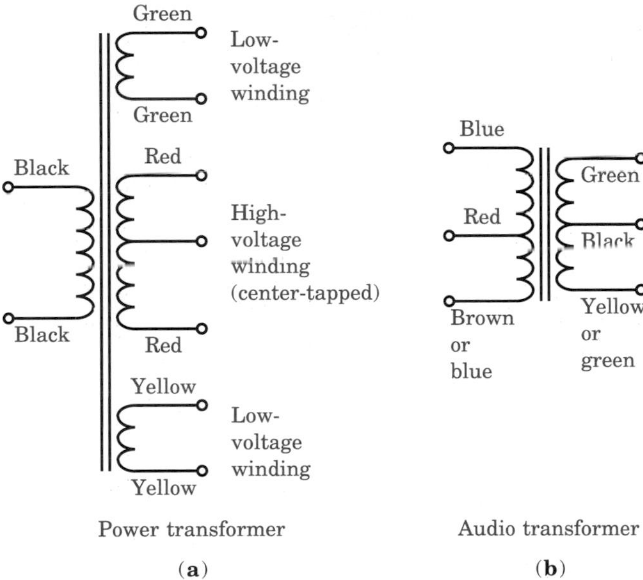

FIGURE 15–11 Typical power and audio transformer color-coding systems

Audio Transformers

Audio transformers are similar to power transformers in their construction. Some typical features include:

1. Smaller and lighter than most power transformers.
2. An example of typical lead color-coding is shown in Figure 15–11b.

RF Transformers and Coils

Because the flux is changing *very rapidly,* rf transformers or coils, Figures 15–12 and 15–13, have different features compared to power or audio transformers. Some features include:

1. Cores are air core, or powdered iron-core, depending on frequency and application. Air-core coils have virtually zero hysteresis or eddy-current losses. Powdered iron-cores are made by grinding metal materials and nonconductive materials into granules, which are pressed into a cylindrical-shaped “slug.” Each metal particle is therefore insulated, greatly reducing losses such as eddy current. See Figure 15–12a for examples of this coil.

(a)

Open lead style, clamp fixing

Printed circuit style

A

Terminal pins
0.08 dia. × 0.16 long
located on 0.1 in. grid

B

Threaded
bushes

(b)

FIGURE 15–12 Examples of various types coils and cores. (a) Powdered iron-core type coils; (b) Examples of toroidal cores (Photo courtesy of Miller Division of Bell Industries)

(a)

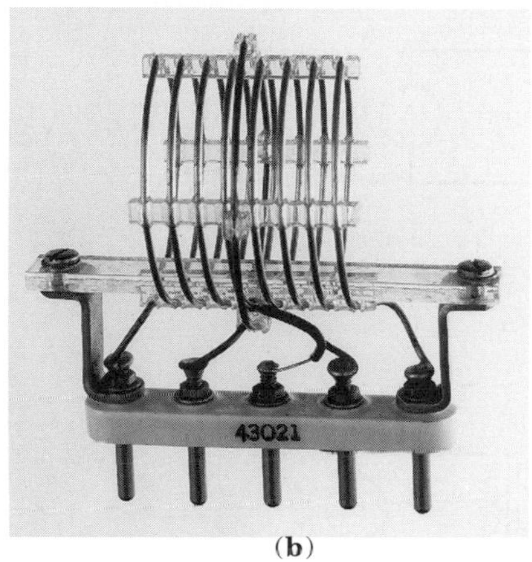

(b)

FIGURE 15–13 Examples of air-core rf coils and transformers (Courtesy of James Miller Electronics)

Also, nonconductive ferrite materials can be used in rf coils or transformers. As you recall, ferrite materials are special magnetic materials that provide easy paths for magnetic flux but are nonconductive. Often, toroid-type cores are used in radio-frequency circuits and lower-frequency applications, Figure 15–12b.

2. Rf coils and transformers typically have very low inductance values. Generally, these values are in the microhenry or millihenry ranges.
3. Figure 15–13 shows examples of air-core coils found in transmitter applications.

TROUBLESHOOTING HINTS

Transformer problems are virtually the same problems as those described in the chapter on inductance. Two basic problems are opens or shorts. The primary test used to detect these defects is the ohmmeter test.

Open Winding(s)

With an open winding, the obvious result is zero secondary output voltage, even when source voltage is applied to the primary. When the primary is *not* open and only one secondary of a multiple-secondary transformer is open, only the open winding shows zero output. To check these conditions, remove the power and use the ohmmeter.

Refer to Figure 15–14a to see an open winding. In this case, the ohmmeter registers an infinite resistance reading, indicating the winding under test has an open. Both the primary and secondary windings should be checked! Normal readings are the specified dc resistance values for each winding.

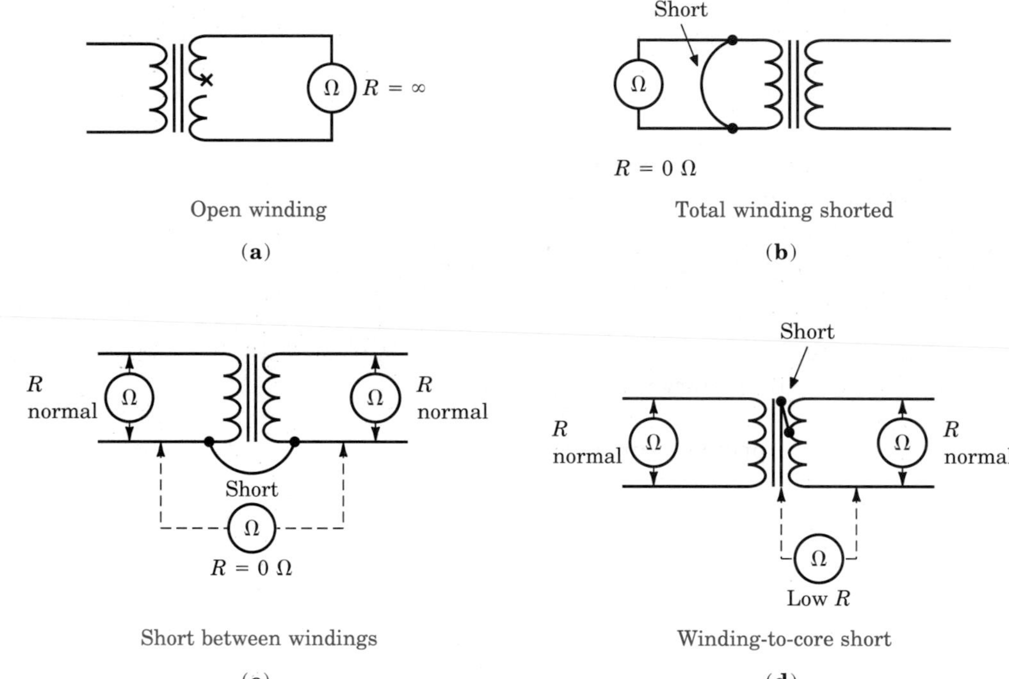

FIGURE 15–14 Typical transformer troubles

Shorted Winding(s)

Shorted turns or a shorted winding can cause a fuse in the primary circuit to blow (open) due to excessive current. Another possible clue to this condition is if the transformer operates at a hotter than normal temperature. Also, another clue of shorted turns is lower than normal output voltage. Actually, partially shorted windings or windings with very low *normal* resistances are hard to check only by an ohmmeter. Therefore, watch for these clues when troubleshooting transformer problems.

However, the ohmmeter is useful to detect certain types of transformer short conditions. Figure 15–14b shows a total-winding short condition. In this case, the ohmmeter reads virtually zero ohms when it should be reading the normal winding resistance value, if there were no short.

Figure 15–14c illustrates the condition when the primary and secondary windings are shorted together. In this case, each winding resistance might be normal; however, there would be a low resistance reading between windings. When conditions are normal, there are infinite ohms between primary and secondary windings.

Another possible short condition is when one or both windings are shorted to the metal core, Figure 15–14d. Again, if things are normal, an infinite ohms reading will be present between either winding and the core material.

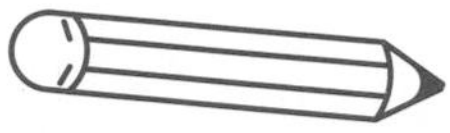

PRACTICAL NOTES

The following discussion describes one practical approach used to determine whether a short in a *power transformer* exists. This approach is a simple test involving the transformer in question and a light bulb. In Figure 15–15, you can see a light bulb is placed in series with the transformer primary winding. When a short is present in *either* the primary or secondary, the lamp glows brightly due to the higher than normal primary current. When no short exists, the lamp does *not* glow or may dimly glow.

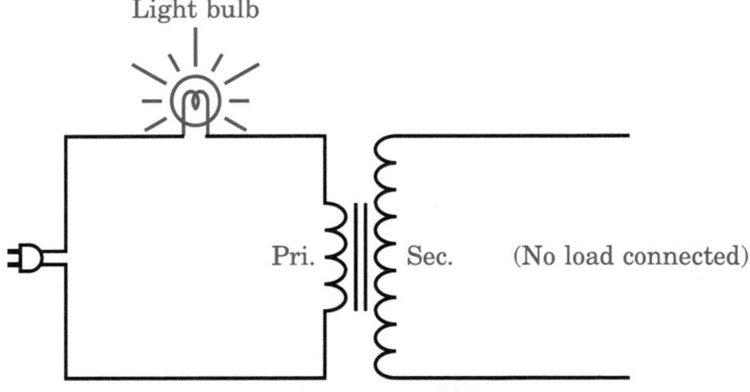

If lamp glows dimly or not at all = no short(s)
If lamp glows brightly = short(s); (throw transformer away!)

NOTE: Use small bulb rating for small transformers; use larger bulb rating for large transformers.

FIGURE 15–15 One way to find shorted turns

The requirements for bulb size and its electrical ratings are:

1. Voltage rating should equal the same voltage rating as the transformer primary.
2. For small transformers, use a small wattage lamp.
3. For large transformers, use a large wattage lamp.

Be aware that choosing the correct bulb size and interpreting the results come only with experience. This test is simply mentioned to show you some innovative ways technicians find to learn *what they need to know,* even in difficult situations.

Practical Transformer Ratings

Power transformers	Audio transformers
Input volts	Primary impedance
Output V, I, and kVA ratings	Secondary impedance
Insulation ratings	Maximum V, I, and P
Size	Size
Mounting dimensions	Mounting dimensions
Weight	

SUMMARY

- When the magnetic flux of two coils link and interact, mutual inductance and inductive coupling occur between the two circuits.
- When mutual inductance is present, current changing in one circuit causes an induced voltage in the other circuit. The symbol for mutual inductance is M or L_M, and the unit of measure is the henry.
- One henry of mutual inductance is present when a change of one ampere per second in one circuit induces one volt in the mutually coupled circuit.
- Expressing the fraction of total flux linking one circuit to another circuit is called coefficient of coupling. The abbreviation for coefficient of coupling is k. NOTE: Multiplying k by 100 yields the percentage of total flux lines linking the circuits.
- The relationship between mutual inductance and coefficient of coupling is shown by the formulas:

$$M = k\sqrt{L_1 \times L_2} \text{ and } k = \frac{M}{\sqrt{L_1 \times L_2}}$$

where M = mutual inductance in henrys
k = coefficient of coupling
L_1 = self-inductance of coil or transformer winding 1
L_2 = self-inductance of coil or transformer winding 2

- Mutual inductance can exist between coils that are not transformer windings but that are close enough for flux linkage to occur. This can affect total inductance for the circuit.
- Transformers are comprised of separate coils or windings, having inductive coupling. The primary winding is connected to an ac source. The load(s) are connected to the secondary winding(s).
- One feature of a two-winding transformer is that the primary and secondary are magnetically coupled, yet electrically isolated from each other. This feature can keep circuits and loads from being directly connected to the power lines, which increases safety.
- One key factor about transformers is the turns ratio. Turns ratio is expressed either by the primary-to-secondary ratio $\left(\frac{N_P}{N_S}\right)$, or secondary-to-primary ratio $\left(\frac{N_S}{N_P}\right)$. The latter approach is used most often.
- The ratio of voltages between primary and secondary is the same as the ratio of turns. That is, $\frac{N_S}{N_P} = \frac{V_S}{V_P}$.
- The current ratio of a transformer is the inverse of the voltage ratio. When voltage is stepped up from primary to secondary, current

is stepped down by a proportionate factor: $\frac{V_S}{V_P} = \frac{I_P}{I_S}$.

- The impedance ratio of a transformer is related to the square of the turns ratio. That is, $\frac{Z_P}{Z_S} = \frac{{N_P}^2}{{N_S}^2}$
- When a transformer has 100% efficiency, the power delivered to the load by the secondary equals the power supplied to the primary by the ac source. The efficiency of iron-core transformers is usually in the 90–98% range.
- Two categories of transformer losses are copper losses and core losses. Copper losses relate to the I^2R losses of the copper wire in the primary and secondary windings. Core losses result from eddy currents being induced in the core by the changing magnetic flux, and the energy wasted due to hysteresis losses. Eddy currents in the core material cause I^2R losses.
- Transformers have various applications at different frequencies. Three transformer categories include power transformers, audio transformers, and rf transformers.
- Power and audio transformers generally have laminated iron-cores, for efficiency's sake. Rf transformers usually use air cores, special powdered iron, or ferrite in their core material.
- Transformers can develop abnormal operations because of opens or shorts. Opens are evident by lack of output. Shorts between turns, from winding-to-winding or from winding-to-core, can be seen by overheating, blown fuses, and so forth.

REVIEW QUESTIONS

1. What is the primary current of a transformer whose source is 120 volts and secondary is supplying 240 V at 100 mA?

2. What is the secondary voltage of a transformer having a secondary-to-primary turns ratio of 5.5:1, if the source voltage is 100 volts?

3. What is the primary-to-secondary turns ratio of a transformer whose primary voltage is 120 volts and secondary voltage is 6 volts?

4. If a transformer has a primary-to-secondary turns ratio of 5:1 and the load resistance connected to the secondary is 8 ohms, what impedance is seen looking in the primary?

5. What is the mutual inductance of two coils having a coefficient of coupling of 0.6, if each coil has a self-inductance of 2 henrys?

6. What is the total inductance of two coils in series having 0.3 k, if each of the coils has a self-inductance of 10 mH? (Assume their fields are aiding.)

7. What is the total inductance of two coils in parallel having 0.5 k, if each coil has a self-inductance of 10 H. (Assume their fields are opposing.)

8. What is the secondary voltage of an ideal transformer whose secondary-to-primary turns ratio is 15 and primary voltage is 12 volts?

9. How many volts-per-turn are in a transformer having a 500-turn primary and a 100-turn secondary, if the primary voltage is 50 volts?

10. A 6:1 step-down transformer has a secondary current of 0.1 A. Assuming an ideal transformer, what is the primary current?

11. What is the number of turns in the primary winding of the transformer shown in Figure 15–16?

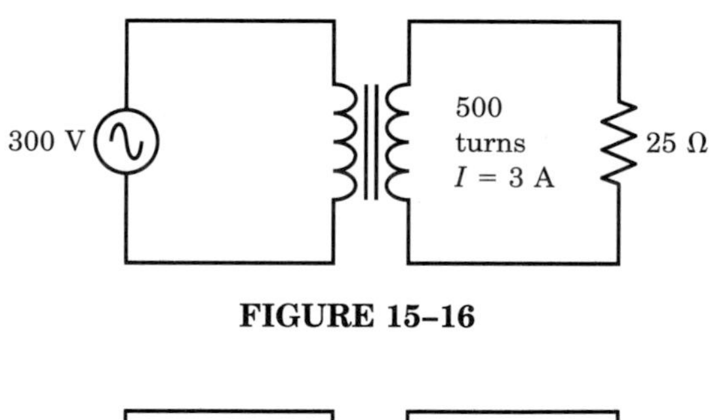

FIGURE 15–16

12. Assuming an ideal transformer, what is the power delivered to R_L in Figure 15–17?

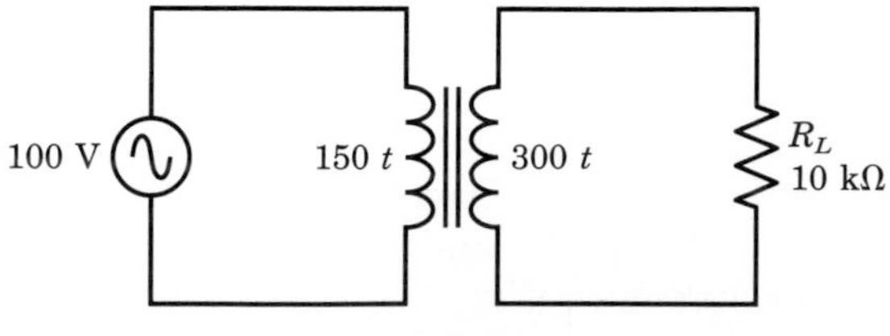

FIGURE 15–17

13. What is the primary-to-secondary turns ratio in Figure 15–18?

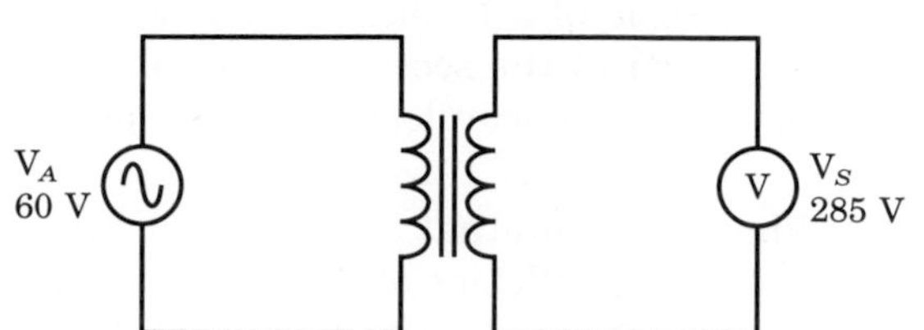

FIGURE 15–18

14. What is the primary current in the transformer circuit in Figure 15–19? (Assume an ideal transformer.)

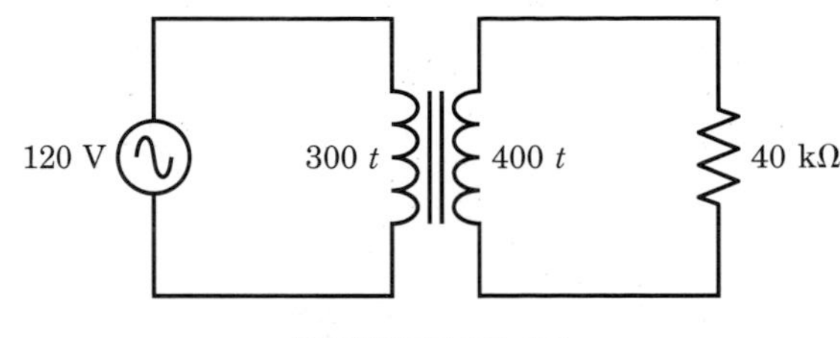

FIGURE 15–19

15. What is the voltage applied to the primary of the circuit in Figure 15–20?

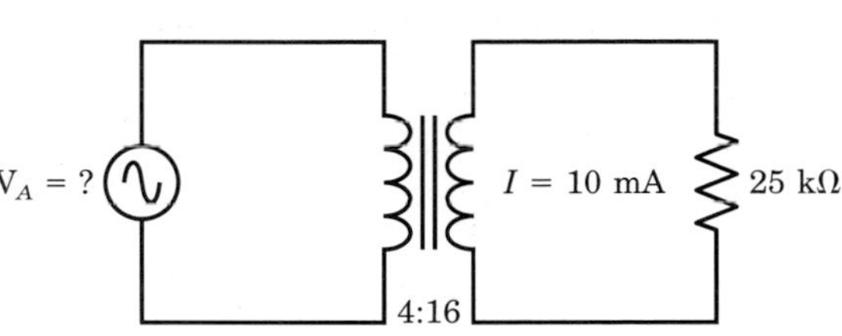

FIGURE 15–20

TEST YOURSELF

1. An amplifier having an output impedance of 5,000 ohms is feeding its output signal to a matching transformer. If the speaker connected to the secondary is a 16-ohms speaker, and the number of turns on the secondary is 200 turns, what number of turns should the primary winding have to provide the best impedance match? (Assume a perfect transformer.)

2. The secondary-to-primary turns ratio of a transformer is 1:8. What is the primary current, if the power delivered to a 200 Ω load resistance is 405 mW?

3. If the impedance ratio (secondary-to-primary) of an ideal transformer is 1:36 and the number of turns on the primary is 1200, how many turns are on the secondary winding?

4. A TV receiver contains a transformer in its power supply. If you unplug the receiver from its 120-volt, 60-Hz wall plug (ac source), turn the on/off switch to the on position, and connect an ohmmeter to each lead of the ac cord's plug, the ohmmeter reads 10 ohms. Using Ohm's law, you think the current supplied to the transformer is 12 amperes, since $I = \frac{V}{R}$, or $\frac{120\text{ V}}{10\ \Omega}$. Explain why the circuit is fused with only a 5-ampere fuse, yet, when the TV receiver is turned on, the fuse does not blow!

PERFORMANCE PROJECT CORRELATION CHART

Suggested performance projects that correlate with topics in this chapter are:

CHAPTER TOPIC	PERFORMANCE PROJECTS	PROJECT NUMBER
Mutual Inductance and Transformer Action	Turns, Voltage and Current Ratios	46
and	and	
Important Transformer Ratios	Turns Ratio(s) versus Impedance Ratios	47

Capacitance

Key Terms

Capacitance

Capacitor

Charging a capacitor

Dielectric

Dielectric constant

Dielectric strength

Discharging a capacitor

Electrostatic (electric) field

RC time constant

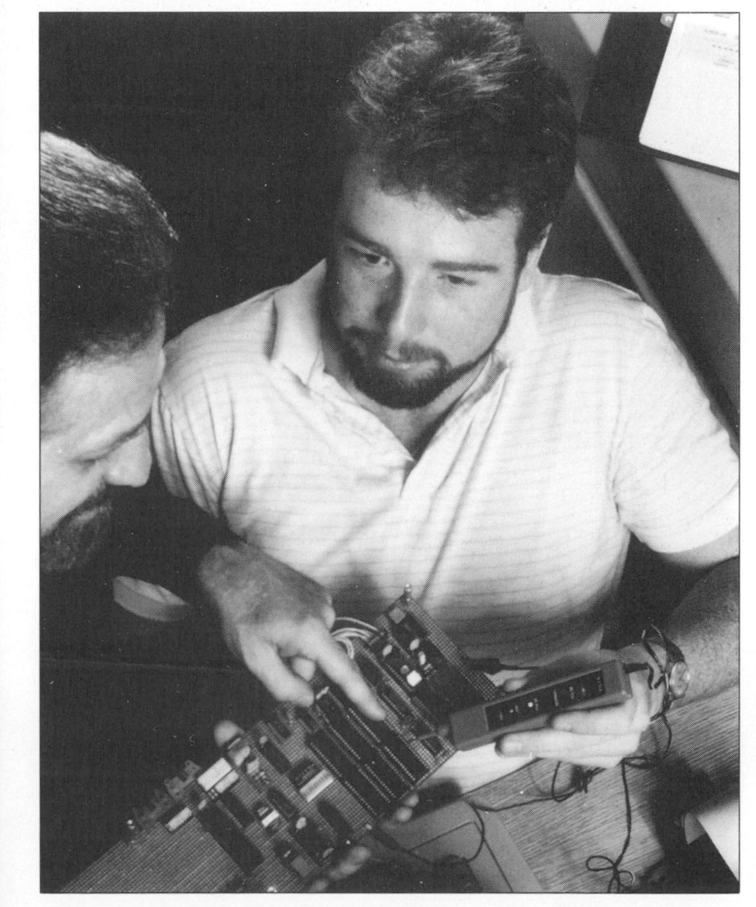

Courtesy of DeVry Institutes

16

Outline

Definition and Description of a Capacitor

The Electrostatic Field

Charging and Discharging Action

The Unit of Capacitance

Energy Stored in a Capacitor's Electrostatic Field

Factors Affecting Capacitance Value

Capacitance Formulas

Total Capacitance in Series and Parallel

The RC Time Constant

Types of Capacitors

Typical Color Codes

Typical Problems and Troubleshooting Techniques

Chapter Preview

Three basic properties of electrical circuits are resistance, inductance, and capacitance. You have studied resistance (opposition to current flow) and inductance (property opposing a *change in current*). In this chapter, you will study capacitance (property opposing a *change in voltage*).

You will learn that capacitors have the unique ability to store electrical energy because of an electric field. Even as current produces a magnetic field, voltage establishes an electric field. This phenomenon and other important dc characteristics of capacitors will be analyzed.

Also, you will study how capacitors charge and discharge; the physical factors affecting capacitance values; some examples of capacitors; how to calculate total capacitance for circuits containing more than one capacitor; examples of capacitor color codes; and some common capacitor problems and troubleshooting techniques.

Objectives

After studying this chapter, you will be able to:

- Define **capacitor, capacitance, dielectric, dielectric constant, electric field,** farad, **RC time constant,** and leakage resistance
- Describe **capacitor charging** action and **discharging** action
- Calculate charge, voltage, **capacitance,** and stored energy, using the appropriate formulas
- Determine total **capacitance** in circuits with more than one **capacitor** (series and parallel)
- Calculate circuit voltages using appropriate **RC time constant** formulas
- List and describe the physical and electrical features of at least four types of **capacitors**
- List typical **capacitor** problems and describe troubleshooting techniques

DEFINITION AND DESCRIPTION OF A CAPACITOR

In *physical* terms, a **capacitor** is an electrical component, generally consisting of two conducting surfaces (often called capacitor plates) that are separated by a nonconductor (called the dielectric), Figure 16–1.

The nonconducting material, or **dielectric,** can be any number of materials. Some examples of dielectric materials are vacuum, air, waxed paper, plastic, glass, and ceramic material.

The schematic symbol for the capacitor plates is one straight line and one curved line, each perpendicular to the lines representing the wires or circuit conductors connected to the two plates, again see Figure 16–1.

In *electrical* terms, a capacitor is an electrical component that stores electrical charge when voltage is applied.

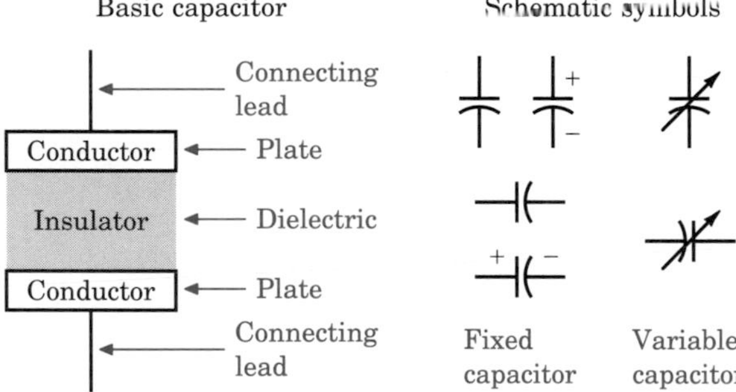

FIGURE 16–1 Basic capacitor and symbols

THE ELECTROSTATIC FIELD

You know that unlike charges attract and like charges repel. Refer to Figure 16–2 as you study the following discussion.

Lines are used to illustrate **electrostatic fields** between charged bodies, much like lines illustrate magnetic flux between magnetic poles. Direction of the electrostatic field lines is usually shown from the positive charged

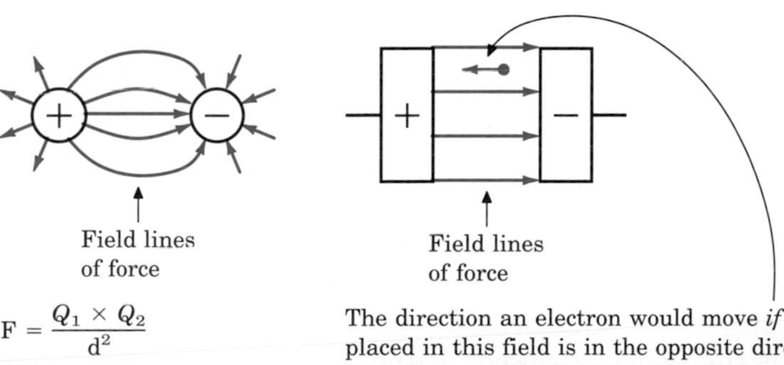

FIGURE 16–2 The electrostatic field

body to the negative charged body. (NOTE: An electron placed in this field travels in the opposite direction from the electrostatic field. In other words, the electron is repelled from the negative body and attracted to the positive body.)

The force between charged bodies is *directly* related to the product of the charges on the bodies and *inversely* related to the square of the distance between them: $F = \frac{Q_1 \times Q_2}{d^2}$ (coulomb's law). This means the closer together capacitor's plates are for a given charge, the stronger the electric field produced. Also, the greater the charge (in coulombs) stored on the plates, the stronger the field produced.

The electrostatic field represents the storage of electric energy. This energy originates with a source and can be returned to a circuit when the source is removed. (This is analogous to energy stored in the buildup of an inductor's magnetic field, which may be returned to the circuit when the magnetic field collapses.)

CHARGING AND DISCHARGING ACTION

Figure 16–3a illustrates a capacitor with no *charge,* that is, no electrostatic field exists between the plates. The plates and dielectric between them are electrically *neutral.*

FIGURE 16–3 Capacitor charging action

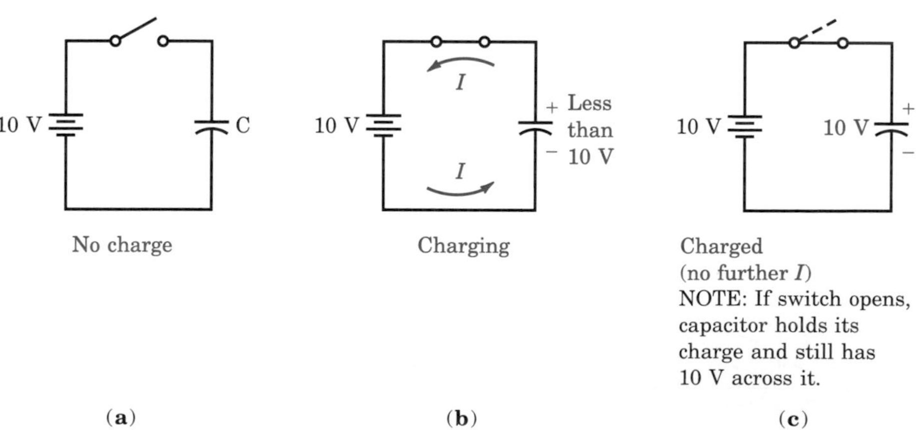

Charging Action

In Figure 16–3b, the circuit switch is closed and electrons move for a short time, **charging the capacitor.** Let's look at this action.

As soon as the switch is closed, neither capacitor plate has an electron excess or deficiency. Also, there is no voltage across the plates or electric field across the dielectric to oppose electron movement to or from the plates.

Electrons leave the negative terminal of the battery and travel to the capacitor plate, which is electrically connected to it. The electrons do not travel through the nonconductive dielectric material between the capacitor plates. Rather, the electrons accumulate on that side of the capacitor, establishing an electron excess and a negative charge at that location. Incidentally, the electron orbital paths in the atoms of the dielectric material (unless it is vacuum) are distorted, since they are rejected by the negative charged plate of the capacitor, Figure 16–4. (NOTE: The distorted orbits represent storage of some electrical energy.)

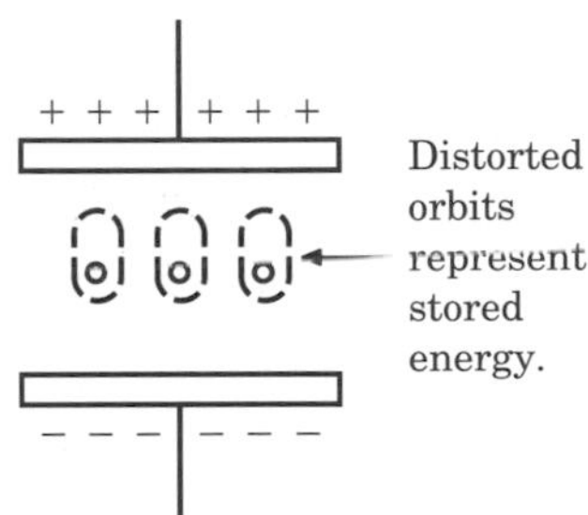

FIGURE 16–4 Distorted orbits represent stored electrical energy.

As electrons accumulate on one side of the capacitor, making that side negative, electrons are being simultaneously attracted to the source's positive terminal from the capacitor's opposite side. Recall in a series circuit, the current is the same through all circuit parts.

Obviously, as an electron surplus builds on one plate and an equal electron deficiency occurs on the other plate, a potential (voltage) difference is established between the capacitor plates.

Notice in Figure 16–3c the voltage polarity established across the capacitor plates is one that *series-opposes* the source voltage. When sufficient *charging current* has flowed causing the capacitor voltage to equal the source voltage, no more current can flow. In other words, the capacitor has *charged* to the source voltage.

Current flowing from the negative source terminal to one capacitor plate and to the positive source terminal from the other capacitor plate is maximum the moment the switch is closed. For a moment, the capacitor acts almost like a short. NOTE: Current does not flow *through* the capacitor by the nonconductive dielectric. Rather, current acts like it has flowed because of the electron movement on either side of the capacitor. Charging current is maximum the first instant the switch is closed and decreases to zero as the capacitor becomes *charged* to the source voltage value.

When the source voltage increases to a higher voltage, current again flows until the capacitor becomes charged to this new voltage level. Once the capacitor has charged to the source voltage, it acts like an open and blocks dc, again see Figure 16–3c.

In summary, maximum charging current flows when voltage is first applied to the capacitor. As the capacitor charges, voltage across the capacitor increases to the source voltage value and charging current decreases. When the capacitor is charged to V applied, no further current flows and the capacitor is holding a charge, which causes a voltage equal to V applied across its plates. The electrical energy has been stored in the electrostatic (dielectric) field of the capacitor.

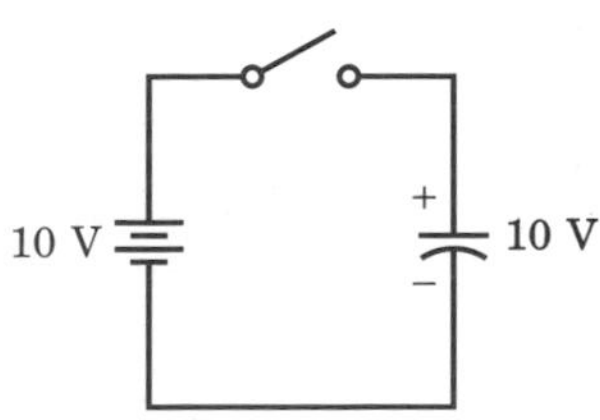

FIGURE 16–5 Capacitors retain charge until it bleeds off either through leakage *R* of dielectric or through some external current path.

When we open the switch in the charged capacitor circuit, Figure 16–5, and we assume the dielectric has infinite resistance to the current, (i.e., no leakage current flows between the plates through the dielectric), the capacitor remains charged *as long* as no external *discharge* path for current is provided.

SAFETY HINTS

Caution: ***Always discharge circuit capacitors*** **after power has been removed and** ***prior*** **to working on circuits containing them!** Technicians have been killed by accidentally becoming the discharge path! By providing an appropriate discharge path to bleed the charge from circuit capacitors prior to working on the circuit, this danger is avoided.

Discharging Action

Figure 16–6 illustrates **discharging a capacitor.** An external path for electron movement has been supplied across the capacitor plates. The excess electrons on the bottom plate quickly move through this path to the top plate, which has an electron deficiency. The charge difference between the plates is neutralized, and the potential difference between the plates falls to zero. When full neutralization occurs and the potential difference between plates is zero, we have completely discharged the capacitor.

IN-PROCESS LEARNING CHECK I

Fill in the blanks as appropriate.

1. A capacitor is an electrical component consisting of two conducting surfaces called ____________ that are separated by a nonconductor called the ____________.
2. A capacitor is a device that stores electrical ____________ when voltage is applied.
3. Electrons ____________ travel through the capacitor dielectric.
4. During charging action, one capacitor plate collects ____________, making that plate ____________. At the same time, the other plate is losing ____________ to become ____________.
5. Once the capacitor has charged to the voltage applied, it acts like an ____________ circuit to dc. When the level of dc applied increases, the capacitor ____________ to reach the new level. When the level of dc applied decreases, the capacitor ____________ to reach the new level.
6. When voltage is first applied to a capacitor, ________ charging current occurs.
7. As the capacitor becomes charged, current ____________ through the circuit in series with the capacitor.

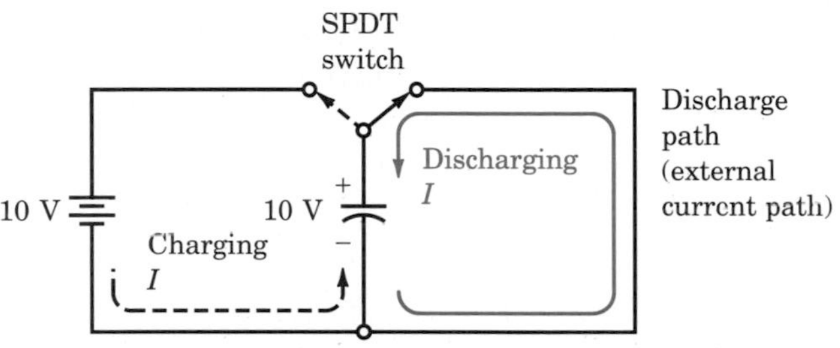

FIGURE 16–6 Discharging a capacitor

THE UNIT OF CAPACITANCE

You have seen that **capacitance** relates to the *capacity* of a *capacitor* to store electrical charge. As you might expect, the more charge a capacitor stores for a given voltage, the larger its capacitance value must be. The unit of capacitance is the farad, named in honor of Michael Faraday. The farad is that amount of capacitance where a charge of one coulomb develops a potential difference of one volt across the capacitor plates, or terminals. The farad is abbreviated as capital F.

From this definition, relationships between capacitance (C), voltage (V), and charge (Q) are:

Formula 16–1	$C = \frac{Q}{V}$

Formula 16–2	$Q = CV$

Formula 16–3	$V = \frac{Q}{C}$

where V = volts
Q = coulombs
C = capacitance in farads

EXAMPLE

When 10 coulombs of charge are stored on a capacitor charged to 2 volts, the capacitance (in farads) is calculated as:

$$C \text{ (farads)} = \frac{Q \text{ (coulombs)}}{V \text{ (volts)}}$$

$$C(F) = \frac{10 \text{ C}}{2 \text{ V}}$$

$$C = 5 \text{ farads}$$

NOTE: The unit of capacitance (C) is the farad (F). The unit of charge (Q) is the coulomb, also abbreviated as "C."

It should be noted that the farad is a huge amount of capacitance. In fact, it would take a room-sized capacitor to have that much capacitance. Practical capacitors are usually in the range of millionths of a farad (μF) or millionths of a millionth of a farad, formerly called micro-micro farad ($\mu\mu$F). This value is now called picofarad, pF. Using powers of ten, these values are expressed as:

One microfarad = 1 μF = 1×10^{-6} F
One picofarad = 1 pF = 1×10^{-12} F

PRACTICE PROBLEM I

Answers are in Appendix B.

1. What is the capacitance of a capacitor that develops 25 volts across its plates when storing a 100-μC charge? (Express answer in microfarads.)
2. What is the charge on a 10-μF capacitor when the capacitor is charged to 250 volts? (Express answer in μC.)
3. What voltage develops across the plates of a 2-μF capacitor with a charge of 50 μC?

ENERGY STORED IN CAPACITOR'S ELECTROSTATIC FIELD

It was stated earlier that electrical energy, supplied by the source during the capacitor charge, is stored in the electrostatic field of the charged capacitor. This energy is returned to the circuit when the capacitor discharges.

For inductors, the energy (joules) stored in the magnetic field is related to the inductance and the current creating that field ($W = \frac{1}{2}LI^2$). However, for capacitors, the joules of stored energy are related to the capacitance and the voltage, or potential difference between capacitor plates created by the electrostatic field. The formula for energy (in joules) is:

Formula 16–4	$\text{Energy} = \frac{1}{2} CV^2$

EXAMPLE
How many joules are stored in a 5-μF capacitor that is charged to 250 volts?
Answer:

$$\text{Energy (J)} = \frac{1}{2} CV^2$$

$$\text{Energy} = \frac{5 \times 10^{-6} \times (250)^2}{2} = \frac{0.3125}{2} = 0.15625 \text{ J}$$

PRACTICE PROBLEM II

How much electrical energy is stored in a 10-μF capacitor that is charged to 100 volts?

FACTORS AFFECTING CAPACITANCE VALUE

Capacitance value is related to how much charge is stored per given voltage. The physical factors influencing the capacitance of parallel-plate capacitors are size of plate area; thickness of dielectric material (spacing between plates); and type of dielectric material, Figure 16–7. Let's examine these factors.

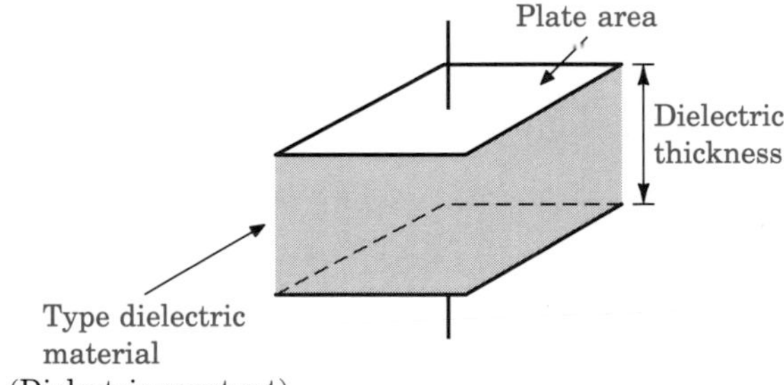

FIGURE 10–7 Physical factors affecting capacitance value

- **Plate area** Obviously, the larger the area of the plates facing each other, the greater the number of electrons (and charge) that can be stored. Therefore in Figure 16–7, C is directly related to plate area. For example, when the plate area doubles, there is room for twice as many electric lines of force with the same potential difference between plates. Thus, the amount of charge the capacitor holds per given voltage has doubled and the capacitance has doubled. The area of the plates directly facing each other equals the area of one plate. Using SI units, this area is expressed in square meters.

- **Distance between plates (dielectric thickness)** From previous discussions, it is logical to assume the further apart the plates, the less the electric field strength established for a given charge and potential difference between plates. Thus, the less the electrical energy or charge stored in this electric field. In other words, the *capacitance value is inverse to the spacing between plates.* Conversely, the closer the plates (the thinner the dielectric material) are, the stronger the electric field intensity per given voltage and the higher the capacitance value. In SI units, this distance (thickness of dielectric) is expressed in meters.

- **Type of Dielectric Material** Different dielectric materials exhibit different abilities to concentrate electric flux lines. This property is sometimes called *permittivity,* which is analogous to conductivity of conductors. Vacuum (or air) has a permittivity of one. The relative permittivity of a given dielectric material is the ratio of the absolute permittivity (ϵ_o) of the given dielectric to the absolute permittivity of free space (vacuum) (ϵ_v). This relative permittivity is called the **dielectric constant** (k) of the material.

Dielectric Constants

Refer to the chart in Figure 16–8 for examples of the average dielectric constant of various materials. The ratio reflecting the dielectric constant is shown as:

Formula 16–5	$k = \dfrac{\epsilon_o \text{ (absolute permittivity of dielectric material)}}{\epsilon_v \text{ (absolute permittivity of vacuum)}}$

(NOTE: Sometimes the symbol K_ϵ is used in lieu of k to represent dielectric constant.)

The higher the dielectric constant is, the greater the density of electric flux lines established for any given plate area and spacing between plates. Therefore, capacitance is directly related to the dielectric constant. For example, when the average dielectric constant (k or K_ϵ) of a given material is five, the capacitance is five times as great with this dielectric material compared to a dielectric of air or vacuum.

FIGURE 16–8 Sample dielectric constants

Kind of Material	Approximate k
Air	1.0
Glass	8.0
Waxed paper	3.5
Mica	6.0
Ceramic	100.0 +

NOTE: Values vary depending on quality, grade, environmental conditions, and frequency.

Dielectric Strength

The electric field intensity is increased by increasing the voltage across a dielectric. As we continue increasing the potential difference across the capacitor plates, eventually a point is reached where the electron orbits in the dielectric material are stressed (distorted in their orbital paths), causing electrons to be torn from their orbits. At this time, the dielectric material breaks down and typically punctures or arcs, becoming a conductor. The material is no longer functioning as a nonconducting dielectric material (unless it is air or vacuum, which are self-healing). The breakdown value relates to the **dielectric strength** of the material. The practical aspect of dielectric strength is called the *breakdown voltage.* The breakdown voltage depends on the kind of material and its thickness. Figure 16–9 lists some materials and their approximate average dielectric strength ratings in terms of volts per mil (thousandths of an inch) breakdown voltage ratings. Be aware these numbers will vary, depending on manufacture, frequency of operation, and so on.

Kind of Material	Approximate Dielectric Strength (Volts per 0.001 inch)
Air	80
Glass	200 + (up to 2kV)
Waxed paper	1200
Mica	2000
Ceramic	500 + (up to ≈ 1kV)

FIGURE 16–9
Examples of dielectric strengths

CAPACITANCE FORMULAS

As you have learned, capacitance depends on the surface dimensions of plates facing each other, and the spacing and the dielectric constant of the material between plates. Using the appropriate factors to compute capacitance in farads using the SI units, the formula for capacitance of a parallel-plate capacitor with air (or vacuum) as the dielectric is:

Formula 16–6	$C = \frac{8.85\ A}{10^{12} s}$

where C = capacitance in farads
A = area of the plates facing each other, in square meters
s = spacing between plates, in meters

NOTE: If the plates have different dimensions, *use the smaller plate area for "A"* in the formula.

To calculate capacitance when dielectric materials other than air or vacuum are used, use this formula and insert the dielectric constant (k) in the formula. Thus:

Formula 16–7	$C = \frac{8.85kA}{10^{12} s}$

Again, C = capacitance in farads
A = area of the plates facing each other in square meters
k = the dielectric constant
s = the spacing between the parallel plates in meters

For your information, a practical formula used to determine the capacitance in *microfarads* (μF) of a multiplate capacitors is in Figure 16–10. NOTE: This formula uses inches and square inches, rather than meters and square meters.

FIGURE 16–10
Multi-plate capacitor

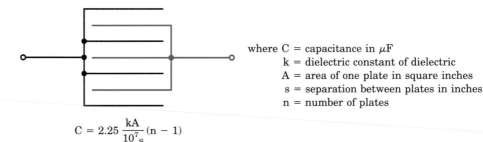

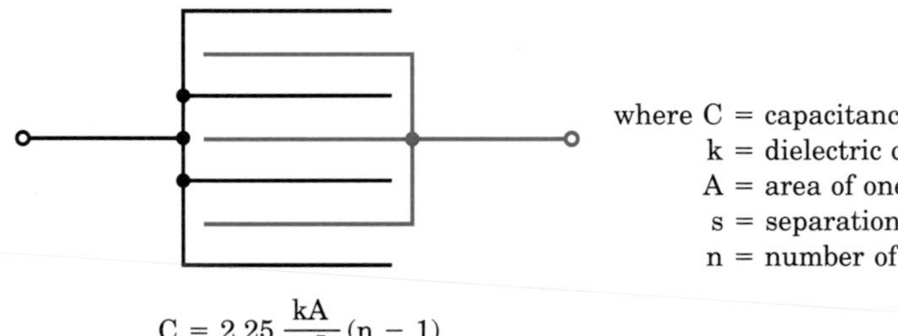

where C = capacitance in μF
k = dielectric constant of dielectric
A = area of one plate in square inches
s = separation between plates in inches
n = number of plates

$$C = 2.25\frac{kA}{10^7 s}(n-1)$$

Formula 16–8	$C = 2.25\frac{kA}{10^7 s}(n-1)$

where C = capacitance in μF
k = dielectric constant in material between plates
A = area of one plate in square inches
s = separation between plate surfaces in inches
n = number of plates

EXAMPLE

1. A two-plate capacitor has a dielectric with a dielectric constant of 3. The plates are each 0.01 square meters in area and spaced 0.001 meter apart. What is the capacitance?
   Answer:

$$C = \frac{8.85kA}{10^{12}s} = \frac{8.85 \times 3 \times 0.01}{10^{12} \times 0.001}$$

$$C = \frac{0.2655}{1 \times 10^9} = 0.0000000002655 \text{ farads}$$

$$C = 265.5 \text{ pF}$$

2. What is the capacitance in μF of a capacitor if the plate area of one plate is 2 square inches, and the dielectric has a k of 100 and is 0.005 inches thick? (Assume a two-plate capacitor.)
   Answer:

$$C = 2.25\frac{kA}{10^7 s}(n-1)$$

$$C = \frac{2.25 \times 100 \times 2}{10^7 \times 0.005} \times (2-1) = \frac{450}{50{,}000} \times 1 = 0.009\ \mu\text{F}$$

PRACTICE PROBLEM III

1. A capacitor has four plates of equal size. Each plate has an area of 1 square inch. This capacitor uses a dielectric having a k of 10. The dielectric thickness between each pair of plates is 0.1 inch. What is the capacitance value in μF?
2. Each plate of a two-plate capacitor has an area of 0.2 square meters. The plates are uniformly spaced 0.005 meters apart and the dielectric is air. What is the capacitance?

TOTAL CAPACITANCE IN SERIES AND PARALLEL

Generally, technicians do not compute capacitance from physical data, as we have done. We did this to help you understand capacitance. (In fact, capacitance checkers exist so you can measure, rather than compute, capacitance). As a technician, however, you may have to determine circuit parameters with capacitors connected in series or parallel. The knowledge you gain will help you perform such required circuit analysis.

Capacitance of Series Capacitors

Refer to Figure 16–11 as you study the following comments.

1. Connecting two capacitors in series increases the dielectric thickness, while the plate areas connected to the source remain the same, Figure 16–11. From our previous discussions, you can surmise that connecting capacitors in series *decreases* the resultant total capacitance.

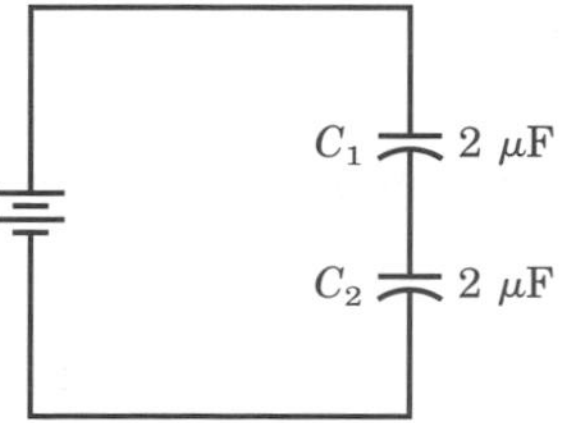

Series capacitors:

$$C_T = \frac{C_1 \times C_2}{C_1 + C_2}$$

For equal value C's:

$$C_T = \frac{C_1}{N}$$

Generalized formula:

$$C_T = \frac{1}{\frac{1}{C_1} + \frac{1}{C_2} \cdots + \frac{1}{C_n}}$$

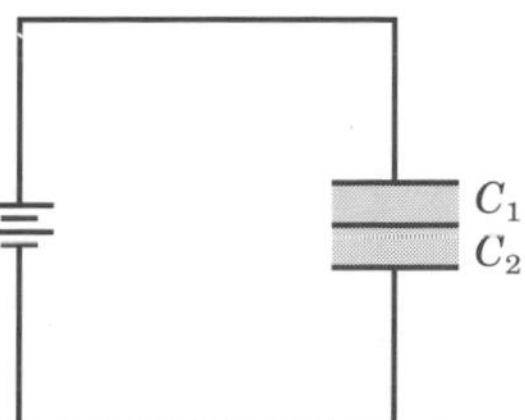

Equivalent:
Dielectric thicknesses add and effective plate area remains the same, therefore $C \downarrow$.

FIGURE 16–11 Total capacitance of series capacitors

2. Finding total capacitance of *capacitors in series* is analogous to finding total resistance of *resistors in parallel.*
3. The formulas to determine total capacitance (C_T) of series-connected capacitors are as follows:
   a. For two capacitors in series:

Formula 16–9	$C_T = \dfrac{C_1 \times C_2}{C_1 + C_2}$

   b. For equal-value capacitors in series:

Formula 16–10	$C_T = \dfrac{C}{N}$

   where C is the value of one of the equal-value capacitors
   N is the number of capacitors in series

   c. The general formula for any number of capacitors in series:

Formula 16–11	$C_T = \dfrac{1}{\dfrac{1}{C_1} + \dfrac{1}{C_2} + \dfrac{1}{C_3} \ldots + \dfrac{1}{C_n}}$

Voltage Distribution Related to Series Capacitors

Because the current throughout a series circuit is the same at any point, capacitors connected in series have the same number of coulombs (C) of charge (Q). The potential difference between plates of equal-value capacitors for a given charge is equal. But for different-value capacitors, potential difference between plates is not equal. Recall the $V = \dfrac{Q}{C}$ formula, where V is directly related to Q and inversely related to C. This indicates in a series circuit (since the Q is the same), the voltage across any given capacitor is

PRACTICAL NOTES

Due to the large tolerances typical of electrolytics, the voltage distribution may not "track" the theoretical values computed by using the "marked values" on the capacitors. If the capacitors' values are precisely as marked, the resulting voltage distribution will be as indicated in the preceding discussion.

inversely proportional to its C value. For example, if we have a 5-μF capacitor and a 10-μF capacitor in series, the voltage across the 5-μF capacitor is two times the voltage across the 10-μF capacitor, Figure 16–12. To obtain capacitances in the ranges described (i.e., multi-microfarad), the electrolytic capacitor is typically used. You will study about this capacitor later in the chapter. However, you should take note of the preceding comments and following examples about voltage distribution.

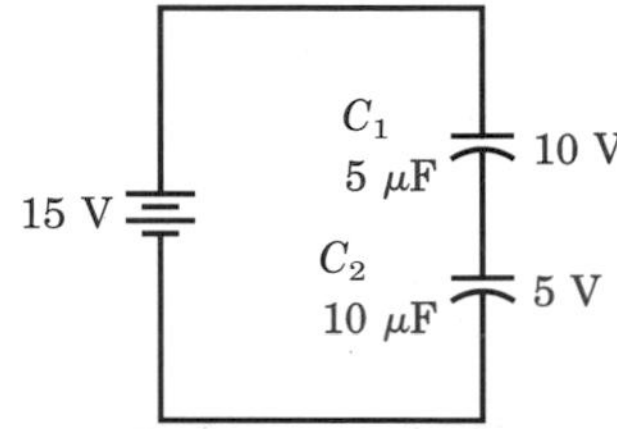

FIGURE 16–12 Voltage distribution across series capacitors

EXAMPLE

1. What is the total capacitance of the circuit in Figure 16–13?
   Answer:

$$C_T = \frac{C_1 \times C_2}{C_1 + C_2} = \frac{3 \times 6}{3 + 6} = \frac{18}{9} = 2\ \mu F$$

2. What is the total capacitance of three 10-μF capacitors connected in series?
   Answer:

$$C_T = \frac{C}{N} = \frac{10\ \mu F}{3} = 3.33\ \mu F$$

3. When the voltage applied to the circuit in Figure 16–13 is 90 volts, what are the V_{C_1} and V_{C_2} values?
   Answer:
   Since the voltage distribution is *inverse* to the capacitance values, V_{C_1} equals twice the value of V_{C_2}, or two-thirds of the applied voltage. Thus, $V_{C_1} = \frac{2}{3} \times 90 = 60$ volts, and $V_{C_2} = \frac{1}{3} \times 90 = 30$ volts.

   An alternate solution is to find Q by using the total capacitance value in the formula $Q = C_T \times V_T$. Then solve for each capacitor voltage value using $V = \frac{Q}{C}$. Therefore, since the total capacitance is 2 μF:

$$Q = 2 \times 10^{-6} \times 90;\ Q = 180\ \mu C.$$

$$V_{C_1} = \frac{180\ \mu C}{3\ \mu F};\ V_{C_1} = 60\ V$$

$$V_{C_2} = \frac{180\ \mu C}{6\ \mu F};\ V_{C_2} = 30\ V.$$

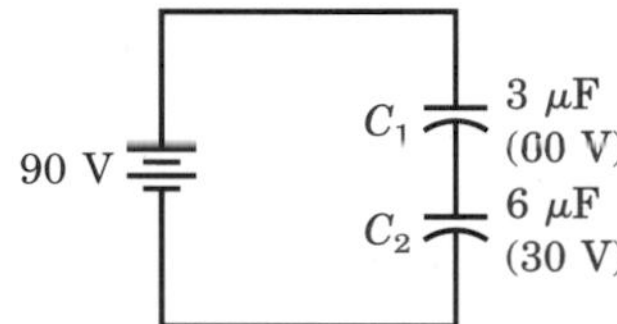

$$C_T = \frac{(3 \times 10^{-6}) \times (6 \times 10^{-6})}{(3 \times 10^{-6}) + (6 \times 10^{-6})}$$

$$C_T = \frac{18}{9} = 2\ \mu F$$

$$V_{C_1} = \frac{2}{3} \times V_T$$

$$V_{C_1} = 60\ V$$

$$V_{C_2} = \frac{1}{3} \times V_T$$

$$V_{C_2} = 30\ V$$

$$Q_T = V_T \times C_T$$

$$Q_T = 90\ V \times 2\ \mu F = 180\ \mu C$$

FIGURE 16–13

PRACTICE PROBLEM IV

1. What is the total capacitance of the circuit in Figure 16–14?

C1 5 μF
100 V
C2 20 μF

$C_T = ?$
$V_{C_1} = ?$
$V_{C_2} = ?$

FIGURE 16–14

2. What is the total capacitance of four 20-μF capacitors connected in series?
3. What are the voltages across each capacitor in Figure 16–14?

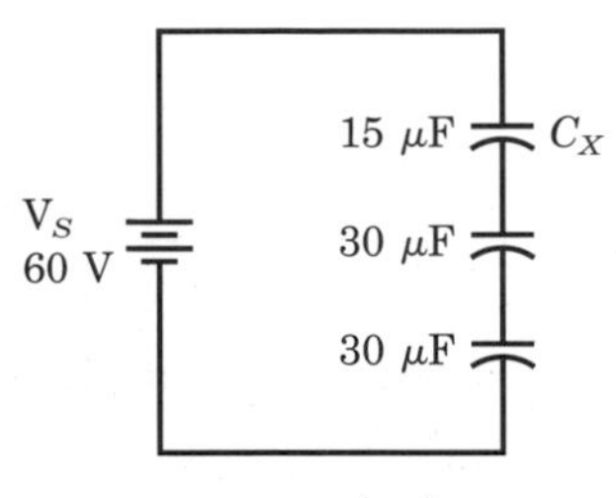

$$V_X = V_S \left(\frac{C_T}{C_X}\right)$$

$$V_X = 60 \left(\frac{7.5}{15}\right)$$

$$V_X = 60 \times 0.5$$

$$V_X = 30 \text{ volts}$$

FIGURE 16–15 Series capacitors voltage formula

FINDING VOLTAGE WHEN THREE OR MORE CAPACITORS ARE IN SERIES

A handy formula, Figure 16–15, used to find voltage when three or more capacitors are in series is:

Formula 16–12	$V_X = V_S \dfrac{C_T}{C_X}$

where V_X = voltage across capacitor "x"
V_S = the dc applied voltage
C_T = Total series capacitance
C_X = value of capacitor "x"

Capacitance of Parallel Capacitors

Refer to Figure 16–16 as you study the following information. Since the top and bottom plates of all the parallel capacitors are electrically connected to each other, the combined plate area equals the sum of the individual capacitor plate areas. However, the dielectric thicknesses do not change and are not additive.

FIGURE 16–16 Capacitance of parallel capacitors

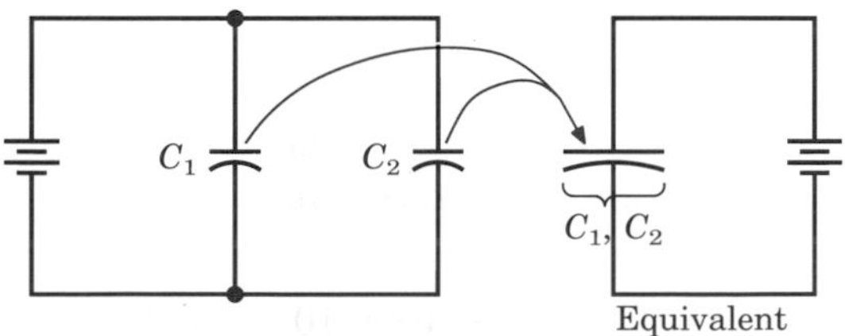

When plate areas are combined, dielectric thickness remains unchanged.

Increasing the effective plate area while keeping the dielectric thickness(es) the same results in the total capacitance of the parallel-connected capacitors equaling the sum of the individual capacitances. In other words, *capacitances in parallel add like resistances add in series.* The formula for total capacitance (C_T) of parallel-connected capacitors is:

Formula 16–13	$C_T = C_1 + C_2 \ldots + C_n$

Charge Distribution Related to Parallel Capacitors

Because the voltage across components connected in parallel is the same, the charge (in coulombs) contained by each capacitor in a parallel-capacitor circuit is *directly* related to each capacitance value. That is, the larger the capacitance value is, the greater is the charge for a given voltage, Q = CV. For example, a 10-μF, 20-μF and 30-μF capacitor are in parallel, and the voltage applied to this circuit is 300 volts. Thus, the charge (Q) on the 10-μF capacitor is half that on the 20-μF capacitor and one-third that on the 30-μF capacitor, (Q = CV). The Q on the 10-μF capacitor equals 10 μF × 300 V = 3,000 μC. The charge on the 20-μF capacitor equals 20 μF × 300 V = 6,000 μC. The charge on the 30-μF capacitor equals 30 μF × 300 V = 9,000 μC.

NOTE: Since total charge equals the sum of the charges, it can also be computed from the $Q_T = C_T \times V_T$ formula. Therefore, total charge = 60 μF × 300 V, so Q_T = 18,000 μC, Figure 16–17.

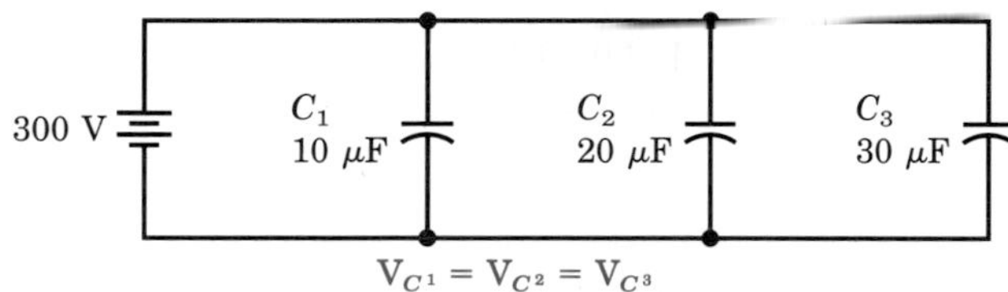

$V_{C1} = V_{C2} = V_{C3}$

$Q_{C_1} = C_1 \times V = 10\ \mu F \times 300\ V = 3000\ \mu C$
$Q_{C_2} = C_2 \times V = 20\ \mu F \times 300\ V = 6000\ \mu C$
$Q_{C_3} = C_3 \times V = 30\ \mu F \times 300\ V = 9000\ \mu C$

Charge on parallel capacitors is directly related to each one's capacitance value.

FIGURE 16–17 Charge distribution on parallel capacitors

EXAMPLE

1. What is the total capacitance of a parallel circuit containing one 15-μF capacitor, two 10-μF capacitors, and one 3-μF capacitor?
   Answer:

   $$C_T = C_1 + C_2 + C_3 + C_4 = 15 + 10 + 10 + 3\ \mu F = 38\ \mu F$$

2. What is the charge on the 3-μF capacitor, if the voltage applied to the circuit in Question 1 is 50 volts?
   Answer:

   $$Q = CV = 3 \times 10^{-6} \times 50 = 150\ \mu C$$

PRACTICE PROBLEM V

1. What is the total capacitance of five 12-μF capacitors connected in parallel?

2. If V_T equals 60 V, what is the charge on each capacitor in Problem 1? What is the total charge on all capacitors?

THE RC TIME CONSTANT

At the beginning of the chapter, we said a capacitor opposes a *change* in voltage. You have seen how charge current is required to establish a potential difference between the capacitor plates. Obviously, the rate of charge for a given capacitor is directly related to the charge current, $V = \frac{Q}{C}$. That is, the higher the current is, the quicker the capacitor is charged to V applied.

You also know current allowed to flow is *inversely* related to the resistance in the current path. Therefore, the larger the resistance value in the charge (or discharge) path of the capacitor is, the longer the time required for voltage to reach V applied across the capacitor during charge, or to reach zero volts during discharge.

The larger the capacitance (C) value is, the more charge required to develop a given potential difference between its plates, $V = \frac{Q}{C}$. Because a capacitor prevents an instantaneous voltage change across its plates (if there is any resistance in the charge path or discharge path), a capacitor is said to oppose a *change* in voltage across itself.

Recall when inductors were discussed in a previous chapter, there was an $\frac{L}{R}$ time constant relating the time required to change current through the coil. You learned that *five* $\frac{L}{R}$ time constants are needed for a complete change from one circuit current value to the new level.

In like manner, an **RC time constant** relating the time required to charge (or discharge) a capacitor exists. To completely change from one voltage level to a new level of applied voltage, the capacitor needs *five* RC time constants. *One RC time constant (TC) = R (in ohms) × C (in farads).* That is:

Formula 16–14	$1\ TC = R \times C$

Refer to Figure 16–18 as you study the following information about a typical series RC circuit.

The sum of the dc voltages across the capacitor and the resistor equals the applied voltage (V_T) at all times.

As the capacitor is charged, the voltage across the capacitor is series-opposing the source voltage. Therefore, the voltage across the resistor at any given moment equals the source voltage minus the capacitor voltage. That is, $V_R = V_T - V_C$. Thus, as the capacitor charges to the V applied value, the

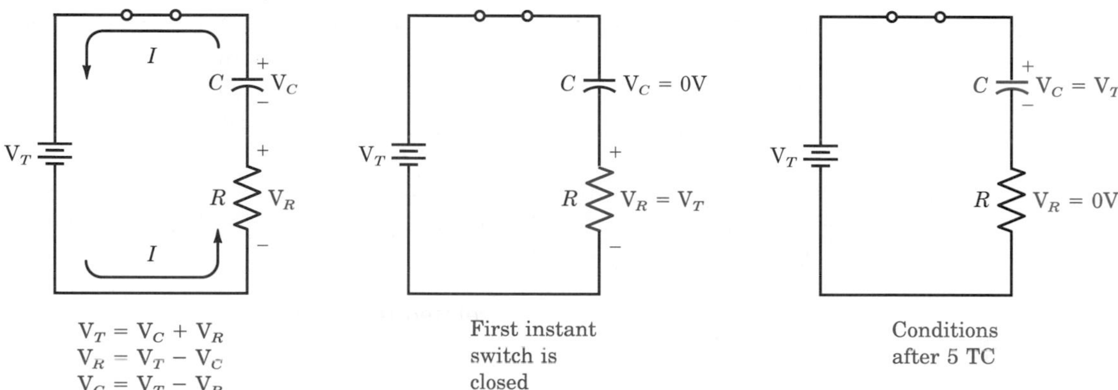

FIGURE 16–18 Voltages in an RC circuit

resistor voltage decreases toward zero volts. When the capacitor is fully charged, V_C equals V applied and zero current flows, and no IR drop occurs across the resistor.

When the switch is initially closed, the capacitor has zero charge and zero voltage. Therefore, the applied voltage is dropped by the resistor (R). Since this is a series circuit, and the current is the same throughout a series circuit, Ohm's law reveals what the charging current is at that moment, $I = \frac{V_R}{R}$. In other words, at that moment, the capacitor behaves like a short; therefore, only the resistor limits current.

As the charging current flows, a series-opposing potential difference across the capacitor develops, and the charging current diminishes at a non-linear rate (toward zero). The IR drop across the resistor drops accordingly. For example, after a certain interval of time, the capacitor is charged to half of V applied. At that time, $\frac{1}{2}$ V applied is across the resistor, and $\frac{1}{2}$ of V applied is across the capacitor.

The percentage of voltage across the capacitor at any given moment relates to the RC time constant, where one time constant (*in seconds*) = R (Ω) × C (farads). Often the Greek lower case letter tau (τ) is used to represent one time constant. Note that 1 TC, or τ (sec) = R (megohms) × C (microfarads). Also, observe that 1 TC, or τ (μ*sec*) = R (Ω) × C (μF).

Refer to Figure 16–19 (curve A) and notice that after one time constant (τ), the capacitor voltage (V_c) equals 63.2% of V applied. After two τ, V_C equals 86.5% of V applied; after three τ, V_C equals 95%; after four τ, V_C equals 98.5%; and after *five* τ, V_C equals over 99% of V applied, therefore it is considered fully charged. Are these percentages and corresponding numbers familiar? Of course! They are the same percentages and numbers used when the $\frac{L}{R}$ time constant was discussed. In other words, the same exponential curves (or time constant chart) is applicable for analyzing voltages in a capacitor circuit as was used to analyze current in an inductor circuit.

Note in Figure 16–19 that curve B illustrates what happens to the resistor voltage as the capacitor is charged. That is, at the first instant V_R equals V applied. As the capacitor charges, V_R decreases toward zero. In effect, the resistor voltage and the circuit current are inverse to the capacitor voltage. As the capacitor voltage increases, the circuit current, and thus the voltage across the resistor decreases.

Using Exponential Formulas to Find Voltage(s)

Values along the exponential curves on the time constant chart in Figure 16–19 are more precisely found by using some simple formulas. The values along the descending exponential curve relate to the base of *natural logarithms,* called epsilon (ϵ), which has a value of 2.718. (Recall this number of natural growth was discussed in an earlier chapter).

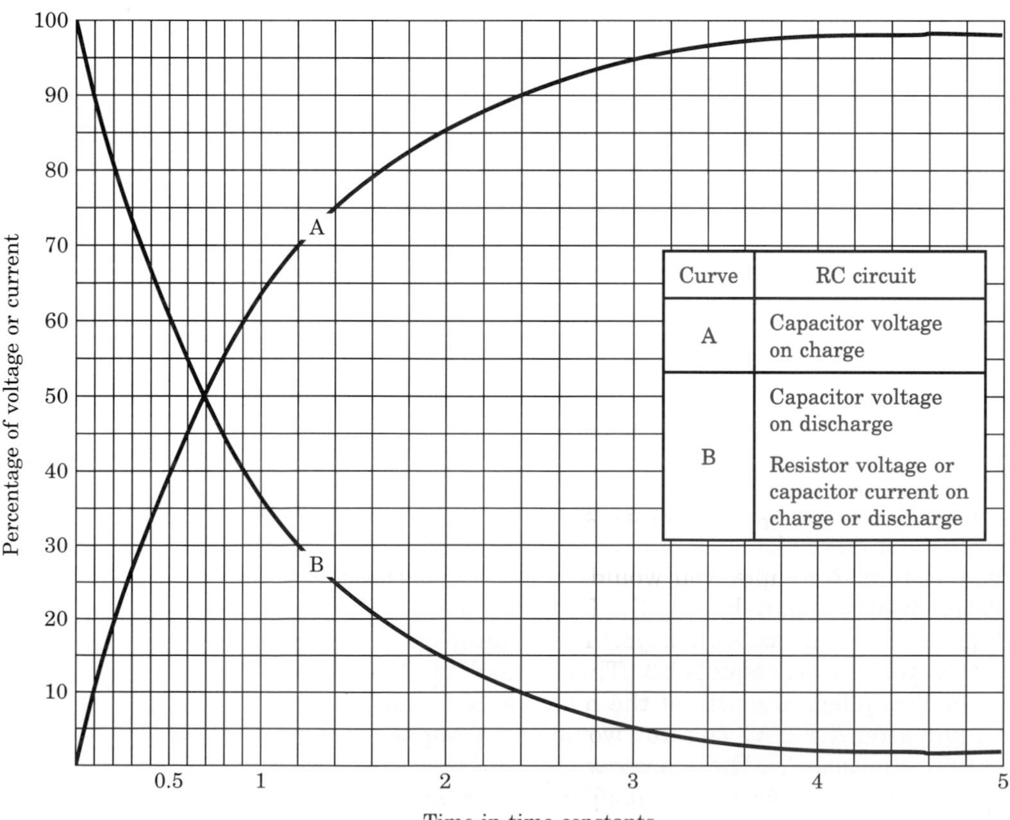

FIGURE 16–19 Time-constant chart

When a capacitor is charging in a series RC circuit, the voltage across the resistor during the first instant is maximum, then decreases exponentially as the capacitor charges to the new voltage level. Thus, the resistor voltage follows the descending exponential curve shown in the time constant chart. The level of V_R for any given point of time relates to the length of time the capacitor has charged relative to the circuit's RC time constant. That is, the time allowed (t) divided by the time of one RC time constant (τ), or $\frac{t}{\tau}$. Obviously, when the t (time allowed) equals five or more τ, then the capacitor has fully charged, no current is flowing and V_R equals zero. When t is less than 5 τ, then V_R equals a value that is found by the formula:

Formula 16–15	$v_R = V \text{ applied} \times \epsilon^{-t/\tau}$

where ϵ = epsilon (or 2.71828)
t = time allowed in seconds
$\tau = R \times C$ (or 1 time constant)

EXAMPLE

For the circuit in Figure 16–20 find V_R after 10,000 μseconds.

$$V_R = V \text{ app} \times \epsilon^{-t/\tau}$$
$$V_R = 50 \times 2.718^{-10{,}000/100{,}000 \times 0.1}$$
$$V_R = 50 \times 2.718^{-1}$$
$$V_R = 50 \times 0.3679$$
$$V_R = 18.395 \text{ volts}$$

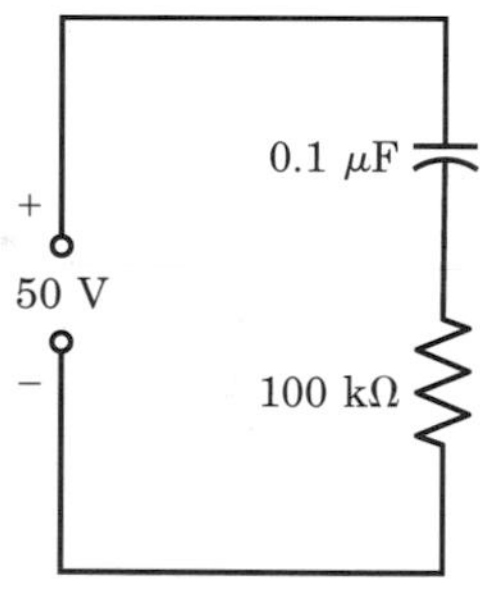

FIGURE 16–20

PRACTICAL NOTES

Using a calculator to find the ϵ value raised to the $\frac{-t}{\tau}$ power is quite easy. In our previous example, you would enter −1 (for the power that 2.718 is raised to); then depress the INV and lnx keys, in that order. (If your calculator has these keys.) This provides the *natural antilogarithm* of the displayed number. This is the equivalent of the epsilon value raised to the power of the number in the display. If your calculator has the e^x key, then you would enter −1, followed by the e^x key. For our example, the answer would show epsilon raised to the negative first power equals 0.367879441 that we would round-off to the 0.3679 as previously shown.

The capacitor voltage is the difference between the resistor voltage and the new voltage applied for any given instant. A formula for v_C is:

Formula 16–16	$v_C = V\ app\ (1 - \epsilon^{-t/\tau})$

NOTE: The small letter v denotes an instantaneous voltage value for given moments in time.

For our example circuit then:

$v_C = 50 \times (1 - 0.3679)$
$v_C = 50 \times 0.6321$
$v_C = 31.605$ volts

To check, add the V_R and the v_C (18.395 + 31.605); the result is 50 volts (our applied voltage). It checks out!

PRACTICE PROBLEM VI

Use either the TC chart or the formulas to solve the following.

1. What is the voltage across the capacitor of the circuit in Figure 16–21 20 μsec after the switch is closed?
2. What is the voltage across the resistor in Figure 16–21, 15 μsec after the switch is closed? What is the charging current value at that moment?

FIGURE 16–21

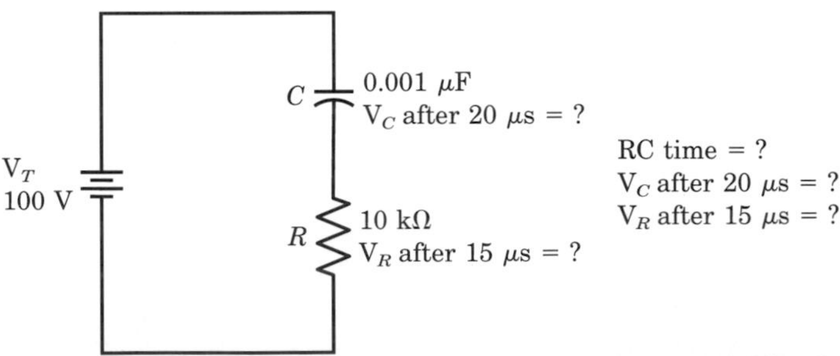

TYPES OF CAPACITORS

General Classification

There are two basic classifications of capacitors—*fixed* and *variable*. These capacitors come in various configurations, sizes, and characteristics. As the two names imply, fixed capacitors have one value of capacitance; variable capacitors have a range of capacitance values that may be adjusted, Figure 16–22. By changing the position of the *rotor* plates with respect to the fixed *stator* plates, Figure 16–22a, the plate areas facing each other are varied. This changes the capacitance value.

(a)

(b)

FIGURE 16–22 Variable capacitors (Photo a courtesy of Amateur Radio Relay League; photo b courtesy of Cornell Dubilier Electronics and Sprague-Goodman Electronics)

When the plates are fully meshed, capacitance is maximum. When the plates are fully unmeshed, capacitance value is minimum. The dielectric for this type of capacitor is air.

Another variable capacitor is shown in Figure 16–22b. This compression-type capacitor has only one movable plate. By turning the screw, the spacing between the movable plate and the fixed plate is varied thus changing the capacitance value. When the movable plate is screwed tight, capacitance is maximum. As the screw is loosened, capacitance decreases.

Capacitors and Their Characteristics

Capacitors are generally referred to in terms of their dielectric material. The following discussion describes several common capacitors and their features.

1. *Paper and Plastic capacitors* use a variety of dielectric materials. Some of these include mylar, polystyrene and polyethylene for the plastic types and waxed or oiled paper for the older, less expensive paper type. Typically, the plates are long strips of tinfoil separated by the dielectric material. The foil and dielectric material are frequently rolled into a cylindrical component, Figure 16–23. It is interesting to note that because of their construction, each plate has two active surfaces. This means to calculate the plate areas, use twice the area of one plate, rather than only the area of one plate. Typical capacitance values for this type capacitor range from about 0.001 μF to several μF.

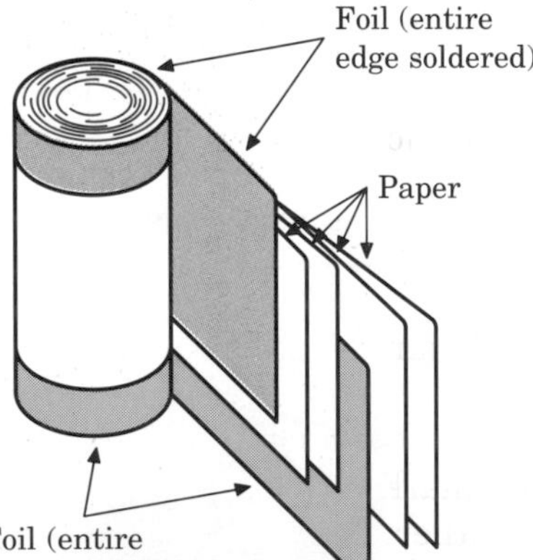

Typical available capacitance values:
Approximately 0.001 to 1.0 μF

Typical working voltage DC (WVDC):
100 to 1500 V

FIGURE 16–23 Paper-type capacitor (Adapted with permission from Duff and Herman, *Alternating Current Fundamentals*, 3rd edition, © 1986 by Delmar Publishers Inc., Albany, New York)

2. *Mica capacitors* use mica for the dielectric material. Because the mica dielectric has a high breakdown voltage, these low-capacitance, high-voltage capacitors are frequently found in high-frequency circuits. Often, their construction is alternate layers of foil with mica that is molded into a plastic case, Figure 16–24. They are compact, moisture-proof and durable. Voltage ratings are in thousands of volts. Typical capacitance values range from about 5 to 50,000 picofarads, depending on voltage ratings.
3. *Ceramic capacitors* are typified by their small size and high-dielectric strength. Ceramic capacitors generally come in the shape of a flat

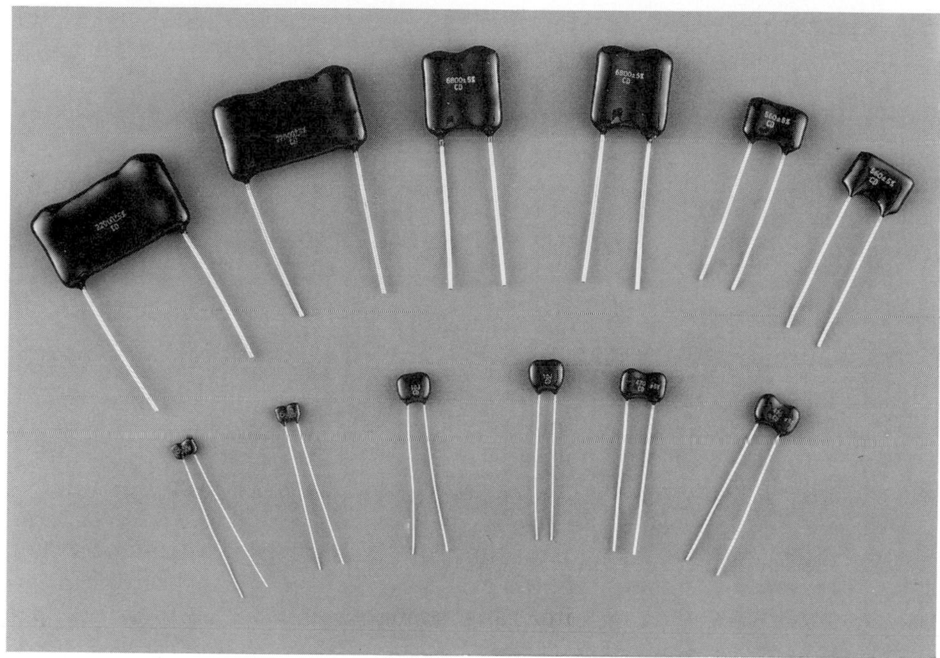

FIGURE 16–24 Mica capacitors (Courtesy of Cornell Dubilier Electronics)

disk (disk-ceramics), or in tubular shapes, Figure 16–25. These capacitors are also compact, moisture-proof and durable. Typical available ranges having 1000 volt ratings are from approximately 5 pF to about 5,000 pF. At lower voltage ratings, higher capacitance values are available.

4. *Electrolytic capacitors* have several prominent characteristics, including:
   a. high capacitance-to-size ratio
   b. polarity sensitivity and terminals marked + and −
   c. allow more leakage current than other types
   d. low cost

Refer to Figure 16–26 for the typical construction of electrolytic can-type capacitors. The external aluminum can or housing is typically the negative plate (or electrode). The positive electrode external contact is generally aluminum foil immersed or in contact with an electrolyte of ammonium borate (or equivalent). If the electrolyte is a borax solution, the capacitor is called a *wet electrolytic.* If the electrolyte is a gauze material saturated with borax solution, the capacitor is called a *dry electrolytic.* To create the dielectric, a dc current is passed through the capacitor. This causes a very thin aluminum oxide film (about 10 microcentimeters thick) to form on the foil surface. This thin oxide film is the dielectric. (NOTE: This type of capacitor is *not* named after its dielectric.) Because of the extremely thin dielectric, capaci-

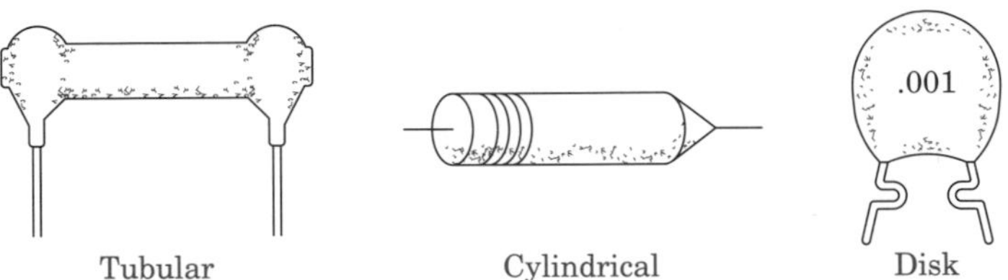

Typical available capacitance values: 1 to 10,000 pF
Typical WVDC: 1,000 to 6,000 V

(a)

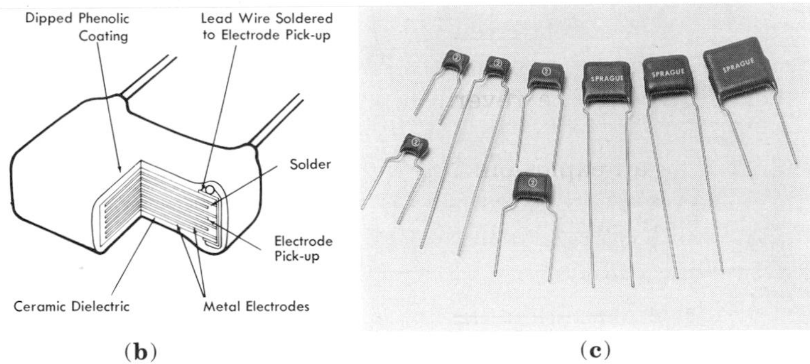

(b) (c)

FIGURE 16–25 Ceramic capacitors (Photos courtesy of Sprague-Goodman Electronics)

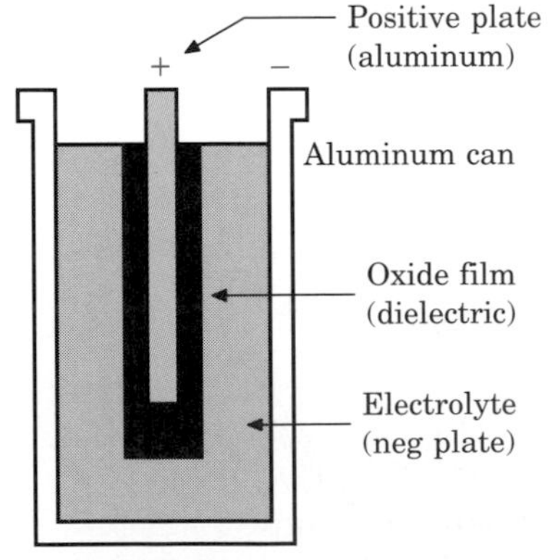

Typical available capacitance values: 10 to 1,000 μF

Typical WVDC: 5 to 500 V

(a)

(b)

FIGURE 16–26 Aluminum type electrolytic capacitors (Photo courtesy of Sprague-Goodman Electronics)

tance value per given plate area is very high. Also, voltage breakdown levels are relatively low. Furthermore, since the dielectric is very thin, some leakage current occurs. Typically, this leakage is a fraction of a milliampere for each microfarad of capacitance.

SAFETY HINTS

A **very important** fact about electrolytics is you must ***observe polarity*** when connecting them into a circuit. If current is passed in the wrong direction through the capacitor, the chemical action that created the dielectric layer will be reversed and will destroy the capacitor. A shorted capacitor is actually created. Gas may build up in the component, causing an explosion! Depending on the voltage ratings, common values of electrolytic capacitors range from 2 μF to several hundred μF. Usual voltage ratings for this capacitor range from 10 volts to about 500 volts.

The main advantage of the electrolytic capacitor is the large capacitance-per-size factor. Two obvious disadvantages are the polarity, which must be observed, and the higher leakage current feature. Also, in many capacitors of this type, the electrolyte can dry out with age and depreciate the capacitor quality or render it useless.

5. *Tantalum capacitors,* Figure 16–27, are in the electrolytic capacitor family. They use tantalum instead of aluminum. One important qual-

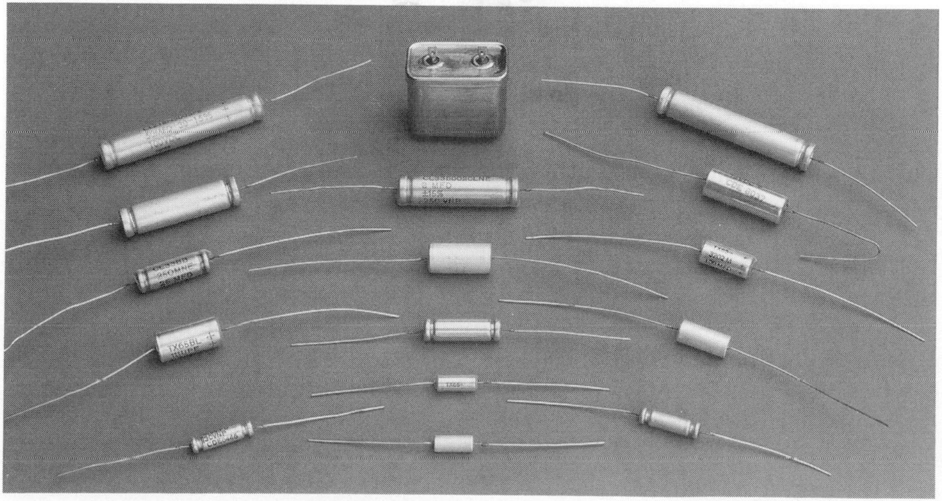

FIGURE 16–27 Tantalum type electrolytic capacitors (Courtesy of Cornell Dubilier Electronics)

ity of tantalum capacitors is they provide very high capacities in small-sized capacitors. They have lower leakage current than the older electrolytics. Also, they do not dry out as fast, thus have a longer shelf-life. Because they are generally only manufactured with low-voltage ratings, they are used in low-voltage semiconductor circuitry. Tantalum capacitors are expensive.

Ratings

There are several important factors that should be observed when selecting or replacing capacitors in various applications. Some of these include:

- Physical size and mounting characteristics
- Capacitance value (in μF or pF)
- Capacitance tolerance (in percentage of rated value)
- Voltage ratings (e.g., working volts dc or WVDC)
- Safe temperature operating range
- Temperature coefficient (for temperature compensating ceramic capacitors used in tuned rf circuits)
- Power factor of the capacitor (expressing the loss characteristics of dielectric used in capacitor)
- Inductance characteristics (when used at high frequencies)

Taking into consideration these factors, Figure 16–28 shows examples of typical ratings for various types of capacitors.

TYPICAL COLOR CODES

Capacitors are color-coded with various systems. It is beyond the scope of our discussion to learn all of these systems. However, you should be aware that they exist.

Figure 16–29 displays some common methods of color-coding for mica, ceramic disc, ceramic tubular, and paper capacitors. Appendix A gives a more detailed look at color codes, including the capacitor color codes. As in other color-coding systems, the position and color of the dots or bands conveys the desired information. (NOTE: Electrolytic capacitors are generally large enough so that the information is printed on their body, rather than using color codes. Also, ceramic capacitors often have the critical information printed on them.)

TYPICAL RANGE OF RATINGS FOR VARIOUS CAPACITORS (Shown in standard electronic parts catalogs)					
CAPACITOR TYPE	CAPACITANCE RATINGS	VOLTAGE RATINGS	TOLERANCE RATINGS	OPER. TEMPS	TEMP. COEFF.
Paper	0.001–1.0 μF	100–1500 V	±10% (common)	−40 to +85°C	N/A
Mica	5–50,000 pF	600 to several kV	±1–5%	−55 to +125°C	N/A
Ceramic	1–10,000 pF	1kV–6kV	±10–20%	−55 to +85°C	N or P 0–750
Electro-lytic	10–1,000 μF	5–500 V	−10–+50%	−40 to +85°C	N/A

NOTE: Temperature coefficient ratings are generally given for ceramic-type capacitors that are designed to increase or decrease in capacitance with temperature change. The rating is normally given in "parts-per-million per degree Celsius." For example, if a capacitor decreases 500 ppm per degree Celsius, it would be labeled "N500."

FIGURE 16–28 Summary of typical ratings

TYPICAL PROBLEMS AND TROUBLESHOOTING TECHNIQUES

Capacitors develop problems more frequently than resistors or inductors. As with the other components, troubles may be related to a shorted capacitor,

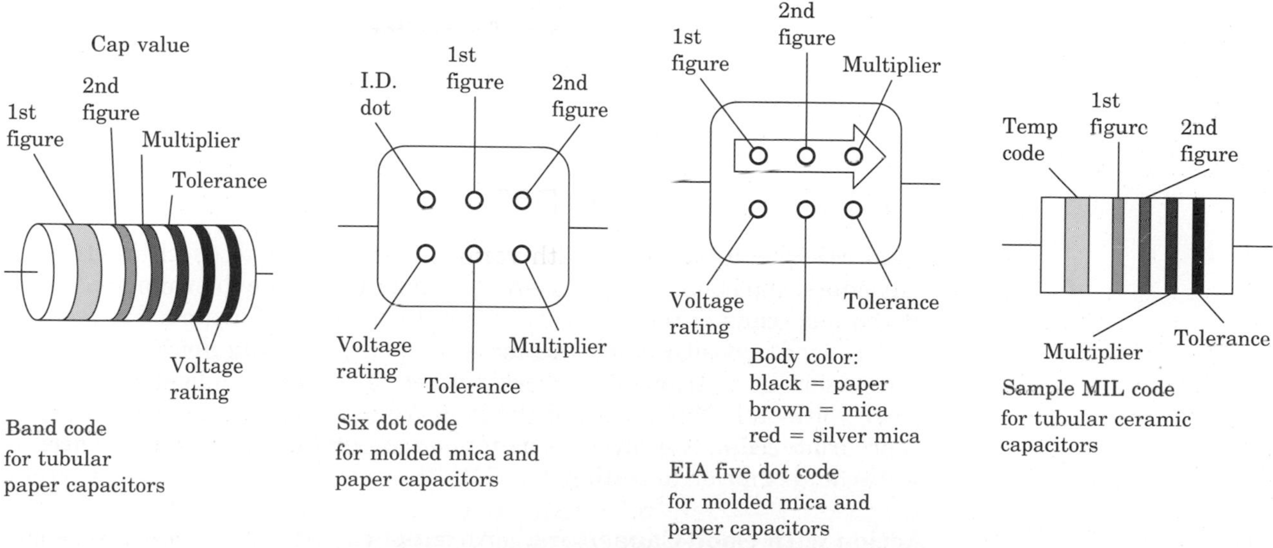

FIGURE 16–29 Typical color-coding systems

an open capacitor, or degenerated performance due to excessive leakage, age, and other causes.

Causes of Open Capacitors

One reason a capacitor acts similar to an open is that a broken connection occurs between an external lead and a capacitor plate. In fact, this situation can occur with any capacitor! In electrolytic capacitors, the electrolyte resistance can drastically increase due to drying out, age, or operation at higher than normal temperatures.

Causes of Shorted Capacitors

Shorts can occur due to external leads touching or plates shorted internally. Internal shorts are most common with paper capacitors and electrolytic capacitors. Common causes for the dielectric material breaking down are age and usage. Again, operating under high temperature conditions accelerates this aging process. In paper electrolytic capacitors, the dielectric material sometimes deteriorates with age, which decreases leakage resistance. The resulting increase in leakage current is equivalent to a *partial short* condition.

Troubleshooting Techniques

Two general methods for testing capacitors are *in-circuit* testing and *out-of-circuit* testing. There are capacitance testers that measure capacitance values, leakage characteristics, and so on; however, the ohmmeter is often useful for making out-of-circuit general-condition tests. Because of the capacitor charging action, there typically is a current surge when a voltage is first applied to the capacitor, and the current decreases to zero (or to leakage-current level) when the capacitor is fully charged. Often, this action is monitored with the ohmmeter.

Using the Ohmmeter for General Tests

General Precautions. When using the ohmmeter to check electrolytic capacitors, you should *observe polarity*. First, determine the voltage polarity at the test leads of the ohmmeter when in the ohms function. Frequently, the red lead is positive and the black lead is negative *but not* for all brands and models of instruments—check with the instrument's literature or another voltmeter. Next, observe the polarity markings on the capacitor, as appropriate. Also, to protect the meter, *make sure the capacitor to be checked is discharged* prior to testing.

Action with Good Capacitors. An out-of-circuit test is where one capacitor lead is disconnected from the circuit, or the capacitor is physically removed from the circuit. When checking a capacitor, use a high range (e.g.,

R × 1 meg selector position) and connect the ohmmeter leads across the capacitor, Figure 16–30. As shown, the ohmmeter pointer initially deflects close to the zero ohms position. As the capacitor charges to the voltage provided by the ohmmeter battery, the pointer moves across the scale toward the infinite resistance scale position, and eventually rests at a very high resistance. For small-value capacitors, this value can be close to infinite. For electrolytic capacitors, the leakage resistance by this method varies from about 0.25 to 1 megohm, depending on the capacitor value and condition. When the R reading is less than this, the capacitor probably needs to be replaced.

For small-value capacitors, the changing deflection action is very quick and little pointer movement is seen. For larger values of capacitors, the charging action and resultant deflection are slower due to the larger RC time constant involved.

Open Capacitors. When an ohmmeter is connected across an open capacitor, the pointer does not move (or an infinite Ω reading is shown), and no charging action is observed. Of course, the technician should be aware of the

FIGURE 16–30 Checking capacitors with an ohmmeter

With leads first connected, Ω reading low

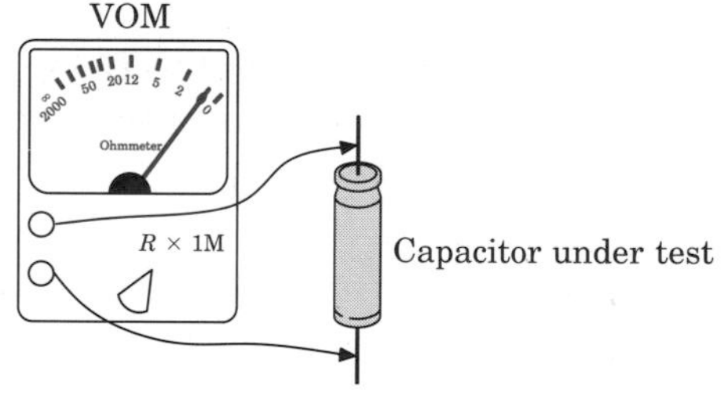

As capacitor charges, Ω reading increases

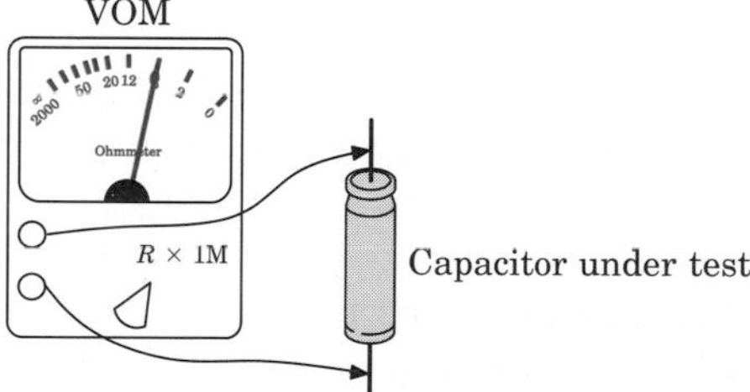

After good capacitor is charged, Ω reading stable at high R value

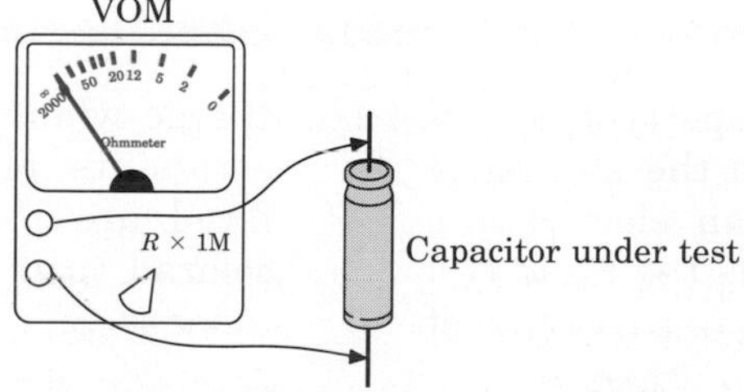

capacitor type and value being checked. For example, small-value capacitors (such as mica capacitors) may not provide an observable deflection, while large-value capacitors should provide some discernable charging action.

Shorted Capacitors. When an ohmmeter is connected across a shorted capacitor, the pointer moves to zero ohms on the scale and remains there. Sometimes, a variable capacitor short is remedied by locating the plate(s) that touch. In this case, simply bend the plates apart. Fixed capacitors indicating a shorted condition must be replaced.

Excessive Leakage Indication. When an ohmmeter is connected across a capacitor having excessively high leakage current (excessively low leakage resistance), the obvious result is a lower-than-normal resistance reading when the pointer finally comes to rest. The capacitor shows a charging action; however, the leakage resistance is lower than it should be.

Replacing Electrolytic Capacitors. Electrolytic capacitors definitely deteriorate with age. Their electrolyte may dry out, causing an increase in resistance and degeneration of operation. Also, the dielectric resistance greatly decreases from constant use causing excessive leakage current and reduced effectiveness in operation. When selecting an electrolytic capacitor for replacement, pay attention to the age of the capacitor—shelf-life is an important consideration.

Closing Comments

As you continue in your studies of electronics, you will find that even the small amount of *stray capacitance* between wires, wires and the chassis, and between sections or parts of certain components affect circuit operation. You will learn that every component has resistance, capacitance, and inductance in one form or another. It is beyond the scope of this chapter to discuss this topic. However, you should be aware that the basic electrical properties of capacitance, inductance, and resistance are present in various forms in every circuit. The frequency where these components are expected to operate determines how critical these *stray* or *distributed* R, L, and C values are to circuit operation.

SUMMARY

- Capacitance is the circuit property that opposes a voltage *change* due to the storing of electrical energy (charge) in an electrostatic field. The unit of capacitance is the farad (F). One farad of capacitance stores one coulomb of charge when one volt is applied or is present across its plates. Practical subunits of the farad are the microfarad (μF) and the picofarad ($\mu\mu$F or pF). The symbol for capacitance is C.

CHAPTER CHALLENGE

Following the SIMPLER troubleshooting sequence, find the problem with this circuit. As you follow the sequence, record your responses to each step on a separate sheet of paper.

CHALLENGE CIRCUIT 9

SYMPTOMS Gather, verify, and analyze symptom information. (Look at the "Starting Point Information.")

IDENTIFY Identify initial suspect area for the location of the trouble.

MAKE Make a decision about "What type of test to make" and "Where to make it." (To simulate making a decision about each test you want to make, select the desired test from the "TEST" column listing below.)

PERFORM Perform the test. (Look up the result of the test in Appendix C. Use the number in the "RESULTS" column to find the test result in the Appendix.)

LOCATE Locate and define a new "narrower" area in which to continue troubleshooting.

EXAMINE Examine available information and again determine "what type of test" and "where."

REPEAT Repeat the preceding analysis and testing steps until the trouble is found. What would you do to restore this circuit to normal operation? When you have solved the problem, compare your results with those shown in the color insert.

Starting Point Information

1. Circuit diagram
2. V_{C_2} is lower than it should be

Test	Results in Appendix C
V_{C_1}	(116)
V_{C_2}	(15)
V_{C_3}	(94)

- A capacitor is a device that possesses the property of capacitance. Capacitance is present when there are two conductors separated by an insulator (nonconductor or dielectric). Capacitance can be lumped as in the capacitor component, or distributed as the capacitance between wires and a chassis.
- Capacitors come in various sizes and shapes and in various capacitance values. Capacitors are often named after the type of dielectric used in their construction, e.g., paper, mica, ceramic, and so on. Electrolytic capacitors are an exception and are *not* named after their aluminum oxide dielectric. This capacitor is also the only basic category that is polarity sensitive.
- Factors affecting the capacitance amount include conductive plate areas, spacing between the plates (dielectric thickness), and the type of dielectric material. The formula is $C = \frac{8.85kA}{10^{12}s}$. The greater the plate areas, the greater the capacitance value. The greater the spacing between plates (the thicker the dielectric) is, the less the capacitance value. Also, the higher the dielectric constant (k) is, the higher the capacitance value.
- The capacitance, charge, and voltage relationships for a capacitor are expressed by the formulas $Q = CV$, $C = \frac{Q}{V}$, and $V = \frac{Q}{C}$, where Q is charge in coulombs; C is capacitance in farads; and V is voltage across the capacitor in volts.
- Dielectric constant (k or K_ϵ) is the comparative charge that different materials store relative to that charge stored in vacuum or air. That is, vacuum and air have a reference k of one.
- When capacitors charge, they are accumulating electrons on their negative plate(s) and losing electrons from their positive plate(s). As this happens, the electron orbits in the dielectric material are stressed, or distorted. This stored energy is returned to the circuit when the capacitor discharges. When the capacitor discharges, the excess electrons on the negative plate(s) travel to the positive plate(s), eventually neutralizing the charge and the potential difference between plates.
- Leakage resistance is related to the resistance of the dielectric material to current flow. If the dielectric were a perfect nonconductor, the leakage resistance would be infinite. Most capacitors have very high leakage resistance, unless their voltage ratings are exceeded. Electrolytic capacitors have a lower leakage resistance than the other types due to the thinness and type of dielectric.
- When capacitors are connected in series, the total capacitance is less than the least capacitance in series. That is, capacitances in series add like resistances in parallel.
- When capacitors are in parallel, total capacitance equals the sum of all the individual capacitances. That is, capacitances in parallel add like resistances in series, where $C_T = C_1 + C_2 + \ldots C_n$.
- Capacitors take time to charge and discharge. The time depends on the amount of capacitance storing the charge and the amount of resistance limiting the charging or discharging current. One RC time constant equals R (in ohms) times C (in farads). Five RC time constants are required for the capacitor to completely charge or discharge to a new voltage level.
- Basic types of capacitors are paper, mica, ceramic, and electrolytic types that come in a variety of combinations. Some of these variations are epoxy-dipped tantalum capacitors, epoxy-coated polyester film capacitors, dipped-mica capacitors, polystyrene film capacitors, dipped-tubular capacitors, and so on.
- Capacitors have various color codes. Electronic Industries Association (EIA) and MIL (mili-

tary) color codes are common. Tubular capacitors often use a banded color-code system (typically six bands). Older types of mica capacitors are frequently color coded with either a five-dot, or six-dot system. Ceramic capacitors may sometimes be color coded with MIL color codes, using three, four, or five color dots or bands.

- Typical problems of capacitors include opens, shorts, or degeneration in operation due to low leakage resistance, or high electrolyte resistance.
- An ohmmeter check of a good capacitor will indicate a charging action occurring when the ohmmeter is first connected to the capacitor. The capacitor is charging to the ohmmeter's internal voltage source value. This is displayed by the ohmmeter pointer initially moving to a lower resistance point on the scale then moving to a final resting point at very high resistance once the capacitor has charged.
- Typical R values after charging are 100 or more megohms. Electrolytic capacitors have a lower leakage resistance in the range of $\frac{1}{2}$ to 1 megohm.

REVIEW QUESTIONS

1. The capacitor plate area doubles while the dielectric thickness is halved. What is the relationship of the new capacitance value to the original capacitance value? (Assume the same dielectric material.)

2. What charge is a 100-pF capacitor storing, if it is charged to 100 volts?

3. What is the capacitor value that stores 200 pC when charged to 50 volts?

4. What voltage is present across a 10-μF capacitor having a charge of 2000 μC?

5. What is the total capacitance of a circuit containing a 100-pF, 400-pF and 1000-pF capacitor connected in series? If the capacitors were connected in parallel, what is the total capacitance?

6. A 10-μF, 20-μF, and 50-μF capacitor are connected in series across a 160-volt source. What voltages appear across each capacitor after it is fully charged? What value of charge is on each capacitor?

7. A 10-μF, 20-μF and 50-μF capacitor are connected in parallel across a 100-volt source. What is the charge on each capacitor? What is the total capacitance and total charge for the circuit?

8. Using the $V_x = V_S \frac{C_T}{C_X}$ formula, determine the voltage across the 20-μF capacitor in a series circuit containing a 20-μF, 40-μF and 60-μF capacitor across a dc source of 100 volts. (Show all work.)

9. Use the appropriate exponential formula(s) and find the V_R value in a series R-C circuit after 1.2 τ if V applied is 200 volts. What is the V_C value? (Use a calculator with the e^x or l_{nx} keys, if possible.)

TEST YOURSELF

1. What is the total capacitance of the circuit in Figure 16–31?

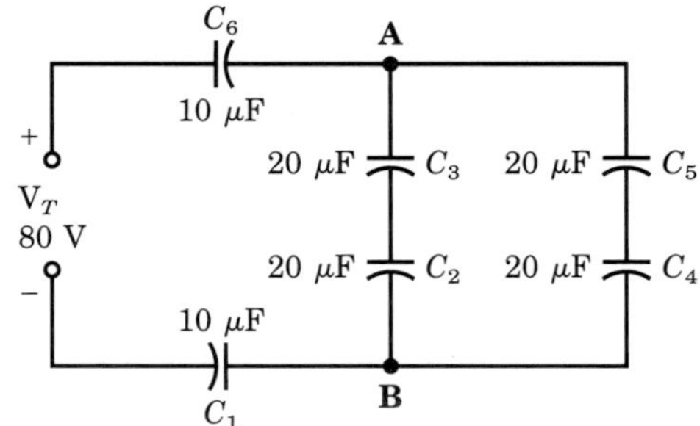

FIGURE 16–31

2. What is the voltage across C_5 in Question 1?
3. What is the charge on C_6 in Question 1?
4. Assume each capacitor in Question 1 above has a breakdown voltage rating of 300 volts. What voltage should not be exceeded between points A and B?
5. For the voltage described between points A and B in Question 4, what is the V_T value that causes that condition to exist?
6. What is the total energy stored in the circuit of Question 1 above with 80-volts V applied?
7. Assume a 15-μF capacitor is charged from a 150-volt source and another 15-μF capacitor is charged from a separate 300-volt source. If the capacitors are disconnected from their sources after being charged and connected in parallel with positive plate to positive plate and negative plate to negative plate, what is the resultant *voltage* and *charge* on each capacitor?
8. For the conditions described in Question 7, what is the total stored energy after the capacitors are connected?

PERFORMANCE PROJECT CORRELATION CHART

Suggested performance projects that correlate with topics in this chapter are:

CHAPTER TOPIC	PERFORMANCE PROJECT	PROJECT NUMBER
Charging and discharging action	Charge and Discharge Action and RC Time	48
The RC Time Constant		
Total Capacitance in Series and Parallel	Total Capacitance in Series and Parallel	49

Capacitive Reactance in AC

Key Terms

Capacitive reactance (X_C)

Rate of change of voltage

Reactance

Courtesy of DeVry Institutes

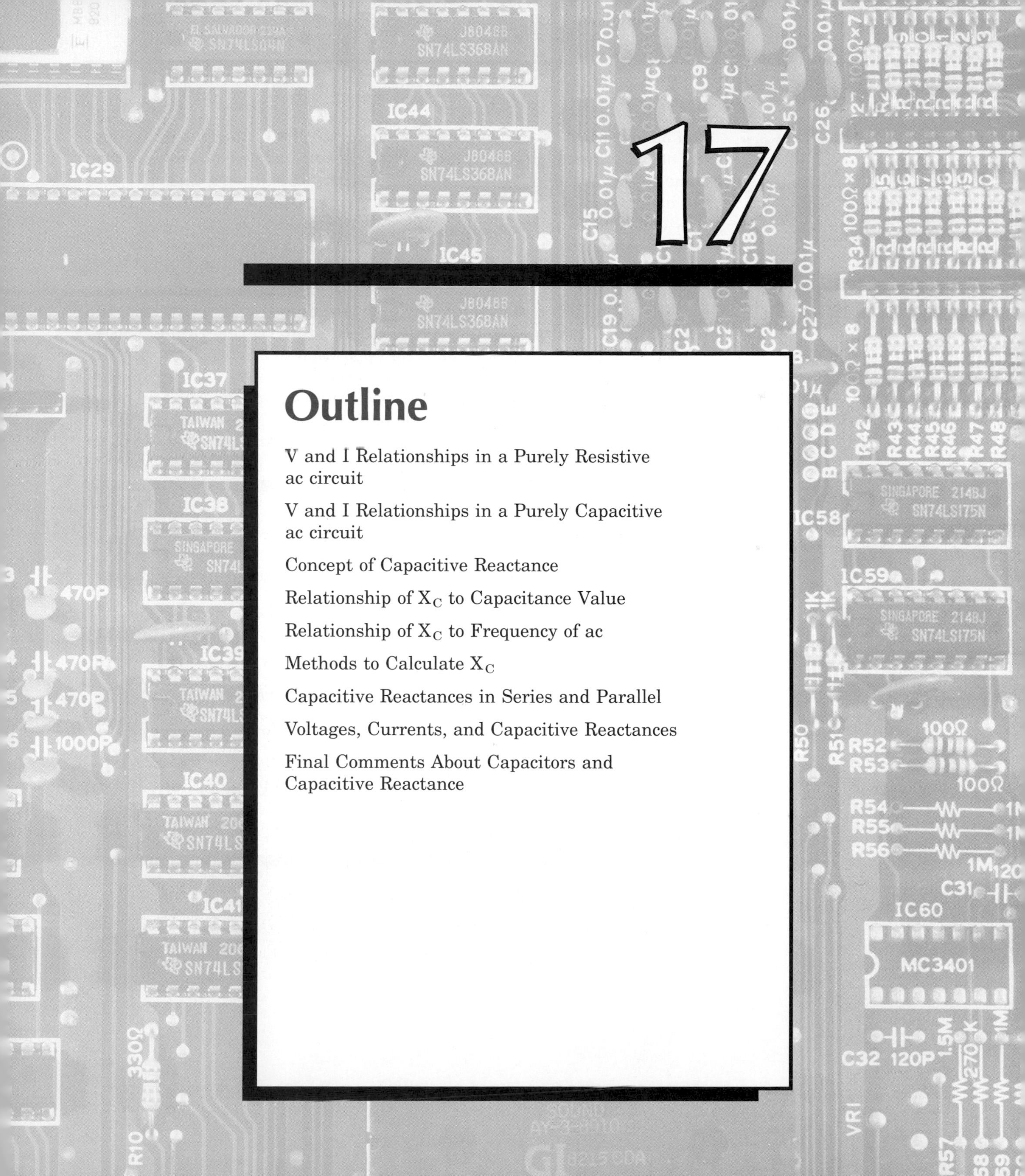

17

Outline

V and I Relationships in a Purely Resistive ac circuit

V and I Relationships in a Purely Capacitive ac circuit

Concept of Capacitive Reactance

Relationship of X_C to Capacitance Value

Relationship of X_C to Frequency of ac

Methods to Calculate X_C

Capacitive Reactances in Series and Parallel

Voltages, Currents, and Capacitive Reactances

Final Comments About Capacitors and Capacitive Reactance

Chapter Preview

You have learned that a capacitor allows current flow, provided it is charging or discharging. However, a capacitor does not allow dc current to continue once charged to the source voltage. Also, you learned a capacitor does not instantly charge or discharge to a new voltage level; it requires current flow and time.

This characteristic of allowing ac current due to the charging and discharging current will be studied in this chapter. You will learn there is an inverse relationship between the amount of *opposition* a capacitor gives to ac current and the circuit frequency and capacitance values involved. Capacitor opposition to ac current is called **capacitive reactance.** (Recall inductive reactance describes inductor opposition to ac current.)

In this chapter, you will examine capacitance as it relates to the capacitor action in a sine-wave ac circuit. You will study the phase relationship of voltage and current for capacitive components; learn to compute capacitive reactance; discover some applications of this characteristic; and learn other practical information about this important electrical parameter.

Objectives

After studying this chapter, you will be able to:

- Illustrate V-I relationships for purely resistive and purely capacitive circuits
- Explain **capacitive reactance**
- Use Ohm's law to solve for X_C value(s)
- Use capacitive reactance formula to solve for X_C value(s)
- Use X_C formula to solve for unknown C and f values
- Use Ohm's law and **reactance** formulas to determine circuit reactances, voltages, and currents for series- and parallel-connected capacitors
- List two practical applications for X_C

V AND I RELATIONSHIPS IN A PURELY RESISTIVE AC CIRCUIT

As a quick review, look at Figure 17–1. Note that V and I are in-phase in a purely resistive ac circuit.

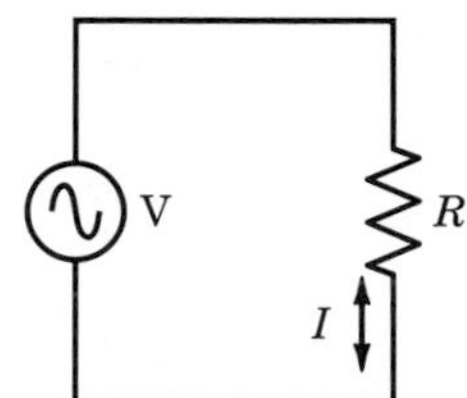

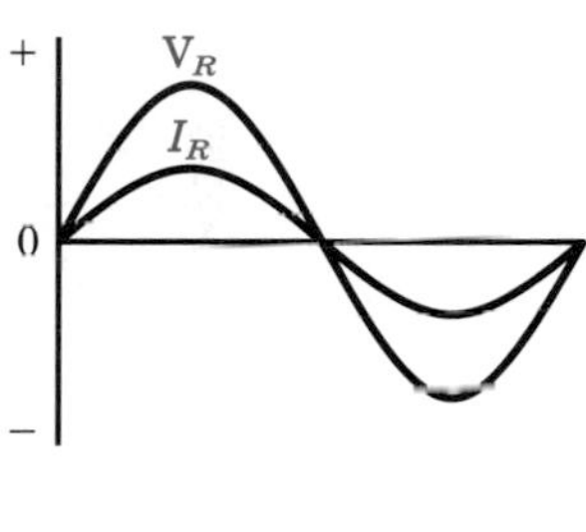

V and I are in-phase

FIGURE 17–1 V and *I* phase relationships in a purely resistive circuit

V AND I RELATIONSHIPS IN A PURELY CAPACITIVE AC CIRCUIT

Observe Figure 17–2 as you study the following statements about the relationship of V and I in a purely capacitive ac circuit.

1. To develop a voltage (potential difference or p.d.) across capacitor plates, charge must move. That is, charging current must flow first.
2. When a sine-wave voltage is applied across a capacitor, which is effectively in parallel with the source terminals, voltage across the capacitor must be the same as the applied voltage.
3. For the voltage across the capacitor to change, there must be alternate charging and discharging action. This implies charging and discharging current flow. The value of instantaneous capacitor current (i_C) directly relates to the amount of capacitance and the rate of change of voltage. (This principle fits well with the concept learned earlier where Q = CV.)
4. The formula that expresses the direct relationship of i_C to C and rate of change of voltage is:

Formula 17–1	$i_C = C\frac{dv}{dt}$

where $\frac{dv}{dt}$ indicates change in voltage divided by amount of time that that change occurs, or the rate of change of voltage.

For example, if the voltage across a 100-pF capacitor changes by 100 volts in 100 μsec, then:

$$i_C = C\frac{dv}{dt}$$

$$i_C = 100 \times 10^{-12} \times \frac{100}{100 \times 10^{-6}}$$

$$i_C = \frac{10{,}000 \times 10^{-12}}{100 \times 10^{-6}}$$

$$i_C = 100 \times 10^{-6}, \text{ or } 100\ \mu\text{A}$$

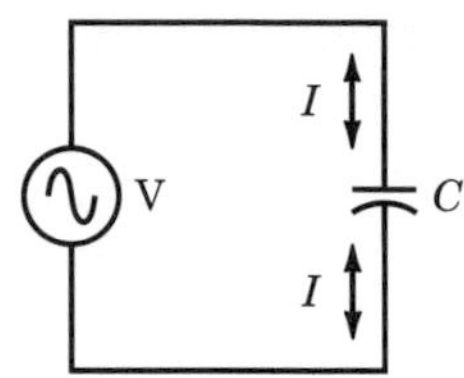

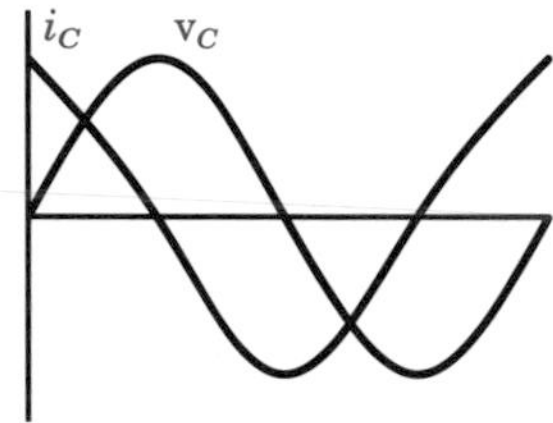

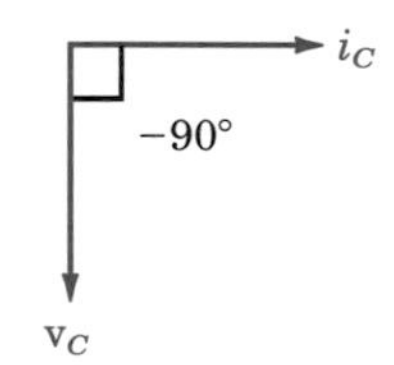

FIGURE 17–2 V and *I* phase relationships in a purely capacitive circuit

5. Since the amount of charging (or discharging) current directly relates to the rate of change of voltage across the capacitor, capacitor current, i_C, is maximum when voltage rate of change is maximum and minimum when voltage rate of change is minimum.
6. Since **rate of change of voltage** is maximum near the zero points of the sine wave and minimum at the maximum points on the sine wave, current is maximum when voltage is minimum and vice-versa. In other words, I is 90° out-of-phase with V.
7. To illustrate these facts, Figure 17–2 shows that capacitor current *leads* the voltage across the capacitor by 90°. Remember our phrase mentioned in an earlier chapter—*"Eli, the ice man." Eli* represents voltage or emf (E) that comes before (leads) current (I) in an inductor (L) circuit. Obviously, the word "ice" represents that *current (I) leads voltage (E) in a capacitive (C) circuit.* If this phrase helps you remember phase relationships of voltage and current for inductors and capacitors, good! But if you understand the concepts related to inductors opposing a current change and capacitors opposing a voltage change, you will not need to remember this phrase!

CONCEPT OF CAPACITIVE REACTANCE

As you have just seen, a sine-wave current develops in a capacitor circuit by the charge and discharge action of the capacitor when a sine-wave voltage is applied. This sine-wave capacitor current leads the sine-wave capacitor voltage by 90°.

In a purely capacitive circuit, the factor that limits the ac current is called capacitive reactance. The symbol used to represent this quantity is "X_C". It is measured in ohms. As you would anticipate, the higher the capacitive reactance, the lower the amount of current produced for a given voltage. That is, $I_C = \frac{V_C}{X_C}$ which is simply an application of Ohm's law.

What factors affect this current-limiting trait called capacitive reactance? It is apparent that for a given voltage the larger the amount of charge moved per unit time, the higher the current and the lower the value of X_C produced and vice versa. Let's examine the relationship of X_C to the two factors that affect this movement of charge, or amount of current for a given voltage. These two factors are *capacitance* and *frequency.*

RELATIONSHIP OF X_C TO CAPACITANCE VALUE

The larger the capacitor, the greater the charge accumulated for a given V across its plates $\left(V = \frac{Q}{C}\right)$. To match a given V applied, a higher charging

IN-PROCESS LEARNING CHECK I

Fill in the blanks as appropriate.

1. For a capacitor to develop a potential difference between its plates, there must be a __________ current.
2. For the potential difference between capacitor plates to decrease (once it is charged), there must be a __________ current.
3. When ac is applied to capacitor plates, the capacitor will alternately __________ and __________.
4. The value of instantaneous capacitor current directly relates to the value of __________ and __________ rate of change.
5. The amount of charging or discharging current is maximum when the rate of change of voltage is __________. For sine-wave voltage, this occurs when the sine wave is near __________ points.
6. Capacitor current __________ the voltage across the capacitor by __________ degrees. The expression that helps remember this relationship is "Eli, the __________ man."

current is required in a high C value circuit than is required with a smaller capacitance. This implies a lower opposition to ac current or a lower capacitive reactance (X_C). This means *X_C is inversely proportional to C.* In other words, the larger the C, the lower the X_C present; and the smaller the C, the higher the X_C present. Refer to Figure 17–3a to observe this relationship.

RELATIONSHIP OF X_C TO FREQUENCY OF AC

If the V_C is tracking the source V, as in the circuit of Figure 17–3b, then the higher the frequency and the less available time there is for the required amount of charge (Q) to move to and away from the respective capacitor plates during charge and discharge action. As you know, it takes a given amount of Q to achieve a specific voltage across a given capacitor $\left(V = \frac{Q}{C}\right)$.

The higher the frequency of applied voltage, the faster is the rate of change of voltage for any given voltage level. As you learned earlier, the i_C directly relates to the voltage rate of change $\left(\frac{dV}{dt}\right)$. Since higher frequencies produce a greater charge and discharge current for any given voltage level, as frequency increases opposition to current decreases. More current for a given V applied indicates there is *less* opposition, meaning capacitive reac-

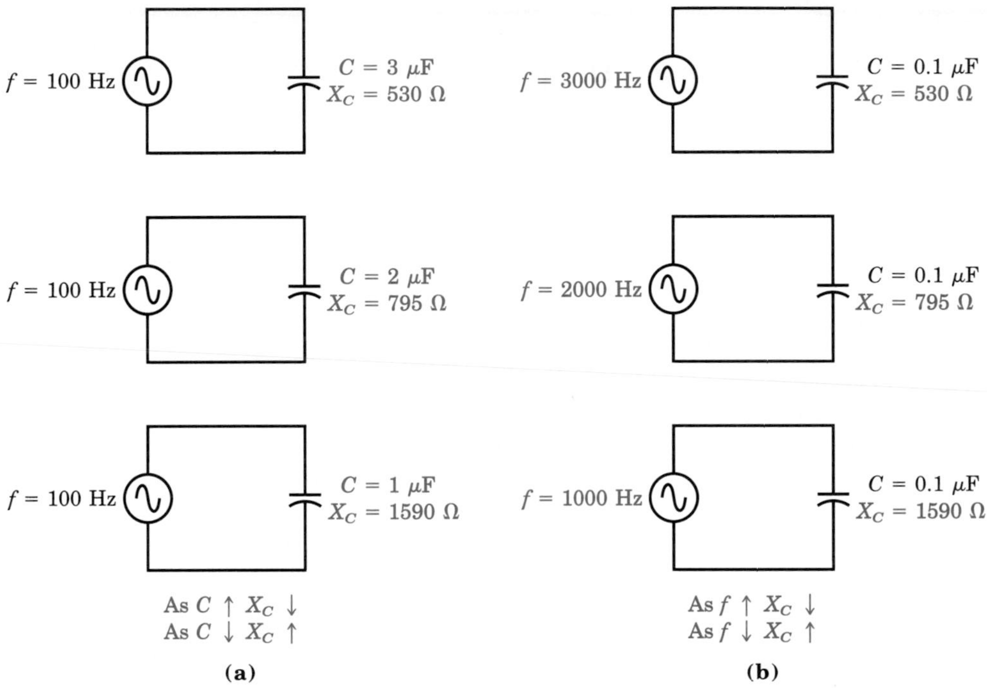

FIGURE 17–3 Relationship of X_C to capacitance and frequency

tance (X_C) decreases as frequency increases. That is, *X_C is inversely proportional to f.* In other words, the higher the frequency, the lower the X_C present; and the lower the frequency, the greater the X_C present.

METHODS TO CALCULATE X_C

Ohm's Law Method

Ohm's law is one practical method to calculate X_C, where $X_C = \frac{V_C}{I_C}$. For example, if the effective ac voltage applied to a capacitive circuit is 100 volts, and the I_C is 1 ampere (rms), capacitive reactance (X_C) must equal 100 ohms. NOTE: Unless told otherwise, assume effective (rms) values of ac voltage and current in your measurements and computations relating to V_C, I_C, and X_C. As an electronics student and technician, you will frequently make such measurements and calculations. Remember, the Ohm's law formula can also be transposed to yield $V_C = I_C \times X_C$, and $I_C = \frac{V_C}{X_C}$.

The Basic X_C Formula

From the previous discussions, it is easy to understand the concepts illustrated by the formula to calculate X_C when only C and f are known. The formula states that:

Formula 17–2	$X_C \text{ (ohms)} = \dfrac{1}{2\pi fC}$

where f is frequency in hertz
C is capacitance in farads

Notice that X_C is *inversely* proportional to both f and C. That is, if either f or C (or both) are increased, the resulting larger product in the denominator divided into one results in a smaller value of X_C. In fact, if either f or C is doubled, X_C is half. If either f or C is halved, X_C doubles, and so forth, Figure 17–4.

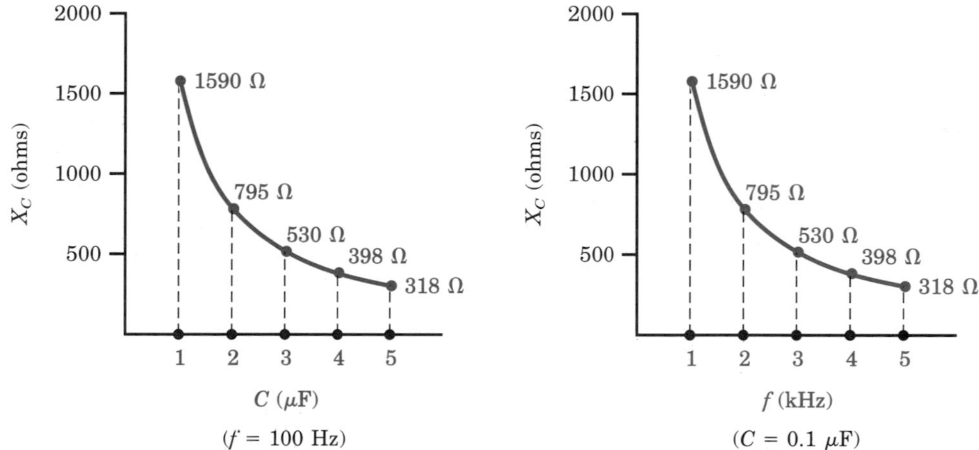

FIGURE 17–4 Examples of the inverse relationship of X_C to C and to f

Stated another way, larger C means more charging and discharging current flows to charge to and discharge from a given voltage. Thus, a larger C results in a lower X_C. Since a higher f means the voltage rate of change is faster, more charging and discharging current must flow in the time allowed by the higher frequency; hence, a higher f results in a lower X_C.

Recall the "$2\pi f$" in the formula relates to angular velocity which relates to sine-wave frequency. The X_C formula is the second formula where you have found this angular velocity expression—the other formula was for inductive reactance where $X_L = 2\pi fL$.

Prior to using the capacitive reactance formula, let's simplify it. Since the one in the numerator and the 2π in the denominator will not be changing (called "constants") when we divide 1 by 2π (6.28), this simplifies the formula. That is, $\frac{1}{6.28} = 0.159$. Therefore, we can restate the X_C formula as follows:

Formula 17–3	$X_C = \frac{0.159}{fC}$

where X_C is capacitive reactance in ohms
f is frequency in hertz
C is capacitance in farads

(Remember this version—you will use it the most!)

PRACTICAL NOTES

Since many calculators have the reciprocal function, it is easy to solve for X_C using either of the approaches described (Formulas 17–2 or 17–3). Use the approach that best suits your needs!

EXAMPLE

1. What is the capacitive reactance of a 2-μF capacitor operating at a frequency 120 Hz?
   Answer:
   $$X_C = \frac{0.159}{fC} = \frac{0.159}{120 \times (2 \times 10^{-6})} = \frac{159{,}000}{240} = 662.5\ \Omega$$

2. The frequency of applied voltage doubles while the C remains the same in a capacitive circuit. What happens to the circuit X_C?
   Answer:
   $X_C = \frac{0.159}{fC}$, where X_C for stated conditions $= \frac{0.159}{(2f)C}$. Therefore, the denominator's product is twice as large as that divided into the unchanged 0.159 numerator. That is, the X_C is half of the original frequency X_C.

3. In a capacitive circuit, if X_C quadruples while the frequency of the applied voltage remains the same, what must have happened to the C value in the circuit?

Answer:
Since X_C is inversely proportional to f and C and f remains unchanged, C must decrease to one-fourth its original value. That is, 4 × original $X_C = \frac{0.159}{f(0.25C)}$.

PRACTICE PROBLEM I

Answers are in Appendix B.

1. What is the capacitive reactance of a 5-μF capacitor operating at a frequency of 200 Hz?
2. What happens to the X_C in a circuit when f doubles and the capacitance value is halved at the same time?

Rearranging the X_C Formula to Find f and C

By rearranging the X_C equation, we can solve for frequency if both C and X_C are known, or solve for C if both f and X_C are known. When the equation is transposed, the new equations are as follows:

Formula 17–4	$f = \frac{0.159}{X_C C}$	and	Formula 17–5	$C = \frac{0.159}{f X_C}$

Try these formulas on the following practice problems.

PRACTICE PROBLEM II

1. What is the frequency of operation of a capacitive circuit where the total X_C is 1,000 ohms and the total capacitance is 500 pF?
2. What is the total circuit capacitance of a circuit exhibiting an X_C of 3,180 ohms at a frequency of 2 kHz?

CAPACITIVE REACTANCES IN SERIES AND PARALLEL

In the last chapter, you learned that capacitances in series add like resistances in parallel, and capacitances in parallel add like resistances in series. It is interesting that when we consider capacitive *reactances* in series and parallel, the converse is true.

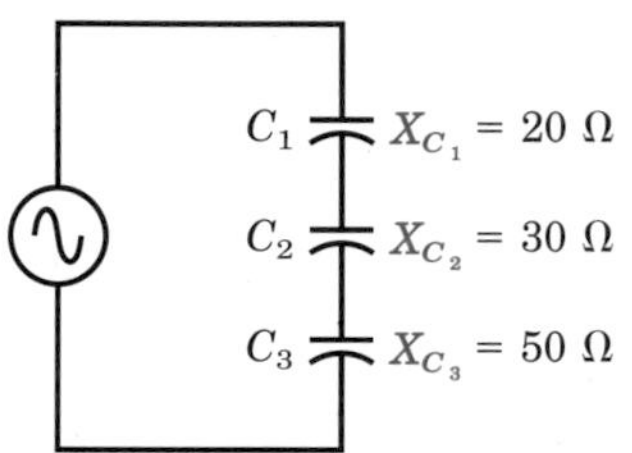

$X_{CT} = X_{C_1} + X_{C_2} + X_{C_3}$
$X_{CT} = 100\ \Omega$

(X_C in series add like resistances in series)

FIGURE 17–5 X_C in series

Capacitive Reactances in Series

Since capacitive reactance (X_C) is an opposition to current flow, measured in ohms, capacitive reactances (current oppositions) in series add like resistances in series. That is, $X_{CT} = X_{C_1} + X_{C_2} + X_{C_3} \ldots + X_{C_n}$, Figure 17–5.

EXAMPLE

What is the total X_C of a circuit containing three series-connected 3-μF capacitors operating at a frequency of 100 Hz?

Answer:

Approach 1: Find total C. Use the X_C formula. Thus,

$$C = \frac{1}{3\text{rd}}\ 3\ \mu\text{F, or } 1\ \mu\text{F}$$

$$X_{CT} = \frac{0.159}{fC} = \frac{0.159}{100 \times 1 \times 10^{-6}}$$

$$X_{CT} = \frac{159{,}000}{100} = 1{,}590\ \Omega$$

Approach 2: Find X_C of a 3 μF C, and multiply by 3. Thus,

$$X_C = \frac{0.159}{fC} = \frac{0.159}{100 \times 3 \times 10^{-6}}$$

$$X_C = \frac{159{,}000}{300} = 530\ \Omega$$

$$X_{CT} = 3 \times 530\ \Omega = 1{,}590\ \Omega$$

PRACTICE PROBLEM III

What is the total X_C of the circuit in Figure 17–6?

FIGURE 17–6

$f = 1$ kHz
C_1 2 μF
C_2 4 μF
C_3 $X_{C_3} = 200\ \Omega$
C_4 $X_{C_4} = 300\ \Omega$

Capacitive Reactances in Parallel

Since capacitive reactance is an opposition to ac current flow, when we supply more current paths by connecting X_Cs in parallel, the resulting total opposition decreases just as it does when we parallel resistances.

Useful formulas to solve for total capacitive reactance of capacitors in parallel are as follows:

- General formula for X_Cs in parallel:

Formula 17–6	$X_{CT} = \dfrac{1}{\dfrac{1}{X_{C_1}} + \dfrac{1}{X_{C_2}} + \dfrac{1}{X_{C_3}} \ldots + \dfrac{1}{X_{C_n}}}$

- Formula for two X_Cs in parallel:

Formula 17–7	$X_{CT} = \dfrac{X_{C_1} \times X_{C_2}}{X_{C_1} + X_{C_2}}$

- Formula for equal-value X_Cs in parallel:

Formula 17–8	$X_{CT} = \dfrac{X_{C_1}}{N}$

where N = number of equal-value X_Cs in parallel

EXAMPLE

What is the total X_C of a circuit containing three parallel-connected capacitor branches with values of 2 μF, 3 μF, and 5 μF respectively? Assume the circuit is operating at a frequency of 400 Hz.

Answer:

Two possible approaches are:

1. Find total C, then use X_C formula to find X_{CT}
2. Find X_C of each capacitor using X_C formula, then use the general formula to solve for X_{CT}.

Of these two approaches, approach 1 is easiest for these circumstances. Therefore,

$$C_T = C_1 + C_2 + C_3$$
$$C_T = 2\ \mu F + 3\ \mu F + 5\ \mu F = 10\ \mu F$$
$$X_{CT} = \frac{0.159}{f C_T} = \frac{0.159}{400 \times 10 \times 10^{-6}} = \frac{159{,}000}{4{,}000} = 39.75\ \Omega$$

PRACTICE PROBLEM IV

1. What is the total capacitive reactance of a circuit containing three parallel-connected capacitors of 1 μF, 2 μF, and 3 μF, respectively. (Assume an operating frequency of 1 kHz.)
2. What is the total capacitive reactance of two parallel-connected capacitors where one capacitance is 5 μF and the

other capacitance is 20 μF? (Circuit is operating at a frequency of 500 Hz.)

3. What is the total capacitive reactance of four 1600-ohm capacitive reactances connected in parallel?

VOLTAGES, CURRENTS, AND CAPACITIVE REACTANCES

Our computations have concentrated on finding the capacitive reactances in capacitive circuits. Let's discuss voltage distribution throughout series capacitive circuits and current distribution throughout parallel capacitive circuits.

Since capacitive reactances represent opposition to ac current (in ohms) when dealing with V, I, and ohms of opposition, we treat X_C like we treat resistances in resistive circuits. That is, voltage drops are calculated by using $I \times X_C$; and currents are calculated by using $\frac{V}{X_C}$, and so on.

Refer to Figure 17–7 to see series circuit parameters and computations illustrated. Then refer to Figure 17–8 and observe how parallel circuit parameters and computations are made. After studying these figures, try the following practice problems.

PRACTICE PROBLEM V

1. For the circuit in Figure 17–9, find X_{C_1}; X_{C_2}; X_{C_3}; X_{CT}; C_2; C_3; V_2; and I_T.
2. For the circuit in Figure 17–10, find C_1; X_{C_1}; X_{C_2}; X_{C_3}; I_2 and I_T.

FINAL COMMENTS ABOUT CAPACITORS AND CAPACITIVE REACTANCE

Having studied capacitive reactance, you can more clearly see how capacitors are used. As you proceed in your studies and career, you should *keep in mind* the inverse relationship of capacitance and capacitive reactance. Also, you remember capacitances in series add like resistances in parallel. Keeping these ideas in mind will help prevent errors when analyzing circuits containing both inductances and capacitances.

A list of similarities and differences between capacitive, inductive and resistive circuits is shown in Figure 17–11. Review what you have learned about the operating features of these three electronic components. It is important that you thoroughly understand each component to analyze circuits that combine two or more of these components.

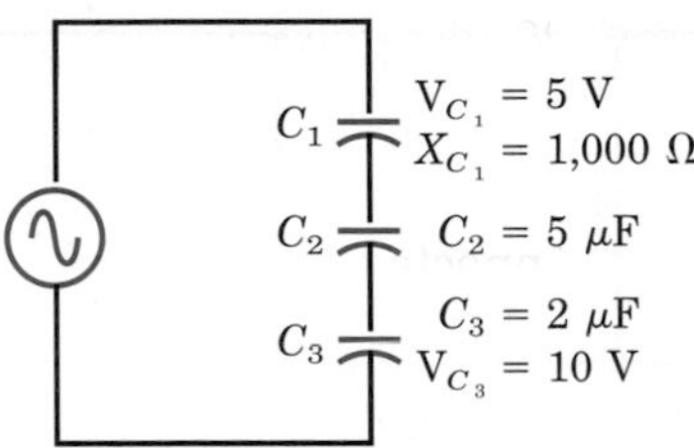

FIGURE 17–7
Series capacitive circuit analysis

Find f; C_1; I; X_{C_2}; X_{C_3}; X_{CT}; V_{C_2}; and V_A.

(Since we know two things about C_1, let's start there.)

$$I = \frac{V_{C_1}}{X_{C_1}} = \frac{5}{1{,}000} = 0.005 \text{ A, or } I = 5 \text{ mA.}$$

Since V_{C_3} is double V_{C_1}, its X_C must be double that of C_1 since they have the same I and $V = I \times X_C$. Thus, $X_{C_3} = 2{,}000\ \Omega$.

Since V_{C_1} is half V_{C_3}, C_1 must have twice the capacitance and half the X_C of C_3. Thus, $C_1 = 4\ \mu$F.

Since C_2 is 2.5 times the C value of C_3 and X_C is inversely related to C, X_{C_2} must equal $\frac{X_{C_3}}{2.5}$ or $\frac{2{,}000}{2.5}$. Thus, $X_{C_2} = 800\ \Omega$.

$$V_{C_2} = I \times X_{C_2} = 5 \text{ mA} \times 800\ \Omega$$
$$V_{C_2} = 4 \text{ V}$$

$$X_{CT} = X_{C_1} + X_{C_2} + X_{C_3} = 1{,}000 + 800 + 2{,}000$$
$$X_{CT} = 3{,}800\ \Omega$$

$$C_T = \frac{1}{\left(\frac{1}{4} + \frac{1}{5} + \frac{1}{2}\right)}$$
$$C_T = 1.05\ \mu\text{F}$$

$$f = \frac{0.159}{C_T X_{CT}} = \frac{0.159}{1.05 \times 10^{-6} \times 3{,}800}$$
$$f = \frac{159{,}000}{3{,}990}$$
$$f = 39.8 \text{ Hz or about 40 Hz}$$

$$V_A = I \times X_{CT} = 5 \text{ mA} \times 3.8 \text{ k}\Omega$$
$$V_A = 19 \text{ V}$$
$$(\text{Also, } V_1 + V_2 + V_3 = 19 \text{ V})$$

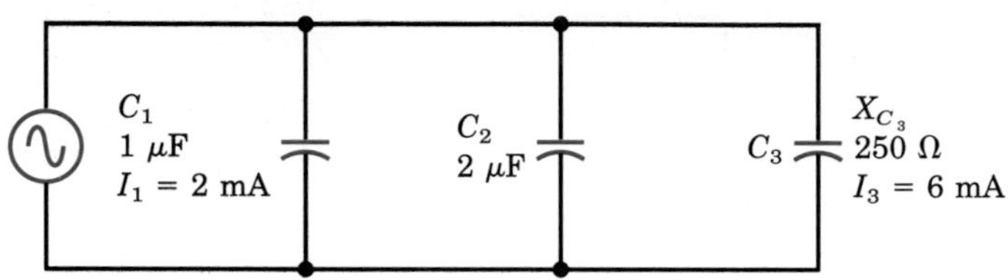

FIGURE 17–8 Parallel capacitive circuit analysis

Find V_A; f; X_{C_2}; X_{C_1}; I_2; I_T.

Starting with information regarding C_3,

$$V_A = V_3 = I_3 \times X_{C_3}$$
$$V = 6\text{ mA} \times 0.25\text{ k}\Omega$$
$$V_A = 1.5\text{V}$$

Since I_3 is triple I_1, X_{C_3} must be $\frac{1}{3}$rd X_{C_1}. Therefore, C_3 must be triple C_1.

$$C_3 = 3\ \mu\text{F}$$

$$X_{C_1} = 3 \times X_{C_3} = 3 \times 250\ \Omega$$
$$X_{C_1} = 750\ \Omega$$

$$X_{C_2} = \frac{1}{2} X_{C_1} = \frac{1}{2} \text{ of } 750\ \Omega$$
$$X_{C_2} = 375\ \Omega$$

$$I_2 = \frac{V_2}{X_{C_2}} = \frac{1.5\text{ V}}{375\ \Omega}$$
$$I_2 = 4\text{ mA}$$

$$I_T = I_1 + I_2 + I_3 = 2\text{ mA} + 4\text{ mA} + 6\text{ mA}$$
$$I_T = 12\text{ mA}$$

f is found by using any one of the branches information or by using the total C and X_C parameters. That is,

$$f = \frac{0.159}{CX_C} = \frac{0.159}{(1 \times 10^{-6} \times 750)}$$
$$f = \frac{159{,}000}{750}$$
$$f = 212\text{ Hz}$$

FIGURE 17–9

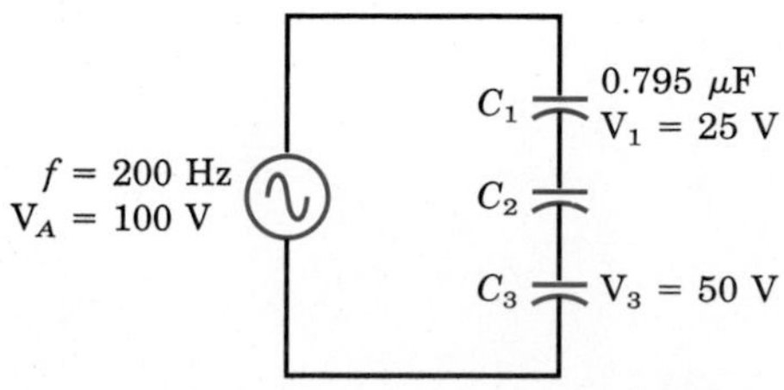

Find X_{C_1}; X_{C_2}; X_{C_3}; X_{CT}; C_2; C_3; V_2; and I_T.

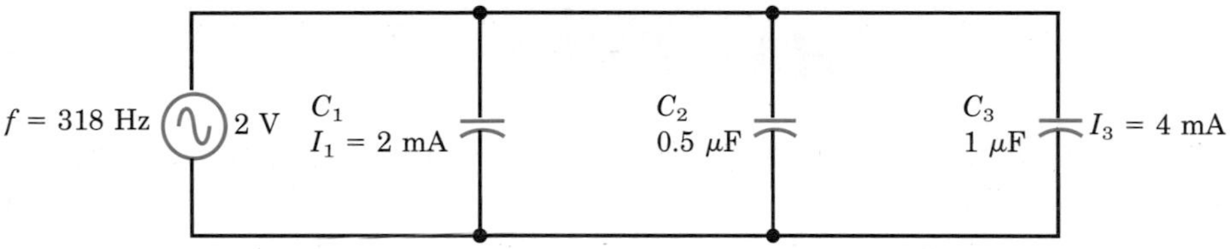

FIGURE 17–10

Find C_1; X_{C_1}; X_{C_2}; X_{C_3}; I_2; and I_T.

Parameter Involved	Capacitive Circuit	Inductive Circuit	Resistive Circuit
Opposition to I	X_C in ohms	X_L in ohms	R in ohms
As f increases	X_C decreases	X_L increases	R no change
As f decreases	X_C increases	X_L decreases	R no change
As C increases	X_C decreases	not applicable	not applicable
As C decreases	X_C increases	not applicable	not applicable
As L increases	not applicable	X_L increases	not applicable
As L decreases	not applicable	X_L decreases	not applicable
Voltage =	$I \times X_C$	$I \times X_L$	$I \times R$
Phase of V and I	I leads V by 90°	V leads I by 90°	V and I in phase
To calculate X	$X_C = \frac{0.159}{fC}$	$X_L = 2\pi fL$	not applicable
Series-connected	$C_T <$ least C	$L_T =$ sum of Ls	$R_T =$ sum of Rs
Parallel-connected	$C_T =$ sum of Cs	$L_T <$ least L	$R_T <$ least R
Series-connected	$X_{CT} =$ sum of X_Cs	$X_{LT} =$ sum of X_Ls	$R_T =$ sum of Rs
Parallel-connected	$X_{CT} <$ least X_C	$X_{LT} <$ least X_L	$R_T <$ least R

FIGURE 17–11 Similarities and differences of C, L, and R circuit characteristics

CHAPTER CHALLENGE

Following the SIMPLER troubleshooting sequence, find the problem with this circuit. As you follow the sequence, record your responses to each step on a separate sheet of paper.

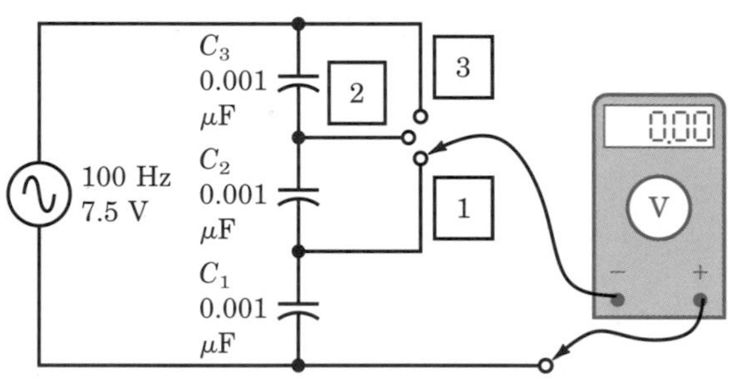

CHALLENGE CIRCUIT 10

SYMPTOMS Gather, verify, and analyze symptom information. (Look at the "Starting Point Information.")

IDENTIFY Identify initial suspect area for the location of the trouble.

MAKE Make a decision about "What type of test to make" and "Where to make it." (To simulate making a decision about each test you want to make, select the desired test from the "TEST" column listing below.)

PERFORM Perform the test. (Look up the result of the test in Appendix C. Use the number in the "RESULTS" column to find the test result in the Appendix.)

LOCATE Locate and define a new "narrower" area in which to continue troubleshooting.

EXAMINE Examine available information and again determine "what type of test" and "where."

REPEAT Repeat the preceding analysis and testing steps until the trouble is found. What would you do to restore this circuit to normal operation? When you have solved the problem, compare your results with those shown in the color insert.

Starting Point Information

1. Circuit diagram
2. When testing at test point "1," voltmeter reading is zero volts.
3. When testing at test point "2," voltmeter reading is 7.5 volts.
4. When testing at test point "3," voltmeter reading is 7.5 volts.

Test	Results in Appendix C
V_{C_1}	(28)
V_{C_2}	(67)
V_{C_3}	(102)
V_T	(79)
"Bridging" with known good 0.001 μF capacitor:	
Bridge C_1	(71)
Bridge C_2	(24)
Bridge C_3	(111)

SUMMARY

- Capacitive reactance is the opposition a capacitor displays to ac or pulsating dc current. The symbol is X_C and it is measured in ohms. Capacitive reactance is calculated by the formula $X_C = \frac{1}{2\pi fC}$, or $X_C = \frac{0.159}{fC}$, where f is frequency in hertz and C is capacitance in farads.
- Factors influencing the amount of capacitive reactance (X_C) in a circuit are the amount of capacitance (C) and the frequency (f). Voltage does *not* affect capacitive reactance. As C *or* f increases, X_C decreases. As C *or* f decreases, X_C increases proportionately.
- In a purely capacitive circuit, current leads voltage by 90°. (Charging current must flow before a potential difference develops across the capacitor plates.)
- Capacitive reactance is calculated by using Ohm's law or the capacitive reactance formula, where $X_C = \frac{V_C}{I_C}$, or $X_C = \frac{0.159}{fC}$.
- The capacitive reactance formula can be rearranged to find either f or C, if all other factors are known. For example, $f = \frac{0.159}{CX_C}$, and $C = \frac{0.159}{fX_C}$.
- Capacitive reactances in series add like resistances in series. That is, $X_{CT} = X_{C_1} + X_{C_2} + X_{C_3} \ldots + X_{C_n}$.
- Capacitive reactances in parallel add like resistances in parallel. The reciprocal formula, product-over-the-sum formula, or equal-value parallel formulas are used, as appropriate, to find total capacitive reactance of parallel-connected capacitors.
- All three variations of Ohm's law may be used with relation to capacitor parameters. That is, $I_C = \frac{V_C}{X_C}$; $V_C = IX_C$; and $X_C = \frac{V_C}{I_C}$.
- Capacitance and capacitive reactance exhibit converse characteristics. In a *series* circuit, the *smallest* C has the largest voltage drop and the largest C the least voltage drop. In this same series circuit, the largest X_C drops the most voltage (because the smallest C has the highest X_C), and so forth. In *parallel* circuits, the largest C holds the largest amount of charge (Q) and the smallest C has the least Q.

REVIEW QUESTIONS

1. What is the capacitive reactance of a 0.00025-μF capacitor operating at a frequency of 5 kHz?
2. What capacitance value exhibits 250 ohms of reactance at a frequency of 1500 Hz?
3. The frequency of applied voltage to a given circuit triples while the capacitance re-

mains the same. What happens to the value of capacitive reactance?

4. Series-circuit capacitors are all exchanged for Cs, each having twice the capacitance of the capacitor it replaces. What happens to total circuit X_C and total C?

5. For the circuit in Figure 17–12 find C_T, X_{CT}, V_A, C_1, C_2, and X_{C_2}.

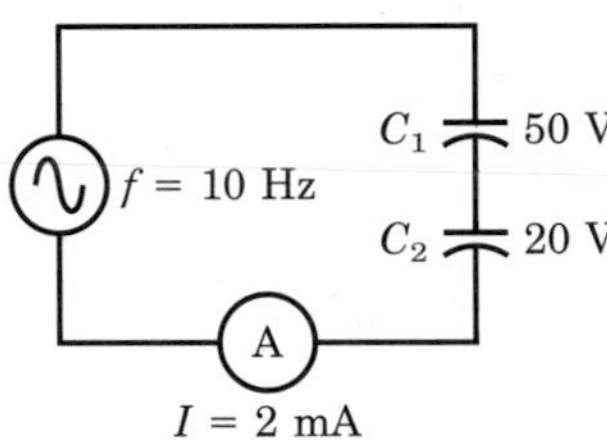

FIGURE 17–12

6. For the circuit in Figure 17–12, if C_1 and C_2 were equal in value, what would be the new voltages on C_1 and C_2?

7. Assume the frequency doubles in the circuit in Figure 17–12. Answer the following statements with "I" for increase, "D" for decrease, and "RTS" for remain the same.
   a. C_1 will ____________
   b. C_2 will ____________
   c. C_T will ____________
   d. X_{C_1} will ____________
   e. X_{C_2} will ____________
   f. X_{CT} will ____________
   g. Current will ____________
   h. V_{C_1} will ____________
   i. V_{C_2} will ____________
   j. V_A will ____________

8. For the circuit in Figure 17–13 find C_T, I_1, I_2, X_{CT}, X_{C_1}, X_{C_2}, and I_T.

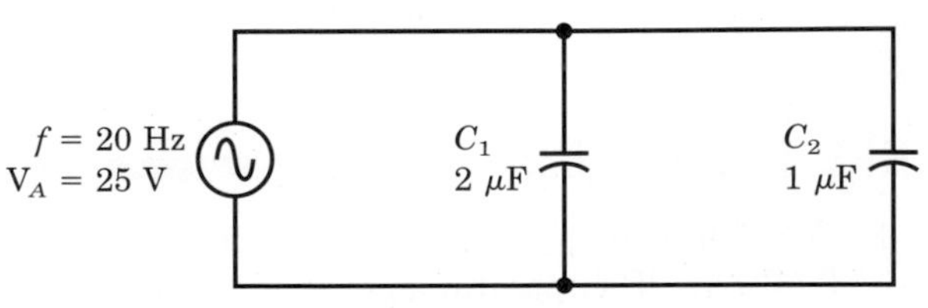

FIGURE 17–13

9. For the circuit in Figure 17–13, if C_2 is made equal to C_1, what is the I_T?

10. Assume the frequency doubles in the circuit in Figure 17–13. Answer the following statements with "I" for increase, "D" for decrease, and "RTS" for remain the same.
   a. C_1 will ____________
   b. C_2 will ____________
   c. C_T will ____________

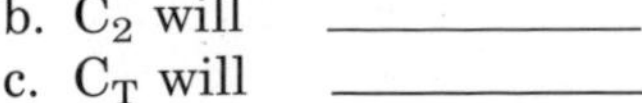

d. X_{C_1} will ____________
e. X_{C_2} will ____________
f. X_{CT} will ____________
g. Current will ____________
h. V_{C_1} will ____________
i. V_{C_2} will ____________
j. V_A will ____________

Refer to Figure 17-13 and answer questions 11–15:

11. What is the X_C of C_1?

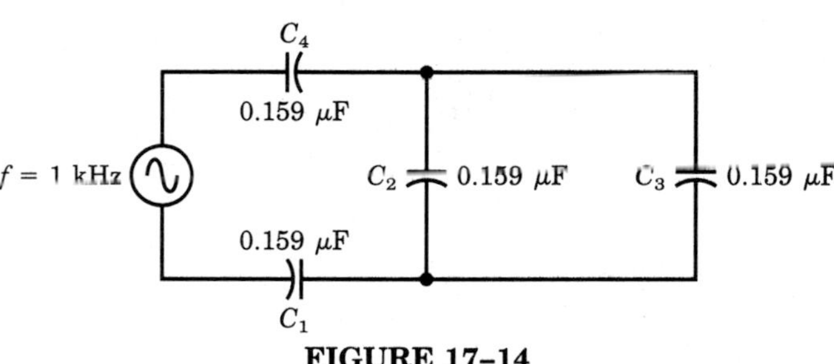

FIGURE 17–14

12. What is the total circuit X_{CT}?

13. What is the value of C_T?

14. If f were changed to 5 kHz, what would be the new value of circuit X_{CT}?

15. If C_2 were replaced by a larger capacitance value capacitor, would the voltage across C_4 increase, decrease, or remain the same?

TEST YOURSELF

1. Draw a capacitive voltage divider circuit where the following conditions would be found: (Label the diagram with all component and parameter values.)
   a. Three capacitors are used.
   b. The voltage across the middle capacitor will be double that found across the capacitors on either side of it.
   c. The frequency of operation is 2 kHz.
   d. The capacitor with maximum X_C has a reactance value of 79.5 Ω.
   e. $V_A = 159$ V

2. Draw a circuit configuration where the total capacitance is 0.66 μF, using six 1-μF capacitors.

PERFORMANCE PROJECT CORRELATION CHART

Suggested performance projects that correlate with topics in this chapter are:

CHAPTER TOPIC	PERFORMANCE PROJECT	PROJECT NUMBER
V & I Relationships in a Purely Capacitive ac Circuit	Capacitance Opposing a Change in Voltage	50
Relationship of X_C to Capacitance Value and Relationship of X_C to Frequency of ac	X_C Related to Capacitance and Frequency	51
Methods to Calculate X_C	The X_C Formula	52

RC Circuits in AC

Key Terms

Current leading voltage

Long time-constant circuit

Short time-constant circuit

Courtesy of International Business Machines Corporation

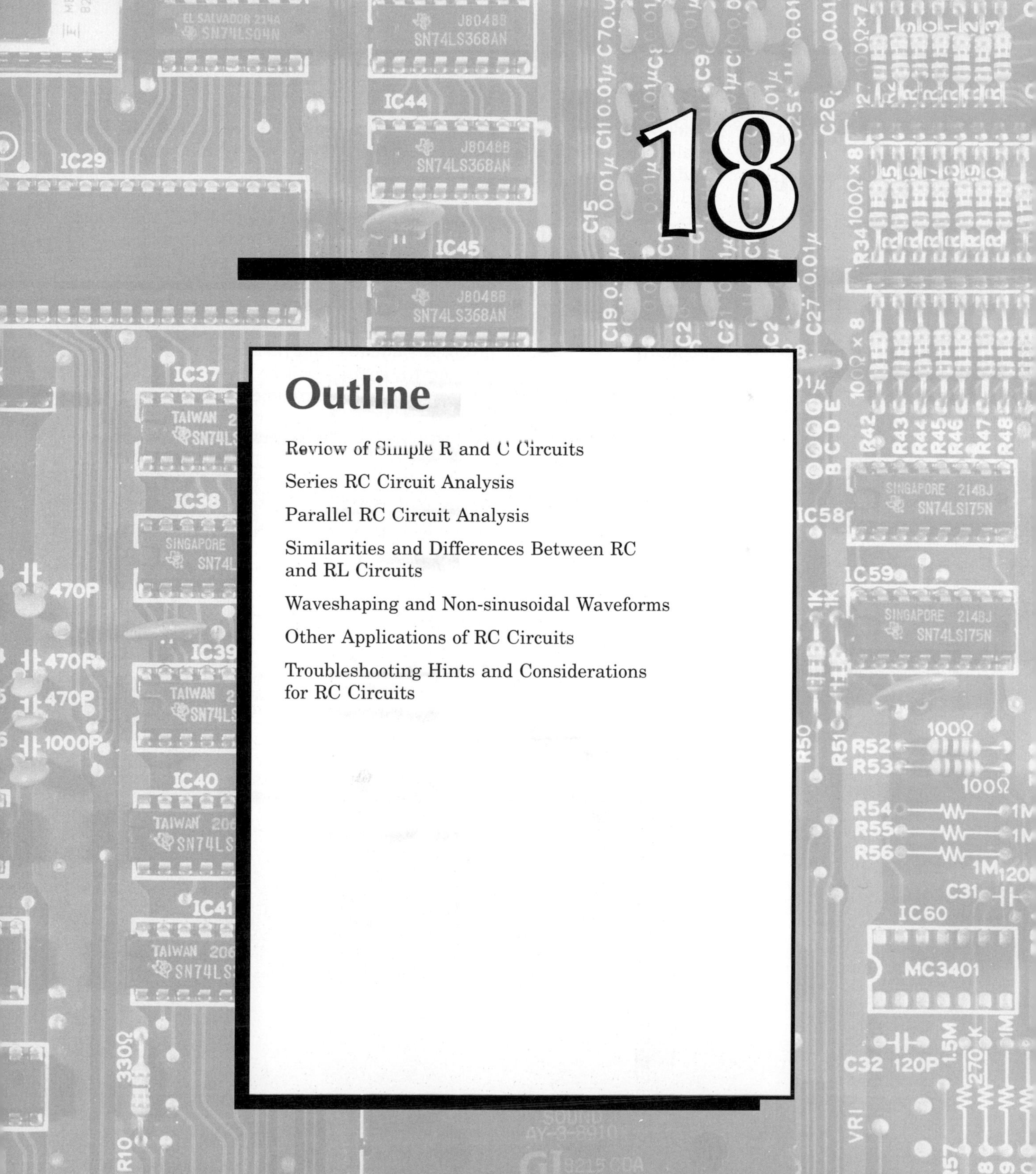

18

Outline

Review of Simple R and C Circuits

Series RC Circuit Analysis

Parallel RC Circuit Analysis

Similarities and Differences Between RC and RL Circuits

Waveshaping and Non-sinusoidal Waveforms

Other Applications of RC Circuits

Troubleshooting Hints and Considerations for RC Circuits

Chapter Preview

Seldom does the technician find capacitors alone in a circuit. They are generally used with other components, such as resistors, inductors, transistors, and so on, to perform some useful function.

In this chapter, we will investigate simple circuits containing both resistors and capacitors. Understanding RC circuits is foundational to dealing with practical circuits and systems, such as amplifiers, receivers, transmitters, computers, power supplies, and other systems and subsystems.

You will learn to analyze both series and parallel combinations of resistance and capacitance that operate under sine-wave ac conditions. An overview of how these components are used for waveshaping in non-sinusoidal applications will also be given. A summary and review of the similarities and differences between RC and RL circuits will be provided to help you remember these important facts. Finally, some troubleshooting hints for RC circuits will be given.

Objectives

After studying this chapter, you will be able to:

- Draw or describe operation of simple R and C circuits
- Analyze appropriate series and parallel RC circuit parameters using Pythagorean theorem
- Use vector analysis to analyze series and parallel RC circuit parameters
- List differences between RC and RL circuits
- Predict output waveform(s) from a waveshaping network
- List two practical applications for RC circuits
- Apply troubleshooting hints to help identify problems

REVIEW OF SIMPLE R AND C CIRCUITS

Although you have studied purely resistive and capacitive circuits, review Figure 18–1 as a quick refresher.

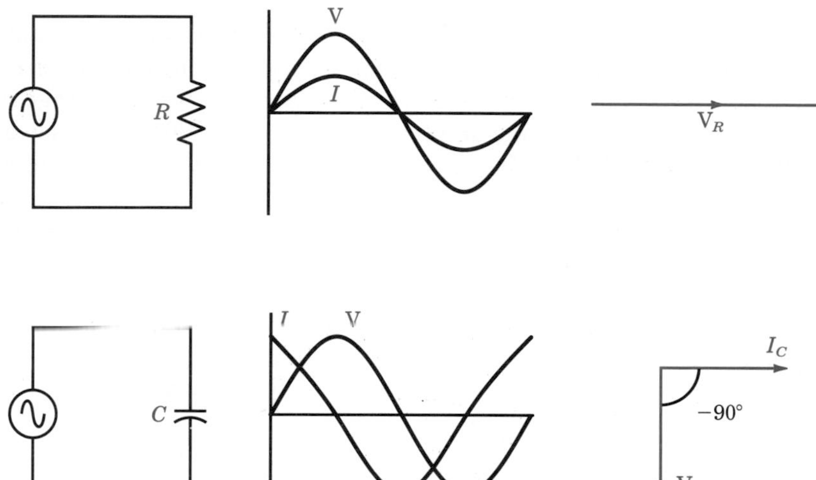

FIGURE 18–1 Voltage-current relationships in purely *R* and purely *C* circuits

SERIES RC CIRCUIT ANALYSIS

As Figure 18–1 has just revealed, current and voltage are in-phase for purely resistive ac circuits. Circuit **current leads voltage** by 90° in purely capacitive circuits. Therefore, when resistance and capacitance are combined, the overall difference in angle between circuit voltage and current is an angular difference between 0° and 90°, Figure 18–2. The actual angle

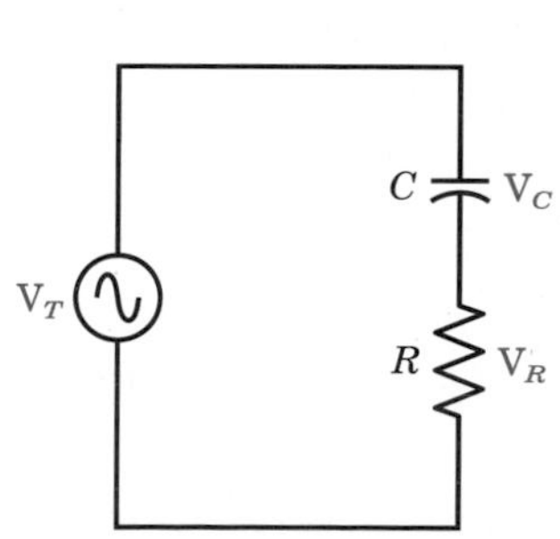

$V_R = I \times R$

$V_C = I \times X_C$

$V_T = I \times Z$ (and vector result of V_R and V_C)

$$I = \frac{V_T}{Z}$$

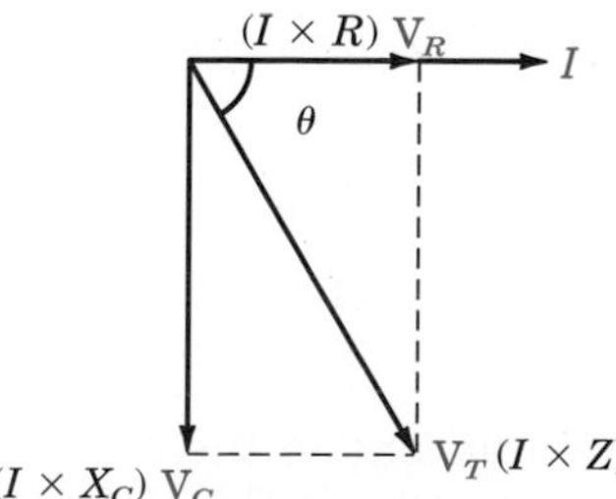

Angular difference between V_T and I (θ) is greater than 0° but less than 90°.

FIGURE 18–2 Positional information for phasors in series RC circuits

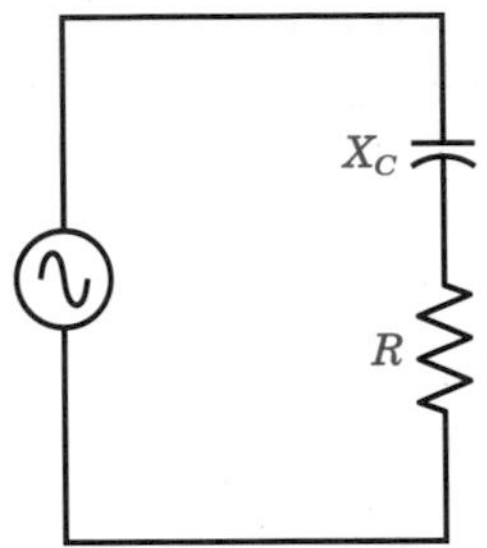

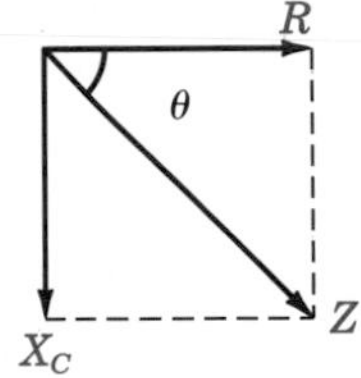

FIGURE 18–3 Impedance diagram for a series RC circuit

depends on the relative values of the components' electrical parameters and the frequency of operation.

Positional Information for Phasors

In series circuits, the current is used as the zero degree reference vector for V-I vector diagrams because *current is common throughout a series circuit.* The V-I vector diagram, therefore, shows the resultant circuit voltage lagging behind the current by some angle between 0° and −90°, again see Figure 18–2. Remember, this is in the fourth quadrant of the coordinate system.

In *series* RC circuits (and *only* for *series* circuits), we can also develop an impedance diagram as well as a V-I vector diagram, Figure 18–3. Since the resistive voltage drop (I × R) vector must indicate the resistor voltage is in-phase with the circuit current, the impedance diagram line representing R is shown at the 0° position. Because the capacitor voltage ($I \times X_C$) lags the capacitor (and circuit) current by 90°, the X_C on the impedance diagram is shown downward at the −90° position.

Summarizing for *series* RC circuits, I and V_R are shown at 0° and V_C is shown at −90° in V-I vector diagrams. The R is shown at 0° and X_C at −90° for impedance diagrams. The angle between V_T and I (in V-I vector diagrams) and between Z and R (in impedance diagrams) is the value of the circuit phase angle. With this information, let's proceed with some analysis techniques.

Solving Magnitude Values by the Pythagorean Theorem

As in the chapter on RL circuits, we can use the Pythagorean theorem to solve for magnitudes of voltages, currents, and circuit impedance in series RC circuits.

Recall the Pythagorean theorem deals with right triangles and the relationship of triangle sides. The hypotenuse (c) relates to the other sides by the formula $c^2 = a^2 + b^2$, or $c = \sqrt{a^2 + b^2}$. Observe Figure 18–4 as you study the following statements to see how this theorem is applied in phasor addition for V-I vector and impedance diagrams for a series RC circuit.

Solving For V_T in the V-I Vector Diagram

1. The hypotenuse of the right triangle in the V-I vector diagram is V_T, or the vector resultant of combining V_R and V_C.
2. The side adjacent to the angle of interest (the phase angle) is V_R. The side opposite the angle is V_C.
3. Substituting into the Pythagorean formula, the formula to solve for the magnitude of the hypotenuse, or V_T, is:

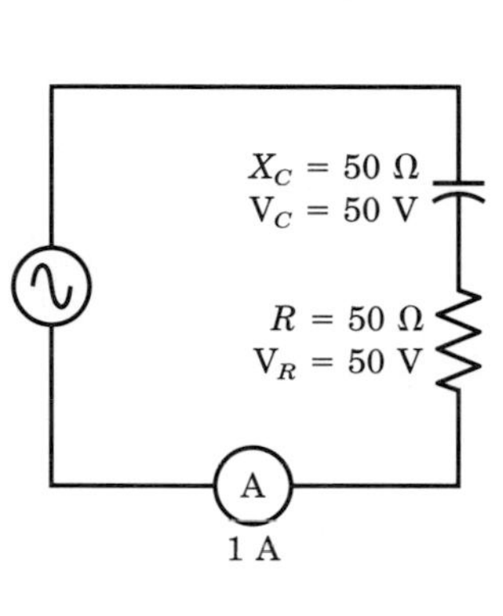

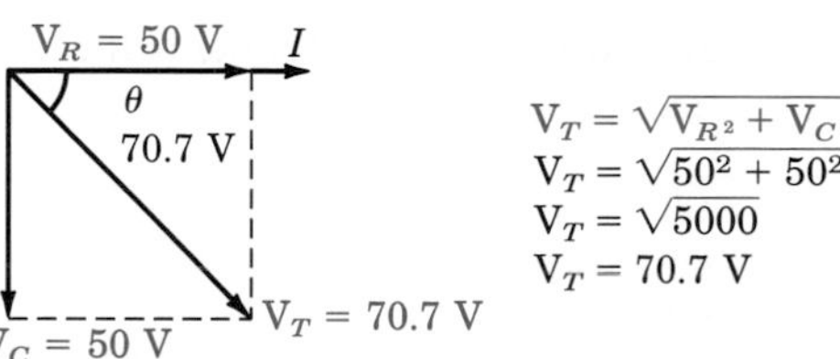

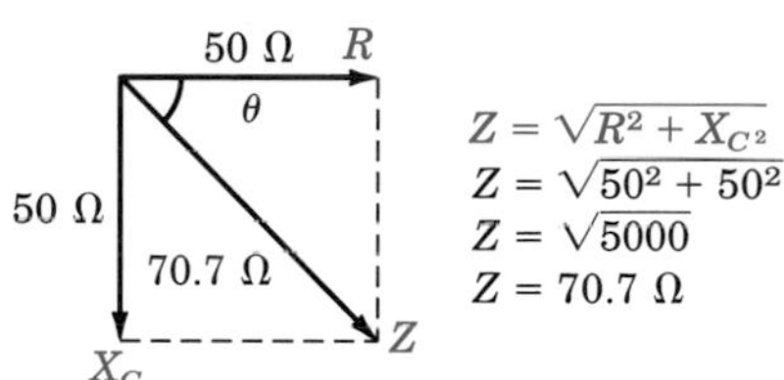

FIGURE 18–4
Using the Pythagorean theorem approach

Formula 18–1	$V_T = \sqrt{V_R{}^2 + V_C{}^2}$

EXAMPLE

Thus, in Figure 18–4, $V_T = \sqrt{50^2 + 50^2} = \sqrt{2500 + 2500} = \sqrt{5000} = 70.7$ volts.

Solving for Z in the Impedance Diagram

1. The hypotenuse of the right triangle in the impedance diagram is Z (impedance).
2. The side adjacent to the angle representing the phase angle is R. The side opposite the angle represents X_C.
3. Substituting into the Pythagorean formula, the formula to solve for the magnitude of the hypotenuse or Z, is:

Formula 18–2	$Z = \sqrt{R^2 + X_C{}^2}$

EXAMPLE

Thus, in Figure 18–4, $Z = \sqrt{50^2 + 50^2} = \sqrt{2500 + 2500} = \sqrt{5000} = 70.7\ \Omega$.

Try applying the Pythagorean formula to the following problems.

PRACTICE PROBLEM I

Answers are in Appendix B.

1. For the circuit in Figure 18–5, use the Pythagorean approach and solve for V_T. Draw the V-I vector diagram and label all elements. (NOTE: You may want to try the

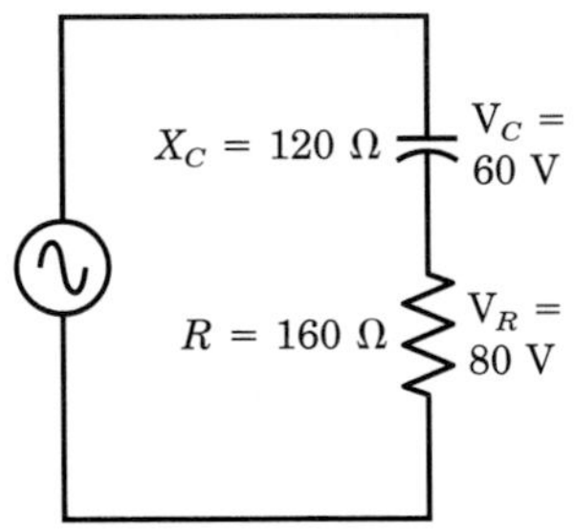

FIGURE 18–5 Pythagorean approach series RC circuit practice problem

drawing-to-scale approach learned in an earlier chapter to check your results.)

2. For the circuit in Figure 18–5, use the Pythagorean approach and solve for Z. Draw and label an appropriate impedance diagram.

PRACTICAL NOTES

Reminder: It is a good habit to sketch diagrams for any problem you are solving even if you don't draw them precisely. By sketching, you can approximate to see if the answers you get are logical.

Solving for Phase Angle Using Trigonometry

Recall from the RL circuits chapter that we can use simple trig functions to solve for the angles of a right triangle. Remember that $\sin\theta = \frac{\text{opp}}{\text{hyp}}$, $\cos\theta = \frac{\text{adj}}{\text{hyp}}$, and $\tan\theta = \frac{\text{opp}}{\text{adj}}$. Since we have already identified which sides are the side adjacent, side opposite, and the hypotenuse in the series RC circuit V-I vector and impedance diagrams, substitute into the trig formulas to solve for angle information.

Note the trig formula that does not require knowing the magnitude of the hypotenuse to solve for the angle information is the tangent function. Therefore, this function is often used.

Solving the Angle from the V-I Vector Diagram

EXAMPLE

Refer again to Figure 18–4. Since we know all three sides, we can use any basic trig formula. However, let's use the tangent function.

1. To find the phase angle, or the angle between V_T and I, the V_C and V_R values can be used.
2. Substitute these values into the tangent formula. Thus,

$$\text{Tan}\ \theta = \frac{\text{opp}}{\text{adj}}$$

$$\text{Tan}\ \theta = \frac{V_C}{V_R} = \frac{50}{50} = 1.$$

3. Using a trig table or a calculator, we find the angle whose tangent equals 1 is 45°.
4. Therefore, the phase angle is 45°. This makes sense since both legs of the triangle are equal in length. If you draw this diagram to scale and use a protractor to measure the angle, the result should verify a 45° angle.

Solving the Angle from the Impedance Diagram

EXAMPLE

Refer again to Figure 18–4 and use the tangent function.

1. The angle between Z and R is the same value as the phase angle.
2. Using the tangent function to find this angle requires the X_C and R values.
3. Substitute into the tangent formula. Thus,

$$\text{Tan } \theta = \frac{\text{opp}}{\text{adj}}$$

$$\text{Tan } \theta = \frac{X_C}{R} = \frac{50}{50} = 1.$$

4. The angle whose tangent equals 1 is 45°. Therefore, the phase angle is 45°.

Apply these principles to solve the following problems.

PRACTICE PROBLEM II

1. Refer to Figure 18–6. Draw a V-I vector diagram for the circuit shown, and find the phase angle using a basic trig function. (Round-off to the nearest volt.)
2. Draw an impedance diagram for the circuit in Figure 18–6. Then solve for the value of θ, using an appropriate trig function.

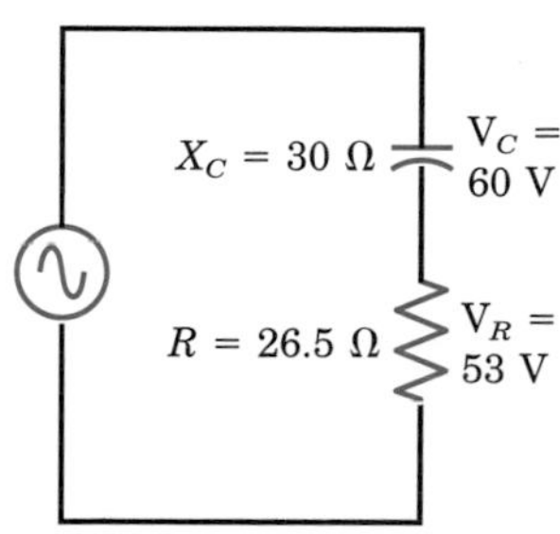

FIGURE 18–6
Solving for θ in a series RC circuit

Summary of Series RC Circuits

Figure 18–7 shows examples of series RC circuit parameters for various circuit conditions. These examples help reinforce the following generalizations about series RC circuits.

1. Ohm's law formulas can be used at any time to solve for individual component or total circuit voltages, currents, resistances, reactances,

Line Numbers	Frequency (Hz)	Capacitance (μF)	X_C (Ohms) $\left(\frac{0.159}{fC}\right)$	R (Ohms)	Z (Ohms)	I (Amperes) $\left(\frac{VT}{Z}\right)$	V_C ($I \times X_C$)	V_R ($I \times R$)	VT (Volts) ($I \times Z$)	Phase Angle (Degrees)
1	100	2	795	7950	7989.65	.025A	19.90	199.01	200	−5.71
2	100	1	1590	7950	8107.44	.0246A	39.22	196.12	200	−11.31
3	100	1	1590	3975	4281.21	.0467A	74.28	185.70	200	−21.80
4	10	1	15900	3975	16389.34	.0122A	194.03	48.51	200	−75.96

NOTICE THE FOLLOWING FACTS ABOUT THE CHANGES:

Compare results between lines 1 and 2. The only change was decreasing value of C to half. Notice X_C doubled; Z increased; I decreased; V_C increased; V_R decreased; and θ increased.

Compare results between lines 2 and 3. The change was decreasing R to half. Notice Z decreased; I increased; V_C increased; V_R decreased; and θ increased.

Compare results between lines 3 and 4. The change was decreasing f to one-tenth f. Notice X_C increased ten times; Z increased; I decreased; V_C increased; V_R decreased; and θ increased greatly.

FIGURE 18–7 Sample series RC circuit parameters chart

and impedances, as long as the appropriate parameter values are substituted into the formulas. That is, $V_C = I_C \times X_C$, $Z = \frac{V_T}{I_T}$, and so forth.

2. The circuit voltage lags the circuit current by some angle between 0° and −90°. Conversely, the circuit current leads the circuit voltage by that same angular difference.
3. Since current is common throughout series circuits, current is the reference vector in phasor diagrams of V and I.
4. The larger the R value is with respect to the X_C value, the more resistive the circuit is. That is, the phase angle is smaller or closer to 0°. Conversely, the larger the X_C value is with respect to the circuit R value, the greater is the phase angle and the closer it is to −90°.
5. Impedance of a series RC circuit is greater than either the R value alone or the X_C value alone but *not* as great as their arithmetic sum. Impedance equals the vector sum of R and X_C.
6. When either R or X_C increases in value, circuit impedance increases and circuit current decreases.
7. Total voltage of a series RC circuit is greater than either the resistive voltage drops or the reactive voltages alone but *not* as great as their arithmetic sum. The voltages must be added vectorially to find the resulting V_T value.
8. For series RC circuits, both V-I vector and impedance diagrams are used. The quantities used in the V-I vector diagram are the circuit current as a reference vector at 0°; V_R (also at the 0° reference posi-

tion); and V_C at $-90°$. V_T (or V_A) is plotted as the vector resulting from the resistive and reactive voltage vectors. For the impedance diagram, R is the reference vector and X_C is plotted downward at the $-90°$ position. Circuit impedance (Z) is the vector sum resulting from R and X_C.

IN-PROCESS LEARNING CHECK I

Fill in the blanks as appropriate.

1. In series RC circuits, the __________ is the reference vector, shown at 0°.
2. In the V-I vector diagram of a series RC circuit, the circuit voltage is shown __________ the circuit current.
3. In series RC circuits, an impedance diagram __________ be drawn.
4. Using the Pythagorean approach, the formula to find V_T in a series RC circuit is __________.
5. Using the Pythagorean approach, the formula to find Z in a series RC circuit is __________.
6. The trig function used to solve for phase angle when you do not know the hypotenuse value is the __________ function.

PARALLEL RC CIRCUIT ANALYSIS

In parallel RC circuits, the voltage is common to all parallel branches. Therefore, *voltage* is the reference vector for the V-I vector diagrams, Figure 18–8a. Again, you *will not* draw impedance diagrams for *parallel* circuits.

In drawing a V-I vector diagram, branch currents are plotted relative to the voltage reference vector. I_R is plotted at the zero degree position and is the side adjacent to the angle of interest. I_C is plotted at the $+90°$ position because it leads the voltage by 90° and is the side opposite the angle for computational purposes. Again, refer to Figure 18–8a.

Solving for I_T Using the Pythagorean Approach

For parallel RC circuits, the Pythagorean formula to solve for I_T becomes:

Formula 18–3	$I_T = \sqrt{I_R^2 + I_C^2}$

EXAMPLE

Use the quantities in Figure 18–8b, and substitute into the above formula. Thus,

$$I_T = \sqrt{I_R^2 + I_C^2} = \sqrt{15^2 + 20^2} = \sqrt{625} = 25 \text{ A}$$

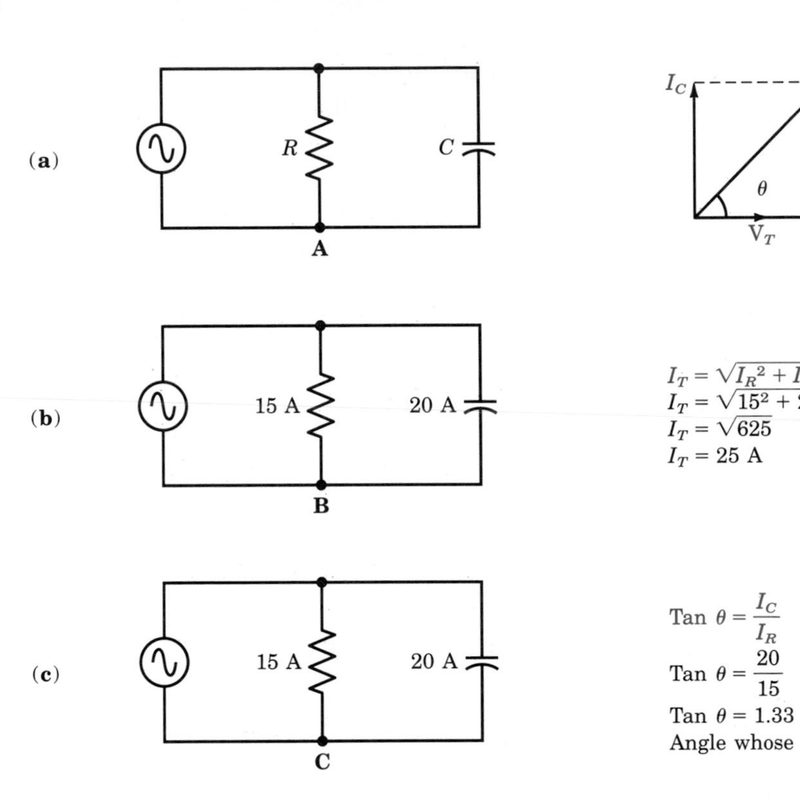

$I_T = \sqrt{I_R^2 + I_C^2}$
$I_T = \sqrt{15^2 + 20^2}$
$I_T = \sqrt{625}$
$I_T = 25$ A

$\text{Tan } \theta = \frac{I_C}{I_R}$

$\text{Tan } \theta = \frac{20}{15}$

Tan $\theta = 1.33$

Angle whose tan = 1.33 is 53.1°

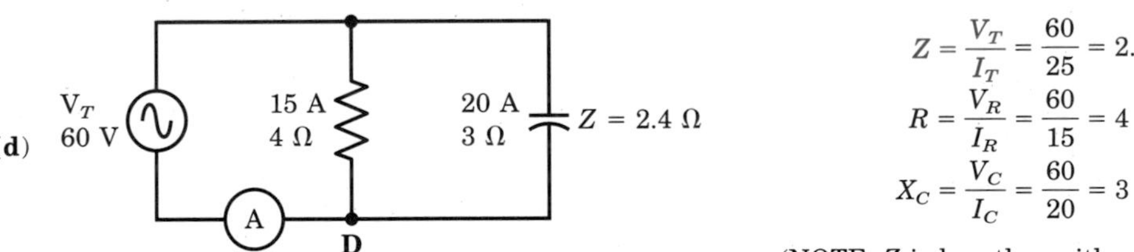

$$Z = \frac{V_T}{I_T} = \frac{60}{25} = 2.4\ \Omega$$

$$R = \frac{V_R}{I_R} = \frac{60}{15} = 4\ \Omega$$

$$X_C = \frac{V_C}{I_C} = \frac{60}{20} = 3\ \Omega$$

(NOTE: Z is less than either of the branch ohms.)

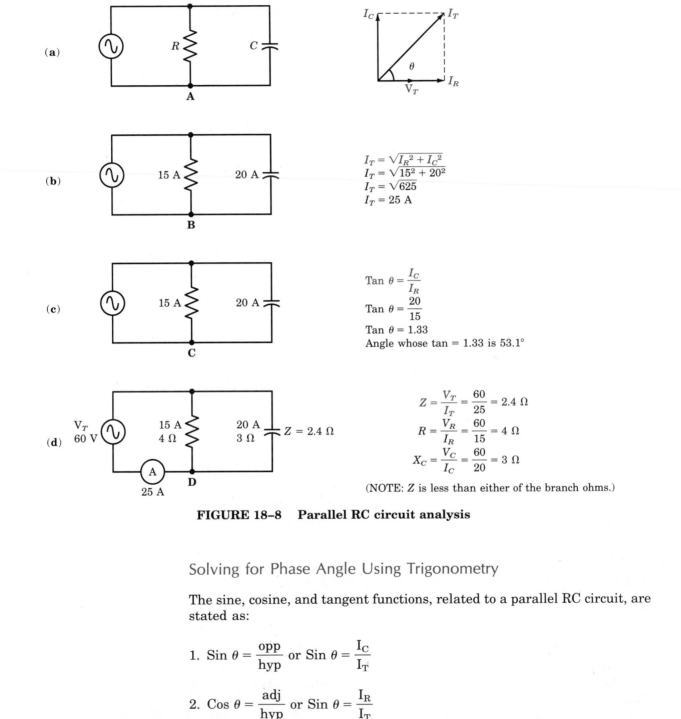

FIGURE 18–8 Parallel RC circuit analysis

Solving for Phase Angle Using Trigonometry

The sine, cosine, and tangent functions, related to a parallel RC circuit, are stated as:

1. $\text{Sin } \theta = \frac{\text{opp}}{\text{hyp}}$ or $\text{Sin } \theta = \frac{I_C}{I_T}$

2. $\text{Cos } \theta = \frac{\text{adj}}{\text{hyp}}$ or $\text{Sin } \theta = \frac{I_R}{I_T}$

3. $\text{Tan } \theta = \frac{\text{opp}}{\text{adj}}$ or $\text{Tan } \theta = \frac{I_C}{I_R}$

Again, we will use the tangent function.

EXAMPLE

Using the parameters shown in Figure 18–8c, the phase angle is computed as follows:

$$\text{Tan } \theta = \frac{\text{opp}}{\text{adj}}$$

$$\text{Tan } \theta = \frac{I_C}{I_R} = \frac{20}{15} = 1.33.$$

The angle whose tangent equals 1.33 is 53.1°; therefore, the phase angle is 53.1°.

Looking at the diagram, this makes sense because the current through the reactive branch is greater than the current through the resistive branch. The circuit is acting more reactive than resistive. In other words, the angle is greater than 45°.

Solving for Impedance in a Parallel RC Circuit

An impedance diagram for parallel circuits is *not* used because the circuit impedance is less than any one branch opposition. It is not the hypotenuse value or the vector sum of the branch oppositions as it is in the series configuration. Neither is it equal to the product over the sum or the results obtained from any of the other parallel resistance formulas used for purely resistive parallel circuits. NOTE: If the resistance and reactive branch oppositions of a two-branch parallel RC circuit are equal, the impedance is 70.7% of one branch's ohmic value, not half of one branch like it is for a purely resistive circuit with two equal-R branches.

To solve for impedance of a parallel RC circuit, we simply use Ohm's law, dividing the applied voltage by the vector resultant total current. For Figure 18–8d, $Z = \frac{V_T}{I_T} = \frac{60}{25} = 2.4\ \Omega$. Since the voltage and branch currents are known, we can also use Ohm's law to solve for the R and X_C values. Thus,

$$R = \frac{V_R}{I_R} = \frac{60}{15} = 4\ \Omega$$

$$X_C = \frac{V_C}{I_C} = \frac{60}{20} = 3\ \Omega$$

PRACTICE PROBLEM III

Refer to Figure 18–9 and solve the following problems.

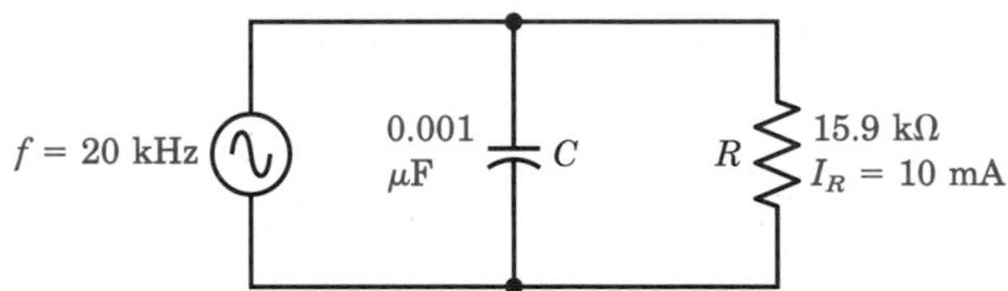

FIGURE 18–9
Parallel RC circuit

1. Use the capacitive reactance formula, Ohm's law, and the Pythagorean approach, as appropriate, to solve for I_T.
2. Draw the V-I vector diagram for the circuit and determine the phase angle, using appropriate trig formula(s). Label all the elements in the diagram.
3. Determine the circuit impedance.

IN-PROCESS LEARNING CHECK II

1. In parallel RC circuits, __________ is the reference vector in V-I vector diagram.
2. In parallel RC circuits, current through the resistor(s) is plotted __________ with the circuit current vector.
3. In parallel RC circuits, current through capacitor branch(es) is plotted at __________ with respect to circuit applied voltage.
4. The Pythagorean formula to find circuit current in parallel RC circuits is __________.
5. Is the Pythagorean formula used to find circuit voltage in a parallel RC circuit? __________
6. Can the tangent function be used to find phase angle in a parallel RC circuit? __________
7. Can an impedance diagram be used to find circuit impedance in a parallel RC circuit? __________

Summary of Parallel RC Circuits

Figure 18–10 provides several examples of parallel RC circuit parameters that illustrate the following facts. Study and understand these general principles before proceeding in the chapter.

Line Numbers	Frequency (Hz)	VT (Volts)	Capacitance (μF)	X_C (Ohms) $\left(\frac{0.159}{fC}\right)$	R (Ohms)	I_C (Amps) $\left(\frac{VT}{X_C}\right)$	I_R (Amps) $\left(\frac{VT}{R}\right)$	I_T (Amps) $I_T = I_R^2 + I_C^2$	Z (Ohms) $\left(\frac{VT}{I_T}\right)$	Phase Angle (Degrees)
1	100	100	2	795	5000	.1258	.0200	.1274	785.14	80.97
2	100	100	1	1590	5000	.0629	.0200	.0660	1515.23	72.36
3	100	100	1	1590	2500	.0629	.0400	.0745	1341.64	57.54
4	10	100	1	15900	2500	.0063	.0400	.0405	2469.66	8.94
5	10	100	1	15900	15900	.0063	.0063	.0089	11243.00	45.00
6	10	200	1	15900	15900	.0126	.0126	.0178	11243.00	45.00

NOTICE THE FOLLOWING FACTS ABOUT THE CHANGES:

Compare results between lines 1 and 2. The only change was to decrease C to half. Notice X_C doubled; I_C halved; I_R stayed the same; I_T decreased; and θ decreased.

Compare results between lines 2 and 3. The only change was to decrease R to half. Notice X_C stayed the same; I_C stayed the same; I_R doubled; I_T increased; and θ decreased.

Compare results between lines 3 and 4. The only change was to decrease f to one-tenth original value. Notice X_C increased 10 times; I_C decreased to one-tenth previous value; I_R stayed the same; I_T decreased; and θ decreased greatly.

Compare results between lines 4 and 5. The only change was to make $R = X_C$. Notice X_C stayed the same; I_C stayed the same; I_R decreased greatly; I_T decreased greatly; and $\theta = 45°$ (since I_C and I_R are equal).

Compare results between lines 5 and 6. The only change was to double V_T. Notice all currents doubled; but Z and the phase angle remained unchanged. Changing V applied does not change X_C, R, Z, or phase angle.

FIGURE 18–10 Sample parallel RC circuit parameters chart

1. Ohm's law formulas are used at any time to solve for individual component or overall circuit parameters, providing you substitute proper quantities into the formulas.
2. The applied voltage (V_T) is the reference vector in phasor diagrams of parallel RC circuits, since V is common to all branches.
3. The total circuit current leads the circuit voltage by some angle between 0° and +90°.
4. The *smaller* the R value with respect to the X_C value, the more resistive the circuit is. That is, the closer to 0° the phase angle is. Conversely, the larger the R value is with respect to the X_C value, the more capacitive the circuit is.
5. Impedance in a parallel RC circuit is less than the ohmic opposition of any one branch and must be found through Ohm's law techniques. Impedance for a parallel RC circuit can *not* be found by drawing an impedance diagram.
6. When either R or X_C increases, impedance increases and circuit current decreases.
7. Total current in a parallel RC circuit is greater than any one branch current. However, total current is *not* as great as the sum of the

branch currents. The branch currents must be added vectorially to find the resulting I_T value.

8. For a V-I vector diagram in a parallel RC circuit, voltage applied is the reference vector; I_R is plotted at 0° and I_C is plotted at +90°. I_T is plotted as the vector resulting from the resistive and reactive branch currents.

SIMILARITIES AND DIFFERENCES BETWEEN RC AND RL CIRCUITS

So that you can see and remember the key points about RL and RC circuits, Figure 18–11 provides an important list of similarities and differences. If you understand the facts listed, you will find the following chapters about circuits containing all three quantities of resistance, inductance, and capacitance easy to understand.

Item Description	**Series RC Circuit**	**Parallel RC Circuit**	**Series RL Circuit**	**Parallel RL Circuit**
	$V_T = \sqrt{V_R^2 + V_C^2}$	$V_T = I_T \times Z$	$V_T = \sqrt{V_R^2 + V_L^2}$	$V_T = I_T \times Z$
	$I_T = I_R = I_C = \frac{V_T}{Z}$	$I_T = \sqrt{I_R^2 + I_C^2}$	$I_T = I_R = I_L = \frac{V_T}{Z}$	$I_T = \sqrt{I_R^2 + I_L^2}$
Formulas	$Z = \sqrt{R^2 + X_C^2}$	$Z = \frac{V_T}{I_T}$	$Z = \sqrt{R^2 + X_L^2}$	$Z = \frac{V_T}{I_T}$
	$\text{Tan } \theta = \frac{X_C}{R}$ or $\frac{V_C}{V_R}$	$\text{Tan } \theta = \frac{I_C}{I_R}$	$\text{Tan } \theta = \frac{X_L}{R}$ or $\frac{V_L}{V_R}$	$\text{Tan } \theta = \frac{I_L}{I_R}$
	$X_C = \frac{.159}{fC}$ or $\frac{V_C}{I_C}$	$X_C = \frac{.159}{fC}$ or $\frac{V_C}{I_C}$	$X_L = 2\pi fL$ or $\frac{V_L}{I_L}$	$X_L = 2\pi fL$ or $\frac{V_L}{I_L}$
As f↑	X_C↓ Z↓ I↑ θ↓	X_C↓ Z↓ I↑ θ↑	X_L↑ Z↑ I↓ θ↑	X_L↑ Z↑ I↓ θ↓
As C↑	X_C↓ Z↓ I↑ θ↓	X_C↓ Z↓ I↑ θ↑	N/A	N/A
As L↑	N/A	N/A	X_L↑ Z↑ I↓ θ↑	X_L↑ Z↑ I↓ θ↓
As R↑	Z↑ I↓ θ↓	Z↑ I↓ θ↑	Z↑ I↓ θ↓	Z↑ I↓ θ↑

FIGURE 18–11 Similarities and differences between RL and RC circuits

WAVESHAPING AND NON-SINUSOIDAL WAVEFORMS

Many electronic circuits depend on the ability to change non-sinusoidal circuit waveforms (such as square or rectangular waveforms) into different shapes for different purposes. For example, in radar systems it is not uncommon to need sharp or pointed waveforms to act as *trigger pulses* for

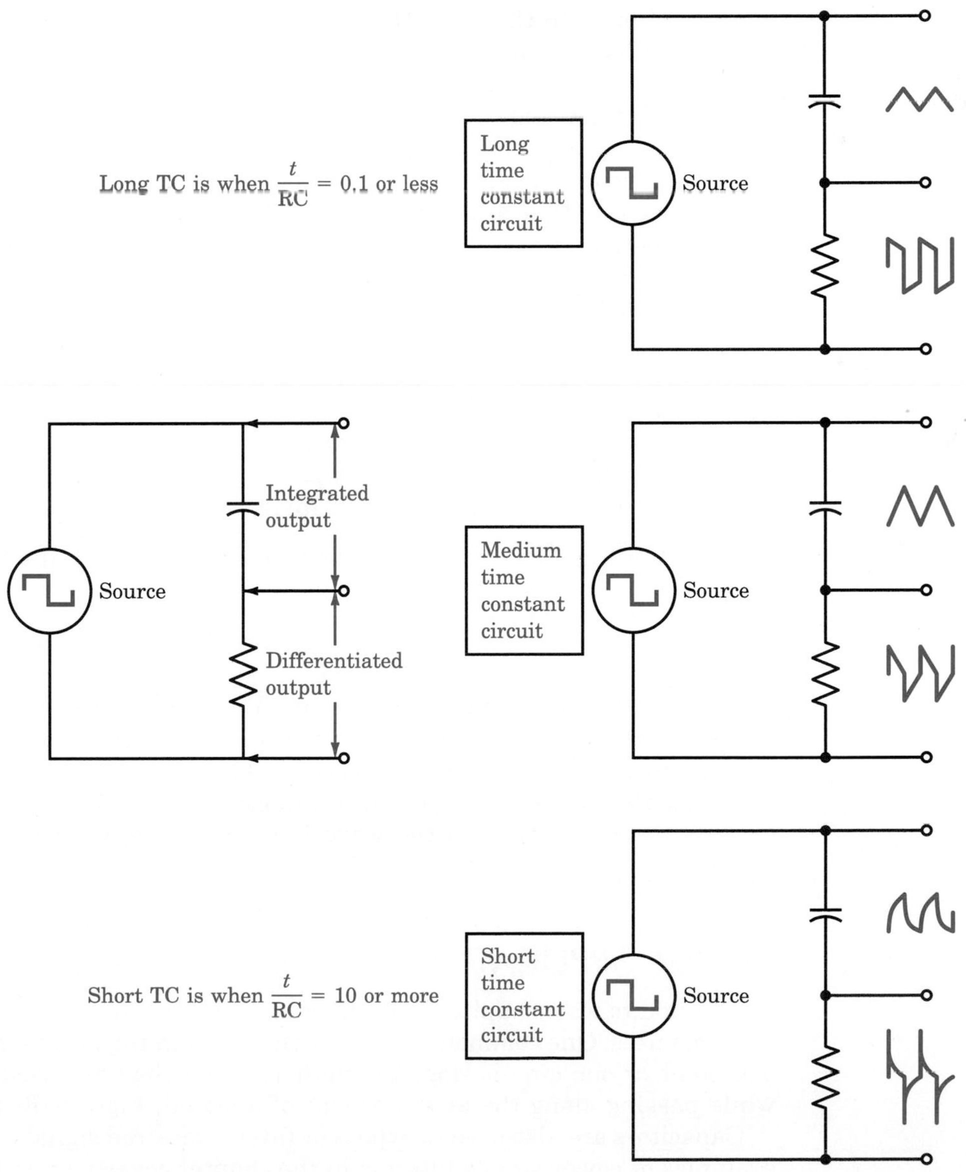

FIGURE 18–12 Examples of RC circuit waveshaping

circuits where timing action is performed. It is beyond the scope of this chapter to thoroughly discuss *waveshaping*. However, you should be familiar with the concepts.

In Figure 18–12, a square wave input fed to a series RC network is shaped to be somewhat sharp or peaked across the resistor. This output is called the *differentiated output*. By definition, a differentiating circuit is one whose output relates to the rate of change of voltage or current of the input.

On the other hand, across the capacitor an output called an *integrated output* is taken. In this case, the waveform is rounded-off or flattened-out. This seems logical since a capacitor opposes a voltage change. Therefore, the voltage can't make sudden or "sharp" changes across the capacitor. By definition, an integrating circuit output relates to the integral of the input signal.

The amount of change of the input signal waveshape at the output relates to the amount of time allowed for the capacitor to charge or discharge during one alternation of the input signal. If the capacitor has more than ample time (over five time constants) to make its voltage change, the circuit is a **short time-constant circuit.** By definition, a short time-constant circuit is one where the time allowed (t) is 10 or more times one RC time constant. That is, $\frac{t}{RC} = 10$ or more. RC time is short compared with t.

Conversely, if the RC time constant is such that the time allowed by the input signal alternation is short compared to the time needed for complete change across the RC network, it is a **long time-constant circuit.** That is, $\frac{t}{RC} = 0.1$ or less. RC time constant is long compared with the time allowed to charge or discharge.

Refer again to Figure 18–12. Notice both the differentiated and integrated waveforms present across the resistor and capacitor for long, medium, and short time constant conditions. You will see practical applications of such circuits as you study a variety of waveshaping circuit applications.

In Figures 18–13 and 18–14, you can again see how the RC time of resistive-capacitive networks is useful in integrating or differentiating signals, causing desired change in the waveshape of square or rectangular wave signals.

OTHER APPLICATIONS OF RC CIRCUITS

In addition to waveshaping and timing applications, RC circuits have other valuable uses. One common function is that of coupling desired signals from one point or one circuit stage to another. Often they are used to block dc while passing along the ac component of a signal, Figure 18–15.

Capacitors are also used to bypass or filter undesired signals. You will see examples of power supply filtering in the chapter covering rectifier circuits.

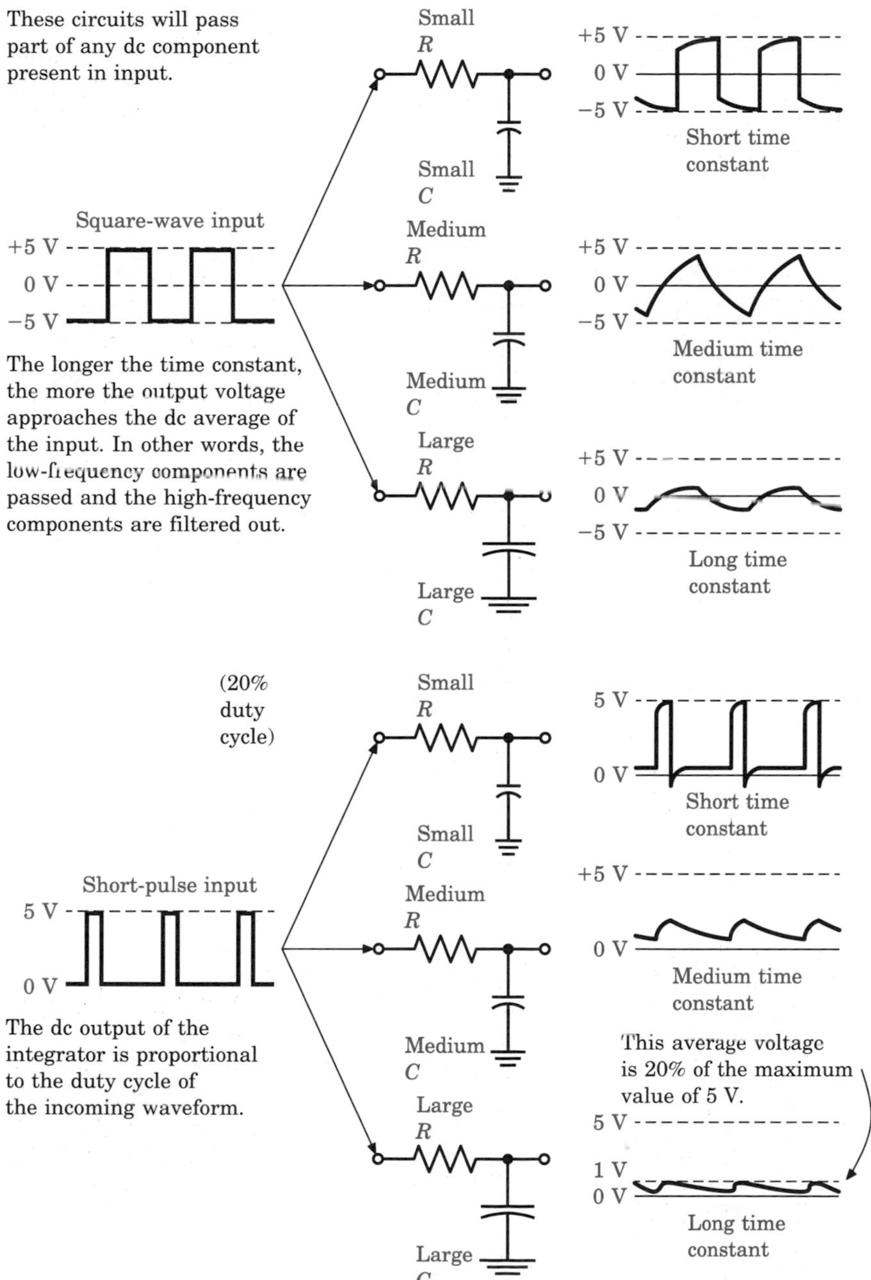

FIGURE 18–13 Waveshaping—RC integrator circuits

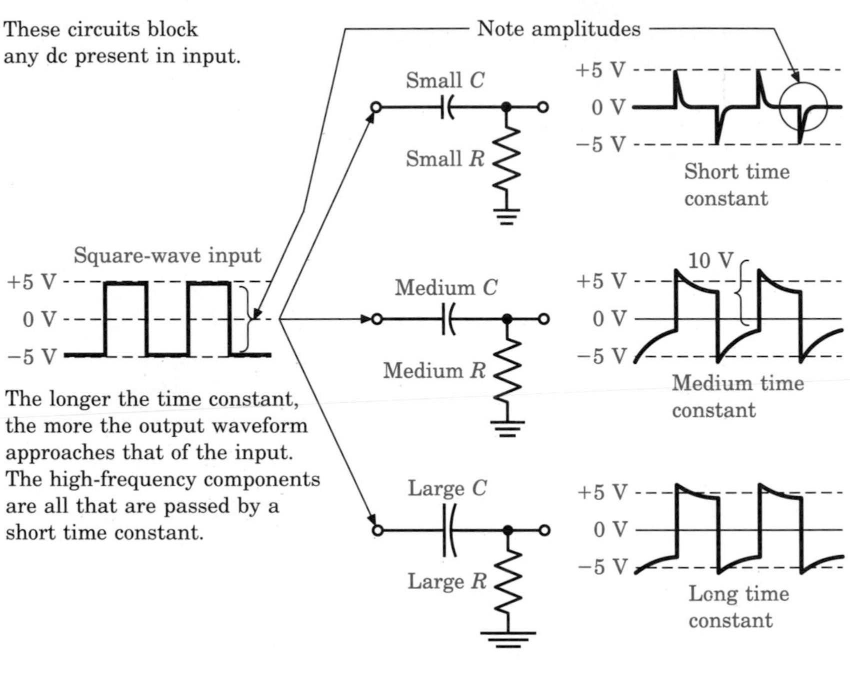
These circuits block any dc present in input.
Note amplitudes
Small C
Small R
+5 V
0 V
−5 V
Short time constant
Square-wave input
+5 V
0 V
−5 V
Medium C
Medium R
10 V
Medium time constant
The longer the time constant, the more the output waveform approaches that of the input. The high-frequency components are all that are passed by a short time constant.
Large C
Large R
Long time constant

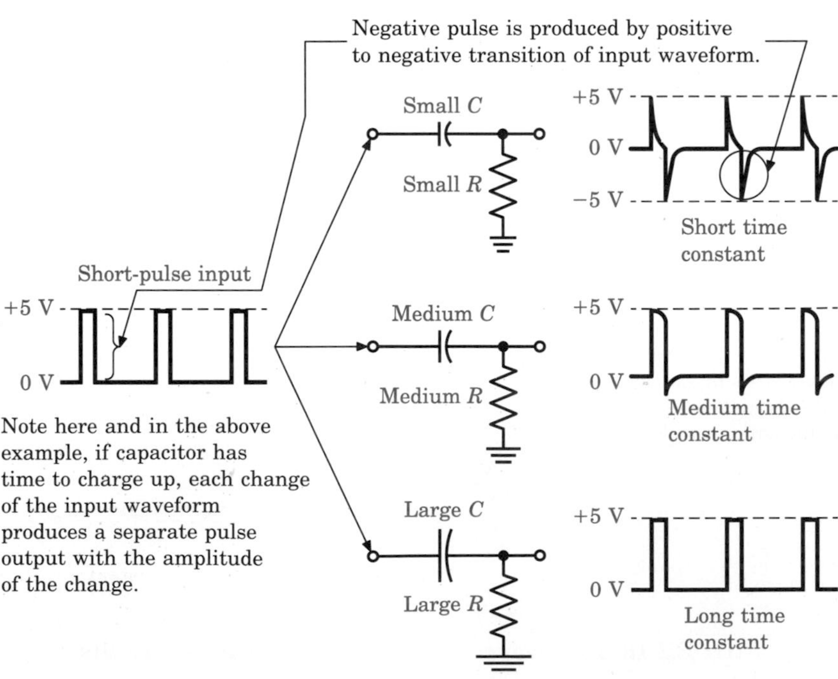
Negative pulse is produced by positive to negative transition of input waveform.
Small C
Small R
+5 V
0 V
−5 V
Short time constant
Short-pulse input
+5 V
0 V
Medium C
Medium R
Medium time constant
Note here and in the above example, if capacitor has time to charge up, each change of the input waveform produces a separate pulse output with the amplitude of the change.
Large C
Large R
Long time constant

FIGURE 18–14 Waveshaping—RC differentiator circuits

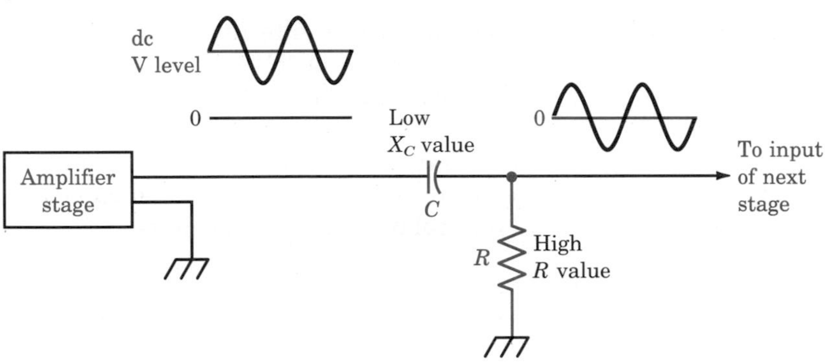

FIGURE 18–15 Example of an RC coupling network

Also, capacitor networks are used for ac and dc voltage dividers, Figure 18–16. Recall, voltage divides inverse to the capacitance values when capacitors are connected in series. That is, the smallest C has the highest V and the largest C has the smallest V.

By now, it is obvious that capacitors are found in almost every electronic circuit and system you will study or work on as a technician. The importance of learning about them and their operation cannot be overemphasized.

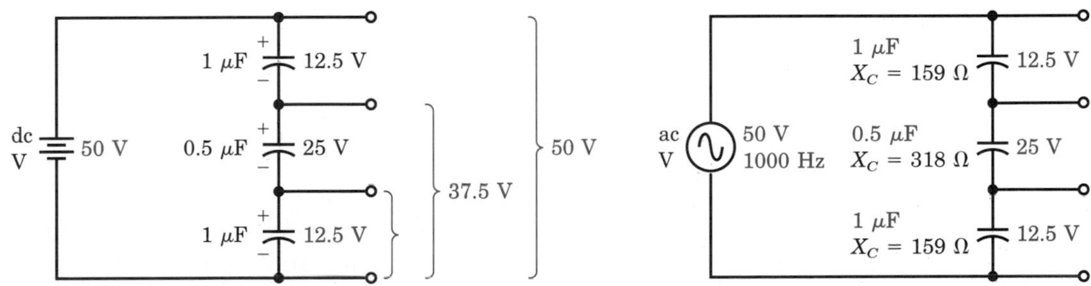

FIGURE 18–16 Capacitor voltage dividers

TROUBLESHOOTING HINTS AND CONSIDERATIONS FOR RC CIRCUITS

Based on information supplied earlier regarding typical troubles with capacitors, one can surmise some of the problems that can occur for the various applications where capacitors are used. Some examples follow.

When used as a filtering device, for example, in power supplies or in decoupling circuits, either an excessive leakage condition or excessive internal resistance condition will prevent proper filtering action. As you will learn, in such systems as amplifiers, radio or TV receivers a range from excessive hum to high-pitched squeals can be heard when such problems exist.

The only cure for this situation is to isolate the bad capacitor and *replace it!* Sometimes, bridging a degenerated capacitor with a good capacitor displays sufficient difference to help isolate the bad capacitor.

SAFETY HINTS

1. To avoid shock, don't make the connections with the circuit on!
2. If a capacitor is used in a *timing circuit,* one symptom of a possible bad C (or R) is an improper time period.
3. If the capacitor is used in a waveshaping function, obvious clues are noted if an oscilloscope is used to analyze the normality or abnormality of the waveform(s).
4. If a capacitor is used in a *coupling circuit* where it is supposed to block dc and pass ac, the presence of measurable dc across the network resistor in series with the coupling capacitor shows the capacitor is probably leaking dc.

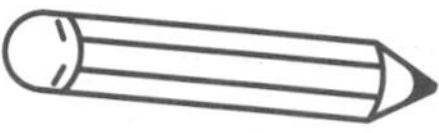

PRACTICAL NOTES

The primary tools used to help find defective capacitors in these various functions are the oscilloscope, digital multimeters, and capacitor testers.

Another useful device is a component *substitution box* that enables selection of an appropriate value component. This component is temporarily substituted in the suspected area of trouble to see if the problem disappears with the substitute part.

As you study power supplies, amplifiers, oscillators, pulse and digital circuits, you will gain more familiarity with common symptoms and troubleshooting techniques associated with RC circuits. Perhaps these few comments will whet your appetite to learn more.

SUMMARY

- Circuit current and circuit voltage are in-phase in purely resistive ac circuits. They are 90° out-of-phase in purely capacitive circuits. They are somewhere between 0° and 90° out-of-phase with each other in circuits containing both resistance and capacitance.

- To calculate total circuit voltage in a series RC circuit, vectorially add the resistive and capacitive voltages, as opposed to adding them with simple arithmetic.

- Since current is common throughout any series circuit, the reference vector for phasor diagrams of series RC circuits is the circuit current. The various circuit voltages (V_C and V_R) are then plotted with respect to the reference vector. V_C is shown lagging the current by 90°, and V_R is in-phase with the current.

- The reference vector for phasor diagrams of parallel RC circuits is the circuit voltage. The various parallel branch currents are plotted with respect to the reference vector. I_C is shown leading the voltage by 90°, and I_R is in-phase with the circuit voltage.

- Impedance is the opposition shown to ac current by the combination of resistance and reactance.

- Impedance diagrams can only be used for *series* circuits. They should not be used for analyzing parallel RC (or RL) circuits.

- The Pythagorean theorem approach is used to solve for the magnitudes of various phasors. In series RC circuits, use the formula to calculate Z and V_T. For parallel RC circuits, use the formula to calculate I_T.

- Trig functions are used to solve for both phase angles and magnitudes of various sides. One commonly used trig function is the tangent function, used for both series and parallel resistive and reactive circuits.

- In series RC circuits, the larger the C is the smaller the X_C present. And the less the capacitance controls the circuit parameters. In series circuits, the smaller the X_C is with respect to the R, the less the $I \times X_C$ drop is compared to the $I \times R$ drop. Hence, the less capacitive the circuit acts, and the smaller the phase angle will be

- In parallel RC circuits, the larger the C is the smaller the X_C present. And the more the capacitance controls the circuit parameters. In parallel circuits, the smaller the X_C is with respect to the R, the higher is the current through the capacitive branch. Hence, the more capacitive the circuit acts, and the larger the phase angle will be.

- Any frequency change changes the X_C. Therefore, frequency change affects circuit current, component voltages, impedance, and phase angle for series RC circuits. Frequency changes also affect branch currents, impedance, and phase angle in parallel RC circuits.

- Capacitors and RC circuits have various applications, including filtering, coupling, voltage-dividing, and waveshaping. These uses take advantage of the capacitor's sensitivity to frequency, ability to block dc while passing ac and to store charge.

- Because electrolytic capacitors age faster than other components, such as resistors or inductors, periodically replace them to assure optimum operation of the equipment where they are used.

REVIEW QUESTIONS

Answer questions 1 through 5 in reference to Figure 18–17.

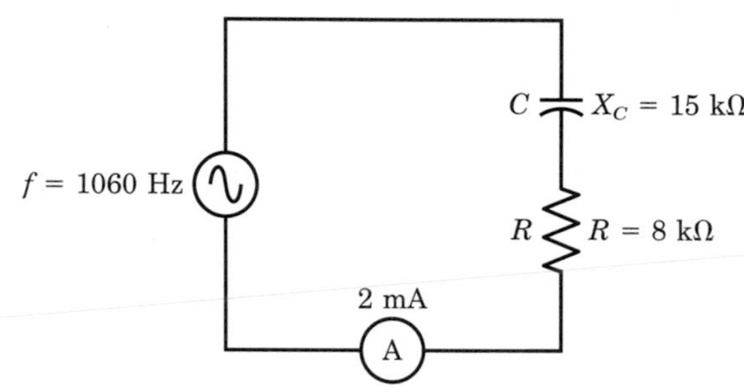

FIGURE 18–17

1. What is the circuit impedance?
2. What are the V_C, V_R, and V_T values?
3. What is the C value?
4. Is circuit voltage leading or lagging circuit current? By how much?
5. Answer the following with "I" for increase, "D" for decrease and "RTS" for remain the same.
   If the frequency of the circuit in Figure 18–17 doubles and component values remain the same:
   a. R will
   b. X_C will
   c. Z will
   d. I_T will
   e. θ will
   f. V_C will
   g. V_R will
   h. V_T will

Answer questions 6 through 10 in reference to Figure 18–18.

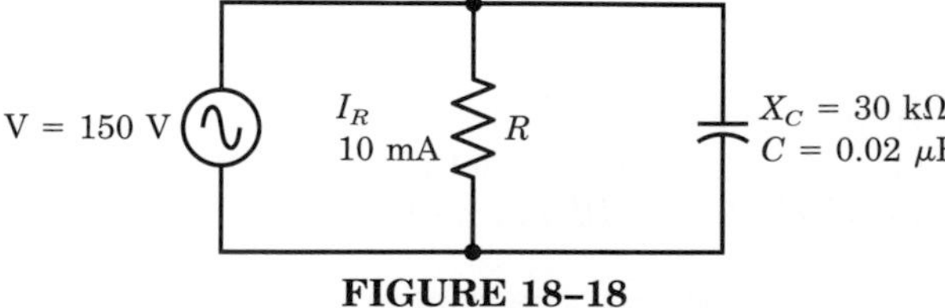

FIGURE 18–18

6. What are the I_C and I_T values?
7. What is the R value?
8. What is the Z value?
9. What is the frequency of operation?
10. Answer the following with "I" for increase, "D" for decrease, and "RTS" for remain the same.
    If the frequency of the circuit in Figure

18–18 doubles and component values remain the same:
a. R will
b. X_C will
c. Z will
d. I_T will
e. θ will
f. I_C will
g. I_R will
h. V_T will

Answer questions 11 through 15 in reference to Figure 18–19.

11. Find R_T

12. Find X_{CT}

13. Find Z

14. Find θ

15. If C_1 equals 0.1 μF, what is the frequency of the source voltage?

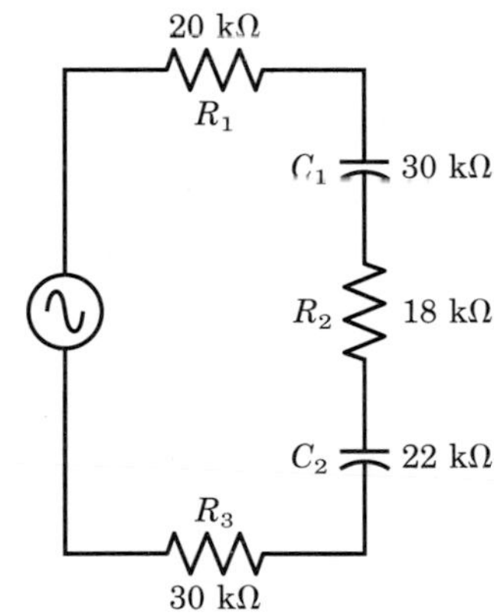

FIGURE 18–19

Answer questions 16 through 20 in reference to Figure 18–20.

16. Find I_R

17. Find I_C

18. Find I_T

19. Find θ

20. Find Z_T

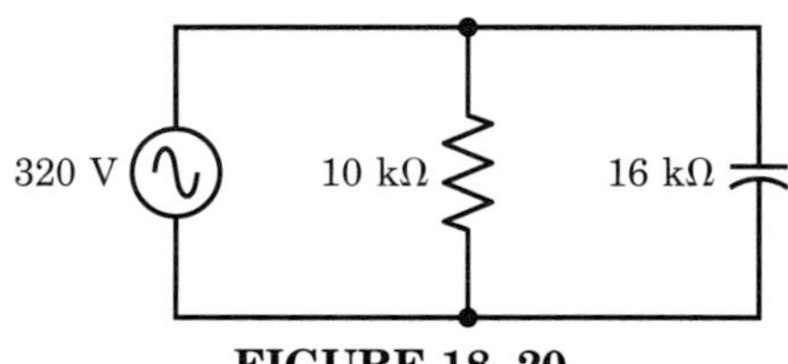

FIGURE 18–20

TEST YOURSELF

1. A two-component series RC circuit is operated at a frequency of 3975 Hz. The circuit is comprised of a 0.04-μF capacitor and an undisclosed value resistance. The circuit impedance is 1250 Ω. The applied

voltage is 25 V. Determine the following circuit parameters: R, X_C, I, V_R, V_C, and θ.

2. A two-branch parallel RC circuit has a total current of five milliamperes. The circuit's capacitive reactance is 2 kΩ. The applied voltage is 5 volts. Determine the following circuit parameters: R, I_C, I_R, Z, and θ.

PERFORMANCE PROJECT CORRELATION CHART

Suggested performance projects that correlate with topics in this chapter are:

CHAPTER TOPIC	PERFORMANCE PROJECT	PROJECT NUMBER
Series RC Circuit Analysis	V, I, R, Z, and θ Relationships in a Series RC Circuit	53
Parallel RC Circuit Analysis	V, I, R, Z, and θ Relationships in a Parallel RC Circuit	54

After completing the above projects, perform the Summary checkout at the end of the performance projects section.

RLC Circuit Analysis

Key Terms

Admittance (Y)

Apparent power (S)

Complex numbers

Imaginary numbers

Polar form notation

Power factor (p.f.)

Reactive power (Q)

Real numbers

Rectangular form notation

Susceptance (B)

True (or real) power (P)

Voltampere-reactive (VAR)

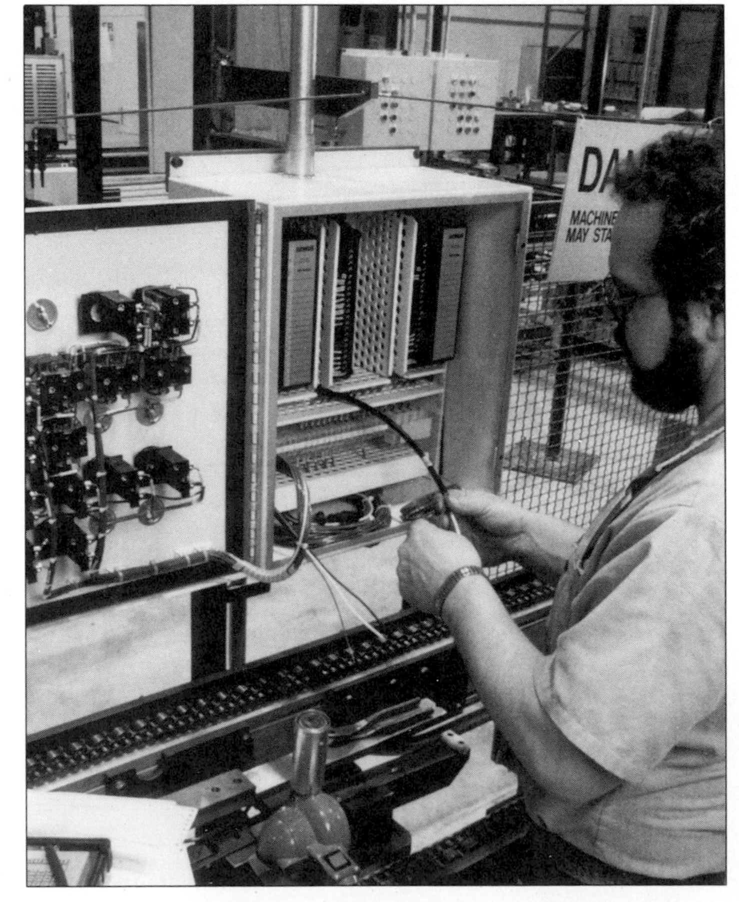

Courtesy of GE Fanuc Automation

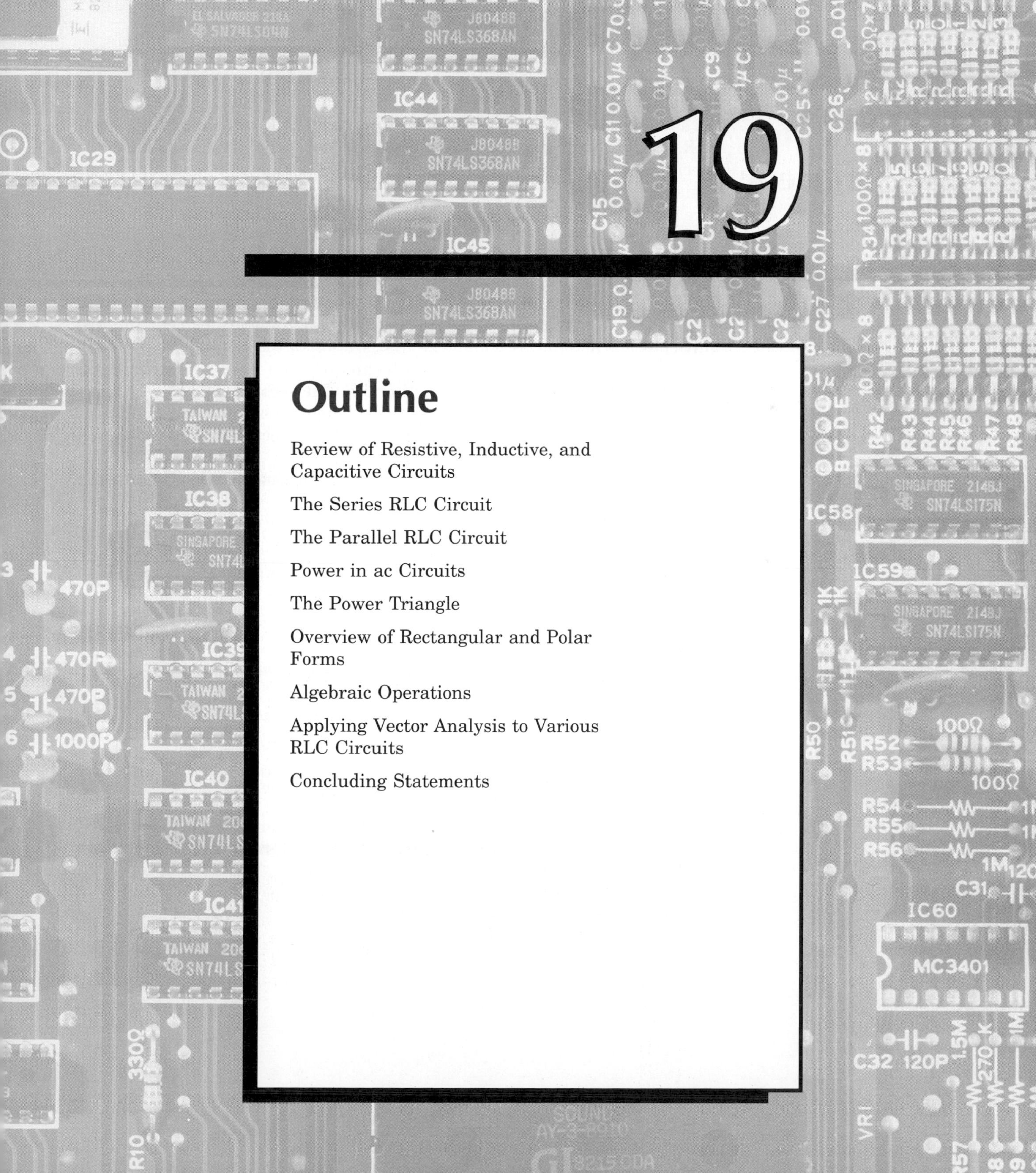

19

Outline

Review of Resistive, Inductive, and Capacitive Circuits

The Series RLC Circuit

The Parallel RLC Circuit

Power in ac Circuits

The Power Triangle

Overview of Rectangular and Polar Forms

Algebraic Operations

Applying Vector Analysis to Various RLC Circuits

Concluding Statements

Chapter Preview

As a technician, you will work with many circuits that contain combinations of R, L, and C. You have already investigated the property of inductance in ac circuits, in pure inductive circuits, and in circuits with inductance and resistance. Likewise, you have seen how capacitance behaves in pure capacitive circuits and in circuits with both capacitance and resistance. Furthermore, you have learned to use the Pythagorean theorem and simple trigonometric functions to analyze these circuits. In this chapter, you will analyze circuits with various combinations of R, L, and C.

The outline of this chapter accommodates either programs that teach vector algebra and its manipulation of complex numbers or only the approaches learned to this point. The first section of the chapter will present RLC circuits and the basic analysis techniques you have been using and provide a discussion of power in ac circuits. The second section introduces rectangular and polar form vector analysis, sometimes called vector algebra. The last portion provides an opportunity to apply vector algebra to various configurations of RLC circuits.

Objectives

After studying this chapter, you will be able to:

- Solve RLC circuit problems using Pythagorean approach and trig functions
- Define and illustrate ac circuit parameters using both **rectangular** and **polar form notation**
- Define **real numbers** and **imaginary numbers**
- Define real power, **apparent power, power factor,** and **voltampere-reactive**
- Calculate values of real power, apparent power, and power factor, and draw the power triangle
- Analyze RLC circuits and state results in rectangular and polar forms

BASIC RLC CIRCUIT ANALYSIS AND POWER IN AC CIRCUITS

Review of Resistive Circuits

In studying circuits containing only resistances, you learned resistance in ac is treated the same as resistance in dc. Key facts about series and parallel resistive circuits are in Figure 19–1. Take a few minutes to review these facts.

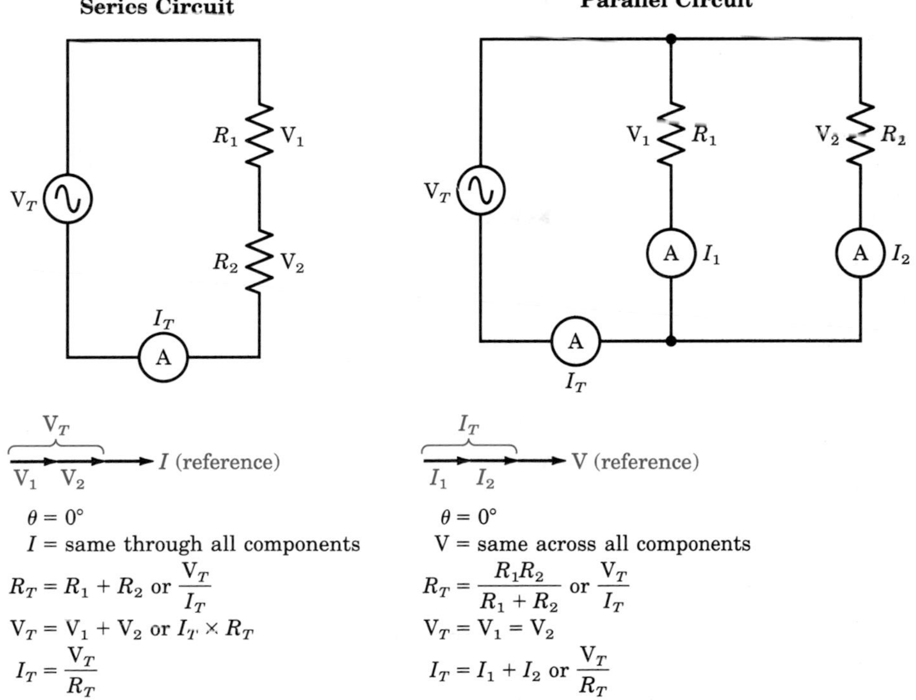

FIGURE 19–1 Pure resistive circuits

Review of Inductive Circuits

As you know, inductance opposes a change in current. This feature causes the voltage to lead current by 90° in purely inductive circuits and by some angle between 0° and 90° in inductive circuits with both L and R. Examine Figures 19–2 and 19–3 as you review inductive circuits. Remember in vector or phasor diagrams of circuit parameters V_L is plotted upward at 90° for *series* circuits with I as the reference vector. V_R is plotted at the 0° position. For the impedance diagram, X_L is plotted upward at 90°, and R is at the 0° position.

For *parallel* circuits, I_L is plotted downward at −90°. When resistance exists, I_R is plotted at the 0° position with circuit V, which is the reference vector.

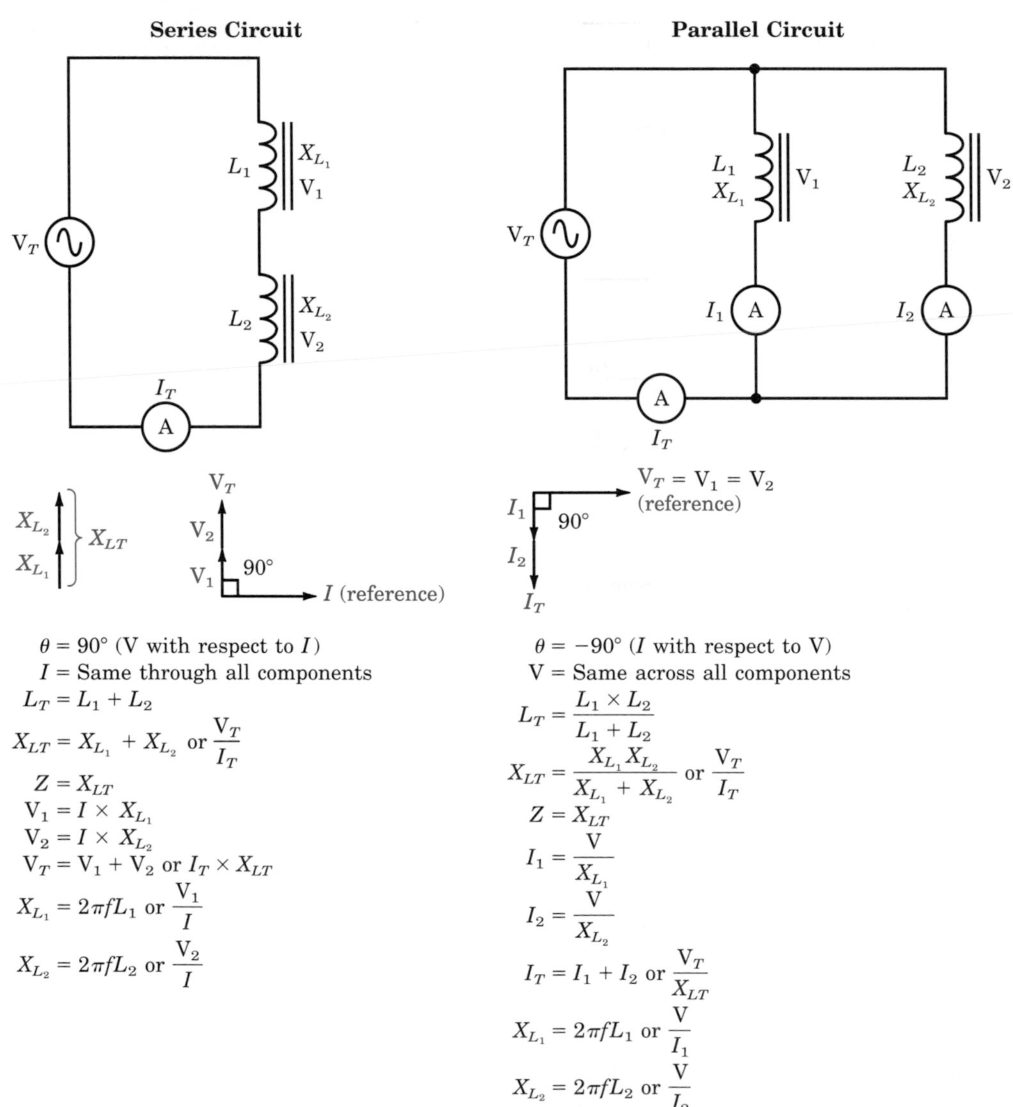

FIGURE 19–2 Pure inductive circuits

Review of Capacitive Circuits

Since it takes time and potential difference on a capacitor to build up a charge, capacitance opposes a voltage change. This feature causes current to lead voltage by 90° in pure capacitive circuits and by some angle less than 90° in circuits with both C and R.

Refer to Figures 19–4 and 19–5. Notice for series circuit V-I vector diagrams, V_C is plotted downward at −90°. For Z diagrams, X_C is plotted down-

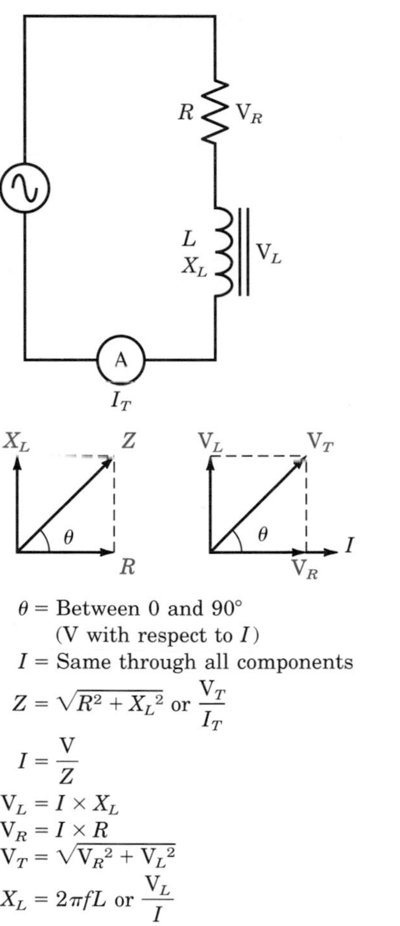

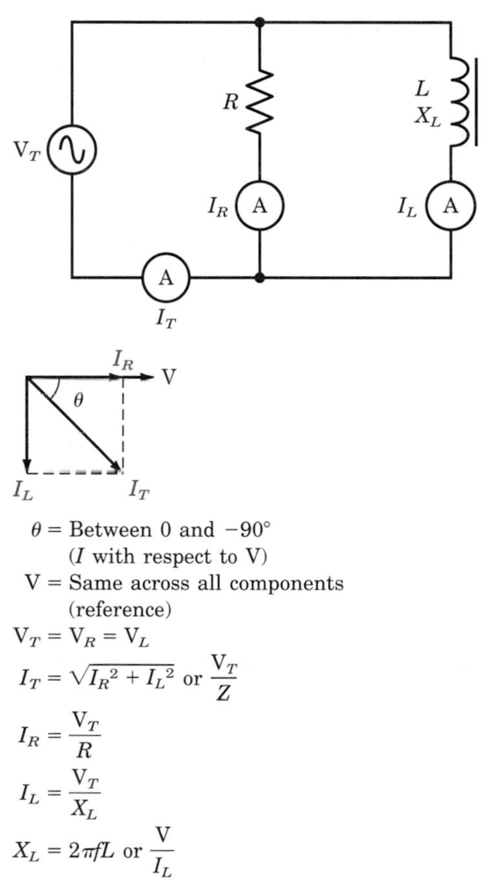

FIGURE 19–3
RL circuits

ward at −90°. When resistance exists, V_R is plotted at the 0° position with the circuit I, which is the reference vector. X_C is plotted downward and R is plotted at 0° for series circuit impedance diagrams.

For parallel circuit V-I diagrams, I_C is plotted upward at 90°, and I_R is plotted at the 0° position along with circuit V, which is the reference vector. Remember all of these facts as we move to the analysis of circuits containing all three components—R, L, and C.

THE SERIES RLC CIRCUIT

A Simple Series LC Circuit Without Resistance

Let's look at what happens when we combine inductive reactance (X_L) and capacitive reactance (X_C) in a simple series circuit. Refer to Figure 19–6 on page 616 as you study the following statements.

FIGURE 19–4
Pure capacitive circuits

Series Circuit

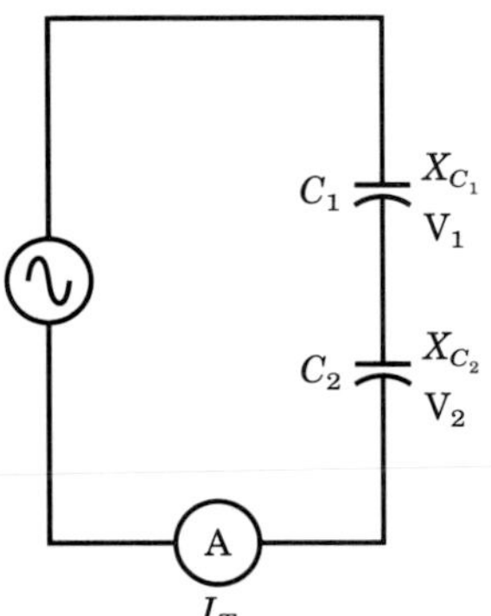

X_{C_1} X_{C_2} X_{CT}

V_{C_1} V_{C_2} V_T −90° I (reference)

$$\theta = -90° \text{ (V with respect to } I)$$
$$I = \text{Same throughout circuit}$$
$$C_T = \frac{C_1 \times C_2}{C_1 + C_2}$$
$$X_{CT} = X_{C_1} + X_{C_2} \text{ or } \frac{V_T}{I_T}$$
$$Z = X_{CT}$$
$$V_1 = I \times X_{C_1}$$
$$V_2 = I \times X_{C_2}$$
$$V_T = V_1 + V_2 \text{ or } I \times X_{CT}$$
$$X_{C_1} = \frac{0.159}{fC_1}$$
$$X_{C_2} = \frac{0.159}{fC_2}$$

Parallel Circuit

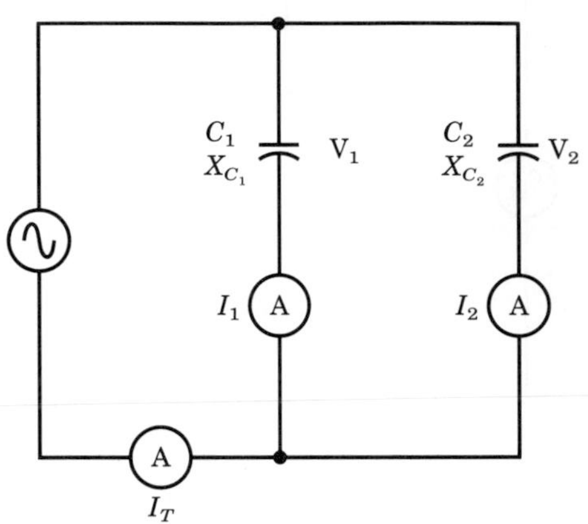

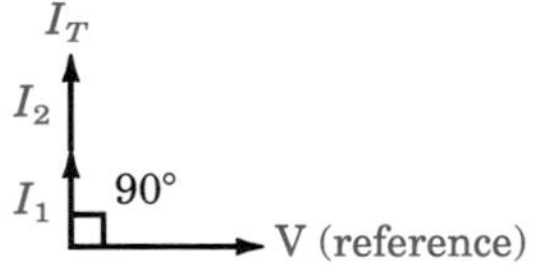

$$\theta = 90° \text{ (} I \text{ with respect to V)}$$
$$V = \text{Same across all components}$$
$$C_T = C_1 + C_2$$
$$X_{CT} = \frac{X_{C_1} X_{C_2}}{X_{C_1} + X_{C_2}} \text{ or } \frac{V_T}{I_T}$$
$$Z = X_{CT}$$
$$I_T = I_1 + I_2 \text{ or } \frac{V_T}{X_{CT}}$$
$$I_1 = \frac{V}{X_{C_1}}$$
$$I_2 = \frac{V}{X_{C_2}}$$
$$X_{C_1} = \frac{0.159}{fC_1} \text{ or } \frac{V}{I_1}$$
$$X_{C_2} = \frac{0.159}{fC_2} \text{ or } \frac{V}{I_2}$$

1. Since X_L is plotted upward and X_C is plotted downward, they *cancel* each other.
2. The result of combining them is that any remaining, or *net* reactance is the difference between the two reactances. That is, if X_L is greater than X_C, the net reactance is some value of X_L. If X_C is greater than X_L, the net circuit reactance is X_C.
3. Apply the same rationale to voltages. Since the voltages are plotted 180° out-of-phase with each other and calculated by multiplying the common current (I) times each reactance, the resulting voltage of the

FIGURE 19–5
RC circuits

Series Circuit

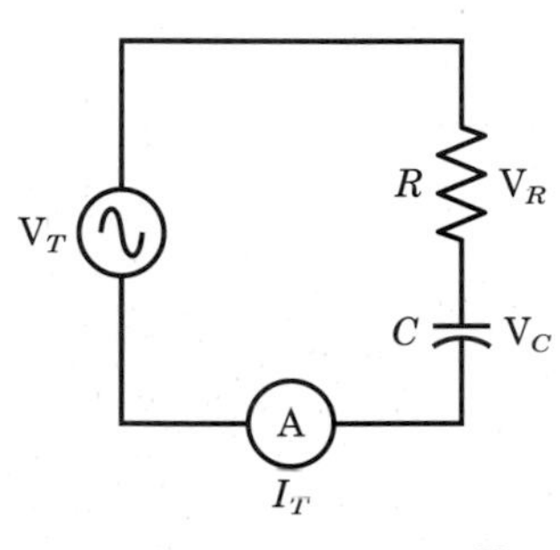

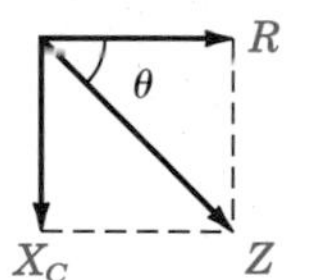

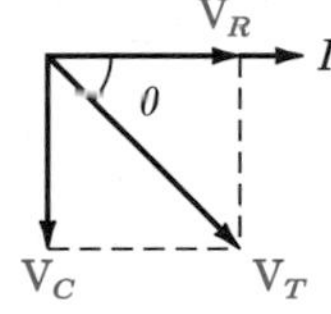

θ = Between 0 and −90°
(V with respect to I)
I = Same throughout circuit

$$Z = \sqrt{R^2 + X_C{}^2} \text{ or } \frac{V_T}{I_T}$$

$$I_T = \frac{V_T}{Z}$$

$$V_C = I \times X_C$$

$$V_R = I \times R$$

$$V_T = \sqrt{V_R{}^2 + V_C{}^2}$$

$$X_C = \frac{0.159}{fC} \text{ or } \frac{V_C}{I}$$

Parallel Circuit

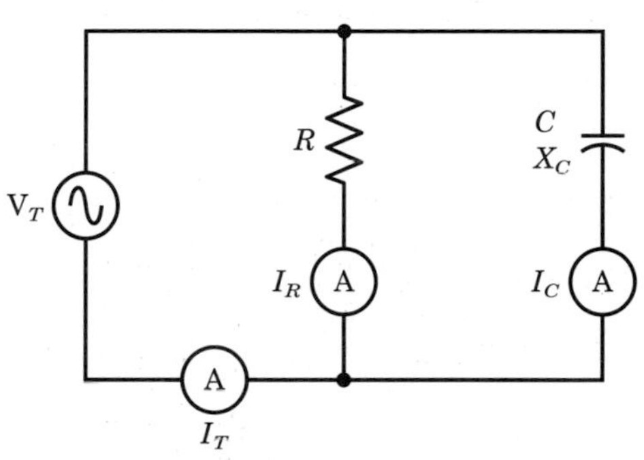

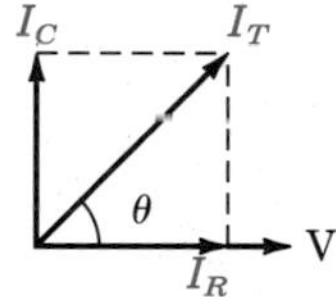

θ = Between 0 and 90°
(I with respect to V)
V = Same across all components

$$V_T = V_R = V_C$$

$$I_T = \sqrt{I_R{}^2 + I_C{}^2}$$

$$I_R = \frac{V_T}{R}$$

$$I_C = \frac{V_T}{X_C}$$

$$X_C = \frac{0.159}{fC}$$

$$Z = \frac{V_T}{I_T}$$

two opposite-reactive voltages is the difference between them. In other words, V_{XT} = largest V_X − smaller V_X.

4. Since the resulting reactance ohms are less than either reactance (due to the cancelling effect), the current is higher than it would be with either reactance alone. This means the $I \times X_L$ or $I \times X_C$ voltages are higher than the V applied! Obviously, since the opposite reactive voltages also cancel each other, the net voltage equals V applied.

Series RLC Circuit Analysis

Refer to Figure 19–7 and note the RLC circuit analysis as you study the following procedures.

FIGURE 19–6 Series LC circuit without resistance

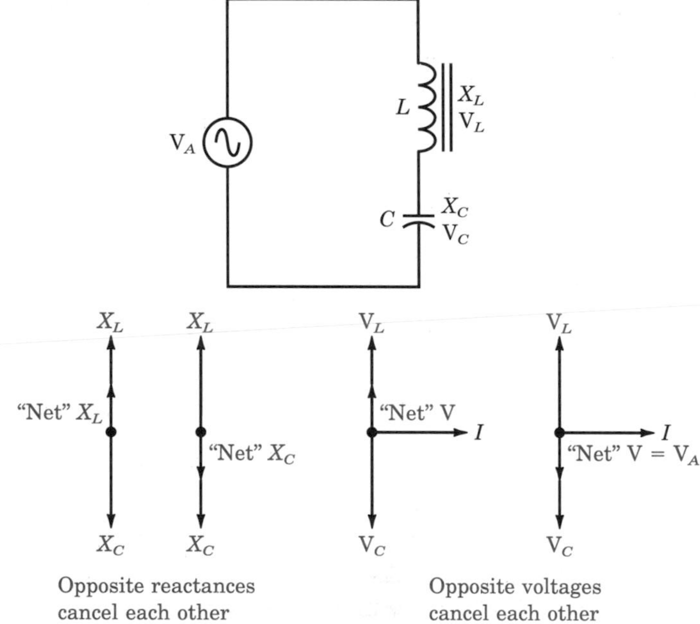

Computing Impedance by Pythagorean Approach

Using the concepts previously presented, the impedance must equal the vector sum of R and the remaining reactance. The formula is $Z = \sqrt{R^2 + (X_L - X_C)^2}$ when X_L is greater than X_C. If X_C is greater than X_L, then $Z = \sqrt{R^2 + (X_C - X_L)^2}$. In essence,

Formula 19–1	$Z = \sqrt{R^2 + X^2}$

where net reactance (X) depends on the relative reactance values.

In Figure 19–7, $R = 60\ \Omega$, $X_L = 160\ \Omega$, and $X_C = 80\ \Omega$. The net reactance is $80\ \Omega$ of inductive reactance in series with $60\ \Omega$ of resistance. Using the Pythagorean formula to find Z:

$$Z = \sqrt{R^2 + X^2}$$
$$Z = \sqrt{60^2 + 80^2}$$
$$Z = \sqrt{3600 + 6400}$$
$$Z = \sqrt{10{,}000}$$
$$Z = 100\ \Omega$$

Computing Current by Ohm's Law

In series circuits, current is common throughout the circuit. Circuit current is found through Ohm's law where $I = \frac{V}{Z}$. Since the V_A is given as 100 V, then $I = \frac{V}{Z} = \frac{100\text{ V}}{100\ \Omega} = 1\text{ A}$.

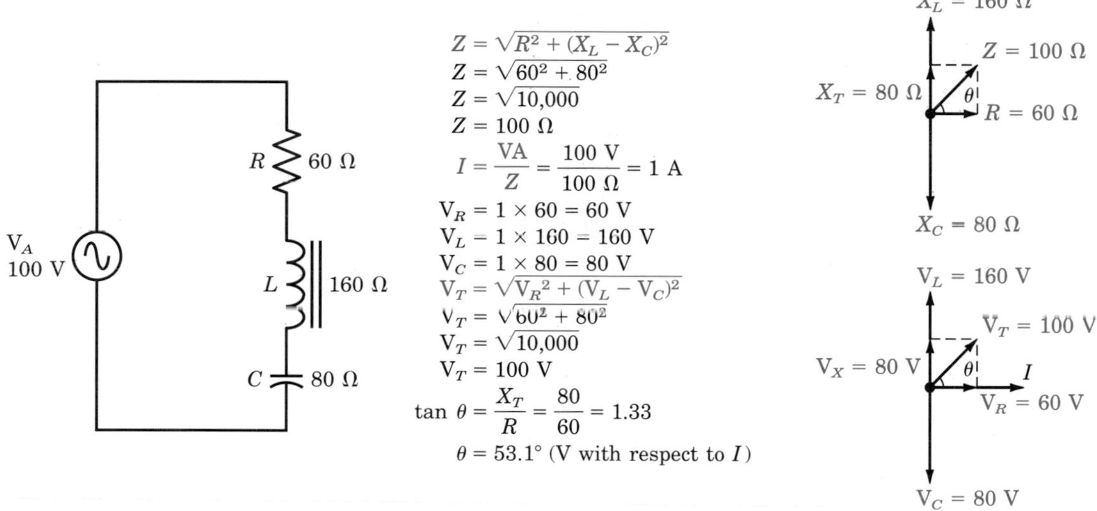

FIGURE 19–7 Series RLC circuit

Computing Individual Voltage by Ohm's Law

The voltage across the resistor = IR, across the inductor = IX_L, and across the capacitor = IX_C.

$V_R = 1\ A \times 60\ \Omega = 60$ volts
$V_L = 1\ A \times 160\ \Omega = 160$ volts
$V_C = 1\ A \times 80\ \Omega = 80$ volts

Computing Total Voltage by Pythagorean Approach

Since the resistive and reactive voltages are out-of-phase, we must again use vector summation. As you know, total voltage is found by the formula $V_T = \sqrt{V_R^2 + (V_L - V_C)^2}$, if inductor voltage is greater than capacitor voltage. If V_C is greater than V_L, the formula is $V_T = \sqrt{V_R^2 + (V_C - V_L)^2}$. Simplifying, the formula becomes

Formula 19–2	$V_T = \sqrt{V_R^2 + V_X^2}$

where V_X is *net reactive voltage.* Thus, in Figure 19–7:

$$V_T = \sqrt{60^2 + 80^2}$$
$$V_T = \sqrt{3600 + 6400}$$
$$V_T = \sqrt{10{,}000}$$
$$V_T = 100\text{ V}$$

Computing Phase Angle from Impedance and Voltage Data

Using impedance data, the formula is:

Formula 19–3	$\text{Tangent of } \theta = \dfrac{X \text{ (net reactance)}}{R}$

$$\text{Tan } \theta = \frac{80}{60}$$
$$\text{Tan } \theta = 1.33$$

The angle whose tangent = 1.33 is 53.1°. Since the net reactance is inductive reactance (plotted upward), the angle is +53.1°. NOTE: If the net reactance were capacitive reactance (plotted downward) of the same value, the phase angle would be −53.1° rather than +53.1°.

Using voltage data, the formula is:

Formula 19–4	$\text{Tangent of } \theta = \dfrac{V_X}{V_R}$

$$\text{Tan } \theta = \frac{80}{60}$$
$$\text{Tan } \theta = 1.33$$

The angle whose tangent = 1.33 is 53.1°.

Since the net reactive voltage is inductive voltage, which leads current (the reference vector), the angle is +53.1°. NOTE: If the net reactive voltage were V_C (plotted downward) of the same value, the phase angle would be −53.1° rather than +53.1°.

PRACTICAL NOTES

With more than one resistor or one reactance of each kind in the circuit, combine resistances to one total R value; inductive reactances to one total X_L value; and capacitive reactances to one total X_C value. Then solve the problem with the standard procedures to solve series RLC circuits.

EXAMPLE

Refer to the series RLC circuit in Figure 19–8 and calculate Z, V_T, V_{L_1}, V_{C_1}, V_{R_1}, and θ.

Answer:

In our example circuit, $R_1 = 20\ \Omega$ and $R_2 = 10\ \Omega$, so circuit $R = 30\ \Omega$.
$X_{L_1} = 50\ \Omega$ and $X_{L_2} = 40\ \Omega$, so circuit $X_L = 90\ \Omega$.
$X_{C_1} = 80\ \Omega$ and $X_{C_2} = 40\ \Omega$, so circuit $X_C = 120\ \Omega$.
Net reactance = $120\ \Omega\ (X_C) - 90\ \Omega\ (X_L) = 30\ \Omega$ of capacitive reactance

$$Z = \sqrt{R^2 + X^2} = \sqrt{30^2 + 30^2} = \sqrt{900 + 900} = \sqrt{1800} = 42.42\ \Omega$$

I is given as 1 ampere.
$V_R = IR = 1 \times 30 = 30$ V
$V_L = IX_L = 1 \times 90 = 90$ V
$V_C = IX_C = 1 \times 120 = 120$ V

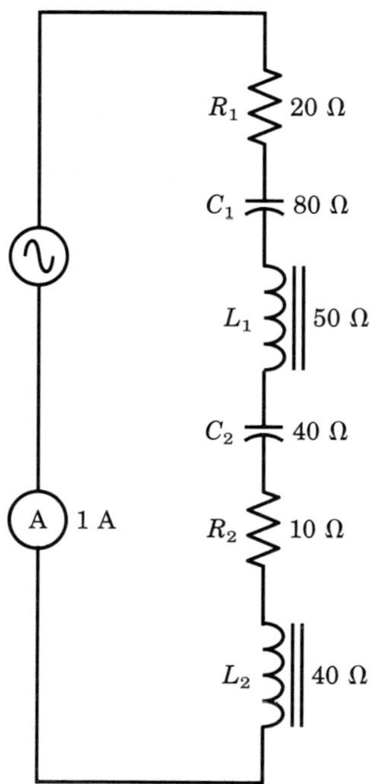

FIGURE 19–8
Series RLC circuit sample problem

$R_T = 30\ \Omega$
$X_{LT} = 90\ \Omega$
$X_{CT} = 120\ \Omega$
X_C "NET" = 30 Ω
$Z = \sqrt{R^2 + X^2} = 42.42\ \Omega$
$I = 1$ A
$V_T = \sqrt{V_R^2 + V_X^2} = 42.42$ V

$$\tan\theta = \frac{V_X}{V_R} = \frac{30}{30} = 1$$

$\theta = -45°$ (V with respect to I)

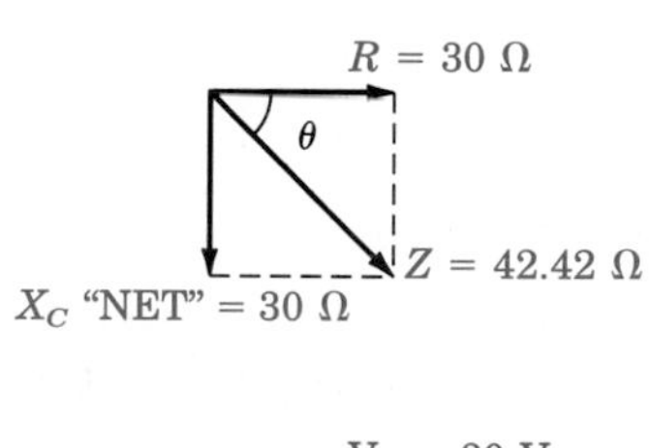

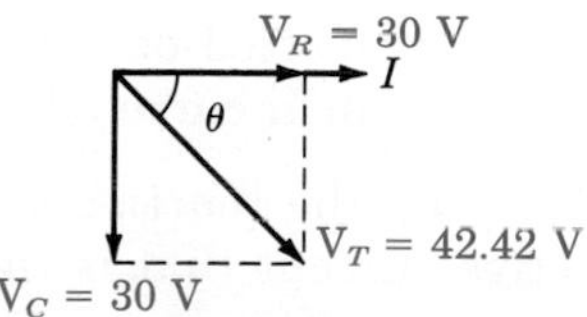

$$V_T = \sqrt{V_R{}^2 + V_X{}^2} = \sqrt{30^2 + 30^2} = \sqrt{900 + 900} = \sqrt{1800} = 42.42 \text{ V}$$
$$V_{L_1} = IX_{L_1} = 1 \text{ A} \times 50\ \Omega = 50 \text{ V}$$
$$V_{C_1} = IX_{C_1} = 1 \text{ A} \times 80\ \Omega = 80 \text{ V}$$
$$V_{R_1} = IR_1 = 1 \text{ A} \times 20\ \Omega = 20 \text{ V}$$
$$\text{Tan } \theta = \frac{X}{R}, \text{ or } \frac{V_X}{V_R}$$
$$\text{Tan } \theta = \frac{30}{30} = 1$$

The angle whose tangent = 1 is 45°.

Now that you have seen how this is done, try the following practice problem.

PRACTICE PROBLEM I

Answers are in Appendix B.

Refer to Figure 19–9 and calculate R, X_{CT}, X_L, Z, V_T, V_{C_2}, V_L, V_R, and θ.

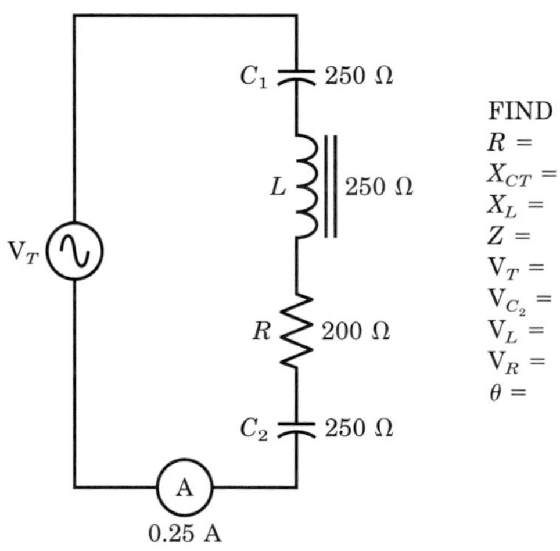

FIGURE 19–9 Series RLC circuit practice problem

THE PARALLEL RLC CIRCUIT

A Simple LC Parallel Circuit

Refer to Figure 19–10 as you study the following discussion on combining inductive and capacitive reactances in parallel.

1. Since I_L is plotted downward and I_C is plotted upward, obviously the reactive branch currents *cancel* each other.
2. Combining the reactive branch currents produces the resulting reactive current that is the difference between the opposite-reactance

IN-PROCESS LEARNING CHECK I

Fill in the blanks as appropriate.

1. When both capacitive and inductive reactance exist in the same circuit, the net reactance is ___________.
2. When plotting reactances for a series circuit containing L and C, the X_L is plotted ___________ from the reference, and the X_C is plotted ___________ from the reference.
3. In a series circuit with both capacitive and inductive reactance, the capacitive voltage is plotted ___________ degrees out of phase with the inductive voltage.
4. If inductive reactance is greater than capacitive reactance in a series circuit, the formula to find the circuit impedance is ___________.
5. In a series RLC circuit, current is found by ___________.
6. True or False. Total applied voltage in a series RLC circuit is found by the Pythagorean approach or by adding the voltage drops around the circuit.
7. If the frequency applied to a series RLC circuit decreases while the voltage applied remains the same, the component(s) whose voltage will increase is/are the ___________.

branch currents. That is, the resulting reactive current (I_X) is the larger reactive branch current minus the smaller reactive branch current.

3. In a parallel circuit, the voltage is the same across all branches, that is, V applied.

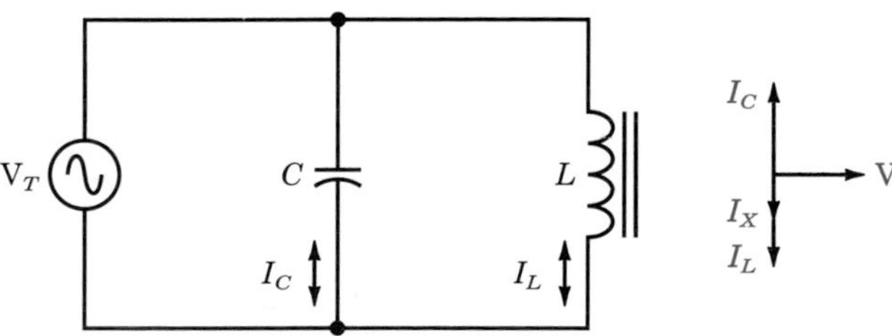

FIGURE 19–10 Parallel LC circuit without resistance

Parallel RLC Circuit Analysis

As was just stated, the *opposite-direction phasors* cancelling effect is also applicable to parallel circuits. But this cancellation effect is different between series and parallel circuits because of the circuit parameters involved. In series circuits, the current is common and the reactive voltage opposite-direction phasors cancel. However, in parallel circuits, the voltage is common and the reactive branch currents are opposite-direction phasors. Look at Figure 19–11 as you study the following discussion.

FIGURE 19–11
Parallel RLC circuit

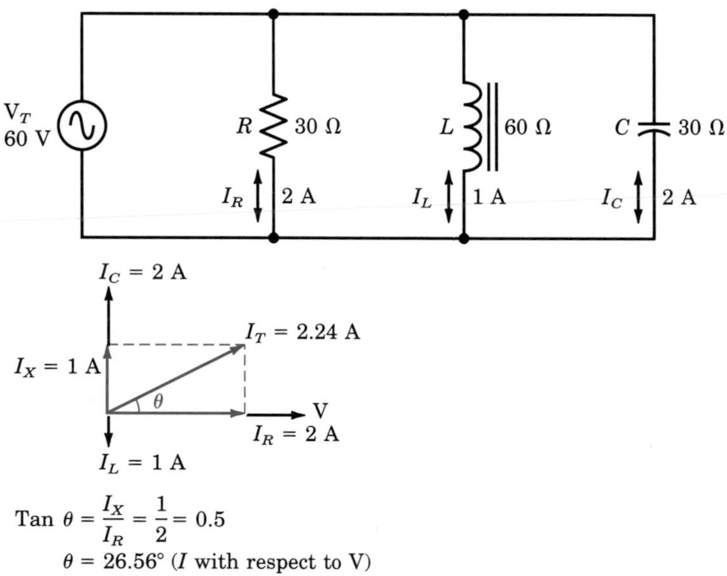

1. I_T is the vector sum of the circuit resistive and reactive branch currents.
2. The resulting reactive branch current value used to calculate total current is found by subtracting the smaller reactive branch current value from the larger reactive branch current value. This value (I_X) is then used to calculate I_T.
3. The general formula for I_T is $I_T = \sqrt{I_R^2 + (I_L - I_C)^2}$, where I_L is greater than I_C. If I_C is greater than I_L, the formula becomes $I_T = \sqrt{I_R^2 + (I_C - I_L)^2}$. Using the *net reactive current,* the formula becomes

Formula 19–5	$I_T = \sqrt{I_R^2 + I_X^2}$

Let's apply this knowledge and Ohm's law to solve a problem.

EXAMPLE

Refer again to the circuit in Figure 19–11 and find I_R, I_L, I_C, I_X, I_T, Z, and θ.

Answer: Use Ohm's law to solve for each branch current:

$$I_R = \frac{V_T}{R} = \frac{60}{30} = 2\text{ A}$$

$$I_L = \frac{V_T}{X_L} = \frac{60}{60} = 1\ A$$

$$I_C = \frac{V_T}{X_C} = \frac{60}{30} = 2\ A$$

Subtract the smaller reactive current from the larger to solve for I_X:

$$I_X = I_C - I_L = 2 - 1 = 1\ A$$

Use the Pythagorean approach to solve for I_T:

$$I_T = \sqrt{I_R^2 + I_X^2} = \sqrt{2^2 + 1^2} = \sqrt{4 + 1} = \sqrt{5} = 2.24\ A$$

Use Ohm's law to solve for Z:

$$Z = \frac{V_T}{I_T} = \text{approximately } 26.8\ \Omega$$

Use the tangent function to solve for the phase angle:

$$\text{Tan}\ \theta = \frac{I_X}{I_R} = \frac{1}{2} = 0.5$$

The angle whose tangent = 0.5 is 26.56°. Since the net reactive current is capacitive, current leads the reference circuit voltage and the phase angle (I_T with respect to V_T) is +26.56°.

Try solving the following practice problem.

PRACTICE PROBLEM II

Refer to Figure 19–12 and solve for I_R, I_L, I_C, I_X, I_T, Z, and θ.

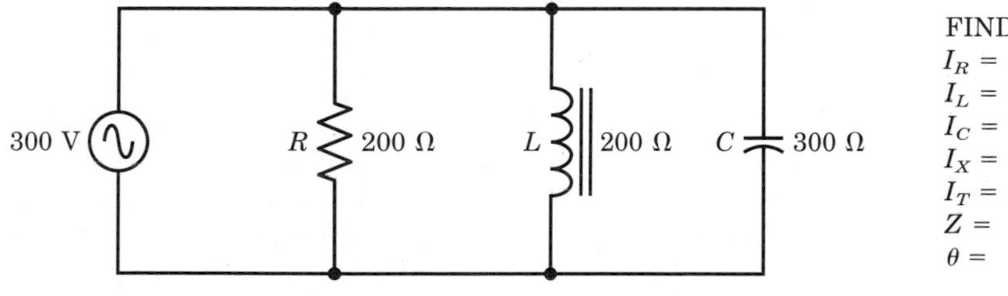

FIGURE 19–12 Parallel RLC circuit practice problem

POWER IN AC CIRCUITS

Background

In dc circuits, you know how to find the power expended by a circuit through the resistive power dissipation. That is, I^2R, $\frac{V^2}{R}$, and $V \times I$.

IN-PROCESS LEARNING CHECK II

1. In parallel RLC circuits, capacitive and inductive branch currents are ________ out-of-phase with each other.
2. The formula to find total current in a parallel RLC circuit when I_C is greater than I_L is ________.
3. True or False. The quickest way to find the circuit impedance of a parallel RLC circuit is to use an impedance diagram.
4. If the frequency applied to a parallel RLC circuit increases while the applied voltage remains the same, the branch current(s) that will increase is/are the ________ branch(es).

In ac circuits with both resistance and reactance, the apparent power supplied by the source is not the real or true power dissipated in the form of heat. Obviously, this is because the reactances take power from the circuit during one portion of the cycle and return it to the circuit during another portion of the cycle. As you know, circuits with reactances cause the voltage and current to be out-of-phase with each other. Therefore, the product of the **out-of-phase** circuit v and i does not indicate the actual circuit power dissipation.

Power in a Purely Resistive Circuit

Look at Figure 19–13. The power is dissipated during both half cycles. During the positive alternation, the power graph is represented by the *in-phase* $+v \times +i$ products. During the negative alternation, the power graph represents the product of the *in-phase* $-v \times -i$ values, which still yields a positive result. In other words, the resistor dissipates power during both alternations, and, over the whole cycle, an average power dissipation equal to the $v \times i$ product results. It can be mathematically shown that the average power equals the effective (rms) voltage times the effective (rms) current.

Power in Purely Reactive Circuits

Phase differences between voltage and current for purely inductive or capacitive circuits cause the power graphs, Figure 19–14. Because reactances take power from the circuit for one-quarter cycle and return it the next quarter cycle, you see a power graph of twice the source frequency. Notice, too, the average power over the complete cycle is zero, since the reactive instantaneous e times i products are positive half the time and negative half the time. This fact implies that pure inductive or capacitive reactances do

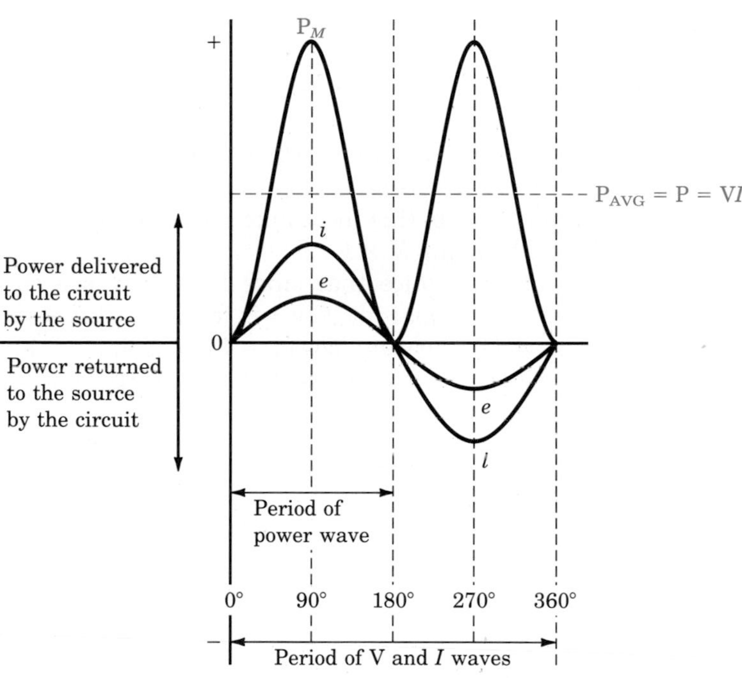

FIGURE 19–13 Instantaneous power graphs in a purely resistive circuit

not dissipate power. In reality, inductors (due to their wire resistance) have some resistive component, which dissipates some power in the coil. Good capacitors have very low effective resistance and typically dissipate almost zero power.

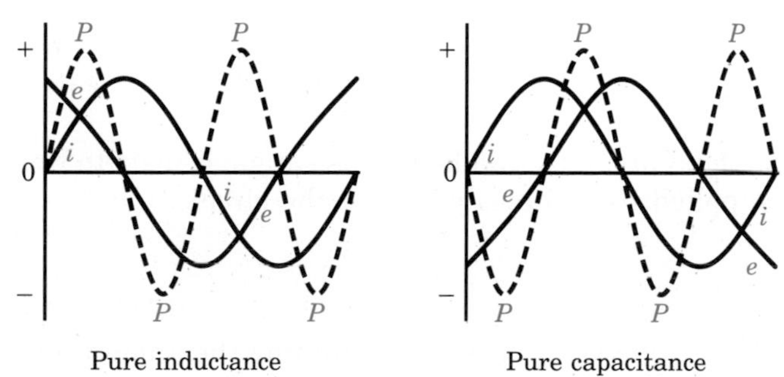

FIGURE 19–14 Instantaneous power graphs in purely reactive circuits

Combining Resistive and Reactive Power Features

View Figure 19–15 and note the curves for v × i products in an ac circuit with both resistance and reactance indicate power dissipation over one cycle. That is, the graphs resulting from positive v × i products and those resulting from negative v × i products are not equal. There is power dissipated. The amount of power represents the *true power* dissipated by the resistive share of the circuit.

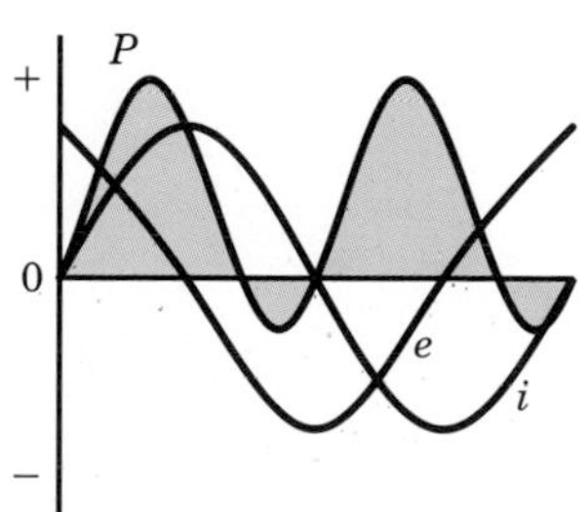

FIGURE 19–15 Instantaneous power graphs—*R* and *X* combined

Apparent Power

To someone not familiar with our preceding discussion, it would appear that by taking a voltmeter reading and a current meter reading for a given ac circuit, the power is found by multiplying their readings (V × I). But the meters are *not* indicating the phase difference between the two quantities. Therefore, what is computed from this technique is *apparent power. Apparent power* is found by multiplying V × I.

Refer to Figure 19–16 and notice that since R and X_L are equal, the circuit phase angle is 45°. The apparent power is found by the product of V applied times I. In this case, apparent power = 141.5 volts × 1 ampere = 141.5 VA. NOTE: Apparent power is expressed as **volt-amperes,** not watts! Also, note the abbreviation for apparent power is "*S*".

FIGURE 19–16 Power and power factor in a series ac circuit

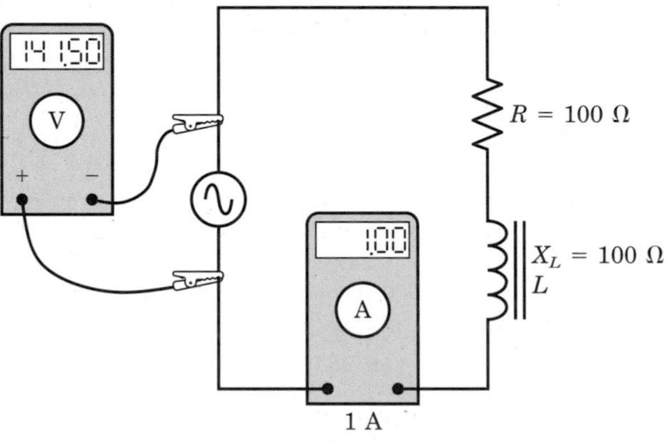

$$\text{p.f.} = \cos\theta = \cos 45° = 0.707$$

$$\text{p.f.} = \frac{R}{Z} \text{ or } \frac{V_R}{V_T}$$

$$S = VI = 141.5 \text{ VA}$$

$$P = VI\cos\theta = I^2R = 100 \text{ W}$$

$$Q = VI\sin\theta = 100 \text{ VAR}$$

With these thoughts, let's find how phase angle is used to find the real or true power supplied to and dissipated by the circuit.

True Power

True power dissipation is determined by finding the amount of power dissipated by the resistance in the circuit. If we assume a perfect inductor in Figure 19–16, the power dissipated by the resistor equals the true power dissipation. Because the I_R and V_R are in-phase with each other, the formula $V_R \times I_R$ is used. Therefore, true power = 100 V × 1 A = 100 W.

The I^2R Method

The preference is to say that true power equals I^2R, where R represents all circuit resistive quantities. This also includes resistor components and resistance values in reactive components. These R values are easily measured. It is not as easy (for example) to separate reactive and resistive voltage components in an inductor by measuring V_L.

The Power Factor Method

The power factor method considers the phase angle between current and voltage, created by the reactance. The value of the circuit *power factor* (p.f.) reflects how resistive or reactive the circuit is acting. A power factor of one indicates a pure resistive circuit. A power factor of zero represents a pure reactive circuit. Obviously, a power factor between zero and one indicates the circuit has both resistance and reactance, and true power is less than apparent power. Power factor is equal to the cosine of the phase angle.

Formula 19–6	Power factor (p.f.) = $\cos\theta$

Then for our sample circuit in Figure 19–16:

Formula 19–7	Apparent Power (S) = $V \times I$

$S = 141.5 \times 1 = 141.5$ volt-amperes

Formula 19–8	True Power (P) = I^2R

$P = 1^2 \times 100 = 100$ watts OR

Formula 19–9	P = Apparent power × the power factor

$P = 141.5 \times \cos\theta$, where $\cos 45°$
$P = 141.5 \times 0.707 = 100$ watts

Power Factor Formulas. Power factor really expresses the ratio of true power to apparent power (which is the same ratio expressed by the $\cos\theta$).

Formula 19–10a	$\text{p.f.} = \dfrac{\text{True Power (P)}}{\text{Apparent Power (S)}}$

$$\frac{VI \times \text{Cos}\,\theta}{VI} \text{ OR } \frac{R}{Z}\text{(Series)} = \frac{I_R}{I_T}\text{ (Parallel)}$$

Formula 19–10b	p.f. = $\text{Cos}\,\theta$

EXAMPLE

Recall in *series circuits* the $\cos\theta = \dfrac{R}{Z}\left(or\ \dfrac{V_R}{V_T}\right)$. For Figure 19–16, our series circuit example:

$$Z = \frac{V_T}{I_T} = \frac{141.5 \text{ V}}{1 \text{ A}} = 141.5\ \Omega.$$

$$\text{Cos } \theta = \frac{R}{Z} = \frac{100}{141.5} = 0.707. \text{ Therefore, p.f.} = 0.707.$$

(Incidentally, the angle whose cos = 0.707 is 45°, as you know.)

In *parallel circuits* the *power factor* = cos $\theta = \frac{I_R}{I_T}$. Refer to Figure 19–17 and notice the calculations for power factor using the resistive and total current values. In this case, the voltage applied is 100 volts. Thus:

$$I_R = \frac{100}{100} = 1 \text{ A}$$

$$I_L = \frac{100}{100} = 1 \text{ A}$$

$$I_T = \sqrt{I_R^2 + I_L^2} = \sqrt{1^2 + 1^2} = \sqrt{2} = 1.414.$$

$$\text{p.f.} = \frac{I_R}{I_T} = \frac{1}{1.414} = 0.707.$$

Again, in this case $\theta = 45°$.

FIGURE 19–17 Power factor in a parallel ac circuit

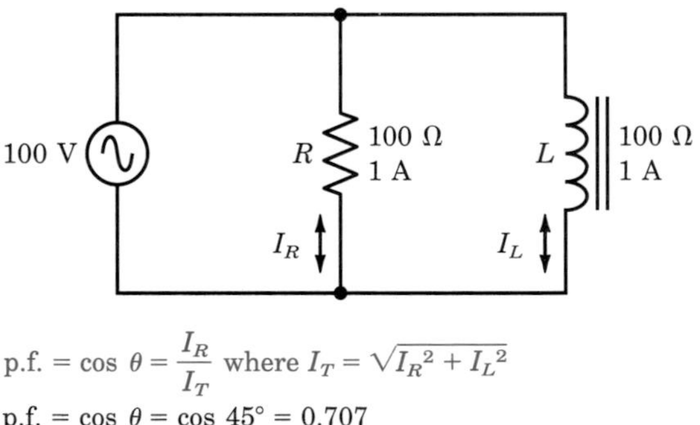

p.f. = cos $\theta = \frac{I_R}{I_T}$ where $I_T = \sqrt{I_R^2 + I_L^2}$

p.f. = cos θ = cos 45° = 0.707

The Power Triangle. The relationship between true and apparent power is sometimes represented visually by a *power triangle* diagram.

Figure 19–18a is a power triangle for a series RL circuit where R = X_L. **Apparent power** (S) in volt-amperes is plotted on a line at an angle equal to θ from the x-axis. Its length, representing magnitude or value, is V × I. Notice the apparent power is the result of two components:

1. The **true power** (P), sometimes called *resistive power,* is plotted on the x-axis (horizontal) and has a magnitude of VI × cos θ watts.
2. The **reactive power** (Q), often called **voltampere-reactive** (VAR), is plotted on the y-axis (vertical) and has a magnitude of VI × sin θ.

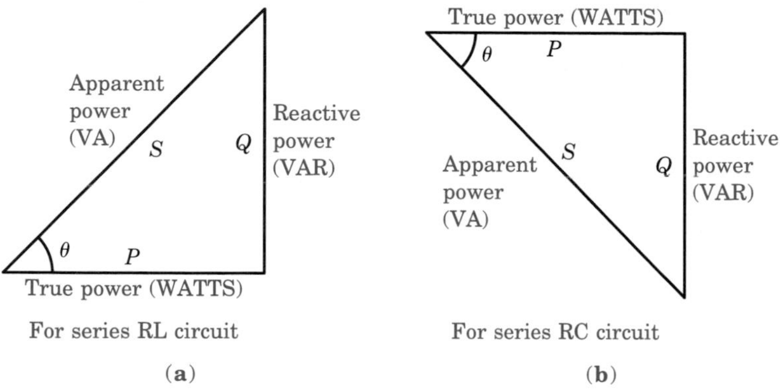

FIGURE 19–18 The power triangle

Figure 19–18b is a power triangle for a circuit with resistance and capacitive reactance. From your knowledge of the Pythagorean theorem and right-triangle analysis, you can see apparent power is computed from the formula:

Formula 19–11	$S = \sqrt{P^2 + Q^2}$

where S = apparent power
P = true power
Q = net reactive power or voltampere-reactive (VAR).

(NOTE: Power triangles are often used to analyze ac power distribution circuits.)

PRACTICAL NOTES

This formula is applicable to both series and parallel circuits. To summarize:

1. Knowing only the value of voltage or current does not indicate the amount of power actually being dissipated by the circuit.
2. A power triangle (different from V–I vector diagram) is used to diagram the various types of power in the ac circuit. Because the power waveform is a different frequency sinusoidal representation compared to the V and I quantities, we cannot plot the power values on the same vector diagram with V and I. Thus, the right-triangle (Pythagorean) approach displays the various types of power in the ac circuit.
3. The vertical element in the power triangle is the reactive power. The horizontal element in the triangle is the true or resistive power. Therefore, the hypotenuse of the triangle is the apparent power, which is the result of reactive and resistive power.

IN-PROCESS LEARNING CHECK III

1. Power dissipated by a perfect capacitor over one whole cycle of input is ____________.
2. The formula for apparent power is ____________, and the unit expressing apparent power is ____________.
3. True power in RLC circuits is that power dissipated by the ____________.
4. The formula for power factor is ____________.
5. If the power factor is one, the circuit is purely ____________.
6. The lower the power factor, the more ____________ the circuit is acting.
7. In the power triangle, true power is shown on the ____________ axis, reactive power (VAR) is shown on the ____________ axis, and apparent power is the ____________ of the right triangle.

RECTANGULAR AND POLAR VECTOR ANALYSIS

Overview of Rectangular and Polar Forms

In studying dc circuits, you learned that adding, subtracting, multiplying, and dividing electrical quantities was all that was needed to analyze circuits.

When you studied ac circuits, you learned that out-of-phase quantities need a different treatment. You learned several approaches—graphic analysis (plotting and measuring vectors or phasors and angles); the Pythagorean theorem (to find resulting magnitudes or values); and simple trigonometry (to find related angles). Also, you learned trigonometry is used to find magnitude information.

In this section, you will learn about algebraic vector analysis that involves both rectangular and polar forms to analyze vector quantities. You will learn how the j operator is used in complex numbers to distinguish the vertical vector from other components. Actually, these techniques are only a small extension of what you already know and use!

Phasors

You know that phasors represent sinusoidal ac quantities when they are the *same* frequency. Of course, they must be the same frequency so that when using phasor drawings the quantities compared maintain the same phase differential throughout the ac cycle. The normal direction of rotation of the constant-magnitude rotating radius vector (phasor) is counterclockwise from the 0° position (reference point). The relative position of the phasor shows its relationship to the reference (0°) position and to other phasors in the diagram. The phasor length indicates the magnitude, or relative value of the quantity represented, Figure 19–19. In ac, this is the effective (rms) value of V, I, and so forth.

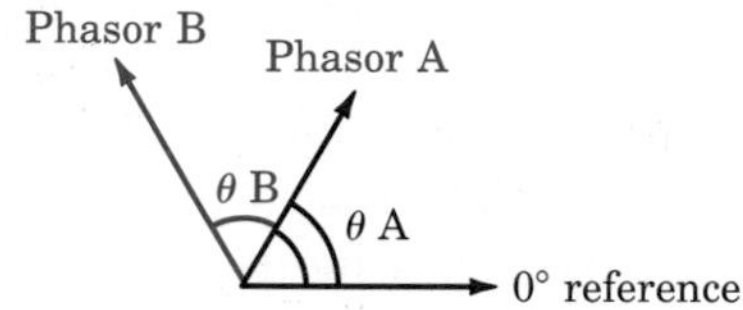

FIGURE 19–19 Phasor illustration

Rectangular Notation

In **rectangular form notation,** the magnitude and direction of a vector relates to both the x-axis (horizontal) and the y-axis (vertical) of the rectangular coordinate system, Figure 19–20. One way to find the location and magnitude of a given vector on this coordinate system is by **complex numbers.** A complex number locates a positional point with respect to two axes.

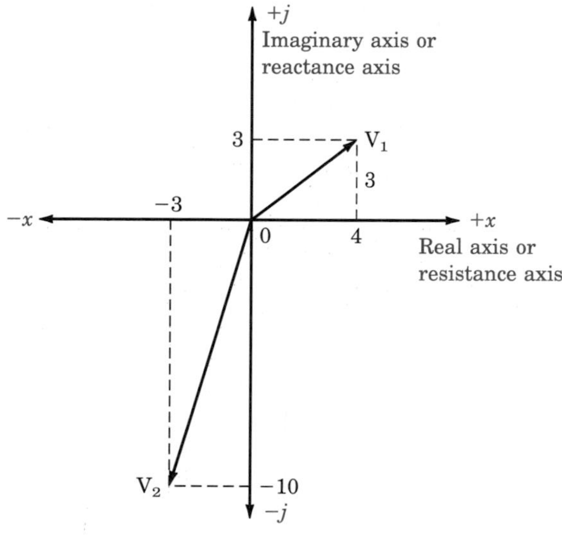

FIGURE 19–20 Illustration of rectangular notation form

A complex number has two elements—**real numbers** and **imaginary numbers.** The first element, real term, indicates the horizontal component on the x-axis. The second element, imaginary term, indicates the vertical component on the y-axis. For example, if it is stated that a specific vector = 4 + j3, this means the point at the end of that vector is located by projecting upward from 4 on the horizontal axis until a point even with 3 on the +j (vertical) axis is reached. (Incidentally, in this case, the length of this vector from the origin is 5, and its angle from the reference plane is 36.9°.)

Notice in Figure 19–20 the horizontal axis is also called the x-axis, the real axis, or the resistance axis. The vertical axis is also called the imaginary axis, or the reactance axis. The j operator designates values on that axis. The +j indicates an upward direction, while −j signifies a downward direction. When a number in a formula is preceded by the letter "j", the number is a vertical (or y-axis) component.

Since the j operator is so important in the rectangular form of vector algebra, let's define the j operator more specifically:

$$j = \sqrt{-1}$$
$$j^2 = (\sqrt{-1})(\sqrt{-1}) = -1$$
$$j^3 = (j^2)(j) = (-1)(j) = -j$$
$$j^4 = (j^2)(j^2) = (-1)(-1) = +1$$

Look at Figure 19–21 to relate these equations to the rectangular coordinate format. Notice the various powers of the j operator rotate a phasor as follows:

- +j is a 90° operator, rotating a phasor 90° in the counterclockwise (CCW) direction, or +90 degrees.
- −j rotates a phasor 90° in the clockwise (CW) direction, or −90°.
- j^2 rotates the phasor 180°.
- j^3 rotates the phasor to the +270° position, or −90°.
- j^4 is +1, rotating the vector back to the starting point (0°).

FIGURE 19–21 The *j* operator

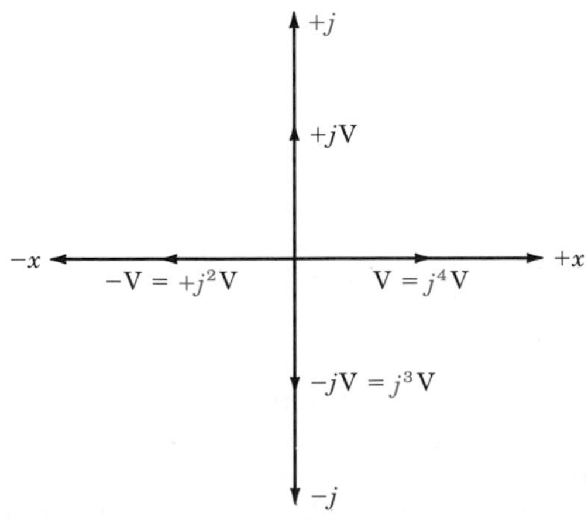

Effect of operators j, j^2, j^3, and j^4

EXAMPLE

If the value of V_T for a series RL circuit (with a perfect inductor) is 30 + j40, it might be illustrated as in Figure 19–22. Observe V_T is 50 volts and θ is 53.1°. Suppose that V_T, expressed in rectangular form, is 4 + j3. What is the value of V_T and the angle? This should be familiar to you.
Answer: 5 volts at an angle of 36.9°.

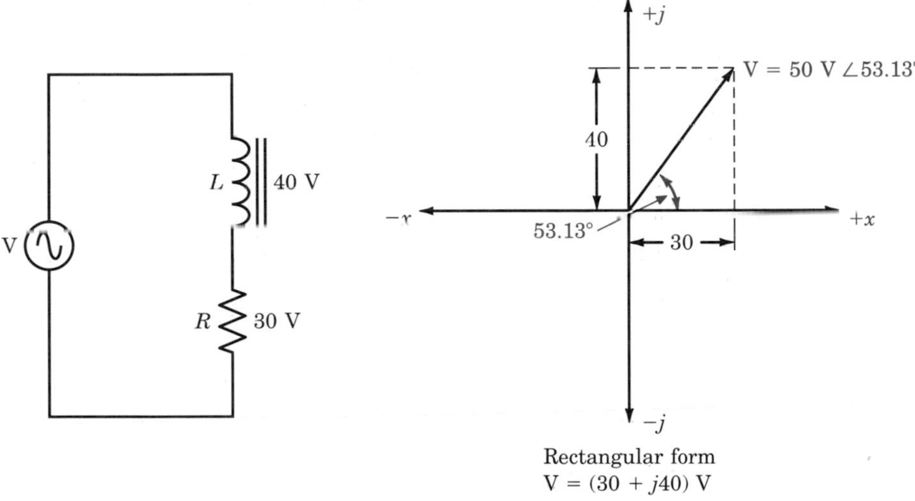

FIGURE 19–22
Example of rectangular notation

Polar Notation

Refer to Figure 19–23 to see how vectors are represented on polar coordinates.

Whereas the rectangular form expresses vectors by the horizontal and vertical axes, the **polar form notation** expresses the length and angle of the resulting vector to define a vector. Previously, we stated one value of V_T is 30 + j40 in rectangular notation. This same V_T is expressed in polar form as 50 V ∠53.1°. (The resulting vector of the other two vectors described in the rectangular form.)

The advantage for using polar form notation is that the values are the same as those measured by instruments such as voltmeters, ammeters, and so on. That is, in our example above, a voltmeter would actually read 50 volts for V_T.

Converting From Rectangular to Polar Form

To convert a complex number in rectangular form to polar form simply find the vector resulting from the real term, j term, and angle.

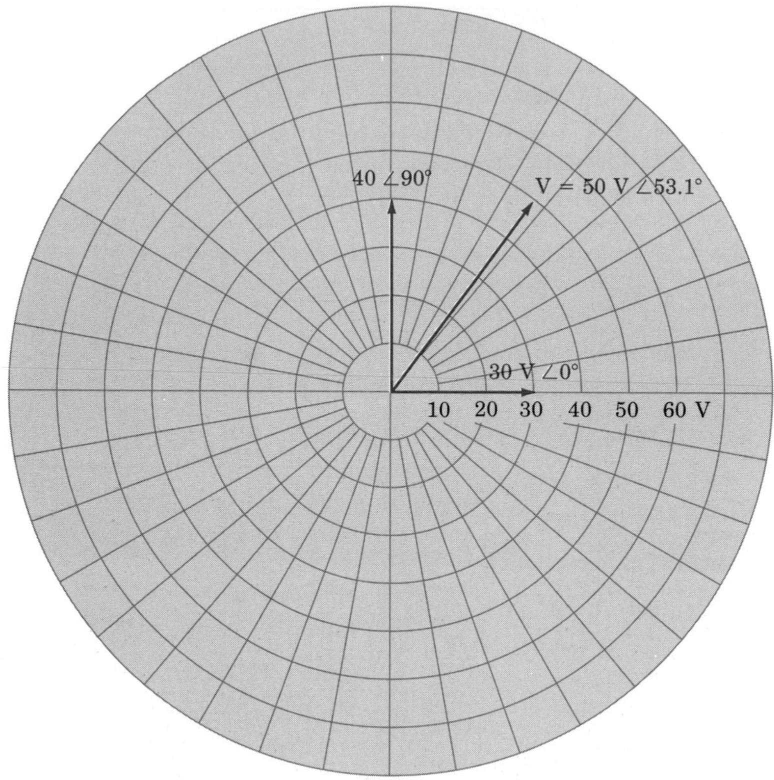

FIGURE 19–23 Illustration of vectors represented on polar coordinate system

---

EXAMPLE

In a series RL circuit, what is the polar notation for a value expressed in rectangular notation as 60 V +j70 V?

Answer:

Use phasor addition to find the magnitude:

$$V_T = \sqrt{V_R^2 + V_L^2} = \sqrt{60^2 + 70^2} = \sqrt{8500} = 92.2 \text{ V}$$

Use the tangent function to solve for the angle:

$$\text{Tan } \theta = \frac{V_L}{V_R} = \frac{70}{60} = 1.67$$

The angle whose tangent = 1.67 is 59.1°.

Polar Form = 92.2 V ∠59.1°

You get the idea! To convert, solve for the magnitude by vectorially adding the real and the j terms; then solve for the angle, using the arc-tangent, as appropriate. The resulting magnitude and angle are then combined to express the polar notation for the vector. Several examples of rectangular and polar notation are:

Rectangular		*Polar*
3 + j4	=	5∠53.1°
4 − j3	=	5∠−36.9°
60 + j70	=	92.2∠59.1°

Now you try a conversion.

PRACTICE PROBLEM III

Convert 20 +j30 to polar form.

Converting From Polar to Rectangular Form

Later in the chapter, you will see the rectangular form is most useful to *add and subtract* vector values and the polar form is useful to *multiply and divide* vector values. For this reason, it is convenient to know how to convert from rectangular to polar form and from polar to rectangular form, as appropriate.

Recall that trig is used to solve for the sides of a right triangle, if the angle and one side are known. Let's put this fact to use.

Since the polar form expresses the hypotenuse of the right triangle plus the angle, we can solve for the other two sides of the triangle by using the appropriate trig functions. Obviously, knowing the sides allows us to express the real and j terms appropriately.

Refer to Figure 19–24 and note the horizontal component, i.e., the resistive or real term (R in series circuit impedance diagrams), V_R in series circuits, and I_R in parallel circuits is found by:

$$R = Z \cos \theta$$
$$V_R = V_T \cos \theta$$
$$I_R = I_T \cos \theta$$

Series circuit

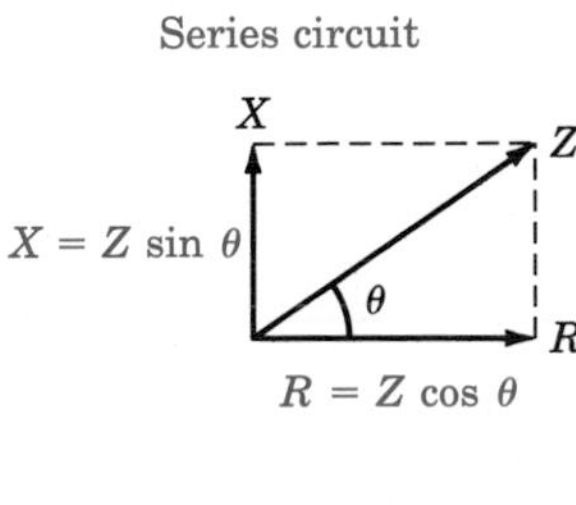

Parallel circuit

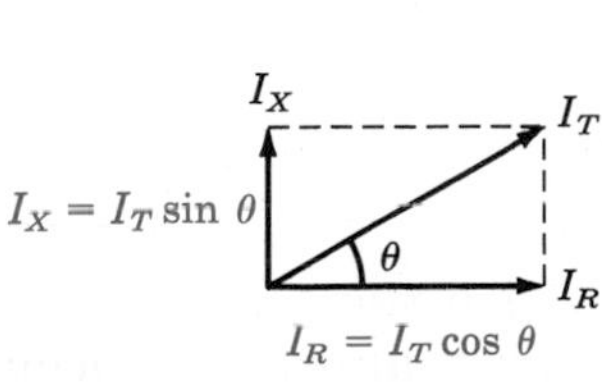

V_X V_T

$V_X = V_T \sin \theta$ θ V_R

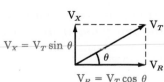

$V_R = V_T \cos \theta$

FIGURE 19–24
Converting from polar to rectangular form

Recall $\cos\theta = \frac{\text{adj}}{\text{hyp}}$; therefore, adj = hyp × cos θ.

In like manner, use trig to solve for the j (imaginary) term, (i.e., X in series impedance diagrams, V_X in series circuits, and I_X in parallel circuits). Referring to Figure 19–24 again, the reactive term(s) are found by:

$$X = Z \sin\theta$$
$$V_X = V_T \sin\theta$$
$$I_X = I_T \sin\theta$$

Recall $\sin\theta = \frac{\text{opp}}{\text{hyp}}$. Therefore, opp = hyp × sin θ.

Once the real and j terms are known, we can express the answer in rectangular form.

Therefore, the general conversion formulas are expressed as:

Formula 19–12	$Z \angle\theta° = Z\cos\theta + jZ\sin\theta$

for a series circuit impedance diagram

Formula 19–13	$V_T \angle\theta° = V_T\cos\theta + jV_T\sin\theta$

for a series circuit V-I vector diagram

Formula 19–14	$I_T \angle\theta° = I_T\cos\theta + jI_T\sin\theta$

for a parallel circuit V-I vector diagram

EXAMPLE

Convert 45 ∠30° to rectangular form.

Answer:

$$45\angle 30° = 45(\cos 30°) + j45(\sin 30°)$$
$$45\angle 30° = 45 \times 0.866 + j45 \times 0.5$$
$$45\angle 30° = 38.97 + j22.5$$

Check this answer by drawing a sketch.

PRACTICE PROBLEM IV

Convert 66 ∠60° to rectangular form.

Now that you are familiar with rectangular and polar notation and their relationships, we will show how these expressions are used mathematically.

ALGEBRAIC OPERATIONS

Addition (Rectangular form)

Basic rule: Add in-phase (resistive) terms and add out-of-phase (reactive) "j" terms.

EXAMPLE
Find the sum of V_1 and V_2.

Example 1:
$$V_1 = 15 + j20$$
$$V_2 = 20 - j10$$
$$V_1 + V_2 = 35 + j10$$

Example 2:
$$V_1 = 20 + j20$$
$$V_2 = 30 + j10$$
$$V_1 + V_2 = 50 + j30$$

Subtraction (Rectangular form)

Basic rule: Change the sign of the subtrahend and add.

EXAMPLE
Subtract branch current I_2 from branch current I_1, where

$$I_1 = 5 - j6 \text{ A}$$
$$I_2 = 4 + j7 \text{ A}$$

Change the sign of branch current I_2 and add:

$$5 - j6$$
$$-4 - j7$$
$$I_1 - I_2 = 1 - j13$$

EXAMPLE
When given total current and one branch current (I_1), find the other branch current (I_2).

$I_T = I_1 + I_2$; therefore, $I_2 = I_T - I_1$.

where

$$I_T = 13 - j5.5 \text{ mA}$$
$$I_1 = 6 - j4.0 \text{ mA}$$

Change the sign of branch current I, and add:

$$13 - j5.5$$
$$-6 + j4.0$$
$$I_2 = 7 - j1.5$$

PRACTICAL NOTES

If addition or subtraction are needed to analyze a problem and the terms are given in polar form, convert them to rectangular form because you cannot use polar form to add or subtract.

On the other hand, even though multiplication and division are possible in rectangular form, polar form is easier for multiplying and dividing. If you know the rectangular form, you will first convert it to the polar form to multiply or divide. For example, $V_T = I_T \times Z_T$.

Multiplication (Polar form)

Basic rule: Multiply the two magnitudes and add the angles algebraically.

EXAMPLE

If a current of $5 \angle 30°$ A flows through a Z of $20 \angle -15°\ \Omega$, what is the voltage?

$$V = IZ$$
$$V = (5 \angle 30°) \times (20 \angle -15°)$$
$$V = 100 \angle 15°\ \text{V}$$

Division (Polar form)

Basic rule: Divide one magnitude by the other and subtract the angles algebraically.

EXAMPLE

Find the current in a circuit if the voltage = $35 \angle 60°$ and the impedance = $140 \angle -20°$.

$$I = \frac{V}{Z}$$
$$I = \frac{35 \angle 60°}{140 \angle -20°}$$
$$I = 0.25 \angle 80°\ \text{A}$$

Raising to Powers (Polar form)

Basic rule: Raise the magnitude to the power indicated and multiply the angle by that power.

EXAMPLE

If $I = 3 \angle 20°$, then $I^2 = 9 \angle 40°$.

Taking a Root (Polar form)

Basic rule: Take the root of the magnitude and divide the angle by the root.

EXAMPLE

If $I^2 = 100 \angle 60°$, then $I = 10 \angle 30°$.

PRACTICE PROBLEM V

1. Add $20 + j5$ and $35 - j6$
2. Subtract $20 + j5$ from $35 - j6$
3. Multiply $5 \angle 20° \times 25 \angle -15°$
4. Divide $5 \angle 20°$ by $25 \angle -15°$
5. Find I^2 if $I = 12 \angle 25°$
6. Find I if $I^2 = 49 \angle 66°$

APPLICATION OF RECTANGULAR AND POLAR ANALYSIS

Sample Series RLC Circuit Analysis and Applying Vector Analysis

Refer to Figure 19–25 as you study the following discussion on finding various circuit parameters. Note if addition and subtraction are needed, the calculations use rectangular notation; if multiplication or division are involved, the calculations use polar notation.

1. Find Z_T:

Polar Form:	Rectangular Form: $20 + j10$
	Z_T = vector sum of impedances
	$Z_T = \sqrt{R_T^2 + X_T^2}$
	$Z_T = \sqrt{20^2 + 10^2}$
	$Z_T = \sqrt{500}$
$Z = 22.36 \angle 26.56°$	$Z_T = 22.36\ \Omega$

2. Find I_T (Assume I_T is at 0° since this is a series circuit.):

Polar Form:	Rectangular Form:
$I_T = \dfrac{V_T}{Z_T}$	
$I_T = \dfrac{60.00 \angle 26.56}{22.36 \angle 26.56}$	
$I_T = 2.68 \angle 0°$ A	$I_T = 2.68 + j0$ A

FIGURE 19–25
Sample RLC circuit analysis

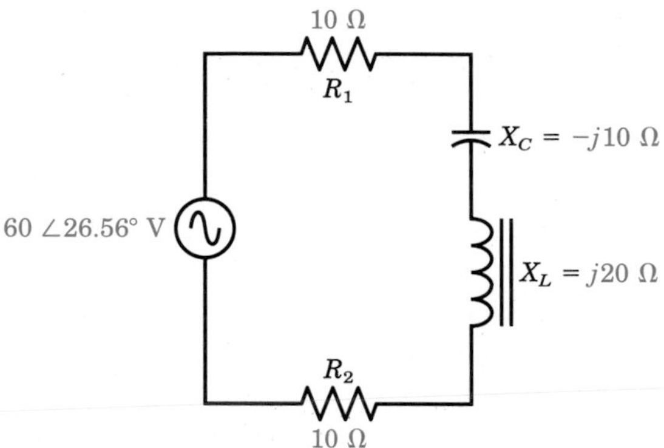

3. Find individual voltages:
   a. $V_{R_1} = I \times R_1 = 2.68\ \angle 0° \times 10\ \angle 0° = 26.8\ \angle 0°\ V$
   b. $V_{R_2} = I \times R_2 = 2.68\ \angle 0° \times 10\ \angle 0° = 26.8\ \angle 0°\ V$
   c. $V_L = I \times X_L = 2.68\ \angle 0° \times 20\ \angle 90° = 53.6\ \angle 90°\ V$
   d. $V_C = I \times X_C = 2.68\ \angle 0° \times 10\ \angle -90° = 26.8\ \angle -90°\ V$
4. Verify total voltage via sum of individual drops. That is, convert each V from polar to rectangular form, then add:

Polar Form:	Rectangular Form:
$V_{R_1} = 26.8\ \angle 0°$	a. $26.8 + j0$
$V_{R_2} = 26.8\ \angle 0°$	b. $26.8 + j0$
$V_L = 53.6\ \angle 90°$	c. $0.0 + j53.6$
$V_C = 26.8\ \angle -90°$	d. $0.0 - j26.8$

Total Voltage = $53.6 + j26.8$ or $60\ \angle 26.56°$
V_T = approximately $60\ \angle 26.56°$ V (polar) or $53.6 + j26.8$ (rectangular)

Sample Parallel RLC Circuit Analysis

Remember for parallel circuits in dc, it is easier to use the conductance of branches or add branch currents, rather than combining reciprocal resistances. Recall conductance is the reciprocal of resistance. That is, $G = \frac{1}{R}$ siemens.

In ac circuits, we have to deal with pure resistance *and* reactance *and* the combination of resistance and reactance (impedance). The comparable term for conductance of a pure reactance is called **susceptance,** where susceptance $= \frac{1}{\pm X}$. The symbol for susceptance is "B". The reciprocal of impedance is called **admittance,** which is equal to $\frac{1}{Z}$. Thus:

- For pure resistance, conductance (G) $= \frac{1}{R}$ S
- For pure reactance, susceptance (B) $= \frac{1}{\pm X}$ S
- For combined R and X admittance (Y) $= \frac{1}{Z}$ S

The phase angle designation for B or Y is opposite that expressed for X or Z, since it is a reciprocal relationship. This means reactance for an inductive branch is $+jX_L$, and inductive susceptance is $-jB_L$. Also, capacitive reactance is $-jX_C$ and capacitive susceptance is jB_C.

Formula 19–15	Inductive susceptance: $-jB_L = \frac{1}{jX_L} = \frac{1}{2\pi fL}$

Formula 19–16	Capacitive susceptance: $jB_C = \frac{1}{jX_C} = 2\pi fC$

Notice the inversion of the X_L and X_C formulas. Important admittance relationships are:

$$Y = G \pm jB = G + j(B_C - B_L)$$

Formula 19–17	$Y = \sqrt{G^2 + B^2}$

$$\theta = \text{angle whose tangent} = \frac{B}{G}$$

NOTE: $Y = \frac{I}{V}$

$$I = V \times Y$$

$$V = \frac{I}{Y}$$

EXAMPLE

Refer to Figure 19–26. Use the admittance approach and find the total current of a parallel circuit operating at 60 Hz with the following parameters: R = 500 Ω; L = 1 H; C = 4 μF; and V = 100 V.

Answers:

$$G = \frac{1}{R} = \frac{1}{500} = 0.002 \text{ S}$$

$$-jB_L = \frac{1}{2\pi fL} = \frac{1}{6.28 \times 60 \times 1} = -j0.00265 \text{ S}$$

$$jB_C = 2\pi fC = 6.28 \times 60 \times (4 \times 10^{-6}) = j0.00151 \text{ S}$$

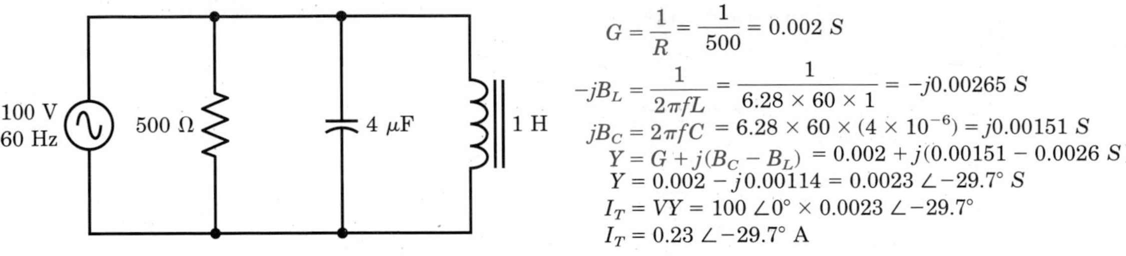

FIGURE 19–26 Sample admittance calculations

$$Y = G + j(B_C - B_L) = 0.002 + j(0.00151 - 0.00265)$$
$$Y = 0.002 - j0.00114 = 0.0023\angle -29.7°\ S$$

$$I_T = VY = 100\ \angle 0° \times 0.0023\ \angle -29.7° = 0.23\ \angle -29.7°\ \text{ampere}$$

You can calculate the parameters using rectangular forms to add and subtract and polar forms to multiply and divide. When using these notations, it is a good idea to make appropriate conversions so each significant parameter is expressed in both forms. In this way, you use the form most applicable to the calculation you are performing.

Sample Combination Circuit Calculations

Refer to Figure 19–27 and note the following calculations.

1. Find Z_T.

   Since in a parallel circuit $Z_T = \frac{Z_1 Z_2}{Z_1 + Z_2}$, we need to know both the rectangular and polar forms of each branch impedance. These forms are a good starting point.

FIGURE 19–27 Sample combination circuit calculations

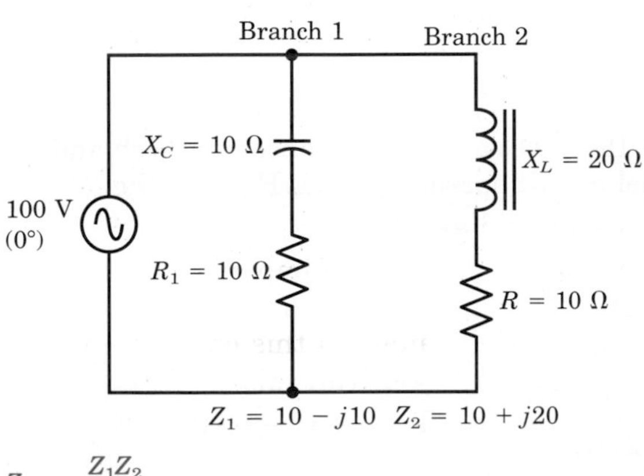

$$Z_T = \frac{Z_1 Z_2}{Z_1 + Z_2}$$

$Z_1 = 10 - j10$ (rectangular) $14.14\angle -45°$ (polar)

$Z_2 = 10 + j20$ (rectangular) $22.36\ \angle 63.4°$ (polar)

$$Z_T = \frac{Z_1 + Z_2}{Z_1 + Z_2} = \frac{14.14\ \angle -45° \times 22.36\ \angle 63.4°}{10 - j10 + 10 + j20} = \frac{316.17\ \angle 18.4°}{20 + j10}$$

$$Z_T = \frac{316.17\ \angle 18.4°}{22.36\ \angle 26.56°} = 14.13\ \Omega\ \angle -8.16°$$

2. Assume 100 volts V applied at 0° and find the current.

$$\text{Branch 1 current} = \frac{V}{Z_1}$$

$$\text{Branch 1 current} = \frac{100\ \angle 0°}{14.14\ \angle -45°} = 7.07\ \angle 45°\ \text{A}$$

$$\text{Branch 2 current} = \frac{V}{Z_2}$$

$$\text{Branch 2 current} = \frac{100\ \angle 0°}{22.36\ \angle 63.4°} = 4.47\ \angle -63.4°\ \text{A}$$

3. To find I_T, convert each branch current to rectangular form and add the branch currents.

$I_1 = 7.07\ \angle 45°$ (polar $5 + j5$ (rectangular)

$I_2 = \underline{4.47\ \angle -63.4°}$ (polar) $\underline{2 - j4}$ (rectangular)

$I_T = 7.07\ \angle 8.13°$ A (polar) $7 + j1$ A (rectangular)

4. Individual component voltages are calculated using $I \times R$, $I \times X_L$, and $I \times X_C$, as appropriate. You can practice rectangular and polar notation and appropriate conversion techniques by doing this. If you do this, remember to sketch vector diagram(s), as appropriate, to check the logic of your results.

PRACTICAL NOTES

Although the calculations introduced in this section may not have to be routinely performed by technicians, having this knowledge will give you a better understanding of important practical concepts. For example, in radio and TV transmission systems, many principles relate to the knowledge gained in this chapter. Also, the ability to control power factors in power transmission is important. When you have mastered these concepts, your understanding of the electronics field will be thorough.

CHAPTER CHALLENGE

Following the SIMPLER troubleshooting sequence, find the problem with this circuit. As you follow the sequence, record your responses to each step on a separate sheet of paper.

CHALLENGE CIRCUIT 11

SYMPTOMS Gather, verify, and analyze symptom information. (Look at the "Starting Point Information.")

IDENTIFY Identify initial suspect area for the location of the trouble.

MAKE Make a decision about "What type of test to make" and "Where to make it." (To simulate making a decision about each test you want to make, select the desired test from the "TEST" column listing below.)

PERFORM Perform the test. (Look up the result of the test in Appendix C. Use the number in the "RESULTS" column to find the test result in the Appendix.)

LOCATE Locate and define a new "narrower" area in which to continue troubleshooting.

EXAMINE Examine available information and again determine "what type of test" and "where."

REPEAT Repeat the preceding analysis and testing steps until the trouble is found. What would you do to restore this circuit to normal operation? When you have solved the problem, compare your results with those shown in the color insert.

Starting Point Information

1. Circuit diagram
2. V_{R_3} = 0.057 V

Test	Results in Appendix C
$V_{A\text{-}B}$	(78)
$V_{B\text{-}C}$	(38)
$V_{C\text{-}D}$	(96)
$R_{A\text{-}B}$	(104)
$R_{B\text{-}C}$	(63)
$R_{C\text{-}D}$	(32)

SUMMARY

- Vectors in opposite directions cancel each other. If one has greater magnitude than the other, there is a net resulting vector equal to the larger value minus the smaller value.
- Power in ac circuits with both resistive and reactive components is described in three ways:
  1. Apparent power (S)—applied voltage times the circuit current, expressed in volt-amperes (VA) and plotted as the hypotenuse of the power triangle.
  2. Reactive power (Q) or voltampere-reactive (VAR)—equal to VI × sin θ, and plotted on the vertical axis.
  3. True power (P) in watts—equal to VI cos θ, and plotted on the horizontal axis.

  Because of the right-triangle relationship: $S = \sqrt{P^2 + Q^2}$.
- Vector quantities are analyzed in several ways—graphic means, algebraic expressions based on rectangular coordinates, and expressions based on polar coordinates.
- Separating a vector quantity into its components is easily accomplished by simple trigonometric functions. These functions are sine, cosine, and tangent.
- Complex numbers express vector quantities relative to the rectangular notation system. The complex number has a term that defines the horizontal (real or resistive) component and a j term that expresses the reactive component or out-of-phase element. The general term for impedance is $R \pm jX$, for applied voltage is $V_R \pm V_X$, and for total current in a parallel circuit is $I_R \pm jI_X$.
- Vectors expressed algebraically in rectangular coordinates are added, subtracted, multiplied, and divided. However, this form of notation is most easily used to add and subtract. To add, like terms are collected (i.e., resistive terms collected and j terms collected), and added algebraically. To subtract, change the sign of the subtrahend and add algebraically.
- Vectors expressed in polar form *cannot* be added or subtracted algebraically. However, this form is easily used to multiply, divide, and find square roots. To multiply polar expressions, multiply magnitudes and add the angles algebraically. To divide polar expressions, divide the magnitudes and subtract the angles algebraically. To raise a polar expression to a power, raise the magnitude by the given power and multiply the angle by the power. To take a root, take the root of the magnitude value and divide the angle by the root value.
- Because polar form facilitates multiplication and division and rectangular form expedites addition and subtraction, it is a good idea to convert from one form to the other.
- When two branches are in parallel with each having resistive and reactive components, the resultant Z is found by the formula $Z = \frac{Z_1 Z_2}{(Z_1 + Z_2)}$. It is convenient to use the polar form for the numerator (for multiplying) and the rectangular form for the denominator (for adding). Then, convert the denominator to the polar form to allow division into the numerator to get the final answer.
- The ease that current flows through a resistance is called conductance, where $G = \frac{1}{R}$. A comparable term for reactance is called susceptance, where $B = \frac{1}{\pm X}$. When resistance and reactance are combined in a circuit, the term for ease of current flow is admittance (Y). Admittance is the reciprocal of impedance, where $Y = \frac{1}{Z}$. The unit of measure for all these terms is siemens (S).

▶ Some important relationships between admittance, susceptance, and conductance are $Y = G \pm jB$, $Y = \sqrt{G^2 + B^2}$, and $\theta = \frac{B}{G}$.

REVIEW QUESTIONS

(Be sure to make appropriate vector diagram sketches to help clarify your thinking.)

1. Refer to Figure 19–28a. Find Z, I, V_C, V_L, V_R, and θ using the Pythagorean theorem, Ohm's law, and appropriate trig functions. Draw impedance and V-I vector diagrams, labelling all parts, as appropriate.

2. Refer to Figure 19–28b. Find Z, I_C, I_L, I_R, I_T, and θ using the Pythagorean theorem, Ohm's law, and appropriate trig functions. Draw an appropriate V-I vector diagram and label all parts, as appropriate.

3. Calculate true power, apparent power, and reactive power for the circuit of Question 1. What is the power factor?

4. Draw a power triangle for the circuit in Question 1.

5. A series circuit has a 20-Ω resistance and a 40-Ω inductive reactance. How do you state the circuit Z in rectangular form? If the inductive reactance is changed to an equal value of capacitive reactance, how do you state Z?

6. A series RLC circuit is composed of 10-Ω resistance, 20-Ω inductive reactance and 30-Ω capacitive reactance. How is the circuit Z stated in rectangular form?

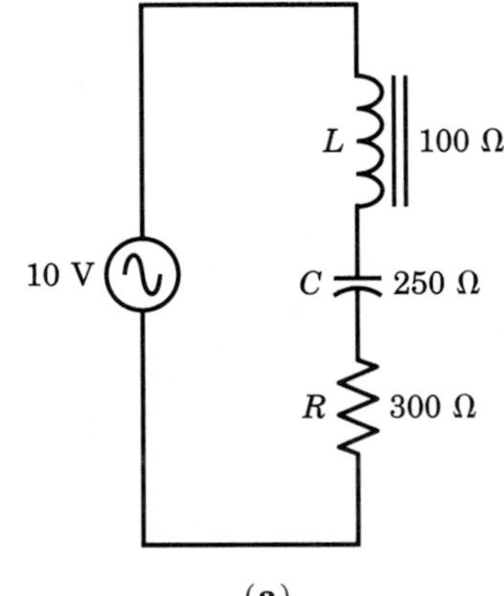

(a)

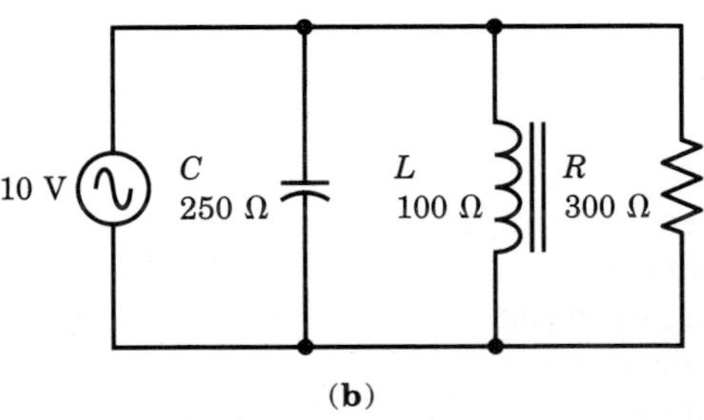

(b)

FIGURE 19–28

7. A parallel circuit with a 20-volt source has a resistive branch of 20 Ω and a capacitive reactance branch of 20 Ω. What is the total current stated in polar form?

8. Convert 30 + j40 to polar form.

9. Convert 14.14 $\angle 45°$ to rectangular form.

10. A series RLC circuit has a 10 V, 1 kHz source, an R of 8 kΩ, an X_L of 6 kΩ, and an X_C of 4 kΩ. What is the total circuit current? (Express in polar form.) What is the true power?

11. What is the total impedance of a series RLC circuit that has a 20-kHz ac source, C = 100 pF, L = 0.159 H, and R = 15 kΩ? (Express in polar form and in rectangular form.)

12. A series RLC circuit has 100 kΩ of resistance, a capacitor with 0.0005 μF of capacity and an inductor with 1.5 H of inductance. What is the value of voltage drop across the inductor, if the applied ac voltage is 100 volts at a frequency of 10 kHz? (Express in polar form.)

13. If the frequency of the applied voltage increases, what happens to V_C?

14. What is the true power dissipated by a series RLC circuit with 100 volts applied, if R = 30 Ω, X_L = 40 Ω, and X_C = 10 Ω?

15. A parallel RC circuit has a capacitor with 2 ohms of X_C and a resistor with 4 ohms of resistance. What is the impedance? (Express in polar form.)

16. A parallel RC circuit has a power factor of 0.5 and I_T = 5 A, f = 1 kHz, and R = 60 Ω. What is the value of C?

17. What is the total impedance of a parallel RLC circuit comprised of $R = 6\ \Omega$, $X_L = 12\ \Omega$, and $X_C = 24\ \Omega$?

TEST YOURSELF

1. A parallel RLC circuit has 400-Ω resistance, 500-Ω inductive reactance branch, and 800-Ω capacitive reactance branch. What is the circuit admittance value?
2. What is the Z and θ of the circuit in Question 1? (Assume a voltage applied of 800 volts.)
3. What value of X_C produces a $\theta = 0°$?
4. For the circuit in question 1, what happens to the following parameters if frequency doubles? (Answer with “I” for increase, “D” for decrease, and “RTS” for remain the same.)
   a. I_R will
   b. I_C will
   c. I_L will
   d. I_T will
   e. Z_T will
   f. θ will

Series and Parallel Resonance

Key Terms

Bandpass

Bandpass filter

Bandstop filter

Bandwidth

High-pass filter

$\frac{L}{C}$ Ratio

Low-pass filter

Q factor

Resonance

Selectivity

Skirts

Courtesy of Motorola, Inc.

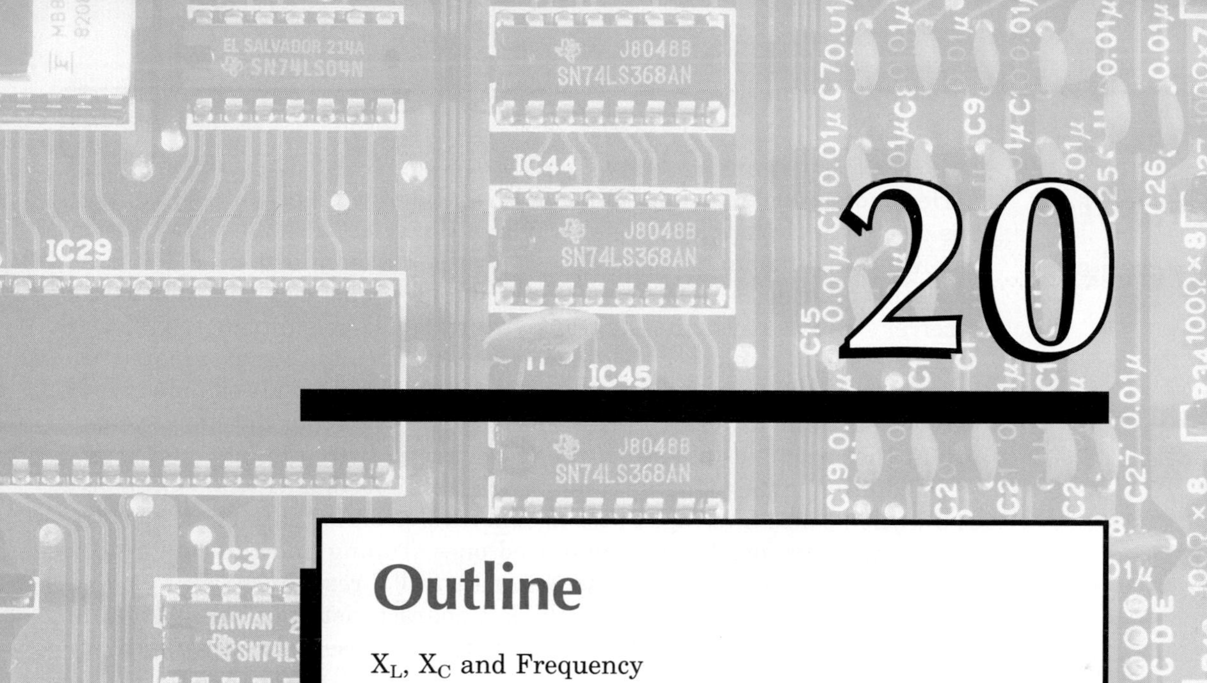

20

Outline

X_L, X_C and Frequency

Series Resonance Characteristics

The Resonant Frequency Formula

Some Resonance Curves

Q and Resonant Rise of Voltage

Parallel Resonance Characteristics

Parallel Resonance Formulas

Effect of a Coupled Load on the Tuned Circuit

Q and Resonant Rise of Impedance

Selectivity, Bandwidth and Bandpass

Measurements Related to Resonant Circuits

Filter Applications of Non-Resonant and Resonant RLC Circuits

Chapter Preview

Series and parallel resonant circuits provide valuable capabilities in numerous electronic circuits and systems. When you select a radio or TV station, you are using the unique capability of a resonant circuit to select a narrow group of frequencies from a wide spectrum of frequencies that impact the receiver's antenna system. The transmitter that is transmitting the selected signals also has a number of "tuned" resonant circuits that assure only the desired frequencies are being transmitted and that spurious signals are not being transmitted. Both the transmitting and receiving antenna systems are further examples of special tuned or resonant circuits.

In addition to the valuable tuning application, resonance allows a related application called *filtering*. Filtering can involve either passing desired signals or voltages, or blocking undesired ones. Tuning and filtering are the most frequent applications for series and parallel resonance.

In this chapter, you will study the key characteristics of both series and parallel resonant circuits. Your knowledge of these circuits will be frequently utilized throughout your career.

Objectives

After studying this chapter, you will be able to:

- List the key characteristics of series and parallel resonant circuits
- Calculate the resonant frequency of circuits
- Calculate **L** or **C** values needed for **resonance** at a given f_r
- Calculate **Q** factor for series and parallel resonant circuits
- Determine **bandwidth** and **bandpass** of resonant circuits
- Draw circuit diagrams for three types of **filters**

X_L, X_C AND FREQUENCY

Previously you learned that as frequency increases, X_L increases proportionately. The formula for inductive reactance expresses this by $X_L = 2\pi fL$. You also learned that as frequency increases X_C decreases in inverse proportion, expressed in the formula $X_C = \frac{1}{2\pi fC}$. Conversely, as frequency decreases, X_L decreases and X_C increases. In other words, X_L and X_C change in opposite directions as frequency changes.

To review, refer to Figure 20–1 and note the relationships between X_L, X_C and frequency. Notice X_C is not illustrated at frequencies approaching zero Hz. This is because X_C is approaching infinite ohms at these frequencies, thus is difficult to illustrate on such a graph.

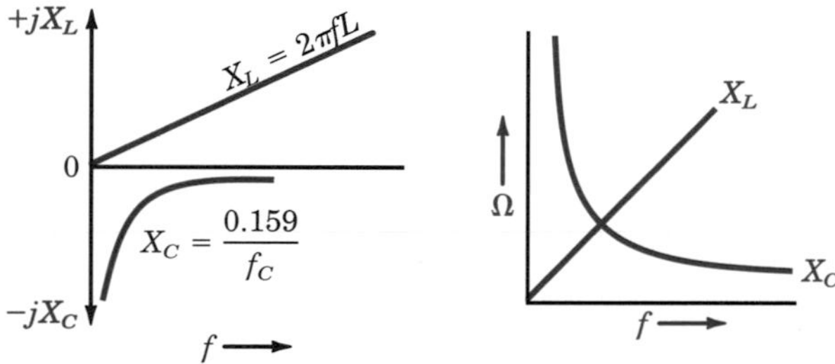

FIGURE 20–1 Relationships between X_L, X_C and frequency

SERIES RESONANCE CHARACTERISTICS

The implication of these statements about the relationships of X_L, X_C and frequency is that for a given set of L and C values, there is a specific frequency where the *absolute* magnitudes of X_L and X_C are equal. Look at Figure 20–2 and notice that for this series RLC circuit, there is one frequency where this occurs. In a series circuit containing L and C, the condition when the magnitude of X_L equals the magnitude of X_C is called **resonance** and the frequency where this occurs is the *resonant frequency*. At

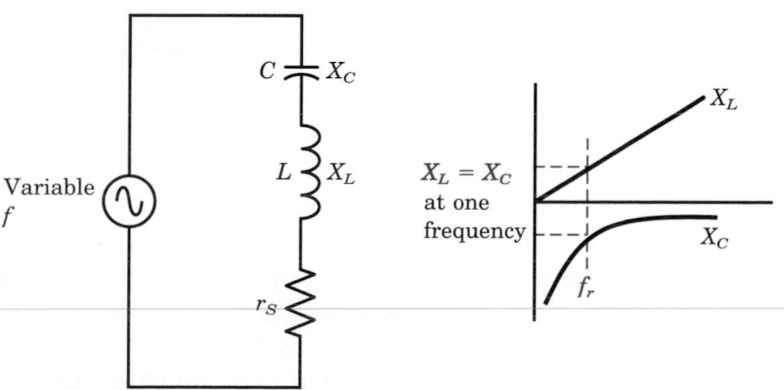

FIGURE 20–2 $X_L = X_C$ at resonant frequency (f_r)

resonance, some interesting observations can be made about circuit electrical parameters.

Refer to Figure 20–3 as you study the following list of features.

FIGURE 20–3 Sample series circuit parameters at resonance

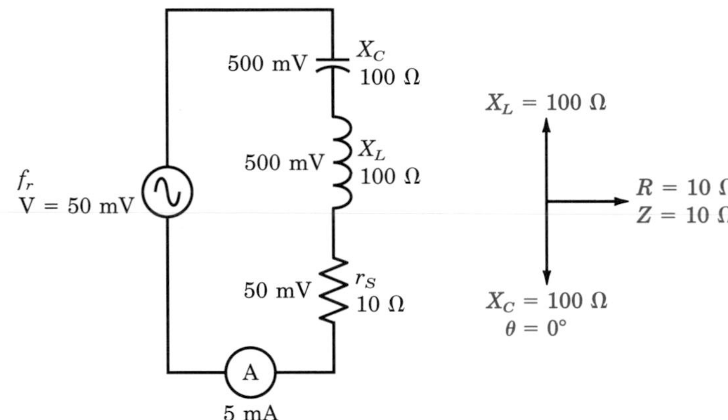

The characteristics of a series RLC circuit at resonance are:

1. $X_L = X_C$ (In this case, 100 Ω = 100 Ω.)
2. Z = minimum and equals R in circuit. (In this case, 10 Ω.)
3. I = maximum and equals $\frac{V}{R}$. (In this case $\frac{50 \text{ mV}}{10\ \Omega} = 5$ mA.)
4. V_C and V_L are maximum. ($I \times X_C = 500$ mV; $I \times X_L = 500$ mV.)
5. $\theta = 0°$ (Since circuit is acting purely resistive.)

Let's examine these statements.

1. When X_L equals X_C, the reactances cancel, leaving zero Ω of net reactance in the circuit. The remaining opposition to current, under these conditions, is any resistance in the circuit. Generally, most of this resistance is the resistance in the inductor windings (unless we have purposely added an additional discrete resistor component). The coil resistance is shown in Figure 20–3 as r_s, indicating series resistance.
2. Because the reactances cancel each other, the Z is minimum and equals the R in the circuit.
3. When Z is minimum (Z = R), I is maximum. That is, $I = \frac{V}{R}$.
4. V_C and V_L are maximum since I is limited only by the R value and is therefore maximum. The voltage across the capacitor equals $I \times X_C$, and the voltage across the inductor equals $I \times X_L$. As noted in our sample circuit, the voltage across the coil and the capacitor is 10

times V applied. There is a *resonant rise* in voltage across the reactive components. We will examine this fact again later in the chapter.

5. Again, because the reactances cancel each other, the circuit acts purely resistive; therefore, phase angle (θ) is zero.

IN-PROCESS LEARNING CHECK I

Fill in the blanks as appropriate.

1. In circuits containing both inductance and capacitance, when the source frequency increases, inductive reactance ____________ and capacitive reactance ____________.
2. In a series RLC (or LC) circuit, when X_L equals X_C, the circuit is said to be at ____________.
3. In a series RLC (or LC) circuit, when X_L equals X_C, the circuit impedance is ____________; the circuit current is ____________; and the phase angle between applied voltage and current is ____________ degrees.
4. In a series RLC (or LC) circuit, when X_L equals X_C, the voltage across the inductor or the capacitor is ____________.

Since you now know some of the series resonance effects, you may wonder how the frequency of resonance is determined.

Let's deviate from examining circuit parameters to look at the basic formulas that allow you to determine the resonant frequency for a circuit with given component values, or to find the appropriate component values for a desired resonant frequency.

THE RESONANT FREQUENCY FORMULA

The Basic Formula

In our previous discussions, a fixed value of L and C were assumed, while f was changed until X_L equaled X_C. With fixed L and C values, there is only one possible resonant frequency. In many practical applications, either the C or L value or both can be varied to achieve resonance at various frequencies. An example is tuning in different radio stations on a receiver. Also, different $\frac{L}{C}$ **ratios** for any given frequency can be obtained. You can see that a technician should know how to determine the resonant frequency of any given set of LC parameters. Let's look briefly at the derivation and formula for frequency of resonance. Starting with resonance $X_L = X_C$, where L is in henrys, C is in farads, and f is in Hz, the derivation is $X_L = X_C$.

$$2\pi fL = \frac{1}{2\pi fC}$$

$$2\pi f_r L = \frac{1}{2\pi f_r C}$$

$$2\pi L(f_r)^2 = \frac{1}{2\pi C}$$

$$f_r^2 = \frac{1}{(2\pi)^2 LC}$$

Formula 20–1	$f_r = \frac{1}{2\pi\sqrt{LC}}$ OR $\frac{0.159}{\sqrt{LC}}$

EXAMPLE

What is the resonant frequency of a series LC circuit having a 200-μH inductor and a 400-pF capacitor?

Answer:

$$f_r = \frac{0.159}{\sqrt{LC}}$$

$$f_r = \frac{0.159}{\sqrt{200 \times 10^{-6} \times 400 \times 10^{-12}}}$$

$$f_r = \frac{0.159}{\sqrt{80{,}000 \times 10^{-18}}}$$

$$f_r = \frac{0.159}{283 \times 10^{-9}} = 561.84 \text{ kHz}$$

Now you try one!

PRACTICE PROBLEM I

Answers are in Appendix B.

What is the resonant frequency of a series LC circuit containing an inductor with an inductance of 250 μH and a capacitor with a capacitance of 500 pF?

Variations of the f_r Formula. As you can see from the basic resonant frequency formula, the L and C values determine the frequency of resonance. That is, the larger the LC product is, the lower is the f_r, and vice versa, Figure 20–4a. Note also in Figure 20–4b how various combinations of L and C can be resonant at one particular frequency (in this case, 1 MHz).

It is frequently useful to find the L value needed with an existing C value or the C value needed with an existing L value to make the circuit resonant

(a)

L (μh)	C (pF)	f_r (MHz)
100	100	1.59
100	200	1.13
400	400	0.398
400	10	2.52

$\left(\text{Using } f_r = \frac{0.159}{\sqrt{LC}}\right)$

Different LC products cause different f_rs

(b)

L (μh)	C (pF)	f_r (MHz)
500	51	1
400	63.75	1
300	85	1
200	127.5	1

$\left(\text{Using } f_r = \frac{0.159}{\sqrt{LC}}\right)$

Different values having same LC product yield same f_r

FIGURE 20–4
(a) Values of L and C determine f_r and (b) different combinations of L and C can produce same f_r.

at some specified frequency. By rearranging the resonant frequency formula (i.e., squaring both sides to delete the square root portion), we can find L or C, if the other two variables are known. When this is done, the following useful formulas are derived:

$$L = \frac{1}{(2\pi)^2 f_r^2 C} = \frac{1}{4\pi^2 f_r^2 C} =$$

Formula 20–2	$L = \frac{0.02533}{f_r^2 C}$

$$C = \frac{1}{(2\pi)^2 f_r^2 L} = \frac{1}{4\pi^2 f_r^2 L} =$$

Formula 20–3	$C = \frac{0.02533}{f_r^2 L}$

NOTE: These formulas are for your information—You need not memorize them!

Other Useful Variations. When dealing in radio frequency circuits, it is sometimes inconvenient to manage henrys, farads and hertz, as in the previous formulas. It is often more convenient to deal with microhenrys (μH), picofarads (pF), and megahertz (MHz). The following formulas are based on these convenient units.

$$f^2 = \frac{25{,}330}{LC} \text{ or}$$

Formula 20–4	$L = \frac{25{,}330}{f^2C}$

where L is in μH, C is in pF, f is in MHz
or

Formula 20–5	$C = \frac{25{,}330}{f^2L}$

NOTE: Again, these formulas are for your information, but you may find occasion to use them.

EXAMPLE
A circuit has an L value of 100 μH. What C value will make the circuit resonant frequency equal 1 MHz?
Answer:

$$\text{Using } C = \frac{0.02533}{f_r^{\,2}L}$$

where L is in H, C in F, f in Hz.

$$C = \frac{0.02533}{(1 \times 10^6)^2 \times 100 \times 10^{-6}}$$

$$C = \frac{0.02533}{1 \times 10^{12} \times 100 \times 10^{-6}}$$

$$C = \frac{25{,}330 \times 10^{-12}}{100} = 253.3 \text{ pF}$$

$$\text{Using } C = \frac{25{,}330}{f^2L}$$

where L in μH, C in pF, and f in MHz.

$$C = \frac{25{,}330}{f^2L} = \frac{25{,}330}{100} = 253.3 \text{ pF}$$

Incidentally, you can also use the X_C equals X_L approach. Since you know the frequency and the L, you can calculate X_L using the $X_L = 2\pi fL$ formula. Then, knowing that X_C is that same value at resonance, you use the X_C formula to solve for C. That is, $C = \frac{0.159}{fX_C}$, Figure 20–5. Now you apply the appropriate formula(s) to solve for an unknown L when C and f_r are known.

Using basic units: $L = H$, $C = F$, and $f_r = Hz$
$L = \frac{0.02533}{f_r^2 C}$ and $C = \frac{0.02533}{f_r^2 L}$
Using convenient units: $L = \mu H$, $C = pF$, and $f_r = MHz$
$L = \frac{25{,}330}{f_r^2 C}$ and $C = \frac{25{,}330}{f_r^2 L}$
Using the $X_L = X_C$ approach:
If you know X_C, then $L = \frac{X_C}{2\pi f_r}$
If you know X_L, then $C = \frac{0.159}{f_r X_L}$

FIGURE 20–5 Some useful variations derived from the f_r and reactance formulas

PRACTICE PROBLEM II

What L value is needed with a 100-pF capacitor to make the circuit resonant frequency equal 5 MHz?

SOME RESONANCE CURVES

Figure 20–6 shows the resonance curves (graphic plots) of current, impedance, and phase angle versus frequency for a series RLC circuit.

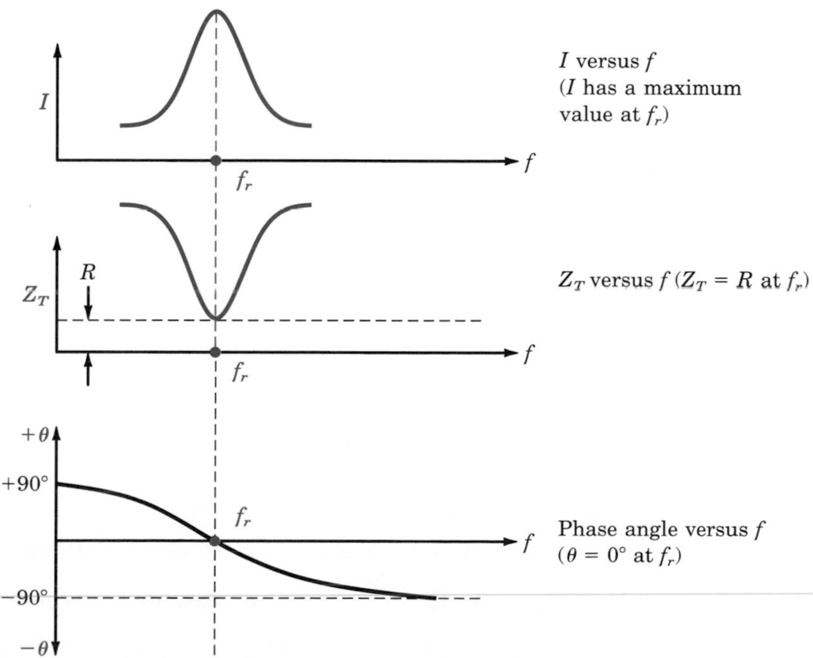

FIGURE 20–6 Plots of I, Z and θ versus frequency

Plot of Current (I) Versus Frequency (f)

When frequency is considerably below the circuit resonant frequency, the current is low compared to current at resonance. This is because $Z = \sqrt{r^2 + (X_C - X_L)^2}$; therefore, circuit impedance (Z) is much higher than it is at resonance. For series RLC circuits below resonance, the net reactance is capacitive reactance. The same pattern is true above resonance, except the net reactance for a series RLC circuit is X_L. (NOTE: For series resonant circuits, plotting current rising to maximum value at resonance is the most significant aspect of the response curves.)

Plot of Impedance (Z) Versus Frequency

As you would expect, impedance is lowest at the resonant frequency, since the reactances cancel each other at resonance. The impedance graph in Figure 20–6 shows that Z equals R (resistance) at resonance. Thus, for any circuit, when Z is low, I is high and when Z is high, I is low.

Plot of Phase Angle Versus Frequency

At resonance, the circuit acts resistively and the phase angle is 0°. (This is because the reactances have cancelled.) Below resonance the circuit acts capacitively, since the net reactance is capacitive reactance. Above resonance, the circuit acts inductively since the net reactance is X_L.

Q AND RESONANT RISE OF VOLTAGE

Q and the Response Characteristic

The shape of a given circuit's resonance response relates to an important factor called the **Q factor,** or figure of merit, Figure 20–7. The higher the Q factor is, the sharper is the response characteristic. This sharp response curve is sometimes described as having steep **skirts,** or as indicating a circuit with high **selectivity** (i.e., one that is very frequency selective). The lower the Q factor is, the broader, or flatter the response curve and the less selective the circuit is.

Q Related to $\frac{X_L}{R}$

The Q factor of a resonant circuit relates to two parameters. The first parameter is associated with the ratio of reactance to resistance at the resonant frequency. In formula form, $Q = \frac{X_L}{r_s}$ or $\frac{X_C}{r_s}$ at resonance, where r_s is the series resistance in the circuit, which often is the coil resistance. In other

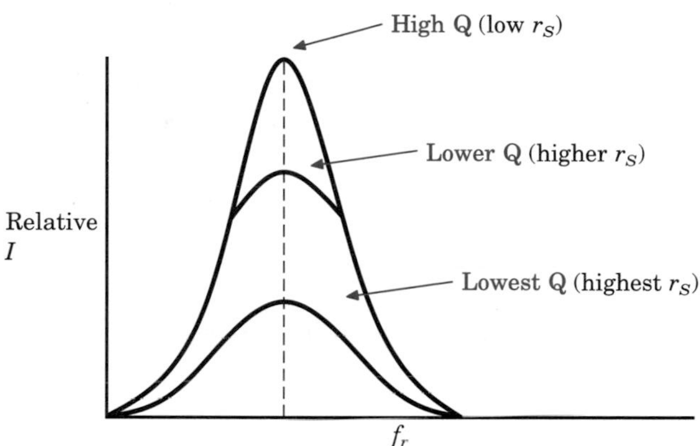

NOTE: r_S value does not change f_r but does affect response.

$$Q = \frac{X_L}{r_S}$$

FIGURE 20–7
Response curves versus circuit Q

words, the Q of the coil (X_L over R) is frequently the same as the Q of the resonant circuit. Obviously, the lower the series circuit resistance is, the higher the Q is and the sharper the resonant circuit response. The higher the r_s is, the lower the Q is and the flatter the response. Note that since I is the same throughout the circuit, the ratio of $\frac{I \times X_L}{I \times R}$, or $\frac{V_L}{V_R}$ is also the same as the Q factor. Likewise, the ratio of reactive power to true power in the circuit also equals the Q factor.

Q Related to $\frac{L}{C}$ Ratio

A second related factor that affects the Q is the ratio of L to C; that is $\frac{L}{C}$. Since both L and C values affect the frequency of resonance, a *given* resonant frequency can be achieved by various LC values, refer again to Figure 20–4b. However, different combinations of LC values have different reactances at resonance. For this reason, it is best to have a *high* $\frac{L}{C}$ ratio to achieve a high X_L at resonance; hence, a high $\frac{X_L}{R}$ or Q factor, Figure 20–8.

Resonant Rise of Voltage

Refer to Figure 20–9 and note the voltages around the series circuit. The term *magnification factor* describes the fact that voltage across each of the reactive components at resonance is Q times V applied. This is because of

FIGURE 20–8
Q related to $\frac{L}{C}$ ratios

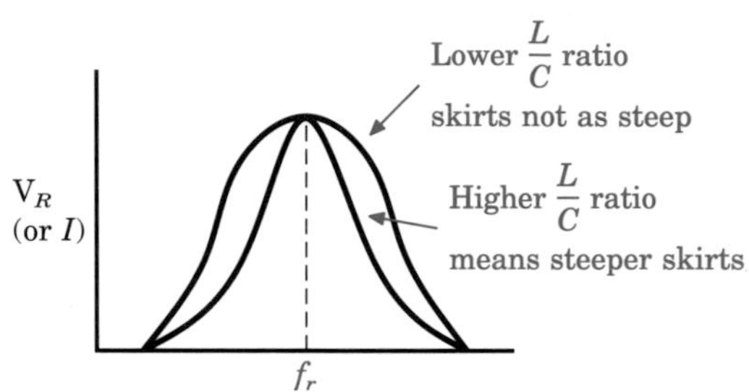

FIGURE 20–9
Magnification factor

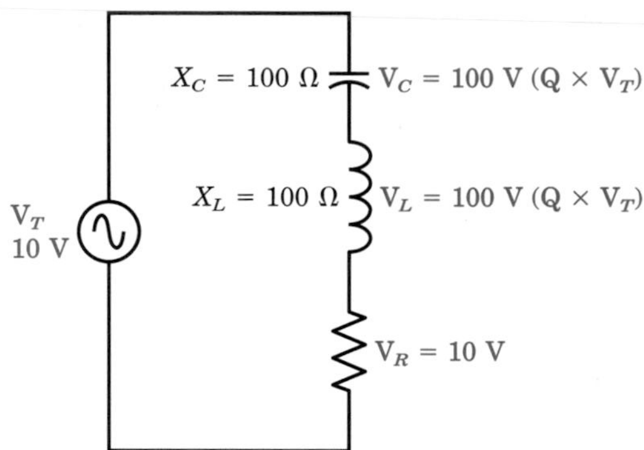

the resonant rise of I at resonance, which is due to the reactances cancelling each other at the resonant frequency. Since I is limited only by R at that frequency, and since each reactive voltage drop is I times X with X being very high, the reactive voltage can be very high. In fact, at resonance, the reactive voltages are Q times V applied; thus, voltage has been magnified (Q times) across the reactive components. In other words, the higher the Q is, the greater the magnification is. As stated earlier, the ratio of reactive voltage to applied voltage at resonance equals the Q. If we measure the voltage across one reactive component and call it V_{out} and call V applied V_{in}, then $Q = \frac{V_{out}}{V_{in}}$.

PRACTICE PROBLEM III

1. What is the Q of the circuit in Figure 20–10?
2. What is the V_C value at resonance, if the applied voltage in Figure 20–10 is 12 volts?
3. If r_s doubles in this circuit, what is the Q value?
4. If r_s is a smaller value than shown in Figure 20–10, are the response curve skirts more or less steep?

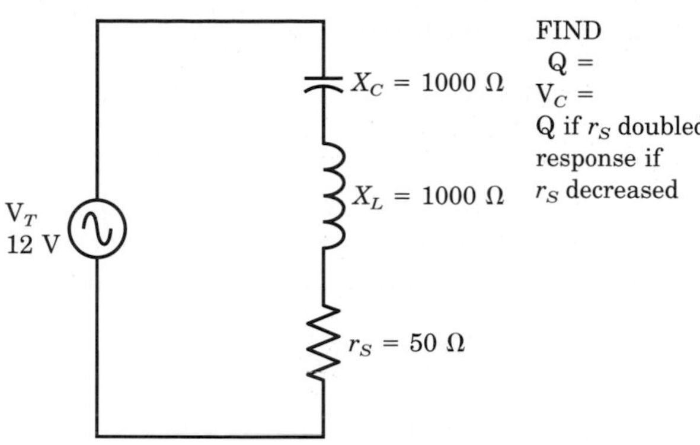

FIGURE 20–10

PARALLEL RESONANCE CHARACTERISTICS

A parallel resonant circuit is a frequently used circuit in receivers, transmitters, and frequency measuring devices. As you might expect, many parallel resonant circuit features are opposite compared to those of a series resonant circuit. Look at Figure 20–11 as you study the following discussion.

Highlighting the basic features of a high-Q parallel resonant circuit:

1. Z = maximum (and is purely resistive)
2. Line current = minimum
3. Current inside *tank* = maximum
4. $X_L = X_C$ (ideal components)
5. $\theta = 0°$ (and p.f. = 1)

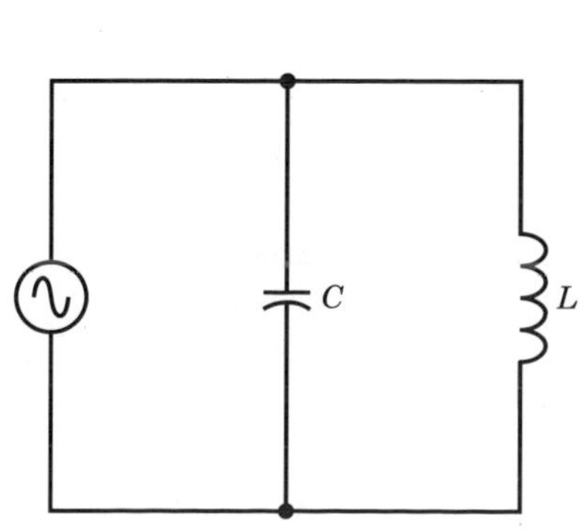

Ideal L and C (no R losses)
At resonance:
Z = maximum (∞)
I line = minimum (0)
$X_L = X_C$
$\theta = 0°$

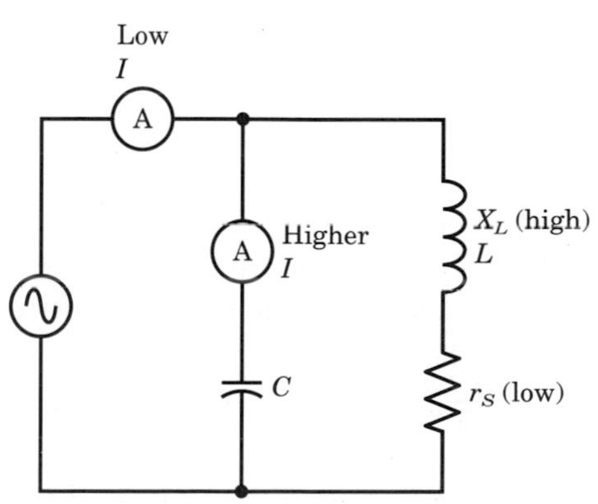

Practical circuit
At resonance:
Z = high (max)
I line = low (not 0)
$X_L = X_C$
$\theta \cong 0°$ when $X_L = X_C$

FIGURE 20–11 Parallel resonant circuit characteristics

Let's examine these statements further.

1. Z is maximum at resonance because the inductive and capacitive branch currents are 180° out-of-phase, thus cancel each other (assuming ideal L and C components). This means the vector sum of the branch currents (or I line) is very small. If the inductor and capacitor are perfect, the Z is infinite. Since practical inductors and capacitors have resistances and losses, there is an effective R involved, Figure 20–11. $\left(\text{For your information, } Z_T = \frac{L}{C_{r_s}}.\right)$
2. Since Z is maximum at resonance, obviously, the line current (or total current supplied by the source) is minimum.
3. Within the LC (tank) circuit, the circulating current can be very high. In fact, the current inside the tank circuit equals Q × I line and the total circuit impedance is $Q \times X_L$.
4. When $X_L = X_C$, the above characteristics are exhibited. In practical parallel resonant circuits having less-than-perfect components, some r_s value associated with the coil exists. This means when $X_L = X_C$, the Z_L of the inductive branch is slightly greater than the X_C of the capacitive branch. In low-Q circuits, this means the frequency needs to be slightly lower to allow more I_L, so the inductive branch current equals the capacitive branch current. For circuits having a Q above 10, however, this effect is negligible.
5. At resonance (for high-Q circuits), circuit current and applied voltage is virtually in-phase.

PRACTICAL NOTES

For low-Q circuits, the condition of unity power factor occurs at a slightly different frequency than when $X_L = X_C$. The frequency of p.f. = 1 is sometimes termed antiresonance, or the antiresonant frequency.

PARALLEL RESONANCE FORMULAS

For Q Greater than 10

When the circuit Q is high (when the r_s is low, or the equivalent R_p is high), the resonant frequency is determined by the LC component values as it is in

IN-PROCESS LEARNING CHECK II

1. In a high-Q parallel resonant circuit, Z is ________.
2. In a high-Q parallel circuit using ideal components, X_L ________ equal X_C at resonance.
3. The power factor of an ideal parallel resonant circuit is ________.
4. Phase angle in an ideal parallel resonant circuit is ________.
5. In a parallel resonant circuit, the current circulating in the tank circuit is ________ than the line current.

series resonant circuits. With Qs greater than 10, the resonant frequency is calculated with the same formula used for series resonance. That is:

$$f_r = \frac{0.159}{\sqrt{LC}}$$

For Q Less than 10

When the circuit Q is low (i.e., when r_s is high or effective R_p is low, or when a reflected impedance from a load coupled to the tuned circuit is low causing an effective low value of equivalent R_p), the resistance, L, and C values enter in the formula for parallel resonance. That is, the resistance has a significant effect on the frequency of resonance so it should be considered. The following formula (only for your information) illustrates that resistance affects resonance in parallel LC circuits.

Formula 20–6	$f_r = \frac{0.159}{\sqrt{LC}} \times \sqrt{1 - \frac{Cr_s^2}{L}}$

In most cases, the Q of parallel tuned circuits is high enough so that the effect of the resistance on the resonant frequency can be generally neglected.

To check your grasp of parallel resonant circuits thus far, try to solve the following problems.

PRACTICE PROBLEM IV

1. Calculate the resonant frequency of a parallel LC circuit with the following parameters: L = 200 μH having a resistance of 10 Ω and C = 20 pF.
2. What is the Q of the circuit in Question 1?

3. What is the impedance at resonance of the circuit in Question 1?
4. Is the resonant frequency of the circuit in Question 1 lower or higher if $r_s = 1\ k\Omega$?

EFFECT OF A COUPLED LOAD ON THE TUNED CIRCUIT

Whenever a parallel resonant circuit delivers energy to a load (i.e., a load is coupled to the tuned circuit), the Q of the tank circuit is affected, Figure 20–12. If the power dissipated by the coupled load is greater than 10 times the power lost in the tank circuit reactive components, the tank impedance is high compared to the load impedance. Thus, for all practical purposes, the impedance of the parallel combination of the tank and load virtually equals the load resistance value. When a tank is heavily loaded by such a resistive load impedance, the Q of the system equals the parallel value of Z tank and R over X, or $Q = \frac{R}{X}$, where R is the parallel load resistance and X is the reactance. (NOTE: This is the inverse of the $\frac{X_L}{R}$ formula used for Q in series resonance and high-Q parallel resonant circuits). For example, if a resistive load of 4000 ohms is connected across a parallel resonant circuit where X_C (or X_L) = 400 ohms, the Q equals $\frac{R}{X_L} = \frac{4000}{400} = 10$.

FIGURE 20–12 Effect of coupling a load to a parallel resonant circuit

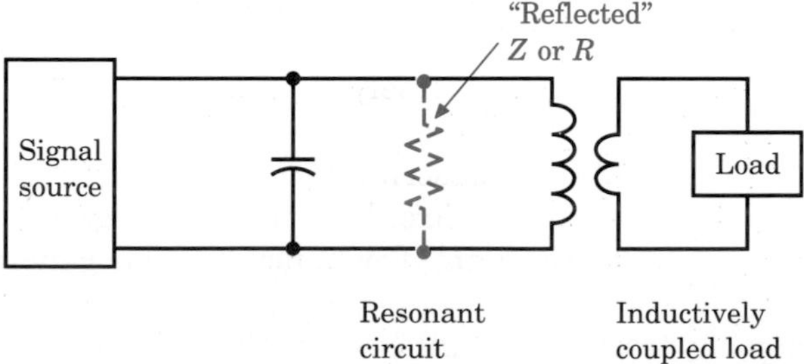

The lower the reflected R value, the lower the Q. If load is "heavy" (i.e., much energy absorbed by load) $Z_T \approx Z$ of load and $Q = \frac{R_p}{X_L}$.

Q AND THE RESONANT RISE OF IMPEDANCE

The preceding discussions show us that the current characteristic is the most meaningful response factor in series resonant circuits, where the impedance characteristic is the most significant response factor in parallel resonant circuits. The response characteristics of a parallel tuned circuit are also related to the Q factor. In Figure 20–13 you can see there is a resonant rise of impedance as the resonant frequency is approached. As we previously stated, the impedance at resonance is Q × X_L. You can also see from this illustration that the higher the Q is, the sharper the response characteristics, and the lower the Q is, the flatter and broader the response.

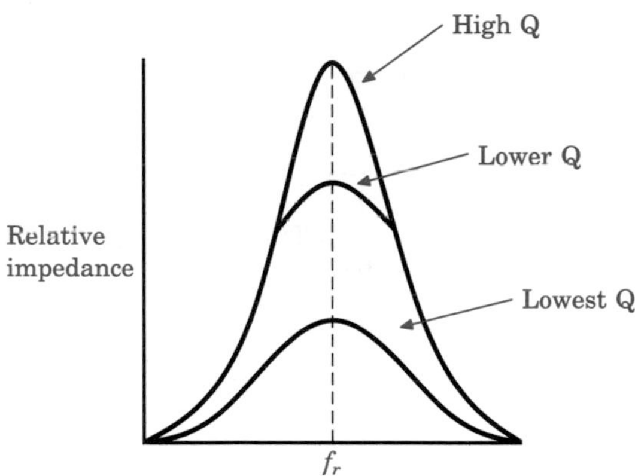

FIGURE 20–13 Resonant rise of *Z* for a parallel resonant circuit

SELECTIVITY, BANDWIDTH AND BANDPASS

Selectivity

The sharper the response curve of a resonant circuit, the more selective it is. This means that only frequencies close to resonance appear near the maximum response portion of the resonant circuit response. This indicates that in series resonant circuits, the current value *falls off* rapidly from its maximum value at frequencies slightly below or above the resonant frequency. In parallel resonant circuits, the impedance *falls off* rapidly from the maximum value obtained at resonance. Conversely, resonant circuits that have less selectivity allow a wider band of frequencies close to their maximum response levels. **Selectivity,** used to describe a resonant circuit, refers to the capability of the circuit to differentiate between frequencies. Refer to Figure 20–14 for examples of various selectivities versus response characteristics.

Bandwidth and Bandpass

You have looked at various response curves and have noticed that none have perfect vertical skirts. Therefore, to define **bandwidth,** some value or level on the response curve is chosen that indicates all frequencies below that level are rejected and frequencies above the chosen value are selected. The accepted level is 70.7% of maximum response level.

For series resonant circuits, we consider those two points on the response curve where current is 70.7% of maximum. For parallel resonant circuits, the points where Z is 70.7% of maximum are used.

FIGURE 20–14 Examples of various selectivities

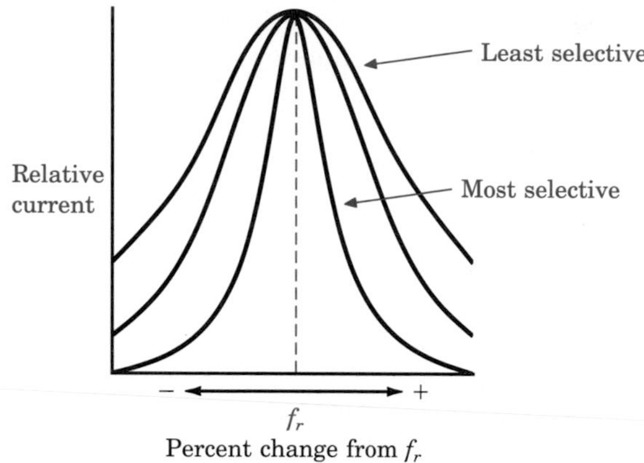

Total bandwidth (BW) is calculated as:

Formula 20–7	$\Delta f = f_2 - f_1$

where: Δf = BW
f_2 = higher frequency 0.707 point
f_1 = lower frequency 0.707 point on response curve

$$\text{Also, } f_1 = f_r - \frac{\Delta f}{2}, \text{ and } f_2 = f_r + \frac{\Delta f}{2}.$$

Notice that if $P = I^2R$ and at f_1 and f_2 the current level is 0.707 of maximum, then at those points on the response curve, $P = (0.707\ I_{max})^2R = 0.5\ P_{max}$. For this reason, these points (f_1 and f_2) are often called *half-power points,* Figure 20–15.

FIGURE 20–15 Bandwidth and bandpass

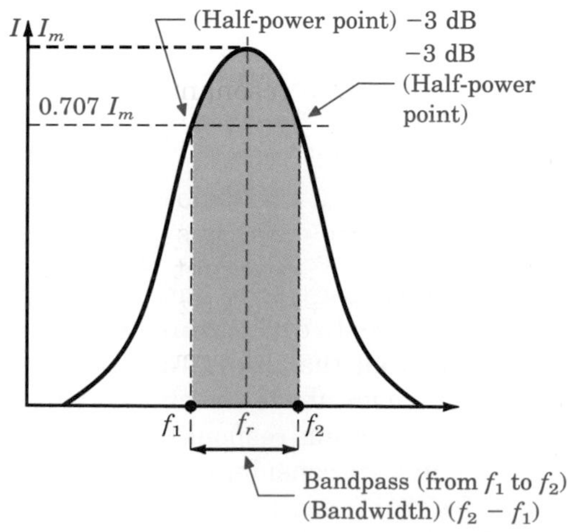

Bandwidth Related to Q

You have learned the Q of a resonant circuit is the primary factor determining the response curve shape. Thus, BW and Q are related. The formula showing this relationship is:

Formula 20–8	$\Delta f = \frac{f_r}{Q}$

Rearranging this formula, we can also say that:

Formula 20–9	$Q = \frac{f_r}{\Delta f}$

Bandpass

As has been discussed, the bandwidth is the width of the band of frequencies where a given resonant circuit responds within specified criteria. For example, a given resonant circuit responds over a 20-kHz bandwidth with levels at or above 70.7% of maximum response level.

On the other hand, **bandpass** is the specific frequencies at the upper and lower limits of the bandwidth. For example, if the resonant frequency is 2000 kHz, and the circuit has a bandwidth of 20 kHz, the bandpass is from 1990 kHz to 2010 kHz, (assuming a symmetrical response). This indicates the specific band of frequencies passed is from 1990 to 2010 kHz, and frequencies outside this limit cause a response that falls below the 70.7% level, or below the half-power points (sometimes expressed as the −3 dB points, as you will learn later). Again, Figure 20–15 illustrates bandwidth and bandpass.

EXAMPLE

1. What is the total bandwidth of a resonant circuit if f_r is 1000 kHz and f_1 is 990 kHz?
   Answer:
   Since one half-power point is 10 kHz below the resonant frequency, it is assumed that if the response curve is symmetrical, the other half-power point is 10 kHz above the resonant frequency. This means the total BW = 20 kHz. That is, $f_1 = 990$ kHz and $f_2 = 1010$ kHz.

2. What is the bandwidth of a resonant circuit if resonant frequency is 5 kHz and Q is 25?
   Answer:
   $\Delta f = \frac{f_r}{Q} = \frac{5000}{20} = 250 \text{ Hz}.$

3. What is the bandpass of the circuit described in Question 2?
Answer:
Bandpass = from 5000 − 125 Hz to 5000 + 125 Hz = 4875 to 5125 Hz.

PRACTICE PROBLEM V

1. In a resonant circuit, f_r is 2500 kHz and Δf is 12 kHz. What are the f_1 and f_2 values?
2. In a resonant circuit, Q is 50 and bandwidth is 10 kHz. What is the resonant frequency and the bandpass?
3. What is the f_1 value in a circuit whose resonant frequency is 3200 kHz and whose bandwidth is 2 kHz?

MEASUREMENTS RELATED TO RESONANT CIRCUITS

Because resonant circuits are so prevalent in electronics, it is worthwhile to know some of their practical measurements. These measurements include measuring resonant frequency; determining Q and magnification factor; determining maximum current (for series resonant circuits) and maximum impedance (for parallel resonant circuits); determining bandwidth and bandpass; measuring the *tuning range* of a tunable resonant circuit (with variable C or L); and determining the tuning ratio. Let's briefly look at each of these measurements.

1. Measuring resonant frequency
   There are several ways to determine resonant frequency. One way is to use a metering device that is coupled to the resonant circuit. Examples are the old *grid-dip meter,* the newer *tunnel-diode dipper,* and various types of frequency meters and frequency counters. Also, the versatile *oscilloscope* is a good device to measure resonant frequency. It is beyond this chapter to teach you how to use these devices, but keep them in mind. Eventually you will get some practical experience using them to measure frequency, Figure 20–16.
2. Determining Q and magnification factor
   In a series resonant circuit, using a high input impedance DVM, or an oscilloscope allows the measurement of voltage across one reactive component (generally, the capacitor). When the circuit is tuned to resonance, the voltage is maximum. This voltage is then compared to the voltage source to determine the Q and the voltage magnification factor. That is, $Q = \frac{V_{out}}{V_{in}}$, Figure 20–17.
3. Determining maximum current (I) for series and maximum impedance (Z) for parallel resonant circuits.

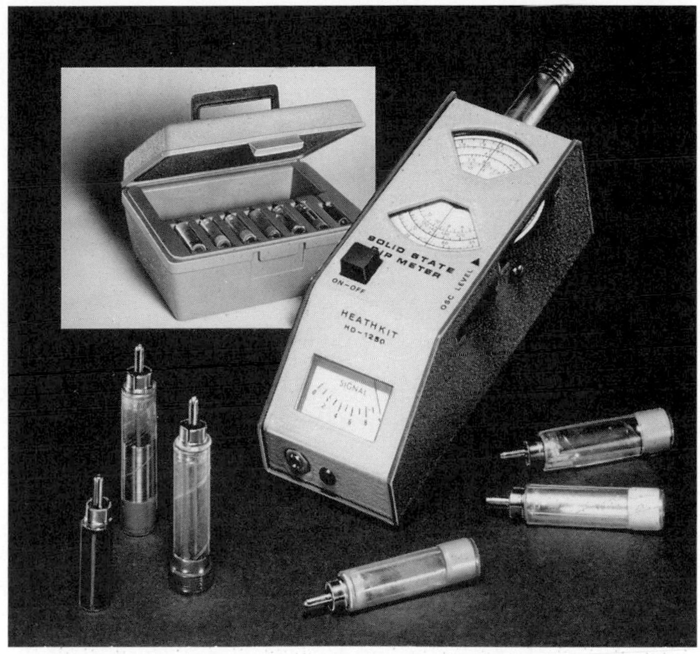

(a)

(b)

FIGURE 20–16 Devices used to determine f_r. (Photo a courtesy of Heath Zenith Electronics; photo b courtesy of Hewlett Packard)

FIGURE 20–17 Determining Q of series resonant circuit

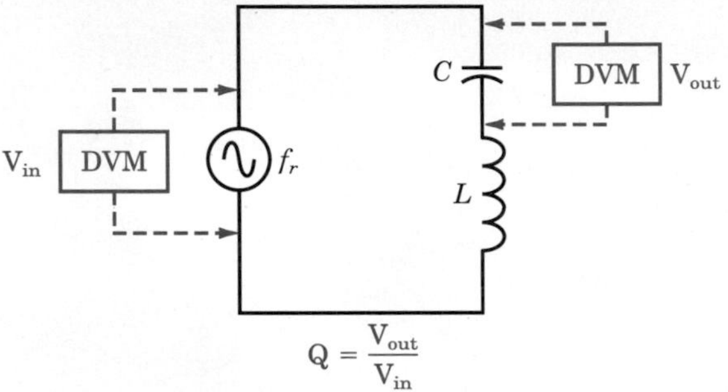

In a series resonant circuit, a current meter or a voltmeter reading across a series resistor is used to determine maximum current. Read the current meter directly, or use Ohm's law where $\frac{V_R}{R} = I$ to determine current when the circuit is tuned for maximum current or resonance, Figure 20–18.

FIGURE 20–18 Determining maximum *I* for a series resonant circuit

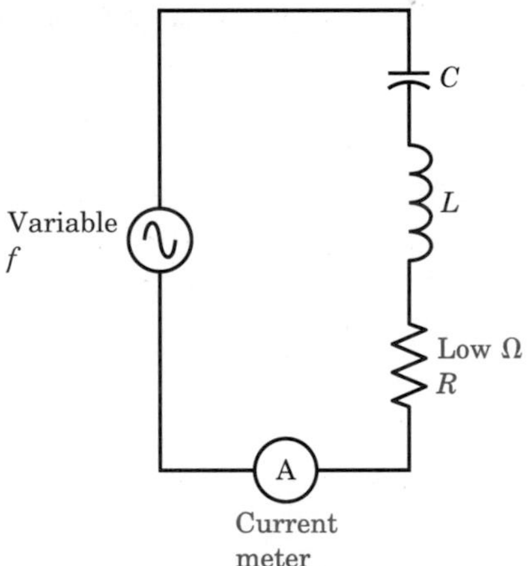

Use a current meter, *or*, insert low *R* value, measure V, and then calculate *I*. $\left(I = \frac{V}{R}\right)$

In a parallel resonant circuit, use a variable R in series with the source and adjust it so that Vs across the tuned circuit and across the variable R are equal (at resonance). Remove the R from the circuit and measure its value. Its value is the same as the Z of the tuned circuit at resonance since the V drops were equal. Since we can calculate X_L from the resonant frequency formula and we know that $Q = \frac{Z_T}{X_L}$, we can determine the Q, Figure 20–19.

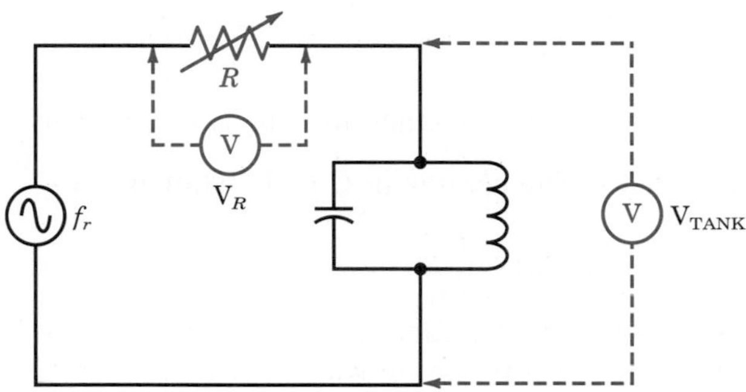

FIGURE 20–19 Determining maximum *Z* for a parallel resonant circuit

4. Determining bandwidth and bandpass
   For a series resonant circuit, tune the circuit through resonance and note the frequencies where current is 70.7% of maximum value. The difference between the lower frequency and the higher frequency is the bandwidth. The actual frequencies at those two points represent the bandpass.

   For a parallel resonant circuit, you can use the Z determining system and find the frequencies where Z is 70.7% of maximum value. Again, the difference between those frequencies is the bandwidth, and the precise lower-to-upper frequencies are the bandpass, Figure 20–20.

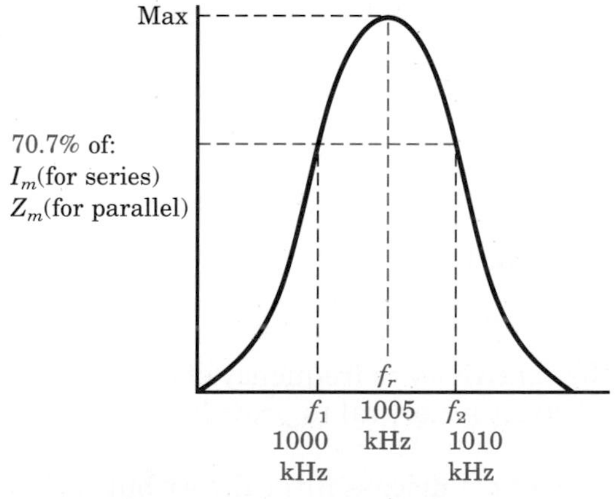

Bandpass = 1000–1010 kHz
Bandwidth = 10 kHz
f_r = 1005 kHz

FIGURE 20–20 Determining bandwidth and bandpass

5. Measuring the tuning range
   If the tuned circuit has a variable C or L, or if both are variable, there are limits to the range of resonant frequencies to which the circuit can be tuned.

   The various resonant frequency measurement devices previously mentioned can check the actual resonant frequencies that are capable of being tuned by the circuit. Simply measure the frequency at both ends of the tuning extremities, and you will know the circuit tuning range.

6. Determining the tuning ratio
   From the resonant frequency formula, you can reason that the frequency change caused by a change in C or L is inversely proportional to the square root of the *change* in C or L. That is, $f_r = \frac{0.159}{\sqrt{LC}}$; therefore, f_r is proportional to $\frac{1}{\sqrt{LC}}$.

   If it is desired to tune through a range of frequencies from 1000 to 6000 kHz, this means this tuning ratio of 6:1 (highest to lowest frequency) requires a change of C or L in the ratio of 36:1. If C is the variable component and its minimum C value is 10 pF, it has to have a maximum C value of 36×10, or 360 pF to tune the desired range.

SAFETY HINTS

When working around equipment with resonant circuits, particularly in transmitter circuits, be aware that it is possible to get bad *rf burns* by touching or getting too near a live circuit or component. In addition, these circuits often have high dc voltages, as well as high radio frequency ac energy—If touched while the circuit is live, injury or death can result!

FILTER APPLICATIONS OF NON-RESONANT AND RESONANT RLC CIRCUITS

Introduction

Since you have learned about the sensitivity to frequency of reactive components and various resonance effects, it is natural to provide an introduction to filters.

This brief section about filters will not discuss filter theory but will introduce some filter networks and their common applications. Two general filter classifications that will be discussed are non-resonant filters and resonant filters. There are numerous ways to classify filters. For example, there are *active filters* (involving some amplification or use of devices such as transistors or integrated circuits, along with RCL components) and *passive filters* (which only use passive components, such as resistors, capacitors, or inductors).

Another way filters are categorized is by the frequency range where they are used. For example, power supply filters typically operate in frequencies of 60 Hz, 120 Hz, 400 Hz, and so forth. Audio filtering networks obviously operate in audio frequency environments. Rf filters are used at frequencies above the audio range in receivers, transmitters, and other rf systems.

Names of filter networks or individual filter components often reflect what they do, for example, low-pass, high-pass, bandpass, bandstop, smoothing, decoupling, and so on. Another common way to refer to filter networks is the way the components are *laid out,* for example, L-shaped, pi-shaped, T-shaped, and so on. With these few classifications in mind, let's look at some filter networks.

A General Principle

Remember this general principle while analyzing any type filter: A device or circuit that has low opposition to a given frequency or band of frequencies and is in series with the output or load passes that frequency or band of frequencies along to the output or load. A device or circuit that has a low opposition to a given frequency or band of frequencies and is in parallel (or shunt) with the output or load bypasses or prevents that frequency or band of frequencies at the output or load. The converse of these two statements is also true. High oppositions in series with the signal path prevent signals from arriving at the load. High oppositions in parallel allow signals at the output or load.

Non-Resonant Filters

Low-Pass Filters. A **low-pass filter** passes along low frequency components of a given signal or waveform, while impeding the passage of higher frequency components. One example of a low-pass filter application is the filter in a power supply, designed to pass the dc component of the rectifier circuit output, while preventing the ripple (ac component) of the rectified voltage or current from being passed to the load, Figure 20–21.

Notice in Figure 20–21 that the L is in series with the output and offers low opposition to low frequencies and high opposition to high frequencies. The C, in parallel with the output (and load R), offers a low opposition path

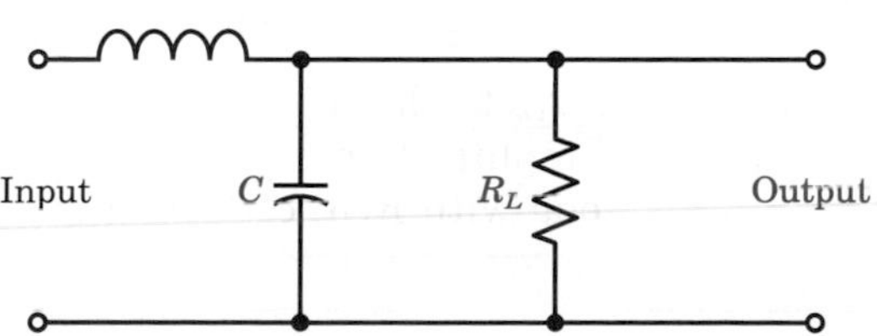

FIGURE 20–21 Example of a low-pass filter

to the higher frequency components of the signal, effectively bypassing or shunting them around the load R. The combination of L and C passes along the lower frequency signal components to the load, while preventing the high frequency components at the load. Thus, the filter is called a low-pass filter. The frequency response of a low-pass filter is shown in Figure 20–22. Various circuit and component configurations that provide low-pass filtering are shown in Figure 20–23.

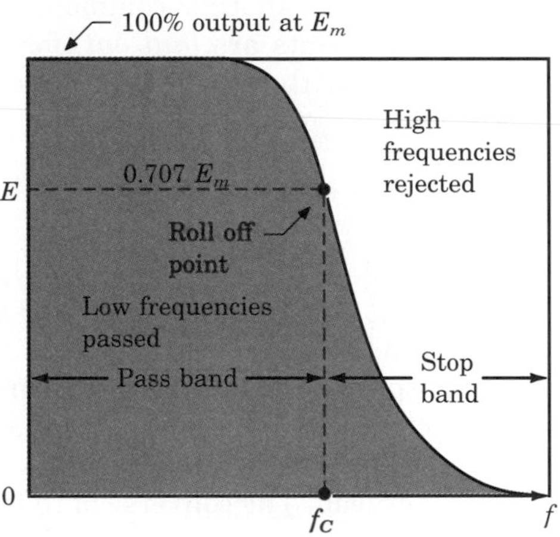

FIGURE 20–22 Frequency response characteristics of a low-pass filter

PRACTICAL NOTES

As you know, points falling below 70.7% or the half-power points of a filter response are out of the band of frequencies that the filter will pass. In low-pass filters, Figures 20–23, the high frequencies are rejected. In high-pass filters, Figures 20–24 and 20–25, the lower frequencies are rejected. The frequency where rejection begins is sometimes called the *cutoff frequency (f_c)*. For a simple RC network, Figure 20–24a, X_C = R at this cutoff frequency. The formula to find the cutoff frequency for this RC network is:

Formula 20–10	$f_c = \frac{1}{2\pi RC}$

To find the cutoff frequency where X_L = R for an RL network similar to the one shown in Figure 20–23a, the formula is:

Formula 20–11	$f_c = \frac{1}{2\pi\left(\frac{L}{R}\right)}$

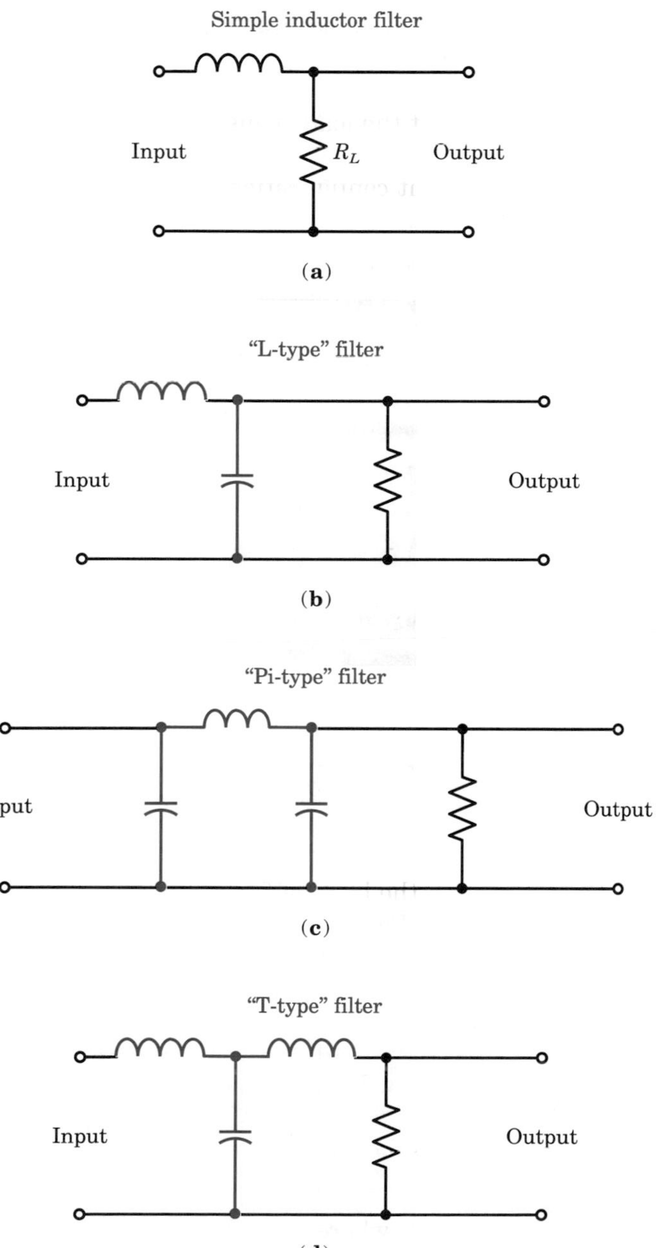

FIGURE 20–23 Examples of circuit configurations for low-pass filters

High-Pass Filters. A **high-pass filter** passes the high frequency components of a signal (or group of signals) to the output or load, while greatly attenuating or preventing the low frequency components from being passed to the output or load. Examples of their applications include use as tone

FIGURE 20–24 Examples of different configurations for high-pass filters

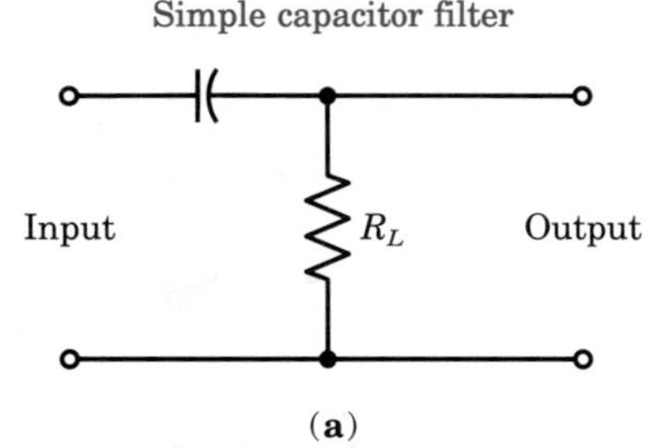

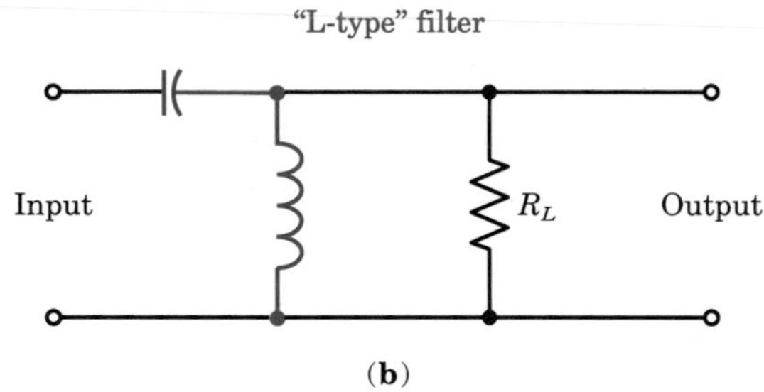

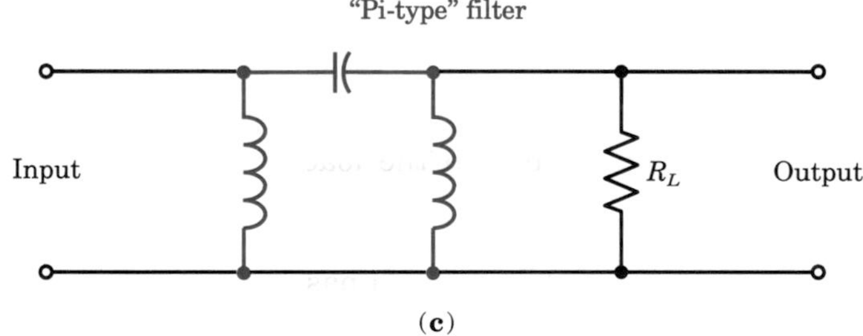

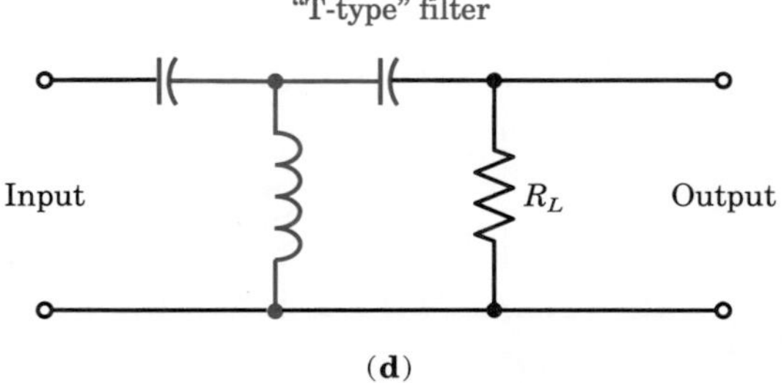

controls in radios and use in the antenna input circuits to receivers to prevent unwanted interference. Figure 20–24 shows several examples of configurations of R, L, and C components used as high-pass filters. Typically, the capacitors, which have a high opposition to low frequencies, are in series

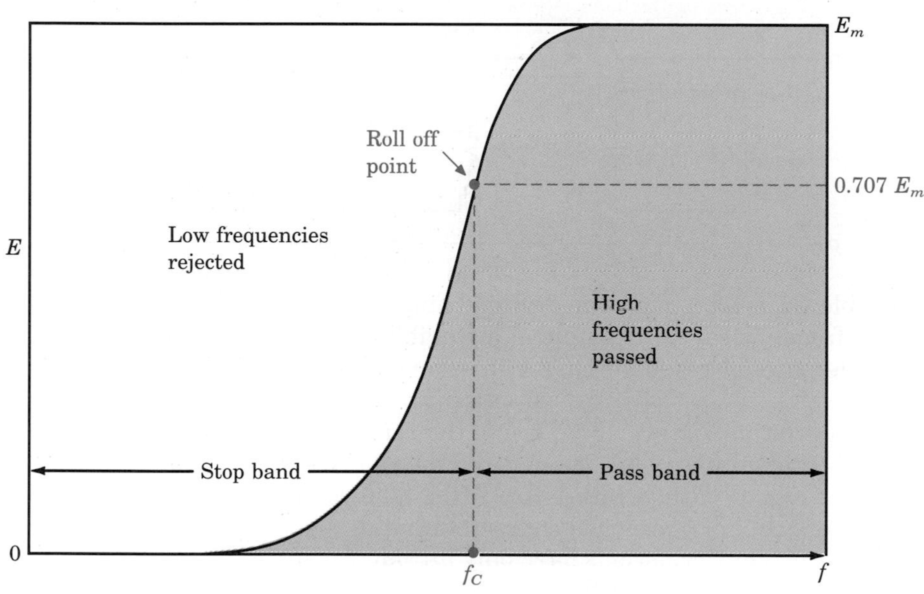

FIGURE 20–25 Typical response characteristic of a high-pass filter

with the output or load. The inductors, which have a low opposition to low frequencies, are in parallel, or shunt with the output, thus effectively bypassing the low frequencies around the load. The result is the high frequency components of the input signal are passed to the load; and the low frequency components are greatly attenuated or prevented from reaching the load. Thus, the filter is called a high-pass filter.

Figure 20–25 shows the typical response characteristic of a high-pass filter.

NOTE: Low-pass and high-pass filter networks can be combined to pass a desired band of frequencies. Those frequencies passed along to the output or load are the frequencies that are not stopped by either filter network. This is one application of *bandpass filtering*. Typically however, this application is often fulfilled by resonant circuits.

Resonant Filters

The Bandpass Filter. Refer to Figure 20–26 and notice a combination of series and parallel resonant circuits are used to pass a desired band of frequencies. The desired output band of frequencies finds little opposition from the series resonant circuits in series with the output path and very high opposition from the parallel resonant circuits. Therefore, the desired band of frequencies appears at the output or load. This tunable filter is tuned with

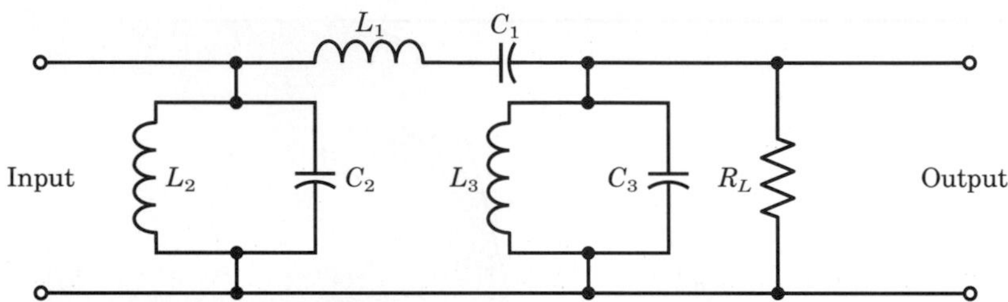

FIGURE 20–26
Resonant-type bandpass filter

variable Cs or Ls to pass one desired band of frequencies while rejecting all other frequencies. An example of a tunable **bandpass filter** is the single tunable parallel resonant circuit that tunes different radio stations.

The Bandstop Filter. By placing the series resonant circuit(s) in shunt with the signal path and one or more parallel resonant circuit(s) in series with the signal path, a **bandstop filter** is developed, Figure 20–27. Also, note the frequency response characteristic.

The previous discussions have only introduced you to filters. We have not attempted to investigate filter design formulas (e.g., for *constant-k* or *m*-

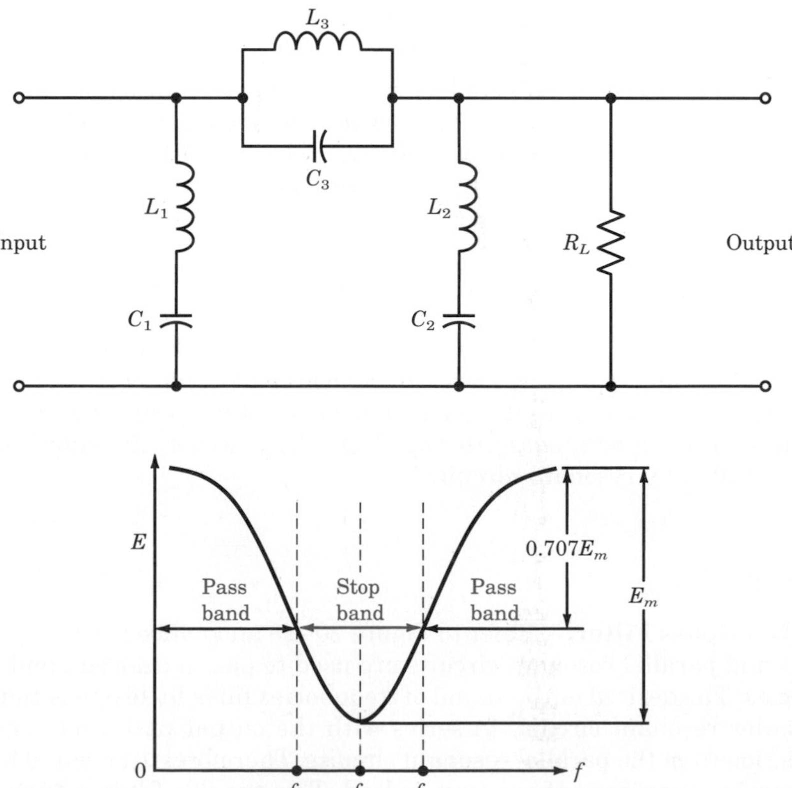

FIGURE 20–27
Resonant-type bandstop filter and response characteristic

derived filters), or the wide spectrum of modern filters, such as YIG (yttrium-iron-garnet) filters, multi-channel analog filters, and computer-controlled filters.

Perhaps your interest in studying filters was increased by this simple overview. Filters have numerous practical uses, from simple power supply filtering to signal processing in such high-technology areas as speech processing and medical instrumentation.

SUMMARY

- X_L and X_C change in opposite directions for a given frequency change. In other words, as f increases, X_L increases and X_C decreases. Also, as f decreases, X_L decreases and X_C increases.

- For any LC combination, there is one frequency where X_L and X_C are equal: the resonant frequency or the frequency of resonance.

- At resonance, series RLC circuits exhibit the characteristics of: $X_L = X_C$, maximum current, minimum Z (and Z = R), θ of 0°, and voltage across each reactance of Q times V applied.

- At resonance, parallel RLC circuits exhibit the characteristics of: $X_L = X_C$, minimum current, maximum Z ($Z = Q \times X_L$), $\theta = 0°$, and current within the tank circuit = Q times I line.

- See Figure 20–28 for comparison of series and parallel resonant circuits.

- The general formula to find the resonant frequency for series LC circuits and for high-Q parallel LC circuits is:

$$f_r = \frac{1}{2\pi\sqrt{LC}} \quad \text{or} \quad \frac{0.159}{\sqrt{LC}}$$

- The Q of a resonant circuit is determined by the circuit resistance and the $\frac{L}{C}$ ratio. The more the circuit resistive losses are, the lower the Q is. The higher the $\frac{L}{C}$ ratio is, the higher the Q is for a given f_r.

- The higher the Q of a resonant circuit is, the greater the magnification factor is for voltage (series resonant circuit) and impedance (parallel resonant circuit).

- The higher the Q of a resonant circuit is, the sharper is its response curve and the more selective it is to a band of frequencies. High Q means narrow bandwidth and narrow bandpass.

- Bandwidth is often defined by the frequencies where the response curve for the circuit is above 70.7% of the maximum level. This is sometimes called the half-power or —3 dB points on the response curve.

- The relationship of resonant frequency, Q, and bandwidth (BW) is shown by: BW (Δf) = $\frac{f_r}{Q}$, or $Q = \frac{fr}{\Delta f}$.

- R, L, and C components can be configured to produce filtering effects. Low-pass filters allow low frequencies to pass to the output, while attenuating high frequencies. High-pass filters pass high frequencies to the output while attenuating the low frequencies.

- Generally, to allow a certain frequency spectrum to pass to the output, a low impedance circuit or component is placed in series with the signal path, and/or a high impedance circuit or component is placed in shunt with the signal path.

FIGURE 20–28 Similarities and differences between series and parallel resonance

	SERIES RESONANCE	PARALLEL RESONANCE
SIMILARITIES	$X_L = X_C$	$X_L = X_C$
	$f_r = \frac{0.159}{\sqrt{LC}}$	$f_r = \frac{0.159}{\sqrt{LC}}$
	$\theta = 0°$	$\theta = 0°$ (High Q circuits)
	p.f. = 1	p.f. = 1 (High Q circuits)
	$\Delta f = \frac{f_r}{Q}$	$f = \frac{f_r}{Q}$
DIFFERENCES	$f_r = \frac{0.159}{\sqrt{LC}}$ no matter circuit R value	f_r can be affected by R
	I = maximum	I line = minimum
	Z = minimum (=R)	Z = maximum
	$Q = \frac{X_L}{r_s}$	$Q = \frac{R}{X_L}$ (low Q circuits)
	Resonant rise of I	Resonant rise of Z
	Resonant rise of reactive V ($Q_X V_A$)	($Z = Q \times X_L$)
	Above f_r = Inductive	Above f_r = Capacitive
	Below f_r = Capacitive	Below f_r = Inductive p.f. = 1 called "antiresonance" . . . can be a different f than maximum Z point.

- ▶ To prevent a certain frequency spectrum at the output of the filter, a high impedance circuit or component is placed in series with the signal path, and/or a low impedance circuit or component is placed in shunt with the signal path.
- ▶ Filters are categorized in various ways. Some classifications are physical layout shape (L, pi, T); usage (bypass, decouple, bandpass, bandstop, smoothing); component types (LC, RC, active, passive); number of sections they have (single-section, two-section); and the frequency spectrum in which they are used (power supply, audio, rf).
- ▶ Multiple-section filters generally provide more filtering than single-section filters.

CHAPTER CHALLENGE

Following the SIMPLER troubleshooting sequence, find the problem with this circuit. As you follow the sequence, record your responses to each step on a separate sheet of paper.

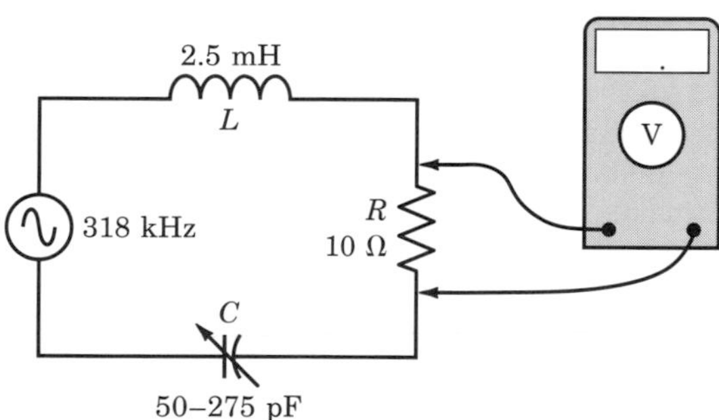

CHALLENGE CIRCUIT 12

SYMPTOMS Gather, verify, and analyze symptom information. (Look at the "Starting Point Information.")

IDENTIFY Identify initial suspect area for the location of the trouble.

MAKE Make a decision about "What type of test to make" and "Where to make it." (To simulate making a decision about each test you want to make, select the desired test from the "TEST" column listing below.)

PERFORM Perform the test. (Look up the result of the test in Appendix C. Use the number in the "RESULTS" column to find the test result in the Appendix.)

LOCATE Locate and define a new "narrower" area in which to continue troubleshooting.

EXAMINE Examine available information and again determine "what type of test" and "where."

REPEAT Repeat the preceding analysis and testing steps until the trouble is found. What would you do to restore this circuit to normal operation? When you have solved the problem, compare your results with those shown in the color insert.

Starting Point Information

1. Circuit diagram
2. V_R is lower than it should be when C is adjusted so that the LC resonance is below the 318 kHz source signal frequency.

Test	Results in Appendix C
V_R (when C set at the position for 318 kHz LC resonance)	(55)
V_R (when C set at a position for LC resonance above 318 kHz)	(85)
V_R (when C set at a position for LC resonance below 318 kHz)	(30)
R_C (C set at a 318 kHz position)	(58)
R_C (C set above 318 kHz position)	(100)
R_C (C set below 318 kHz position)	(72)

NOTE: R_C is resistance measured from stator-to-rotor plates across the capacitor.

CHAPTER CHALLENGE

Following the SIMPLER troubleshooting sequence, find the problem with this circuit. As you follow the sequence, record your responses to each step on a separate sheet of paper.

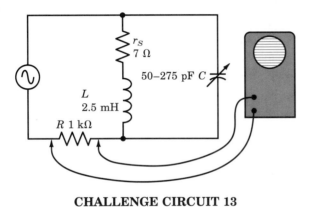

r_S 7 Ω

50–275 pF C

L 2.5 mH

R 1 kΩ

CHALLENGE CIRCUIT 13

Starting Point Information

1. Circuit diagram
2. Using the scope to monitor when V across the 1 kΩ R is minimum (indicating that the parallel LC impedance is maximum), we find that the range of frequencies tuned from maximum C value to minimum C value is slightly higher in frequency than should be expected from the rated values shown on the diagram.

SYMPTOMS Gather, verify, and analyze symptom information. (Look at the "Starting Point Information.")

IDENTIFY Identify initial suspect area for the location of the trouble.

MAKE Make a decision about "What type of test to make" and "Where to make it." (To simulate making a decision about each test you want to make, select the desired test from the "TEST" column listing below.)

PERFORM Perform the test. (Look up the result of the test in Appendix C. Use the number in the "RESULTS" column to find the test result in the Appendix.)

LOCATE Locate and define a new "narrower" area in which to continue troubleshooting.

EXAMINE Examine available information and again determine "what type of test" and "where."

REPEAT Repeat the preceding analysis and testing steps until the trouble is found. What would you do to restore this circuit to normal operation? When you have solved the problem, compare your results with those shown in the color insert.

Test	Results in Appendix C
$R_{inductor}$	(87)
$R_{capacitor}$	(19)
C of capacitor when "fully meshed"	(60)
C of capacitor when "fully unmeshed"	(107)
L of inductor	(41)

REVIEW QUESTIONS

1. State the condition that must exist if a series RLC circuit is at resonance.

2. Disregarding the generator or source impedance, a series LC circuit is comprised of a 100-pF capacitor and a 1-μH inductor having 10-Ω of resistance.
   a. What is the impedance at resonance?
   b. What is the resonant frequency?
   c. What is the Q?
   d. What is the bandwidth?
   e. What is the bandpass?
   f. Below the resonant frequency, does the circuit act inductively or capacitively?

3. The voltage applied to the circuit in Question 2 is 20 volts.
   a. What is the voltage across the capacitor at resonance?
   b. What is the current at resonance?
   c. What is the true power at resonance?
   d. What is the net reactive power?

4. A parallel LC circuit has a capacitive branch with a 30-pF capacitor and an inductive branch with an inductor having 28 μH and a r_s of 30 Ω.
   a. What is the f_r?
   b. What is the Q? (Assume a Q greater than 10.)
   c. What is the bandwidth?
   d. What is the bandpass?
   e. At resonance, what is the Z?
   f. Below resonance, does the circuit act inductively or capacitively?

5. If the voltage applied to the circuit in Question 4 is 20 volts, what is the line current? What is the current through the capacitor branch?

6. Draw a pi-type LC filter that can filter a power supply output. Is this a low-pass or a high-pass filter?

7. Draw a circuit showing a series resonant and a parallel resonant circuit connected so the combination passes a band of frequencies close to the resonant frequencies of the circuits.

8. A series RLC circuit consists of a 100-ohm resistor, a 75-H inductor and a 300-pF capacitor. What is the resonant frequency?

9. A series RLC circuit has a resonant frequency of 40 kHz and the Q is 20. At what frequency will the phase angle be 45°?

10. A series RLC circuit is resonant at 150 kHz, and has R = 100 ohms and C = 25 pF. What is the inductance value?

11. A series RLC circuit is resonant at 21 kHz and has R = 40 ohms and inductance = 12 H. What is the C value?

12. Given a series RLC circuit that is resonant at 1.5 MHz, and knowing that C has a reactance of 500 ohms, what is the R value for the circuit to have a bandwidth of 10 kHz?

13. A series RLC circuit is comprised of a 200-ohm resistor, a 36-mH inductor and a capacitor. What is the voltage across the R at resonance if V applied is 100 volts?

14. What is the resonant frequency of a parallel LC combination having a 200-μH inductor and a 0.0002-μF capacitor?

15. If the frequency of the source in Question 14 is adjusted to a new frequency above the resonance frequency, will I_T lead or lag V_A?

16. A 220-μH inductor has a distributed capacitance of 25 pF and an effective resistance of 80 ohms. What impedance will it exhibit at resonant frequency?

17. If a parallel LC circuit is acting inductively at a frequency of 150 kHz, to tune the circuit to resonance requires which of the following?
    a. Increasing X_C
    b. Decreasing X_L
    c. Decreasing L
    d. Increasing C

TEST YOURSELF

Answer the following questions with "I" for increase, "D" for decrease and "RTS" for remain the same.

1. For a series resonant RLC circuit, if frequency is slightly increased:
    a. X_L will ________
    b. X_C will ________
    c. R will ________
    d. Z will ________
    e. I will ________
    f. θ will ________
    g. V_C will ________
    h. V_L will ________

i. V_R will ____________
j. V_T will ____________
k. BW will ____________
l. Q will ____________
m. I_C will ____________
n. I_L will ____________
o. I_T will ____________

2. For a series resonant RLC circuit, if frequency is slightly decreased:
a. X_L will ____________
b. X_C will ____________
c. R will ____________
d. Z will ____________
e. I will ____________
f. θ will ____________
g. V_C will ____________
h. V_L will ____________
i. V_R will ____________
j. V_T will ____________
k. BW will ____________
l. Q will ____________
m. I_C will ____________
n. I_L will ____________
o. I_T will ____________

3. For a parallel resonant RLC circuit, if frequency is slightly increased:
a. X_L will ____________
b. X_C will ____________
c. R will ____________
d. Z will ____________
e. I will ____________
f. θ will ____________
g. V_C will ____________
h. V_L will ____________
i. V_R will ____________
j. V_T will ____________
k. BW will ____________
l. Q will ____________
m. I_C will ____________
n. I_L will ____________
o. I_T will ____________

4. For a parallel resonant RLC circuit, if frequency is decreased:
a. X_L will ____________
b. X_C will ____________
c. R will ____________
d. Z will ____________
e. I will ____________
f. θ will ____________
g. V_C will ____________
h. V_L will ____________
i. V_R will ____________

j. V_T will ___________
k. BW will ___________
l. Q will ___________
m. I_C will ___________
n. I_L will ___________
o. I_T will ___________

5. Research the terms m-derived and constant-k filters and write the definition of each.

PERFORMANCE PROJECT CORRELATION CHART

Suggested performance projects that correlate with topics in this chapter are:

CHAPTER TOPIC	PERFORMANCE PROJECT	PROJECT NUMBER
X_L, X_C and Frequency	X_L and X_C Relationship to Frequency	55
Series Resonance Characteristics The Resonant Frequency Formula	V, I, R, Z, and θ Relationships when $X_L = X_C$	56
Q and Resonant Rise of Voltage	Q and Voltage in a Series Resonant Circuit	57
Selectivity, Bandwidth and Bandpass	Bandwidth Related to Q	58
Parallel Resonance Characteristics The Resonant Frequency Formula	V, I, R, Z, and θ Relationships when $X_L = X_C$	59
Q and Resonant Rise of Impedance	Q and Impedance in a Parallel Resonant Circuit	60
Selectivity, Bandwidth, and Bandpass	Bandwidth Related to Q	61

PART V

INTRODUCTORY DEVICES AND CIRCUITS

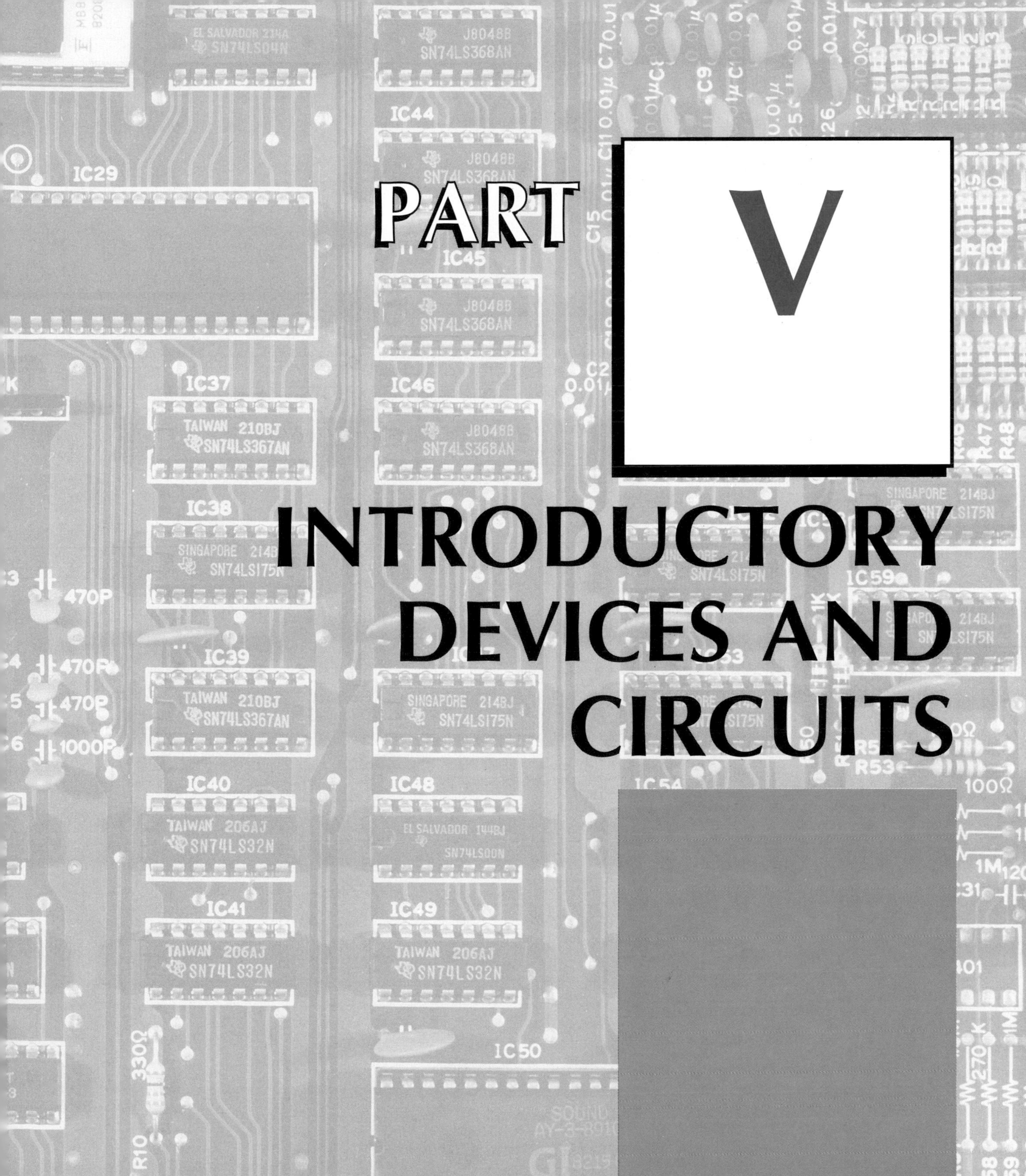

Diodes and Power Supply Circuits

Key Terms

Anode
Bleeder
Cathode
Covalent bonding
Depletion region
Doping
Filter
Forward bias
Majority carriers
Minority carriers
Peak-inverse-voltage (PIV)
Pentavalent materials
Pulsating dc
P-N junction
Rectify (rectification)
Regulator
Reverse bias
Ripple
Semiconductor diode
Trivalent materials
Valence electrons

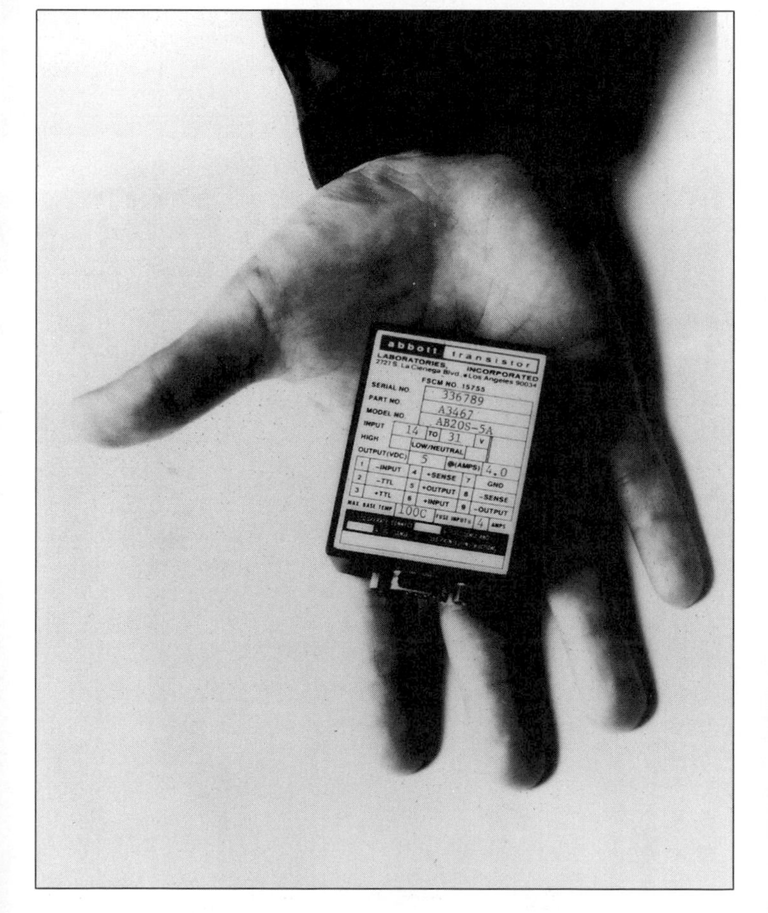

Courtesy of Abbott Transistor Laboratories

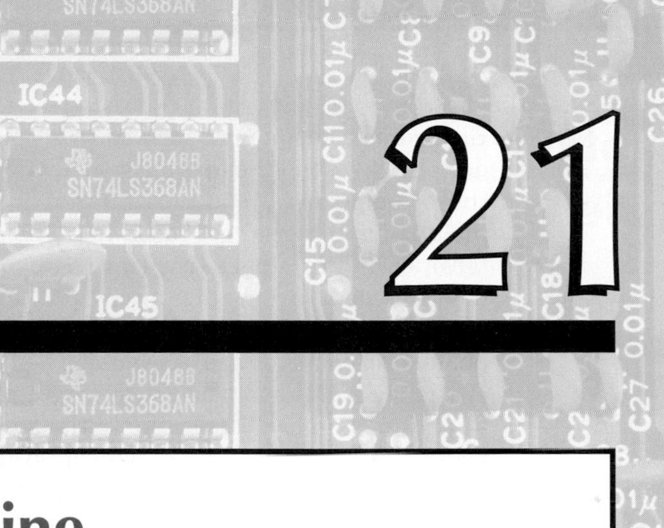

Outline

Semiconductor Materials
The P-N Junction
The Semiconductor Diode
Diode Clippers
The Power Supply System
The Half-Wave Rectifier Circuit
The Full-Wave Rectifier Circuit
The Bridge Rectifier Circuit
Basic Power Supply Filters
Basic Voltage Multiplier Circuits
Troubleshooting Hints

Chapter Preview

Electronic components, such as transistors, and integrated circuits require dc voltages when they perform their functions. Typically, most circuits with these components require regulated dc voltages. Examples of such circuits include amplifier stages (used in audio amplifiers, receivers, and transmitters); oscillator stages (used in receivers, transmitters, and measurement devices); and various digital circuits (used in VCRs and computers).

Generally, the voltage supplied by power companies to our homes and businesses is ac voltage. As you learned earlier, this is because of the higher efficiency of transporting power from the power company to its customers by ac rather than dc. This is due to the step-up or step-down ability of ac voltages by using transformers, which minimize line losses.

Power supplies, which change ac source voltages to appropriate dc voltages, are found in most electronic equipment. Because this is true, circuits associated with power supplies are a critical area that a technician should know about.

In this chapter, you will study semiconductor diodes and the basic concepts of how transformers, diodes, and filter components change ac to usable dc. It is not our intention to thoroughly discuss semiconductor theory, filter design, or regulation networks. This chapter will simply introduce you to some basics of diode action, and concepts of rectification and filtering. This understanding will aid you as a technician when you work with power supplies. You will study semiconductor theory and applications and filter circuits in more detail later in your education.

Objectives

After studying this chapter, you will be able to:

- List key features of **P** and **N** semiconductor materials
- Draw the **semiconductor diode** symbol
- List the important ratings for semiconductor diodes
- Briefly explain clipping action
- List the basic power supply system elements
- Draw the basic half-wave, full-wave, and bridge rectifier circuits
- Determine the unfiltered dc output voltage of specified rectifier circuits
- Briefly explain the power supply filter action
- Identify power supply filter configurations
- Recognize two voltage multiplier circuits
- List three common troubles and related symptoms in power supplies

SEMICONDUCTOR MATERIALS

Conductors, Non-Conductors, and Semiconductors

Recall from early chapters that conductors, non-conductors, and semiconductors are materials that offer different resistance levels to electron movement (current). Remember that conductors, such as copper, gold, and silver, offer little opposition to current flow. Typical specific resistance or resistivity values for these materials range from 10 to 20 Ω per mil-foot at 20°C. (A mil-foot is the specific resistance of a material having a cross-section of 0.001 inch diameter and a length of one foot, at room temperature.) Non-conductors, on the other hand, have very high resistivities. Materials, such as germanium and silicon, have resistances between those of conductors and non-conductors. Thus, these materials are called semiconductors.

A noteworthy difference between conductors and semiconductors is most conductors have a positive temperature coefficient (i.e., resistance increases with rising temperature). Semiconductors usually have a negative temperature coefficient (i.e., resistivity decreases with rising temperature).

Review of Atomic Structure

Referring to Figure 21–1, remember that three primary particles of atoms are protons and neutrons in the nucleus and orbiting electrons outside the nucleus. Also, recall the orbiting electrons are arranged in shells at different energy levels, and the maximum number of electrons in any shell are $2n^2$; where n equals the shell number moving out from the nucleus. These various shell levels are termed the K, L, M, and so on, starting with the innermost shell and moving outward. Also, recall when the outermost ring of electrons is full and the atom is stable, there are eight **valence electrons.** The discussion about semiconductors will focus on this outermost valence-electron ring. With this review, let's briefly look at semiconductor materials

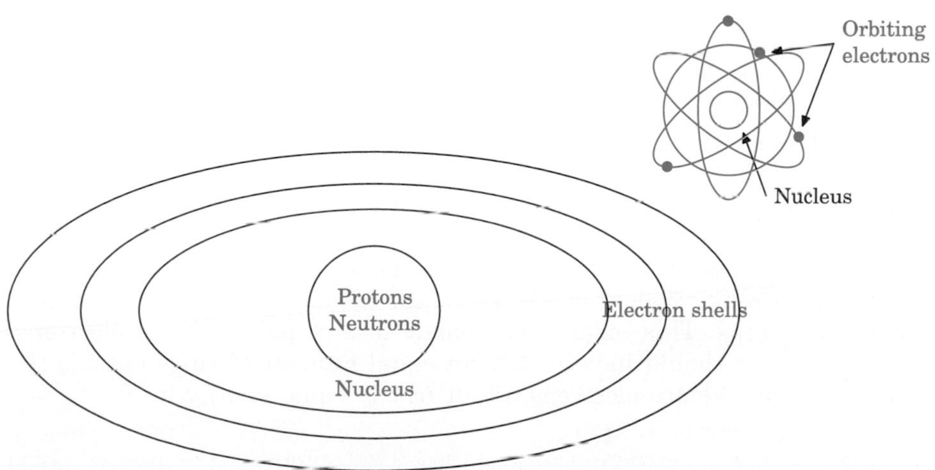

FIGURE 21–1 Basic atomic structure

having four outer ring (valence) electrons. These are sometimes called tetravalent materials, since the prefix "tetra" means four.

Semiconductor Materials

Two common tetravalent materials found in semiconductors are silicon and germanium. Silicon has 14 protons in the nucleus; thus, 14 orbiting electrons arranged in shells of 2, 8, and 4 electrons, respectively. Germanium has 32 orbiting electrons, arranged in shells of 2, 8, 18, and 4 electrons, respectively. These materials have a crystal makeup where their atoms form a lattice structure that forms the solid material, Figure 21–2. Because of their solid makeup, you often hear the term *solid-state devices* used to refer to semiconductor diodes, transistors, and integrated circuits.

FIGURE 21–2 Crystal-lattice structure

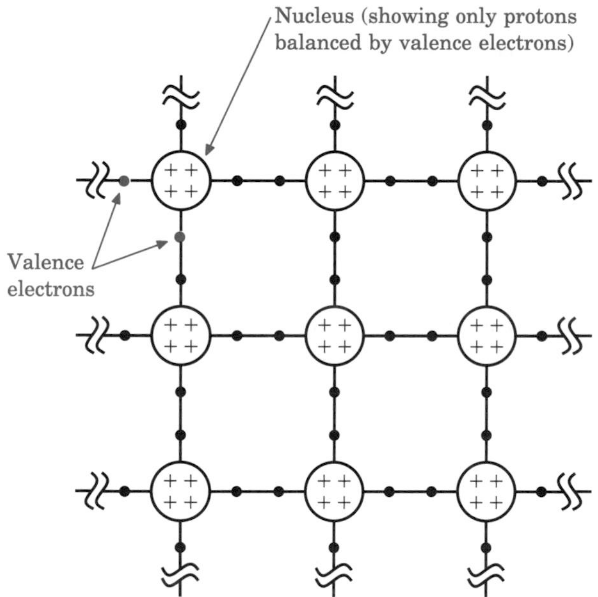

Covalent Bonding

Refer to Figure 21–3 and notice that adjacent atoms of the semiconductor material each have four valence electrons. Adjacent atoms share valence electrons to create a stable eight electron arrangement. **Covalent bonding** is accomplished by this sharing action.

Earlier, we mentioned covalent bonds. There are two primary requirements for forming covalent bonds. First, the atoms involved have 3, 4, or 5 valence electrons. This means they easily lose or gain valence electrons. Also, the atoms should have nearly an equal number of rings outside the nucleus so the valence electrons are at nearly equal energy levels.

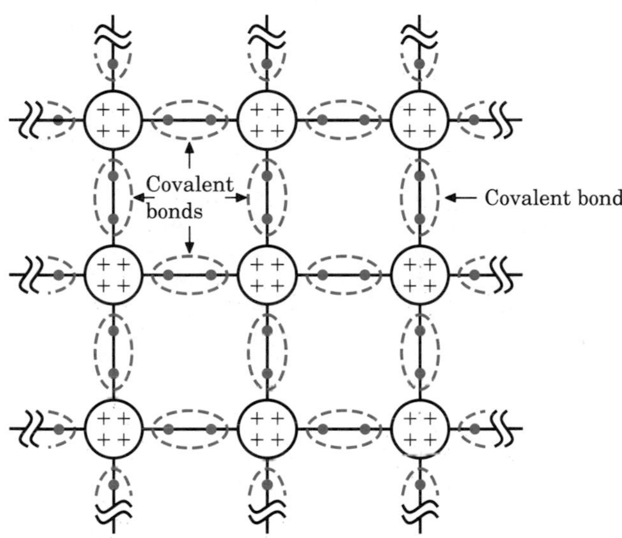

FIGURE 21–3
Covalent bonds

Doping

At room temperatures, pure semiconductor materials have few electrons that are free to travel through the lattice structure. This means that pure semiconductor material is not a good conductor at these temperatures. (As you learned, if temperature increases, the material resistance decreases as heat provides the energy levels needed to break some electrons free.) By purposely introducing certain impurities into the lattice structure of the semiconductor material, we can increase the conductivity of the material. This process of introducing impurities is called **doping** the material.

By doping a semiconductor material with a **pentavalent material** (five valence electrons), such as arsenic, or antimony (commonly used to dope germanium), or phosphorus (frequently used to dope silicon), the conductivity of the material increases. This is because four of the five valence electrons in the doping material atoms form covalent bonds with neighboring semiconductor atoms, with the remaining electron, which is not securely attached, moving easily through the structure, Figure 21–4. NOTE: it takes only about $\frac{1}{200}$ the energy to move this free electron compared to moving an electron from the covalent bond pairs.

N-Type Semiconductor Material

Materials doped with pentavalent materials are known as *N-type* materials. This is because the free charges (**majority** current **carriers**) are negative electrons. That is, current flows through the material as a result of the free electrons donated by the impurity material atoms that dope the semiconductor. Incidentally, when the fifth electron drifts away from the impurity atom

FIGURE 21–4 Doping with a pentavalent material

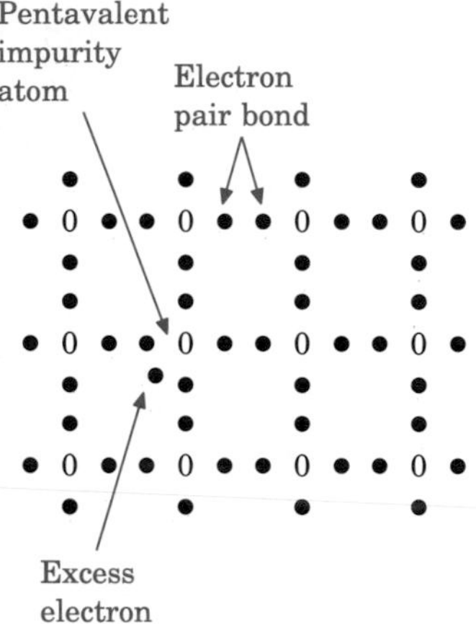

due to thermal or other energy, it leaves behind a pentavalent atom that has four instead of five valence electrons. This means that what remains is a positive ion that has donated an electron. For this reason, the impurity atom in N-type material is sometimes called a *donor* atom.

N-Type Conduction

Figure 21–5 illustrates how the free electrons, supplied by the donor atoms, enable current flow through the doped semiconductor N-type material. The outside voltage source attracts electrons from the material to its positive terminal and injects an equal number of electrons into the other end of the N-type material. In effect, the electrons migrate easily through the lattice network from one end to the other end and on to the source. Thus, current flow is established.

P-Type Semiconductor Material

Figure 21–6 shows that by doping the semiconductor material with a **trivalent material** (three valence electrons, such as gallium or indium for germanium and aluminum or boron for silicon), covalent bonds are formed whereby one location is minus an electron. This positive charged space with an electron absence is called a *hole.* In a silicon semiconductor material doped with boron, for example, there is one hole (electron deficiency, or positive entity) for each aluminum atom in the doped lattice structure. Because the majority current carriers in this material are positive charged entities, the material is known as P-type semiconductor material. The impurity atoms introduced through doping are called *acceptor* atoms, since they accept electrons to fill the holes.

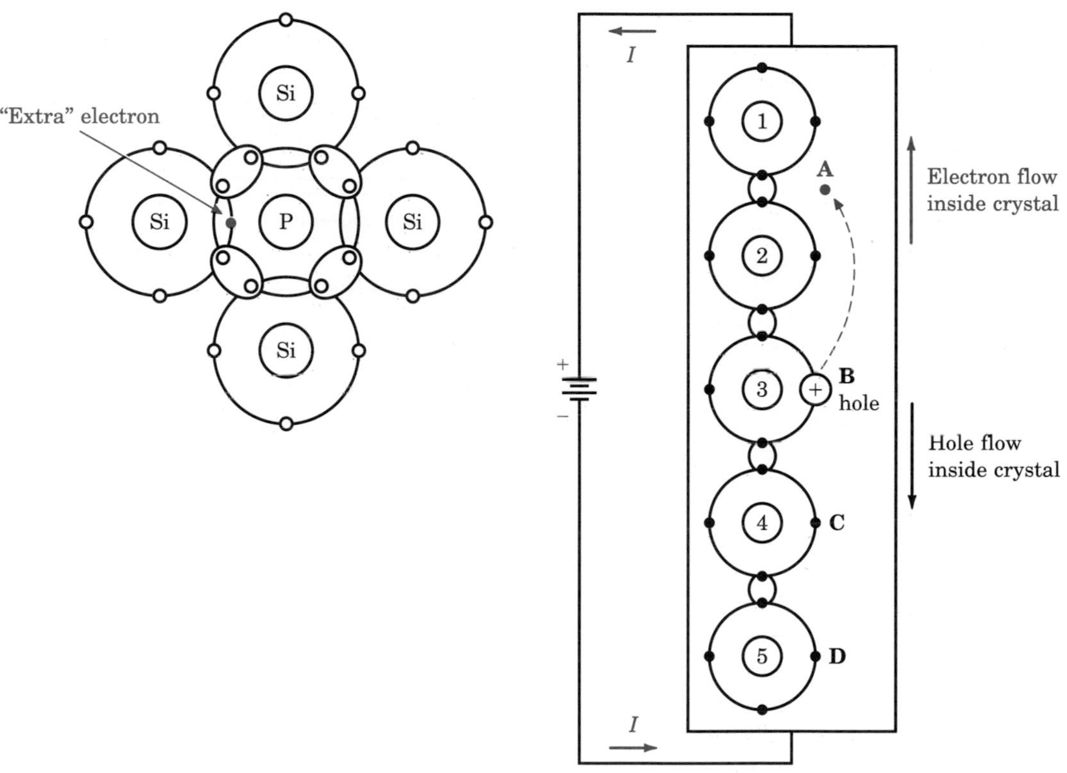

FIGURE 21–5 N-type semiconductor material

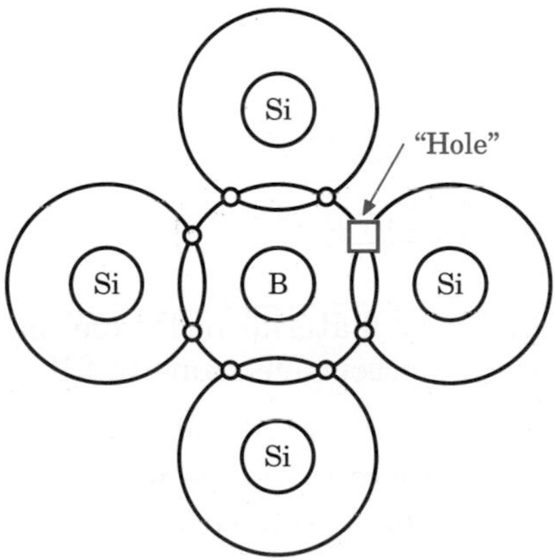

FIGURE 21–6 Trivalent doping with silicon and boron in P-type material

P-Type Conduction

Figure 21–7 illustrates that electrons hop from one hole to another (leaving behind a hole each time they move) that establishes electron flow through

FIGURE 21–7
P-type conduction

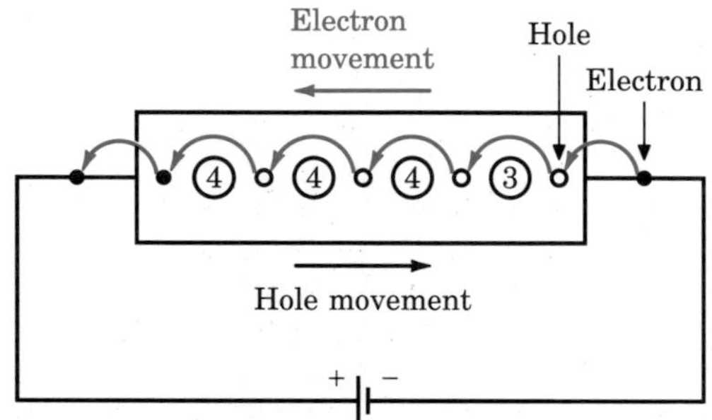

the material. In effect, the holes (majority current carriers in P-type material) move in the opposite direction from the electron movement.

When an electron moves from the negative source side to the lattice structure of the P-type material filling a hole, the structure has a net negative charge. This causes an electron to leave the other end of the lattice network and go to the positive source terminal to neutralize the charge. This process is multiplied and continuous as current flow is established.

Summary of N and P Type Materials

1. Prior to doping (in their pure form) semiconductor materials are sometimes called intrinsic semiconductors. When doped with impurity atoms, they are termed extrinsic semiconductors.
2. N-type material is formed by doping the material with pentavalent materials.
3. The atoms of the impure material in N-type material are called donor atoms since they donate electrons to the structure.
4. The majority current carriers in N-type material are electrons. (The **minority carriers** are holes in this case.)
5. P-type material is formed by doping the material with trivalent materials.
6. The atoms of the impurity material in P-type material are called acceptor atoms, since they accept electrons to fill their holes caused by an electron deficiency.
7. The majority current carriers in P-type material are holes. (The *minority carriers* are electrons in this case.)

THE P-N JUNCTION

What happens if we join an N-type and a P-type material? If this is done, a **P-N** (or N-P) **junction** has been formed. Refer to Figure 21–8 as you study the following discussion.

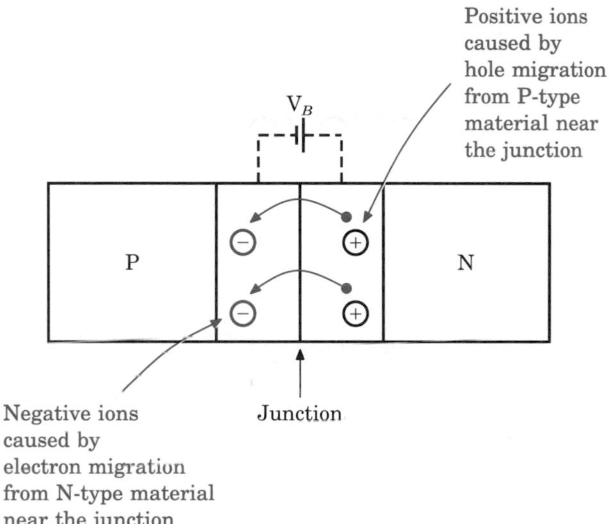

FIGURE 21–8 The P-N junction

Since there are free electron charges in the N-type material near the junction of the P and N material, they are attracted to the positive holes near the junction in the P-type material. Some electrons leave the N-type material to fill holes in the P-type material. In other words, holes from the P-type material are attracted by and diffused into the N-type material. Every electron that leaves the N-type material and goes to the P-type material leaves behind a hole in the N-type material.

The result of these activities is that the N-type material now has an electron deficiency and is positively charged near the P-N junction, and the P-type material has an electron excess and is negatively charged near the junction. The difference of potential near the junction is known as a *barrier potential* or *contact potential,* (V_B). NOTE: This barrier potential decreases if junction temperature increases. That is, this parameter is temperature sensitive!

Current movement over or through the junction *without* external voltage applied is termed *diffusion current.* This diffusion current establishes the barrier potential.

This junction region where the current carriers of both materials have moved is called the **depletion region,** since it has been depleted of some majority current carriers during the diffusion process.

THE SEMICONDUCTOR DIODE

The **semiconductor diode** is a practical application of the P-N junction. Applying external (bias) voltage to the P-N junction causes it to conduct, or not to conduct, current flow. This open or closed current gate concept is very useful in electronics, as you will see. Figure 21–9 shows the schematic sym-

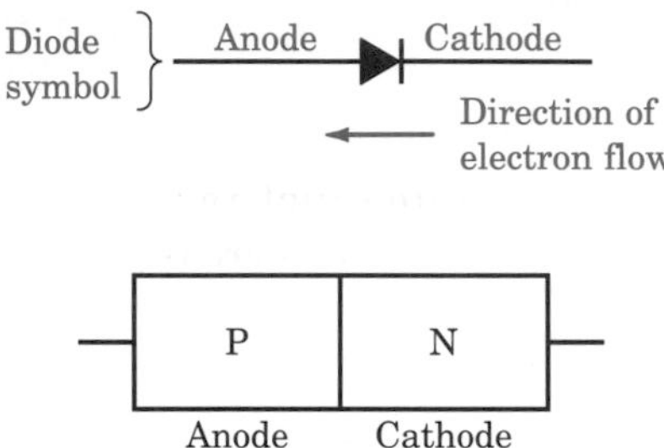

FIGURE 21–9 Diode schematic symbol

bol for a semiconductor diode. Note the **anode** (P-type material) is denoted by an arrow, and the cathode (N-type material) is indicated by a bar. The arrow actually points in a direction opposite to electron current flow through the diode, or in the same direction as hole movement.

Reverse Bias

Reverse bias of a P-N junction diode means we connect the external source negative terminal to the anode (P-type material) and the external source positive terminal to the **(cathode)** N-type material, Figure 21–10. Notice the depletion region is wider because of the source supplying electrons to the P-type material (filling some of its majority carrier holes) and electrons (majority carriers) being attracted from the N-type material to the positive source terminal. This action causes the P-N barrier potential to rise to a value equal to the source voltage, preventing majority carrier current flow through the semiconductor. Only a minute amount of reverse current (leakage current) flows due to the minority carriers. This reverse current is very small until the reverse voltage is increased to a point known as the breakdown or Zener voltage. Prior to reaching this breakdown point, the leakage

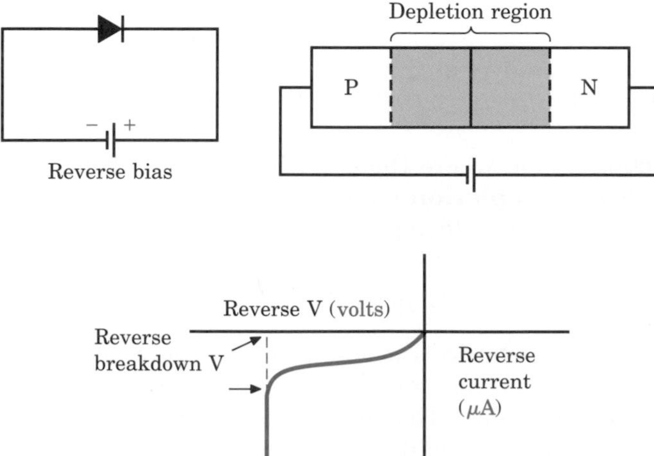

FIGURE 21–10 Reverse-biased diode and graph of *I* versus V characteristic

current in germanium diodes are in the range of 1 mA or less. For silicon type diodes, the reverse leakage current is in microamperes. The symbol for this reverse current is I_{CO}. Of course, temperature affects the amount of leakage current. It has been stated that reverse leakage current approximately doubles for every 10 degrees C increase in temperature. The higher the temperature is, the higher the I_{CO} is. When the breakdown voltage is reached, covalent bonds are broken, and there is an avalanche of carriers created. Thus, reverse current increases dramatically.

FIGURE 21–11 Zener diode symbol

An interesting event occurs when this Zener point is reached. The voltage drop across the diode remains virtually constant over a wide range of reverse current, refer again to Figure 21–10. This feature is advantageous when diodes (called Zener diodes) are designed for voltage regulators, overvoltage protectors, voltage reference devices, and other related devices. Note the schematic symbol for Zener diodes in Figure 21–11.

In summary, a reverse-biased diode does not conduct forward current and acts like an open switch with respect to forward conduction.

Forward Bias

From the previous discussion, you can conclude that **forward bias** occurs when the external source voltage causes the anode (P-type material) to be positive with respect to the cathode (N-type material). Under these circumstances, as bias voltage increases from zero, conduction similar to that shown in Figure 21–12 results. Little current flows until the barrier potential of the diode is nearly overcome. Then, there is a rapid rise in current flow. The typical barrier potential for silicon junction diodes is from 0.6 to 0.7 volts. (For germanium junction diodes, the barrier potential is from about 0.25 to 0.3 volts.) In power supplies, the silicon diode is commonly

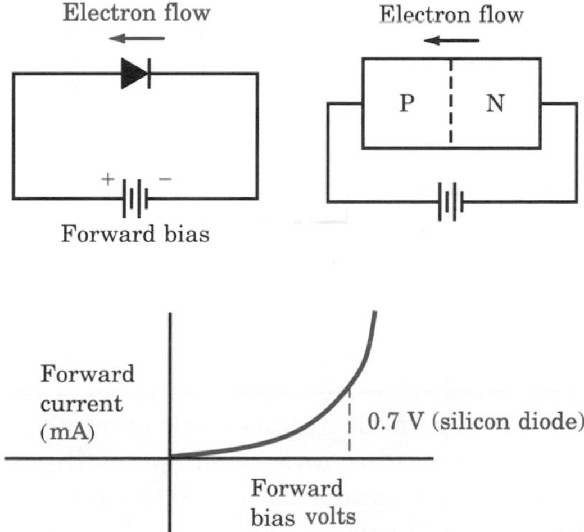

FIGURE 21–12 Forward-biased diode and graph of *I* versus V characteristic

used. As forward bias increases, a point is reached where further voltage increase does not cause a significant change in forward current through the diode. This is called *saturation.*

In summary, when a junction diode is forward biased, it conducts current and acts like a low-resistance closed switch. When reverse biased, the diode acts like an open switch.

IN-PROCESS LEARNING CHECK I

Fill in the blanks as appropriate.

1. An interesting difference between conductors and semiconductors is that most conductors have a __________ temperature coefficient while semiconductors typically have a __________ temperature coefficient. A positive temperature coefficient means that as temperature increases, resistance __________. A negative temperature coefficient means that as temperature increases, resistance __________.
2. Semiconductor materials have four outermost rings, or valence electrons. These materials are sometimes termed __________ materials, where the prefix __________ means four.
3. To increase conductivity of semiconductor materials, a process called __________ is used.
4. Electrons are the majority carriers in __________-type semiconductor materials.
5. When we join a P-type semiconductor material with an N-type, the point where they come together is called a P-N __________.
6. To cause conduction in a semiconductor diode, the diode must be __________ biased.
7. To reverse bias a P-N junction, connect the negative source voltage to the __________-type material and the positive source voltage to the __________-type material. The P-type material is known as the __________ of the diode.

DIODE CLIPPERS

Prior to studying power supply circuits, let's look at a few applications of this effect of diode action. Circuits that remove the extremities of an input wave are sometimes called *clipper* or *limiter* circuits. There are several applications for such circuits in electronics. For example, they are used in waveshaping circuits to get rid of unwanted spikes in waveforms.

As you will see, clipping is similar to the rectifying action you will study later. Because the diode only passes current when it is forward biased, (anode is positive with respect to cathode), it is possible to clip off the undesired portion of a given input waveform. Refer to the series-connected diode clippers in Figure 21–13 and notice it is possible to clip either the positive half-cycle or the negative half-cycle from the input signal by selecting the diode direction in the circuit. In Figure 21–14, you can see the same circuit action but with shunt diode clippers.

Another interesting variation of clipping action is to purposely bias the diodes so they limit the waveform to a value other than zero. This is done by applying (or creating with a resistor) a voltage at one diode element. In effect, the input voltage swing is limited by the amount of ac input wave that causes the anode to be positive with respect to the cathode, Figure 21–15.

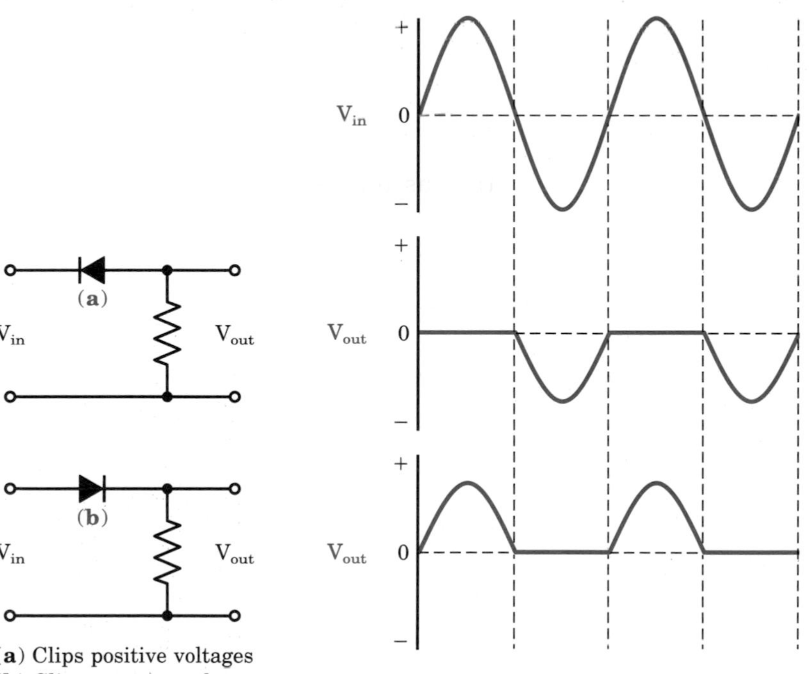

FIGURE 21–13 Series diode clippers

THE POWER SUPPLY SYSTEM

As was stated earlier, the purpose of a power supply system is to convert available ac into usable dc. A block diagram of a typical power supply system is shown in Figure 21–16. As shown in the diagram, the main elements of a power supply are:

1. A *transformer* used to step-up or step-down voltage, as appropriate. (NOTE: A transformer is not present in all power supplies.)
2. A **rectifier** used to change ac to pulsating dc.
3. A **filter** used to filter ac components of pulsating dc and provide smooth dc output.

FIGURE 21–14 Shunt diode clippers

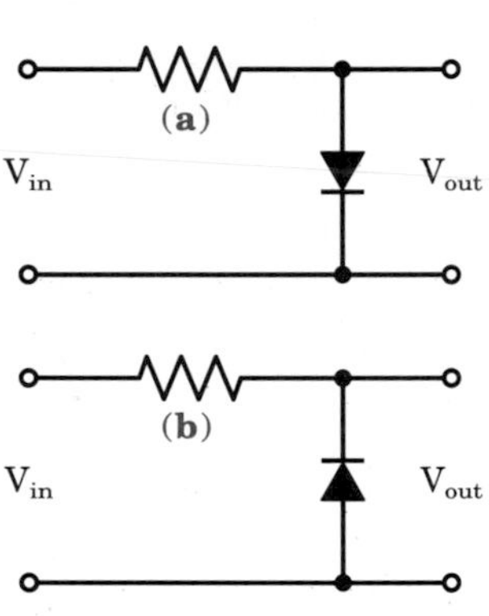

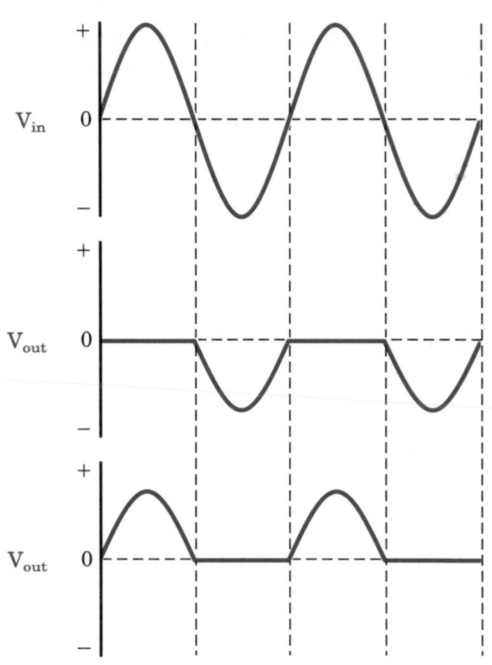

(**a**) Limits or clips positive voltages
(**b**) Limits or clips negative voltages

FIGURE 21–15 Diodes biased to limit to other than zero level

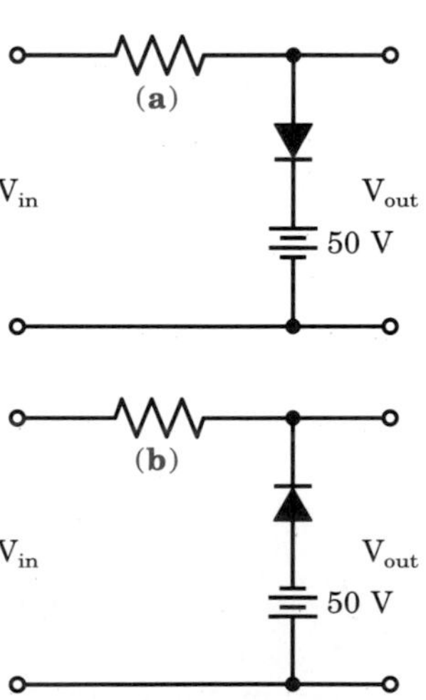

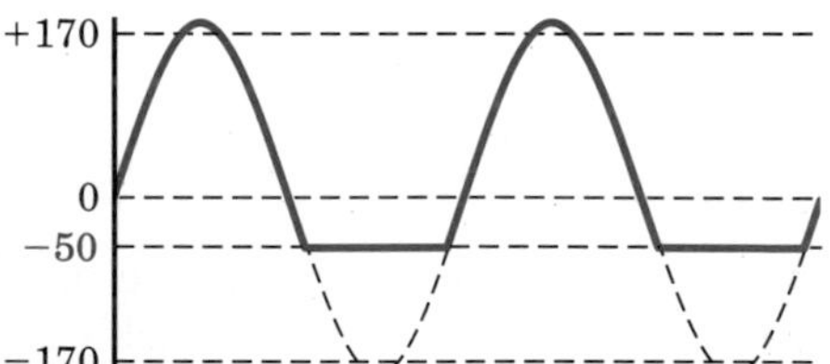

(**a**) Biased positive limiter
(**b**) Biased negative limiter

4. A **bleeder**-divider used to aid filtering and provide safety. (NOTE: This element is not present in all power supplies.)
5. A *regulator* used to provide regulated output V and/or I. (NOTE: A regulator is not needed or present in all power supplies.)

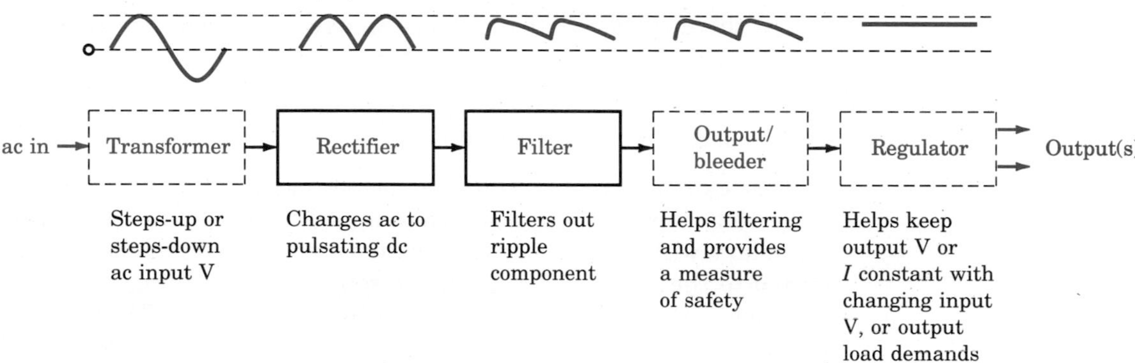

FIGURE 21–16 Power supply system block diagram

You will learn more about these elements when we discuss the various circuit configurations associated with power supplies.

Design Considerations

It is not the intention of this chapter to teach you all the details of power supply design. However, several considerations that power supply designers must regard are of interest. Four common considerations are output voltage required; minimum, average, and peak output current; voltage regulation required over the needed current range; and acceptable ripple voltage limit. These considerations play a part in selecting the components for the various power supply system sections (e.g., transformer, rectifier, filter, and **regulator** components).

THE HALF-WAVE RECTIFIER CIRCUIT

Operation

Observe Figure 21–17 as you study the following discussion of the basic operation of the half-wave (H-W) rectifier circuit.

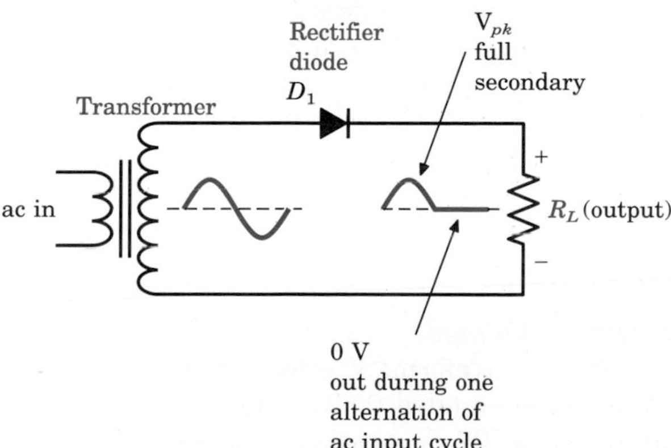

FIGURE 21–17 Half-wave rectifier circuit (without filter)

- Notice the direction that the diode conducts electron flow is from cathode to anode (against the diode symbol arrow).
- The diode (D_1) only conducts when the anode is positive with respect to the cathode (or the cathode is negative with respect to the anode).
- For our example, the ac source is the secondary of a transformer. Assume that during the first alternation of ac, the top of the transformer secondary winding is positive. This means the diode anode is connected to the positive side of the ac source. The winding bottom is negative with respect to the winding top at this time. Therefore, the negative side of the source is connected to the bottom of resistor R_L.
- Since the diode anode is positive with respect to the cathode during the first ac input alternation, the diode conducts current. Electron flow is from the bottom of the transformer secondary winding, through R_L, through the diode (cathode-to-anode), and back to the top of the transformer winding. The voltage polarity developed across R_L is as shown in Figure 21–17. NOTE: The waveform shape and amplitude match the ac voltage waveform of the transformer during this first positive alternation. Only the very small voltage drop of the conducting diode subtracts from this value.
- During the next ac input alternation (the negative alternation), the transformer top is negative and the bottom is positive. This means the anode is negative with respect to the cathode; therefore, the diode *will not* conduct current. Since no current passes through R_L, there is no output voltage during this half-cycle. NOTE: This is indicated by the output waveform.
- This same cycle is repeated for each ac input cycle. During the positive alternation, the diode conducts current, and output voltage develops across R_L. During the negative alternation, the diode does not conduct and no output voltage develops across R_L.
- The unfiltered output of the half-wave rectifier is a **pulsating dc.** For each ac input cycle, there is one dc pulse of output. Since the output voltage is only one polarity, it is a dc voltage. These bumps are sometimes called **ripples.** It is common to say the ripple frequency of a half-wave rectifier output is equal to the frequency of the ac input. If the input is 60 Hz ac, the output dc has a ripple frequency of 60.

Important Characteristics

- *Unfiltered dc Output Voltage*
  Recall for ac voltage waveform, V effective (rms) equaled 0.707 times V peak and V average equaled 0.637 times V peak. Since 0.637 is about nine-tenths of 0.707, V average equaled 0.9 times V effective.

Since the half-wave rectifier system is producing output only during one-half cycle, it is logical that the average dc output is only half of the ac voltage. That is, half of 0.9 = 0.45. Hence, average dc output voltage of the half-wave rectifier circuit equals 0.45 times V effective. Of course, you can also compute the dc output as half of the average ac voltage. That is, half of 0.637 times V peak. This means the unfiltered dc output (half-wave rectifier) is calculated from either of the following formulas:

Formula 21–1	$V_{dc} = 0.45 \times V_{rms}$

Formula 21–2	$V_{dc} = 0.318 \times V_{pk}$

EXAMPLE

What is the unfiltered dc output voltage of a half-wave rectifier circuit, if the ac input voltage is 100-volts peak?

Answer:

$V_{dc} = 0.318 \times V_{pk} = 31.8$ volts

PRACTICE PROBLEM I

Answers are in Appendix B.

What is the unfiltered dc output voltage of a half-wave rectifier circuit, if the ac input is 200-volts rms?

- *Peak-Inverse-Voltage (PIV)*

  You know the voltage drop across a conducting silicon diode is about 0.7 volts. What about the voltage across the diode when it is *not* conducting? More specifically, what is the maximum voltage across the diode when it is not conducting? This value is called the **peak-inverse-voltage (PIV).**

  This voltage appears across the diode when the transformer secondary voltage is at maximum voltage level, with the top of the secondary being negative and the bottom being positive. Since there is no current through R_L when the diode is not conducting (except for negligible reverse leakage current), there is no voltage drop across R_L. This means the top of R_L (and the cathode) are essentially at the same potential as the bottom of the transformer winding. In effect, the voltage across the diode when it is not conducting is the same as the transformer winding voltage; thus PIV = $1.414 \times V_{rms}$.

Formula 21–3	PIV (H-W) = $1.414 \times V_{rms}$

- *Ripple Frequency*

  As you have seen, there is one ripple in the output for each ac cycle.

Formula 21–4	Ripple Frequency (H-W) = ac input frequency

- *Polarity of Output Voltage*
  If the diode is reversed (i.e., the cathode connected to the top of the winding and the anode connected to the top of R_L), the polarity of the output voltage reverses across R_L. Current passes from the top of R_L to the bottom, making the top negative with respect to the bottom. In a half-wave rectifier, to reverse the output polarity, simply reverse the diode cathode and anode connections in the circuit.

PRACTICE PROBLEM II

Refer to Figure 21–18 and find dc output voltage, PIV, output ripple frequency, and output polarity with respect to ground.

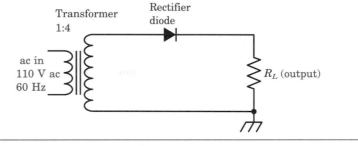

FIGURE 21–18

Transformer
1:4
Rectifier
diode
ac in
110 V ac
60 Hz
R_L (output)

THE FULL-WAVE RECTIFIER CIRCUIT

Operation

Observe Figure 21–19 as you study the following discussion of the basic operation of the full-wave rectifier circuit.

- Observe the transformer secondary is *center-tapped* so a diode and R_L combination is across one half of the transformer secondary when the

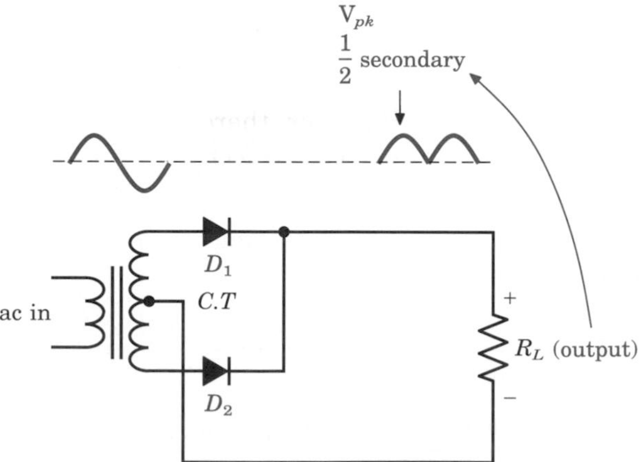

FIGURE 21–19 Full-wave rectifier circuit (without filter)

top diode (D_1) conducts and across the other half of the secondary when the other diode (D_2) conducts. (Both diodes do *not* conduct at the same time due to the voltage polarity from center-tap to each end of the secondary during any given alternation.)

- Let's assume for the first ac alternation, the top of the transformer secondary is positive with respect to the bottom. This means if the peak ac voltage across the whole winding is 100 V, then the top of the secondary is $+50\ V_{pk}$ with respect to the center-tap. Thus, the voltage across R_L is $50\ V_{pk}$ (neglecting the small diode drop across D_1) when the top diode conducts.
- On the next alternation, the voltage across the bottom half of the secondary is of a polarity that causes the bottom diode (D_2) to conduct, and R_L is effectively connected across the bottom half of the secondary. The current direction through R_L will, however, be the same as it was during the first alternation. That is, the top of R_L is positive with respect to the bottom of R_L. Again, the peak voltage is 50 V, or half the secondary peak voltage.
- The full-wave rectifier system provides output pulses on *both* alternations of the ac input. However, the voltage across R_L is only half of the transformer secondary voltage, rather than the whole secondary value.
- Because there are output pulses on both alternations, the ripple frequency of the full-wave rectifier output is twice that of the half-wave rectifier system.

Let's now look at the key parameters of the full-wave rectifier.

Important Characteristics

- *Unfiltered dc Output Voltage*
  For a given transformer total secondary voltage, the unfiltered full-wave rectifier dc output voltage is the same as the half-wave rectifier system because the pulsating dc output is only half the amplitude of the half-wave output pulse. However, there is output on both half cycles, rather than just one half-cycle. (NOTE: One key advantage of the full-wave circuit is that its output is easier to filter since there are many narrow valleys to fill between pulses.) Formulas to compute dc output are:

Formula 21–5	$V_{dc} = 0.9 \times V_{rms}$ of half the secondary

Formula 21–6	$V_{dc} = 0.45 \times V_{rms}$ of the full secondary

Formula 21–7	$V_{dc} = 0.637 \times V_{pk}$ of half the secondary

Formula 21–8	$V_{dc} = 0.318 \times V_{pk}$ of the full secondary

- *Peak-Inverse-Voltage (PIV)*
  The PIV of each diode in the full-wave rectifier circuit equals the peak voltage of the full secondary, *not half.* This is because the anode of the non-conducting diode is directly connected to the negative end (cathode) of the transformer winding during the alternation when it is not conducting. That cathode is at the same potential as the top of R_L, which is connected through the conducting diode to the positive end of the transformer winding. Therefore, the PIV of each diode during its own non-conducting alternation of ac input equals the peak voltage of the full secondary. PIV is calculated as:

Formula 21–9	PIV (F-W) = $1.414 \times V_{rms}$ of full secondary

- *Ripple Frequency*
  With the full-wave rectifier system, there is a ripple (pulse) output for each half-cycle of ac input. This means the output ripple frequency equals two times ac input frequency.

Formula 21–10	Ripple Frequency = $2 \times$ ac input frequency

- *Polarity of Output Voltage*
  Like the half-wave rectifier system, the output polarity can be reversed by reversing the cathode and anode connections to *each* diode. ***Caution:*** Both diodes must be reversed! If only one is reversed, the transformer winding is shorted through the diodes during the half-cycle of the inappropriately connected diodes that both conduct. NOTE: Polarity is also reversed by taking output from the transformer center-tap and making the two connected diode cathodes the ground or common point for the output.

PRACTICE PROBLEM III

Assume a full-wave rectifier circuit is connected to a transformer having a full-secondary rms voltage of 300 volts. Assume the ac source feeding the transformer primary is a 400-Hz ac source. Determine dc output, ripple frequency, and PIV for each diode.

THE BRIDGE RECTIFIER CIRCUIT

Operation

Observe Figure 21–20 as you study the following discussion of the basic operation of the bridge rectifier circuit.

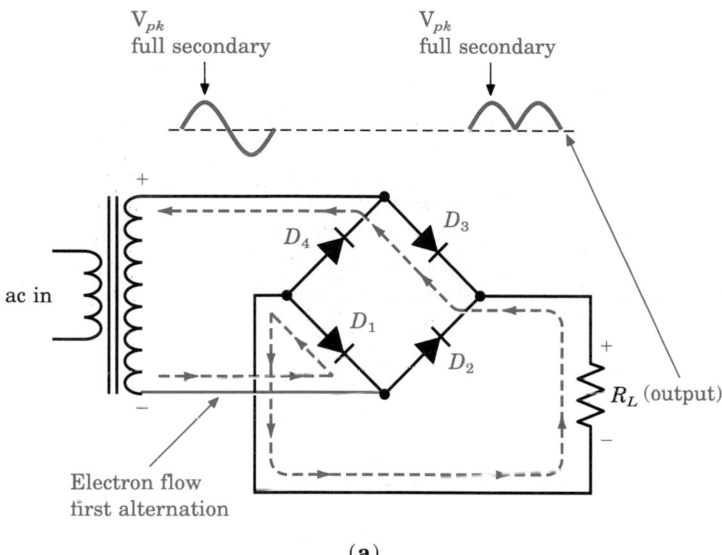

(a)

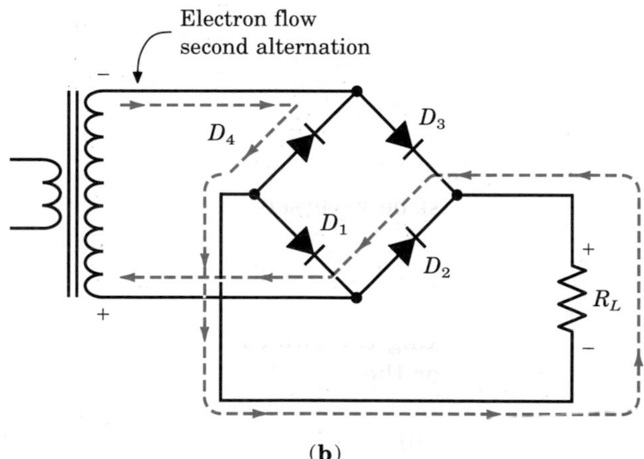

(b)

FIGURE 21–20 Bridge rectifier circuit (without filter)

- Notice that for this circuit, the total transformer secondary is used with no center-tap.
- In the bridge rectifier, which uses four diodes, two diodes conduct on one alternation and the other two diodes conduct on the opposite alternation of the ac input. When the top of the secondary is positive, diodes D_1 and D_3 conduct. When the bottom of the secondary is positive, diodes D_2 and D_4 conduct.
- Tracing the current path during the first alternation, electrons leave the bottom of the winding (negative end), travel through diode D_1, up through R_L, from cathode to anode through D_3, and back to the top

(positive end) of the transformer winding. The resulting voltage across R_L is a positive pulse with a peak value equal to the peak value of the *total* secondary winding (minus the small drops across the two diodes).

- During the second alternation, electrons leave the negative end of the transformer winding (the top in this case), travel through diode D_4, up through R_L, and through D_2 back to the positive side of the winding (the bottom end in this alternation). The resulting voltage across R_L is a positive pulse with a peak value equal to the peak value of the *total* secondary winding (minus the small drops across the two conducting diodes).
- Since there is an output pulse during each alternation of ac input, the output ripple frequency is twice the ac input frequency.

Important Features

- *dc Output Voltage (Unfiltered)*
Since there is an output pulse on both half-cycles of ac input, and since that output amplitude is essentially the same as the *total* secondary voltage, the dc output voltage equals nine-tenths of the effective total secondary voltage. Formulas to calculate the unfiltered dc output of the bridge rectifier circuit are:

Formula 21–11	$V_{dc} = 0.9 \times V_{rms}$ of the full secondary

Formula 21–12	$V_{dc} = 0.637 \times V_{pk}$ of the full secondary

- *Peak-Inverse-Voltage (PIV)* (Each diode)
The PIV of each rectifier equals the peak of the full secondary. For example, when D_3 is not conducting, its anode is connected to the negative end of the transformer winding. At the same time, its cathode is at the same potential as the top of R_L, which is virtually connected to the bottom end of the winding through diode D_2's conduction. This means D_3's anode is at the top of the winding, and its cathode is effectively connected through conducting diode D_2 to the bottom of the winding. The total winding voltage is across D_3 when it is not conducting. This same logic determines PIV for each of the four diodes. Thus:

Formula 21–13	PIV (Bridge) $= 1.414 \times V_{rms}$ of the full secondary

- *Ripple Frequency:*
Since there is an output ripple for each alternation of ac input, the ripple frequency is twice the ac input frequency.

Formula 21–14	Ripple Frequency (Bridge) $= 2 \times$ ac input frequency

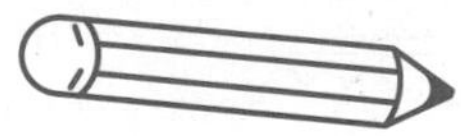

PRACTICAL NOTES

Caution Relating to Grounding

The bridge rectifier circuit dc output terminals cannot be grounded if one input side is unbalanced to ground (i.e., if one input side is grounded).

Special Note About Current

The transformer secondary winding that feeds the bridge rectifier circuit must carry the full load current.

IN-PROCESS LEARNING CHECK II

1. The simplest rectifier circuit is the ____________ rectifier.
2. The ripple frequency of a bridge rectifier is ____________ the ripple frequency of the half-wave rectifier circuit.
3. The circuit having the highest output voltage for a given transformer secondary voltage is the ____________ rectifier.
4. For a given full transformer secondary voltage, the full-wave rectifier circuit unfiltered dc output voltage is ____________ the dc output of a half-wave rectifier that is using the same transformer.
5. To find the dc output (unfiltered) of a full-wave rectifier, multiply the full secondary rms voltage by ____________.

BASIC POWER SUPPLY FILTERS

Purpose

The purpose of a power supply filter is to keep the ripple component from appearing in the output, Figure 21–21.

Some Common Types of Filters

Refer to Figure 21–22 to see some examples of the following filter types.

1. Capacitance (C)
2. Inductance (L)
3. Inductance-capacitance (LC)
   a. Capacitor input

FIGURE 21–21 Examples of rectifier outputs before and after filtering

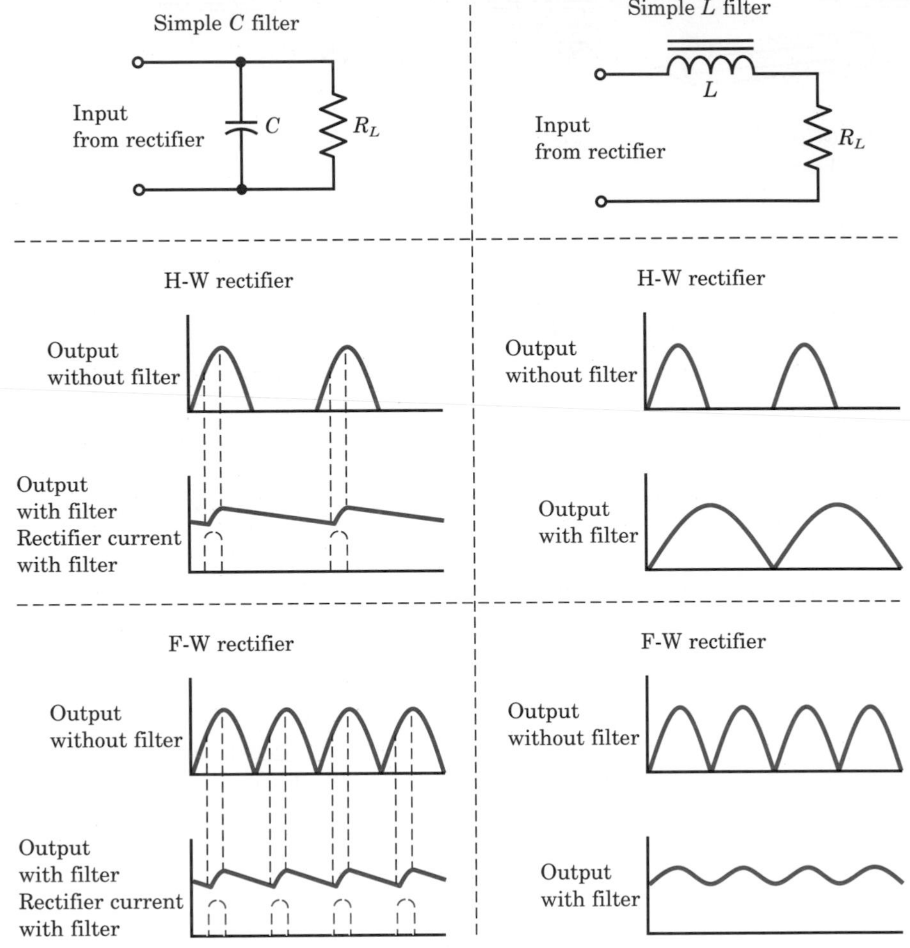

b. Choke input
c. Other (special)
   (1) Brute force
   (2) Resonant

As you can see, capacitors are used in many filter networks. Their filtering action results from much shorter RC charge times compared to discharge times. Charge occurs through the very low resistance conducting diodes, and discharge through the much higher R circuits represented by the power supply loads. This means the capacitors essentially charge to the peak value of the ac voltage applied during diode conduction time and slowly discharge until the next ac cycle reaches a level that causes the diodes to conduct again, recharging the capacitors to near peak value. Since the capacitors charge to the peak value and maintain much of their charge between charging pulses, the output ripple amplitude is greatly decreased (or "smoothed

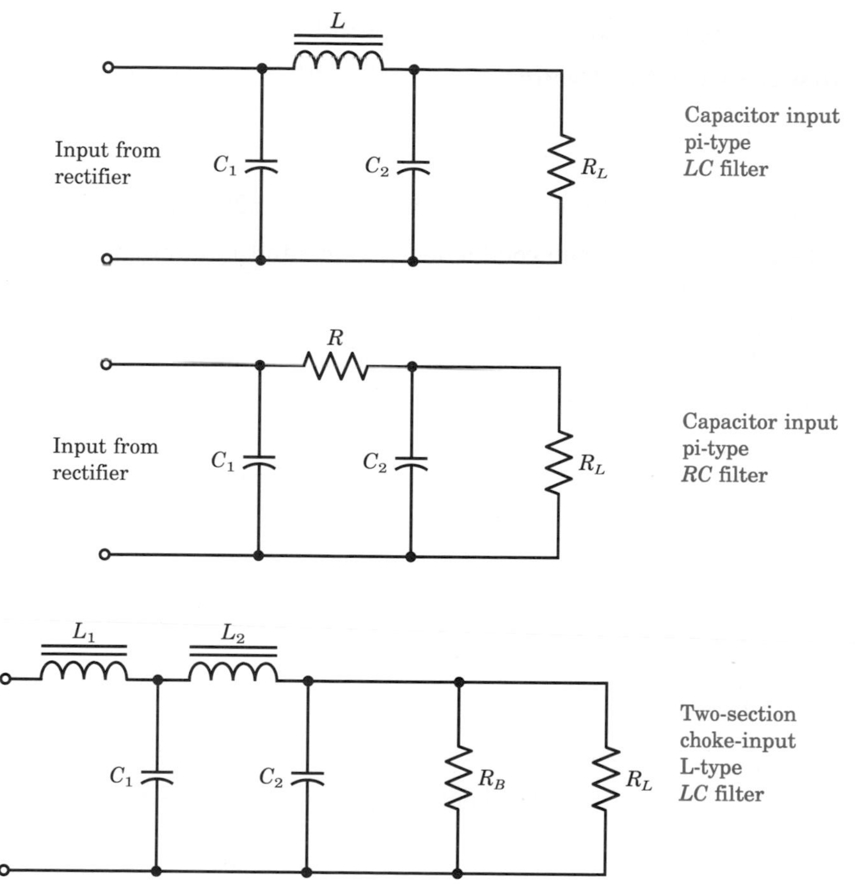

FIGURE 21–22 Some common types of power supply filter networks

out"), and the resulting dc voltage is greatly increased from that of an unfiltered output, refer again to Figure 21–21.

Some Filter Design Considerations

Briefly, some of the considerations for designing filters are:

1. Allowable ripple
2. Allowable regulation
3. Rectifier peak current limits
4. Load current
5. Output voltage

Obviously, these factors affect the values and ratings of components in the power supply system and the filter network.

Effect of Filters on Operation and Output of Rectifier Circuits

Half-Wave Rectifier System.

1. Increases dc output (maximum possible with light load is 1.414 × V_{rms}).
2. Possible PIV with filter capacitor present is 2.828 × V_{rms}.
3. Fills in valleys between conduction periods and minimizes ripple.

Full-Wave Rectifier System.

1. Increases dc output (maximum possible with light load is 1.414 × V_{rms} of *half* the transformer secondary V).
2. Possible PIV with filter capacitor present is 2.828 × V_{rms} of *half* the secondary voltage, or 1.414 × full secondary rms value.

Bridge Rectifier System.

1. Increases dc output (maximum possible with light load is 1.414 × V_{rms} of *full secondary* voltage).
2. PIV per rectifier = 1.414 × V_{rms} of *full secondary* voltage.

Power Supply Output Characteristics and the Bleeder Resistor

Figure 21–23 shows the general output features with capacitor input filters and choke input filters. Notice that below a certain load current level, the output voltage rapidly soars upward when there is a choke input filter system. One way to prevent this is to use a bleeder resistor across the output. A bleeder resistor draws a fixed minimum load, which is enough current to get the output below the knee of the curve. Also, the bleeder resistor performs another useful function. It provides safety by draining the charge from high voltage filter capacitors when the circuit is turned off. This prevents unex-

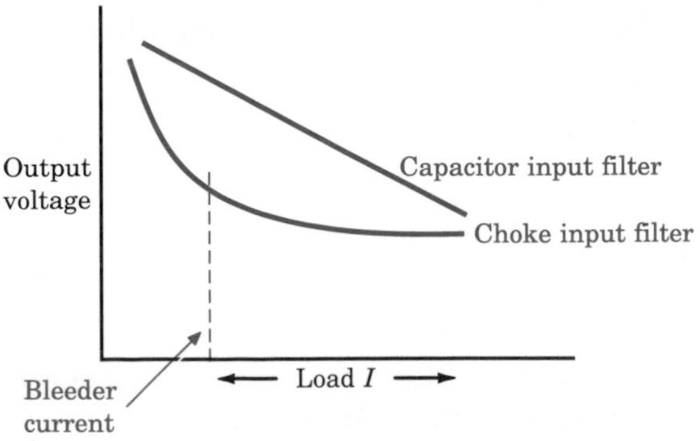

FIGURE 21–23 Output characteristics of capacitor and choke input filters

pected shocks (or death) for someone who might assume the circuit is safe when the circuit is turned off.

SAFETY HINTS

If the bleeder is defective, or there is no bleeder, high voltage can be present for days after the circuit is turned off. The technician should appreciate bleeder resistors, but never assume they are working. *Always* discharge capacitors with an additional shorting method that isolates you from any possible charge on the capacitors.

BASIC VOLTAGE MULTIPLIER CIRCUITS

Full-Wave Doubler Circuit

By arranging the ac input, rectifier diodes, capacitors, and output as shown in Figure 21–24, it is possible to charge the capacitors series-aiding, with each charged to one times the input voltage. Therefore, the output voltage is equal to the sum of the two capacitors' voltages, or approximately *twice* the input voltage (under light loads). This term is full-wave voltage doubler because there are two output pulses for every ac input cycle.

Half-Wave Cascade Voltage Doubler

Again, by arranging the input, rectifier diodes, capacitors, and output as shown in Figure 21–25, it is possible to charge C_2 to approximately twice the ac input voltage. In this case, C_2 gets a pulse of charging current once during each ac input cycle, thus this circuit is considered a half-wave doubler. The reason C_2 charges to 2 times V is because C_1's charge is series aiding with the source in C_2's charging path. Thus, C_2 charges to V_{C_1} plus V source, or approximately 2 times V source.

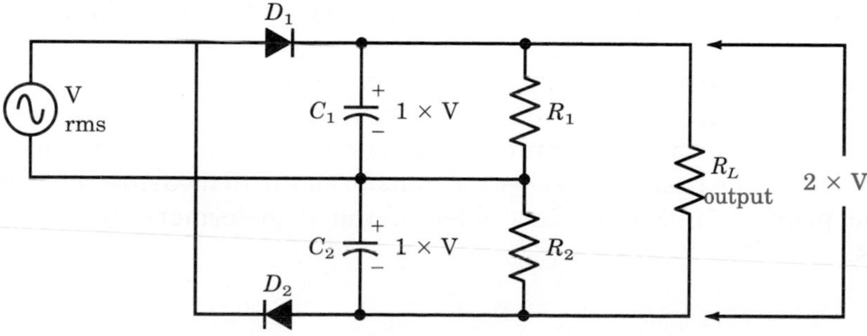

FIGURE 21–24 Full-wave voltage doubler circuit

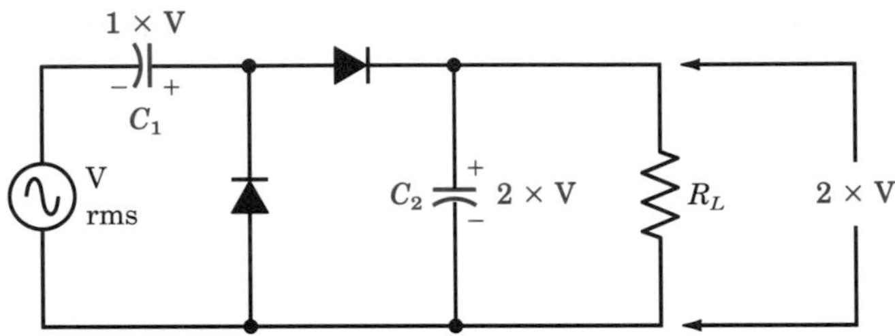

FIGURE 21–25
Half-wave cascade voltage doubler circuit

TROUBLESHOOTING HINTS

First isolate a power supply problem to the general section of trouble by carefully considering the symptoms. Recall the main sections of a typical power supply are the transformer, the rectifiers, the filter network, the output network, and possibly a regulator circuit. Excellent tools to help you with symptoms are the oscilloscope to look at waveforms, and/or a voltmeter to make voltage measurements. Obviously, your *eyes* become a tool as you can see burned parts, blown fuses, and so forth, and to analyze the readings and scope patterns. Your *nose* is a great symptom locator to find overheated parts. If the system you are troubleshooting involves audio output (such as a receiver or record player) your *ears* are a very helpful tool, since you can hear hum and can probably tell if it's 60 Hz or 120 Hz, and you can hear "motorboating".

By noting whether there is excessive ripple in the output and zero or too low an output voltage, you can isolate the section you might want to troubleshoot.

Typical Power Supply Problems

Transformer Section. This can include power sources feeding the primary of the transformer, fuses in the primary and/or secondary, and the transformer itself.

Transformers usually don't go bad by themselves. Usually some external circuit or component connected to the transformer causes too much current through the transformer circuit that either blows a fuse or damages the transformer. The best way to check a transformer is to measure the voltages on the primary and secondary sides of the transformer against known norms.

Rectifiers. A shorted diode places a short across the transformer part of the time, causing the transformer to overheat, frequently blow a fuse, or damage the transformer. Voltage is low at the output of the rectifier system.

The best way to check a rectifier is the following. If the ac input voltage to the rectifier circuit is normal but the circuit output voltage is low, check each diode. Measure the ac across each diode. If the diode is shorted, the ac voltage reading is virtually zero. If the diode is open, the ac voltage reading is too high. Also, with power off, ohmmeter checks should show a very high resistance in one direction and a very low resistance in the other, if the diode is good.

Filter Capacitors in Filter Section. Often, electrolytic capacitors are used. They tend to short frequently, and overheat as a result. Also, electrolytics tend to open by drying out with age.

To check filter capacitors, measure the *ac* voltage across each capacitor. If the capacitor is good, the reading is low. If it is open, the ac voltage is unusually high.

Filter Chokes in the Filter Section. Obviously, since filter chokes are in series with the load current, excessive load current causes too much current through the choke inductor. This is obvious via overheating.

With the power off, the ohmmeter helps to determine if the choke is open or has too low a resistance. (NOTE: Many chokes have low dc resistance, so don't be fooled.) Also, you can usually smell an overheated choke or transformer. If the choke is open, output voltage drops to zero, since the choke is in series with the output.

Bleeder Resistors. Excessive current through the resistor or age causes bleeder resistors to fail.

SAFETY HINTS

Do not feel these resistors. If wire wound resistors are used, they normally are hot enough to burn you badly! Use your eyes and nose to detect overheating. With power off, use an ohmmeter to check for an open or shorted resistor. (Caution: Don't be fooled by sneak parallel paths.)

CHAPTER CHALLENGE

Following the SIMPLER troubleshooting sequence, find the problem with this circuit. As you follow the sequence, record your responses to each step on a separate sheet of paper.

SYMPTOMS Gather, verify, and analyze symptom information. (Look at the "Starting Point Information.")

IDENTIFY Identify initial suspect area for the location of the trouble.

MAKE Make a decision about "What type of test to make" and "Where to make it." (To simulate making a decision about each test you want to make, select the desired test from the "TEST" column listing below.)

PERFORM Perform the test. (Look up the result of the test in Appendix C. Use the number in the "RESULTS" column to find the test result in the Appendix.)

LOCATE Locate and define a new "narrower" area in which to continue troubleshooting.

EXAMINE Examine available information and again determine "what type of test" and "where."

REPEAT Repeat the preceding analysis and testing steps until the trouble is found. What would you do to restore this circuit to normal operation? When you have solved the problem, compare your results with those shown in the color insert.

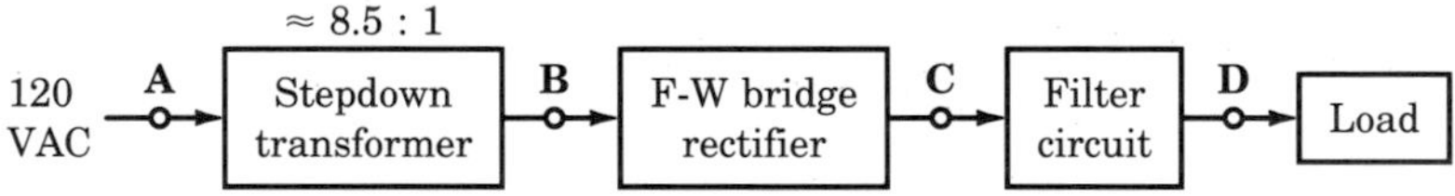

CHALLENGE CIRCUIT 14

Starting Point Information

1. Circuit diagram
2. The output voltage at point D is lower than it should be
3. You are to troubleshoot down to the "block" level

Test — Results in Appendix C

Waveform at point A shows . . . . (74)
Waveform at point B shows . . . . (53)
Waveform at point C shows . . . . (21)
DC voltage at point D measures . . . . (65)

CHAPTER CHALLENGE

Following the SIMPLER troubleshooting sequence, find the problem with this circuit. As you follow the sequence, record your responses to each step on a separate sheet of paper.

SYMPTOMS Gather, verify, and analyze symptom information. (Look at the "Starting Point Information.")

IDENTIFY Identify initial suspect area for the location of the trouble.

MAKE Make a decision about "What type of test to make" and "Where to make it." (To simulate making a decision about each test you want to make, select the desired test from the "TEST" column listing below.)

PERFORM Perform the test. (Look up the result of the test in Appendix C. Use the number in the "RESULTS" column to find the test result in the Appendix.)

LOCATE Locate and define a new "narrower" area in which to continue troubleshooting.

EXAMINE Examine available information and again determine "what type of test" and "where."

REPEAT Repeat the preceding analysis and testing steps until the trouble is found. What would you do to restore this circuit to normal operation? When you have solved the problem, compare your results with those shown in the color insert.

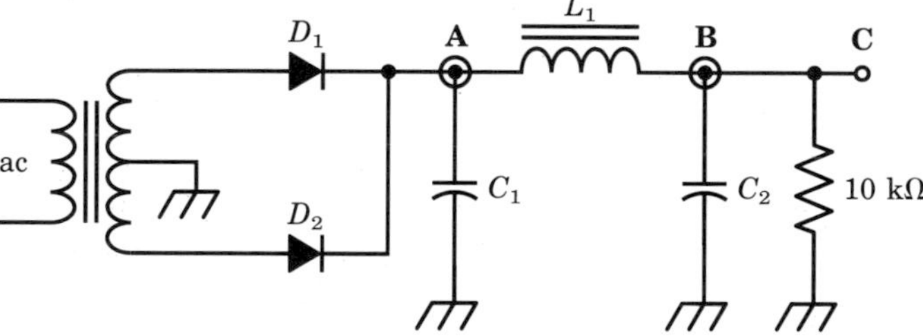

CHALLENGE CIRCUIT 15

Starting Point Information

1. Circuit diagram
2. Output voltage at C is lower than normal. A scope check also indicates higher than normal ac ripple present.

Test	Results in Appendix C
DC voltage at pt C	(69)
AC voltage at pt C	(26)
DC voltage at pt B	(98)
AC voltage at pt B	(6)
DC voltage at pt A	(76)
AC voltage at pt A	(35)
R of L_1	(92)
C_1 R measurements	(45)
C_2 R measurements	(117)

SUMMARY

- Semiconductor material has conductivity falling in a range between metal conductors and non-conducting materials. Another difference between conductors and semiconductor materials is that conductors typically have positive temperature coefficients and semiconductors have negative temperature coefficients.
- Three particles in atoms are protons, neutrons, and electrons. The outer shell or ring of electrons are called the valence electrons.
- Semiconductor materials such as germanium and silicon have four valence electrons in their outer shells. Neighboring atoms of the semiconductor material share electrons in what are called covalent bonds.
- Introducing select impurities into the crystal structure of the semiconductor material is called doping. Doping with pentavalent materials introduces donor atoms into the structure that donate relatively free electrons (N-type material). Doping with trivalent materials introduces acceptor atoms into the structure that easily accept electrons to fill the holes where the covalent bonds lack an electron (P-type material). The majority current carriers in N-type material are electrons; in P-type material, the majority carriers are holes.
- When P- and N-type materials are joined, a P-N junction is formed. Near the junction, a depletion region is created by electrons from the N-type material moving to fill holes in the P-type material, and holes effectively moving the other direction. This creates a charge near the junction called a barrier potential. This potential is negative on the P side of the junction and positive on the N side of the junction.
- The P-N junction creates a diode. (The prefix "di" indicates two parts.) By applying external bias voltage to the diode, it can be made to conduct, or not to conduct. Conduction is caused by forward biasing the P-N junction (i.e., negative connected to the N-type material or cathode and positive connected to the P-type material or anode). When reverse biased, the diode will not conduct and acts like an open switch.
- Diodes are frequently used to rectify ac voltage. That is, to change ac to pulsating dc.
- A typical power supply system takes existing ac voltage and converts it to dc voltage, or voltages of appropriate level(s). Many power supplies use a transformer to step-up or step-down ac voltages; a rectifier system to change ac to pulsating dc; a filter system to smooth out the pulsating dc and eliminate the ac component of the complex wave; an output network (in some cases) that can be a voltage divider network, a bleeder, and so on; and, in some cases, a voltage or current regulator system to keep voltage or current output constant with varying loads.
- Common rectifier circuit configurations are the half-wave rectifier (using one diode); the conventional full-wave rectifier (using two diodes and a center-tapped transformer); and the bridge rectifier (using four diodes). Other special circuit configurations are used that act as voltage multipliers. (Typical circuits are the full-wave doubler and the half-wave cascade doubler.)
- Unfiltered dc output voltages (for a given transformer) can be computed for the various standard circuit configurations as follows:

$$V_{dc}\ (\text{H-W}) = 0.45 \times V_{rms} \text{ or } 0.318 \times V_{pk}$$
$$V_{dc}\ (\text{F-W}) = 0.9 \times V_{rms} \text{ of } 0.5 \text{ secondary or } 0.45 \times V_{rms} \text{ of secondary}$$
$$V_{dc}\ (\text{Bridge}) = 0.9 \times V_{rms} \text{ of full secondary}$$

- Peak-inverse-voltage (PIV) is the maximum voltage across the diode during its non-conducting period. PIVs per rectifier diode (*without filter*) for the common circuit configurations are:

PIV (H-W) = 1.414 × V_{rms} (full secondary)
PIV (F-W) = 1.414 × V_{rms} (full secondary)
PIV (Bridge) = 1.414 × V_{rms} (full secondary)

(NOTE: With capacitor filters, peak inverse voltage can be as great as 2.828 × V_{rms}.)

- Capacitors, inductors, and resistors are often used in filtering networks. Common filters include simple C-types, simple L-types, and more common, combinations of L and C in L-type and pi-type configurations. Filters can be single section or multiple-section. Generally, when there are more sections, filtering is greater.

- Ripple frequency relates to the number of pulses per cycle of ac input. Half-wave rectifiers have ripple frequency outputs equal to the ac input frequency. Full-wave and bridge rectifiers have ripple frequency outputs equal to twice the ac input frequency.

- Oscilloscope waveforms and voltage measurements are two common techniques used to troubleshoot power supplies. Also, the ohmmeter is useful to check fuses, and forward and reverse resistances of diodes.

REVIEW QUESTIONS

1. Explain the term pentavalent.

2. Explain the term intrinsic semiconductor.

3. How many valence electrons does silicon have?

4. Define the term depletion region.

5. What are the majority carriers in N-type material?

6. What are the minority carriers in P-type material?

7. Explain the term barrier potential.

8. Draw the basic diagram of a forward biased P-N junction diode.

9. Draw the basic diagram of a half-wave rectifier.

10. Draw the basic diagram of a typical full-wave rectifier circuit.

11. Draw the basic diagram of a bridge rectifier circuit.

12. Draw the circuit of a full-wave rectifier circuit using a capacitor-input, pi-type filter network feeding a 100 kΩ load. Is the dc output

of this circuit equal to, greater than, or less than the same rectifier without filtering?

13. What is the unfiltered dc output of a full-wave rectifier circuit fed by a 1:2 step-up transformer connected to a primary voltage source of 200-volts rms?

14. What is the PIV of the diode in a half-wave rectifier circuit using a capacitor filter, if the transformer feeding the rectifier circuit is a 1:3 step-up transformer supplied by a 150-volt ac source?

15. If the ripple output of a rectifier-filter system is excessive, is the problem most likely in the transformer, rectifier(s), or filter network? Indicate which component(s) is probably the cause of the problem.

16. Draw a series diode clipper circuit that will clip the negative alternation of the ac input cycle.

17. Draw a shunt diode clipper circuit that will clip the negative alternation of the ac input cycle.

18. Describe what you would do to the circuit to cause it to clip only part of the negative alternation.

19. During the alternation when the diode is not conducting in the clipper circuit, what voltage is observed across the non-conducting diode?

20. Name two possible applications of clipper/limiters.

TEST YOURSELF

1. Draw the circuit of a half-wave rectifier circuit with negative output with respect to ground.
2. Research and write the formula for voltage regulation percentage.
3. Research and write the definition for the term critical inductance.
4. Research and write the formula for percentage ripple.
5. Research and write the definition for the term optimum inductance.
6. Research and list at least four important specifications found in data manuals relating to diodes.

7. A full-wave rectifier has an unfiltered dc output of 100 volts.
   a. What is the ac voltage across the full secondary of the transformer feeding the circuit?
   b. What is the maximum circuit output if a filter is used?
   c. What is the PIV value on each rectifier?
8. If the circuit in Question 7 is a bridge rectifier, rather than a conventional full-wave rectifier, what is the dc output voltage using the same transformer?

PERFORMANCE PROJECT CORRELATION CHART

Suggested performance projects in the Laboratory Manual that correlate with topics in this chapter are:

CHAPTER TOPIC	PERFORMANCE PROJECT	PROJECT NUMBER
The Semiconductor Diode	Forward and Reverse Bias, and I vs V	62
	Rectification	63
The Half-Wave Rectifier Circuit	Average dc and Waveform, Unfiltered	64
	Average dc and Waveform, with C Filter	65
	Effect of Load	66
The Bridge Rectifier Circuit	Average dc and Waveform, Unfiltered	67
	Average dc and Waveform, with C Filter	68
	Effect of Load	69

NOTE: After completing the above projects, perform the "Summary" checkout at the end of the performance projects section in the Laboratory Manual.

Overview of Transistors

Key Terms

Base
Bipolar junction transistor
Collector
Common-base amplifier
Common-collector amplifier
Common-emitter amplifier
Emitter
Field effect transistor (FET)
Light-emitting diode (LED)
Metal-oxide semiconductor (MOS)
NPN transistor
Photodiode
PNP transistor
Silicon-controlled rectifier (SCR)
Source/gate/drain
Thyristor
Triac
Tunnel diode
Varactor diode

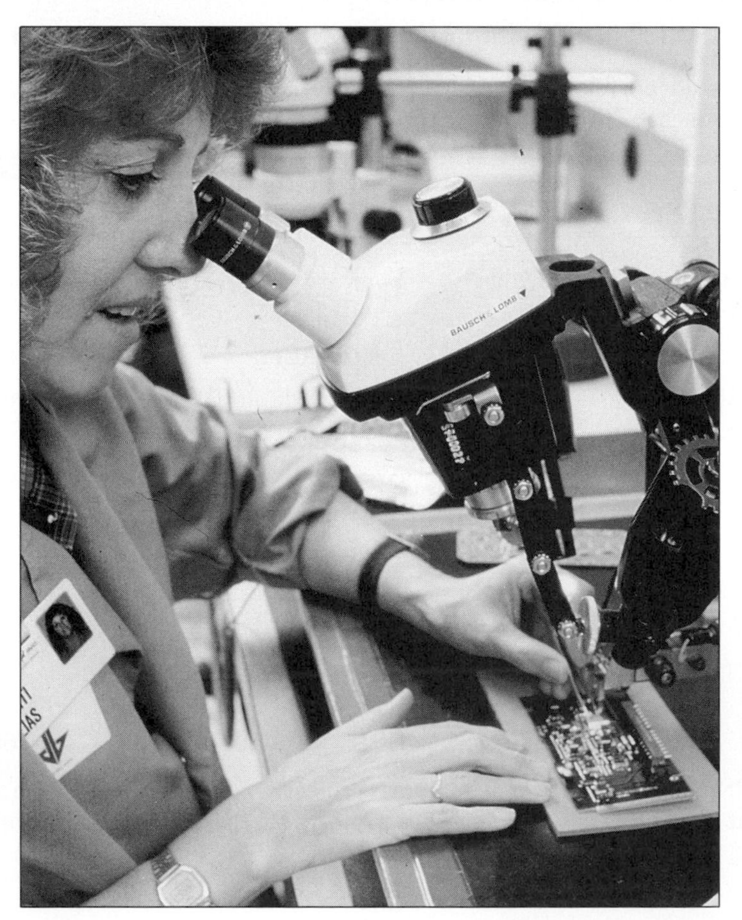

Courtesy of Motorola, Inc.

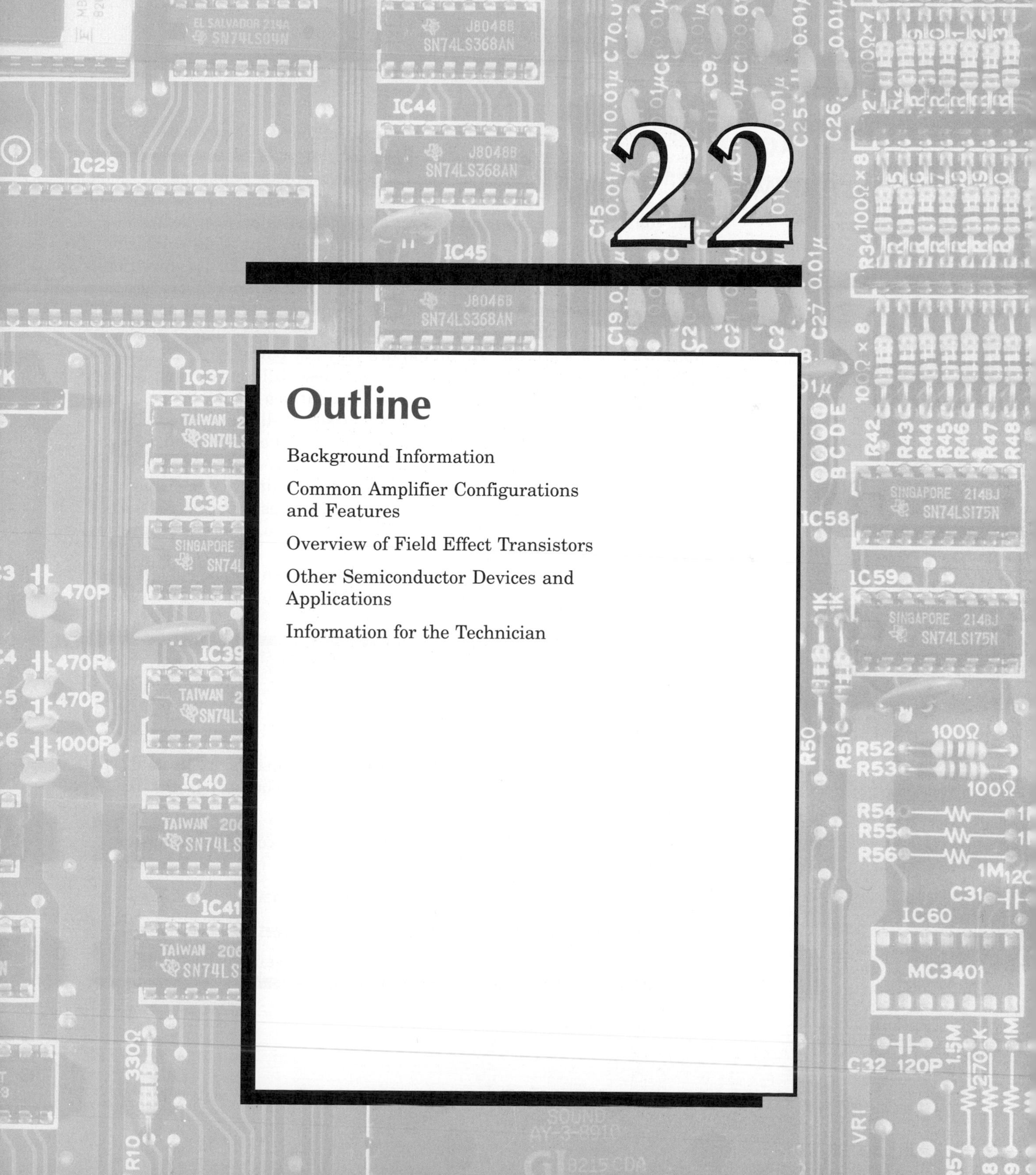

22

Outline

Background Information

Common Amplifier Configurations and Features

Overview of Field Effect Transistors

Other Semiconductor Devices and Applications

Information for the Technician

Chapter Preview

It would be impossible to discuss transistor theory in one chapter. Many books are devoted entirely to that endeavor. For this reason, this chapter will focus on basic features of transistors, introductory information about types, and an overview of typical applications. Perhaps this chapter will whet your appetite for learning more about these important devices.

The transistor was invented in the late 1940s. The term *transistor* is really a contraction of "transfer" and "resistor". Credit for its invention is given to two Bell Laboratories scientists, J. Bardeen and W. H. Brattain. The outstanding features of transistors are apparent when their operational features are compared to vacuum tubes, Figure 22–1. Compared to vacuum tubes, transistors are very small, require very low power for operation, and generate much less heat during operation. These differences exist because vacuum tubes depend on heating a metal element to a point where electrons are emitted from its surface *(thermionic emission),* allowing other elements in the tube to control the emitted electron flow. On the other hand, semiconductor diodes and transistors use solid-state current-carrying mechanisms that do not require high current or high temperatures for operation. Another advantage of transistors is that since much less heating is involved, transistors typically have a much longer life. Also, transistors are less fragile.

In this chapter, you will learn that transistors are a critical subject in any technician's training. Because of their importance, you should look forward to further study as opportunities arise.

Objectives

After studying this chapter, you will be able to:

- List three advantages of transistors compared to vacuum tubes
- Draw symbols for **NPN** and **PNP bipolar** transistors, **MOS** field effect transistors, **SCR thyristors,** and two special semiconductor diode devices
- Describe the input and output characteristics of common transistor **amplifier** stages
- List the advantages of each common type of transistor amplifier stage
- Describe two categories of applications for transistors
- Define several transistor troubles and their preventive measures

(b)

(a)

Item	Transistors	Vacuum Tubes
size	small	large
power	small	large
heat	low	high
voltage sensitivity	very sensitive	very forgiving
durability	rugged	very fragile
cost	many-low	medium to high cost
life expectancy	long	medium

(c)

FIGURE 22–1 Similarities and differences between vacuum tubes and transistors (Photo a courtesy of International Rectifier; photo b courtesy of General Electric Company)

BACKGROUND INFORMATION

As you might surmise from the previous chapter, a forward-biased P-N junction is basically a low-resistance path for current flow. Conversely, a reverse-biased P-N junction represents a high resistance path. If the current is almost the same through a given high resistance as it is through a given low resistance, the I^2R (power) in the high resistance is greater than in the low resistance. This is one basic principle on which power gain is produced by a transistor.

For example, if a signal is introduced in a forward-biased P-N junction and output is taken from a reverse-biased P-N junction with similar current in both output and input circuits, there is *power gain.* In effect, the transistor has transferred the signal from the low resistance (input) to the high resistance (output). Thus, the word **transistor** represents transfer and resistor.

The NPN Transistor

Three sections of semiconductor material are sandwiched together to form a transistor. An **NPN transistor** is one where the outside pieces of the sandwich are N-type material, and the material placed between these two sections is P-type material. See Figure 22–2 as you study the following discussion.

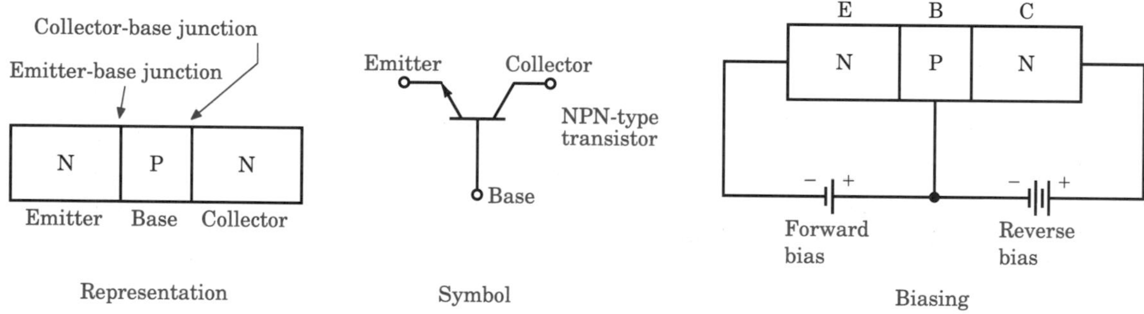

FIGURE 22–2 The NPN transistor

There are two P-N or N-P junctions. Because there are majority and minority carriers and opposite polarity biasing voltages, these devices are sometimes called **bipolar junction transistors.**

The first N-P junction is forward biased and the second junction (middle P to outside N) is reverse biased.

The symbol for the NPN transistor indicates three elements known as **emitter, base,** and **collector.** As shown in Figure 22–2, the emitter-base is the forward-biased, low-resistance input circuit, and the high-resistance output circuit is the collector circuit. The NPN transistor symbol shows the arrow for the emitter "Not-Pointing-iN" (NPN) toward the base. Learn this symbol, it is important!

Electrons in the emitter N-type material are forced toward the junction and gain enough energy to pass through the junction and the very thin base material section, then into the collector region. Once electrons are in the collector region, the strong positive of the collector-base reverse-bias voltage attracts these electrons to the collector source. From this point onward, the circuit has been completed.

About 92 to 99 percent of the emitter current is transferred to the collector circuit. Currents are designated as I_E (emitter current); I_B (base current); and I_C (collector current), Figure 22–3.

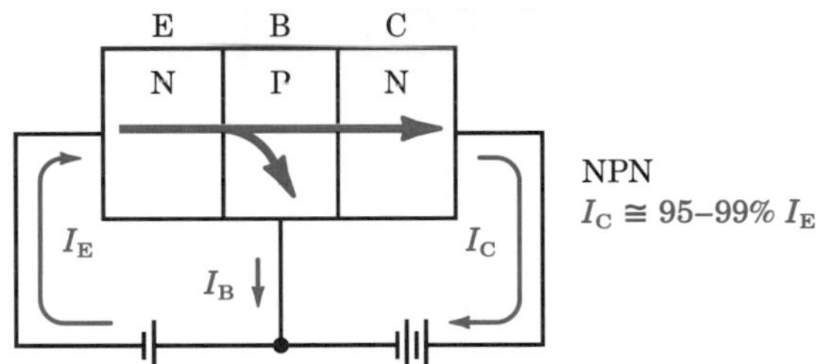

FIGURE 22–3
Currents in a transistor

The amount of current flowing is essentially controlled by the emitter-base current. NOTE: This bipolar transistor is essentially a current-controlled device.

The PNP Transistor

The operational theory for the **PNP transistor** is similar to the NPN transistor with the following exceptions:

1. Bias polarities are reversed to achieve forward biasing of the emitter-base junction and reverse biasing of the collector-base junction.
2. Majority carriers are holes rather than electrons. However, many authors prefer to explain operation in terms of electron movement.
3. The PNP transistor symbol is similar to the NPN. However, the emitter arrow is "Pointing-iN" toward the base, Figure 22–4. Remember that the arrow always points in the opposite direction from electron flow.

TRANSISTOR ALPHA AND BETA PARAMETERS

As you might suspect, the relationship of currents through a transistor provides valuable insight into its operation. We do not intend to discuss transistor parameters or circuit design in depth in this chapter; however, we do want to introduce you to the terms alpha and beta commonly associated with these devices.

The alpha and beta of a transistor refer to specific relationships of transistor element currents to each other. Alpha is abbreviated by the Greek letter alpha (α). Beta is abbreviated using the Greek letter beta (β).

FIGURE 22–4 The PNP transistor

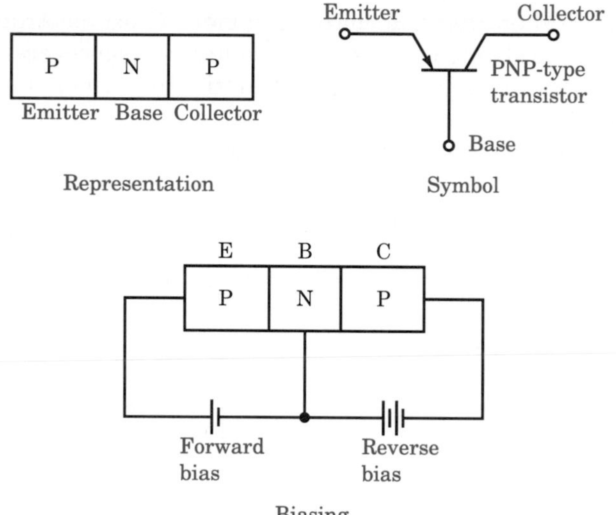

IN-PROCESS LEARNING CHECK I

Fill in the blanks as appropriate.

1. A forward-biased PN junction is a ____________ resistance path for current flow. A reverse-biased PN junction is a ____________ resistance path for current flow.
2. Normally, the forward-biased junction of a junction transistor is the ____________ junction.
3. Normally, the reverse-biased junction of a junction transistor is the ____________ junction.
4. The symbol for a NPN transistor shows the emitter element ____________.
5. The symbol for a PNP transistor shows the emitter element ____________.

In most simplistic terms, alpha is defined as the ratio of collector current to emitter current: dc alpha $(\alpha) = \frac{I_C}{I_E}$. Typical values of alphas range from about 0.96 to 0.99. This indicates that about 96 to 99 percent of the emitter current becomes collector current. The remaining percentage flows as base current.

Beta is defined as the ratio of collector current to base current: dc beta $(\beta) = \frac{I_C}{I_B}$. Typical values of beta range from 50–100 for small-signal type transistors.

Since (for the common-emitter type circuit) the input is considered at the base and the output is considered at the collector, there is a current gain from base to collector. This gain characteristic allows the transistor to be used as an "active" component or element in amplifiers and oscillators and other applications, as you will see in an upcoming chapter in the text.

Some Common Basing or Lead Configuration Schemes

In Figure 22–5 you see some common transistor lead basing layouts (i.e., bottom view of transistor package with lead connections layout). "E" stands for emitter lead, "B" for base lead, and "C" for the collector lead. The transistor outline (TO) numbers identify each specific packaging configuration used by manufacturers.

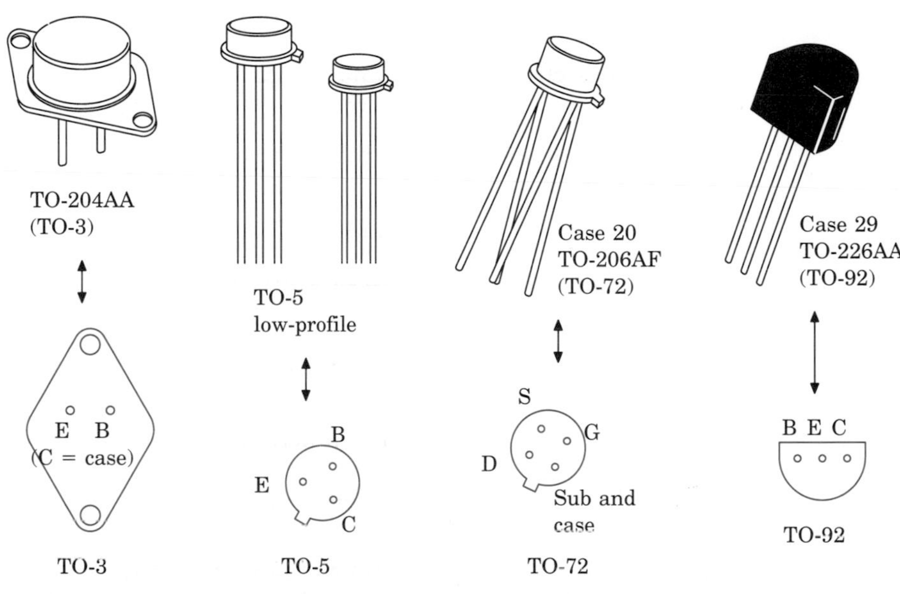

FIGURE 22–5 A sample of some common packaging and basing schemes

COMMON AMPLIFIER CONFIGURATIONS AND CHARACTERISTICS

The Common-Base Amplifier (CB Amplifier)

A simple diagram of a **common-base amplifier** circuit is shown in Figure 22–6. The primary characteristics are:

1. The base is common to both input and output circuits.

FIGURE 22–6 Simplified diagram showing common-base circuit arrangement

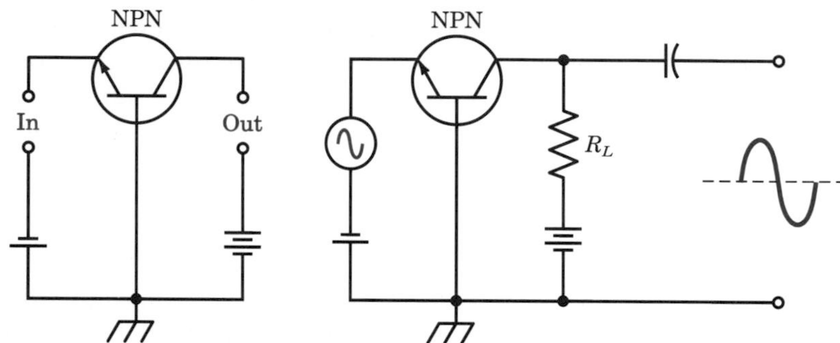

2. The input signal is introduced into the emitter-base circuit, and the output signal is taken from the collector-base circuit.
3. The input circuit is low impedance (forward-biased emitter-base junction). Typically, the impedance is in the range of 1 to 50 ohms.
4. The output circuit is high impedance (approximately 1 kΩ to 1 MΩ). Voltage or power gain is in the 1000 to 1500 range.
5. There is no phase reversal between the input and output signals.

The Common-Emitter Amplifier (CE Amplifier)

A simple diagram of the **common-emitter amplifier** is shown in Figure 22–7. Some basic features are:

1. The emitter is common to both input and output circuits.
2. The input signal is introduced into the emitter-base circuit, and the output signal is taken from the collector-emitter circuit.
3. The input circuit is low impedance (forward-biased emitter-base junction). Typically, the impedance is in the range of 25 to 5000 ohms.
4. The output circuit is medium to high impedance (approximately 50 Ω to 50 kΩ). Power gains range as high as 10,000. This circuit provides

FIGURE 22–7 Simplified diagram showing common-emitter circuit arrangement

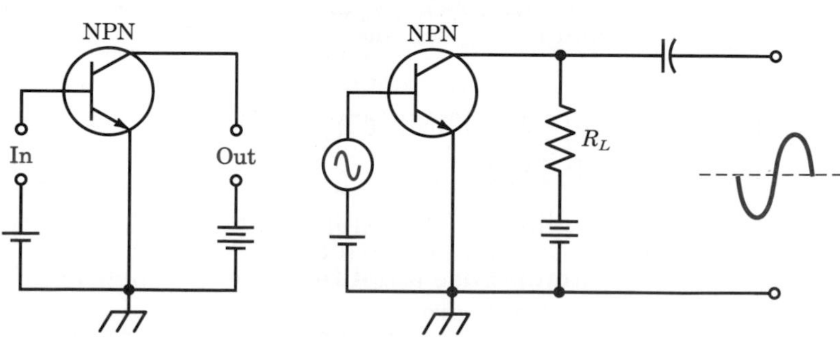

both voltage and current gain. Current gain in CE amplifiers is always greater than unity.

5. There is 180° phase reversal between the input and output signals.

The Common-Collector Amplifier (CC Amplifier)

A simple circuit diagram for a **common-collector amplifier** is shown in Figure 22–8. Several key characteristics are:

1. The input signal is introduced into the base-collector circuit, and the output signal is taken from the emitter-collector circuit.
2. The input impedance is high.
3. The output impedance is relatively low. Voltage gain is *less than* unity. Power gain is also lower than in either the CB or CE amplifiers. The primary application is for impedance matching between circuits connected to its input and output.
4. There is no phase reversal between the input and output signals.

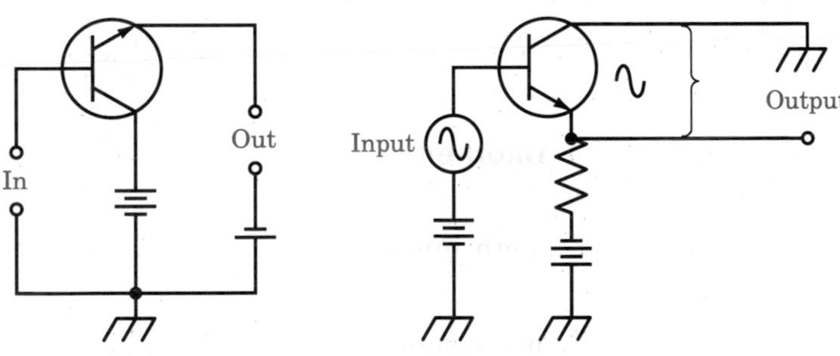

FIGURE 22–8 Simplified diagram showing common-collector circuit arrangement

IN-PROCESS LEARNING CHECK II

1. The transistor amplifier circuit having the base electrode common to both input and output circuits is the ___________ amplifier.
2. The transistor amplifier circuit providing both voltage and current gain is the ___________ amplifier circuit.
3. The transistor amplifier circuit having 180 degrees phase difference between input and output signals is the ___________ amplifier.
4. The transistor amplifier circuit having a voltage gain of less than one (unity) is the ___________ amplifier circuit.
5. A transistor amplifier circuit frequently used for impedance matching is the ___________ amplifier circuit.

OVERVIEW OF FIELD EFFECT TRANSISTORS

The junction transistors we previously discussed are bipolar devices, and conduction depends on two types of charge carriers. The **field effect transistor** (FET), however, is unipolar and depends on one type of charge carrier. Its operation is based primarily on controlling charges. The voltages applied to the gate control the electrical size of the channel where current is conducted. The three electrodes of this device are the **source, gate,** and **drain,** Figure 22–9. The gate is the control electrode. The electrodes are analogous to regular junction transistors in that the source is somewhat equivalent to the emitter, (where current carriers enter the channel). The gate is where input signals are typically applied, and it controls conduction

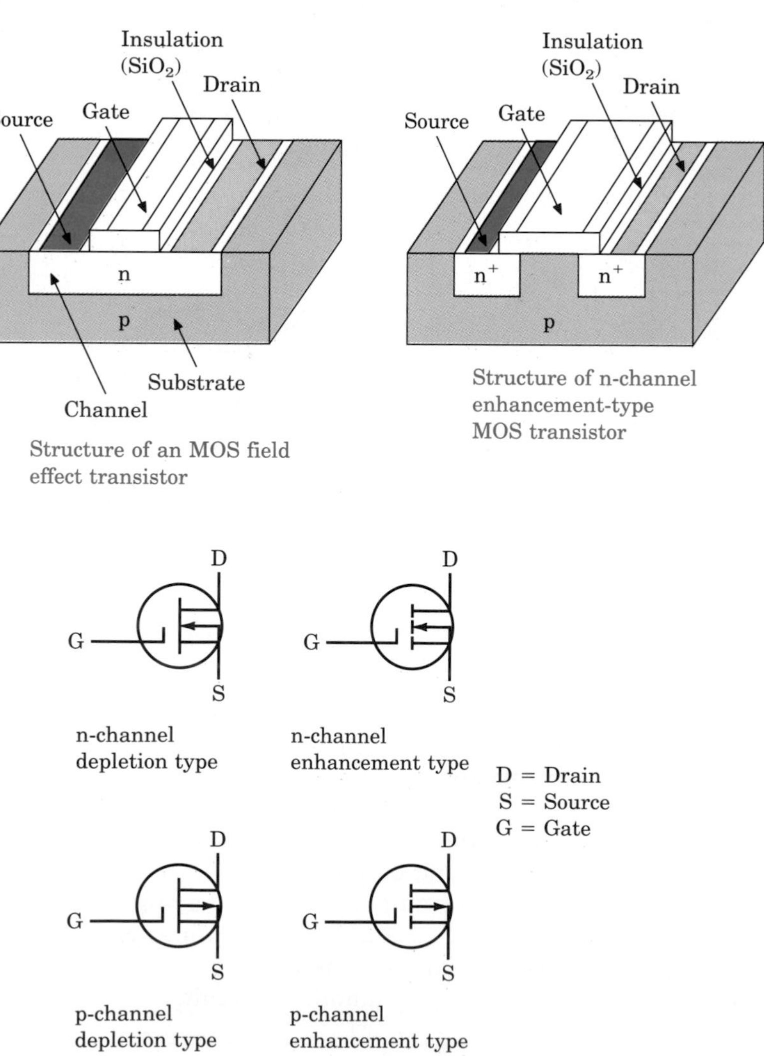

FIGURE 22–9 Metal-oxide-semiconductor field effect transistor (MOSFET) construction arrangements and some related symbols

of the device, which is analogous to the base in a regular transistor. The drain is where current carriers exit the device, roughly equivalent to the collector.

There are various constructions of FETs. Some common types are the JFET (junction), IGFET (insulated-gate) and MOSFET **(metal-oxide-semiconductor).** The MOSFET retains the solid-state device advantages of ruggedness, small size, and low low power requirements, yet provides a very high input impedance (10-15 MΩ). Figure 22–9 shows some typical constructions and related symbols for the MOSFET.

RF amplifiers, other voltage amplifiers, voltage-controlled attenuators, and various switching functions are a few of the typical applications for these FET devices.

OTHER SEMICONDUCTOR DEVICES AND APPLICATIONS

The SCR and the TRIAC (Thyristors)

The **silicon-controlled rectifier** (SCR) is a device that can be switched from on to off, or from off to on by current pulses to the gate electrode. As shown in Figure 22–10, this device is often called a four-layer P-N-P-N device with three principle electrodes—cathode, anode, and gate (control electrode). The SCR is a unidirectional device that controls dc and ac power. It can turn power on or off of a load, or control how much power is applied to the load. Its main disadvantage in controlling ac power is that its unique features can only be used during one alternation of each cycle.

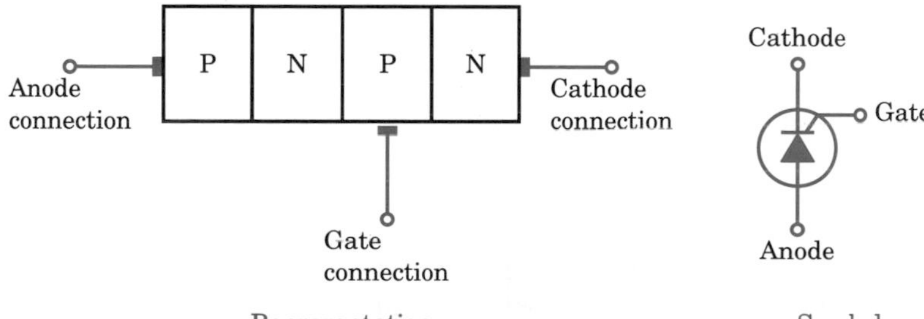

FIGURE 22–10 The silicon-controlled rectifier (SCR)

A device similar to the SCR is the **triac.** See Figure 22–11 for an illustration of the triac terminals, schematic symbol, and so on. The triac (sometimes called a bidirectional triode **thyristor**) is equivalent to having two parallel SCRs connected in opposite directions. This allows the device to control ac power over the entire cycle, not just one alternation.

SCRs and triacs are used in various applications. Examples include controlling the speed of a motor and controlling the light intensity of lamps.

FIGURE 22–11 The triac (bidirectional triode thyristor)

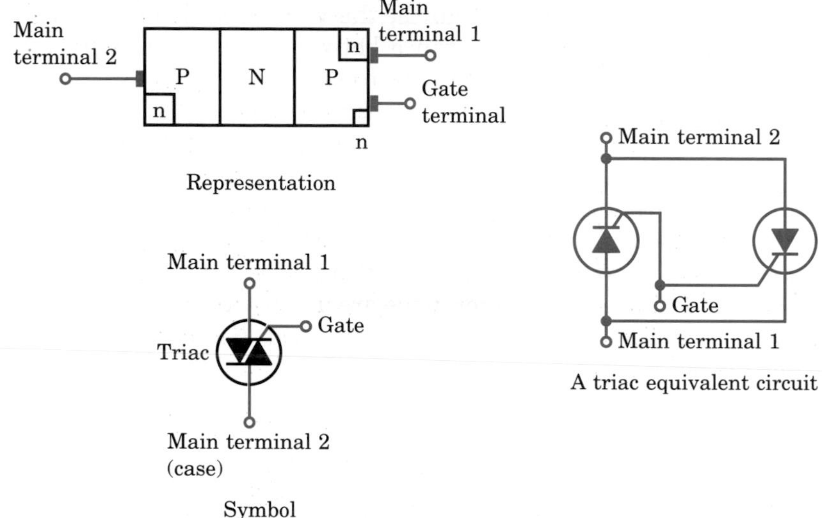

The Tunnel Diode

The **tunnel diode** is a two-terminal diode device manufactured with special characteristics. See Figure 22–12 for the symbols used for this device. Because of the heavy doping, the depletion region is very thin. Even though the barrier voltage is significant due to the heavy doping, electrons are able to tunnel through the thin depletion region barrier voltage. At the biasing voltage levels, which cause this tunneling effect, there is a unique feature exhibited by the diode, called the *negative-resistance* region of the V-I characteristic. This means there is a portion of the diode characteristic curve where an increase in forward voltage causes a decrease in forward current. This feature is advantageous in circuits known as *oscillator circuits,* which you will learn about in the near future. These circuits generate high frequency ac signals. This diode is also sometimes used as an electronic switch. Its key features are it can switch at high speeds, and it takes little power to operate.

FIGURE 22–12 The tunnel diode

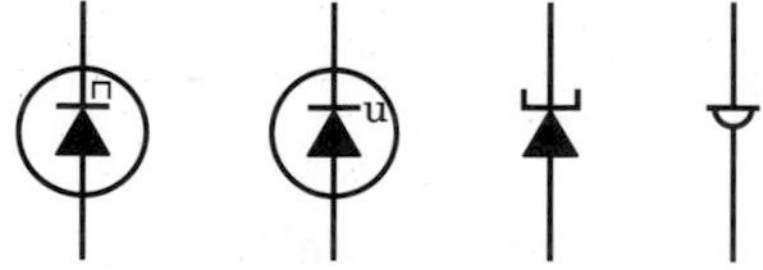

Several schematic symbols used

The Varactor Diode

Varactor diodes act like small variable capacitors. By changing the amount of reverse bias across their P-N junction, the capacitance can be varied. The greater the reverse bias is, the wider the depletion region is. Also, the greater the effective distance between P and N sections is, the smaller the capacitance is. Furthermore, the smaller the reverse bias is, the narrower the depletion region is and the greater the capacitance produced. In effect, the varactor (or varactor diode) is an electronically variable capacitance compared to a variable capacitor whose physical plates must be meshed or unmeshed to change capacitance. These devices are designed with capacitances ranging from 1 pF to 2,000 or more pF. Frequency of operation effects their Q factor. The varactor symbol is shown in Figure 22–13. The unique abilities of the varactor diode are used in various practical applications. Some examples include tuned circuits and other circuits in receivers and transmitters.

Capacitance of junction related to bias voltage

FIGURE 22–13 The varactor diode

The Photodiode Cell

The **photodiode cell** is just one of a family of "optoelectronic" devices. The family includes devices which emit light (such as LED's) and devices that are sensitive to light (such as photodiodes and phototransistors). Sometimes combinations of light emitting and light sensing devices are packaged together to form optocouplers or opto-isolators. You will study this concept later.

The photodiode is a device which performs as a variable resistance whose resistance to current (reverse-bias current) is controlled by light intensity. The stronger the incident light striking the diode junction, the higher the reverse current that will flow. A window, or lens allows light to hit the diode junction area, causing this variable resistance effect. The illustration in Figure 22–14 shows the typical enclosure and schematic symbol for the photodiode cell.

Typical construction

Schematic symbol

FIGURE 22–14 The photodiode cell

The Photocell (Photovoltaic Cell)

Another P-N material, two-terminal device is the photovoltaic cell, sometimes called a solar cell. This device develops a voltage output at its terminals when struck by light. Figure 22–15 shows one schematic symbol for

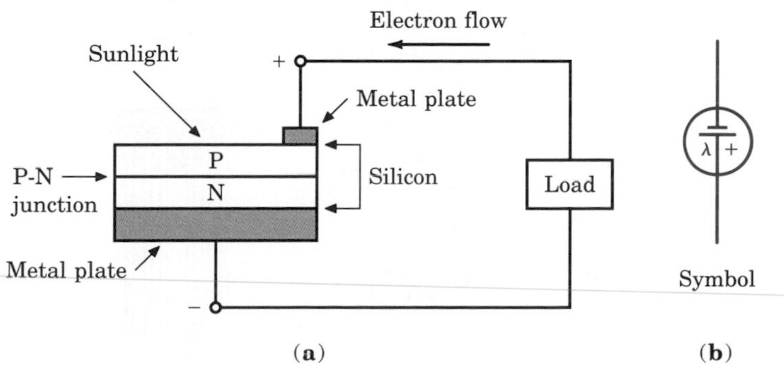

FIGURE 22–15 The photovoltaic (solar) cell
(a) Basic structure
(b) Schematic symbol

this device. Because the efficiency factors are very low (2 to 20%), you usually find these cells in an array that combines outputs of numerous cells. You have probably seen or used units that use these cells, such as the battery-free, solar powered, hand-held calculator.

Symbol

FIGURE 22–16 Schematic symbol for a light-emitting diode (LED)

The Light-Emitting Diode (LED)

The **light-emitting diode** (LED) is a special diode that emits light in the visible spectrum of light frequencies when the diode is forward-biased. Figure 22–16 shows the symbol used for this device. You have probably seen them in a multitude of indicator light applications. Also, they are used in card reading applications, electro-optical switching applications, burglar alarms, and numerous other systems and subsystems.

Opto-Isolators

Sometimes a combination of an LED (used as a light source or transmitter) and a phototransistor (used as a receiver of the transmitted light) are used to provide isolation between circuits or processes. The information is transmitted from the source to the detector (or receiver) by light waves.

Two common applications for opto-isolators include preventing a low impedance load from loading a high impedance source and enabling a low-current source to control a high-current detector.

With this system, it is important that the source emit the light frequency or frequencies where the detector or receiver is sensitive. For this reason, sources and detectors are matched by their light emission frequencies and the receiving device sensitivity to these frequencies. Also, sometimes optical lenses focus the light on the receiver and help prevent unwanted light from activating the detector.

Often the LED source and the phototransistor are contained in the same transistor style package. See the 4N22 diagram shown in Figure 22–17.

The isolation ability of this system makes it useful in many applications.

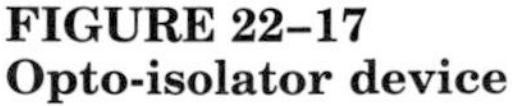

FIGURE 22–17 Opto-isolator device

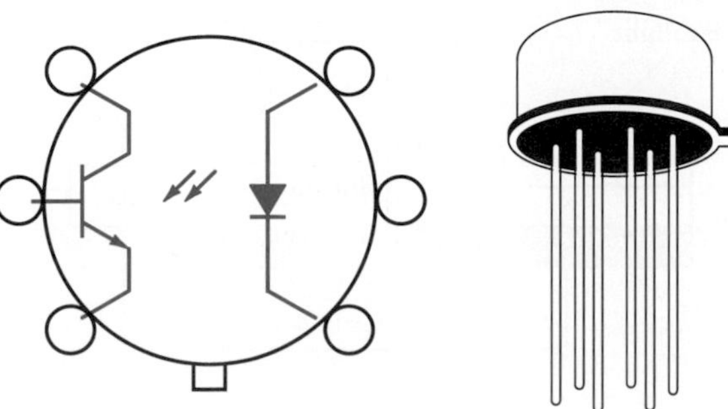

INFORMATION FOR THE TECHNICIAN

PRACTICAL NOTES

Some transistor troubles are caused by physical factors. An example is when a connecting wire lead fails to make contact with a transistor electrode.

Many problems occur in transistors when their electrical ratings are exceeded. Damage results from excessive current through or excessive heating of the materials in the device. (You can damage a semiconductor device when you are soldering it to the circuit or circuit board if you do not appropriately heat-sink the leads you are soldering.) Exceeding the voltage parameters for such devices is easy to do, and they "die silently" with no warning other than abnormal circuit or system operation. That is, there is no smoke or sound involved. Also, wrong polarity voltages applied to the electrodes cause problems. Even static electricity on a technician handling certain types of these devices (e.g., MOS types) can blow out junctions. For that reason, certain grounding precautions are used when handling these sensitive devices. You will learn more about these precautions in the laboratory or on the job.

There are some sophisticated *in-circuit testers* and *out-of-circuit testers,* which test transistors. An ohmmeter can also be used.

Often, transistor troubles are found by checking the forward- and reverse-biased resistances from junction-to-junction. Many times, the problem is seen by an internal short or open. Of course, the ohmmeter supplies the battery voltage for the tests.

NOTE: It helps to know the voltage polarity of the test probes when in the ohmmeter mode. Another *caution* is that currents caused by using ohmmeters on certain ranges can exceed ratings of the transistor being tested. Thus, you can make a transistor go from good to bad instantly.

Refer to Figure 22–18 for some examples of how ohmmeter tests are used. Note that the values shown are only relative and can differ from device to device.

Another technique to check transistors during operation is to check the bias voltages from element-to-element. A bad transistor causes either too much or too little current through the various biasing networks, which can be detected by bias voltage measurements.

In summary, semiconductor devices can be damaged during manufacture, by mishandling during installation or replacement, or by improper currents or voltages. Normal precautions require using heat sinks to drain

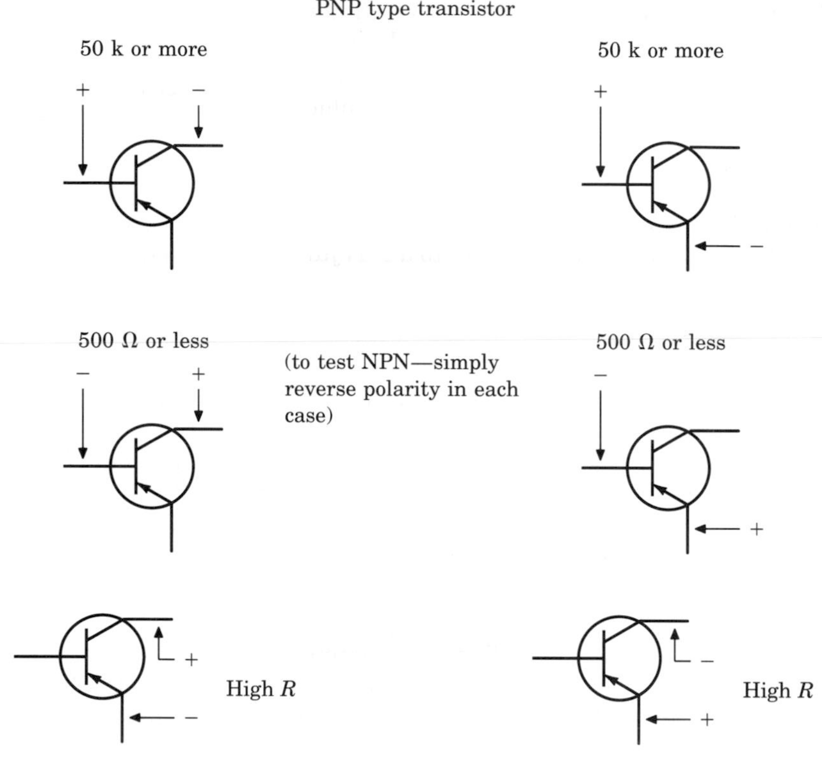

FIGURE 22–18 Ohmmeter testing of transistor junctions (typically on $R \times 10$ range)

heat away from the transistor elements during soldering. Proper precautions about static electricity are also necessary.

If a transistor or semiconductor fails, the device is replaced. But associated circuit components should also be checked prior to replacement to ensure that another malfunctioning part did not initially cause the failure.

Common sense is the common ingredient in preventing problems and in solving them when they occur. In your further studies, you should become familiar with how semiconductor devices and circuits operate, what their normal parameters should be, and the various tools and techniques used to install, test, and replace these remarkable devices.

SUMMARY

- Transistor is derived from the words transfer and resistor.
- A transistor consists of two P-N junctions. Typically, one P-N junction is forward-biased and is a low-resistance path for current; the other junction is reverse-biased and is a high resistance path for current.
- Because the transistor transfers signals from its low-resistance input circuit to its high-resistance output circuit, it exhibits power gain.

- Two basic transistors are the NPN and the PNP. The letters N and P designate the type of semiconductor material in the three sections. In the NPN device, P-type material is placed between two areas of N-type material. In the PNP device, N-type material is placed between two outside areas of P-type material.
- Forward bias is a voltage applied to a P-N junction so the positive polarity is connected to the P-type material and the negative polarity voltage is connected to the N-type material.
- Reverse bias exists when a source's negative polarity voltage is connected to the P-type material, and the positive polarity voltage is connected to the N-type material.
- Three elements of a bipolar transistor are the emitter, base, and collector.
- Three common amplifier circuit configurations are the common-base amplifier, the common-emitter amplifier (most common), and the common-collector amplifier. Each configuration has unique input and output impedances and other features that best accommodate specific electronic applications.
- Field effect transistors, such as the metal-oxide semiconductor (MOSFET), have high input impedances and demand very little power.
- Thyristors, often termed SCRs (silicon-controlled rectifiers), are devices that can be switched on or off with a current pulse to their gate element. The three parts of an SCR are the gate, cathode, and anode.
- SCRs control the percentage of time during each ac cycle of input that a load receives power. Light dimmers, or motor speed controls are examples of their applications. Also, SCRs control dc power to the load.
- A triac is equivalent to two parallel SCRs and controls power over the entire cycle of ac input. In other words, SCRs only control power during one alternation of each ac cycle of input.
- Some special two-element or diode semiconductors include the tunnel diode (used in oscillator circuits); varactor (used as a variable capacitance); photoconductive cell (used as a variable light-dependent resistance); photovoltaic, or solar cell (used to generate voltage/current by exposure to light); and the LED, or light-emitting diode (used to emit light when current flows).
- Semiconductors are quite sensitive. Therefore, proper caution must be used when handling, soldering, desoldering, and applying their operating parameters.
- Semiconductors are tested in various ways. Test equipment is available to test them *in-circuit* and *out-of-circuit*. Also, the ohmmeter, if used properly, is a good tool for general testing. Bias voltages are also good indicators of normal or abnormal circuit operation.

REVIEW QUESTIONS

1. How many PN junctions does a typical junction-type transistor have?
2. For an NPN transistor, what polarity of voltage is applied to the emitter-base junction? To the collector-base junction?
3. For a PNP transistor, what material appears in the middle of the semiconductor materials?

4. Draw the schematic symbols for the NPN and PNP transistors. Label the diagrams, as appropriate.

5. Approximately what percentage of emitter current appears in the collector circuit of typical transistor circuits?

6. Explain forward-biasing with respect to a PN junction.

7. Explain reverse-biasing with respect to a PN junction.

8. Is the base current in a transistor a high or low percentage of the emitter current?

9. Explain the term basing diagram.

10. In the common-emitter amplifier configuration, what transistor electrode is common to both input and output circuits?

11. For the following transistor amplifier configurations, what are the typical input and output impedance characteristics? (NOTE: Use "high", "low", or "medium" to describe.)
    a. Common-emitter: input Z = ____ output Z = ____
    b. Common-base: input Z = ____ output Z = ____
    c. Common-collector: input Z = ____ output Z = ____

12. Which transistor amplifier circuit configuration causes 180° phase shift from input to output?

13. Which transistor amplifier circuit configuration provides both voltage and current gain?

14. Indicate between which junctions or electrodes the input signal is introduced, and between which electrodes the output signal is taken for the following circuit configurations:
    a. Common-emitter amplifier................ input ____ and ____
       output ____ and ____
    b. Common-base amplifier................... input ____ and ____
       output ____ and ____
    c. Common-collector amplifier .... input ____ and ____
       output ____ and ____

15. Are standard bipolar junction transistors current-controlled or voltage-controlled devices? What about FETs?

16. For certain applications, what is the main advantage of a MOSFET device compared to standard bipolar transistors?

17. Explain negative-resistance with respect to a tunnel diode.

18. If more light strikes a photoconductive cell, will its resistance increase or decrease?

19. In practical applications, what electrical property is varied in a varactor when biasing is changed?

20. Explain the term LED and give a primary application.

21. Name two precautions that must be observed in working and using transistors.

22. What common test instrument is used to check the condition of a transistor out-of-circuit?

23. What test is used to check a transistor operation while in the circuit, with the circuit on?

24. What device or technique should be used when installing or removing semiconductor devices by means of soldering or desoldering?

TEST YOURSELF

1. Look up and draw the transistor outlines for the following case/basing configurations: TO-3, TO-5 and TO-92. Label each electrode on your drawing(s).
2. Research and explain the features and use of heat sinks associated with power transistors.
3. Research and briefly explain the operation of the diac device.
4. Briefly explain why leakage current in a transistor increases with temperature.
5. Research and briefly explain the difference between an enhancement and a depletion FET.

PERFORMANCE PROJECT CORRELATION CHART

Suggested performance projects in the Laboratory Manual that correlate with topics in this chapter are:

CHAPTER TOPIC	PERFORMANCE PROJECT	PROJECT NUMBER
Common Amplifier Configurations and Features	Emitter-Base and Collector-Base Biasing	70
	Alpha and Beta	71

Overview of Integrated Circuits and Digital Electronics

Key Terms

Analog (linear) operation

AND gate

Chip

Digital operation

Hybrid integrated circuits

Integrated circuit (IC)

Inverter

Microprocessor

Mode

Monolithic

NAND gate

NOR gate

OR gate

SSI, MSI, LSI, and VLSI

Thin-film process

Courtesy of International Business Machines

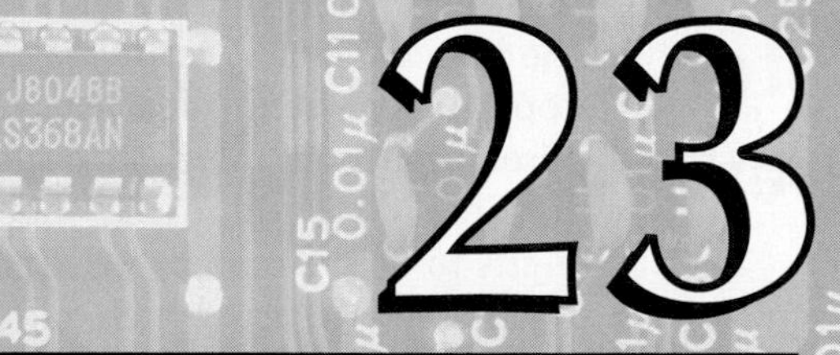

23

Outline

Brief History of Integrated Circuits

Some Classifications of Integrated Circuits

Advantages and Disadvantages of Integrated Circuits

Examples of Analog and Digital Modes of Operation

Integrated Circuits Packaging and Basing

Common Numerical Systems Used With Digital Integrated Circuits

Basic Logic Gates

Other Common Digital Circuits

Memory

The Microprocessor

The Basic Digital Computer

Chapter Preview

The invention of the integrated circuit (IC) in the late 1950s allowed a great step forward in applications of electronic technology. Our trips to the moon were largely enabled through the advances brought about by integrated circuits. Virtually every household appliance, convenience, and luxury seems to be affected by ICs in some way. The ability to package hundreds of circuits in a chip, which may be only 0.1 to 0.3 inches in size, opens up horizons of possibilities limited only by our imaginations. See Figure 23–1 for an example of a semiconductor wafer (used in IC manufacture) where there are many IC chips, each containing thousands of circuits.

These devices are used in both analog electronic circuits (circuits whose outputs vary continuously over a range of values as a function of input values) and digital circuits (circuits whose outputs generally vary or switch between two or more discrete steps or states). An IC used in analog functions is often called a linear IC. Sample uses for linear ICs are in amplifiers, regulators, and other similar applications. The digital IC is probably the most widely used in today's technologies. They are used in various logic, switching, memory, and counting applications.

This chapter will give you only an introduction to these devices. To learn about their important parameters, circuit design considerations, and an analytical approach to the many circuits they contain, other text(s) should be studied. Certainly, you eventually need to study these devices as you prepare to become a technician or engineer.

Objectives

After studying this chapter, you will be able to:

- Classify **integrated circuits** with respect to construction, mode of operation, and packaging
- List three advantages and one disadvantage of integrated circuits
- Identify pin numbers on the most common integrated circuit packages
- Briefly define the difference between **analog** and **digital operations**
- Give two examples each of analog circuit applications and digital circuit applications
- Write the truth table for **OR, AND,** and **NAND gates**
- Write the truth table for an **inverter**
- Explain the difference in operation between RAM and ROM
- Draw a diagram showing the basic building blocks of a digital computer system

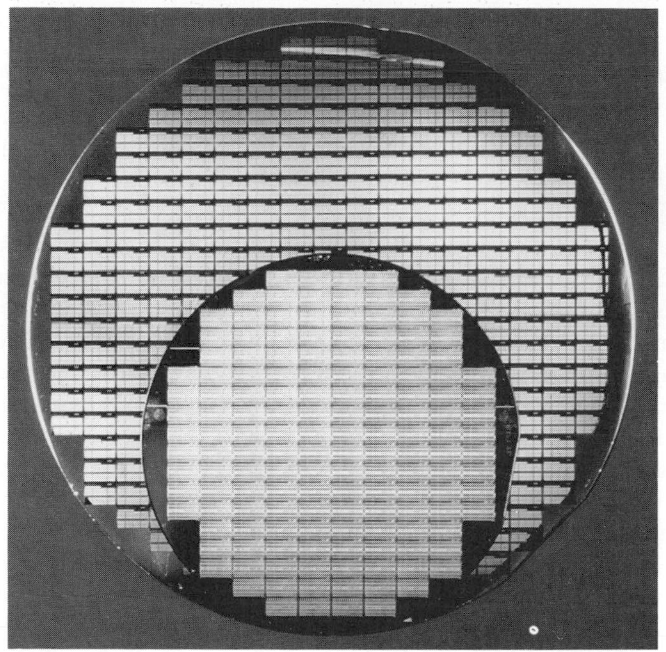

FIGURE 23–1 Semiconductor logic wafer prior to separation of chips (Courtesy of IBM Corporation)

BRIEF HISTORY OF ICs

The transistor was a significant improvement over vacuum tubes. Once this step in technology was made, it became natural to think of putting several semiconductor transistors on one chip of semiconductor material and making appropriate connections between them to form a complete circuit. Furthermore, it was a small step to discover that not only could transistors be formed on these semiconductor chips, but that resistors and capacitors could also be created within the semiconductor chip. Thus the **integrated circuit** (IC) or **microchip** was born.

Jack Kilby, of Texas Instruments, was the first to announce the invention. Six months later, Robert Noyce, of Fairchild Semiconductors, created a similar device. Whereas, Kilby's device was manufactured using germanium semiconductor material, Noyce's device was made from silicon. Now, the silicon IC is one of the most common types.

The obvious advantage of such devices is the miniaturization made possible because of them. For example, computers that used a room full of space are now found with equal computing power housed in a desktop computer.

SOME CLASSIFICATIONS OF ICs

Integrated circuits are classified in various ways. Several classifications are used, according to:

1. Mode
2. Construction (fabrication method)
3. Packaging
4. Number of circuits
5. Family

Two basic IC classes are **linear** (or **analog**) and **digital** integrated circuits. This is a classification based on **mode.**

Another way to classify these devices relates to their *construction* characteristics (monolithic, thin film, or hybrid). **Monolithic** ICs contain electronic elements such as transistors, diodes, resistors and capacitors, all produced or fabricated on a single stone (monolith). This monolithic silicon chip with all these elements is created by selective chemical diffusion processes, Figure 23–2. **Thin-film** ICs use a process of depositing a thin film on the semiconductor substrate, then processing special areas to obtain appropriate integration. **Hybrid** ICs are formed by combining the monolithic and film processes. Hybrid also applies to devices that have a combination of two or more connected units in one package, or units composed of both monolithic and separate transistor devices.

The *packaging* classification relates to the housing of the device and the layout of the connecting terminals. Refer to Figures 23–3 and 23–5 for examples of common housings and pin connections.

Sometimes, digital ICs are classified by the *number* of gates (circuits) in the IC. Typically, *very-large-scale integration* (**VLSI**) are ICs with over 1000 gates on a single chip. *Large-scale integration* (**LSI**) are ICs with more than 100 gates, and *medium-scale integration* (**MSI**) are ICs with less than 100 circuits. Last, *small-scale integration* (**SSI**) are ICs with less than 10 circuits on a single chip.

Last in our classification list is the identification of ICs by their *family* characteristics. Families of ICs are generally distinguished by the transistors, special circuit configurations, or performance features (i.e., speed, power requirements, number of loads they drive, and so on). These families are sometimes called *logic families* when referring to digital integrated circuit families.

One popular family is the *TTL* (transistor-transistor logic) family. The input and output circuitry and the processing sequence through the device (from transistor-to-transistor) give this family its name. Figure 23–4 illustrates a typical **NAND gate** using this circuit format. The TTL family has several sub-families that vary in power and speed features. The *important* identifying factor to aid technicians in knowing that IC devices belong to this family is by the numbering system used. The TTL logic family is in the 7400 series (or 5400 series for military specification versions). TTL logic devices use 5 VDC power supplies and are low-impedance devices, which

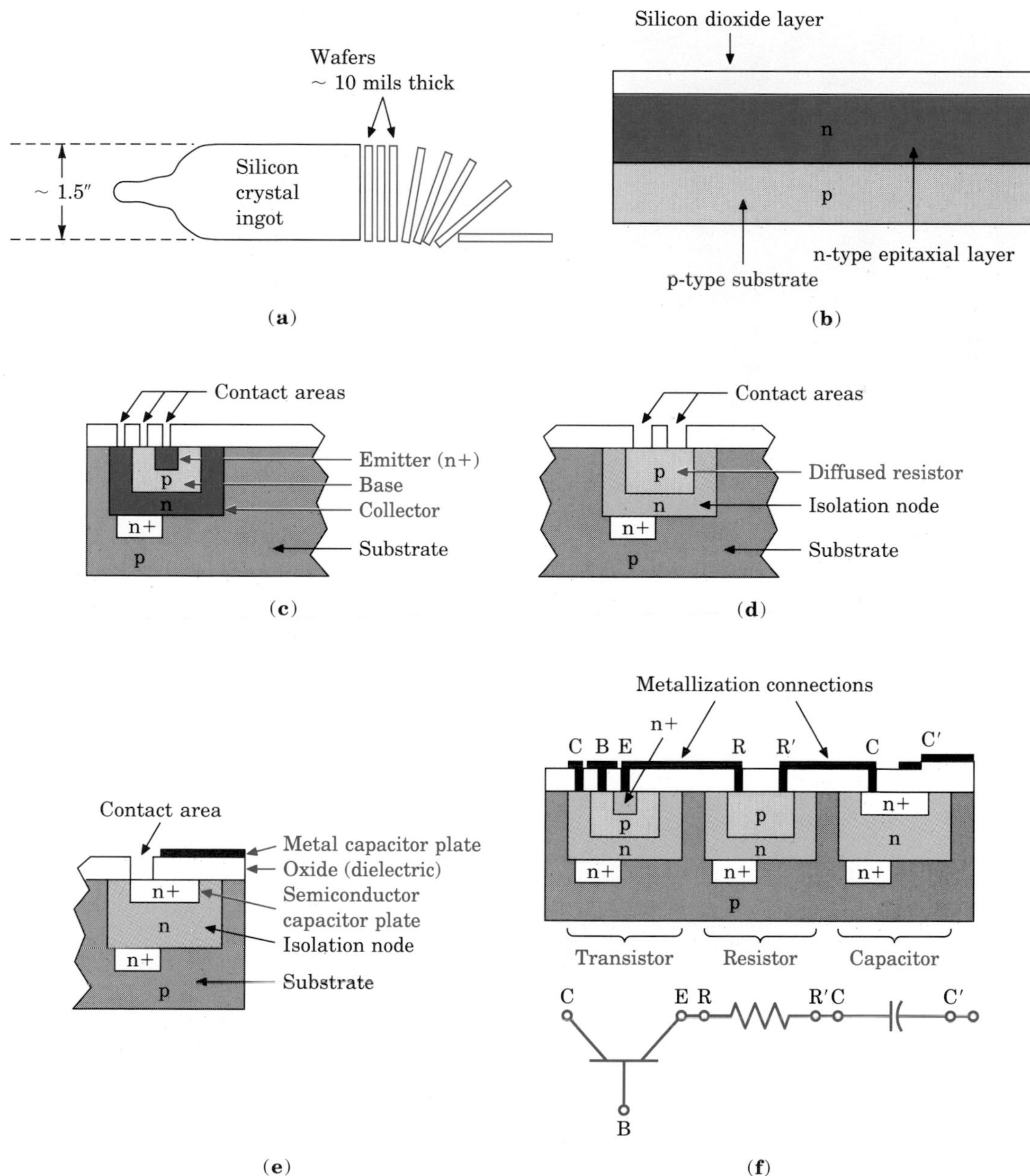

FIGURE 23–2 Select steps in fabrication of an IC chip

require moderate power to operate. They are not very susceptible to damage from static electricity.

Another important IC family is the *CMOS* (complementary symmetry-metal-oxide semiconductor) devices. The identification numbers often used

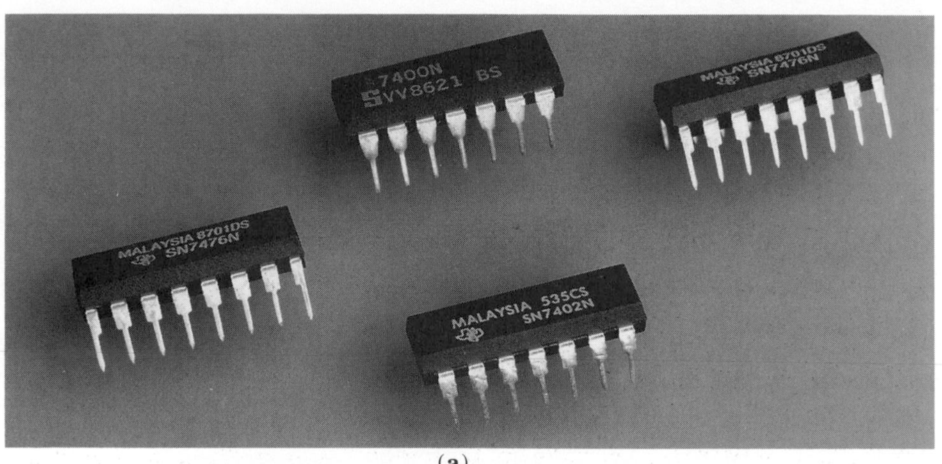

(a)

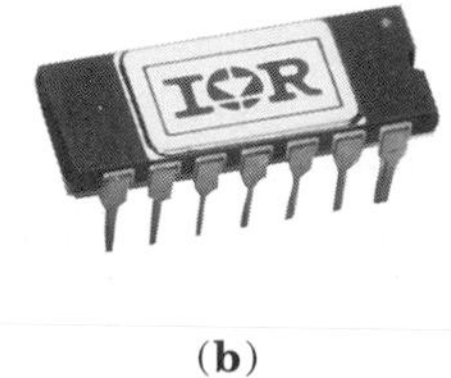

(b)

FIGURE 23–3 Examples of integrated circuit (IC) packages (Photo a courtesy of Texas Instruments; photo b courtesy of International Rectifier)

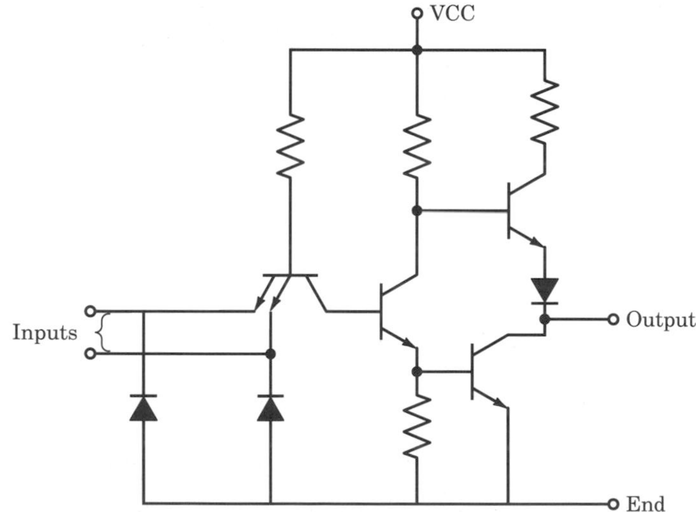

FIGURE 23–4 A transistor-transistor logic (TTL) NAND logic gate circuit

for these devices are in the 4000 series. CMOS devices operate with supply voltages from 3 to 18 volts. Their greatest asset is they require *very low* power for operation. When idling, they draw only nanoamperes from the source. Their main weakness is their susceptibility to damage from static charges. For this reason, they require special handling during shipping, installation, and servicing.

Another notable IC family is the *ECL* (emitter-coupled logic) devices. Their main advantage is their high speed of operation. Their disadvantage is they use considerable power for operation.

The last family is the I^2L (integrated-injection logic) devices. Their name comes from the fact that during operation, a constant current is injected into the bases of the NPN transistors. Because of their circuit design, these devices use very little power. Therefore, these ICs produce little heat, which allows very high density of circuits and more circuits per space.

In summary, the TTL gates are very popular for general use. The CMOS circuits require low power for operation and are inexpensive to produce. The ECL devices provide speed, where needed, but use considerable power to operate. Finally, the I^2L devices allow great density of circuits and require low power for operation.

ADVANTAGES AND DISADVANTAGES OF ICs

Obvious advantages of many ICs are small size; light weight; small space requirements; high circuit complexity; small power requirements; relative immunity to noise for digital devices; and low cost.

Apparent disadvantages of ICs are high degree of controlled manufacturing processes; inability to handle high powers; fragile pins requiring careful handling; susceptibility to damage from static charges (certain devices); and inability to troubleshoot to component levels. (NOTE: The inability to troubleshoot to component levels can also be an advantage due to saving time, and because replacing certain ICs is often more economical than replacing a single component, especially in older systems.)

EXAMPLES OF ANALOG AND DIGITAL MODES OF OPERATION

Recall in the preview of this chapter, we defined analog circuits as devices whose outputs vary continuously over a range and digital circuits as devices whose outputs are typically in one of two conditions.

Examples of analog type operations are light dimmers in your car or home, volume, tone, and color controls on your TV, and certain motor speed controls, such as the speed control on your electric drill.

Examples of digital style operations are a light switch (either on or off), automobile horn, doorbell, and on-off cycles of washers, dryers, and furnaces.

IC PACKAGING AND BASING

Earlier we discussed several common packages used for housing ICs. Now we want to familiarize you with the ways the pins are counted when installing or testing *DIP* (dual-in-line), *can* or *flat-pack* IC devices.

Refer to Figure 23–5 for examples of the *pinout* for common basing arrangements. Notice that in each case, there is a clue such as a marking,

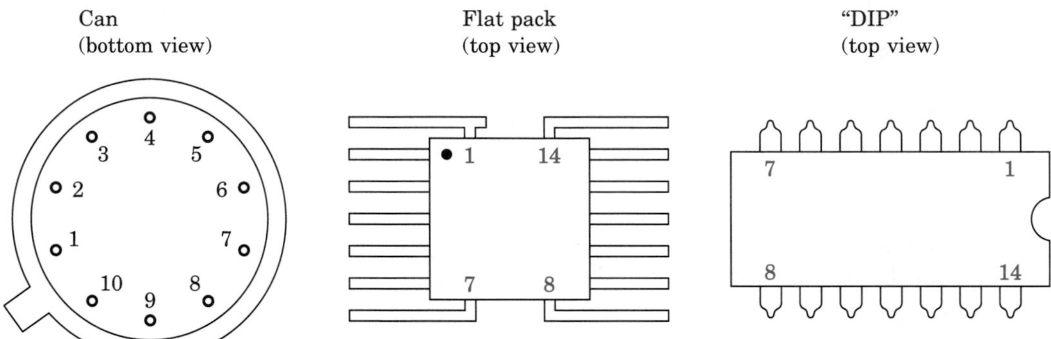

FIGURE 23–5 Examples of IC packaging and pin connections

notch, or indentation on the DIP and flat-pack devices and a tab on the can packages that identifies the location of pin 1. From that point, you count pins around the IC package. When looking at the *top* of the IC, the pins are counted *counterclockwise* from pin 1. When looking at the IC from the *bottom,* the pins are counted in a *clockwise* direction from pin 1. Learn this technique! It is important!

COMMON NUMBER SYSTEMS USED WITH DIGITAL ICs

Without excess detailing of number systems, we do want to introduce those that lend themselves easily to the two-state (on-off, high-low, 1 to 0) operation of digital systems used in digital circuits and computers. Common numerical systems used are the *binary, octal,* and *hexadecimal* systems. You are used to the decimal system, or the base-10 system of numbers (i.e., 1, 10, 100, 1000, and so on).

In any number system, the next place in the number is found by raising the base to the next power. For example, the number 1 in the decimal system equals 10 to the zero power (10^0), 10 equals 10 to the first power (10^1), 100 equals 10 to the second power (10^2), and so forth, Figure 23–6.

FIGURE 23–6 Chart showing significance of digit position in decimal system

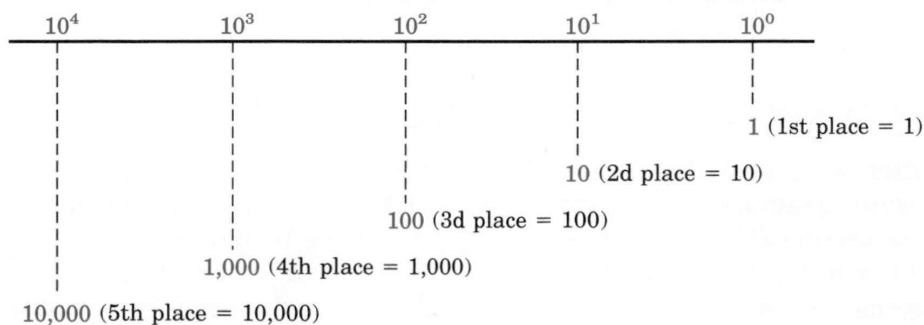

The binary (base 2) system is the basic system used in computers and digital circuits because it correlates with their two-state operation. In the binary system, there are only two digits—0 and 1. (In the decimal system, there are ten digits: 0, 1, 2, 3, 4, 5, 6, 7, 8, 9.) Each place in the decimal system is worth 10. That is, each time a number is moved from one column to the adjacent column to the left, it represents a 10-times factor, (e.g., 1, 10, 100, and so on). Each place in the binary system is worth 2. A 1 in the right-most column equals 2 to the zero power (2^0), or 1; a 1 in the next column to the left equals 2 to the first power, or 2; a 1 in the next column to the left equals 2 to the second power, or four; and so forth. See Figure 23–7 for a comparison between a decimal number and its equivalent binary number.

Because binary numbers can become very long, it is practical to use systems that can be easily transposed to the binary operation. Two such systems are the *octal* system (base 8) and the *hexadecimal* (base 16) system. Refer to Figure 23–8 to see examples of how the various systems correlate.

DECIMAL	BINARY
0	0
1	1
2	10
3	11
4	100
5	101
6	110
7	111
8	1000
9	1001
10	1010
11	1011
12	1100
13	1101
14	1110
15	1111

FIGURE 23–7
Comparison of decimal and binary numbers

BASIC LOGIC GATES

The Two-States or Logic Levels

Two-state logic gates are the heart of digital electronics. These logic gates perform the function of decision-making. The two states of output for these devices are *on* or *off* or output voltage *high* or *low,* or their output representing a binary *1* or *0.* By interconnecting various digital logic gates, some very complex problems can be solved.

DECIMAL	BINARY	OCTAL	HEXADECIMAL
0	0	0	0
1	1	1	1
2	10	2	2
3	11	3	3
4	100	4	4
5	101	5	5
6	110	6	6
7	111	7	7
8	1000	10	8
9	1001	11	9
10	1010	12	A
11	1011	13	B
12	1100	14	C
13	1101	15	D
14	1110	16	E
15	1111	17	F

FIGURE 23–8
Correlation of various number systems

IN-PROCESS LEARNING CHECK I

Fill in the blanks as appropriate.

1. To classify ICs, two modes of operation are ________ and ________ modes.
2. Monolithic and thin film refer to the ________ method of classifying ICs.
3. Flat-pack, can, and DIP describe IC classification in terms of ________.
4. An IC having more than 100 gates or equivalent circuits is categorized as ________.
5. Two very important IC families are the transistor-transistor logic, or ________ family, and the complementary symmetry-metal-oxide semiconductor, or ________ devices.
6. The basic numerical system used in computers and digital circuits is the ________ system.

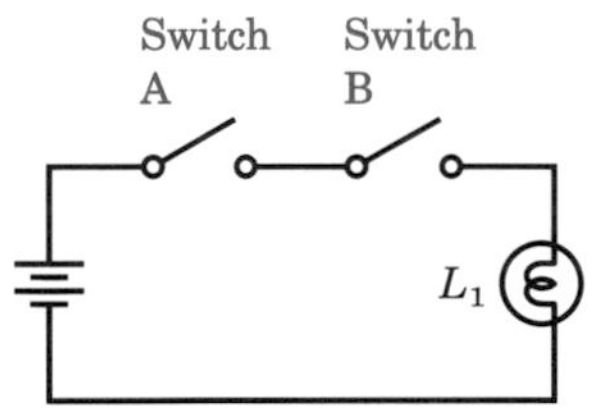

The AND Gate

Switch A AND switch B must be closed in order for light (L_1) to light

FIGURE 23–9 Light switching circuit representing AND logic function

An analogy to an **AND gate** is a light circuit having two series switches, Figure 23–9. Both switch A *AND* switch B must be closed before the output to the bulb is *high,* and the bulb lights. This is equivalent to a *two-input AND* gate where both inputs must be high (logic 1) before the gate output is high (or logic 1). If either or both A or B are not at logic 1 (closed), the bulb does not light.

The symbol representing an AND gate in digital electronics is shown in Figure 23–10.

Also, the *truth table* for the AND gate is shown in Figure 23–10. A truth table indicates the device output for various input conditions. For the two-input AND gate, you can see that both inputs must be at logic level 1 for the output to be at logic level 1. The output for all other input conditions is a logic 0 (or low logic level).

In digital logic systems, *Boolean algebra* is often used. These Boolean expressions often simplify digital circuits, as you will learn in your further studies of digital electronics. We won't take time to teach you this skill. Instead, we will show you the Boolean expressions for the various gates we discuss. In this method of expression, the AND function is shown by a dot (·). For example, to say that high input A AND B equal high output X, you can say that A · B = X. This says that *if* inputs A AND B are at logic 1, then output X is at logic 1 (again, refer to Figure 23–10).

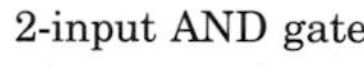

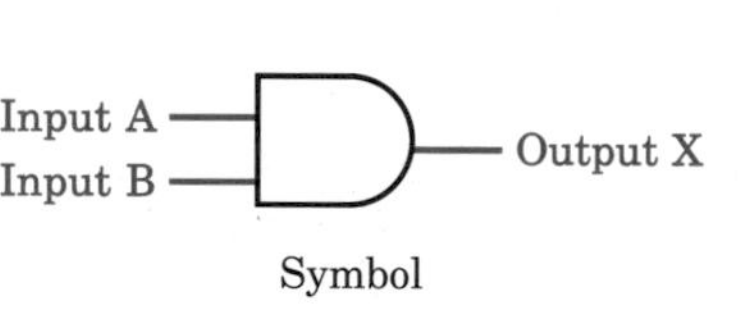

Truth table

Inputs		Output
A	B	X
0	0	0
0	1	0
1	0	0
1	1	1

(Input) (AND) (Input) (Output)

$$A \cdot B = X$$

Boolean expression

If A AND B are high (1) then X will be high (1).

FIGURE 23–10 AND gate symbol, truth table, and Boolean expression

The OR Gate

A two-input **OR gate** is equivalent in operation to a light circuit where the switches are in parallel with each other. This results in a condition where if either switch is closed (representing a logic high condition for a logic gate), the output is high. Figure 23–11 shows the switch circuit, logic symbol, truth table, and Boolean expression for a two-input OR gate. Become familiar with these.

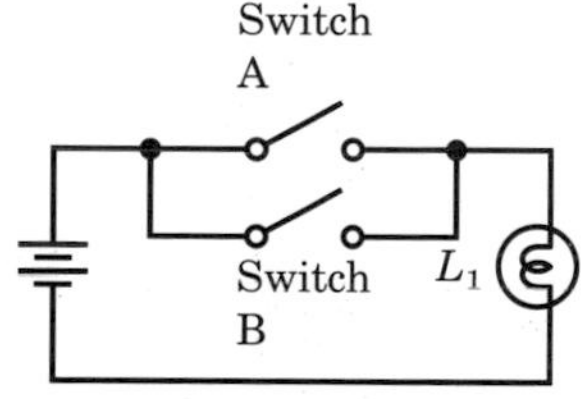

Equivalent circuit

Inputs		Output
A	B	X
0	0	0
0	1	1
1	0	1
1	1	1

Truth table

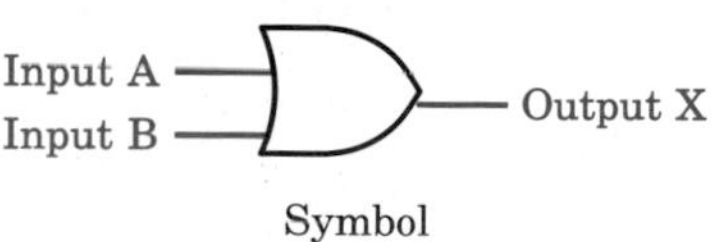

Symbol

(OR)

$$A + B = X$$

Boolean expression

If A OR B is high (1), then X will be high (1).

FIGURE 23–11 OR gate equivalent switch circuit, logic symbol, truth table, and Boolean expression

The Inverter

An important operation in digital electronic circuits is the ability to invert a 0 to a 1 or a 1 to a 0. An **inverter** has one input and one output. It is easy to remember what it does by its name. The logic symbol, truth table, and Boolean expression for an inverter are in Figure 23–12.

FIGURE 23–12 Inverter logic symbol(s), truth table, and Boolean expression

A — In — Out or A — In — Out

Symbols

Input	Output
A	X
0	1
1	0

Truth table

$A = \overline{A}$ ← (Output "not" A)

(Input) (Output)

Boolean expression

(If A = 1, then X = 0
If A = 0, then X = 1)

The NOR Gate

If we invert the output of an **OR gate,** we get a **NOR** (Not-OR) **gate.** The NOR gate is the opposite of the OR gate, or inverted. The symbol, truth table, and Boolean expression for the NOR gate are in Figure 23–13.

FIGURE 23–13 NOR gate symbol, truth table, and Boolean expression

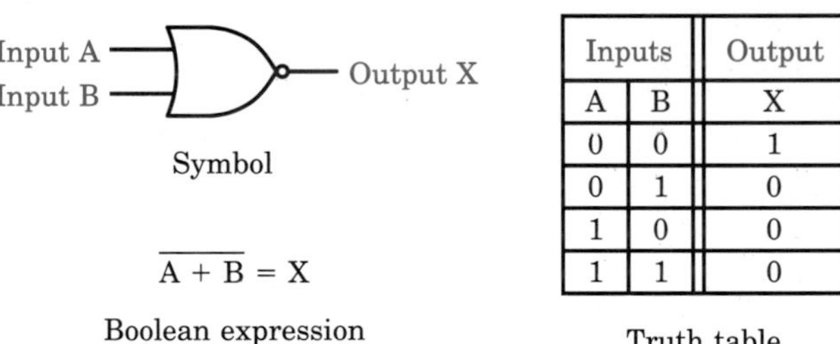

Symbol

$\overline{A + B} = X$

Boolean expression

Inputs		Output
A	B	X
0	0	1
0	1	0
1	0	0
1	1	0

Truth table

The NAND Gate

One frequently used gate in digital electronics is the **NAND** (Not-AND) **gate.** The NAND gate output can be regarded as an inverted AND gate output. That is, its output is the opposite of the AND gate for any given set of input conditions. The logic symbol, truth table, and Boolean expression for the NAND gate is in Figure 23–14.

FIGURE 23–14 NAND gate symbol, truth table, and Boolean expression

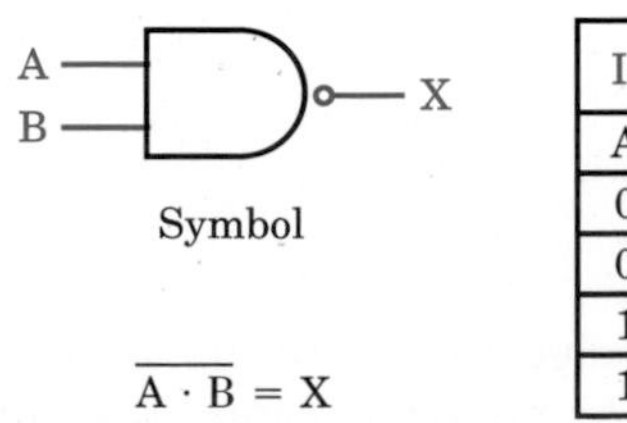

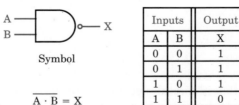

Symbol

$\overline{A \cdot B} = X$

Boolean expression

Inputs		Output
A	B	X
0	0	1
0	1	1
1	0	1
1	1	0

Truth table

The EXCLUSIVE OR Gate

This gate is generally created by a combination of other gates. A two-input, exclusive OR gate functions so that if either input is high, the output is high. *However,* if both inputs are high, the output is low. This is the difference from the regular OR gate. See Figure 23–15 for the logic symbol, truth table, and Boolean expression.

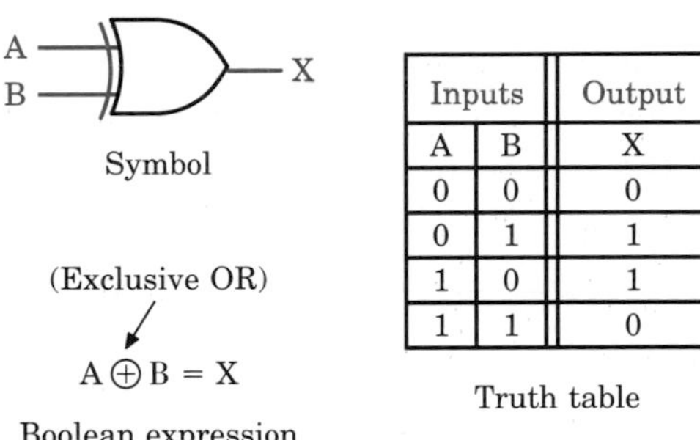

Inputs		Output
A	B	X
0	0	0
0	1	1
1	0	1
1	1	0

FIGURE 23–15 Exclusive OR gate symbol, truth table, and Boolean expression

The EXCLUSIVE NOR Gate

This gate is an *exclusive OR* gate with its output inverted. See Figure 23–16 for the logic symbol, truth table, and Boolean expression.

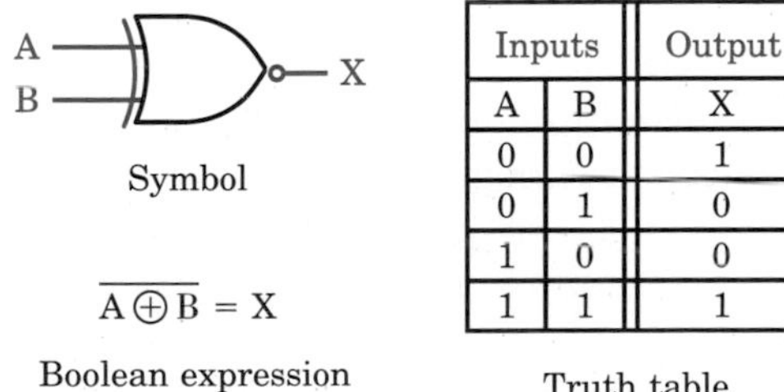

Inputs		Output
A	B	X
0	0	1
0	1	0
1	0	0
1	1	1

FIGURE 23–16 Exclusive NOR gate symbol, truth table, and Boolean expression

Summary of Logic Gates

The various two-state gates described are used alone or in combination to perform both arithmetic and logic functions. Binary addition, binary subtraction, "ANDing," "ORing," and so forth are all possible. By using these various *building blocks,* the digital computer is possible. (NOTE: There are also special three-state gates called tri-state gates that connect and isolate as appropriate, various circuits in digital electronics.)

IN-PROCESS LEARNING CHECK II

1. A two-input logic gate where if either input goes high the output goes high is called a(n) __________ gate.
2. A logic gate whose output is inverted from the input level is called a(n) __________.
3. A two-input logic gate where both inputs must be high to have a high output is called a(n) __________ gate.
4. If we invert the output of an OR gate, we essentially have a __________ gate.
5. A logic gate that is essentially the inversion of an AND gate output is called a(n) __________ gate.

OTHER COMMON DIGITAL CIRCUITS

In this overview chapter, we will not thoroughly discuss the various circuits commonly found in digital electronics. However, we will mention some terms that you should be looking for as you continue your studies.

Flip-Flops

These are bi-stable circuits that store one *bit* of information. That is, it remembers its output state (1 or 0) until the circuit is purposely reset or triggered to its second stable state. In other words, this circuit has two stable states, thus the term flip-flop. In one state, the output is a logic 1, in the other state, the output is a logic 0.

Counters

Counters are circuits that measure input pulses. They are useful in a number of applications for digital electronics. The typical operation is one output pulse for a given number of input pulses, determined by the circuit design.

Shift-Registers

Shift-registers are another storage circuit where binary data are stored until appropriately needed for processing.

MEMORY

We have briefly discussed memory elements. Memories in computers store digital data, either in temporary or permanent form. Now, let's discuss some other memory terminologies you will encounter in digital computers.

Basic Types of Memory

Volatile and Nonvolatile. Memories are classified as either *volatile* or *nonvolatile* storage. When memory is volatile, the data disappear when power is removed from the circuit or when the system is turned off. Obviously, the nonvolatile memory retains data, even when power is turned off, and makes it available, upon request, when the power is restored.

RAM and ROM. *RAM* (random access memory) is memory where data are temporarily stored and recalled when needed (provided the system isn't turned off, or power is not lost). This memory is commonly created by use of flip-flops. With this memory, the user can read what is already stored in the memory or can write new data into the memory. Sometimes, when talking about computers, RAM is called *user* memory because the user writes new data and reads existing data from this memory. The amount of available storage is often referred to in thousands of *bytes* of information, or kilobytes. (NOTE: A bit is a 1 or a 0, a byte is a grouping of bits usually 8 or 16 bits on most home computer systems.)

ROM (read-only memory) is different from RAM memory. ROM memories are typically programmed with data at the factory, and their contents are generally not changed. In other words, the user does not typically write to a ROM. ROM is a nonvolatile memory that retains what is stored, even after the power is turned off. ROMs are manufactured in different ways. Often, they are made of a diode matrix or resistance elements that are left closed or purposely opened between matrix points, providing a matrix location that reads as a 1 or a 0. Special ROMs are available where the technician programs the device, such as the *PROM* (programmable read-only memory). Also, the *EPROM* (erasable-programmable-read-only memory) is one special ROM where changes are made by erasing the existing data with ultraviolet light, then reprogramming, as desired. The PROM is not reprogrammable once the matrix fuses have been burned, whereas the EPROM can be reprogrammed after being erased by ultraviolet light.

THE MICROPROCESSOR

A **microprocessor** can be regarded as a *central processing unit* (CPU) on a single chip. The technology that enables thousands of circuits in a single device in a very minute space has allowed the production of this technological wonder. Microprocessors combine many elements that have been discussed throughout this chapter.

Since these devices are inexpensive, small, require little power, and perform the functions needed in the CPU of computer systems, they are found in full-blown computer systems and in a myriad of other applications. For example, computer control of your automobile's timing and air-fuel mixtures are easily performed, and built-in troubleshooting is now found on many vehicles. Temperature and suspension control are also possible as a result of microprocessors. Obviously, digital readouts for time, gasoline, and speed are all quite common due to microprocessors.

Digital sound systems, digital tuning of radios and TVs, digital synthesizers, automated cameras, and many other high-tech amenities are here to stay.

In industry, digital control of robots, quality-control devices, and automated production lines are all becoming more and more common—All brought about by the technologies which we have been discussing.

Perhaps these facts will motivate you to learn more about microprocessors and related technologies.

THE BASIC DIGITAL COMPUTER

By combining various digital circuits, such as the counters, registers, arithmetic units, memories, clocks, and so on, it is possible to assemble a system called a digital computer. Figure 23–17 shows a very simple block diagram of a microcomputer system. Each element can be briefly explained as follows:

1. Input
   Can be a keyboard, inputs from external memory devices (discs or tapes), and inputs from opto-electronic devices, (card readers).

2. Processing
   Generally, this is the section of a computer that processes the data entered or executes a software program that guides the computer through a number of processing steps. Generally, in microcomputers, the CPU is the microprocessor chip. It contains the on-board storage, arithmetic-logic unit, and other special *registers* to perform arithmetic operations, processing of instructions, creation of timing signals, communication with input, output and memory, and other operations.

3. Memory
   Memory includes various forms of information or data storage. Some operating memory is on-board the CPU. Other memories not present in the CPU include RAM, ROM, discs, tapes, and so forth. Memories store data until the computer needs it to perform an instruction or calculation and other functions.

4. Output
   Computer output goes to a number of places and/or devices. One common device for computer output is the video screen. Output can go to other devices, such as printers, machines or instruments.

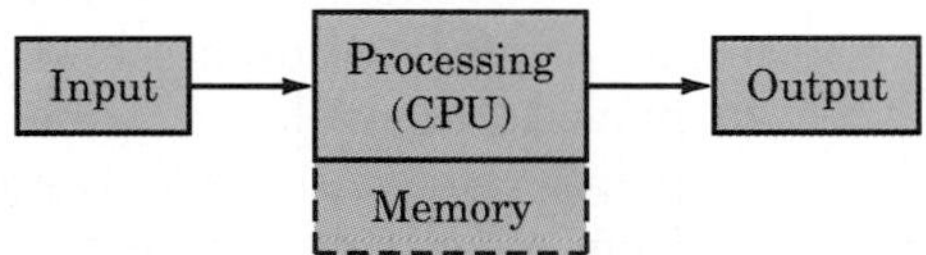

FIGURE 23–17 Simplified diagram of a basic digital computer system

Final Comments About Digital Electronics

The progression of digital electronics has been amazing. As you have read, vacuum-tube computers used to fill rooms and even buildings. The invention of the transistor was a large step toward our present-day technology. Even at that, the use of hard wiring and discrete components did limit applications of digital electronics. Later, the invention of the IC was a giant step forward. The progression has been from small-scale integration (SSI), to medium-scale integration (MSI), to large-scale integration (LSI), to very-large-scale integration (VLSI). This has propelled digital electronics to the forefront of the electronics industry. Almost everything we touch in business or in the home is now affected in some way by digital electronics.

You should give further study to the topics introduced in this chapter. You will want to do this in order to be a technician who is on the cutting edge of technology. In doing so, you will ensure yourself of job security and advancement in this exciting field of electronics.

SUMMARY

- The integrated circuit (IC) has fostered many advances in electronics technology. Its features are very small size and weight, ability to contain many circuits in microminiature space, low power demands, possibly customized operating characteristics, and durability.
- The IC was conceived from the notion that more than one transistor might be created on a given piece of semiconductor material. Eventually, ICs replaced discrete transistors in many circuits and systems.
- ICs are classified by mode of operation, construction, packaging, number of circuits, and family.
- Two modes to define ICs are linear and digital operation. Construction refers to monolithic, thin film, and hybrid. Common package categories include dual-in-line (DIP), can, and flatpack. Categorized by number of circuits, ICs are SSI, MSI, LSI, and VLSI. SSI is small-scale integration (less than 10 circuits); MSI is medium-scale integration (less than 100 circuits);

LSI is large-scale integration (more than 100 circuits), and VLSI is very-large-scale integration (over 1000 circuits). Family refers to their component types, operating parameters, and so on.

- Analog operation involves circuits whose outputs vary continuously over a range and is directly related to the input levels. Volume controls and variable light dimmers are examples of analog operation.
- Digital operation refers to circuits having discrete output levels, often either on or off and in a high or low state. Examples are a light switch circuit and an automobile horn.
- Because of the binary nature of digital circuitry and computers, numerical systems often used with computers are the binary (base 2), octal (base 8), and hexadecimal (base 16) systems. (NOTE: The term *radix* is sometimes used instead of the term *base.*)
- Several basic logic gates include the AND gate; OR gate; inverter; NOR gate, and NAND gate. Each gate is named by the input requirements needed to achieve a given output result.
- Other types of digital circuits are flip-flops, counters, and shift-registers.
- Common memories used in digital electronics are volatile (contents lost when power is removed); nonvolatile (retains contents with power removed); RAM (random access memory from which one can read stored data or write new data), and ROM (read-only memory).
- Mass data storage devices can be memories using diskettes and tapes of various types and characteristics.
- The microprocessor typically performs the function of a central processing unit (CPU). It performs mathematical and logical functions, as well as other housekeeping functions, such as timing and tasks that allow appropriate data in and out of the microprocessor.
- A very basic computer system needs to have input, processing capability, memory, and output.

REVIEW QUESTIONS

1. In what decade was the integrated circuit invented?

2. Define analog operation and give several examples.

3. Define digital operation and give several examples.

4. Draw the symbol, and list the truth table and the Boolean expression for each of the following:
   a. AND gate
   b. OR gate
   c. NAND gate

5. List three advantages of ICs.

6. List two disadvantages of ICs.

7. Draw a 16-pin DIP IC package, both top and bottom view, and number the pins on the drawings.

8. Name one advantage and one disadvantage of CMOS devices.

9. What does the term TTL represent?

10. What is a primary advantage of ECL devices?

11. What is the base of the binary number system?

12. How many bits of information can a single flip-flop circuit store? How many states of operation does this device have?

13. Draw the block diagram for a simple computer and label all blocks.

TEST YOURSELF

1. Research and draw a diagram illustrating the 7400 quadruple two-input positive NAND gate. Show all pin numbers.

2. Research and draw a diagram illustrating the 7404 hex inverter. Show all pin numbers.

3. Research and draw a diagram illustrating the 7476 Dual J-K Master-Slave Flip-Flop (with preset and clear). Show all pin numbers and device lettering, as appropriate.
4. Write both the decimal and binary numbers representing 0 through 15. Align them in two columns so the binary representation for each number is immediately adjacent to the decimal representation for each number.
5. Express the numbers 0 through 15 (decimal) in both the octal and hexadecimal systems. Again, align these numbers in two adjacent columns, as appropriate.

Introduction to Transistor Amplifiers and Oscillators

Key Terms

Clapp oscillator

Class A, B, and C

Colpitts oscillator

Common-base

Common-collector

Common-emitter

Crystal oscillator

Feedback

Hartley oscillator

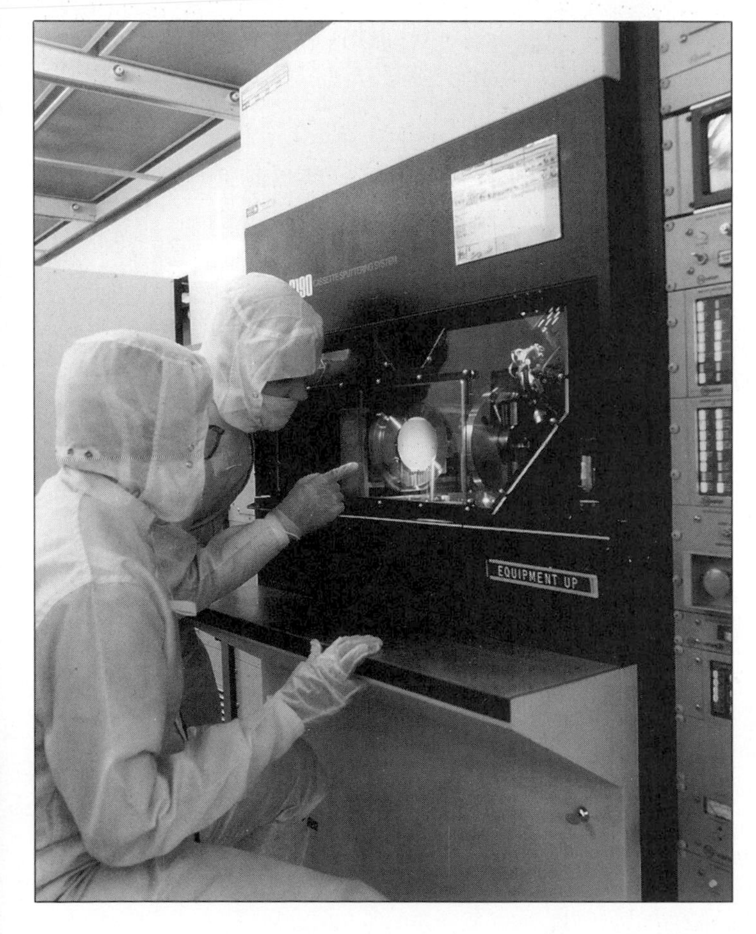

Courtesy of Motorola, Inc.

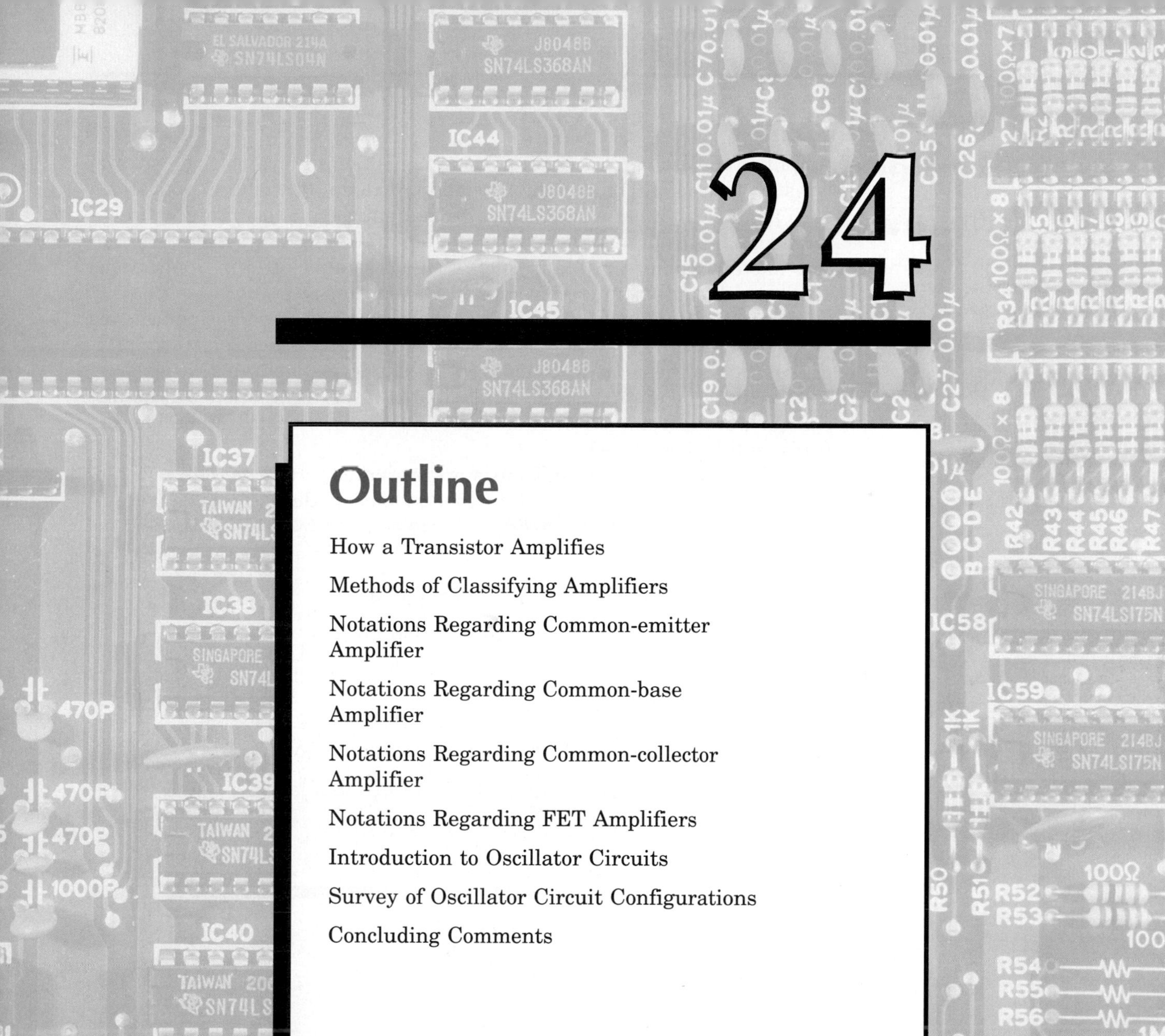

24

Outline

How a Transistor Amplifies

Methods of Classifying Amplifiers

Notations Regarding Common-emitter Amplifier

Notations Regarding Common-base Amplifier

Notations Regarding Common-collector Amplifier

Notations Regarding FET Amplifiers

Introduction to Oscillator Circuits

Survey of Oscillator Circuit Configurations

Concluding Comments

Chapter Preview

Amplifier and oscillator circuits are found in numerous systems and subsystems in electronics. For example, the ability to amplify enables applications such as building a very small signal from a microphone or recording into a large signal that drives the speakers in a stereo amplifier system. Oscillators make possible variable frequency signal sources used in applications such as test equipment, receivers, and transmitters.

This chapter will provide an overview about some key operating features of amplifier and oscillator circuits. You will learn classifications of amplifiers relative to their circuit configuration, bias levels, and applications.

Also, basic requirements for an oscillator stage to function will be discussed. Several common oscillator circuits, along with their notable features, will be introduced.

The circuit applications of transistors presented in this chapter will provide information to help you in further studies of these important areas.

Objectives

After studying this chapter, you will be able to:

- Explain the basic transistor amplification process
- Classify amplifiers by circuit and class of operation
- Identify key features of each type of amplifier
- List typical applications for each classification of amplifier
- Specify the conditions required for sustained oscillation
- Recognize basic **oscillator** circuit configurations
- List typical application(s) for each oscillator

HOW A TRANSISTOR AMPLIFIES

In a previous chapter, you learned that a transistor transfers current from a low resistance input circuit to a high resistance output circuit, thus exhibiting power gain. In actuality, the amplifier stage is not manufacturing power from nothing. Rather, the power source provides the power, but the amplifier's special transfer feature enables amplification. Likewise, by proper selection of transistor elements in the input and output circuits, voltage gain and/or current gain and changes in impedance values from input to output circuits can occur.

An analogy sometimes used to relate transistor operation is that a transistor acts like a variable resistor in a three-resistor voltage divider, Figure 24–1. For our purposes, we'll use the **common-emitter** (CE) amplifier for illustration. The transistor is in series with its collector and emitter resistors. The collector voltage source (V_{CC}) is the voltage source for the voltage divider.

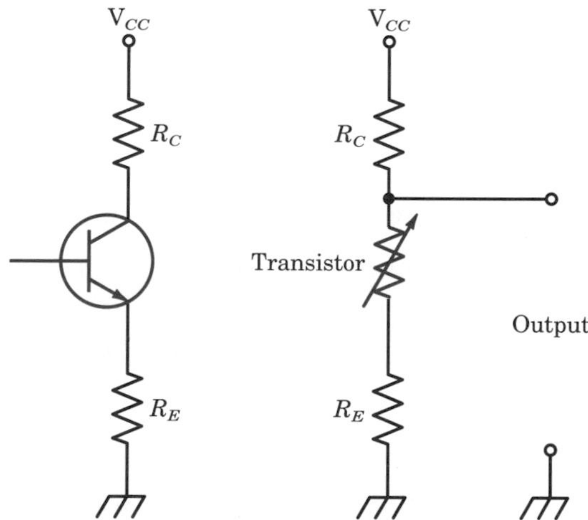

FIGURE 24–1 Analogy of a transistor as a variable resistor

If the transistor operates so that its collector-to-emitter resistance is a very low resistance, the voltage drop across the transistor is low and voltages across the fixed series resistors are high. If the transistor operates so that it acts like an open (very high resistance) between collector and emitter, then virtually all the source voltage is across the transistor.

The bias setting of the base-emitter junction and the input signal to the stage determines where the transistor operates.

For your information, transistor amplifier circuit designers may use one of three levels of approximation, depending on their tasks.

The first level of approximation considers the transistor extremes of operation to have the features of a switch. In other words, completely open (infinite ohms) and completely shorted (zero ohms) are the operational extremes.

A second approximation level relates to the diode actions within the transistor, where 0.7 V is the nominal emitter-base forward-bias requirement for silicon transistors. 0.25 V is the nominal emitter-base forward bias requirement for germanium transistors.

The third approximation level involves detailed mathematical models to analyze transistor operation.

In this chapter we will use the second approximation approach in our explanations.

If a very small voltage change between the base and emitter (the CE amplifier input) causes a large voltage change at the collector, we have voltage amplification. If a small current change at the input of the stage causes a large current change at the output, we have current amplification.

With this background information, let's look at some important classifications and features of transistor amplifiers.

METHODS OF CLASSIFYING AMPLIFIERS

There are several common ways to classify amplifiers. Some of these classifications include:

1. By signal levels involved (small-signal amplifiers and large-signal amplifiers).
2. By class of operation (**Class A, Class B,** and **Class C**).
3. By circuit configuration (**common-emitter, common-base,** and **common-collector**).
4. By type of device used (bipolar transistor and FET).

Classification by Signal Levels

As implied by the name, a small-signal amplifier is designed and operates to effectively handle small-signal levels, Figure 24–2. Examples of small-signal amplifiers are the initial stages in radio and TV receivers. The small signals coming from the antennas are typically in the range of microvolts or millivolts.

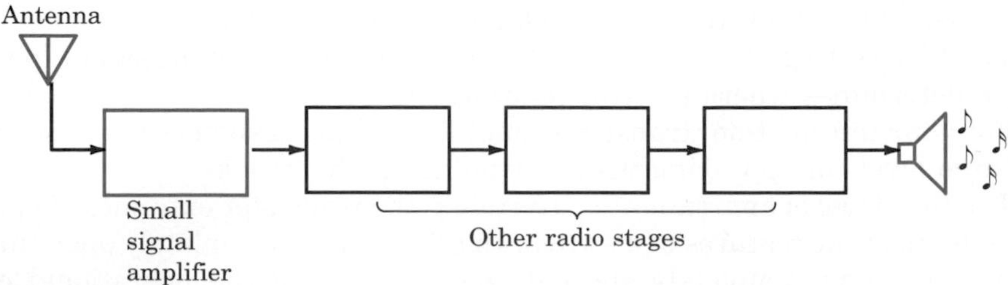

FIGURE 24–2 Small-signal amplifier fed by an antenna signal

Large-signal amplifiers properly handle signals that have significantly larger amplitude or current levels. The final power amplifier stage(s) that drives speakers in audio systems or provides transmitter signals to transmitting antennas are examples of large-signal amplifiers, Figure 24–3.

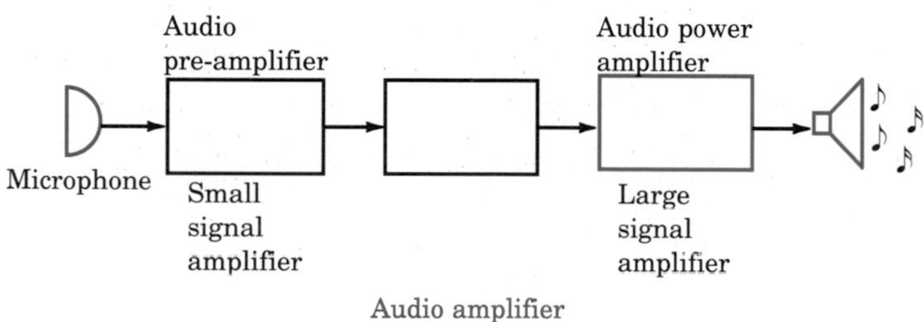

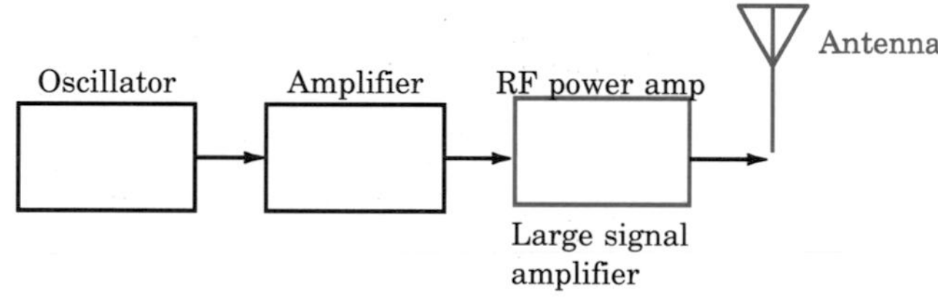

FIGURE 24–3 Power amplifiers

Classification by Class of Operation

Bipolar Transistor Collector Curves. Before discussing the classes of operation for amplifiers, let's briefly look at some important dc (no signal) features of a transistor stage as illustrated on characteristic curves.

In Figure 24–4 you see a circuit where we can set the base current I_B, and then note the change in collector current for various values of V_{CE}. By changing I_B to different values and repeating the process, we can generate a set of curves, termed a family of static collector curves, for the given transistor. This set of curves for the various settings is known as a family of static (dc) collector characteristic curves for the device. Each type transistor has its own characteristics for its curves, depending on its physical and electrical properties. An example of a set of collector characteristic curves for a transistor is shown in Figure 24–5.

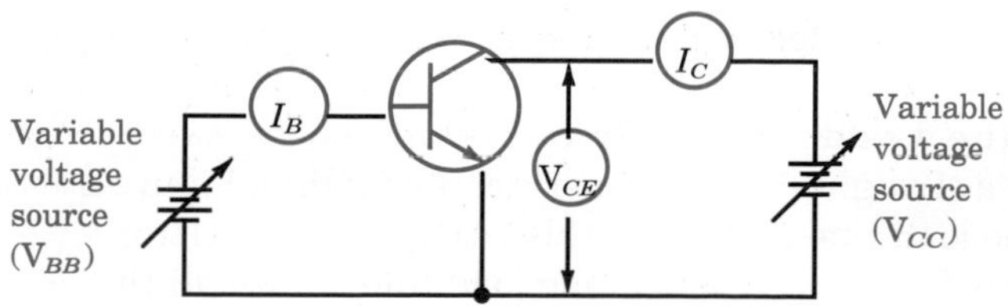

FIGURE 24–4 Transistor circuit to develop collector curves

FIGURE 24–5 Sample family of collector characteristic curves

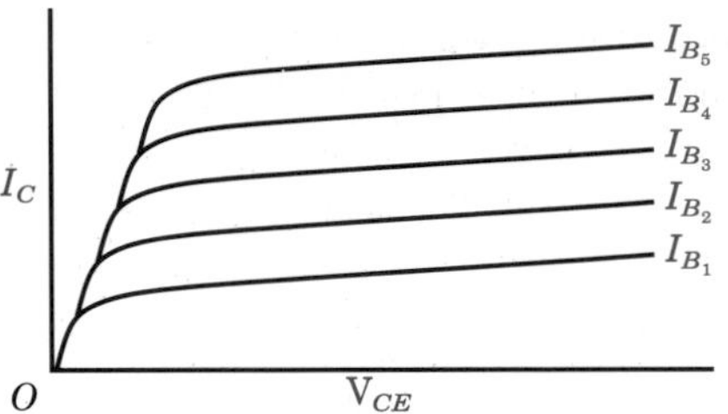

The Load Line. By using these characteristic curves, we can show a complete range of operating points for a transistor amplifier stage by means of a load line superimposed on these curves.

One extreme operational condition is when base current is zero. Under these conditions, the collector current is also zero with a very small amount of collector leakage current. This condition occurs when both the base-emitter and base-collector junctions of the transistor are reverse biased. Under these conditions, the collector-to-emitter transistor voltage drop is maximum due to minimum I × R drop across the resistor(s) in series with the transistor. At that point, V_{CE} virtually equals the source voltage (V_{CC}), and this operating point (and condition) is called cutoff, Figure 24–6.

FIGURE 24–6 Transistor cutoff point shown on characteristic curves

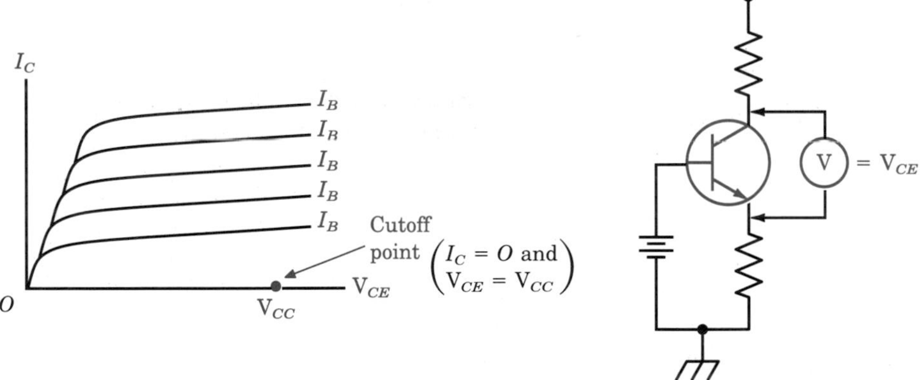

The opposite extreme operational condition occurs when I_C is maximum or the transistor operates at a point termed saturation. This state exists when the base current increases to a point where a further base current increase causes no additional increase in collector current. At this point, the collector-to-emitter voltage (V_{CE}) is minimum due to the maximum I times R drop across the collector and emitter resistors (R_C and R_E), Figure 24–7.

Drawing the Load Line. By connecting these two extreme operating points with a straight line, we construct a load line, Figure 24–8. The significance of the load line is that for this particular transistor with a given V_{CC} and specified R_C and R_E values, the load line shows all the possible operat-

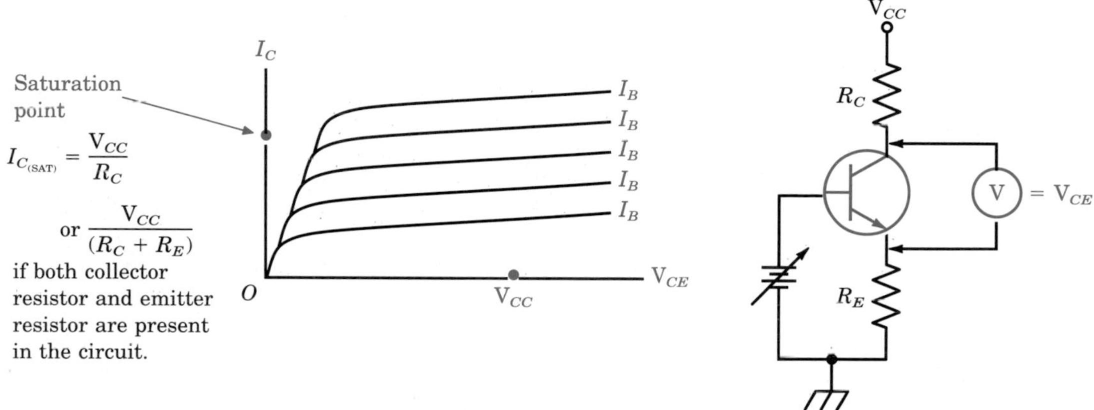

FIGURE 24–7 Transistor saturation point shown on characteristic curves

ing conditions for the transistor stage, including the two extremes of minimum I_C and maximum I_C.

As you will soon see, by setting the base-bias value, we can select an operating point anywhere along this load line. We can then determine all the important stage parameters, for example V_{CE} and I_C for a given I_B. To construct this load line for a given circuit on a specific transistor's curves, perform the following:

1. Mark the cutoff point on the collector curves graph where V_{CE} equals V_{CC}. Refer again to Figure 24–6.
2. Determine the saturation, or maximum I_C point by calculating the collector current required to cause the total V_{CC} voltage value to be dropped across R_E plus R_C (disregarding the very small drop across the transistor). For example, if $V_{CC} = 40$ volts and $R_E + R_C =$ a total of 500 ohms, then maximum I_C equals $\frac{40\text{ V}}{500\ \Omega}$, or 80 mA. Thus, the maximum I_C point is at the juncture of 0 V_{CE} and 80 mA I_C. Refer again to Figure 24–7.
3. Draw the load line by connecting the maximum I_C and cutoff points on the characteristic curves with a straight line. Refer again to Figure 24–8.

Q Point. The Q point or quiescent operating point of an amplifier stage, is determined by the value of dc base bias. The intersection of the base-bias (I_B) line and the load line is called the Q point. By projecting a line down to the V_{CE} axis (x-axis), we can see the V_{CE} value for that circuit operating at that particular operating point. By projecting a line horizontally from the Q point to the I_C axis (y-axis), we can determine the collector current (I_C) value

FIGURE 24–8 A load line shown on characteristic curves

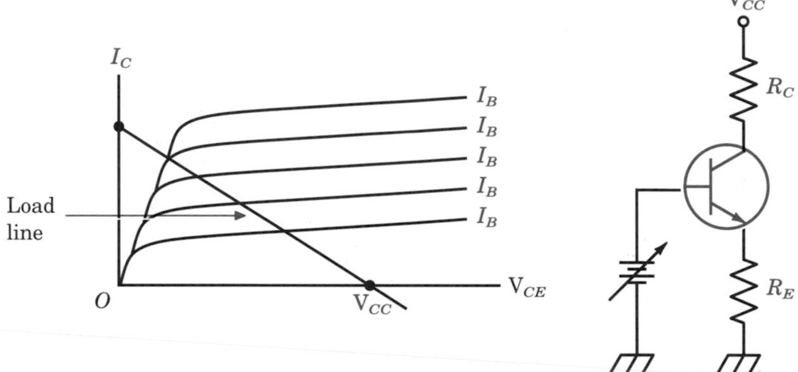

for that circuit operating with that base bias and the circuit's component values.

Class of Operation and Signal Characteristics Versus Q Points

Class A Operation. When the Q point is set at the midpoint of the load line, Figure 24–9, the stage is operating at or near Class A operation. By definition, collector current flows during 360 degrees (or all) of the input signal cycle if the amplifier is operating Class A. Under these conditions, the input signal causes the maximum swing in collector current and V_{CE} before signal distortion occurs, Figure 24–9. Because of the low distortion characteristic, many audio and small-signal amplifier stages operate as Class A amplifiers. Class A stages provide minimum distortion; however, they have low efficiency. In other words, the amount of ac signal power out is small compared to the dc input power supplied by the V_{CC} source.

FIGURE 24–9 Q point set at midpoint on load line

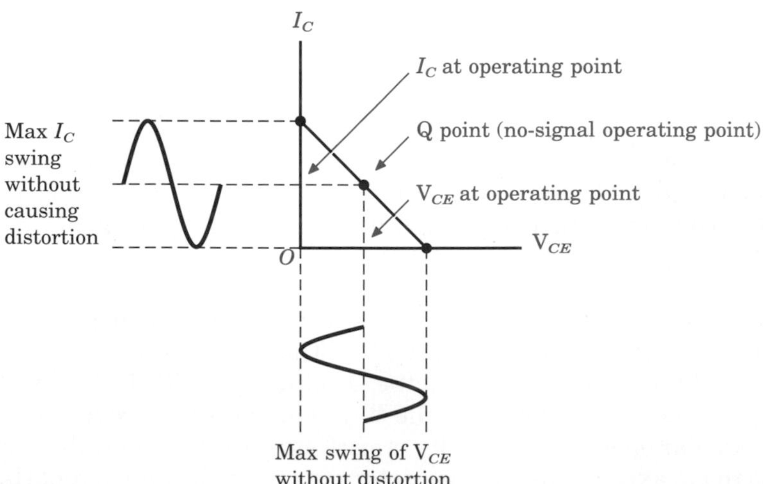

Class B Operation. When the transistor stage is biased so that its Q point is at cutoff, Figure 24–10, the stage is biased for Class B operation. Collector current flows for 180 degrees (or half) the input signal cycle. Because of being biased at cutoff, half of the input cycle biases the transistor stage below cutoff. Thus, the output signal waveform reflects half of the input cycle has been cut off. The output waveform resembles the half-wave rectified signal you studied in an earlier chapter. Again, see Figure 24–10. This class of amplifier is used in radio frequency circuits and can be used in audio amplifiers by using special two-transistor circuit configurations. An example is the push-pull amplifier stage that uses two transistors to conduct on opposite half cycles of the input signal, thus providing a complete 360 degree signal at the output, Figure 24–11.

Class B stages have higher efficiency than Class A stages. However, they distort the signal when used as a single-device circuit.

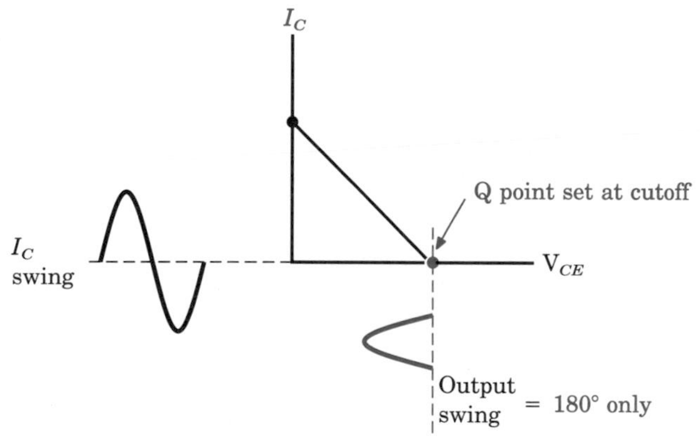

FIGURE 24–10 Q point set at cutoff bias. (NOTE: Class B output resembles half-wave rectified signal.)

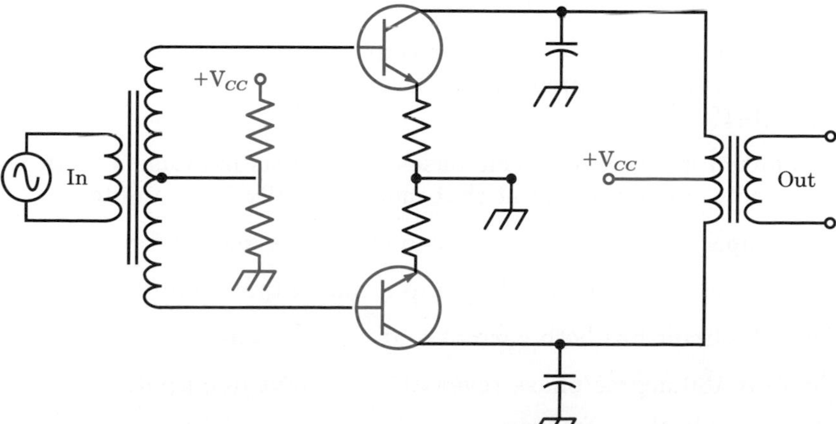

FIGURE 24–11 Basic push-pull stage

Class C Operation. When a transistor amplifier stage is biased below cutoff value, Figure 24–12, collector current flows for less than 180 degrees of the input cycle. Typical conduction time for Class C operation is approximately 120 degrees.

FIGURE 24–12 Class C operated stage

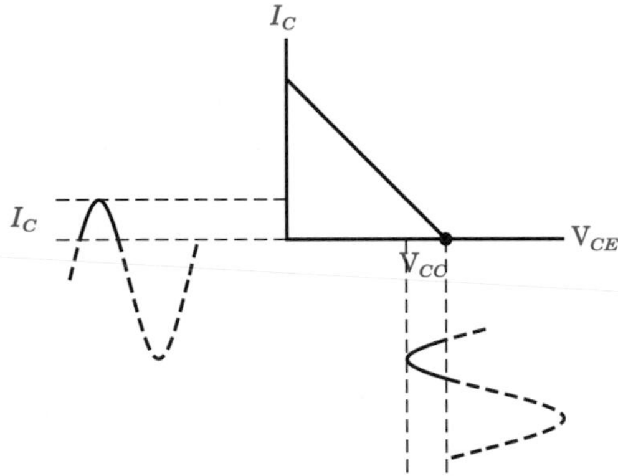

Because Class C operation can be used in radio frequency (RF) circuits where the other circuit components (such as a tuned LC circuit) complete the missing part of the output waveform, you frequently find this operation used in RF power amplifiers. Class C operation has high efficiency and large-signal handling capability.

Classification by Circuit Configuration

Recall in a previous chapter we discussed three common circuit configurations for transistor amplifiers—common-emitter, common-base, and common-collector circuits. Let's look at these configurations.

THE COMMON-EMITTER (CE) AMPLIFIER

Since the common-emitter amplifier is an important circuit, we will study several aspects of this circuit first. Several notable features of the circuit, Figure 24–13, are:

1. A common method to provide base-emitter bias is to use a voltage divider network. We'll look at that more carefully in a minute.
2. Each capacitor in this circuit has one or more special functions.
3. The emitter is common to both input and output circuits.
4. The CE circuit has both current and voltage gain.
5. There is 180 degree phase reversal from input to output.

Now let's study these features.

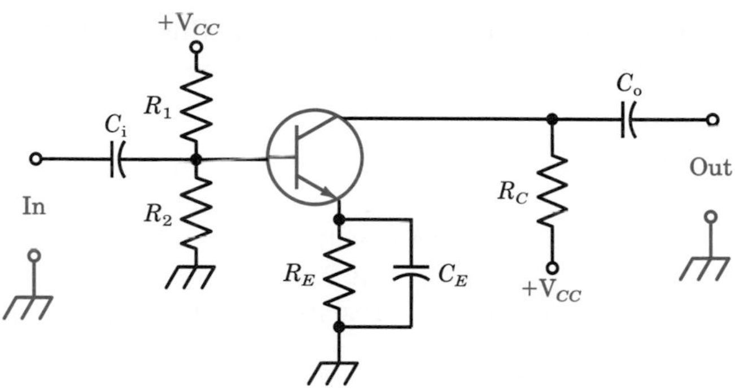

FIGURE 24–13 Common-emitter amplifier circuit

Voltage-Divider Bias Method

As you have previously learned, to forward bias the emitter-base junction of a silicon transistor takes approximately 0.7 V difference of potential between the base and emitter (for silicon semiconductors). Let's use that fact as a starting point.

Some further second approximation level assumptions will be made for an easier analysis, since these assumptions introduce only a small percentage of error. Refer to Figure 24–14 as you study the following discussion.

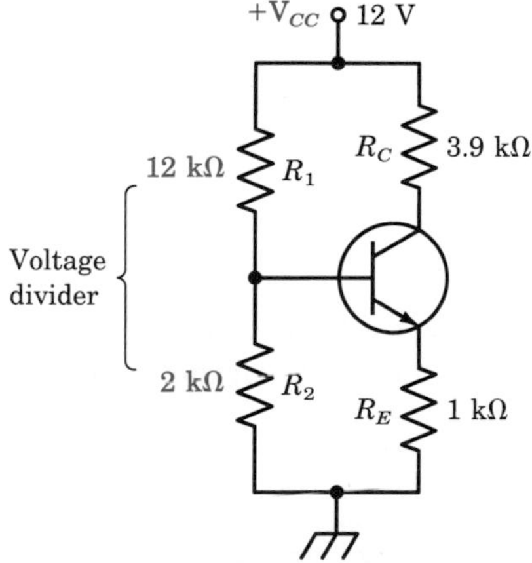

FIGURE 24–14 Voltage-divider bias circuit

1. Assume collector current is virtually equal to emitter current.
2. Assume the base current is much less than the voltage divider current. Therefore, base current has negligible effect on the divider voltages, and base current is not considered.
3. Assume in the divider branch, the sum of the voltage drops across

R_1 and $R_2 = V_{CC}$. Using Ohm's law, current through the divider $= \frac{V_{CC}}{(R_1 + R_2)}$.

4. Assume in the transistor branch, the sum of the emitter-resistor voltage drop, transistor voltage drop (V_{CE}), and collector-resistor voltage drop $= V_{CC}$. This means collector voltage (V_C) $= V_{CC}$ minus the drop across the collector resistor ($I_C \times R_C$). And $V_{CE} = V_{CC}$ minus the voltage dropped across the collector resistor and the emitter resistor.

EXAMPLE

Using the above information and Ohm's law, let's analyze the circuit parameters of Figure 24–14 so you can see how easy it is.

1. Divider current: $I = \frac{V_{CC}}{R_1 + R_2} = \frac{12\text{ V}}{14\text{ k}\Omega} = 0.857\text{ mA}$
2. Base voltage: $V_B = I \times R_2 = 0.857\text{ mA} \times 2\text{ k}\Omega = 1.71\text{ V}$
3. Emitter voltage: $V_E = V_B - 0.7\text{ V} = 1.01\text{ V}$
4. Emitter current: $I_E = \frac{V_E}{R_E} = \frac{1.01\text{ V}}{1\text{ k}\Omega} =$ about 1 mA
5. Collector voltage: $V_C = V_{CC} - (I_C \times R_C) = 12\text{ V} - 3.9\text{ V} = 8.1\text{ V}$
6. Transistor voltage: $V_{CE} = V_C - V_E = 8.1\text{ V} - 1.01\text{ V} = 7.09\text{ V}$

PRACTICE PROBLEM I

Answers are in Appendix B.

Assume that the circuit in Figure 24–14 has $R_1 = 10\text{ k}\Omega$; $R_2 = 2\text{ k}\Omega$; $R_E = 1\text{ k}\Omega$; $R_C = 4.7\text{ k}\Omega$ and $V_{CC} = 10\text{ V}$. Calculate divider current, base voltage, emitter voltage, emitter current, collector voltage, and voltage across the transistor.

PRACTICAL NOTES

The critical parameters for the transistor stage are calculated with three of the above formulas.

1. $V_B = I\text{ (divider)} \times R_2$
2. $V_E = V_B - 0.7\text{ V}$

3. $I_E = \dfrac{V_E}{R_E}$

Using our approximations and knowing these parameters allows you to easily determine all the other parameters of interest.

Function of Capacitors

Referring again to Figure 24–13, each circuit capacitor performs a special function.

The input coupling capacitor, C_i, couples the ac signal from its input side to the transistor input circuit. It passes the ac signal while blocking any dc voltages from the preceding stages.

The emitter bypass capacitor, C_E, has a very low reactance (impedance) at the operating frequencies. As result, it puts the emitter at signal ground yet allows the dc voltage level at the emitter to be maintained above signal ground. Also, C_E prevents unwanted variations of the dc emitter voltage, thus increasing the gain of the stage. The output coupling capacitor, C_O, couples the amplified signal from the collector circuit of the CE amplifier to the following stage(s), as appropriate. Also, C_O blocks dc while passing along the amplified ac signal.

The Emitter is "Common"

Signal input voltage is applied to the base of the transistor. Signal output is taken from the collector circuit. Thus, the emitter is common to both the input and output circuits.

CE Circuit Gain and Phase Reversal

The common-emitter circuit has both current and voltage gain and a 180 degree phase inversion between input and output signals.

Signal variations in the range of a few millivolts or less at the base (input circuit) cause appreciable change in base current that in turn causes milliamperes of change in collector current. As a result of this change in collector current, voltage swings of several volts can occur at the output, depending on the value of the collector resistor.

Gain Calculations

Gain, or voltage amplification (A_V) is calculated by comparing the voltage swing (peak-to-peak change) at the output to the voltage swing at the input. That is:

$$A_V = \frac{V_{out}}{V_{in}}$$

Current gain (A_I) is computed in a similar fashion. That is, output change in current over input change in current, or

$$A_I = \frac{i_{out}}{i_{in}} = \frac{i_C \text{ (pk-to-pk)}}{i_B \text{ (pk-to-pk)}}$$

Power gain is calculated as the voltage gain times the current gain. That is, $A_P = A_V \times A_I$. As you can see, for common-emitter amplifiers having both voltage and current gain, the power gain (product of the gains) can be very high.

IN-PROCESS LEARNING CHECK I

Fill in the blanks as appropriate.

1. Transistor amplifiers are classified according to ______________________________, ______________________________, ______________________________, and ______________________________.
2. Class A amplifiers conduct during __________ degrees of the input cycle; Class B amplifiers conduct for __________ degrees of the input cycle; and Class C amplifiers conduct for approximately __________ degrees of the input cycle.
3. A load line shows all the possible operating points of a given amplifier circuit from __________ to __________.
4. The input signal to a common-emitter amplifier is 4 mV and the output signal (V_C) swings between +4 and +5 volts. What is the voltage gain of the amplifier?
5. A signal causes a common-emitter base current to swing by 0.03 mA and the collector current varies 1.5 mA pk-to-pk. What is the current gain of the amplifier?
6. A common-collector amplifier has a voltage gain of 600 and a current gain of 30. What is the power gain?

THE COMMON-BASE (CB) AMPLIFIER

A typical common-base (CB) amplifier is shown in Figure 24–15. The circuit characteristics are:

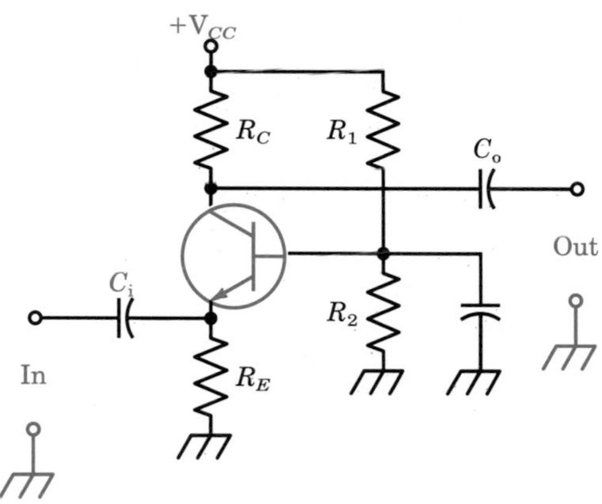

FIGURE 24–15 Common-base amplifier circuit

1. The base is common to both input and output circuits (i.e., at signal ground).
2. The input signal fed is to the emitter and the output signal is taken from the collector circuit.
3. This type of circuit has low input impedance and relatively high output impedance. There is no phase reversal between input and output signals.
4. This type of circuit has a current gain of about one and has fairly high voltage and power gain.

THE COMMON-COLLECTOR (CC) AMPLIFIER

Figure 24–16 illustrates the common-collector amplifier. Some interesting points about this circuit include:

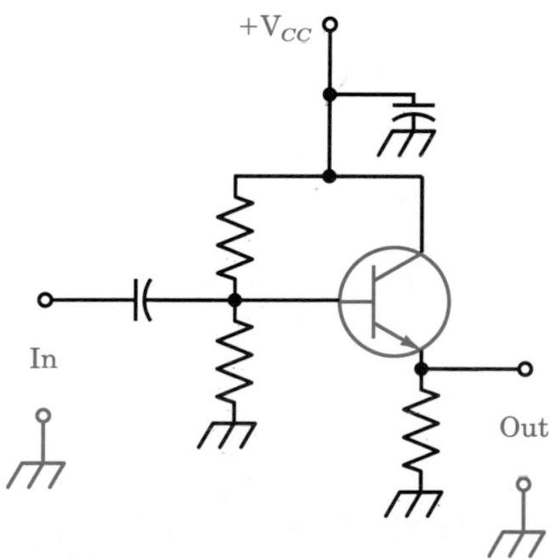

FIGURE 24–16 Common-collector amplifier circuit

1. The collector is essentially at signal ground (although still at $+V_{CC}$ with respect to dc). The resistor where the output signal is generated is the emitter resistor.
2. The input signal is fed to the base and the output is taken across the emitter resistor. Since the output signal is taken at the emitter and that signal follows (essentially tracks) the input signal, this circuit is often termed an emitter follower. There is no phase reversal from input to output signals.
3. The input impedance of a CC is quite high, but the output impedance (across R_E) is low. For this reason, the circuit is often used for isolation and impedance matching purposes.
4. Voltage gain of this type circuit is less than one. Current gain is quite high.

FET AMPLIFIER CONFIGURATIONS

Three configurations of FET amplifiers are common-source (CS), (similar to the CE junction transistor amplifier), Figure 24–17a; common-drain (CD), (similar to the CC or emitter follower), Figure 24–17b; and common-gate (CG), Figure 24–17c.

FIGURE 24–17a Common-source FET amplifier circuit

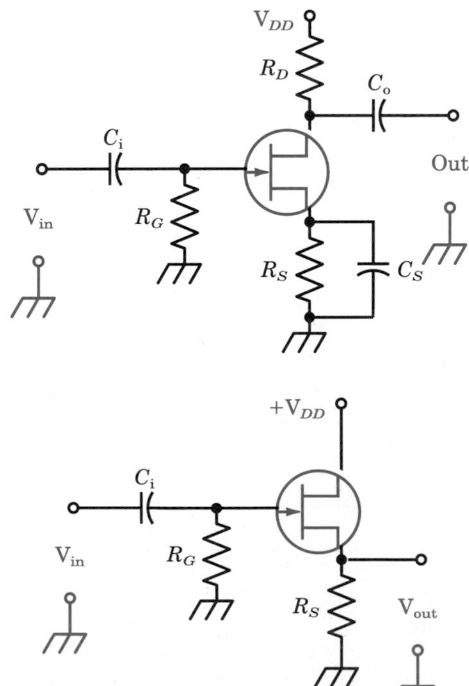

FIGURE 24–17b Common-drain FET amplifier circuit

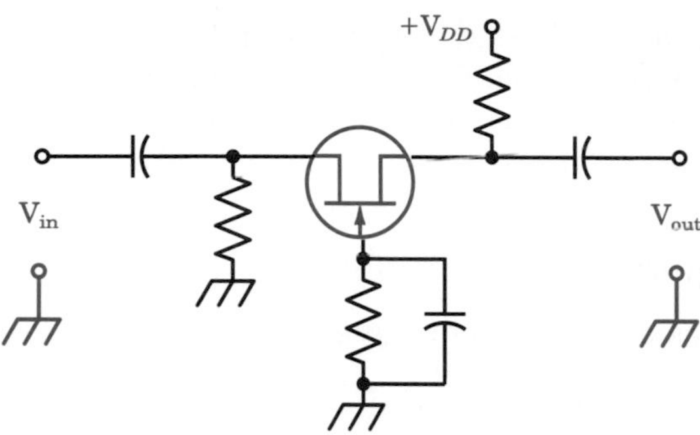

FIGURE 24–17c Common-gate FET amplifier circuit

As in our bipolar transistor configurations, each FET amplifier has unique characteristics. Therefore, each lends itself to certain applications. You will probably thoroughly study these configurations if you take a course in solid-state devices and circuits.

One difference between FET devices and the junction transistors we previously discussed is FET amplifiers are voltage-operated devices, whereas junction transistors are current-operated. That is, in the junction transistor, the base current controls collector current. In FETs, the gate voltage controls the drain current. Due to this fact, an important parameter of the FET amplifier not mentioned with junction transistors is transconductance, abbreviated g_m. This transconductance factor determines the voltage gain of an FET amplifier. A general formula for this parameter is $g_m = \frac{\Delta I_d}{\Delta V_g}$.

Another difference of the CS and CD FET amplifiers is that they have a very high input impedance compared to the CE amplifier that has a relatively low input impedance. You can see some advantages of this type of amplifier when it is used to prevent loading effects on circuits feeding the input of the amplifier stage.

Classification by Type of Device

Two examples of this classification are bipolar transistor amplifier stages and FET amplifier stages.

Sometimes, further details are given about the devices when classifying amplifiers. For example, the additional information might include the precise type of device, such as NPN or PNP transistor, N-channel depletion FET, and so on.

INTRODUCTION TO OSCILLATOR CIRCUITS

Definition of an Oscillator

Someone once said an oscillator is a “hot rod” (or “turbo”) amplifier. Let’s explain that comment.

An oscillator is a circuit that generates and sustains an output signal without an input signal supplied by another circuit or source. As mentioned earlier, this capability enables oscillators to be used as signal sources for many electronic circuit applications.

Conditions Required for Oscillation

There are four basic requirements to enable developing and maintaining oscillation:

1. A dc supply source
2. A frequency-determining circuit or device
3. Amplification
4. **Feedback** in proper phase

NOTE: It is not the intention of this chapter to detail operations or to perform electron chasing specifics. Rather, we will give you identifying features of common oscillator circuits to provide an awareness of these circuits.

We will now examine several common oscillator circuit configurations and briefly identify how these four requirements are met.

The Hartley Oscillator Circuit

Refer to Figure 24–18 as you read the following description of the **Hartley oscillator circuit.**

1. The dc supply is the V_{CC} source.
2. The frequency-determining circuit is comprised of the LC tank circuit, L_{1a}, L_{1b}, and C_1. (Recall the LC circuit resonant frequency formula studied earlier.)

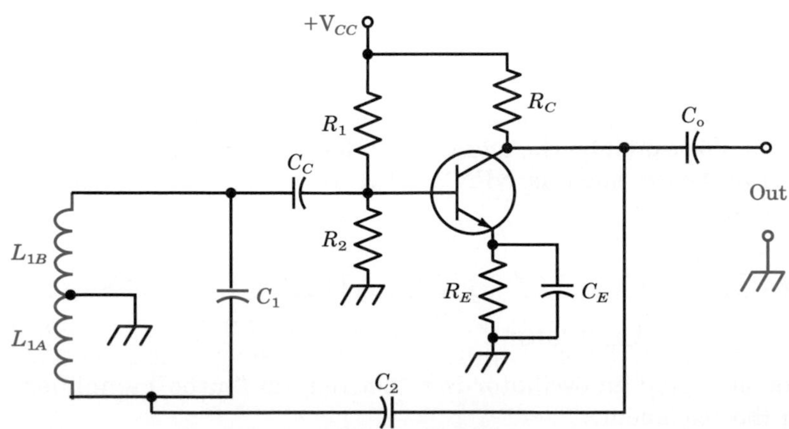

FIGURE 24–18 Hartley oscillator circuit

3. Amplification is supplied by the transistor amplifier circuit.
4. Feedback is provided through the tapped coil in the tank circuit. This tapped coil acts as an ac (rf) voltage divider of the output voltage. The amount and phase of the feedback is appropriate to sustain oscillation.

The Colpitts Oscillator Circuit

Refer to Figure 24–19 as you read the following description of the **Colpitts oscillator circuit.**

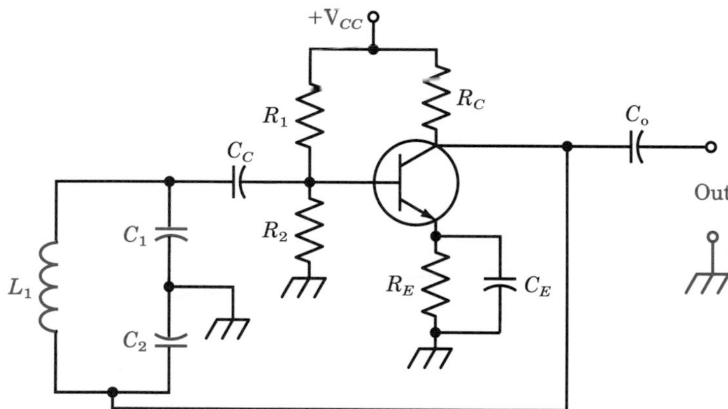

FIGURE 24–19 Colpitts oscillator circuit

1. The dc supply is the V_{CC} source.
2. The frequency-determining circuit is comprised of the LC tank circuit, $C_1 - C_2$, and L_1.
3. Amplification is supplied by the transistor amplifier circuit.
4. Feedback is provided through the rf capacitive voltage divider $\frac{C_1}{C_2}$ in the tank circuit. Again, the amount and phase of the feedback is appropriate to sustain oscillation.

The Clapp Oscillator Circuit

The **Clapp oscillator** circuit, Figure 24–20, is similar to the Colpitts oscillator circuit. The difference is that there is an additional capacitor in series with the coil in the Clapp oscillator circuit. Certain temperature stability and frequency control advantages are gained with this arrangement when using appropriate components.

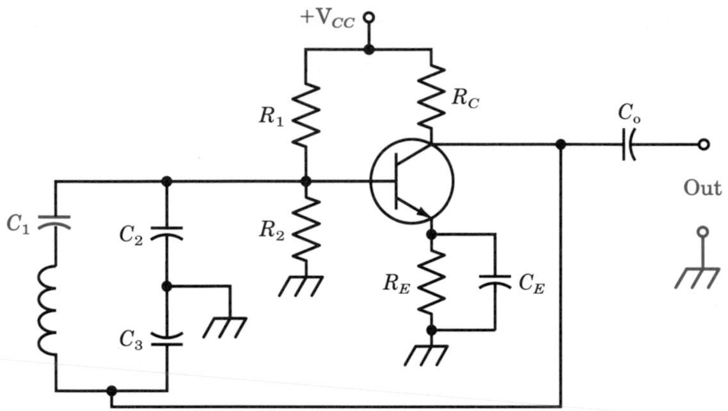

FIGURE 24–20 Clapp oscillator circuit

The Crystal Oscillator Circuit

Figure 24–21 illustrates a **crystal oscillator.** This circuit uses the unique piezoelectric effect of certain crystalline materials. This unique characteristic is that if ac voltage is applied to the crystal, it vibrates at its natural resonant frequency. Conversely, if we mechanically stress the crystal, it generates a voltage.

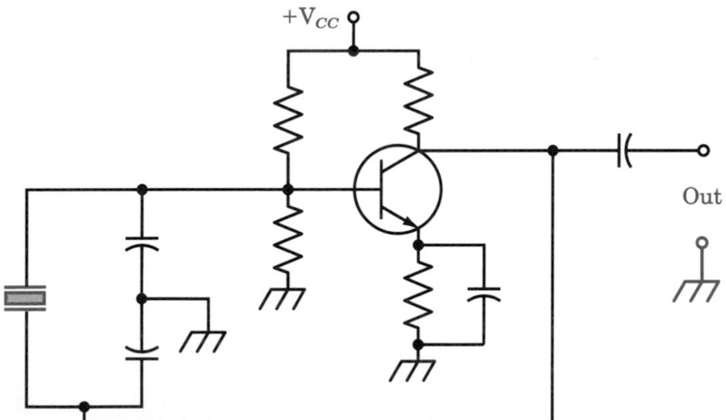

FIGURE 24–21 Crystal oscillator circuit

For this circuit, the crystal determines the resonant frequency. The physical dimensions of the crystal control its natural resonant frequency.

Again, the collector source is the dc voltage needed. The transistor stage provides the amplification, and the remaining components and circuitry enable appropriate feedback to sustain oscillation.

Each amplifier and oscillator discussed has advantages and disadvantages. The application or function for which they are chosen to be used relates to their specific advantages and determines which circuit is selected.

For amplifiers, the frequency of operation, the type of gain, and input and output impedances are factors in selection. For oscillators, the desired frequency of operation, required stability, and other factors influence the choice.

We have only investigated two general applications for transistors. We have not looked at switching and other applications. However, we hope these applications will encourage you to further study these and other transistor circuit applications.

SUMMARY

- Bipolar junction transistors provide amplification because a small current through one junction controls a larger current through another junction.
- Semiconductor amplifiers are classified by signal levels, class of operation, circuit configuration, and type of device used in the circuit.
- Three classes of operation for transistor amplifiers are Class A, Class B, and Class C. In Class A operation, collector current flows at all times, or during 360 degrees of the input signal cycle. In Class B operation, current flows during half of the input signal cycle (180 degrees). In Class C operation, collector current flows for approximately 120 degrees of the input signal cycle.
- Three basic transistor amplifier circuit configurations are common-emitter (CE), common-base (CB), and common-collector (CC) circuits. Each has special features that are used in specific applications.
- The common-emitter circuit provides both current and voltage gain, and the phase of the amplified signal is 180 degrees different than the input signal. That is, the signal is inverted during the amplification process. The common-base amplifier output is in-phase with the input signal. The current gain of the CB amplifier is approximately one. The common-collector circuit features include a gain voltage of less than one and high current gain. It has a high input impedance and low output impedance, causing it to be used frequently for impedance matching or circuit isolation purposes.
- Field effect transistors (FETs) are voltage-operated devices that provide a very high input impedance to prevent loading the circuitry feeding input signals to the FET.
- A family of collector characteristics curves is developed for a transistor by noting the changes in collector current at various collector-to-emitter voltages for various settings of emitter-base current.
- A load line is constructed on collector characteristic curves for a given transistor and circuit by drawing a line between the extremes of cutoff and saturation. In other words, the load line illustrates all possible operating points between those two extremes.
- The Q point, or quiescent operating point, of a transistor amplifier circuit is identified on the load line at the juncture of the base current and the load line for the given circuit. The "no signal" (quiescent dc values) for base current, collector current, and collector-to-emitter voltage are read from the collector curves/load line representation.
- If the Q point for a transistor stage is halfway between cutoff and saturation, the stage is op-

erating as a Class A amplifier. If the Q point is set at cutoff, the stage is operating Class B. If the Q point is set below cutoff, the stage is operating Class C.

- Oscillator circuits use dc sources and unique circuitry to produce ac signals.
- Basic requirements to produce oscillation include a dc source, amplification, a frequency-determining mechanism, and feedback from output to input in proper magnitude and phase.
- Several types of oscillator circuits include Hartley, Colpitts, Clapp, and crystal oscillators. The identifying feature of the Hartley oscillator is its tapped coil in the tank circuit. The capacitive voltage divider in the tank circuit distinguishes the Colpitts circuit. The distinguishing feature of the Clapp oscillator is a special capacitor in series with the tank coil. Crystal oscillators use the piezoelectric effect of special crystalline (quartz) materials and are easily identified because of the crystal element.

REVIEW QUESTIONS

1. Name one application for small-signal and one application for large-signal transistor amplifier stages.
2. Which amplifier class has the lowest efficiency?
3. Which amplifier class has the least distortion?
4. Which amplifier class has the highest efficiency?
5. Where is the Q point set to achieve Class A operation?
6. What amplifier class often uses multiple transistors, particularly in audio amplification applications?
7. Where is the Q point set to achieve Class B operation?
8. Name a common advantage of FET amplifiers compared to bipolar junction transistor amplifiers.

9. A common-emitter circuit has a VCC of 20 volts. What is the approximate quiescent value of V_{CE} if the circuit operates as a Class A amplifier?

10. Briefly list four assumptions frequently used when analyzing a common-emitter circuit that uses voltage-divider bias.

11. For a voltage-divider bias circuit, V_{CC} equals 15 volts, R_C equals 10 kΩ, and R_E equals 1 kΩ. (Assume R_1 = 9.1 kΩ and R_2 = 1 kΩ in the bias divider.) Find the following values:
    a. Divider current
    b. Base voltage
    c. Emitter voltage (Assume a silicon transistor.)
    d. Collector voltage
    e. Collector-to-emitter voltage

12. If you needed a transistor amplifier to provide 180 degrees phase difference between the input and output signals, which circuit configuration would you choose?

13. A transistor amplifier circuit is needed that matches a high impedance at its input side to a low impedance connected to its output side. Which circuit configuration would you choose?

14. For the circuit shown in Figure 24–22, which circuit in Figure 24–23 would complete the circuit for a Hartley oscillator?

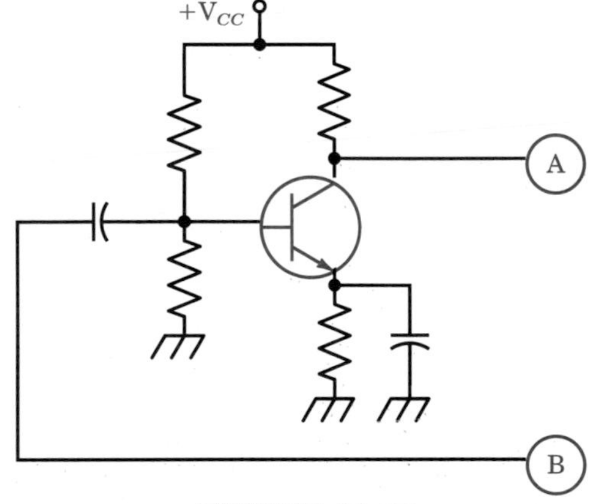

FIGURE 24–22

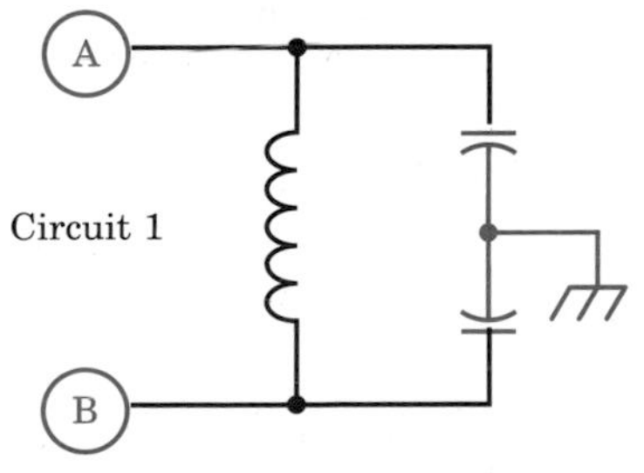

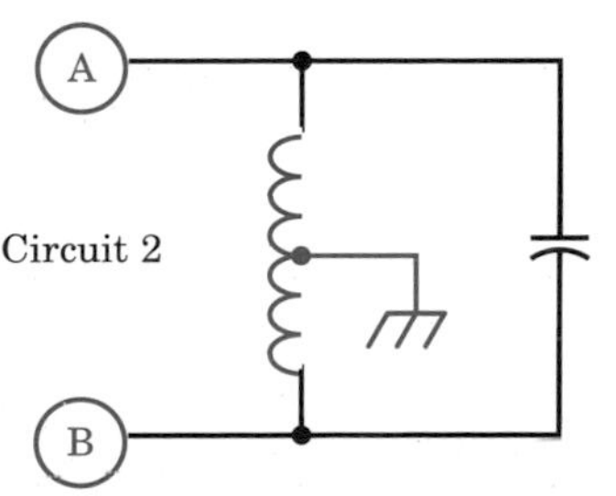

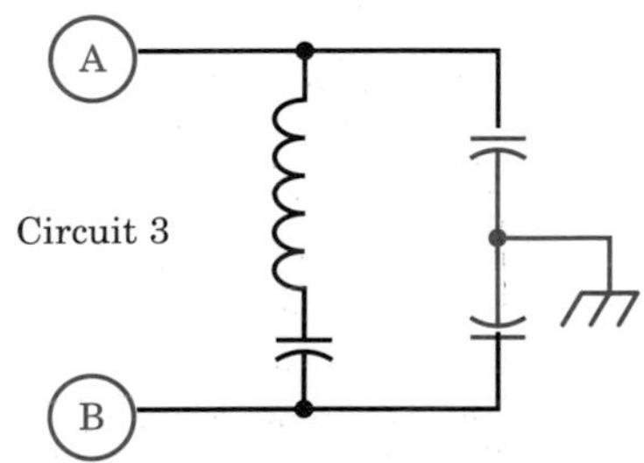

FIGURE 24–23

15. For the circuit shown in Figure 24–22, which circuit in Figure 24–23 would complete the circuit for a Clapp oscillator?

16. For the circuit shown in Figure 24–22, which circuit in Figure 24–23 would complete the circuit for a Colpitts oscillator?

TEST YOURSELF

Refer to the characteristic curves in Figure 24–24 for the following.

FIGURE 24–24

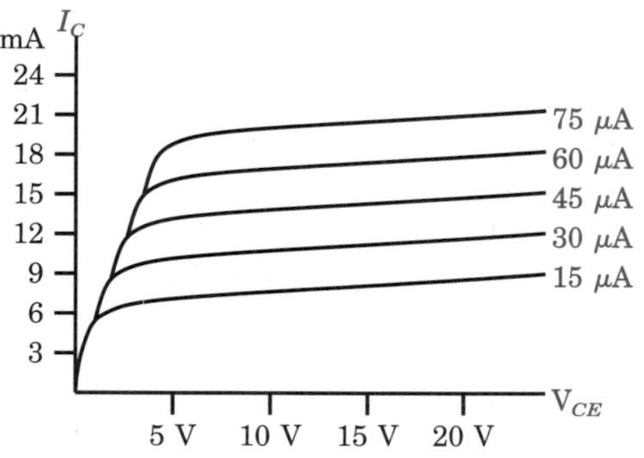

1. From the curves, determine the collector current value for a base current of 60 μA and a V_{CE} of 15 volts.
2. What are the V_{CE} and I_C values at cutoff and saturation if a collector resistor with a value of 1 kΩ and an emitter resistor with a value of 220 Ω were present? (Assume 15 volts V_{CC}.)
3. Using the information from Question 2, what is the I_E value when the Q point is set for Class A operation? What is the V_E value?
4. For the conditions described in Question 3, what are the divider resistor values that will set the Q point for Class A operation?

Introduction to Operational Amplifiers

Key Terms

Common-mode gain

Common-mode input

Differential amplifier

Differential-mode gain

Operational amplifier (Op-amp)

Courtesy of Hewlett-Packard Company

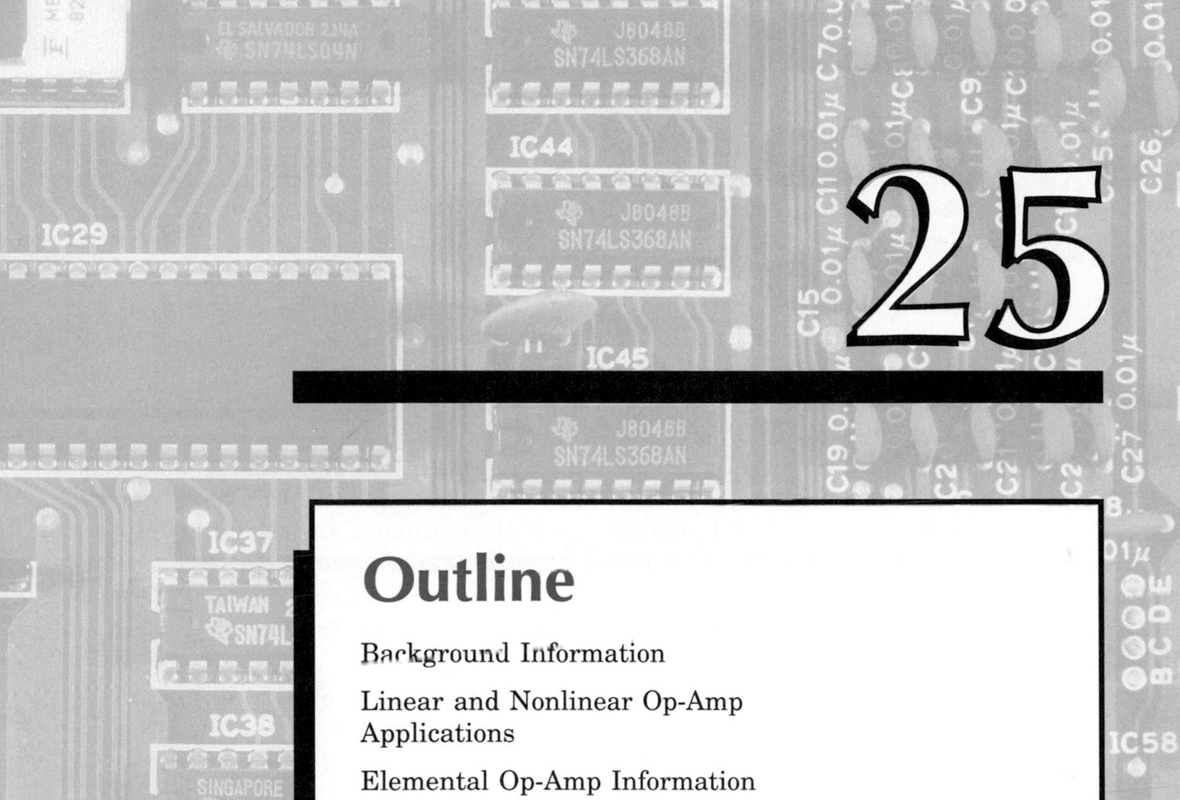

25

Outline

Background Information

Linear and Nonlinear Op-Amp Applications

Elemental Op-Amp Information

Characteristics of the Ideal Op-Amp

An Inverting Amplifier

A Noninverting Amplifier

Basic Op-Amp Parameters

Sample Op-Amp Applications

Concluding Comments

Chapter Preview

Operational amplifiers (op-amps) are very versatile and frequently used. In fact, many linear integrated circuits manufactured are op-amps. For that reason, we want to give you an overview of this subject.

In this chapter you will learn about the basic features of op-amps. A list of linear and nonlinear applications for these devices will be given so you can understand their versatility. Additionally, you will learn that in many applications, a simple ideal op-amp approach can be used for amplifier analysis and design. Also, a brief discussion of some nonideal op-amp parameters will be presented.

This chapter is an introduction to op-amps. For this reason, we recommend that you pursue further studies on this topic to learn more about the operation, design, and applications of these unique and useful devices.

Objectives

After studying this chapter, you will be able to:

- Explain the derivation of the term operational amplifier **(op-amp)**
- Draw op-amp symbol(s)
- Define the term **differential amplifier**
- Draw a block diagram of typical circuits used in op-amps
- List key characteristics of an ideal op-amp
- List important parameters of nonideal op-amps

BACKGROUND INFORMATION

The name **operational amplifiers** comes from the fact that these circuits were originally used in large analog (vacuum-tube type) computers to perform mathematical operations, such as summing, differentiating, and integrating. These early operational amplifier circuits were hard-wired circuits using many discrete (individual) components, such as vacuum tubes, resistors, capacitors, and so on.

Today, the op-amp is a linear-integrated circuit device with the transistors, resistors, diodes and their interconnections all fabricated on a single semiconductor substrate. Some sample packages are shown in Figure 25–1.

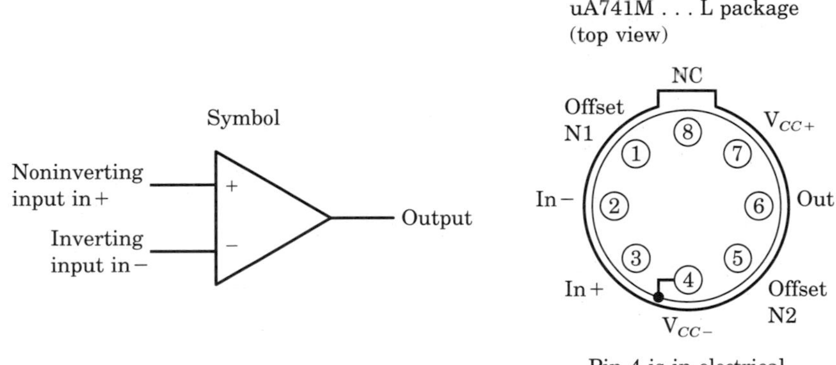

FIGURE 25–1 General-purpose operational amplifier (Adapted with permission from Texas Instruments)

It is easy to fabricate the IC transistors, resistors, and diodes, but more difficult to manufacture meaningful capacitor values on ICs. Therefore, we find the operational amplifier stages are direct coupled. Direct coupled means without coupling capacitors between stages. That is, one stage's output is coupled directly into the next stage's input. Because dc is not blocked between stages, sometimes these amplifiers are referred to as dc amplifiers, as well as direct-coupled amplifiers.

By definition, an op-amp is a very high gain, direct-coupled (dc) amplifier. By controlling and using appropriate feedback from output to input, important amplifier characteristics are controlled by design. Characteristics, such as the amplifier gain and bandwidth features, are controlled by the external components connected to the operational amplifier IC.

To understand the versatility of these devices, let's look at their numerous applications.

LINEAR AND NONLINEAR OP-AMP APPLICATIONS

Linear Applications

Various linear applications (where the output waveform is similar to the input waveform) include inverting amplifiers (where the output is inverted from the input), noninverting amplifiers, voltage followers, summing amplifiers, differential amplifiers, amplifiers for instrumentation, stereo preamplifiers, logarithmic amplifiers, differentiators, integrators, sine-wave oscillators, and active filters.

Nonlinear Applications

Several nonlinear applications for the op-amp (where the output waveform is different from the input waveform) include comparators, bistable/astable/monostable flip-flops (multivibrators), hysteresis oscillators, and other special purpose circuits.

ELEMENTAL OP-AMP INFORMATION

Op-amps can have open-loop voltage gains ranging from several thousand up to about a million. The packaging requires few external connections. Look at the schematic symbols and typical connection points in Figure 25–2.

A simple block diagram of the typical stages in an operational amplifier is shown in Figure 25–3.

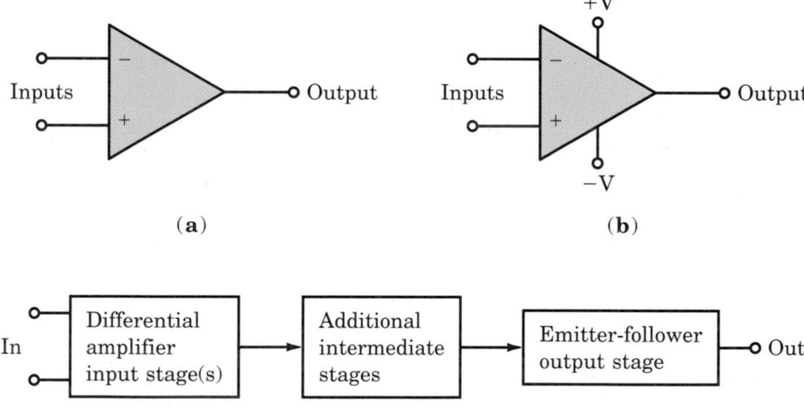

FIGURE 25–2
(a) Simplified symbol
(b) Symbol showing V source connections

FIGURE 25–3
Simplified block diagram of an operational amplifier (op-amp)

As you can see from the preceding two figures, the schematic symbol, device connections, and block diagram for an op-amp are simple. However, the circuitry inside the device contains many circuits and connections shown in the typical IC op-amp diagram in Figure 25–4.

In Figure 25–5, the differential amplifier input stage of the op-amp has both an inverting input (at the negative input terminal) and a noninverting input (at the positive input terminal).

If an input signal is applied to the inverting (negative input) with the noninverting (positive input) grounded, the output signal polarity is opposite the input signal polarity, Figure 25–6a. Conversely, if the input signal

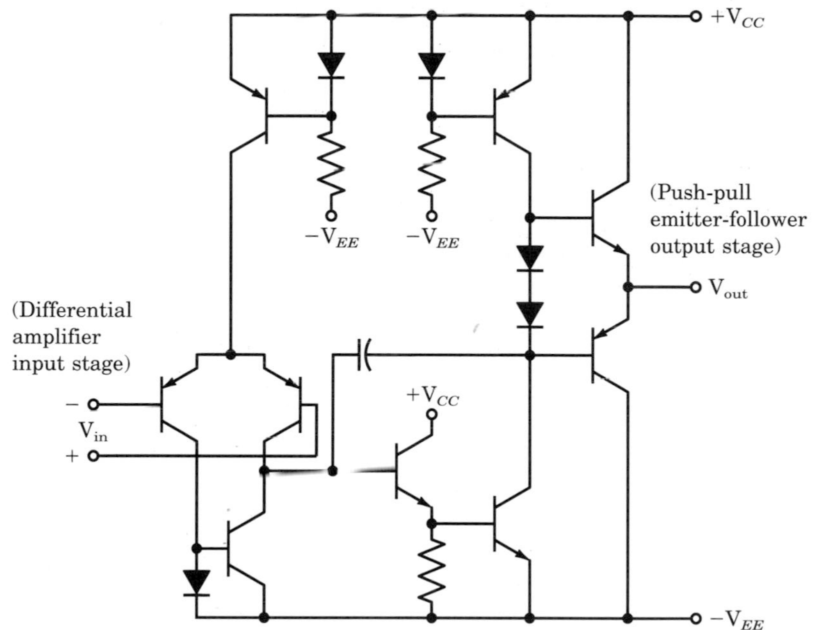

FIGURE 25–4 Typical op-amp circuitry

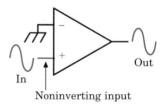

is fed to the positive input terminal while the negative input terminal is grounded, the output signal is the same polarity as the input, Figure 25–6b.

The enormous gain factors of the basic op-amp device (i.e., open-loop gain) make it necessary to control this gain by use of external components. Notice in the op-amp circuit configurations in Figure 25–6, there is an input component, resistor (R_i), and a feedback component resistor (R_f). This feedback component, which feeds some of the output signal back to an input, controls the amplifier gain. When the feedback feeds to the negative input, it reduces gain, which stabilizes the amplifier. Also, it causes the frequency response of the amplifier to broaden.

Inverting input

In

Out

Signal fed to inverting input; output is inverted

In

Out

Noninverting input

Signal fed to noninverting input; output is not inverted

FIGURE 25–5 Inverting and noninverting inputs

R_f R_i In Out

(a)

R_f R_i In Out

(b)

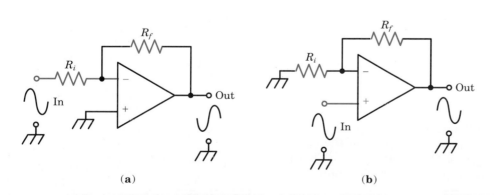

FIGURE 25–6
(a) Inverting op-amp connections
(b) Noninverting op-amp connections

When the feedback component connects from output to inverting input (negative input), the feedback is negative. When the feedback component connects from output to noninverting input (positive input), the feedback is positive. For amplifier applications, we will focus on the negative feedback configuration.

FIGURE 25–7 Differential amplifier

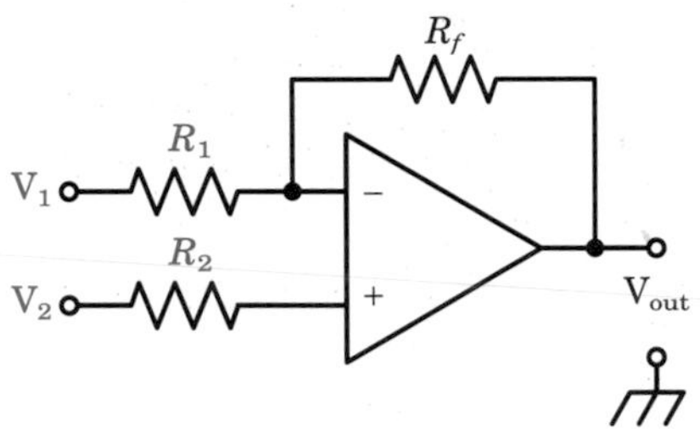

Input resistors (R_1 and R_2) must be equal, then:

$$V_{out} = \frac{R_1}{R_2}(V_1 - V_2)$$

Since the circuit output signals are of opposite polarity for signals fed to the negative input and the positive input—if the same signal feeds both inputs—their outputs cancel each other. This indicates that this amplifier circuit responds only to the *difference* between signals fed to the two inputs. For this reason, this amplifier configuration is called a **differential amplifier,** Figure 25–7.

This differential effect introduces another term called **common-mode rejection ratio.** This is the degree the amplifier rejects (does not respond or produce output) for signals that are common to both inputs. This circuit feature provides useful rejection of undesired signals that can be present at both inputs. For example, undesired signals, such as 60-Hz pickup or other stray noise signals, that may be present at the inputs, Figure 25–8.

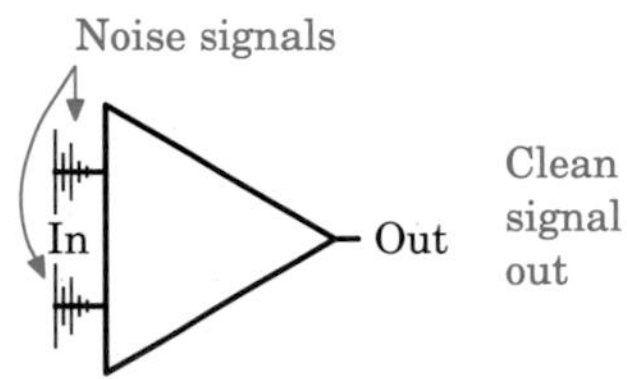

FIGURE 25–8 Noise signals common to both inputs cancel in the output.

IN-PROCESS LEARNING CHECK I

Fill in the blanks as appropriate.

1. An op-amp is a high gain ___________-coupled amplifier.
2. The input stage of an op-amp uses a ___________ amplifier configuration.
3. The name operational amplifier derives from the fact that their early use was to perform mathematical ___________ .
4. The open-loop gain of an op-amp is much ___________ than the closed loop gain.
5. To achieve an output that is the inversion of the input, the input signal is fed to the ___________ input of the op-amp.
6. Input signals to the op-amp can be fed to the ___________ input, the ___________ input, or to both inputs.

CHARACTERISTICS OF THE IDEAL OP-AMP

With this background, let's now look at the characteristics of an ideal op-amp. An ideal op-amp would have infinite voltage gain, infinite bandwidth, infinite input impedance, and zero output impedance.

Obviously, these parameters are not achievable in a practical device. However, they are used to approximate operation when designing certain circuits using op-amps.

In practice, these devices have limitations similar to other semiconductor devices. That is, there are maximum current-carrying capabilities, maximum voltages, which must not exceed power limitations. Also, there is something less than flat-frequency response if amplification of signal frequencies from dc to light is attempted.

PRACTICAL NOTES

Manufacturers produce explicit data sheets on every device they manufacture. These data sheets accurately describe each device's capabilities and limitations. Thus, it is critical that a technician learn how to appropriately find, use, and interpret the data found on these data sheets.

As you can surmise from the previous discussions, real world op-amps cannot meet the absolute ideal parameters of infinite gain and so on. However, it is realistic to assume these devices exhibit high voltage gain, high input impedance, low output impedance, and wide bandwidth amplifying characteristics. Using this information, let's look at some basic op-amp circuits.

AN INVERTING AMPLIFIER

In Figure 25–9 you see an amplifier with input signal applied to the inverting input through the input resistor R_i. Part of the amplified and inverted output feeds back to the input through the feedback resistor R_f. With the positive input grounded, the differential input signal equals the value of the input signal fed to the negative input.

If ideal (infinite input impedance) conditions are assumed, it can be shown that the closed-loop voltage gain $\left(\frac{V_{out}}{V_{in}}\right)$ of this amplifier equals the ratio of the feedback resistor value to the input resistor value. This makes designing for a specific gain quite easy.

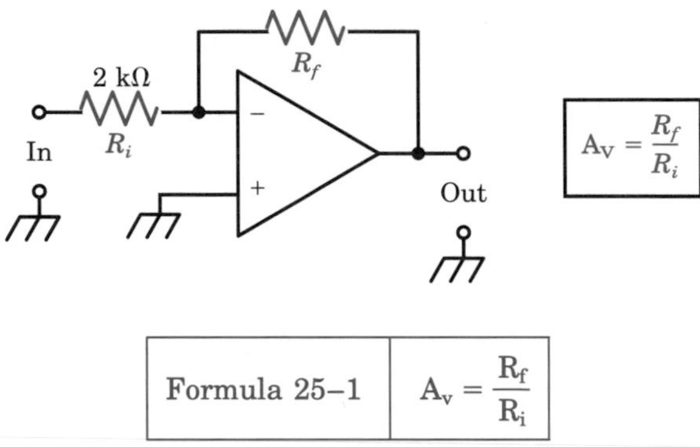

FIGURE 25–9
Inverting amplifier closed-loop gain

Formula 25–1	$A_v = \frac{R_f}{R_i}$

(NOTE: Open-loop gain is the gain of the op-amp device without any external feedback system affecting gain. Open-loop gain values are much higher than the practical closed-loop gain described above.)

EXAMPLE

For the circuit in Figure 25–9, if $R_f = 27\ k\Omega$ and $R_i = 2\ k\Omega$, what is the voltage gain of the amplifier?

Answer:

$$A_v = \frac{R_f}{R_i} = \frac{27\ k\Omega}{2\ k\Omega} = -13.5$$

PRACTICE PROBLEM I

Answers are in Appendix B.

1. For the circuit in Figure 25–9, what R_f value is needed to obtain a gain of -100?
2. Using the R_f value computed in Question 1, what R_i value is needed to achieve a voltage gain of -200?

A NONINVERTING AMPLIFIER

The circuit for a noninverting op-amp amplifier is shown in Figure 25–10. As you can see, the junction of the feedback resistor (R_f) and R_i is still at the negative input terminal. However, the difference is that the input signal feeds to the positive input and R_i is grounded. R_f and R_i form a voltage divider for the feedback voltage. The differential voltage felt by the op-amp is the difference between the input voltage (V_{in}) fed to the positive input, and the feedback voltage felt at the negative input terminal. This differential input is then amplified by the gain of the op-amp, and produces the voltage output. Without discussing the derivation of the noninverting amplifier gain, let us simply state that the gain of the noninverting amplifier

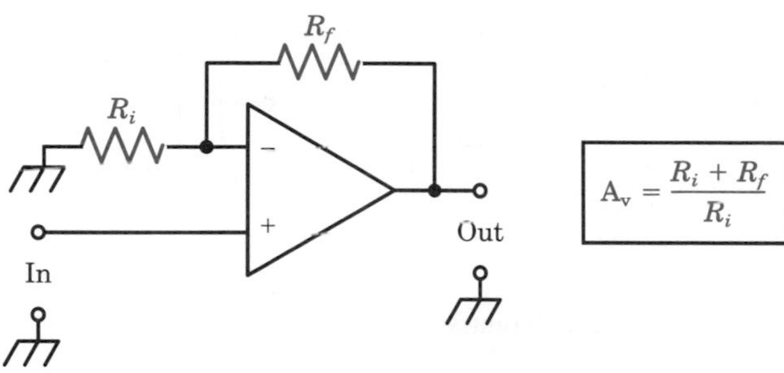

FIGURE 25–10 Noninverting amplifier closed-loop gain

equals the sum of the input and feedback resistances divided by the input resistance.

Formula 25–2	$A_v \text{ (noninverting)} = \frac{R_i + R_f}{R_i}$

EXAMPLE

If the R_i and R_f values for a noninverting op-amp circuit are 1 kΩ and 12 kΩ respectively, what is the voltage gain of the amplifier?

Answer:

$$A \text{ (noninverting)} = \frac{(R_i + R_f)}{R_i} = \frac{(1\ \text{k}\Omega + 12\ \text{k}\Omega)}{1\ \text{k}\Omega} = 13$$

PRACTICE PROBLEM II

What is the voltage gain of a noninverting amplifier with a feedback resistor of 20,000 ohms and a R_i value of 2,000 ohms?

BASIC OP-AMP PARAMETERS

As you have studied, the input circuit of the multi-stage operational amplifier is a differential amplifier. You have also learned some terms used to describe this type of circuit. As a result, you will be somewhat familiar with these terms in your future studies.

Some additional op-amp terms you will also see in additional studies are described in the following section.

Differential Amplifier Related Terms

1. Modes of operation include single-ended, differential, and common mode. Single-ended mode is the operation with one input grounded, Figure 25–11a. Differential (or double-ended) mode is the operation where signals of opposite polarity are fed to the two inputs, Figure

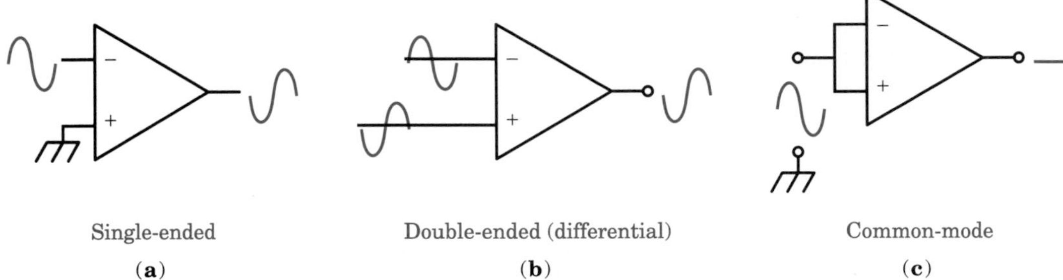

FIGURE 25–11 Modes of operation for differential amplifiers

25–11b. Common mode refers to the operation where the same signal is fed to both inputs, Figure 25–11c.

2. **Differential-mode gain** is the ratio of output voltage to the difference between two signals fed to the two inputs. Usually this gain is very high.
3. **Common-mode gain** is the small (usually less than 1) gain the differential amplifier shows when fed the same signal at both inputs. That is, it is the ratio of the output voltage to the **common-mode input** (input common to the two inputs). Ideally, there would be no output with the same signal fed to both inputs; however, there is a small output in practical amplifiers.

Input Parameters of Interest

1. Input-offset voltage exists since practical op-amps produce some output voltage even with no voltage at the inputs (called output-offset voltage). Thus, an input-offset voltage can be applied to bring the output to zero when there are no signal inputs to the op-amp, Figure 25–12.

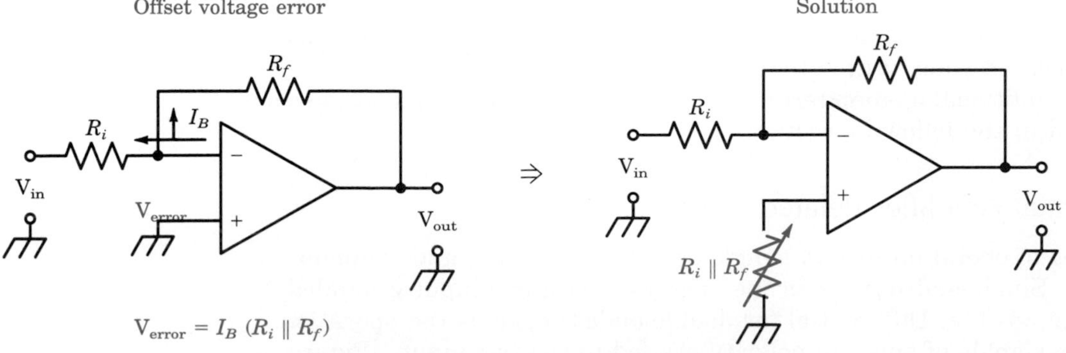

FIGURE 25–12 Offset error correction

2. Input-bias current is the small dc base current needed so that the input junctions of the input stage transistors function properly.
3. Input-offset current is the difference between input bias currents into the two inputs when the output is at zero volts.

Three important considerations for designers are output offset, gain, and frequency compensation. These and other parameters, such as input and output impedance considerations, are important to those who design applications circuits using op-amps. However, these considerations are not necessary for you to study in this overview chapter.

SAMPLE OP-AMP APPLICATIONS

Summing Amplifier

See Figure 25–13a and 25–13b for concept examples of the summing amplifier. As the name implies, the output has a relationship to the sum of the inputs. In Figure 25–13a, you see all the resistors have the same value. In Figure 25–13b, you see the summing effect resulting with resistors of different values in the input and output circuits. Use Ohm's law to compute each input current. The sum of those currents passes through the feedback resistor (R_f), and the product of that current times the feedback resistor value provides the output voltage.

Integrator Circuit

This circuit uses a capacitor as the feedback component, Figure 25–14. It performs the mathematical (calculus) function of integrating. One popular

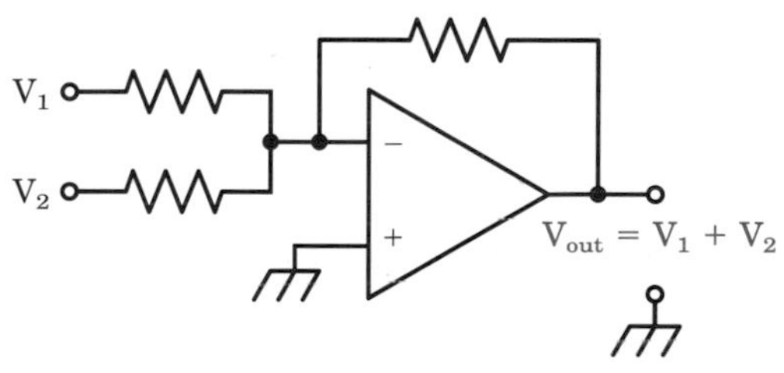

(a) Assuming all Rs equal:
$V_{out} = V_1 + V_2$

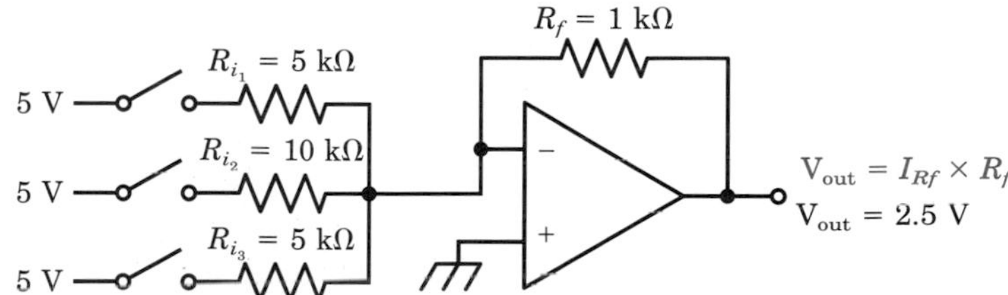

(b) With different R values and with all three switches closed:

$$I_{i_1} = \frac{5\ V}{5\ k\Omega} = 1\ mA$$

$$I_{i_2} = \frac{5\ V}{10\ k\Omega} = 0.5\ mA$$

$$I_{i_3} = \frac{5\ V}{5\ k\Omega} = 1\ mA$$

I through $R_f = 1 + 0.5 + 1 = 2.5\ mA$

$V_{out} = I_{Rf} \times R_f = 2.5\ mA \times 1\ k\Omega = 2.5\ V$

FIGURE 25–13 Examples of summing amplifiers

FIGURE 25–14
Example of op-amp integrator

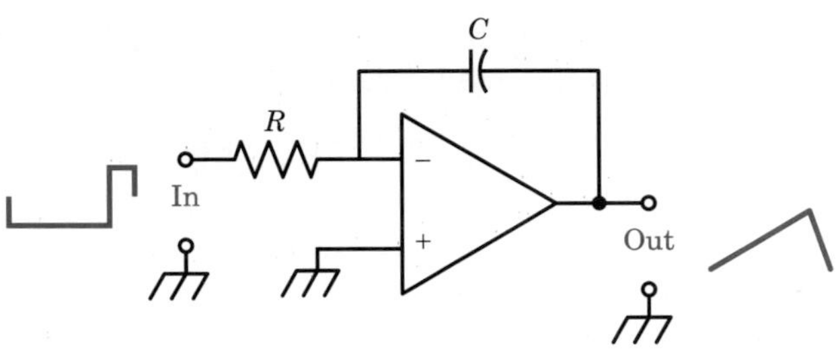

application for this circuit is to produce a ramp-shaped voltage output when its input is a rectangular wave.

Differentiator Circuit

Again, the op-amp performs a mathematical function—the calculus function of differentiating. This means the circuit output is proportional to the derivative of the input voltage. This is somewhat the inverse of integrating. Therefore, one application of this circuit is to produce a rectangular output from a ramp-shaped input voltage, Figure 25–15. Another useful function of this circuit is a detector of pulse edges (such as the leading or trailing edges of a rectangular wave).

FIGURE 25–15
Examples of op-amp differentiators

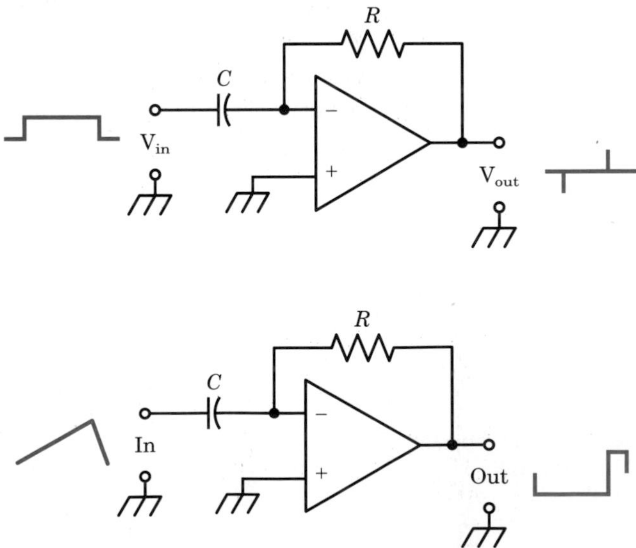

Comparator

Frequently, it is useful to compare voltages to see which is larger. An op-amp circuit that performs this function is shown in Figure 25–16. As you can see, the op-amp has two inputs. With this circuit, if the input to the

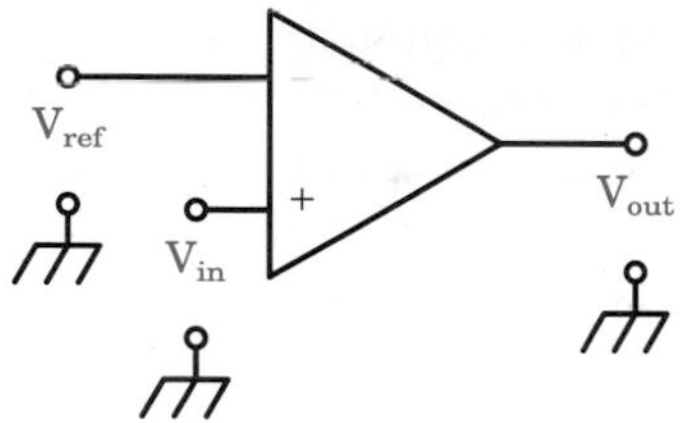
V_{ref}
+
V_{in}
V_{out}
When noninverting V_{in} is greater than inverting V_{ref}, V_{out} is high.
When noninverting V_{in} is less than inverting V_{ref}, V_{out} is low.

FIGURE 25–16 Example of op-amp comparator

noninverting input is greater than the input to the inverting input, the output is high (positive V_{sat}). On the other hand, if the input to the noninverting input is lower than the inverting input, the output is low (negative V_{sat}). For example, if V_{ref} is some dc voltage reference level and a sine-wave signal is fed to the noninverting input, the output is at its high extreme only during the period where the sine-wave is higher than the dc reference value. During the remaining time, the output is at its low extreme. By changing the reference voltage level, the output can be varied from a square wave to a rectangular wave. One practical use of such a circuit is to convert a sine wave to a square wave.

Series-Type Linear Voltage Regulator

Figure 25–17 shows a simple series-type voltage regulator circuit that uses an op-amp as one critical element.

This circuit regulates for any changes in input voltage (V_{in}) or for changes in output load current demand (I_L). It essentially keeps the output voltage constant, or regulates, for these changes.

The voltage divider resistors from the V_{out} line to ground sense changes in output voltage. Since one op-amp differential input is held constant by the Zener diode reference voltage circuit, any change fed from the output through the R-divider network to the other op-amp input causes a differential input between the op-amp inputs. This, in turn, is amplified and con-

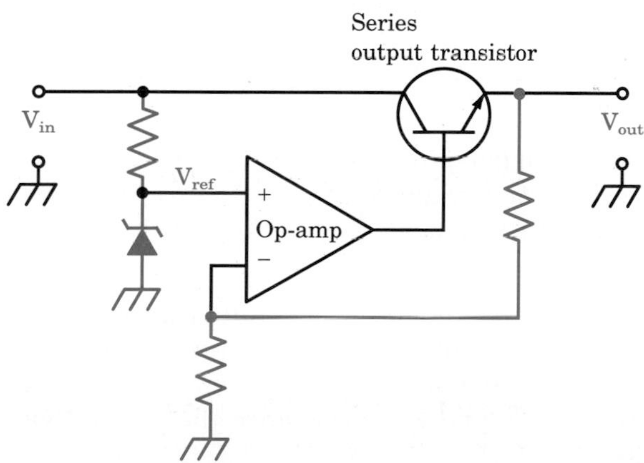
Series
output transistor
V_{in}
V_{ref}
+
Op-amp
−
V_{out}

FIGURE 25–17 Example of op-amp series-type linear voltage regulator

trols the series output transistor conduction, and consequently, its V drop from collector-to-emitter. Obviously, if the output voltage attempts to decrease (due to a decrease in circuit V_{in} or an increase in load demand I_L), we want less drop across the transistor, bringing the output back up to the desired level. Conversely, if the output voltage attempts to increase (due to V input increasing or I load decreasing), we want the transistor voltage drop ($V_{C\text{-}E}$) to increase, bringing the output voltage back down to the desired regulated level.

Bistable Multivibrator (Flip-Flop) Circuit

A bistable multivibrator or flip-flop circuit has two possible stable output voltage states. One state is positive saturation; the other is negative saturation. To change from one state to another state requires an input trigger signal.

Observe Figure 25–18 as you read the following comments.

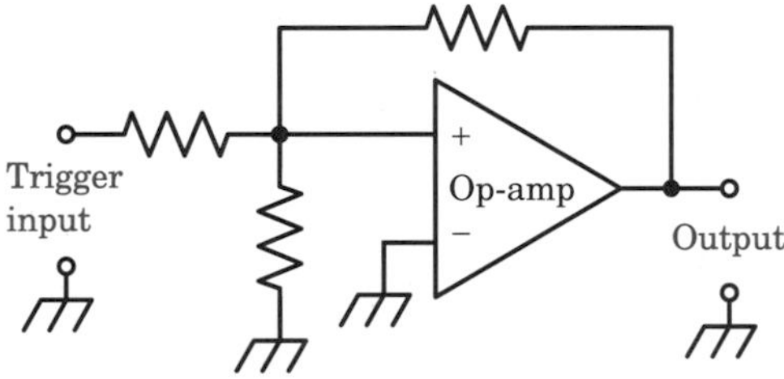

FIGURE 25–18 Example of op-amp dc-coupled multivibrator (flip-flop)

1. The example shows a dc-coupled flip-flop.
2. Notice the feedback is positive feedback, being fed from the output back to the positive input of the op-amp.
3. Because of this regenerative positive feedback, the circuit has only two stable states.
4. If the device is initially in positive saturation, a positive pulse applied to the input has no effect since the output is already at maximum positive saturation.
5. When a negative trigger pulse is fed to the positive input and drives the input negative, the regenerative feedback system drives the op-amp to negative saturation. It remains in that state until a positive trigger reverses the circuit action.

To be a well-rounded technician, you should continue to study these circuits and other practical applications of circuits listed in this introductory chapter. We highly recommend you study them in detail. Op-amps are important to the technician and engineer since they are prevalent devices in today's technology.

SUMMARY

- Operational amplifiers were initially used in analog computers to perform mathematical operations.
- Integrated circuit op-amps are high gain, direct-coupled, dc amplifiers. Op-amps perform various useful functions, therefore they are applied in many electronic systems and subsystems. They are used as inverting and noninverting amplifiers for numerous purposes, including summing, differentiating, and integrating. They are also used in oscillator circuits and nonlinear applications, such as flip-flops and other special circuits.
- The open-loop gain of op-amp devices can be in the hundreds of thousands. The closed-loop gain of an op-amp circuit is much less than the open-loop gain. The closed-loop gain is controlled by the values and types of external components connected to the inputs and output terminals. Using some ideal op-amp assumptions, the closed-loop gain is computed as the ratio of the feedback resistor values compared to the input resistor.
- To produce an output that is inverted from the input, the input signal is connected to the inverting (negative) input of the op-amp. Conversely, to produce an output with the same phase as the input, the input signal is fed to the noninverting (positive) input.
- The single-ended mode of operation for the input differential amplifier used in op-amps is achieved by grounding one input and feeding the signal to the other input.
- The differential amplifier's differential mode is used when the two input signals are opposite in polarity.
- The common-mode operation of a differential amplifier is used when the same signal feeds both inputs.
- Op-amps are used in numerous functions. Some examples include inverting amplifiers, noninverting amplifiers, summing amplifiers, integrators, differentiators, comparators, oscillators, flip-flops, and voltage regulator circuits.

REVIEW QUESTIONS

1. Which amplifier circuit is typically used as the input stage for an operational amplifier?
2. List the key parameters to describe the ideal op-amp.
3. Define open-loop gain.
4. An op-amp used as an inverting amplifier has a feedback resistor of 250 kΩ and an input resistor of 1200 ohms. Compute the closed-loop gain.
5. A noninverting op-amp amplifier has an R_f equal to 25 kΩ and an R_i equal to 1 kΩ. Compute the closed-loop gain.

6. For an op-amp operating in single-ended mode, describe the output signal polarity when the positive input is grounded and the input signal feeds the negative input.

7. For an op-amp in common-mode operation, should the output magnitude be large, small, or medium?

8. To produce integration in an op-amp, what type component should be used as the feedback component?

9. To produce differentiation in an op-amp, what type component should be used as the feedback component?

10. Assuming equal resistors throughout the circuit, write the formula for the output of a summing amplifier having three input voltages.

11. Briefly explain the function of a comparator.

12. For the circuit in Figure 25–17, briefly explain how a condition where V_{out} trying to increase could be compensated for by the regulator circuit. Explain in terms of the coordination between the feedback circuit, the op-amp and the series output transistor.

13. For the circuit in Figure 25–18, would a positive input trigger pulse make the output go high or low?

14. For the circuit in Figure 25–18, if the output is currently in the low state, would a negative trigger pulse at the input cause a change in output states?

15. What two parameters of an op-amp can be somewhat controlled by the feedback and input resistor values?

TEST YOURSELF

1. Draw the schematic diagram of an inverting op-amp circuit having a gain of −500.
2. Find and list the data handbook parameters typically shown for a 741 type of operational amplifier device.
3. Research and draw a circuit of an inverting op-amp circuit using an offset-null adjustment circuit that minimizes the output offset voltage.

APPENDIX A

Color Codes

A. RESISTOR COLOR CODE

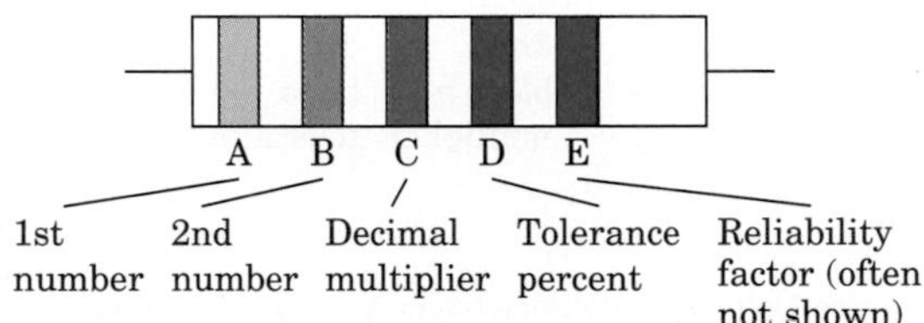

COLOR SIGNIFICANCE CHART
(For resistors and capacitors)

COLOR	NUMBER COLOR REPRESENTS	DECIMAL MULTIPLIER	TOLERANCE PERCENT	VOLTAGE RATING	% CHANGE PER 1,000 HRS OPER.
BLACK	0	1	—	—	—
BROWN	1	10	1*	100*	1%
RED	2	100	2*	200*	0.1%
ORANGE	3	1,000	3*	300*	0.01%
YELLOW	4	10,000	4*	400*	0.001%
GREEN	5	100,000	5*	500*	—
BLUE	6	1,000,000	6*	600*	—
VIOLET	7	10,000,000	7*	700*	—
GRAY	8	100,000,000	8*	800*	—
WHITE	9	1,000,000,000	9*	900*	—
GOLD	—	0.1	5	1,000*	—
SILVER	—	0.01	10	2,000*	—
NO COLOR	—	—	20	500*	—

(*Applicable to capacitors only)

B. CAPACITOR COLOR CODES

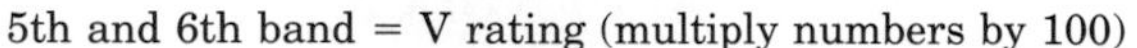

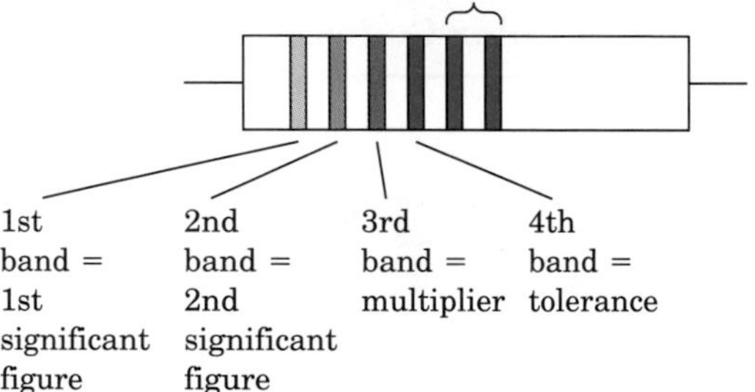

See Color Significance Chart for meaning of colors

See Color Significance Chart for meaning of colors

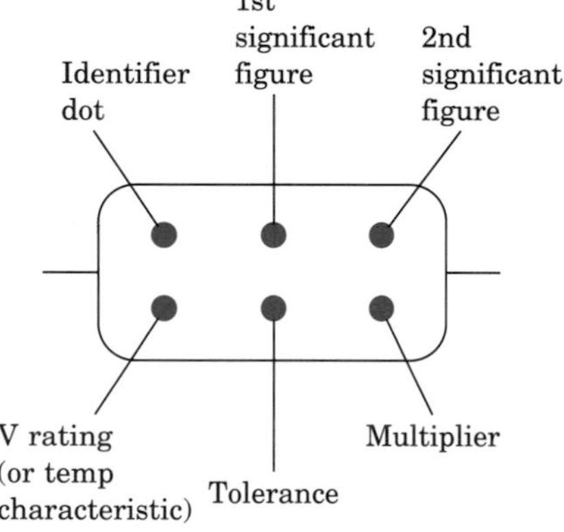

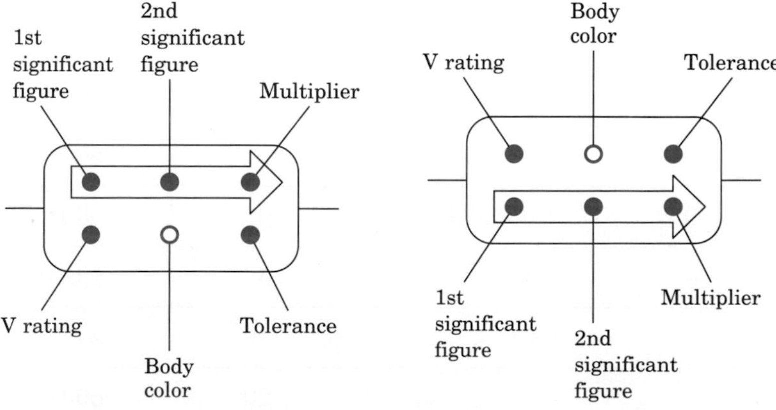

Band color coding system for tubular ceramic capacitors

Band color coding system for tubular ceramic capacitors

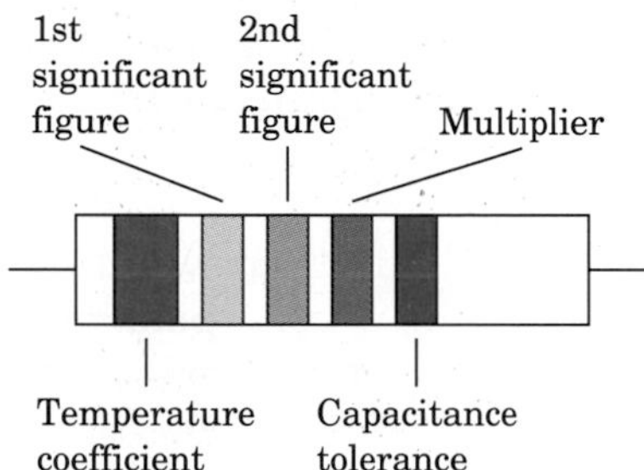

C. SEMICONDUCTOR DIODE COLOR CODE(S)

(NOTE: Prefix "1N . . . " is understood.)

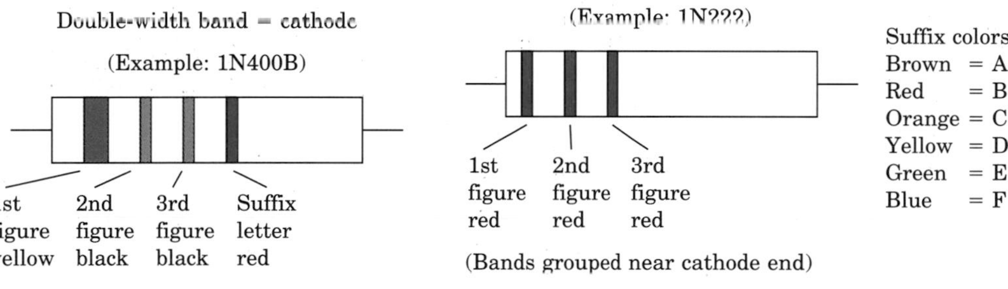

APPENDIX B

Answers to In-Process Learning Checks, Practice Problems, and Odd Answers to Review Questions and Test Yourself

Chapter 1

IN-PROCESS LEARNING CHECKS

I

1. Matter is anything that has ***weight*** and occupies ***space.***
2. Three physical states of matter are ***solid, liquid,*** and ***gas.***
3. Three chemical states of matter are ***elements, compounds,*** and ***mixtures.***
4. The smallest particle that a compound can be divided into and retain its physical properties is the ***molecule.***
5. The smallest particle that an element can be divided into and retain its physical properties is the ***atom.***
6. Three parts of an atom, which interest electronics students, are the ***electron, proton,*** and ***neutron.***
7. The atomic particle having a negative charge is the ***electron.***
8. The atomic particle having a positive charge is the ***proton.***
9. The atomic particle having a neutral charge is the ***neutron.***
10. The ***protons*** and ***neutrons*** are found in the atom's nucleus.
11. The atomic particle that orbits the nucleus of the atom is called the ***electron.***

II

1. An electrical system (circuit) consists of a source, a way to transport the electrical energy, and a ***load.***
2. Static electricity is usually associated with ***nonconductors.***
3. The basic electrical law is that ***unlike*** charges attract each other and ***like*** charges repel each other.
4. A positive sign or a negative sign often shows ***polarity.***
5. If two quantities are *directly* related, as one increases the other will ***increase.***
6. If two quantities are *inversely* related, as one increases the other will ***decrease.***
7. The unit of charge is the ***coulomb.***
8. The unit of current is the ***ampere.***
9. An ampere is an electron flow of one ***coulomb*** per second.
10. If two points have different electrical charge levels, there is a difference of ***potential*** between them.
11. The volt is the unit of ***electromotive force,*** or ***potential*** difference.

REVIEW QUESTIONS

1. Matter is anything that has weight and occupies space whether it is a solid, liquid, or gas.
3. Three chemical states of matter are elements, compounds, and mixtures. Copper is an element, sugar is a compound, and sand and gold dust are a mixture.
5. A compound is a form of matter that can be chemically divided into simpler substances and that has two or more types of atoms. Examples are water (hydrogen and oxygen) and sugar (carbon, hydrogen, and oxygen).
7. I.D. of charges on particles. See page 7.
9. Free electrons are electrons in the atomic structure that can be easily moved or removed from their original atom. These are outer-ring or valence electrons in conductor materials.
11. Chemical, mechanical, light, and heat
13. Unlike charges attract and like charges repel each other.
15. Force of attraction or repulsion will be one-ninth the original force.

17. Unit of measure of electromotive force (emf) is the volt.
19. a. Unit of measure for current is the ampere.
    b. 4 amperes of current flow
21. A closed circuit has an unbroken path for current flow. An open circuit has discontinuity or a broken path; therefore, the circuit does not provide a continuous path for current.

TEST YOURSELF

1. Drawing:

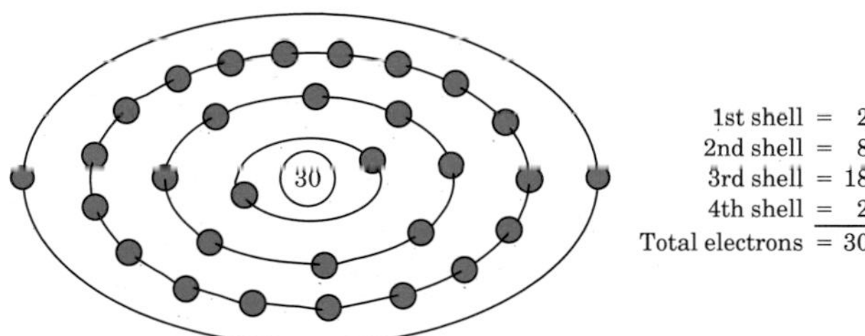

3. Zinc
5. 30
7. Circuit diagram:

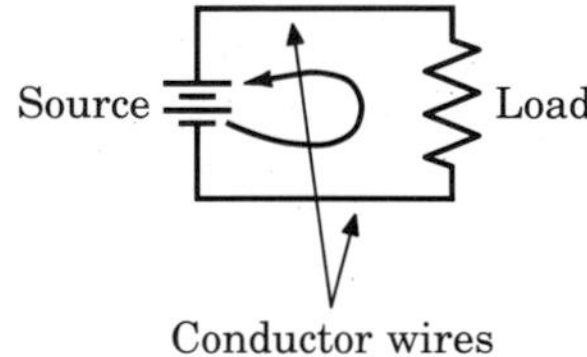

Chapter 2

PRACTICE PROBLEMS

I

1. radio transmitter
2. from oscillator to RF amplifier
3. from modulator to RF amplifier
4. from power supply to speech amplifier

II

1. 5, battery, switch, inductor, capacitor, and resistor
2. 10 k (10,000 ohms)
3. SPST (single-pole, single-throw)
4. battery

III

1. 3
2. 5
3. R_5, 25 kΩ
4. 2
5. 100 volts

IV

$$I = \frac{V}{R}$$

$$I = \frac{150\text{ V}}{10\ \Omega}$$

I = 15 amperes

V

1. P = 40 W; I = 2 amperes
2. I = 1 mA; R = 100 k ohms
3. I = 12 mA; P = 1.44 W (or 1,440 mW)

IN-PROCESS LEARNING CHECKS

I

1. Charge is represented by the letter ***Q.*** The unit of measure is the ***coulomb,*** and the abbreviation is ***C.***
2. The unit of potential difference is the ***volt.*** The symbol is ***V.***
3. The abbreviation for current is ***I.*** The unit of measure for current is the ***ampere,*** and the symbol is ***A.***
4. The abbreviation for resistance is ***R.*** Unit of measure for resistance is the ***ohm,*** and the symbol ***the Greek letter omega, Ω.***
5. Conductance is the ***ease*** with which current flows through a component or circuit. The abbreviation is ***G.*** The unit of measure for conductance is the ***siemens,*** and the symbol is ***S.***
6. 0.0000022 amperes = 2.2 microamperes (μA). Could be expressed as 2.2×10^{-6}A.
7. One-thousandth is represented by the metric prefix ***milli.*** When expressed as a power of ten, it is $\boldsymbol{10^{-3}}$.

8. 10,000 ohms might be expressed as ***10 kΩ*** (kilohms). Expressed as a whole number times a power of ten, ***10 × 10³ ohms.***

II

1. I = 1.5 amperes, or 1500 mA. ($I = \frac{V}{R}$; $I = \frac{15\ V}{10\ \Omega}$ = 1.5 A)
2. V = 30 V. (V = I × R; V = 2 mA × 15 kΩ = 30 V)
3. R = 27 kΩ.
4. a. Current ***decreases.***
   b. Voltage applied is ***higher.***
   c. Circuit resistance is ***halved.***

REVIEW QUESTIONS

1. 0.003 V
3. A micro unit is one-thousandth of a milli unit.
5. Schematic diagram of battery, two resistors, and a SPST switch:

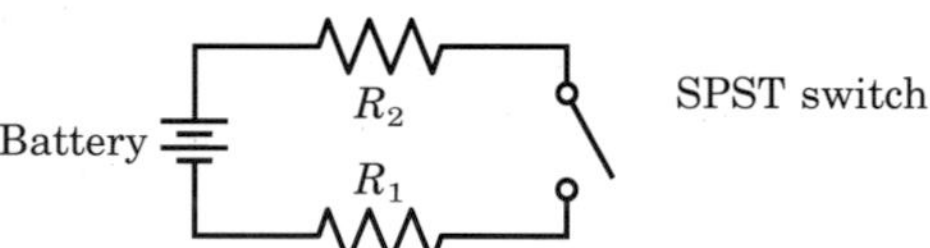

7. Resistor
9. Voltmeter
11. a. $I = \frac{V}{R}$

   $V = I \times R$

   $R = \frac{V}{I}$

   b. $P = V \times I$; $P = I^2R$; $P = \frac{V^2}{R}$
13. New current is four times original current.
15. a. Schematic diagram of V source, current meter, and resistor:

+ A − 2 mA + 30 V − I R = 15 kΩ − Power = 60 mW

   b. $R = \frac{V}{I} = \frac{30\ V}{2\ mA} = 15\ k\Omega$

   $P = V \times I = 30\ V \times 2\ mA = 60\ mW$

   c. See diagram above.
17. Carbon or composition resistors and wirewound resistors. Carbon/composition resistors are composed of carbon or graphite mixed with powdered insulating material. Wirewound resistors are constructed with special resistance wire wrapped around a ceramic or insulator core.
19. Value, power rating, tolerance, and physical size

TEST YOURSELF

1. Block diagram of a simple intercom system:

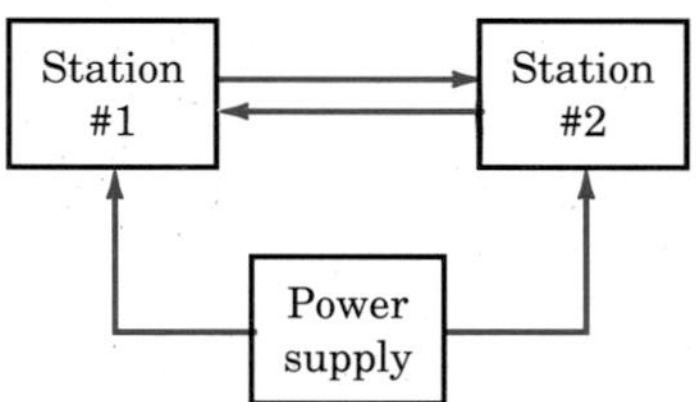

3. a. Schematic diagram of source, resistor, and current meter:

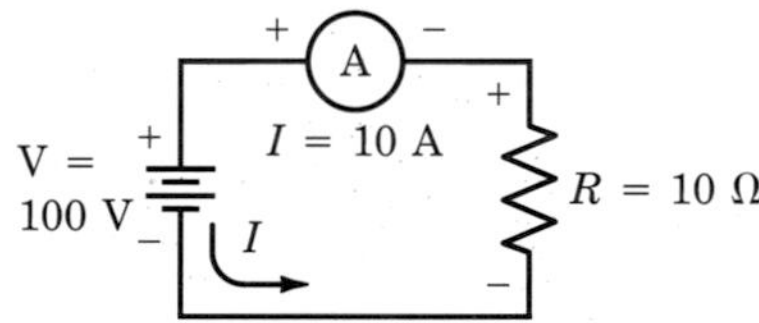

   b. $P = I^2R$

   $1{,}000\ W = I^2 \times 10\ \Omega$

   $I^2 = \frac{P}{R}$

   $I^2 = \frac{1{,}000}{10}$

   $I = \sqrt{\frac{P}{R}}$

   $I = \sqrt{\frac{1000}{10}} = \sqrt{100} = 10\ A$

   c. $V_A = I \times R$

   $= 10\ A \times 10\ \Omega = 100\ V$

   d. Labelled diagram

5. a. 0.01
   b. 12th power (10^{12})
   c. 8.570×10^4 (or 85,700)
   d. 6.17×10^3

Chapter 3

PRACTICE PROBLEMS

I

Total resistance = 74 kΩ

II

$R_1 = 47\ \Omega$; $R_T = 100\ \Omega$

III

$V_A = 411$ V; $I_T = 3$ mA; $V_{R_2} = 81$ V; $V_{R_1} = 300$ V

IV

$V_2 = 30$ V; Voltage Applied = 75 V

V

$P_1 = 90$ W; $P_2 = 90$ W; $P_3 = 90$ W; $P_T = 270$ W; $I_T = 3$ A;
$V_1 = 30$ V; $V_2 = 30$ V; $V_3 = 30$ V; $R_3 = 33.3\%$ of R_T; $P_3 = 33.3\%$ of P_T

VI

1. a. increase; b. decrease; c. decrease; d. increase; e. decrease
2. Yes

VII

1. With the 47 kΩ resistor shorted the voltage drops are:
   10 k resistor drops 14.6 volts
   27 k resistor drops 39.4 volts
   47 k resistor drops 0 volts
   100 k resistor drops 146 volts
2. They would increase.

VIII

Givens: $R_1 = 20$ kΩ; $V_1 = \frac{2}{5}V_A$; both V_2 and $V_3 = \frac{1.5}{5}V_A$; $V_A = 50$ V; $R_2 = 15$ kΩ; $R_3 = 15$ kΩ; $V_2 = 15$ V; $V_3 = 15$ V; $P_T = 50$ mW; $P_1 = 20$ mW; $P_2 = 15$ mW; $P_3 = 15$ mW; I = 1 mA

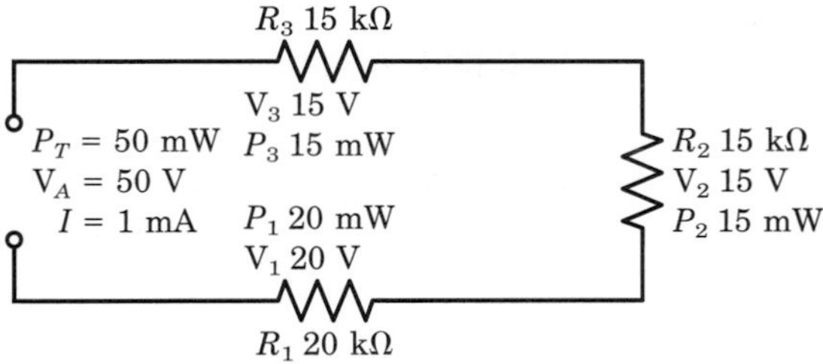

IX

D to C = 94 V; C to B = 54 V; B to A – 40 V; D to B = 148 V

X

1. Minimum power rating should be 2 times (60 V × 30 mA), or 2 × 1800 mW = 3600 mW (3.6 W).
2. R dropping equals $\frac{110 \text{ V}}{30 \text{ mA}} = 3.66$ kΩ. Minimum power rating should be 2 times (110 V × 30 mA), or 2 × 3300 mW = 6600 mW (6.6 W).

CALCULATOR CORNER

$V_T = 40$ V; $P_3 = 40$ mW; $R_1 = 5$ kΩ; $R_2 = 5$ kΩ; $R_3 = 10$ kΩ

IN-PROCESS LEARNING CHECKS

I

1. The primary identifying characteristic of a series circuit is that the ***current*** is the same throughout the circuit.
2. The total resistance in a series circuit must be greater than any one ***resistance*** in the circuit.
3. In a series circuit, the highest value voltage is dropped by the ***largest or highest*** value resistance, and the lowest value voltage is dropped by the ***smallest or lowest*** value resistance.
4. V drop by the other resistor must be ***100 V.***
5. Total resistance ***increases.*** Total current ***decreases.*** Adjacent resistor's voltage drop ***decreases.***
6. Applied voltage is ***130 V.***

REVIEW QUESTIONS

1. $R_T = \frac{10\text{ V}}{5\text{ A}} = 2\ \Omega$; each resistor = 1 Ω

3. Diagram, labeling, and calculations for a series circuit.

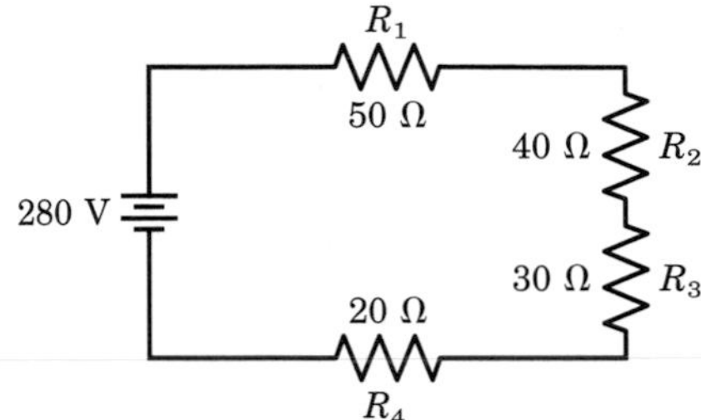

a. $R_T = 50 + 40 + 30 + 20 = 140\ \Omega$
b. V applied = I × R = 2 A × 140 Ω = 280 V
c. $V_1 = 100$ V
$V_2 = 80$ V
$V_3 = 60$ V
$V_4 = 40$ V
d. $P_T = I_T \times V_T = 2\text{ A} \times 280\text{ V} = 560$ W
e. $P_2 = 2\text{ A} \times 80\text{ V} = 160$ W
$P_4 = 2\text{ A} \times 40\text{ V} = 80$ W
f. one seventh V_A dropped by R_4
g. (1) Total resistance increases.
(2) Total current decreases.
(3) V_1 decreases; V_2 decreases; V_4 decreases.
(4) P_T decreases since I decreases and V remains the same.
h. (1) Total resistance decreases.
(2) Total current increases.
(3) V_1 increases; V_3 increases; V_4 increases (due to increased I with Rs remaining the same)

5. Diagram of three sources to acquire 60 volts, if sources are 100 V, 40 V, and 120 V.

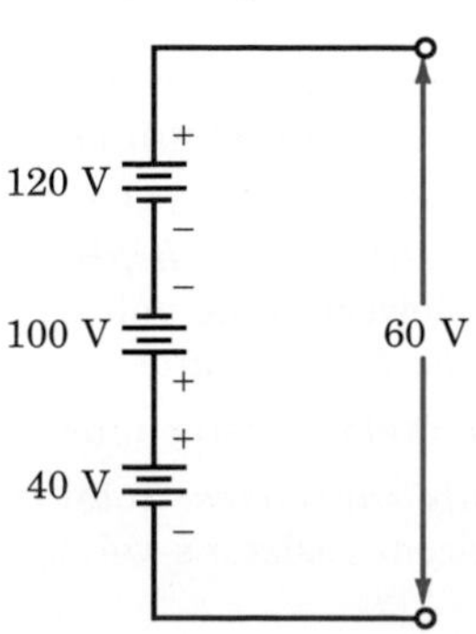

7. Use power-off testing. This prevents damage to the power supply and other components (due to abnormally high currents) and allows measuring resistances with an ohmmeter to troubleshoot.

9. d.

11. Diagram of a three-resistor voltage divider.

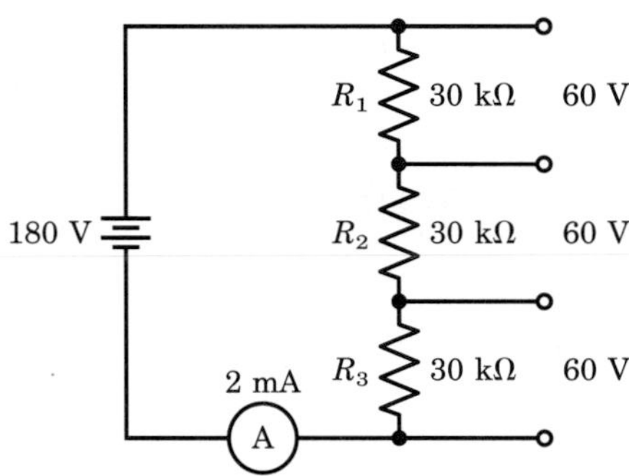

13. V applied = 125 V

15. R_1 is set at 3.81 kΩ
a. $I_T = 4.2$ mA
b. R_1 dissipates 16% of total power
c. $I_T = 4$ mA with R_1 set at middle of its range
d. Voltmeter would indicate 50 V.

17. a. $P_{R_2} = 423$ mW
b. $P_T = 819$ mW
c. $I_T = 3$ mA

TEST YOURSELF

1. Circuit diagram where $V_3 = 3\ V_1$; $V_2 = 2\ V_1$ and $V_A = 60$ V.

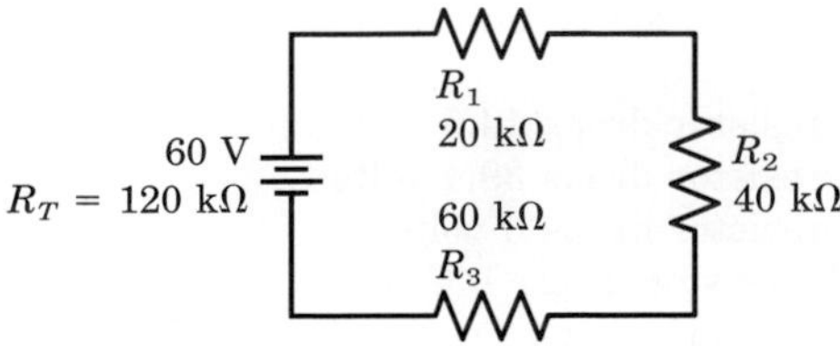

a. $R_1 = 20$ kΩ
b. $V_1 = 10$ V
$V_2 = 20$ V
$V_3 = 30$ V
c. I = 0.5 mA; $P_T = 30$ mW

Chapter 4

PRACTICE PROBLEMS

I

$V_{R_1} = V_{R_2} = 2 \text{ mA} \times 27 \text{ k}\Omega = 54 \text{ V}$
$V_A = V_{R_1} = V_{R_2} = 54 \text{ V}$

II

$I_T = I_1 + I_2 + I_3 + I_4 + I_5$
The lowest R value branch passes the most current.
The highest R value branch passes the least current.

III

$V_2 = 50$ V; $V_1 = 50$ V; $V_A = 50$ V; $I_1 = 5$ A; $I_T = 6$ A

IV

$R_1 = 15$ kΩ; $I_2 = 2.5$ mA; $I_3 = 2.5$ mA; $R_2 = 30$ kΩ

V

$V_{R_1} = 30$ V; $V_T = 30$ V; If $R_2 = 10$ kΩ, then $R_T = 1.66$ kΩ

VI

$R_T = 5.35$ kΩ

VII

Total resistance = 25 Ω

VIII

$R_T = 12$ Ω

IX

$R_T = 28.6$ Ω

X

$R_2 = 100$ Ω

XI

$P_T = 2250$ mW or 2.25 W

XII

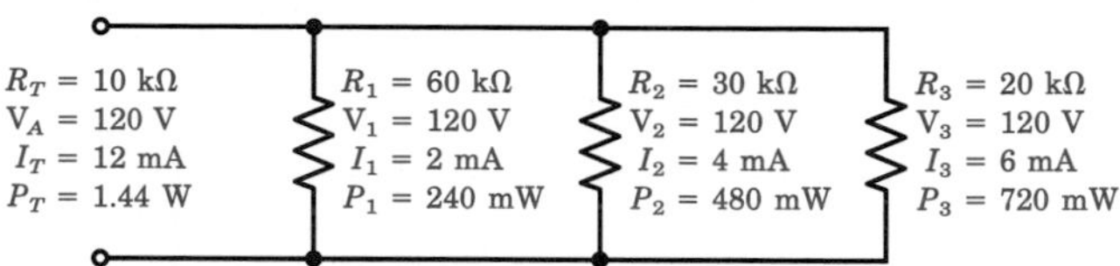

XIII

$I_1 = 1.85$ mA (through 27 kΩ resistor)
$I_2 = 1.07$ mA (through 47 kΩ resistor)
$I_3 = 0.50$ mA (through 100 kΩ resistor)

XIV

$I_1 = 1.65$ mA
$I_2 - 0.35$ mA

REVIEW QUESTIONS

1. $R_T = 17.15$ kΩ; product-over-the-sum method
3. $I_2 = 30$ mA; Kirchhoff's current law
5. a. I_1 will RTS
   I_2 will D
   I_3 will RTS
   I_T will D
   P_T will D
   V_1 will RTS
   V_T will RTS

   b. I_1 will D
   I_2 will D
   I_3 will I
   I_T will I
7. $I_1 = 12$ mA
   $I_2 = 6$ mA (since I branch is inverse to R branch).
9. $P_{R_1} = 0.5 \text{ A} \times 50 \text{ V} = 25 \text{ W}$
11. a. $M_1 = 2.13$ mA
    b. $M_2 = 37$ mA
    c. $M_3 = 39.13$ mA
    d. $M_4 = 10$ mA
    e. $M_5 = 49.13$ mA
    f. $P_{R_1} = 213$ mW
    g. $P_{R_2} = 3.7$ W
    h. $P_{R_3} = 1$ W
    i. $P_T = 4.91$ W
    j. New $P_T = 1.21$ W
13. $R_X = 31.59$ kΩ
15. a. R_4
    b. $P_{R_4} = 71.7$ mW

TEST YOURSELF

1. Circuit diagram of a three-branch parallel circuit.

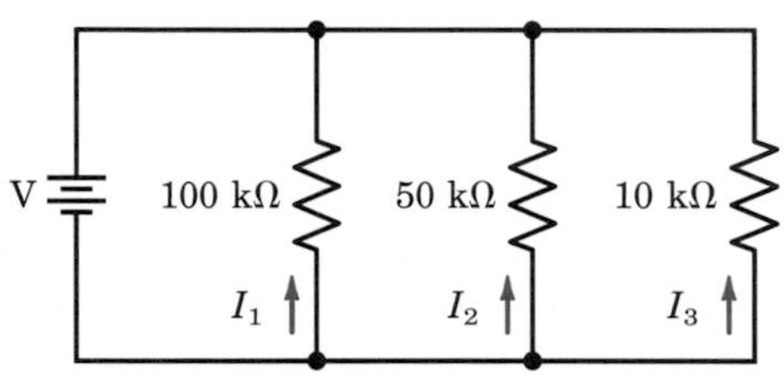

V

1. V_{RL} = 150 V, P_{RL} = 300 mW
2. R_L = 75 kΩ
3. R_2 = 50 kΩ
4. The change, as specified, means R_2 has decreased.
5. The change, as specified, means R_1 has increased.
6. V_{R_2} equals 166.66 V.
7. If R_1 opens, V_{RL} drops to zero volts.
   If R_2 shorts, V_{RL} drops to zero volts.

Chapter 5

PRACTICE PROBLEMS

I

1. The only component carrying total current is R_7.
2. The only single components in parallel with each other are R_2 and R_3.
3. The only single components that pass the same current are R_1, R_4 and R_5.

II

I_T = 2 A; I_1 = 1.33 A; I_2 = 0.67 A; I_3 = 0.33 A

III

voltage across: R_1 = 30 V; R_2 = 100 V; R_3 = 25 V; R_4 = 50 V; R_5 = 50 V; R_6 = 25 V.
current through: R_1 = 3 A; R_2 = 0.5 A; R_3 = 2.5 A; R_4 = 0.5 A; R_5 = 2 A; and R_6 = 2.5 A.

IV

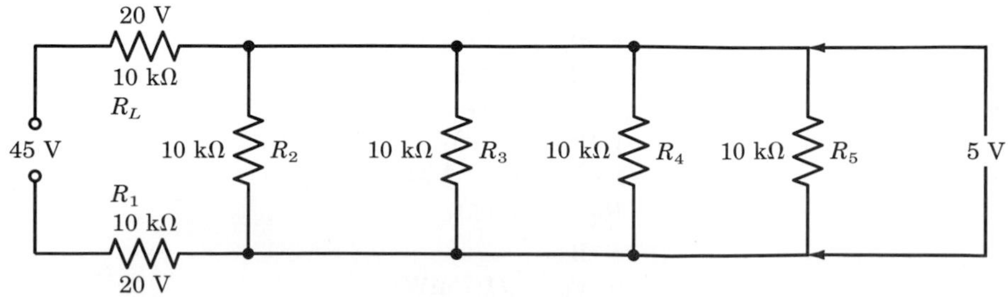

IN-PROCESS LEARNING CHECKS

I

For Figure 5–10a: R_T = 10 Ω
For Figure 5–10b: R_T = 120 Ω

II

I_1 = 0.4 mA; I_T = 1 mA; V_T = 85 V; R_T = 85 kΩ; P_3 = 36 mW; and the 100 kΩ resistor (R_3) is dissipating the most power.

III

Figure 5–19:
The defective component is R_5. It has drastically increased in value from 3.3 kΩ to 17.3 kΩ.

Figure 5–20:
(NOTE: Total circuit current *should be* 2 mA, meaning R_3 should drop 2 mA × 47 kΩ volts, or 94 volts. Instead it is dropping 121 − 17.6 = 103.4 volts, which is higher than it should be.)
Possible bad components could include R_1 or R_2 decreased in value, or R_3 increased in value.

Figure 5–21:
If R_2 increases in value, V_{R_1} increases and V_{R_3} decreases. (This is due to total circuit resistance increasing that decreases the total current through R_3, resulting in R_3 dropping less voltage. Kirchhoff's voltage law would indicate, therefore, that the voltage across parallel resistors R_1 and R_2 have to increase so the loop voltages equal V applied.)

ANSWERS TO SPECIAL "THINKING EXERCISE"

If R_2 opened:
V_T remains the same.
V_1 decreases due to lower total current through R_1.
V_2 increases because V_T stays the same, and V_1 and V_4 decrease due to lower total current through them.
V_3 increases for the same reason V_2 increases.
V_4 decreases due to lower I_T caused by higher R_T.
I_T decreases due to higher R_T with same applied voltage.
I through R_1 decreases due to higher circuit total R.
I through R_2 decreases to zero due to its opening.
I through R_3 increases due to higher V across it.
I through R_4 decreases due to lower I_T through it.
P_T decreases due to lower I_T with same V_T.
P_3 increases due to higher I through same R value.
P_4 decreases due to lower current through same R value.

REVIEW QUESTIONS

1. $R_T = 22\ \Omega$
3. R_T will D
   R_4 will RTS
   V_1 will I
   V_3 will D
   P_T will I
5. Minimum voltage is 13 V.
   Maximum voltage is 18 V.
7. $R_4 = 6\ k\Omega$ (so that branches voltage equals 80 V). $I_2 = 10$ mA (since $I_T = 20$ mA and I_1 can be calculated as 10 mA, due to V across the 8 kΩ resistor having to be 80 V. That is, the 100 V applied minus the 20 V drop across the 1 kΩ resistor that has I_T passing throught it).
9. Light gets dimmer, due to the additional current that passes through the resistor in series with the source. This drops the voltage across the remaining sections of the circuit.
11. 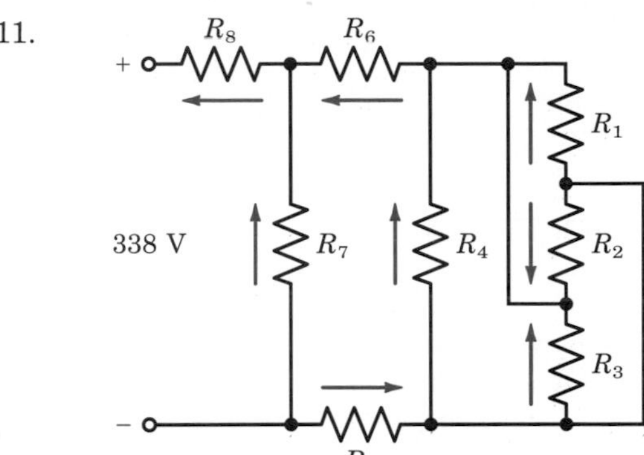

   a. $R_T = 169\ k\Omega$
   b. $I_T = 2$ mA
   c. $V_{R_1} = 14$ V
   d. $V_{R_2} = 14$ V
   e. $V_{R_3} = 14$ V
   f. $V_{R_4} = 14$ V
   g. $V_{R_5} = 62$ V
   h. $V_{R_6} = 62$ V
   i. $V_{R_7} = 138$ V
   j. $V_{R_8} = 200$ V

TEST YOURSELF

1. $R_1 = 0.5\ k\Omega$
   $R_2 = 0.5\ k\Omega$
   $R_3 = 0.75\ k\Omega$
   $R_4 = 6\ k\Omega$
   (NOTE: Current through R_4 = 10 mA; through R_3 = 20 mA; through R_2 = 30 mA; through R_1 = 60 mA. R_1 drops 30 V with 60 mA through it, therefore = 0.5 kΩ.
   R_2 drops 15 V with 30 mA through it, therefore = 0.5 kΩ; R_3 drops 15 V with 20 mA through it, therefore = 0.75 kΩ; and R_4 drops 60 V with 10 mA through it, therefore = 6 kΩ.)

Chapter 6

PRACTICE PROBLEMS

I

I = 4 A
$V_L = 16$ V

P_T = 80 W
P_L = 64 W
Efficiency = 80%

II

1. I through R_1 = 0 mA
   V across R_1 with respect to point A = 0 V
   I through R_2 = 7.5 mA
   V across R_2 with respect to point A = +75 V
   I through R_3 = 7.5 mA
   V across R_3 with respect to point A = −75 V
2. I through R_1 = 10 mA
   V across R_1 with respect to point A = +100 V
   I through R_2 = 12.5 mA
   V across R_2 with respect to point A = −125 V
   I through R_3 = 2.5 mA
   V across R_3 with respect to point A = +25 V

III

1. If R_L = 175 Ω, I_L = 0.25 A and V_L = 43.75 V (Figure 6–8)
2. If I_L = 125 mA, then R_L = 375 Ω (Figure 6–8)
3. In Figure 6–9, R_{TH} = 4 kΩ and V_{TH} = 50 V. If R_L = 16 kΩ, then V_L = 40 V and I_L = 2.5 mA

IV

If R_L is 60 Ω, I_L = 0.588 A and V_L = 35.28 V

V

1. If R_L = 100 Ω in Figure 6–14A: I_L = 0.4 A; V_L = 40 V
2. If R_L changes from 25 Ω to 50 Ω, I_L decreases (from 1 A to 0.66 A), and V_L increases (from 25 V to 33.5 V).

IN-PROCESS LEARNING CHECKS

I

1. For maximum power transfer to occur between source and load, the load resistance should be ***equal to*** the source resistance.
2. The higher the efficiency of power transfer from source to load, the ***greater*** the percentage of total power is dissipated by the load.
3. Maximum power transfer occurs at ***50%*** efficiency.
4. If the load resistance is less than the source resistance, efficiency is ***smaller than*** the efficiency at maximum power transfer.
5 To analyze a circuit having two sources, the superposition theorem indicates that Ohm's law ***can*** be used.
6. What is the key method in using the superposition theorem to analyze a circuit with more than one source? ***Noting the direction of current flow and polarity of voltage drops.***
7. Using the superposition theorem, if the sources are considered "voltage" sources, are these sources considered shorted or opened during the analysis process? ***Shorted.***
8. When using the superposition theorem when determining the final result of your analysis, the calculated parameters are combined, or superimposed, ***algebraically.***

REVIEW QUESTIONS

1. Matching the impedance of an antenna system to the impedance of a transmitter output stage, and matching the output impedance of a audio amplifier to the impedance of a speaker.
3. V_{R_1} = 20 V
   V_{R_2} = 20 V
5. V_{TH} = 30 V
   R_{TH} = 10 kΩ
   I_L = 0.857 mA
   V_L = 21.43 V
7.

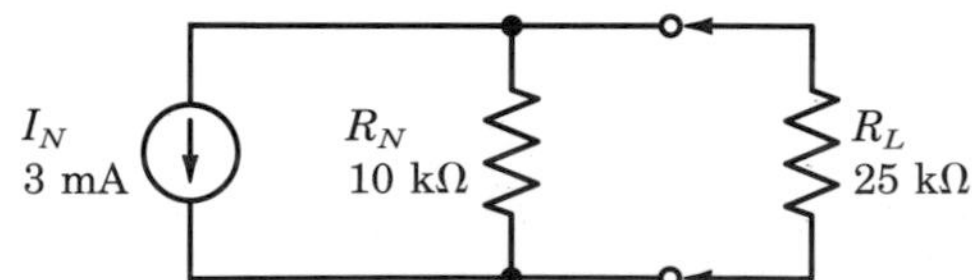

9. To "Thevenize" a circuit:
   1) open or remove R_L.
   2) determine open circuit V at points where R_L is to be connected. (This value is V_{TH}.)
   3) determine resistance looking toward source from R_L connection points, assuming the source is shorted.
   4) draw the Thevenin equivalent circuit with V_{TH} as source and R_{TH} in series with source.

5) calculate I_L and V_L for given value(s) of R_L using two-resistor series circuit analysis techniques.

11. Decreased
13. $R_{TH} = 30\ \Omega$; $V_{TH} = 36\ V$
15. $R_{TH} = 10\ \Omega$; $V_{TH} = I_{TH} \times R_{TH} = 3\ A \times 10\ \Omega = 30\ V$
17. In circuits with multiple sources
19. Norton's Theorem

TEST YOURSELF

1. Student practices using immediate mode BASIC to solve problems.

Chapter 7

PRACTICE PROBLEMS

I

1. $A = 1.5\ V - \frac{1.25\ V}{0.1\ A} = \frac{0.25\ V}{0.1\ A} = 2.5\ \Omega$
2. Since we defined useful life voltage between 1.4 V and 1.6 V, this cell is *bad.*

II

1. It will deliver 7.5 amperes for 20 hours $\left(\frac{150\ \text{ampere-hours}}{20\ \text{time in hours}}\right)$.
2. It will deliver 18.75 amperes for 8 hours $\left(\frac{150\ \text{ampere-hours}}{8\ \text{time in hours}}\right)$.

IN-PROCESS LEARNING CHECKS

I

1. The carbon-zinc type dry cell is composed of a negative electrode made of ***zinc*** and a positive electrode made of ***carbon.*** The electrolyte solution in this type cell is ***ammonium chloride,*** called sal ***ammoniac.***
2. The process which causes hydrogen bubbles to form on the positive electrode, and which is detrimental to the cell's operation is called ***polarization.***
3. Manganese dioxide is often added to the cell to prevent the process mentioned in question 2. This agent is sometimes known as a ***depolarizer*** agent.
4. Another problem that can occur in dry cells is that of little batteries being formed due to impurities suspended in the cell's electrolyte near the zinc electrode. This type activity is known as ***local action.*** To reduce this undesired activity, mercury is put on the zinc by a process called ***amalgamation.***
5. The shelf life of dry cells is ***hurt*** by storing the cells in hot conditions.
6. Nominal cell terminal voltage of dry cells is about ***1.5*** V.

II

1. The lead-acid battery comprises ***secondary*** cells which ***are*** rechargeable.
2. Lead peroxide makes up the ***positive*** plates of a lead-acid battery. Spongy lead makes up the ***negative*** plates of a lead-acid battery. The electrolyte of a lead-acid battery is dilute ***sulphuric acid.***
3. The normal closed-circuit voltage per cell in a lead-acid battery is approximately ***two*** volts.
4. Does the electrolyte in a lead-acid battery become more or less dilute as the battery discharges? ***more dilute.***
5. Does the specific gravity of the electrolyte in a lead-acid battery increase or decrease as the battery becomes more discharged? ***decrease.*** A fully discharged battery has a specific gravity of approximately ***1.150 or less.***
6. What method is used to rate the current delivering capabilities of a lead-acid battery? ***The ampere-hour rating.***
7. Why are sparks and/or flame dangerous near a charging battery? ***Because battery releases volatile hydrogen gas while being charged.***
8. Two types of tests that can be used to check the condition of a lead-acid battery are: a ***hydrometer (specific gravity)*** test and the short-term ***heavy load*** test.

REVIEW QUESTIONS

1. Positive ion is an atom that has lost electron(s).
3. Amalgamation is the process of coating the inside of the zinc electrode with mercury in a carbon-zinc cell in order to minimize local action.
5. Manganese dioxide
7. Mercury cell
9. 100 ampere-hours

TEST YOURSELF

1. The source's positive terminal is connected to the battery's positive terminal and the source's negative terminal is connected to the battery's negative terminal. This reverses the chemical action in the battery and recharges it.
3. a. Ni-cad cells are in the range of 110–1200 mA per hr.
   b. Alkaline cells are in the range of 150–750 mA (for 10 hrs).
5. AAA (carbon-zinc): V = 1.5 V; Current = not given in catalog; mA/hr = not given in catalog
   AA (carbon-zinc): V = 1.5 V; Current = not given in catalog; mA/hr = not given in catalog
   A (carbon-zinc): not a standard type
   C (carbon-zinc): V = 1.5 V; Current = not given in catalog; mA/hr = not given in catalog
   D (carbon-zinc): V = 1.5 V; Current = not given in catalog; mA/hr = not given in catalog
7. Current delivering capability primarily determined by size of active area of the electrodes.
9. Some applications of lithium cells are in computers, cameras, watches and calculators.

Chapter 8

PRACTICE PROBLEMS

I

II

Figure 8–13a = Move away from each other.
Figure 8–13b = Move toward one another.
Figure 8–13c = Move away from each other.

III

Figure 8–17a = N at right end, S at left end; Figure 8–17b = N at left end, S at right end; and Figure 8–17c = N at left end, S at right end.

IN-PROCESS LEARNING CHECKS

I

1. Small magnets that are suspended and free to move align in a ***North*** and ***South*** direction.
2. Materials that lose their magnetism after the magnetizing force is removed are called ***temporary magnets.*** Materials that retain their magnetism after the magnetizing force is removed are called ***permanent magnets.***
3. A wire carrying dc current ***does*** establish a magnetic field. Once a magnetic field is established, if the current level doesn't change, the field is ***stationary.***
4. A law of magnetism is like poles ***repel*** each other and unlike poles ***attract*** each other.
5. A maxwell represents ***one*** line of force.
6. A weber represents 10^8 lines of force.
7. Lines of force are ***continuous.***
8. Lines of force related to magnets exit the ***North*** pole of a magnet and enter the ***South*** pole.
9. Nonmagnetic materials do ***not*** stop the flow of magnetic flux lines through themselves.
10. Yes, magnetism ***can*** be induced from one magnetic object to another object, if it is a magnetic material.

II

1. For a given coil dimension and core material, what two factors primarily affect the strength of an electromagnet? ***Number of turns*** and ***amount of current.***
2. The left-hand rule for determining the polarity of electromagnetics states that when the fingers of your left hand point ***in the same direction*** as the current passing through the coil, the thumb points toward the ***North*** pole of the electromagnet.
3. Adjacent current-carrying conductors which are carrying current in the same direction tend to ***attract each other.***

4. If you grasp a current-carrying conductor so that the thumb of your left hand is in the direction of current through the conductor, the fingers "curled around the conductor" ***will indicate*** the direction of the magnetic field around the conductor.
5. When representing an end view of a current conductor pictorially, it is common to show current coming out of the paper via a ***dot.***

III

1. A "B-H" curve is also known as a ***magnetization*** curve.
2. B stands for flux ***density.***
3. H stands for magnetizing ***intensity.***
4. The point where increasing current through a coil causes no further significant increase in flux density is called ***saturation.***
5. The larger the area inside a "hysteresis loop" the ***larger*** the magnetic losses represented.

REVIEW QUESTIONS

1. Magnetism is the property causing certain materials to develop attraction or repulsion when the molecular structure has certain alignments. Also, magnetism causes induced voltage or current into conductors when there is relative motion between the magnet and the conductor. (Magnetism is the property associated with the setting up of a magnetic field around a current-carrying conductor, called electromagnetism).
FLUX = Magnetic lines of force. *Field* = Area of magnetic influence of a magnet. *Pole* = Either the N-seeking or S-seeking end of a magnet.
Magnetic Polarity = Identification of the N-seeking or S-seeking ends of a magnet; relative direction of a magnet's flux.
3. Like poles repel and unlike poles attract.
5. North and South poles are indicated.

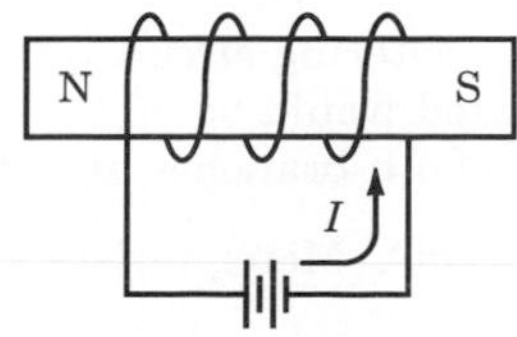

7. Permeability
9. Using SI units and terms:
Flux – Weber (lines/10^8)
Flux Density – Webers/sq meter (Teslas or T)
11. Permeability – μ (mu) Teslas/(AT/meter)
Relative Permeability – μ_r (flux w/core material/flux w/vacuum core)
13. Magnetomotive force is the force by which a magnetic field is produced and relates to the ampere-turns involved in an electromagnet. Magnetic field intensity adds the dimension of ampere-turns per a given length of the electromagnetic coil. In the SI system, field intensity is AT/meter.
15. Reluctance will triple.
17. Flux density will increase.
19. A magnetic shield effectively "shunts" the magnetic flux around the object being shielded due to the shield's providing an easy path for flux through itself.

TEST YOURSELF

1. Hysteresis loop showing saturation, residual magnetism, and coercive force areas is shown.

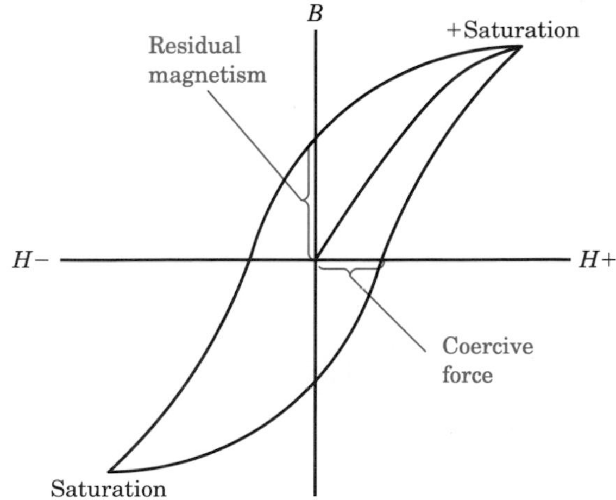

3. a. $\frac{10^8}{10^6} = 100$ lines of flux
b. Teslas = webers per square meter. Since there are 100 lines in 0.005 m^2, there are $\frac{1}{0.005} \times$ 100 lines in a square meter, or lines = 200 ×

100 = 20,000 number of webers per square meter (teslas) = $\frac{20{,}000}{100{,}000{,}000}$ = 0.0002 T (or 0.2 mT or 200 μT)

Chapter 9

PRACTICE PROBLEMS

I

$$R_s = \frac{V_m}{I_s}$$

$$R_s = \frac{100 \text{ mV}}{99 \text{ mA}}$$

$$R_s = 1.01\ \Omega$$

II

$$R_{mult} = \frac{V\ (\text{range})}{I\ (\text{fs})} - R_m$$

$$R_{mult} = \frac{250}{50 \times 10^{-6}} - 2{,}000$$

$$R_{mult} = 5{,}000{,}000 - 2{,}000$$

$$R_{mult} = 4{,}998{,}000 \text{ or } 4.998\ \text{M}\Omega$$

III

1. Ohms-per-volt rating of a 20-microampere movement is $\frac{1}{\text{fs}} = \frac{1}{20 \times 10^{-6}}$ = 50,000 ohms per volt.
2. Full-scale current rating of a meter rated at 2,000-ohms per volt is $\frac{1}{2{,}000}$ = 0.5 mA movement.

IN-PROCESS LEARNING CHECKS

I

1. Three essential elements in an analog meter movement include a ***moving*** element, an ***indicating*** means, and some form of ***damping.***
2. The purpose of the permanent magnet in a d'Arsonval meter movement is to establish a fixed ***magnetic field.***
3. The movable element in a moving coil meter moves due to the interaction of the fixed ***magnetic field*** and the magnetic field of the movable coil when current passes through the coil.
4. Electrical conduction of current in a movable meter movement is accomplished through the ***springs,*** which also provides a desired ***mechanical*** force against which the movable coil must act.
5. The element in a moving-coil movement producing electrical damping is the ***coil frame.***

REVIEW QUESTIONS

1. Full-scale current rating = 2 mA

   R meter = $\frac{200 \text{ mV}}{2 \text{ mA}}$ = 100 Ω

3. A shorting-type switch is where the movable contact contacts the next contact position before leaving the previous contact position. This switch is frequently used in multiple-range current meters to assure the meter is never unshunted and therefore has some measure of protection from overcurrent.
5. Circuit for measuring specified currents is shown.

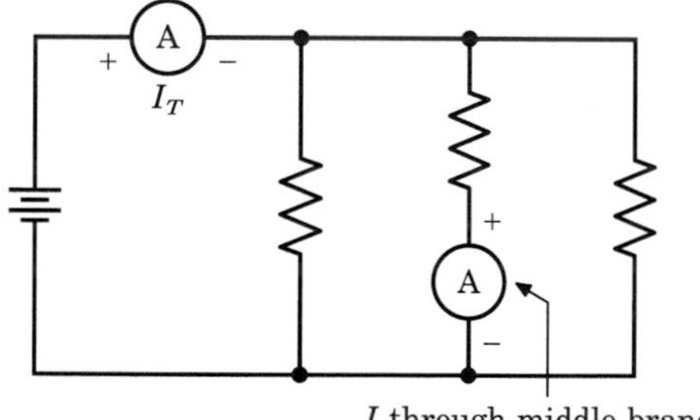

7. Ohms-per-volt sensitivity = $\frac{1}{(\text{fs})} = \frac{1}{40\ \mu\text{A}} = \frac{25\ \text{k}\Omega}{\text{V}}$

   $R_m + R_{mult}$ = 500 × 25 kΩ = 12,500 kΩ, or 12.5 MΩ
9. The resulting metering circuit resistance for the conditions stated would be 50 Ω. (Shunt R = meter R since both branches are to carry 1 mA for a total of 2 mA. Meter resistance = $\frac{V_m}{I_m}$ or $\frac{100 \text{ mV}}{1 \text{ mA}}$ = 100 Ω.)

11. A voltmeter should have a high resistance to prevent appreciably changing circuit conditions when it is connected in parallel with the portion of the circuit where the measurement is being made.
13. Midscale = 1500 Ω
    Quarter scale = 4500 Ω
15. With voltmeter checks, it is not necessary to break the circuit and insert the meter in series, as is required for direct current meter measurements.
17. Damping is the designed prevention of excess mechanical pointer oscillation above and below an indicated meter reading in a meter movement.
19. It loads the circuit under test much less, thus giving a truer picture of the circuit operation; it is easier to read without error.

TEST YOURSELF

1. Diagram of a movement with Ayrton shunt is shown.

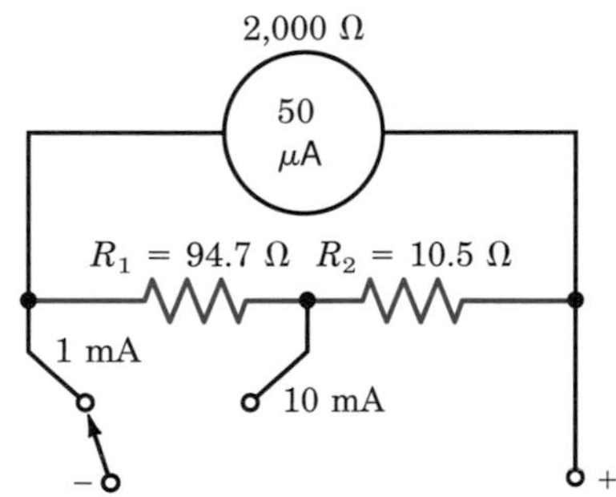

In 10 mA position:

$I_m = 50\ \mu A$

$I_s = 9{,}950\ \mu A$

$50\ \mu A \times (R_1 + R_m) = 9{,}950\ \mu A \times (105.26 - R_1)$

$50\ R_1 + 100{,}000 = 1{,}047{,}337 - 9{,}950\ R_1$

$10{,}000\ R_1 = 947{,}337$

$R_1 = 94.7337\ \Omega$

In 1 mA position:

$I_s = 1\ mA - 50\ \mu A = 950\ \mu A$

$$I_m = \frac{50}{950} = \frac{1}{19} \text{ of } I_T$$

$$I_s = \frac{18}{19} \text{ of } I_T$$

$0.1\ V = 950\ \mu A \times R_s$

$$R_s = \frac{0.1}{950 \times 10^{-6}} = 105.26\ \Omega$$

$R_1 + R_2 = 105.26\ \Omega$

$R_2 = 105.26 - 94.7337 = 10.5263\ \Omega$

3. 49.75 V

Chapter 10

PRACTICE PROBLEMS

I

1. The sine of 35° = 0.5735.
2. Cycle 1 ends and cycle 2 begins at point G.
3. Maximum rate of change for the waveform shown occurs at points J, N, R, and T.

II

1.

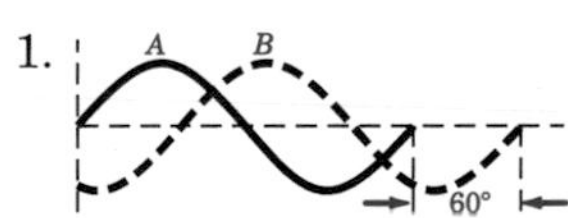

2.

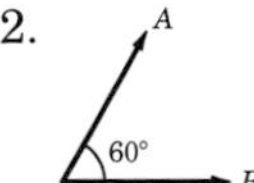

III

1. 10 V
2. 9 V
3. 90%
4. No
5. Yes
6. a. 120.8 V
   b. 171 V
   c. 120.8 V
   d. 171 V
   e. 108.9 V

IV

1. a. $V_1 = 50$ V
   b. $V_2 = 50$ V
   c. $I_{pk} = 7.07$ mA
   d. $P_{pk} = 999.7$ mW
   e. $P_1 = 250$ mW

2. a. $I_1 = 10$ mA
   b. $I_2 = 10$ mA
   c. $I_{pk} = 28.28$ mA
   d. $V_{pk\text{-}pk} = 282.8$ V
3. a. 66.7 V
   b. 33.3 V
   c. 33.3 V
   d. 443.55 mW

IN-PROCESS LEARNING CHECKS

I

1. The basic difference is that dc (direct current) is in one direction and of one polarity. Alternating current (ac) periodically changes direction and polarity.
2. The reference, or zero degree position is horizontally to the right when describing angular motion.
3. The y-axis is the vertical axis in the coordinate system.
4. The second quadrant in the coordinate system is that quadrant between 90 and 180 degrees.
5. A vector represents a given quantity's magnitude and direction with respect to location in space.
6. A phasor represents a rotating vector's relative position with respect to time.

II

1. 10-kHz
2. 0.0025 seconds, or 2.5 msec
3. 50 μsec
4. Amplitude
5. 45° and 135°; 225° and 315°
6. 1 and 2
7. 0.067 μsec
8. 40-Hz
9. As frequency increases, T ***decreases.***
10. The longer a given signal's period, the ***longer*** the time for each alternation.

REVIEW QUESTIONS

1. Differences between ac and dc:
   a. dc is of one polarity and ac changes polarity periodically
   b. dc is of one amplitude and ac is constantly changing amplitude
   c. dc is unidirectional and ac is bidirectional.
3. $f = \frac{1}{t} = \frac{1}{0.2\ \mu\text{sec}} = 5$ MHz
5. A 10 V rms, 1 MHz ac signal is illustrated as follows.

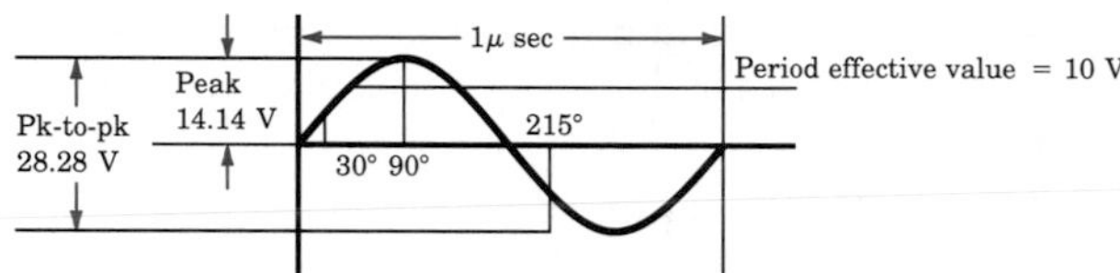

7. 400 Hz.
9. Pk-to-pk voltage is approximately 63 volts. $(P = \frac{V^2}{R}$; $50\text{ mW} = \frac{V^2}{10\text{ k}\Omega}$; $500 = V^2$; $22.36 = V$ rms; rms $\times 2.828 =$ pk-to-pk V; $22.36 \times 2.828 = 63.24$ V)
11. If the rms value of an ac voltage doubles, the peak-to-peak value must double.
13. Frequency has been reduced to one-third the original frequency.
15. Time is one half the time of a period, where $T = \frac{1}{f} = \frac{1}{500} = 0.002$ seconds. Therefore, one alternation equals 0.001 seconds, or 1 millisecond.

TEST YOURSELF

1. Labeled waveform diagram showing angles and amplitudes is shown. (NOTE: Cosine wave is leading the sine wave by 90°.)

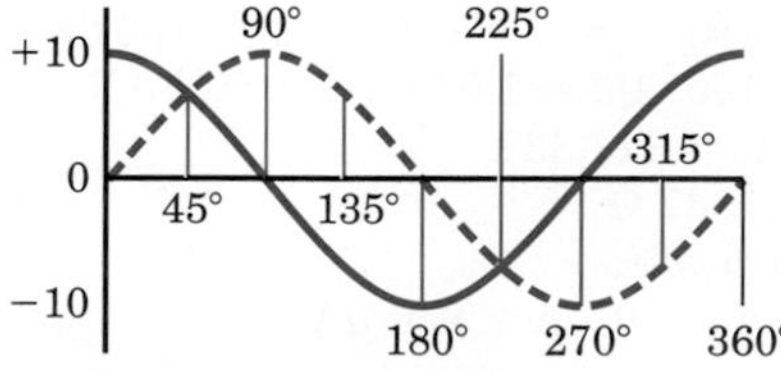

   a. see diagram
   b. sine wave at 45° = +7.07 units
      sine wave at 90° = +10 units

c. cosine wave at 0° = +10 units
   cosine wave at 45° = +7.07 units
   cosine wave at 90° = 0 units
d. cosine leading sine by 90°

3. Diagram of the three-resistor series-parallel circuit is shown.

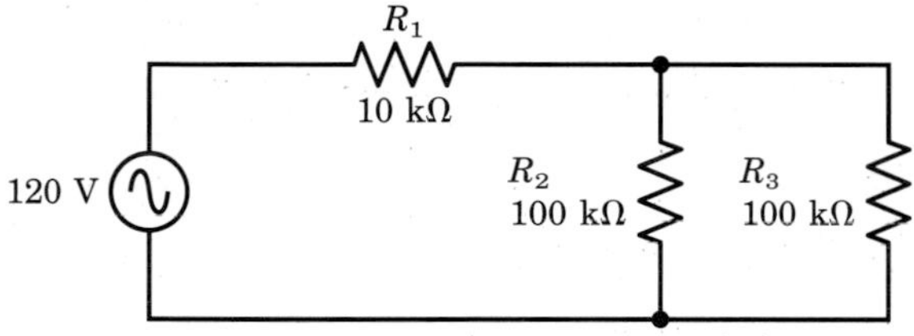

a. $V_{R_1} = 20$ V
b. $I_{R_1} = 2$ mA
c. $P_{R_1} = 40$ mW
d. $I_{R_2 = 1 \text{ mA}}$
e. $V_3 = 100$ V
f. $\theta = 0°$

Chapter 11

PRACTICE PROBLEMS

I

When the vertical frequency is four times the horizontal frequency, the waveform shows four cycles of the given waveform.

II

3.5 V (rms: 5 V peak voltage)

III

72 degrees per major division

IN-PROCESS LEARNING CHECKS

I

1. The part of the oscilloscope producing the visual display is the ***cathode-ray tube.***
2. The scope control that influences the brightness of the display is the ***intensity*** control.
3. A waveform is moved up or down on the screen by using the ***vertical position*** control.
4. A waveform is moved left or right on the screen by using the ***horizontal position*** control.
5. For a signal fed to the scope's vertical input, the controls that adjust the number of cycles seen on the screen are the horizontal ***time/cm*** and the horizontal ***time variable*** controls.
6. The control(s) that help keep the waveform from moving or jiggling on the display are associated with the ***synchronization*** circuitry.

REVIEW QUESTIONS

1. Name is cathode-ray tube and abbreviation is CRT.
3. To protect and add life to coating on CRT screen
5. vertical input terminal(s)
7. "Ext Trig" terminal
9. vertical (or y-axis)
11. a. 30 V pk-pk
    b. T = 0.004 s
    c. horizontal time/cm and/or time variable
    d. 0.0004 cm/s
    e. rms = 10.6 V; pk = 15 V; pk-pk = 30 V

TEST YOURSELF

1. Diagram B: vertical position and horizontal gain controls
   Diagram C: vertical position and horizontal gain controls
   Diagram D: vertical gain and horizontal position controls
   Diagram E: horizontal gain control
   Diagram F: vertical gain control
   Diagram G: horizontal sweep frequency controls (horizontal time/cm and time variable)
   Diagram H: vertical gain control
   Diagram I: vertical and horizontal gain controls
   Diagram J: horizontal sweep frequency controls (horizontal time/cm and time variable)
   Diagram K: horizontal gain control
   Diagram L: vertical position, vertical and horizontal gain controls
   Diagram M: vertical gain control

Chapter 12

PRACTICE PROBLEMS

I

Since 1 weber $= 10^8$ flux lines and cutting $\frac{1 \text{ Wb}}{\text{sec}}$ induces 1 V, then cutting 2×10^8 flux lines (2 Wb) in 0.5 sec = 4 volts

II

$$V_L = -L \frac{di}{dt}$$

$$V_L = -5 \left(\frac{3}{1}\right) = -15 \text{ V}$$

III

10 mH + 6 mH + 100 μH = 16.1 mH total inductance

IV

Three equal value inductors (15 H) in parallel have a total equivalent inductance $= \frac{1}{3}$ of one of them. L_T = 5 henrys.

V

a. dc current through coil = 5 A

b. energy $= \frac{LI^2}{2} \ldots = \frac{1(25)}{2} = 12.5$ joules

VI

Two seconds = approximately 2.4 TC.
After 2.4 TC, V_R is about 91% of V applied, or 9.1 volts.
After 2.4 TC, V_L is about 9% of V applied, or 0.9 volts.

IN-PROCESS LEARNING CHECKS

I

1. All other factors remaining the same, when an inductor's number of turns increases four times, the inductance value ***increases*** by a factor of ***sixteen times.***
2. All other factors remaining the same, when an inductor's diameter triples, the inductance value ***increases*** by a factor of ***nine times.*** (NOTE: A is directly proportional to d^2.)
3. All other factors remaining the same, when an inductor's length increases, the inductance value ***decreases.***
4. When the core of an air-core inductor is replaced by material having a permeability of ten, the inductance value ***increases*** by a factor of ***ten times.***

REVIEW QUESTIONS

1. Inductance (or self-inductance) is the property of a circuit that opposes a change in current flow.
3. a. number of turns on the coil
   b. length of the coil
   c. cross-sectional area of the coil
   d. relative permeability of the core material
5. Amount of emf induced in a circuit is related to the number of magnetic flux lines cut or linked per unit time.
7. Inductance will decrease.
9. Total inductance will increase.
11. Inductance will increase to 400 mH, (L varies as the square of the turns).
    Expressed in μH = 400,000 μH, or 0.4 henrys.
13. New inductance is $500 \times 100\ \mu$H = 0.05 H.
15. $L_T = L_1 + L_2 + L_3 = 0.1 \text{ H} + 0.2 \text{ H} + 0.001 \text{ H} =$ 0.301 H, or 301 mH
17. $1 \text{ TC} = \frac{L}{R} = 250 \times \frac{10^{-3}}{100} = 0.0025$ sec, or 2.5 ms.
    It will take 5 times 2.5 ms, or 12.5 ms for current to complete change from one level to another.
19. Tapping a coil at midpoint yields an inductance of one-fourth the value of the total inductance of the coil. In the case of the 0.4 H coil, L equals about 0.1 H at the 250 turn tap.

TEST YOURSELF

1. Since the inductance change is proportional to the square of the turns, $\frac{L_1}{L_2} = \frac{N_1^2}{N_2^2}$; $\frac{4}{6} = \frac{1000^2}{N_2^2}$; $N_2 =$ 1224.74 turns.

3. Because the voltage-per-turn is less than the varnish insulation breakdown voltage. (The voltage per turn in this case is approximately 2 V per turn.)

Chapter 13

PRACTICE PROBLEMS

I

1. 3 kΩ
2. 333.3 Ω

II

1. 6280 Ω
2. 79.6 mH, or approximately 80 mH
3. 238.85 Hz, or approximately 239 Hz

III

1. 5 kΩ
2. $X_L = 5024\ \Omega$, so $I \times X_L = 3\text{ mA} \times 5025\ \Omega =$ approximately 15 volts
3. 40 mA

IN-PROCESS LEARNING CHECKS

I

1. When current increases through an inductor, the cemf ***hinders*** the current increase.
2. When current decreases through an inductor, the cemf ***hinders*** the current decrease.
3. In a pure inductor, the ***voltage*** leads the ***current*** by 90 degrees.
4. What memory aid helps to remember the relationship described in Question 3? ***"Eli"*** the ice man.
5. The opposition that an inductor shows to ac is termed ***inductive reactance.***
6. The opposition that an inductor shows to ac ***increases*** as inductance increases.
7. The opposition that an inductor shows to ac ***decreases*** as frequency decreases.
8. X_L is ***directly*** related to inductance value.
9. X_L is ***directly*** related to frequency.

REVIEW QUESTIONS

1. $L = \dfrac{X_L}{2\pi f} = \dfrac{376.8}{6.28} \times 100 = \dfrac{376.8}{628} = 0.6\text{ H}$
   $X_L = 4.5 \times 376.8\ \Omega = 1695.6\ \Omega$
3. X_L is six times greater than the original value.
5. Circuit I will increase, since X_L will decrease and V is assumed to remain the same.
7. Q will decrease since $Q = \dfrac{X_L}{R}$. X_L will decrease while R stays the same.
9. Total inductance $= \dfrac{2 \times 6}{2 + 6} = \dfrac{12}{8} = 1.5\text{ H}$

   Frequency $= \dfrac{X_L}{2\pi L} = 39.8\text{ Hz}$
11. $Q = \dfrac{X_L}{R} = \dfrac{6.28 \times 200 \times 1}{10} = \dfrac{1256}{10} = 125.6$

    Q will increase, since $Q = \dfrac{X_L}{R}$ and X_L triples while R only doubles.
    Q is not an appropriate consideration for inductors in dc circuits. X_L is reactance to ac due to the changing current inducing back-emf.
13. $V_T = 45\text{ V}$
    $V_{L_1} = 15\text{ V}$
    $I_T = 0.75\text{ mA}$

TEST YOURSELF

1. Total inductance is 3 H

   (NOTE: L_1 computes to be 2 H via the $L = \dfrac{X_L}{2\pi f}$ formula.)

Chapter 14

PRACTICE PROBLEMS

I

5 pounds at an angle of 36.9° from horizontal.

II

$V_T = \sqrt{V_R^2 + V_L^2} = \sqrt{25 + 25} = \sqrt{50} = 7.07\text{ V}$

III

1. $\text{Cos } \theta = \frac{\text{adj}}{\text{hyp}} = \frac{5}{10} = 60°$
2. Tangent of 45° = 1.0
3. Sine of 30° = 0.5

IV

V applied = 150 volts

V

1. $\theta = 33.69°$ (Tan $\theta = \frac{\text{opp}}{\text{adj}} = \frac{20}{30} = 0.6666$. Angle whose tangent = 0.6666 is 33.69°.)
2. Tangent function
3. Angle would have been greater (56.3°).

VI

$V_R = 35.35$ volts (Cos $\theta = \frac{\text{adj}}{\text{hyp}}$ and cos 45° = .7071.

Therefore, $0.7071 = \frac{\text{adj}}{50} = 35.35$).

VII

1.

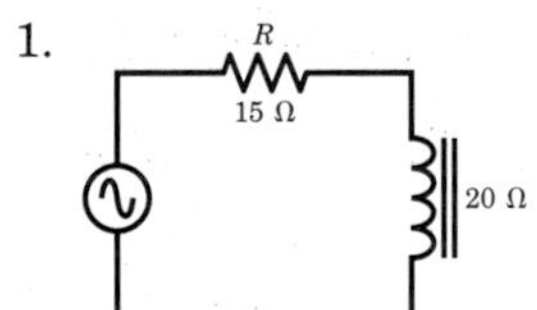

2.

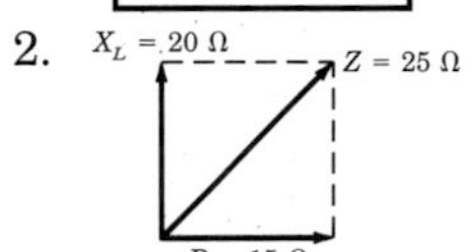

3. $Z = \sqrt{15^2 + 20^2}$

$Z = \sqrt{625}$

$Z = 25\ \Omega$

4. $\text{Tan } \theta = \frac{X_L}{R}$

$\text{Tan } \theta = \frac{20}{15}$

Tan θ = 1.33
Angle = 53.1°

VIII

1. $I_T = \sqrt{I_R^2 + I_L^2}$

$I_T = \sqrt{5^2 + 3^2}$

$I_T = \sqrt{34}$

$I_T = 5.83$ A

2. $Z = \frac{V_T}{I_T}$

$Z = \frac{300}{5.83}$

$Z = 51.45\ \Omega$

3. $\theta = 30.9°$ (Tan $\theta = \frac{\text{opp}}{\text{adj}} = \frac{3\text{ A}}{5\text{ A}} = 0.6$. Angle whose tangent equals 0.6 is 30.9°.)

IN-PROCESS LEARNING CHECKS

I

1. In a purely resistive ac circuit, the circuit voltage and current are ***in***-phase.
2. In a purely inductive ac circuit, the current through the inductance and the voltage across the inductance are ***90*** degrees out-of-phase. In this case, the ***voltage*** leads the ***current.***
3. A quantity expressing both magnitude and direction is a ***vector*** quantity.
4. The length of the vector expresses the ***magnitude.***

REVIEW QUESTIONS

1. a. 36 Ω
   b. $Z = \sqrt{R^2 + X_L^2} = \sqrt{20^2 + 30^2} = \sqrt{1300} =$ approximately 36 Ω
   c. $\text{Cos } \theta = \frac{R}{Z} = \frac{20}{36} = 0.555$. Angle whose cosine equals 0.555 is 56.3°; thus, angle between R and Z = 56.3°.
3. Diagrams for the specified series RL circuit are shown.

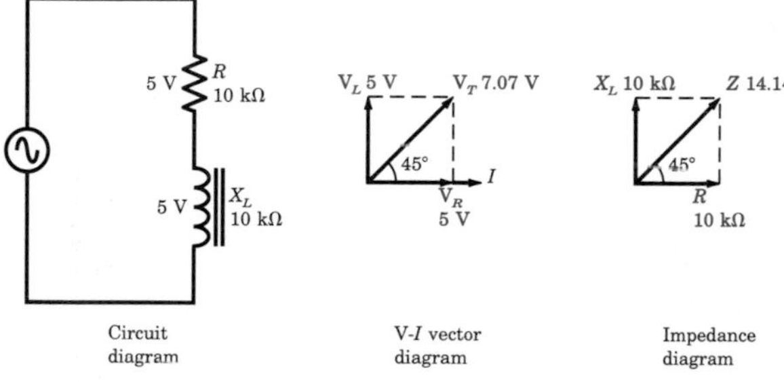

Since $\theta = 45°$, then R must = X_L, and V_R must = V_L.

$Z = \sqrt{R^2 + X_L^2}$

$Z = \sqrt{100 \times 10^6}$

$Z = 14.14 \times 10^3 = 14.14\ k\Omega$

5. a. R will RTS
   b. L will RTS
   c. X_L will I
   d. Z will I
   e. θ will D
   f. I_T will D
   g. V_T will RTS
7. a. R will I
   b. L will RTS
   c. X_L will RTS
   d. Z will I
   e. θ will D
   f. I_T will D
   g. V_T will RTS
9. a. R will I
   b. L will RTS
   c. X_L will RTS
   d. Z will I
   e. θ will I
   f. I_T will D
   g. V_T will RTS
11. a. Z = vector sum of 10 kΩ R and approximate 20 kΩ X_L = 22.36 kΩ
    b. $V_R = I \times R = \left(\frac{67\ V}{22.36\ k\Omega}\right) \times 10\ k\Omega =$ approximately 3 mA × 10 kΩ = 30 V
    c. θ = approximately 63.43°
13. a. I = 5 mA
    b. Z = 40 kΩ
    c. V_L = 105.4 V
    d. X_L = 21.08 kΩ
    e. L = 55 H

TEST YOURSELF

1. Diagrams and calculations for the specified series RL circuit are shown.

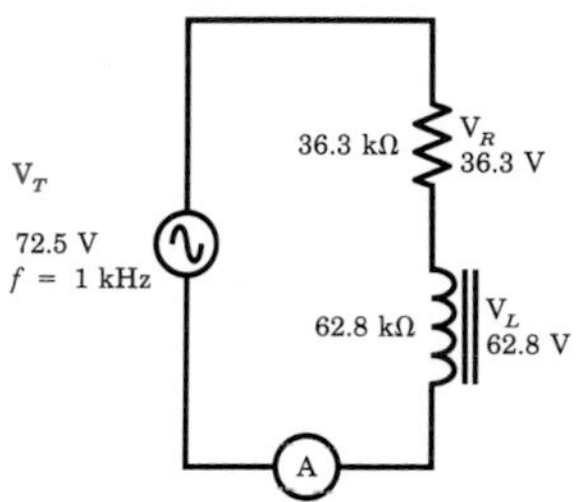

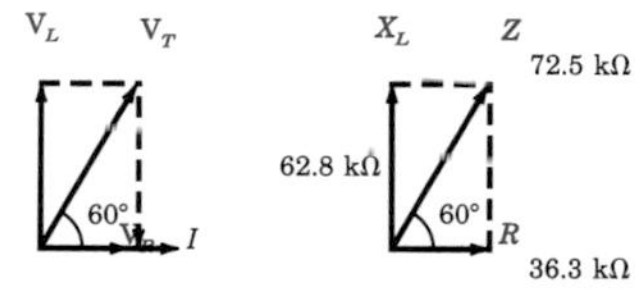

$X_L = 2\pi fL = 6.28 \times 1 \times 10^3 \times 10 = 62.8\ k\Omega$

$\text{Sin}\ \theta = \frac{X_L}{Z}$; $\sin 60° = \frac{62.8\ k\Omega}{Z}$; $0.866 = \frac{62.8\ k\Omega}{Z}$;

$0.866\ Z = 62.8\ k\Omega$; $Z = \frac{62.8\ k\Omega}{0.866} = 72.517\ k\Omega$

$\text{Tan}\ \theta = \frac{\text{opp}}{\text{adj}} = \frac{X_L}{R}$; $\tan 60° = \frac{62.8\ k\Omega}{R}$;

$1.73 = \frac{62.8\ k\Omega}{R}$

$1.73\ R = 62.8\ k\Omega$; $R = \frac{62.8\ k\Omega}{1.73} = 36.3\ k\Omega$

$I = \frac{V_T}{Z}$; $I = \frac{72.5\ V}{72.5\ k\Omega} = 1\ mA$

$V_R = I \times R = 1\ mA \times 36.3\ k\Omega = 36.3\ V$

$V_L = I \times X_L = 1\ mA \times 62.8\ k\Omega = 62.8\ V$

3. The circuit diagram for the specified circuit is shown. I_T = approximately $\frac{140\ \text{volts}}{14\ k\Omega} = 10\ mA$

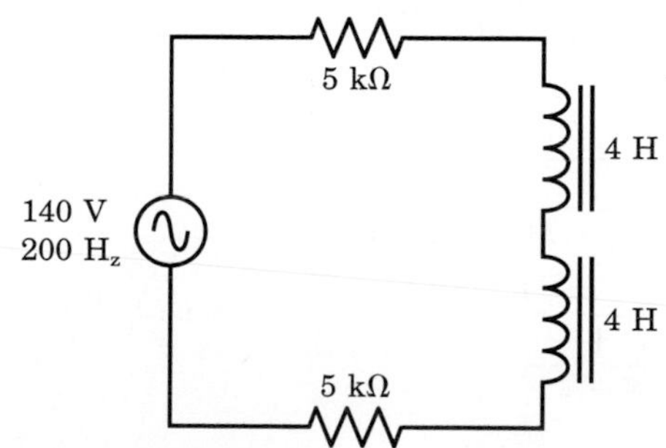

$Z = \sqrt{R^2 + X_L^2}$
$X_L = 2\pi fL = 6.28 \times 200 \times 4 = 5.024\ k\Omega$ (or 5 kΩ) each inductor
$R = 5\ k\Omega + 5\ k\Omega = 10\ k\Omega$
$Z = \sqrt{200 \times 10^6} = 14.14\ k\Omega$ (or 14 kΩ)
$X_{LT} \cong 10\ k\Omega$

Chapter 15

PRACTICE PROBLEMS

I

$M = k\sqrt{L_1 \times L_2}$

$M = 0.95\sqrt{10 \times 15}$

$M = 0.95\sqrt{150}$

$M = 0.95 \times 12.25 = 11.63\ H$

II

$L_T = L_1 + L_2 - 2M$
$L_T = 10\ H + 10\ H - (2 \times 5\ H)$
$L_T = 20\ H - 10\ H = 10\ H$

III

1. $L_T = \dfrac{1}{\dfrac{1}{L_1 + M} + \dfrac{1}{L_2 + M}}$

$L_T = \dfrac{1}{\dfrac{1}{8+4} + \dfrac{1}{8+4}}$

$L_T = \dfrac{1}{\dfrac{1}{12} + \dfrac{1}{12}}$

$L_T = \dfrac{1}{\dfrac{1}{6}} = 1 \times \dfrac{6}{1} = 6\ H$

2. $L_T = \dfrac{1}{\dfrac{1}{L_1 - M} + \dfrac{1}{L_2 - M}}$

$L_T = \dfrac{1}{\dfrac{1}{8-4} + \dfrac{1}{8-4}}$

$L_T = \dfrac{1}{\dfrac{1}{4} + \dfrac{1}{4}}$

$L_T = \dfrac{1}{\dfrac{1}{2}} = 1 \times \dfrac{2}{1} = 2\ H$

IV

1. $\dfrac{N_s}{N_P} = \dfrac{V_s}{V_P}$; s-p turns ratio $= \dfrac{300}{100}$ or 3:1
2. Turns ratio equals voltage ratio. Since the transformer steps-up the voltage six times, the secondary must have six times the turns of the primary. Therefore, the s-p turns ratio is 6:1.
3. Since the s-p turns ratio is 5:1, the secondary has five times the voltage of the primary, or 5 × 50 volts = 250 volts.
4. Voltage ratio with s-p turns ratio of 1:4 is 1:4 (secondary-to-primary). This is a step-down transformer, since the primary voltage is four times that of the secondary.

V

Since the impedance is transformed in relation to the square of the turns ratio, when the secondary has twice the number of turns of the primary, the impedance at the primary looks like one-fourth the impedance across the secondary, or $\dfrac{2000}{4} = 500\ \Omega$.

VI

1. The *primary-to-secondary* impedance ratio is related to the square of the turns relationship of primary-to-secondary. Since the primary has 5 times as many turns as the secondary, the primary impedance is 5^2 times greater than secondary. Therefore the p-s Z ratio is 25:1.
2. The impedance looking in the primary is $(16)^2 \times 4\ \Omega = 256 \times 4 = 1024\ \Omega$.

IN-PROCESS LEARNING CHECKS

I

1. Producing voltage via a changing magnetic field is called electromagnetic ***induction.***

2. Inducing voltage in one circuit by varying current in another circuit is called ***mutual inductance.***
3. The fractional amount of the total flux that links two circuits is called the ***coefficient*** of coupling, which is represented by the letter ***k.*** When 100% of the flux links the two circuits, the ***coefficient*** of coupling has a value of ***1.***
4. The closer coils are, the ***higher*** the coupling factor produced. Compared to parallel coils, perpendicular coils have a ***lower*** degree of coupling.

REVIEW QUESTIONS

1. $\frac{V_s}{V_P} = \frac{I_P}{I_s}$

$\frac{240\text{ V}}{120\text{ V}} = \frac{I_P}{100\text{ mA}}$

$I_P = 200\text{ mA}$

3. $\frac{N_P}{N_s} = \frac{V_P}{V_s}; \frac{N_P}{N_s} = \frac{120\text{ V}}{6\text{ V}}; \frac{N_P}{N_s} = 20:1$

5. $M = k\sqrt{L_1 \times L_2} = 0.6\sqrt{2 \times 2} = 0.6 \times 2 = 1.2\text{ H}$

7. $L_T = \frac{1}{\frac{1}{L_1 - M} + \frac{1}{L_2 - M}}$

where $M = k\sqrt{L_1 \times L_2} = 0.5\sqrt{10 \times 10} = 0.5 \times 10 = 5\text{ H}$

$L_T = \frac{1}{\frac{1}{10 - 5} + \frac{1}{10 - 5}}$

$L_T = \frac{1}{0.4} = 2.5\text{ H}$

9. $\frac{50}{500} = 0.1$ volts-per-turn

11. 2000 turns

13. 1:4.75

15. 62.5 V

TEST YOURSELF

1. $\frac{N_P}{N_s} = \sqrt{\frac{Z_P}{Z^s}}$

$\frac{N_P}{200} = \sqrt{\frac{5000}{16}}$

$N_P = 200 \times 17.67 =$ approximately 3535.5

3. If the impedance ratio (s-p) of an ideal transformer is 1:36, the turns ratio (s-p) is the square root of 1:36, or 1:6. This means the primary has six times as many turns as the secondary. Since the number of turns on the primary is 1200, the $N_s = \frac{1200}{6} = 200$ turns.

Chapter 16

PRACTICE PROBLEMS

I

1. $C = \frac{Q}{V} = \frac{100 \times 10^{-6}}{25} = 4 \times 10^{-6} = 4\ \mu\text{F}$
2. $Q = CV = 10 \times 10^{-6} \times 250 = 2500\ \mu\text{C}$
3. $V = \frac{Q}{C} = \frac{50 \times 10^{-6}}{2 \times 10^{-6}} = 25\text{ V}$

II

$E = \frac{1}{2}CV^2 = \frac{10 \times 10^{-6} \times (100)^2}{2} = \frac{100{,}000 \times 10^{-6}}{2} = 50{,}000\ \mu\text{J}$

III

1. $C = \frac{2.25\text{ kA}}{10^7{}_s} \times (n - 1) = \frac{2.25 \times 10 \times 1}{10^7 \times 0.1}$

$C = \frac{22.5}{10^6} \times (3) = 22.5 \times 10^{-6} \times 3 = 67.5\ \mu\text{F}$

2. $C = \frac{8.85\text{ kA}}{10^{12}{}_s}\ C = \frac{8.85 \times 1 \times 0.2}{10^{12} \times 0.005} = \frac{1.77 \times 10^{-12}}{0.005} =$ 354 pF

IV

1. $C_T = \frac{C_1 \times C_2}{C_1 + C_2} = \frac{5 \times 10^{-6} \times 20 \times 10^{-6}}{25 \times 10^{-6}} = \frac{100 \times 10^{-12}}{25 \times 10^{-6}} = 4 \times 10^{-6}$, or $4\ \mu\text{F}$
2. Total capacitance of four 20-μF capacitors in series is equal to one of the equal value capacitors divided by 4. Thus, $\frac{20}{4} = 5\ \mu\text{F}$, $C_T = \frac{C_1}{C_N}$.

3. V_{C_1} = four-fifths V_T, or 80 volts. V_{C_2} = one-fifth V_T, or 20 volts.

   $(Q_T = C_T \times V_T; = 4\ \mu F \times 100V = 400\ \mu C)$

   $$(V_{C_1} = \frac{Q}{C_1}; = \frac{400\ \mu C}{5\ \mu F} = 80\ V)$$

   $$(V_{C_2} = \frac{Q}{C_2}; = \frac{400\ \mu C}{20\ \mu F} = 20\ V)$$

V

1. C_T = sum of individual capacitances; therefore = 60 μF.
2. $Q_{C_1} = C_1 \times V = 12\ \mu F \times 60\ V = 720\ \mu C$.
   $Q_{C_1} = Q_{C_2} = Q_{C_3} = Q_{C_4} = Q_{C_5}$
   Q_T = sum of all the charges = $5 \times 720\ \mu C$ = 3600 μC.

VI

1. One time constant for circuit = $0.001 \times 10^{-6} \times 10 \times 10^3$
   1 TC = 0.01 milliseconds (or 10 microseconds)
   Time allowed for charge = 20 μsec, or 2 TC
   After 2 TC, capacitor is charged to 86.5% of V applied.
   In this case, V_C = 86.5% of 100 volts, or 86.5 V, after 20 μsec.
2. The voltage across the resistor 15 μsec after the switch is closed will be equal to the percentage of V applied that V_R drops to after 1.5 TC. (15 μsec = 1.5 × the 10 μsec TC). According to the TC chart, the resistor voltage is approximately 23% of V applied, or 23 volts.

IN-PROCESS LEARNING CHECKS

I

1. A capacitor is an electrical component consisting of two conducting surfaces called ***plates*** that are separated by a nonconductor called the ***dielectric.***
2. A capacitor is a device that stores electrical ***charge*** when voltage is applied.
3. Electrons ***do not*** travel through the capacitor dielectric.
4. During charging action, one capacitor plate collects ***electrons,*** making that plate ***negative.*** At the same time, the other plate is losing ***electrons*** to become ***positive.***
5. Once the capacitor has charged to the voltage applied, it acts like an ***open*** circuit to dc. When the level of dc applied voltage increases, the capacitor ***charges*** to reach the new level. When the level of dc applied decreases, the capacitor ***discharges*** to reach the new level.
6. When voltage is first applied to a capacitor ***maximum*** charge current occurs.
7. As the capacitor becomes charged, current ***decreases*** through the circuit in series with the capacitor.

REVIEW QUESTIONS

1. New capacitance is four times the original capacitance, since C is directly related to the area of the plates facing each other and inversely related to the dielectric thickness.

3. $C = \frac{Q}{V} = \frac{200\ pC}{50} = 4\ pF$

5. $$C_T\ (pF) = \frac{1}{\frac{1}{C_1} + \frac{1}{C_2} + \frac{1}{C_3}} = \frac{1}{\frac{1}{100} + \frac{1}{400} + \frac{1}{1000}}$$

   $$C_T = \frac{1}{0.01 + 0.0025 + 0.001} = \frac{1}{0.0135} = 74.07\ pF$$

   $C_T = 100\ pF + 400\ pF + 1000\ pF = 1500\ pF$

7. $Q_{C_1} = C_1 \times V_1 = 10\ \mu F \times 100\ V = 1000\ \mu C$
   $Q_{C_2} = C_2 \times V_2 = 20\ \mu F \times 100\ V = 2000\ \mu C$
   $Q_{C_3} = C_3 \times V_3 = 50\ \mu F \times 100\ V = 5000\ \mu C$
   $C_T = C_1 + C_2 + C_3 = 80\ \mu F$
   $Q_T = Q_1 + Q_2 + Q_3 = 8000\ \mu C$

9. $V_R = V\ \text{applied} \times \epsilon^{-t/\tau} = 200 \times 2.718^{-1.2} = 60.23\ V$
   $V_C = V_A - V_R$, or approximately $200 - 60 = 140\ V$

TEST YOURSELF

1. Total capacitance of circuit shown is calculated using knowledge that capacitors in series add like resistors in parallel, and capacitors in parallel add like resistors in series. Thus, total capacitance of the two branches (points A to B) = 20 μF. This 20 μF is in series with C_6's 10 μF and C_1's 10 μF. Simplifying further, the two 10 μFs in series are equivalent to 5 μF. This value is in series with

the 20 μF of the two branches. $\frac{20 \times 5}{20 + 5} = 4\ \mu F$ total capacitance.

3. Q_{C_6} = same as Q_T, or 320 μC
5. Voltage applied is 3000 volts. (Since the voltage across C_1 and across C_6 are each double that across points A and B, so 1200 V + 1200 V + 600 V = 3000 V.) (NOTE: This cannot be actually done, since the voltage ratings of C_1 and C_6 were described as being 300 V each. Therefore, it would not be possible to have 1200 volts across each without their breaking down. But, for purposes of achieving 600 volts across points A and B, it takes 3000 volts applied, as shown.)
7. Charge on first capacitor (Q = C × V) = 15 μF × 150 V = 2250 μC. Charge on second capacitor = 15 μF × 300 V = 4500 μC. Total charge to be distributed on total capacitance is 6750 μC. Since capacitors are connected in parallel, total C = 30 μF. $V = \frac{Q}{C} = \frac{6750\ \mu C}{30\ \mu F} = 225$ volts.

Chapter 17

PRACTICE PROBLEMS

I

1. $X_C = \frac{0.159}{200 \times 5 \times 10^{-6}} = \frac{159{,}000}{1000} = 159\ \Omega$
2. X_C remains the same if f is doubled and C is halved simultaneously.

II

1. $f - \frac{0.159}{CX_C} = \frac{0.159}{500 \times 10^{-12} \times 1000}$

   $f = \frac{159{,}000 \times 10^6}{500 \times 10^3} = 318 \times 10^3$, or 318 kHz
2. $C = \frac{0.159}{fX_C} = \frac{0.159}{2000 \times 3180}$

   $C = \frac{0.159}{6360000} = .000000025$ F, or 0.025 μF

III

$X_{C_1} = \frac{0.159}{fC} = \frac{0.159}{1000 \times 2 \times 10^{-6}}$

$X_{C_1} = \frac{159{,}000}{2000} = 79.5\ \Omega$

$X_{C_2} = \frac{0.159}{fC} = \frac{0.159}{1000 \times 4 \times 10^{-6}}$

$X_{C_2} = \frac{159{,}000}{4000} = 39.75\ \Omega$

$X_{CT} = X_{C_1} + X_{C_2} + X_{C_3} + X_{C_4} = 79.5 + 39.75 + 200 + 300 = 619.25\ \Omega$

IV

1. $X_{CT} = \frac{0.159}{fC_T} = \frac{0.159}{1000 \times 6 \times 10^{-6}}$

   $X_{CT} = \frac{159{,}000}{6000} = 26.5\ \Omega$
2. $X_{CT} = \frac{0.159}{fC_T} = \frac{0.159}{500 \times 25 \times 10^{-6}}$

   $X_{CT} = \frac{159{,}000}{12500} = 12.72\ \Omega$
3. $X_{CT} = \frac{1600}{4} = 400\ \Omega$

V

1. $X_{C_1} = 1{,}000\ \Omega$; $X_{C_2} = 1{,}000\ \Omega$; $X_{C_3} = 2{,}000\ \Omega$; $X_{CT} = 4{,}000\ \Omega$;
   $C_2 = 0.795\ \mu F$; $C_3 = 0.3975\ \mu F$; $V_2 = 25$ V; $I_T = 25$ mA
2. $C_1 = 0.5\ \mu F$; $X_{C_1} = 1{,}000\ \Omega$; $X_{C_2} = 1{,}000\ \Omega$; $X_{C_3} = 500\ \Omega$;
   $I_2 = 2$ mA; $I_T = 8$ mA

IN-PROCESS LEARNING CHECKS

I

1. For a capacitor to develop a potential difference between its plates, there must be a ***charging*** current.
2. For the potential difference between capacitor plates to decrease (once it is charged), there must be a ***discharging*** current.
3. When ac is applied to capacitor plates, the capacitor will alternately ***charge*** and ***discharge.***
4. The value of instantaneous capacitor current directly relates to the value of ***capacitance*** and the ***voltage*** rate of change.

5. The amount of charging or discharging current is maximum when the voltage rate of change is ***maximum.*** For sine-wave voltage, this occurs when the sine wave is near ***zero*** points.
6. Capacitor current ***leads*** the voltage across the capacitor by ***90*** degrees. The expression that helps remember this relationship is Eli the ***ice*** man.

REVIEW QUESTIONS

1. $X_C = \frac{0.159}{fC} = \frac{0.159}{250} \times 10^{-12} \times 5 \times 10^3 =$
   $\frac{0.159}{1250} \times 10^{-9} = \frac{159000000}{1250} = 127.2\ k\Omega$
3. X_C decreases to one-third its original value, since X_C is inverse to frequency $\left(X_C = \frac{0.159}{fC}\right)$.
5. For the circuit shown:
   $C_T = \frac{C_1 \times C_2}{C_1 + C_2} = \frac{0.636 \times 1.59}{0.636 + 1.59} = 0.454\ \mu F$
   $X_{CT} = X_{C_1} + X_{C_2} = 25\ k\Omega + 10\ k\Omega = 35\ k\Omega$
   $V_A = 70\ V$ (sum of V_{C_1} and V_{C_2})
   $C_1 = \frac{0.159}{10\ Hz \times X_C}$; where $X_C = \frac{V}{I} = \frac{50\ V}{2\ mA} =$
   $25\ k\Omega$. Thus, $C_1 = \frac{0.159}{10 \times 25\ k\Omega} = \frac{0.159}{250 \times 10^3} =$
   $0.636\ \mu F$.
   $C_2 = \frac{0.159}{10\ Hz \times X_C}$, where $X_C = \frac{V}{I} = \frac{20\ V}{2\ mA} =$
   $10\ k\Omega$. Thus, $C_2 = \frac{0.159}{10 \times 10\ k\Omega} = \frac{0.159}{100} \times 10^3 =$
   $1.59\ \mu F$.
7. a. C_1 will RTS
   b. C_2 will RTS
   c. C_T will RTS
   d. X_{C_1} will D
   e. X_{C_2} will D
   f. X_{CT} will D
   g. I will I
   h. V_{C_1} will RTS
   i. V_{C_2} will RTS
   j. V_A will RTS
9. Total current increases to 12.58 mA, if $C_2 = C_1$ and all other factors remain the same.
11. X_C of $C_1 = 1000\ \Omega$
13. $C_T = 0.063\ \mu F$
15. increase

TEST YOURSELF

1. Capacitive voltage divider with the parameters specified is shown.

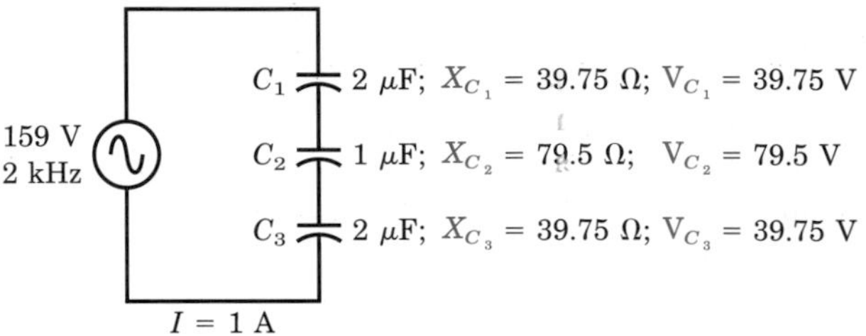

Chapter 18

PRACTICE PROBLEMS

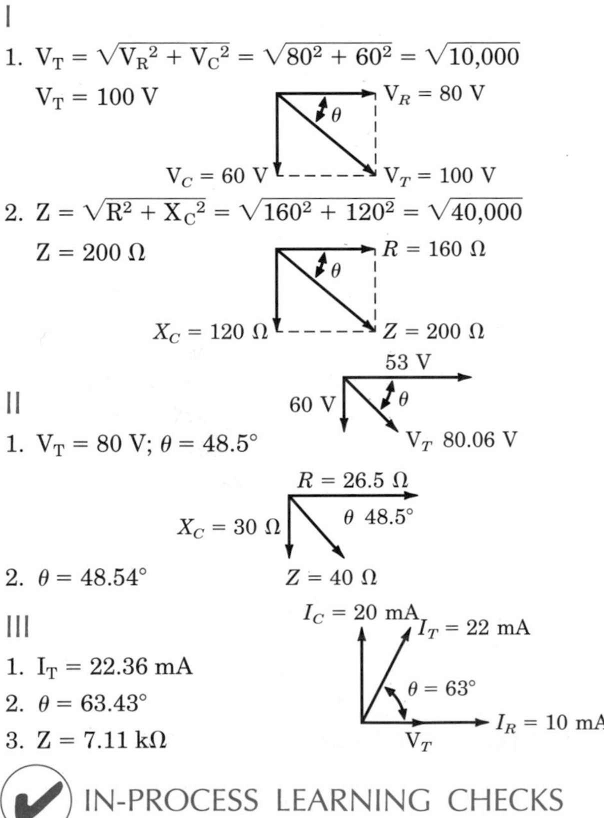

I

1. $V_T = \sqrt{V_R^2 + V_C^2} = \sqrt{80^2 + 60^2} = \sqrt{10{,}000}$
   $V_T = 100\ V$
2. $Z = \sqrt{R^2 + X_C^2} = \sqrt{160^2 + 120^2} = \sqrt{40{,}000}$
   $Z = 200\ \Omega$

II

1. $V_T = 80\ V;\ \theta = 48.5°$
2. $\theta = 48.54°$

III

1. $I_T = 22.36\ mA$
2. $\theta = 63.43°$
3. $Z = 7.11\ k\Omega$

IN-PROCESS LEARNING CHECKS

I

1. In series RC circuits, the ***current*** is the reference vector, shown at 0°.

2. In the V-I vector diagram of a series RC circuit, the circuit voltage is shown ***lagging*** the circuit current.
3. In series RC circuits, an impedance diagram ***can*** be drawn.
4. Using the Pythagorean approach, the formula to find V_T in a series RC circuit is $V_T = \sqrt{V_R^2 + V_C^2}$.
5. Using the Pythagorean approach, the formula to find Z in a series RC circuit is $Z = \sqrt{R^2 + X_C^2}$.
6. The trig function used to solve for phase angle when you do not know the hypotenuse value is the ***tangent*** function.

II

1. In parallel RC circuits, ***voltage*** is the reference vector in V-I vector diagram.
2. In parallel RC circuits, current through the resistor(s) is plotted ***out-of-phase*** with the circuit current vector.
3. In parallel RC circuits, current through capacitor branch(es) is plotted at ***+90 degrees*** with respect to circuit applied voltage.
4. The Pythagorean formula to find circuit current in parallel RC circuits is $I_T = \sqrt{I_R^2 + I_C^2}$.
5. ***No.***
6. ***Yes.***
7. ***No.***

REVIEW QUESTIONS

1. $Z = \sqrt{R^2 + X_C^2} = \sqrt{64 \times 10^6 + 225 \times 10^6} = \sqrt{289 \times 10^6} = 17\ k\Omega$
3. $C = \frac{0.159}{fX_C} = \frac{0.159}{1060 \times 15\ k\Omega} = \frac{0.159}{15.9 \times 10^6} = 0.01\ \mu F$
5. a. R will RTS
   b. X_C will D
   c. Z will D
   d. I_T will I
   e. θ will D
   f. V_C will D
   g. V_R will I
   h. V_T will RTS
7. $R = \frac{V_R}{I} = \frac{150\ V}{10\ mA} = 15\ k\Omega$
9. $f = \frac{0.159}{CX_C} = \frac{0.159}{0.02 \times 10^{-6} \times 30 \times 10^3} = \frac{0.159}{0.6 \times 10^{-3}} = 265\ Hz$
11. $R_T = 20\ k\Omega + 18\ k\Omega + 30\ k\Omega = 68\ k\Omega$
13. $Z = \sqrt{R^2 + X_C^2} = \sqrt{(68 \times 10^3)^2 + (52 \times 10^3)^2} = \sqrt{7328 \times 10^6} = 85.6\ k\Omega$
15. $f = \frac{0.159}{CX_C} = \frac{0.159}{0.1 \times 10^{-6} \times 30 \times 10^3} = \frac{0.159}{0.003} =$ 53 Hz
17. $I_C = \frac{320\ V}{16\ k\Omega} = 20\ mA$
19. $\text{Tan}\ \theta = \frac{I_C}{I_R} = \frac{20}{32} = 0.625$. Angle whose tangent equals 0.625 is 32°. $\theta = 32°$

TEST YOURSELF

1. $R = 750\ \Omega$
   $X_C = 1{,}000\ \Omega$
   $I = 20\ mA$
   $V_R = 15\ V$
   $V_C = 20\ V$
   $\theta = -53.1°$
   $X_C = \frac{0.159}{3975 \times 0.04 \times 10^{-6}} = 1{,}000\ \Omega$
   $\text{Tan}\ \theta = \frac{opp}{adj} = \frac{-X_C}{R} = \frac{-1{,}000}{750} = -1.33$
   θ = the angle whose tan = -1.33
   $\theta = -53.1°$ (V_T lags reference vector for series circuits, I_T.)
   $R = Z\cos\theta = 1250 \times \cos 53.1° = 1250 \times 0.6 = 750\Omega$

Chapter 19

PRACTICE PROBLEMS

I

$R = 200\ \Omega$
$X_{CT} = 500\ \Omega$
$X_L = 250\ \Omega$
$Z = \sqrt{R^2 + X^2} = \sqrt{200^2 + 250^2} = \sqrt{102500} = 320.15\ \Omega$

$V_T = I \times Z = 0.25 \times 320.15 = 80.04$ V
$V_{C_2} = I \times X_{C_2} = 0.25 \times 250 = 62.5$ V
$V_l = I \times X_L = 0.25 \times 250 = 62.5$ V
$V_R = I \times R = 0.25 \times 200 = 50$ V

$$\theta = \frac{\text{angle whose tangent equals X(net)}}{\text{R}}$$

$\text{Tan } \theta = \frac{250}{200} = 1.25$; therefore, $\theta = -51.34°$ (V with respect to I)

II

$$I_R = \frac{V}{R} = \frac{300}{200} = 1.5 \text{ A}$$

$$I_L = \frac{V}{X_L} = \frac{300}{200} = 1.5 \text{ A}$$

$$I_C = \frac{V}{X_C} = \frac{300}{300} = 1 \text{ A}$$

$$I_X = I_L - I_C = 1.5 - 1 = 0.5 \text{ A}$$

$$I_T = \sqrt{I_R^2 + I_X^2} = \sqrt{1.5^2 + 0.5^2} = \sqrt{2.5} = 1.58 \text{ A}$$

$$Z = \frac{V}{I} = \frac{300}{1.58} = 189.8\ \Omega$$

$$\theta = \text{angle whose tan equals } \frac{I_X}{I_R} = \frac{0.5}{1.5} = 0.333$$

Therefore, $\theta = -18.43°$ (I with respect to V)

III

20 + j30 in rectangular form converts to 36.05 $\angle 56.31°$ in polar form.

IV

66 $\angle 60°$ in polar form converts to 33 + j57.15 in rectangular form.

V

1. $\begin{array}{r} 20 + j5 \\ \underline{35 - j6} \\ 55 - j1 \end{array}$
2. $\begin{array}{l} 35 - j6 \\ \underline{20 + j5 \text{ (change sign)}} \end{array} \quad \begin{array}{r} 35 - j6 \\ \underline{-20 - j5} \\ 15 - j11 \end{array}$
3. $5 \angle 20° \times 25 \angle -15° = 125 \angle 5°$
4. $\dfrac{5 \angle 20°}{25 \angle -15°} = 0.2 \angle 35°$
5. $I = 12 \angle 25°$, then $I^2 = 144 \angle 50°$
6. If $I^2 = 49 \angle 66°$, then $I = 7 \angle 33°$

IN-PROCESS LEARNING CHECKS

I

1. When both capacitive and inductive reactance exist in the same circuit, the net reactance is ***the difference between their reactances.***
2. When plotting reactances for a series circuit containing L and C, the X_L is plotted ***upward*** from the reference, and the X_C is plotted ***downward*** from the reference.
3. In a series circuit with both capacitive and inductive reactance, the capacitive voltage is plotted ***180*** degrees out-of-phase with the inductive voltage.
4. If inductive reactance is greater than capacitive reactance in a series circuit, the formula to find the circuit impedance is $\boldsymbol{Z = \sqrt{R^2 + (X_L - X_C)^2}}$.
5. In a series RLC circuit, current is found by $\boldsymbol{\frac{V}{Z}}$.
6. ***False.***
7. If the frequency applied to a series RLC circuit decreases while the voltage applied remains the same, the component(s) whose voltage will increase is/are the ***capacitor(s).***

II

1. In parallel RLC circuit, capacitive and inductive branch currents are ***180 degrees*** out-of-phase with each other.
2. The formula to find total current in a parallel RLC when I_C is greater than I_L is $\boldsymbol{I_T = \sqrt{I_R^2 + (I_C - I_L)^2}}$.
3. ***False.***
4. If the frequency applied to a parallel RLC circuit increases while the applied voltage remains the same, the branch current(s) that will increase is/are the ***capacitive*** branch(es).

III

1. Power dissipated by a perfect capacitor over one whole cycle of input is ***zero.***
2. The formula for apparent power is $S = V \times I$, and the unit expressing apparent power is ***volt-amperes.***
3. True power in RLC circuits is dissipated by the ***resistances.***
4. The formula for power factor is ***p.f. = cos θ.***
5. If the power factor is one, the circuit is purely ***resistive.***
6. The lower the power factor, the more ***reactive*** the circuit is acting.
7. In the power triangle, true power is on the ***horizontal*** axis, reactive power (VAR) is on the ***vertical*** axis, and apparent power is the ***hypotenuse*** of the right triangle.

REVIEW QUESTIONS

1. $Z = \sqrt{R^2 + X_T{}^2} = \sqrt{300^2 + 150^2} = \sqrt{112500} = 335.4\ \Omega$

   $I = \frac{V_T}{Z} = \frac{10\text{ V}}{0.3354\text{ k}\Omega} = 29.8\text{ mA}$

   $V_C = I \times X_C = 29.8\text{ mA} \times 250\ \Omega = 7.45\text{ V}$

   $V_L = I \times X_L = 29.8\text{ mA} \times 100\ \Omega = 2.98\text{ V}$

   $V_R = I \times R = 29.8\text{ mA} \times 300\ \Omega = 8.94\text{ V}$

   $\text{Cos } \theta = \frac{\text{adj}}{\text{hyp}} = \frac{R}{Z} = \frac{300}{335.4} = 0.89$

   θ = angle whose cos = 0.89 = 26.56° (I_T will lead V_T.)

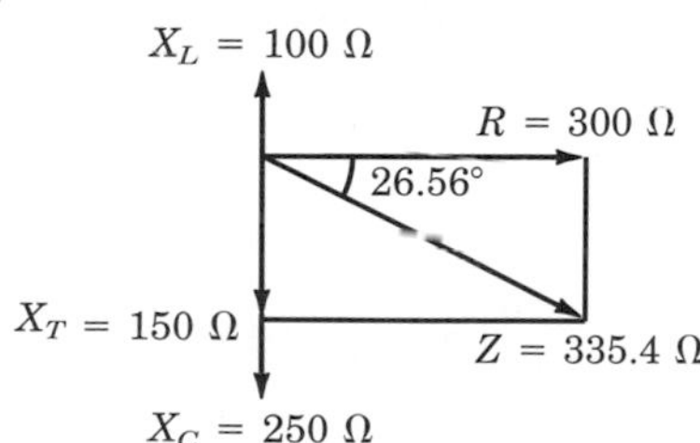

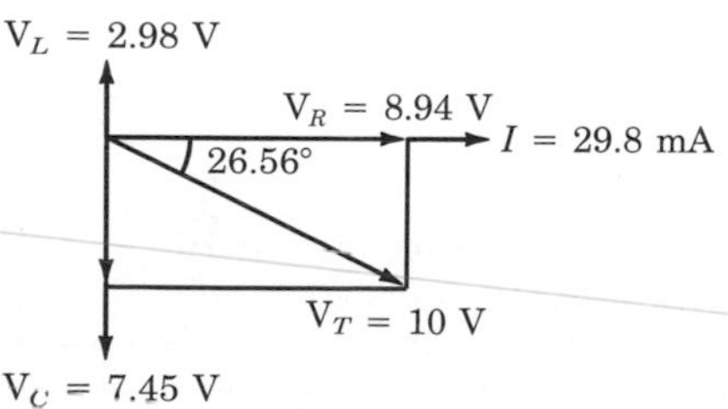

3. True power = $I^2R = 0.0298^2 \times 300 = 0.266$ W

   Apparent power = V × I = 10 V × 0.0298 mA = 0.298 VA

   Reactive power = V × I × sin θ = 10 V × 0.0298 × 0.447 = 0.133 VAR

   $\text{Power factor} = \frac{\text{true power}}{\text{apparent power}} = \frac{0.266}{0.298} = 0.89$

   (or p.f. = Cos θ = cos 26.56° = 0.89)

5. Z = 20 + j40

   If the circuit were an RC circuit, it would be stated as

   Z = 20 − j40

7. $I_R = \frac{20}{20} = 1\text{ A}$

   $I_C = \frac{20}{20} = 1\text{ A}$

   I_T = 1.414 A at an angle of 45°

   $I_T = 1.414\ \angle 45°$

9. Converting 14.14 ∠45° to rectangular form:

   10 + j10

   That is, R = Z × cos θ = 14.14 × cos 45° = 10, since the angle is 45°; R and X must be equal; thus $Z = R + jX_L$

   14.14 = 10 + j10

11. $Z = 61.43\ \angle -75.96°$ kΩ (polar form)

    Z = 14900 − j59600 (rectangular form)

13. V_C will decrease.

15. $Z = 1.79\ \angle -63.4°$ ohms

17. $Z = 5.82\ \angle 14°$ ohms

TEST YOURSELF

1. $G = \frac{1}{400} = 0.00250$

   $-jB_L = \frac{1}{500} = 0.00200$

   $jB_C = \frac{1}{800} = 0.00125$

   $Y = G + j(B_C - B_L)$

   $Y = 0.0025 + (0.00125 - 0.002)$

   $Y = 0.0025 - 0.00075 = 0.00175$ siemens

3. 500 Ω

Chapter 20

PRACTICE PROBLEMS

I

$f_r = \frac{0.159}{\sqrt{LC}} = 450.4$ kHz

II

$L = \frac{25{,}330}{f^2C} = 10.13\ \mu H$

III

1. $Q = \frac{X_L}{R} = \frac{1000}{50} = 20$
2. $V_C = I \times X_C$

   $I = \frac{V_T}{Z} = \frac{12}{50} = 0.24$ A

   $V_C = 0.24\ A \times 1000\ \Omega = 240$ V
3. If r_s doubles, $Q = \frac{1000}{100} = 10$.
4. If r_s is made smaller, the skirts are steeper.

IV

1. $f_r = \frac{0.159}{\sqrt{LC}} = 2.514$ MHz
2. $Q = \frac{X_L}{R} = 315.8$
3. $Z = Q \times X_L$ = approximately 997 kΩ
4. If r_s is 1 kΩ, the f_r is lower.

V

1. $f_1 = 2494$ kHz and $f_2 = 2506$ kHz
2. $f_r = 500$ kHz and bandpass = from 495 kHz to 505 kHz
3. $f_1 = 3200\ kHz - \frac{2\ kHz}{2} = 3200\ kHz - 1\ kHz =$ 3199 kHz

IN-PROCESS LEARNING CHECKS

I

1. In circuits containing both inductance and capacitance, when the source frequency increases, inductive reactance ***increases*** and capacitive reactance ***decreases.***
2. In a series RLC (or LC) circuit, when X_L equals X_C, the circuit is said to be at ***resonance.***
3. In a series RLC (or LC) circuit, when X_L equals X_C, the circuit impedance is ***minimum;*** the circuit current is ***maximum;*** and the phase angle between applied voltage and current is ***zero*** degrees.
4. In a series RLC (or LC) circuit, when X_L equals X_C, the voltage across the inductor or the capacitor is ***maximum.***

II

1. In a high-Q parallel resonant circuit, Z is ***maximum.***
2. In a high-Q parallel circuit using ideal components, X_L ***does*** equal X_C at resonance.
3. The power factor of an ideal parallel resonant circuit is ***1.***
4. Phase angle in an ideal parallel resonant circuit is ***0 degrees.***
5. In a parallel resonant circuit, the current circulating in the tank circuit is ***higher*** than the line current.

REVIEW QUESTIONS

1. $X_L = X_C$
3. a. $V_C = Q \times V_A = 9.98 \times 20$; $V_C = 199.6$ V.
   b. I at resonance $= \frac{V_A}{R} = \frac{20\ V}{10\ \Omega} = 2$ A
   c. True power $= I^2R = 2^2 \times 10\ \Omega = 40$ W
   d. Net reactive power is 0 VAR.
5. I line $= \frac{V}{Z} = \frac{20\ V}{31.018\ k\Omega} = 0.6447$ mA

   $I_C = Q \times$ I line $= 32.155 \times 0.6447$ mA $= 20.73$ mA
7. Circuit is shown.

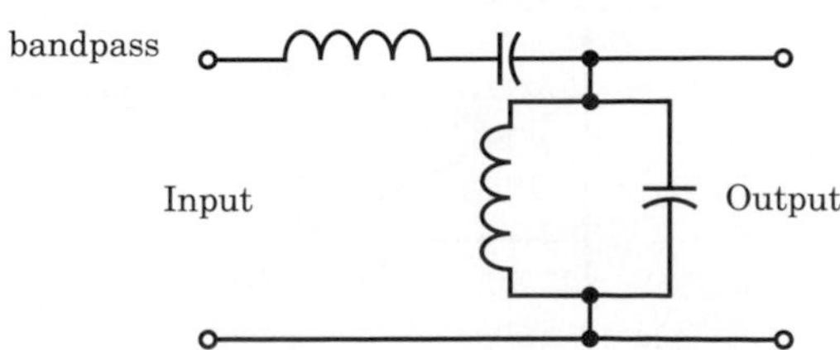

9. Phase angle will be 45° at 39 kHz and/or 41 kHz. (Start solution with $f = \frac{f_r}{Q}$, 45° phase angle occurs at a half power point.)
11. $C = \frac{0.02533}{f_r{}^2 L} = 4.79$ pF
13. At resonance, voltage across R = V applied or 100 V.
15. I_T will lead V_A due to higher capacitive branch current and lower inductive branch current. I_T would be capacitive.
17. To tune to resonance, increase C (choice d). (Since it was originally acting inductively, this indicates current through inductive branch was greater than current through capacitive branch. By increasing C, X_C decreases and I_C increases. Changed by the proper amount, resonance is reached.)

TEST YOURSELF

1. a. X_L will I
   b. X_C will D
   c. R will RTS
   d. Z will I
   e. I will D
   f. θ will I
   g. V_C will D
   h. V_L will I
   i. V_R will D
   j. V_T will RTS
   k. BW will RTS
   l. Q will RTS (close to the same)
   m. I_C will D
   n. I_L will D
   o. I_T will D
3. a. X_L will I
   b. X_C will D
   c. R will RTS
   d. Z will D
   e. I will I
   f. θ will I
   g. V_C will RTS
   h. V_L will RTS
   i. V_R will RTS
   j. V_T will RTS
   k. BW will RTS
   l. Q will RTS
   m. I_C will I
   n. I_L will D
   o. I_T will I
5. The m-derived filter is a modified version of a constant-k filter. Its characteristics are such that the "m" factor relates to the ratio of the filter cutoff frequency to the frequency of infinite attenuation. In general, m ranges in values between 0 and 1. (Typically around 0.6.)
   The constant-k filter is one where the product of the impedances in the filter "arms" (typically a capacitive element and an inductive element) are constant at any frequency. The "k" represents whatever this "constant" factor is, (e.g., $Z_1 \times Z_2 = k^2$).
   The purpose for such filter design parameters as the constant-k and m-derived is to have filters whose input and output impedances are relatively constant over the frequency range of expected operation, or the passband. That is, to maintain their impedance match between output (load) and input (source) connected to the filter network.

Chapter 21

PRACTICE PROBLEMS

I

$V_{dc} = 0.45 \times V_{rms} = 0.45 \times 200 = 90$ V

II

$V_{dc} = 0.45 \times V_{rms} = 0.45 \times 440 = 198$ V
PIV = $1.414 \times V_{rms} = 1.414 \times 440 = 622.16$ V
Ripple frequency = same as ac input = 60 pps (pulses per second)
Polarity of output = positive with respect to ground reference

III

$V_{dc} = 0.9 \times V_{rms}$ of half the secondary = $0.9 \times 150 =$ 135 V
Ripple frequency = 2 × ac input frequency = 2 × 400 = 800 pps
PIV = $1.414 \times V_{rms}$ full secondary = $1.414 \times 300 =$ 424.2 V

IN-PROCESS LEARNING CHECKS

I

1. An interesting difference between conductors and semiconductors is that most conductors have a ***positive*** temperature coefficient while semiconductors typically have a ***negative*** temperature coefficient. A positive temperature coefficient means that as temperature increases, resistance ***increases.*** A negative temperature coefficient means that as temperature increases, resistance ***decreases.***
2. Semiconductor materials have four outermost ring electrons or valence electrons. These materials are sometimes termed ***tetravalent*** materials, where the prefix ***tetra*** means four.
3. To increase conductivity of semiconductor materials, a process called ***doping*** is used.
4. Electrons are the majority carriers in ***N***-type semiconductor materials.
5. When we join a P-type semiconductor material with an N-type, the point where they come together is called a P-N ***junction.***
6. To cause conduction in a semiconductor diode, the diode must be ***forward*** biased.
7. To reverse bias a P-N junction means to connect the negative source voltage to the ***P***-type material and the positive side of the source voltage to the ***N***-type material. In this case, the P-type material is known as the ***anode*** of the diode.

II

1. The simplest rectifier circuit is the ***half-wave*** rectifier.
2. The ripple frequency of a bridge rectifier is ***two times*** the ripple frequency of the half-wave rectifier circuit.
3. The circuit having the highest output voltage for a given transformer secondary voltage is the ***bridge*** rectifier.
4. For a given full transformer secondary voltage, the full-wave rectifier circuit unfiltered dc output voltage is ***equal to*** the dc output of a half-wave rectifier that is using the same transformer.
5. To find the dc output (unfiltered) of a full-wave rectifier, multiply the full secondary rms voltage by ***0.45.***

REVIEW QUESTIONS

1. The term pentavalent indicates a material whose atoms have five valence electrons. Examples are arsenic and antimony. Pentavalent materials are used to dope semiconductor material to make it N-type semiconductor material.
3. Four
5. Electrons
7. Barrier potential refers to the reverse-bias potential voltage developed as a result of some majority carriers of the P and N materials diffusing across the junction, leaving a net positive charge near the junction on the N-type material side and a net negative charge near the junction on the P-type material side.
9. Diagram of a half-wave rectifier circuit is shown.

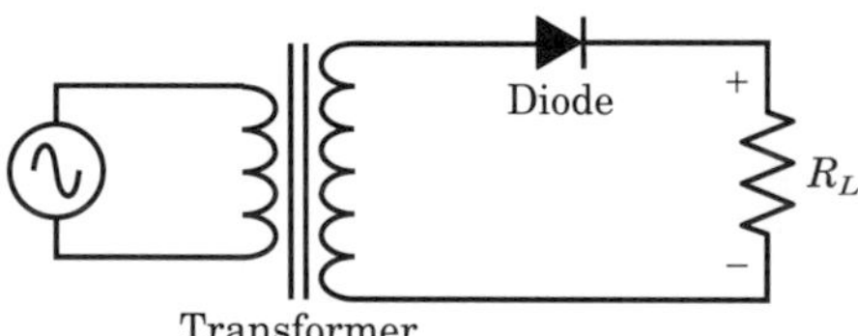

11. Diagram of a bridge rectifier circuit is shown.

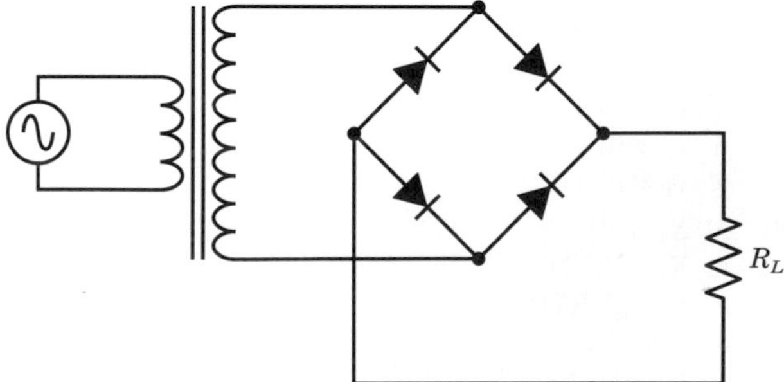

13. $V_{dc} = 0.9 \times V_{rms}$ of half the secondary
Since the transformer is step-up, primary voltage = 200 volts rms and full secondary voltage = 400 volts rms.
$$V_{dc} = 0.9 \times \left(\frac{400\text{ V}}{2}\right) = 0.9 \times 200 = 180\text{ V}$$

15. Most likely cause is the filtering. In most instances, the problem would be a deteriorated condition of the capacitor element(s).
17. Diagram of a shunt diode clipper is shown.

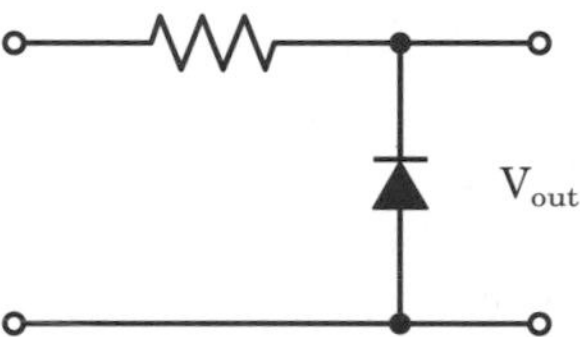

19. ac applied voltage for that alternation. (Reaching a maximum level of V-peak of the ac.)

TEST YOURSELF

1. The circuit of a H-W rectifier circuit is shown.

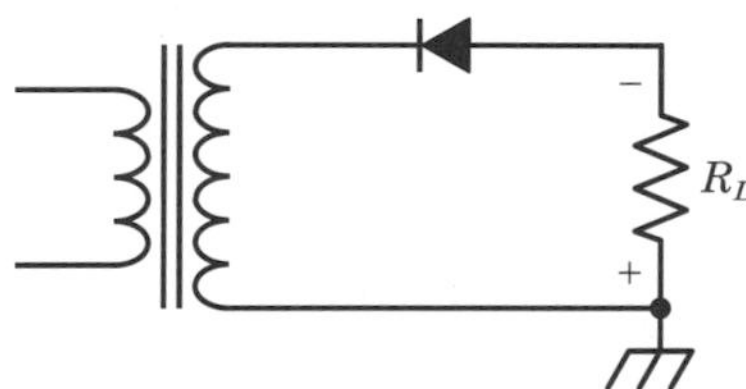

3. Critical inductance is the value of inductance required in the first choke of a filter network in order to keep the output voltage from going above the average value of the rectified wave at the input of the choke when the load current is small (i.e., minimum load current).
5. Optimum inductance is that value of the first choke where peak rectifier current is kept from exceeding the dc load current by more than 10% at maximum load levels (i.e., when load current is large).
7. a. Voltage across the full secondary of a transformer is calculated as shown:

   Since output $= 0.9 \times \frac{1}{2}$ secondary voltage,

   $100 = 0.9 \times \left(\frac{\text{secondary V}}{2}\right)$

   $200 = 0.9 \times \text{sec}$

   Secondary $V = \frac{200}{0.9} = 222.22$ volts

   b. Maximum possible output would be $1.414 \times V_{rms}$ of one-half the secondary, V max out $= 1.414 \times 111.11 = 157$ volts

   c. PIV on each rectifier would be $2.828 \times \frac{1}{2}$ secondary, or $1.414 \times$ full secondary rms voltage. Thus, PIV = 314 volts

Chapter 22

IN-PROCESS LEARNING CHECKS

I

1. A forward-biased PN junction is a ***low*** resistance path for current flow. A reverse-biased PN junction is a ***high*** resistance path for current flow.
2. Normally, the forward-biased junction of a junction transistor is the ***emitter-base*** junction.
3. Normally, the reverse-biased junction of a junction transistor is the ***collector-base*** junction.
4. The symbol for a NPN transistor shows the emitter element ***pointing away from the base.***
5. The symbol for a PNP transistor shows the emitter element ***pointing toward the base.***

II

1. The transistor amplifier circuit having the base electrode common to both input and output circuits is the ***common-base*** amplifier.
2. The transistor amplifier circuit configuration providing both voltage and current gain is the ***common-emitter*** amplifier circuit.
3. The transistor amplifier circuit having 180 degrees phase difference between input and output signals is the ***common-emitter*** amplifier.
4. The transistor amplifier circuit having a voltage gain of less than one (unity) is the ***common-collector*** amplifier circuit.
5. A transistor amplifier circuit frequently used for impedance matching is the ***common-collector*** amplifier circuit.

REVIEW QUESTIONS

1. Two
3. N-type material.
5. Approximately 96–99%

7. A reverse-biased PN junction is one where the P-type material is connected to the negative side of the source and the N-type material is connected to the positive side of the source.
9. A basing diagram is an outline sketch of a device, showing electrical elements of the device to which each "pin" is connected and the physical layout, or configuration, of the connection points relative to the packaging of the device.
11. Typical impedance characteristics of the various amplifier configurations are as shown below:
    a. Common-emitter: Input Z = low; Ouput Z = low to medium
    b. Common-base: Input Z = low; Output Z = fairly high
    c. Common Collector: Input Z = high; Output Z = low
13. Common-emitter circuit
15. Standard bipolar junction transistors are thought of as "current-operated" devices; whereas, FETs are thought of as "voltage-operated" devices.
17. Negative-resistance feature of a tunnel diode is that portion of its operating range where an increase in forward bias voltage causes a decrease in current through the device.
19. Capacitance
21. a. Use heat sinks when installing or removing devices, as appropriate.
    b. Use precautions relating to static electricity when working with certain types of transistor devices, such as MOSFETS.
23. Testing in circuit can be done by measuring bias voltages to see if they are normal and/or by use of special in-circuit test equipment.

TEST YOURSELF

1. Outline drawings for the TO–3, TO–5, and TO–92 transistor packages are shown.

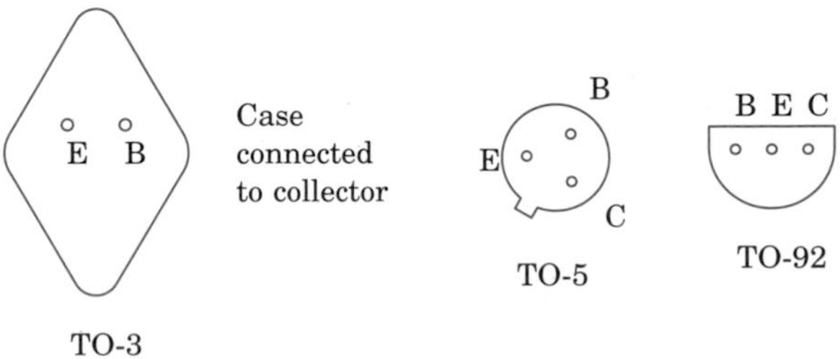

3. A diac is a two-electrode, three-layer bidirectional avalanche diode that can be switched on or off for either polarity of applied voltage. The bidirectional characteristic results from fairly equal doping of the two diode materials.
5. In the enhancement type, the gate must be forward-biased to produce active carriers and permit conduction through the channel. In the depletion type, charge carriers are present in the channel, and the channel is conductive when no bias voltage is applied to the gate.

Chapter 23

IN-PROCESS LEARNING CHECKS

I

1. To classify ICs, two modes of operation are ***linear*** and ***digital*** modes.
2. Monolithic and thin film refer to the ***fabrication/construction*** method of classifying ICs.
3. Flat-pack, can, and DIP describe IC classification in terms of ***packaging.***
4. An IC having more than 100 gates or equivalent circuits is categorized as ***LSI.***
5. Two very important IC families are the transistor-transistor logic, or ***TTL*** family, and the complementary symmetry-metal-oxide semiconductor, or ***CMOS*** devices.
6. The basic numerical system used in computers and digital circuits is the ***binary*** system.

II

1. A two-input logic gate where if either input goes high the output goes high is called a(n) ***OR*** gate.
2. A logic gate whose output is inverted from the input level is called a(n) ***inverter.***
3. A two-input logic gate where both inputs must be high to have a high output is called a(n) ***AND*** gate.
4. If we invert the output of an OR gate, we essentially have a ***NOR*** gate.
5. A logic gate that is essentially the inversion of an AND gate output is called a(n) ***NAND*** gate.

REVIEW QUESTIONS

1. During the 1950s
3. Digital operation refers to devices or circuits whose outputs vary or switch between discrete steps or levels. Examples are light switch (on or off states), doorbell, washers, dryers, furnaces, logic circuits, and switching circuits.
5. Some advantages of ICs include: small size and small space requirements, low power requirements, high complexity of circuitry possible, immunity to noise, and low cost.
7. Top and bottom views of a 16-pin DIP IC package are shown.

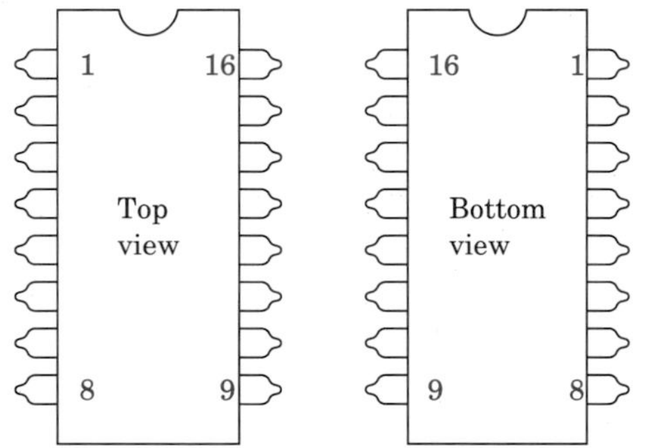

9. TTL means "Transistor-Transistor-Logic"
11. The "base" of the binary system is "2"
13. Block diagram of a simple computer is shown.

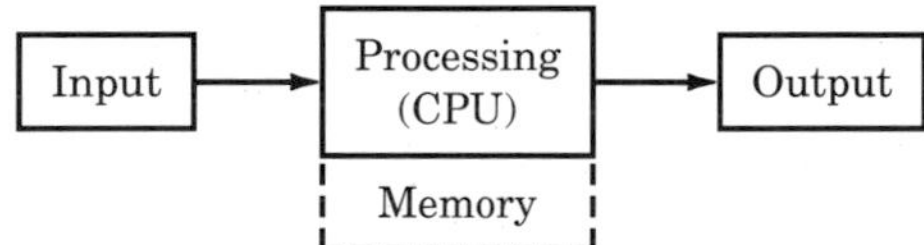

TEST YOURSELF

1. Diagram showing a 7400, quadruple, 2-input positive NAND gate is shown.

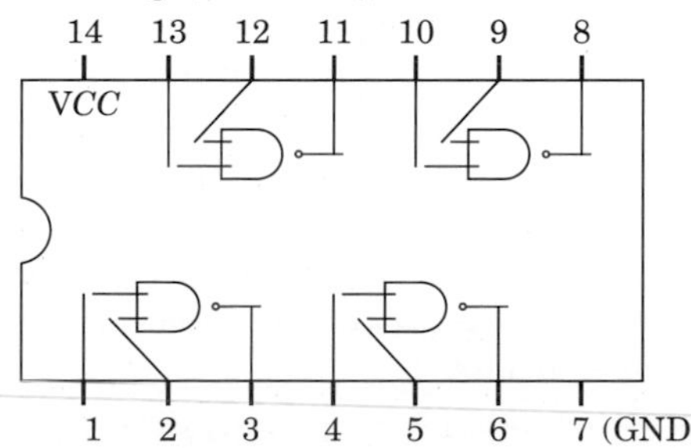

3. Diagram is shown.

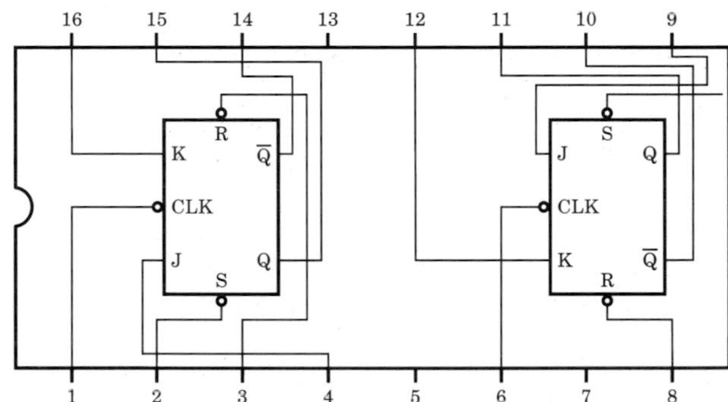

5. Chart showing the numbers 0 through 15 (decimal) in both octal and hexadecimal number systems is shown.

OCTAL	DECIMAL	HEXADECIMAL
0	0	0
1	1	1
2	2	2
3	3	3
4	4	4
5	5	5
6	6	6
7	7	7
10	8	8
11	9	9
12	10	A
13	11	B
14	12	C
15	13	D
16	14	E
17	15	F

Chapter 24

PRACTICE PROBLEMS

I

$$\text{Divider current: } I = \frac{V_{CC}}{R_1 + R_2} = \frac{10\text{ V}}{12\text{ k}\Omega} = 0.83\text{ mA}$$

Base voltage: $V_B = V_{R_2} = I$ divider $\times R_2 = 0.83$ mA $\times$ 2 kΩ = 1.66 V

Emitter voltage: $V_E = V_B - 0.7$ V = 1.66 − 0.7 = 0.96 V

Emitter current: $I_E = \frac{V_E}{R_E} = \frac{0.96\ V}{1\ k\Omega} = 0.96$ mA

Collector voltage: $V_C = V_{CC} - (I_C \times R_C) = 10$ V − (0.96 mA × 4.7 kΩ) = 10 V − 4.5 V = 5.5 V

Voltage across transistor: $V_{CE} = V_C - V_E = 5.5$ V − 0.96 V = approximately 4.5 V

IN-PROCESS LEARNING CHECKS

I

1. Transistor amplifiers are classified according to ***signal levels, class of operation, circuit configuration,*** and ***type of device used.***
2. Class A amplifiers conduct during ***360*** degrees of the input cycle; Class B amplifiers conduct for ***180*** degrees of the input cycle; and Class C amplifiers conduct for approximately ***120*** degrees of the input cycle.
3. A load line shows all the possible operating points of a given amplifier circuit from ***cutoff*** to ***saturation.***
4. The input signal to a common-emitter amplifier is 4 mV and the output signal (V_C) swings between +4 and +5 volts. What is the voltage gain of the amplifier? ***250***
5. A signal causes a common-emitter base current to swing by 0.03 mA and the collector current varies 1.5 mA pk-to-pk. What is the current gain of the amplifier? ***50***
6. A common-emitter amplifier has a voltage gain of 600 and a current gain of 30. What is the power gain? ***18,000***

REVIEW QUESTIONS

1. Small-signal transistor amplifier stages may be used in audio amplifier pre-amplifier stages and in radio receiver "front-end" radio-frequency (rf) amplifier stages; large-signal transistor amplifier stages in audio power amplifier stages and high-level transmitter stages.
3. Class A
5. The Q point is generally set at the mid-point of the loadline for Class A operation.
7. The Q point is set at cutoff for Class B operation.
9. Approximately 10 volts
11. a. $I = \frac{V_{CC}}{R_1 + R_2} = \frac{15\ V}{10.1\ k\Omega} = 1.49$ mA
    b. $V_B = I \times R_2 = 1.49$ mA × 1 kΩ = 1.49 V
    c. $V_E = V_B - 0.7$ V = 1.49 − 0.7 = 0.79 V (and $I_E = 0.79$ mA)
    d. $V_C = V_{CC} - (I_C \times R_C) = 15 - (0.79$ mA × 10 kΩ) = 15 − 7.9 = 7.1 V
    e. $V_{CE} = V_C - V_E = 7.1 - 0.79 = 6.31$ V
13. Common-collector amplifier
15. Circuit 3

TEST YOURSELF

1. Approximately 17 to 18 mA
3. I_E = about 6 mA; $V_E = I_E \times R_E = 6$ mA × 220 Ω = 1.32 V

Chapter 25

PRACTICE PROBLEMS

I

1. 200 kΩ
2. 1 kΩ

II

Gain = 11

IN-PROCESS LEARNING CHECKS

1. An op-amp is a high gain ***direct-coupled*** amplifier.
2. The input stage of an op-amp uses a ***differential*** amplifier configuration.
3. The name operational amplifier derives from the fact that their early use was to perform mathematical ***operations.***
4. The open-loop gain of an op-amp is much ***higher*** than the closed loop gain.

5. To achieve an output that is the inversion of the input, the input signal is fed to the ***negative*** input of the op-amp.
6. Input signals to the op-amp can be fed to the ***inverting*** input, ***noninverting*** input, or to both inputs.

REVIEW QUESTIONS

1. Differential amplifier circuit
3. Open-loop gain is the gain of the op-amp device without any external feedback system affecting gain.
5. $\frac{R_i + R_f}{R_i} = \frac{26\ k\Omega}{1\ k\Omega}$ = a gain of 26
7. Output signal should be small.
9. For differentiation, use a resistor as the feedback component.
11. A comparator is often used to compare voltages in order to see which one (which of the two input signals) is larger.
13. If the device were NOT already in positive saturation, it would cause the output to go "high." If the device were in positive saturation, the device would not change states.
15. The two parameters are gain and frequency response.

TEST YOURSELF

1. The circuit of an inverting op-amp having a gain of 500 is as shown below.

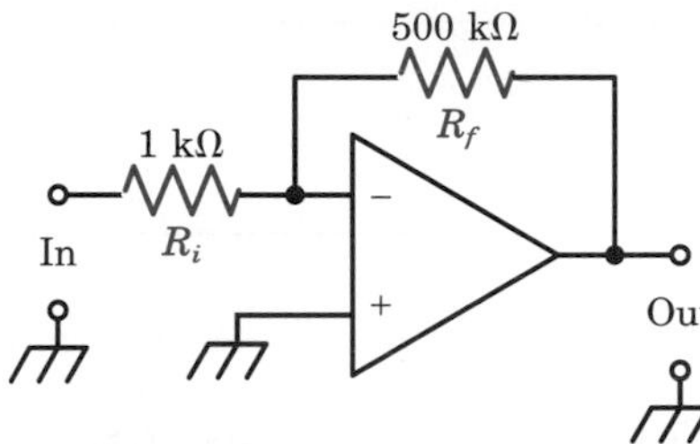

3. The circuit of an inverting op-amp circuit using an offset-null adjustment circuit that minimizes output offset voltage is as shown below.

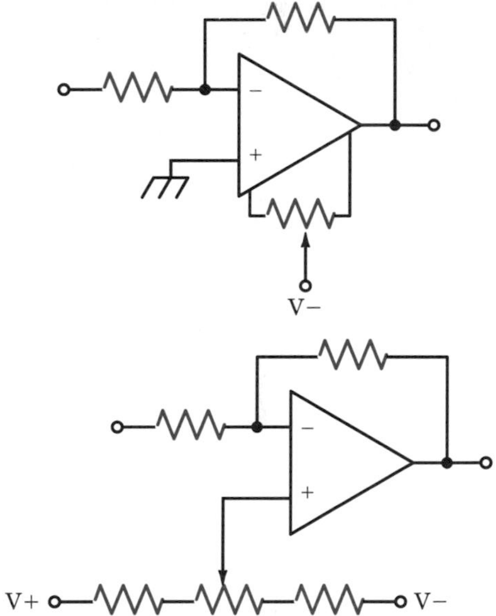

Circuits for device without internal nulling provisions.

APPENDIX C

Chapter Challenge Test Results

Find the number listed next to the test you chose and record the result.

1. 1.4 V
2. signal normal
3. 2 kΩ
4. 2 V
5. 1.5 V
6. high
7. 1.5 V
8. 0 V
9. 0 V
10. 3 V
11. 0 V
12. 7 V
13. ∞ Ω
14. 0 Ω
15. ≈0 V
16. 1 kΩ
17. 400 Ω
18. 1.5 V
19. ∞ Ω
20. 1.5 V
21.
22. 10 V
23. no signal
24. circuit operates normally
25. 10 kΩ
26. slightly high
27. 0 V
28. 0 V
29. 0 Ω
30. greatly below max value
31. 5 V
32. 400 Ω
33. ∞ Ω
34. 6.4 V
35. noticeably high
36. 4 kΩ
37. 10 kΩ
38. 3.3 V
39. 14 V
40. 0 V
41. ≈1.5 mH
42. 3.0 V
43. 7 V
44. 1.5 V
45. high
46. 0 Ω
47. 0 V
48. 2.3 V
49. 14 V
50. 0.5 V
51. 1.5 V
52. 10 V
53. (14 VAC)
54. 10 kΩ
55. max value
56. 10 V
57. signal normal
58. ∞ Ω
59. 1.5 kΩ
60. 275 pF
61. 14 V
62. 1 kΩ
63. ∞ Ω
64. 1.5 V
65. ≈6 VDC
66. 0 V
67. 7.5 V
68. 10 kΩ
69. low
70. signal normal
71. no change in operation
72. 0 Ω
73. 1.5 V
74. (120 VAC)
75. 10 V
76. slightly low
77. 5.9 V
78. 3.1 V
79. 7.5 V
80. 7 V
81. 2 V
82. 0 V
83. 10 V
84. 0 Ω
85. slightly below max value
86. 1.5 V
87. 5 Ω
88. 14 V
89. 2.3 V
90. ∞ Ω
91. 0 V
92. normal
93. 10 kΩ
94. ≈5 V
95. 10 V
96. ≈11 V
97. 0 V
98. low
99. 1.5 V
100. ∞ Ω
101. 0 V
102. 0 V
103. 0 V
104. ≈5 kΩ
105. 0 V
106. no signal
107. 50 pF
108. ∞ Ω
109. 12 kΩ
110. 0 V
111. no change in operation
112. 1.5 V
113. signal normal
114. 297 Ω
115. 0 V
116. ≈5 V
117. within normal range

APPENDIX D

Copper-Wire Table

Wire Size A.W.G. (B&S)	Diam. in Mils[1]	Circular Mil Area	Turns per Linear Inch (25.4 mm)[2] Enamel	S.C.E.	D.C.C.	Cont.-duty current[3] single wire in open air	Cont.-duty current[3] wires or cables in conduits or bundles	Feet per Pound (0.45 kg) Bare	Ohms per 1000 ft. 25° C	Current Carrying Capacity[4] at 700 C.M. per Amp.	Diam. in mm.	Nearest British S.W.G. No.
1	289.3	83690	—	—	—	—	—	3.947	.1264	119.6	7.348	1
2	257.6	66370	—	—	—	—	—	4.977	.1593	94.8	6.544	3
3	229.4	52640	—	—	—	—	—	6.276	.2009	75.2	5.827	4
4	204.3	41740	—	—	—	—	—	7.914	.2533	59.6	5.189	5
5	181.9	33100	—	—	—	—	—	9.980	.3195	47.3	4.621	7
6	162.0	26250	—	—	—	—	—	12.58	.4028	37.5	4.115	8
7	144.3	20820	—	—	—	—	—	15.87	.5080	29.7	3.665	9
8	128.5	16510	7.6	—	7.1	73	46	20.01	.6405	23.6	3.264	10
9	114.4	13090	8.6	—	7.8	—	—	25.23	.8077	18.7	2.906	11
10	101.9	10380	9.6	9.1	8.9	55	33	31.82	1.018	14.8	2.588	12
11	90.7	8234	10.7	—	9.8	—	—	40.12	1.284	11.8	2.305	13
12	80.8	6530	12.0	11.3	10.9	41	23	50.59	1.619	9.33	2.053	14
13	72.0	5178	13.5	—	12.8	—	—	63.80	2.042	7.40	1.828	15
14	64.1	4107	15.0	14.0	13.8	32	17	80.44	2.575	5.87	1.628	16
15	57.1	3257	16.8	—	14.7	—	—	101.4	3.247	4.65	1.450	17
16	50.8	2583	18.9	17.3	16.4	22	13	127.9	4.094	3.69	1.291	18
17	45.3	2048	21.2	—	18.1	—	—	161.3	5.163	2.93	1.150	18
18	40.3	1624	23.6	21.2	19.8	16	10	203.4	6.510	2.32	1.024	19
19	35.9	1288	26.4	—	21.8	—	—	256.5	8.210	1.84	.912	20
20	32.0	1022	29.4	25.8	23.8	11	7.5	323.4	10.35	1.46	.812	21
21	28.5	810	33.1	—	26.0	—	—	407.8	13.05	1.16	.723	22
22	25.3	642	37.0	31.3	30.0	—	5	514.2	16.46	.918	.644	23
23	22.6	510	41.3	—	37.6	—	—	648.4	20.76	.728	.573	24
24	20.1	404	46.3	37.6	35.6	—	—	817.7	26.17	.577	.511	25
25	17.9	320	51.7	—	38.6	—	—	1031	33.00	.458	.455	26
26	15.9	254	58.0	46.1	41.8	—	—	1300	41.62	.363	.405	27
27	14.2	202	64.9	—	45.0	—	—	1639	52.48	.288	.361	29
28	12.6	160	72.7	54.6	48.5	—	—	2067	66.17	.228	.321	30
29	11.3	127	81.6	—	51.8	—	—	2607	83.44	.181	.286	31
30	10.0	101	90.5	64.1	55.5	—	—	3287	105.2	.144	.255	33
31	8.9	80	101	—	59.2	—	—	4145	132.7	.114	.227	34
32	8.0	63	113	74.1	61.6	—	—	5227	167.3	.090	.202	36
33	7.1	50	127	—	66.3	—	—	6591	211.0	.072	.180	37
34	6.3	40	143	86.2	70.0	—	—	8310	266.0	.057	.160	38
35	5.6	32	158	—	73.5	—	—	10480	335	.045	.143	38–39
36	5.0	25	175	103.1	77.0	—	—	13210	423	.036	.127	39–40
37	4.5	20	198	—	80.3	—	—	16660	533	.028	.113	41
38	4.0	16	224	116.3	83.6	—	—	21010	673	.022	.101	42
39	3.5	12	248	—	86.6	—	—	26500	848	.018	.090	43
40	3.1	10	282	131.6	89.7	—	—	33410	1070	.014	.080	44

[1] A mil is 0.001 inch. A circular mil is a square mil $\times \frac{\pi}{4}$. The circular mil (c.m.) area of a wire is the square of the mil diameter.

[2] Figures given are approximate only; insulation thickness varies with manufacturer.

[3] Max. wire temp. of 212°F (100°C) and max. ambient temp. of 135°F (57°C).

[4] 700 circular mils per ampere is a satisfactory design figure for small transformers, but values from 500 to 1000 c.m. are commonly used.

APPENDIX E

Useful Conversion Factors

To Convert	Into	Multiply by
ampere-turns/cm	amp-turns/inch	2.54
ampere-turns/inch	amp-turns/cm	0.3937
ampere-turns/inch	amp-turns/meter	39.37
ampere-turns/meter	amp-turns/inch	0.0254
Centigrade	Farenheit	$\left(C^\circ \times \frac{9}{5}\right) + 32$
centimeters	feet	3.281×10^{-2}
centimeters	inches	0.3937
centimeters/sec	feet/sec	0.03281
circular mils	sq mils	0.7854
circumference	radians	6.283
circular mils	sq inches	7.854×10^{-7}
coulombs	faradays	1.036×10^{5}
degrees	radians	0.01745
dynes	joules/cm	10^{-7}
dynes	joules/meter (newtons)	10^{-5}
ergs	foot-pounds	7.367×10^{-6}
ergs	joules	10^{-7}
farads	microfarads	10^{6}
faradays	ampere-hours	26.8
faradays	coulombs	9.649×10^{4}
feet	centimeters	30.48
foot-pounds	ergs	1.356×10^{7}
gausses	lines/sq inch	6.452
gausses	webers/sq meter	10^{-4}
gilberts	ampere-turns	0.7958
grams	dynes	980.7
grams	pounds	2.205×10^{-3}
henries	millihenries	1,000
horsepower	foot-lbs/sec	550
horsepower	kilowatts	0.7457
horsepower	watts	745.7
inches	centimeters	2.54
inches	millimeters	25.4
joules	ergs	10^{7}
joules	watt-hrs	2.778×10^{-4}
kilolines	maxwells	1,000
kilowatts	foot-lbs/sec	737.6
kilowatts	horsepower	1.341
kilowatts	watts	1,000
kilowatt-hrs	joules	3.6×10^{6}
lines/sq inch	webers/sq meter	1550×10^{-5}
maxwells	kilolines	0.001
maxwells	webers	10^{-8}
microfarad	farads	10^{-6}
microns	meters	1×10^{-6}
millihenries	henries	0.001
mils	inches	0.001
nepers	decibels	8.686
ohms	megohms	10^{-6}
quadrants	degrees	90
quadrants	radians	1.571
radians	degrees	57.3
square inches	sq cms	6.452
temperature (°F)−32	temperature (°C)	$\frac{5}{9}$
watts	horsepower	1.341×10^{-3}
watts (Abs.)	joules/sec	1
webers	maxwells	10^{8}
webers/sq meter	webers/sq inch	6.452×10^{-4}
yards	meters	0.9144

APPENDIX F
Schematic Symbols

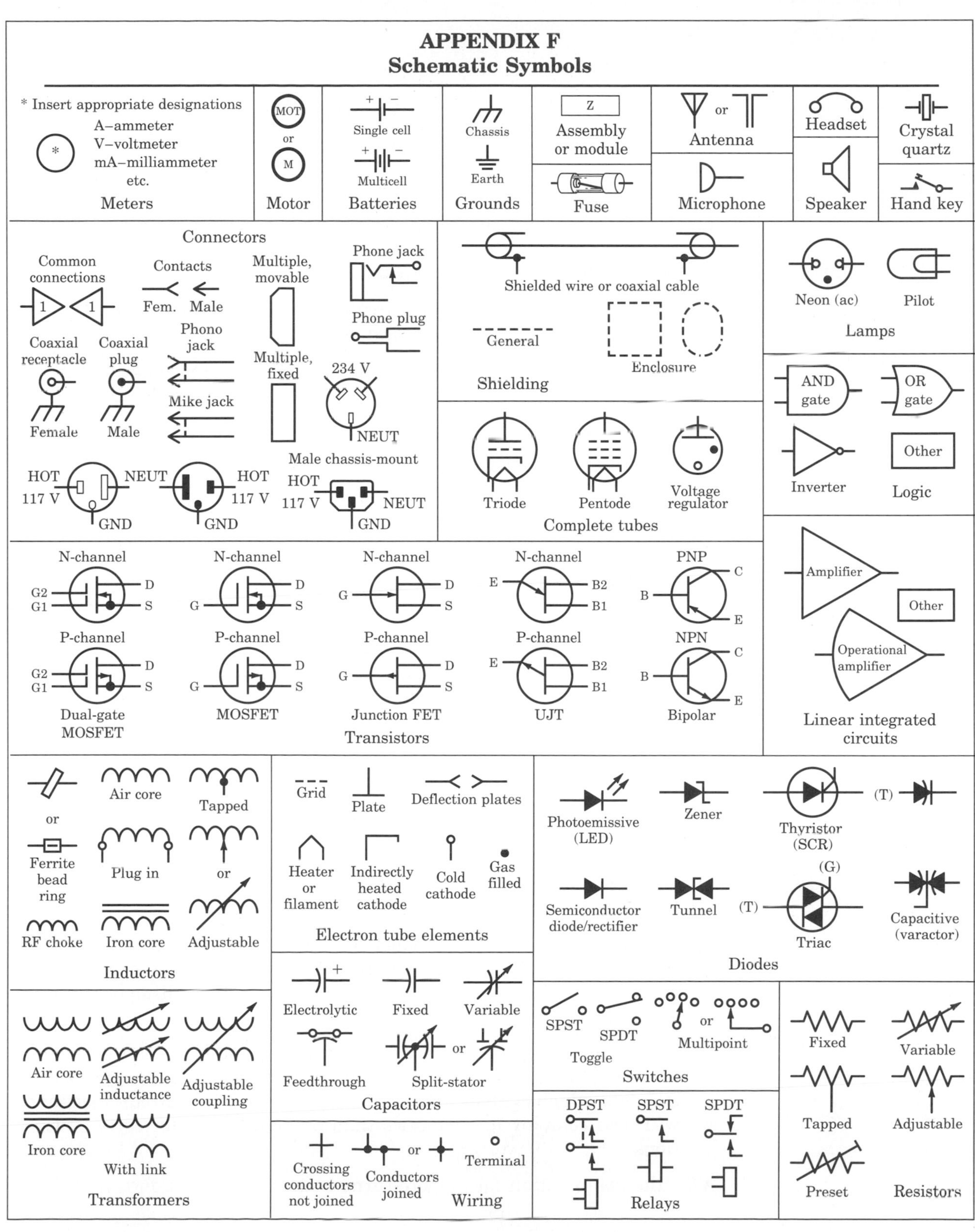

APPENDIX G

Glossary

ac abbreviation for alternating current. The letters "ac" are also used as a prefix, or modifier to designate voltages, waveforms, and so forth that periodically alternate in polarity.

admittance (Y) the ability of a circuit with both resistance and reactance to pass ac current; the reciprocal of impedance.

alternation one-half of a cycle. Alternations are often identified as the positive alternation or the negative alternation when dealing with ac quantities.

amalgamation in carbon-zinc dry cell manufacturing, coating the inside of the zinc electrode with a thin layer of mercury to help prevent local action.

ampere (A) the basic unit of current flow. An ampere of current flow represents electron movement at a rate of one coulomb per second; that amount of current flowing through one ohm of resistance with one volt applied.

ampere-hour rating a method of rating cells or batteries based on the amount of current they supply over a specified period with specified conditions.

analog operation circuit operation where the outputs vary continuous over a range as a direct result of the changing input levels; also called linear operation.

AND gate a digital logic gate where both inputs must be high for the output to be high.

anode the diode part that needs to be more positive than the cathode so the diode will conduct current. Its symbol is a triangle or arrow.

apparent power (S) the product of V and I in an ac circuit, without regard to phase difference between V and I. The unit of measure is volt-amperes. In the power triangle it is the vector resulting from apparent and reactive power.

assumed voltage method a method of finding total resistance in parallel circuits by assuming a circuit applied voltage value to find branch currents, which enables finding total current and total circuit resistance with less complex mathematics than other methods for certain situations.

AT the abbreviation for ampere-turns or magnetomotive force.

atom the basic building block of matter composed of different types of particles. Major atom particles are the electron, proton, and neutron.

attenuator a controllable voltage divider network in oscilloscopes that help adjust the amplitude of input signals to a useable size.

autotransformer a special single-winding transformer where part of the winding is in both the primary and secondary circuit. These transformers can be made to step-up or step-down, depending on where the winding of the primary source and the secondary load are connected.

average value (for one-half cycle) in ac sinusoidal values, the average height of the curve above the zero axis, expressed as 0.637 times maximum value. (NOTE: The average height over an *entire* cycle is zero; however, the average height of one alternation is as defined above.)

B the symbol representing flux density.

backoff ohmmeter scale the reverse (right-to-left) scale characteristic of series-type ohmmeters. Zero ohms is indicated on the scale's right end and infinite ohms on the scale's left end. Also, this scale is a nonlinear scale. That is, calibration marks and values are not evenly spaced across the scale.

bandpass the band of frequencies that passes through a filter with minimal attenuation or degradation; often designated by specific frequencies, where the lower frequency is the frequency at the low-frequency half-power point, and the upper frequency is the frequency at the upper half-power point on the resonance curve; also called pass-band.

bandpass filter a filter that passes a selected band of frequencies with little loss but greatly attenuates frequencies either above or below the selected band.

bandstop filter a filter that stops (greatly attenuates) a selected band of frequencies while allowing all other frequencies to pass with little attenuation; also called band-reject filter.

bandwidth the frequency limits where a specified response level is shown by a frequency-sensitive circuit, such as a resonant circuit. The cutoff level is defined at 70.7% of peak response, sometimes called the half-power points on the response curve.

base the center of the NPN or PNP sandwich in a BJT that separates the emitter and the collector transistor regions. The emitter-base junction is forward-biased and shows low impedance to current flow. The collector-base junction is reverse-biased and displays high resistance to current flow.

battery a dc voltage source containing a combination of cells, connected to produce higher voltage or current than a single cell produces alone.

bilateral resistance any resistance having equal resistance in either direction; that is, R is the same for current passing either way through the component.

bipolar junction transistor (BJT) a three-terminal, bipolar semiconductor containing both majority and minority current carriers. The device has alternate layers of P- and N-type semiconductor materials (which form two PN junction diodes). Examples are the NPN and the PNP transistor.

bleeder the resistive component assuring the minimum current drawn from a power supply system is the output voltage to achieve reasonable regulation over the range of output currents demanded from the supply; serves to discharge the filter capacitors when the power supply is turned off, which adds safety for people working on electronic equipment fed by the power supply. This resistive component also called bleeder resistor.

bleeder current the fixed current through a resistive system that helps keep voltage constant under varying load conditions.

bleeder resistor the resistor connected in parallel with power supply to draw a bleeder current.

block diagrams graphic illustrations of systems or subsystems by means of blocks (or boxes) that contain information about the function of each block. These blocks are connected by lines that show direction of flow (signal, fluid, electrical energy, etc.) and how the blocks interact. Block diagrams can illustrate any kind of system: electrical, electronic, hydraulic, mechanical, and so forth. "Flow" generally moves from left to right through the diagram but is not a requirement.

bridge circuit a special series-parallel circuit usually composed of two, two-resistor branches. Depending on the relative values of these resistors, the bridge output (between the center points of each branch) may be balanced or unbalanced.

bridging or shunting connecting a component, device, or circuit in parallel with an existing component, device, or circuit.

capacitance the ability to store electrical energy in the form of an electrostatic field.

capacitive reactance (X_C) the opposition a capacitor gives to ac or pulsating dc current. The symbol used to represent capacitive reactance is X_C.

capacitor a device consisting of two or more conductors, separated by nonconductor(s). When charged, a capacitor stores electrical energy in the form of an electrostatic field and blocks dc current.

cathode the diode part that needs to be negative with respect to the anode so the diode will conduct current. Its symbol is a bar with a perpendicular connecting lead. (This bar symbol is also perpendicular to the anode arrow symbol.)

cell (relates to voltage sources) a single (stand-alone) element or unit within a combination of similar units that convert chemical energy into electrical energy. The chemical action within a cell produces dc voltage at its output terminals or contacts.

cemf (counter-emf or back-emf) the voltage induced in an inductance that opposes a current change and is present due to current changes causing changing flux linkages with the conductor(s) making up the inductor.

charging a capacitor moving electrons from one plate of a capacitor and moving electrons off the other plate; resultant charge movement is called charging current.

chip a semiconductor substrate where active devices (transistors) and passive devices (resistors and capacitors) form a complete circuit.

circuit a combination of elements or components that are connected to provide paths for current flow to perform some useful function.

Clapp oscillator a circuit that resembles a modified Colpitts oscillator circuit. Compared with the Colpitts oscillator, the Clapp circuit uses an additional capacitor placed in series with the inductor in the frequency-determining circuit.

class A, B, and C in reference to amplifier operation, the classifications that denote the operating range of the circuit. Class A operates halfway between cutoff and saturation. Class B operates at cutoff. Class C operates at below cutoff bias.

coefficient of coupling (k) the amount or degree of coupling between two circuits. In inductively coupled circuits, it is expressed as the fractional amount of the total flux in one circuit that links the other circuit. If all the flux links, the coefficient of coupling is 1. The symbol for coefficient of coupling is k. When k is multiplied by 100, the percentage of total flux linking the circuits is expressed.

collector the transistor electrode that receives current carriers that have transversed from the emitter through the base region; the electrode that removes carriers from the base-collector junction.

Colpitts oscillator a circuit that generates ac signals with dc voltages applied; typified by the capacitive voltage divider used in its frequency-determining and feedback circuits.

common-base relating to transistor amplifiers, the circuit configuration where the base is common to both input and output circuits.

common-base amplifier (CB) a transistor amplifier circuit configuration where the base electrode is common to both input and output circuits with respect to signals; sometimes called a grounded-base amplifier.

common-collector the transistor amplifier circuit configuration where the collector is common for input and output circuits.

common-collector amplifier (CC) a transistor amplifier circuit configuration where the collector electrode is common to both input and output circuits with respect to signals; sometimes called an emitter follower.

common-emitter in reference to transistor amplifiers, the circuit configuration where the emitter is common to both input and output circuits.

common-emitter amplifier (CE) a transistor amplifier circuit configuration where the emitter electrode is common to both input and output circuits with respect to signals; sometimes called a grounded-emitter amplifier.

common-mode gain comparison (ratio) of the output voltage of a differential amplifier to the common-mode input voltage. Ideally, the output should be zero when the same input is fed to both differential amplifier inputs.

common-mode input an input voltage to a differential amplifier that is common to both its inputs.

common-mode rejection ratio (CMRR) a measure expressing the amount of rejection a differential amplifier shows to common-mode inputs; the ratio between the differential-mode gain and the common-mode gain equals the CMRR.

complex numbers term that expresses a combination of a *real* and an *imaginary* term which must be added vectorially or with phasors. The real term is either a positive or negative number displayed on the horizontal axis of the rectangular coordinate system. The imaginary term, *j operator,* expresses a value on the vertical or j-axis. The value of the real term is expressed first. For example, 5 + j10 means plus 5 units (to the right) on the horizontal axis and up 10 units on the vertical axis.

compound a form of matter that can be chemically divided into simpler substances and that has two or more types of atoms.

conductance the ease that current flows through a component or circuit; the opposite of resistance.

conductor a material that has many free electrons due to its atoms' outer ring having less than 4 electrons, which is less than the 8 needed for chemical and/or electrical stability.

constant current source a current source whose output current is constant even with varying load resistances. Its output voltage varies to enable constant output current, as appropriate.

constant voltage source a voltage source where output voltage (voltage applied to circuitry connected to source output terminals) remains virtually constant even under varying current demand (load) situations.

coordinate system in ac circuit analysis, a system to indicate phase angles within four, 90° quadrants. The vertical component is the "Y" axis. The horizontal component is the "X" axis.

coordinates the vertical and horizontal axes that describe the position and magnitude of a vector.

copper losses the I^2R losses in the copper wire of the transformer windings. Since current passes through the wire and the wire has resistance to current flow, an I^2R loss is created in the form of heat, rather than this energy being a useful part of energy transfer from primary source to secondary load.

core soft iron piece to help strengthen magnetic field; helps produce a uniform air gap between the permanent magnet's pole pieces and itself.

core losses energy dissipated in the form of heat by the transformer core that subtracts from the energy transferred from primary to secondary; basically comprised of eddy-current losses (due to spurious induced currents in the core material creating an I^2R loss) and hysteresis losses, (due to energy used in constantly changing the polarity of the magnetic domain in the core material).

cosine a trigonometric function of an angle of a right triangle; the ratio of the adjacent side to the hypotenuse.

cosine function a trigonometric function used in ac circuit analysis; relationship of a specific angle of a right triangle to the side adjacent to that angle and the hypotenuse where $\cos\theta = \frac{\text{adjacent}}{\text{hypotenuse}}$.

coulomb the basic unit of charge; the amount of electrical charge represented by 6.25×10^{18} electrons.

covalent bonding the pairing of adjacent atom electrons to create a more stable atomic structure.

CRT (cathode-ray tube) the oscilloscope component where electrons strike a luminescent screen material causing light to be emitted. Waveforms are displayed on an oscilloscope's CRT.

crystal oscillator an oscillator circuit that uses a special element (usually quartz-crystal material and holder) for the primary frequency-determining element in the circuit.

current (I) the progressive movement of electrons through a conductor. Current is measured in Amperes.

current divider circuit any parallel circuit divides total current through its branches in inverse relationship to its branch R values.

current leading voltage sometimes used to describe circuits where the circuit current leads the circuit voltage, or the circuit voltage lags the circuit current.

current ratio the relationship of the current values of secondary and primary, or vice versa. In an ideal transformer, the current ratio is the *inverse* of the voltage and turns ratios; $\frac{I_P}{I_S} = \frac{N_S}{N_P} = \frac{V_S}{V_P}$.

cycle a complete series of values of a periodic quantity that recur over a period of time.

damping (electrical) the process of preventing an instrument's pointer from oscillating above and below the measured value. Electrical damping is due to the magnetic field produced by induced current opposing the motion that induced it.

depletion region the region near the junction of a P-N junction where the current carriers are depleted due to the combining effect of holes and electrons.

depolarizer in a dry cell, the material to help prevent or lessen polarization. A common depolarizing material is manganese dioxide.

dielectric nonconductive material separating capacitor plates.

dielectric constant the relative permittivity of a given material compared to the permittivity of vacuum or air; the relative ability of a given material to store electrostatic energy (per given volume) to that of vacuum or air.

dielectric strength the highest voltage a given dielectric material tolerates before electrically breaking down or rupturing; rated in volts per thousandths of an inch, V per mil.

differential amplifier an amplifier circuit configuration having two inputs and one output; produces an output that is related to the difference (or differential) between the two inputs.

differential-mode gain the gain of a differential amplifier when operated in the differential mode; the ratio of its output voltage to the differential-mode (difference between the two input voltages) input voltage.

digital operation circuit operation where the output varies in discrete steps or in an on-off, high-low, 1-0 category of operation.

discharging a capacitor balancing (or neutralizing) the charges on capacitor plates by allowing excess electrons on the negative charged plate to move through a circuit to the positive charged plate. When opposite capacitor plates have the same charge, the voltage between them is zero and the capacitor is fully discharged.

doping introducing specific impure atoms into a semiconductor crystal (lattice) structure to aid current conduction through the material.

dropping resistor a resistor connected in series with a given load that drops the circuit applied voltage to the level required by the load when passing the rated load current through it. (Recall a *voltage drop* is the difference in voltage between two points caused by a loss of pressure (or loss of emf) as current flows through a component offering opposition to current flow).

effective value (rms) the value of ac voltage or current that produces the same heating effect in a resistor as that produced by a dc voltage or current of equal value. In ac, this value is 0.707 times maximum value.

efficiency percentage, or ratio, of useful output (such as power) delivered compared to amount of input required to deliver that output. For example, in electrical circuits, the amount of power input to a device, component stage, or system compared to the amount of power delivered to the load. Generic formula is Efficiency (%) = $\left(\frac{P_{out}}{P_{in}}\right) \times 100$

electrical charge an excess or deficiency of electrons compared to the normal neutral condition of having an equal number of electrons and protons in each atom of the material.

electrode generally, cne of the electrical terminals that is a conducting path for electrons in or out of a cell, battery, or other current generating device.

electrolyte a liquid, paste, or substance through which electrical conduction occurs due to chemical processes of the electrolyte and other materials immersed or exposed to it.

electromagnetism the property of a conductor to produce a magnetic field when current is passing through it.

electron the negatively charged particle in an atom orbiting the atom's nucleus.

electrostatic (electric) field the region surrounding electrically charged bodies; area where an electrically charged particle experiences force.

element a form of matter that cannot be chemically divided into simpler substances and that has only one type of atom.

emitter the transistor electrode that injects current carriers into the adjacent base region.

energy the ability to do work, or that which is expended in doing work. The basic unit of energy is the erg, where 980 ergs = 1 gram cm of work. A common unit in electricity/electronics is the joule which equals 10^7 ergs.

ϵ the Greek letter epsilon used to signify the mathematical value of 2.71828. This number, raised to various exponents that relate to time constants, determines electrical circuit parameter values in circuits where exponential changes occur.

equation a mathematical statement of equality between two quantities or sets of quantities. Example, $A = B + C$.

equivalent resistance the total resistance of two or more resistances in parallel.

Faraday's law the law states the amount of induced voltage depends on the rate of cutting magnetic flux lines. In other words, greater number of flux lines cut per unit time produces greater induced voltage.

feedback with signal amplifying or signal generating circuits, that small portion of the output signal that is coupled back to the input circuit of the stage.

field effect transistor (FET) a unipolar transistor that depends on the effect of an electric field on charges to control current flow.

filter in power supplies, the component or combination of components that reduce the ac component of the pulsating dc output of the rectifier prior to the output fed to a load.

flux relating to magnetism; the magnetic lines of force.

flux density (B) the magnetic flux per unit cross-sectional area; unit of measure is the tesla (T) representing 1 Wb per square meter.

focus control a scope control that adjusts the narrowness of the electron stream striking the screen to create a clear and sharp presentation.

force in electrical charges, the amount of attraction or repulsion between charges. In electromotive force, the potential difference in volts.

forward bias the external voltage applied to a semiconductor junction that causes forward conduction, for example, positive to P-type material and negative to N-type material.

frame/movable coil the parts in a moving-coil meter movement that move in response to the interaction between the fixed magnetic field and the field of the current-carrying, movable coil.

free electrons the electrons in the outer ring of conductor materials that are easily moved from their original atom. Sometimes termed valence electrons.

frequency (f) the number of cycles occurring in a unit of time. Generally, frequency is the number of cycles per second, and Hertz (Hz) is cycles per second. In formulas, frequency is expressed as f, where $f = \frac{1}{T}$.

full-scale current the value of current that causes full-scale, or full-range deflection of the pointer or indicator device in a current meter movement; the maximum current that should pass through the moving-coil in a current meter movement.

gain controls oscilloscope controls that adjust the horizontal and vertical size of the image displayed.

ground reference a common reference point for electrical/electronic circuits; a common line or conducting surface in electrical circuits used as the point where electrical measurements are made; a common line or conductor where many components make connection to one side of a power source. Typically, ground reference is the chassis, or a common printed circuit "bus" (or line) called chassis ground. In power line circuits, may also be "earth ground."

H the symbol representing magnetic field intensity.

Hartley oscillator a type of circuit that generates ac signals with dc voltages applied; typified by the special tapped coil used in its frequency-determining and feedback circuits.

henry the basic unit of inductance, abbreviated "H". One henry of inductance is that amount of inductance having one volt induced cemf when the rate of current change is one ampere per second.

Hertz (Hz) the unit of frequency expressing a frequency of one cycle per second.

high-pass filter a filter that passes frequencies *above* a specified cutoff frequency with little attenuation. Frequencies below the cutoff are greatly attenuated.

horizontal (X) amplifier the scope amplifier circuitry that amplifies signals fed to the horizontal deflection circuitry.

hybrid circuits generally, circuits comprised of components created by more than one technique, such as thin-film, discrete transistors, and so on.

I_m or I_{fs} full-scale current value of the meter movement.

imaginary numbers complex numbers; the imaginary part of a complex number that is a real number multiplied by the square root of minus one.

impedance the total opposition an ac circuit displays to current at a given frequency; results from both resistance and reactance. Symbol representing impedance is "Z" and unit of measure is the ohm.

impedance ratio the relationship of impedances for the transformer windings; directly related to the *square* of the turns ratio; $\frac{Z_P}{Z_S} = \frac{{N_P}^2}{{N_S}^2}$. Reflected impedance is the impedance reflected back to the primary from the secondary. This value depends on the square of the turns ratio and value of the load impedance connected across the secondary.

inductance (self-inductance) the property of a circuit to oppose a change in current flow.

induction the ability to induce voltage in a conductor; ability to induce magnetic properties from one object to another by magnetically linking the objects.

inductive reactance the opposition an inductance gives to ac or pulsating current.

in-line typically means in series with the source.

instantaneous value the value of a sinusoidal quantity at a specific moment, expressed with lower case letters, such as "e" or "v" for instantaneous voltage and "i" for instantaneous current.

insulator a material in which the atoms' outermost ring electrons are tightly bound and not free to move from atom to atom, as in a conductor. Typically, these materials have close to the 8 outer shell electrons required for chemical/electrical stability.

integrated circuit (IC) a combination of electronic circuit elements. In relation to semiconductor ICs, this term generally indicates that all of these components are fabricated on a single substrate or piece of semiconductor material.

intensity control a scope control that adjusts the number of electrons striking the screen per unit time; thus controlling the brightness of the screen display.

inverter a digital logic gate where the output is the opposite of the input. If the input is low, the output is high. If the input is high, the output is low.

ion atoms that have gained or lost electrons and are no longer electrically balanced, or neutral atoms. Atoms that lose electrons become positive ions. Atoms that gain electrons become negative ions.

I_s current through the shunt resistance.

isolation transformer a 1:1 turns ratio transformer providing electrical isolation between circuits connected to its primary and secondary windings.

joule the unit of energy. One joule of energy is the amount of energy moving one coulomb of charge between two points with a potential difference of one volt of emf; the work performed by a force of one newton acting through a distance of one meter equals one joule of energy. Also, 3.6×10^6 joules = 1 kilowatt hour.

Kirchhoff's current law the value of current entering a point must equal the value of current leaving that same point in the circuit.

Kirchhoff's voltage law the arithmetic sum of voltage drops around a closed loop equals V applied; the algebraic sum of voltages around the entire closed loop, including the source, equals zero.

L, L_T L is the abbreviation representing inductance or inductor. L_T is the abbreviation for total inductance. L_1, L_2, and so on represent specific individual inductances.

$\frac{L}{C}$ ratio the ratio of inductance to capacitance in a LC resonant circuit. Generally, the higher this ratio is, the higher is the Q of the circuit and the more selective its response.

leading in phase in ac circuits the electrical quantity that first reaches its maximum positive point is specified leading the electrical quantity that reaches that point later.

LED (light-emitting diode) a semiconductor that radiates light when current passes through it; a P-N junction device emitting light when forward biased.

left-hand rules related to magnetism; when the thumb and fingers of the left hand determine the direction of magnetic fields of current-carrying conductors to find the polarity of an electromagnet's poles.

Lenz's law the law states the direction of an induced voltage or current opposes the change causing it.

linear network a circuit whose electrical behavior does not change with different voltage or current values.

lines in magnetic circuits, lines represent flux.

lines of force (flux) imaginary lines representing direction and location of the magnetic influence of a magnetic field relative to a magnet.

Lissajous pattern(s) scope patterns on the CRT screen when sine-wave signals are simultaneously applied to the horizontal and vertical deflection systems. These patterns are useful to determine phase and frequency relationships between the two signals.

load the amount of current or power drain required from the source by a component, device, or circuit.

loaded voltage divider a network of resistors designed to create various voltage levels of output from one source voltage; where parallel loads, which demand current, are connected.

loading effect changing the electrical parameters of an existing circuit when a load component, device, or circuit is connected. With voltmeters, the effect of the meter circuit's resistance when in parallel with the portion of the tested circuit; thus, causing a change in the circuit's operation and some change in the electrical quantity values throughout the circuit.

local action in carbon-zinc cells, the formation of multiple little local cells created by zinc impurities that are suspended in electrolyte near the zinc electrode. Local action causes the cell to be eaten away and degenerates the cell's chemical capability, even when the cell is not in use.

long time-constant circuit when the circuit conditions are such that the $\frac{t}{RC}$ ratio = 0.1 (or less). This means the RC time required to charge or discharge the capacitor is long compared with the time allowed during each alternation or cycle.

low-pass filter a filter that permits all frequencies *below* a specified cutoff frequency to pass with little attenuation. Frequencies above the cutoff frequency are attenuated greatly.

$\frac{L}{R}$ **time constant** the time required for an exponentially changing quantity to change by 63.2% of the total value of change to be achieved. For a circuit containing inductance, it is the time (in seconds) required for the current to acquire a level that is 63.2% of its final value. One time constant (in seconds) in a series RL or inductor circuit equals $\frac{L}{R}$, where L is in henrys and R is in ohms.

magnetic field the field of influence or area where magnetic effects are observed surrounding magnets; a region of space where magnetic effects are observed.

magnetic field intensity (H) the amount of magnetomotive force per unit length. In the MKSA system, it is the ampere-turns per meter (AT/meter).

magnetic polarity the relative direction of a magnet's flux with respect to its ends or poles; defined as the North-seeking pole and the South-seeking pole of a magnet.

magnetism the property causing forces of attraction or repulsion in certain materials; property causing voltage to be induced in nearby conductors when there is relative motion between the magnetic object and the conductor.

magnetomotive force (mmf or F) the force causing magnetic lines of force through a medium, thus establishing a magnetic field; developed by current through a coil, where, the amount of mmf relates to the amount of current and the number of turns on the coil (F = AT).

magnitude the comparative size or amount of one quantity with respect to another quantity of the same type.

majority carriers the electrons or holes in extrinsic semiconductor materials that enable current flow. In N-type materials, the majority carriers are electrons. In P-type materials, the majority carriers are holes.

matter anything that has weight and occupies space, whether it is a solid, liquid, or gas.

maximum power transfer theorem a theorem that states when the resistance (or impedance) of the load is properly matched to the internal resistance of the power source, maximum power transfer can take place between the source and the load.

Maxwell (Mx) one line of force, or one flux line.

meter shunt (current meter) the parallel resistance used with current meter movements to allow them to measure currents higher than their full-scale current rating; a percentage of the total current being measured is shunted, or bypassed around the meter by the meter shunt.

microprocessor typically, a VLSI chip that acts as the central processing unit (CPU) for a computer system.

minority carriers the electrons in P-type material and the holes in N-type material; called minority since there are fewer electrons in P-type material than holes and fewer holes in N-type material than electrons.

mode related to semiconductor ICs; the type of operation for which the IC is designed. The linear mode is commonly used for amplifiers. The digital mode is often used for gates.

molecule the smallest particle of a compound that resembles the compound substance itself.

monolithic one piece as in a monolithic IC formed on one semiconductor substrate (or one stone).

MOS abbreviation for metal-oxide semiconductor; relates to the fabrication method.

moving-coil (d'Arsonval) movement a type of meter movement. A coil is suspended within a permanent magnetic field and caused to move when current passes through it. Movement is due to the interaction of its magnetic field with the permanent magnet's field.

μ_r the symbol for relative permeability.

multimeter an instrument designed to measure several types of electrical quantities, for example, current, voltage, and resistance; commonly found in the form of the VOM and digital multimeter (DMM).

multiple-source circuit a circuit containing more than one voltage or current source.

multiplier resistor(s) (voltmeter) the current-limiting series resistance(s) used with current meter movements, enabling them to be used as voltmeters that measure voltages much greater than the basic movement's voltage drop at full-scale current. The value of the multiplier resistance is that its resistance (plus the meter's resistance) limits current to the full-scale value at the desired voltage-range level.

mutual inductance (M or L_M) the property that enables a current change in one conductor or coil to induce a voltage in another conductor or coil, and vice versa. This induced voltage results from magnetic flux of one conductor or coil linking the other conductor or coil. The symbol for mutual inductance is M or L_M and the unit of measure is the henry.

NAND gate an AND gate with inverted output. That is, if both inputs are high, the output is low. For all other input conditions, the output is high.

network a combination of electrically connected components.

neutron a particle in the nucleus of the atom that displays no charge. Its mass is approximately equal to the proton.

NOR gate an OR gate with inverted output. That is, if either or both inputs are high, the output is low. Only when both inputs are low is output high.

Norton's theorem a theorem regarding circuit networks stating that the network can be replaced (and/or analyzed) by using an equivalent circuit consisting of a single "constant-current" source (called I_N) and a single shunt resistance (called R_N).

NPN transistor a transistor made of P material sandwiched between two outside N material areas. Then, going from end-to-end, the transistor is an NPN type. The symbol for the NPN transistor shows the emitter arrow going away from the base.

ohm (Ω) the basic unit of resistance; the amount of electrical resistance limiting the current to one ampere with one volt applied.

Ohm's law a mathematical statement describing the relationships among current, voltage, and resistance in electrical circuits. Common equations are $V = I \times R$, $I = \frac{V}{R}$, and $R = \frac{V}{I}$.

ohms-per-volt rating a voltmeter rating indicating how many ohms of resistance are needed to limit the current through the meter to its full-scale value when one volt is applied to the meter circuit.

op-amp abbreviation for operational amplifier; a versatile amplifier circuit with very high gain, high input impedance, and low output impedance.

open circuit any break in the current path that is undesired, such as a broken wire or component, or designed, such as open switch contacts in a lighting circuit.

OR gate a digital logic gate where if either input is high, or if both inputs are high, the output is high.

oscilloscope a device that visually displays signal waveforms so time and amplitude parameters are easily determined.

parallel branch a single current path within a circuit having two or more current paths where each of the paths are connected to the same voltage points; a single current path within a parallel circuit.

parallel circuit a circuit with two or more paths for current flow where all components are connected between the same voltage points.

peak-inverse-voltage (PIV) the maximum voltage across a diode in the reverse (non-conducting) direction. The maximum voltage that a diode tolerates in this direction is the peak-inverse-voltage rating.

peak-to-peak value (pk-pk) the difference between the positive peak value and the negative peak value of a periodic waveform; computed for the sine wave as 2.828 times effective value.

peak value (V_{pk} or maximum value, V_{max}) the maximum positive, negative value, or amplitude that a sinusoidal quantity reaches, sometimes referred to as peak value. It is calculated as 1.414 times effective value.

pentavalent materials materials that have five valence electrons; used in doping semiconductor material to produce N-type materials where electrons are the majority current carriers.

period (T) the time required for one complete cycle. In formulas, it is expressed as T, where $T = \frac{1}{f}$.

permeability (mu or μ) the ability of a material to pass, conduct, or concentrate magnetic flux.

phase angle (θ) the difference in time or angular degrees between ac electrical quantities (or periodic functions) when the two quantities reach a certain point in their periodic function. This difference is expressed by time difference (the fraction of a period involved), or by angular difference (expressed as degrees or radians). (NOTE: A radian = 57.3°. $2\pi \times$ a radian = 360°). An example is the difference in phase between circuit voltage and circuit current in ac circuits that contain reactances, such as inductive reactance.

phasor a quantity that expresses position relative to time.

photodiode cell a device where resistance between terminals is changed by light striking its light-sensitive element; a light-controlled variable resistance. Typically, the more light striking it, the lower its terminal resistance is.

P-N junction the location in a semiconductor where the P and N type material join; where the transition from one type material to the other type occurs.

PNP transistor a transistor where the middle section is N-type material and the two outside sections are composed of P-type material, thus called a PNP transistor. The symbol for the PNP transistor shows the emitter arrow going toward the base.

pointer the element in a meter that provides the indicating function of how much movement has taken place by the movable coil.

polar form notation expressing circuit parameters with respect to polar coordinates; expressing a point with respect to distance and angle (direction) from a fixed reference point on the polar axis. For example, $25 \angle 45°$.

polarity in an electrical circuit, a means of designating differences in the electrical charge condition of two points, or a means of designating direction of current flow. In a magnetic system, a means of designating the magnetic differences between two points or locations.

polarization relating to cells or batteries, the undesired formation of hydrogen bubbles around the positive electrode that degenerates the cell's operation.

pole pieces special meter pieces used to help linearize the magnetic field in a moving-coil movement.

position control(s) (vertical and horizontal) scope controls that move the display vertically or horizontally; thus centering the display on the screen.

potential in electrical circuits, the potential, or ability, to move electrons. In electromotive force, the difference in charge levels at two points. This potential difference is measured in volts.

power the rate of doing work. In mechanical terms, a horsepower equals work done at a rate of 550 foot pounds per second. In terms of electrical power, it is electrical work at a rate of one joule per second.

power factor (p.f.) the ratio of true power to apparent power. It is equal to the cosine of the phase angle (θ).

powers of ten working with small or large numbers can be simplified by converting the unit to a more manageable number times a power of ten. An example is 0.000006 amperes expressed as 6×10^{-6} amperes. Some of the frequently used powers of ten are 10^{-12} for pico (p) units, 10^{-6} for micro (μ) units, 10^{-3} for milli units, 10^{3} for kilo units, and 10^{6} for mega units.

product-over-the-sum method a method of finding total resistance in parallel circuits by using two resistances at a time in the formula $R_T = \frac{R_1 \times R_2}{R_1 + R_2}$.

proton the positively charged particle in an atom located in the nucleus of the atom. Its weight is approximately 1836 times that of the electron, and approximately the same weight as the neutron.

pulsating dc a varying voltage that is one polarity, as opposed to reversing polarity as ac does.

Pythagorean theorem a theorem expressing the mathematical relationship of the sides of a right triangle, where the square of the hypotenuse is equal to the sum of the squares of the other two sides. That is, $c^2 = a^2 + b^2$. Equation used in electronics is $c = \sqrt{a^2 + b^2}$.

Q (figure of merit) a number indicating the ratio of the amount of stored energy in the inductor magnetic field to the amount of energy dissipated. Found by using the ratio of the inductance's inductive reactance (X_L) to the resistance of the inductor (R). That is, $Q = \frac{X_L}{R}$.

Q factor the relative quality or figure of merit of a given component or circuit; often related to the ratio of $\frac{X_L}{R}$ of the inductor in the resonant circuit. Other circuit losses can also enter into the effective Q of the circuit, such as the loading that a load coupled to the resonant circuit causes and the effect of R introduced in the circuit. The less the losses are, the higher the Q is. The sharpness of the resonant circuit response curve is directly related to the Q of the circuit.

quadrants segments of the coordinate system representing 90° angles, starting at a 0° reference plane.

range select (switch) the switch used with meter movements that connects or disconnects various shunt resistances values (for current or ohmmeter measurements), or multiplier resistances for the voltage measurement mode.

RC time constant the time, in seconds, represented by the product of R in ohms and C in farads; time required for a capacitor to charge or discharge 63.2% of the change in voltage level applied. Five (5) RC time constants are needed for a capacitor to completely charge or discharge to a changed voltage level.

reactance the opposition a capacitor or inductor gives to alternating current or pulsating direct current.

reactive power (Q) the vertical component of the ac power triangle that equals VI sin θ; expressed as VAR (voltampere-reactive).

real numbers in math, any rational or irrational number.

rectangular form notation expressing a set of parameters with the complex number system. For example, $R \pm jX$ indicates a combination of resistance and reactance. Thus, $5 + j10$ means 5 ohms of resistance in series with 10 ohms of inductive reactance.

rectify (rectification) the process of changing ac to pulsating dc.

regulator the component or circuitry to help keep output voltage and/or current constant with varying input voltages and/or with varying loads.

relative permeability (μ_r) comparing a material's permeability with air or a vacuum; relative permeability is not a constant, since it varies for a given material depending on the degree of magnetic field intensity present; relative permeability $(\mu_r) = \dfrac{\text{flux density with given core}}{\text{flux density with vacuum core}}$

reluctance (R) the opposition of a given path to the flow of magnetic lines of force through it.

residual magnetism the magnetism remaining in the core material of an electromagnet after the magnetizing force is removed.

resistance (R) in an electrical circuit, the opposition to electron movement or current flow. Its value is affected by many factors, including the dimensions and material of the conductors, the physical makeup and types of the components in the circuit, and temperature.

resistor an electrical component that provides resistance in ohms and that controls or limits current or distributes or divides voltage.

resistor color code a system to display the ohmic value and tolerance of a resistor by means of colored stripes or bands around the body of the resistor.

resonance a special circuit condition existing when opposite types of reactances in a circuit are equal and have cancelling effects (i.e., $X_L = X_C$).

reverse bias the external voltage applied to a semiconductor junction that causes the semiconductor not to conduct, for example, negative to P-type material and positive to N-type material.

ripple the ac component in a pulsating dc signal, such as the output of a power supply.

R_m meter movement resistance value.

R_{mult} resistance value of the resistor used that enables a basic current meter to measure voltage by its current-limiting effect.

saturation in magnetism, the condition when increasing the magnetizing force does not increase the flux density of a magnetic material.

scalar a quantity expressing only quantity, such as a "real number."

scale the face of a meter containing calibration marks that interpret the pointer's position in appropriate units of measure.

schematic diagrams graphic illustrations of circuits or subcircuits and/or systems or subsystems. Schematics are more detailed than block diagrams and give specific information regarding how components are connected. Schematics show the types, values, and ratings of component parts, and in many cases show the electrical values of voltage, current, and so forth at various points throughout the circuit.

SCR abbreviation for silicon-controlled rectifier; often used to control the time during an ac cycle the load receives power from the power source. A four-layer, P-N-P-N device where conduction is initiated by a voltage to the gate. Conduction doesn't cease until anode voltage is reversed, removed, or greatly reduced although the initiating signal has been removed.

selectivity referring to resonant circuits, selectivity indicates the sharpness of the response curve. A narrow and more pointed response curve results in a higher degree of frequency selectiveness or selectivity.

semiconductor a material with 4 valence (outermost ring) electrons. It is neither a good conductor, nor a good insulator.

semiconductor diode A semiconductor P-N junction that passes current easily in one direction while not passing current easily in the other direction.

sensitivity rating a means of rating meters, or meter movements to indicate how much current is required to cause full-scale deflection; how much voltage is dropped by the meter movement at full-scale current value. Also, voltmeters are rated by the ohms-per-volt required to limit current to full-scale value. (This is inversely related to the current sensitivity rating. That is, the reciprocal of the full-scale current value yields the ohms-per-volt rating.

series circuit a circuit in which components are connected in tandem or in a string providing only one path for current flow.

series-parallel circuit a circuit comprised of a combination of both series-connected and parallel-connected components and/or subcircuits.

series-type ohmmeter circuit the ohmmeter circuit where the meter, its zero-adjust resistance, and its internal voltage source are in series. When the test leads are connected across an unknown resistance to be measured, that resistance is placed in series with the meter circuit.

short circuit generally, an undesired very low resistance path across two points in a circuit. It may be across one component, several components, or across the entire circuit. There are designed shorts in circuits, such as purposeful "jumpers" or closed switch contacts.

short time-constant circuit describes a circuit where the frequency is such that the capacitor has at least ten times one RC time constant to charge or discharge. A short time-constant circuit is sometimes described as one where: $\frac{t}{RC} = 10$ (or greater).

siemans the unit used to measure or quantify conductance; the reciprocal of the ohm $\left(G = \frac{1}{R}\right)$.

sine a trigonometric function of an angle of a right triangle; the ratio of the opposite side to the hypotenuse.

sine function a trigonometric function used in ac circuit analysis; expresses the relationship of a specified angle of a right triangle to the side opposite that angle and the hypotenuse where $\sin \theta = \frac{\text{opposite}}{\text{hypotenuse}}$.

sine wave a wave or waveform expressed in terms of the sine function relative to time.

sinusoidal quantity a quantity that varies as a function of the sine or cosine of an angle.

skirts term typically used to describe the portions of a resonance response curve that descend down from the peak response point. The steepness of the skirts reflects the selectivity of the circuit.

source in electrical circuits, the source is the device supplying the circuit applied voltage, or the electromotive force causing current to flow through the circuit. Sources range from simple flashlight cells to complex electronic circuits.

source/gate/drain the electrodes in a field effect transistor device; gate controls current flow between the source and drain.

springs the elements in a moving-coil meter movement that conduct current to the movable coil and provide a mechanical force against which the moving coil must operate.

SSI, MSI, LSI, and VLSI SSI is small-scale integration (less than 10 gates on a single chip). MSI is medium-scale integration (less than 100 gates on a single chip). LSI is large-scale integration (more than 100 gates on a single chip). VLSI is very-large-scale integration (over 1000 gates on a single chip).

steady-state condition the condition where circuit values and conditions are stable or constant; opposite of transient conditions, such as switch-on or switch-off conditions, when values are changing from one state to another.

stops pegs that limit the left and right movement of the meter's pointer and related moving elements.

subcircuit a part or portion of a larger circuit.

superposition theorem superimposing one set of values on another to create a set of values different from either set of the superimposed values alone; the algebraic sum of the superimposed sets of values.

susceptance (B) the relative ease that ac current passes through a pure reactance; the reciprocal of reactance; the imaginary component of admittance.

sweep circuits circuits in an oscilloscope that cause the horizontal trace and the rapid retrace of the electron beam across the CRT.

sweep frequency control(s) scope controls that set the number of times the horizontal trace occurs per unit time. Frequently, this control is calibrated in time/cm to measure the time per centimeter traced on the display. (NOTE: The formula $f = \frac{1}{T}$ is then used to determine frequency.)

switches devices for making or breaking continuity in a circuit.

synchronization the process of bringing two entities into time coherence. In a scope, the circuitry and process that cause the beginning of the horizontal trace to be synchronized with the vertical input signal.

synchronization control(s) scope controls that control the timing of the horizontal sweep in synchronization with the signal(s) being tested.

tangent function a trigonometric function used in ac circuit analysis; expresses the relationship of a specified angle of a right triangle to the side opposite and side adjacent to that angle where $\tan \theta = \frac{\text{opposite}}{\text{adjacent}}$.

TC the abbreviation often used to represent time constant.

tesla a unit of flux density representing one weber per square meter in the MKSA system of units.

Thevenin's theorem a theorem used to simplify analysis of circuit networks because the network can be replaced by a simplified equivalent circuit consisting of a single voltage source (called V_{TH}) and a single series resistance (called R_{TH}).

thin-film process a process where thin layers of conductive or nonconductive materials are used to form integrated circuits.

thyristor a P-N-P-N semiconductor switch with special characteristics.

transformers devices that transfer energy from one circuit to another through electromagnetic (mutual) induction.

triac a special semiconductor device, sometimes called a 5-layer N-P-N-P-N device, that controls power over an entire cycle, rather than half the cycle, like the SCR device; equivalent of two SCRs in parallel connected in opposite directions but having a common gate.

trivalent materials materials that have three valence electrons; used in doping semiconductor material to produce P-type materials where holes are the majority current carriers.

true (or real) power (P) the power expended in the ac circuit resistive parts that equals VI cos θ, or apparent power $\times$ cos θ. The unit of measure is watts.

tunnel diode a special PN junction diode that has been doped to exhibit high-speed switching capability and a special negative-resistance over part of its operating range; sometimes used in oscillator and amplifier circuits.

turns ratio the number of turns on the separate windings of a transformer; expressed as the number of turns on the secondary versus the number of turns on the primary $\left(\frac{N_S}{N_P}\right)$; frequently stated as the ratio of primary-to-secondary turns.

universal (Ayrton or ring) shunt a method of connecting and selectively tapping resistances so various series and shunt resistance values are connected to the meter circuit, thus affording multiple current ranges. Circuitry assures that the meter is never without some shunt resistance value; thus, providing some measure of meter protection.

valence electrons the electrons in the outer shell of an atom.

varactor diode a silicon, voltage-controlled semiconductor capacitor where capacitance is varied by changing the junction bias; used in various applications where voltage-controlled, variable capacitance can be advantageous.

vector a quantity that illustrates both magnitude and direction. A vector's direction represents position relative to space.

vertical (Y) amplifier the scope amplifier circuitry that amplifies signals fed to the vertical input jack(s).

vertical volts/cm control(s) scope controls that adjust the amplification or attenuation of the input signal fed to the vertical deflection system. Generally, this control has positions that provide calibrated vertical deflections in ranges of millivolts per centimeter or volts per centimeter; thus enabling measurement of waveform voltages.

volt (V) the basic unit of potential difference or electromotive force (emf); the amount of emf causing a current flow of one ampere through a resistance of one ohm.

voltage the amount of emf available to move a current.

voltage divider action the dividing of a specific source voltage into various levels of voltages by means of series resistors. The distribution of voltages at various points in the circuit is dependent on what proportion each resistor is of the total resistance in the circuit.

voltage rate of change $\left(\frac{dV}{dt}\right)$ the amount of voltage level change per unit time. For example, a rate of change of one volt per second, 10 volts per μ second, and so forth.

voltage ratio in an ideal transformer with 100% coupling (k = 1), the voltage ratio is the same as the turns ratio, since the voltage induced in each turn of the secondary and the primary windings is the same for a given rate of flux change; $\frac{V_S}{V_P} = \frac{N_S}{N_P}$.

voltampere-reactive (VAR) the unit of reactive power in contrast to real power in watts; 1 VAR is equal to one reactive volt-ampere.

watt the basic unit of electrical power. The basic formula for electrical power is P (in watts) equals V (in volts) times I (in amperes).

weber the unit of magnetic flux indicating 10^8 maxwells (flux lines).

Wheatstone bridge circuit a special application of the bridge circuit commonly used to measure unknown resistances. It utilizes a sensitive current or voltage meter to determine when the bridge is balanced.

work the expenditure of energy, where work = force × distance. The basic unit of work is the foot poundal. NOTE: At sea level, 32.16 foot poundals = one foot pound.

zero-adjust control the adjustable series resistor in an ohmmeter used to regulate current through the meter movement so that with the test probes across zero ohms (or touching each other), there is full-scale current through the meter, and the pointer is located above the zero-ohms mark on the scale.

Index

Abbreviations:
basic electrical units, 39–40
Absolute permeability, 274
ac. *See* Alternating current
Acceptor atoms, 696
Ac circuit:
analyses of, 461–66
power in, 623–29
V and I relationships in purely capacitive circuit, 563–64
V and I relationships in purely inductive circuit, 435–37
V and I relationships in purely resistive circuit, 435, 563
Active filters, 674
Admittance, 640–42
Air-core coils, 511
Algebraic operations, 637–39
Alkaline cell, 249
Alpha, 731–32
Alternating current (ac), 348
angular motion and, 350–51
compared with dc, 349
coordinate system and, 351–52
parallel RC circuit analysis, 591–96
period and frequency, 358–60
phase relationships, 360–63
power transmission and, 349–50
purely resistive ac circuit, 367–68
rectangular wave, 369–70
sawtooth wave, 369–70
series RC circuit analysis, 585–91
sine wave current and voltage values, 363–67
sine wave descriptors, 357–58
sine wave generation, 352–57
square wave, 369–70
See also Capacitive reactance; Inductive reactance
Alternation, 357–58
Amalgamation, 238
Ampere, 25, 39
Amplifiers:
bipolar transistor collector curves, 775, 776
circuit configuration, 780–87
Class A operation, 778
Class B operation, 779
Class C operation, 780
classification methods, 774–80
common-base (CB) amplifier, 784–85
common-collector (CC) amplifier, 785–86
common-emitter (CE) amplifier, 780–84
drawing load line, 776–78
FET amplifier, 786–87
load line, 776
oscillator circuits, 787–91
Q point, 778
signal levels, 774–75
transistor amplification, 773–74
type of device, 787
See also Op-amps
Amplitude, 436
Analog circuits, 750, 755
AND gate, 758–59
Angle:
phase, 360–63, 481
right triangle, 462–66
of rotation, 350–51, 354–55
Angle of rotation, 350–51, 354–55
Angular motion, 350–51
Angular velocity, 439–40
Anode, 700
Apparent power, 626
Applied voltage, 86
Arcs, 424
Assumed voltage method, total resistance, 138–39
Atomic number, 10
Atomic shells, 11–12
Atomic weight, 10
Atoms, 5, 693–94
atomic number and weight, 10
atomic shells, 11–12
electron theory, 12–13
energies changing electrical balance of, 14
ions, 13
model of explaining electron theory, 8–10
particles, 8
structure of, 8–12
Audio frequencies, 482, 483
Audio transformer, 511
Autotransformer, 510
Average value, 364–65
Ayrton shunt. *See* Universal shunt

Back-emf. *See* Counter-emf
Backoff ohmmeter scale, 325
Bandpass, 669–70
determining, 673
Bandpass filter, 678–80
Bandpass filtering, 677
Bandstop filter, 680
Bandwidth, 667–68
determining, 673
related to Q, 669
Bar magnets, 260–63
Barrier potential, 699
Base, 730
Batteries, 234, 250–51
examples of, 249
lead-acid battery, 239–48
See also Cells
Battery ratings, 243
Beta, 731–32
B (flux density), 273–74
B-H curve, 276–77
Binary system, 756–57
Bipolar junction transistors, 730
Bipolar transistor collector curves, 775–76
Bistable multivibrator (flip-flop) circuit, 810
Bleeder current, 189
Bleeder-divider, 704
Bleeder resistor, 189, 716–17, 719

Block diagrams, 43, 46–47, 705, 800
Block or module level, troubleshooting, 113, 115
Bohr, Niels, 8
Boolean algebra, 759
Branch current, 478–81
Branch power dissipation, 140–41
Breakdown voltage, 533
Bridge circuits, 195–96
  See also Wheatstone bridge circuit
Bridge rectifier circuit, 710–12
  effect of filters on, 716
  peak-inverse-voltage, 712
  ripple frequency, 712
  unfiltered dc output voltage, 712

Calibrated scale, 301, 302
Calibrating voltage, 389–90
  measuring voltages greater or less than, 392–93
Capacitance:
  factors affecting value of, 531–33
  formulas, 533–35
  of parallel capacitors, 538–40
  of series capacitors, 535–36, 538
  unit of (farad), 529–30
Capacitive reactance, 562, 564, 572
  methods to calculate X_C, 566–69
  in parallel, 570–71, 574
  in series, 570, 573
  voltages and currents and, 572
  X_C and capacitance value, 564–65
  X_C and frequency of ac, 565–66
Capacitor:
  capacitance of parallel capacitors, 538
  ceramic, 546–47, 548
  charge distribution on parallel capacitors, 539
  charging action, 526–28
  color codes, 550, 551
  defined, 525
  discharging action, 528, 529
  electrolytic, 547–49
  electrostatic field, 525–26
  energy stored in electrostatic field of, 530–31
  fixed, 544
  mica, 546
  paper and plastic, 546
  problems with, 551–52
  ratings for, 550, 551
  RC time constant, 540–44
  tantalum, 549–50
  total capacitance in series, 535–36
  troubleshooting techniques, 552–54
  variable, 544–46
  voltage and series capacitors, 536–37
  voltage when three or more capacitors are in series, 538
Carbon-zinc cell. *See* Dry cell
Cathode, 700
Cathode-ray tube (CRT), 380, 382
C circuit, 585
Cells, 234, 250–51
  chemical action in, 235–37
  dry cell, 237–39
  other types of primary and secondary cells, 248–49
  wet cell, 239–48
  See also Batteries
Central processing unit (CPU), 764
Ceramic capacitors, 546–48
Charge, 39
Charged conditions, 241
Charges in motion:
  ampere (unit of current), 25
  current flow and, 24
  example of current flow, 24–25
  formula relating charge movement, time, and current, 25–26
Charging action, 526–27
Charging conditions, 242
Chassis ground, 103–4, 188
Circuit, 14
  basic, 29–30
  closed, 28
  open, 28–29
  total resistance in, 82–84
  See also ac circuit; Parallel circuit; Parallel RL circuit; RL circuit; RLC circuit; Series circuit; Series-parallel circuit; Series RL circuit
Clamp-on current probe, 330, 331
Clapp oscillator circuit, 789, 790
Class A operation, 778
Class B operation, 779
Class C operation, 780
Clipper (limiter) circuits, 702
Closed circuit, 28
Coefficient of coupling, 495–96
  and formulas for k and M, 500
Collapsing field, 424
Collector, 730
Color code, resistors:
  color values chart, 68
  gold and silver colors, special use of, 68–69
  significance of band positions, 68
  significance of colors, 67–68
Color codes, capacitors, 550, 551
Colpitts oscillator circuit, 789
Common-base (CB) amplifier, 733–34, 784–85
Common-collector (CC) amplifier, 735, 785–86
Common-emitter (CE) amplifier, 734–35, 773–74, 780–84
Common-mode gain, 806
Common-mode input, 806
Common-mode rejection ratio, 801–802
Comparator, 808–809
Complex numbers, 631
Component level, troubleshooting, 113, 116
Compounds, 6
Conductance, 39
Conductance method, total resistance, 134–35
Conductors, 14
  compared with semiconductors, 693
  current-carrying, 264–67
Constant current method, 244–45
Constant current source, 223
Constant voltage method, 244–45

Constant voltage source, 219
Contact potential, 699
Continuity checks, 335–36
Coordinate system, 351–52, 465, 631–32
Copper loss, 508
Core loss, 279, 508
Core materials, 275
  B-H curve, 276–77
  diamagnetic, 291
  ferrites, 291
  hysteresis loop, 277–79
  paramagnetic, 291
Cosine, 464
Coulomb, 20, 39
Coulomb, Charles, 19–20
Counterbalance, 301
Counter-emf (cemf), 410, 412
Counters, 763
Coupling:
  coefficient of, 495–96
  inductive, 495
Coupling capacitor, 598, 601
Covalent bonding, 694, 695
CRT (cathode-ray tube), 380, 382
Crystal oscillator circuit, 790
Current, 24–25, 39
  ac sine wave values, 363–67
  ampere (unit of current), 25
  bleeder, 189
  build-up and decay of in inductor's magnetic field, 421–24
  direction of current flow, 58
  measurement of, 334–35, 393
  in parallel circuits, 128–31
  relationship of to resistance with voltage unchanged, 49–51
  relationship of to voltage with resistance constant, 49
  in series-parallel circuits, 173–75
  in transistors, 730–31
  voltage checks to determine, 333–34
Current-carrying conductors:
  direction of field around, 264–65
  force between parallel conductors, 266
  left-hand rule for, 265–66
Current-divider circuits, 149
  formula for parallel circuit, any number of branches, 150
  formula for parallel circuit, two branches, 150–53
Current meters, 314–15
  connection of, 302–303
  measurement techniques, 334–35
  sensitivity of, 304–305
  trade-offs, 305–306
  troubleshooting, 332
Current meter shunts:
  basic shunt, 306
  multiple-range meter circuit, 309–10
  parameters for calculating resistance value, 306–307
  R value calculation methods, 307–309
  universal (Ayrton or ring) shunt, 310–14
Current ratio, 503–504
Cutoff frequency, 679
Cycle, 355, 357–58

Damping function, 302
d'Arsonval (moving coil) movement, 299–301
  advantages and disadvantages of, 302
  operation of, 301–302
dc. *See* Direct current
Demagnetizing coil, 279–80
Depletion region, 699
Depolarizing agent, 238
Design:
  parallel circuit to specifications, 148–49
  power supply, 705
  series circuit to specifications, 98–100
  series-parallel circuit to specifications, 185–87
Diagrams:
  block, 43, 46–47, 705, 800
  impedance, 473–75, 587, 589
  phasor, 457, 458
  power triangle, 629
  schematic, 47–48
  V-I vector, 586, 588
Diamagnetic materials, 291
Dielectric, 525
  constants, 532
  strength, 532–33
  thickness of, 531
  type of material, 531–32
Differential amplifier, 801, 802, 805–806
Differential-mode gain, 806
Differentiated output, 598
Differentiator circuit, 808
Diffusion current, 699
Digital circuits, 750, 755
Digital computer, 764–65
Diode clippers, 702–703
Direct current (dc), 348
  compared with ac, 349
  power transmission and, 349–50
  unfiltered output voltage, 706–707, 709, 712
  voltage measurement with oscilloscope, 392
Discharged conditions, 241
Discharging action, 528–29
Discharging conditions, 241
Divide-and-conquer approach, troubleshooting, 96–97
DMM (digital multimeter), 330
Donor atom, 696
Doping, 695
Drain, 736
Drawing-to-scale approach, 474
Dry cell:
  construction of, 237
  local action in, 238
  operating characteristics of, 238
  polarization in, 237–38
  shelf life of, 238
Dry electrolytic, 547

Earth ground, 188
Eddy current losses, 508
Edison cell, 249
Effective value (rms), 365
Efficiency factor, 212
Electrical balance, energies that change, 14

Electrical charges, law of, 17–18
Electrical circuit. *See* Circuit
Electrical potential:
  defined, 21
  methods for establishing and maintaining, 22–23
  volts (unit of potential difference), 21–22
Electrical shock, 70
Electrical system, 15–16
Electrical units, basic:
  charge, 39
  conductance, 39–40
  current, 39
  potential, 39
  resistance, 39
Electrodes, 235–36
Electrolyte, 235–36
Electrolytic capacitors, 547–49, 554
Electromagnetism, 269–70
  field around multiple-turn coil, 267
  field direction around current-carrying conductor, 264–65
  field strength of electromagnet, 268–69
  force between parallel current-carrying conductors, 266
  left-hand rule for current-carrying conductors, 265–66
  left-hand rule to determine polarity, 267–68
Electromotive force (emf), 21, 236–37
  induced, 281–84
  self-induced, 410
Electron beam, 385, 386
Electronic symbols, 44–45
Electrons, 8–10
  valence, 12
Electron theory, 12–13
Electrostatic fields, 525–26
Elements, 6
Emitter, 730
Energy, 59
  in capacitor's electrostatic field, 530
EPROM, 764
Erg, 59
Error:
  meter loading effects, 321–22, 337
  parallax, 337–38
EXCLUSIVE NOR gate, 761, 762
EXCLUSIVE OR gate, 761, 762
Exponential curve, 421

Farad, 529–30
Faraday, Michael, 259, 529
Faraday's law, 284, 410–11
Ferrites, 291
Ferromagnetic materials, 291
FET amplifier, 786–87
Field effect transistors (FET), 736–37
Fields of force, 20–21
Figure of merit, 447
Filter capacitors, 719
Filter chokes, 719
Filtering, 652
Filters, 681, 703
  active, 674
  frequency range of, 675
  general principle for, 675
  networks of, 675
  non-resonant, 675–78
  passive, 674
  resonant, 678–80
  See also Power supply filters
Fixed capacitor, 544
Flip-flops, 763, 810
Flux density (B), 260, 273–74, 290, 291
Flux lines. *See* Lines of force
Flux linkages, 495–96
Focus controls, 380
Foot poundal, 59
Forward bias, 701–702
Frequency (f), 358
  Hertz (Hz), 359
  plot of current versus, 660
  plot of impedance versus, 660
  plot of phase angle versus, 660
  relationship with period, 360
  resonant, 653
  X_C and, 565–66
  X_L and, 439–40
Frequency ratios, measurement of, 395–99
Full-scale current, 305
Full-wave doubler circuit, 717
Full-wave rectifier circuit, 708–709
  effect of filters on, 716
  peak-inverse-voltage, 710
  polarity of output voltage, 710
  ripple frequency, 710
  unfiltered dc output voltage, 709

Gain calculations, 783–84
Gain control, 385
Gas, 5–6
Gate, 736
Gaussmeter, 291
Generator, motor effect, 286–87
Generator action, 281–83
Generator effect, 287–88
Ground reference, 188, 333

Half-power points, 668
Half-wave cascade voltage doubler, 717
Half-wave rectifier circuit, 705–706
  effect of filters on, 716
  peak-inverse-voltage, 707
  polarity of output voltage, 708
  ripple frequency, 707
  unfiltered dc output voltage, 706–707
Hall effect, 290–91
Hall, E.H., 290
Hartley oscillator circuit, 788–89
henry (H), 411
Henry, Joseph, 259, 411
Hertz, Heinrich, 359
Hertz (Hz), 359
Hexadecimal system, 757
High-pass filters, 676–78
High voltage probe, 330, 331
H (magnetic field intensity), 273–74
Hole, 696

Horizontal sweep frequency
controls, 383–84
Horizontal sweep signal, 385
Horsepower, 59–60
Hydrometer test, 247
Hypotenuse, 463, 586–87
Hysteresis loop, 277–79
Hysteresis loss, 508

I^2R method, 626–27
ICs. *See* Integrated circuits
IGFET (insulated-gate), 737
Imaginary numbers, 631
Impedance:
defined, 472
in parallel RC circuits, 593–94
in series RLC circuit, 616
in series RL circuits, 472–75
Impedance matching, 211
Impedance ratio, 506–508
In-circuit test, 336, 552, 741
Induced magnetism, 263–64
Inductance, 408
defined, 409
factors that determine, 413–14
generator action, 281–83
L/R time constant, 419–24
motor action, 280–81
mutual, 415, 495–500
in parallel, 415–17
self-inductance, 410–12, 495
in series, 415
See also Inductors
Inductive coupling, 495
Inductive reactance in ac, 437,
444–46
calculating X_L, 440–43
quality factor (Q), 446–47
relationship of X_L to frequency
of ac, 439–40
relationship of X_L to inductance
value, 438–39
in series and parallel, 443–44
Inductors, 424–26, 434
applications for, 482–84
characteristics of, 409–10
collapsing field of, 421–24
energy stored in magnetic field
of, 417–19
in parallel, 415–17
physical properties affecting
inductance, 413–14
in series, 415
troubleshooting, 426
See also Inductance; Inductive
reactance
Instantaneous values, 367
Instruments, 298
Insulator materials, 14
Integrated circuits (ICs):
advantages and disadvantages
of, 755
classifications of, 751–55
counters, 763
digital, 752
digital computer, 764–65
examples of analog and digital
modes of operation, 755
flip-flops, 763
history of, 751
hybrid, 752
linear (analog), 752
logic families, 752–55
logic gates, 757–62
LSI, 752
memory, 763–64
microprocessor, 764
monolithic, 752
MSI, 752
number systems used with
digital ICs, 756–57
packaging and basing of, 755–56
shift-registers, 763
SSI, 752
thin film, 752
VLSI, 752
Integrated output, 598
Integrator circuit, 807–808
Intensity controls, 380
Internal resistance, 211–14, 250,
305
International System (SI), 271–72
Inverse proportionality concept,
308–309, 312
Inverter, 760
Inverting amplifier, 803–804
Ions, 13
Isolation transformer, 510
Isotopes, 10
IT:
solving for using Pythagorean
approach, 591–92

Jewel bearings, 300, 301
JFET (junction), 737
j operator, 632
Joule, 59, 418

k, formula for, 500
Keeper bar, 263
Kelvin, Lord, 298
Kirchhoff's current law, 129–31
Kirchhoff's voltage law, 86–88

Laminations, 508
Law of electrical charges, 17–18
L circuit, 458–59
L/C ratios, 655–59
Lead-acid battery, 239
battery ratings, 243
care of, 247–48
charged conditions, 241
charging conditions, 242
charging methods, 244–45
construction of, 240–41
discharged conditions, 241
discharging conditions, 241
safety hints, 246
testing methods, 245, 247
Lead basing layouts, 733
Leading voltage, 436–37
Leakage flux, 288–89, 496
LeClanche cell. *See* Dry cell
Left-hand rules:
current-carrying conductors,
265–66
polarity of electromagnets, 267–
68
Lenz's law, 284–86, 411
Light-emitting diode (LED), 740
Like poles, 262
Lines of force, 260, 261
rules concerning, 262–63, 410–
11
Liquid, 5–6
Lissajous pattern technique, 394–
99

Lithium cell, 249
Load, 188
Load current, 188
Loaded voltage dividers, 187
  potentiometer varying voltage output, 193–94
  terminology, 188–89
  three-element divider, multiple loads, 190–93
  two-element divider, 190
Loading effect, 321–22, 337
Load line, 776–78
Local action, 238
Logic gates, 757–62
Long cables, continuity checks of, 335–36
Long time-constant circuit, 598
Low-pass filters, 675–76
L/R time constant, 419–24

M, formula for, 500
Magnetic core piece, 299, 301
Magnetic field, 259, 260
  collapsing, 424
  of inductor, 417–19
Magnetic field intensity (H), 273–74
Magnetic flux, 260
Magnetic linkage, 495
Magnetic shield, 289–90
Magnetic units, 271–72
Magnetism:
  B-H curve, 276–77
  classifications of materials, 291
  early history, 259
  and electricity, 259
  elemental electromagnetism, 264–70
  Faraday's law, 284
  fundamental laws, rules and terms, 260–64
  Hall effect, 290–91
  hysteresis loop, 277–80
  induction and related effects, 280–83
  Lenz's law and reciprocal effects of motors and generators, 284–88
  magnetic shields, 289–90
  magnetic units, terms, symbols, and formulas, 270–75
  pole pieces linearizing a field, 290
  toroidal coil form, 288–89
Magnetomotive force, 274
Magnets:
  permanent, 259, 263, 299, 301
  temporary, 259
Magnification factor, 661–62
  determining, 670
Magnitude:
  Pythagorean theorem in determining, 463, 586
  trig functions in determining, 463–66
  vectors in determining, 459–61
Matter:
  chemical states of defined, 6–7
  composition of, 7–8
  physical states of defined, 5–6
Maximum current, determining, 670, 672
Maximum power transfer theorem:
  defined, 211
  efficiency factor formula, 212–13
  practice problem, 214–15
Maxwell (Mx), 260
Memory:
  RAM and ROM, 763–64
  volatile and nonvolatile, 763
Mercury cell, 249
Meters, 298, 337–38
  basic movement requirements, 299
  current meters, 302–306
  current meter shunts, 306–15
  d'Arsonval (moving coil) movement, 299–302
  ohmmeters, 324–27
  troubleshooting, 332–37
  voltmeters, 315–23
  VOMs, DMMs, and related devices, 327–32
Meter scales, 319
Meter shunt, 305
Metric system, 40–43
  application of metric prefixes and powers of ten, 54–55
Mho, 40
Mica capacitors, 546, 547
Microchip, 751
Microprocessor, 764
Mixtures, 6–7
Molecules, 7–8
MOSFET (metal-oxide-semiconductor), 737
Motion, charges in. *See* Charges in motion
Motor, generator effect, 287–88
Motor action, 280–81, 301
Motor effect, 286–87
Movable coil, 300–302
Moving-coil movement. *See* d'Arsonval (moving coil) movement
Multimeters:
  clamp-on current probe, 330–31
  digital (DMM), 330
  high voltage probe for, 330, 331
  VOM, 328–29
Multiple-range meter circuit, 309
Multiple-source circuit, 215
Multiple-turn coil, field around, 267
Multiplier band, 68
Multiplier resistor, 315
  approximating multiplier values, 318
  calculating multiplier values, 316–18
Mutual inductance, 415, 495–500
  and formulas for k and M, 500

NAND gate, 760–61
Negative charge, 19
Negative peak, 357, 358
Negative-resistance region, 738
Net reactive current, 622
Networks, simplification of, 219–22
Neutrons, 8–10
Nickel-cadmium battery, 249
Non-conductors, 693
Noninverting amplifier, 804–805
Nonrechargeable primary cells and batteries, 249

Non-sinusoidal waveforms, 597
NOR gate, 760, 761
Norton's theorem, 222
  compared with Thevenin equivalent circuit, 224–26
  conversions with Thevenin equivalent parameters, 227
  defined, 223
NPN transistor, 730–31
N-type conduction, 696
N-type semiconductor material, 695–96
Number systems, 756–57, 758

Octal system, 757
Oersted, Hans, 259, 264
Ohm, 27–28, 39
Ohm, Georg Simon, 27, 49
Ohmmeters:
  adjusting, using, and reading, 324–27
  for capacitor checks, 552–54
  measurement techniques, 335–36
  series-type, 324
  sneak paths, 336–37
  for transistor checks, 741
  troubleshooting, 332
Ohm's law:
  application of metric prefixes and powers of ten, 54–55
  combined with power formula, 61–65
  computing current in series RLC circuit by, 616–17
  computing individual voltage in series RLC circuit by, 617
  multiplier resistance value, 316–18
  rearranging to find unknown quantity, 52–54
  relationship of current to resistance with voltage unchanged, 49–52
  relationship of current to voltage with resistance constant, 49
  relationships between electrical quantities, 49–65
  shunt resistance calculations, 307–308
  total resistance formula, 82–84, 132–33, 169–72
  voltage in series-parallel circuits, 176–77
Ohms-per-volt rating, 319–21
Op-amps, 799
  basic information about, 800–802
  basic parameters, 805–807
  ideal op-amp characteristics, 803
  inverting amplifier, 803–804
  linear applications, 799–800
  noninverting amplifier, 804–805
  nonlinear applications, 800
  sample applications, 807–10
Open capacitors, 552–54
Open circuit, 28–29
  continuity checks to find, 335
  in parallel circuits, 142–43
  in series circuits, 91–93
  in series-parallel circuits, 180–81
  troubleshooting hints, 93, 144, 181–82
  voltage measurements to find, 334
Open coil, 426
Open winding(s), 513
Operational amplifiers. *See* Op-amps
Opposite-direction phasors, 620–22
Opto-isolator, 740
OR gate, 759, 760
Oscillator circuits, 738
  Clapp oscillator circuit, 789, 790
  Colpitts oscillator circuit, 789
  conditions required for oscillation, 788
  crystal oscillator circuit, 790
  defined, 787–88
  Hartley oscillator circuit, 788–89
Oscilloscope, 379, 400
  CRT, 380, 382
  horizontal and vertical signals combined to view waveform, 386–89
  horizontal section, 385
  horizontal sweep frequency control(s), 383–84
  intensity and focus controls, 380
  key sections of, 380–86
  measuring frequency ratios with, 395–99
  measuring voltage and current with, 389–93
  for phase comparisons, 393–95
  position controls, 382–83
  synchronization controls, 385–86
  vertical section, 385
Out-of-circuit test, 552–53, 741
Outside toward the source approach, total resistance, 169–70
Overlaying technique, 393–95

Paper and plastic capacitors, 546
Parallax error, 337–38
Parallel circuit:
  capacitive reactances in, 570–72, 574
  compared with series circuit, 147
  current dividers, 149–53
  current in, 128–29
  defined, 127
  designing to specifications, 148–49
  features of, 127–28
  Kirchhoff's current law, 129–31
  opens in and troubleshooting hints, 142–44
  power in, 140–42
  resistance in, 132–40
  shorts in and troubleshooting hints, 144–46
  sources in, 149
  voltage in, 128
Parallel coils:
  total inductance with aiding fields, 499
  total inductance with opposing fields, 499–500
Parallel RC circuit, 594–96
  solving for impedance in, 593–94
  solving for IT using Pythagorean approach, 591
  solving for phase angle using trig, 592–93

Parallel resonance, 663–67
Parallel RL circuit, 477–82
Parallel RLC circuit:
  circuit analysis, 621–23
  simple circuit, 620–21
Paramagnetic materials, 291
Passive filters, 674
Peak-inverse-voltage (PIV), 707, 710, 712
Peak-to-peak values, 357, 358, 363
Peak value, 363
Pentavalent material, 695
Period (T), 358
  relationship with frequency, 360
Permanent magnets, 259
  in d'Arsonval movement, 299, 301
  practical points about, 263
Permeability factors, 274–75
Permittivity, 532
Phase angle, 360, 436, 481
  computing from impedance and voltage data, 618–20
  solving from impedance diagram, 589
  solving for using trig, 588, 592–93
  solving for from V-I vector diagram, 588–89
Phase comparisons, 393
  Lissajous pattern technique, 395
  overlaying technique, 394–95
Phase relationships, 360–63
Phasor, 350, 630–31
  opposite-direction, 620–22
  positional information for in RC circuits, 585, 586
Photocell (photovoltaic cell), 739–40
Photodiode cell, 739
Pin connections, ICs, 755–56
Plate area, capacitor, 531
P-N junction, 698–99, 729–30
PNP transistor, 731
Pointer, 301, 302, 324
Polar form, 633
  converting from rectangular form to, 633–35
  converting to rectangular form from, 635–36
Polarity, 18–19, 58–59, 104–105, 436
  left-hand rule for, 267–68
  for moving-coil movement, 301–302
  of output voltage, 708, 710
Polarization, 237–38
Pole pieces, 290, 299, 301
Position controls, 382–83
Positive charge, 19
Positive peak, 357, 358
Potential, 39
Potential difference, 21
Potentiometer, 189
  variable-voltage-divider capability, 193–94
Powdered iron-core coils, 511, 512
Power:
  in ac circuits, 623–29
  applications of formulas for, 62–65
  formula for, 60
  individual component calculations, 90
  mechanical, 59
  in parallel circuits, 140–42
  in series circuits, 89–91
  in series-parallel circuits, 178–80
  total circuit calculations, 90–91
  variations of power formula, 61
Power cords, continuity checks of, 335–36
Power factor method, 627–28
Power frequencies, 482
Powers of ten:
  addition of, 56
  application of metric prefixes and, 54–55
  basic rules for, 55–58
  division using, 56
  moving decimal places using, 55
  multiplication and division combined and, 56
  multiplication using, 55–56
  subtraction of, 57
Power supply filters:
  design considerations, 715
  effect of on operation and output of rectifier circuits, 716
  output characteristics and the bleeder resistor, 716–17
  types of, 713–15
Power supply system:
  bridge rectifier circuit, 710–12
  design considerations, 705
  filters, 713–17
  full-wave rectifier circuit, 708–10
  half-wave rectifier circuit, 705–708
  main elements of, 703–705
  problems in, 718–19
  voltage multiplier circuits, 717
Power transformers, 509–11
Power transmission, ac advantages over dc in, 349–50
Power triangle, 628–29
Primary cells:
  defined, 234
  dry cell, 237–39
  examples of, 249
Primary windings, 497, 498
Product-over-the-sum method, total resistance, 135–37
Protons, 8–10
P-type conduction, 697–98
P-type semiconductor material, 696
Pulsating dc, 706
Purely capacitive ac circuit, 585, 612–13
  V and I relationships in, 563–64
Purely inductive ac circuit, 611–12
  V and I relationships in, 435–37
Purely reactive ac circuit:
  power in, 624–25
Purely resistive ac circuit, 367–68, 585, 611
  power in, 624
  V and I relationships in, 435, 563
Pythagorean theorem, 463
  computing impedance in series RLC circuit, 616
  computing total voltage in series RLC circuit, 617–18
  solving magnitude values by, 586

using to analyze voltage in series RL circuits, 467–68

Q factor, 660
determining, 670
related to L/C ratio, 661
related to XL/R, 660–61
resonant rise of impedance and, 666–67
Q point, 778
Quadrants, 351
Quality factor (Q), 446–47

Radio frequencies, 482, 484
RAM, 763–64
Range selector switch, 318
Ratings, capacitors, 550, 551
RC circuit:
applications of, 598, 601
compared with RL circuits, 596
parallel circuit analysis, 591–96
series circuit analysis, 585–91
timing applications, 598
troubleshooting, 601–602
waveshaping, 597–98, 599–600
R circuit, 457–58, 585
RC time constant, 540–44
Reactive power, 628
Real numbers, 631
Recall conductance, 640
Rechargeable secondary cells and batteries, 249
Reciprocal method, total resistance, 133–34
Rectangular and polar vector analysis:
algebraic operations, 637–39
application of, 639–43
converting from polar to rectangular form, 635–36
converting from rectangular to polar form, 633–35
phasors, 630–31
polar notation, 633
rectangular notation, 631–33
Rectangular form, 631–33
converting from polar form to, 635–36
converting to polar form from, 633–35
Rectangular wave, 369–70
Rectifier, 703
problems with, 719
Reduce and redraw approach, total resistance, 170–72
Reference points, 18–19, 103–104, 188, 333
Regulator, 704
Relative permeability, 275, 413
Reluctance, 270, 275
Resistance, 39
defined, 27
internal, 211–14, 250, 305
in meter shunts, 306–309
in parallel circuit, 132–40
in series circuit, 81–84
in series-parallel circuit, 169–72
unknown, finding using Wheatstone bridge circuit, 196–98
Resistive power, 628
Resistors:
bleeder, 189
color coding of, 67–69
construction of, 66–67
multiplier, 315
series-dropping, 105–106
types of, 65–66
uses for, 65
Resonance, 653
Resonance curves, 659–60
Resonant circuits:
coupled load, effect on tuned circuit, 666
filter applications, non-resonant and resonant RLC circuits, 674–81
measurements related to, 670–74
parallel resonance characteristics, 663–64
parallel resonance formulas, 664–66
Q and resonant rise of impedance, 666–67
Q and resonant rise of voltage, 660–63
resonance curves, 659–60
resonant frequency formula, 655–59
selectivity, bandwidth and bandpass, 667–70
series resonance characteristics, 653–55
Resonant frequency, 653
measuring, 670
Resonant frequency formula, 655–59
Reverse bias, 700–701
RF transformers and coils, 511–13
Right triangle problems, 462–66
Ring shunt. *See* Universal shunt
Ripple frequency, 707, 710, 712
Ripples, 706
RL, 188
RLC circuit:
algebraic operations, 637–39
applying vector analysis, 639–43
basic circuit analysis, 611–13
current by Ohm's law, 616–17
filter applications, 674–81
impedance by Pythagorean approach, 616
individual voltage by Ohm's law, 617
parallel circuit analysis, 620–23
phase angle from impedance and voltage data, 618–20
power in ac circuits, 623–29
rectangular and polar vector analysis, 630–43
series circuit analysis, 615–20
series circuit without resistance, 613–15
total voltage by Pythagorean approach, 617–18
RL circuits:
applications for, 482–84
compared with RC circuits, 596
ROM, 763–64
Rotating vector, 350–51, 439–40
Rowland's law, 273
R (shunt) value, 306–309

Safety hints, 69–71
battery charging, 246
capacitors, 528, 549
measurement instruments, 319
power supply circuits, 717, 719

Safety hints (*Cont.*)
  RC circuits, 602
  resonant circuits, 674
  transformers, 510
Saturation, 702
Sawtooth wave, 369–70
Scalars, 459
Schematic diagrams, 47–48
Scopes. *See* Oscilloscope
Secondary cells:
  defined, 234
  examples of, 249
  wet cell, 239–48
Secondary windings, 497, 498
Selectivity, 660, 667
Self-inductance, 410, 411–12, 495
Semiconductor diode, 692, 699
  forward bias of, 701–702
  reverse bias of, 700–701
Semiconductor materials, 694, 698
  N-type, 695–96
  P-type, 696, 697
Semiconductors, 14
  compared with conductors, 693
Sensitivity, 304–305
Series circuit:
  applied voltage, finding value of, 86
  capacitive reactances in, 570, 573
  characteristics of, 81
  compared with parallel circuit, 147
  defined, 81
  designing to specifications, 98–100
  divide-and-conquer troubleshooting approach, 96–97
  individual component power calculations, 90
  individual component voltages, 84–86
  Kirchhoff's voltage law and, 86–88
  opens and troubleshooting hints, 91–93
  power in, 89–91
  reference point(s) in, 103–105
  resistance in, 81–84
  series-dropping resistor and, 105–106
  shorts and troubleshooting hints, 93–96
  special applications, 100–106
  total circuit power calculations, 90–91
  total resistance formula, 82
  total resistance using Ohm's law, 82–83
  voltage divider action using, 102–103
  voltage in, 84–88
  voltage sources in, 100–102
Series coils:
  total inductance with aiding fields, 499
  total inductance with opposing fields, 499
Series-dropping resistor, 105–106
Series-parallel circuit:
  analysis and recognition of, 166–69
  characteristics of, 165–66
  current in, 173–75
  defined, 165
  designing to specifications, 185–87
  loaded voltage dividers, 187–95
  opens in and troubleshooting hints, 180–82
  power in, 178–80
  shorts in and troubleshooting hints, 183–85
  total resistance in, 169–72
  voltage in, 176–78
  Wheatstone bridge circuit, 195–98
Series RC circuit, 585, 589–91
  positional information for phasors, 585, 586
  solving for phase angle from impedance diagram, 589
  solving for phase angle using trig, 588
  solving for phase angle using V-I vector diagram, 588–89
  solving for Z in the impedance diagram, 587
  solving magnitude values by Pythagorean theorem, 586
Series resonance, 653–63
Series RLC circuit:
  applying vector analysis to, 639–40
  circuit analysis, 615–20
  computing current, 616–17
  computing impedance, 616
  computing individual voltage, 617
  computing phase angle, 618–20
  computing total voltage, 617–18
  simple circuit without resistance, 613–15
Series RL circuit:
  analyzing impedance in, 472–75
  fundamental analysis of, 466–77
  Pythagorean theorem to analyze voltage in, 467–68
  trig to solve for voltages in, 470–72
Series-type linear voltage regulator, 809–10
Series-type ohmmeter circuits, 324
Shelf life, 238, 250–51
Shift-registers, 763
Short circuit:
  ohmmeter checks to find, 335
  in parallel circuits, 144–46
  in series circuits, 93–95
  in series-parallel circuits, 183
  troubleshooting hints, 95–96, 146, 184–85
  voltage measurements to find, 334
Shorted capacitors, 552, 554
Shorted turns, 426
Shorted winding(s), 514–15
Shorting-type switch, 309–10
Short-term heavy-load test, 247
Short time-constant circuit, 598
Siemens, 39–40
Siemens, Ernest von, 39
Signal levels, 774–75
Signals:
  comparing phase of with oscilloscope, 393–95
  measuring frequency ratios with oscilloscope, 395–99

measuring frequency ratios with oscilloscope, 395–99
Silicon-controlled rectifier (SCR), 737
SIMPLER troubleshooting sequence, 107–109, 114
block or module level, 113, 115
component level of, 113, 116
example of, 109–13
Sine, 353–54, 463–64
Sine wave, 352–57, 564
current and voltage values, 363–67
descriptors of, 357–58
in-phase, 361
out-of-phase, 360–61
period and frequency, 358–60
Skirts, 660
Sneak paths, 336–37
Solar cell, 249
Solid, 5
Solid-state devices, 694
Source, 736
Spindle, 300, 301
Split technique, troubleshooting, 96–97
Springs, 301
Square wave, 369–70
Static electricity:
Coulomb's formula and, 19–20
coulomb (unit of charge), 20
defined, 17
fields of force and, 20–21
law of electrical charges and, 17–18
polarity and reference points, 18–19
Step-down transformer, 501
Step-up transformer, 501
Stops, 301
Sulfation, 241
Summing amplifier, 807
Superposition theorem:
defined, 215
practice problem, 218
reversed polarities, 217–18
verification of, 215–16
Support frame (movable coil), 300, 302
Susceptance, 640–41
Symbols:
basic electrical units, 39–40
capacitors, 525
electronic, 44–45
magnetism, 260
Synchronization controls, 385–86

Tangent, 464
Tantalum capacitors, 549–50
Temporary magnets, 259
Terminology:
ac sine wave current and voltage values, 363–67
loaded voltage dividers, 188–89
magnetism, 260
Thermionic emission, 728
Thevenin's theorem:
compared with Norton equivalent circuit, 223–26
conversions with Norton equivalent parameters, 227
defined, 219–22
Three-element voltage divider, 190–93
Time constant, 420
Timing applications, 598
Tolerance band, 68
Toroidal cores, 288–89, 512, 513
Total current value, 334, 478
Total impedance, 478
Total inductance, 415–17
Total resistance, 82–84, 132–40, 169–72
Transformation ratio, 500–501
Transformers, 408, 703
audio transformers, 511
autotransformer, 510
current ratio, 503–504
impedance ratio, 506–508
losses of, 508–509
mutual inductance between windings, 496–98
power transformers, 509–11
primary windings, 497–98
problems with, 718
RF transformers and coils, 511–13
secondary windings, 497–98
step-down, 501
step-up, 501
transformation ratio, 500–501
troubleshooting, 513–15
turns ratio, 501, 502
voltage ratio, 501–502
Transistors, 728
alpha and beta parameters, 731–33
amplification by, 773–74
common amplifiers, 733–35
common-base (CB) amplifier, 733–34
common basing or lead configuration schemes, 733
common-collector (CC) amplifier, 735
common-emitter (CE) amplifier, 734–35
compared with vacuum tubes, 729
field effect transistors, 736–37
light-emitting diode (LED), 740
NPN transistor, 730–31
opto-isolators, 740
photocell (photovoltaic cell), 739–40
photodiode cell, 739
problems with and testing of, 741–42
SCR and TRIAC (thyristors), 737, 738
tunnel diode, 738
varactor diode, 738–39
TRIAC (thyristor), 737, 738
Trigger pulses, 597
Trigonometric functions, 463–66
for angle information in series RL circuits, 468–70
solving for phase angle, 588, 592–93
for voltages in series RL circuits, 470–72
Trivalent material, 696
Troubleshooting:
block or module level, 113, 115
capacitors, 552–54
component level, 113, 116
divide-and-conquer approach, 96–97

Troubleshooting (*Cont.*)
  inductors, 426
  measuring instruments, 332–37
  open circuit, 93, 144
  power supply system, 718
  RC circuits, 601–602
  short circuit, 95–96, 146
  SIMPLER sequence for, 107–14
  transformers, 513–15
True power, 626
Truth tables, 759–62
Tuned circuits, 408
  effect of coupled load on, 666
Tuning, 652
Tuning range, 673–74
Tuning ratio, determining, 674
Tunnel diode, 738
Turns ratio, 501, 502
Two-element voltage divider, 190

Universal (Ayrton or ring) shunt, 310–11
  10-mA range, 314
  100-mA range, 314
  1000-mA range, 314
  shunt calculations, 312–14
Universal Time Constant (UTC) chart, 422
Unknown quantity, finding, 52–54
Unlike poles, 262

Valence electrons, 12, 693–94
V and I relationships:
  in pure inductive ac circuit, 435–37
  in pure resistive ac circuit, 435
Varactor diode, 738–39
Variable capacitor, 544–46
Variable controls, 385
Vector resultant, 473
Vectors, 350
  plotting and measuring, 459–61
  rectangular and polar forms analysis, 630–36
Vertical attenuator and amplifier, 385
Vertical section, 385
Vertical volts/cm, 385
Volt, 21, 39
Volta, Alessandro, 21, 235
Voltage, 21, 58–59
  ac measurement with oscilloscope, 389–91
  ac sine wave, 352–57, 363–67
  applied voltage, finding value of, 86
  chassis ground reference point, 103–105
  dc measurement with oscilloscope, 392
  dividing of, 102–103
  exponential formulas to find, 542–44
  individual component voltages, 84–86
  Kirchhoff's voltage law, 86–88
  in parallel circuits, 128
  resonant rise of, 661–63
  in series circuits, 84–89
  series-connected sources, 100–102
  series-dropping resistor, 105–106
  in series-parallel circuits, 176–78
  in series RL circuits, 467–68, 470–72
Voltage checks, 333–34
Voltage divider, 102, 187, 601
  See also Loaded voltage dividers
Voltage-divider bias method, 781–82
Voltage multiplier circuits, 717
Voltage ratio, 501, 502
Voltaic cell, 235
Voltampere-reactive (VAR), 628
Voltmeters, 323
  approximating multiplier values, 318
  connecting to circuit being tested, 316
  loading effect, 321–22
  measurement techniques, 333–34
  multiplier value calculation, 316–18
  ohms-per-volt rating, 319–21
  purpose of multiplier resistor, 315
  range selector switch, 318–19
  troubleshooting, 332
VOM (volt-ohm-milliammeter), 328–29
VT, solving for in V-I vector diagram, 586–87

Watt, 59
Watt, James, 59
Waveforms:
  combining horizontal and vertical signals to view, 386–89
  linear, 799–800
  nonlinear, 800
  non-sinusoidal, 597
  rectangular wave, 369–70
  sawtooth wave, 369–70
  sine wave, 352–57
  square wave, 369–70
Waveshaping, 597–98, 599–600
Weber (Wb), 260
Wet cell, 239–48
Wet electrolytic, 547
Wheatstone bridge circuit, 196–98
Work, 59

X_C:
  basic formula for, 567–69
  Ohm's law method to calculate, 566
  relationship of to capacitance value, 564–65
  relationship of to frequency of ac, 565–66
X_C and frequency, 653
X_L, 437
  calculating using inductive reactance formula, 440–42
  calculating using Ohm's law approach, 442–43
  relationship of to frequency of ac, 439–40
  relationship of to inductance value, 438–39
X_L and frequency, 653

Z, solving for in impedance diagram, 587–88
Zener diodes, 701
Zero-adjust control, 324
Zinc-chloride cell, 249